MW00846446

PRINCIPLES OF INORGANIC CHEMISTRY

PRINCIPLES OF INORGANIC CHEMISTRY

Brian W. Pfennig

WILEY

Published by John Wiley & Sons, Inc., Hoboken, New Jersey
Published simultaneously in Canada

Library of Congress Cataloging-in-Publication Data:

Pfennig, Brian William.
Principles of inorganic chemistry / Brian W. Pfennig.
pages cm
Includes bibliographical references and index.
ISBN 978-1-118-85910-0 (cloth)
1. Chemistry, Inorganic–Textbooks. 2. Chemistry, Inorganic–Study and teaching (Higher) 3. Chemistry, Inorganic–Study and teaching (Graduate) I. Title.
QD151.3.P46 2015
546–dc23

2014043250

Cover image :Courtesy of the author
Typeset in 10/12pt GillSans by Laserwords Private Limited, Chennai, India.

Printed in the United States of America

10 9 8 7 6 5 4 3 2 1

1 2015

Contents

Preface

This book was written as a result of the perceived need of mine and several other colleagues for a more advanced physical inorganic text with a strong emphasis on group theory and its applications. Many of the inorganic textbooks on the market are either disjointed—with one chapter completely unrelated to the next—or encyclopedic, so that the student of inorganic chemistry is left to wonder if the only way to master the field is to memorize a large body of facts. While there is certainly some merit to a descriptive approach, this text will focus on a more principles-based pedagogy, teaching students how to rationalize the structure and reactivity of inorganic compounds—rather than relying on rote memorization.

After many years of teaching the inorganic course without a suitable text, I decided to write my own. Beginning in the summer of 2006, I drew on a variety of different sources and tried to pull together bits and pieces from different texts and reference books, finishing a first draft (containing 10 chapters) in August, 2007. I used this version of the text as supplementary reading for a few years before taking up the task of writing again in earnest in 2012, subdividing and expanding the upon original 10 chapters to the current 19, adding references and more colorful illustrations, and including problems at the ends of each chapter.

The book was written with my students in mind. I am a teacher first and a scientist second. I make no claims about my limited knowledge of this incredibly expansive field. My main contribution has been to collect material from various sources and to organize and present it in a pedagogically coherent manner so that my students can understand and appreciate the principles underlying such a diverse and interesting subject as inorganic chemistry.

The book is organized in a logical progression. Chapter 1 provides a basic introduction to the composition of matter and the experiments that led to the development of the periodic table. Chapter 2 then examines the structure and reactivity of the nucleus. Chapter 3 follows with a basic primer on wave-particle duality and some of the fundamentals of quantum mechanics. Chapter 4 discusses the solutions to the Schrödinger equation for the hydrogen atom, the Pauli principle, the shapes of the orbitals, polyelectronic wave functions, shielding, and the quantum mechanical basis for the underlying structure of the periodic table. Chapter 5 concludes this section of the text by examining the various periodic trends that influence the physical and chemical properties of the elements. Chapter 6 then begins a series of chapters relating to chemical bonding by reviewing the basics of Lewis structures, resonance, and formal charge. Chapter 7 is devoted to the molecular geometries of molecules and includes not only a more extensive treatment of the VSEPR model than most other textbooks but it also presents the ligand close-packing model as a complementary model for the prediction of molecular geometries. Symmetry and group theory are introduced in detail in Chapter 8 and will reappear as a recurring theme throughout the remainder of the text. Unlike most inorganic textbooks on the market, ample coverage is given to representations of groups, reducing representations, direct products, the projection operator, and applications of group theory. Chapter 9 focuses on one of the applications of group theory to the vibrational spectroscopy of molecules, showing how symmetry coordinates can be used to approximate the normal modes of vibration of small molecules. The selection rules for IR and Raman spectroscopy are discussed and the chapter closes with a brief introduction to resonance Raman spectroscopy. The next three chapters focus on the three different types of chemical bonding: covalent, metallic, and ionic bonding. Chapter 10 examines the valence bond and molecular orbital models, which expands upon the application of group theory to chemical problems. Chapter 11 then delves into metallic bonding, beginning with a primer on crystallography before exploring the free electron model and band theory of solids. Chapter 12 is focused on ionic bonding—lattice enthalpies, the Born–Haber cycle, and Pauling's rules for the rationalization of ionic solids. It also has extensive coverage of the silicates and zeolites. The structure of solids is reviewed in greater detail in Chapter 13, which explores the interface between the different types of chemical bonding in both solids and discrete molecules. Switching gears for a while from structure and bonding to chemical reactivity, Chapter 14 introduces the two major types of chemical reactions: acid–base reactions and oxidation–reduction reactions. In addition to the usual coverage of hard–soft acid–base theory, this chapter also examines a more general overview of chemical reactivity that is based on the different topologies of the MOs involved in chemical transformations. This chapter also serves as a bridge to the transition metals. Chapter 15 presents an introduction to coordination compounds and their thermodynamic and magnetic properties. Chapter 16 examines the structure, bonding, and electronic spectroscopy of coordination compounds, making extensive use of group theory. Chapter 17 investigates the reactions of coordination compounds in detail, including a section on inorganic photochemistry. Finally, the text closes with two chapters on organometallic chemistry: Chapter 18 looks at the different types of bonding in organometallics from an MO point of view, while Chapter 19 presents of a survey of organometallic reaction mechanisms, catalysis, and organometallic photochemistry and then concludes with connections to main group chemistry using the isolobal analogy. Throughout the textbook, there is a continual building on earlier material, especially as it relates to group theory and MOT, which serve as the underlying themes for the majority of the book.

This text was originally written for undergraduate students taking an advanced inorganic chemistry course at the undergraduate level, although it is equally suitable as a graduate-level text. I have written the book with the more capable and intellectually curious students in my undergraduate courses in mind. The prose is rather informal and directly challenges the student to examine each new experimental observation in the context of previously introduced principles of inorganic chemistry. Students should appreciate the ample number of solved sample problems interwoven throughout the body of the text and the clear, annotated figures and illustrations. The end-of-chapter problems are designed to invoke an active wrangling with the material and to force students to examine the data from several different points of view. While the text is very physical in emphasis, it is not overly mathematical and thorough derivations are provided for the more important physical relationships. It is my hope that students will not only enjoy using this textbook in their classes but will read and reread it again as a valuable reference book throughout the remainder of their chemical careers.

While this book provides a thorough introduction to physical inorganic chemistry, the field is too vast to include every possible topic; and it is therefore somewhat limited in its scope. The usual group by group descriptive chemistry of the elements, for example, is completely lacking, as are chapters on bioinorganic chemistry or inorganic materials chemistry. However, it is my belief that what it lacks in breadth is more than compensated for by its depth and pedagogical organization. Nonetheless, I eagerly welcome any comments, criticisms, and corrections and have opened a dedicated e-mail account for just such a purpose at pfennigtext@hotmail.com. I look forward to hearing your suggestions.

BRIAN W. PFENNIG

Lancaster, PA
June, 2014

Acknowledgments

This book would not have been possible without the generous contributions of others. I am especially indebted to my teachers and mentors over the years who always inspired in me a curiosity for the wonders of science, including Al Bieber, Dave Smith, Bill Birdsall, Jim Scheirer, Andy Bocarsly, Mark Thompson, Jeff Schwartz, Tom Spiro, Don McClure, Bob Cava, and Tom Meyer. In addition, I thank some of the many colleagues who have contributed to my knowledge of inorganic chemistry, including Ranjit Kumble, Jim McCusker, Dave Thompson, Claude Yoder, Jim Spencer, Rick Schaeffer, John Chesick, Marianne Begemann, Andrew Price, and Amanda Reig. I also thank Reid Wickham at Pearson (Prentice-Hall) for her encouragement and advice with respect to getting published and to Anita Lekwhani at John Wiley & Sons, Inc. for giving me that chance. Thank you all for believing in me and for your encouragement.

There is little original content in this inorganic text that cannot be found elsewhere. My only real contribution has been to crystallize the content of many other authors and to organize it in a way that hopefully makes sense to the student. I have therefore drawn heavily on the following inorganic texts: *Inorganic Chemistry* (Miessler and Tarr), *Inorganic Chemistry* (Huheey, Keiter, and Keiter), *Chemical Applications of Group Theory* (Cotton), *Molecular Symmetry and Group Theory* (Carter), *Symmetry and Spectroscopy* (Harris and Bertulucci), *Problems in Molecular Orbital Theory* (Albright and Burdett), *Chemical Bonding and Molecular Geometry* (Gillespie and Hargittai), *Ligand Field Theory* (Figgis and Hitchman), *Physical Chemistry* (McQuarrie and Simon), *Elements of Quantum Theory* (Bockhoff), *Introduction to Crystallography* (Sands), and *Organometallic Chemistry* (Spessard and Miessler).

In addition, I am grateful to a number of people who have assisted me in the preparation of my manuscript, especially to the many people who have reviewed sample chapters of the textbook or who have generously provided permission to use their figures. I am especially indebted to Lori Blatt at Blatt Communications for producing many of the amazing illustrations in the text and to Aubrey Paris for her invaluable assistance with proofing the final manuscript.

I would be remiss if I failed to acknowledge the contributions of my students, both past and present, in giving me the inspiration and perseverance necessary to write a volume of this magnitude. I am especially indebted to the intellectual interactions I have had with Dave Watson, Jamie Cohen, Jenny Lockard, Mike Norris, Aaron Peters, and Aubrey Paris over the years. Lastly, I would like to acknowledge the most important people in my life, without whose undying support and tolerance I would never have been able to complete this work—my family. I am especially grateful to my wonderful parents who instilled in me the values of a good education, hard work, and integrity; to my wife Jessica for her unwavering faith in me; and to my incredibly talented daughter Rachel, who more than anyone has suffered from lack of my attention as I struggled to complete this work.

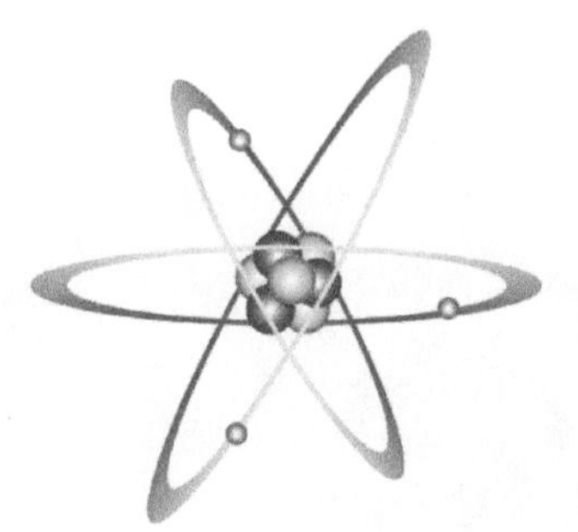

The Composition of Matter 1

"Everything existing in the universe is the fruit of chance and necessity."
—*Democritus*

1.1 EARLY DESCRIPTIONS OF MATTER

Chemistry has been defined as the study of matter and its interconversions. Thus, in a sense, chemistry is a study of the physical world in which we live. But how much do we really know about the fundamental structure of matter and its relationship to the larger macroscopic world? I have in my rock collection, which I have had since I was a boy, a sample of the mineral cinnabar, which is several centimeters across and weighs about 10 g. Cinnabar is a reddish granular solid with a density about eight times that of water and the chemical composition mercuric sulfide. Now suppose that some primal instinct suddenly overcame me and I were inclined to demolish this precious talisman from my childhood. I could take a hammer to it and smash it into a billion little pieces. Choosing the smallest of these chunks, I could further disintegrate the material in a mortar and pestle, grinding it into ever finer and finer grains until I was left with nothing but a red powder (in fact, this powder is known as *vermilion* and has been used as a red pigment in artwork dating back to the fourteenth century). Having satisfied my destructive tendencies, I would nonetheless still have exactly the same material that I started with—that is, it would have precisely the same chemical and physical properties as the original. I might therefore wonder to myself if there is some inherent limitation as to how finely I can divide the substance or if this is simply limited by the tools at my disposal. With the proper equipment, would I be able to continue dividing the compound into smaller and smaller pieces until ultimately I obtained the *unit cell*, or smallest basic building block of the crystalline structure of HgS, as shown in Figure 1.1? For that matter (no pun intended), is there a way for me to separate out the two different types of atoms in the substance?

If *matter* is defined as anything that has mass and is perceptible to the senses, at what point does it become impossible (or at the very least impractical) for me to continue to measure the mass of the individual grains or for them to no longer be perceptible to my senses (even if placed under an optical microscope)? The ancient philosopher Democritus (ca 460–370 BC) was one of the first to propose that matter is constructed of tiny indivisible particles known as *atomos* (or *atoms*), the different varieties (sizes, shapes, masses, etc.) of which form the fundamental building blocks of the natural world. In other words, there should be some lower limit as to how

Principles of Inorganic Chemistry, First Edition. Brian W. Pfennig.

(a)

(b)

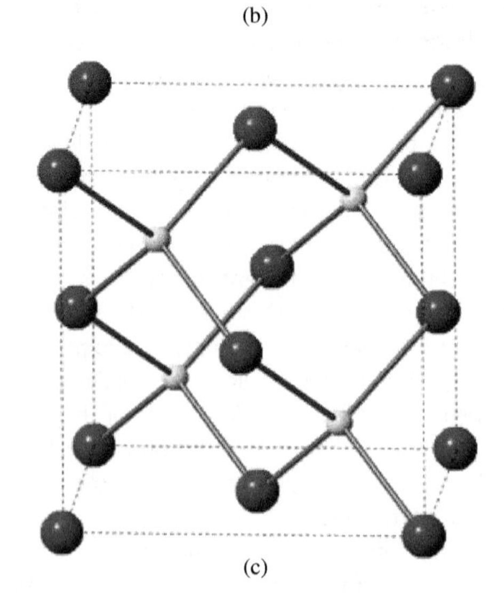
(c)

FIGURE 1.1
Three examples of the same chemical material ranging from the macroscopic to the atomic scale: (a) the mineral cinnabar, (b) vermilion powder, and (c) the unit cell of mercuric sulfide. [Vermilion pigment photo courtesy of Kremer Pigments, Inc.]

finely I can continue to carve up my little chunk of cinnabar. As far back as the Middle Ages, the alchemists learned that one could decompose a sample of HgS by heating it up in a crucible. At temperatures above 580 °C, the heat drives off the sulfur and leaves behind a pool of silvery liquid mercury. Eventually, I could break the molecule itself apart into its individual atoms, but then I could go no further.

Or could I? In the late 1800s, scientists discovered that if they constructed a hollow glass tube with an anode in one end and a cathode in the other and pumped out as much of the air as they could, an electrical discharge between the two electrodes could produce a faint glow within the tube. Later, *cathode ray tubes*, as they became to be known, were more sophisticated and contained a phosphorescent coating in one end of the tube. William Crookes demonstrated that the rays were emitted from the cathode and that they traveled in straight lines and could not bend around objects in their path. A while later, Julius Plücker was able to show that a magnet applied to the exterior of the cathode ray tube could change the position of the phosphorescence. Physicists knew that the cathode ray carried a negative charge (in physics, the cathode is the negatively charged electrode and because the beam originated from the cathode, it must therefore be negatively charged). However, they did not know whether the charge and the ray could be separated from one another. In 1897, Joseph J. Thomson finally resolved the issue by demonstrating that both the beam and the charged particles could be bent by an electrical field that was applied perpendicular to the path of the beam, as shown in Figure 1.2. By systematically varying the electric field strength and measuring the angle of deflection, Thomson was able to determine the charge-to-mass (e/m) ratio of the particles, which he called *corpuscles* and which are now known as *electrons*. Thomson measured the e/m ratio as -1.76×10^8 C/g, a value that was at least a thousand times larger than the one expected on the basis of the known atomic weights of even the lightest of atoms, indicating that the negatively charged electrons must be much smaller in size than a typical atom. In other words, the atom was not indivisible, and could itself be broken down into smaller components, with the electron being one of these subatomic particles. As a result of his discovery, Thomson proposed the so-called plum pudding model of the atom, where the atom consisted of one or more of these tiny electrons distributed in a sea of positive charge, like raisins randomly dispersed in a gelatinous pudding. Thomson was later awarded the 1906 Nobel Prize in physics for his discovery of the electron and his work on the electrical conductivity of gases.

In 1909, Robert Millikan and his graduate student Harvey Fletcher determined the charge on the electron using the apparatus shown in Figure 1.3. An atomizer from a perfume bottle was used to spray a special kind of oil droplet having a low vapor pressure into a sealed chamber. At the bottom of the chamber were two parallel circular plates. The upper one of these plates was the anode and it had a hole drilled into the center of it through which the oil droplets could fall under the influence of gravity. The apparatus was equipped with a microscope so that Millikan could observe the rate of fall of the individual droplets. Some of the droplets became charged as a result of friction with the tip of the nozzle, having lost one or more of their electrons to become positively charged cations. When Millikan applied a potential difference between the two plates at the bottom of the apparatus, the positively charged droplets were repelled by the anode and reached an equilibrium

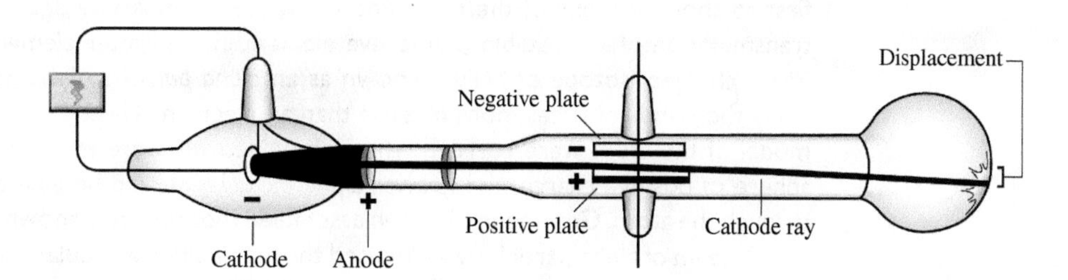

FIGURE 1.2
Schematic diagram of a cathode ray tube similar to the one used in J. J. Thomson's discovery of the electron. [Blatt Communications.]

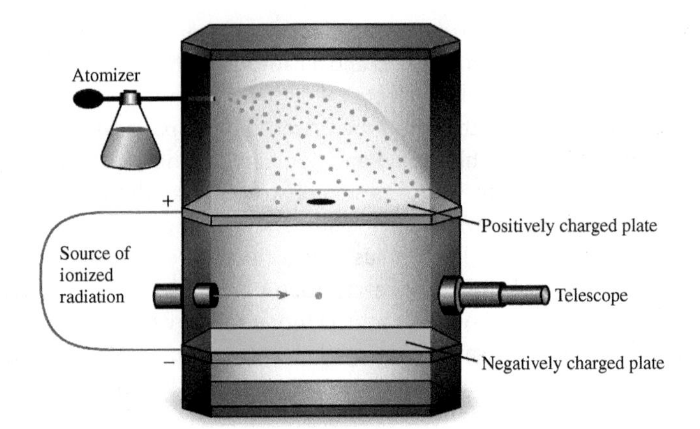

FIGURE 1.3
Schematic diagram of the Millikan oil drop experiment to determine the charge of the electron. [Blatt Communications.]

state where the Coulombic repulsion of like charges and the effect of gravity were exactly balanced, so that appropriately charged particles essentially floated there in space inside the container. By systematically varying the potential difference applied between the two metal plates and counting the number of particles that fell through the opening in a given period of time, Millikan was able to determine that each of the charged particles was some integral multiple of the electronic charge, which he determined to be -1.592×10^{-19} C, a measurement that is fairly close to the modern value for the charge on an electron ($-1.60217733 \times 10^{-19}$ C). Using this new value of e along with Thomson's *e/m* ratio, Millikan was able to determine the mass of a single electron as 9.11×10^{-28} g. The remarkable thing about the mass of the electron was that it was 1837 times smaller than the mass of a single hydrogen atom. Another notable feature of Millikan's work is that it very clearly demonstrated that the electronic charge was quantized as opposed to a continuous value. The differences in the charges on the oil droplets were always some integral multiple of the value of the electronic charge e. Millikan's work was not without controversy, however, as it was later discovered that some of his initial data (and Fletcher's name) were excluded from his 1913 publication. Some modern physicists have viewed this as a potential example of pathological science. Nevertheless, Millikan won the 1923 Nobel Prize in physics for this work.

Also in 1909, one of J. J. Thomson's students, Ernest Rutherford, working with Hans Geiger and a young graduate student by the name of Ernest Marsden, performed his famous "gold foil experiment" in order to test the validity of the plum pudding model of the atom. Rutherford was already quite famous by this time, having won the 1908 Nobel Prize in chemistry for his studies on radioactivity. The fact that certain compounds (particularly those of uranium) underwent spontaneous radioactive decay was discovered by Antoine Henri Becquerel in 1896. Rutherford was the first to show that one of the three known types of radioactive decay involved the transmutation of an unstable radioactive element into a lighter element and a positively charged isotope of helium known as an *alpha particle*. Alpha particles were many thousands of times more massive than an electron. Thus, if the plum pudding model of the atom were correct, where the electrons were evenly dispersed in a sphere of positive charge, the heavier alpha particles should be able to blow right through the atom. Geiger and Marsden assembled the apparatus shown in Figure 1.4.

A beam of alpha particles was focused through a slit in a circular screen that had a phosphorescent coating of ZnS on its interior surface. When an energetic alpha particle struck the phosphorescent screen, it would be observed as a flash of light. In the center of the apparatus was mounted a very thin piece of metal foil (although it is often referred to as the *gold foil experiment*, it was in fact a piece of platinum foil, not gold, which was used). While the majority of alpha particles struck the screen

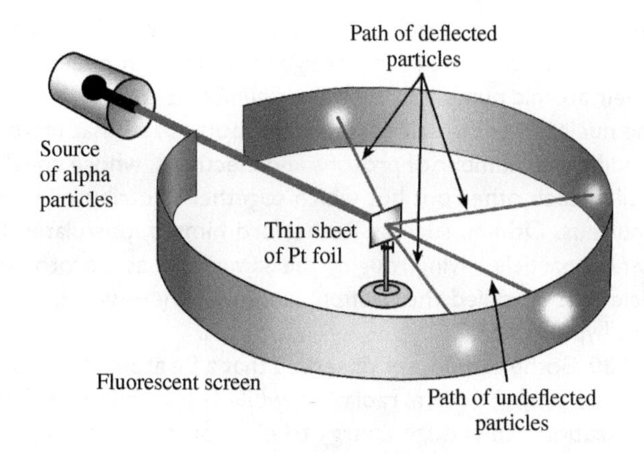

FIGURE 1.4
Schematic diagram of the Geiger–Marsden experiment, also known as *Rutherford's gold foil experiment*. [Blatt Communications.]

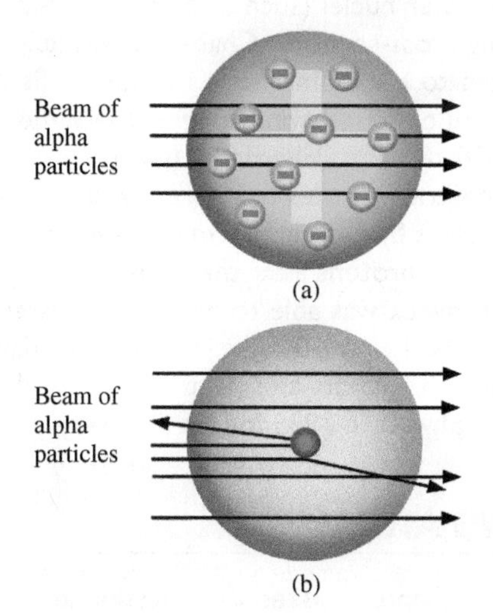

FIGURE 1.5
Atomic view of the gold foil experiment. If the plum pudding model of the atom were accurate, a beam of massive alpha particles would penetrate right through the atom with little or no deflections (a). The observation that some of the alpha particles were deflected backward implied that the positive charge in the atom must be confined to a highly dense region inside the atom known as the *nucleus* (b). [Blatt Communications.]

immediately behind the piece of metal foil as expected, much to the amazement of the researchers, a number of alpha particles were also deflected and scattered at other angles. In fact, some of the particles even deflected backward from the target. In his own words, Rutherford was said to have exclaimed: "It was quite the most incredible event that has ever happened to me in my life. It was almost as incredible as if you fired a 15-inch shell at a piece of tissue paper and it came back and hit you. On consideration, I realized that this scattering backwards must be the result of a single collision, and when I made calculations I saw that it was impossible to get anything of that order of magnitude unless you took a system in which the greater part of the mass of an atom was concentrated in a minute nucleus." Further calculations showed that the diameter of the nucleus was about five orders of magnitude smaller than that of the atom. This led to the rather remarkable conclusion that matter is mostly empty space—with the very lightweight electrons orbiting around an incredibly dense and positively charged nucleus, as shown in Figure 1.5. As a matter of fact, 99.99999999% of the atom is devoid of all matter entirely! On the atomic scale, solidity has no meaning. The reason that a macroscopic object "feels" at all hard to us is because the atom contains a huge amount of repulsive energy, so that whenever we try to "push" on it, there is a whole lot of energy pushing right back.

It wasn't until 1932 that the final piece of the atomic puzzle was put into place. After 4 years as a POW in Germany during World War I, James Chadwick returned to England to work with his former mentor Ernest Rutherford, who had taken over J. J. Thomson's position as Cavendish Professor at Cambridge University. It was not long before Rutherford appointed Chadwick as the assistant director of the nuclear physics lab. In the years immediately following Rutherford's discovery that the nucleus contained *protons*, which existed in the nucleus and whose charges were equal in magnitude to the electronic charge but with the opposite sign, it was widely known that the nuclei of most atoms weighed more than could be explained on the basis of their atomic numbers (the *atomic number* is the same as the number of protons in the nucleus). Some scientists even hypothesized that maybe the nucleus contained an additional number of protons and electrons, whose equal but opposite charges cancelled each other out but which together contributed to the increased mass of the nucleus. Others, such as Rutherford himself, postulated the existence of an entirely new particle having roughly the same mass as a proton but no charge at all, a particle that he called the *neutron*. However, there was no direct evidence supporting this hypothesis.

Around 1930, Bothe and Becker observed that a Be atom bombarded with alpha particles produced a ray of neutral radiation, while Curie and Joliot showed that this new form of radiation had enough energy to eject protons from a piece of paraffin wax. By bombarding heavier nuclei (such as N, O, and Ar) with this radiation and calculating the resulting cross-sections, Chadwick was able to prove that the rays could not be attributed to electromagnetic radiation. His results were, however, consistent with a neutral particle having roughly the same mass as the proton. In his next experiment, Chadwick bombarded a boron atom with alpha particles and allowed the resulting neutral particles to interact with nitrogen. He also measured the velocity of the neutrons by allowing them to interact with hydrogen atoms and measuring the speed of the protons after the collision. Coupling the results of each of his experiments, Chadwick was able to prove the existence of the neutron and to determine its mass to be 1.67×10^{-27} kg. The modern-day values for the charges and masses of the electron, proton, and neutron are listed in Table 1.1. Chadwick won the Nobel Prize in physics in 1935 for his discovery of the neutron.

1.2 VISUALIZING ATOMS

At the beginning of this chapter, I asked the question at what point can we divide matter into such small pieces that it is no longer perceptible to the senses. In a sense, this is a philosophical question and the answer depends on what we mean as being perceptible to the senses. Does it literally mean that we can see the individual components with our naked eye, and for that matter, what are the molecular characteristics of vision that cause an object to be seen or not seen? How many photons of light does it take to excite the rod and cone cells in our eyes and cause them to fire neurons down the optic nerve to the brain? The concept of perceptibility is somewhat vague. Is it fair to say that we still see the object when it is multiplied under an optical microscope? What if an electron microscope is used instead? Today, we have "pictures" of individual atoms, such as those shown in Figure 1.6, made by a scanning tunneling microscope (STM) and we can manipulate individual atoms on

TABLE 1.1 Summary of the properties of subatomic particles.

Particle	Mass (kg)	Mass (amu)	Charge (C)
Electron	$9.10938291 \times 10^{-31}$	0.00054857990946	$-1.602176565 \times 10^{-19}$
Proton	$1.672621777 \times 10^{-27}$	1.007276466812	$1.602176565 \times 10^{-19}$
Neutron	$1.674927351 \times 10^{-27}$	1.00866491600	0

Source: *The NIST Reference on Constants, Units, and Uncertainty* (http://physics.nist.gov, accessed Nov 3, 2013).

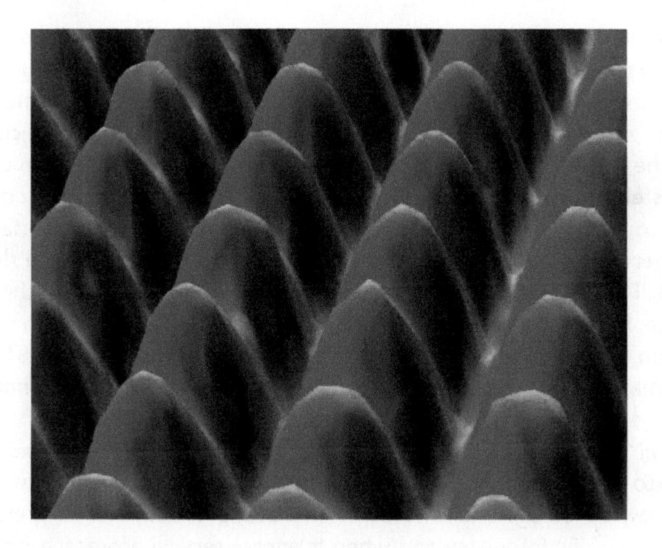

FIGURE 1.6
Scanning tunneling microscopy of the surface of the (110) face of a nickel crystal. [Image originally created by IBM Corporation.]

a surface in order to create new chemical bonds at the molecular level using atomic force microscopy (AFM).

But are we really capable of actually seeing an individual atom? Technically speaking, we cannot see anything smaller than the shortest wavelength of light with which we irradiate it. The shortest wavelength that a human eye can observe is about 400 nm, or 4×10^{-7} m. As the diameter of an atom is on the scale of 10^{-11} m and the diameter of a typical nucleus is even smaller at 10^{-15} m, it is therefore impossible for us to actually see an atom. However, we do have ways of visualizing atoms. A scanning tunneling microscope, like the one shown in Figure 1.7, works by moving an exceptionally sharp piezoelectric tip (often only one atom thick at its point) across the surface of a conductive solid, such as a piece of crystalline nickel in an evacuated

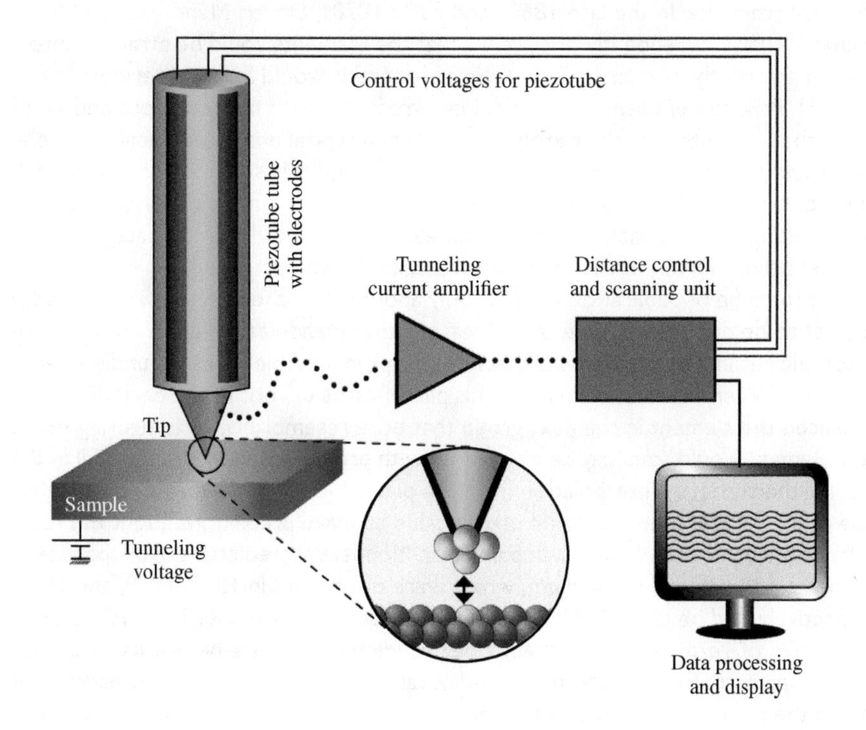

FIGURE 1.7
Schematic diagram of a scanning tunneling microscope (STM). [Blatt Communications.]

chamber. When a small voltage is applied to the tip of the STM, a tunneling current develops whenever the tip is close to the surface of a Ni atom. This tunneling current is proportional to the distance between the tip of the probe and the atoms on the surface of the crystal. By adjusting the STM so that the tunneling current is a constant, the tip will move up and down as it crosses the surface of the crystal and encounters electron density around the nuclei of the nickel atoms. A computer is then used to map out the three-dimensional contour of the nickel surface and to color it different shades of blue in this case, depending on the distance that the tip has moved. The STM can also be used to pick up atoms and to move them around on a surface. In fact, the scientists who invented the STM (Gerd Binnig and Heinrich Rohrer, both of whom shared the 1986 Nobel Prize in physics) used an STM to spell out the name of their sponsoring company IBM by moving around 35 individual Xe atoms affixed to a Ni surface.

The AFM, which has a smaller resolution than the STM, has the advantage of being able to visualize nonconductive surfaces. It functions using a cantilever with a very narrow tip on the end. Instead of interacting directly with the electrons, it vibrates at a specific frequency and when it encounters an atom, the frequency of the vibration changes, allowing one to map out the contour of the surface.

1.3 THE PERIODIC TABLE

While chemistry is the study of matter and its interconversions, inorganic chemistry is that subdiscipline of chemistry which deals with the physical properties and chemistry of all the elements, with the singular exclusion of carbon. An *element* is defined by the number of protons in its nucleus. There are 90 naturally occurring elements (all of the elements up to and including atomic number 92, with the exception of Tc (atomic number 43) and Pm (atomic number 61)). However, if all of the man-made elements are included, a total of 118 elements are currently known to exist. It has long been known that many of the elements had similar valences and chemical reactivity. In the late 1860s and early 1870s, Dmitri Mendeleev and Julius Lothar Meyer independently discovered that the elements could be arranged into a table in an orderly manner such that their properties would follow a periodic law. In his book *Principles of Chemistry*, Mendeleev wrote: "I began to look about and write down the elements with their atomic weights and typical properties, analogous elements and like atomic weights on separate cards, and this soon convinced me that the properties of elements are in periodic dependence upon their atomic weights." His resulting periodic table organized the elements into eight broad categories (or *Gruppe*) according to increasing atomic mass, as shown in Figure 1.8.

At the time of publication in 1871, only about half of the elements known today had yet to be discovered. One of the reasons that Mendeleev's version of the periodic table became so popular was that he left gaps in his table for as yet undiscovered elements. When the next element on his pile of cards did not fit the periodic trend, he placed the element in the next group that bore resemblance to it, figuring that a new element would someday be discovered with properties appropriate to fill in the gap. Furthermore, by interpolation from the properties of those elements on either side of the gaps, Mendeleev could use his table to make predictions about the reactivity of the unknown elements. In particular, Mendeleev predicted the properties of gallium, scandium, and germanium, which were discovered in 1875, 1879, and 1886, respectively, and he did so with incredible accuracy. For example, Table 1.2 lists the properties of germanium that Mendeleev predicted 15 years before its discovery and compares them with the modern-day values. It is this predictive capacity that makes the periodic table one of the most powerful tools in chemistry. Mendeleev's periodic table was organized according to increasing mass. With the discovery of

Reibco	Gruppo I. — R^1O	Gruppo II. — RO	Gruppo III. — R^1O^3	Gruppo IV. RH^4 RO^1	Gruppo V. RH^2 R^1O^5	Gruppo VI. RH^2 RO^3	Gruppo VII. RH R^2O^7	Gruppo VIII. — RO^4
1	II=1							
2	Li=7	Bo=9.4	B=11	C=12	N=14	O=16	F=19	
3	Na=23	Mg=24	Al=27,8	Si=28	P=31	S=32	Cl=35,5	
4	K=39	Ca=40	—=44	Ti=48	V=51	Cr=52	Mn=55	Fo=56, Co=59, Ni=59, Cu=63.
5	(Cu=63)	Zn=65	—=68	—=72	As=75	So=78	Br=80	
6	Rb=85	Sr=87	?Yt=88	Zr=90	Nb=94	Mo=96	—=100	Ru=104, Rh=104, Pd=106, Ag=108.
7	(Ag=108)	Cd=112	In=113	Sn=118	Sb=122	To=125	J=127	
8	Cs=183	Ba=187	?Di=188	?Co=140	—	—	—	— — — —
9	(—)	—	—	—	—	—	—	
10	—	—	?Er=178	?La=180	Ta=182	W=184	—	Os=195, Ir=197, Pt=198, Au=199.
11	(Ag=199)	Hg=200	Tl=204	Pb=207	Bi=208	—	—	
12	—	—	—	Th=231	—	U=240	—	— — — —

FIGURE 1.8
Dmitri Mendeleev's periodic table (1871).

TABLE 1.2 Properties of the element germanium (eka-silicon) as predicted by Mendeleev in 1871 and the experimental values measured after its discovery in 1886.

Physical and Chemical Properties	Predicted	Actual
Atomic mass (amu)	72	72.3
Density (g/cm^3)	5.5	5.47
Specific heat (J/g °C)	0.31	0.32
Atomic volume (cm^3/mol)	13	13.5
Formula of oxide	RO_2	GeO_2
Oxide density (g/cm^3)	4.7	4.70
Formula of chloride	RCl_4	$GeCl_4$
Boiling point of chloride (°C)	<100	86
Density of chloride (g/cm^3)	1.9	1.84

the nucleus in the early 1900s, the modern form of the periodic table is instead organized according to increasing atomic number. Furthermore, as we shall see in a later chapter, the different blocks of groups in the periodic table quite naturally reflect the quantum nature of atomic structure.

1.4 THE STANDARD MODEL

As an atom is the smallest particle of an element that retains the essential chemical properties of that substance, one might argue that atoms are the fundamental building blocks of matter. However, as we have already seen, the atom itself is not indivisible, as Democritus believed. As early as the 1930s, it was recognized that there were other fundamental particles of matter besides the proton, the neutron, and the electron. The muon was discovered by Carl Anderson and Seth Nedermeyer in 1936. Anderson was studying some of the properties of cosmic radiation when he noticed a new type of negatively charged particle that was deflected by a magnetic field to a lesser extent than was the electron. The *muon* has the same charge as the electron, but it has a mass that is about 200 times larger, which explains why it was not deflected as much as an electron. Muons are not very stable particles, however; they have a mean lifetime of only 2.197×10^{-6} s. Muons occur when cosmic radiation interacts with matter and are also generated in large quantities in modern-day particle accelerators. As it turns out, however, the muon represents just one strange beast in a whole zoo of subatomic particles that include hadrons, baryons, neutrinos, mesons, pions, quarks, and gluons—to name just a few, begging the question of just how divisible is matter and what (if anything) is fundamental?

The *standard model* of particle physics was developed in the 1970s following experimental verification of quarks. The standard model incorporates the theory of general relativity and quantum mechanics in its formulation. According to the standard model, there are a total of 61 elementary particles, but ordinary matter is composed of only six types (or flavors) of leptons and six types of quarks. Leptons and quarks are themselves examples of fermions, or particles that have a spin quantum number of ½ and obey the Pauli exclusion principle. It is the various combinations of these fundamental particles that make up all of the larger particles, such as protons and neutrons. Thus, for example, a proton is composed of two up quarks and one down quark (pronounced in such a way that it rhymes with the word "cork"). Electrons, muons, and neutrinos are all examples of leptons. Both leptons and quarks can be further categorized into one of three different generations, as shown in Figure 1.9. First-generation particles, such as the electron and the up and down quarks that make up protons and neutrons, are stable, whereas second- and third-generation particles exist for only brief periods of time following their generation. Furthermore, each of the 12 fundamental particles has a corresponding antiparticle. An *antiparticle* has the same mass as a fundamental particle, but exactly the opposite electrical charge. The antiparticle of the electron, for instance, is the positron, which has a mass of roughly 9.109×10^{-31} kg like the electron, but an electrical charge of $+1.602 \times 10^{-19}$ C or +1e. Whenever a particle and its antiparticle collide, they annihilate each other and create energy. In addition to the 12 fundamental particles and their antiparticles, there are also force-carrying particles, such as

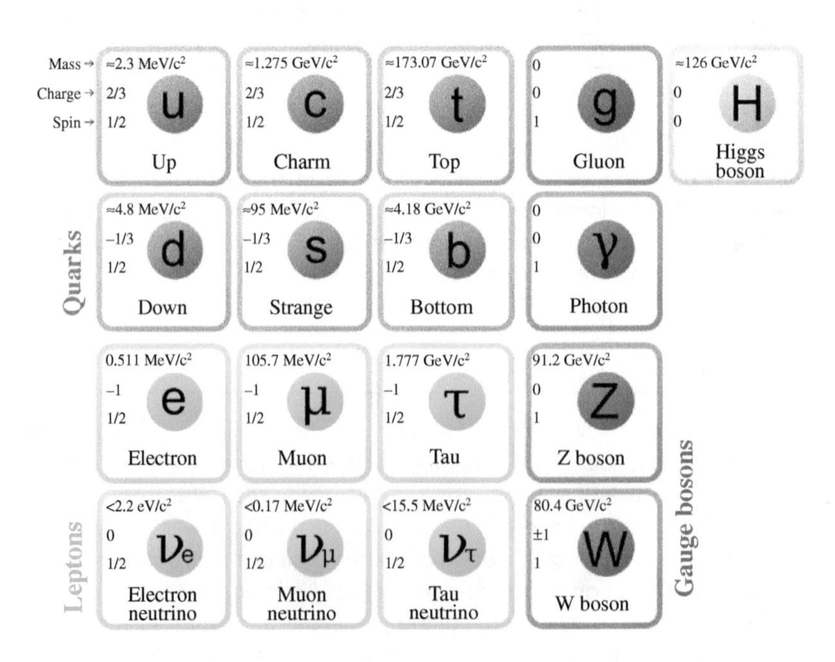

FIGURE 1.9
The 12 fundamental particles (leptons in green and quarks in purple) and the force-carrying particles (in red) that comprise the standard model of particle physics. The newly discovered Higgs boson, which explains why some particles have mass, is shown at the upper right. [Attributed to MissMJ under the Creative Commons Attribution 3.0 Unported license (accessed October 17, 2013).]

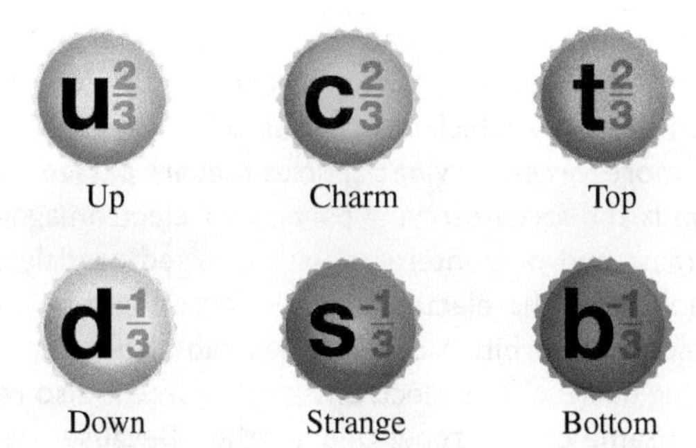

FIGURE 1.10
Cartoon representation of the six different flavors of quarks (arranged into pairs by their generations). The numbers inside each quark represent their respective charges. [Blatt Communications.]

the photon, which carries the electromagnetic force. Collectively, the 12 fundamental particles of matter are known as *fermions* because they all have a spin of ½, while the force-carrying particles are called *bosons* and have integral spin. The different types of particles in the standard model are illustrated in Figure 1.9.

There are four types of fundamental forces in the universe, arranged here in order of increasing relative strength: (i) gravity, which affects anything with mass; (ii) the weak force, which affects all particles; (iii) electromagnetism, which affects anything with charge; and (iv) the strong force, which only affects quarks. There are six *quarks*, as shown in Figure 1.10, and they are arranged as pairs of particles into three generations. The first quark in each pair has a spin of +2/3, while the second one has a spin of −1/3.

Quarks also carry what is known as *color charge*, which is what causes them to interact with the strong force. Color charges can be represented as red, blue, or green, by analogy with the RGB additive color model, although this is really just a nonmathematical way of representing their quantum states. Like colors tend to repel one another and opposite colors attract. Because of a phenomenon known as *color confinement*, an individual quark has never been directly observed because quarks are always bound together by gluons to form *hadrons*, or combinations of quarks. *Baryons* consist of a triplet of quarks, as shown in Figure 1.11. Protons and neutrons are examples of baryons that form the basic building blocks of the nucleus. *Mesons*, such as the kaon and pion, are composed of a pair of particles: a quark and an antiquark.

Unlike quarks, which always appear together in composite particles, the leptons are solitary creatures and prefer to exist on their own. Furthermore, the leptons do not carry color charge and they are not influenced by the strong force. The electron, muon, and tau are all negatively charged particles (with a charge of -1.602×10^{-19} C), differing only in their masses. Neutrinos, on the other hand, have no charge and are particularly difficult to detect. The electron neutrino has an extremely small mass and can pass through ordinary matter. The heavier leptons (the muon and the tau) are not found in ordinary matter because they decay very quickly into lighter leptons, whereas electrons and the three kinds of neutrinos are stable.

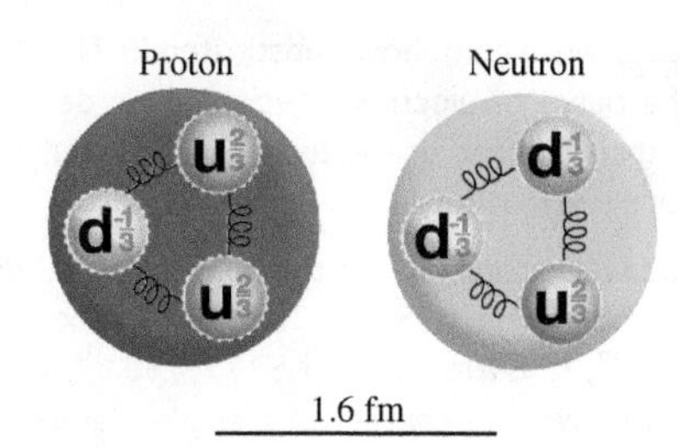

FIGURE 1.11
Representation of a proton, which is made from two up and one down quarks, and a neutron, which is made from one up and two down quarks. The diameter of the proton and neutron are roughly drawn to scale; however, the quarks are about 1000 times smaller than a proton or a neutron. [Blatt Communications.]

Well, now that we know what matter is made of, we might ask ourselves the question of what it is that holds it together. Each of the four fundamental forces (with the exception of gravity, which has not yet fully been explained by the standard model) has one or more force-carrying particles that are passed between particles of matter. The *photon* is the force-carrying particle of electromagnetic radiation. The photon has zero mass and only interacts with charged particles, such as protons, electrons, and muons. It is the electromagnetic force that holds atoms together in molecules—the electrons orbiting one nucleus can also be attracted to the protons in a neighboring nucleus. The electromagnetic force is also responsible for why particles having the same charge repel one another. Because they are all positively charged, one might wonder how it is that more than one proton can exist within the very small confines of the nucleus. The explanation for this conundrum is that protons are made up of quarks. The quarks are held together in triplets in the proton by the strong force because they have color charge. Likewise, it is the residual strong force, where a quark on one proton is attracted to a quark on another proton or neutron, which holds the protons and neutrons together inside the nucleus. The force-carrying particle for the strong force is the *gluon*. Quarks absorb and emit gluons very rapidly within a hadron, and so it is impossible to isolate an individual quark. The weak force is responsible for an unstable heavier quark or lepton disintegrating into two or more lighter quarks or leptons. The *weak force* is carried by three different force-carrying particles: the W^+, W^-, and Z bosons. The W^+ and W^- particles are charged, whereas the Z particle is neutral. The standard model also predicts the presence of the Higgs boson, popularly known as the *god particle*, which is responsible for explaining why the fundamental particles have mass. Recently, scientists working at the LHC (Large Hadron Collider) particle accelerator have finally discovered evidence suggesting the existence of the elusive Higgs boson. In fact, Peter Higgs, after which the Higgs boson was named, shared the 2013 Nobel Prize in physics for his contributions in the area of theoretical particle physics. The particles that comprise the standard model of particle physics are to date the most fundamental building blocks of matter. Despite its incredible successes, the standard model has yet to accurately describe the behavior of gravity or why there are more particles in the universe than antiparticles and why the universe contains so much dark matter and dark energy. Physicists continue to search for a grand unified theory of everything, and one is therefore left to wonder whether anything at all is truly fundamental. In the following chapter, we examine some further properties of the nucleus and show how matter and energy themselves can be interconverted.

EXERCISES

1.1. In Thomson's cathode ray tube experiment, the electron beam will not be deflected unless an external electric or magnetic field has been applied. What does this result imply about the force of gravity on the electrons (and hence about the mass of an electron)?

1.2. If a beam of protons were somehow substituted in Thomson's cathode ray tube experiment instead of a beam of electrons, would their deflection by an electrical field be larger or smaller than that for an electron? Explain your answer. What would happen if a beam of neutrons were used?

1.3. The following data were obtained for the charges on oil droplets in a replication of the Millikan oil drop experiment: 1.5547×10^{-19}, 4.6192×10^{-19}, 3.1417×10^{-19}, 3.0817×10^{-19}, 1.5723×10^{-19}, 1.5646×10^{-19}, 1.5420×10^{-19}, and 1.5547×10^{-19} C. Use these data to calculate the average charge on a single electron. Explain how you arrived at your result.

1.4. An alpha particle is the same as a helium-4 nucleus: it contains two protons and two neutrons in the nucleus. Given that the radius of an alpha particle is approximately 2.6 fm, calculate the density of an alpha particle in units of grams per cubic centimeter.

1.5. Given that the mass of an average linebacker at Ursinus College is 250 lbs and the radius of a pea is 0.50 cm, calculate the number of linebackers that would be required to be stuffed into the volume of a pea in order to obtain the same density as an alpha particle.

1.6. Given that the radius of the helium-4 nucleus is approximately 2.6 fm, the classical electron radius is 2.8 fm, and the calculated atomic radius of ^{4}He is 31 pm, calculate the percentage of the space in a helium-4 atom that is actually occupied by the particles.

1.7. Explain the similarities and differences between scanning tunneling microscopy and atomic force microscopy.

1.8. At the time when Mendeleev formulated the periodic table in 1871, the element gallium had yet to be discovered, and Mendeleev simply left a gap in his periodic table for it. By interpolating data from the elements that surround gallium in the periodic table, predict the following information about gallium and then compare your predictions to the actual values: its atomic mass, its density, its specific heat, its atomic volume, its melting point, the molecular formula for its oxide, the density of its oxide, the molecular formula for its chloride, and the density of its chloride.

1.9. Which of the following particles will interact with an electromagnetic field? (a) An electron, (b) an up quark, (c) an electron neutrino, (d) a proton, (e) a positron, (f) a muon, (g) a pion.

1.10. Explain why it is that electrons traveling in the same region of space will always repel one another, but protons can exist in close proximity with each other in the interior of the nucleus.

BIBLIOGRAPHY

1. Atkins P, Jones L, Laverman L. *Chemical Principles: The Quest For Insight*. 6th ed. New York: W. H. Freeman and Company; 2013.
2. McMurry J, Fay RC. *Chemistry*. 4th ed. Upper Saddle River, NJ: Pearson Education, Inc; 2004.
3. Nave, C. R. *HyperPhysics*. http://hyperphysics.phy-astr.gsu.edu/hbase/hframe.html (accessed Oct 10, 2013).
4. Schaffner, P. and the Particle Data Group at Lawrence Berkeley National Laboratory, *The Particle Adventure: The Fundamentals of Matter and Force*. http://www.particleadventure.org/index.html (accessed July 2, 2012).
5. Segrè E. *From X-Rays to Quarks: Modern Physicists and Their Discoveries*. New York: W. H. Freeman and Company; 1980.

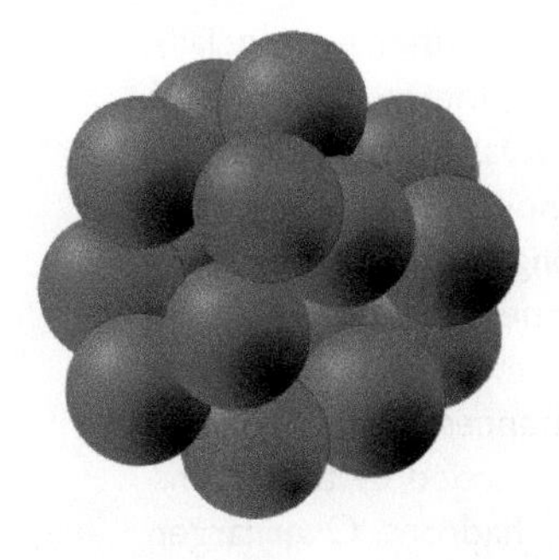

The Structure of the Nucleus | 2

"If, as I have reason to believe, I have disintegrated the nucleus of the atom, this is of greater significance than the war."

—*Ernest Rutherford*

2.1 THE NUCLEUS

The defining characteristic of any element is given by the composition of its nucleus. The nucleus of an atom is composed of the *nucleons* (protons and neutrons), such that an element is given the symbol ${}^{A}_{Z}X$, where Z is the atomic number (or number of protons), A is the mass number (also known as the *nucleon number*), and X is the one- or two-letter abbreviation for the element. A *nuclide* is defined as a nucleus having a specific mass number A. Most elements exist as multiple *isotopes*, which differ only in the number of neutrons present in the nucleus. It is important to recognize that while the different isotopes of an element have many of the same chemical properties (e.g., react with other elements to form the same stoichiometry of compounds), they often have very different physical properties. Thus, for example, while cobalt-59 (${}^{59}Co$) is a stable isotope and is considered one of the elements essential to human life, its slightly heavier isotope cobalt-60 (${}^{60}Co$) is highly unstable and releases the destructive gamma rays used in cancer radiation therapy. Further, while "heavy water" or deuterium oxide (D_2O or ${}^{2}H_2O$) is not radioactive, the larger atomic mass of the deuterium isotope significantly increases the strength of a hydrogen bond to oxygen, which slows the rates of many important biochemical reactions and can (in sufficient quantities) lead to death.

The nucleus of an atom is restricted to a very small radius (typically on the order of 10^{-14}–10^{-15} m). As the majority of an atom's mass is located in a highly confined space, the density of a nucleus is exceptionally large (approximately 10^{14} g/cm^3). In fact, it was the presence of a very dense nucleus in the Geiger–Marsden experiment that led to the unexpected observation that some of the alpha particles were deflected backward toward the source instead of passing directly through the thin foil. At first glance, this result should be surprising to you, given that the protons in

The nucleus. [Attributed to Marekich, reproduced from http://en.wikipedia.org/wiki/Atomic_nucleus (accessed October 17, 2013).]

Principles of Inorganic Chemistry, First Edition. Brian W. Pfennig.

a nucleus are positively charged and should therefore repel one another—especially at short distances. It was not until the 1970s when the strong interaction, one of the four fundamental forces of nature that comprise the standard model of particle physics, was discovered. The *strong force* is, as its name implies, the strongest of these fundamental forces. It is approximately 10^2 times stronger than the electromagnetic force, which is what causes the protons to repel one another, 10^6 times stronger than the weak force, and 10^{39} times more powerful than the gravitational force. However, the strong force acts only over very short distances, typically on the order of 10^{-15} m. The strong interaction is the force that is carried by the gluons and holds quarks having unlike color charges together to form hadrons. Over larger distances, it is the residual strong force that is responsible for holding the protons and neutrons together in the nucleus of an atom.

2.2 NUCLEAR BINDING ENERGIES

The *nuclear binding energy* is a measure of how strongly the nucleons are held together in the nucleus by the strong force. In one sense, it is analogous to the bond dissociation energy, which measures how strongly atoms are held together in a molecule. The nuclear binding energy (ΔE) can be calculated from Equation (2.1), where Δm is the mass defect and c is the speed of light in vacuum (2.99792458×10^8 m/s):

$$\Delta E = (\Delta m)c^2 \tag{2.1}$$

According to Einstein's theory of relativity, matter and energy are interchangeable. It is for this reason that the masses of subatomic particles are often listed with energy units of MeV/c^2, as shown in Table 2.1. Solving Equation (2.1) for c^2 using appropriate units, one obtains the useful equality that $c^2 = 931.494$ MeV/amu. Because it always takes energy to split a nucleus apart into its isolated nucleon components, the mass of an atom or a nuclide is always less than the sum of its parts. The *mass defect* of the particle is therefore defined as the difference in mass between all the subatomic particles that comprise the atom or nuclide and the mass of the isotope itself.

Example 2-1. Calculate the nuclear binding energy of an alpha particle if its mass is 4.00151 amu.

Solution. An alpha particle is a helium-4 nucleus. The sum of the masses of two neutrons (2 * 1.008665 amu) and two protons (2 * 1.007276 amu) is 4.03188 amu. The mass defect is therefore 4.03188 – 4.00151 = 0.03037 amu. Given that 1 amu = 1.6605×10^{-27} kg and the speed of light in a vacuum is 2.9979×10^8 m/s:

$$\Delta E = (\Delta m)c^2 = (0.03037\,\text{amu}) \left(\frac{1.6605 \times 10^{-27}\,\text{kg}}{1\,\text{amu}} \right) (2.9979 \times 10^8\,\text{m/s})^2$$

$$\Delta E = 4.532 \times 10^{-12}\,\text{J}(6.022 \times 10^{23}\,\text{mol}^{-1}) = 2.729 \times 10^{12}\,\text{J/mol}$$

As 1 eV = 96485 J/mol, $E = 2.829 \times 10^7$ eV or 28.29 MeV. It is more useful, however, to compare the binding energy of one nucleus with that of another in terms of MeV/nucleon. Therefore, the binding energy of an alpha particle is 28.29 MeV/4 nucleons or 7.072 MeV/nucleon. Alternatively, the nuclear binding energy can be directly calculated in MeV using the values in the right-most column of Table 2.1 as follows:

$$\Delta E = (2 * 938.272 + 2 * 939.565) - (4.00151 * 931.494) = 28.29\,\text{MeV}$$

$$\Delta E = 28.29\ \text{MeV}/4 = 7.073\ \text{MeV/nucleon}$$

TABLE 2.1 The masses of subatomic particles in different units.

Particle	Mass (kg)	Mass (amu)	Mass (MeV/c^2)
Proton, m_p	1.67262×10^{-27}	1.00728	938.272
Neutron, m_n	1.67493×10^{-27}	1.00867	939.565
Electron, m_e	9.10938×10^{-31}	5.48580×10^{-4}	0.510999
Atomic mass unit, u	1.66054×10^{-27}	1	931.494

Example 2-2. Calculate the nuclear binding energy of a carbon-12 atom.

Solution. Carbon-12 consists of six protons, six neutrons, and six electrons and weighs exactly 12.0000 amu.

$$\Delta E = (6 * 938.272 + 6 * 939.565 + 6 * 0.510999) - (12.0000 * 931.494)$$
$$= 92.16\ \text{MeV}$$
$$\Delta E = 92.16\ \text{MeV}/12 = 7.680\ \text{MeV/nucleon}$$

In nuclear chemistry, the entropy is usually zero (except in the interiors of stars) and therefore the nuclear binding energy can be used as a measure of the stability of a particular nucleus. Because each isotope of an atom has a different nuclear binding energy, some isotopes will be more stable than others. Figure 2.1 shows the nuclear binding energy curve (per nucleon) as a function of the mass number. In general, elements having mass numbers around 60 have the largest binding energies per nucleon. Isotopes having these mass numbers belong to Fe and Ni, which explains the prevalence of these elements in planetary cores. The maximum in the nuclear binding energy curve occurs for ^{56}Fe, which helps to justify its overall cosmic abundance. Iron is believed to be the 10th most prevalent element in the universe, as shown in Figure 2.2, and it is the 4th most abundant in the earth's crust.

2.3 NUCLEAR REACTIONS: FUSION AND FISSION

The transmutation of the elements has long been the goal of the alchemists. In 1917, Ernest Rutherford was the first person to realize that dream. Rutherford converted nitrogen-14 into oxygen-17 and a proton by bombarding a sample of ^{14}N with a stream of alpha particles, according to the nuclear reaction shown in Equation (2.2):

$$^{14}_{7}N + ^{4}_{2}He \rightarrow ^{17}_{8}O + ^{1}_{1}H \tag{2.2}$$

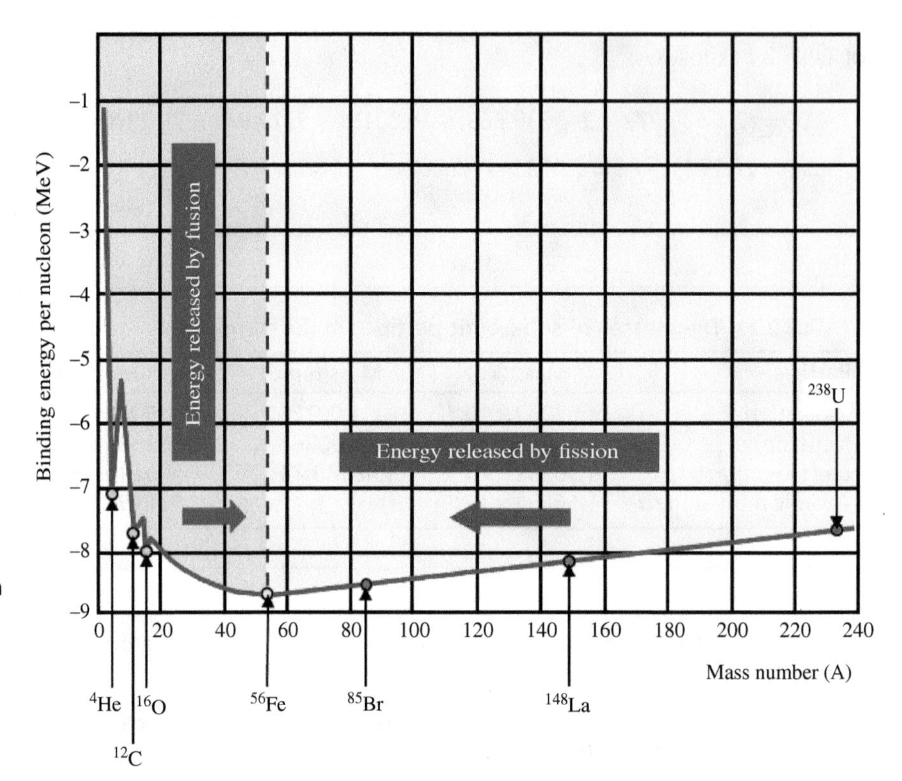

FIGURE 2.1
Nuclear binding energy curve plotting the average binding energy per nucleon as a function of the mass number *A*. The maximum binding energy occurs for the "iron group" of isotopes having mass numbers between 56 and 60. [© Keith Gibbs, www.schoolphysics.co.uk.]

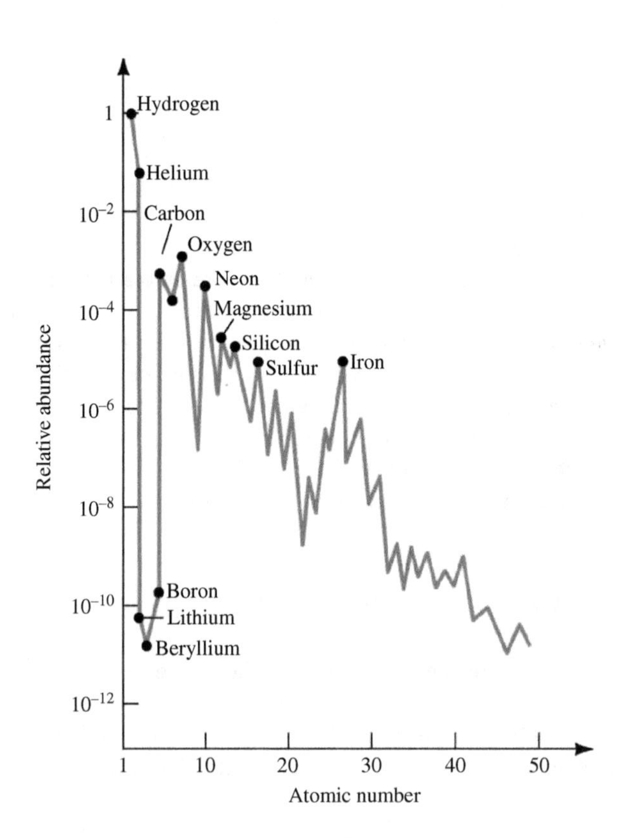

FIGURE 2.2
Relative cosmic abundances of the elements, as compared to that of hydrogen. [Reproduced by permission from *Astronomy Today*, Chaisson and McMillan, 8th ed., Pearson, 2014.]

Whenever writing a nuclear equation, the sum of the atomic numbers of the reactants must equal the sum of the atomic numbers of the products and the sums of the mass numbers on each side of the nuclear equation must also be equal. This does not, however, imply that conservation of mass must apply. Because each isotope has a unique nuclear binding energy, some mass may be lost or gained in the form of energy during a nuclear reaction. The energetics of nuclear reactions are measured in terms of Q, which can be calculated from Equation (2.3), where the masses of the individual nuclides are recorded in MeV/c^2, as in Table 2.1. If the sign of Q for a nuclear equation is positive, the reaction is said to be exothermic. By contrast, if the sign of Q is negative, the nuclear reaction is endothermic and it will require kinetic energy in order to proceed:

$$Q = -\Delta H = \sum_{\text{reactants}} \text{masses} - \sum_{\text{products}} \text{masses} \tag{2.3}$$

Example 2-3. Given that the masses of the isotopes in Equation (2.2) are 14.00307, 4.00151, 16.99913, and 1.00728 amu for ^{14}N, an alpha particle, ^{17}O, and a proton, respectively, calculate Q for the nuclear reaction given by Equation (2.2). Is the reaction endothermic or exothermic?

Solution

$$Q = [(14.00307 + 4.00151) - (16.99913 + 1.00728)\text{amu}] * 931.494 \text{ MeV/amu}$$
$$= -1.7046 \text{ MeV}$$

Because Q is negative, the reaction is endothermic.

Nuclear fusion occurs when two or more small nuclei are joined together to form a larger nucleus. Typically, when two smaller nuclei fuse together, a tremendous amount of energy is released—many orders of magnitude larger than the energy released in an ordinary chemical reaction. Thus, for example, because the average nuclear binding energy per nucleon (Figure 2.1) is much larger for He than it is for H, a self-sustaining nuclear fusion reactor would be a fantastic source of energy. One such example of a typical nuclear reaction occurring in a fusion reactor is given by Equation (2.4) and is illustrated by the diagram shown in Figure 2.3:

$$^{2}_{1}H + ^{3}_{1}H \rightarrow ^{4}_{2}He + ^{1}_{0}n \tag{2.4}$$

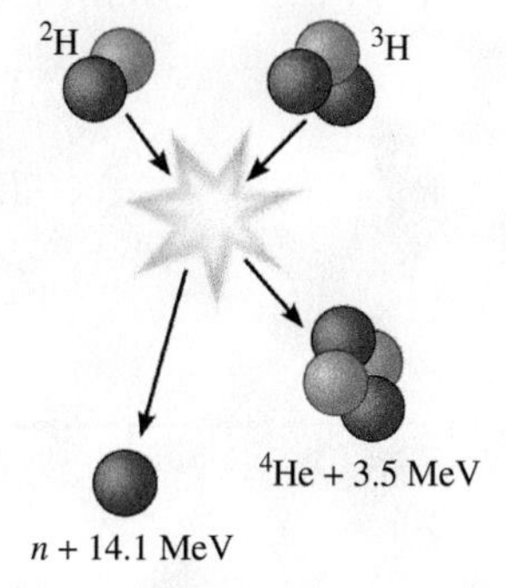

FIGURE 2.3
Illustration of the nuclear fusion reaction given by Equation (2.4). [Reproduced from http://en.wikipedia.org/wiki/Nuclear_fusion (accessed October 17, 2013).]

Example 2-4. Given that the masses of deuterium, tritium, helium-4, and a neutron are 2.01410, 3.01605, 4.00260, and 1.00867 amu, respectively, prove that the total energy released by the fusion reaction illustrated in Figure 2.3 and given by Equation (2.4) is 17.6 MeV.

Solution

$$Q = [(2.01410 + 3.01605) - (4.00260 + 1.00867)\text{amu}] * 931.494\ \text{MeV/amu}$$
$$= 17.587\ \text{MeV}$$

Nuclear fission occurs when a heavier nucleus splits apart to form lighter (or daughter) nuclei. Fission processes can also release tremendous amounts of energy, as illustrated by the use of atomic weapons. In his famous letter to President Franklin D. Roosevelt in August 1939, Albert Einstein, acting on the request of Leo Szilard, informed the president of the possibility that scientists in Nazi Germany were working on a powerful new weapon based on nuclear fission reactions. Shortly thereafter, incredible financial and R&D resources were poured into the super-secret Manhattan Project in an effort to produce a viable nuclear weapon. As a result of these efforts, the first atomic bomb, known simply as *The Gadget*, was detonated near the desert town of Alamogordo, NM, on July 16, 1945 (Figure 2.4). Only several weeks later, the first atomic bombs used in combat were dropped on the Japanese cities of Hiroshima and Nagasaki on August 6 and 9, respectively. These weapons were credited with ending World War II and saving the lives of the many American soldiers, which would have been required for a ground invasion. The basic fission reaction used in the first nuclear weapons is shown by Equation (2.5):

$$^{235}_{92}\text{U} + ^{1}_{0}\text{n} \rightarrow ^{140}_{56}\text{Ba} + ^{93}_{36}\text{Kr} + 3^{1}_{0}\text{n} \tag{2.5}$$

FIGURE 2.4
The Trinity test of the first atomic bomb in the desert near Alamogordo, NM. [Photo credit: US Department of Energy.]

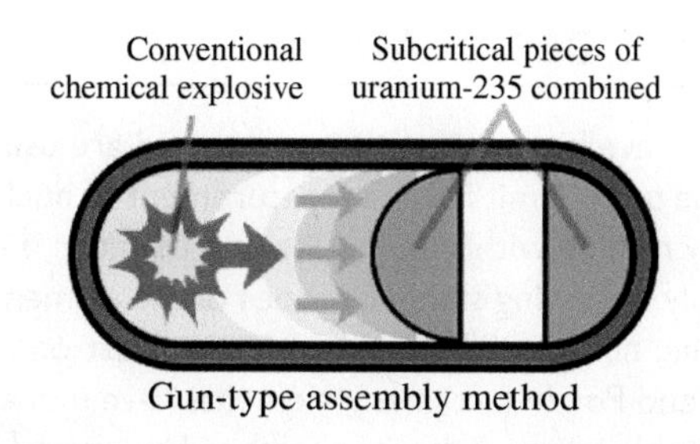

FIGURE 2.5
Illustration of the way two subcritical pieces of ^{235}U are combined in a nuclear weapon to initiate the self-sustaining fission reaction shown by Equation (2.5). [Reproduced from http://en.wikipedia.org/wiki/Fission_bomb#Fission_weapons (accessed November 30, 2013).]

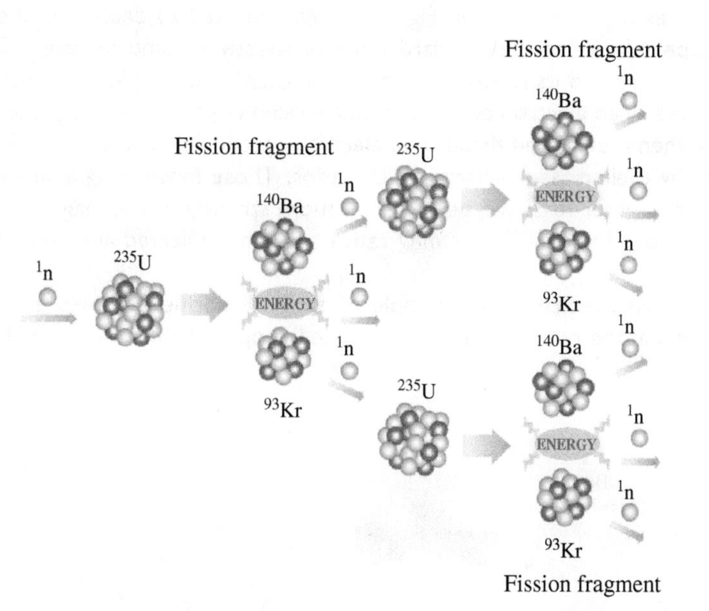

FIGURE 2.6
Self-sustaining chain reaction from the fission of ^{235}U once a supercritical mass of uranium has been assembled. [Blatt Communications.]

In order for the reaction to be self-sustaining, a supercritical mass of at least 3% (enriched) ^{235}U must be assembled using a conventional explosive, as shown in Figure 2.5. At this high of a concentration, the neutrons produced by the fission of the uranium-235 isotope have a large enough cross section and sufficient kinetic energy to initiate the fission of a neighboring ^{235}U nucleus, leading to a chain reaction, as shown in Figure 2.6. The earliest atomic bombs had a total energy equivalent to 18 kton of TNT. Modern hydrogen bombs typically have a plutonium core and use the energy generated from the initial fission reaction to initiate a fusion reaction of hydrogen nuclei, further enhancing the destructive output. As a result, modern atomic weapons have a frighteningly large destructive capacity of approximately 1.2 Mton of energy.

Example 2-5. Given the following masses, calculate the energy released by the fission reaction illustrated by Equation (3.5): ^{235}U (235.0439 amu), ^{140}Ba (139.9106 amu), ^{93}Kr (92.9313 amu), and a neutron (1.00867 amu).

Solution

$$Q = [(235.0439 + 1.00867) - (139.9106 + 92.9313 + 3 * 1.00867)\text{amu}]$$
$$* 931.494\ \text{MeV/amu} = 172.0\ \text{MeV}$$

2.4 RADIOACTIVE DECAY AND THE BAND OF STABILITY

Most fission reactions have high activation barriers and are usually very slow. Using today's technology, the upper limit for the measurement of nuclear lifetimes is about 10^{20} years, so that any nuclide with a longer lifetime than this is considered as stable. There are 266 naturally occurring stable isotopes of the elements. Every element up to and including atomic number 83 (bismuth) has at least one stable isotope, with the exceptions of Tc and Pm. In fact, many elements have more than one stable isotope. The masses of stable isotopes can be measured by several different techniques, but they are most commonly measured using mass spectrometry. A *mass spectrometer*, such as the one shown in Figure 2.7, can be used to determine the mass of an isotope relative to the standard value of exactly 12 amu for the ^{12}C isotope and can also record its relative abundance. A volatilized sample of the substance is bombarded by an electron beam to form a stream of positively charged ions. These ions are then accelerated through an electromagnetic field and separated from one another by their mass-to-charge (m/Q) ratios. Those ions having a larger positive charge or a lighter mass will be deflected more strongly by the magnetic field. The separated ions having different m/Q ratios are then collected and counted by the detector.

The mass spectrum for a sample of atomic chlorine is shown in Figure 2.8, indicating that the two most abundant stable isotopes of chlorine are ^{35}Cl and ^{37}Cl.

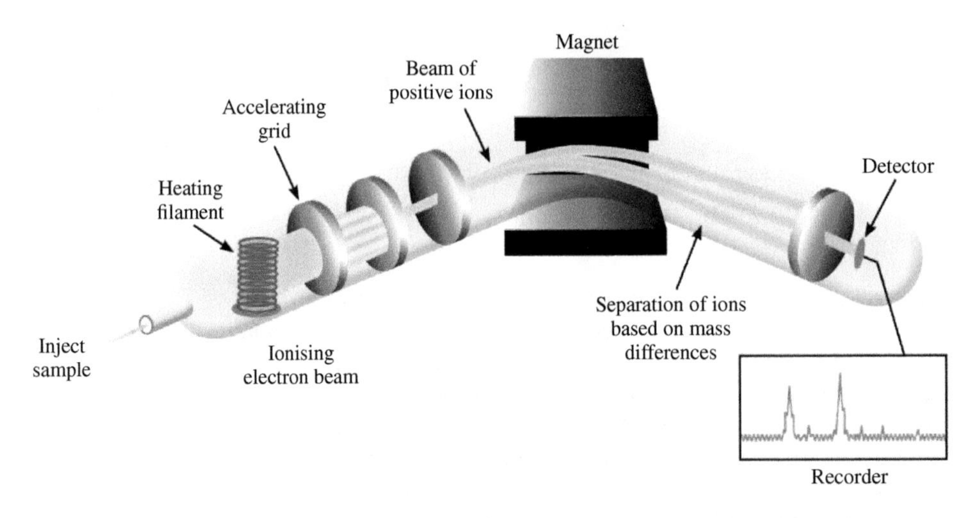

FIGURE 2.7
Schematic diagram of a mass spectrometer. [Blatt Communications.]

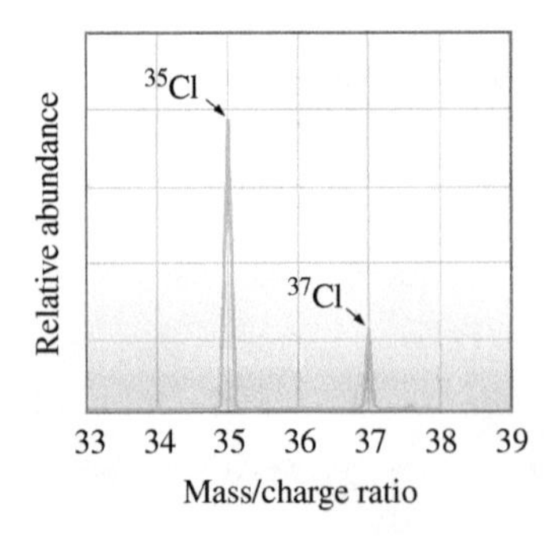

FIGURE 2.8
Mass spectrum for the two isotopes of chlorine. [Blatt Communications.]

The atomic mass of an element as listed in the periodic table is a weighted average of all the known isotopes having the same atomic number, as shown by Equation (2.6), where f_i is the decimal percentage of the naturally occurring abundance of an isotope and m_i is its atomic mass and the sum is for all the known isotopes having atomic number Z:

$$\begin{bmatrix}\text{atomic mass on}\\ \text{periodic table}\end{bmatrix} = \sum_i f_i m_i \tag{2.6}$$

The masses and natural abundances of every known isotope have been tabulated by the National Nuclear Data Center in a booklet known as the *Nuclear Wallet Cards*. An electronic version of the nuclear wallet cards can be found at http://www.nndc.bnl.gov/wallet/wccurrent.html. The mass excess ($\Delta = M - A$) in this table is given in units of megaelectronvolts and needs to be converted into amu for all practical calculations (recall that 1 amu = 931.494 MeV).

Example 2-6. Given the mass spectrum for the sample of chlorine shown in Figure 2.8 and the data in the *Nuclear Wallet Cards*, calculate the weighted average atomic mass of a chlorine atom.

Solution. The two stable isotopes of chlorine have the following natural abundances and masses:

^{35}Cl	75.76%	−29.013 MeV	34.969 amu*
^{37}Cl	24.24%	−31.761 MeV	36.966 amu

*Calculated as follows: $M = \Delta + A = -29.013\text{ MeV}\left(\frac{1\text{ amu}}{931.494\text{ MeV}}\right) + 35\text{ amu} = 34.969\text{ amu}$

Therefore, the weighted average calculated using Equation (2.6) is

$$\text{atomic mass} = 0.7576(34.969\text{ amu}) + 0.2424(36.966\text{ amu}) = 35.453\text{ amu}$$

Example 2-7. Silicon exists as three stable isotopes ($^{28}Si = 27.977$ amu, $^{29}Si = 28.976$ amu, and $^{30}Si = 29.974$ amu). Given that the atomic mass of Si on the periodic table is 28.086 amu and the natural abundance of ^{28}Si is 92.23%, calculate the natural abundances of ^{29}Si and ^{30}Si, respectively.

Solution. The total decimal percentage of each isotope must equal unity. Letting x be the decimal percentage of ^{29}Si and y be the decimal percentage of ^{30}Si, then $0.9223 + x + y = 1$ or $y = 1 - 0.9223 - x = 0.0777 - x$.

Substituting these values into Equation (2.6), we obtain the following equation:

$$28.086\text{ amu} = 0.9223\,(27.977\text{ amu}) + x(28.976\text{ amu}) + (0.0777 - x)\,(29.974\text{ amu})$$

Solving for x, we get $x = 0.0463$ and $y = 0.0314$. Thus, the percentages of ^{29}Si and ^{30}Si are 4.63 and 3.14%, respectively.

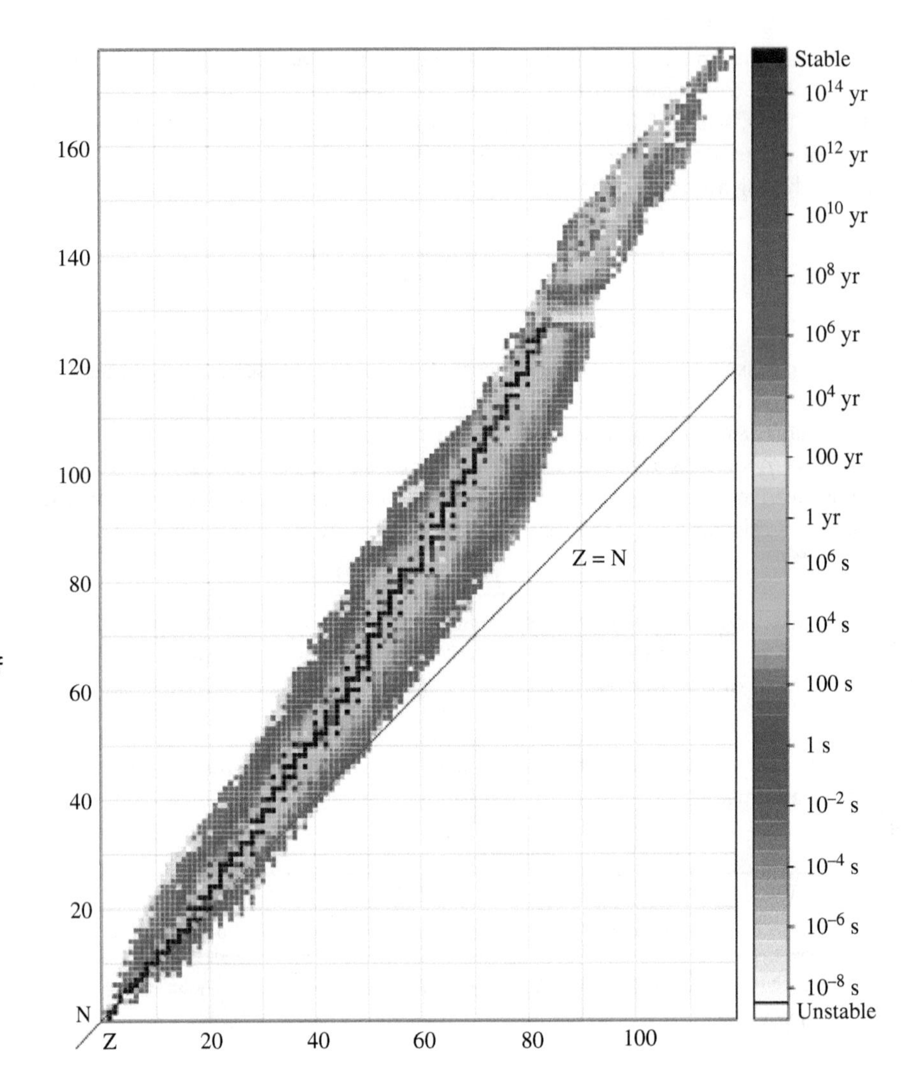

FIGURE 2.9
Plot of the different isotopes of the elements as the N/Z ratio. The solid line is for $N/Z = 1$ and the dark band of stability rises away from this line as the number of neutrons in the nucleus increases. Elements on either side of the band of stability will undergo spontaneous radioactive decay. [Reproduced from http://en.wikipedia.org/wiki/Table_of_nuclides_(complete) (accessed November 30, 2013).]

When the 266 known stable isotopes are plotted as N (number of neutrons) versus Z (number of protons), they form a band of stable nuclides on the positive side of the line $A = 2Z$, as shown in Figure 2.9, because the increasing percent composition of neutrons helps dilute the Coulombic repulsion of the positively charged protons in the nucleus. The largest stable isotope on this graph occurs for ^{208}Pb. A maximum occurs because the strong force can act only over very short ranges to hold the nucleons together and the radii of the heavier elements eventually becomes larger than this threshold. The band of stability can be compared to an island rising up out of an ocean ridge. Those isotopes having the largest nuclear binding energies will have the highest elevations and those beneath a certain threshold nuclear binding energy will occur below sea level and will therefore be unstable. Thus, the isotopes on either side of the dark band of stability in Figure 2.9 are underwater by this analogy and will always undergo spontaneous radioactive decay to form a more stable nuclide. The concept of *radioactivity* was discovered in 1896 by Henri Becquerel, although the term itself was actually first coined by Marie Curie. This pair (along with Madam Curie's husband Pierre) shared the 1903 Nobel Prize in physics for their work on radioactivity.

The different types of radioactive decays are listed in Table 2.2, according to the type of particle or radiation that is emitted. Although it is not strictly a form of

TABLE 2.2 Types of radioactive decay.

Type	Penetration	Speed	Particle	Example
α-Decay	Not very far, but severe	~0.10c	$-{}^{4}_{2}He^{2+}$	${}^{226}_{88}Ra \rightarrow {}^{222}_{86}Rn^{2-} + {}^{4}_{2}He^{2+}$
β-Decay	Moderately far	<0.90c	$-{}^{0}_{-1}e$	${}^{231}_{53}I \rightarrow {}^{231}_{52}Xe + {}^{0}_{-1}e$
γ-Emission	Very far	c	–Photon	${}^{60}_{27}Co^{*} \rightarrow {}^{60}_{27}Co + \gamma$
Positron emission	Moderately far	<0.90c	$-{}^{0}_{1}e$	${}^{11}_{6}C \rightarrow {}^{11}_{5}B + {}^{0}_{1}e$
Proton emission	Low to moderate	~0.10c	$-{}^{1}_{1}H^{+}$	${}^{53}_{27}Co \rightarrow {}^{52}_{26}Fe + {}^{1}_{1}H$
Neutron emission	Very far	<0.10c	$-{}^{1}_{0}n$	${}^{137}_{53}I \rightarrow {}^{136}_{53}I + {}^{1}_{0}n$
Electron capture*	NA	NA	$+{}^{0}_{-1}e$	${}^{44}_{22}Ti + {}^{0}_{-1}e \rightarrow {}^{44}_{21}Sc$

*Electron capture is actually a nuclear reaction and not a genuine type of radioactive decay. Therefore, it does not follow first-order kinetics.

radioactive decay, electron capture has also been included in the table. The only way to increase the atomic number Z is by beta decay, or the emission of an electron. Thus, any nuclide having a higher mass number than the band of stability can decrease its N/Z ratio by an increase in Z. On the other hand, isotopes below the band of stability can decrease Z by one of several different methods: positron emission (which is most common for low atomic numbers), alpha decay (which is more typical for larger atomic numbers), and neutron emission (which is rare). Extremely large unstable isotopes can also undergo fission in order to split apart into smaller, more stable daughter nuclides. The specific type of radioactive decay that an unstable nucleus will undergo is listed in the *Nuclear Wallet Cards*.

All radioactive decays occur with first-order kinetics, with the exception of electron capture, which is a two-particle collision. The differential rate law for radioactive decay is given by Equation (2.7). After integration, an alternative and more useful form of the rate law is shown by Equation (2.8). The *half-life* of radioactive decay is defined as the length of time it takes for the number of unstable nuclides to decrease to exactly one-half of their original value. The half-life, τ can be calculated using Equation (2.9), where k is the first-order rate constant:

$$\text{rate} = -\frac{dN}{dt} = kN \tag{2.7}$$

$$\frac{N}{N_0} = e^{-kt} \tag{2.8}$$

$$\tau = \frac{\ln(2)}{k} \tag{2.9}$$

Example 2-8. Radioactive iodine is used to image the thyroid gland. Typically, a saline solution of $Na^{131}I$ is administered to the patient by an IV drip. Predict the most likely type of radioactive decay for this nuclide and calculate Q for the reaction. Given that the half-life of ^{131}I is 8.025 days, what percentage of the isotope will have decayed during the 2.0-h procedure?

Solution. According to the *Nuclear Wallet Cards*, the only stable isotope of iodine is ^{127}I. Therefore, ^{131}I lies to the higher side of the band of stability and will need to increase Z in order to become a stable isotope. The only form of radioactive decay that increases Z is the emission of a beta particle. Using the principles of conservation of mass number and conservation of atomic number during a nuclear reaction, the nuclear equation for beta decay is

$$^{131}_{53}I \rightarrow {}^{0}_{-1}e + {}^{131}_{54}Xe$$

The mass excesses of these particles given by the *Nuclear Wallet Cards* are as follows: $\Delta = -87.442$ MeV for iodine-131 and -88.413 MeV for xenon-131. Using the formula that $M = \Delta + A$ and the conversion that 1 amu = 931.494 MeV, the masses of each isotope can be calculated as follows:

$$^{131}_{53}I \quad 131\ \text{amu}\left(\frac{931.494\ \text{MeV}}{1\ \text{amu}}\right) - 87.442\ \text{MeV}$$

$$= 121,938.272\ \text{MeV}\ (\text{or}\ 130.906\ \text{amu})$$

$$^{131}_{54}Xe \quad 131\ \text{amu}\left(\frac{931.494\ \text{MeV}}{1\ \text{amu}}\right) - 88.413\ \text{MeV}$$

$$= 121,937.301\ \text{MeV}\ (\text{or}\ 130.905\ \text{amu})$$

$$Q = 121,938.272 - (0.510999 + 121,937.301) = 0.460\ \text{MeV}$$

The reaction is exothermic, as might be expected for a spontaneous process (given that the entropy is zero). The kinetics of the process are first-order, so that the rate constant k can be calculated from Equation (2.9):

$$k = \frac{\ln(2)}{\tau} = \frac{0.6931}{8.025\ \text{days}} = 0.08637\ \text{days}^{-1}$$

Next, Equation (2.8) can be used to calculate the ratio of ^{131}I remaining compared with the initial amount after 2 h (or 0.083 days):

$$\frac{N}{N_0} = e^{-(0.08637\ \text{days}^{-1})(0.083\ \text{days})} = 0.9928$$

Thus, 99.28% of the original ^{131}I is remaining and only 0.72% has decayed.

Example 2-9. The half-life of lead-214 is 26.8 min. Assuming that the sample is initially 100% ^{214}Pb, use a spreadsheet to calculate the percentage of ^{214}Pb remaining as a function of time every 10 min for a total of 100 min. Then graph these data as % ^{214}Pb versus time.

Solution. Using Equation (2.9), the rate constant k was calculated as 0.0259 min^{-1}. Using Equation (2.8), the percentage of ^{214}Pb remaining was calculated every 10 min for a total period of 100 min using the spreadsheet. The results are plotted in the graph below the table.

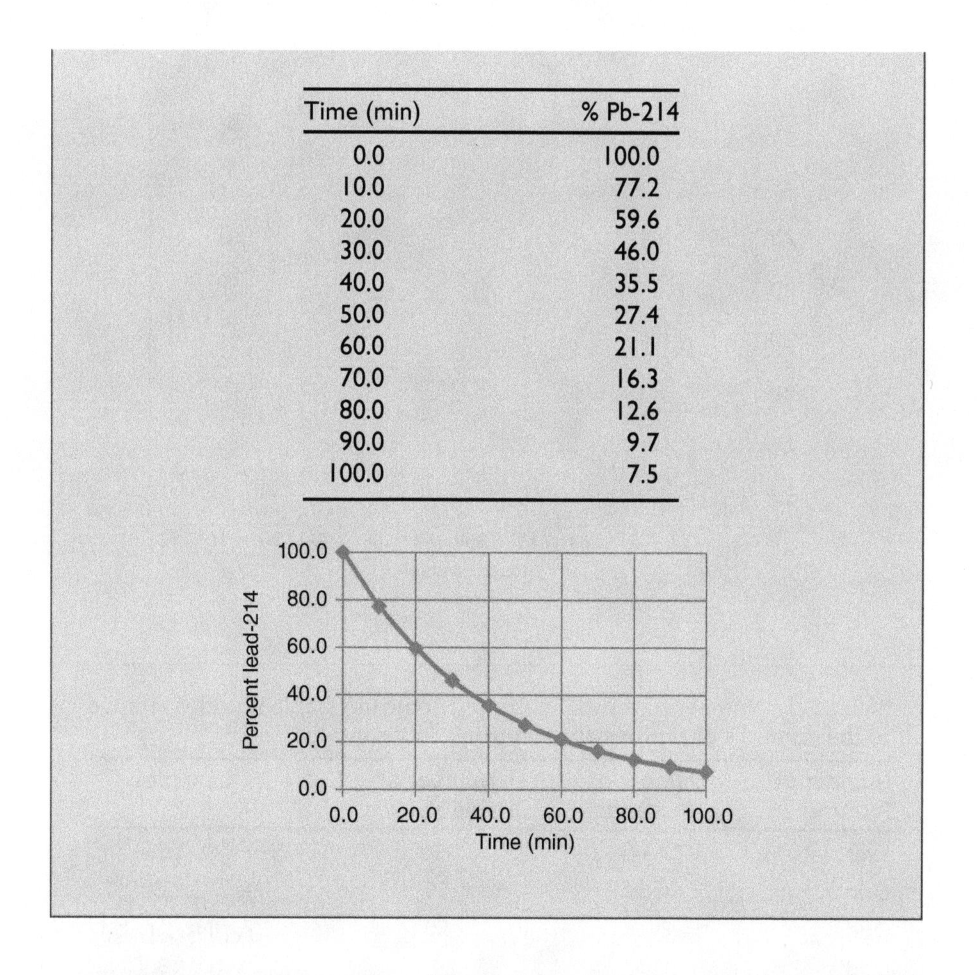

Time (min)	% Pb-214
0.0	100.0
10.0	77.2
20.0	59.6
30.0	46.0
40.0	35.5
50.0	27.4
60.0	21.1
70.0	16.3
80.0	12.6
90.0	9.7
100.0	7.5

As the largest stable isotope is ^{208}Pb, any nuclei heavier than this will undergo spontaneous radioactive decay to form a smaller, more stable nucleus. However, this process does not always occur in a single step. More commonly, a number of nuclear reactions occur until a stable isotope is formed through what is known as a *radioactive decay series*. Thus, for example, ^{238}U emits an alpha particle to form ^{234}Th, which in turn undergoes two successive beta decays to form the unstable ^{214}U nucleus. Following a series of five successive alpha decays, the isotope ^{214}Pb is formed. Finally, the stable ^{208}Pb nucleus is reached following two more beta decays, an alpha decay, two further beta decays, and then one final alpha decay. The entire sequence of steps is shown in Figure 2.10. All of the steps in the series take place at different rates until the stable nucleus is achieved in the end. Similar decay series are known for many other heavier elements.

2.5 THE SHELL MODEL OF THE NUCLEUS

Table 2.3 shows the number of stable isotopes as a function of the even or odd nature of the number of nucleons. Of the 266 naturally occurring stable isotopes, more than half have an even number of both protons and neutrons. By comparison, there are only four stable isotopes having both an odd number of protons and neutrons. Furthermore, even a cursory inspection of the cosmic abundances of the elements shown in Figure 2.2 makes it clear that there are more elements having an even atomic number than there are those with an odd atomic number. In fact, there even seem to be certain "magic numbers" of nucleons that are consistent with the most stable nuclei. The nuclear magic numbers are analogous to the common observation

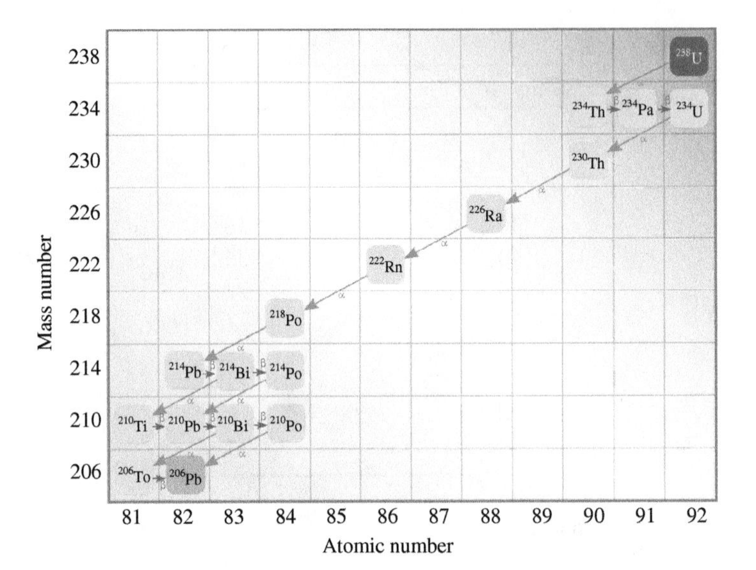

FIGURE 2.10
The radioactive decay series for ^{238}U to ^{208}Pb. [Blatt Communications.]

TABLE 2.3 Number of the 266 naturally occurring stable nuclides relative to the numbers of protons and neutrons they contain.

Number of Protons	Number of Neutrons	Number of Stable Nuclides	Examples
Even	Even	157	$^{4}_{2}He$, $^{16}_{8}O$, $^{40}_{20}Ca$, $^{208}_{82}Pb$
Even	Odd	55	$^{9}_{4}Be$, $^{13}_{6}C$, $^{29}_{14}Si$, $^{47}_{22}Ti$
Odd	Even	50	$^{19}_{9}F$, $^{23}_{11}Na$, $^{89}_{39}Y$, $^{127}_{53}I$
Odd	Odd	4	$^{2}_{1}H$, $^{6}_{3}Li$, $^{10}_{5}B$, $^{14}_{7}N$

that certain elements on the periodic table (the noble gases) are especially stable (atomic numbers 2, 10, 18, 36 54, and 86), except that the magic numbers of nucleons are different: 2, 8, 20, 28, 50, 82, and 126. Those nuclei that have double magic numbers are especially stable. Examples of double magic number nuclei include ^{4}He (2p, 2n), ^{16}O (8p, 8n), ^{40}Ca (20p, 20n), and ^{208}Pb (82p, 126n). Clearly, there must be some underlying phenomenon responsible for this rather unusual observation.

Just as it is now understood that the underlying reason for the stability of the noble gases has to do with the quantum nature of the electrons in atoms (e.g., Bohr's shell model of the atom in which the energy levels are quantized), the shell model of the nucleus states that the different energy levels of the nucleus are also quantized and that it is this quantization that leads to the enhanced stability of those nuclei having the nuclear magic numbers listed earlier. The main difference between the two models has to do with the fact that the electrons in an atom repel each other through the electromagnetic force (Coulomb's law), while the nucleons are attracted to one another through the strong force. As a first approximation, the structure of the nucleus can be approximated using the harmonic oscillator model (see Chapter 3 for more details), where the attractive forces of the nucleons in Figure 2.11 are given by the potential energy described in Equation (2.10), where r is the distance between two nucleons and R is the size of the corresponding square well potential. The solutions to the harmonic oscillator problem are governed by four quantum numbers: υ, the principal quantum number (which takes values of 1, 2, 3, ...), l, the angular

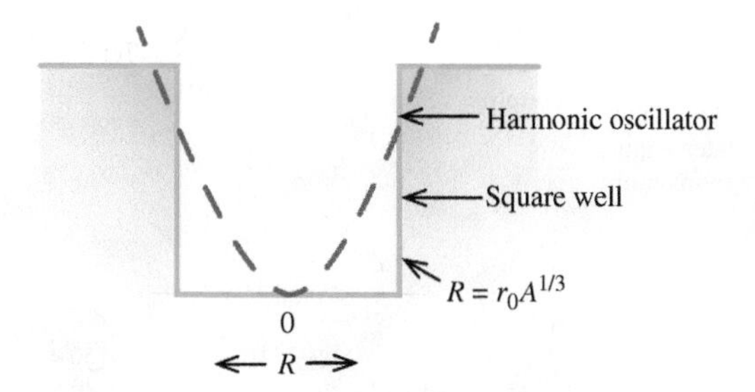

FIGURE 2.11
Harmonic oscillator versus square well potential energy model for the nucleus. [Blatt Communications.]

momentum quantum number (which takes values of 0, 1, 2, 3, … and have the corresponding designations s ($l=0$), *p* ($l=1$), *d* ($l=2$), *f* ($l=3$), *g* ($l=4$), etc.), m_l, the magnetic substate (which can take $2l+1$ possible values of $\pm l$), and s, the spin state (which has values of $\pm 1/2$). The energies $E_{\upsilon l}$ corresponding to the allowed quantum states are given by Equation (2.11), where $\hbar = h/2\pi$ and ω is the angular frequency.

$$V(r) = -V_0\left[1 - \frac{r^2}{R^2}\right] \quad (2.10)$$

$$E_{\nu l} = [2(\nu - 1) + l]\hbar\omega \quad (2.11)$$

Because the nucleons are fermions, the *Pauli principle* must also apply; that is to say that no two nucleons of the same type (proton or neutron) can have the same identical set of all four quantum numbers. Because protons have an electric charge quantum number of +1 while neutrons have a charge quantum number of 0, it *is*, however, possible for a proton and a neutron to have the same set of the quantum numbers υ, *l*, m_l, and s. Even so, the resulting set of nuclear orbitals still could not explain the observed set of nuclear magic numbers. This problem was later solved independently in 1949 by Maria Göppert Mayer and Hans Jensen, who shared the 1963 Nobel Prize in physics for their work on the nuclear shell model. It was noted that because the nucleons were in such close proximity with each other, the orbital angular momentum and spin angular momentum would couple with one another. The closest macroscopic analogy would be the way that the moon influences the tides on earth and how earth's gravity influences the moon's rotation. Therefore, a new quantum number *j*, which represents the total angular momentum, must be introduced, as shown in Equation (2.12):

$$j = l + s = l \pm 1/2 \quad (2.12)$$

The new quantum mechanical notation for the different energy levels that includes the *spin–orbit coupling* term is given by υl_j and has a degeneracy of $2j+1$ nucleons. Assuming that all the nucleons are arranged pairwise into each shell beginning with the lowest energy level, the rearranged energy level diagram, shown in Figure 2.12, can be used to explain the experimentally observed magic numbers.

Evidence supporting the shell model of the nucleus includes the enhanced cosmic abundance of nuclei containing the magic numbers of protons or neutrons (as shown in Figure 2.13), the stability of isotopes at the end of a radioactive decay series having magic numbers of protons or neutrons, the nuclear binding energy for a neutron being a maximum for one of the magic numbers and dropping off significantly with the addition of the next neutron, and the electric quadruple moments being approximately zero for those isotopes having a magic number. All of these results are indicative of the added stability of closed nuclear shells.

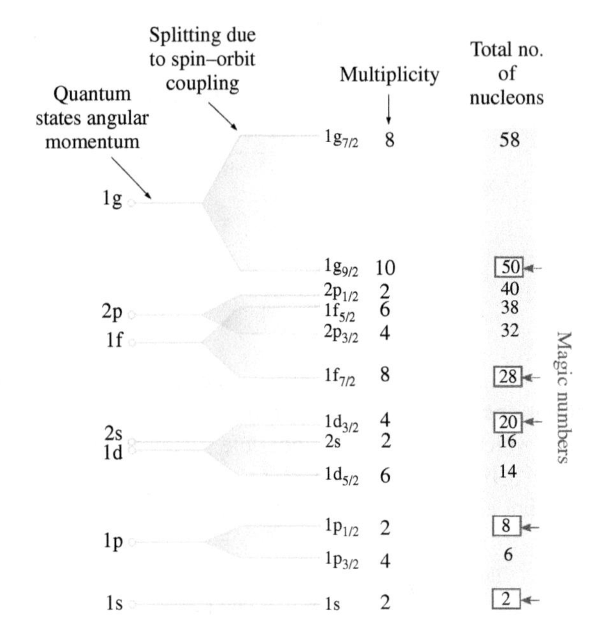

FIGURE 2.12
Shell model of the nucleus including strong spin–orbit coupling. [Blatt Communications.]

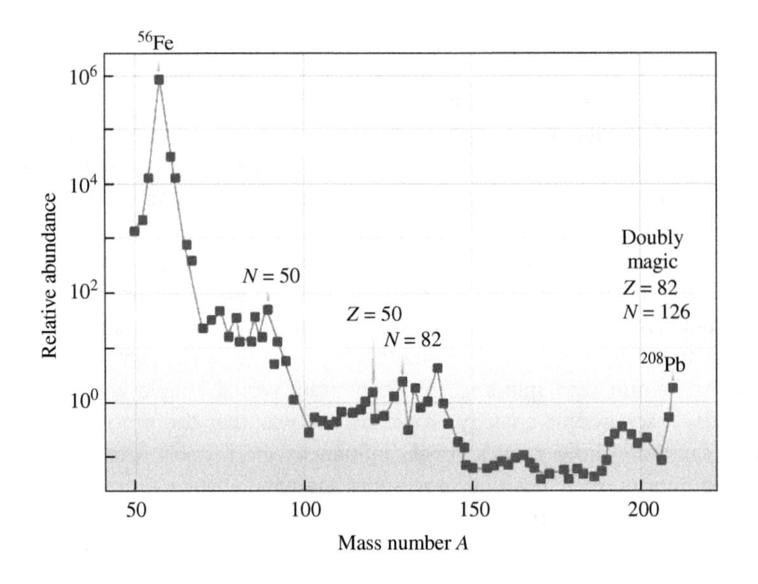

FIGURE 2.13
Cosmic abundances of the elements, showing how many of the more prevalent elements have magic numbers of protons or neutrons. [Blatt Communications.]

Example 2-10. Show all of the different energy levels for a $2d$ nuclear shell. Also indicate the degeneracy of each state.

Solution. $\upsilon = 2$, $l = 2$ for $2d$. The possible values of j are 5/2 and 3/2. There are $2j + 1$ degenerate states associated with each energy level. Thus, the degeneracy of the $2d_{5/2}$ energy level is (6) and the degeneracy of the $2d_{3/2}$ energy level is (4). The $2d$ shell as an aggregate can therefore hold a total of 10 nucleons.

2.6 THE ORIGIN OF THE ELEMENTS

The famous physicist Carl Sagan once said that "we are star stuff" and he was speaking literally. All of the naturally occurring elements have their origins in the stars.

In the beginning, approximately 13.7 billion years ago, all of the known matter and energy in the universe was concentrated into a single dense region of space known as the *ylem* and having a temperature of about 10^{13} K. As there were no walls to contain it, the ylem exploded in a cosmic genesis event known as the *Big Bang*, and the universe has been expanding ever since. There is a variety of evidence supporting the theory of a Big Bang. First of all, the electronic spectra from distant galaxies has been shown to be Doppler-shifted to the red (shifted to longer wavelengths), which indicates that they are moving away from our solar system. By correlating the magnitude of the red shift, the age of the universe can be calculated to be approximately 13.5–13.9 billion years. Furthermore, there exists a universal cosmic microwave background radiation from every corner of the universe that is believed to be the remnant of the initial Big Bang event and that has a black-body radiation temperature of exactly 2.725 K. Finally, the ratio of He : H in the universe is essentially constant at 0.23 and there is little variation in the isotopic ratios of these two elements across the entire universe, indicating that the majority of hydrogen and helium was created as a result of a single cosmic event.

The evolution of the universe has taken place in stages. The first stage following the Big Bang was the creation of the elementary particles. Initially, matter and energy existed in a sort of cosmic soup where they could easily interconvert with one another. As the universe expanded, however, it began to cool to a temperature between 10^{10} and 10^{12} K and the first protons and neutrons were formed. This stage in the evolution of the universe, which occurred between 10^{-6} and 1 s after the Big Bang, is known as *baryogenesis*. At this point in time, protons could be converted into neutrons and positrons by bombardment with antineutrinos and neutrons could be converted back into protons and electrons by the interaction with a neutrino, as shown in Equations (2.13) and (2.14). A gamma ray could also spontaneously split apart to form both a positron and an electron in a process known as *pair production*, as shown in Equation (2.15):

$$^{1}_{1}\mathrm{H} + \overline{\nu} \rightarrow {}^{1}_{0}\mathrm{n} + {}^{0}_{1}\mathrm{e} \quad (2.13)$$

$$^{1}_{0}\mathrm{n} + \nu \rightarrow {}^{1}_{1}\mathrm{H} + {}^{0}_{-1}\mathrm{e} \quad (2.14)$$

$$\gamma \rightarrow {}^{0}_{1}\mathrm{e} + {}^{0}_{-1}\mathrm{e} \quad (2.15)$$

In the first few moments after the Big Bang, the energy density was sufficiently large that Equations (2.13) and (2.14) occurred at roughly the same rate, so that the numbers of protons and neutrons were initially equal. However, as the universe cooled as a result of its expansion, the rate of proton formation began to dominate. This occurred because the reaction given by Equation (2.13) requires slightly more energy than the one in Equation (2.14) due to the larger mass of the neutron. As a result, by the time that the neutrinos and anti-neutrinos stopped interacting with the other elementary particles (about 1 s after the Big Bang), the composition of the universe consisted of about 87% protons and 13% neutrons.

At first, the energy density of the universe was still sufficiently strong that the proton and neutron collisions would immediately break apart again. The next phase in the evolution of the universe, known as the *Big Bang nucleosynthesis* (*BBN*), began after the universe had cooled even further to a temperature around 10^9 K, roughly 3 min after the initial expansion so that the density of the universe was approximately 0.1 g/cm^3. In this nucleosynthesis phase, the first nucleons were combined together to form heavier nuclei. Before this point in time, the next larger nucleus (^{2}H) was unable to form because of its low nuclear binding energy. The BBN, which occurred in the series of fusion reactions shown by Equations (2.16)–(2.19), is therefore responsible for the creation of most of the hydrogen and helium in the universe.

$$^{1}_{1}H + ^{1}_{0}n \rightarrow ^{2}_{1}H + \gamma \quad (2.16)$$

$$^{2}_{1}H + ^{2}_{1}H \rightarrow ^{3}_{2}He + ^{1}_{0}n \quad (2.17)$$

$$^{2}_{1}H + ^{2}_{1}H \rightarrow ^{3}_{1}H + ^{1}_{1}H \quad (2.18)$$

$$^{3}_{2}He + ^{1}_{0}n \rightarrow ^{4}_{2}He + \gamma \quad (2.19)$$

Because there were initially less neutrons than protons, the former acted as a limiting reagent, so that by the time the BBN had concluded, the universe consisted of about 25% He, 74% H, and 1% D, by mass. Trace amounts of Li and Be were also produced by the reactions given in Equations (2.20) and (2.21). The synthesis of the heavier elements was significantly reduced by the stability of the double magic number ^{4}He nucleus. The BBN continued from about 3–20 min following the Big Bang. The brevity of the BBN also helped prevent the formation of the heavier elements. As a result, no nuclei heavier than ^{8}Be were formed by the BBN:

$$^{3}_{2}He + ^{4}_{2}He \rightarrow ^{7}_{4}Be \quad (2.20)$$

$$^{3}_{1}H + ^{4}_{2}He \rightarrow ^{7}_{3}Li \quad (2.21)$$

The BBN was followed by a cooling phase, which allowed the neutrons to decay into protons. The temperature of the universe at this point in time was no longer high enough for fusion reactions to occur and it largely consisted of matter. Approximately 379,000 years after the Big Bang, the universe had gradually cooled to a temperature of about 3000 K. This was finally cool enough for electrons to cling to the nuclides to form the first neutral atoms. This is sometimes known as the *chemistry phase* in the evolution of the universe.

Not much of anything happened again until approximately 10^9 years after the Big Bang when the temperature of the universe had cooled even further to about 20 K. At this point in time, clouds of interstellar matter began to coalesce under the influence of gravity. Gradually, the density of these clouds increased until the first protostars began to form. Approximately 90% of the stars in the universe at this point in time were stars having sizes and compositions similar to our sun; stars that are classified as *Main Sequence stars*. This type of star was the first kind to form following the Big Bang. For this reason, the earliest stars are sometimes referred to as *first-generation stars*. As the gravitational forces began to contract matter together in the first stars, the increase in density in their cores caused the temperatures to rise. At a certain critical point, the increasing temperatures ionized the hydrogen and helium nuclei. Furthermore, the presence of neutrons was no longer observed, as they were during the BBN. When the pressure inside a star was sufficiently large, the interior temperature became hot enough for nuclear fusion to occur. Once the density increased to $\sim$100 g/cm^3 and the internal temperature of the star reached at least 4×10^6 K, the gases were ignited and *stellar nucleosynthesis* was initiated. At temperatures greater than 10^7 K, helium-4 can be produced by a process known as *hydrogen burning*, given by Equations (2.22)–(2.24). This process is also known as the *proton–proton chain*; it is depicted in Figure 2.14. The first step in this process (the formation of D) is rate-limiting because it produces a lepton and an anti-lepton and it is therefore governed by the weak force. As a result, Main Sequence stars can burn for many billions of years:

$$^{1}_{1}H + ^{1}_{1}H \rightarrow ^{2}_{1}H + ^{0}_{-1}e + \nu \quad (2.22)$$

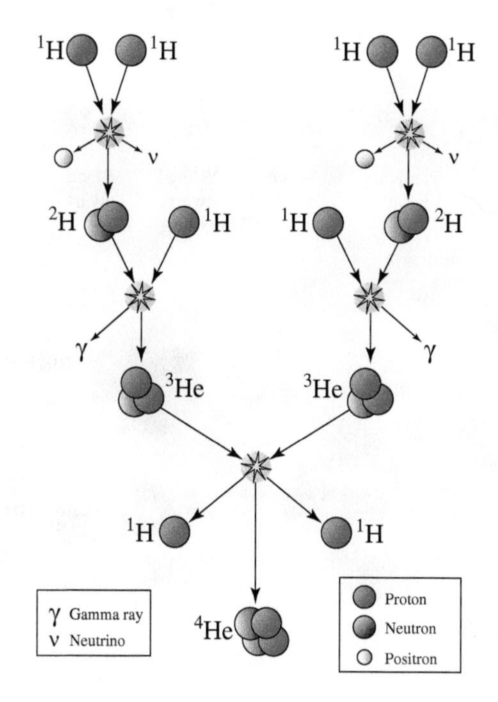

FIGURE 2.14
Schematic diagram of the proton–proton chain in stellar nucleosynthesis. [Attributed to Borb, reproduced from http://en.wikipedia.org/wiki/Proton%E2%80%93proton_chain_reaction (accessed October 17, 2013).]

$$^{2}_{1}\mathrm{H} + {}^{1}_{1}\mathrm{H} \rightarrow {}^{3}_{2}\mathrm{He} + \gamma \tag{2.23}$$

$$^{3}_{2}\mathrm{He} + {}^{3}_{2}\mathrm{He} \rightarrow {}^{4}_{2}\mathrm{He} + {}^{1}_{1}\mathrm{H} + {}^{1}_{1}\mathrm{H} \tag{2.24}$$

A star the size of our sun will continue to burn hydrogen until most of its core is consumed. As the star contracts, if its mass is greater than 10^{30} kg, eventually the more dense helium that accumulates in the core of the star increases its internal temperature to the point where *helium burning* can occur. This heating of the core to temperatures greater than 2×10^{7} K and a density of 105 g/cm^3 causes an expansion of the star's outer surface, creating a *Red Giant star*, the next step in the stellar evolution sequence depicted in Figure 2.15.

As shown by Equations (2.25) and (2.26), two ^{4}He nuclei can fuse together to form ^{8}Be, which in turn can fuse with another ^{4}He nucleus to make a ^{12}C nuclide. Because the net reaction involves the fusion of three alpha particles, the first two steps in this mechanism are often called the *triple alpha process* (Figure 2.16). Because this is a three-body process, the reaction is slow, allowing Red Giant stars to continue to burn for $10^7 - 10^8$ years. Our sun will eventually become a Red Giant, swelling in size to engulf all of the inner planets.

$$^{4}_{2}\mathrm{He} + {}^{4}_{2}\mathrm{He} \rightarrow {}^{8}_{4}\mathrm{Be} + \gamma \tag{2.25}$$

$$^{8}_{4}\mathrm{Be} + {}^{4}_{2}\mathrm{He} \rightarrow {}^{12}_{6}\mathrm{C} + \gamma \tag{2.26}$$

The ^{8}Be nucleus formed by Equation (2.25) is metastable, with a half-life of about 10^{-16} s. Thus, it might seem unlikely that the reaction given by Equation (2.26) can occur with much probability. However, by some fortuitous chance, the energy of the ^{8}Be ground state has almost the same exact value as the energy of two alpha particles and the reaction given by Equation (2.26) generates about the same amount of energy as an excited state of ^{12}C. If it were not for the coincidence of these resonant processes, there probably would not be enough ^{12}C in the universe to support the existence of carbon-based life forms. Furthermore, the reaction of ^{4}He with ^{12}C to

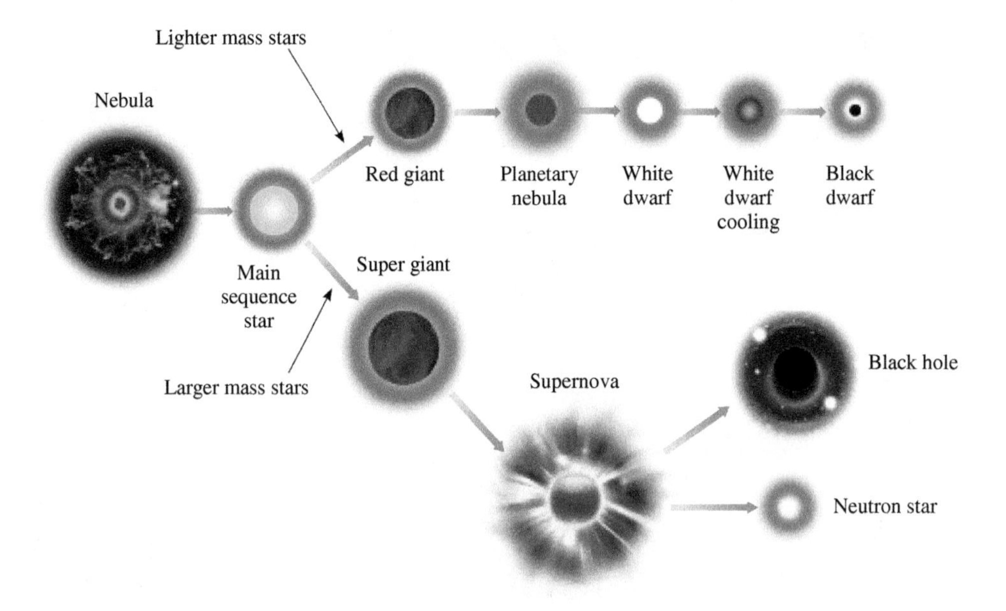

FIGURE 2.15
Life cycle of a star (stellar evolution).

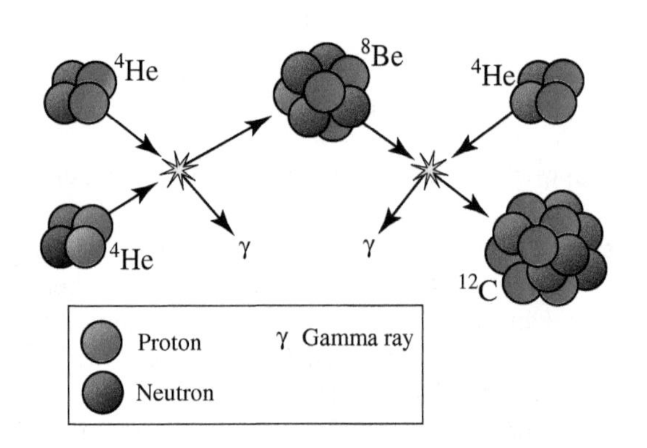

FIGURE 2.16
Schematic diagram illustrating the triple alpha process that occurs during stellar nucleosynthesis. [Attributed to Borb, reproduced from https://en.wikipedia.org/wiki/Triple-alpha_process (accessed October 17, 2013)]

generate ^{16}O, as shown in Equation (2.27), is a very slow process. However, once it occurs, both this reaction and the one given by Equation (2.28) to form ^{20}Ne are exothermic, which accounts for the cosmic abundances of these two elements.

$$^{12}_{6}C + ^{4}_{2}He \rightarrow ^{16}_{8}O + \gamma \qquad (2.27)$$

$$^{16}_{8}O + ^{4}_{2}He \rightarrow ^{20}_{10}Ne + \gamma \qquad (2.28)$$

In more massive stars (>1.3 times the mass of our sun), which have even higher internal temperatures ($>1.7 \times 10^7$ K), the carbon–nitrogen–oxygen (CNO) cycle can provide an alternative synthesis of ^4He, as shown by Equations (2.29)–(2.34). The *CNO cycle* is catalytic for the production of ^4He, as shown in Equation (2.35). This is the dominant source of energy in stars that are larger than our sun. A schematic diagram of the CNO cycle is shown in Figure 2.17:

$$^{12}_{6}C + ^{1}_{1}H \rightarrow ^{13}_{7}N + \gamma \quad 1.95\text{ MeV} \qquad (2.29)$$

$$^{13}_{7}N \rightarrow ^{13}_{6}C + ^{0}_{1}e + \nu \quad 2.22\text{ MeV} \qquad (2.30)$$

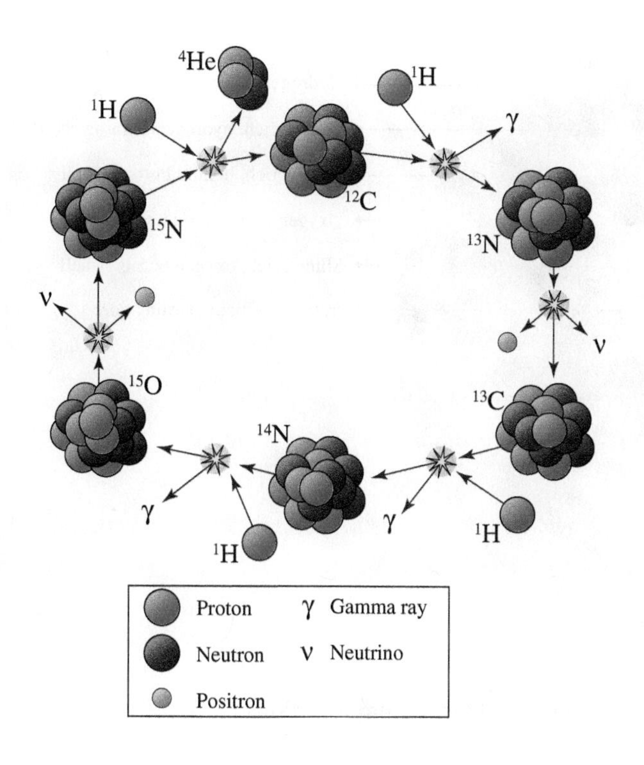

FIGURE 2.17
Schematic diagram of the CNO cycle. [Attributed to Borb, reproduced from https://en.wikipedia.org/wiki/CNO_cycle (accessed October 17, 2013).]

$$^{13}_{6}C + {}^{1}_{1}H \rightarrow {}^{14}_{7}N + \gamma \qquad 7.54\ \text{MeV} \tag{2.31}$$

$$^{14}_{7}N + {}^{1}_{1}H \rightarrow {}^{15}_{8}O + \gamma \qquad 7.35\ \text{MeV} \tag{2.32}$$

$$^{15}_{8}O \rightarrow {}^{15}_{7}N + {}^{0}_{1}e + \nu \qquad 2.75\ \text{MeV} \tag{2.33}$$

$$^{15}_{7}N + {}^{1}_{1}H \rightarrow {}^{4}_{2}He + {}^{12}_{6}C \qquad 4.96\ \text{MeV} \tag{2.34}$$

$$4\ {}^{1}_{1}H \rightarrow {}^{4}_{2}He + 2\,{}^{0}_{1}e + 2\nu + 3\gamma \qquad 26.77\ \text{MeV} \tag{2.35}$$

As these more massive stars consumed their hydrogen and helium to make some of the richer elements, these heavier atoms would gravitate to the interior of the star (Figure 2.18), creating an even more dense and hot environment. Depending on the size of the star, helium burning could be followed by carbon burning, which in turn could be followed by neon burning, oxygen burning, and silicon burning. As shown in Figure 2.1, those elements with the largest nuclear binding energies occurred for Fe and Ni isotopes having mass numbers around 56. The fusion reactions leading to these nuclides were therefore exothermic, sustaining the life cycle of the star and preventing gravitational collapse. In this manner, all of the elements up to ^{56}Ni could be formed. In the interiors of stars, however, ^{56}Ni undergoes two successive positron emissions to produce ^{56}Fe, which makes up most of the core in the interior of stars. No elements more massive than these can be produced by stellar nucleosynthesis because any further fusion processes would necessarily be endothermic.

All of the elements heavier than ^{56}Fe can only be formed in super massive stars as a result of nuclear fission, followed by neutron capture (smaller stars will eventually burn themselves out to form White Dwarfs, as shown in Figure 2.15). In these larger Red Supergiant stars, which have lifetimes of several million years, the heavier elements are formed gradually over thousands of years by the *s-process* (where the abbreviation s signifies slow). The s-process increases A by a series of neutron

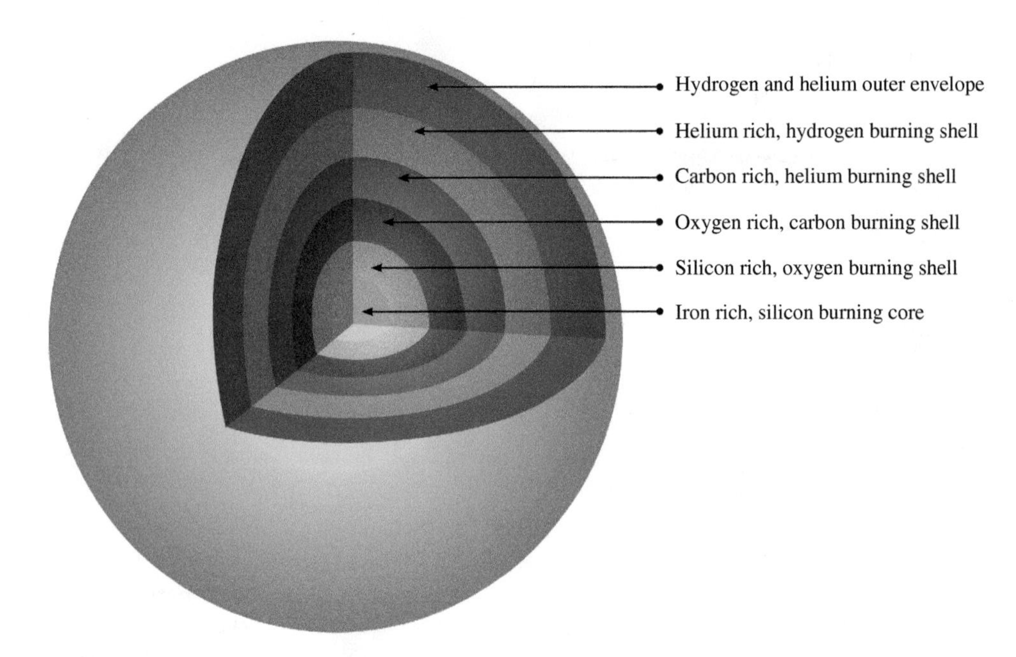

FIGURE 2.18
Schematic diagram showing the interior of a massive star. [Blatt Communications.]

captures, followed by beta decay. Two of the primary sources of the neutrons used in the subsequent neutron capture derive from the collisions of alpha particles with ^{13}C and ^{22}Ne, as shown in Equations (2.36) and (2.37):

$$^{13}_{6}C + ^{4}_{2}He \rightarrow ^{16}_{8}O + ^{1}_{0}n \tag{2.36}$$

$$^{22}_{10}Ne + ^{4}_{2}He \rightarrow ^{25}_{12}Mg + ^{1}_{0}n \tag{2.37}$$

An example of s-process reaction is given by Equation (2.38). The stable isotope ^{56}Fe slowly increases its mass number by three consecutive neutron captures to form the unstable nuclide ^{59}Fe. Because the neutrons are captured slowly, any unstable nucleus will have time to decay, which is exactly what happens in this case as ^{59}Fe undergoes beta decay to form the stable ^{59}Co nucleus.

$$^{56}_{26}Fe + 3\,^{1}_{0}n \rightarrow ^{59}_{26}Fe \rightarrow ^{59}_{27}Co + ^{0}_{-1}e \tag{2.38}$$

The s-process can start with any stable seed nucleus, as shown in Figure 2.19. In each case, neutron capture occurs to increase A and then beta decay occurs whenever an unstable nucleus is encountered. The largest isotope that can be produced by the s-process is ^{209}Bi, which is the heaviest stable nucleus on the band of stability.

The second process by which nuclei heavier than ^{56}Fe can be formed is known as the *r-process*. In many of the Red Supergiants, the star eventually runs out of nuclear fuel and the core collapses under the influence of gravity. When this happens, it is a very rapid process. Once the collapse reaches a certain critical nuclear density ($\sim 5 \times 10^{14}$ g/cm^3), any further collapse initiates a recoil and the star explodes as a *supernova*, emitting tremendous amounts of energy and large bursts of neutrons and neutrinos. Under these extreme conditions of high neutron flux, neutron capture occurs on a faster timescale than beta decay, so that much heavier nuclei can be formed before the nucleus undergoes spontaneous beta decay. Eventually, however, a point is reached where the nuclear binding energy is so small that the nucleus

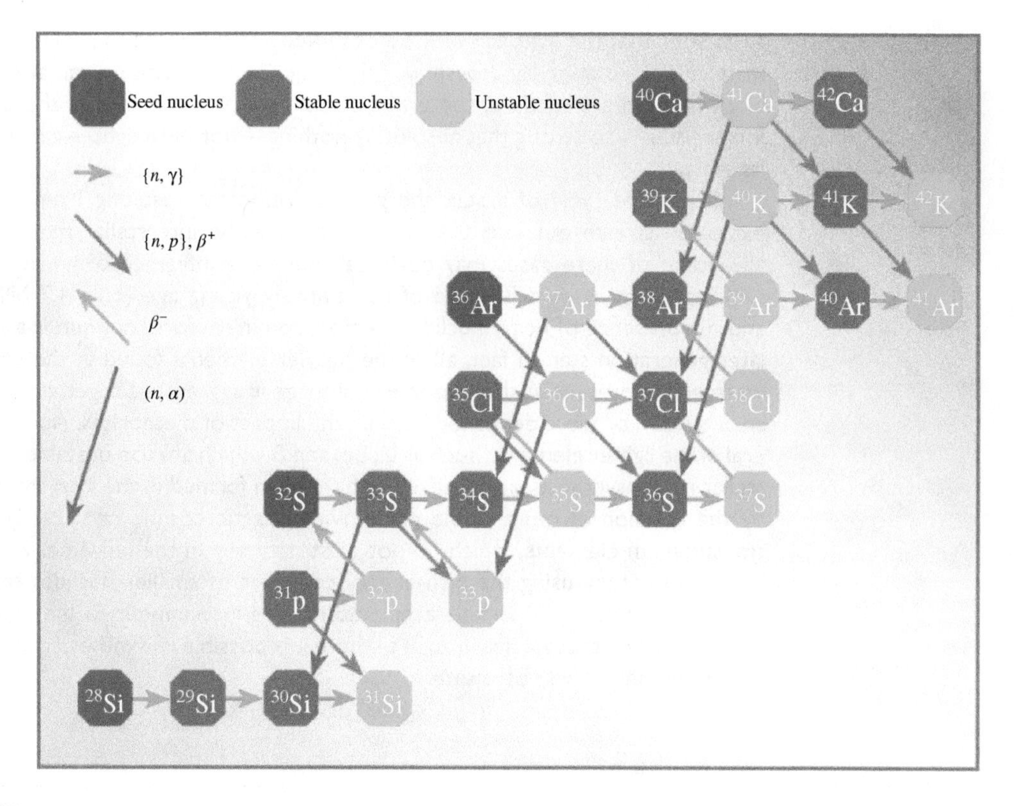

FIGURE 2.19
Illustration of the *s*-process. A stable seed nucleus (red) increases its mass number by a series of neutron captures until an unstable nucleus (green) is encountered. Whenever an unstable nucleus is made, it immediately undergoes beta decay to form the next stable nucleus (violet) in the series. [Blatt Communications.]

can no longer absorb any more neutrons and the mass buildup is truncated. At this point, the nucleus undergoes a series of beta decays until a stable nuclide results. An example of the *r*-process is shown by Equation (2.39):

$$^{116}_{48}\text{Cd} + 14\,^{1}_{0}\text{n} \rightarrow \,^{130}_{48}\text{Cd} \rightarrow\rightarrow\rightarrow\rightarrow \,^{130}_{52}\text{Te} + 4\,^{0}_{-1}\text{e} \tag{2.39}$$

Example 2-11. Show how ^{197}Au can be created from ^{175}Lu using the *r*-process.

Solution. All of the gold in the universe was originally created using the *r*-process in the supernovae explosions.

$$^{175}_{71}\text{Lu} + 22\,^{1}_{0}\text{n} \rightarrow \,^{197}_{71}\text{Lu} \rightarrow\rightarrow\rightarrow\rightarrow\rightarrow\rightarrow\rightarrow\rightarrow \,^{197}_{79}\text{Au} + 8\,^{0}_{-1}\text{e}$$

The ultimate fate of the star following a supernova depends on the exact conditions. A *neutron star* is a very dense and extremely hot star that is composed entirely of neutrons and which is prevented from undergoing further collapse by the Pauli exclusion principle, which states that no two neutrons can have the same exact quantum state. Because of a neutron star's enormous density, it has an extremely large gravitational pull. The massive star that's at the center of the Crab nebula, whose explosion was observed by the Chinese in 1054 AD, is an example of a rapidly rotating neutron star known as a *pulsar*. A pulsar emits electromagnetic radiation as it spins such that the light can only be observed when it faces the earth, in the same manner as a rotating lighthouse works. A supernova explosion can also result in a black hole. In a *black hole*, the gravitational field of the object at the core of the former star is so strong that absolutely nothing—not even light—can escape from its grasp.

In the life cycle of a star, the gaseous remnants resulting from a supernova explosion stream out into the cosmos forming the interstellar medium. Eventually, some of these gases may coalesce under the influence of gravity to form a future-generation star. Because of its relatively young age (only 4.7 billion years) and the presence of heavy nuclides such as iron in its core, our sun is a second- or later-generation star. In fact, all of the heavier elements found in the earth's crust were either present in the interstellar dust as it aggregated together to form our solar system or were deposited here by the impact of meteorites. Additionally, several of the lighter elements, such as Li, Be, and B, which are too unstable to survive in stellar nucleosynthesis, are believed to have been formed in the interstellar medium by the collision of more stable nuclei with galactic cosmic rays. Lastly, all of the transuranium elements, which do not exist naturally in the universe, were created in the laboratory using the high-energy collisions of smaller nuclides that exist in nuclear reactors or in particle accelerators. It is a testament to the power of the human intellect that we live in an age where it is possible to synthesize elements that Nature herself could not create.

EXERCISES

2.1. Calculate the nuclear binding energy per nucleon for ^{56}Fe in units of MeV.

2.2. If the nuclear binding energy of a single atom of ^{55}Mn is 7.727238×10^{-11} J, calculate the atomic mass of ^{55}Mn.

2.3. Calculate the binding energy (in MeV and J/mol) and binding energy per nucleon for ^{136}Xe (135.9072 amu), ^{208}Pb (207.9766 amu), and ^{28}Si (27.97693 amu). Which nucleus is the most stable?

2.4. Using data from the *Nuclear Wallet Cards* for the two stable isotopes of copper, calculate the chemical atomic mass of Cu from the weighted average of these isotopes.

2.5. In 1989, Stanley Pons and Martin Fleischmann startled the scientific community with their claims that they could achieve fusion under ordinary experimental conditions (so-called cold fusion). One of the nuclear reactions they claimed to have achieved is as follows. Calculate Q (or the amount of energy) that is produced by this reaction.

$$^{2}_{1}\text{H} + ^{2}_{1}\text{H} \rightarrow ^{3}_{2}\text{He} + ^{1}_{0}\text{n}$$

2.6. Lead can exist as four stable isotopes: ^{204}Pb, ^{206}Pb, ^{207}Pb, and ^{208}Pb. Which of these isotopes of lead has the greatest cosmic abundance? Rationalize your answer. Lead-205 is unstable and undergoes electron capture. Write the nuclear reaction for this process. Lead-206 is produced from ^{222}Rn by a radioactive decay chain. Show the nuclear reactions for each step of the chain.

2.7. Radioactive ^{222}Rn is known to accumulate in the basements of buildings and to constitute a health risk.

a. What is the parent primary natural radionuclide that produces the ^{222}Rn?

b. Write the reaction for the nuclear decay of ^{222}Rn.

c. Why is it that none of the intermediate (or the parent) radionuclides are considered as a health risk?

2.8. Radiocarbon dating is a widely used technique to determine the age of organic matter. In a living organism, the ratio of ^{14}C : ^{12}C is assumed to be a constant because the system is at equilibrium with the atmosphere. However, once the organism has died, the carbon is trapped in the material as organic matter and any ^{14}C undergoes beta decay with a half-life of 5,730 years. If the beta activity of a recently felled tree is 0.25 Bq and the radioactivity of a piece of the Shroud of Turin is 0.23 Bq, estimate the age of the Shroud of Turin.

2.9. Predict the other fission product that occurs in the following nuclear reaction:

$$^{235}_{92}U + ^{1}_{0}n \rightarrow ^{137}_{52}Te + \ ? + 2^{1}_{0}n$$

2.10. Our sun radiates energy at a rate of 3.9×10^{26} J/s. Calculate the rate of mass loss of the sun in units of kilogram per second.

2.11. Early forms of the periodic table left a gap between molybdenum and ruthenium. The missing element, technetium, does not have any stable isotopes. The longest-lived isotope of technetium is ^{98}Tc. Predict the radioactive decay products for this isotope and explain why it does not exist naturally on earth.

2.12. Calculate the amount of energy released during each step of the proton–proton chain.

2.13. Predict the first stable isotope when ^{98}Mo is used as a seed nucleus in the s-process.

BIBLIOGRAPHY

1. Burbidge EM, Burbidge GR, Fowler WA, Hoyle F. Synthesis of the Elements in Stars. *Reviews of Modern Physics* 1957;29:547.
2. Friedlander G, Kennedy JW, Macias ES, Miller JM. *Nuclear and Radiochemistry*. 3rd ed. New York: John Wiley & Sons; 1981.
3. Greenwood NN, Earnshaw A. *Chemistry of the Elements*. 2nd ed. Oxford: Butterworth-Heinemann; 1997.
4. Housecroft CE, Sharpe AG. *Inorganic Chemistry*. 3rd ed. Essex: Pearson Education Limited; 2008.
5. Nave, R. *HyperPhysics*. http://hyperphysics.phy-astr.gsu.edu/hbase/hframe.html (accessed Oct 10, 2013).
6. Tuli JK. *Nuclear Wallet Cards*. 7th ed. New York: U.S. Department of Energy; 2005.
7. Viola, V. E.; deSouza, R. T. *Chemistry C460 eText*. http://courses.chem.indiana.edu/c460/lecture_notes.asp (accessed July 10, 2012).

A Brief Review of Quantum Theory 3

"Anyone who is not shocked by quantum theory has not understood it."

—*Niels Bohr*

3.1 THE WAVELIKE PROPERTIES OF LIGHT

It is a relatively simple matter to observe the macroscopic properties of matter—that is to say, we can measure the mass of a mole of particles or the volume of a gas or the density of a solid. However, if we want to observe a single atom or something on the subatomic level, our powers of observation are limited. At this microscopic level, we are forced to use a form of *spectroscopy*, which deals with the interaction of light with matter. The wavelike properties of light have been well-documented, beginning with the pioneering work of Christiaan Huygens in 1678. In contrast to Newton's corpuscular theory of light, Huygens "wavelets" could better explain certain properties of light, such as reflection, refraction, and diffraction. For mathematical purposes, light can be considered as a sine wave, oscillating in a plane parallel to its direction of travel. The *wavelength* (λ) of light, measured in meters (m) is defined as the distance between successive crests on the wave, while the *frequency* (υ) is the number of times a crest passes a fixed point per unit time (s^{-1}). Thus, the product of the wavelength and the frequency will be equal to its velocity (m/s), as shown by Equation (3.1). One of the most important properties of waves is the principle of *superposition*. Whenever two waves are travelling at the same velocity and in the same direction, the amplitudes of the two waves can be added together to create a new wave front. If the original waves have the same amplitude initially and are exactly in phase with one another, *constructive interference* will occur, leading to a new wave front having a magnitude that is twice that of either of the originals, as shown in Figure 3.1(a). If, on the other hand, the original waves are exactly out of phase with one another and have identical (but opposite) amplitudes, then they will exactly cancel each other out in a process known as *destructive interference*, which is shown in Figure 3.1(b). Of course, it is also possible to mathematically add the sine waves from more than two waves

Quantum corral. [Image originally created by IBM Corporation.]

Principles of Inorganic Chemistry, First Edition. Brian W. Pfennig.

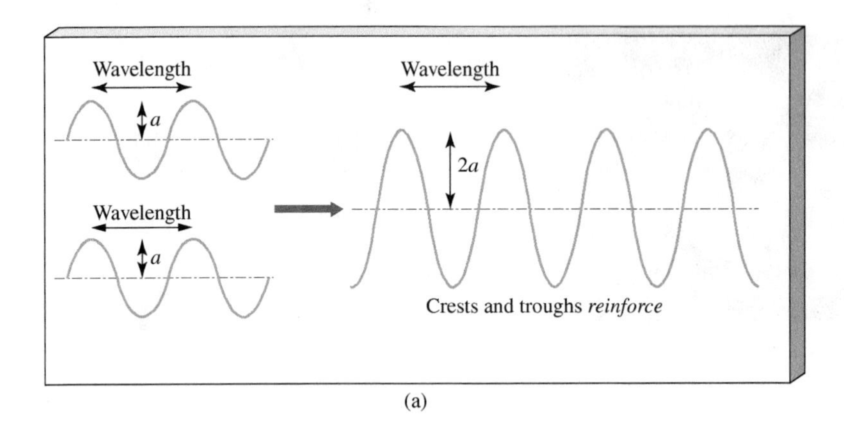

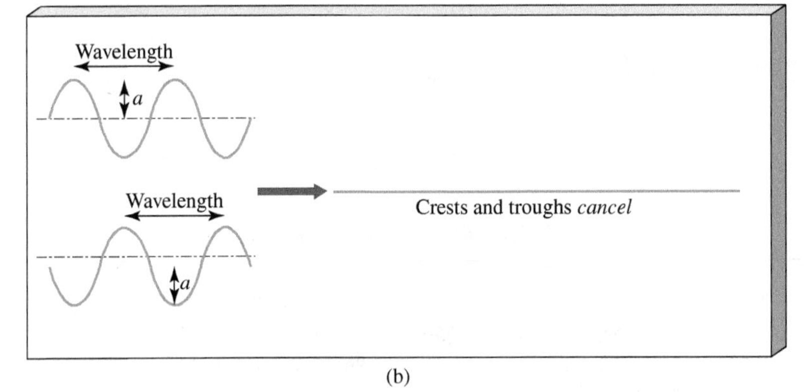

FIGURE 3.1
(a) Constructive and (b) destructive interference of two in-phase light waves. [Reproduced by permission from Astronomy Today, McMillan and Chaisson, 2nd ed., Prentice Hall, 1997.]

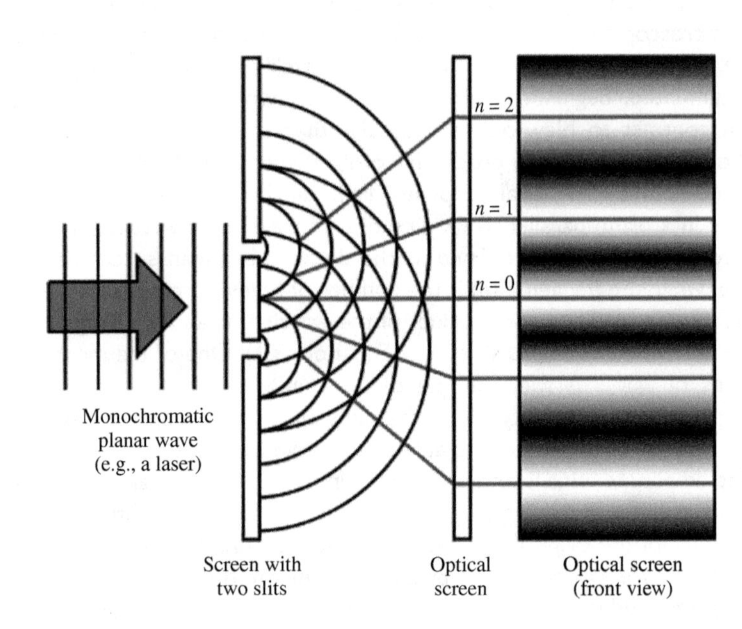

FIGURE 3.2
The interference (or diffraction) pattern that results when two monochromatic beams of light pass through a screen containing two narrow slits and are then observed on a parallel optical screen some distance away from the light source. This experiment, which is known as *Thomas Young's double slit experiment*, demonstrates the wavelike nature of light. [Reproduced from http://en.wikibooks.org/wiki/High_School_Chemistry/The_Dual_Nature_of_Light (accessed November 30, 2013).]

together at any given time, just as it is also possible to add together waves that are not in phase with one another or which have differing initial amplitudes.

$$v = \upsilon\lambda \tag{3.1}$$

In 1801, Thomas Young performed his famous *double slit experiment*, finally proving the wavelike nature of light. A diagram of this experiment is shown in Figure 3.2. A beam of monochromatic light (such as that from a LASER beam) is passed through two narrow openings. As the coherent (in-phase) light waves pass through the holes, they spread out to form a semicircular pattern of propagating wavefronts. This phenomenon is completely analogous to the way that the wake from a passing boat might spread out around a rock in the middle of an otherwise calm reservoir. In the figure, the black concentric circles represent the crests of the wave at any given point in time. Where the two wavefronts are in phase with one another, constructive interference occurs, leading to a larger amplitude (or bright light fringes) on an optical screen that is parallel to the screen containing the two slits. Destructive interference, on the other hand, occurs when the waves are exactly out of phase, causing the amplitudes to cancel with each other and leading to *nodes* (or dark fringes) on the optical screen.

In 1873, building on the work of others, James Clerk Maxwell published his comprehensive theory of electromagnetic radiation in *A Treatise on Electricity and Magnetism*. In his book, Maxwell argued that light can be thought of as a transverse wave that consists of perpendicularly oscillating electric (E) and magnetic (B) fields that each lie perpendicular to the direction of travel, as shown in Figure 3.3. Light travelling in a vacuum travels at the speed of light, which is defined as follows: $c \equiv 2.99792458 \times 10^8$ m/s. Substituting c for v in Equation (3.1) yields the more familiar Equation (3.2). The different types or colors of *electromagnetic radiation* can be classified according to either their wavelengths or their frequencies, as shown in Figure 3.4.

$$c = \upsilon\lambda = 2.99792458 \times 10^8 \text{ m/s} \qquad (3.2)$$

Example 3-1. The standard garage door opener typically operates at a frequency of about 400 Hz (1 Hz = 1 s^{-1}). Calculate the wavelength for this frequency of light and identify to which region of the electromagnetic spectrum it belongs.

Solution. Solving Equation (3.2) for the wavelength, one obtains:

$$\lambda = \frac{c}{\nu} = \frac{2.99792458 \times 10^8 \text{ m/s}}{400(1/\text{s})} = 0.750 \text{ m}$$

This wavelength falls in the microwave region of the electromagnetic spectrum.

At Ursinus College, where I teach, there is a training drill that the football team uses called the *ropes exercise*. With one end of a heavy rope affixed to the wall, the football player swings the other end of the rope up and down, creating a transverse wave in the rope, as shown in Figure 3.5. Initially, the wave travels from the football

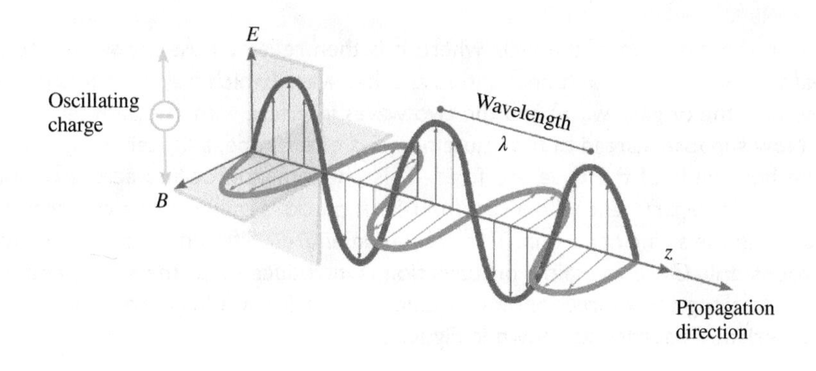

FIGURE 3.3
Maxwell's depiction of a light wave as perpendicularly oscillating electric (E) and magnetic (B) fields. [Blatt Communications.]

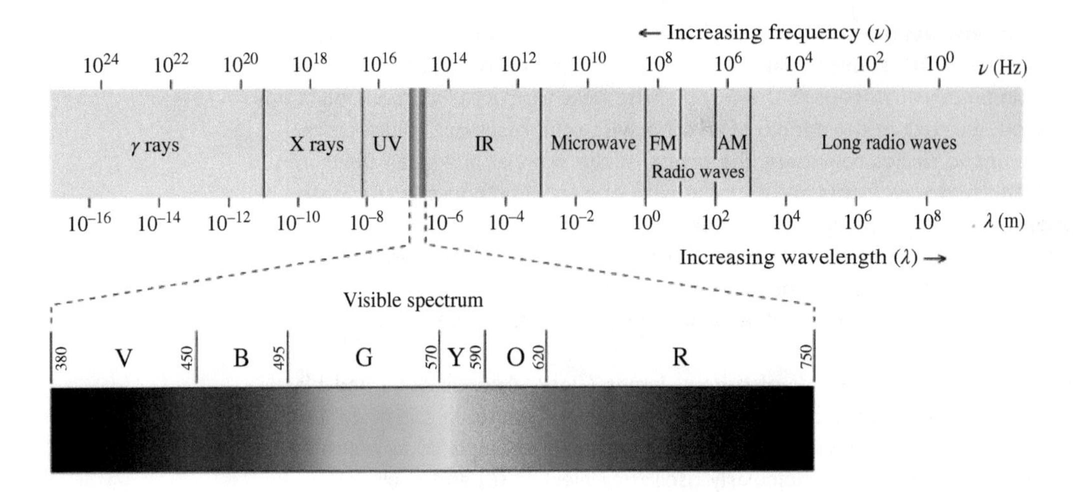

FIGURE 3.4
The electromagnetic spectrum, showing an expansion of the visible region in color. [Attributed to Philip Ronan under the Creative Commons Attribution-Share Alike 3.0 Unported license (accessed October 17, 2013).]

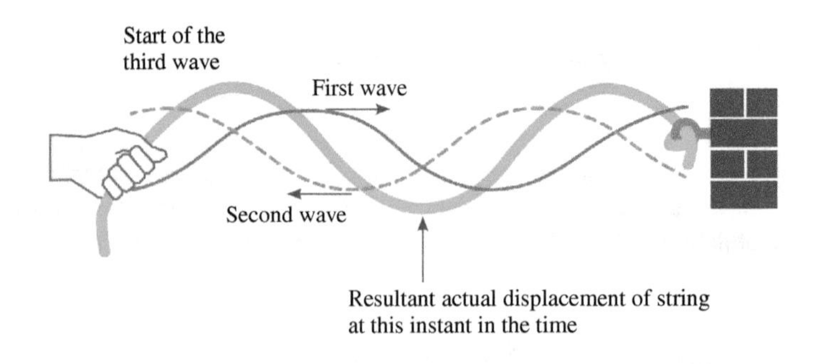

FIGURE 3.5
Schematic diagram of the ropes exercise, where one end of the rope is fixed while the football player initiates a transverse wave in the rope by an up-and-down motion. [Blatt Communications.]

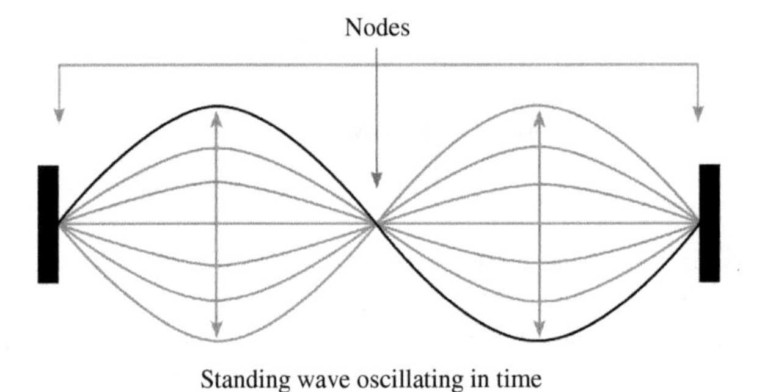

FIGURE 3.6
An example of a standing wave such as the one produced when a guitar string is plucked. [Blatt Communications.]

player in the direction of the wall, where it is then reflected. As the wave returns toward the football player, it encounters another wave (which may or may not be in phase with the original wave) and the two waves interfere with one another.

Now suppose instead of fixing just one end of the rope, a lighter string is used where both ends of the rope are fixed. This phenomenon occurs across campus in the music department whenever a musician plucks a guitar string or creates a vibration in the string of a violin. This creates an entirely different kind of wave than the ropes drill. Once the initial perturbation is introduced into the string, creating amplitude along the *y*-direction, for instance, the string will begin to oscillate at a characteristic frequency, as shown in Figure 3.6.

In order to sustain the note and to create music, only those wavelengths that have an amplitude of zero at both ends of the fixed string can constructively interfere with one another to create what is known as a *standing wave*. A *standing wave* (also called a *stationary wave*) is a wave that exists in a fixed position. Although the standing wave appears to be stationary, it is actually oscillating up and down in place as a function of time. Therefore, the amplitude of the wave (y) is a function of its position (x) and time (t), as shown in Equation (3.3). This is the classical wave equation in one-dimension (see Appendix A for a derivation).

$$\frac{\partial^2 y}{\partial x^2} = \frac{1}{v^2}\frac{\partial^2 y}{\partial t^2} \tag{3.3}$$

The wave equation is a linear, second-order partial differential equation. While the solution to this type of differential equation is not particularly difficult, it might present a challenge for the average student taking inorganic chemistry. Therefore, a general solution is presented and it is left as an exercise to demonstrate the validity of this result. One possible solution to the one-dimensional wave equation is the sine wave given by Equation (3.4), where A is the maximum amplitude of the wave.

$$y(x,t) = A\sin\left[2\pi\left(\frac{x}{\lambda} - \nu t\right)\right] \tag{3.4}$$

Example 3-2. Prove that Equation (3.4) is a solution to the one-dimensional classical wave equation given in Equation (3.3).

Solution. Taking the first partial derivative of y with respect to x and t yields the following:

$$\frac{\partial y}{\partial x} = \frac{2\pi}{\lambda}A\cos\left[2\pi\left(\frac{x}{\lambda} - \nu t\right)\right]$$

$$\frac{\partial y}{\partial t} = -2\pi\nu A\cos\left[2\pi\left(\frac{x}{\lambda} - \nu t\right)\right]$$

Now, taking the second partial derivative of y with respect to x and t yields:

$$\frac{\partial^2 y}{\partial x^2} = -\left(\frac{2\pi}{\lambda}\right)^2 A\sin\left[2\pi\left(\frac{x}{\lambda} - \nu t\right)\right] = -\left(\frac{2\pi}{\lambda}\right)^2 y(x,t)$$

$$\frac{\partial^2 y}{\partial t^2} = -(2\pi\nu)^2 A\sin\left[2\pi\left(\frac{x}{\lambda} - \nu t\right)\right] = -(2\pi\nu)^2 y(x,t)$$

Operators such as this that yield back the original function times a constant are called *eigenfunctions* and the corresponding constants are known as the *eigenvalues*. You will find that most of the operators in quantum mechanics will be of this type. Substituting the second partial derivatives into Equation (3.3) and cancelling the signs yields

$$\left(\frac{2\pi}{\lambda}\right)^2 y(x,t) = \frac{1}{v^2}(2\pi\nu)^2 y(x,t)$$

which reduces to

$$\frac{1}{\lambda^2} = \frac{\nu^2}{v^2}$$

Taking the square root of both sides yields the equality given by Equation (3.1).

By introducing the constraint that the amplitude of the wave must be zero at both ends of the string in order to have constructive interference, we have imposed a set of *boundary conditions* on the mathematical problem, namely, that $y(0, t) = y(L, t) = 0$ at any value of t, where L is the length of the string. Because this must be true for any value of t, we can arbitrarily set $t = 0$ to simplify the math, so that $A \sin(0) = A \sin(2\pi L/\lambda) = 0$. Ignoring the trivial solution that $A = 0$ and recognizing that $\sin(n\pi) = 0$ whenever n is an integer, we obtain the result that $\lambda=2L/n$. As this example demonstrates, the concept of *quantization* (integral values of n) arises naturally whenever we force the ends of the string to be fixed in place. Several of the standing wave solutions for different values of n are depicted in Figure 3.7. One feature that is common to all of the solutions is the presence of nodes (where the amplitude is always zero) and the presence of antinodes (where the amplitude reaches its maximum value).

The lowest frequency occurs when $n = 1$ and is called the *fundamental*. Doubling the frequency corresponds to raising the pitch by an octave. Those solutions having values of $n > 1$ are known as the *overtones*. As mentioned previously, one important property of waves is the concept of superposition. Mathematically, it can be shown that any periodic function that is subject to the same boundary conditions can be represented by some linear combination of the fundamental and its overtone frequencies, as shown in Figure 3.8. In fact, this type of mathematical analysis is known as a *Fourier series*. Thus, while the note middle-A on a clarinet, violin, and piano all have the same fundamental frequency of 440 Hz, the sound (or timbre) that the different instruments produce will be distinct, as shown in Figure 3.9.

If electromagnetic radiation can be considered a wave, then what exactly is it that is doing the waving? Imagine a cork bobbing up and down on the surface of a reservoir as a result of the wave from a passing boat. To a first approximation, the cork will not change its longitudinal or latitudinal position. It will simply oscillate up and down in position as the wavefront passes by. Now, imagine the opposite scenario, where a cork is forced to oscillate in a periodic manner on the surface of a smooth body of water. The periodic motion of the cork will lead to the generation of waves in the liquid. By analogy, light can be produced by an oscillating charge, such as an electron (or for that matter any charged particle) being accelerated back and forth between two poles (e.g., in a radio transmission tower).

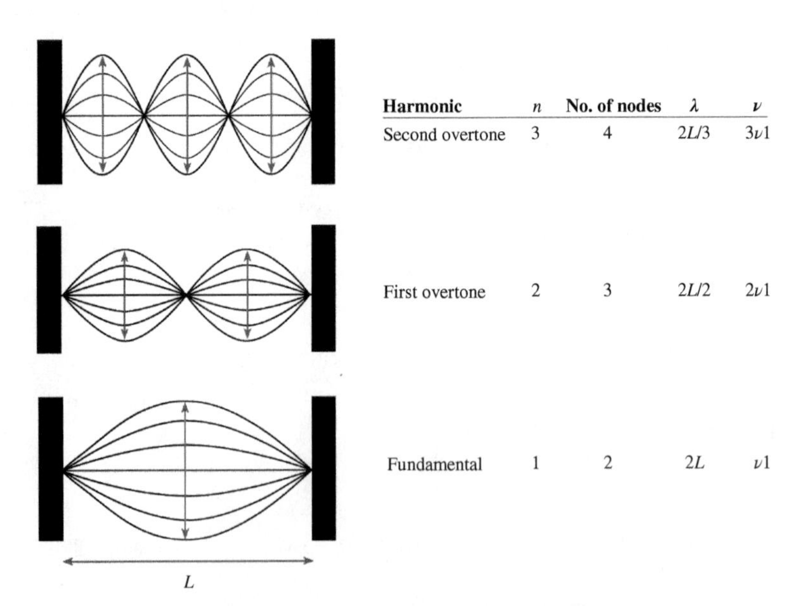

FIGURE 3.7
Several solutions to the wave equation subject to the boundary conditions for a standing wave that $y(0,t) = y(L,t) = 0$. [Blatt Communications.]

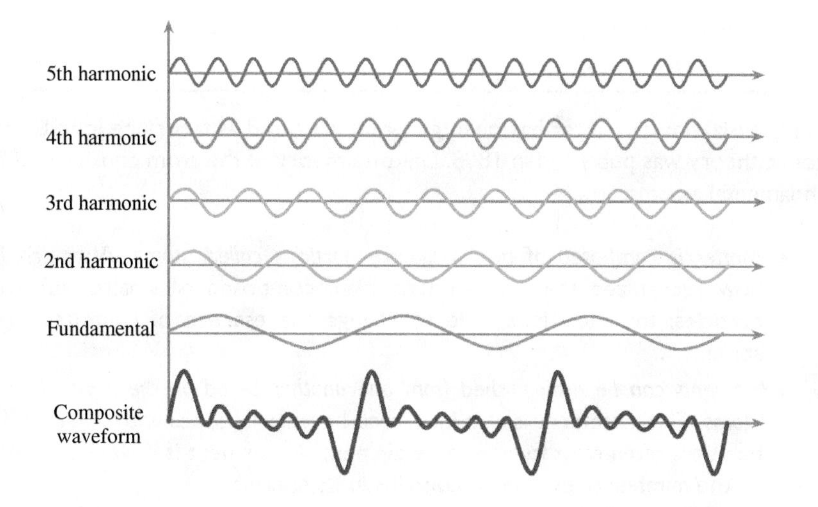

FIGURE 3.8
A composite periodic waveform can be constructed from a linear combination of the fundamental and its overtones (or partials). [Blatt Communications.]

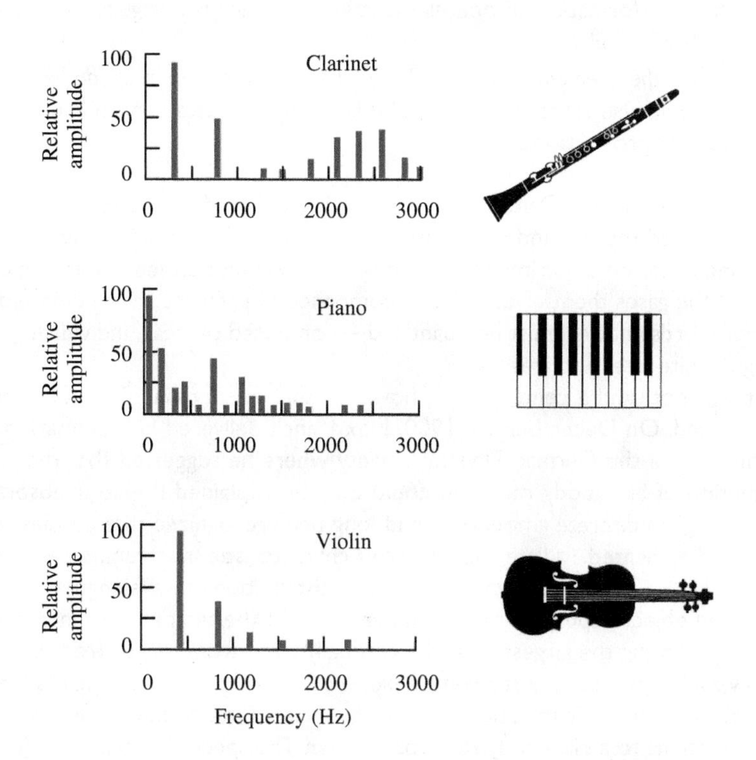

FIGURE 3.9
Illustration of how the timbre of different musical instruments playing the same note can produce distinctly different sounds depending on the exact combination of the fundamental and its overtones. [Blatt Communications.]

Now suppose that a cork was floating peacefully on the surface of a perfectly smooth body of water. Then along comes the wake from a passing motorboat. What will happen? The cork will begin to oscillate up and down in a periodic manner as the wavefronts from the motorboat pass by in the water underneath it. In a similar manner, the interaction of light with matter can cause oscillations of the charged particles within the substance. Low-frequency radiation, such as infrared light, can cause the positions of the atoms in molecules to vibrate, whereas high-frequency radiation, such as ultraviolet light, can effect electronic transitions or even ionization. Therefore, the different regions of the electromagnetic spectrum can be used to probe various aspects of a molecule's fundamental structure. This serves as the basis for the field of atomic and molecular spectroscopy.

3.2 PROBLEMS WITH THE CLASSICAL MODEL OF THE ATOM

The quantization of matter has been generally accepted as fact since John Dalton's atomic theory was published in 1808. Dalton's theory of the atom consisted of four fundamental axioms:

- *Matter is composed of tiny, indivisible particles called atoms.* Although it is now recognized that the atom is itself composed of smaller subatomic particles, this fact does little to change the essence of Dalton's original statement.
- *Elements can be distinguished from one another based on the masses of their atoms.* This statement is also incorrect, because isotopes with different mass numbers often exist for the same element. An element is now characterized by the number of protons it contains in its nucleus.
- *Atoms can neither be created, nor destroyed.* This assertion is also false, because the transformation of one element into another is a regular occurrence in nuclear chemistry.
- *When the elements combine in a chemical reaction, they do so in small, whole-number ratios.* This final point is simply a restatement of Proust's law of definite proportions.

The importance of Dalton's work is that it was based on the scientific method. Dalton realized that air and evaporated water consisted of fundamentally different gases; and it was his experiments using mixtures of gases that led him to the conclusion that the gases themselves must be composed of particles having different sizes. In other words, matter must be quantized—composed of small, individual particles having definite sizes and masses.

It was not until a century later, however, that energy was first considered to be quantized. On December 14, 1900, Max Planck delivered his landmark address at a meeting of the German Physical Society where he suggested that the spectral distribution of blackbody radiation could only be explained if matter absorbs and emits energy in discrete amounts. It had long been recognized that certain metallic objects, when heated, radiated light having a characteristic spectrum, such as the one shown in Figure 3.10. Furthermore, the exact distribution of wavelengths radiated by the heated object depended on its temperature. As the temperature increases, the wavelength having the largest intensity gradually shifts from longer (red) to shorter (blue) wavelengths. Strictly speaking, a *blackbody* is an idealized object that can perfectly absorb and emit radiation of all frequencies. One of the closest real-world approximations to a blackbody radiator is a star. The spectral distribution (intensity vs wavelength profile) of a star is analogous to that of an ideal blackbody, with the wavelength of its maximum intensity decreasing with increasing temperature. This explains why a cooler star (~3000 K) has a red appearance, while the hottest of stars (>5000 K) are blue, as shown in Figure 3.10.

The distribution curve for a blackbody that is predicted by classical physics is also shown in Figure 3.10. This curve is based on the *Rayleigh–Jeans law*, which is given by Equation (3.5), where ρ is the energy density per frequency (proportional to the intensity axis shown in Figure 3.10), k_B is the Boltzmann constant ($k_B = R/N_A = 1.381 \times 10^{-23}$ J/K), and T is the absolute temperature (K). At long wavelengths, there is a fairly good agreement between the experimental and predicted distribution curves for a blackbody having $T = 5000$ K. However, at shorter wavelengths, the classical prediction clearly deviates from observation. Because this deviation begins to occur at wavelengths that fall in the ultraviolet range of the electromagnetic spectrum, this

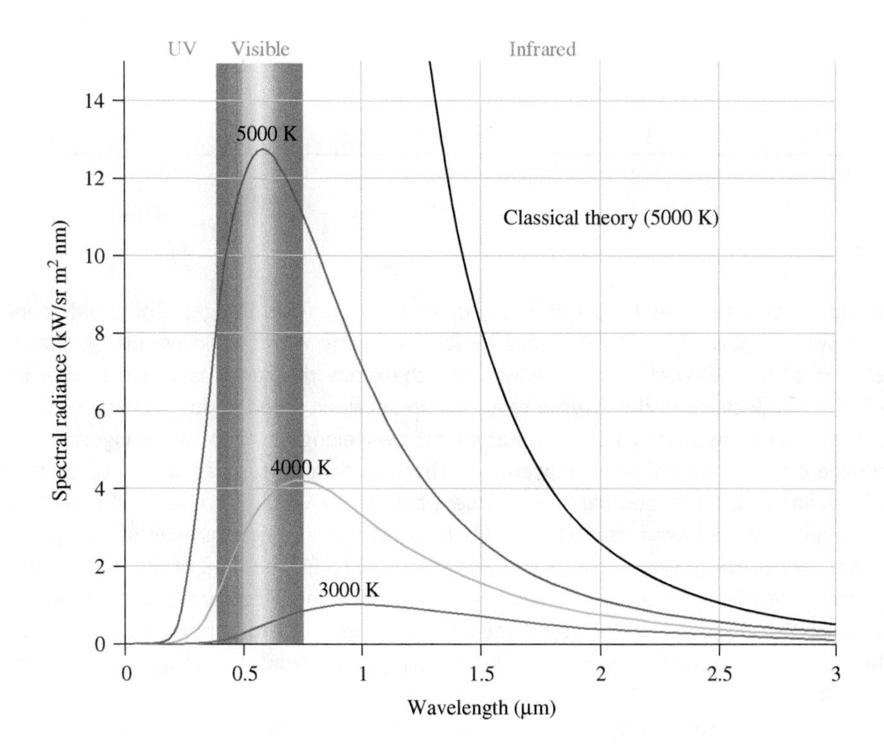

FIGURE 3.10 The spectral distribution of radiation as a function of wavelength for three different temperature stars. A star is one of the closest approximations to a perfect blackbody radiator. [Reproduced from http://en.wikipedia.org/wiki/Black-body_radiation (accessed October 17, 2013).]

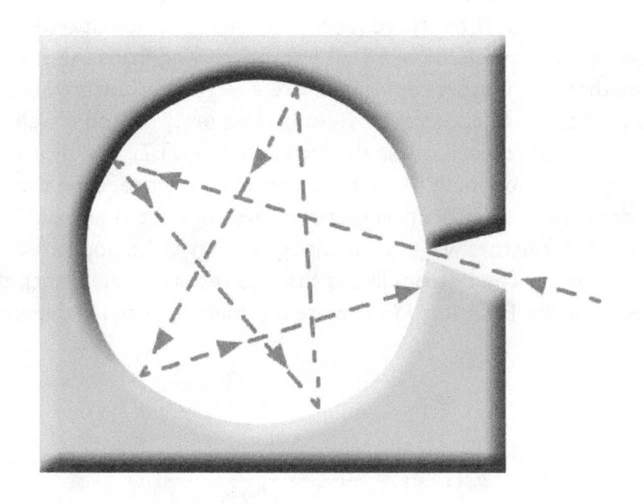

FIGURE 3.11 Analogy of a perfect blackbody with a hollow metal cavity having a tiny pinhole. [Blatt Communications.]

shortcoming of classical theory is often referred to as the *UV catastrophe*.

$$\rho_\nu(T)\,d\nu = \frac{8\pi\nu^2 k_B T}{c^3}\,d\nu \tag{3.5}$$

According to classical mechanics, the absorption of heat by a metal object leads to an increase in the kinetic energy of its constituent atoms. As the energy is absorbed, it causes the electrons and charged nuclei in the solid to oscillate back and forth with characteristic frequencies. To a first approximation, an idealized blackbody will behave like a hollow cavity with a small pinhole in it, as the one depicted in Figure 3.11. When the object is heated or absorbs electromagnetic radiation, the total energy is forced to equilibrate inside the cavity before any radiation can be emitted through the pinhole in the form of light. As was shown in Section 3.1, the boundary conditions for constructive interference of standing waves in a cavity

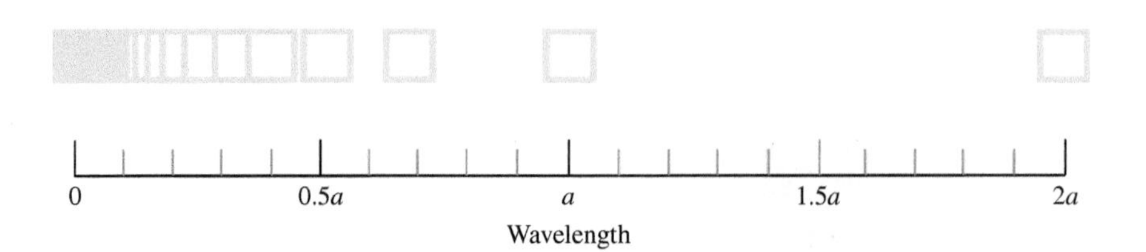

FIGURE 3.12
Illustration of how the number of allowed oscillators increases as the wavelength decreases. [Blatt Communications.]

of diameter L requires that $\lambda = 2L/n$, where n is a positive integer. Thus, only certain wavelengths of oscillations may exist as standing waves. The wavelengths and shapes of the allowed standing waves are therefore the same as those shown in Figure 3.7. Because of the inverse relationship between λ and n, the probability density of allowed oscillations increases at shorter wavelengths, as shown in Figure 3.12, where $a = L$. Classical physics predicted that the total energy absorbed by all the vibrations of the charged particles inside a blackbody would be equally distributed over all of the allowed oscillators. This is known as the *equipartition of energy*. In other words, there were no special rules imparted to make some of the oscillations more favorable than others. Thus, the intensity of the electromagnetic radiation emitted through the pinhole would necessarily increase at shorter and shorter wavelengths, simply because there is a higher density of allowed wavelengths at shorter wavelengths.

Planck overcame this limitation by postulating that the energies of the oscillations could only exist in discrete amounts, with the energy (E) of a given oscillation being directly proportional to its frequency, according to Equation (3.6), where h is Planck's constant ($h = 6.6261 \times 10^{-34}$ J s). Borrowing from the statistical work of Boltzmann, those oscillators requiring more energy to vibrate would necessarily have a lower probability of occurrence. Because frequency (υ) and wavelength (λ) are inversely proportional to each other through Equation (3.2), the shorter the wavelength, the less probable would be its occurrence. Therefore, even though there is still a higher density of oscillators having short wavelengths, the probability that these short-wavelength oscillators will absorb enough energy to be populated will eventually begin to decrease. The bottom line is that the results of the Planck distribution law, which is given by Equation (3.7), perfectly match those of the experimental curve!

$$E = h\upsilon \tag{3.6}$$

$$\rho_\nu(T)\,d\nu = \frac{8\pi h k_B T}{c^3}\frac{\nu^3 d\nu}{e^{h\nu/k_B T} - 1} \tag{3.7}$$

The essential mathematical substitution made by Planck was the recognition that the energy differences between the allowed oscillations in the solid were quantized, as shown in Figure 3.13. As a result, there were certain rules that had to be introduced in order to govern how much of the total energy a specific oscillator was able to absorb. Another way of thinking about the distribution problem is to consider an old-fashioned coin scramble. Suppose that there is a total of \$9.00 worth of coins (a certain total amount of absorbed energy) to distribute among three very eager children (three of the allowed standing waves or oscillators in Figure 3.13). There are \$3.00 worth of quarters (12 quarters), \$3.00 worth of nickels (60 nickels), and \$3.00 worth of pennies (300 pennies). In the absence of any special rules (no quantization of energy), each of the children will pick up exactly 4 quarters, 20 nickels, and 100 pennies for a total of \$3.00 each. This is the classical analogy to the equipartition of energy—each of the children (oscillators) acquired the same total amount of money (energy). Next, the coin scramble is repeated; but as the organizer, Planck imposes

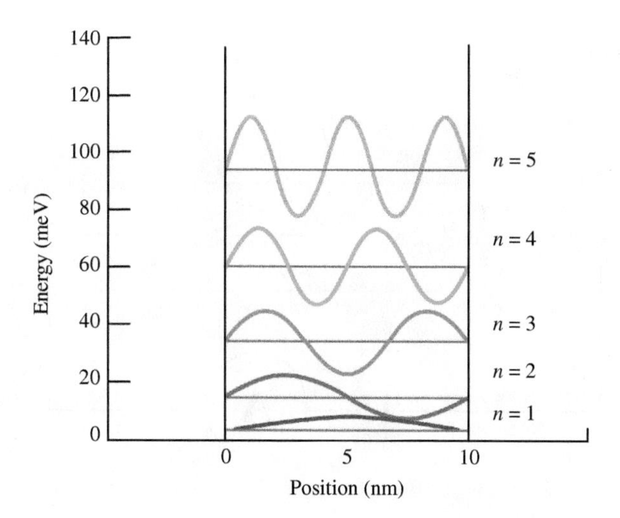

FIGURE 3.13
Energy level diagram for the different allowed vibrations of an oscillating string, showing how the energy increases with n. According to Boltzmann's statistical population theorem, low-frequency oscillations will be populated (or occur) more often than higher frequency ones. [Blatt Communications.]

some rules on the game: the first child is not picky and can pick up any denomination of currency, but the second child is a little more discerning and refuses to bend over to pick up any pennies, and finally the third child comes from an exceedingly snobbish background and will only stoop to pick up quarters.

The final results of this second coin scramble are given in Table 3.1. Notice that the rules (or quantum restrictions) that Planck placed on the coin scramble led to an unequal distribution of wealth among the three participants. The restriction that the third child could only acquire the largest denomination of currency led him/her to a smaller take of the total amount of money, just as those oscillators that required a larger quantum jump were less populated in the equilibrium distribution of energy. As it turns out, these are the oscillations with smaller wavelengths; and therefore the intensity profile of Planck's distribution curve follows the Rayleigh–Jeans law at longer wavelengths, but then curves back downward as the wavelength gets shorter.

At first, the implications of the quantization of energy were not fully appreciated. Planck himself was reluctant to accept the concept, convinced that it was merely a mathematical coincidence and that a more satisfactory classical description would eventually replace it. Quoting Banesh Hoffmann in *The Strange Story of the Quantum*, the scientists "could but make the best of it, and went around with woebegone faces sadly complaining that on Mondays, Wednesdays, and Fridays they must look on light as a wave: on Tuesday, Thursdays, and Saturdays as a particle. On Sundays, they simply prayed."

It was not until 5 years later (in 1905) when Albert Einstein explained the photoelectric effect that the concept of quantization looked as if it had any real staying

TABLE 3.1 Results of the coin scramble with and without Planck's special rules.

	Equal Chance (No Rules)			Planck's Rules		
Distribution of coins:	#1	#2	#3	#1	#2	#3
Pennies	100	100	100	300	0	0
Nickels	20	20	20	30	30	0
Quarters	4	4	4	4	4	4
Total amount ($)	3.00	3.00	3.00	5.50	2.50	1.00

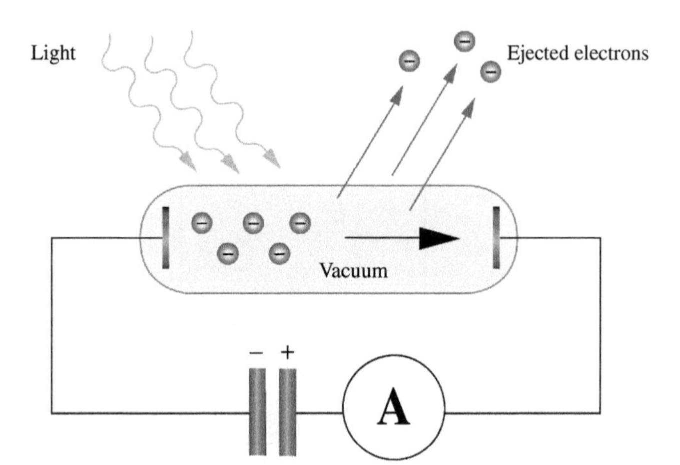

FIGURE 3.14
The photoelectric effect: light striking a metal surface in an evacuated chamber can eject electrons from the metal if the frequency of the light is greater than the work function for the metal. [Blatt Communications.]

power. In 1887, Heinrich Hertz had observed that UV light can eject electrons from a metal plate in a partially evacuated tube (Figure 3.14). There were three major observations associated with this phenomenon. (i) While the kinetic energy of the ejected electrons depended on the frequency, the number of ejected electrons was dependent on the intensity of the light. (ii) There was a certain threshold frequency, characteristic of the metal, below which no electrons at all would be ejected, regardless of the intensity of the light, as shown in Figure 3.15. (iii) There was no lag time between the time the light struck the metal plate and the time that the electrons were ejected. Each of these observations was in direct conflict with the predictions of classical theory. Because the intensity of light (according to Maxwell's equations) is proportional to the square of the amplitude of its electric field, the classical prediction was that the kinetic energy of the electrons should depend on the intensity of the light, not on its frequency. Furthermore, before an electron could be ejected, it was expected that a sufficient amount of energy must be accumulated in order to overcome the surface potential and yet no lag time was observed. Lastly, there was no classical explanation for the threshold energy.

Only by assuming that light behaved as packets of energy did could Einstein explain the experimental data. Einstein originally called these *quanta*, while the modern term for them is *photons*. Furthermore, the energy of each photon was directly related to its frequency through the same proportionality constant, h, that Planck had used in his explanation of blackbody radiation. Only those photons having sufficient energy to overcome the work function (ϕ) of the metal (or its ionization energy) would be capable of ejecting an electron. The kinetic energy (KE) of the ejected electron was therefore dependent on the frequency of the photon, as shown by Equation (3.8). The number of ejected electrons, in turn, depended on the intensity

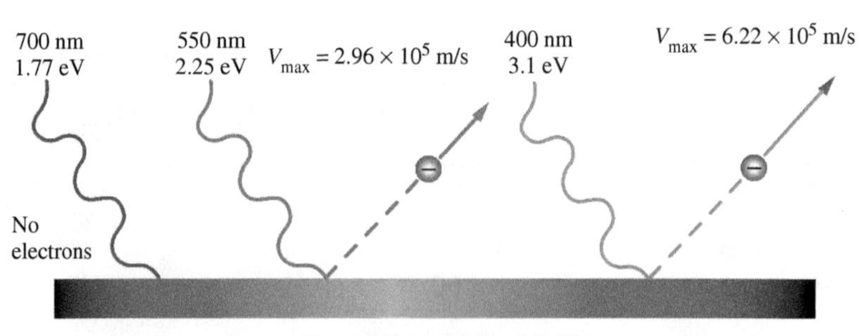

FIGURE 3.15
The photoelectric effect illustrated for potassium metal in a partially evacuated tube. The work function required for the ejection of electrons from a potassium surface is 2.0 eV. Therefore, a photon of red light ($h\nu$ = 1.77 eV) does not have sufficient energy to eject an electron. Photons of green or blue light, which have energies greater than 2 eV, can both eject electrons; but the kinetic energies of the ejected electrons will differ. [Blatt Communications.]

of light, or the number of photons striking the metal surface. Thus, for every photon striking the surface (as long as $h\nu > \phi$), one electron would be ejected. In other words, energy (in the form of electromagnetic radiation) can be considered to have particle-like properties.

$$KE = h\nu - \phi \tag{3.8}$$

Example 3-3. Potassium has a work function, $\phi = 2.0$ eV. Calculate the velocity of an electron ejected from potassium using blue light with a wavelength of 400 nm.

Solution. The energy of a photon of blue light is directly proportional to its frequency. Combination of Equations (3.6) and (3.2) yields

$$E = \frac{hc}{\lambda} = \frac{(6.6261 \times 10^{-34}\,\text{J s})(2.998 \times 10^{8}\,\text{m/s})}{(400\,\text{nm})(1\,\text{m}/10^{9}\,\text{nm})} = 4.97 \times 10^{-19}\,\text{J}$$

Given that 1 eV = 96,485 J/mol and $N_A = 6.022 \times 10^{23}$ mol^{-1}:

$$E = 4.97 \times 10^{-19}\,\text{J}\left(\frac{1\,\text{eV}}{96,485\,\text{J/mol}}\right)(6.02 \times 10^{23}\,\text{mol}^{-1}) = 3.10\,\text{eV}$$

Using Equation (3.8), the kinetic energy of the ejected electron is

$$KE = h\nu - \phi = 3.10\,\text{eV} - 2.0\,\text{eV} = 1.1\,\text{eV}$$

$$KE = 1.1\,\text{eV}\left(\frac{96,485\,\text{J/mol}}{1\,\text{eV}}\right)\left(\frac{1\,\text{mol}}{6.022 \times 10^{23}}\right) = 1.8 \times 10^{-19}\,\text{J}$$

Because KE = (1/2)mv^2, rearrangement to solve for v yields

$$v = \sqrt{\frac{2KE}{m_e}} = \sqrt{\frac{2(1.8 \times 10^{-19}\,\text{J})}{9.11 \times 0^{-31}\,\text{kg}}} = 6.2 \times 10^{5}\,\text{m/s}$$

The first model of the atom to include the concept of quantization was proposed by Niels Bohr in 1913. Before this time, the prevailing notion was that the atom consisted of a dense, positively charged nucleus, about which the electrons orbited much like the planets orbit around the sun. One of the inherent contradictions with the *planetary model* was that the negatively charged electrons should undergo a constant acceleration by virtue of their electrostatic attraction to the nucleus. At the same time, Maxwell's theory of electromagnetism requires that any charged particle undergoing acceleration must continuously emit light. In fact, classical calculations predicted that the electron in a hydrogen atom should rapidly collapse into the nucleus in a matter of several nanoseconds. This, of course, does not occur. Furthermore, the gradual loss of energy as the electron's orbit spirals in closer and closer to the nucleus should lead to an emission spectrum that resembles a continuum of many different wavelengths.

In fact, it had been known for several decades that when certain metal cations are heated in a Bunsen burner, they will emit light of a characteristic color. These are the common flame tests of a qualitative general chemistry laboratory and that form the basis for the different colors in fireworks. For instance, Li^+ and

Sr^{2+} are red, Na^+ is yellow-orange, and Ba^{2+} is green. Instead of a continuum, as would have been predicted by classical theory, each ion emitted light with a characteristic fingerprint of narrow wavelengths (or lines) when passed through a prism. The line spectrum of hydrogen was well established at the time and is shown in Figure 3.16. In 1885, a Swiss math teacher named Johann Balmer noticed the mathematical pattern in the spectral lines of hydrogen in the visible spectrum and derived an empirical equation to explain them. Then, in 1888, the Swedish physicist Johannes Rydberg generalized the Balmer equation to include several other series of lines that were discovered in the UV and near-IR regions of the hydrogen line spectrum (shown in Figure 3.17). The Rydberg formula, shown in Equation (3.9), can be used to calculate the wavenumber ($1/\lambda$) of any line in the emission spectrum of hydrogen, where R_H is the Rydberg constant and n_f and n_i are both positive integers (with $n_f < n_i$). Each series of lines converges at short wavelengths on what is known as the *series limit*. Because the series limit occurs at the shortest wavelength in each series (or the largest wavenumber), the value of n_i for the

FIGURE 3.16
The line spectrum of hydrogen in the visible region. [Reproduced from http://en.wikipedia.org/wiki/Hydrogen_spectral_series (accessed November 30, 2013).]

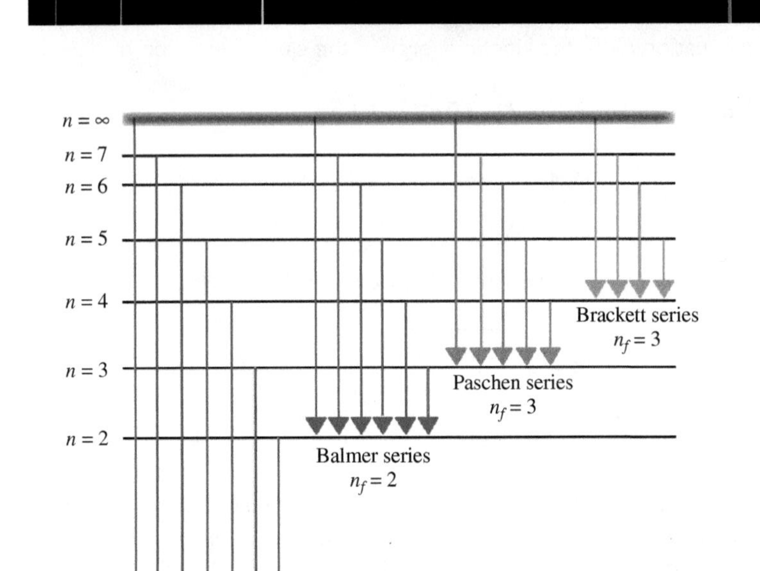

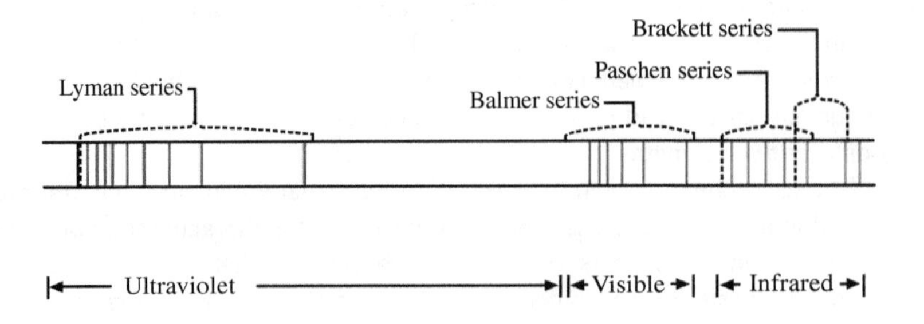

FIGURE 3.17
The complete line spectrum of hydrogen from the ultraviolet to the infrared. Transitions from the $n_i = \infty$ energy level to the lowest energy level in each series correspond with the series limit, where the individual energy levels all start to blur together. [Blatt Communications.]

series limit is always taken as infinity. While mathematically convenient, the Rydberg equation was entirely an empirical equation in desperate need of a more theoretical explanation.

$$\bar{\nu} = \frac{1}{\lambda} = R_H \left(\frac{1}{n_f^2} - \frac{1}{n_i^2} \right), \text{ where } R_H = 109,678\,\text{cm}^{-1} \text{ and } n_f < n_i \tag{3.9}$$

Example 3-4. Calculate the wavelengths (nm) of the five lines shown in Figure 3.16 for the line spectrum of hydrogen if $n_f = 2$ for the Balmer series.

Solution. Using Equation (3.9) with $n_f = 2$ and $n_i = 3$, the wavenumber of the red line in the Balmer series is determined to be 1.52×10^4 cm^{-1}:

$$\bar{\nu} = \frac{1}{\lambda} = 109,678\,\text{cm}^{-1} \left(\frac{1}{2^2} - \frac{1}{3^2} \right) = 1.52 \times 10^4\,\text{cm}^{-1}$$

The wavelength of this line is obtained by taking the reciprocal of the wavenumber and then converting from cm into nm, as follows:

$$\lambda = \frac{1}{\bar{\nu}} = \frac{1}{1.52 \times 10^4\,\text{cm}^{-1}} = 6.56 \times 10^{-5}\,\text{cm} \left(\frac{1\,\text{m}}{100\,\text{cm}} \right) \left(\frac{10^9\,\text{nm}}{1\,\text{m}} \right) = 656\,\text{nm}$$

Using $n_i = 4$, 5, 6, and 7, the resulting wavelengths for the remaining lines are 486, 434, 410, and 397 nm, respectively.

Example 3-5. The wavelength of the series limit for the Pfund series of lines in the hydrogen spectrum is 2.28 μm. Determine n_f for the Pfund series and then calculate the longest wavelength line for this series.

Solution. Because $n_i = \infty$ for the series limit, the second term in parentheses in Equation (3.9) approaches zero.

$$\frac{1}{\lambda} = \frac{109,678\,\text{cm}^{-1}}{n_f^2} = \frac{1}{2.28\,\mu\text{m}} \left(\frac{10^6\mu\text{m}}{1\,\text{m}} \right) \left(\frac{1\,\text{m}}{100\,\text{cm}} \right) = 4390\,\text{cm}^{-1}$$

$$n_f^2 = \frac{109,678\,\text{cm}^{-1}}{4390\,\text{cm}^{-1}} = 25 \text{ so } n_f = 5$$

3.3 THE BOHR MODEL OF THE ATOM

In the summer of 1912, Niels Bohr wrote to his older brother Harald: "Perhaps I have found out a little about the structure of atoms." His revolutionary model of the atom was published the following year. In an attempt to develop a theoretical model of the hydrogen atom that was consistent with the lines predicted by the Rydberg formula, Bohr proposed the following:

- The electron moves around the nucleus in circular orbits, where the centripetal and centrifugal forces are exactly balanced. The centrifugal force results from the electrostatic attraction of the negatively charged electron for the positively charged proton in the nucleus and can be calculated according to Coulomb's law. The electron's centripetal motion is governed by Newton's second law. The two forces are set equal in Equation (3.10), where F is the force, e is the electronic charge = 1.602×10^{-19} C, r is the radius, $4\pi\varepsilon_0$ is the permittivity of free space = 1.113×10^{-12} C^2/Nm^2, and r is the radius of the orbit.

$$F = \frac{e^2}{4\pi\varepsilon_0 r^2} = ma = \frac{mv^2}{r}$$

(Coulomb's law) (Newton's 2nd law) (3.10)

- *The stationary state assumption*: In a given orbit, the total energy (kinetic + potential) will be a constant. The kinetic energy KE is equal to $(1/2)mv^2$, while the potential energy V can be obtained by integrating Coulomb's law with respect to distance. While this assumption was in opposition to the classical prediction that the electron will spiral into the nucleus, it was necessary in order to explain the experimental observations.

$$E = KE + V = \frac{1}{2}mv^2 + \int \frac{e^2}{4\pi\varepsilon_0^2 r^2}dr = \frac{1}{2}mv^2 - \frac{e^2}{4\pi\varepsilon_0^2 r} \tag{3.11}$$

Given the equality in Equation (3.11), followed by substitution from Equation (3.10) yields

$$E = \frac{1}{2}mv^2 - mv^2 = -\frac{e^2}{8\pi\varepsilon_0 r} \tag{3.12}$$

- *The quantum restriction postulate*: Only certain quantized orbits will be allowed. These orbits are restricted to the condition where the angular momentum (l) is an integral multiple of $h/2\pi$:

$$l = mvr = \frac{nh}{2\pi}, \quad \text{where } n = 1, 2, 3, \ldots \tag{3.13}$$

Solving Equation (3.12) for mvr and setting it equal to Equation (3.13) yields

$$mvr = \frac{e^2}{4\pi\varepsilon_0 v} = \frac{nh}{2\pi} \tag{3.14}$$

Solving Equation (3.14) for v and substituting for E in Equation (3.12) yields

$$v = \frac{2\pi e^2}{4\pi\varepsilon_0 nh} \tag{3.15}$$

$$E = -\frac{1}{2}mv^2 = -\frac{2\pi^2 me^4}{(4\pi\varepsilon_0)^2 n^2 h^2} \tag{3.16}$$

- When an electron jumps from a higher to a lower energy orbit, the energy difference (ΔE) will be emitted as a photon:

$$E = E_i - E_f = \frac{2\pi^2 me^4}{(4\pi\varepsilon_0)^2 h^2}\left(\frac{1}{n_f^2} - \frac{1}{n_i^2}\right) \tag{3.17}$$

Substituting e = 1.602×10^{-19} C, $m_e = 9.109 \times 10^{-31}$ kg, $4\pi\varepsilon_0 = 1.113 \times 10^{-10}$ C^2/J·m, and $h = 6.626 \times 10^{-34}$ J·s, one obtains a value of 2.178×10^{-18} J for R_∞.

- It is impossible to describe the electron when it is in between the stationary state orbits.

Example 3-6. Prove that Equation (3.17) is just another way of stating the Rydberg formula given in Equation (3.9).

Solution. Combining Equation (3.17) with the Planck formula $\Delta E = h\nu$:

$$\frac{1}{\lambda} = \frac{\nu}{c} = \frac{\Delta E}{hc} = \frac{2.178 \times 10^{-18}\,\mathrm{J}}{(6.626 \times 10^{-34}\,\mathrm{J \cdot s})(2.998 \times 10^{8}\,\mathrm{m/s})}\left(\frac{1}{n_f^2} - \frac{1}{n_i^2}\right)$$

$$\frac{1}{\lambda} = 1.096 \times 10^{7}\,\frac{1}{\mathrm{m}}\left(\frac{1\,\mathrm{m}}{100\,\mathrm{cm}}\right) = 1.096 \times 10^{5}\,\mathrm{cm}^{-1}$$

This result is consistent with the Rydberg formula to the number of significant figures presented.

Using the circular orbits of the Bohr model, the meaning of the subscripts in the quantum numbers n_f and n_i now become obvious: n_i is the initial Bohr orbit (the one farther from the nucleus, or the excited state) and n_f is the final Bohr orbit, as shown in Figure 3.18. The following values of n_f correspond with the listed series: $n_f = 1$ (Lyman), $n_f = 2$ (Balmer), $n_f = 3$ (Paschen), $n_f = 4$ (Brackett), and $n_f = 5$ (Pfund). Thus, for example, the first four lines for the Balmer series originate in the $n_i = 3$, 4, 5, and 6 energy levels and terminate in the $n_f = 2$ level, as shown in Figure 3.18.

The actual radius of a Bohr orbit can be calculated by solving Equation (3.12) for r and substituting the value of E given by Equation (3.17), as shown in Equation (3.18). Defining the first Bohr radius ($n = 1$) as a_0, the other allowed orbital radii therefore go as $a_0\, n^2$, as shown in Figure 3.19.

$$r = -\frac{e^2}{8\pi\varepsilon_0 E} = \frac{4\pi\varepsilon_0 n^2 h^2}{4\pi^2 m e^2} \tag{3.18}$$

Example 3-7. Prove that $a_0 = 52.9$ pm.

Solution. According to Equation (3.18), $a_0 = \frac{(1.113\times10^{-10}\,\mathrm{C^2/Jm})(6.626\times10^{-34}\,\mathrm{J \cdot s})^2}{4(3.1416)^2(9.109\times10^{-31}\,\mathrm{kg})(1.602\times10^{-19}\,\mathrm{C})^2}$ $= 5.29 \times 10^{-11}\,\mathrm{m} = 52.9$ pm

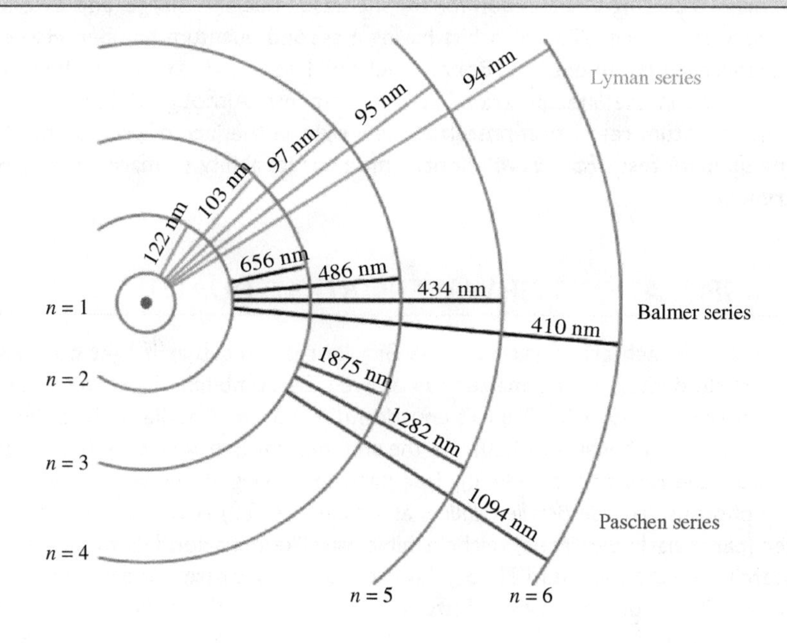

FIGURE 3.18
According to the Bohr model of the atom, the lines in the emission spectrum of hydrogen result from transitions between stationary orbits that exhibit quantization of angular momentum. The first several transitions and their corresponding wavelengths are shown for each series of lines in the line spectrum of hydrogen. [Attributed to Szdori, reproduced from http://en.wikipedia.org/wiki/Hydrogen_spectral_series (accessed November 30, 2013).]

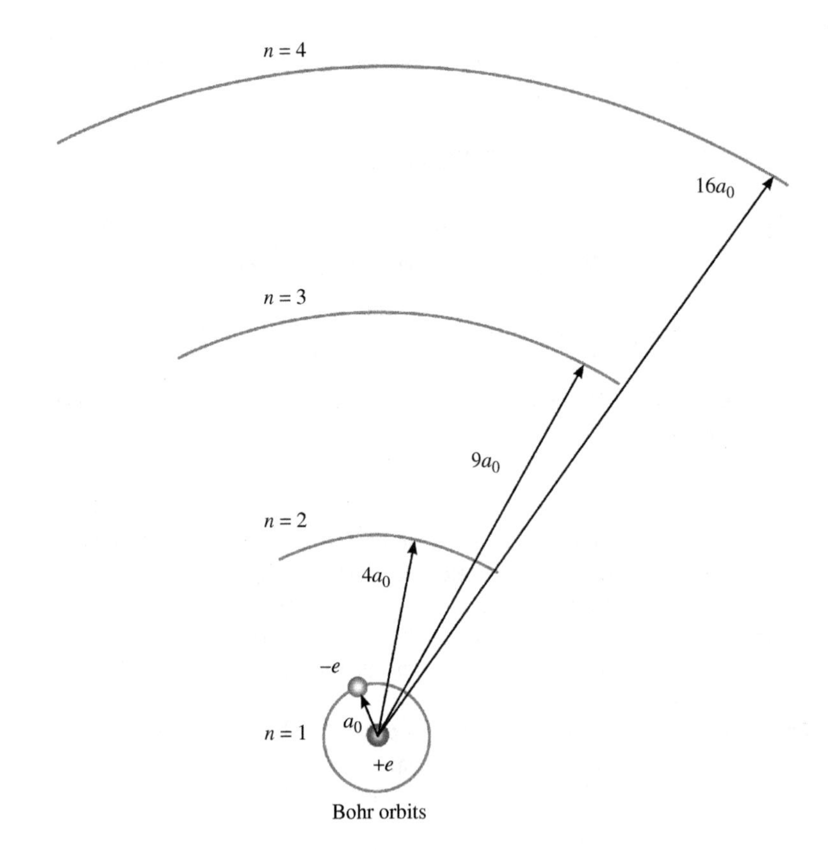

FIGURE 3.19
The allowed orbital radii in the Bohr model of the atom. [Blatt Communications.]

The Bohr model of the atom was well received by the scientific community. Although the theory could only explain the line spectrum of hydrogen and none of the other elements, it is nevertheless important because it was the first theoretical model of the atom to incorporate the concept of quantization. Various modifications were proposed by Sommerfeld and others in an attempt to reconcile the Bohr model with elements other than hydrogen. They did this by accounting for the fact that the electron and nucleus each revolve around a common center of mass, including a correction for the increased nuclear charge, and extending the theory to include elliptical orbits having a second quantum number. However, despite these improvements, the Bohr model could never fully explain the fine structure observed in the line spectra of certain elements. Although Bohr's stationary state and quantum restriction postulates flew right in the face of classical mechanics, the ultimate test of any revolutionary model is its ability to marry theory with experiment.

3.4 IMPLICATIONS OF WAVE-PARTICLE DUALITY

In 1924, the French aristocrat Louis de Broglie reasoned that if light can exhibit wave-particle duality, maybe matter can as well. By combining Einstein's equation for relativity (Eq. 3.19) with Planck's equation (Eq. 3.6), de Broglie derived the simple relationship in Equation (3.20). As the momentum is $p = mv$, de Broglie then generalized the result to include any free particle moving at any velocity v, instead of only photons. The *de Broglie relation*, as Equation (3.20) is called, predicted that matter (particularly electrons) might exhibit wavelike behavior. His thesis, entitled "Researches on the Quantum Theory," was reviewed by Einstein and was one of the shortest doctoral dissertations in history. The wave-particle duality of matter was so revolutionary that Nobel laureate Max von Laue was quoted as saying, "If that turns out to be true, I'll quit physics."

$$E = mc^2 \tag{3.19}$$

$$\lambda = \frac{hc}{E} = \frac{h}{mc} = \frac{h}{mv} = \frac{h}{p} \tag{3.20}$$

Nonetheless, experimental confirmation was not far behind. Only 2 years later, Davisson and Germer were the first to observe the wavelike properties of matter. Upon firing a beam of electrons at a nickel crystal, the researchers observed a diffraction pattern emerging from the other side of the crystal. Feynman likened the experiment to a modern-day double-slit experiment, where the bullets from a machine gun, fired one at a time through two closely spaced holes somehow managed to interfere with each other. The only plausible explanation was if one assumed that the electrons had wave-particle duality. Incidentally, Davisson shared the 1937 Nobel Prize in physics with George Paget Thomson, who independently discovered the phenomenon of electron diffraction. In one of the strangest ironies in the history of science, G. P. Thomson received the Nobel Prize for his experiments proving that the electron behaved like a wave, while his father J. J. Thomson won the 1906 Nobel Prize for proving that the electron was a particle. De Broglie won his own Nobel Prize in physics in 1929.

Example 3-8. Calculate the de Broglie wavelength of an electron traveling at one-tenth the speed of light. Do the same for a Nolan Ryan fastball traveling at 100 mph. Comment on the results.

Solution. The de Broglie wavelength of an electron can be calculated from Equation (3.20):

$$\lambda = \frac{h}{mv} = \frac{(6.626 \times 10^{-34}\ \text{J}\cdot\text{s})}{(9.109 \times 10^{-31}\ \text{kg})(2.998 \times 10^{8}\ \text{m/s})} = 2.43 \times 10^{-11}\ \text{m}$$

For a 142 g baseball traveling at roughly 100 mph:

$$v = \frac{100\ \text{mi}}{\text{h}}\left(\frac{1.609\ \text{km}}{1\ \text{mi}}\right)\left(\frac{10^3\ \text{m}}{1\ \text{km}}\right)\left(\frac{1\ \text{h}}{3600\ \text{s}}\right) = 44.7\ \text{m/s}$$

$$\lambda = \frac{h}{mv} = \frac{(6.626 \times 10^{-34}\ \text{J}\cdot\text{s})}{(0.142\ \text{kg})(44.7\ \text{m/s})} = 1.04 \times 10^{-34}\ \text{m}$$

The wavelength of an electron is larger than its radius, while the wavelength of the baseball is insignificant compared with the size of the ball. Therefore, the wavelike nature of matter is only important for extremely small objects, such as atoms, nucleons, and electrons.

The de Broglie relation provided some additional footing for the quantum restriction postulate in the Bohr model of the atom. By assuming that the electron exhibits wavelike properties, only those circular orbits for which the circumference ($2\pi r$) is equal to an integral multiple of the wavelength ($n\lambda$) will form a standing wave, as shown in Figure 3.20. Thus, the quantization of angular momentum that Bohr applied in the third postulate (Eq. 3.13) of his derivation follows naturally from a wave-mechanical description of the electron, according to Equation (3.21). Hence, the allowed Bohr orbits can be considered circular standing waves, such as the ones depicted in this figure. The corresponding shapes of the vibrations shown in Figures 3.7 and 3.13 can be obtained by snipping each circular orbit

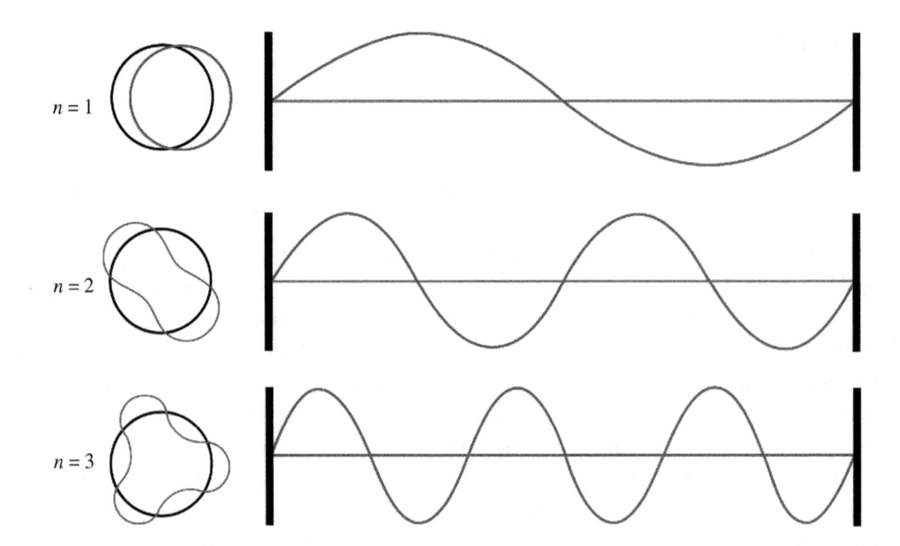

FIGURE 3.20
The first three (n = 1, 2, 3) de Broglie waves superimposed on the Bohr model of the atom. [Blatt Communications.]

and stretching the resulting string out between two fixed points in space. The wavelike properties of electrons form the basis of electron microscopes. Using these instruments, exceedingly short wavelengths of light can be used to probe the sample by controlling the velocity of the electron with an applied voltage.

$$2\pi r = n\lambda = \frac{nh}{mv}$$
$$l = mvr = \frac{nh}{2\pi} \tag{3.21}$$

In 1925, when he was only 23 years of age, Werner Heisenberg published his principle of indeterminacy, more commonly known as the *Heisenberg uncertainty principle*. According to the *Heisenberg uncertainty principle*, the wave-particle duality of matter places an inherent limitation on one's ability to simultaneously measure both the position and the momentum (and hence the velocity) of an electron. In order to precisely measure the position of an electron, one would need to bombard it with a photon having a very short wavelength. Because shorter wavelengths imply higher energies, the collision of the photon with the electron would impart a large uncertainty in the momentum of the electron. In mathematical terms, the result is expressed by Equation (3.22).

$$\Delta x \Delta p \geq \frac{h}{4\pi} \tag{3.22}$$

An appropriate analogy involves trying to photograph the headlights of a car in a busy intersection at night. If a very short shutter speed is used, the position of the car (or at least its leading edge) can be determined with a high degree of precision. However, any calculation of the car's momentum using this photograph will have a large amount of error. On the other hand, one can measure the car's momentum by using a longer shutter speed and holding the aperture of the camera open for a specified period of time. The resulting photograph will show a blurring of the headlights, so that their exact position cannot be ascertained. It is important to recognize that the Heisenberg principle has nothing at all to do with the limitations of the equipment—in this case, the camera. Rather, it is a fundamental property of the measurement itself.

Suppose that we were to add two sine waves having slightly different frequencies together. The two waves will constructively interfere with each other at certain

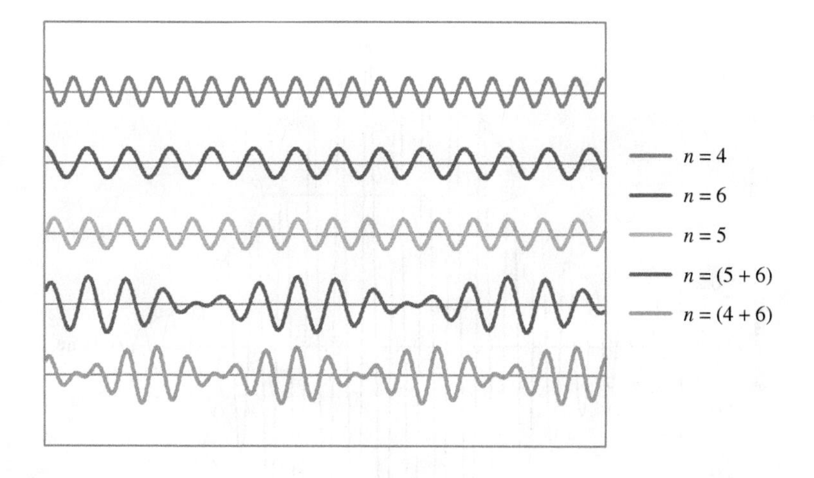

FIGURE 3.21
The addition of two (or more) standing waves leads to a beat pattern, where the width of the pulse is inversely related to the difference in frequencies used in the linear combination. The standing waves for $n = 4$, $n = 6$, and $n = 5$ are shown at the top of the illustration. Addition of the standing waves $n = 5$ and $n = 6$ yields a beat pattern having an uncertainty in x of 1.0, while addition of the standing waves $n = 4$ and $n = 6$ yields a beat pattern having an uncertainty in x of 0.5. In general, Δx for any superposition of two standing waves will be inversely proportional to Δn. Thus, in order to localize the wave function to a narrow region in space, the difference in energy between the standing waves in the superposition must increase, leading to a larger uncertainty in the momentum.

points in time and then destructively interfere at different times, creating what is known as a *beat pattern*, as shown in Figure 3.21.

The beat pattern consists of what is known as a *pilot wave* or a *wave packet*. As indicated in the diagram, one of the ways to localize the wave packet into a narrow region of space is to combine two waves that have vastly different wavenumbers (the wavenumber $1/\lambda$ is directly related to the momentum by a factor of h). Thus, in order to minimize the uncertainty in the position, a wider range of wavelengths (or a greater uncertainty in momentum) is required, as illustrated in Figure 3.22. This is simply a qualitative restatement of the Heisenberg uncertainty principle.

Further, it can be shown that the larger the number of sine waves added together, such that they are all arranged to be in phase with one another at time $t = 0$, the more defined the wave packet will become. As shown in Figure 3.23, the superposition of many sine waves produces a pulse of radiation having a measurable pulse width. Suppose that we add together a very large number of sine waves chosen from a distribution of frequencies that is $\pm5\%$ of the center frequency υ_0, such that $\Delta\upsilon = \pm0.05\upsilon_0$. If we define the uncertainty in time as the half-width at half-maximum of the wave packet, then the uncertainty for this particular group of sine waves will be $\Delta t = 5$, because about five cycles will occur within this timeframe (half the length of the arrow). Thus, the product of the uncertainties is given by Equation (3.23). Substitution of Equation (3.6) into Equation (3.23) yields

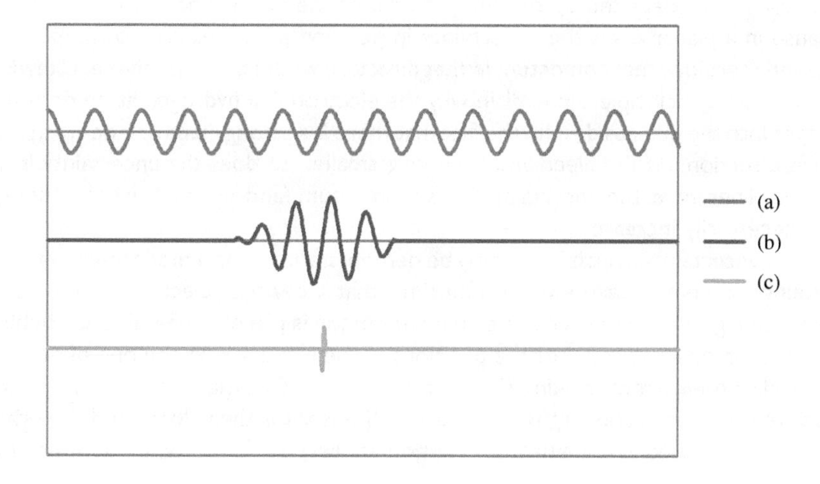

FIGURE 3.22
Illustration of the Heisenberg uncertainty principle for a wave packet. (a) A standing wave having $\Delta p_x \to 0$ has $\Delta x \to \infty$; (b) the superposition of standing waves having a nonzero but finite Δp_x has an intermediate value of Δx as the wave packet becomes defined in space; and (c) the superposition of standing waves where $\Delta p_x \to \infty$ causes $\Delta x \to 0$ and localizes the wave packet.

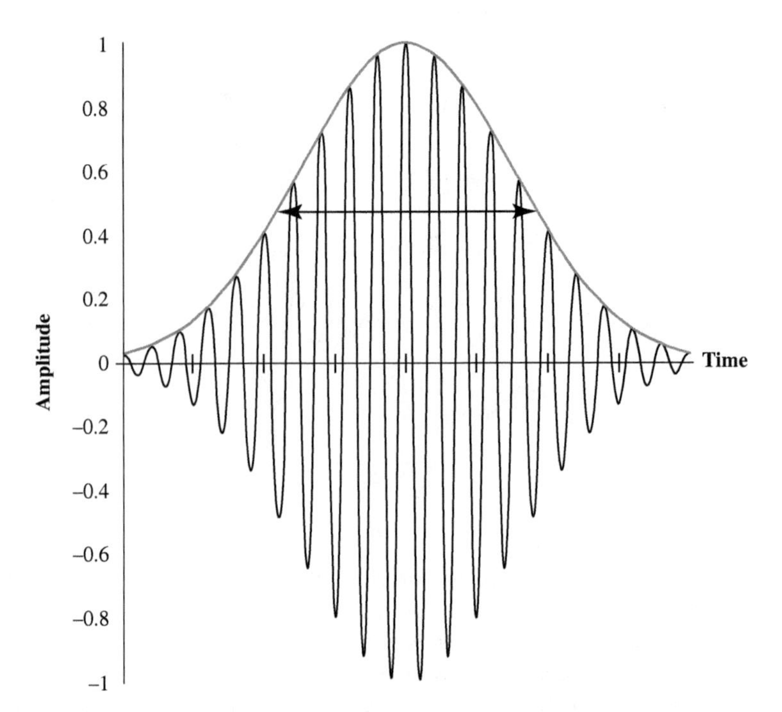

FIGURE 3.23
The superposition of a large number of sine waves having a distribution of frequencies such that $\Delta\upsilon = \pm 0.05\upsilon_0$. The arrow represents the full-width at half-maximum; half of its length represents the uncertainty in time of the wave packet. [Reproduced by permission from Warren, W. *The Physical Basis of Chemistry*, page 114, Copyright Elsevier (1994).]

Equation (3.24).

$$\Delta\nu\,\Delta t \geq 1/4 \tag{3.23}$$

$$\Delta E\,\Delta t \geq h/4 \tag{3.24}$$

Because the form of each sine wave was defined by Equation (3.4), the quantity x/λ must behave in the same manner as υt. Thus, $\Delta x\,\Delta(1/\lambda) \geq 1/4$. Finally, substitution of Equation (3.20) affords Equation (3.25), which is nearly identical to the Heisenberg uncertainty relationship in Equation (3.22), the main difference being in the exact manner in which the uncertainty is defined.

$$\Delta x\,\Delta p \geq h/4 \tag{3.25}$$

The Heisenberg uncertainty principle explains several interesting features of atoms. For instance, electrons cannot exist in planar orbits around the nucleus, as is so commonly depicted by the Bohr model of the atom. The reason for this is because in a planar orbit the uncertainty in position perpendicular to the plane is zero and therefore the momentum in that direction would become infinite. Likewise, the uncertainty principle can explain why the electron in a hydrogen atom does not collapse into the nucleus despite the fact that there is a strong electrostatic attraction in that direction. As the electron's orbit gets smaller, so does the uncertainty in its position. Therefore, the uncertainty in its momentum (and also in its kinetic energy) must necessarily increase.

The uncertainty principle can also be demonstrated using a modern-day version of Young's double-slit experiment. Consider that a beam of electrons is fired at a screen having two narrow slits. A suitable detector is placed some distance behind the screen in order to monitor the positions of the electrons. When one of the slits is closed, the electrons striking the detector form a Gaussian distribution exactly opposite the open slit, as might be expected. If this slit is then closed and the other one opened, a second Gaussian distribution appears at the detector opposite the

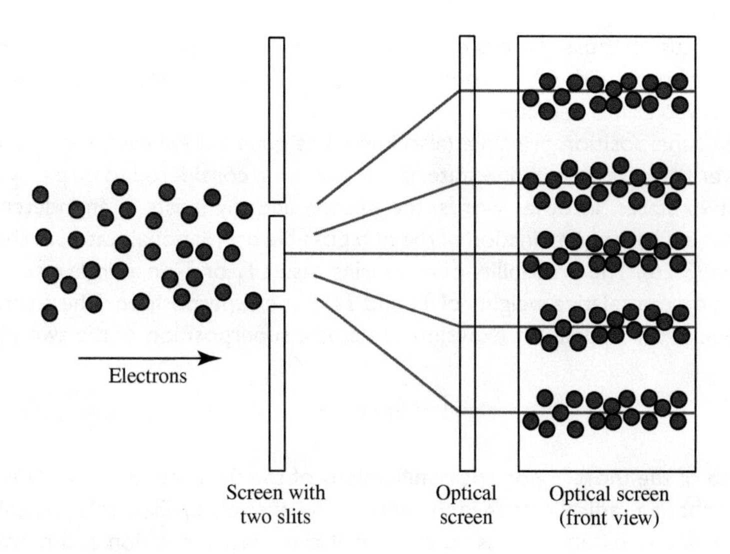

FIGURE 3.24
Modern-day version of the double-slit experiment using a beam of electrons. Even when the electrons are fired one at a time through the screen with the slits, an interference pattern develops on the optical screen. [Reproduced from http://commons.wikimedia.org/wiki/File:Two-Slit_Experiment_Electrons.svg (accessed December 1, 2013).]

second slit. However, when a large number of electrons are fired at the screen with both of the slits open, instead of just the sum of two Gaussians, the detector records an interference pattern similar to the one shown in Figure 3.2. This is a remarkable observation in and of itself. Somehow the electrons passing through the two open slits have interfered with one another. Now suppose that the electron gun is modulated so that it releases just a single electron at a time. At first, the pattern observed at the detector appears completely random. But if we wait long enough for a larger number of electrons to pass through one or the other slit (*one at a time*), an astounding result is observed—the exact same interference pattern develops at the detector, as shown in Figure 3.24. The only conclusion that can be drawn is that the electron somehow interfered with itself as it passed through the barrier with the two slits in it.

The key to understanding this seeming paradox is put forward in the *superposition principle*, which is entirely a quantum mechanical concept. Consider the apparatus shown in Figure 3.25. A beam of photons enters from the left and strikes a beam splitter. Exactly one-half of the time the photon will be directed to the left along translation state 1 toward a mirror, while the other half of the time it will pass directly through the beam splitter along translation state 2. Regardless of the pathway, the total distance that the photon travels before it returns to the beam splitter and is sent to a detector is exactly the same. Furthermore, we can adjust the intensity of the light beam so that only one photon at a time will pass through

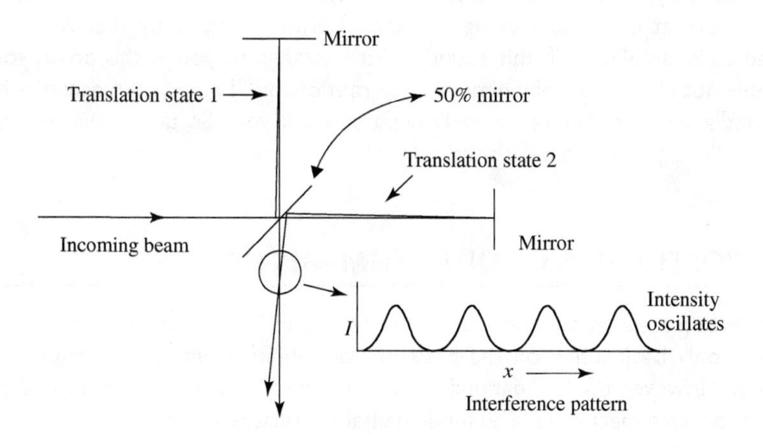

FIGURE 3.25
Schematic of an optical interference apparatus. [© Michael D Fayer, *Elements of Quantum Mechanics*, 2001, by permission of Oxford University Press, USA.]

the apparatus. Because there is only one photon in the apparatus at a time, there is no other photon with which to interfere. Nevertheless, an interference pattern is still observed at the detector.

The superposition principle (also known as *quantum entanglement*) states that whenever the photon is in one state, it can always be considered to be partly in *each* of the two states. In other words, the photon actually exists in an indeterminate state that is a linear combination of the two possible translational states, as shown by Equation (3.26). The probability of measuring result T_1 or T_2 in a *single measurement* depends on the relative weights of T_1 and T_2 in the superposition. The electrons in the modern-day double-slit experiment act as a superposition of the two different slits.

$$T = c_1 T_1 + c_2 T_2 \tag{3.26}$$

One of the most important ramifications of the uncertainty principle is that it brought about a radical change in the philosophy of science. Classical mechanics was deterministic in nature; that is to say that if the precise position and momentum of a particle or a collection of particles were known, Newton's laws could be used (at least in principle) to determine all the future behavior of the particle(s). The uncertainty principle, however, tells us that there is an inherent limitation to how accurately we can measure the two quantities simultaneously. Any observation of an extremely small object (one whose wavelength is on the same magnitude or larger than the particle itself) necessarily effects a nonnegligible disturbance to the system, and thereby it influences the results. Einstein never liked the statistical nature of quantum mechanics, saying "God does not play dice with the universe." Nonetheless, the quantum mechanical model is a statistical one.

It is for this reason that the Bohr model of the atom is correct only for one-electron systems such as hydrogen. For any multielectron system, such as helium, if we chose to focus on one electron and to calculate the electrostatic field that it felt as a result of its interactions with the nucleus and with the other electron, we would encounter a situation known as the *electron correlation problem*. While we might be able to locate the position of the other electron fairly precisely, because of the uncertainty in its momentum we would have no way of predicting its future behavior. The bottom line is that we no longer have a deterministic model of the atom. As strange as it might seem, given the inherent limitations of our observations, the best that we can do is to have a statistical view of the subatomic world. We can say, for instance, that there will be a 50% chance that the second electron will occupy a certain region of space. Hence, in the forthcoming discussion of quantum mechanics, we change our terminology from electron "orbits" to "orbitals," or regions of space where there is simply a strong likelihood of finding an electron. By virtue of its existence as a standing wave, the electron can exist "everywhere at once" (excluding the nodes) with a probability that is somehow related to its amplitude. If this sounds a little strange to you at this point, you are probably not alone. As Bohr himself once cautioned: "If quantum mechanics hasn't profoundly shocked you, you haven't understood it yet." So take a moment to ask yourself, "Are you profoundly shocked yet?"

3.5 POSTULATES OF QUANTUM MECHANICS

The quantum mechanical model of atomic structure is based on a set of postulates that can only be justified on the basis of their ability to rationalize experimental behavior. However, the foundations of quantum theory have their origins in the field of classical wave mechanics. The fundamental postulates are as follows:

- *Postulate 1*: The state of a particle is completely described by a wave function Ψ, such that all the possible information that can be measured about the particle (its position, momentum, energy, etc.) is contained in Ψ. The *Born interpretation* of the wave function, which itself derives from the fact that the intensity of light is proportional to the square of its amplitude, states that $\Psi^*\Psi$ represents the *probability density* that the particle will exist in the volume element $d\tau$ at point (x, y, z). Recall that the Heisenberg uncertainty principle states that we cannot measure the exact position of the particle—only a statistical probability can be obtained. The quantity Ψ^* is the complex conjugate of the wave function Ψ. Thus, for example, if Ψ is the complex function $x + iy$, Ψ^* will equal $x - iy$. The derivative $\partial\tau$ is an infinitely small, three-dimensional volume element. Because $\Psi^*\Psi$ represents a probability density, Ψ and Ψ^* must both be single-valued, continuous, finite, and smoothly varying. Furthermore, the integral of $\Psi^*\Psi$ over all space must be unity, as shown in Equation (3.27), because the total probability of finding the particle somewhere has to be 100%. Wave functions that satisfy Equation (3.27) are said to be *normalized*.

$$\int_{-\infty}^{\infty} \Psi^*\Psi \, \partial\tau = 1 \tag{3.27}$$

- *Postulate 2*: For every physically observable variable in classical mechanics, there exists a corresponding linear, Hermitian operator in quantum mechanics. Examples are shown in Table 3.2, where the ^ symbol indicates a quantum mechanical operator and $\hbar = h/2\pi$. A *Hermitian operator* is one which satisfies Equation (3.28).

$$\int \Psi_i{}^*\hat{A}\Psi_j \partial\tau = \int \Psi_j{}^*\hat{A}\Psi_i \, \partial\tau \tag{3.28}$$

Many quantum mechanical operators will define an eigenvalue equation. An *eigenfunction* is any function that when operated on yields back the original function times a constant. That constant is known as the *eigenvalue*. The requirement that each operator be Hermitian therefore guarantees that the eigenvalues will always be real numbers. Any two eigenfunctions that have the same eigenvalue are said to be *degenerate* and will possess the same energy. As a result of the requirement that quantum mechanical operators be linear, any linear combination of degenerate wave functions will also be an acceptable solution. Furthermore, any well-behaved function (one subject to the same restrictions as Ψ was in Postulate 1) can be expanded as a linear combination of eigenfunctions. This property is analogous to the superposition of standing waves. Eigenfunctions that are not degenerate will be *orthogonal* to each other, according to Equation (3.29).

$$\int \Psi_i{}^*\Psi_j \partial\tau = 0, \quad i \neq j \tag{3.29}$$

- *Postulate 3*: In any measurement where an exact solution can be obtained, the only values that will ever be observed are the eigenvalues a_n that satisfy the eigenvalue equation given by Equation (3.30). This postulate is the one that is responsible for the quantization of energy.

$$\hat{A}\Psi_n = a_n\Psi_n \tag{3.30}$$

Although measurements must always yield an eigenvalue, the state does not initially have to be an eigenstate of $\hat{A}$. Any arbitrary state can be expanded

TABLE 3.2 Classical mechanical observables and their quantum mechanical operators.

Observable		Operator	
Type	Symbol	Symbol	Operation
Position	x	$\widehat{X}$	Multiply by x
Position vector	$\mathbf{r}$	$\widehat{R}$	Multiply by r
Momentum	p_x	$\widehat{P}$	$-i\hbar\frac{\partial}{\partial x}$
Momentum vector	$\mathbf{p}$	$\widehat{P}$	$-i\hbar\left(\mathbf{i}\frac{\partial}{\partial x}+\mathbf{j}\frac{\partial}{\partial y}+\mathbf{k}\frac{\partial}{\partial z}\right)$
Kinetic energy	K_x	$\widehat{K}_x$	$-\frac{\hbar^2}{2m}\frac{\partial^2}{\partial x^2}$
	K	$\widehat{K}$	$-\frac{\hbar^2}{2m}\left(\frac{\partial^2}{\partial x^2}+\frac{\partial^2}{\partial y^2}+\frac{\partial^2}{\partial z^2}\right) = -\frac{\hbar^2}{2m}\nabla^2$
Potential energy	$V(x)$	$\widehat{V}(x)$	Multiply by $V(x)$
	$V(x,y,z)$	$\widehat{V}(x,y,z)$	Multiply by $V(x,y,z)$
Total energy	E	$\widehat{H}$	$-\frac{\hbar^2}{2m}\nabla^2+V(x,y,z)$
Angular momentum	L_x	$\widehat{L}_x$	$=-i\hbar\left(y\frac{\partial}{\partial z}-z\frac{\partial}{\partial y}\right)$
	L_y	$\widehat{L}_y$	$=-i\hbar\left(z\frac{\partial}{\partial x}-x\frac{\partial}{\partial z}\right)$
	L_z	$\widehat{L}_z$	$=-i\hbar\left(x\frac{\partial}{\partial y}-y\frac{\partial}{\partial x}\right)$

as a linear combination in the complete set of eigenfunctions, as shown in Equation (3.31).

$$\Psi = \sum_{i}^{n} c_i \Psi_i \tag{3.31}$$

- *Postulate 4*: If Ψ is not an eigenfunction of the quantum mechanical operator, then a series of measurements on identical systems of particles will yield a distribution of results, such that Equation (3.32) will describe the average (or "expectation") value of the observable (assuming the wave function is normalized).

$$< a >= \int_{-\infty}^{\infty} \Psi^* \widehat{A} \Psi \, \partial\tau \tag{3.32}$$

- *Postulate 5*: The wave function of a particle evolves in time according to the time-dependent Schrödinger equation, given by Equation (3.33).

$$-\frac{\hbar^2}{2m}\nabla^2\Psi + V\Psi = \widehat{H}\Psi = -i\hbar\frac{\partial\Psi}{\partial t} \tag{3.33}$$

3.6 THE SCHRÖDINGER EQUATION

Quantum mechanics is a model that is based entirely on postulates that explain the observations associated with atomic and subatomic particles. As such, the Schrödinger equation cannot be derived from first principles. However, what follows is a rationale for the Schrödinger equation. The classical wave equation in one-dimension was given by Equation (3.3) and is reproduced in Equation (3.34) with the substitution of Ψ for y.

$$\frac{\partial^2\Psi}{\partial x^2} = \frac{1}{v^2}\frac{\partial^2\Psi}{\partial t^2} \tag{3.34}$$

A general solution to the Schrödinger equation is given by Equation (3.35).

$$\Psi(x,t) = A\,e^{2\pi i\left(\frac{x}{\lambda} - \nu t\right)} \tag{3.35}$$

Example 3-9. Prove that Equation (3.35) is a solution to Equation (3.34).

Solution. Taking the first partial derivative of Ψ with respect to x and t gives the following two equations:

$$\frac{\partial\Psi}{\partial x} = \frac{2\pi i}{\lambda}Ae^{2\pi i\left(\frac{x}{\lambda} - \nu t\right)}$$

$$\frac{\partial\Psi}{\partial t} = -2\pi i\nu Ae^{2\pi i\left(\frac{x}{\lambda} - \nu t\right)}$$

Taking the second partial derivative with respect to each independent variable yields

$$\frac{\partial^2\Psi}{\partial x^2} = -\left(\frac{2\pi}{\lambda}\right)^2 Ae^{2\pi i\left(\frac{x}{\lambda} - \nu t\right)} = -\left(\frac{2\pi}{\lambda}\right)^2\Psi$$

$$\frac{\partial^2\Psi}{\partial t^2} = -(2\pi\nu)^2 Ae^{2\pi i\left(\frac{x}{\lambda} - \nu t\right)} = -(2\pi\nu)^2\Psi$$

After substituting into Equation (3.34):

$$-\left(\frac{2\pi}{\lambda}\right)^2\Psi = -\frac{(2\pi\nu)^2}{v^2}\Psi$$

which yields the following after taking the square root of both sides and cancelling:

$$\frac{1}{\lambda} = \frac{\nu}{v}$$

Because this is just a reformulation of Equation (3.1), Equation (3.35) is an acceptable solution to Equation (3.34). In fact, it is the most general solution.

Substituting the de Broglie relation given by Equation (3.20) into the second partial derivative of Equation (3.35) with respect to x yields Equation (3.36).

$$\frac{\partial^2\Psi}{\partial x^2} = -\frac{4\pi^2}{\lambda^2}\Psi = -\frac{4\pi^2 m^2 v^2}{h^2}\Psi \tag{3.36}$$

Given that the kinetic energy $K = (1/2)mv^2$ and the total energy $E = K + V$, where V is the potential energy, Equation (3.36) can be rewritten as

$$\frac{\partial^2\Psi}{\partial x^2} = -\frac{8\pi^2 mK}{h^2}\Psi = -\frac{8\pi^2 m}{h^2}(E\Psi - V\Psi) \tag{3.37}$$

Solving for the first partial derivative of Equation (3.35) with respect to time and substituting the Planck formula $E = h\nu$, we obtain Equation (3.38):

$$\frac{\partial\Psi}{\partial t} = -2\pi i\nu\Psi = -\frac{2\pi iE}{h}\Psi \tag{3.38}$$

Solving both Equations (3.37) and (3.38) for $E\Psi$ and setting them equal yields Equation (3.39):

$$E\Psi = -\frac{h^2}{8\pi^2 m}\frac{\partial^2\Psi}{\partial x^2} + V\Psi = -\frac{h}{2\pi i}\frac{\partial\Psi}{\partial t}\left(\frac{i}{i}\right) = \frac{ih}{2\pi}\frac{\partial\Psi}{\partial t} \tag{3.39}$$

Finally, some simple rearrangement yields the time-dependent Schrödinger equation in one-dimension, given by Equation (3.40).

$$-\frac{h^2}{8\pi^2 m}\frac{\partial^2\Psi}{\partial x^2} + V\Psi = \frac{ih}{2\pi}\frac{\partial\Psi}{\partial t} \tag{3.40}$$

Generalizing from one-dimension to three-dimensions and substituting del-squared as the operator yields Equation (3.41), the time-dependent Schrödinger equation in three-dimensions, which is the same as Equation (3.33).

$$-\frac{\hbar^2}{2m}\nabla^2\Psi + V\Psi = H\Psi = i\hbar\frac{\partial\Psi}{\partial t},$$
$$\text{where } \nabla^2 = \frac{\partial^2}{\partial x^2} + \frac{\partial^2}{\partial y^2} + \frac{\partial^2}{\partial z^2} \tag{3.41}$$

For stationary states, analogous to Bohr's stationary orbits where the potential energy is independent of time, $\Psi(x, y, z, t)$ can be factored into a time-dependent term $\phi(t)$ and a time-independent term $\psi(x, y, z)$, as shown in Equation (3.42).

$$\Psi(x, y, z, t) = \psi(x, y, z)\phi(t) \tag{3.42}$$

Taking Equation (3.41) and dividing both sides by $\psi\phi$ yields Equation (3.43).

$$-\frac{\hbar^2}{2m}\frac{\nabla^2\psi\phi}{\psi\phi} + \frac{V\psi\phi}{\psi\phi} = i\hbar\frac{\psi}{\psi\phi}\frac{d\phi}{dt} \tag{3.43}$$

Because the del-squared operator acts only on position coordinates, both it and the potential energy term on the left-hand side of Equation (3.43) are independent of time and the $\phi(t)$s cancel on this side. Likewise, because the operator on the right-hand side of Equation (3.43) is independent of the coordinates, the $\psi(x, y, z)$s cancel on the right-hand side. Setting both sides equal to the separation constant (a scalar quantity) yields Equation (3.44).

$$-\frac{\hbar^2}{2m}\frac{\nabla^2\psi}{\psi} + \frac{V\psi}{\psi} = \frac{i\hbar}{\phi}\frac{d\phi}{dt} = E \tag{3.44}$$

Rearrangement of Equation (3.44) and substitution of the symbol for the Hamiltonian operator yields the familiar time-independent form of the Schrödinger equation shown in Equation (3.45).

$$-\frac{\hbar^2}{2m}\nabla^2\psi + V\psi = \widehat{H}\psi = E\psi = \frac{i\hbar}{\phi}\frac{d\phi}{dt} \tag{3.45}$$

The time-independent Schrödinger equation in one-dimension is a linear, second-order differential equation having constant coefficients. A general method for solving this type of differential equation is to rearrange the equation into a quadratic of the form shown in Equation (3.46), where y'' and y' are the second and first derivatives, respectively, with respect to x.

$$y'' + py' + qy = 0 \tag{3.46}$$

Because we are looking for a function whose first and second derivatives yield back the function times a constant, one logical solution would be an exponential function, such as $y = e^{sx}$. In this case, $y' = se^{sx}$ and $y'' = s^2e^{sx}$, so that substitution into Equation (3.46) yields Equation (3.47). When both sides of Equation (3.47) are divided by e^{sx}, the result is known as the *auxiliary equation* and is shown in (3.44).

$$s^2e^{sx} + pse^{sx} + qe^{sx} = 0 \tag{3.47}$$

$$s^2 + ps + q = 0 \tag{3.48}$$

Because Equation (3.48) is a quadratic, there are two independent solutions to the auxiliary equation. Thus, the most general solution to the differential equation given by Equation (3.46) is a linear combination, as shown in Equation (3.49), where a and b are weighting constants.

$$y(x) = a\,e^{s_1x} + b\,e^{s_2x} \tag{3.49}$$

In this context, the one-dimensional Schrödinger equation can be rewritten as Equation (3.50).

$$\frac{d^2\psi}{dx^2} + \frac{2m(E-V)\psi}{\hbar^2} = \frac{d^2\psi}{dx^2} + \beta^2\psi = 0 \tag{3.50}$$

After substitution, the auxiliary equation becomes:

$$(s^2 + \beta^2) = 0 \tag{3.51}$$

Therefore, $s^2 = -\beta^2$, or $s = \pm i\beta$ and the general solution to time-independent Schrödinger equation is given by Equation (3.52).

$$\psi(x) = a\,e^{i\beta x} + b\,e^{-i\beta x} \tag{3.52}$$

Using Euler's formula that $e^{i\beta x} = \cos(\beta x) + i\sin(\beta x)$, an alternative form of Equation (3.52) is Equation (3.53).

$$\psi(x) = A\,\cos(\beta x) + B\,\sin(\beta x) \tag{3.53}$$

Whether it is more convenient to use the exponential form given in Equation (3.52) or the trigonometric form given in Equation (3.53) will depend on the specific nature of the problem at hand.

One final point about wave functions is the requirement that they be normalized. Depending on the particular problem, the solution to Equation (3.53) might yield an eigenfunction where the integral of $\psi^*\psi \, d\tau$ over all space is not equal to one. In this case, an appropriate constant c must be found such that the wave function can be normalized. The normalization process is shown in Equations (3.54)–(3.56).

$$\int_{-\infty}^{\infty} \psi^*\psi \, \partial\tau = \int_{-\infty}^{\infty} |\psi|^2 \partial\tau = N \tag{3.54}$$

$$\int_{-\infty}^{\infty} |c\psi|^2 \partial\tau = c^2 N = 1 \tag{3.55}$$

$$c = 1/\sqrt{N} \tag{3.56}$$

3.7 THE PARTICLE IN A BOX PROBLEM

Ultimately, our goal is to solve the Schrödinger equation in three-dimensions for the electron in the hydrogen atom. This electron is subject to a potential energy term that involves a Coulombic attraction toward the nucleus. However, the solution to this differential equation is not a trivial one. We therefore choose a somewhat similar, but simpler, problem—the particle in a box—to demonstrate the procedure and to illustrate some of the principles of quantum mechanics. We then extrapolate those results to the hydrogen atom in a later chapter.

Consider an electron that is trapped in a one-dimensional box of length a, such as the one shown in Figure 3.26. Inside the box, the electron experiences zero potential energy. However, the walls of the box are infinitely steep, so that the potential energy of an electron is infinite outside the box.

The wave function is zero everywhere outside the box. Inside the box, the time-independent Schrödinger equation in one-dimension reduces to Equation (3.57), which has the trigonometric solution given by Equation (3.53), where $\beta = \sqrt{2mE}/\hbar$.

$$\frac{d^2\psi}{dx^2} + \frac{2m}{\hbar^2} E\psi = 0 \tag{3.57}$$

The boundary conditions are such that the wave function approaches zero at both walls of the box. In other words, $\psi(0) = \psi(a) = 0$. The condition that $\psi(0) = 0$ implies that $A = 0$, because $B \sin(0) = 0$. The condition that $\psi(a) = 0$ therefore implies that $B \sin(\beta a) = 0$. There are two conditions under which this will be true: if $B = 0$ or if $\sin(\beta a) = 0$. If $B = 0$, then the wave function would never exist, because A is also zero. Therefore, $\sin(\beta a)$ must equal zero; and this will only be true when $\beta a = n\pi$, where n is any integer. Solving Equation (3.58) for the energy yields the results given by Equation (3.59). Because the energy must always be greater than or equal to zero, n must equal a non-negative integer. Furthermore, if $n = 0$, E would also equal zero, and the two roots of the auxiliary equation would be identical. Thus, $n = 0$ is not an acceptable solution and the quantum number, n, must take on a positive value.

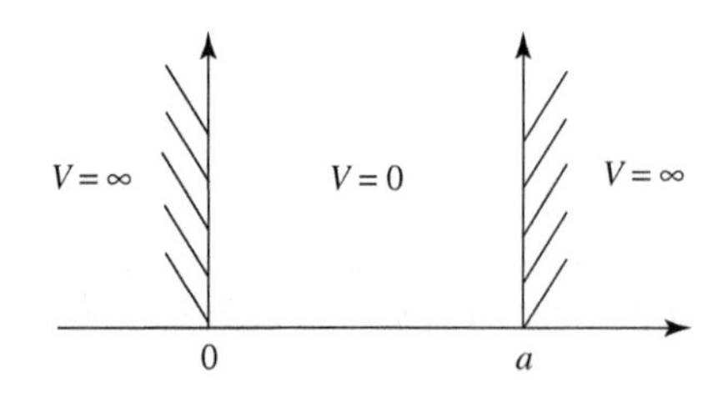

FIGURE 3.26
Sketch of the one-dimensional particle in a box problem.

After substitution for A and β into Equation (3.53), the acceptable solutions to the particle in a box problem are given in Equation (3.60).

$$\beta = \frac{\sqrt{2mE}}{\hbar} = \frac{n\pi}{a} \tag{3.58}$$

$$E = \frac{n^2\hbar^2\pi^2}{2ma^2} = \frac{n^2h^2}{8ma^2}, \quad \text{where } n = 1, 2, 3, \ldots \tag{3.59}$$

$$\psi(x) = B \sin\left(\frac{n\pi x}{a}\right) \tag{3.60}$$

Example 3-10. Use the normalization procedure discussed in the previous section to prove that the value of B in Equation (3.60) is $(2/a)^{1/2}$.

Solution. In the process of normalization, the integral of $|\psi|^2\, d\tau$ over all space is determined to be N. By substitution of Equation (3.60) for ψ, the following integral must be evaluated:

$$B^2 \int_0^a \sin^2\left(\frac{n\pi x}{a}\right) dx = N$$

Consultation of an integral table yields the following analytic solution:

$$\int_0^a \sin^2(kx)\, dx = \frac{a}{2} - \frac{1}{4k} \sin(2ka)$$

Because $\sin(2kx) = \sin(2n\pi x/a) = \sin(n'\pi) = 0$ for all n', the original integral reduces to $B^2\,(a/2) = N$. Because N must equal 1 in order for the wave function in Equation (2.60) to be normalized, B must therefore equal $(2/a)^{1/2}$. Thus, the normalized solutions to the one-dimensional particle in a box problem are

$$\psi(x) = \sqrt{\frac{2}{a}} \sin\left(\frac{n\pi x}{a}\right)$$

The acceptable solutions to the one-dimensional particle in a box problem are sketched in Figure 3.27(a) for the first several quantum numbers. The *Born interpretation* of the wave function states that the product $\psi^*\psi$ represents the probability density of finding the electron in a finite region of space. Because the Born interpretation of the wave function is $|\psi|^2$, this function is shown in Figure 3.27(b).

Three features are immediately evident:

1. The energy is quantized and it increases as the square of n.
2. For all $n > 1$, there are $n-1$ nodal regions of space where there is zero probability that the electron will exist.
3. When n is infinite, the shape of the wave function approaches that of a straight line. The peaks and valleys of the standing wave all blur together into a single continuum. Hence, the quantum mechanical model approximates the classical one at very large values of n, a property which is known as the *correspondence principle*. In order for any new theory of the atom to be valid, it must not only explain and predict new behavior but it must also incorporate all the experimental evidence that preceded it.

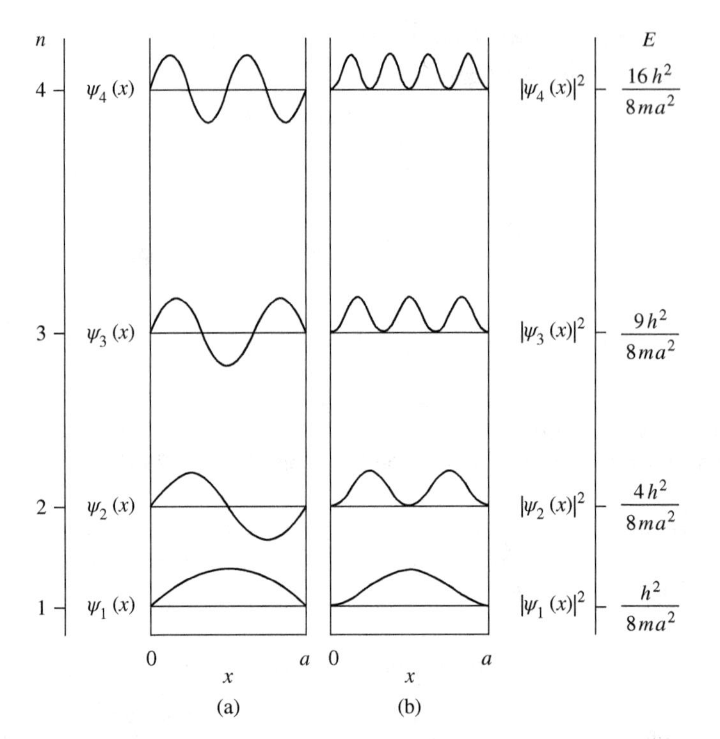

FIGURE 3.27
The first of several solutions to the particle in a box problem showing (a) $\psi(x)$ and (b) $|\psi(x)|^2$, along with their corresponding energies. [Copyright University Science Books, Mill Valley, CA. Used with permission. All rights reserved. McQuarrie, D. A.; Simon, J. D. Physical Chemistry: A Molecular Approach, 1997.]

Example 3-11. Prove that $\psi(x) = \sqrt{\frac{2}{a}} \sin\left(\frac{n\pi x}{a}\right)$ is an acceptable solution to the Schrödinger equation.

Solution. The time-independent Schrödinger equation in one-dimension is given by Equation (3.45):

$$-\frac{\hbar^2}{2m}\frac{d^2\psi}{dx^2} + V\psi = \widehat{H}\psi = E\psi$$

Taking the first and second derivatives of Equation (3.45) yields

$$\frac{d\psi}{dx} = \left(\frac{n\pi}{a}\right)\sqrt{\frac{2}{a}}\cos\left(\frac{n\pi x}{a}\right)$$

$$\frac{d^2\psi}{dx^2} = -\left(\frac{n\pi}{a}\right)^2\sqrt{\frac{2}{a}}\sin\left(\frac{n\pi x}{a}\right) = -\frac{n^2\pi^2}{a^2}\psi$$

Because $V = 0$ inside the box, substitution of the second derivative into the Schrödinger equation yields

$$-\frac{\hbar^2}{2m}\frac{n^2\pi^2}{a^2}\psi = E\psi$$

Substitution for $\hbar$ and division of both sides by ψ gives the expected result:

$$E = \frac{n^2h^2}{8ma^2}$$

Example 3-12. Using Postulate 4, show that the average position <x> of the electron in the one-dimensional particle in a box problem lies exactly in the center of the box.

Solution. The expectation value of position, <x> is given by Equation (3.32):

$$\langle x \rangle = \int_0^a \psi * x\psi \, d\tau = \frac{2}{a}\int_0^a x \sin^2\left(\frac{n\pi x}{a}\right) dx$$

Using a table of integrals:

$$\int_0^x x \sin^2(kx)dx = \frac{x^2}{4} - \frac{x \sin(2kx)}{4k} - \frac{\cos(2kx)}{8k^2}$$

Because $k = n\pi/a$, $\sin(2ka) = \sin(2n\pi) = \sin(0) = 0$ and $\cos(2ka) = \cos(2n\pi)$ $= \cos(0) = 1$. Therefore, the integrand reduces to

$$\int_0^x x \sin^2(kx)dx = \frac{x^2}{4} - \frac{1}{8k^2} - \left(\frac{0}{4} - \frac{1}{8k^2}\right) = \frac{a^2}{4}$$

Substitution into the expectation value for x yields

$$\langle x \rangle = \frac{2}{a}\left(\frac{a^2}{4}\right) = \frac{a}{2} \text{ or in the exact middle of the box}$$

Because the hydrogen atom is a three-dimensional problem, let us next consider what happens when the particle is confined to a rectangular parallelepiped having the dimensions a, b, and c, as shown in Figure 3.28.

Inside the box, where the potential energy is zero, the time-independent Schrödinger equation in three-dimension is given by Equation (3.45). Separation of the variables yields Equation (3.61):

$$\psi(x, y, z) = X(x)\,Y(y)\,Z(z) \tag{3.61}$$

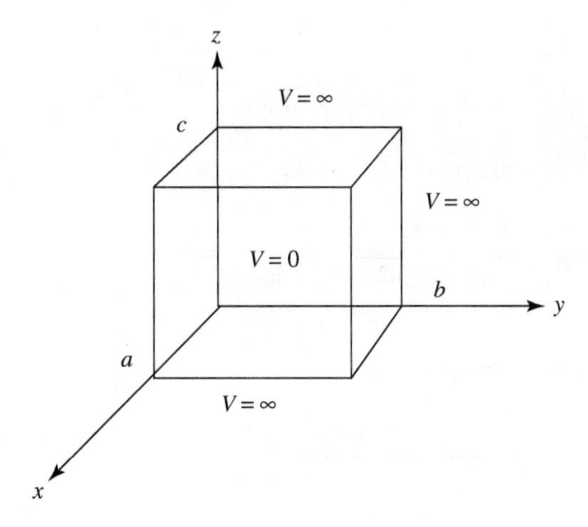

FIGURE 3.28
The particle in a three-dimensional box problem, where the box has dimensions of a, b, and c.

Substitution into Equation (3.45), followed by division by $X(x)Y(y)Z(z)$ gives Equation (3.62):

$$-\frac{\hbar^2}{2m}\frac{1}{X(x)}\frac{d^2X}{dx^2}-\frac{\hbar^2}{2m}\frac{1}{Y(y)}\frac{d^2Y}{dy^2}\frac{\hbar^2}{2m}\frac{1}{Z(z)}\frac{d^2Z}{dz^2}=E \tag{3.62}$$

Because each term on the left-hand side of Equation (3.62) is a function of a single variable, all three terms must be equal to a constant in order for the equation to be true at all values of x, y, and z. Thus, Equation (3.62) can be simplified as Equation (3.63):

$$E_x+E_y+E_z=E \tag{3.63}$$

The boundary conditions are $X(0) = X(a) = Y(0) = Y(b) = Z(0) = Z(c) = 0$. Each term in Equation (3.62) therefore reduces to the same form as the particle in a one-dimensional box, such that Equations (3.64) and (3.65) result:

$$\psi(x,\,y,\,z)=A\sin\left(\frac{n\pi x}{a}\right)B\sin\left(\frac{n\pi x}{b}\right)C\sin\left(\frac{n\pi x}{c}\right) \tag{3.64}$$

$$E=\frac{\hbar^2}{2m}\left(\frac{n_x^2}{a^2}+\frac{n_y^2}{b^2}+\frac{n_z^2}{c^2}\right) \tag{3.65}$$

Normalization of Equation (3.65) yields a value of $(8/abc)^{1/2}$ for the conglomerate constant ABC.

The first several energy levels for the three-dimensional particle in a box problem (where $a = b = c$) are shown in Figure 3.29, where the energy axis has units of $h^2/8\,ma^2$. Many of the energy levels are degenerate—having more than one acceptable set of quantum numbers at the same energy. For example, there are three ways to have the sum $n_x^2 + n_y^2 + n_z^2 = 6$: $n_x, n_y, n_z = 2, 1, 1; 1, 2, 1;$ and $1, 1, 2$. As a result, all three wave functions will have the same energy, where the $E = 6(h^2/8\,ma^2)$ energy level is said to be triply degenerate. The degeneracy is removed, however, if the symmetry of the box is lowered, so that it is no longer a cube: $a \neq b \neq c$. Consider the case where $b = (1/a)^{1/2}$ and $c = (1/a)^{1/3}$. The energies of the 2, 1, 1; 1, 2, 1; and 1, 1, 2 sets of quantum numbers are now $7(h^2/8\,ma^2)$, $8(h^2/8\,ma^2)$, and $9(h^2/8\,ma^2)$. This is the first of many examples in this textbook where we will find that the symmetry of an object greatly influences its physical and chemical properties.

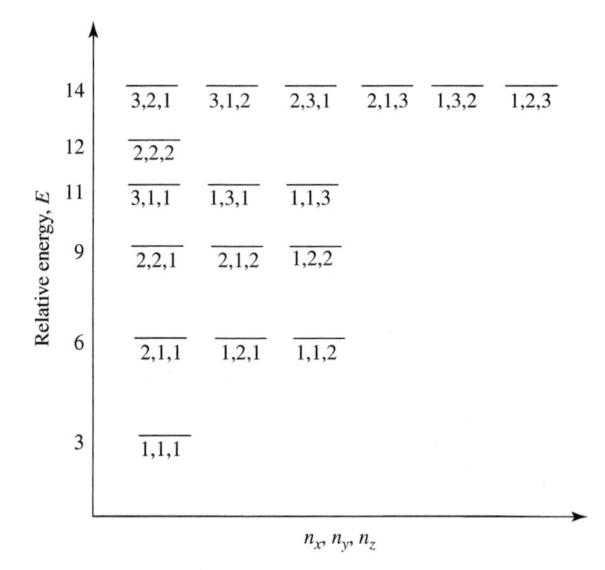

FIGURE 3.29
The relative energies (in units of $h^2/8\,ma^2$) for the three-dimensional particle in a box having sides with lengths $a = b = c$. The energy level labels list the three quantum numbers n_x, n_y, and n_z in that order. Notice that many of the energy levels are degenerate.

3.8 THE HARMONIC OSCILLATOR PROBLEM

A second common problem that occurs in chemistry is given by the harmonic oscillator model. Consider a block of mass m attached to a spring, as shown in Figure 3.30. If the other end of the spring is attached to a fixed point and the mass is displaced from its equilibrium position, its motion will oscillate in time. The harmonic oscillator problem is important in quantum chemistry because, to a certain extent, it models the behavior of nucleons in the nucleus, the motions of atoms in metallic solids, and the vibrations in polyatomic molecules. The periodic behavior of a harmonic oscillator can also be described by the angular motion of a pendulum swinging in the xy plane, as shown in Figure 3.31. The pendulum swings through an arc of length θ with an angular velocity ω that is equal to 2π radians/sec, such that Equation (3.66) describes the relationship between frequency and the angular velocity.

$$\nu = \frac{2\pi}{\omega} \tag{3.66}$$

Using Hooke's law, the restoring force for the mass on the spring in Figure 3.30 is given by Equation (3.67). Taking the integral of the force with respect to position yields the potential energy in Equation (3.68). Setting the force equal to Newton's second law yields Equation (3.69), which rearranged becomes Equation (3.70). This is a second-order differential equation with constant coefficients and it can therefore

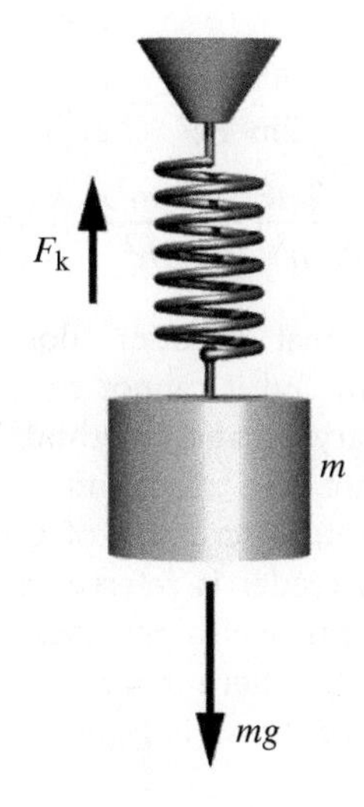

FIGURE 3.30
Harmonic oscillator model employing a mass m affixed to a spring having a force constant k. [Attributed to Svjo, reproduced from http://en.wikipedia.org /wiki/Hooke's_law (accessed December 1, 2013).]

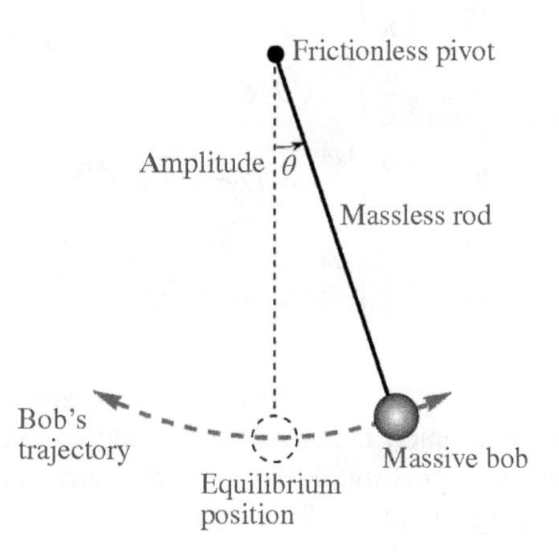

FIGURE 3.31
Harmonic motion of a pendulum. [Reproduced from http://en.wikipedia.org/wiki /Pendulum (accessed Decemebr 1, 2013).]

be solved using the auxiliary equation method described in the previous section. The general solution to Equation (3.70) in trigonometric form is given by Equation (3.53). The boundary condition that $x(0) = L$, where L is the maximum distance that the mass was stretched from its equilibrium position implies that the second term in Equation (3.53) is zero and $A = L$ and therefore the solution to Equation (3.70) reduces to Equation (3.71).

$$F = -kx \tag{3.67}$$

$$V(x) = -\int F dx = -\int kx dx = \frac{1}{2}kx^2 \tag{3.68}$$

$$F(x) = -kx = m\frac{d^2x}{dt^2} \tag{3.69}$$

$$\frac{d^2x}{dt^2} + \frac{k}{mx} = 0 \tag{3.70}$$

$$x(t) = A\cos(\beta t) = L\cos(\omega t),$$

$$\text{where } \omega = \sqrt{\frac{k}{m}} \tag{3.71}$$

Substituting the classical potential energy into the Schrödinger equation in one-dimension gives Equation (3.72). Following rearrangement and substitution of $\omega^2 = k/m$, Equation (3.73) results.

$$-\frac{\hbar^2}{2m}\frac{d^2\psi}{dx^2} + \frac{1}{2}kx^2\psi = E\psi \tag{3.72}$$

$$\frac{d^2\psi}{dx^2} + \left(\frac{2mE}{\hbar^2} - \frac{\omega^2 m^2 x^2}{\hbar^2}\right)\psi = 0 \tag{3.73}$$

This second-order differential equation does not have constant coefficients (because of the x^2 term) and it cannot be solved in the same manner as Equation (3.57) using the auxiliary equation method. The solutions to the quantum harmonic oscillator problem are not trivial and will be presented here without proof. For a rigorous, but lucid, description of the calculus involved in solving Equation (3.73), the interested reader is referred to Fayer's *Elements of Quantum Mechanics*. The first few solutions, which are based on Hermite polynomials, are given by Equations (3.74)–(3.77), where $\alpha = m\omega/\hbar$, $y = \sqrt{\alpha}x$, and graphs of the corresponding wave functions are shown in Figure 3.32.

$$\Psi_0 = \left(\frac{\alpha}{\pi}\right)^{1/4} e^{-y^2/2} \tag{3.74}$$

$$\Psi_1 = \left(\frac{\alpha}{\pi}\right)^{1/4} \sqrt{2}y e^{-y^2/2} \tag{3.75}$$

$$\Psi_2 = \left(\frac{\alpha}{\pi}\right)^{1/4} \frac{1}{\sqrt{2}}(2y^2 - 1)e^{-y^2/2} \tag{3.76}$$

$$\Psi_3 = \left(\frac{\alpha}{\pi}\right)^{1/4} \frac{1}{\sqrt{3}}(2y^3 - 3y)e^{-y^2/2} \tag{3.77}$$

There are several very important features to note in Figure 3.32. First of all, according to classical mechanics, the mass on the end of the spring can have any energy, while the quantum mechanical harmonic oscillator must have the discrete energy levels given by Equation (3.78). Second, the potential energy of the mass will

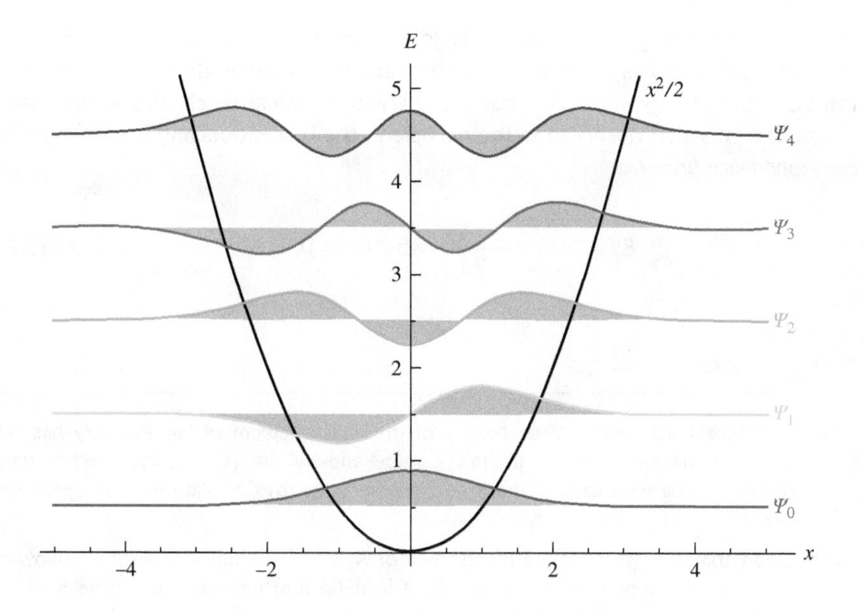

FIGURE 3.32
Solutions to the one-dimensional Schrödinger equation for the harmonic oscillator model. [Figure created by Oscar Castillo-Felisola using Sagemath. This figure is licensed under a Creative Commons Attribution-ShareAlike 3.0 Unported License.]

oscillate back and forth along the bottom of the parabola in Figure 3.32 in much the same way as the bob on the end of the pendulum will swing back and forth through its arc. Using classical mechanics, the oscillator will be moving faster in the center of its arc and therefore it will spend the least amount of time at the bottom of the potential energy well. The exact opposite is observed for the quantum mechanical harmonic oscillator—the probability density is greatest in the $v = 0$ (lowest energy) level at the exact center of the potential well. Third, while it is entirely possible for the classical oscillator to have zero energy (in fact this occurs every time the mass is restored to its equilibrium position), the quantum mechanical harmonic oscillator cannot be zero. This is due to the fact that if the potential energy were exactly zero, the position of the mass would be well defined to be at the center of the potential and the momentum would be well defined as exactly zero. This would violate the Heisenberg uncertainty principle. Fourth, it should be noted that the wave functions extend beyond the classical barrier for the potential energy (in other words, they leak outside of the parabola into the classically forbidden zone). In the limit of a very large v, as shown in Figure 3.33, the quantum harmonic oscillator spends more of

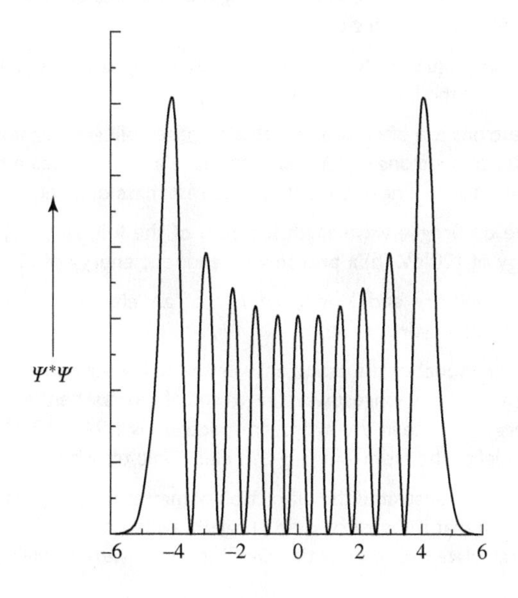

FIGURE 3.33
The probability function $\psi^*\psi$ for the harmonic oscillator with $v = 10$, showing how the potential energy for the quantum mechanical harmonic oscillator approaches that of the classical harmonic oscillator for very large values of v. [© Michael D Fayer, *Elements of Quantum Mechanics*, 2001, by permission of Oxford University Press, USA.]

its time near the edges of the parabola, in agreement with the classical observation that the oscillator has a higher probability at the extremes than in the center of the potential energy well. This correlation with classical mechanics at very large values of n was also observed in the particle in the box model and is known as the *correspondence principle*.

$$E_v = \hbar\omega\left(v + \frac{1}{2}\right), \quad \text{where } v = 0, 1, 2, \ldots \tag{3.78}$$

EXERCISES

3.1. Fluorescent light bulbs often consist of the line spectrum of Hg. Mercury has two strong atomic emission lines in the UV at 254 and 366 nm. (a) Calculate the frequency corresponding with each of these two wavelengths. (b) Calculate the energy of each line in units of joules.

3.2. Given that the (first) ionization energy of K is 419 kJ/mol, answer the following: (a) What is the maximum wavelength of light (in nm) necessary to ionize K metal? (b) Calculate the velocity of the ejected electron when a photon of UV light with $\lambda =$ 235 nm strikes the surface of K metal in an evacuated chamber.

3.3. Prove that $y = Ae^{2\pi i\upsilon tx} + Be^{-2\pi i\upsilon tx}$ is a solution to the one-dimensional wave equation given by Equation (3.3).

3.4. Indicate which of the following are eigenfunctions: (a) d^2/dx^2 (cos ax), (b) d/dt (e^{int}), (c) d/dy ($y^2 - 2y$), (d) $\partial/\partial y$ (x^2 e^{6y}), (e) d/dx (sin ax). For each eigenfunction, determine the eigenvalue.

3.5. At what wavelength does the maximum in the cosmic background radiation from the Big Bang occur given that the average temperature of the universe is 2.725 K? In which region of the electromagnetic spectrum does this occur?

3.6. Given that the (first) ionization energy of Cs is 376 kJ/mol, answer the following: (a) What is the maximum wavelength of light (in nm) necessary to ionize Cs metal? (b) Calculate the velocity of the ejected electron when a photon of UV light with $\lambda = 235$ nm strikes the surface of Cs metal in an evacuated chamber.

3.7. The Paschen series of lines in the line spectrum of hydrogen occur in the near-IR. (a) Calculate the wavelength (in nm) of the series limit for the Paschen series of lines in the line spectrum of hydrogen. (b) The frequency of one line in the Paschen series of hydrogen is 2.34×10^{14} Hz. Using the Bohr model of the atom with its circular orbits, sketch this specific electronic transition.

3.8. Use the Rydberg equation to calculate the wavelengths of the first three lines in the Brackett series of the line spectrum of hydrogen.

3.9. Beams of neutrons are often used to obtain images of lightweight atoms in molecules. What velocity of neutrons is necessary to make a neutron beam having a wavelength of 0.0150 pm? (You will need to look up the rest mass of a neutron.)

3.10. Calculate the de Broglie wavelength for each of the following: (a) an electron with a kinetic energy of 120 eV, (b) a proton with a kinetic energy of 120 eV.

3.11. Calculate the velocity and kinetic energy of an electron in the first Bohr orbit ($a_0 = 52.9$ pm) of a hydrogen atom.

3.12. Using the Bohr model of the atom, answer the following questions: (a) Calculate the wavelength (nm) for the longest wavelength line of the Paschen series ($n_f = 3$). (b) Given that the energy of a line in the hydrogen spectrum is 1.94×10^{-18} J, draw a picture of the Bohr model of the atom showing this electronic transition.

3.13. Scientists have demonstrated the diffraction of matter for particles as large as a Buckyball (C_{60}). Given that the observed de Broglie wavelength for a beam of Buckyballs was 0.0025 nm, calculate the velocity that the Buckyballs were travelling.

3.14. One of the shortest LASER pulses ever generated was 100 as with an uncertainty of 12 as. Using the Heisenberg uncertainty principle, calculate the uncertainty in the frequency of the LASER pulse.

3.15. If it were possible to locate the position of an electron to within 10 pm, what would be the uncertainty in v?

3.16. Calculate the energy difference (in units of cm^{-1}) between the $n = 2$ energy level and the $n = 1$ energy level for the particle in a box model with length 1.0 nm.

3.17. To a first approximation, the six *pi* electrons in the molecule hexatriene, $CH_2{=}CH{-}CH{=}CH{-}CH{=}CH_2$, can be considered using the particle in a box model, where the length of the box is equal to the distance between the two end C atoms (or 867 pm). Calculate the energies for the first four energy levels using this model. Given that each energy level can hold two electrons in the ground state, the highest filled energy level will have $n = 3$ and the lowest unfilled energy level will have $n = 4$. Calculate the wavelength of an electronic transition between the $n = 3$ and the $n = 4$ energy levels.

3.18. The Bohr radius for the $n = 1$ level in hydrogen atom is 52.9 pm. Assuming that $a = 2r$ for the one-dimensional particle in a box model, calculate the energies for the first three quantum levels in units of megajoules per mole.

3.19. What is the degeneracy of the fifth lowest energy level for a particle in a box with lengths $a = b = 1.0$ nm and $c = 2.0$ nm?

3.20. Given that the force constant for a harmonic oscillator is 250 N/m, calculate the fundamental vibrational frequency and the zero-point energy for this vibrational mode.

BIBLIOGRAPHY

1. Adamson, A. W. *A Textbook of Physical Chemistry*, 3rd ed., Academic Press, Orlando, FL, 1986.
2. Bockhoff, F. J. *Elements of Quantum Theory*, Addison-Wesley Publishing Company, Inc., Reading, MA, 1969.
3. Eyring, H.; Walter, J.; Kimball, G. E. *Quantum Chemistry*, John Wiley & Sons, Inc., New York, 1944.
4. Fayer, M. D. *Elements of Quantum Mechanics*, Oxford University Press, New York, 2001.
5. Hoffmann, B. *The Strange Story of the Quantum*, Dover Publications, Inc., New York, 1959.
6. Levine, I. N. *Quantum Chemistry*, 2nd ed., Allyn and Bacon, Inc., Boston, MA, 1974.
7. McQuarrie, D. A.; Simon, J. D. *Physical Chemistry: A Molecular Approach*, University Science Books, Sausalito, CA, 1997.
8. Pauling, L.; Wilson, E. B. *Introduction to Quantum Mechanics with Applications to Chemistry*, McGraw-Hill Book Company, New York, 1935.
9. Warren, W. S. *The Physical Basis of Chemistry*, Academic Press, Inc., San Diego, CA, 1994.

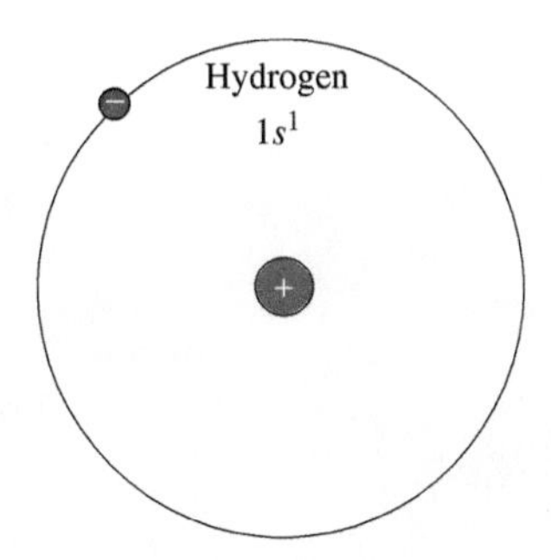

Atomic Structure | 4

"That's like saying you're the most important electron in the hydrogen atom. 'Cause you see, there's only one electron in a hydrogen atom."

—*Sheldon Cooper (Big Bang Theory)*

4.1 THE HYDROGEN ATOM

The goal of this chapter is to solve the Schrödinger equation for the one electron in the hydrogen atom. This electron experiences an electrostatic attraction for the nucleus which is distance dependent and which has a potential energy term given by Coulomb's law. The time-independent Schrödinger equation for the hydrogen atom is given in Equation (4.1), where Z is the atomic number ($Z = 1$ for H) and e is the charge on an electron. Because of the spherical symmetry for the potential energy term, it is more convenient to switch from Cartesian to polar coordinates, as shown in Figure 4.1. The conversions are given in the figure caption.

$$-\frac{\hbar^2}{2m}\left(\frac{\partial^2}{\partial x^2}+\frac{\partial^2}{\partial y^2}+\frac{\partial^2}{\partial z^2}\right)\psi-\frac{Ze^2}{4\pi\varepsilon_0 r}=E\psi \tag{4.1}$$

After substituting the polar coordinates for Cartesian coordinates and a very lengthy application of the chain rule, Equation (4.1) becomes Equation (4.2) in spherical polar coordinates.

$$-\frac{\hbar^2}{2m}\left[\frac{1}{r^2}\frac{\partial}{\partial r}\left(r^2\frac{\partial\psi}{\partial r}\right)+\frac{1}{r^2\sin\theta}\frac{\partial}{\partial\theta}\left(\sin\theta\frac{\partial\psi}{\partial\theta}\right)+\frac{1}{r^2\sin^2\theta}\frac{\partial^2\psi}{\partial\phi^2}\right]-\frac{Ze^2}{4\pi\varepsilon_0 r}\psi=E\psi \tag{4.2}$$

This equation can be separated into a radial part and an angular part, such that the wave function ψ can be taken as the product of a radial function R and an angular function Y, as shown in Equation (4.3). Multiplication of both sides of Equation (4.2) by $2mr^2$, followed by the method of separation of variables yields the slightly more manageable form of the Schrödinger equation for the hydrogen atom given in Equation (4.4), where β is the separation constant and we have incorporated $\hbar$ in with β.

$$\psi(r,\theta,\phi)=R(r)\,Y(\theta,\phi) \tag{4.3}$$

Principles of Inorganic Chemistry, First Edition. Brian W. Pfennig.

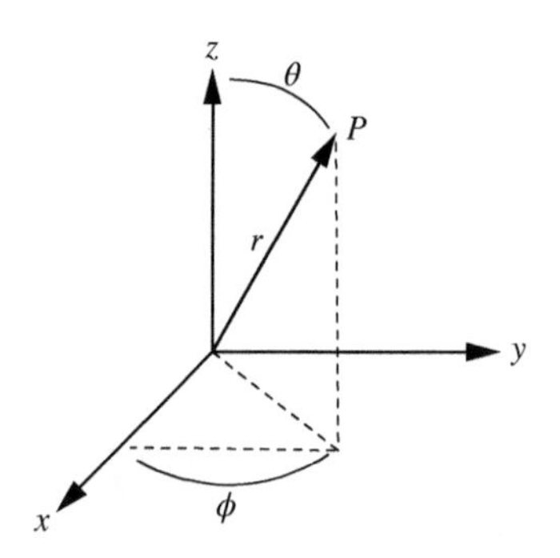

FIGURE 4.1
The point P is expressed in Cartesian units as $P(x, y, z)$ and in spherical polar coordinates as $P(r, \theta, \phi)$, where $x = r\sin\theta\cos\phi$, $y = r\sin\theta\sin\phi$, and $z = r\cos\theta$.

$$\frac{1}{R(r)}\left[\frac{d}{dR}\left(r^2\frac{dR}{dr}\right)+\frac{2mr^2}{\hbar^2}\left(\frac{e^2}{4\pi\varepsilon_0 r}+E\right)R(r)\right]=$$
$$-Y(\theta,\phi)\left[\frac{1}{\sin\theta}\frac{\partial}{\partial\theta}\left(\sin\theta\frac{\partial Y}{\partial\theta}\right)+\frac{1}{\sin^2\theta}\frac{\partial^2 Y}{\partial\phi^2}\right]=\beta \qquad (4.4)$$

Neither of the differential equations in Equation (4.4) is particularly easy to solve. Owing to the separation of variables, however, each solution will consist of two parts: a radial wave function $R(r)$ and an angular wave function $Y(\theta,\phi)$. The solutions to the angular part are referred to as the *spherical harmonics*, a fairly common type of function in a wide variety of physical problems. Furthermore, because there are three variables, the solutions to Equation (4.4) will be subject to three different quantum numbers (Q.N.): n, l, and m_l. The three quantum numbers can only assume the values given and have the designations listed at the right. Thus, for instance, when $n = 3$, l can equal 0, 1, or 2. If $l = 2$, m_l can be −2, −1, 0, 1, or 2.

$n = 1, 2, 3, \ldots$	Principal Q.N.
$l = 0, 1, 2, \ldots, n-1$	Azimuthal Q.N.
$m_l = 0, \pm 1, \pm 2, \ldots, \pm l$	Magnetic Q.N.

The solutions to the Schrödinger equation for the hydrogen atom are shown in Table 4.1 for the first few sets of quantum numbers. By analogy to the Bohr model of the hydrogen atom, each set of three quantum numbers specifies a particular orbital, instead of an orbit. An *orbital* is nothing more than one of the allowed wave function solutions to the Schrödinger equation for an electron in the hydrogen atom. Each orbital is given a symbol, such as the $2p_z$ orbital, where the numeral indicates the value of the principal quantum number, the letter indicates the value of the azimuthal quantum number ($l = 0$, 1, 2, and 3 correspond with the letters s, p, d, and f), and the subscript has to do with the magnetic quantum number. To a first approximation, the principal quantum number determines an orbital's size, the azimuthal quantum number is reflective of its shape, and the magnetic quantum number indicates its relative orientation in space.

4.1.1 The Radial Wave Functions

The radial wave functions all show an exponential decay as the radius increases. The exponential decay is slower with increasing n because the denominator in the exponential term contains a factor of na_0. Thus, the average radius (or size) of an orbital also increases with n. For $n > 1$, the radial functions all have at least one radial

TABLE 4.1 Mathematical forms of the radial wave functions, $R(r)$, and angular wave functions, $Y(\theta, \phi)$, for the hydrogen atom for the first few sets of allowed quantum numbers, where $a_0 = 52.9$ pm, the Bohr radius.

n, l, m_l	$R(r)$	$Y(\theta, \phi)$	Orbital
1, 0, 0	$2\left(\frac{Z}{a_0}\right)^{3/2} e^{-\frac{Zr}{a_0}}$	$\left(\frac{1}{4\pi}\right)^{1/2}$	1s
2, 0, 0	$\left(\frac{Z}{2a_0}\right)^{3/2}\left(2-\frac{Zr}{a_0}\right) e^{-\frac{Zr}{2a_0}}$	$\left(\frac{1}{4\pi}\right)^{1/2}$	2s
2, 1, 0	$\frac{1}{\sqrt{3}}\left(\frac{Z}{2a_0}\right)^{3/2}\left(\frac{Zr}{a_0}\right) e^{-\frac{Zr}{2a_0}}$	$\left(\frac{3}{4\pi}\right)^{1/2} \cos\theta$	$2p_0$
2, 1, 1	$\frac{1}{\sqrt{3}}\left(\frac{Z}{2a_0}\right)^{3/2}\left(\frac{Zr}{a_0}\right) e^{-\frac{Zr}{2a_0}}$	$\left(\frac{3}{8\pi}\right)^{1/2} \sin\theta e^{i\phi}$	$2p_{+1}$
2, 1, −1	$\frac{1}{\sqrt{3}}\left(\frac{Z}{2a_0}\right)^{3/2}\left(\frac{Zr}{a_0}\right) e^{-\frac{Zr}{2a_0}}$	$\left(\frac{3}{8\pi}\right)^{1/2} \sin\theta\, e^{-i\phi}$	$2p_{-1}$
3, 0, 0	$\frac{2}{27}\left(\frac{Z}{3a_0}\right)^{3/2}\left(27-\frac{18Zr}{a_0}+\frac{2Z^2r^2}{a_0^2}\right) e^{-\frac{Zr}{3a_0}}$	$\left(\frac{1}{4\pi}\right)^{1/2}$	3s
3, 1, 0	$\frac{1}{81\sqrt{3}}\left(\frac{2Z}{a_0}\right)^{3/2}\left(6-\frac{Zr}{a_0}\right)\frac{Zr}{a_0} e^{-\frac{Zr}{3a_0}}$	$\left(\frac{3}{4\pi}\right)^{1/2} \cos\theta$	$3p_0$
3, 1, 1	$\frac{1}{81\sqrt{3}}\left(\frac{2Z}{a_0}\right)^{3/2}\left(6-\frac{Zr}{a_0}\right)\frac{Zr}{a_0} e^{-\frac{Zr}{3a_0}}$	$\left(\frac{3}{8\pi}\right)^{1/2} \sin\theta e^{i\phi}$	$3p_{+1}$
3, 1, −1	$\frac{1}{81\sqrt{3}}\left(\frac{2Z}{a_0}\right)^{3/2}\left(6-\frac{Zr}{a_0}\right)\frac{Zr}{a_0} e^{-\frac{Zr}{3a_0}}$	$\left(\frac{3}{8\pi}\right)^{1/2} \sin\theta\, e^{-i\phi}$	$3p_{-1}$
3, 2, 0	$\frac{1}{81\sqrt{15}}\left(\frac{2Z}{a_0}\right)^{3/2}\left(\frac{Zr}{a_0}\right)^2 e^{-\frac{Zr}{3a_0}}$	$\left(\frac{5}{16\pi}\right)^{1/2} (3\cos^2\theta - 1)$	$3d_0$
3, 2, 1	$\frac{1}{81\sqrt{15}}\left(\frac{2Z}{a_0}\right)^{3/2}\left(\frac{Zr}{a_0}\right)^2 e^{-\frac{Zr}{3a_0}}$	$\left(\frac{15}{8\pi}\right)^{1/2} \sin\theta \cos\theta\, e^{i\phi}$	$3d_{+1}$
3, 2, −1	$\frac{1}{81\sqrt{15}}\left(\frac{2Z}{a_0}\right)^{3/2}\left(\frac{Zr}{a_0}\right)^2 e^{-\frac{Zr}{3a_0}}$	$\left(\frac{15}{8\pi}\right)^{1/2} \sin\theta \cos\theta\, e^{-i\phi}$	$3d_{-1}$
3, 2, 2	$\frac{1}{81\sqrt{15}}\left(\frac{2Z}{a_0}\right)^{3/2}\left(\frac{Zr}{a_0}\right)^2 e^{-\frac{Zr}{3a_0}}$	$\left(\frac{15}{32\pi}\right)^{1/2} \sin^2\theta e^{2i\phi}$	$3d_{+2}$
3, 2, −2	$\frac{1}{81\sqrt{15}}\left(\frac{2Z}{a_0}\right)^{3/2}\left(\frac{Zr}{a_0}\right)^2 e^{-\frac{Zr}{3a_0}}$	$\left(\frac{15}{32\pi}\right)^{1/2} \sin^2\theta e^{-2i\phi}$	$3d_{-2}$

node. For the 2s orbital, for instance, which has the quantum numbers $n = 2$, $l = 0$, $m_l = 0$, the radial node occurs when $r = 2a_0/Z$ because this makes the second term in the parentheses zero ($a_0 = 52.9$ pm, the Bohr radius). A general rule is that there are $n - l - 1$ radial nodes. Thus, a 2s orbital has one radial node, while a 2p orbital has no radial nodes.

When a wave function or a product of wave functions is integrated over all space, the volume element in Cartesian coordinates is $d\tau = dx\,dy\,dz$. In polar coordinates,

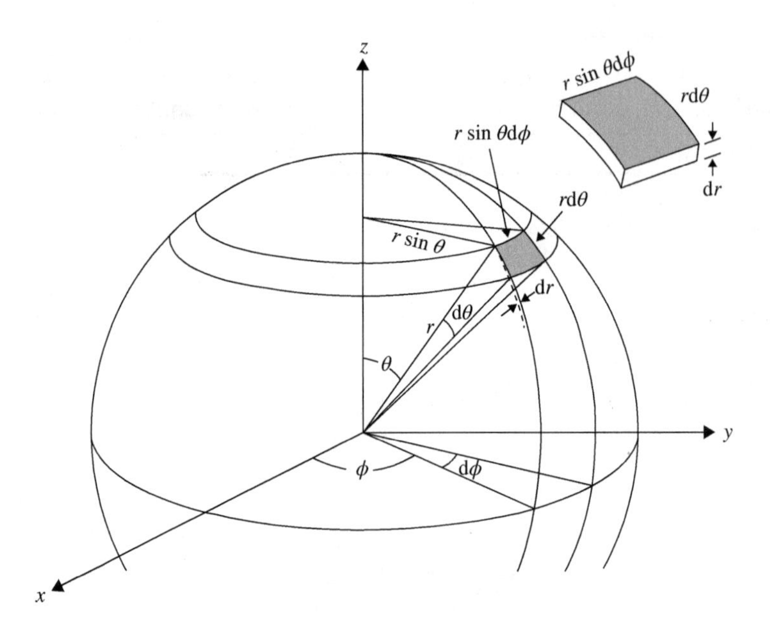

FIGURE 4.2
Definition of the volume element $d\tau$ in polar coordinates: $d\tau = dV = r^2 \sin\theta \, dr \, d^2\theta \, d\phi$. [© University Science Books, Mill Valley, CA. Used with permission. All rights reserved. McQuarrie, D. A.; Simon, J. D. *Physical Chemistry: A Molecular Approach*, 1997.]

however, the volume element becomes $d\tau = dV = r^2 \sin\theta \, dr \, d\theta \, d\phi$, as shown in Figure 4.2.

A more useful quantity than the radial wave function is the radial distribution function, also called the *radial probability function. The radial distribution function* is the probability that the electron will exist in a thin volume element dV at a distance r from the nucleus. One way of visualizing this is to think of the volume element as a thin spherical shell, similar to one of the layers in an onion skin, existing at a distance r away from the nucleus. The volume element dV shown in Figure 4.2 represents a fraction of this "onion skin." Because the probability of finding an electron in a given region of space goes as the square of the wave function (the Born interpretation), the radial distribution function is equal to $R(r)^2 \, dV$. The volume of a sphere is $V = (4/3)\pi r^3$, and therefore $dV/dr = 4\pi r^2$. Following substitution, the radial distribution function is defined as $4\pi r^2 R(r)^2 \, dr$. Plots of the radial distribution function for the first several types of orbitals in the hydrogen atom are shown in Figure 4.3. The presence of the radial nodes is clearly indicated on the diagram. A second noteworthy feature is that the probability of the electron being close to the nucleus for a given value of n decreases in the order $s > p > d > f$. In other words, the s orbital "penetrates" the nucleus better than a p orbital having the same principal quantum number. This fact is of utmost importance in the forthcoming section of shielding and influences a large number of an element's chemical properties.

Example 4-1. Use Equation (3.18) to prove that the Bohr radius a_0 has a value of 52.9 pm when $n = 1$.

Solution. Using Equation (3.18) and solving for r yields:

$$r = \frac{4\pi\varepsilon_0 n^2 h^2}{4\pi^2 m e^2} = \frac{(1.113 \times 10^{-10}\,\mathrm{C^2/Jm})(6.626 \times 10^{-34}\,\mathrm{J\,s})^2}{4(3.1416)^2(9.109 \times 10^{-31}\,\mathrm{kg})(1.602 \times 10^{-19}\,\mathrm{C})^2}$$

$$= 5.29 \times 10^{-11}\,\mathrm{m} = 52.9\,\mathrm{pm}$$

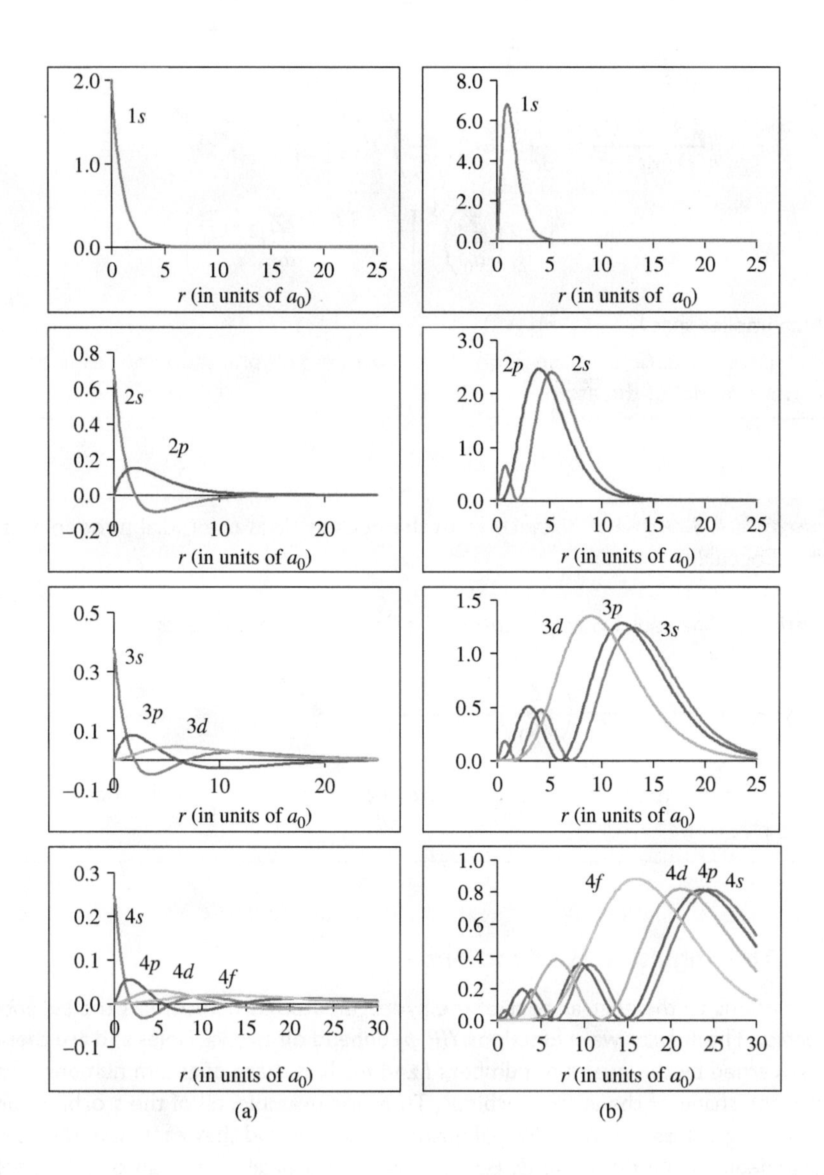

FIGURE 4.3
The radial function $R(r)$ (a) and the radial distribution function (b) for several types of orbitals in the hydrogen atom. The y-scale varies from one orbital to the next.

Example 4-2. Show that the most probable radius for an electron in the 1s orbital of hydrogen is equal to the Bohr radius, a_0.

Solution. The most probable radius can be obtained from the highest peak in the radial distribution function, because this function is a measure of the probability of finding an electron in a volume element at a certain distance from the nucleus. Because the radial distribution function for a 1s orbital has a single peak, the radius at which this peak occurs can be calculated by taking the first derivative of the function with respect to r and setting it equal to zero. For a 1s orbital, $R(r)$ and the first derivative of the radial probability function are

$$R(r) = 2\left(\frac{Z}{a_0}\right)^{3/2} e^{-\frac{Zr}{a_0}}$$

$$\frac{d(4\pi^2 r^2 R^2)}{dr} = \frac{d}{dr}\left[16\pi^2\left(\frac{Z}{a_0}\right)^3 r^2 e^{-\frac{2Zr}{a_0}}\right]$$

$$= 16\pi^2\left(\frac{Z}{a_0}\right)^3\left[2re^{-\frac{2Zr}{a_0}} - \frac{2Z}{a_0}r^2 e^{-\frac{2Zr}{a_0}}\right] = 0$$

which implies that $[2r - \frac{2Z}{a_0}r^2] = 0$

Hence, $r = a_o/Z$. Because $Z = 1$ for H, $r = a_0 = 52.9$ pm, the same result as in the Bohr model of the atom.

Example 4-3. At what distance from the nucleus does the radial node in a 2*s* orbital occur?

Solution. The radial wave function for a 2*s* orbital is given here:

$$\left(\frac{Z}{2a_0}\right)^{3/2}\left(2 - \frac{Zr}{a_0}\right)^{3/2} e^{-\frac{Zr}{2a_0}}$$

Setting the middle term equal to zero, we find that $2 = \frac{Zr}{a_0}$ and hence $r = \frac{2a_0}{Z}$ = 106 pm.

4.1.2 The Angular Wave Functions

The solutions to the angular part of the hydrogen atom are known as the *spherical harmonics*. The angular wave functions $Y(\theta,\phi)$ depend on two variables and are therefore governed by the quantum numbers l and m_l. It is these quantum numbers that dictate the shape of the atomic orbitals. Thus, for example, all of the s orbitals are spherical regardless of their principal quantum number and they each have the same angular dependence on θ and ϕ. Because $l = 0$ for s orbitals, m_l can only be zero; therefore, only one type of s orbital exists for any given value of *n*. By contrast, there are three different kinds of *p* orbitals because m_l can take values of -1, 0, or $+1$. The shapes of these orbitals are shown in Figure 4.4, where the orbital boundary indicates a 90% or greater probability of finding the electron in the enclosed region of space. When $m_l = 0$, the angular wave function is real, and the shape of the orbital consists of two lobes that lie along the z coordinate axis. Because its lobes lie along the z-axis, the p_0 orbital is also known as the p_z *orbital*. The sign of the wave function before squaring is also indicated in the figure. Notice that one of the lobes is positive while the other is negative and therefore an angular nodal surface exists in

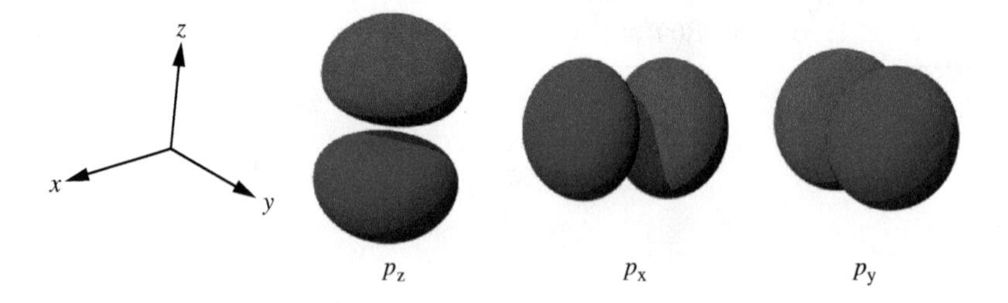

FIGURE 4.4
Shapes of the three *p* orbitals. [Modified from http://en.wikipedia.org/wiki/Atomic_orbital (accessed November 30, 2013).]

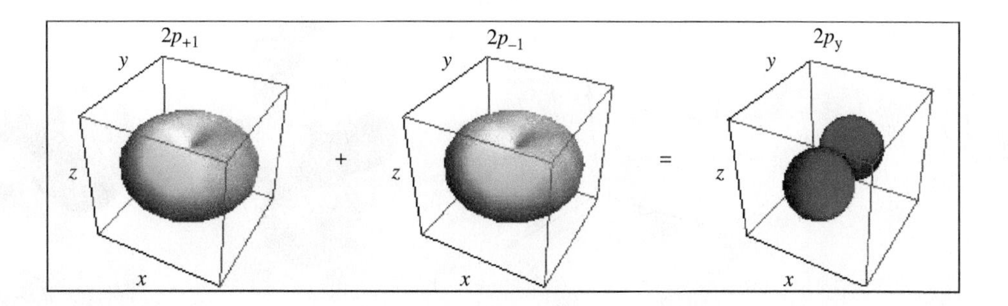

FIGURE 4.5
Illustration of how a linear combination of the $2p_{+1}$ and $2p_{-1}$ orbitals can be used to construct the more familiar $2p_y$ orbital. [Images by Lisa M. Goss. Used by permission.]

the *xy* plane. As a rule, every orbital will exhibit *l* angular nodal surfaces, which can be either planar or conical in shape. Thus, all three of the *p* orbitals will have a single angular nodal plane. Likewise, each *d* orbital will have two angular nodal surfaces and each *f* orbital will have three.

Although the angular wave functions of orbitals having $|m_l| > 0$ all contain imaginary components in their exponential terms, the product Y^*Y is real and can therefore be plotted. The angular dependence of the 2*p* orbitals having $m_l = -1$ and $m_l = +1$ are identical. Both orbitals take on the shape of a donut with the z-axis passing through the center of the donut hole. The only difference between the two orbitals is that the electron is moving in opposite directions in each of them. Because of the requirement that all quantum mechanical operators be linear and Hermitian, any linear combination of two degenerate wave functions will also be an acceptable solution to the Schrödinger equation. When the positive linear combination $Y(1,1) + Y(1,-1)$ is taken, as shown in Equation (4.6), where the numbers in parentheses refer to *l* and m_l, respectively, the equation for the familiar p_y orbital is obtained after normalization, as shown in Figure 4.5. When the negative linear combination $Y(1,1) - Y(1,-1)$ is taken, as shown in Equation (4.7), the equation for the p_x orbital results after normalization. This process is also known as the *hybridization* of atomic orbitals. As shown in Figure 4.4, the p_x hybrid orbital has its lobes lying along the *x*-axis, while the p_y hybrid has its lobes lying along the *y*-axis. According to quantum theory, the hybrid orbitals must also be orthogonal to one another. These particular linear combinations ensure that all three *p* orbitals will have the same shape with their lobes pointing along the three orthogonal Cartesian axes.

$$Y(1,1)^* Y(1,1) = Y(1,-1)^* Y(1,-1) = \frac{3}{8\pi}\sin^2\theta \tag{4.5}$$

Normalization means that the integral of Y^*Y over all space must equal unity. Because each of the original wave functions $Y(1,1)$ and $Y(1,-1)$ are normalized and have integrals of one, the integral of the positive linear combination must equal two. Therefore, $N = 2$ in the normalization equation and the normalizing coefficient is $c = 1/2^{1/2}$.

$$p_x = \frac{1}{\sqrt{2}}(Y(1,1) + Y(1,-1)) = \frac{1}{\sqrt{2}}\left(\frac{3}{8\pi}\right)^{1/2}[\sin\theta\, e^{i\phi} + \sin\theta\, e^{-i\phi}] =$$
$$\frac{1}{\sqrt{2}}\left(\frac{3}{8\pi}\right)^{1/2}\sin\theta\,[\cos\phi + i\sin\phi + \cos\phi - i\sin\phi] = \frac{1}{\sqrt{2}}\left(\frac{3}{8\pi}\right)^{1/2}\sin\theta\cos\phi \tag{4.6}$$

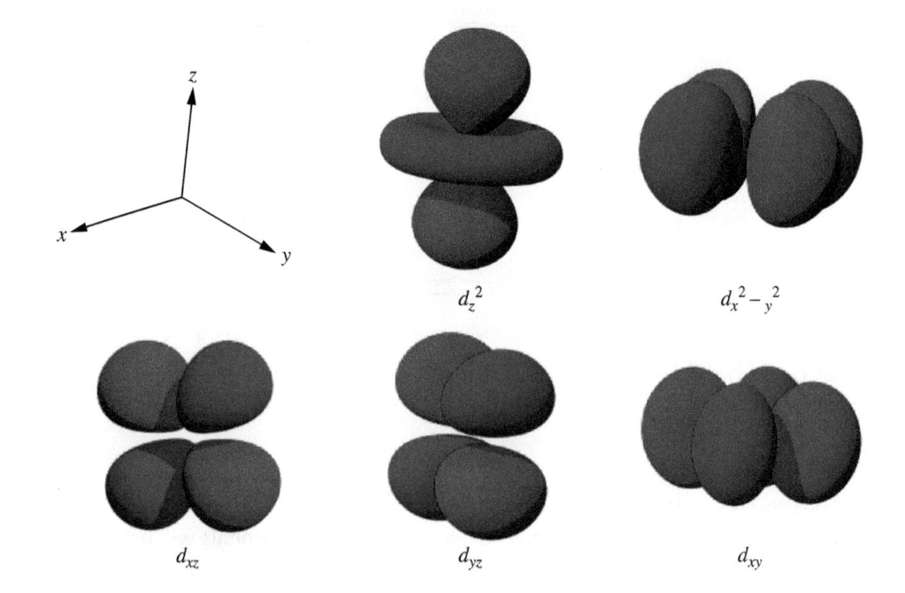

FIGURE 4.6
Shapes of the five *d* orbitals. [Modified from http://en.wikipedia.org/wiki/Atomic_orbital (accessed November 30, 2013).]

For normalization of the negative linear combination, the i factors out in the determination of the normalizing coefficient because of the complex conjugate in Equation (3.54), such that:

$$p_y = \frac{-i}{\sqrt{2}}(Y(1,1) - Y(1,-1)) = -\frac{i}{\sqrt{2}}\left(\frac{3}{8\pi}\right)^{1/2}[\sin\theta\, e^{i\phi} - \sin\theta\, e^{-i\phi}] =$$

$$-\frac{i}{\sqrt{2}}\left(\frac{3}{8\pi}\right)^{1/2}\sin\theta\,[\cos\phi + i\sin\phi - \cos\phi + i\sin\phi] = \frac{1}{\sqrt{2}}\left(\frac{3}{8\pi}\right)^{1/2}\sin\theta\,\sin\phi \tag{4.7}$$

The five different kinds of *d* orbitals are shown in Figure 4.6.

When $m_l = 0$, the d_{z^2} orbital results. This orbital has two lobes of the same sign pointing along the z-axis, with a donut-shaped lobe of the opposite sign in the *xy* plane. Notice that there are two conical nodes, each beginning at the origin and pointing in a different direction along the z-axis. When $|m_l| = 1$, the product Y^*Y yields a probability function containing two donuts centered on the z-axis, where one lies above the *xy* plane and the other lies beneath it. When $|m_l| = 2$, the product Y^*Y looks similar to a single hollow donut lying in the *xy* plane, as shown in Figure 4.7. Taking the positive and negative linear combinations of $Y(2,1)$ with $Y(2,-1)$ yields the d_{xz} and the d_{yz} orbitals shown in Figure 4.6. Both of these orbitals contain four lobes (as in a four-leaf clover) of alternating sign of the wave function and lying in the *xz* and *yz* planes, respectively. Each of the four lobes lies between the coordinate axes. Likewise, linear combinations of the $Y(2,2)$ and $Y(2,-2)$ wave functions yield the d_{xy} and $d_{x^2-y^2}$ orbitals shown in Figure 4.6. Both orbitals lie in the *xy* plane and have the same alternating four-leaf clover shape. However, the d_{xy} orbital has its lobes pointing between the coordinate axes, while the $d_{x^2-y^2}$ orbital's lobes lie squarely on the coordinate axes. The names and shapes of the five *d*-orbitals are especially important in the field of coordination chemistry and should be memorized by the student at this time.

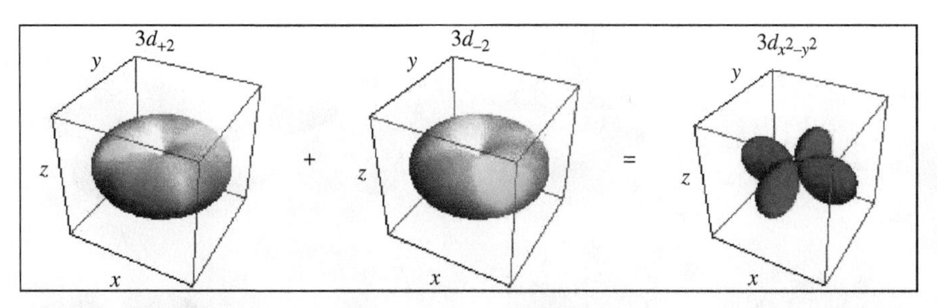

FIGURE 4.7
Illustration of how a linear combination of the $3d_{+2}$ and $3d_{-2}$ orbitals can be used to construct the more familiar $3d_{x^2-y^2}$ orbital. [Images by Lisa M. Goss. Used by permission.]

Example 4-4. Write the mathematical form of the angular part of the wave function for the d_{xz} and d_{yz} orbitals by taking the positive and negative linear combinations of $Y(2,1)$ with $Y(2,-1)$, respectively.

Solution. The positive linear combination (d_{xz}) of $Y(2,1)$ with $Y(2,-1)$ is given by

$$d_{xz} = \frac{1}{\sqrt{2}}(Y(2,1) + Y(2,-1)) = \frac{1}{\sqrt{2}}\left(\frac{15}{8\pi}\right)^{1/2}$$
$$[\sin\theta\cos\theta(\cos\phi + i\sin\phi + \cos\phi - i\sin\phi)]$$
$$= \left(\frac{15}{4\pi}\right)^{1/2}\sin\theta\cos\theta\cos\phi$$

The negative linear combination (d_{yz}) of $Y(2,1)$ with $Y(2,-1)$ is given by

$$d_{yz} = -\frac{i}{\sqrt{2}}(Y(2,1) - Y(2,-1)) = -\frac{i}{\sqrt{2}}\left(\frac{15}{8\pi}\right)^{1/2}$$
$$[\sin\theta\cos\theta(\cos\phi + i\sin\phi - \cos\phi + i\sin\phi)]$$
$$= \left(\frac{15}{4\pi}\right)^{1/2}\sin\theta\cos\theta\sin\phi$$

The shapes of the seven f orbitals are shown in Figure 4.8. When $m_l = 0$, the f_{z^3} orbital results, which has lobes of opposite sign along the z-axis and two donuts encircling that axis that also have opposite signs. For $|m_l| = 1$, the positive and negative linear combinations $Y(3,1) \pm Y(3,-1)$ yield the f_{xz^2} and f_{yz^2} orbitals, which have six lobes each lying in the *xz* and *yz* planes, respectively. For $|m_l| = 2$, the two hybrids are the f_{xyz} and $f_{z(x^2-y^2)}$ orbitals. Both of these orbitals have eight lobes forming a cubic shape, with the former lying between the *x*- and *y*-axes and the latter lying on the coordinate axes. Lastly, for $|m_l| = 3$, the hybrid orbitals are the $f_{x(x^2-3y^2)}$ and $f_{y(3x^2-y^2)}$ orbitals, which have six lobes each lying in the *xy* plane. The shapes of the seven *f*-orbitals are typically unimportant in chemical bonding and do not need to be memorized.

Several features common to all of the atomic orbitals are as follows:

- For any given value of *l*, the summation of the electron density probabilities for the complete set of orbitals will be a sphere. This is known as *Unsöld's theorem*. Thus, for example, the sum of the electron density for the $2p_x$, $2p_y$, and $2p_z$ orbitals is a sphere, as is the case for the lone 2*s* orbital.

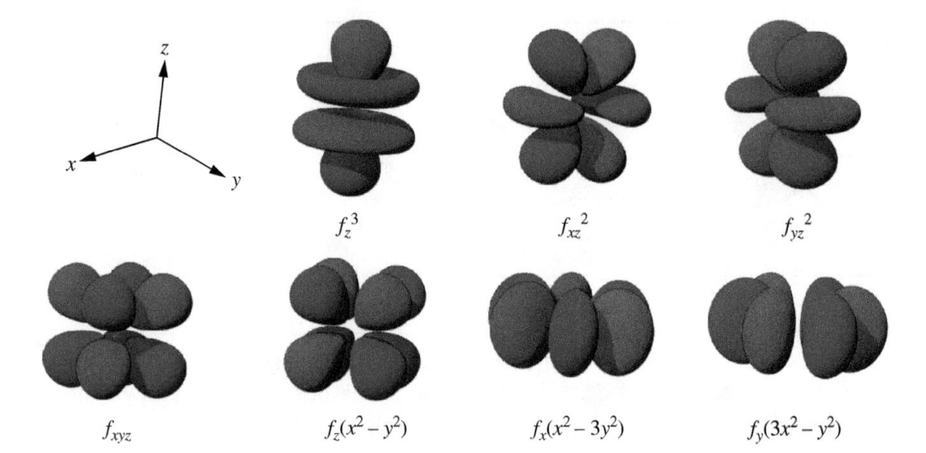

FIGURE 4.8
Shapes of the seven *f* orbitals, with $|m_l|$ increasing from left to right in the diagram. [Modified from http://en.wikipedia.org/wiki/Atomic_orbital (accessed November 30, 2013).]

- The absolute value of m_l is the number of angular nodes that present themselves when the orbital is viewed from either direction along the z-axis. Thus, a p_z orbital ($m_l = 0$), when viewed from the "top" of the z-axis, will present with a positive lobe. The negative lobe that lies beneath this will not be observed from this vantage point. However, the p_x and p_y orbitals ($|m_l| = 1$) will each present one angular nodal plane when viewed along the z-axis.
- The symmetry of the atomic orbitals with respect to inversion alternates in a regular pattern. The inversion operation means that if one takes any point (x, y, z) back through the origin an equal distance to point ($-x$, $-y$, $-z$), the probability density will be identical in magnitude and sign. All *s* and *d* orbitals are symmetric (or *gerade*) with respect to inversion. All *p* and *f* orbitals are antisymmetric (or *ungerade*) with respect to inversion (they have the opposite sign). More generally, whenever *l* is odd, the atomic orbitals will be *gerade* and whenever *l* is even, they will be *ungerade*.

Example 4-5. Prove that the total probability density for the three 2*p* orbitals is a sphere.

Solution. The sum of the squares of the angular components of the wave functions for the three 2*p* orbitals is

$$\left(\frac{3}{4\pi}\right)[\cos^2\theta + \sin^2\theta\cos^2\phi + \sin^2\theta\sin^2\phi] = \left(\frac{3}{4\pi}\right)[\cos^2\theta + \sin^2\theta(\cos^2\phi + \sin^2\phi)]$$

We can ignore the coefficient in front just as we ignored the radial part of the wave function because these just relate to the radius of the orbital summation and have nothing to do with its shape. We can also use the trigonometric identity that $\cos^2(a) + \sin^2(a) = 1$. Thus, the equation reduces to

$$[\cos^2\theta + \sin^2\theta(1))] = [\cos^2\theta + \sin^2\theta] = 1$$

Because the total angular component is a constant, there is no net angular dependence and the overall shape is that of a sphere.

4.2 POLYELECTRONIC ATOMS

Whenever two or more electrons are present in an atom or ion, an exact solution to the Schrödinger equation cannot be obtained because of the *electron correlation problem*. Consider any given electron in a polyelectronic atom. The electrostatic field experienced by this electron cannot be known exactly because of the Heisenberg uncertainty principle, which states that the exact position of other electrons cannot be measured precisely. In order to circumvent this problem, an approximation method is used. The most common approximation technique is known as the *self-consistent field (SCF) method*. Using this procedure, a reasonable wave function is used as a first approximation for all but one of the electrons. Then, the force field felt by this one electron is calculated to obtain a wave function for the electron. Next, a second electron is chosen and the wave function just obtained for the first electron is used in the calculation of the force field that the second electron experiences. This method is repeated for each of the electrons in an iterative process until the force field for each electron begins to converge to a single value; or in other words, when a SCF results. This field is then used to calculate the approximate wave function solutions (known as the *Hartree–Fock equations*) to the polyelectronic Schrödinger equation.

The following is an example of the SCF method in practice in the treatment of the two electrons present in a helium atom. Because helium has two electrons orbiting a +2 nucleus, it presents itself as the three-body problem shown in Figure 4.9, where the nucleus is presumed to be at rest and therefore sits at the origin of the coordinate system. The Hamiltonian for the helium atom includes three potential energy terms: an attractive force between electron 1 and the nucleus (r_1), an attractive force between electron 2 and the nucleus (r_2), and the electron–electron repulsion between the two electrons (r_{12}), as shown in Equation (4.8).

$$\hat{H} = \left[-\frac{\hbar^2}{2m}\nabla_1^2 - \frac{\hbar^2}{2m}\nabla_2^2 - \frac{Ze^2}{4\pi\varepsilon_0 r_1} - \frac{Ze^2}{4\pi\varepsilon_0 r_2} + \frac{e^2}{4\pi\varepsilon_0 r_{12}}\right] \tag{4.8}$$

We allow the overall wave function to be a product of two wave functions, one for each individual electron, as shown in Equation (4.9). The effective Hamiltonian for electron 1 can then be calculated using Equation (4.10), where V^{eff} is the effective potential energy that electron 1 feels with respect to electron 2 and is given by Equation (4.11).

$$\psi(r_1, r_2) = \phi(r_1)\phi(r_2) \tag{4.9}$$

$$\hat{H}_1^{\text{eff}}(r_1) = \left[-\frac{\hbar^2}{2m}\nabla_1^2 - \frac{Ze^2}{4\pi\varepsilon_0 r_1} + V_1^{\text{eff}}\right] \tag{4.10}$$

$$V_1^{\text{eff}}(r_1) = \int \phi * (r_2)\frac{e^2}{4\pi\varepsilon_0 r_{12}}\phi(r_2)dr_2 \tag{4.11}$$

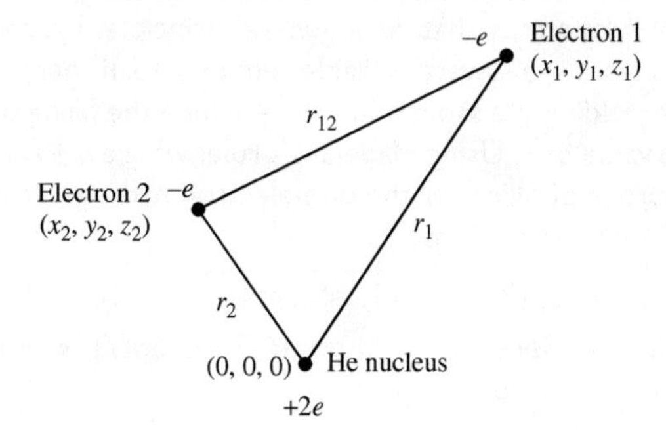

FIGURE 4.9
Definition of the different potential energy interactions between the two electrons and the +2 nucleus in a helium atom.

There are, of course, two additional equations corresponding to Equations (4.10) and (4.11) for the second electron. We begin by assuming a reasonable wave function (such as a 1*s* hydrogen wave function) for the second electron $\phi(r_2)$ and then using this wave function to evaluate the effective potential energy that electron 1 experiences according to Equation (4.11). This will allow a suitable effective Hamiltonian to be calculated for electron 1 using Equation (4.10). Next, we solve the Schrödinger equation using our effective Hamiltonian for electron 1, as shown by Equation (4.12)

$$\hat{H}_1^{\text{eff}}(r_1)\phi(r_1) = E_1\,\phi(r_1) \tag{4.12}$$

We then substitute the resulting value for $\phi(r_1)$ into the effective potential energy equation for electron 2. This value is then used in the equation that corresponds to Equation (4.10) to determine the effective Hamiltonian for electron 2. Then the Schrödinger equation corresponding to Equation (4.12) for electron 2 is solved in order to determine a new value for $\phi(r_2)$. The whole process is repeated in an iterative manner until the wave functions for $\phi(r_1)$ and $\phi(r_2)$ no longer change with time. We call this a SCF and the two resulting wave functions obtained by this method are known as the *Hartree–Fock orbitals*.

It is fortunate that whenever the SCF method is employed for any element or ion having more than one electron, the resulting wave functions always tend to resemble those of the hydrogen atom. Hence, the probability electron densities for the other elements can be compared to the *hydrogenic* orbital shapes. However, these wave functions are not identical to those of hydrogen, and the following differences should be noted:

- All of the hydrogenic orbitals contract as Z increases.
- Unlike the hydrogen atom, where the energies of the orbitals depend only on the principal quantum number, the energies of the hydrogenic orbitals also depend on the magnitude of l. For any given value of n, the energies of the hydrogenic orbitals for a polyelectronic atom increase in the order: $s < p < d < f$. The lowered degeneracy is a direct result of the differing degrees of penetration of the nucleus that the electrons in these orbitals exhibit. Comparing the radial distribution functions for a 3*d*, 3*p*, and 3*s* electron in Figure 4.3, it is apparent that an electron in the 3*s* orbital will have a higher probability of residing closer to the nucleus than will a 3*p* or a 3*d* electron. Thus, an electron in a 3*d* orbital of iron, for instance, will be somewhat shielded (or screened) from feeling the full effect of the +26 charge of the nucleus by any electrons that have a higher probability of lying closer to the nucleus. Therefore, the *effective nuclear charge* that the 3*d* electron feels will be less than that of a 3*p* or a 3*s* electron. Consequently, a 3*d* electron will be held less tightly by the nucleus than a 3*s* or 3*p* electron and will therefore lie at higher energy, as shown in Figure 4.10.
- As a result of the different energies of the *s*, *p*, *d*, and *f* hydrogenic orbitals having the same value of *n*, some overlapping of orbital energies between different principal quantum numbers occurs. *Madelung's rule* (also known as *Klechkowski's rule*) states that, as a general principle, hydrogenic orbital filling proceeds from the lowest available sum of $n + l$. If there is more than one combination yielding the same value of $n + l$, then the filling will occur first for the smallest value of n. Using Madelung's rule, where $n + l$ is shown in parentheses, the order of filling for the one-electron hydrogenic orbitals (ignoring electron–electron repulsions) is

  $1s(1) < 2s(2) < 2p(3) < 3s(3) < 3p(4) < 4s(4) < 3d(5) < 4p(5) < 5s(5) < 4d(6) < 5p(6) < 6s(6) < 4f(7) < 5d(7) < 6p(7) < 7s(7) < 5f(8) < 6d(8) < 7p(8)$

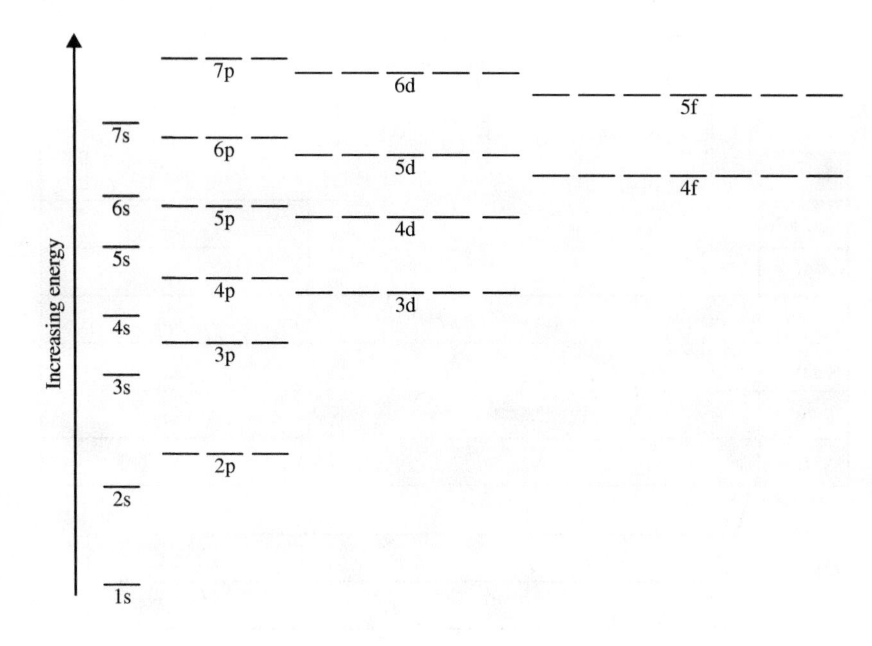

FIGURE 4.10 Energy ordering of the hydrogenic orbitals in a polyelectronic atom. [Reproduced from http://en.wikibooks.org/wiki/High_School_Chemistry/Families_on_the_Periodic_Table (accessed November 30, 2013).]

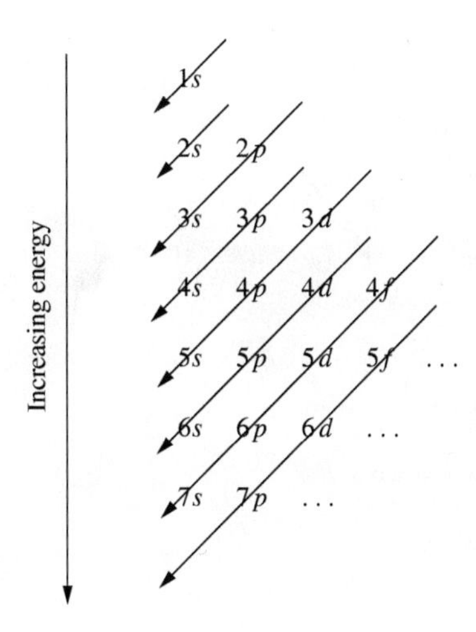

FIGURE 4.11 Moeller's rubric for the general order of orbital filling using the one-electron hydrogenic orbitals.

The same ordering is obtained if one employs *Moeller's rubric*, which is shown in Figure 4.11. This general ordering of the hydrogenic orbitals neatly mimics the pattern of the periodic properties of the elements that were first discovered in the late 1860s by Dmitri Mendeleev and J. Lothar Meyer. A version of the periodic table showing its correlation with orbital filling is shown in Figure 4.12.

4.3 ELECTRON SPIN AND THE PAULI PRINCIPLE

A fourth quantum number, called the *spin quantum number*, is required when the relativistic effects of electronic motion are taken into consideration. The concept of *electron spin* was first postulated by Goudsmit and Uhlenbeck in 1925 in order to explain the fine structure (or splitting) of the line spectra of several of the alkali metals. For example, the yellowish glow of many incandescent lights found in large city parking lots is actually due to two very closely spaced lines in the emission spectrum of Na. The two sodium D-lines, which arise from a transition between 3*p* and 3*s* hydrogenic orbitals, have wavelengths of 588.9950 and 589.5924 nm. The

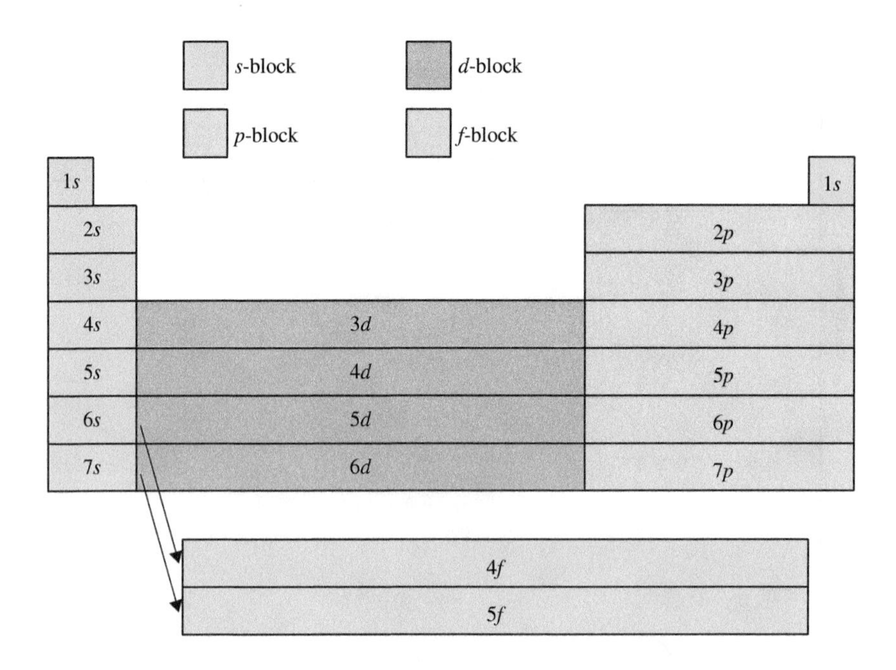

FIGURE 4.12
Periodic table showing the correlation with Moeller's rubric for orbital filling.

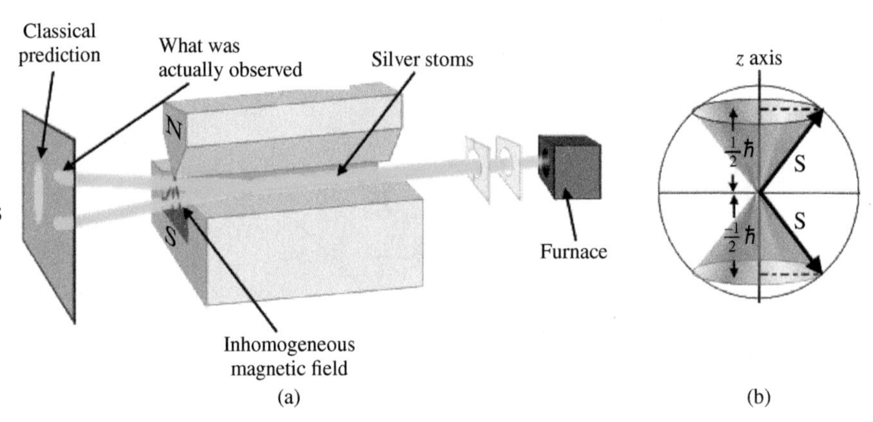

FIGURE 4.13
(a,b) Schematic diagram of the Stern–Gerlach experiment, showing how a beam of Ag atoms can be split by an inhomogeneous magnetic field into two different trajectories as a result of the different spin states they possess. [(a,b) Attributed to Theresa Knott, reproduced from http://en.wikipedia.org/wiki/Stern%E2%80%93Gerlach_experiment (accessed November 30, 2013).]

Stern–Gerlach experiment, depicted in Figure 4.13, provided the first experimental confirmation for the quantization of electron spin. In their 1922 experiment, Otto Stern and Walther Gerlach showed that a beam of silver atoms could be split into two beams by passing it through an inhomogeneous magnetic field. In addition to electrons, many nuclei also exhibit spin, forming the basis of nuclear magnetic resonance (NMR) and electron spin resonance (ESR) analytical techniques.

In 1928, Paul Dirac developed a relativistic theory of quantum mechanics from which the concept of spin arose naturally. The inclusion of a fourth variable (time) required the presence of a fourth quantum number. According to Dirac's derivation, an electron possesses both orbital (L) and spin (S) angular momentum. The total angular momentum (J) is a linear combination of the two, as shown in Equation (4.13).

$$\hat{J} = \hat{L} + \hat{S} = -i\hbar\frac{\partial}{\partial\phi} \pm \frac{\hbar}{2} \tag{4.13}$$

The spin angular momentum vector S can take values of $\pm m_s\ \hbar/2$, where m_s can take values of $+1/2$ (α) or $-1/2$ (β), depending on whether it aligns against or with the external magnetic field, respectively. The usual classical picture of electron spin, where the electron can be considered as a top spinning on its axis either in the clockwise or the counterclockwise direction, is only useful as a conceptual tool. In actuality, spin is strictly a quantum mechanical phenomenon and it has no classical analogy.

The introduction of a fourth quantum number to account for the spin angular momentum of an electron also necessitates the introduction of a sixth fundamental postulate of quantum mechanics.

Postulate 6: The total wave function must be antisymmetric with respect to the interchange of all coordinates of one fermion with those of another. The *Pauli exclusion principle*, which states that no two electrons within an atom can have the same set of quantum numbers, is a direct result of this antisymmetry principle.

Another way of thinking about the Pauli exclusion principle is to give each electron in an atom its own electronic (e-mail) address. For instance, at one point in time, my e-mail address was brpfennig@vaxsar.vassar.edu. One could think of my e-mail address as consisting of four quantum numbers. The principal quantum number here would be the .edu domain, which identifies this electron (me) as belonging to the realm of educational institutions. The second quantum number in my e-mail address gets a little more specific and identifies me at Vassar College, where I used to teach. The third quantum number is even more specific, sending my junk mail to the Vaxsar server (clever, huh?) at the college. Finally, the mail is delivered specifically to me. Just as there are no two people in the world having the exact same email address, there are no two electrons within an atom that can have the same set of all four quantum numbers: n, l, m_l, and m_s.

Returning to the helium atom, suppose that electrons 1 and 2 occupy the two states a and b, respectively. The wave function in Equation (4.14) would be unacceptable because the two electrons are distinguishable upon interchange. Taking linear combinations of the product of the two states, however, provides two acceptable wave functions because now the electrons become indistinguishable upon interchange, as shown in Equations (4.15) and (4.16). The former of these equations is symmetric with respect to electron exchange (it yields the same mathematical expression), while the latter is antisymmetric.

$$\psi = \psi_a(1)\psi_b(2) \neq \psi_a(2)\psi_b(1) \tag{4.14}$$

$$\psi_+ = \frac{1}{\sqrt{2}}[\psi_a(1)\psi_b(2) + \psi_a(2)\psi_b(1)] \tag{4.15}$$

$$\psi_- = \frac{1}{\sqrt{2}}[\psi_a(1)\psi_b(2) - \psi_a(2)\psi_b(1)] \tag{4.16}$$

For the same pair of electrons, the spins of the two electrons could be both positive $[\alpha(1)\alpha(2)]$, both negative $[\beta(1)\beta(2)]$, or some linear combination of the two: $1/\sqrt{2}[\alpha(1)\beta(2) + \alpha(2)\beta(1)]$ or $1/\sqrt{2}[\alpha(1)\beta(2) - \alpha(2)\beta(1)]$. The former three spin functions are symmetric with respect to interchange, while the latter is antisymmetric. Because it is the total wave function (spatial plus spin) that must be antisymmetric according to the Pauli principle, of the eight possible combinations between the two spatial wave functions given in Equations (4.15) and (4.16) with the four spin wave functions, only the following are antisymmetric overall:

$$\psi_+ = \frac{1}{2}[\psi_a(1)\psi_b(2) + \psi_a(2)\psi_b(1)][\alpha(1)\beta(2) - \alpha(2)\beta(1)] \tag{4.17}$$

$$\psi_- = \frac{1}{\sqrt{2}}[\psi_a(1)\psi_b(2) - \psi_a(2)\psi_b(1)][\alpha(1)\alpha(2)] \tag{4.18}$$

$$\psi_- = \frac{1}{\sqrt{2}}[\psi_a(1)\psi_b(2) - \psi_a(2)\psi_b(1)][\beta(1)\beta(2)] \tag{4.19}$$

$$\psi_- = \frac{1}{\sqrt{2}}[\psi_a(1)\psi_b(2) - \psi_a(2)\psi_b(1)][\alpha(1)\beta(2) + \alpha(2)\beta(1)] \quad (4.20)$$

Any two electrons occupying the same orbital would have a symmetric spatial wave function. Therefore, their spin wave functions must be antisymmetric, as is the case for Equation (4.17). In other words, the Pauli exclusion principle states that no two electrons in the same atom may have all four quantum numbers the same. Each electron in an atom must possess a unique set of quantum numbers. As a result, every hydrogenic orbital in a polyelectronic atom can hold at most two electrons, and then if and only if their electron spins are opposite. Hence, the sets of s, *p*, *d*, and *f* orbitals for a given value of *n* can hold a maximum of 2, 6, 10, and 14 electrons, as suggested by the blocks of elements shown in Figure 4.12.

4.4 ELECTRON CONFIGURATIONS AND THE PERIODIC TABLE

The *Aufbau* (or "building up") *principle* uses one-electron (hydrogenic) atomic orbital energies to predict the electron configurations of polyelectronic atoms. The electrons are placed into the orbitals one at a time to form the lowest energy configuration that s consistent with the Pauli exclusion principle and Hund's rule of maximum multiplicity. *Hund's rule* requires that the electrons be placed in a degenerate set of orbitals in such a way as to maximize the spin multiplicity. The *spin multiplicity* is defined as $2S + 1$, where S is the sum of the m_s values for all of the electrons. Because $S = 0$ for any combination of paired electrons, the spin multiplicity is solely determined by the number of unpaired electrons. Using the Aufbau principle, along with Madelung's rule, the electron configuration of *N* is $1s^2 2s^2 2p^3$. Hund's rule implies that the three electrons in the 2*p* orbitals all have identical spins and are unpaired. Any atom that has unpaired electrons will be *paramagnetic* and will be strongly attracted by a magnetic field. Atoms that have all their electrons paired are *diamagnetic* and are weakly repelled by a magnetic field. The electron configurations for the 118 elements are shown in Table 4.2. In order to avoid having to write out lengthy descriptions for the entire electron configuration of the heavier elements, a short-hand method is used whereby a set of square brackets containing the symbol for the most recent noble gas is used to abbreviate the electron configuration.

The astute reader will notice that there are quite a few exceptions to the electron configurations predicted by the Aufbau principle. This is because the hydrogenic orbitals only have meaning for one-electron systems. Whenever there is more than one electron, electron–electron repulsions must also be considered. Consider the case of the two electrons discussed, electrons 1 and 2, occupying the states a and b. There were four allowed wave functions for this particular case, given by Equations (4.17)–(4.20). The energies of the unperturbed wave functions are shown in Figure 4.14 at left. Because both of the electrons are negatively charged, there exists a coulombic repulsion between them whenever the electrons are in different regions of space. As a result of this repulsion, which is given by the Coulomb integral *j*, the total energy of the system will be raised as shown in Figure 4.14, center.

When the two electrons occupy the same region of space (the overlap region), the energies of the four different wave functions are no longer degenerate. When the electrons are all paired, $S = 0$ and the spin multiplicity is 1, forming a singlet state, *S*, which is singly degenerate. The wave function that describes this state is the one given by Equation (4.17), where the spatial portion of the total wave function is symmetric with respect to interchange. The spatially symmetric state tends to bring the electrons together into the same region of space. Therefore, the energy of the singlet state will be raised even further in Figure 4.14 as a result of the larger

TABLE 4.2 Electron configurations for the elements in terms of the one-electron, hydrogenic orbitals.

Symbol	Z	Configuration	Symbol	Z	Configuration
H	1	$1s^1$	Nd	60	$[\mathrm{Xe}]6s^24f^4$
He	2	$1s^2$	Pm	61	$[\mathrm{Xe}]6s^24f^5$
Li	3	$[\mathrm{He}]2s^1$	Sm	62	$[\mathrm{Xe}]6s^24f^6$
Be	4	$[\mathrm{He}]2s^2$	Eu	63	$[\mathrm{Xe}]6s^24f^7$
B	5	$[\mathrm{He}]2s^22p^1$	Gd*	64	$[\mathrm{Xe}]6s^24f^75d^1$
C	6	$[\mathrm{He}]2s^22p^2$	Tb	65	$[\mathrm{Xe}]6s^24f^9$
N	7	$[\mathrm{He}]2s^22p^3$	Dy	66	$[\mathrm{Xe}]6s^24f^{10}$
O	8	$[\mathrm{He}]2s^22p^4$	Ho	67	$[\mathrm{Xe}]6s^24f^{11}$
F	9	$[\mathrm{He}]2s^22p^5$	Er	68	$[\mathrm{Xe}]6s^24f^{12}$
Ne	10	$[\mathrm{He}]2s^22p^6$	Tm	69	$[\mathrm{Xe}]6s^24f^{13}$
Na	11	$[\mathrm{Ne}]3s^1$	Yb	70	$[\mathrm{Xe}]6s^24f^{14}$
Mg	12	$[\mathrm{Ne}]3s^2$	Lu	71	$[\mathrm{Xe}]6s^24f^{14}5d^1$
Al	13	$[\mathrm{Ne}]3s^23p^1$	Hf	72	$[\mathrm{Xe}]6s^24f^{14}5d^2$
Si	14	$[\mathrm{Ne}]3s^23p^2$	Ta	73	$[\mathrm{Xe}]6s^24f^{14}5d^3$
P	15	$[\mathrm{Ne}]3s^23p^3$	W	74	$[\mathrm{Xe}]6s^24f^{14}5d^4$
S	16	$[\mathrm{Ne}]3s^23p^4$	Re	75	$[\mathrm{Xe}]6s^24f^{14}5d^5$
Cl	17	$[\mathrm{Ne}]3s^23p^5$	Os	76	$[\mathrm{Xe}]6s^24f^{14}5d^6$
Ar	18	$[\mathrm{Ne}]3s^23p^6$	Ir	77	$[\mathrm{Xe}]6s^24f^{14}5d^7$
K	19	$[\mathrm{Ar}]4s^1$	Pt*	78	$[\mathrm{Xe}]6s^14f^{14}5d^9$
Ca	20	$[\mathrm{Ar}]4s^2$	Au*	79	$[\mathrm{Xe}]6s^14f^{14}5d^{10}$
Sc	21	$[\mathrm{Ar}]4s^23d^1$	Hg	80	$[\mathrm{Xe}]6s^24f^{14}5d^{10}$
Ti	22	$[\mathrm{Ar}]4s^23d^2$	Tl	81	$[\mathrm{Xe}]6s^24f^{14}5d^{10}6p^1$
V	23	$[\mathrm{Ar}]4s^23d^3$	Pb	82	$[\mathrm{Xe}]6s^24f^{14}5d^{10}6p^2$
Cr*	24	$[\mathrm{Ar}]4s^13d^5$	Bi	83	$[\mathrm{Xe}]6s^24f^{14}5d^{10}6p^3$
Mn	25	$[\mathrm{Ar}]4s^23d^5$	Po	84	$[\mathrm{Xe}]6s^24f^{14}5d^{10}6p^4$
Fe	26	$[\mathrm{Ar}]4s^23d^6$	At	85	$[\mathrm{Xe}]6s^24f^{14}5d^{10}6p^5$
Co	27	$[\mathrm{Ar}]4s^23d^7$	Rn	86	$[\mathrm{Xe}]6s^24f^{14}5d^{10}6p^6$
Ni	28	$[\mathrm{Ar}]4s^23d^8$	Fr	87	$[\mathrm{Rn}]7s^1$
Cu*	29	$[\mathrm{Ar}]4s^13d^{10}$	Ra	88	$[\mathrm{Rn}]7s^2$
Zn	30	$[\mathrm{Ar}]4s^23d^{10}$	Ac*	89	$[\mathrm{Rn}]7s^26d^1$
Ga	31	$[\mathrm{Ar}]4s^23d^{10}4p^1$	Th*	90	$[\mathrm{Rn}]7s^26d^2$
Ge	32	$[\mathrm{Ar}]4s^23d^{10}4p^2$	Pa*	91	$[\mathrm{Rn}]7s^25f^26d^1$
As	33	$[\mathrm{Ar}]4s^23d^{10}4p^3$	U*	92	$[\mathrm{Rn}]7s^25f^36d^1$
Se	34	$[\mathrm{Ar}]4s^23d^{10}4p^4$	Np*	93	$[\mathrm{Rn}]7s^25f^46d^1$
Br	35	$[\mathrm{Ar}]4s^23d^{10}4p^5$	Pu	94	$[\mathrm{Rn}]7s^25f^6$
Kr	36	$[\mathrm{Ar}]4s^23d^{10}4p^6$	Am	95	$[\mathrm{Rn}]7s^25f^7$
Rb	37	$[\mathrm{Kr}]5s^1$	Cm*	96	$[\mathrm{Rn}]7s^25f^76d^1$
Sr	38	$[\mathrm{Kr}]5s^2$	Bk	97	$[\mathrm{Rn}]7s^25f^9$
Y	39	$[\mathrm{Kr}]5s^24d^1$	Cf*	98	$[\mathrm{Rn}]7s^25f^96d^1$
Zr	40	$[\mathrm{Kr}]5s^24d^2$	Es	99	$[\mathrm{Rn}]7s^25f^{11}$
Nb*	41	$[\mathrm{Kr}]5s^14d^4$	Fm	100	$[\mathrm{Rn}]7s^25f^{12}$
Mo*	42	$[\mathrm{Kr}]5s^14d^5$	Md	101	$[\mathrm{Rn}]7s^25f^{13}$
Tc	43	$[\mathrm{Kr}]5s^24d^5$	No	102	$([\mathrm{Rn}]7s^25f^{14})$
Ru*	44	$[\mathrm{Kr}]5s^14d^7$	Lr*	103	$[\mathrm{Rn}]7s^25f^{14}7p^1$
Rh*	45	$[\mathrm{Kr}]5s^14d^8$	Rf	104	$[\mathrm{Rn}]7s^25f^{14}6d^2$
Pd*	46	$[\mathrm{Kr}]4d^{10}$	Db	105	$([\mathrm{Rn}]7s^25f^{14}6d^3)$
Ag*	47	$[\mathrm{Kr}]5s^14d^{10}$	Sg	106	$([\mathrm{Rn}]7s^25f^{14}6d^4)$
Cd	48	$[\mathrm{Kr}]5s^24d^{10}$	Bh	107	$([\mathrm{Rn}]7s^25f^{14}6d^5)$
In	49	$[\mathrm{Kr}]5s^24d^{10}5p^1$	Hs	108	$([\mathrm{Rn}]7s^25f^{14}6d^6)$
Sn	50	$[\mathrm{Kr}]5s^24d^{10}5p^2$	Mt	109	$([\mathrm{Rn}]7s^25f^{14}6d^7)$
Sb	51	$[\mathrm{Kr}]5s^24d^{10}5p^3$	Ds*	110	$([\mathrm{Rn}]7s^15f^{14}6d^9)$
Te	52	$[\mathrm{Kr}]5s^24d^{10}5p^4$	Rg*	111	$([\mathrm{Rn}]7s^15f^{14}6d^{10})$
I	53	$[\mathrm{Kr}]5s^24d^{10}5p^5$	Cn	112	$([\mathrm{Rn}]7s^25f^{14}6d^{10})$
Xe	54	$[\mathrm{Kr}]5s^24d^{10}5p^6$	Uut	113	$([\mathrm{Rn}]7s^25f^{14}6d^{10}7p^1)$

(continued)

TABLE 4.2 (*Continued*)

Symbol	Z	Configuration	Symbol	Z	Configuration
Cs	55	$[Xe]6s^1$	Fl	114	$([Rn]7s^2 5f^{14} 6d^{10} 7p^2)$
Ba	56	$[Xe]6s^2$	Uup	115	$([Rn]7s^2 5f^{14} 6d^{10} 7p^3)$
La*	57	$[Xe]6s^2 5d^1$	Lv	116	$([Rn]7s^2 5f^{14} 6d^{10} 7p^4)$
Ce*	58	$[Xe]6s^2 4f^1 5d^1$	Uus	117	$([Rn]7s^2 5f^{14} 6d^{10} 7p^5)$
Pr	59	$[Xe]6s^2 4f^3$	Uuo	118	$([Rn]7s^2 5f^{14} 6d^{10} 7p^6)$

Exceptions to the Aufbau principle are noted with an asterisk. Values in parentheses are calculated, not experimental.

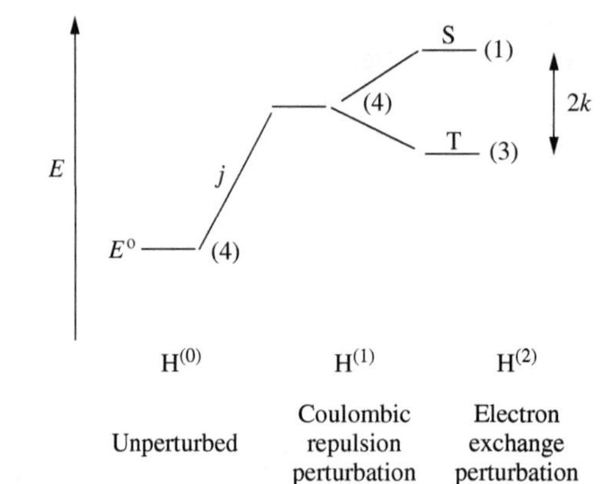

FIGURE 4.14
Energy diagram for a two-electron, two-state system, showing the changes in energy and the degeneracies (in parentheses) that result from the electron–electron repulsions.

electron–electron repulsion. The remaining three wave functions, which possess an antisymmetric spatial component, form a triplet state, T, which is triply degenerate. In the antisymmetric spatial wave function, the electrons tend to avoid each other. This natural tendency of the electrons to stay away from each other decreases the amount of electron–electron repulsion. Therefore, the triplet state will be lower in energy than the singlet state, as shown in Figure 4.14 at the right. This observation is the basis for Hund's rule of maximum multiplicity, which would also predict that the triplet state should be more stable than the singlet state. The energy difference between the two states is given by twice the exchange integral k. The magnitude of the exchange integral, in turn, is proportional to $N(N-1)$, where N is the number of unpaired electrons. For those electron configurations that are exceptions to the Aufbau principle, the exception is usually a result of the increased exchange energy that results when there are a large number of unpaired electrons. For example, the electron configuration of Cr is $[Ar]4s^1 3d^5$ instead of the predicted $[Ar]4s^2 3d^4$. The added stability of the former electron configuration over the latter is due in large part to the larger number of unpaired electrons that it possesses.

4.5 ATOMIC TERM SYMBOLS

The electron configurations of polyelectronic atoms given by the one-electron hydrogenic orbitals are an incomplete description of the ways that the electrons can occupy these orbitals. When the electron configuration of Fe, for instance, is said to be $[Ar]4s^2 3d^6$ using Madelung's rule, this ignores the contribution that the electron–electron repulsions make to the one-electron hydrogenic orbital energies. For the d^6 configuration in Fe, there are actually 210 different ways that the six electrons can occupy the five d orbitals. Some of these 210 microstates will have the same values of j and k and therefore have the same energy, while other combinations will have different energies. Each energy level describes a *state*, or a *term*. It is more realistic for us to think about the electron configurations of polyelectronic atoms in terms of the energies of their terms instead of the one-electron hydrogenic orbital energies. The energies of these terms depend on three factors: (i) the average energy of the one-electron hydrogenic orbitals from which they are derived, $E°$; (ii) the magnitude of the Coulomb integral j; and (iii) the magnitude of the exchange integral k.

The following method can be used to extract the *term symbols* (which are used to describe electronic transitions) from all of the different microstates that the electrons in an atom can assume. Consider the carbon atom, for example. Carbon has the electron configuration $1s^2 2s^2 2p^2$. The occupation of the filled $1s$ and $2s$ orbitals is unambiguous. According to the Pauli exclusion principle, there is only one way that the electrons can occupy each of these singly degenerate orbitals: one with $m_s = +1/2$ and the other with $m_s = -1/2$. However, there are 15 different ways to place two electrons into the triply degenerate $2p$ subshell. These 15 possibilities are called *microstates* and are shown in Figure 4.15, where the 15 sets are p orbitals arranged vertically. The m_l quantum number for each p orbital is listed at the left in the diagram.

In general, the number of Pauli-allowed microstates (M.S.) can be calculated using the formula given by Equation (4.21), where n_o is the number of degenerate orbitals and n_e is the number of electrons to be placed in those orbitals.

$$\#\text{M.S.} = \frac{(2n_o)!}{n_e!(2n_o - n_e)!} \tag{4.21}$$

Applying this formula to the carbon atom ($n_o = 3$, $n_e = 2$), 15 Pauli-allowed microstates are predicted. For the $3d^2$ configuration in V^{3+} (g), there are 45 possible microstates ($n_o = 5$, $n_e = 2$). In the case of electron configurations such as $[Ar]4s^1 3d^5$, the probabilities are multiplicative. The number of allowed microstates for the s^1 configuration is 2, while that for the d^5 configuration is 252. Thus, there are 504 possible microstates for a gaseous Cr atom!

As a result of electron–electron repulsions, not all of the microstates for a given electron configuration will necessarily have the same energy. In the case of carbon, for instance, it is logical to conclude that the microstates where both electrons are paired in the same orbital will have a greater coulombic repulsion than those that are unpaired. Thus, our goal is to take the Pauli-allowed microstates and to extract from them all those combinations that have the same energy. These are collectively referred to as *states*. Each *state*, or collection of microstates, has its own energy and can be assigned a symbol, known as a *term symbol* that characterizes some of the properties of the state (or *term*). In general, there are two formalisms for extracting term symbols from the microstates: Russell–Saunders (RS) coupling and *jj* coupling. Each method is discussed individually.

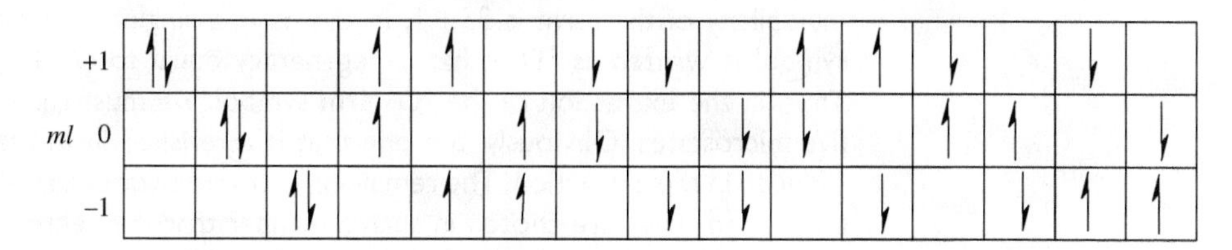

FIGURE 4.15
The 15 ways (microstates) that two electrons can occupy the three $2p$ orbitals.

4.5.1 Extracting Term Symbols Using Russell–Saunders Coupling

Every electron has both orbital (L) and spin (S) angular momentum. The RS or LS coupling scheme, which is generally valid for the lighter elements ($Z < 30$), provides a mechanism whereby the orbital angular momenta l of the individual electrons couple together to produce a total orbital angular momentum L and the individual spin angular momenta s couple together to yield a total spin angular momentum S, as shown in Equation (4.22):

$$L = \sum_i l_i \text{ and } S = \sum_i s_i \tag{4.22}$$

Just as the orbital angular momentum l can take on $2l + 1$ components having $m_l = -l, -l + 1, \cdots, l - 1, l$, the total orbital angular momentum L will also have $2L + 1$ components having $M_L = -L, -L + 1, \ldots, L - 1, L$. Similarly, the states described by $L = 0, 1, 2, 3$, and so on, will have symbols reflecting the shape of their orbital angular momenta, namely S, P, D, and F. It is common practice to use lower case symbols to represent "one-electron" orbitals and upper case symbols for polyelectronic states. For a given value of S, there will be $2S + 1$ spin states having values of M_S ranging from $-S$ to $+S$ in integral steps. The value of $2S + 1$ is known as the *spin multiplicity*. States having $2S + 1 = 1$ are called *singlets*, while those having a spin multiplicity of three are known as *triplets*.

The procedure for extracting term symbols from the microstates using the RS coupling scheme is as follows:

- Determine the number of allowed microstates and sketch a microstate table similar to the one shown earlier for carbon.
- Add a row across the top of the microstate table that calculates the M_L as the sum of the individual m_l's. Add a similar row across the bottom that calculates M_S as the sum of the individual m_s's. The microstate table for carbon should now look similar to the one in Figure 4.16:
- Count the number of microstates that have the same values of M_L and M_S and organize them into a new table similar to the one shown in Figure 4.17. Note that this form of the microstate table will always be symmetric about its center, a fact that can be used to simplify the amount of work involved in the table's construction.
- Remove the first term symbol from the chart by choosing the maximum value of M_L. If there is more than one entry having the same value, then maximize the value of M_S as a secondary consideration. In this case, the entry with the largest value of M_L is indicated by an asterisk in Figure 4.18. This entry has $M_L = 2$ and $M_S = 0$. These are the maximum values for the total orbital and total spin angular momentum components, and they therefore correspond with $L = 2$ and $S = 0$. The extracted term is a D-term, because $L = 2$ has the same orbital angular momentum as a hydrogenic *d*-orbital ($l = 2$). The spin multiplicity of the term is $2S + 1$, in this case, a singlet. The singlet-D term symbol is written as ^1D. It has a degeneracy equal to $(2L + 1)(2S + 1) = 5$. Thus, in the extraction of the ^1D term symbol, we must remove a total of five microstates. Obviously, the one that is asterisked in the table must be included in the extraction. The remaining four microstates to be removed are underlined. They are chosen in such a manner that the center of symmetry in the table following extraction will be preserved.
- After removal of the five degenerate microstates that comprise the singlet-D term, the microstate table is reduced as follows (Figure 4.18(b)).

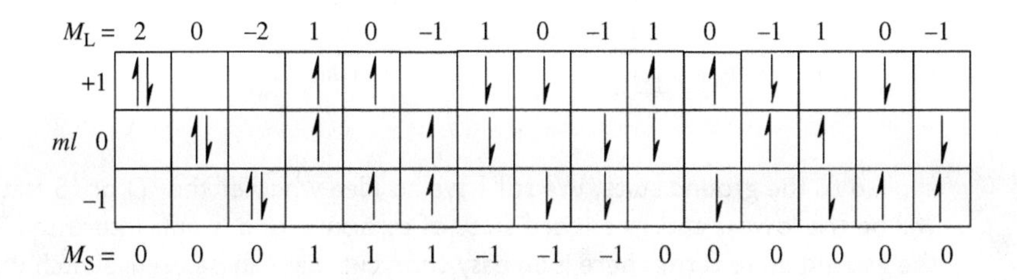

FIGURE 4.16 Microstate table for carbon, with the M_L and M_S values tabulated across the top and bottom, respectively.

M_L \ M_S	1	0	−1
2	0	1	0
1	1	2	1
0	1	3	1
−1	1	2	1
−2	0	1	0

FIGURE 4.17 Table for a p^2 electron configuration showing the numbers of microstates having the corresponding values of M_L and M_S.

- The extraction process is repeated until all of the microstates have been assigned to term symbols. The next extraction removes a term with $L = 1$ and $S = 1$, or a triplet-P term. The degeneracy of the ^{3}P state is $(2L + 1)(2S + 1) = 9$. Thus, nine microstates must be removed from the table in a symmetrical manner. The table now reduces to Figure 4.18(c). The final term has $L = 0$ and $S = 0$, or ^{1}S, and is singly degenerate.

Using the RS coupling scheme, it was determined that there are three different energy levels for the $1s^2 2s^2 2p^2$ electron configuration of carbon, having the degeneracies listed in parentheses: ^{1}D (5), ^{3}P (9), and ^{1}S (1). As required by the formula given in Equation (4.21), there are a total of 15 Pauli-allowed microstates. However, these 15 configurations exist in states that have three separate energy levels.

- We can use Hund's rules to determine the ground-state term. First, we choose the term with the maximum spin multiplicity and then (if necessary) the term with the largest total orbital angular momentum L. In the case of the C atom, the ground-state term is the ^{3}P state. Hund's rules can only be

(a) ^{1}D

M_L \ M_S	1	0	−1
2	0	1*	0
1	1	2	1
0	1	3	1
−1	1	2	1
−2	0	1	0

(b) ^{3}P

M_L \ M_S	1	0	−1
2	0	0	0
1	1	1	1
0	1	2	1
−1	1	1	1
−2	0	1	0

(c) ^{1}S

M_L \ M_S	1	0	−1
2	0	0	0
1	0	0	0
0	0	1*	0
−1	0	0	0
−2	0	0	0

FIGURE 4.18 (a–c) Extracting the term symbols from the p^2 microstate table.

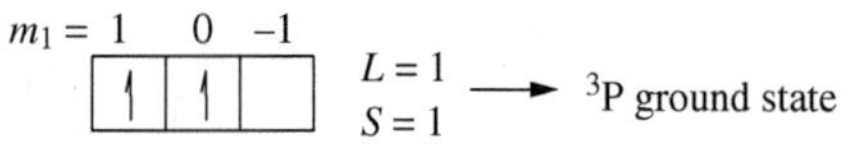

$m_l = 1 \quad 0 \quad -1$

$L = 1$, $S = 1$ → ^{3}P ground state

FIGURE 4.19
The shortcut method for determining the ground-state term symbol for a p^2 electron configuration.

applied to the ground state. We still have no idea whether the ^{1}D or ^{1}S state will be the lowest energy excited state of carbon. If one is only interested in the ground-state term, there is an easy shortcut that can be used. Sketch the electron configuration by filling in the orbital "boxes" such that the electrons fill in unpaired first beginning with the orbital having the largest value of m_l. Then, it is a simple matter to calculate L and S from the individual angular momenta in order to determine the ground-state term. The example for C is shown in Figure 4.19.

Example 4-6. Determine the ground-state term symbol for the d^7 electron configuration.

Solution. Using the given figure, the sum of the m_l values is 3 such that $L = 3$ (*F*) and $S = 3/2$ so the spin multiplicity is 4. Thus, the ground-state term symbol is ^{4}F and has a degeneracy of $(2L + 1)(2S + 1)$ or 28.

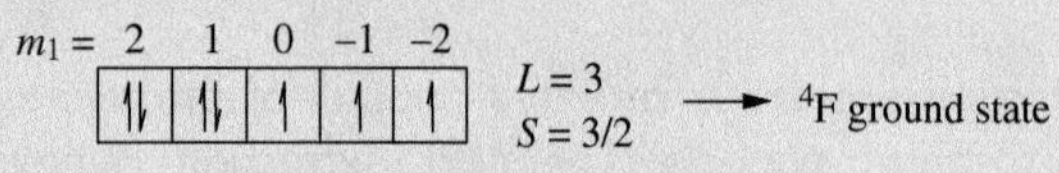

$m_l = 2 \quad 1 \quad 0 \quad -1 \quad -2$

$L = 3$, $S = 3/2$ → ^{4}F ground state

The term symbols for a variety of possible electron configurations are shown in Table 4.3. For each entry, the ground-state term symbol is listed first. Note that the completely filled subshells s^2, p^6, and d^{10} are all spherically symmetric, as required by Unsöld's theorem and have the ^{1}S term. Half-filled subshells (such as p^3 or d^5 are also spherically symmetric. The term symbols for the pairs: p^1 and p^5, p^2 and p^4, d^2 and d^8, and so on, are identical to one another because the p^5 electron configuration, for instance, can be viewed equivalently as a p^1 (hole) configuration.

4.5.2 Extracting Term Symbols Using *jj* Coupling

For the heavier elements ($Z \geq 30$), the total orbital L and spin S angular momentum quantum numbers are no longer valid. Instead, the orbital l and spin s angular momentum of each individual electron couple together first, as shown by Equation (4.23), to produce a new quantum number j. The directional components of j are given by the quantum number m and range from $j \to -j$ in integral increments, as shown in Equation (4.24).

$$j_i = l_i \pm s_i \tag{4.23}$$

$$m_i = -j_i, -j_i + 1, \ldots, j_i - 1, j_i \tag{4.24}$$

The individual j's then couple together to yield the total angular momentum J, according to Equation (4.25). The components of J can range from $-J$ to $+J$, as shown by Equation (4.26).

$$J = \sum_i j_i \tag{4.25}$$

$$M = \sum_i m_i = -J, -J + 1, \ldots, J - 1, J \tag{4.26}$$

TABLE 4.3 Term symbols for common electron configurations.

Equivalent electrons	
s^2, p^6, d^{10}	^1S
p^1, p^5	^2P
p^2, p^4	^3P, ^1D, ^1S
p^3	^4S, ^2D, ^2P
d^1, d^9	^2D
d^2, d^8	^3F, ^3P, ^1G, ^1D, ^1S
d^3, d^7	^4F, ^4P, ^2H, ^2G, ^2F, ^2D, ^2D, ^2P
d^4, d^6	^5D, ^3H, ^3G, ^3F, ^3F, ^3D, ^3P, ^3P, ^1I, ^1G, ^1G, ^1F, ^1D, ^1D, ^1S, ^1S
d^5	^6S, ^4G, ^4F, ^4D, ^4P, ^2I, ^2H, ^2G, ^2G, ^2F, ^2F, ^2D, ^2D, ^2D, ^2P, ^2S
Nonequivalent electrons	
s s	^1S, ^3S
s p	^1P, ^3P
s d	^1D, ^3D
p p	^3D, ^1D, ^3P, ^1P, ^3S, ^1S
p d	^3F, ^1F, ^3D, ^1D, ^3P, ^1P
d d	^3G, ^1G, ^3F, ^1F, ^3D, ^1D, ^3P, ^1P, ^3S, ^1S

Consider the p^2 electron configuration of lead (earlier we had used C for the p^2 configuration, but carbon's atomic number is less than 30, so here we will use Pb instead). Because $l = 1$ for a p-electron and $s = \pm 1/2$, $j = 3/2$ or $1/2$. When the values of jj' are 3/2, 3/2, the combinations of mm′ quantum numbers possible are shown in Table 4.4. For $j = 3/2$, $m = 3/2$, $1/2$, $-1/2$, or $-3/2$. Certain combinations, such as (3/2, 3/2) are Pauli-excluded because $j = j'$ and $m = m'$. Likewise, only one combination of pairs such as (3/2, 1/2), (1/2, 3/2) is allowed because of configurational exclusion. For this particular set of jj', there are six possible microstates, as shown in Table 4.4.

In order to extract the terms from these microstates, one must first maximize M and then select $2J + 1$ microstates to remove. Thus, the first term to be extracted will have $J = 2$ and is fivefold degenerate ($M = 2$, 1, 0, -1, and -2). This leaves one remaining microstate in the table, having $J = 0$ and where $M = 0$ (singly-degenerate). Consequently, there are two terms from this table: $J = 2$ (5) and $J = 0$ (1), where the degeneracy of each term is shown in parentheses. For the combination $jj' = 3/2$, 1/2, eight combinations of mm′ are allowed, as shown in Table 4.5. In this case, m can equal m', because $j \neq j'$. Extraction of the terms from Table 4.5 yields: $J = 2$ (5), $J = 1$ (3). For $jj' = 1/2$, 1/2, there is only one possible microstate ($m = 1/2$, $m' = -1/2$). Because $j = j'$, the Pauli exclusion principle states that $m \neq m'$. Hence, the combinations mm′ = 1/2, 1/2 or $-1/2$, $-1/2$ are not allowed. The combination $-1/2$, 1/2 is

TABLE 4.4 Microstates having $jj' = 3/2, 3/2$.

m_1	m_2	M
3/2	1/2	2
3/2	−1/2	1
3/2	−3/2	0
1/2	−1/2	0
1/2	−3/2	−1
−1/2	−3/2	−2

TABLE 4.5 Microstates having $jj' = 3/2, 1/2$.

m_1	m_2	M
3/2	1/2	2
3/2	−1/2	1
1/2	1/2	1
1/2	−1/2	0
−1/2	1/2	0
−1/2	−1/2	−1
−3/2	1/2	−1
−3/2	−1/2	−2

configurationally excluded. Extraction of the final term yields $J = 0$ (1). In summary, applying the *jj* coupling scheme to the p^2 configuration of Pb yields a total of 15 possible microstates (6 from $jj' = 3/2, 3/2$; 8 from $jj' = 3/2, 1/2$; and 1 from $jj' = 1/2, 1/2$) and five terms ($J = 2, J = 0, J = 2, J = 1$, and $J = 0$).

4.5.3 Correlation Between RS (LS) Coupling and *jj* Coupling

For elements having atomic numbers that are intermediate between the extremes of RS coupling and *jj* coupling, there is a one-to-one correspondence between the terms. For every RS term symbol, one can introduce a spin-orbit perturbation that further splits these terms into different energies by introducing the quantum number J. For intermediate cases, $J = |L + S| \rightarrow |L - S|$. Consider the p^2 electron configuration as an example. In this case, the RS term symbols are ^{3}P (9), ^{1}D (5), and ^{1}S (1). For the ^{3}P term, where $L = 1$ and $S = 1$, J can take values of 2, 1, or 0. Thus, the ^{3}P term will split into three different energy levels, each having a degeneracy of $2J + 1$, as follows: $^3P_2$ (5), $^3P_1$ (3), and $^3P_0$ (1). The revised RS term symbol lists the quantum number J as a subscript following the term. As before, the number in parentheses represents the degeneracy of the modified term. Note that the total number of microstates both before and after spin-orbit coupling is the same (in this case, 9). Likewise, the ^{1}D (5) term will become $^1D_2$ (5) as a result of the spin-orbit perturbation and the ^{1}S (1) term will become $^1S_0$ (1). Figure 4.20 shows the one-to-one correspondence between terms in the RS coupling scheme and those in the *jj* coupling scheme for the p^2 electron configuration.

In progressing from the RS term symbol to the spin-orbit modified terms, a *barycenter* of energy must be maintained. In other words, the total energy of the microstates before and after the perturbation must be equivalent. The energies of

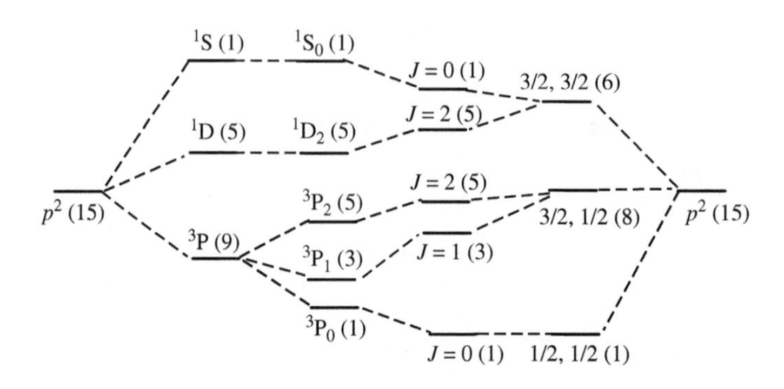

FIGURE 4.20
There is a one-to-one correspondence between terms in the RS coupling scheme and terms in the *jj* coupling scheme.

the spin-orbit perturbed states (relative to each unperturbed RS term) are given by Equation (4.27), where λ is the spin-orbit coupling constant:

$$E_{S.O.} = (\lambda/2)[J(J+1) - L(L+1) - S(S+1)] \quad (4.27)$$

For the ^3P term, the energy of the perturbed $^3P_2$, $^3P_1$, and $^3P_0$ states are $+\lambda$, $-\lambda$, and -2λ, respectively, according to the given equation. Although the magnitude of λ is a function of the element, it always increases with increasing atomic number. Hence, the degree of spin-orbit coupling will increase with Z. Because the $^3P_2$ state is fivefold degenerate, it is destabilized with respect to the unperturbed term by $5(+1\lambda)$, or $+5\lambda$. The triply degenerate $^3P_1$ state and the singly degenerate $^3P_0$ state are stabilized by $3(-1\lambda)$ and $1(-2\lambda)$, respectively. Thus, the total energy of the $^3P_2$, $^3P_1$, and $^3P_0$ terms are $+5(+1\lambda)+3(-1\lambda)+1(-2\lambda)=0$, relative to the unperturbed ^3P state. Note that the total energy is conserved following the spin-orbit perturbation. The very important concept of term symbols will be revisited in a later section of this text when the spectroscopy of coordination compounds is discussed.

4.6 SHIELDING AND EFFECTIVE NUCLEAR CHARGE

One of the largest differences between hydrogen and polyelectronic atoms is that the nuclear charge that an outer electron feels in a polyelectronic atom will be reduced by the electron–electron repulsions of all the inner electrons. Thus, the nuclear charge Z should be replaced by the effective nuclear charge Z^*. As mentioned previously, for a given value of n, the ability of an electron to penetrate the nucleus decreases in the order $s > p > d > f$. Therefore, the ns and np electrons are particularly good at shielding (or screening) the nd or nf electrons from feeling the full effect of the nuclear charge. The effective nuclear charge is equal to the difference between the nuclear charge Z and a shielding parameter σ, as shown in Equation (4.28). The shielding parameter depends on the electron configuration of the atom and the electron of interest.

$$Z^* = Z - \sigma \quad (4.28)$$

In 1930, Slater developed a set of empirical rules for calculating the magnitude of the shielding parameter:

- The electron configuration is written in groups with increasing values of n according to the paradigm given here:

$$(1s)(2s, 2p)(3s, 3p)(3d)(4s, 4p)(4d)(4f)(5s, 5p)(5d)(5f) \ldots$$

- The electron of interest and any electrons that lie to the right of that electron in the paradigm will contribute zero to the shielding parameter.
- For s and p electrons, any other electron in the same group contributes 0.35 each (exception: if the electron of interest lies in the $1s$ orbital, any other $1s$ electron will contribute only 0.30). Electrons in the $n-1$ group contribute 0.85 each and all other groups lying to the left contribute 1.00 each to the shielding parameter.
- For d or f electrons, any electron in the same group contributes 0.35 each and all other electrons to the left contribute 1.00 each. This is a result of the fact that the s and p orbitals are better screeners than are d and f orbitals.

Example 4-7. Calculate the effective nuclear charge that the outermost electron in K would feel if: (a) it was placed into the 4*s* orbital, and (b) it was instead placed into a 3*d* orbital.

Solution. The electron configuration of the first 18 electrons in K using the paradigm is $(1s)^2(2s,\ 2p)^8(3s,\ 3p)^8$.

(a) If the electron of interest is placed into a 4*s* orbital, the shielding parameter and effective nuclear charge calculated using Slater's rules are

$$\sigma = 0(0.35) + 8(0.85) + 10(1.00) = 16.8$$

$$Z^* = Z - \sigma = 19.0 - 16.8 = 2.2$$

(b) If the electron of interest is placed into a 3*d* orbital, the shielding parameter and effective nuclear charge calculated using Slater's rules are

$$\sigma = 0(0.35) + 18(1.00) = 18.0$$

$$Z^* = Z - \sigma = 19.0 - 18.0 = 1.0$$

Because of the larger value of Z^*, Slater's rules predict correctly that the 19th electron of K will fill the 4*s* orbital instead of the 3*d* orbital.

To a first approximation, the ionization energy (I.E.) of an atom is equal to the negative of the energy of the electron (E_{el}) that is being removed. This is known as *Koopman's theorem*. Because electron energies are always negative, the ionization energy will always be positive. The I.E. can be calculated from Equation (4.29), where Z^* is the effective nuclear charge calculated using Slater's rules and *n* is the principal quantum number for the outermost electron.

$$\text{I.E.} = -E_{el} = -2.179 \times 10^{-18}\,\text{J}(6.022 \times 10^{23}\,\text{mol}^{-1})\frac{Z^*}{n^2} \quad (4.29)$$

Example 4-8. Calculate the effective nuclear charge for: (a) a 3*d* electron in Ni and (b) for a 4*s* electron in Ni. Which electron will have the smaller first ionization energy?

Solution. The electron configuration for Ni using the Slater paradigm is $(1s)^2(2s,\ 2p)^8(3s,\ 3p)^8(3d)^8(4s,\ 4p)^2$.

(a) If the electron of interest is a 3*d* electron, the shielding parameter and effective nuclear charge calculated using Slater's rules are

$$\sigma = 7(0.35) + 18(1.00) = 20.45$$

$$Z^* = Z - \sigma = 28.00 - 20.45 = 7.55$$

(b) If the electron of interest is a 4*s* electron, the shielding parameter and effective nuclear charge calculated using Slater's rules are

$$\sigma = 1(0.35) + 16(0.85) + 10(1.00) = 23.95$$

$$Z^* = Z - \sigma = 28.00 - 23.95 = 4.05$$

Slater's rules predict that the $3d$ electron will experience a stronger effective nuclear charge than will the $4s$ electron. Therefore, the $4s$ electron will have the smaller first ionization energy. Thus, the electron configuration of Ni^+ is $[Ar]4s^1 3d^8$, not $[Ar]4s^2 3d^7$. Likewise, the electron configuration of Ni^{2+} is $[Ar]4s^0 3d^8$. It is a general property of the transition metals that the hydrogenic ns orbitals typically fill before the $(n-1)d$ orbitals fill; and the ns orbitals also ionize (or empty) before the $(n-1)d$ orbitals do.

Example 4-9. Calculate the theoretical ionization energy for a $2p$ electron in the N atom using Equation (4.29). Then, calculate the ionization energy if the effects of shielding had not been considered (using just Z instead of Z^*). Compare both answers to the experimental ionization energy of 1.4 MJ/mol.

Solution. The electron configuration of N using the Slater paradigm is $(1s)^2 (2s, 2p)^5$. The electron of interest is a $2p$ electron. The shielding parameter and effective nuclear charge can be calculated using Slater's rules:

$$\sigma = 4(0.35) + 2(0.85) = 3.1$$

$$Z^* = Z - \sigma = 7.0 - 3.1 = 3.9$$

The first ionization energy using Equation (4.29) is

$$\text{I.E.} = -E_{el} = -2.179 \times 10^{-18}\,\text{J}(6.022 \times 10^{23}\,\text{mol}^{-1})\frac{3.9}{2^2}\left(\frac{1\,\text{MJ}}{10^6\,\text{J}}\right) = 1.3\ \text{MJ/mol}$$

Ignoring shielding, one obtains:

$$\text{I.E.} = -E_{el} = -2.179 \times 10^{-18}\,\text{J}(6.022 \times 10^{23}\,\text{mol}^{-1})\frac{7}{2^2}\left(\frac{1\,\text{MJ}}{10^6\,\text{J}}\right) = 2.3\ \text{MJ/mol}$$

Clearly, the answer that includes shielding is closer to the experimental value.

The periodic properties of the elements, including the first ionization energy, are discussed in the next chapter. Almost all of these properties will ultimately depend on three simple factors: (i) the electron configuration of the atom, (ii) the principal quantum number of the outermost electron, and (iii) the effective nuclear charge. Therefore, a quantitative understanding of Slater's rules is an important tool in the prediction of the chemical reactivity of the elements.

EXERCISES

4.1. How far from the hydrogen nucleus will the radial node for an electron in the $3p$ orbital reside?

4.2. How many radial nodes and how many angular nodal planes will a $4f$ orbital possess?

4.3. Write the mathematical form of the angular part of the wave function for the d_{xy} and $d_{x^2-y^2}$ orbitals by taking the positive and negative linear combinations of $Y(2,2)$ with $Y(2,-2)$, respectively.

4.4. Write the complete electron configurations (no shorthand notation) for each of the following: (a) Fe, (b) Fe^{3+}, (c) O^-, (d) Pb, (e) Mn^{2+}, (f) Sc, (g) Zn^{2+}, (h) Tl, (i) Pu, and (j) Re^+.

4.5. Which atoms or ions in Problem 4.4 are paramagnetic and which are diamagnetic?

4.6. Determine the term symbols for the N atom. Show all work. (a) Determine the number of possible microstates. (b) Write out all the possible combinations of the electrons in a microstates table. (c) Extract the term symbols and determine the degeneracy of each term. (d) Determine the ground-state term symbol using Hund's rule of maximum multiplicity.

4.7. Determine the spin–orbit splitting for each of the RS term symbols for the P atom and use Equation (4.27) to calculate their energies relative to the barycenter of each term.

4.8. Determine the ground-state term symbol for each of the following: (a) O, (b) Ni, (c) Fe^{2+}, (d) Co, and (e) Pu.

4.9. Using Slater's rules, calculate Z^* for: (a) $2p$ electron in F, (b) $4s$ electron in Cu, and (c) $3d$ electron in Cu.

4.10. When Cu is oxidized by one electron, will the electron be removed from the $4s$ or the $3d$ orbital first? Explain your answer.

4.11. The stabilization of a half-filled d-subshell is greater than that for a half-filled p-subshell. Explain why.

4.12. Calculate the first ionization energy for P in units of megajoule per mole.

BIBLIOGRAPHY

1. Adamson AW. *A Textbook of Physical Chemistry*. 3rd ed. Orlando, FL: Academic Press; 1986.
2. Bockhoff FJ. *Elements of Quantum Theory*. Reading, MA: Addison-Wesley Publishing Company, Inc.; 1969.
3. David, C. W. "The Laplacian in Spherical Polar Coordinates," *Chemistry Education Materials*, Paper 34, 2007. http://digitalcommons.uconn.edu/chem_educ/34/ (accessed October 19, 2013).
4. Eyring H, Walter J, Kimball GE. *Quantum Chemistry*. New York: John Wiley & Sons, Inc.; 1944.
5. Fayer MD. *Elements of Quantum Mechanics*. New York: Oxford University Press; 2001.
6. Hoffmann B. *The Strange Story of the Quantum*. New York: Dover Publications, Inc.; 1959.
7. Housecroft CE, Sharpe AG. *Inorganic Chemistry*, 3rd ed. Essex, England: Pearson Education Limited; 2008.
8. Huheey JE, Keiter EA, Keiter RL. *Inorganic Chemistry: Principles of Structure and Reactivity*. 4th ed. New York: Harper Collins College Publishers; 1993.
9. Levine IN. *Quantum Chemistry*. 2nd ed. Boston, MA: Allyn and Bacon, Inc.; 1974.
10. McQuarrie DA, Simon JD. *Physical Chemistry: A Molecular Approach*. Sausalito, CA: University Science Books; 1997.
11. Miessler GL, Tarr DA. *Inorganic Chemistry*. 4th ed. Upper Saddle River, NJ: Pearson Education Inc.; 2011.
12. Pauling L, Wilson EB. *Introduction to Quantum Mechanics with Applications to Chemistry*. New York: McGraw-Hill Book Company; 1935.

Periodic Properties of the Elements | 5

"The Periodic Table is Nature's Rosetta Stone."

—*Rudy Baum*

5.1 THE MODERN PERIODIC TABLE

The modern form of the periodic table is so iconic that almost anyone would recognize the image above simply by looking at its general shape. Nowadays, it is almost impossible to believe that there was ever a time when the periodic nature of the elements was not yet understood. The first primitive form of the periodic table was developed in 1829 by Johann Döbereiner, who noticed a repetitive pattern in the chemical properties of some of the elements, lumping them together in groups of three that he called *triads*. The elements chlorine, bromine, and iodine, for instance, formed a triad. A few years later, in 1865, the Englishman John Newlands recognized that if the elements were listed in order of their increasing atomic weights, they could be arranged into what he called the *law of octaves* (or groups of eight) according to their chemical properties. However, it was not until 1869, when the Russian teacher Dmitri Mendeleev gave a presentation to the Russian Chemical Society called "A Dependence between the Properties of the Atomic Weights of the Elements" that the idea of periodic trends among the various elements began to take hold. Mendeleev wrote out the physical properties of all the known elements on cards and arranged them into groups on the basis of trends in a game that his friends called *Patience*. In Mendeleev's day, only a third of the elements known today had been discovered. The genius of Mendeleev's work was that when he arranged the elements in order of increasing atomic mass and the next element did not fit the properties of the next group, he simply left a gap in his deck of cards, assuming that a new element would someday be discovered having the appropriate pattern of trends. It is the predictive capability of Mendeleev's periodic table that set his above those of his contemporaries.

The modern version of the periodic table, of course, is based on quantum mechanics and the relative energies of the hydrogenic orbitals. When Moeller's rubric is applied and the orbitals are arranged according to increasing energy, the *Aufbau* (or "building up") *principle* states that the electrons will fill those orbitals having the lowest energies first. Further, the Pauli exclusion principle states that no more than two electrons can occupy the same orbital and thus each orbital can hold only two electrons. When the elements are arranged in order of increasing atomic number, their physical and chemical properties follow the same periodic pattern as the

Principles of Inorganic Chemistry, First Edition. Brian W. Pfennig.

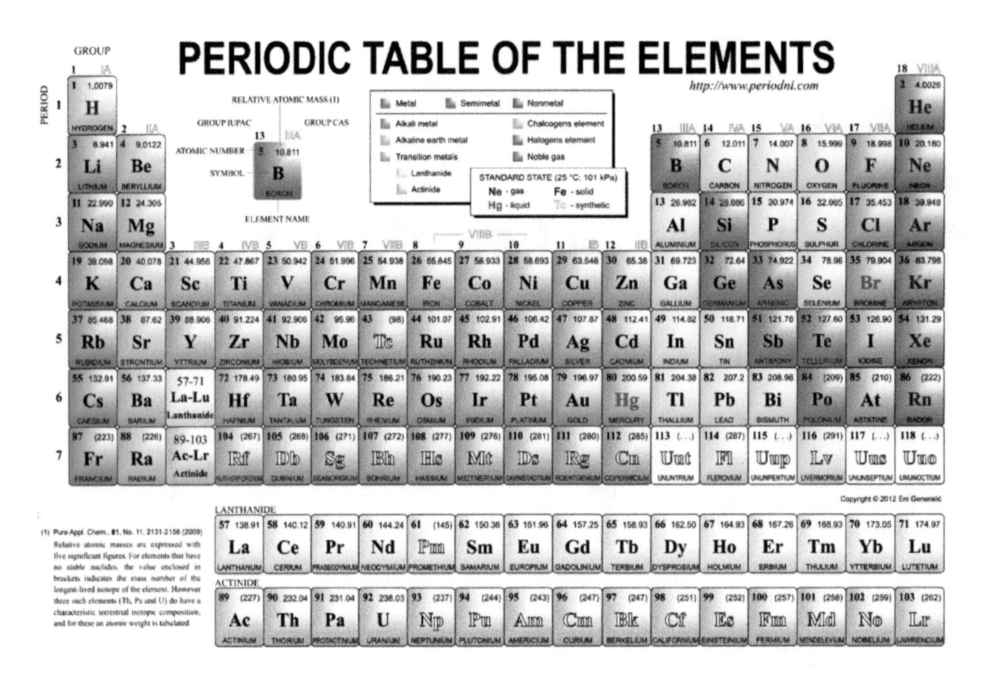

FIGURE 5.1
Modern version of the periodic table. [©E. Generalic, http://www.periodni.com, used with permission.]

one predicted based on this ordering of the hydrogenic orbitals. A modern version of the periodic table of the elements is shown in Figure 5.1.

The modern periodic table has 18 columns or groups. It is divided into the metals, which lie on the left-hand side of the red staircase of elements, and the nonmetals, which lie to the right of this staircase. The metals typically lose their valence (or outermost) electrons easily, are malleable and ductile, and are good conductors of heat and electricity. The nonmetals, on the other hand, gain or share their electrons easily, are brittle, lack luster, and are usually poor conductors of heat or electricity. Those elements lying along the staircase (highlighted in red in Figure 5.1) are known as the *metalloids* and have intermediate character between the metals and the nonmetals and are often semiconductors. A number of the different groups have specific names: Group 1 contains the alkali metals, Group 2 contains the alkaline earths, Group 15 contains the pnictogens, Group 16 contains the chalcogens, Group 17 contains the halogens, and Group 18 contains the noble or the inert gases. Furthermore, the *d*-block elements are known as the *transition metals*, the 4*f*-block is called the *lanthanide series*, and the 5*f*-block is called the *actinides*. A long version of the periodic table, which shows the accurate quantum mechanical placement of the lanthanides and actinides more clearly, is shown in Figure 5.2.

Mendeleev's periodic table was organized according to increasing mass. In some cases, the order of the elements had to be reversed in order to fit the periodic trends more neatly—thus, for instance, Te fell before I even though the atomic mass of Te is heavier. With the discovery of the nucleus in the early 1900s, the modern form of the periodic table is instead organized according to increasing atomic number. We owe this improvement to Henry Moseley, who bombarded the elements with electrons in order to observe their characteristic X-ray fluorescence. The resulting X-ray spectra fit a formula, known as *Moseley's law* that assigned a unique integral number to each element (now known as the *atomic number*). Tragically, Moseley was killed during World War I in the Battle of Gallipoli at the age of 27 and was thus ineligible to win the Nobel Prize he so assuredly deserved.

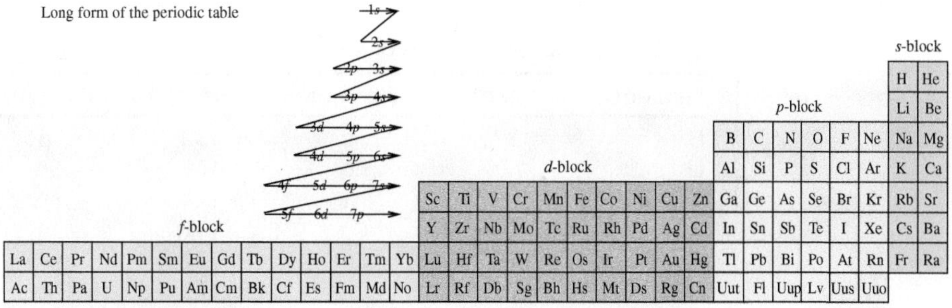

Long form of the periodic table

FIGURE 5.2
The long version of the periodic table.

In the sections that follow, we will explore some of the more common periodic trends in the physical and chemical properties of the elements. Although it may seem like a daunting task to memorize the descriptive properties of each of the elements, this text aims to highlight the underlying principles that serve as the architecture of the periodic table. Most of these generalizations, including the exceptions to the rule, can be readily understood in terms of only three parameters: (i) the principal quantum number *n*, (ii) the effective nuclear charge Z^*, and (iii) the element's electron configuration. A fourth property, relativity, also plays a critical role for some of the heavier elements. Most of the periodic trends that follow can be easily rationalized using the above parameters as a basis.

5.2 RADIUS

As a general rule, the radius of an atom will increase down and to the left in the periodic table. The radius of an atom depends on its electron probability. Because of the inherent uncertainty in defining where the end of one atom's sphere of influence over the electron density ends and the next begins, the radius of an atom can be defined in a number of different ways, and the magnitude of an atom's radius will vary from compound to compound. Furthermore, even for the same molecule, the value might depend on the experimental technique employed. For instance, electron diffraction measures the distance between two nuclei, whereas X-ray crystallography measures the distance between peaks of maximum electron density. Other molecules cannot be crystallized and the radius can only be measured in the gas phase, typically using microwave spectroscopy.

The *covalent radius* is defined as exactly one-half the internuclear separation between covalently bonded atoms in a predominantly covalent molecule containing a single bond. This definition can be problematic because not every atom can form a pure covalently bonded molecule containing only single bonds. The lighter noble gas atoms, for instance, do not even form bonds. Selected covalent radii are listed in Table 5.1.

The term *atomic radius* is also sometimes employed to describe the covalent radius of an atom. Atomic radii are based on the internuclear distances measured in a large number of covalently bonded heteronuclear compounds. The covalent radii for N, O, and F (measured using N_2H_4, H_2O_2, and F_2, respectively) are abnormally long because the bonding in these particular compounds is unusually weak. While the covalent radius is half the distance between the nuclei within a single molecule, the *van der Waals radius* is defined as half the distance between the nuclei of two atoms in neighboring molecules. For metal atoms, the *metallic radius* is defined as one-half the distance between neighboring atoms in the extended metallic solid. The metallic radius will depend on the coordination number (or number of nearest

TABLE 5.1 Covalent radii for selected elements.

Element	*r* (pm)	Element	*r* (pm)	Element	*r* (pm)
H	37	Fe	132	In	142
Li	128	Co	126	Sn	139
Be	96	Ni	124	Sb	139
B	84	Cu	132	Te	138
C	76	Zn	122	I	139
N	71	Ga	122	Xe	140
O	66	Ge	120	Cs	244
F	57	As	119	Ba	215
Na	166	Se	120	La	187
Mg	141	Br	120	Hf	175
Al	121	Kr	116	Ta	170
Si	111	Rb	220	W	162
P	107	Sr	195	Re	151
S	105	Y	190	Os	144
Cl	102	Zr	175	Ir	141
Ar	106	Nb	164	Pt	136
K	203	Mo	154	Au	136
Ca	176	Tc	147	Hg	132
Sc	170	Ru	146	Tl	145
Ti	160	Rh	142	Pb	146
V	153	Pd	139	Bi	148
Cr	139	Ag	145	Po	140
Mn	139	Cd	144	At	150

neighbors) of the metal in its crystalline lattice. Metals that exist in more than one type of allotrope will therefore have different metallic radii. Finally, for ionic solids, the *ionic radius* is defined by an electron density contour map, such as the one shown in Figure 5.3, showing where the sphere of influence of one ion ends and the other begins in that intermediate region between oppositely charged ions in an ionic solid. Selected ionic radii (Shannon's data) are listed in Table 5.2.

The different definitions of the radius are required because of the uncertain nature of exactly where the electron probability density should be cut off in defining the size of an atom. Because the different definitions depend on the type of bonding and the coordination number of the element, their values can be dramatically different from one molecule to another. For sodium, the metallic, covalent, and ionic (six-coordinate) radii are 186, 166, and 116 pm, respectively. For the neutral atom, the metallic radius is the largest because it has the largest coordination number. Sodium crystallizes in a body-centered cubic lattice, where each sodium atom is surrounded by eight nearest neighbor Na atoms. Therefore, to a first approximation, the electron density of each atom in metallic Na is shared over all eight of its neighbors (and to a lesser extent over its more distant neighbors). As a result, the distance between atoms will be larger than in the case of covalently bonded Na, where the electron density is shared with only one other Na atom. There is only so much electron density to go around, so the more atoms that are bonded to the Na, the greater its radius will be. It is for this reason that the metallic radius of Na (186 pm) is larger than its covalent radius (166 pm). Because the Na^+ ion contains one less electron than the neutral Na atom, there will be less electron–electron repulsions and less screening of the outer-most electrons. Therefore, the ionic radius (118 pm) of Na^+ is shorter than that of the neutral atom. For anions, the larger number of electrons increases the ionic radius because of increased electron–electron repulsions. Thus,

FIGURE 5.3
Definition of the ionic radius as viewed from a hard-spheres ionic crystalline lattice (a) and an electron density contour map (b). [Reproduced from Kittel, C. Introduction to Solid State Physics, 6th ed., John Wiley & Sons, Inc: New York, 1986. This material is reproduced with permission of John Wiley & Sons, Inc.]

while the covalent radius of Cl is 102 pm, the ionic radius of Cl^- (six-coordinate) is 181 pm. In fact, it is a universal fact that the radius of a cation will always be shorter than that for the neutral atom and the radius of an anion will always be larger than that for the neutral atom.

Across a row or series, the principal quantum number remains essentially the same, whereas the effective nuclear charge is increasing as more and more protons are added to the nucleus. This larger value of Z^* exerts a stronger pull on the electrons, shrinking the atomic radius. Down a column or group in the periodic table, Z^* varies only slightly, while n is increasing. As n increases, the outer-most electron will be added to an orbital having a larger average radial probability. Thus, the atomic radius increases down a column. The periodic trends in atomic radii are illustrated in Figure 5.4. While the trend holds for most of the main group elements, there are a few exceptions. One notable exception (known as the *lanthanide contraction*) is the smaller-than-expected radii of the lanthanides and actinides that result from poor shielding of the f orbitals.

An obvious corollary to the periodic trend for atomic radius is the atomic volume. Because the volume of an atom goes as the cube of the radius, those elements having large atomic radii will have especially large atomic volumes. In fact, Julius Meyer's original graph of atomic volumes, shown in Figure 5.5, makes a strong case for the concept of periodicity. The largest atomic volumes occur for the alkali metals. Because each of these metals has a valence of one, Meyer had already organized them into the same group of the periodic table. While the strength of Mendeleev's

TABLE 5.2 Shannon's ionic radii (pm) for selected ions, according to their coordination number.

Element	2	3	4	6	8
Li^{+}			59	76	92
Be^{2+}		16	27	45	
B^{3+}			11	27	
N^{3-}				146	
O^{2-}	135	136	138	140	
F^{-}	129	130	131	133	
Na^{+}			99	102	118
Mg^{2+}			57	72	89
Al^{3+}			39	54	
P^{5+}			17	38	
S^{6+}			26	43	
S^{2-}				184	
Cl^{-}				181	
K^{+}			137	138	151
Ca^{2+}				100	112
Ga^{3+}			47	62	
Ge^{4+}			39	53	
As^{5+}			34	46	
Se^{6+}			28	42	
Se^{2-}				198	
Br^{-}				196	
Rb^{+}				152	161
Sr^{2+}				118	126
In^{3+}			62	80	92
Sn^{4+}			55	69	81
Sb^{5+}				60	
Te^{6+}			43	56	
Te^{2-}				221	
I^{-}				220	
Cs^{+}				167	178
Ba^{2+}				135	142
Tl^{3+}			75	89	98
Pb^{4+}			65	78	94
Bi^{5+}				76	

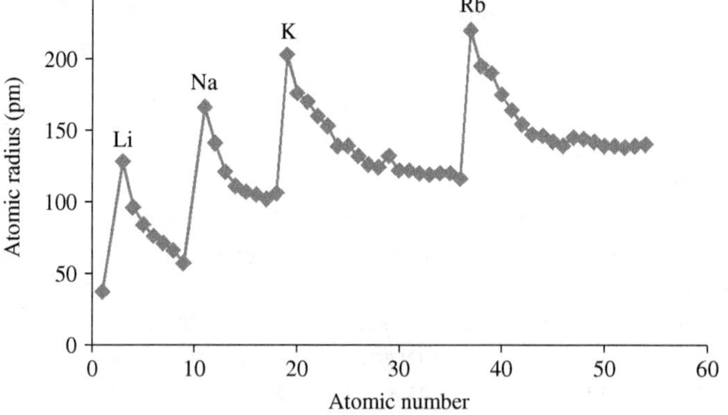

FIGURE 5.4
Covalent radii as a function of atomic number. The peaks occur for the alkali metals.

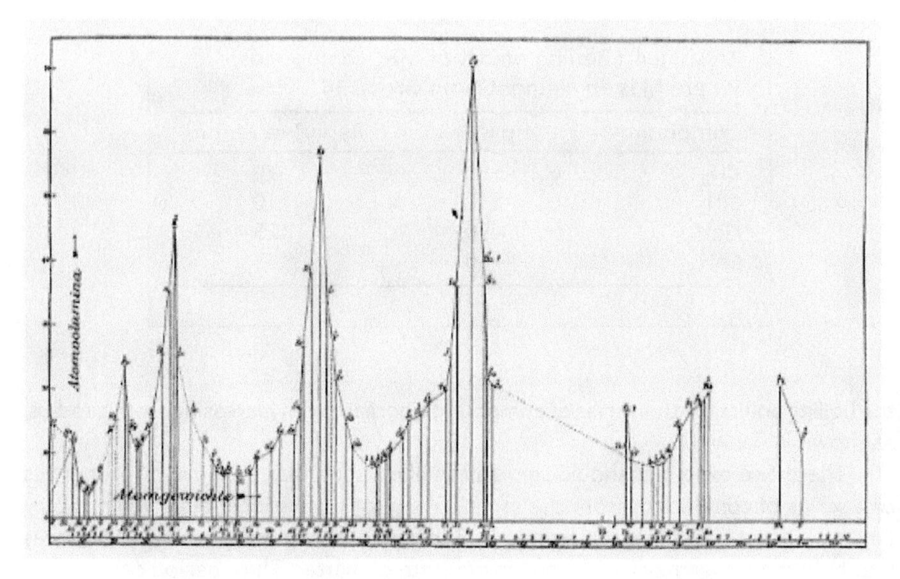

FIGURE 5.5
Julius Lothar Meyer's original plot of atomic volumes versus increasing atomic mass. The peaks occur for the alkali metals.

work was in its predictive capability, Meyer's research more clearly emphasized the periodicity of elemental properties.

There are a number of other physical properties of the elements that are either directly or indirectly related to the periodic trend for radius. Consider, for example, the melting points of the lithium halides. Because the bonding in these salts is primarily ionic in nature because lithium is a metal and the halogens are nonmetals, the strength of the cation–anion attractive force is expected to have an inverse dependence on the ionic radius of the halide, according to Coulomb's law of electrostatic attraction. The melting point of each crystalline solid depends on the strength of the electrostatic forces holding the ions together. Thus, the melting points for this series of ionic compounds are expected to increase as the ionic radius of the anion decreases, as shown in Table 5.3. One should be careful not to carry such periodic trends to the extreme, however. One of the assumptions in the earlier example is that each of the ionic solids crystallizes in the same type of lattice and therefore has the same number of nearest neighbor ions of the opposite charge. A second assumption is that the solids can be treated as exclusively ionic in nature. In fact, the bonding in lithium iodide has a large amount of covalent character.

There is also a radial trend in the MH_4 boiling points, where M is an element from Group 14. In this case, the bonding is primarily covalent and the species are molecular solids, so that the boiling point depends on the degree of intermolecular forces holding the molecules together in the liquid state. Because the strength of a London dispersion force increases with the polarizability of the element's electron cloud and the polarizability in turn depends on the diffuseness (or size) of the atom,

TABLE 5.3 Melting points of some ionic solids.

Ionic Solid	mp (°C)	Halide Radius (pm)
LiF	842	119
LiCl	614	167
LiBr	547	182
LiI	450	206

TABLE 5.4 Boiling points of MH_4 compounds, where M is an element from Group 14.

Compound	bp (°C)	Radius of M (pm)
CH_4	−161	70
SiH_4	−112	110
GeH_4	−90	125
SnH_4	−50	145

the boiling points for this series of compounds increase with increasing atomic radius, as shown in Table 5.4.

These two examples should serve as an illustration that the physical properties of a series of compounds from the same column of the periodic table will not only depend on the trend in their atomic radii but also on the different type of forces that hold them together in the condensed state of matter. Thus, periodicity is more than a simple set of rules or trends to memorize. It is actually an exercise in rational thought that uses the periodic table as its scaffold.

The *lattice energy* of an ionic solid is defined as the amount of energy gained when the gaseous ions are brought together from a separation distance of infinity to form the ionic solid, as illustrated in Equation (5.1):

$$M^+(g) + X^-(g) \rightarrow MX(s) \qquad U_0 \tag{5.1}$$

Because the electrostatic attraction between gaseous ions is expected to follow a similar relationship to Coulomb's law, the magnitude of the lattice energy should show an inverse dependence on the distance between the ions in the crystalline solid. This interionic separation will, in turn, depend on the sum of the ionic radii for the two ions. Assuming that each of the magnesium halides crystallizes in the same type of unit cell and that the bonding is primarily ionic in nature, one would therefore expect the magnitude of the lattice energy to increase as the anion is changed from I^- to Br^- to Cl^- to F^-. This is indeed the case. The lattice energies for MgI_2, $MgBr_2$, $MgCl_2$, and MgF_2 are −2327, −2440, −2526, and −2957 kJ/mol, respectively.

The size of an ion can also influence the *hydration enthalpies* of ionic compounds in water. The smaller the ionic radius, the higher the charge density (Z/r) of the ion will be. Ions with large charge densities can form stronger ion-dipole forces with the polar water molecule. Therefore, the enthalpy of hydration is expected to become more negative (more favorable from a thermodynamic standpoint) as the ionic radius decreases. The hydration enthalpies for the alkali metals, alkaline earths, and halogens are shown in Table 5.5, illustrating the expected periodic trend.

TABLE 5.5 Hydration enthalpies for selected ions (kJ/mol).

Cation	ΔH_{hyd}	Cation	ΔH_{hyd}	Anion	ΔH_{hyd}
Li^+	−520				
Na^+	−405	Mg^{2+}	−1920	F^-	−506
K^+	−321	Ca^{2+}	−1650	Cl^-	−364
Rb^+	−300	Sr^{2+}	−1480	Br^-	−337
Cs^+	−277	Ba^{2+}	−1360	I^-	−296

The solubility of the ions in water, on the other hand, is a tradeoff between the hydration enthalpy and the lattice energy. The solubility product, K_{sp}, for an ionic compound dissolving in water can be calculated from the sum of the hydration enthalpies for the ions and the reverse of the lattice energy for the ionic solid, as shown in Equations (5.2)–(5.5):

$$M^+(g) \rightarrow M^+(aq) \qquad \Delta H^\circ_{hydr}(M^+) \tag{5.2}$$

$$X^-(g) \rightarrow X^-(aq) \qquad \Delta H^\circ_{hydr}(X^-) \tag{5.3}$$

$$MX(s) \rightarrow M^+(g) + X^-(g) \qquad -U_0 \tag{5.4}$$

$$MX(s) \rightarrow M^+(aq) + X^-(aq) \qquad \Delta H^\circ_{soln}(\propto K_{sp}) \tag{5.5}$$

As the ionic radii decrease, the negative lattice energy term in Equation (5.4) becomes more positive. At the same time, the smaller ionic radii increase the magnitude of the hydration enthalpies in Equations (5.3) and (5.4). Thus, the radial trends tend to cancel each other. If the magnitude of the hydration enthalpies is the dominating factor, as is the case for the alkaline earth sulfates, then the solubility will increase as the ionic radius of the cation decreases, as shown in Table 5.6. On the other hand, if the lattice energy term is more important, as is the case for the alkaline earth hydroxides, then the solubility will decrease as the ionic radius of the cation decreases.

Example 5-1. Given that the lattice energy of CsF is −724 kJ/mol, use the hydration enthalpies in Table 5.3 to calculate the enthalpy of solution for CsF.

Solution. As shown in Equations (5.2)–(5.5), the enthalpy of solution is the sum of the individual hydration enthalpies for the ions minus the lattice energy of the salt. The sum is given by −277 − 506 + 724 = −59 kJ/mol. A graphical representation of this thermodynamic cycle is shown below.

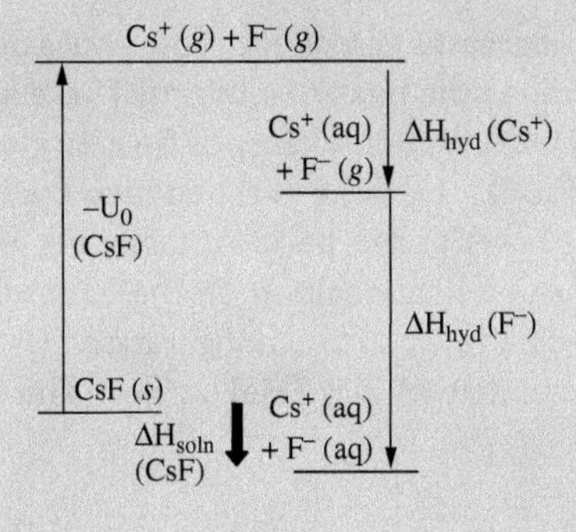

TABLE 5.6 Solubility product (K_{sp}) for selected compounds of the alkaline earths.

Compound	K_{sp}	Compound	K_{sp}
$MgSO_4$	Soluble	$Mg(OH)_2$	9×10^{-12}
$CaSO_4$	6×10^{-5}	$Ca(OH)_2$	1×10^{-6}
$SrSO_4$	3×10^{-7}	$Sr(OH)_2$	3×10^{-4}
$BaSO_4$	2×10^{-9}	$Ba(OH)_2$	5×10^{-3}

5.3 IONIZATION ENERGY

The ionization energy is defined as the amount of energy required to remove an electron from an atom in the gas phase to a distance of infinity away from the nucleus. The first ionization energy for an atom is shown in Equation (5.6) and is the one that removes the outer-most (or valence) electron from the electrostatic attraction of the nucleus. The second ionization energy is the additional amount of energy required to remove the second electron, and so on. Because it will always take energy to remove an electron, ionization energies are always positive. Theoretical ionization energies can be calculated from the negative of the energy of an electron using Koopman's theorem. Using the Bohr model, the first ionization energy can be calculated using Equation (5.7), where the Rydberg constant is expressed as 2.179×10^{-18} J. Experimentally, ionization energies are measured using photoelectron spectroscopy (PES). This technique employs either soft X-ray or vacuum UV radiation to knock the core and valence electrons, respectively, out of their atomic orbitals. The kinetic energies of the ejected electrons are then measured, and the ionization energy (or work function) is calculated according to Einstein's equation for the photoelectric effect, Equation (3.8):

$$\mathrm{M}\,(g) \rightarrow \mathrm{M}^{+}(g) + e^{-}(\infty) \quad \text{I.E.} \tag{5.6}$$

$$\text{I.E.} = -E_{el} = -R_{\infty} N_A \left(\frac{Z^*}{n^2}\right) \tag{5.7}$$

The general periodic trend for the first ionization energy is that it increases up and to the right in the periodic table. Progressing down a column, n is increasing, while Z^* remains essentially constant. Owing to the inverse dependence of the ionization energy on n^2 given by Equation (5.7), the first ionization energy is expected to decrease down a column. Moving across a series, the principal quantum number is constant while Z^* is increasing. As a result, the first ionization energy increases across a row. A plot of the first ionization energy as a function of atomic number is shown in Figure 5.6.

As a general rule, I.E. increases across each row until a noble gas atom is reached and then it drops off rapidly as the next row begins. There are some notable exceptions. For instance, the first ionization energy of Be is larger than that of B and that of N is greater than that for O. The same exceptions to the rule occur for the group congeners of subsequent rows in the periodic table: Mg > Al, P > S, Ca > Ga, and As > Se. The exceptions can be rationalized on the basis of the third major factor that affects periodic trends: the electron configuration. The electron configuration of Be is [He] $2s^2$, whereas that of B is [He] $2s^2 2p^1$. The removal of the valence

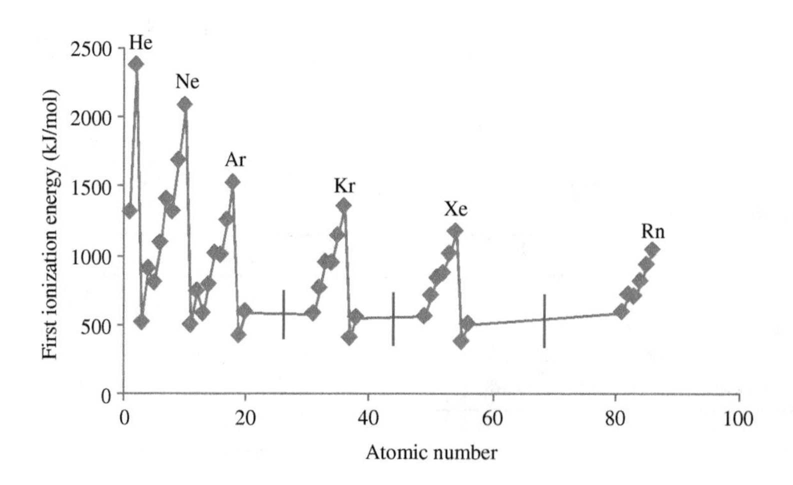

FIGURE 5.6
A plot of the first ionization energies for selected elements as a function of increasing atomic number. The peaks occur for the noble gases.

electron in B leaves the filled 2*s* subshell, which is a fairly stable electron configuration. Removal of a valence electron in Be, on the other hand, has to occur from a filled 2*s* subshell. Likewise, the electron configuration of N is [He] $2s^2 2p^3$, whereas that of O is [He] $2s^2 2p^4$. It is easier to remove the valence electron in O (despite its greater effective nuclear charge) because it yields a half-filled 2*p* subshell that is stabilized as a result of its large exchange energy. Removal of a 2*p* electron in N from this half-filled subshell decreases the magnitude of the exchange energy and is therefore less favorable. Exceptions to the general trend also occur at the end of the row of transition metals. The first ionization energy for Zn is considerably larger than that for Ga. For one, the valence electron in Zn is in a 3*d* subshell ($n = 3$) instead of the 4*p* level ($n = 4$). More significantly, however, the electron configuration of Zn is [Ar] $4s^2 3d^{10}$ as compared with that of Ga, which is [Ar] $4s^2 3d^{10} 4p^1$. The valence electron in Zn would need to be removed from a stable filled 3*d* subshell, whereas removal of the 4*p* electron from Ga leads to a stable electron configuration. The numerous exceptions to the general trend highlight the fact that it is far better to understand the factors that influence the periodic trends than it is to memorize simple generalizations.

For any given element, the second and third ionization energies are always larger than the first ionization energy. Table 5.7 lists the first several ionization energies for the more common elements.

An atom's ionization energy will strongly affect its chemical reactivity. The noble gases, which have the largest first ionization energies of all the elements, are generally unreactive. Hence, they are also known as the *inert gases*. The first noble gas compound was not discovered until 1962 when Neil Bartlett synthesized the salt $Xe[PtF_6]$. Other noble gas compounds include XeF_2, XeF_4, XeF_6, XeF_8^{2-}, $XeOF_4$, XeO_3, KrF^+, and KrF_2. The heavier noble gases have lower first ionization energies and are therefore more reactive than the lighter ones. They also have more room to accept ligands around them because of their larger atomic radii. On the other

TABLE 5.7 Ionization energies (kJ/mol) for selected elements.

Element	1st IE	2nd IE	3rd IE	Element	1st IE	2nd IE	3rd IE
H	1312.0			Ge	762.1	1537	2735
He	2372.3	5250.4		As	947.0	1798	
Li	513.3	7298.0		Se	940.9	2044	
Be	899.4	1757.1		Br	1139.9	2104	
B	800.6	2427		Kr	1350.7	2350	
C	1086.2	2352		Rb	403.0	2632	
N	1402.3	2856.1		Sr	549.5	1064.2	
O	1313.9	3388.2		In	558.3	1820.6	2704
F	1681	3374		Sn	708.6	1411.8	2943.0
Ne	2080.6	3952.2		Sb	833.7	1794	2443
Na	495.8	4562.4		Te	869.2	1795	
Mg	737.7	1450.7		I	1008.4	1845.9	
Al	577.4	1816.6	2744.6	Xe	1170.4	2046	
Si	786.5	1577.1		Cs	375.5	2420	
P	1011.7	1903.2	2912	Ba	502.8	965.1	
S	999.6	2251		Tl	589.3	1971.0	2878
Cl	1251.1	2297		Pb	715.5	1450.4	3081.5
Ar	1520.4	2665.2		Bi	703.2	1610	2466
K	418.8	3051.4		Po	812		
Ca	589.7	1145		At	930		
Ga	578.8	1979	2963	Rn	1037		

end of the spectrum, the alkali metals, as a group, have the lowest first ionization energies and are extremely reactive. Oxidation by one electron occurs quite readily and leads to a stable electron configuration that is isoelectronic with the noble gases. Therefore, the second and third ionization energies of the alkali metals are extremely large. Both K and Na will burst into flame if enough solid is added to water. Sodium is oxidized by water according to Equation (5.8) to make sodium hydroxide and hydrogen gas. The reaction is very exothermic and the hydrogen will ignite if sufficient quantities are used. The alkali metals will also ionize in liquid ammonia to make a blue solution containing solvated electrons, as shown in Equation (5.9). They will also react readily with Lewis bases such as OH^- or SO_4^{2-} and atoms from the right-hand side of the periodic table to make a wide variety of ionic salts.

$$2\,Na\,(s) + 2\,H_2O\,(l) \rightarrow 2\,NaOH\,(aq) + H_2(g) \quad (5.8)$$

$$Na\,(s) + x\,NH_3(l) \rightarrow Na^+(solv) + e^-(NH_3)_x \quad (5.9)$$

It is the ionization energy that is largely responsible for the staircase-shaped, metal–nonmetal line that divides the periodic table into two parts, as shown in Figure 5.7. The general periodic trend for ionization energy is that it increases up and to the right in the periodic table. Thus, the metal–nonmetal line is diagonal in nature. Metals lie to the left of the staircase, whereas nonmetals lie to its right. Those elements that actually border the staircase are called the *metalloids* because they often exhibit properties that are intermediate between those of a nonmetal and those of a true metal. Because of the low first ionization energies of the metals, these atoms hold their valence electrons only loosely and are easily oxidized. As a result, metallic bonding can be thought of as a crystalline array of metal atoms that have each lost one or more valence electrons to become cations that are surrounded by a "sea" of mobile electrons. The lower the ionization energy, the more readily the metal atom will lose its valence electron(s). The resistivity

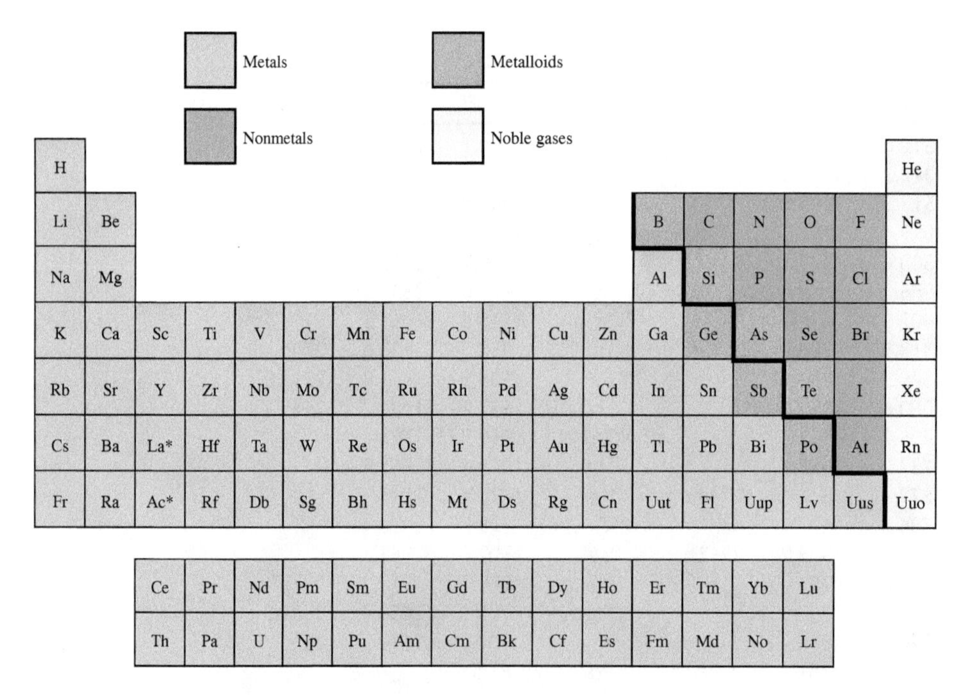

FIGURE 5.7
Representation of the periodic table showing the division line between the metals and the nonmetals.

of a metal, however, has more to do with an atom's size than does its ionization energy. The larger the metallic radius, the more diffuse its electron cloud will be. As metallic bonding occurs as a result of the overlap of atomic orbitals to form delocalized molecular orbitals over the extended structure, there will be less overlap for more diffuse orbitals than for more compact ones. Thus, although the first ionization energy of Na is larger than that of Cs, sodium is the more conductive of the two.

Example 5-2. Use Equation (5.7) to calculate the first ionization energy for each of the following: (a) Na, (b) F, and (c) Kr.

Solution. Using Slater's rules, the values of Z^* for the valence electron Na, F, and Kr are 2.2, 5.2, and 8.3, respectively.

(a) Na $\text{I.E.} = 2.179 \times 10^{-18}\,\text{J}(6.022 \times 10^{23}\,\text{mol}^{-1})\left(\frac{1\,\text{kJ}}{10^3\,\text{J}}\right)\left(\frac{2.2}{3^2}\right) = 320\,\text{kJ/mol}$

(b) F $\text{I.E.} = 2.179 \times 10^{-18}\,\text{J}(6.022 \times 10^{23}\,\text{mol}^{-1})\left(\frac{1\,\text{MJ}}{10^6\,\text{J}}\right)\left(\frac{5.2}{2^2}\right) = 1.7\,\text{MJ/mol}$

(c) Kr $\text{I.E.} = 2.179 \times 10^{-18}\,\text{J}(6.022 \times 10^{23}\,\text{mol}^{-1})\left(\frac{1\,\text{kJ}}{10^3\,\text{J}}\right)\left(\frac{8.3}{4^2}\right) = 680\,\text{kJ/mol}$

5.4 ELECTRON AFFINITY

When an electron is added to an atom in the gas phase from a distance of infinity, the energy of this process is known as the *electron gain enthalpy* and is defined in Equation (5.10). Electron gain enthalpies are usually negative because energy is typically released as the electron comes under the influence of the nucleus. For historical reasons, electron affinities are almost always reported as positive numbers. Therefore, electron affinity (E.A.) has the same magnitude but the opposite sign of the electron gain enthalpy.

$$\text{M}(g) + e^-(\infty) \rightarrow \text{M}^-(g) \quad \Delta H^\circ_{eg} = -\text{E.A.} \tag{5.10}$$

The periodic trend for electron affinities is that they generally increase up and to the right in the periodic table. As Z^* increases across a series, the larger nuclear charge exerts a stronger attraction on the electron, resulting in a large value for the E.A.. Progressing down a column, the electron is added to a quantum level that is, on average, farther from the nucleus (larger n). Thus, that electron will experience a smaller nuclear attraction, and the atom will have a low E.A. The electron affinities for a number of the main group elements are shown in Table 5.8.

There are quite a few exceptions to the general periodic trend. For instance, in the second series, the E.A. for N is less than that for C and the E.A. for Ne is considerably less than that for F (it is even negative!). As was the case for the ionization energies, exceptions to the trend can usually be explained on the basis of electron configurations. The electron configuration of N contains a half-filled $2p$ subshell. The addition of an electron would decrease the exchange energy and is therefore less favorable than in the case of C, where one more electron will yield the more stable electron configuration. Likewise, the noble gases already have filled np subshells. To decrease their valence would require the addition of an electron into an $(n+1)s$ subshell, which is not a very favorable process. Thus, the noble gases all have negative values for their electron affinities. Because the noble gases have large ionization energies and small electron affinities, they are largely inert. The halogens,

TABLE 5.8 Electron affinities (kJ/mol) for selected elements. For O and S, both the first and second E.A.s are listed.

Element	E.A.	Element	E.A.	Element	E.A.
H	72.8	P	71.7	In	34
He	−21	S	200/−532	Sn	121
Li	59.8	Cl	348.7	Sb	101
Be	≤0	Ar	−35	Te	190.2
B	23	K	48.3	I	295.3
C	122.5	Ca	2.4	Xe	−41
N	−7	Ga	36	Cs	45.5
O	141/−844	Ge	116	Ba	14.0
F	322	As	77	Tl	30
Ne	−29	Se	195.0	Pb	35.2
Na	52.9	Br	324.5	Bi	101
Mg	≤0	Kr	−39	Po	186
Al	44	Rb	46.9	At	270
Si	133.6	Sr	5.0	Rn	−41

on the other hand, have very large electron affinities and are quite reactive. Fluorine gas, for example, will react with almost any other element.

Vertical exceptions within a group also occur. The E.A. for F is less than that for Cl, for example. The smaller-than-expected E.A. for fluorine can be rationalized because of fluorine's extremely small radius. The addition of an electron to its valence shell would therefore increase the magnitude of the electron–electron repulsions. Consequently, the E.A. for fluorine is somewhat less than that for chlorine. Additionally, the bond dissociation enthalpy for F_2 (155 kJ/mol) is considerably less than that expected based on the other members of its group. For comparison, Cl_2, Br_2, and I_2 have bond dissociation enthalpies of 242, 193, and 151 kJ/mol. Other anomalies occur for N and O, whose electron affinities are also less than the group trend would have predicted. By analogy with the F–F bond strength, the N–N and O–O bonds are likewise weaker than those for P–P or S–S. In fact, both the hydrazine (N–N) and peroxide (O–O) classes of compounds are particularly reactive. Hydrazine, N_2H_4, was once used as a rocket fuel, and many peroxides are potentially explosive.

5.5 THE UNIQUENESS PRINCIPLE

In many ways, the anomalies mentioned earlier for the second series elements, such as their smaller-than-expected electron affinities and bond dissociation enthalpies, are more of the rule than the exception. In general, the physical properties of the second series elements are not always representative of their groups, a characteristic that is often referred to as the *uniqueness principle*. Consider, for example, the melting points of the chlorides shown in Table 5.9. The melting points of the second series compounds are very different from the remainder of the group.

One of the proposed reasons for the uniqueness of the second series elements is that their extremely small sizes ensure larger-than-average charge densities (Z/r). The compact and concentrated nature of the charge for second series elements exerts such a strong polarizing effect on the electron cloud of the anion that the orbitals overlap to form a significant percentage of covalent character to the bonding. Because covalent bonds are more directional in nature than ionic bonds, it does

TABLE 5.9 Melting points of selected chlorides, illustrating the second series anomaly known as the *uniqueness principle*.

Compound	mp (°C)	Compound	mp (°C)	Compound	mp (°C)
LiCl	610	$BeCl_2$	415	BCl_3	−107
NaCl	801	$MgCl_2$	714	$AlCl_3$	193
KCl	771	$CaCl_2$	782		
RbCl	718	$SrCl_2$	874	$InCl_3$	583

TABLE 5.10 Boiling points of the nonmetal hydrides, illustrating the uniqueness principle for the second series elements.

Compound	bp (°C)	Compound	bp (°C)	Compound	bp (°C)
NH_3	−33	H_2O	100	HF	20
PH_3	−87	H_2S	−65	HCl	−85
AsH_3	−60	H_2Se	−45	HBr	−69
SbH_3	−25	H_2Te	−15	HI	−35

not take as much energy to break the ions apart from each other in a covalent bond. Hence, the melting points of the more covalent second series compounds are considerably lower than those for their heavier group congeners.

A similar effect is observed for the boiling points of the nonmetal hydrides, shown in Table 5.10. As a rule, the boiling points increase down a group as a result of the stronger London dispersion forces associated with increased polarizability. The second series elements, however, exhibit a striking exception to the trend, bearing the highest boiling point of any compound in the same group. The larger charge densities of the second row elements (N, O, F) help them form strong hydrogen bonding interactions, which in turn elevates their boiling points.

Another significant difference between the series 2 elements and their heavier group members is that the former are more likely to form multiple bonds. For example, C=C, C≡C, N≡N, C=O, C≡O, O=O, N=O, and N≡O multiple bonds are prevalent, while examples of multiples with Si, P, and S are almost nonexistent (except for P=O and S=O). Nitrogen exists as a triple-bonded diatomic, whereas phosphorous, on the other hand, forms a tetrahedron of singly bonded P_4. Similarly, oxygen exists naturally as a doubly bonded diatomic, but sulfur forms a singly bonded S_8 ring in its natural form. The greater incidence of multiple bonding in the second series elements results from the smaller size of their atomic orbitals, which allows for an enhanced overlap of the *p* orbitals to form pi bonds. The larger and more diffuse *p* orbitals of the third and fourth series elements decrease the amount of overlap so that pi bonds between these atoms will only occur under special circumstances.

A third characteristic of the uniqueness principle is the lack of availability of low-lying *d* orbitals for participation in bonding. Hence, the second series elements cannot violate the octet rule in the formation of compounds. Consider, for example, the mixed halogens with fluorine as the central atom: F_2 and ClF, and BrF compared with those of chlorine, bromine, or iodine, which take higher coordination numbers: ClF_3, ClF_5, BrF_3, BrF_5, IF_3, IF_5, and IF_7. Similarly, the carbon atom in CF_4 is sp^3 hybridized, while the Si atom in SiF_6^{2-} is d^2sp^3 hybridized. The extent to which the *d* orbitals actually participate in the bonding in the so-called hypervalent compounds is a matter of contention. However, the lack of low-lying *d* orbitals available for mixing in the second series elements restricts their compounds to the usual group valence.

The decreased solubilities in water of lithium salts are also a direct result of the uniqueness principle. The large charge-to-size ratio of the lithium ion exerts a strong polarizing effect on the anion and increases the degree of covalent character in the bonding. As a result, compounds such as LiOH, LiF, Li_2CO_3, and Li_3PO_4 are not very soluble in water. Instead, they are more soluble in slightly less polar solvents, such as methanol or ethanol. The anomalous solubility of Li^+ salts is mimicked by those of Mg^{2+}, which leads naturally into a discussion of the diagonal effect.

5.6 DIAGONAL PROPERTIES

Diagonal relationships are commonly observed between elements from the second and third series. This periodic trend is especially true for the following pairs of elements: Li/Mg, Be/Al, and B/Si. While vertical periodic trends are still predominant, some properties match better along a diagonal. These diagonal periodic trends are no doubt related to the fact that the radius of an atom increases down and to the left in the periodic table, whereas I.E. and E.A increase up and to the right. The diagonal nature of the metal–nonmetal line has already been discussed.

The solubility of many Li^+ and Mg^{2+} salts in alcohols is but one example of the way that these two elements are similar. The perchlorate salts of both ions are very hygroscopic. Lithium and magnesium both form organometallic compounds that are soluble in organic solvents because of the covalent nature of their bonding. Organo-lithium reagents, such as LiI, and the Grignards, RMgX, both serve as activating agents in organometallic synthesis. The six-coordinate ionic radius of Li^+ (76 pm) is much closer in size to Mg^{2+} (72 pm) than it is to Na^+ (102 pm). Lithium's unusually high enthalpy of vaporization (135 kJ/mol) as compared with its group congeners is also closer to that of Mg (132 kJ/mol) than it is to Na (98 kJ/mol). Both Li and Mg are unique in that they are the only two elements that will react directly with dinitrogen to make a metal nitride.

Beryllium and aluminum also have a number of physical and chemical properties in common. Both metals form compounds that are largely covalent in their bonding and therefore more soluble in organic solvents than other members of their groups. Beryllium and aluminum both form amphoteric oxides, compounds that can act either as acids or bases. The standard reduction potential of Be (−1.85 V) is more similar to that of Al (−1.66 V) than it is to that of Mg (−2.37 V). Their compounds are numerous and usually more stable than those for their other group congeners. Both form stable compounds with oxygen- and nitrogen-based ligands. Beryllium chloride forms polymeric chains of chloride-bridged Be^{2+} ions that are sp^3 hybridized, as shown in Figure 5.8, whereas $AlCl_3$ exists as a dimer having a similar hybridization.

The diagonal elements boron and silicon also have similar properties. The chlorides of both elements form network covalent solids that act as Lewis bases toward water, reacting according to Equations (5.11) and (5.12):

$$BCl_3 + 3\,H_2O \rightarrow B(OH)_3 + 3\,HCl \tag{5.11}$$

$$SiCl_4 + 4\,H_2O \rightarrow Si(OH)_4 + 4\,HCl \tag{5.12}$$

The hydrides of B and Si are volatile and flammable compounds. The boranes, as the boron hydrides are termed, form many interesting hydrogen-bridged structures that are seldom observed with other main group elements. Both boron and silicon form numerous and complex oxygen-containing structures known as the *borates* and *silicates*. Their ions (B^{3+} and Si^{4+}) also have similar charge densities. Boron is a nonmetal with respect to its chemical reactivity, but it does have semi-metallic properties. Silicon is a semi-conductor and is used in the manufacture of computer chips. Both of these elements lie along the diagonal staircase that separates the metals from the nonmetals.

5.7 THE METAL–NONMETAL LINE

The metal–nonmetal line is shown in Figure 5.7. It runs diagonally along a staircase from just left of B to just left of At. Boron is considered to be a nonmetal, whereas Al and Po are metals. The remainder of the elements along this line exhibit both metallic and nonmetallic properties. These elements are called the *metalloids* and include Si, Ge, As, Sb, Te, and At. The metals form infinite lattices with high coordination numbers. Because of their low ionization energies, the metals actually exist in the crystal structure as cations that are surrounded by a mobile sea of electrons. The electrons are free to flow over the entire structure through a delocalized band of molecular orbitals. This is what gives the metals many of their characteristic properties, such as their high electrical and thermal conductivities and their metallic luster. Nonmetals, on the other hand, are poor thermal conductors and electrical insulators. Most metallic solids are malleable (they can be shaped by hammering) and ductile (they can be drawn into wires), while most nonmetals are brittle. Metal atoms tend to form cations, are good reducing agents, and can react with water or acids to form hydrogen gas. The nonmetals tend to form anions, are good oxidizing agents, and never liberate hydrogen from acids.

The electrical conductivity, σ, of a metal is measured in $1/\Omega{\cdot}m$ and is the current density divided by the electric field strength. The resistivity, ρ, is the inverse of the conductivity. The less tightly the metal atom holds onto its electrons and the greater the overlap between its valence orbitals, the smaller the resistivity will be and the more conductive the metal. Therefore, metals with low first ionization energies and small atomic radii will make the best conductors. Among the transition metals, silver is the most conductive, with a room-temperature resistivity of $1.6 \times 10^{-8}\ \Omega{\cdot}m$.

One of the most significant differences between the metals and the nonmetals besides their electrical conductivity is the acid–base properties of their oxides. Metal oxides are basic and react accordingly, as demonstrated by the reactions shown in Equations (5.13)–(5.15). Nonmetal oxides, on the other hand, act as acids, as shown in Equations (5.16)–(5.18). The oxides of the metalloids are amphoteric and can act either as acids or as bases, as shown by Equations (5.19) and (5.20):

Basic oxides: (Li, Na, Mg, K, Ca, Rb, Sr, Cs, Ba, Tl)

$$CaO\ (s) + H_2O\ (l) \rightarrow Ca^{2+}(aq) + 2\,OH^-(aq) \tag{5.13}$$

$$Na_2O\ (s) + H_2O\ (l) \rightarrow 2\,Na^+(aq) + 2\,OH^-(aq) \tag{5.14}$$

$$BaO\ (s) + 2\,H_3O^+(aq) \rightarrow Ba^{2+}(aq) + 3\,H_2O\ (l) \tag{5.15}$$

Acidic oxides: (B, C, N, O, F, Si, P, S, Cl, Se, Br, Te, I, Po, At)

$$SiO_2(s) + H_2O\ (l) \rightarrow H_2SiO_3(aq) \tag{5.16}$$

$$SO_3(g) + H_2O\ (l) \rightarrow H_2SO_4(aq) \tag{5.17}$$

$$CO_2(g) + 2\,OH^-(aq) \rightarrow CO_3^{\ 2-}(aq) + H_2O \tag{5.18}$$

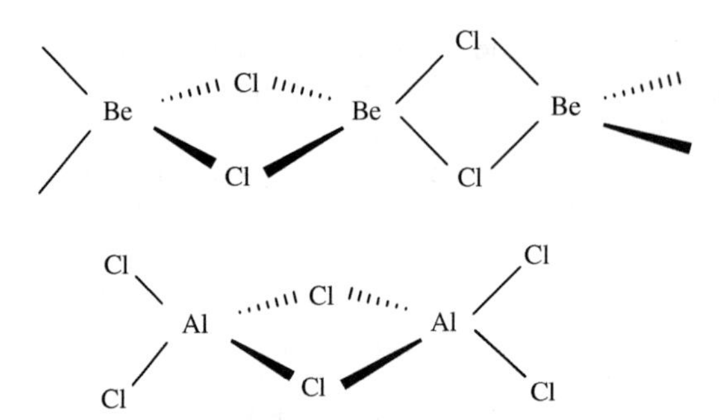

FIGURE 5.8
Chemical structures of $BeCl_2$ and $AlCl_3$, showing how both metal ions have tetrahedral coordinations.

Amphoteric oxides: (Be, Al, Ga, Ge, As, In, Sn, Sb, Pb, Bi)

$$Al_2O_3(s) + 6\,H_3O^+(aq) \rightarrow 2\,Al^{3+}(aq) + 9\,H_2O\,(l) \quad (5.19)$$

$$Al_2O_3(s) + 2\,OH^-(aq) + 3\,H_2O\,(l) \rightarrow 2\,[Al(OH)_4]^-(aq) \quad (5.20)$$

With few exceptions, the metal oxides are ionic solids and react with water to form aqueous ions, the nonmetal oxides are network covalent solids that react with water to make covalent compounds, and the amphoteric oxides of the metalloids form oligomeric polar–covalent solids. Similar relationships hold for the hydrides and fluorides of each element, with the metal forming an ionic solid and the nonmetal forming a network covalent solid, although the actual demarcation line varies somewhat depending on the anion.

5.8 STANDARD REDUCTION POTENTIALS

One of the few periodic trends of the metals not to show a strong diagonal effect is the standard reduction potential. In fact, this trend follows more of a horizontal rule. The standard reduction potential, $E°$, is defined in Equation (5.21). The standard reduction potential for the normal hydrogen electrode (N.H.E.), or the half-reaction shown in Equation (5.22), is given a value of zero. Metal atoms with $E°$'s more negative than the N.H.E. are easier to oxidize and harder to reduce. Metal atoms with $E°$'s more positive than the N.H.E. are easier to reduce and harder to oxidize:

$$M^{n+}(aq) + n\,e^- \rightarrow M^0(s) \quad E° \quad (5.21)$$

$$2\,H^+(aq,\ 1\,M) + 2\,e^- \rightarrow H_2(g,\ 1\,atm) \quad E° \equiv 0\,V \quad (5.22)$$

The standard reduction potential can be calculated using the thermodynamic cycle shown in Equations (5.23)–(5.26):

$$M^+\,(aq) \rightarrow M^+\,(g) - \Delta H_{hydr}\,(-) \quad (5.23)$$

$$M^+\,(g) + e^- \rightarrow M\,(g) \quad -\,I.E.\,(+) \quad (5.24)$$

$$M\,(g) \rightarrow M\,(s) \quad -\,\Delta H_{subl}\,(+) \quad (5.25)$$

$$M^+\,(aq) + e^- \rightarrow M\,(s) \quad \Delta H_{redn}\alpha - E° \quad (5.26)$$

The periodic trend within a group is a balance between two contributing properties. On one hand, as one proceeds down the group, the decrease in ionization energy makes it easier to oxidize the metal, leading to a negative value for the standard reduction potential. On the other hand, the decrease in the charge density (Z/r) down a column leads to a less negative enthalpy of hydration for the metal ion,

TABLE 5.11 Standard reduction potentials for selected elements at 25 °C.

Half-reaction	$E°$ (V vs NHE)
Li^+ (aq) + e^- → Li (*s*)	−3.05
K^+ (aq) + e^- → K (*s*)	−2.93
Ba^{2+} (aq) + $2e^-$ → Ba (*s*)	−2.90
Sr^{2+} (aq) + $2e^-$ → Sr (*s*)	−2.97
Ca^{2+} (aq) + $2e^-$ → Ca (*s*)	−2.87
Na^+ (aq) + e^- → Na (*s*)	−2.71
Mg^{2+} (aq) + $2e^-$ → Mg (*s*)	−2.37
Sc^{3+} (aq) + $3e^-$ → Sc (*s*)	−2.08
Be^{2+} (aq) + $2e^-$ → Be (*s*)	−1.85
Al^{3+} (aq) + $3e^-$ → Al (*s*)	−1.66
Ti^{2+} (aq) + $2e^-$ → Ti (*s*)	−1.63
V^{2+} (aq) + $2e^-$ → V (*s*)	−1.19
Mn^{2+} (aq) + $2e^-$ → Mn (*s*)	−1.18
Cr^{3+} (aq) + $3e^-$ → Cr (*s*)	−0.74
Fe^{2+} (aq) + $2e^-$ → Fe (*s*)	−0.44
Tl^+ (aq) + $2e^-$ → Tl (*s*)	−0.33
Co^{2+} (aq) + $2e^-$ → Co (*s*)	−0.28
Ni^{2+} (aq) + $2e^-$ → Ni (*s*)	−0.25
Sn^{2+} (aq) + $2e^-$ → Sn (*s*)	−0.14
Pb^{2+} (aq) + $2e^-$ → Pb (*s*)	−0.13
$2H^+$ (aq) + $2e^-$ → H_2 (*g*)	0.00
Cu^{2+} (aq) + $2e^-$ → Cu (*s*)	0.34
I_2 (*s*) + $2e^-$ → 2 I^- (aq)	0.54
Ag^+ (aq) + e^- → Ag (*s*)	0.80
Br_2 (*l*) + $2e^-$ → 2 Br^- (aq)	1.07
Cl_2 (*g*) + $2e^-$ → 2 Cl^- (aq)	1.36
F_2 (*g*) + $2e^-$ → 2 F^- (aq)	2.87

shifting $E°$ more positive. As a result of the opposing trends, there is little variation in the standard reduction potential within a group. There is, however, a fairly steady trend in the $E°$ as one proceeds across a series. Metals on the left-hand side of the periodic table have the lowest ionization energies and the most negative values of $E°$. The standard reduction potentials of the fourth series elements are as follows (with units of V vs N.H.E.): K^+/K (−2.93), Ca^{2+}/Ca (−2.87), Sc^{3+}/Sc (−2.08), Ti^{2+}/Ti (−1.63), V^{2+}/V (−1.19), Cr^{3+}/Cr (−0.74), Fe^{2+}/Fe (−0.44), Co^{2+}/Co (−0.28), Ni^{2+}/Ni (−0.25), Cu^{2+}/Cu (+0.34), and $Br_2/2\ Br^-$ (+1.07) (Table 5.11).

Many of the transition metals can take more than one oxidation state. Whenever this occurs, it is useful to show the relationship between the standard reduction potentials graphically using a *Latimer diagram*. Consider the stepwise reduction potentials for Cu^+ and Cu^{2+} shown by Equations (5.27) and (5.28). In order to determine the *skip potential* for the two-electron reduction of Cu^{2+} shown by Equation (5.29), we cannot simply add the two stepwise potentials together to get the final result.

$$Cu^+(aq) + e^- \rightarrow Cu\,(s) \qquad E° = 0.520\text{ V} \tag{5.27}$$

$$Cu^{2+}(aq) + e^- \rightarrow Cu^+(aq) \qquad E° = 0.159\text{ V} \tag{5.28}$$

$$Cu^{2+}(aq) + e^- \rightarrow Cu\,(s) \qquad E° = ??? \tag{5.29}$$

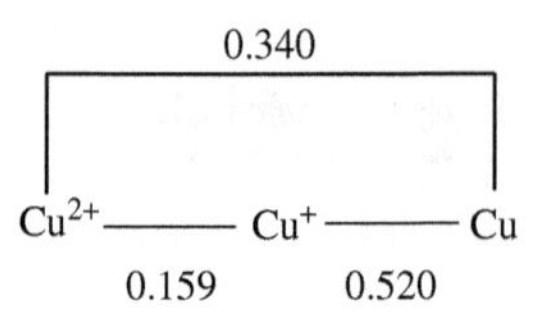

FIGURE 5.9
Latimer diagram for copper.

The reason for this becomes apparent when the equation relating the standard reduction potential to the Gibbs free energy is considered, as shown by Equation (5.30), where *F* is Faraday's constant (96,485 C/mol). While the value of *n* in Equation (5.30) is 1 for Equations (5.27) and (5.28), it is 2 for Equation (5.29):

$$G^{\circ} = -n\,F\,E^{\circ} \tag{5.30}$$

The Gibbs free energies are additive, as shown by Equations (5.31)–(5.33), and these can be used to deduce the skip potential for Equation (5.29):

$$Cu^{+}(aq) + e^{-} \rightarrow Cu\,(s) \quad G^{\circ} = -0.520F \tag{5.31}$$

$$Cu^{2+}(aq) + e^{-} \rightarrow Cu^{+}(aq) \quad G^{\circ} = -0.159F \tag{5.32}$$

$$Cu^{2+}(aq) + e^{-} \rightarrow Cu\,(s) \quad G^{\circ} = -0.679F = -2FE^{\circ} \tag{5.33}$$

Using the relationship given in Equation (5.33), the skip potential for Equation (5.29) is calculated as 0.340 V. The Latimer diagram for Cu shown in Figure 5.9 summarizes all three potentials in a succinct manner. Arrows are usually omitted from the Latimer diagram because it is assumed that all of the potentials are given for the reduction half-reactions.

It is also sometimes useful to illustrate the Gibbs free energy relationships on a *Frost diagram*. The Frost diagram for Cu is shown in Figure 5.10. The Frost diagram is essentially a plot of the Gibbs free energy (in units of nE°) versus the oxidation state of the metal. The Gibbs free energy for the metal in its elemental form is assigned a value of 0 on the *y*-axis. From Equation (5.31), the Gibbs free energy difference between Cu(0) and Cu(I) is 0.520. As the Gibbs free energy difference between Cu(I) and Cu(II) is an additional 0.159, the total change in free energy upon going from Cu(0) to Cu(II) is 0.679. Most metals are powerful reducing agents, but Cu is one of the few exceptions because its oxidized forms have a higher free energy than the element. As a result, Cu is not normally oxidized by H^{+} unless an oxidizing anion is used. Thus, for instance, HCl will not oxidize copper, but HNO_3 will. Furthermore,

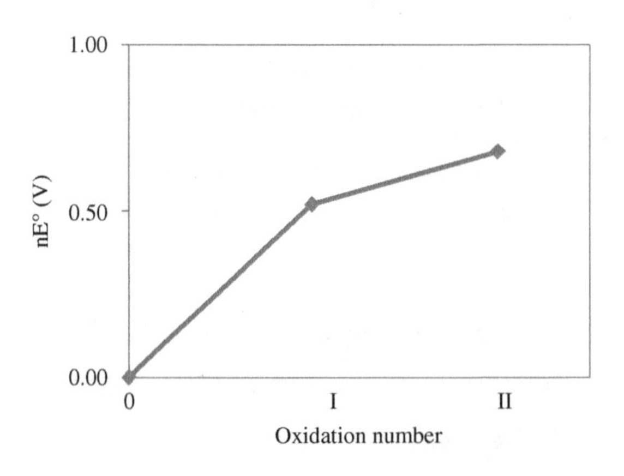

FIGURE 5.10
Frost diagram for copper.

the Frost diagram predicts that when HNO_3 oxidizes Cu metal, the resulting oxidation state will be Cu(II) and not Cu(I). The Cu(I) intermediate is unstable and can disproportionate into Cu(0) and Cu(II), as shown by Equations (5.34)–(5.36):

$$Cu^{+}(aq) + e^{-} \rightarrow Cu\,(s) \qquad G^{\circ} = -0.520\ F \tag{5.34}$$

$$Cu^{+}(aq) \rightarrow Cu^{2+}(aq) + e^{-} \qquad G^{\circ} = +0.159\ F \tag{5.35}$$

$$2\,Cu^{+}(aq) \rightarrow Cu\,(s) + Cu^{2+}(aq) \quad G^{\circ} = -0.361\ F \tag{5.36}$$

The usefulness of Frost diagrams becomes more apparent as the complexity of the different oxidation numbers increases. Consider the Frost diagram for Mn shown in Figure 5.11. Manganese can take a wide variety of oxidation states and its derivatives also have an acid–base dependence.

Those compounds lying high in the Frost diagram are good oxidizing agents relative to those lying lower on the diagram. Thus, the permanganate ion is the strongest oxidizing agent. Furthermore, the slope of the line determines the relative strength of the compound as an oxidizing agent. Permanganate is a stronger oxidizer in the presence of acid than it is in the presence of base. Those oxidation states that lie above the line connecting their two neighbors in the Frost diagram are unstable with respect to disproportionation, as is the case for MnO_4^{3-} in basic solution. In a similar manner, those species that are likely to undergo comproportionation are those that lie on either side of a substance that sits below the lines connecting its two neighbors. For example, $Mn(OH)_2$ and MnO_2 can react with each other to form two equivalents of Mn_2O_3 in basic solution. Substances lying on the lower region of the graph have low free energies and are therefore less reactive.

5.9 THE INERT-PAIR EFFECT

One of the more unusual periodic trends is known as the *inert-pair effect*. The elements that immediately follow the 4*d* and 5*d* transition series are considerably less reactive than their group properties might have predicted and they also prefer oxidation states that are two lower than the usual group valence. Consider, for example, the Group IIIA elements. The lighter elements, B, Al, and Ga take only the 3+ oxidation state predicted by their group valence. However, the heavier elements, In and Tl, take both the 3+ and 1+ oxidation states. The same trend is observed for the Group IVA elements. Again, the lighter elements C, Si, and Ge take the 4+ valence predicted by their period. However, Sn and Pb can exist in either the 4+ or the 2+ oxidation state.

The relative stability of the lower oxidation states for the fifth and sixth series post-transition elements (In, Sn, Sb, Tl, Pb, Bi, and Po) has been attributed to the higher-than-expected ionization energies for these seven elements. Consequently, it costs these elements relatively more energy to achieve a higher oxidation state than their lighter group congeners. At the same time, less energy is gained as a result of bond formation because they also exhibit lower-than-predicted bond dissociation enthalpies. Weaker bond enthalpies are expected for the heavier elements as a result of the diffuse nature of orbital overlap with increasing orbital size. The larger-than-expected ionization energies for the post-transition series Group IIIA elements result from the fact that the valence *s* electrons are not shielded from the nucleus very effectively by the intervening *d* electrons. As the two *s* electrons are both held tighter by the nucleus in the post-transition series elements, the ionization energies for these two electrons are unusually large. The end result is that these elements prefer oxidation states that are two lower than their typical group valence. Relativistic effects, which are discussed in the following section, also contribute to the higher-than-expected ionization energies of the valence *s* electrons.

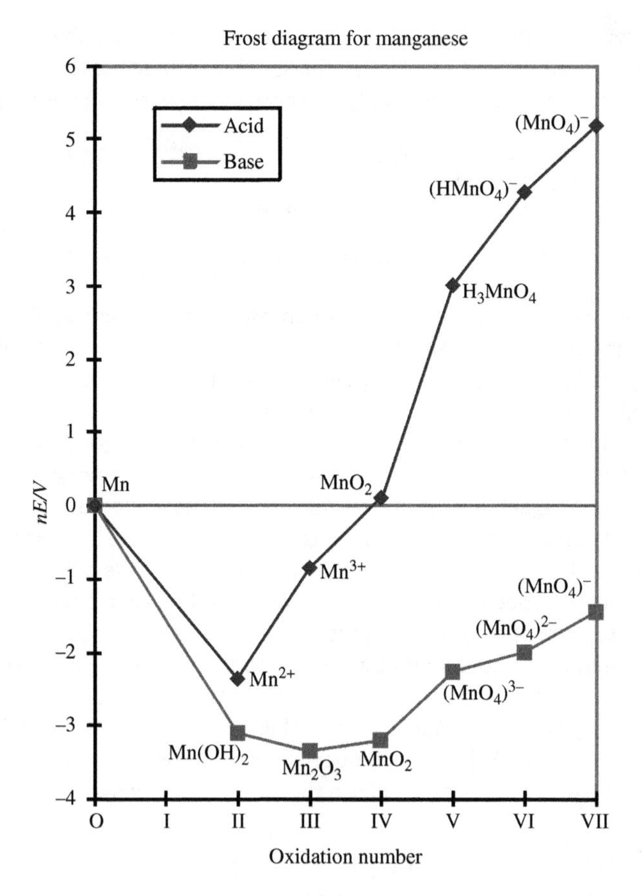

FIGURE 5.11
Frost diagram for manganese under acid and base conditions. [Developed by René T. Boeré, University of Lethbridge, Alberta, Canada, copyright (2000, 2013). Used with permission.]

5.10 RELATIVISTIC EFFECTS

For most applications of chemistry, the effect of relativity on electronic motion can largely be ignored. Einstein's theory of relativity states that the mass of a particle becomes infinite as its velocity approaches the speed of light. For the smaller elements, Schrödinger's nonrelativistic model is perfectly appropriate, as the velocities of the electrons in the lighter elements rarely approach the speed of light. However, the larger mass of the heavier elements exerts a stronger electrostatic pull on the inner-most electrons, sometimes accelerating them to speeds in excess of 10^8 m/s. According to Einstein's theory of relatively, particles that are traveling at velocities approaching the speed of light are more massive than they are at rest, as shown by Equation (5.37), where m_{rel} is the relativistic mass of the electron and m_0 is its rest mass. As a result of their relativistic heavier mass, the accelerated electrons have a smaller average radius, as indicated by Equation (5.38), which results from the derivation of the Bohr model of the atom:

$$m_{rel} = \frac{m_0}{\sqrt{1 - (v/c)^2}} \tag{5.37}$$

$$r = \frac{4\pi\varepsilon_0 n^2 \hbar^2}{m_e Z e^2} \tag{5.38}$$

Because the s and p electrons penetrate the nucleus better than the d or f electrons do, the s and p electrons are accelerated to a greater extent. Thus, the *s* and *p* orbitals effectively contract as relativistic effects become more prominent. Furthermore, as the *s* and *p* electrons migrate closer to the nucleus, this enhances their ability to shield the *d* and *f* electrons from experiencing the full attraction of the nucleus. Thus, the *d* and *f* orbitals expand as a result of relativity. The combined effect is to lower the energies of the *s* and *p* electrons, as shown in Figure 5.12, while raising the energy of the *d* and *f* electrons. The magnitude of the effect increases roughly as the square of the atomic number. Thus, the heavier elements experience considerably larger relativistic effects than do the lighter ones. For a number of elements heavier than Pt, the magnitude of the effect is even comparable to the strength of a chemical bond! The relativistic effect enhances the inert-pair effect mentioned in the preceding section. The smaller radius of the 6*s* orbital in Tl, for instance, causes the 6*s* electrons to be held more tightly than they would have been in the absence of relativity and raises their ionization energies. Hence, Tl will commonly prefer the lower 1+ oxidation number over its 3+ state.

However, unlike the inert-pair effect, any of the heavier elements can experience the relativistic effect. As an example, gold is the least reactive of the coinage metals. This explains why it was one of the first elements to be discovered by ancient civilizations (about 10,000 years ago). Gold exists in nature in its elemental form and is only rarely found in minerals, such as calverite, $AuTe_2$, and sylvanite, $AuAgTe_4$. The electron configuration of Au is [Xe] $4f^{14}5d^{10}6s^1$. As a result of the relativistic effect, the 6*s* electron has an unusually large ionization energy (890 kJ/mol, compared with 746 kJ/mol for Cu and 741 kJ/mol for Ag) and is extremely difficult to oxidize. Hence, gold will not rust in air, nor will it react with the oxidizing acid HNO_3. In fact, in order to dissolve gold, a combination of one part nitric acid with three parts HCl must be employed. Nitric acid acts as the oxidizer, while the Cl^- ion helps to solubilize the gold as the complex ion $AuCl_4^-$, according to Equations (5.39) and (5.40). The mixture is known as *aqua regia* because it is the only acid that can dissolve the "royal" metals Au and Pt:

$$\mathrm{Au\ (s) + 6\,H^+\ (aq) + 3\,NO_3^-\ (aq) \rightarrow Au_3^+\ (aq) + 3\,NO_2(g) + 3\,H_2O\ (\mathit{l})} \tag{5.39}$$

$$\mathrm{Au_3^+\ (aq) + 4\,Cl^-\ (aq) \rightarrow [AuCl_4]^-\ (aq)} \tag{5.40}$$

As a result of relativistic orbital contraction, the atomic radius of Au is less than that expected on the basis of its periodic trends: Cu (135 pm), Ag (160 pm), and Au (135 pm). Because of the small size of its half-filled 6*s* orbital, the E.A. of Au (223 kJ/mol) is considerably larger than that for Ag (126 kJ/mol) or Cu (118 kJ/mol). In fact, the E.A. of Au is so large that gold exists as the Au^- anion in the compound cesium auride (CsAu).

At the same time that the 6*s* orbital shrinks and is therefore stabilized by the relativistic effect, the 5*d* orbital expands and is destabilized. Thus, the energy gap between the 5*d* and 6*s* orbitals is fairly small, and gold reflects light in the low-energy region of the visible spectrum. This is why gold has a distinctive yellowish color, unlike the silvery reflectance of most of the other metals. Gold is also a fairly soft metal. Pure gold, which is 24 carats, is seldom used in the manufacture of jewelry because it is so soft that ordinary wear and tear will damage the metal. Instead, an alloy of 18k or 14k gold is generally used. A typical piece of 18k jewelry consists of 75.0% Au, 12.5% Ag, and 12.5% Cu, by mass. The presence of the other metals helps harden the alloy.

The unusual differences between Au and Hg, which are neighbors on the periodic table, can also be ascribed to the relativistic energies of the orbitals shown in Figure 5.12. Gold has a melting point of 1064 °C, whereas Hg is a liquid at

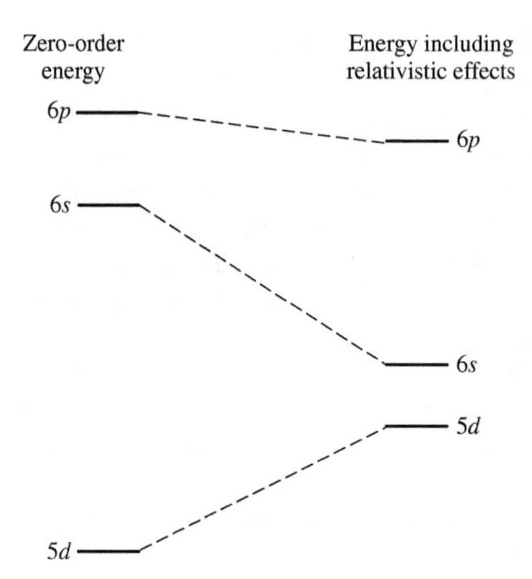

FIGURE 5.12
Valence orbital energies for Au and Hg in the absence of relativistic effects and with relativistic effects included.

room temperature and has a melting point of −39 °C. In many ways, Au acts as a pseudo-halogen, as evidenced by its ability to form the auride anion. The low-lying 6*s* orbital in Au is singly occupied. Therefore, like the halogens, gold can exist as the diatomic Au_2. In fact, Au_2 has a gas-phase bond dissociation energy (BDE) of 221 kJ/mol, which is even larger than that for I_2 (151 kJ/mol) and only slightly smaller than that of Cl_2 (242 kJ/mol). Mercury, on the other hand, has two electrons in the low-lying 6*s* orbital and it behaves as a pseudo-noble gas. Mercury(0) cannot form a diatomic like Au(0) because its electrons are all paired. However, oxidation of Hg by one electron to form Hg(I), which is isoelectronic with Au(0), leads to a stable diatomic cation, Hg_2^{2+}. By analogy with the noble gases, Hg(0) is fairly unreactive. Its unusually low melting point can be ascribed to the fact that Hg atoms can only aggregate together in the condensed phases as a result of weak van der Waals forces. Mercury is unique among the transition metals because it exists in the gas phase almost exclusively as monomeric Hg just like the noble gases do. In comparison with H, whose relativistic mass is 1.00003 times its rest mass and where its 1*s* electron has an average velocity of 0.0073*c*, the relativistic mass of a 1*s* electron in Hg is 1.23 times its rest mass and its average velocity is 0.58*c*, so that its size shrinks by a factor of nearly a quarter.

The spin–orbit coupling of polyelectronic terms is also a consequence of the relativistic effect. For elements with $Z \geq 30$, the *L* and *S* quantum numbers in the Russell–Saunders (*LS* coupling) scheme are no longer valid. Instead, the spin (*s*) and orbital (*l*) angular momentum for each electron couple together first to give a new quantum number *j*. The total angular momentum *J* is given as the sum of the individual *j* values for each of the electrons in a process known as *jj coupling*. This procedure was described in Section 4.5. From a phenomenological point of view, the relativistic approach considers the electron as stationary with the nucleus orbiting it. Because the nucleus is a charged particle, it creates a magnetic moment. The spin and orbital magnetic moments can couple together to mix those states that have the same total angular momenta. The magnitude of spin–orbit coupling depends on the metal, its oxidation state, its ground state term symbol (the values of *J*, *L*, and *S*), and its ligand field, according to Equation (4.27), where λ is the spin–orbit coupling constant. Because the value of λ increases in proportion to Z^4/n^3, it becomes increasingly important for the heavier atoms. For example, the value of λ for the ground state Cu(II) free ion is $-830\ \text{cm}^{-1}$, whereas $\lambda = -1600\ \text{cm}^{-1}$ for Rh(III) and $-4000\ \text{cm}^{-1}$ for Ir(III).

One of the most important consequences of spin–orbit coupling is the mixing of terms in electronic spectroscopy. Electronic transitions between states of different spin multiplicities, such as singlet-to-triplet transitions, are forbidden by the spin selection rule. This explains why phosphorescence ($\Delta S \neq 0$) is a slower process than fluorescence ($\Delta S = 0$). The coupling of the spin and orbital angular momentum relaxes this restraint and allows the different spin states to mix with each other. Hence, the rate of intersystem crossing between states of different multiplicity increases with increasing atomic number. Spin–orbit coupling can also affect the degree of spin–spin coupling in NMR spectroscopy.

5.11 ELECTRONEGATIVITY

The final periodic trend that we shall discuss in this chapter is the important concept of electronegativity. As defined by Linus Pauling, *electronegativity* is the ability of an atom in a molecule to attract the shared electrons closer to itself. Unlike most of the other periodic laws, such as radius, ionization energy, and E.A., electronegativity has no meaning outside the context of a chemical bond. Thus, there is no absolute scale for electronegativity. The Pauling scale of electronegativity was developed in 1932 and is still the most widely used. Pauling arbitrarily assigned the electronegativity of fluorine as 4.00. The electronegativities of the other elements are based on the difference between the experimental bond dissociation energy for a heteronuclear diatomic and the average of the bond dissociation energies for the corresponding homonuclear diatomics. This difference, given by the symbol Δ in Equations (5.41) and (5.42), is known as the *ionic resonance energy*, which is proportional to the percent ionic character in the bonding. The abbreviation BDE stands for the bond dissociation energy. A correction must be made if the diatomic molecule is multiply bonded, as is the case for N_2 or O_2. The Pauling electronegativities of selected elements are listed in Table 5.12, along the corresponding values for more modern definitions of electronegativity:

$$\Delta = \mathrm{BDE}(A-B) - \frac{\mathrm{BDE}(A-A) + \mathrm{BDE}(B-B)}{2} \tag{5.41}$$

$$\Delta\chi = \chi_A - \chi_B = 0.102\sqrt{\Delta} \tag{5.42}$$

Example 5-3. Calculate the electronegativity of Cl given that the bond dissociation energies for Cl_2, F_2, and ClF are 242, 158, and 255 kJ/mol, respectively.

Solution. The ionic resonance energy, Δ, is calculated from Equation (5.41) as follows:

$$\Delta = 255 - \frac{242 + 158}{2} = 55\ \mathrm{kJ/mol}$$

The difference in electronegativity between F and Cl can be calculated from Equation (5.42):

$$\Delta\chi = \chi_F - \chi_{Cl} = 0.102\sqrt{55} = 0.76$$

As the electronegativity of F defined by Pauling is 4.00, the calculated value of χ for Cl is 3.24. The actual values are listed in Table 5.12, where the modern value of χ for F has been corrected to 3.98 instead of 4.00. The tabulated value of 3.16 for Cl is reasonably close to the calculated value of 3.24.

TABLE 5.12 Electronegativities of selected elements using the revised Pauling scale, the Mulliken–Jaffe scale, and Allen's spectroscopic configuration energies. The latter scales have been converted to the revised Pauling scale. The Mulliken–Jaffe values are listed for the sp^3 valence hybridization, unless otherwise indicated.

Element	χ (Pauling)	χ_{MJ} (Mulliken)	χ_{spec} (Allen)
H	2.20	2.25 (*s*)	2.30
Li	0.98	0.97 (*s*)	0.91
Be	1.57	1.54 (*sp*)	1.58
B	2.04	2.04 (sp^2)	2.05
C	2.55	2.48	2.54
N	3.04	3.04	3.07
O	3.44	3.68	3.61
F	3.98	3.91 (14% *s*)	4.19
Na	0.93	0.91 (*s*)	0.87
Mg	1.31	1.37 (*sp*)	1.29
Al	1.61	1.83 (sp^2)	1.61
Si	1.90	2.28	1.92
P	2.19	2.41	2.25
S	2.58	2.86	2.59
Cl	3.16	3.10 (14% *s*)	2.87
K	0.82	0.73 (*s*)	0.73
Ca	1.00	1.08 (*sp*)	1.03
Ga	1.81	2.01 (sp^2)	1.76
Ge	2.01	2.33	1.99
As	2.18	2.38	2.21
Se	2.55	2.79	2.42
Br	2.96	2.95 (14% *s*)	2.68
Kr	3.00	3.31	2.97
Rb	0.82	0.82 (*s*)	0.71
Sr	0.95	1.00 (*sp*)	0.96
In	1.78	1.76 (sp^2)	1.66
Sn	1.96	2.21	1.82
Sb	2.05	2.22	1.98
Te	2.10	2.57	2.16
I	2.66	2.95	2.36
Xe	2.60	3.01	2.58

Alternative electronegativity scales have been developed by others through the years. The Mulliken–Jaffe scale defines electronegativity on the basis of atomic parameters, namely the average of the valence ionization energy and E.A., as shown in Equation (5.43). The *v* subscripts indicate that the valence configuration must be used. Thus, for example, boron forms three equivalent bonds by hybridizing its valence 2*s* and 2*p* orbitals to make sp^2 hybrid orbitals. The theoretical ionization energy and E.A. for the sp^2 hybrid orbitals must be used in Equation (5.43) in place of those for an unhybridized 2*p* orbital. Hence, the original Pauling definition of electronegativity as a property of an atom within a molecule still rings true. Mulliken–Jaffe electronegativities are typically expressed in units of electron volts. However, the Pauling scale has been so firmly entrenched that more often than not the Mulliken–Jaffe scale is converted to the Pauling scale through a simple conversion factor. One advantage of the Mulliken–Jaffe scale is that it correctly predicts the periodic trend for electronegativity. Because the periodic trends for both first ionization energy and E.A. increase up and to the right in the periodic table, the average of these two values should also follow this trend. Indeed, the periodic trend for electronegativity is that it increases toward the upper right-hand corner of the periodic table. When comparing the electronegativities of the elements, there is a very clear demarcation line between the metals and the nonmetals. The metals have only a small difference in energy between their *s* and *p* orbitals, whereas the nonmetals have a much larger gap.

$$\chi_{MJ} = \frac{\text{I.E.}_v + \text{E.A.}_v}{2} \quad (5.43)$$

More recently, Lee Allen has developed an electronegativity scale that is based on spectroscopic data, namely the average valence electron energies obtained by gas-phase atomic spectroscopy. Allen's spectroscopic electronegativity (χ_{spec}) is based on the configuration energy (C.E.), given by Equation (5.44). In this equation, ε_p and ε_s are the "one-electron" energies of the valence *p* and *s* orbitals, respectively, and *m* and *n* are the numbers of *p* and *s* electrons occupying those orbitals. More specifically, the energies ε_p and ε_s are calculated from the average of the $(2L+1)(2S+1)$ multiplets of the ground state with those of the singly *s* and singly *p* ionized configurations. The Allen approach has the advantage that it is not based on molecular properties and can be calculated solely from the spectroscopic properties of each atom. Because the formula in Equation (5.44) ignores contributions from the *d* and *f* electrons, it is not at all useful for the transition metals, lanthanides, or actinides. This is not a problem as the chemical properties of the transition metals are dominated by ligand field effects, rather than electronegativities. As was the case for the Mulliken–Jaffe scale, configuration energies can also be converted to the Pauling scale by multiplying the C.E. (in kJ/mol) by 1.75×10^{-3} mol/kJ. The data in Table 5.12 compare the electronegativities of selected atoms using all three definitions. The electronegativities of the lighter noble gases are not included, as these elements tend not to form too many compounds.

$$\chi_{spec} \propto \text{C.E.} = \frac{m\varepsilon_p + n\varepsilon_s}{m+n} \quad (5.44)$$

Other definitions have also been proposed through the years. The Allred–Rochow scale is based on the electrostatic force exerted by the nucleus on the valence electrons, whereas the Sanderson scale represents a ratio of the experimental electron density to that predicted by interpolation of the electron densities of the two nearest noble gas atoms. Sanderson is also responsible for the concept of electronegativity equalization. When two or more atoms having different initial electronegativities combine to make a bond, they adjust to have the same intermediate electronegativity in the molecule. The electronegativities of ions and of groups of atoms, such as CF_3, NH_2, or COOH, have also been calculated using the principle of electronegativity neutrality. For example, the group electronegativity of CF_3 (3.47) is greater than that of CH_3 (2.31) as a result of the inductive effect of the F atoms. Likewise, the calculated group electronegativity of the electron-withdrawing group COOH (3.04) is greater than that of the electron-donating group NH_2 (2.47).

Example 5-4. Calculate χ_{spec} for N given that $\varepsilon_p = 1270$ kJ/mol and $\varepsilon_s = 2465$ kJ/mol.

Solution. The electron configuration of N is [He] $2s^22p^3$. The C.E. of N can therefore be calculated from Equation (5.44):

$$\text{C.E. (N)} = \frac{3(1270\ \text{kJ/mol}) + 2(2465\ \text{kJ/mol})}{5} = 1750\,\text{kJ/mol}$$

The spectroscopic electronegativity of N is given by

$$\chi_{spec}(\text{N}) = 1.75 \times 10^{-3}\,\text{mol/kJ}\,(1750\,\text{kJ/mol}) = 3.06$$

The electronegativity of N on the revised Pauling scale is 3.04, in excellent agreement with the calculation.

EXERCISES

5.1. Explain why the covalent radius of K is considerably smaller than its metallic radius. How are the two different radii defined?

5.2. The atomic radius of Cl is 102 pm. Explain why the ionic radius of Cl in the perchlorate ion is 41 pm, whereas the ionic radius of Cl in the chloride ion is 181 pm. Compare both values to the atomic radius.

5.3. The ionic radius of Na^+ depends on its coordination number. Rationalize the following trend: 99 pm (four-coordinate), 102 pm (six-coordinate), and 118 pm (eight-coordinate).

5.4. Use ionic radii to rationalize why H_2O has a bent molecular geometry while Li_2O is linear.

5.5. Although Li^+ has a much smaller ionic radius than K^+ in the gas phase, the solvated radii of the two ions is reversed: Li^+ (aq) = 382 pm, whereas K^+ (aq) = 328 pm. Explain these results in terms of the enthalpy of hydration and the numbers of water molecules that surround each ion in aqueous solution.

5.6. Why is Lu^{3+} smaller than Y^{3+}?

5.7. For each pair, circle the species with the larger radius and provide reasoning for your choice: (a) Mg or Mg^{2+}, (b) O or O^{2-}, (c) K^+ or Cl^-, (d) P^{3-} or S^{2-}, (e) N or P, (f) Na or Mg, and (g) Na^+ or Mg^{2+}.

5.8. Arrange the following in increasing order of first ionization energy: F, Na, Al, P, S, and Cl.

5.9. Which of the alkali metals is the most reactive with water, and why?

5.10. Explain why Kr and Xe form more types of noble gas compounds than the lighter noble gases.

5.11. Calculate the first ionization energy for the valence electron in O in MJ/mol.

5.12. Match the following electron configurations with the appropriate first ionization energy:

a. $1s^2\ 2s^2\ 2p^6\ 3s^2\ 3p^6\ 4s^2\ 3d^{10}\ 4p^6\ 5s^1$ (i) 1356 kJ/mol

b. $1s^2\ 2s^2\ 2p^6\ 3s^2\ 3p^6\ 4s^2$ (ii) 595 kJ/mol

c. $1s^2\ 2s^2\ 2p^6\ 3s^2\ 3p^6\ 4s^2\ 3d^{10}\ 4p^6$ (iii) 409 kJ/mol

5.13. One allotrope of carbon is graphite, which is an electrical conductor, is shown below. Predict whether silicon has an allotrope having a structure that is analogous to that for graphite. Explain your answer in detail.

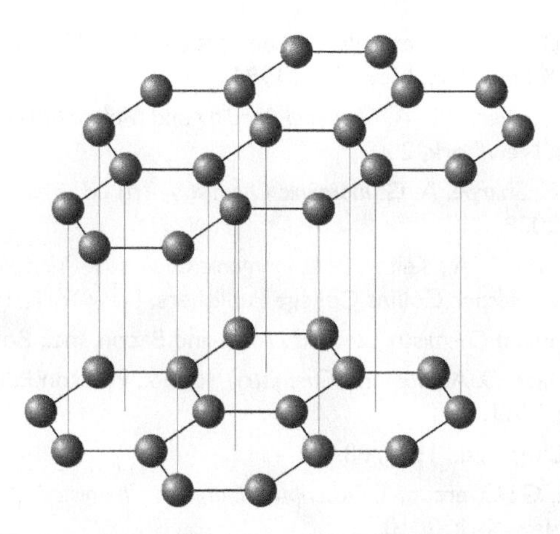

[E. Generalic, http://www.periodni.com, used with permission.]

5.14. When lead is exposed to chlorine gas, the product of the reaction is $PbCl_2$. Only under more extreme conditions will $PbCl_4$ form. However, when chlorine reacts with carbon, the product of the reaction is CCl_4. Explain these differences.

5.15. When PbO_2 reacts with water, it yields the hypothetical hydrated product: $Pb(OH)_4$. On the other hand, CO_2 reacts with water to make carbonic acid, H_2CO_3. Explain the differences in reactivity between these two molecules in as much detail as possible.

5.16. Given the standard reduction potentials below, calculate the standard reduction potential for the Au^{3+}/Au^{+} redox couple and sketch the Latimer diagram for gold.

$$Au^{+}(aq) + e^{-} \rightarrow Au\,(s) \quad E^{\circ} = 1.69\text{ V}$$

$$Au^{3+}(aq) + 3e^{-} \rightarrow Au\,(s) \quad E^{\circ} = 1.40\text{ V}$$

Given the information above, sketch the Frost diagram for gold and predict the product when Au (s) is reacted with H^{+} (aq).

5.17. Given the Latimer diagram shown below for lead, calculate the Gibbs free energy for the disproportionation reaction shown below. Also, fill in the missing skip potential.

$$\overset{??}{\overbrace{Pb^{4+} \xrightarrow[1.67\text{ V}]{} Pb^{2+} \xrightarrow[-0.13\text{ V}]{} Pb}}$$

Disproportionation reaction: $2\,Pb^{2+}\,(aq) \rightarrow Pb\,(s) + Pb^{4+}\,(aq)$

5.18. Given the following bond dissociation energies, calculate the Pauling electronegativity of Br using Equations (5.41) and (5.42): Cl–Cl (242 kJ/mol), Br–Br (183 kJ/mol), and Br–Cl (219 kJ/mol). The Pauling electronegativity of Cl is 3.16.

5.19. Calculate χ_{spec} for O given that $\varepsilon_p = 1529$ kJ/mol and $\varepsilon_s = 3124$ kJ/mol.

BIBLIOGRAPHY

1. Atkins, P.; de Paula, J. *Physical Chemistry*, 7th ed., W. H. Freeman and Company, New York, 2002.
2. Cordero, B.; Gómez, V.; Platero-Prats, A. E.; Revés, M.; Echeverría, J.; Cremades, E.; Barragán, F.; Alvarez, S. Covalent radii revisited, Dalton Trans., 2008 [DOI: 10.1039/b801115j].

3. Douglas, B.; McDaniel, D.; Alexander, J. *Concepts and Models of Inorganic Chemistry*, 3rd ed., John Wiley & Sons, Inc., New York, 1994.
4. Gillespie, R. J.; Popelier, P. L. A. *Chemical Bonding and Molecular Geometry*, Oxford University Press, New York, 2001.
5. Housecroft, C. E.; Sharpe, A. G. *Inorganic Chemistry*, 3rd ed., Pearson Education Limited, Essex, England, 2008.
6. Huheey, J. E.; Keiter, E. A.; Keiter, R. L. *Inorganic Chemistry: Principles of Structure and Reactivity*, 4th ed., Harper Collins College Publishers, New York, 1993.
7. Levine, I. N. *Quantum Chemistry*, 2nd ed., Allyn and Bacon, Inc., Boston, MA, 1974.
8. Miessler, G. L.; Tarr, D. A. *Inorganic Chemistry*, 4th ed., Pearson Education Inc., Upper Saddle River, NJ, 2011.
9. Norrby, L. J. J. Chem. Ed., 1991, 68, 110–113.
10. Rayner-Canham, G.; Overton, T. *Descriptive Inorganic Chemistry*, 5th ed., W. H. Freeman and Company, New York, 2010.
11. Rodgers, G. E. *Descriptive Inorganic, Coordination, and Solid-State Chemistry*, 3rd ed., Brooks/Cole, Cengage Learning, Belmont, CA, 2012.
12. Shannon, R. D. Acta Crystallographica, 1976, A32, 751–767.
13. Yam, V. W.-W.; Cheng, E. C.-C. Chem. Soc. Rev. 2008, 37, 1806–1813.

An Introduction to Chemical Bonding 6

"A bond does not really exist at all—it is a most convenient fiction."

—*Charles Coulson*

6.1 THE BONDING IN MOLECULAR HYDROGEN

One might begin a discussion of chemical bonding by asking the question: *why*? Why are atoms not satisfied being by themselves—like individual islands in the stream? What is the underlying social nature of particular kinds of atoms that makes them come together and form molecules? The answer lies in their quantum structure. Among all the atoms in the periodic table, the noble gases are the most reclusive. Noble gas atoms, particularly the lighter ones, tend not to associate with others—even with others of their same kind, which is the reason that the noble gases have also been called the *inert gases*. The unusual stability of the noble gases arises from the fact that they have completely filled quantum shells. The electron configuration of Ne, for instance, is $1s^2 2s^2 2p^6$, with the $n = 1$ and $n = 2$ quantum levels completely filled with electrons. As a result, the ionization energy for Ne is significantly larger than that for its neighbors (Table 5.7). It takes a lot of energy to remove an electron from the outer shell of neon. The electron affinity of Ne is also less than that of its neighbors (Table 5.8). In fact, it actually costs energy for Ne(g) to accept an additional electron (the electron gain enthalpy is +29 kJ/mol). As a result of its large IE and small EA, Ne is unlikely to become an ion or to react with other atoms. This answers the question of why the element neon is not found in any molecule, but why do other atoms such as carbon love to form bonds?

The underlying reason that all bonds form is so that the atoms can achieve a lower energy configuration. Just as most people need each other for emotional stability, most atoms need to come together to form a more stable electronic configuration. Consider the H–H bond in dihydrogen, the simplest chemical bond that exists. Figure 6.1 shows the potential energy diagram for the diatomic molecule H_2 as a function of the internuclear separation, r. The energy at infinite separation has been set to zero. As the two nuclei approach one another, the nucleus of H_A begins to polarize the electron cloud of H_B, and vice versa. As a result of the

Van Arkel triangle of bonding. [Copyright Peter Wothers, used by permission.]

Principles of Inorganic Chemistry, First Edition. Brian W. Pfennig.

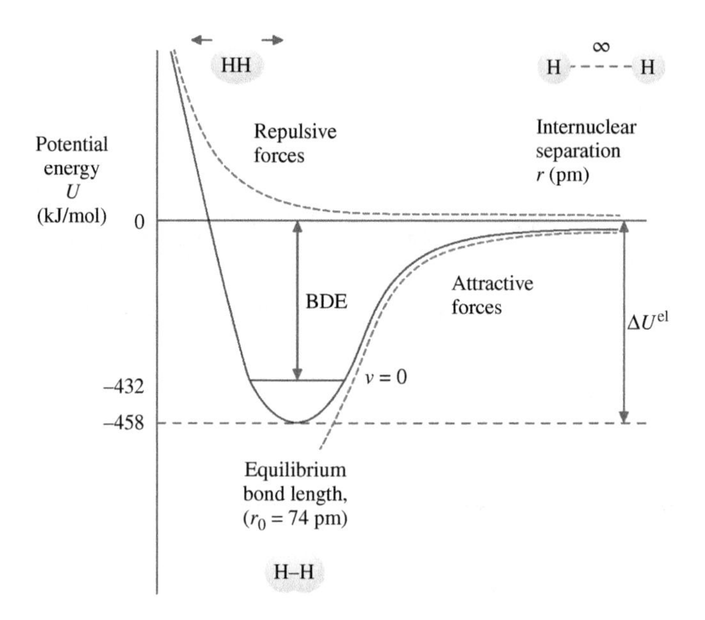

FIGURE 6.1
A potential energy diagram for the bonding in H_2 as a function of the internuclear separation, *r*.

attractive forces, the energy decreases until it reaches a minimum at r_0. As the two nuclei continue to approach each other, eventually nuclear–nuclear and electron–electron repulsions dominate and the overall energy rises abruptly. The minimum electronic internal energy, called ΔU_{el} (458 kJ/mol in this case), is a good first approximation of the bond dissociation energy (BDE), as shown in the diagram. (All molecules have a zero-point energy due to molecular vibration that is higher than the minimum of the curve, and thus the actual BDE will be less than the value shown in the diagram.) The internuclear separation at this minimum energy is known as the *equilibrium bond length*. For H_2, $r_0 = 74$ pm and BDE = 432 kJ/mol. In other words, the two moles of H atoms gain a total of 432 kJ whenever they form a mole of H_2 gas. At $r = r_0$, the electron clouds of the two H atoms overlap, so that the two electrons can be shared between the nuclei. This sharing of electrons between two or more nuclei is known as *covalent bonding*. In the case of H_2, the sharing allows each H atom to achieve the coveted electron configuration of its neighboring noble gas, He, which has a total of two electrons in its valence shell.

6.2 LEWIS STRUCTURES

In his landmark 1916 paper entitled "The Atom and the Molecule," the great American chemist Gilbert Newton Lewis described the nature of chemical bonding in terms of the sharing of valence electrons. The number of *valence electrons* is defined as the number of electrons that an atom contains outside of the last closed noble gas shell and ignoring any filled *d*- or *f*-subshells. Lewis noted that most stable molecules have an even number of valence electrons in their shells, an observation that he dubbed the "rule of two." Because most of the early period elements contained a maximum of eight valence electrons in their compounds, he also postulated a "rule of eight," which Irving Langmuir later referred to as the *octet rule*. As a result of the "rule of eight," Lewis originally visualized the valence shells of atoms as cubes with the electrons occupying positions at their corners, as shown in Figure 6.2. Inner shells consisted of smaller cubes inscribed within the larger valence cube. Atoms could form bonds by sharing electrons along the edges of their cubes—in order to fill a complete "octet" in their valence shell.

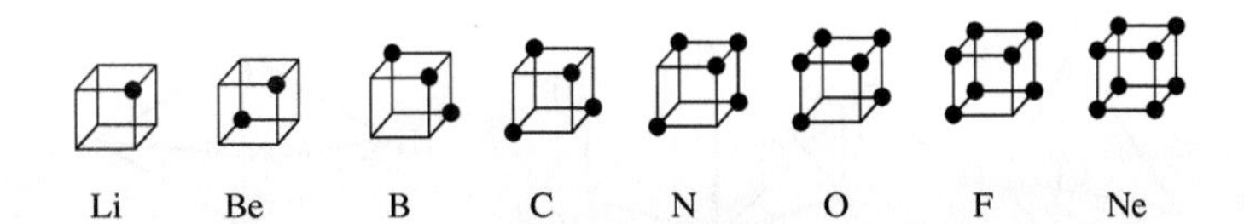

FIGURE 6.2
G. N. Lewis' early depiction of the valence shell of the period 2 elements.

In the same year that Lewis developed his theory of covalent bonds, Walther Kossel independently developed a parallel theory of ionic bonding. Noting that an alkali metal such as Na could achieve a stable noble gas configuration if it lost an electron and a halogen such as Cl could achieve a noble gas configuration by gaining an additional electron, Kossel proposed that the bonding in a molecule such as NaCl consisted of a pair of oppositely charged ions that were electrostatically attracted to one another: Na^+, Cl^-. Evidence supporting the nature of ionic bonding came from the fact that molten salts, such as NaCl, are electrical conductors because the ions are free to migrate in the liquid phase and that they can dissolve in solvents having a high dielectric constant to form solutions that also conduct electricity. The electrostatic forces in ionic solids act in all directions, so that ionic bonding is said to be nondirectional in nature.

The charge on an ion (or the charge that an atom assumes in a polyatomic ion or neutral molecule) is known as its *oxidation number*. For metals, the oxidation number is typically the same as the group number. Thus, for instance, Na is +1 and Mg is +2. For nonmetals, it is usually the negative of the number of electrons the atom must gain in order to achieve a noble gas configuration. Hence, O is −2 (except in peroxides) and F is −1. Elements in their naturally occurring forms (such as H_2, Cu, or S_8) have an oxidation number of zero. For any neutral molecule, the sum of the oxidation numbers on all of the individual atoms must equal zero. Typically, the more electronegative atom is assigned the more negative oxidation number. Thus, for example, the oxidation numbers of K, Mn, and O in $KMnO_4$ are 1, 7, and −2, respectively, so that the overall sum of the oxidation numbers $(1 + 7 + 4 * -2)$ is zero. Some atoms, particularly the transition metals, can take more than one oxidation number, depending on the compound in which they occur. Thus, for instance, the oxidation number of Mn is 4 in MnO_2, but it is 2 in $MnCl_2$.

The Lewis structures for a number of atoms, ions, and molecules are shown in Figure 6.3. Pairs of electrons (following the "rule of two") can be represented either as a pair of dots or as a single dashed line. Because most carbon-containing compounds have tetrahedral geometries, Lewis envisioned four pairs of electrons at the corners of a tetrahedron in these molecules in order to satisfy the "rule of eight," as shown in Figure 6.4. In some cases, multiple bonds are required to satisfy the "rule of eight." For example, ethylene contains two pairs of electrons between the C atoms to form a C=C double bond. Lewis envisioned ethylene as two tetrahedrons that shared an edge in common between the C atoms, so that the two pairs of electrons in the C=C double bond were bent away from each other like the banana-shaped bonds you might have seen in a general chemistry textbook. Thus, he was correctly able to explain why all six atoms in ethylene lie in the same plane. Likewise, the bonding in ethyne requires three pairs of electrons to be shared between the C atoms. In this case, Lewis envisioned the two carbon tetrahedra

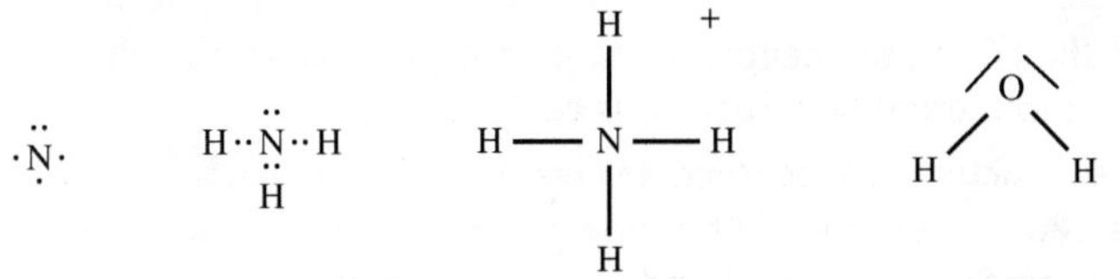

·N· H··N··H H H—N—H + H O H

FIGURE 6.3
Lewis structures for N, NH_3, NH_4^+, and H_2O.

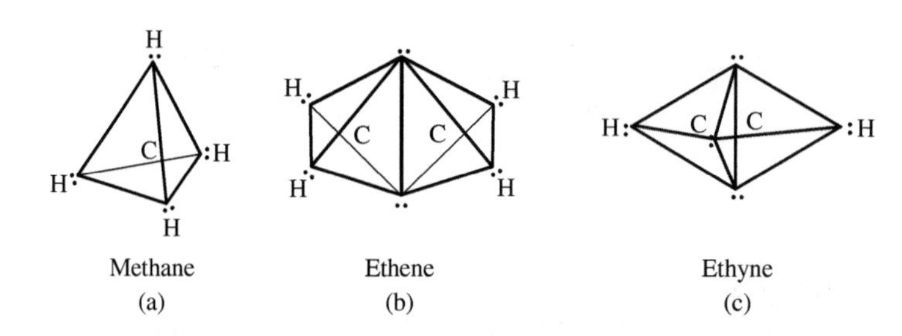

FIGURE 6.4
Lewis structures for (a) methane, (b) ethylene, and (c) ethyne, where the electrons are positioned at the corners of a tetrahedron for each carbon atom.

as sharing a face in common, which correctly predicts the linear geometry of this molecule.

While the precise definition of a Lewis structure has changed somewhat over the years, the essence of the concept is virtually the same. Lewis structures remain one of the single most powerful pedagogical tools in the understanding of chemical bonding. Modern Lewis structures for simple molecules can be constructed from the following set of rules.

Rules for drawing Lewis structures:

- Determine if the bonding is primarily ionic or covalent based on the difference in electronegativities between the elements. The larger the electronegativity difference, the more ionic the bonding. Typically, metals that combine with nonmetals are considered to be ionic, whereas nonmetals combining with nonmetals are principally covalent. If the bonding is ionic, treat each of the ions or polyatomic ions separately.
- Determine the total number of valence electrons by summing the valence electrons on each atom and adding one electron for each negative charge on an anion or subtracting one electron for each positive charge on a cation.
- Position the atoms so that there is a central atom (usually the least electronegative atom, but never H) surrounded by outer atoms called the *ligands*. This arrangement is commonly referred to as the *skeletal structure*.
- Determine a temporary electron distribution by connecting each ligand to the central atom with a single bond (a pair of electrons), arranging lone pairs of electrons around each ligand to satisfy the "rule of eight" (or "rule of two" if the ligand is H) and then placing any remaining electron on the central atom. This is known as the *provisional structure*.
- Calculate the formal charge (FC) on the central atom, as shown in Equation (6.1). The FC essentially compares the number of valence electrons that the free atom contains to the number of electrons that can be apportioned to that atom in the molecule. Any lone-pair electrons belong exclusively to that particular atom in a molecule, whereas any bonding-pair electrons are distributed equally between the atoms in the bond (hence the factor of one-half in the third term of Equation (6.1)):

$$FC = \begin{pmatrix} \text{valence} \\ e^- \text{ in the} \\ \text{free atom} \end{pmatrix} - \begin{pmatrix} \text{no. of lone} \\ \text{pair } e^- \text{ in} \\ \text{the molecule} \end{pmatrix} - \frac{1}{2}\begin{pmatrix} \text{no. of bonding} \\ \text{pair } e^- \text{ in} \\ \text{the molecule} \end{pmatrix} \quad (6.1)$$

  If the FC on the central atom is zero or equal to the charge of the polyatomic ion, then the structure is correct as it is.
- If the structure is incorrect, calculate the FC on each ligand. To obtain the best Lewis structure, shift a lone pair of electrons from the ligand with the most negative FC into a bonding position in order to form a multiple bond.

For Series 2 elements, continue this process until the central atom has a full octet. For all elements higher than Series 2, continue this process until the central atom has an FC of zero.

Example 6-1. Using the above-mentioned rules, sketch the best Lewis structure for (a) CH_2O, (b) PCl_5, (c) BH_3, and (d) CN^-.

Solution. (a) CH_2O has $4+2(1)+6=12$ valence electrons. Carbon is the central atom because it is the least electronegative of C and O (recall that H can never be the central atom). The provisional structure is shown at left, below. The FCs are calculated as follows: (C) $=4-0-0.5(6)=1$, (O) $=6-6-0.5(2)=-1$, and (H) $=1-0-0.5\ (2)=0$. Because the FC on C is not zero, a multiple bond needs to be made between C and the element with the most negative FC (in this case, O). The revised and correct structure of CH_2O is shown at right. In this structure, the central C atom has a complete octet.

Provisional structure

Correct structure

(b) PCl_5 has $5+5(7)=40$ valence electrons. Phosphorous is the central atom because it is the least electronegative of the two. The provisional structure is shown below. The FCs are calculated as follows: (P) $=5-0-0.5(10)=0$ and (Cl) $=7-6-0.5(2)=0$. This structure is therefore already correct. Notice that the central atom has more than eight valence electrons around it. This violation of the "octet rule" is allowed for the central atom if it is an element beyond Series 2. While the traditional explanation for *hypervalent molecules* such as PCl_5 is that they can use their low-lying *d* orbitals in the bonding, modern quantum mechanical calculations show only a minimal contribution from the 3*d* orbitals. Hypervalent molecules typically occur when there is a large difference in electronegativity between the central atom and the ligand, so that the actual bonding is polar covalent and the valence electrons are not shared equally between the two atoms. Lewis himself regarded the "rule of two" as more fundamental than the "rule of eight." It was his contemporary, Irving Langmuir, who coined the phrase "octet rule" and who perhaps overemphasized the importance of this rule. Although Langmuir was a gifted lecturer and did much to popularize Lewis's ideas, the two scientists were often at loggerheads with one another over who deserved the most credit for the discovery of the covalent bond. In fact, when Lewis died in his Berkeley laboratory in the spring of 1946 while working with liquid HCN, some senior members of the faculty believed that it might have been a suicide. Lewis had just returned to the lab following a rather depressing luncheon he had with his nemesis Langmuir, who was visiting the campus that day to receive an honorary degree. While Langmuir had won the 1932 Nobel Prize in chemistry for his work on surface chemistry, Lewis—whose contributions to science included an explanation of covalent bonding, his definition of Lewis acids and bases, and numerous contributions in thermodynamics—never won the prize himself.

(c) BH_3 has a total of six valence electrons. Boron is the central atom. The provisional structure has single bonds between the B and each H atom, as shown below. The FC on B is $3 - 0 - 0.5(6) = 0$ and on H it is $1 - 0 - 0.5(2) = 0$. Therefore, this is already the best Lewis structure, even though B has less than a full octet. BH_3 is considered an electron-deficient molecule because it has less than the optimum eight valence electrons around the central atom. As a result, BH_3 is a good Lewis acid (electron pair acceptor) because the addition of a pair of electrons to form a donor–acceptor compound would now satisfy the boron's "rule of eight."

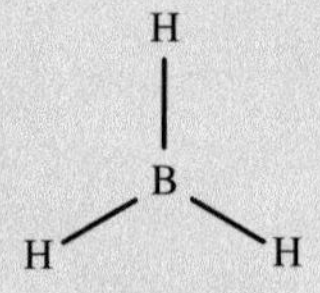

(d) CN^- has $4 + 5 + 1 = 10$ valence electrons. Carbon is the central atom. The provisional structure is shown below, at left. The FCs at this point are $(C) = 4 - 2 - 0.5(2) = 1$ and $(N) = 5 - 6 - 0.5(2) = -2$. By shifting a lone pair on N into a bonding position to make a C=N double bond, the FCs become $(C) = 4 - 2 - 0.5(4) = 0$ and $(N) = 5 - 4 - 0.5(4) = -1$. However, this structure still would not satisfy the "rule of eight" for C. Thus, we must shift another lone pair on N into a bonding position to form the triply bonded Lewis structure shown below, at right. The FCs are now $(C) = 4 - 2 - 0.5(6) = -1$ and $(N) = 5 - 2 - 0.5(6) = 0$. Although the more electronegative atom usually gets the more negative FC, the fact that a proton reacts with the cyanide ion to make HCN and not CNH lends credence to the correct placement of the negative FC on cyanide as lying on the C atom.

Provisional structure

Correct structure

6.3 COVALENT BOND LENGTHS AND BOND DISSOCIATION ENERGIES

As a rule, the more electron pairs holding the atoms together in a chemical bond, the shorter the bond length and the larger the BDE. The same is also true of human relationships: the more you share with your significant other, the closer you will be to each other and the stronger the bonding between you. Table 6.1 lists some typical

TABLE 6.1 Average bond lengths and bond dissociation energies (BDEs) for selected types of covalent bonds.

Bond Type	Average Bond Length (pm)	Average BDE (kJ/mol)	Bond Type	Average Bond Length (pm)	Average BDE (kJ/mol)
H–H	74	432	N–N	145	167
C–H	109	435	N–O	140	201
N–H	101	386	N–F	136	283
O–H	96	459	P–P	221	201
Si–H	148	318	P–O	163	335
P–H	144	322	P–F	154	490
S–H	134	363	P–Cl	203	326
H–F	92	565	O–O	148	142
H–Cl	127	428	O–F	142	190
H–Br	141	362	S–F	156	284
H–I	161	295	C=C	134	602
F–F	142	155	C≡C	120	835
Cl–Cl	199	240	C=N	129	615
Br–Br	228	190	C≡N	116	887
I–I	267	148	C=O	120	799
C–C	154	346	C≡O	113	1072
C–N	147	305	N=N	125	418
C–O	143	358	N≡N	110	942
C–F	135	485	N=O	121	607
C–S	182	272	O=O	121	494
C–Cl	177	327	P=O	150	544
C–Br	194	285	S=O	143	522

bond lengths and BDEs for a variety of covalently bonded molecules. It is important to recognize that the values in the table are the averages of those for many different compounds and that there is indeed quite a bit of variation in these values from one molecule to the next. Thus, for instance, the BDE for the C–H bond in Br_3CH is 380 kJ/mol, whereas the same exact type of bond in CH_4 has a BDE of 435 kJ/mol. According to Table 6.1, a typical C–H bond will require about 411 kJ/mol of energy on average to break.

The exact BDE depends on a variety of factors, including the degree of orbital overlap, the size (or diffuseness) of the overlapping orbitals, the number of electron pairs between the atoms, crystal packing effects, and the percent ionic character (which is related to the electronegativity difference between the two atoms). For instance, the bond lengths in the homonuclear diatomics F_2, Cl_2, Br_2, and I_2 increase down the column as the radius of the individual atoms increases. The BDEs for the same series of compounds decreases down the column as the orbitals become more diffuse (fluorine being the exception due to the uniqueness principle). For multiply bonded atoms, the average bond length decreases and the BDE increases as the number of shared electrons increases. Thus, C–C, C=C, and C≡C have average bond lengths of 154, 134, and 120 pm, respectively, whereas their respective average BDEs are 346, 602, and 835 kJ/mol. The BDEs for the triply bonded series C≡C, C≡N, and C≡O are 835, 887, and 1072 kJ/mol, respectively. This increase across the series is due to the decreasing atomic radius of the heteroatom and the larger percent ionic character in the bonding. As we will see later, the Lewis model, which emphasizes the localization of electron pairs between two and only two atoms, is an oversimplification. In reality, electrons are not always paired; furthermore, it is possible for one or more electrons to be delocalized over more than just two atomic nuclei.

6.4 RESONANCE

Often, it is possible to draw more than one valid Lewis structure for a molecule. In such cases, the true bonding can be considered as a *resonance hybrid* of the different canonical forms. Consider, for example, the bonding in ozone, O_3. Ozone contains a total of 18 valence electrons. Following the rules for Lewis structures, either of the two canonical forms shown in Figure 6.5 is possible. Notice that each *canonical structure* contains one O–O single bond and one O=O double bond. One might therefore expect to see two different bond lengths and two different BDEs in O_3. A typical O–O single bond has a bond length of 148 pm and a bond energy of 142 kJ/mol, whereas the corresponding values for a typical O=O double bond are 121 pm and 494 kJ/mol, respectively. Spectroscopic measurements on ozone indicate, however, that both of the O–O bond lengths are 128 pm, a value that is intermediate in length between that of a single and a double bond. Furthermore, the total BDE is 605 kJ/mol, somewhat less than the sum of 616 kJ/mol expected on the basis of one O–O single and one O=O double bond.

Thus, the true bonding in ozone is a resonance hybrid of the two contributing canonical forms. Neither structure is correct in and of itself, nor are the two structures oscillating back and forth with one another. The real bonding is a 50/50 hybrid of the two canonical forms, where each bond length and each BDE is intermediate between an O–O single bond and an O=O double bond. One can make the analogy to mixing the primary colors yellow and blue to make green. The resulting green paint is neither yellow nor blue. It is also not sometimes yellow and other times blue. It is an equal mixture (or hybrid) of the two primary colors.

When the canonical forms all contribute equally to the bonding, they are called *equivalent canonical forms*; when they contribute unequally, they are nonequivalent canonical forms. Therefore, the structures shown in Figure 6.5 for ozone represent equivalent canonical forms. They are related to each other by a symmetry element—in this case, a twofold rotational axis that passes through the central oxygen. As a result, each canonical structure contributes exactly 50% to the resonance hybrid. The double-headed arrow is used to indicate the concept of resonance. The carbonate ion, shown in Figure 6.6, is another example of equivalent canonical forms; only in this case, each canonical structure contributes 33.3% to the resonance hybrid.

The resonance structures shown in Figures 6.5 and 6.6 are equivalent canonical forms because they are related to each other through a symmetry element.

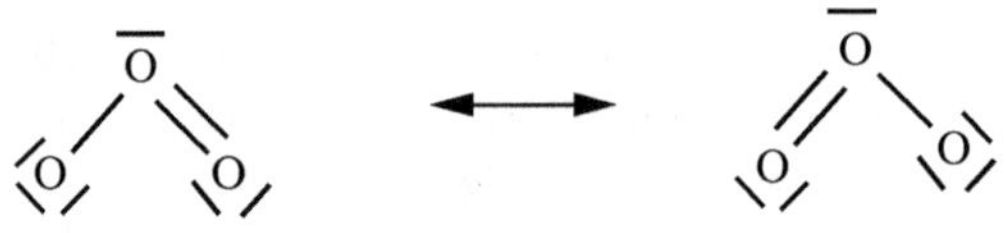

FIGURE 6.5
The two equivalent canonical forms of ozone. The true bonding is a resonance hybrid of the two, with a 50% contribution from each.

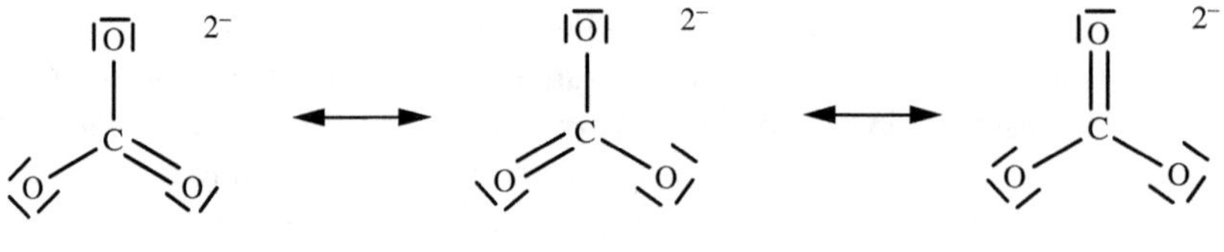

FIGURE 6.6
The three equivalent canonical forms for the carbonate ion. Each contributes 33.3% to the resonance hybrid.

For example, rotation of the ozone molecule by 180° around an imaginary axis that bisects the O–O–O bond angle leads to an equivalent configuration for the molecule. In each case, there is one O–O single bond and one O=O double bond in the contributing canonical structures; the only difference between them is the placement of the single and double bonds. Likewise, in the carbonate ion, rotation of each subsequent canonical structure by 120° or 240° leads to a structure that is identical to the original. As we see in a later chapter, symmetry plays an important role in the structure and physical properties of molecules.

The canonical structures of nonequivalent resonance forms, on the other hand, are not related by a symmetry element. Consider the three possible Lewis structures for SCN^- shown in Figure 6.7. There is no rotational axis about which the canonical forms can be interchanged to form an equivalent configuration. In the case of nonequivalent canonical structures, the weighting is unequal and some of the canonical structures will make a larger contribution to the resonance hybrid than will others.

While it is impossible to determine the exact percent contribution of each canonical form *a priori*, the FCs on each atom can be used to determine a relative weighting. The canonical form that contributes most to the resonance hybrid will be (in no particular order of importance) the one which

- has the least number of atoms with nonzero FCs,
- favors the nonzero FCs to be as close to zero as possible,
- places the negative FCs on the more electronegative atoms,
- maximizes close +/– (attractive) FCs in the structure, and
- minimizes close +/+and –/– (repulsive) FCs in the structure.

This set of rules essentially states that the electronic distribution around each atom in the molecule will tend to be the same as the electronic configuration of the free atom. Furthermore, as electronegativity is defined as the ability of an atom in a molecule to attract the shared electrons closer to itself, those atoms having larger electronegativities will be the ones that tend to acquire a negative FC.

Of the three feasible Lewis structures for the thiocyanate ion shown in Figure 6.7, structure A is the smallest contributor to the resonance hybrid because it has the largest number of nonzero FCs and a fairly large value for the negative FC (–2). Structures B and C are better, as they have only a single nonzero FC. Of these two, structure B will be the largest contributor to the resonance hybrid because it places the negative FC on the more electronegative N atom.

A comparison of the experimental bond lengths for SCN^- with the typical bond lengths shown in Table 6.2 supports this assessment. The S–C bond length in thiocyanate is 165 pm, which is intermediate between a typical S–C single and double bond. The experimental C–N bond length adds credence to the suggestion that structure B contributes more to the actual bonding than structure C does. At the same time, the C–N bond length in SCN^- is 117 pm, which falls between a typical C=N and C≡N bond. This is exactly the expected result if canonical forms B and C are the dominant resonance contributors. Furthermore, protonation of the thiocyanate ion occurs on the nitrogen end. Because the FC on N is –1 in the dominant

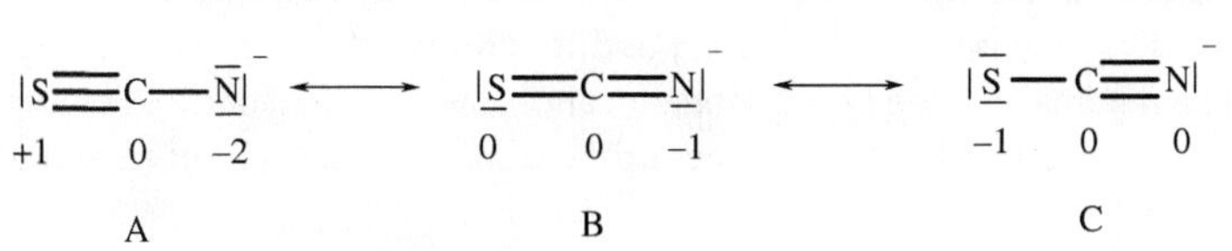

|S≡C—N̄|⁻ ⟷ |S̄=C=N̄|⁻ ⟷ |S̄—C≡N|⁻

+1 0 –2   0 0 –1   –1 0 0

A   B   C

FIGURE 6.7
Three nonequivalent canonical forms for the thiocyanate ion. Canonical structure B makes the largest contribution to the resonance hybrid, whereas canonical form A contributes the least.

TABLE 6.2 Average bond lengths for the S–C and C–N bonds in a variety of compounds.

Bond Type	Average Length (pm)	Bond Type	Average Length (pm)
S–C	182	C–N	147
S=C	160	C=N	129
		C≡N	116

canonical structure, it is entirely logical that the proton will prefer to bind to the N-terminus. If canonical form C had been the dominant contributor, then the proton would have attached to the sulfur end of the thiocyanate ion.

Example 6-2. Sketch any reasonable canonical forms of N_2O and identify their relative contributions to the overall resonance hybrid. The experimental bond lengths for the N–N and N–O bonds in N_2O are 113 and 119 pm, respectively.

Table of typical bond lengths:

Bond Type	Average Length (pm)	Bond Type	Average Length (pm)
N–N	145	N–O	140
N=N	125	N=O	121
N≡N	110		

Solution. The following Lewis structures are valid canonical forms:

(a)		(b)		(c)
0 +1 −1		−1 +1 0		−2 +1 +1
|N≡N—O|	⟷	|N=N=O	⟷	|N—N≡O|

Given the typical bond lengths in the table, the N–N bond length in N_2O is closest to that for a typical N≡N triple bond, whereas the N–O bond length is closest to that for a typical N=O double bond. Of the three canonical forms, the experimental bond lengths suggest that structures A and B contribute a larger percentage to the resonance hybrid than does structure C. This is the same as the result predicted using the rules for FCs. Structures A and B have their FCs closest to zero while also maximizing the +/− attractive forces on neighboring atoms. Structure C has three nonzero FCs, a large (−2) FC, and two +/+ repulsive charges right next to each other; and its contribution to the resonance hybrid will be minimal. Of the two dominant canonical forms, Structure A is probably the larger contributor because it places the negative FC on the more electronegative O atom.

6.5 POLAR COVALENT BONDING

A pure covalent bond, where the electrons are shared equally between the two nuclei, only occurs in homonuclear diatomic molecules, such as H_2, F_2, and O_2. For any heteronuclear molecule, there is always a certain percent ionic character to the bonding due to the difference in electronegativity between the atoms. The recognition that there was a gray area between covalent and ionic bonding led Anton Eduard van Arkel and J. A. A. Ketelaar to develop triangles of bonding in the 1940s, such as the one shown in Figure 6.8.

The original 1941 van Arkel triangle of bonding recognized that there were three main types of chemical bonding, each of which was placed at one of the corners of the triangle. The *y*-axis in Figure 6.8 is proportional to the difference in electronegativity of the atoms involved in the bonding, whereas the *x*-axis represents the average electronegativity of the atoms in the bond. Metallic bonding occurs when there is a negligible or zero difference in electronegativity and the average electronegativity of the atoms is small. Metallic bonding occurs because metals having low ionization energies lose their valence electrons to form cations, where the ionized electrons can be delocalized over the entire metallic solid. Metallic bonding occurs between metal atoms of the same type or between two or more different metals (in which case they are called *alloys*). Metals are characterized by their strength, luster, ductility, malleability, and thermal and electrical conductivity. Metallic bonds are nondirectional, which causes them to have high melting points and boiling points.

As the average electronegativity increases across a row of the periodic table, there is a gradual transition from metallic to covalent bonding as one approaches the metal-to-nonmetal transition. Covalent bonding typically occurs between nonmetal atoms from the right-hand side of the periodic table, where there is a negligible or zero difference in electronegativity between the atoms. In a covalent bond, the valence electrons are shared between atoms. These types of bonds are typically directional in nature because the bonding pair or pairs of electrons are localized between the atoms in the bond. As a result, covalent compounds typically have much lower melting points and boiling points than do metallic bonds.

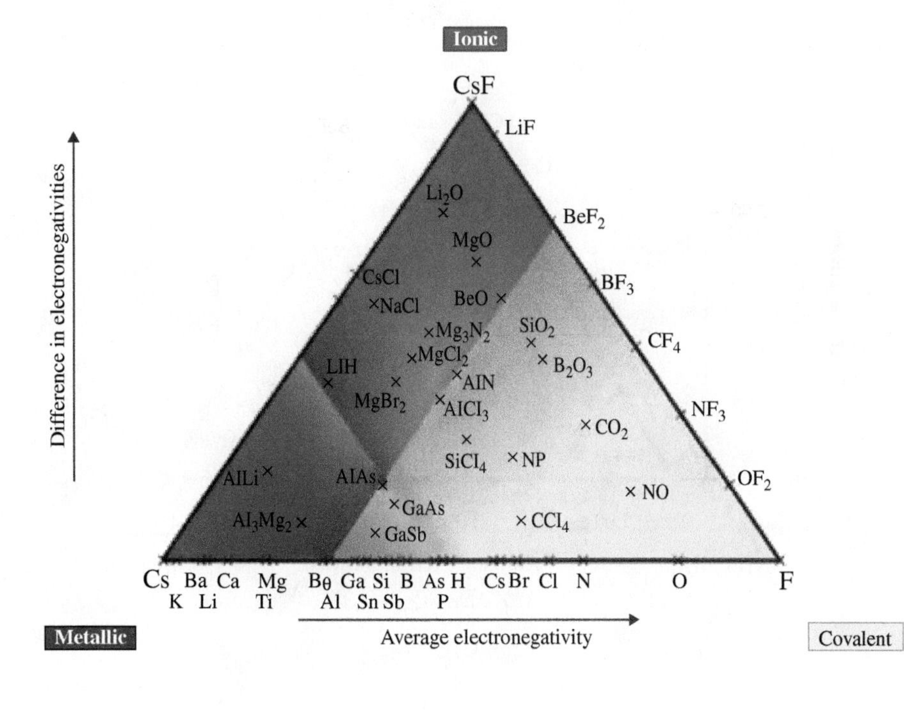

FIGURE 6.8
A typical van Arkel–Ketelaar triangle of bonding based on electronegativity. [Copyright Peter Wothers, used by permission.]

Finally, ionic bonds are formed between elements from opposite sides of the periodic table (typically between a metal and a nonmetal), where there is a large difference in electronegativity between the atoms. Because metals have very low IEs and nonmetals have large EAs, an ionic bond is characterized by the transfer of one or more electrons from one atom to another to form a cation–anion pair. As a general rule, the nondirectional nature of the electrostatic attraction between the ions leads to fairly high melting and boiling points. Most ionic solids are insulators because the ions are fixed in place in the crystalline lattice; however, they become electrical conductors when molten or dissolved in aqueous solution.

Modern triangles of bonding, such as the one shown in Figure 6.8, are quantitative in nature. The x-axis is the average electronegativity of the atoms involved in the bonding. Because electronegativity increases across a row of the periodic table, metallic bonding is therefore on the left-hand side of the triangle of bonding, whereas covalent bonding lies to the right. The y-axis is the difference in electronegativity between the two atoms. Pure metallic and covalent bonds occur at the base of the triangle, where there is zero difference in electronegativity between the atoms. Ionic bonding lies at the apex, where the largest difference of electronegativity occurs. The quantitative Norman triangle of bonding, which uses Lee Allen's electronegativity as its basis, is shown in Figure 6.9.

Polar covalent molecules lie along the right edge of the triangle of bonding along the line that connects covalent and ionic bonding. Because of the difference in electronegativity between the central atom and the ligands, the bonding electrons are shared unequally between the atoms. By definition, the more electronegative atom in the bond attracts the shared electrons closer to itself, thereby developing a partial negative charge. In a diatomic molecule, the more electropositive atom develops a partial positive charge of the same magnitude as a result of this deficiency in electron density. Partial charges are usually represented by the symbol δ, as shown in Figure 6.10, and are sometimes referred to as the *atomic charges (ACs)*.

ACs can be calculated from quantum mechanics by partitioning the wave function or on the basis of calculated electron densities. The AC on an atom in a polar covalent molecule is always intermediate between its oxidation number (which

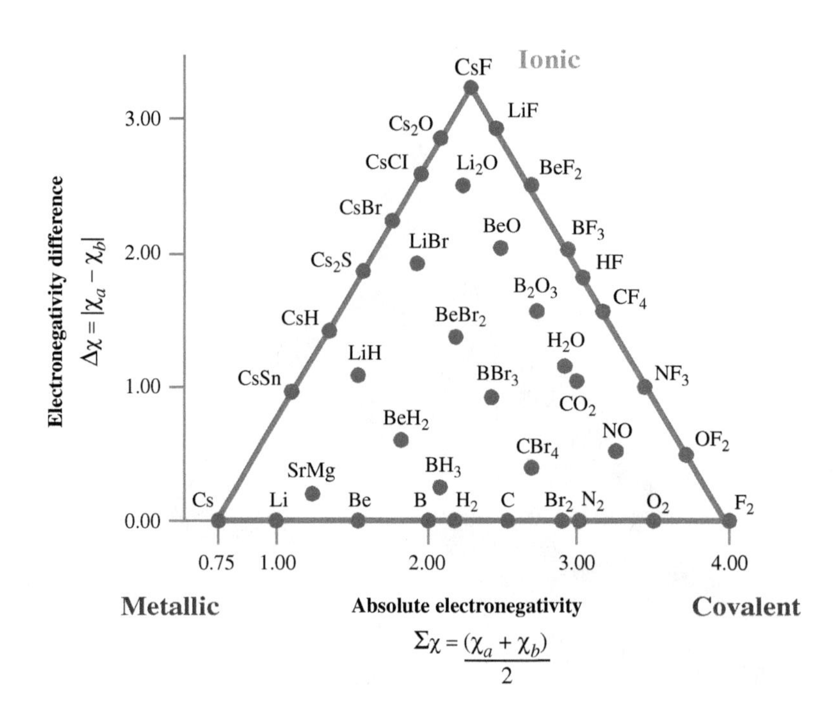

FIGURE 6.9
Norman's quantitative triangle of bonding. [From the *Chemogenesis Web Book*, © Mark R Leach, meta-synthesis.]

assumes that the bonding is 100% ionic) and its FC (which assumes that the bonding is 100% covalent). The *percent ionic character* of the bond can be estimated using a formula developed by Linus Pauling and given by Equation (6.2):

$$\% \text{ ionic character} = \frac{0.33(\chi_A - \chi_B)^2}{1 + 0.33(\chi_A - \chi_B)^2} * 100\% \tag{6.2}$$

Example 6-3. Using the revised Pauling electronegativities from Table 5.12 and applying Equation (6.2), calculate the percent ionic character for each of the following molecules: (a) HF, (b) HI, (c) LiF, and (d) $BeCl_2$.

Solution. Applying Equation (6.2) to HF, for example, yields the following result:

$$\% \text{ ionic character} = \frac{0.33(3.98 - 2.20)^2}{1 + 0.33(3.98 - 2.20)^2} * 100\% = 51.1\%$$

The calculated percent ionic character for each example is as follows: (a) HF 51.1%, (b) HI 6.53%, (c) LiF 74.8%, and (d) 45.5%. The larger percent ionic character in HF than in HI is part of the reason why HF is a weak acid while HI is a strong acid. The larger ACs in HF help strengthen the bonding so that it will not as easily dissociate a proton. Such calculations raise the interesting question as to where to draw the "cutoff" line between an ionic bond and a covalent bond in a polar covalent molecule. At what level of percent ionic character do we dispense with representing a molecule with a line representing the covalent bond (as in Li–F or Cl–Be–Cl) and instead write the molecule as ionic (as Li^+F^- or $Cl^-Be^{2+}Cl^-$)? Naturally, there is no one right answer to this question—it is more of a judgment based on chemical intuition. For example, most chemists would probably write LiF as the ionic Li^+F^- but $BeCl_2$ as the covalent Cl–Be–Cl.

It has long been recognized that polar covalent molecules have larger BDEs than the average BDE of the corresponding homonuclear compounds. In fact, this difference in energy, known as the *ionic resonance energy*, was the basis for Linus Pauling's original electronegativity scale, as demonstrated by Equations (5.41) and (5.42). This is because the electrostatic attraction of the partial charges helps strengthen the covalent bonding. For the same reason, the bond lengths of polar covalent bonds are also shorter than the sum of their covalent radii. Comparing the isoelectronic N_2 and CO species with each other, while both molecules are triply bonded, the BDE of CO (1070 kJ/mol) is considerably larger than that of N_2 (946 kJ/mol). This difference is largely a result of the increased percent ionic character in carbon monoxide.

Of course, it is also possible to initiate a discussion of polar covalent bonding from the opposite apex of the triangle of bonding, beginning with the ionic point of view and introducing covalent character into the mix. The sketches in Figure 6.11 help illustrate the transition between a purely ionic and a polar covalent bond. In a strictly ionic bond, the ions can be considered as hard spheres whose electron clouds cannot interpenetrate with one another. When the cation and anion have identical ionic radii, as shown in Figure 6.11(a), the hard-spheres, ionic model is often

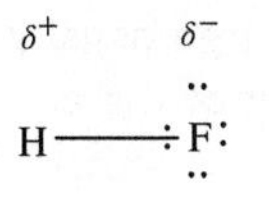

δ^+ δ^-

H—⟶F:

FIGURE 6.10
Polar covalent bonding in the HF molecule.

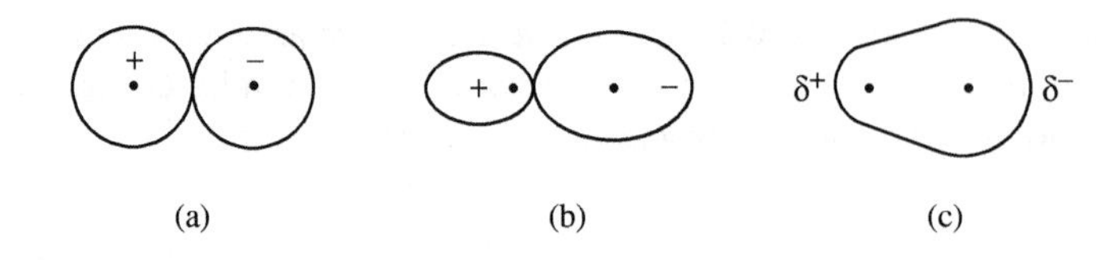

FIGURE 6.11
Fajans' classification of the transition between ionic and covalent bonding: (a) ions that exist with no polarization whatsoever and which can be treated as charged hard-spheres in a purely ionic manner, (b) the mutually polarized ion pair that results when the electron clouds interact with the opposite nucleus, and (c) sufficient polarization of the electron clouds so that they overlap to form a polar covalent bond.

a good first approximation. As the sizes of the two ions begin to differ, however, the smaller size of the cation causes an increase in its charge density, which enhances the cation's ability to polarize the electron cloud of the larger and more diffuse anion, as shown in Figure 6.11(b). Eventually, the degree of polarization increases to the point where there is significant overlap of the electron clouds of the two atoms, as shown in Figure 6.11(c). This sharing of electron density between the two nuclei is the hallmark of a covalent bond. Because the electron density is shared between the nuclei in Figure 6.11(c), the concept of individual ions is no longer relevant. Instead, the bonding is best described as polar covalent in nature, where a partial positive charge resides on the nucleus that was previously classified as the cation and a partial negative charge lies on the nucleus that was previously classified as an anion.

In 1923, Kasimir Fajans established a qualitative set of rules (known as *Fajans' rules*) that allows one to predict the relative degree of covalent character in this intermediate arena of bonding. The covalency of the bond will increase with the degree of polarization. This, in turn, will increase with (i) cations having large charge densities, (ii) anions having diffuse electron clouds, and (iii) cations that have partially filled *d* orbitals.

Small, highly charged cations (such as Al^{3+} or Be^{2+}) have large charge densities (Z/r) and can better polarize the electron clouds of their neighbors. These types of ions are sometimes referred to as *hard* ions because the nucleus holds its electrons rather tightly and acts to a certain degree like a point positive charge. The electron clouds of anions that have a large negative charge and a large radius are highly polarizable. These types of ions are considered "soft" because their electron clouds are rather squishy and easily distorted. The more negative the charge, the more electrons the anion will possess and the less tightly the nucleus will be able to hold onto them. Likewise, the larger the radius of the anion, the farther away from the nucleus's influence the average electron will be. When a hard cation is positioned in the vicinity of a soft anion, the highly concentrated positive charge of the cation polarizes the electrons in the loosely held electron cloud of the anion so that they spend more of their time (on average) lying on the side of the cation, as shown in Figure 6.11(b). The surplus of electron density in the anion that is oriented toward the cation, in turn, polarizes the electron cloud of the cation, repelling the cation's electrons to the side opposite the anion. The degree of this mutual polarization increases even further with the polarizability of the anion. In extreme cases, the polarization is so large that the electron clouds overlap, as shown in Figure 6.11(c). The electrons are now shared (albeit unequally) between the two nuclei, so that it is impossible to distinguish which electrons belong to the cation and which belong to the anion. Thus, the bonding is primarily covalent in nature.

A third factor that increases the covalency is the specific electron configurations of the cations. If two cations have identical charge densities, then the cation containing a partially filled *d*-subshell will be the more polarizing. This is because the electrons in *d* orbitals are very poor at screening the positive charge of the cation from the electron cloud of the anion. Consider, for example, a comparison of Hg^{2+} ($r = 116$ pm) and Ca^{2+} ($r = 114$ pm). The charge densities of these two cations are essentially equal. If anything, the charge density of Ca^{2+} is slightly larger. However, the Hg^{2+} ion is the more polarizing of the two. This is because the electron configuration of Hg^{2+} is [Xe] $5d^{10}$, whereas that of Ca^{2+} is [Ne] $3s^2 3p^6$. The *s* and *p* orbitals are much better screeners than are the *d* orbitals. Thus, the nuclear charge of Hg^{2+} will be felt more strongly by the electron cloud of the anion than will the nuclear charge of Ca^{2+}.

The effects of increasing covalency on the physical properties of ionic solids are numerous. Consider, for example, the melting points of the lithium halides (845, 605, 550, and 449 °C for LiF, LiCl, LiBr, and LiI, respectively). The larger, more polarizable the anion is, the more covalent the bonding will be and the lower the melting point. The iodide ion is much more polarizable than fluoride—hence, the nearly twofold difference in their melting points. A similar trend exists for the series of bromides NaBr, $MgBr_2$, and $AlBr_3$ (with melting points of 747, 700, and 97.5 °C, respectively). The larger the charge density of the cation, the more polarized the electron cloud of the bromide will become, lowering the molecule's melting point. The degree of covalency also influences the solubility in aqueous solution. Silver fluoride, for instance, is freely soluble in water, whereas the K_{sp}s for AgCl, AgBr, and AgI continue to decrease as the degree of covalent character in the bonding increases (2×10^{-10}, 5×10^{-13}, and 8×10^{-17}, respectively). The more polarizable the anion, the more covalent the bonding, and the less likely the substance will dissociate into ions in an aqueous solution.

EXERCISES

6.1. What is the oxidation number of each atom in each of the following: (a) NH_3, (b) NO_2, (c) NO_3^-, (d) SO_3^{2-}, (e) SO_4^{2-}, (f) $Na_2S_2O_3$, (g) $XeOF_4$, (h) Cr_2O_3, (i) $Cr_2O_7^{2-}$, and (j) HOCl?

6.2. Sketch the best Lewis structure for each of the following molecules or ions: (a) PF_6^-, (b) HCP, (c) BCl_3, (d) PCl_3, (e) NO^+, (f) CO_2, (g) SO_2, (h) $CHCl_3$, (i) BrF_3, and (j) NOF.

6.3. Sketch any feasible canonical structures for each of the following molecules or ions: (a) N_3^-, (b) SO_3^{2-}, (c) $SOCl_2$, (d) $XeOF_2$, (e) NO_2^-, (f) SO_2Cl_2, (g) SCN^-, (h) POF_3, (i) SNF_3, And (j) $IO_2F_2^-$. Indicate the FCs on each atom in every canonical structure. Then circle the canonical structure that is the largest contributor to the resonance hybrid. If there are equivalent canonical forms, circle all of them.

6.4. The fulminate ion, CNO^-, where N is the central atom, is highly unstable. In fact, metal fulminates are explosive and have been used in percussion caps. Sketch all the possible resonance structures for the fulminate ion, determine the FCs on each atom in every canonical structure, and discuss why you think the fulminate ion is so unstable.

6.5. Which of the following isomers of N_2CO is the most stable: NOCN, ONNC, or ONCN? Use FCs to explain your answer.

6.6. Explain why PCl_5 is a stable molecule, while NCl_5 is not.

6.7. Lee Allen has defined the following formula for calculating the AC of an atom in a molecule that takes into account the electronegativity differences between the ions in the way that it apportions the bonding pair electrons. He calls these ACs Lewis–Langmuir charges. Using the formula below, calculate the AC of H in HF, HCl, HBr, and HI and then compare these values to the percent ionic character calculated using Equation (6.2). Is there a linear correlation between the two? [Reference: Allen, L. C. *J. Am. Chem. Soc.*, (1989), *111*, 9115]

$$\text{AC} = \begin{pmatrix}\text{valence}\\ \text{e}^- \text{ in the}\\ \text{free atom}\end{pmatrix} - \begin{pmatrix}\text{no. of lone}\\ \text{pair e}^- \text{ in}\\ \text{the molecule}\end{pmatrix} - 2 * \sum_{\text{all bonds}} \frac{\chi_A}{\chi_A + \chi_B}$$

6.8. Using the formula in Problem 7, calculate the ACs for each atom in each of the following molecules or ions: (a) CO_2, (b) ClF, (c) CH_2O, (d) HCN, and (e) CH_2Cl_2.

6.9. Provide a reasonable explanation for the relative BDEs of C≡C, C≡N, and C≡O in Table 6.1.

6.10. Using the revised Pauling electronegativities from Table 5.12 and applying Equation (6.2), calculate the percent ionic character for each of the following molecules: (a) KCl, (b) BCl_3, (c) H_2O, (d) OF_2, and (e) SF_6.

6.11. Explain why the melting point of CuCl is 430 °C, while that of NaCl is nearly twice as large at 801 °C.

6.12. Predict whether $HgCl_2$ or $CaCl_2$ will have the higher melting point and explain your answer.

6.13. For each pair, predict which compound will be expected to have the more covalent bonding and explain your answer: (a) $FeCl_2$ or $FeCl_3$, (b) $FeCl_3$ or $AlCl_3$, and (c) $AlCl_3$ or BCl_3.

BIBLIOGRAPHY

1. Gillespie RJ, Popelier PLA. *Chemical Bonding and Molecular Geometry*. New York: Oxford University Press; 2001.
2. Huheey JE, Keiter EA, Keiter RL. *Inorganic Chemistry: Principles of Structure and Reactivity*. 4th ed. New York: Harper Collins College Publishers; 1993.
3. McQuarrie DA, Simon JD. *Physical Chemistry: A Molecular Approach*. Sausalito, CA: University Science Books; 1997.
4. Miessler GL, Tarr DA. *Inorganic Chemistry*. 4th ed. Upper Saddle River, NJ: Pearson Education Inc.; 2011.

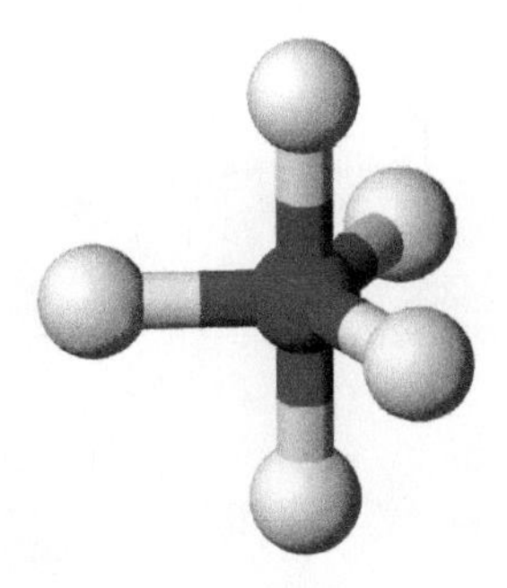

Molecular Geometry | 7

"It is structure that we look for whenever we try to understand anything."
—Linus Pauling

7.1 THE VSEPR MODEL

One of the most difficult tasks for students of chemistry to grasp is the visualization of the atomic world. Suppose that a student was to consume some mysterious bottle of liquid like a modern-day Alice in Wonderland that could shrink her down to the molecular level. How might these connections that hold the atoms together in molecules then appear to her? Would they look like balls on the ends of spokes, like the illustration at the beginning of this chapter, or would they be more bloated-looking like a CPK space-filling model of van der Waals interactions? (Incidentally, the C, P, and K in CPK models stand for Robert Corey, Linus Pauling, and Walter Koltun, who pioneered their use in 1952). Because we cannot visualize a chemical bond (at least not directly), it is a particularly challenging endeavor to imagine what a chemical bond might actually look like. As the theoretical chemist Charles Coulson once said: "One is almost tempted to say … at last I can almost see a bond. But that will never be, for a bond does not really exist at all: it is a most convenient fiction." The Lewis–Langmuir model presented in the previous chapter allows us to predict in two dimensions the number of bonding and nonbonding electrons in molecules containing predominantly covalent, directional bonding. However, we are usually more interested in their molecular geometries, for it is the shape of molecules that is relevant to such important applications as enzyme–substrate binding, host–guest chemistry, stereochemistry and chirality, catalysis, and rational drug design. Thus, the goal of this chapter is to move beyond the "Flat Stanley" world of Lewis structures to the stereoscopic world of 3D.

Although he was better known for his work on solutions, for which he won the very first Nobel Prize in chemistry in 1901, Jacobus Henricus van't Hoff also made important contributions in the field of molecular structure. As early as 1874, van't Hoff argued that the existence of certain isomers (optical isomers such as those shown in Figure 7.1(a)) proved that the valences in a carbon atom were

Trigonal bipyramidal electron geometry. [Reproduced from http://en.wikibooks.org/wiki/High_School_Chemistry/Families_on_the_Periodic_Table (accessed December 22, 2013).]

Principles of Inorganic Chemistry, First Edition. Brian W. Pfennig.

FIGURE 7.1
The optical isomers in (a) a tetrahedral C atom represented as nonsuperimposable mirror images of one another and (b) the optical isomers of the $[Co(en)_3]^{3+}$ ion.

oriented toward the corners of a tetrahedron. In a similar manner, in 1911, the father of modern coordination chemistry, Alfred Werner, demonstrated that certain ethylenediamine (en) coordination complexes of cobalt had an octahedral molecular geometry by separating the two optical isomers of $[Co(en)_3]Cl_3$, which are shown in Figure 7.1(b).

However, it was not until the 1940s when Sidgwick and Powell performed a systematic survey of the structures of molecules that it was proposed that molecular geometry could be predicted on the basis of the number of bonding and nonbonding pairs of electrons in the valence shell of the central atom. If the central atom had two pairs of valence electrons, the molecular geometry was linear; for three pairs, it was trigonal planar; for four pairs, it was tetrahedral; for five pairs, it was trigonal bipyramidal; and for six pairs, it was octahedral. This first formal model of molecular geometry was extended further by Gillespie and Nyholm in 1957 into what has become known as the *valence shell electron pair repulsion* (*VSEPR*) model.

The VSEPR model is an electronic model used to predict the three-dimensional shapes of small molecules. The basic premise of the model is that the electron pairs around the central atom of a molecule will adopt an arrangement that keeps them as far apart from one another as possible and that it is this repulsion of electron pairs that leads to the observed molecular geometry. However, it is important to note that the electronic repulsion is more a result of the Pauli principle than it is a Coulombic repulsion of the negatively charged electrons. The overall wave function in a polyelectronic atom can be separated into an orbital component and a spin component. According to the Pauli principle, whenever any two electrons in an atom are interchanged with one another, the total wave function must be antisymmetric. Because electrons having the same spin wave function must therefore occupy different spatial wave functions in order to satisfy the Pauli principle, any electrons having the same spin state will tend to avoid one another. In a free atom, the α set of electrons and the β set of electrons are like "two worlds that do not see each other in terms of the Pauli principle" (Gillespie, Popelier). For a free atom or ion containing a total of eight valence electrons (e.g., Ne, F^-, and O^{2-}), the four α electrons will (on average) occupy a tetrahedral shape so that they can be as far apart from one another as possible, as shown in Figure 7.2(a). A similar tetrahedron of the four β spin electrons will also exist, as shown in Figure 7.2(b), but it will be

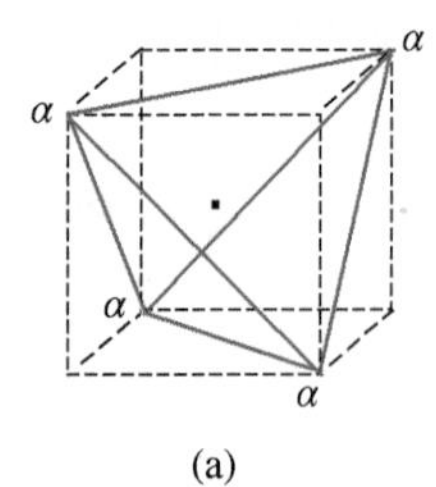

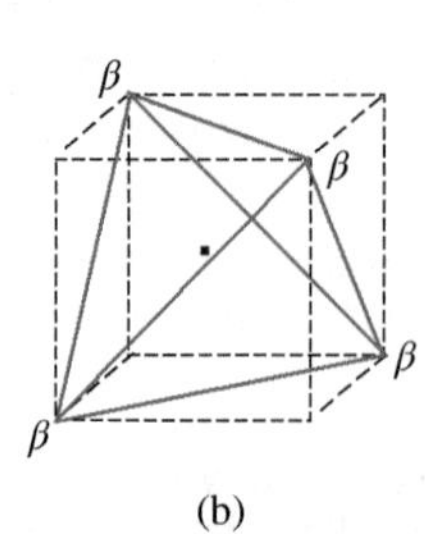

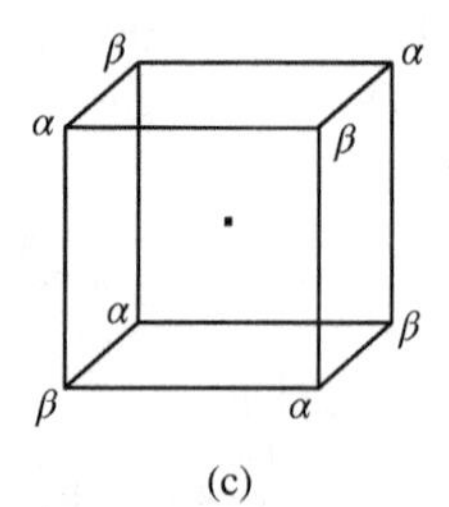

FIGURE 7.2
(a) The four α-spin electrons in a free atom or ion having the same electron configuration as [Ne] will occupy the corners of a tetrahedron as a result of the Pauli principle; (b) a similar tetrahedron will exist for the four β-spin electrons; and (c) the overall average electronic distribution of the eight electrons will position the two tetrahedra so that they occupy opposite corners of a cube.

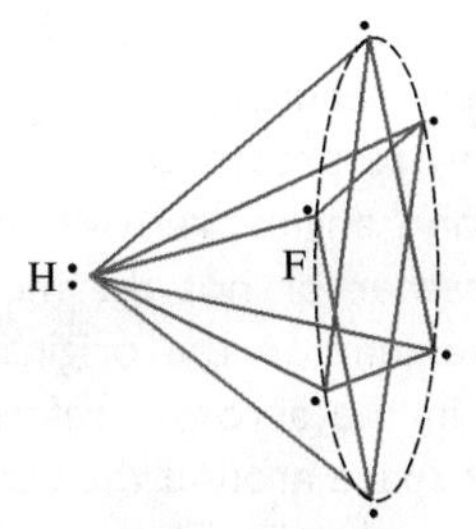

FIGURE 7.3
The average electron distribution for the eight valence electrons in the HF molecule. Electrons having α-spin are shown in the diagram at the corners of the red tetrahedron, whereas the β-spin electrons are shown at the corners of the blue tetrahedron. Notice that the distribution of the six nonbonding electrons on F forms a ring-shaped plane of electron density and avoids the formation of electron pairs.

oriented in a different direction. Because the electrostatic repulsion of electrons will keep the tetrahedral α and tetrahedral β sets of electrons apart from each other, the most probable relative arrangement for all eight electrons is the one shown in Figure 7.2(c), where the α and β electrons occupy alternating corners of a cube. In this closed shell, the overall electron density is spherical, as required by Unsöld's theorem; however, on average, the electrons will distribute themselves in such a way as to minimize electron–electron repulsions. In other words, contrary to what the Lewis structures might seem to imply, the electrons are not actually paired in a free atom or a monatomic ion.

Electron pairing only occurs in molecules as a result of their attraction to the nuclei. Whenever the nucleus of the central atom attracts one of the ligand's electrons and the nucleus of the ligand attracts an electron on the central atom, electrons having opposite spin states can be brought together as a bonding pair. Thus, it is the presence of attractive forces to the nuclei that is actually responsible for the formation of the bonding electron pairs in molecules. If this occurs in a linear molecule, such as HF, one of the apexes from the α spin tetrahedron and one of the apexes from the β spin tetrahedron will be brought into coincidence to form the bonding pair, as shown in Figure 7.3. The remaining six electrons on the F atom will remain (on average) as far apart from each other as possible. In other words, the six valence electrons on the F atom in H–F do not exist as lone pairs of electrons. Instead, they form a ring-shaped plane of electron density centered just beyond the F nucleus.

For nonlinear molecules, such as H_2O, NH_3, and CH_4, whenever two or more of the electrons on the central atom are paired with an electron from each H atom, the α and β spin tetrahedra are brought into coincidence with each other in such a way as to share one of their edges in common, as shown in Figure 7.4. This brings the α and β spin tetrahedra together so that any of the remaining nonbonding electrons on the central atom will necessarily exist as lone pairs of electrons, giving these molecules their characteristic tetrahedral electron geometry. Thus, in any nonlinear molecule that satisfies the octet rule, wherever there is any region of space having a high likelihood of finding a pair of electrons having opposite spin, there will necessarily be a low probability of finding any other electron in the same region of space. It is this consequence of the Pauli principle that causes the pairs of electrons arranged around the central atom to behave *as if* they were repelling one another.

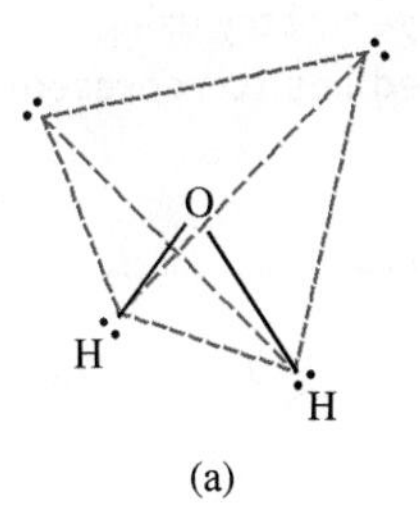

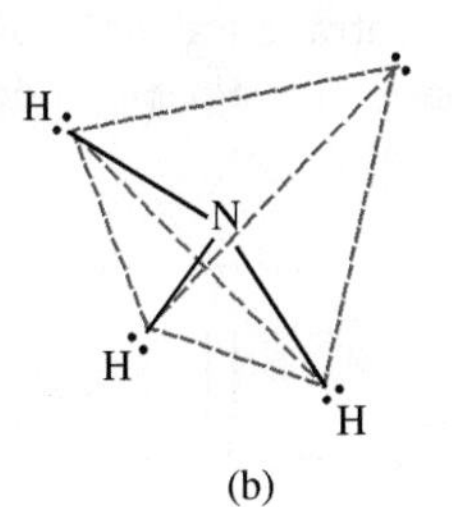

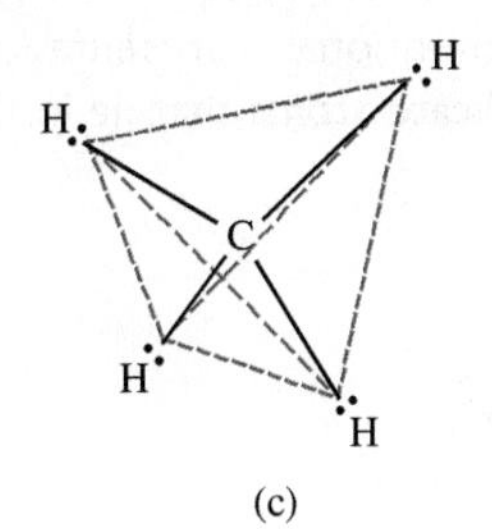

FIGURE 7.4
In nonlinear molecules that obey the octet rule, whenever two or more pairs of electrons are localized as a result of their attractive interactions with the nuclei, the α-spin and β-spin tetrahedra will be brought into coincidence with each other, localizing any nonbonding electrons as lone pairs on the central atom.

The VSEPR model can be used to predict the three-dimensional shapes of molecules based on the apparent repulsive interaction of their valence electrons. The model can be used not only to predict the basic geometry of a molecule but also to qualitatively explain bond angles, relative bond lengths, site preferences in certain geometries, and whether or not the molecule has a dipole moment. A modern pedagogical improvement on the original VSEPR theory is known as *VSEPD theory*, where the "D" in the acronym refers to an electron domain. An *electron domain* is any region of space around the central atom in which there is a high probability of finding electron density. The geometry of the molecule can then be predicted on the basis of the repulsion of electron domains and the relative magnitudes of these repulsions.

Rules for determining the geometry of a molecule using VSEPD theory

- Sketch the best Lewis structure or canonical form(s) of the molecule according to the rules given in Section 6.2. The most appropriate Lewis structure for thionyl chloride is shown in Figure 7.5.
- Count the total number of electron domains around the central atom and determine the electron geometry. In the case of $SOCl_2$, there are four electron domains: an S=O double bond, a lone pair, and two S–Cl single bonds. Although there are a total of five pairs of electrons around the central S atom, there are only four electron domains because the two pairs of electrons comprising the S=O double bond are localized to the same general region of space (between the S atom and the O atom) and therefore act together as a single domain.
- Once the number of domains around the central atom has been ascertained, the basic *electron geometry* can be determined by recognizing that the electron domains will adopt an arrangement that keeps them as far apart from one another as possible.

If there are only two domains around the central atom, the best possible configuration (the one with the lowest energy) will be the one where the two domains are pointed directly away from one another at an angle of 180°. This will result in a linear arrangement of the electron domains, as shown in Table 7.1. When there are three domains, the optimum arrangement is a trigonal planar geometry, in which all three domains lie in the same plane with an angle of 120° between them. The three H atoms in trigonal planar BH_3, for example, would form the apexes of an equilateral triangle.

When there are four domains, the best three-dimensional arrangement is tetrahedral. A tetrahedron is the Platonic solid formed by connecting four equilateral triangles into a pyramidal shape, as shown in Figure 7.6(a). Another way of visualizing a tetrahedron is to place the central atom in the center of a cube with the ligands occupying alternating corners of the cube, as shown in Figure 7.6(b). Because it is difficult to represent a three-dimensional structure on a two-dimensional sheet of paper, certain conventional representations are required. Figure 7.6(c) depicts the conventional representation of a tetrahedral molecular geometry using solid lines to indicate atoms that lie in the plane of the paper, a dotted line to represent that the

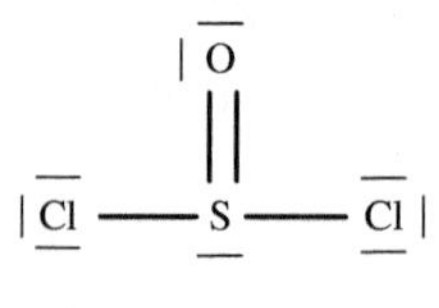

FIGURE 7.5
The best Lewis structure for thionyl chloride is the one that has all the formal charges equal to zero.

TABLE 7.1 Common electron and molecular geometries derived using VSEPD theory.

Domains	# BP	# LP	Electron Geometry	Molecular Geometry	Shape	Sample
2	2	0	Linear	Linear	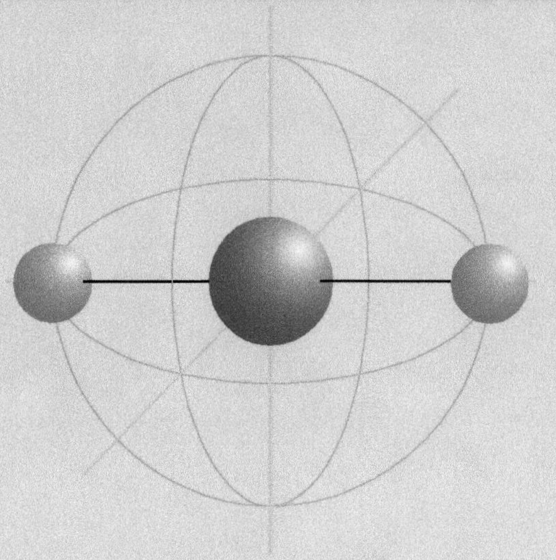	$BeCl_2$
3	3	0	Trigonal planar	Trigonal planar	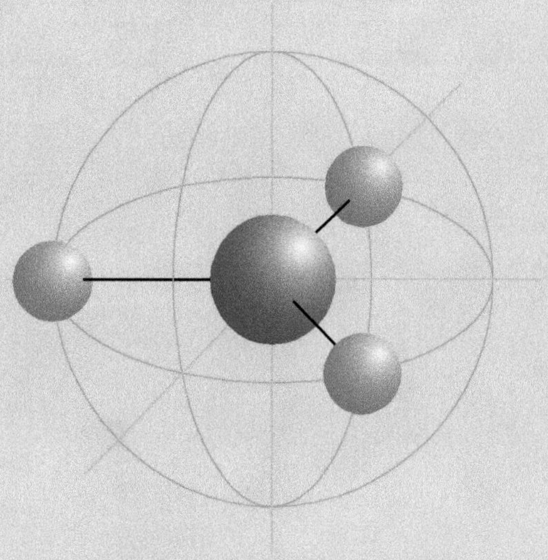	BF_3
3	2	1	Trigonal planar	Bent	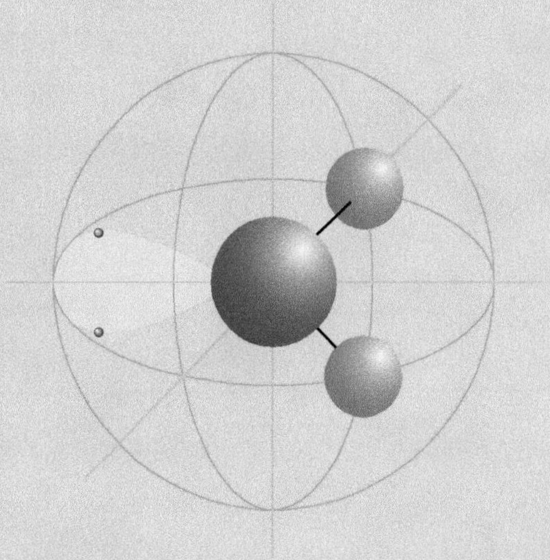	SO_2

(continued)

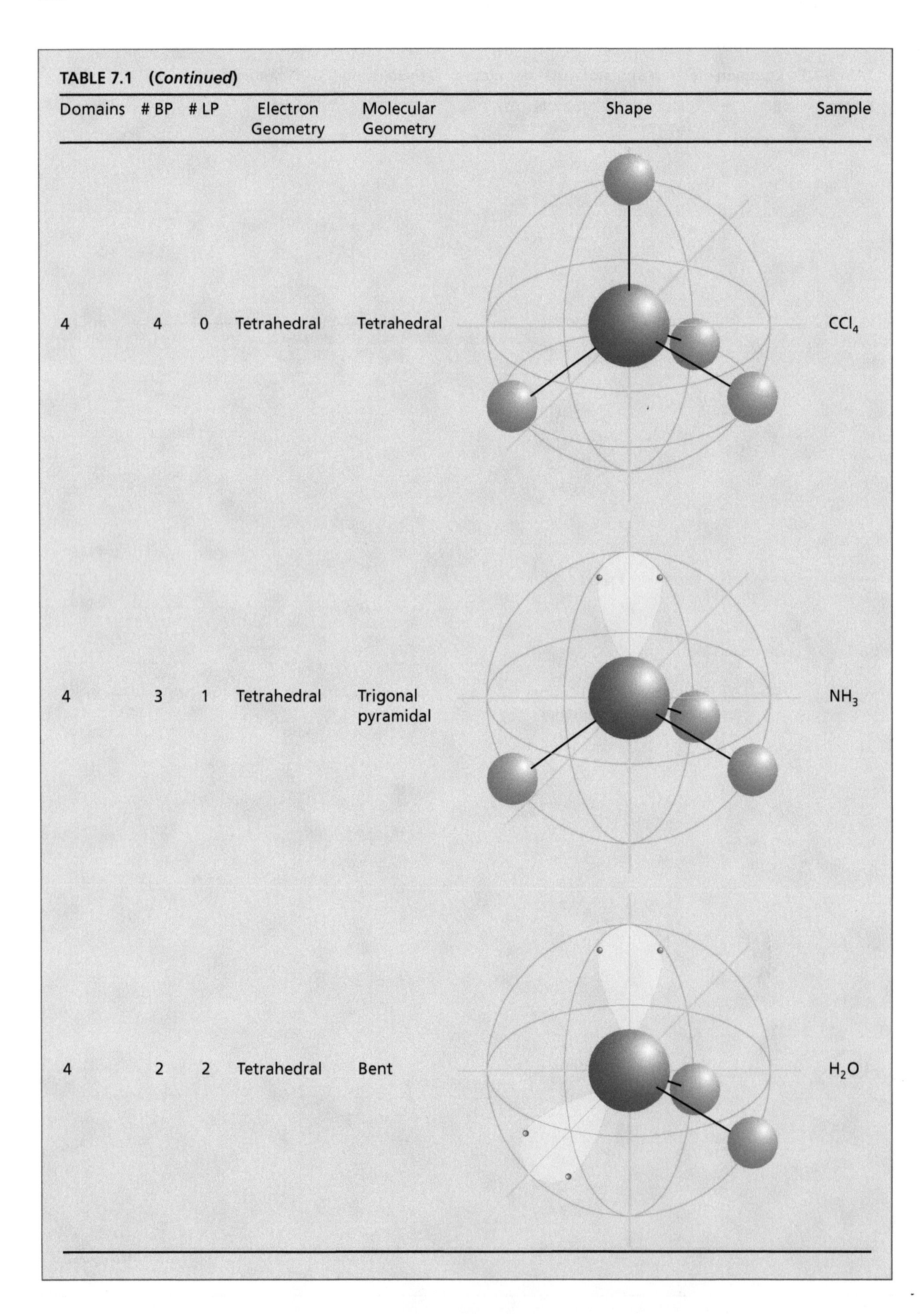

TABLE 7.1 (*Continued*)

Domains	# BP	# LP	Electron Geometry	Molecular Geometry	Shape	Sample
4	4	0	Tetrahedral	Tetrahedral		CCl_4
4	3	1	Tetrahedral	Trigonal pyramidal		NH_3
4	2	2	Tetrahedral	Bent		H_2O

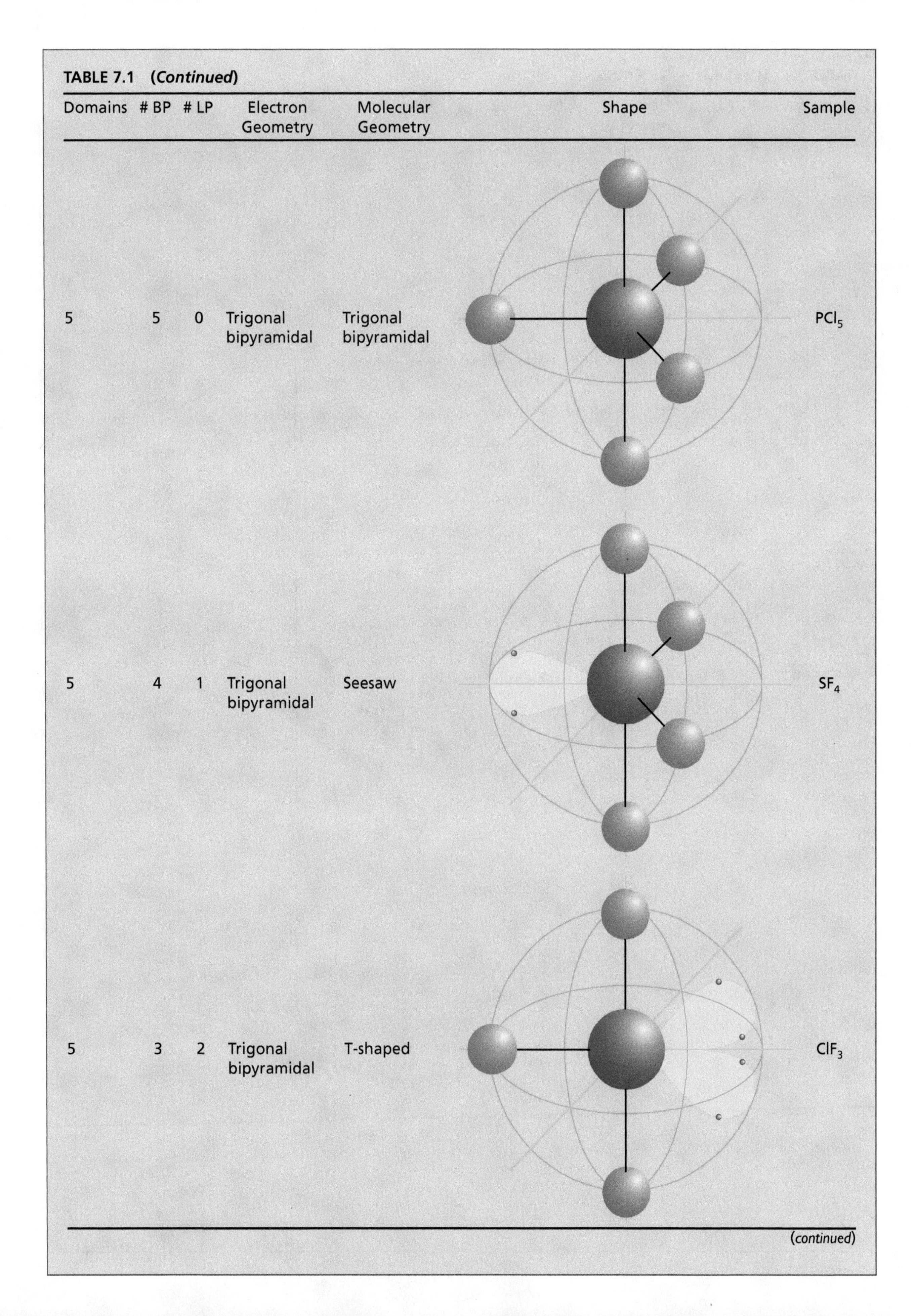

TABLE 7.1 (*Continued*)

Domains	# BP	# LP	Electron Geometry	Molecular Geometry	Shape	Sample
5	5	0	Trigonal bipyramidal	Trigonal bipyramidal		PCl_5
5	4	1	Trigonal bipyramidal	Seesaw		SF_4
5	3	2	Trigonal bipyramidal	T-shaped		ClF_3

(*continued*)

TABLE 7.1 (*Continued*)

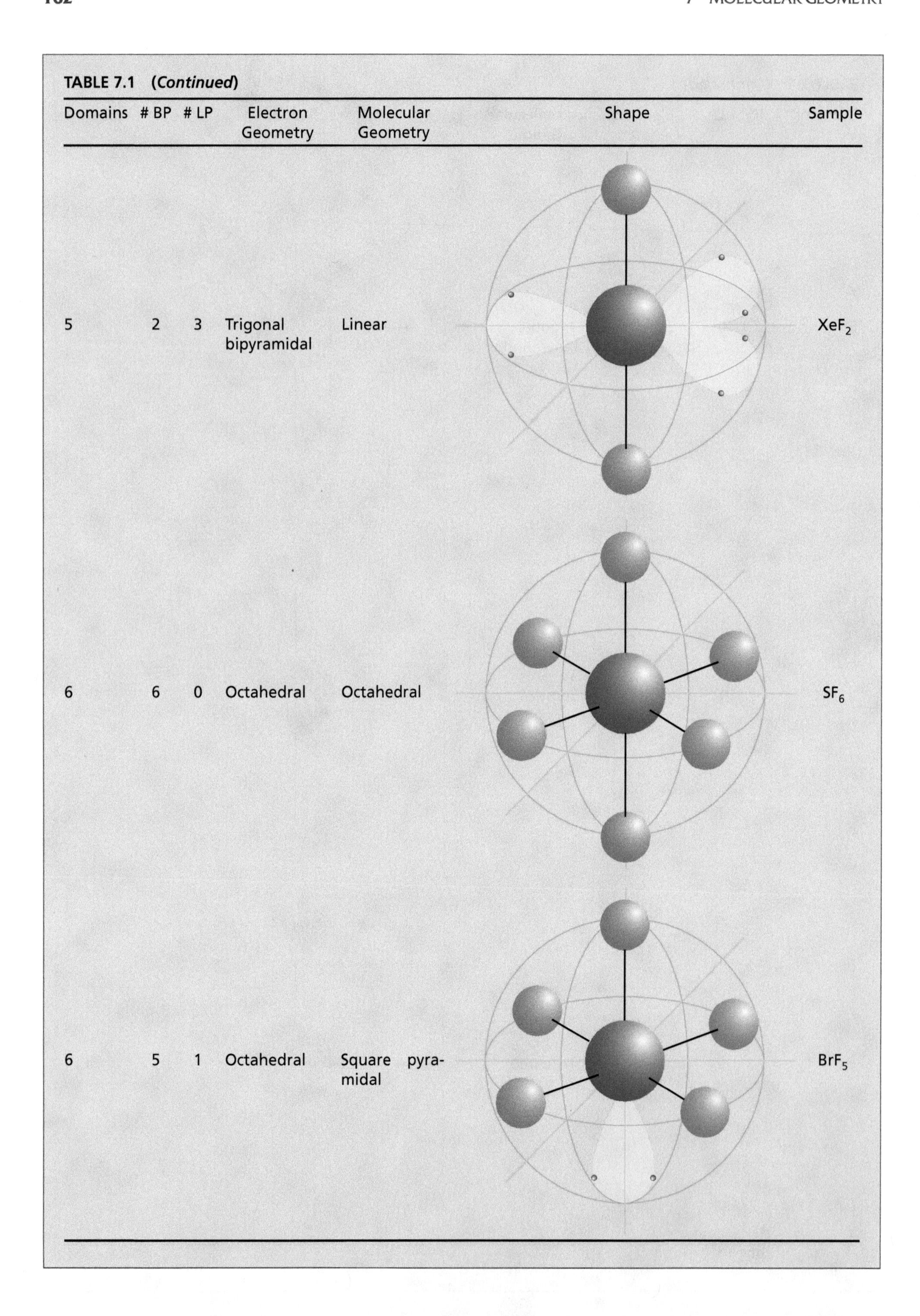

Domains	# BP	# LP	Electron Geometry	Molecular Geometry	Shape	Sample
5	2	3	Trigonal bipyramidal	Linear		XeF_2
6	6	0	Octahedral	Octahedral		SF_6
6	5	1	Octahedral	Square pyramidal		BrF_5

TABLE 7.1 (*Continued*)

Domains	# BP	# LP	Electron Geometry	Molecular Geometry	Shape	Sample
6	4	2	Octahedral	Square planar	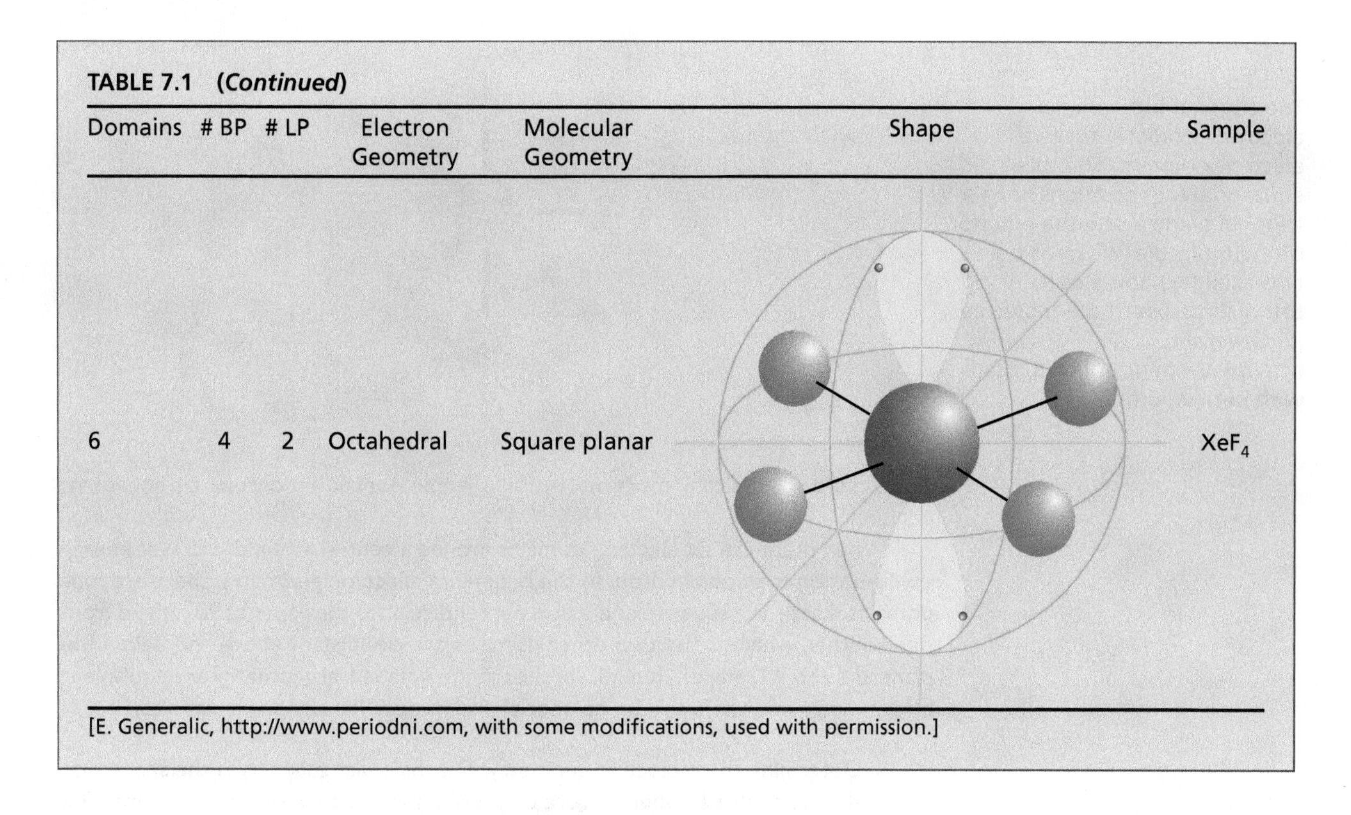	XeF_4

[E. Generalic, http://www.periodni.com, with some modifications, used with permission.]

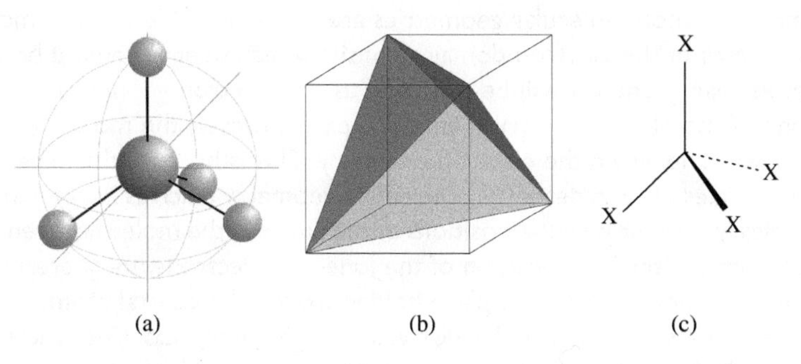

FIGURE 7.6
Different representations of a tetrahedral electron geometry: (a) the Platonic solid having four faces composed of equilateral triangles, (b) as the shape inscribed by connecting opposite corners of a cube, and (c) in the conventional representation used to draw tetrahedral molecules. [(a) E. Generalic, http://www.periodni.com, used with permission. (b) Diagram by Mathew Crawford, copyright Daedalus Education.]

bond is going behind the plane of the paper, and a dark wedge to illustrate that the bond is in front of the plane of the paper. All of the angles in a tetrahedral electron geometry are 109.5°, making each of the four sites in a tetrahedron geometrically equivalent.

This is not, however, the case in a trigonal bipyramidal geometry, which is the best arrangement for five electron domains. The trigonal bipyramidal geometry, shown in Figure 7.7 using the same convention, has three domains at 120° that lie in a trigonal plane around the equator of the molecule (atoms in these positions are called *equatorial*) and two domains that lie at the north and south poles of the molecule's threefold rotational axis (atoms in these positions are called *axial*). The two axial sites lie at 180° to each other and at 90° to each of the three domains in the trigonal plane. The trigonal bipyramid is the only common electron geometry where the different sites are not all equivalent by symmetry. It is for this reason that there is sometimes a preference for one or the other site in a trigonal bipyramidal electron geometry, based on the relative sizes of the domains. A larger-sized domain will prefer an equatorial site because this minimizes the electron–electron repulsions the best. Equatorial sites have only two nearest neighbor domains at 90° angles to them, whereas an axial site will be orthogonal to three other neighbors.

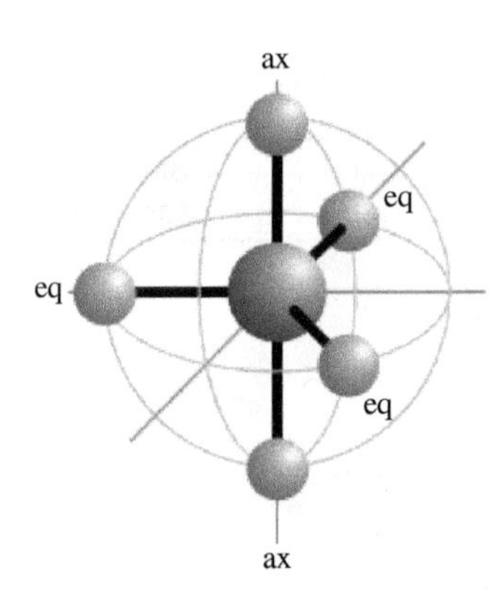

FIGURE 7.7
The trigonal bipyramidal electron geometry for five electron domains. The three equatorial (eq) positions lie in a trigonal plane along the equator of the molecule, whereas the two axial (ax) domains lie along the vertical axis of the molecule. [E. Generalic, http://www.periodni.com, used with permission.]

Thus, there is physically more room for a larger domain to occupy an equatorial position.

When there are six electron domains around a central atom, the lowest energy configuration is an octahedron. In the octahedral electron geometry, there are four domains that lie in a square plane that is perpendicular to the paper at 90° angles from one another, whereas the two other domains are equidistant above and below the plane and at a 90° angle. Thus, all six sites in the octahedral geometry are equivalent by symmetry and there will be no special preferences for any particular site.

- Determine the molecular geometry. The *molecular geometry* is the shape that the molecule has when you are only looking at the positions of the atoms. The most common molecular geometries are shown in Table 7.1. For molecules where all of the electron domains consist of equivalent chemical bonds, the molecular geometry will be the same as the electron geometry. Whenever one of the electron domains is a lone pair, however, the molecular geometry will differ from the electron geometry. This is because the experimental techniques used to determine molecular geometry (such as X-ray diffraction) typically register only the positions of the nuclei in the molecule. Even though we cannot "see" the position of the lone-pair electrons, their presence in a lone-pair domain acts as a place holder around the central atom, occupying one of the sites in the molecule's electron geometry. Consider the example SF_4, which has the Lewis structure shown in Figure 7.8(a), with a total of five electron domains, one of which is a lone-pair domain. The electron geometry of SF_4 is trigonal bipyramidal, but its molecular geometry is seesaw because the nuclei form the shape of a seesaw (standing on its end).
- Determine the effects of the different-sized domains.

Notice that the lone pair of electrons in SF_4 in Figure 7.8(b) was placed in an equatorial position, as opposed to an axial one. This was not an arbitrary decision. The reason for this is that the central atom will hold a lone pair of electrons (which "belong" exclusively to that atom) closer to itself than it will hold a

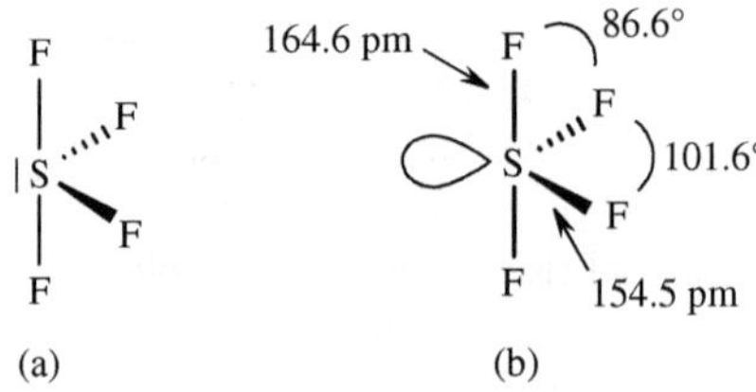

FIGURE 7.8
(a) The Lewis structure of SF_4, ignoring the nonbonding electrons on the F atoms, and (b) the molecular geometry of SF_4, where the lone-pair electrons are simply shown as a lobe.

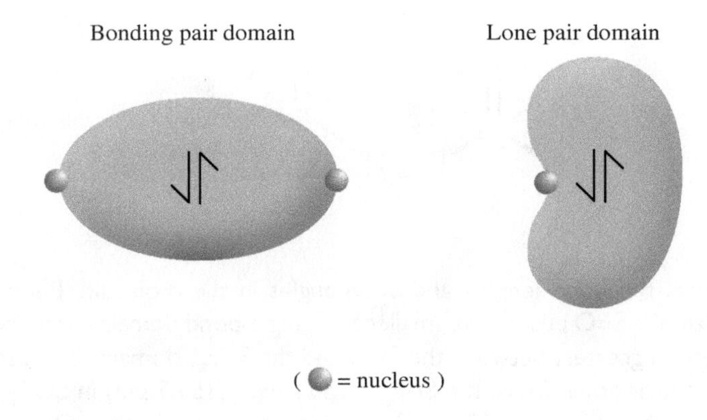

FIGURE 7.9
The relative angular volumes of a bonding pair electron domain versus a lone-pair (nonbonding pair) electron domain [Blatt Communications].

bonding pair of electrons (which it has to share with another element). Thus, the lone-pair domain occupies a larger percentage of the space (or takes up more angular volume) around the central atom than does a bonding pair domain, as shown in Figure 7.9.

In the trigonal bipyramidal electron geometry, which has two nonequivalent sites (equatorial and axial), the larger domains prefer to occupy the equatorial positions where there is more room to accommodate them. As mentioned previously, the equatorial domains have more room in them because they have only two close 90° interactions, whereas an axial domain has three close 90° interactions. Thus, the larger lone-pair domain in SF_4 will occupy any of the three equivalent equatorial positions, leading to a seesaw molecular geometry. The site preference for larger domains to go equatorial also has an effect on the bond lengths and bond angles in molecules, as is the case for SF_4. The larger lone-pair domain will occupy more angular volume around the central atom, which has the effect of pushing the smaller domains away from itself. Thus, the F_{eq}–S–F_{eq} bond angle decreases from its ideal value of 120° to 101.6° and the F_{ax}–S–F_{eq} bond angle decreases from 90° to 86.6°. The larger lone-pair domain also provides a justification for why the S–F_{ax} bond length is longer (164.6 pm) than the S–F_{eq} bond length (154.5 pm), as there is a more direct lone pair to S–F_{ax} bonding pair repulsion than there is a lone pair to S–F_{eq} bonding pair repulsion.

Just as the larger lone-pair domain in SF_4 preferred to occupy an equatorial site in the trigonal bipyramidal electron geometry, the same will also be true for doubly bonded domains. Because a double bond contains four electrons confined to the space between two atoms, a double bond domain will take up more space around the central atom than will a single-bond domain. The Lewis structure of SOF_4, along with its three-dimensional shape, is shown in Figure 7.10. The molecule has a total of five electron domains: one S=O domain and four S–F domains. Repulsion of the electron domains leads to a trigonal bipyramidal electron and molecular geometry. However, because the double bond domain occupies more space around the S atom than do the single-bond domains, the S=O will prefer one of the equatorial sites, where there is more room to accommodate its larger size. The larger S=O domain

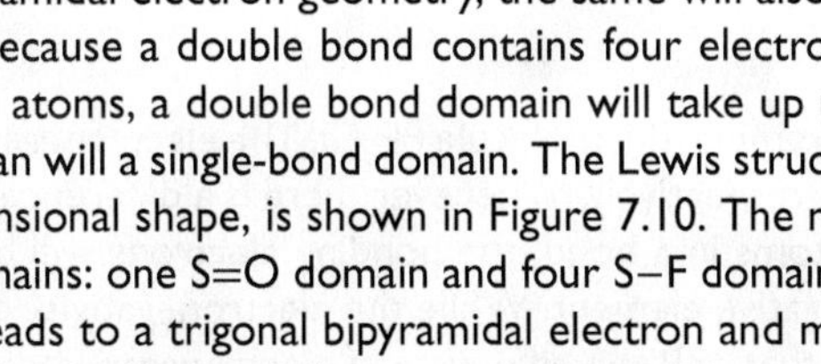

159.5 pm 112.8° 97.7° 153.8 pm

(a) (b)

FIGURE 7.10
(a) The Lewis structure of SOF_4, ignoring the nonbonding electrons on the O and F atoms, and (b) the molecular geometry of SOF_4.

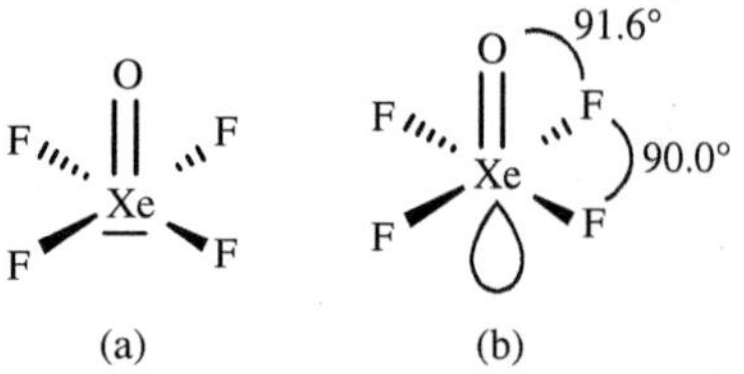

FIGURE 7.11
(a) The Lewis structure of $XeOF_4$, ignoring the nonbonding electrons on the O and F atoms, and (b) the molecular geometry of $XeOF_4$.

will also affect the bond lengths and bond angles in the molecule. Because of its imposing size, the S=O pushes the smaller S–F single-bond domains away from itself. The repulsion is greatest between the S=O and the $S–F_{ax}$ domains because they lie at ~90° to one another. Thus, the $S–F_{ax}$ bond length (159.5 pm) in SOF_4 is longer than the $S–F_{eq}$ bond length (153.8 pm). Likewise, the experimental $O=S–F_{ax}$ bond angle (97.7°) is greater than the usual 90° axial-central atom-equatorial angle and the $O=S–F_{eq}$ bond angle is larger (at 123.6°) than its usual 120°. The $F_{eq}–S–F_{eq}$ bond angle must therefore be less than 120° (the experimental value is 112.8°), whereas the $F_{ax}–S–F_{eq}$ bond angle must be less than 90° (it is 82.3°).

For any electron geometry (at least for those listed in Table 7.1) other than the trigonal bipyramidal geometry, all of the sites are equivalent by symmetry and therefore there is no particular preference for where a larger domain will reside. However, when there is more than one large domain, the larger domains will arrange themselves to be as far apart from each other as possible. Consider the example shown in Figure 7.11 for $XeOF_4$. There are a total of six electron domains around the central Xe atom in $XeOF_4$, giving it an octahedral electron geometry. Four of these domains are Xe–F single-bond domains, whereas the remaining domains are a Xe=O double bond and a lone pair on Xe. The double-bond and lone-pair domains will both be larger than a single-bond domain and will therefore orient themselves as far apart from one another as possible. Thus, while in general there is no particular site preference in an octahedral geometry, the two larger domains in this specific example must occupy sites that are 180° from each other. This is also the reason why the two lone pairs in a square planar molecular geometry are directly opposite one another.

- Determine the effect of the electronegativity of the ligands. Even in molecules where there are no multiple-bond or lone-pair domains, deviations from the ideal bond angles can exist as a result of electronegativity differences between the ligands.

Consider, for example, the molecule PF_4Cl. The electronegativities of P, F, and Cl are 2.2, 4.0, and 3.2, respectively. Whenever there is a difference in electronegativity between the two atoms in a bond, the bonding electrons will be polarized toward the more electronegative element. While the electronegativity of F and Cl are both more negative than that of P, the difference in electronegativity is greater between P and F than it is between P and Cl. Thus, the bonding pair electrons in a P–Cl bond will lie closer to the P atom than they will in a P–F bond, as shown in Figure 7.12(a). This implies that a P–Cl bonding pair domain will occupy a greater angular volume

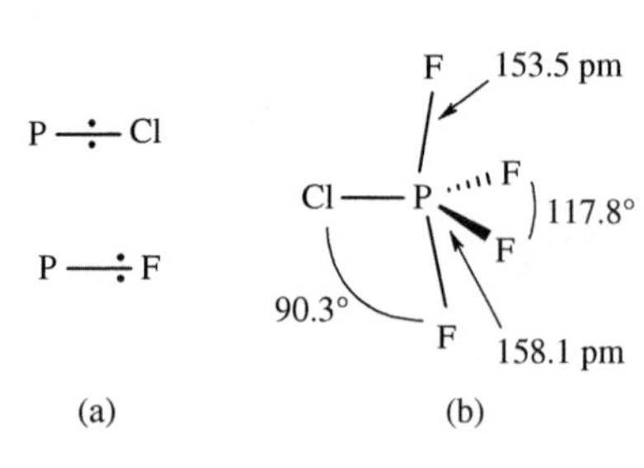

FIGURE 7.12
(a) A comparison of the relative positions of the bonding electrons in a P–Cl bond and in a P–F bond. The difference results from the fact that F is more electronegative than Cl. (b) The molecular structure of PF_4Cl. The $P–F_{ax}$ bond lengths are longer than the $P–F_{eq}$ bond lengths because they lie closer to the larger P–Cl domain. Likewise, the $Cl_{eq}–P–F_{ax}$ bond angle is larger than 90° and the $F_{eq}–P–F_{eq}$ bond angle is smaller than 120°.

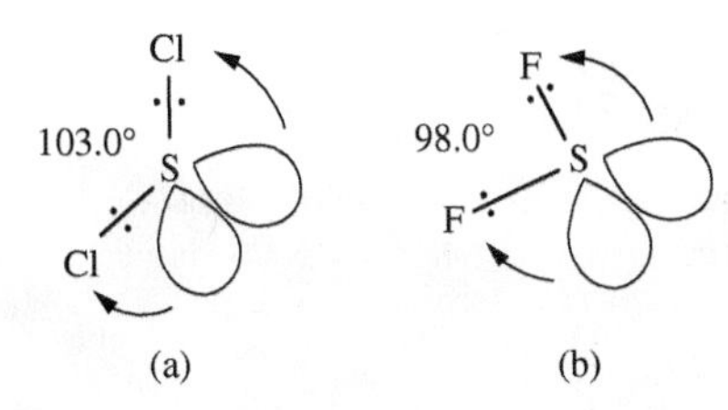

FIGURE 7.13
(a,b) The bond angle in SF_2 can close more than the bond angle in SCl_2 because the bonding electrons lie farther from the central atom in SF_2.

around the central P atom than will a P–F bonding pair domain. Furthermore, the atomic radius of the Cl atom is also larger than the F atom, thereby compounding the effect. As a result, the larger P–Cl domain will prefer to occupy an equatorial position in PF_4Cl, where there is more available space to accommodate it. The larger size of the P–Cl domain will also decrease the F_{eq}–P–F_{eq} bond angle from its ideal value of 120° to 117.8° because the two smaller P–F domains in the trigonal plane will be pushed closer to one another. In terms of the bond lengths, the P–F_{ax} bond length (158.1 pm) is larger than the P–F_{eq} bond length (153.5 pm) because the axial fluorines lie closer to the larger P–Cl domain than do the equatorial fluorines, as shown in Figure 7.12(b).

Electronegativity differences can also be used to rationalize why the bond angle in the bent SF_2 molecule (98.0°) is smaller than the corresponding angle in SCl_2 (103.0°), as shown in Figure 7.13. Both molecules have tetrahedral electron geometries with two lone pairs. Repulsion of these lone-pair domains leads to the bond angle in each species being less than the expected 109.5° for a perfect tetrahedron. The bonding pair of electrons in SF_2 is held farther from the S atom than it is in SCl_2 due to the larger electronegativity of F than Cl. Thus, the S–Cl bonding pair will take up more angular volume around the central S atom in SCl_2 and the Cl–S–Cl bond angle will not be able to squeeze as far closed as it can in SF_2.

Example 7-1. Predict the electron and molecular geometry for each of the following: (a) N_3^-, (b) ICl_4^-, (c) PF_3, (d) POF_3, and (e) BrF_3. Sketch the three-dimensional shape of each species using the proper convention and predict the approximate bond angles.

Solution. The best Lewis structures for each species are shown below at left. The formal charges are listed next to each atom. The three-dimensional shapes of each molecule or ion and their molecular geometries are listed to the right of the Lewis structures.

For N_3^-, there are two domains, forcing a linear electron geometry. As both domains are bonding, the molecular geometry will also be linear and the bond angle will be 180°. The experimental bond angle is 180° with a bond length of 118 pm, consistent with that of a typical N=N double bond.

The ICl_4^- ion has six electron domains around iodine, giving it an octahedral electron geometry. Four of these domains are I–Cl single bonds and the other two are lone-pair domains. Because the lone-pair electrons are held exclusively by the iodine, while the bonding electrons are shared, the two lone pairs will occupy more angular volume around the I atom and will want to be as far apart from one another as possible. Thus, the molecular geometry of ICl_4^- is square planar, and the Cl–I–Cl bond angles are expected to be 90°. The experimental data confirm this prediction.

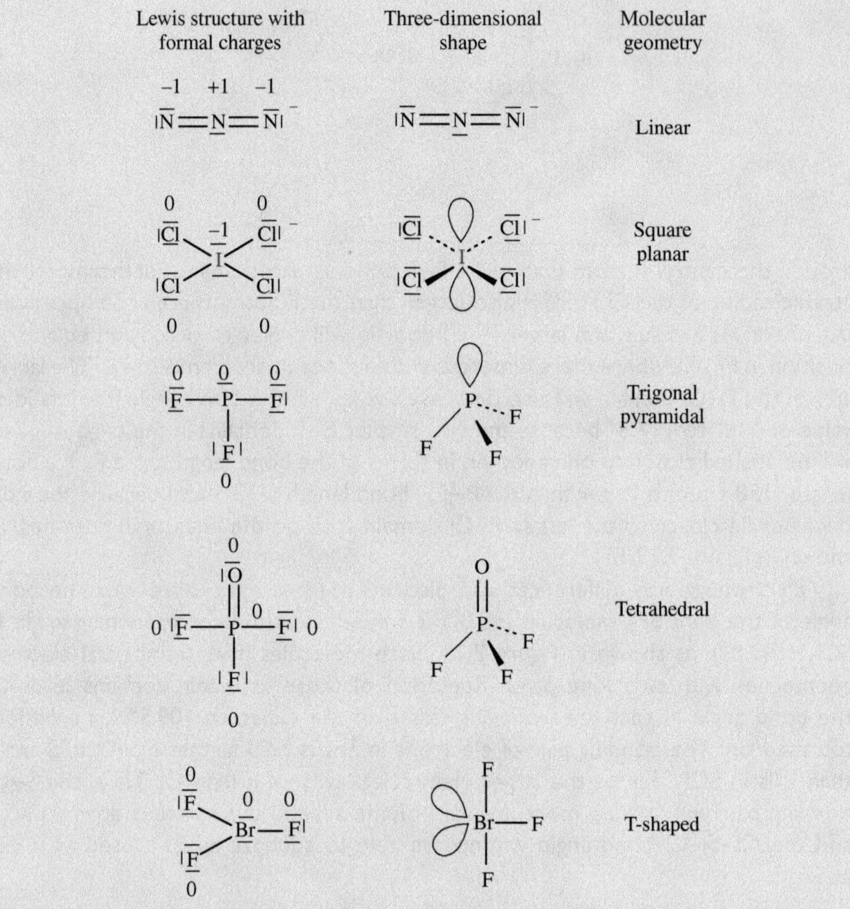

Lewis structure with formal charges	Three-dimensional shape	Molecular geometry
-1 $+1$ -1 $[N{=}N{=}N]^-$	$[N{=}N{=}N]^-$	Linear
$[ICl_4]^-$ (Cl: 0, I: -1)		Square planar
PF_3 (all 0)		Trigonal pyramidal
POF_3 (all 0)		Tetrahedral
BrF_3 (all 0)		T-shaped

In PF_3, there are four domains, forcing a tetrahedral electron geometry. One of these domains is a lone pair, making the molecular geometry trigonal pyramidal. The F–P–F bond angles are expected to be less than the usual 109.5° for a tetrahedral molecule because the lone pair occupies more space than the bonding pairs. The experimental bond angle in PF_3 is 97.7°, which is considerably smaller than 109.5°. Owing to the larger electronegativity of F, the P–F single bonds are highly polarized toward the fluorine atoms. Thus, the bond angle can collapse quite a bit before the P–F bonding electrons will begin to repel one another.

In the POF_3 molecule, there are also four domains, which lead to the prediction of both a tetrahedral electron and a molecular geometry. The larger P=O double bond domain will repel the smaller P–F single-bond domains away from itself, collapsing the F–P–F bond angles from the expected 109.5°. The repulsion between P=O and P–F will not be as large as it was between the lone pair and the P–F single bonds in PF_3 because the four bonding electrons in the P=O double bond are polarized toward the more electronegative O atom. The experimental F–P–F bond angle of 101.3° in POF_3 supports this qualitative prediction. Because of the weaker repulsion in POF_3 than in PF_3, the P–F bond lengths in POF_3 are also shorter (152.4 pm) than those in PF_3 (157.0 pm).

Lastly, the BrF_3 molecule has five electron domains, predicting a trigonal bipyramidal electron geometry. The two lone-pair domains will occupy more space around the central Br atom than will the shared Br–F single-bond domains. The lone pairs will therefore prefer two of the equatorial sites in the trigonal bipyramid, where they will have more room to expand. This leads to a T-shaped molecular geometry. The equatorial lone-pair domains will repel the two $Br{-}F_{ax}$ domains more than they will the $Br{-}F_{eq}$ domain, as the former are closer to the lone pairs. Thus, the $Br{-}F_{ax}$ bond length is expected to be larger than the $Br{-}F_{eq}$ bond length. The experimental values of 181 and 173 pm, respectively, again support this qualitative prediction. Because of the greater repulsion between the lone-pair and bonding-pair domains than between two bonding pair domains, the $F_{ax}{-}Br{-}F_{eq}$ bond angle should be less than 90°; the experimental value is 85°.

Example 7-2. There are two different Br–F bond lengths in BrF_5: 170 and 177 pm. There are also two different Br–F bond lengths in BrF_3: 173 and 181 pm. Sketch the three-dimensional structure of each molecule and label the appropriate bond lengths for each Br–F bond on your diagram. Also, explain why the experimental bond lengths are longer in BrF_3 than they are in BrF_5.

Solution. The molecular geometries of BrF_5 and BrF_3 are square pyramidal and T-shaped, respectively, as shown in the following diagram.

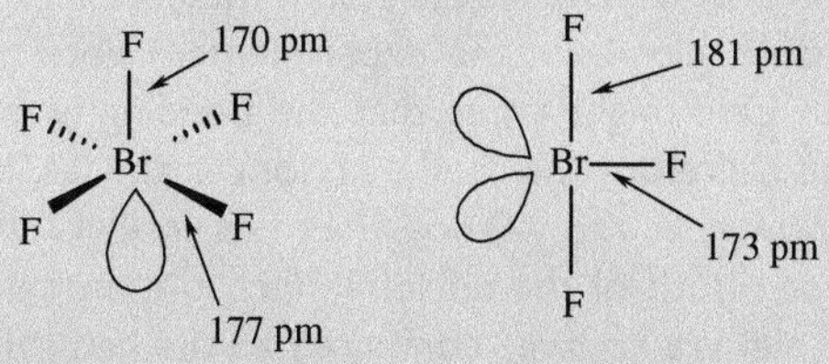

170 pm
177 pm
Square pyramidal

181 pm
173 pm
T-shaped

In both cases, the lone-pair domains occupy more space around the central Br atom and push away the smaller Br–F single-bond domains. The repulsion between a lone pair and a bonding pair is stronger the closer they are to one another. Thus, the four equatorial Br–F bond lengths in BrF_5 will be longer than the Br–F axial bond length. Likewise, the $Br{-}F_{ax}$ bond lengths in BrF_3 will be longer than the $Br{-}F_{eq}$ bond length in this molecule. Both corresponding Br–F single bonds in BrF_3 will be longer than those in BrF_5 because there are two lone pairs in the BrF_3 molecule versus only one lone pair in BrF_5.

Example 7-3. Explain why the bond angle in H_2O is 104.5°, whereas the same angle in H_2S is only 92.1°.

Solution. The two molecules are isomorphous; both have a tetrahedral electron geometry and a bent molecular geometry. In both cases, the two lone-pair domains will occupy the largest angular volume around the central atom, collapsing the bond angle to be less than the usual 109.5° for a perfect tetrahedron. As the lone pairs repel one another, the bond angle decreases until the bonding electrons are close enough to each other to counterbalance the LP–LP repulsion. The electronegativities of H, O, and S are 2.20, 3.44, and 2.58, respectively. Thus, the O–H single bond is highly polarized toward the O atom, whereas the H–S bond is only slightly polarized toward sulfur atom. The bonding electrons in H_2S therefore lie farther away from the central atom than those in H_2O, allowing the bond angle to collapse more in H_2S before the bonding pairs will begin to repel each other. Thus, H_2S has a noticeably smaller bond angle than H_2O. The larger size of the S atom in H_2S further exaggerates this effect, as the H–S single bond is also longer than the O–H single bond, allowing the bonding pairs in H_2S to lie farther from the central atom than those in H_2O.

104.5° 92.1°

For all of its simplicity, the VSEPR model is an incredibly powerful one. The theory is simply based on the Pauli principle, which leads to an apparent repulsion of pairs of electrons occupying differently sized domains around the central atom. Yet, it can not only predict the molecular geometry of most small molecules but can also make qualitative comparisons between bond angles and bond lengths and determine which ligand will occupy which site in a trigonal bipyramidal electron geometry. Like many theories, however, the model is not without its shortcomings. For instance, the best Lewis structure for Li_2O predicts a bent molecular geometry like the isoelectronic H_2O molecule. However, the experimentally determined geometry for lithium oxide is linear. One might argue that the bonding is more ionic in nature than it is covalent, and therefore the model no longer applies. However, molecular orbital calculations show that the valence electrons in Li_2O are shared to a reasonable degree between both of the nuclei. In fact, the model fails for a different reason: the lithium nuclei are so large compared to the central O atom that there is a steric (ligand–ligand) repulsion between the two Li atoms, which forces the molecule into a linear molecular geometry. Another exception occurs for the BrF_6^- ion, which has seven electron domains. VSEPR theory would predict this ion to have a pentagonal bipyramidal electron geometry with significant bond angle deviations due to the larger size of the lone-pair domain. Infrared spectroscopy of BrF_6^-, however, indicates that the molecular geometry is very close to that of a perfect octahedron. An exception also occurs for $As(AuPH_3)_4^+$, which is square pyramidal, despite the fact that the isoelectronic complex cation $N(AuPH_3)_4^+$ is tetrahedral like NH_4^+.

Furthermore, the VSEPR model fails rather often when the central atom is a transition metal. Thus, for example, $NiCl_4^{2-}$ is tetrahedral, whereas isoelectronic $PtCl_4^-$ is square planar. Similarly, $CuCl_5^{3-}$ is trigonal bipyramidal but $Ni(CN)_5^{3-}$ is square pyramidal. Perhaps most disturbing are instances when a simple change of the ligand leads to an entirely different molecular geometry. While $NiCl_4^{2-}$ is tetrahedral, $Ni(CN)_4^{2-}$ is square planar. There are even certain coordination compounds that have one type of molecular geometry at high temperatures and an entirely different one at lower temperatures. Some of the reasons why the VSEPR model does not always predict the correct molecular geometry include crystal packing effects in the solid state, the sizes of the ligands, the degree of percent ionic character in the bonding, and ligand field effects involving the *d* orbitals.

7.2 THE LIGAND CLOSE-PACKING MODEL

As we saw in the previous section, the VSEPR model is a particularly powerful tool in the prediction of the three-dimensional shapes of molecules. However, it does not *always* predict the correct molecular geometry. In this section, we

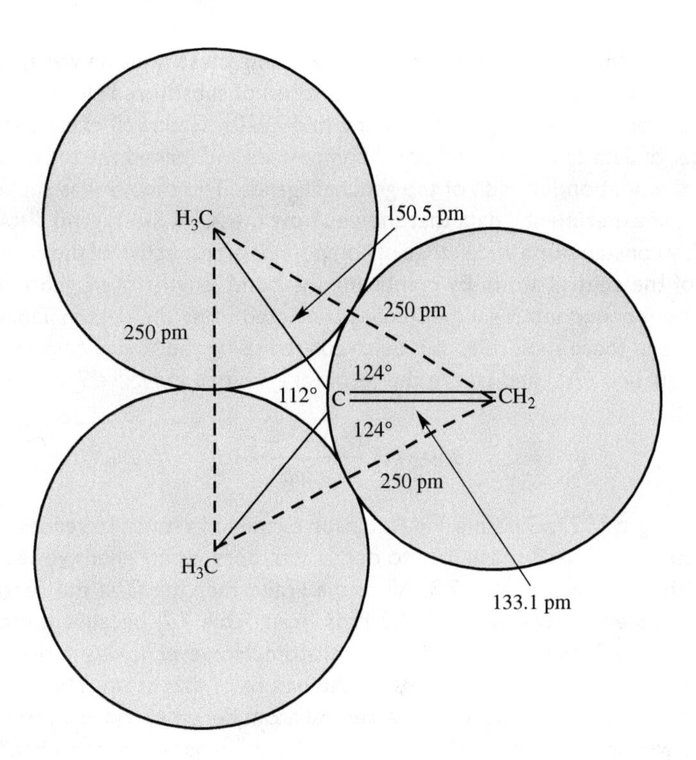

FIGURE 7.14
Bond lengths and bond angles for 2-methylpropene, showing how the terminal C atoms form a nearly perfect equilateral triangle.

discuss an alternative model known as the *ligand close-packing* (*LCP*) model. While VSEPR theory is an electronic model, the LCP model is a steric one. It puts forth the assumption that molecular geometries are dictated by the sizes of the ligands themselves and the repulsive interactions between those ligands. This theory originated with the work of Bartell and Bonham in 1960, who noticed that the terminal C atoms in 2-methylpropene form a near perfect equilateral triangle, as shown in Figure 7.14. These results suggested to the authors that the shapes of molecules could be predicted by allowing the ligands to approach each other as closely as possible around the central atom to form a ligand close-packed structure.

The concept of the close-packing of ions in ionic solids had already been widely accepted, so it was not too much of a stretch to apply the same basic idea to the ligands in covalently bonded species. Bartell assigned a nonbonded radius to each of

TABLE 7.2 Bartell and Glidewell's 1,3-radii for ligands attached to carbon.

Atom	$r_{1,3}$ (pm)
H	92
C	125
N	114
O	113
F	108
Si	155
P	145
S	145
Cl	144

the different ligands (X), as shown in Table 7.2. Using these radii, he was then able to predict the interligand (X–X) distances in a series of substituted alkenes ($X_2C=C$) and substituted ketones ($X_2C=O$). In the mid-1970s, Glidewell expanded Bartell's original set of data to include additional compounds and coined the term *1,3-radii* to describe the nonbonded radii of the geminal ligands. The theory was supported by a large set of experimental data that showed how the X–X interligand distances are remarkably constant for a wide range of compounds, irrespective of the coordination number of the central atom. By combining the bond lengths of neighboring atoms with the nonbonded interligand distances predicted using the data in Table 7.2, the bond angles in these molecules can be calculated using the law of cosines, which is given by Equation (7.1) and where the terms are defined in Figure 7.15:

$$\cos(\gamma) = \frac{a^2 + b^2 - c^2}{2ab} \tag{7.1}$$

Beginning in 1997, Gillespie and Robinson examined a much larger set of molecular structures and used these data to derive the more comprehensive set of ligand radii, which are listed in Table 7.3. All of the ligand radii for C as the central atom are self-consistent with Glidewell's 1,3-radii from Table 7.2, because these are the values that were derived for C as the central atom. However, it was noticed that the magnitude of the ligand radii depends on the nature of that atom. This observation is largely due to a charge effect. As the central atom becomes more electropositive, the ligand withdraws more of the bonding electron density closer to itself. As the partial negative charge on the ligand begins to increase, this has the effect of increasing the ligand radius. In the furthest extreme where there is a very large difference in electronegativity between the central atom and the ligand, the ligand radius would eventually approach the ionic radius for the corresponding ion. Table 7.4 lists the interligand distances for a variety of different molecules, illustrating how the sum of the ligand radii in Table 7.3 can be used to predict the nonbonded interligand distances in molecules.

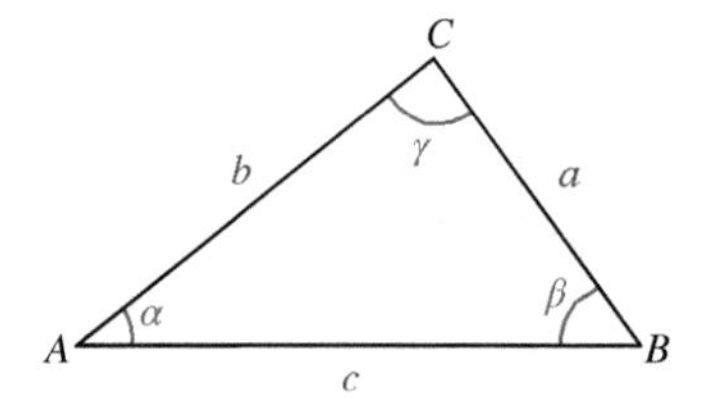

FIGURE 7.15
If the bond lengths *CA* and *CB* are known and the interligand distance *AB* can be predicted from the 1,3-radii listed in Table 7.2, then the bond angle γ can be calculated using Equation (7.1). [Reproduced from http://en.wikipedia.org/wiki/Law_of_cosines (accessed December 15, 2013).]

TABLE 7.3 Gillespie's ligand radii (pm) as a function of the central atom.

	Central Atom			
Ligand	Be	B	C	N
H		110	90	82
C		137	125	120
N	144	124	119	
O	133	119	114	
F	128	113	108	108
Cl	168	151	144	142

TABLE 7.4 Data supporting the LCP model in the prediction of molecular structure.

Compound	C–F	C–Cl	Angle	F···Cl
$FCCl_3$	133	176	109.3	253
F_2CCl_2	134.5	174.4	109.5	253
F_3CCl	132.8	175.1	110.4	254
F_2CHCl	135.0	174.7	110.1	255
FClC=O	133.4	172.5	108.8	250
Predicted	133	178	107.4	252
Compound	*C–F*	*C–O*	*Angle*	*O···F*
$(CF_3)_2O$	132.7	136.9	110.2	221
CF_3O^-	139.2	122.7	116.2	223
CF_3OF	131.9	139.5	109.6	222
Predicted	133	142	107.6	222
Compound	*C–F*	*C=O*	*Angle*	*O···F*
$F_2C=O$	131.7	117.0	126.2	222
$FCH_3C=O$	134.8	118.1	121.7	221
FClC=O	133.4	117.3	123.7	221
FBrC=O	131.7	117.1	125.7	222
Predicted	133	120*	122.6	222
Compound	*C–F*	*C–C*	*Angle*	*C···F*
$F_3C–CF_3$	132.6	154.5	109.8	234
$(CF_3)_3CH$	133.6	156.6	110.9	237
$(CF_3)_3CCl$	133.3	154.4	111.0	237
H_3CCOF	136.2	150.5	110.5	236
Predicted	133	152	109.5	233
Compound	C–F	C=C	*Angle*	*C···F*
$F_2C=CF_2$	131.9	131.1	123.8	232
$F_2C=CCl_2$	131.5	134.5	124.0	235
$F_2C=CH_2$	131.6	132.4	125.2	234
Predicted	133	134*	121.5	233
Compound	*C–C*	*C–Cl*	*Angle*	*C···Cl*
$(CH_3)_2CCl_2$	152.3	179.9	108.9	271
CH_3CH_2Cl	152.8	174.6	110.7	274
CH_3COCl	150.8	179.8	112.2	275
OClCCOCl	153.6	174.6	111.7	272
Predicted	152	178	109.0	269

All distances are measured in picometer. The predicted bond lengths are from the covalent radii in Table 5.1 (except as noted), nonbonded interligand distances are from Table 7.3, and the predicted bond angles were calculated using Equation (7.1).
*http://www.wiredchemist.com/chemistry/data/bond_energies_lengths.html.

Example 7-4. Using covalent radii from Table 5.1 and the ligand radii from Table 7.3, predict the bond angles in each of the following molecules or ions: (a) BeF_4^{2-}, (b) $Cl_3C\text{-}CCl_3$, and (c) NF_4^+.

Solution. (a) The covalent radii of Be and F are 96 and 57 pm, respectively. Thus, the predicted Be–F bond length is 153 pm. The interligand F···F distance calculated from the ligand radius for F with Be as the central atom (2 × 128 pm) is predicted to be 256 pm. Using Equation (7.1), with $a = b = 153$ pm and $c = 256$ pm, the bond angle is predicted to be 113.6°. The actual Be–F bond length is 155.4 pm and the bond angle is 109.5°. (b) The covalent radii of C and Cl are 76 and 102 pm, respectively. Thus, the predicted C–Cl bond length is 178 pm. The interligand Cl···Cl distance calculated from the ligand radius for Cl with C as the central atom (2 × 144 pm) is predicted to be 288 pm. Using Equation (7.1), with $a = b = 178$ pm and $c = 288$ pm, the bond angle is predicted to be 108.0°. The actual C–C bond length is 176.9 pm and the bond angle is 108.9°. (c) The covalent radii of N and F are 71 and 57 pm, respectively. Thus, the predicted N–F bond length is 128 pm. The interligand F···F distance calculated from the ligand radius for F with N as the central atom (2 × 108 pm) is predicted to be 216 pm. Using Equation (7.1), with $a = b = 128$ pm and $c = 216$ pm, the bond angle is predicted to be 115.1°. The actual N–F bond length is 130 pm and the bond angle is 109.5°.

Up until this point, the LCP model has proven useful in the prediction of molecular geometries for molecules having Period 2 central atoms and no lone pairs. Thus, the question arises as to whether the model can also be extended to include a broader range of compounds. If a central atom contains lone-pair electrons in addition to its bonding pair electrons, its electron cloud will be polarized toward both the bonding-pair and the lone-pair regions. In this case, the lone pairs on the central atom act as pseudoligands, spreading out as much as possible and pushing the bonded ligands into a close-packed structure. For example, an AX_3E (where E is a lone pair on the central atom A and X is a bonded ligand) will have the same interligand distances as in an AX_4 geometry, resulting in an identical molecular geometry as the one predicted by VSEPR theory. As a result, the LCP model can also correctly predict the molecular geometries of molecules, but it has the advantage of a more quantitative prediction of the bond angles using Equation (7.1), as shown in the following example.

Example 7-5. Given the molecular geometries of H_2O, HOF, and OF_2 shown below, use the LCP model to predict the bond angle in HOF.

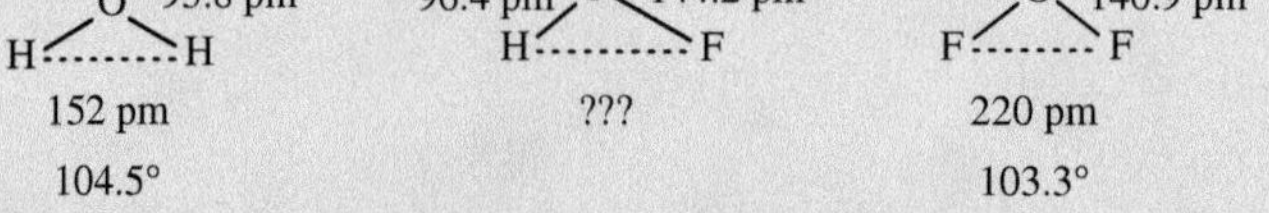

H_2O	HOF	OF_2
O–H 95.8 pm	O–H 96.4 pm, O–F 144.2 pm	O–F 140.9 pm
H···H 152 pm	H···F ???	F···F 220 pm
104.5°		103.3°

Solution. The ligand radius of H with O as the central atom is 0.5(152 pm) = 76 pm and the ligand radius of F with O as the central atom is 0.5(220 pm) = 110 pm. Thus, the predicted H···F nonbonded distance in HOF is 76 + 110 = 186 pm. Using $a = 96.4$, $b = 144.2$, and $c = 186$ in Equation (7.1) yields a predicted HOF bond angle of 99.3°. The actual interligand distance in HOF is 183 pm and the experimental bond angle is 97.2°. In either case, the angle is smaller than the one predicted on the basis of VSEPR theory. VSEPR theory would correctly predict that each of the bond angles is less than the ideal 109.5° for a tetrahedral electron geometry. It would also qualitatively predict that the bond angle in OF_2 would be smaller than the one in H_2O because the bonding electrons are held closer to the central O atom in the H_2O molecule due to O being the more electronegative atom. However, for HOF, the VSEPR model would predict that the bond angle should be intermediate between the observed 103.3° bond angle for OF_2 and the 104.5° bond angle for H_2O, which is clearly not the case! On the other hand, the LCP model, with its assumption of nearly constant interligand distances, can be used to explain the smaller HOF bond angle.

7.3 A COMPARISON OF THE VSEPR AND LCP MODELS

Both the VSEPR and the LCP models can be used in a complementary manner to predict the general three-dimensional shapes of molecules. In one sense, the two models are essentially the same, except for the replacement of BP–BP repulsions in the VSEPR model with ligand–ligand repulsions and the concept of a ligand radius in the LCP model. However, there are also some significant differences between the two models. In addition to being able to predict the interligand distances in molecules that allow for the quantitative determination of bond angles, the LCP model can also explain the molecular shapes of certain molecules that VSEPR cannot. For instance, the LCP model correctly predicts that Li_2O will be linear, unlike the water molecule, because the Li ligand is not electronegative enough to fully localize the electron pairs on O into a tetrahedron. Thus, the geometry in Li_2O will be dictated by the ligand–ligand interactions and the molecule will close-pack in the linear geometry. Likewise, while the VSEPR model would predict a bond angle in $N(CF_3)_3$ of less than 109.5°, the actual bond C–N–C bond angle in this molecule is 117.9°. This result is closer to the 120° bond angle predicted on the basis of the close-packing of three bulky ligands around the central atom than it is to the 109.5° bond angle based on the repulsion of four electron pairs.

One of the criticisms of the LCP model, however, is that it is more or less an exercise in applying empirically derived values. Furthermore, the quantitative predictions that it makes become a little dicier for central atoms beyond the Period 2 elements. For these molecules, the ligands are not squeezed as tightly together in the molecule and their electron clouds will also be "softer." Furthermore, because of the larger radius of the central atom, close-packing of the ligands will only occur for very large values of the coordination number, where the ligands are necessarily more crowded. The combination of these two factors implies that the LCP model, which takes a hard-spheres approach, will not work quite as well as it did for the Period 2 elements. Despite these considerations, Gillespie has tabulated values for Period 3 and higher central atoms and ligand–ligand interactions remain an important factor in determining the overall geometry of these molecules.

To summarize, the VSEPR model is intuitively simple in its formulation and it can be used to predict not only the molecular geometry but also the subtle differences in bond lengths and bond angles. The LCP model, on the other hand, is more general and allows for a more quantitative prediction of the bond angles; however, it has the disadvantage of having to memorize the empirically based ligand radii for each of the different ligands and central atoms. By comparison with the van Arkel triangle of bonding, VSEPR theory can be regarded as approaching the determination of molecular geometries from a purely covalent perspective. It then introduces the effect of polar covalent bonding on the bond angles by considering the positions of the bonding pair electrons in the bonding electron domain as a subsequent perturbation. The VSEPR model works best when the electrons are well localized into bonding and nonbonding pairs. On the other hand, the LCP model can be considered a purely ionic model. It functions best when the ligands can be regarded as hard-spheres whose electron clouds do not interpenetrate with one another. It accounts for polar covalent bonding by changing the size of the ligand radii depending on the electronegativity of the central atom to which the ligands are attached. Thus, the LCP model works especially well for small, electropositive Period 2 central atoms having strongly electronegative ligands (such as O or F), where the electron density on the ligands is closer to that of a free ion. It also works well with larger ligands (such as Cl or CX_3), where ligand–ligand interactions are likely to take precedence over electron pair repulsions. Thus, in a sense, the two models are entirely complementary to each other, and a more complete understanding of molecular geometry can only be achieved when each approach is applied appropriately to the molecule under consideration.

EXERCISES

7.1. Use the VSEPR model to predict the molecular shapes and indicate the qualitative bond angles for each of the following molecules or ions:

a. ClF_3
b. NO_2^+
c. IF_5
d. ClO_2F
e. I_3^-
f. CCl_4
g. XeO_4
h. SF_6
i. $XeOF_4$
j. SNF_3
k. XeF_4
l. XeO_3
m. SF_2
n. KrF_2
o. SO_2
p. HOCl
q. CO_2
r. CCl_3^+
s. SO_4^{2-}
t. NOF

7.2. For each pair of molecules, predict which will have the larger bond angle and rationalize your answer:

a. NH_3 or NF_3
b. PF_3 or NF_3
c. NH_3or BH_3
d. SO_2 or SO_3
e. The smallest equatorial angle in PF_4Cl or PF_3Cl_2

7.3. For each of the following pairs, predict which will have the larger bond length and rationalize your answer:

a. $Br–F_{ax}$ in BrF_3 or in BrF_5
b. $P–Cl_{ax}$ or $P–Cl_{eq}$ in PCl_5
c. $X–F_{ax}$ in ClF_5 or in BrF_5
d. C–C in H_2CCH_2 or in HCCH
e. O–F in HOF or in OF_2

7.4. Use the data from Tables 5.1 and 7.3 to calculate all of the nonbonded distances and bond angles in each of the following molecules:

a. $CClF_3$
b. NH_2^-
c. BCl_2F

7.5. Rationalize why the C–F bond length in CF_3^+ is 123.5 pm, whereas the C–F bond length in $F_2C{=}O$ is 131.7 pm. What effect does the longer C–F bond length in $F_2C{=}O$ have on the F–C–F bond angle? Explain.

7.6. The calculated atomic charges on F for the series of compounds BeF_2, BF_3, and CF_4 are −0.88, −0.81, and −0.61, respectively. Explain how these atomic charges reinforce the notion that the ligand radius of F depends on the nature of the central atom. Why is the ligand radius for F larger when it is attached to Be than when it is attached to B or C?

7.7. Use the LCP model to rationalize the following trend in the bond angle around the central atom: LiOH = 180°; $Be(OH)_2$ = 134.5°; $B(OH)_3$ = 112.8°; $C(OH)_4$ = 106.9°; $N(OH)_3$ = 102.6°; $O(OH)_2$ = 98.7°; and $F(OH)_2$ = 98.6°.

BIBLIOGRAPHY

1. Bartell, L. S. Coord. Chem. Rev., 2000, 197, 37–49.
2. Gillespie, R. J.; Hargittai, I. *The VSEPR Model of Molecular Geometry*, Allyn and Bacon, Needham Heights, MA, 1991.
3. Gillespie, R. J.; Popelier, P. L. A. *Chemical Bonding and Molecular Geometry*, Oxford University Press, New York, 2001.
4. Housecroft, C. E.; Sharpe, A. G. *Inorganic Chemistry*, 3rd ed., Pearson Education Limited, Essex, England, 2008.
5. Huheey, J. E.; Keiter, E. A.; Keiter, R. L. *Inorganic Chemistry: Principles of Structure and Reactivity*, 4th ed., Harper Collins College Publishers, New York, 1993.
6. Miessler, G. L.; Tarr, D. A. *Inorganic Chemistry*, 4th ed., Pearson Education Inc., Upper Saddle River, NJ, 2011.

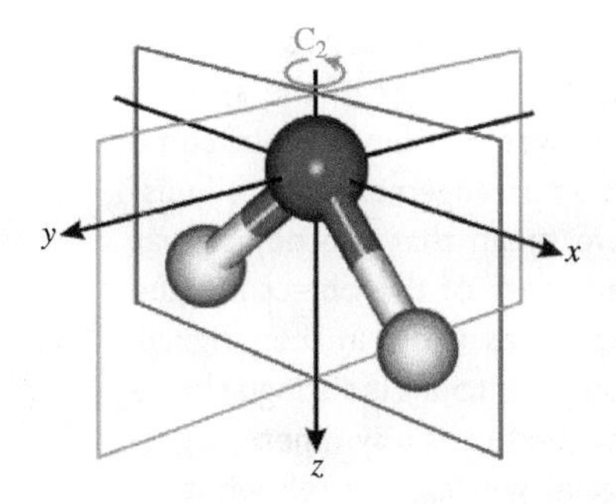

Molecular Symmetry 8

"Symmetry plays a large role in what is often called physical intuition."
—*R. M. Hochstrasser*

8.1 SYMMETRY ELEMENTS AND SYMMETRY OPERATIONS

In the previous chapter, we saw that molecular geometry was a consequence of the tradeoff between electronic effects (the electron–electron repulsions that result from the Pauli principle) and steric effects (the nuclear–nuclear repulsions between the ligands on the central atom). In this chapter, we are concerned with the determination of molecular symmetry. While molecular geometry is concerned with the shapes of molecules, molecular symmetry has to do with the spatial relationships between atoms in molecules. As we shall see, it is the three-dimensional shape of a molecule that dictates its molecular symmetry and we can use a mathematical description of symmetry properties, known as *group theory*, to describe the structure, bonding, and spectroscopy of molecules.

In our discussion of the particle in a box problem in Chapter 3, we have already seen one of the ways in which the symmetry of an object affects its chemical properties. When the symmetry of the three-dimensional box was lowered from a cube to form a parallelepiped, some of the degeneracy in the quantum mechanical energy levels was lost as a result of the symmetry breaking. The same property is also true of molecules—the more symmetric the molecule, the greater its degeneracy, and the simpler the construction of its energy levels. In large molecules having a high degree of symmetry, we can therefore take advantage of the symmetrical relationships of the different atoms in molecules to simplify the mathematical solutions to the wave equation.

The first step in the application of symmetry to molecular properties is therefore to recognize and organize all of the symmetry elements that the molecule possesses. A *symmetry element* is an imaginary point, line, or plane in the molecule about which a symmetry operation is performed. An *operator* is a symbol that tells you to do something to whatever follows it. Thus, for example, the Hamiltonian operator is the sum of the partial differential equations relating to the kinetic and

Water molecule with symmetry elements. [M. Chaplin, http://wwwl.lsbu.ac.uk/water/hoorb.html (accessed December 16, 2013).]

Principles of Inorganic Chemistry, First Edition. Brian W. Pfennig.

potential energy of a system. When we apply the Hamiltonian operator to the wave function of an atom, the solutions correspond to the total energy of the system. A *symmetry operation* is a geometrical operation that moves an object about some symmetry element in a way that brings the object into an arrangement that is indistinguishable from the original. By indistinguishable, we mean that the object was returned to an equivalent position, one in which every part of the object has the same orientation and the same relative position in space as it did in the original. One way of determining whether the object was brought into an indistinguishable arrangement is to close your eyes while somebody else performs a symmetry operation on the object. If, when you open your eyes again, you cannot tell whether or not the other person has actually done anything to the object, then the object has been brought into an equivalent arrangement and it is indistinguishable from the original.

There are two basic categories of symmetry operations: translational symmetry operations and point symmetry operations. *Translational symmetry* results from the movement of an object in space by a certain distance and in a certain direction. Examples of translational symmetry are the artificial ducks that you might encounter in a shooting gallery. The ducks move in one continuous line at a constant velocity across the front of the target. If you were to close your eyes for just the right amount of time for a duck to move an integral number of places down the line and you then reopened your eyes, you would not be able to tell whether any of the ducks had moved. This type of symmetry occurs within the extended lattices of crystalline solids and (in part) leads to the classification of solids into space groups. *Point symmetry*, on the other hand, occurs in molecules where there is at least one point in space that remains unchanged with respect to any symmetry operation. Snowflakes are an excellent example of point symmetry. Each snowflake has a sixfold rotational axis that passes directly through the center of the snowflake. If you were to close your eyes while some fictional Jack Frost rotated the snowflake by any integral multiple of 60° and you then reopened your eyes, the snowflake would appear to be in a position that was indistinguishable from the original. In the case of point symmetry, all of the symmetry elements that the object possesses must have at least one point in common. This point occurs at the exact center or at the origin of the object.

In this chapter, we are only concerned about the point symmetry of molecules. We use the point symmetry operations present in molecules to classify them into molecular point groups. There are five kinds of point symmetry elements that a molecule can possess, and therefore, there are also five kinds of point symmetry operations.

8.1.1 Identity, E

The first and easiest type of point symmetry operation to identify in a molecule is known as the *identity operation*. The identity operation, given the symbol *E*, tells you to do exactly nothing at all to the molecule. Therefore, every molecule—even one that is lacking any symmetry at all—has one and only one identity operation. The identity operation is only required because it is necessary to satisfy the criteria for a mathematical group: namely, that the group contains at least one member that commutes with every other member to leave them unchanged. Thus, for example, the identity operation for the multiplication of a set of real numbers is 1 because multiplication of any other number by 1 gives back the original number. We will show in a later section that the complete set of symmetry operations for any molecule must always satisfy the following four requirements for a mathematical group: identity, closure, associativity, and reciprocity.

8.1.2 Proper Rotation, C_n

The second type of point symmetry operation is the proper rotation. A proper rotation is a symmetry operation that occurs around a line known as the *proper rotational axis* in order to bring the object into an equivalent configuration. For example, a snowflake has a sixfold proper rotational axis that runs perpendicular to the plane of the snowflake and that passes directly through its center. Proper rotations are abbreviated by the symbol C_n, where n is equal to rotation around the axis by $2\pi/n$ radians (or $360°/n$). The benzene molecule also has a C_6 proper rotational axis that passes directly through the donut hole in its center, as shown in Figure 8.1(a). Rotation of the molecule by 60° moves each C atom into its neighbor's original position. Notice that the principal axis does not necessarily need to pass through an atom. The C_6 rotational axis can also serve as a C_3 axis, as rotation by 120° also leads to an equivalent configuration. We say that the C_3 rotational axis is collinear with the C_6 rotational axis. In fact, the C_3 operation is identical to performing two sequential C_6 operations on the molecule. In the benzene molecule, there is also a C_2 rotational axis that is collinear with the aforementioned C_6 and C_3 axes and that corresponds with rotation of the molecule by 180°.

This example serves to illustrate one of the fundamental properties of symmetry operations—that they can be multiplied together in much the same manner as the set of real numbers can be multiplied (with the exception that the symmetry operations of a molecule do not necessarily commute and therefore the order of multiplication matters). Two C_6 rotations performed in succession are identical to one C_3 rotation about the same axis. Similarly, three C_6 rotations are equivalent to one C_2 rotation. The symmetry operations are said to multiply together, as shown in Equations (8.1) and (8.2), where (by convention) the last operation in the product is the one that is always performed first.

$$C_6C_6 = C_6^2 = C_3 \tag{8.1}$$

$$C_6C_6C_6 = C_6^3 = C_2 \tag{8.2}$$

There are also an additional six C_2 proper rotational axes in the benzene molecule. If one were to look at benzene from an edge-on vantage point and look directly down the line that connects one of the C atoms with the imaginary center of the molecule and with a C atom directly opposite the first one, the entire benzene ring could be rotated by 180° (or flipped over like a pancake) to form an equivalent configuration. Thus, there are three C_2' rotational axes that are each perpendicular to the C_6 axis of the molecule and which each contain two of the six C atoms, as shown in Figure 8.1(b). There are also three C_2'' rotational axes that also lie in the plane of the molecule but which bisect the C–C bonds, as shown in Figure 8.1(c). The rotational axis with the largest value of n is called the *principal axis* of rotation. In this case, the principal axis is the C_6 rotational axis. Some molecules with higher symmetry (such as a tetrahedral molecule) may actually contain more than one principal axis. By convention, the z coordinate axis is always taken as the principal axis. The direction of rotation around the principal

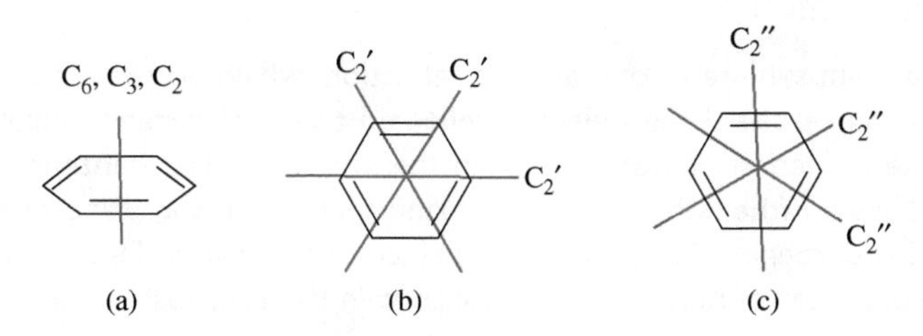

FIGURE 8.1
Proper rotational axes in the benzene molecule: (a) the principal axis C_6 (which also has collinear C_3 and C_2 axes), (b) three C_2' axes that are perpendicular to the principal axis and that each contain two C atoms, and (c) three C_2'' axes that are perpendicular to the principal axis but bisect the C–C bonds.

axis is arbitrary. Some textbooks prefer to use clockwise rotations, whereas others define the rotation in the counterclockwise direction. The choice is entirely up to the author's preference, for it has no real significance (as long as it is always applied in a self-consistent manner) because it is the result of the operation itself that brings the object into an equivalent position.

Example 8-1. Identify all of the proper rotational axes in the square planar $PtCl_4^{2-}$ ion and indicate which is the principal axis. Then show that the following multiplications of symmetry operations are valid: $C_4\,C_4 = (C_4)^2 = C_2$ and $(C_4)^4 = E$.

Solution. The principal axis is the one with the largest value of *n*, which in this case is the C_4 axis.

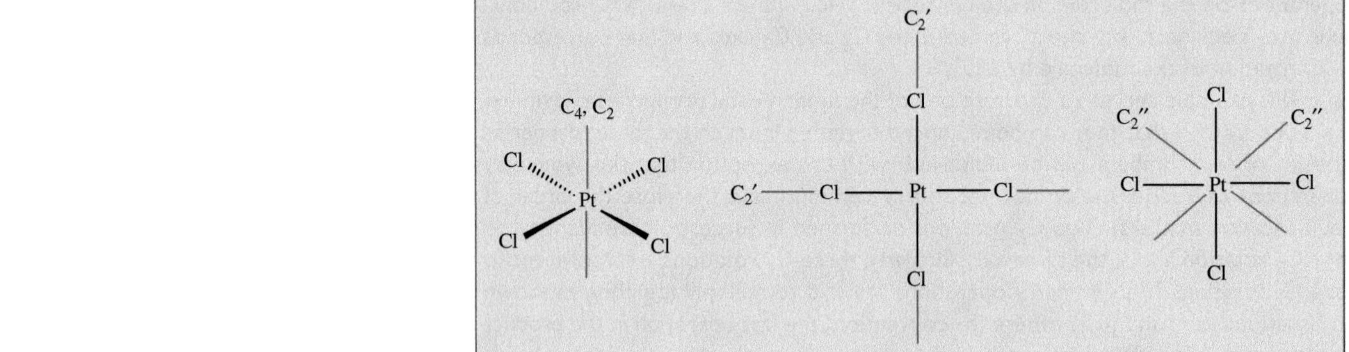

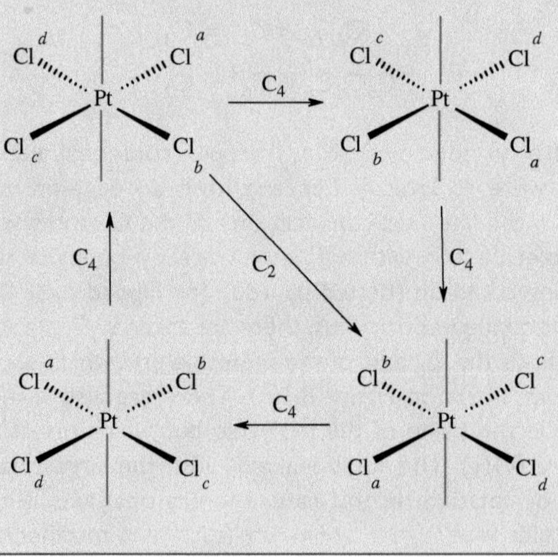

8.1.3 Reflection, σ

The third point symmetry operation is reflection, which occurs when an equivalent configuration results if the object is reflected an equal distance through a mirror plane. The reflection operation is given the symbol σ. The benzene molecule, for instance, has a rather obvious mirror plane that lies in the plane of the molecule itself and is perpendicular to the principal axis, as shown in Figure 8.2(a). Benzene also exhibits a set of mirror planes that contain the principal axis and pass through

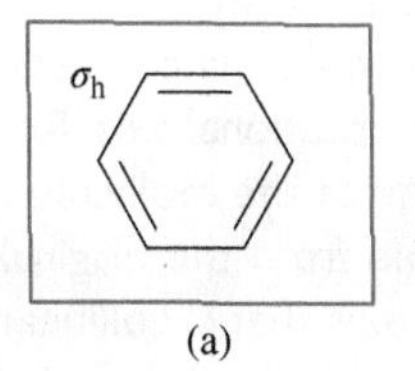

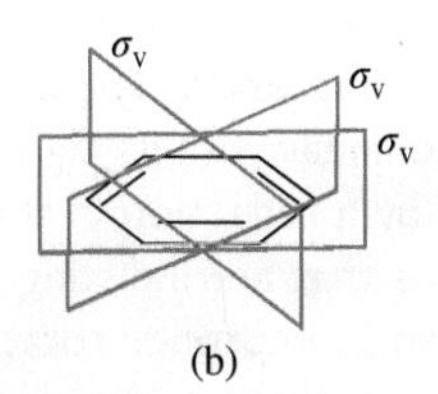

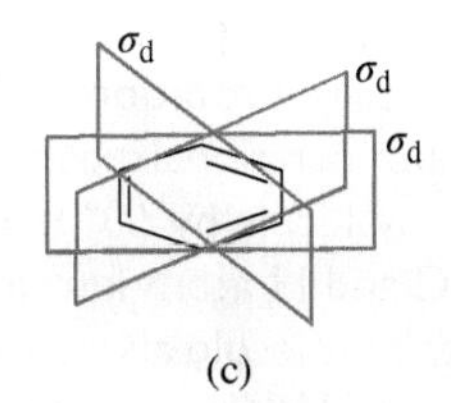

FIGURE 8.2
The different types of mirror plane reflections in the benzene molecule: (a) horizontal mirror plane, (b) vertical mirror planes, and (c) dihedral mirror planes.

two of the C atoms, while simultaneously containing one of the three C_2' axes, as shown in Figure 8.2(b). Finally, there exist three more mirror planes in the molecule that contain the principal axis, bisect a C–C bond, and contain one of the three C_2'' axes, as shown in Figure 8.2(c). Whenever there is more than one mirror plane in a molecule, the mirror planes are divided into classes on the basis of geometrical considerations and given subscripts in order to distinguish between them. A horizontal mirror plane (σ_h) is always perpendicular to the principal axis. Therefore, there can only ever be one horizontal mirror plane in any given molecule. The mirror plane shown in Figure 8.2(a) is a horizontal mirror plane. Vertical mirror planes (σ_v), on the other hand, always contain the principal axis. In the case of molecules such as benzene, where there are two different types of vertical mirror planes, a further distinction between the classes is required. The mirror planes in Figure 8.2(b) that pass through the C atoms and contain as many atoms as possible are designated as σ_v, whereas the mirror planes in Figure 8.2(c) that bisect the C–C bonds are known as *dihedral mirror planes* and are given the symbol σ_d. A dihedral mirror plane is simply a subset of a vertical mirror plane. A more rigorous definition of the difference between vertical and dihedral mirror planes is that a vertical mirror plane will contain more atoms than a dihedral one, or (if they contain the same number of atoms) a vertical mirror plane will contain one of the three Cartesian axes (the *x*-, *y*-, or *z*-axes).

8.1.4 Inversion, *i*

The fourth point symmetry operation is *inversion*. The inversion operation, given the symbol *i*, takes every point (x, y, z) and converts it into an equivalent configuration at point $(-x, -y, -z)$. Thus, there can only be one inversion center and it can only occur in the exact center of the molecule. The inversion center in benzene is shown by an imaginary black point that lies in the center of the molecule shown in Figure 8.3. Notice that the inversion center need not contain an atom. Because the labels (a–f) on the C atoms are imaginary, inversion still brings the molecule into a configuration that is equivalent to the original.

8.1.5 Improper Rotation, S_n

The fifth type of point symmetry operation is an improper rotation, given the symbol S_n. Improper rotations occur around an improper rotational axis by rotating the molecule by $2\pi/n$ radians and then reflecting the molecule through a mirror

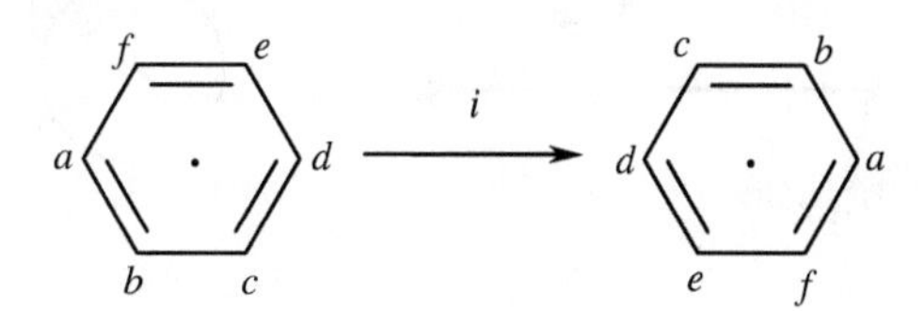

FIGURE 8.3
The inversion center in benzene lies in the plane of the molecule directly in the center of the ring.

plane that is perpendicular to the improper rotational axis. Thus, an improper rotation is really a rotation–reflection operation. The benzene molecule contains an S_6 improper rotational axis that is collinear with its C_6 proper rotational axis. Rotation of the molecule by 60° followed by a reflection in the plane of the molecule brings each C and H atom into positions that are indistinguishable from the original. The benzene molecule also contains an S_3 improper rotational axis that is collinear with its S_6 axis. While it is often the case that an improper rotational axis is coincident with a proper rotational axis of the same order, this does not necessarily need to be the case. Another example of an S_6 rotation is given in Figure 8.4. Notice that the first step in performing the S_6 operation (rotation by 60° in this case) does not return the object to an equivalent configuration in this instance. It is only after both steps (rotation and then reflection) that the object is brought back to a position that is indistinguishable from the original. Thus, the S_6 operation is in fact a unique and distinct symmetry operation. Although it performs the same function as the product σ_h C_3 (recall that in a product the second operation is always the one that is performed first), the object need not contain a C_3 proper rotational axis. Also note the following trivial improper rotations: $S_1 = \sigma$ and $S_2 = i$.

In the staggered configuration of ethane, an S_6 axis exists, even though the molecule lacks a C_6 proper rotational axis, as shown in Figure 8.5. The S_6 improper rotational axis is, however, collinear with the lower-order C_3 proper rotational axis in this molecule.

Tetrahedral molecules contain three S_4 axes, as shown in Figure 8.6(a). If the central atom is placed in the center of an imaginary cube, with the four terminal atoms at opposite corners of the cube, then the three S_4 axes will pass through the centers of the faces of the cube. Figure 8.6(b) illustrates the combination rotation–reflection for one of the S_4 rotational axes in CCl_4, where the Cl ligands have been labeled with imaginary subscripts in order to demonstrate the transformation. Because the subscripts are only present for bookkeeping purposes and are not real, a single S_4 rotation moves the molecule into an equivalent configuration.

Table 8.1 summarizes the five different types of point symmetry elements and their corresponding symmetry operations.

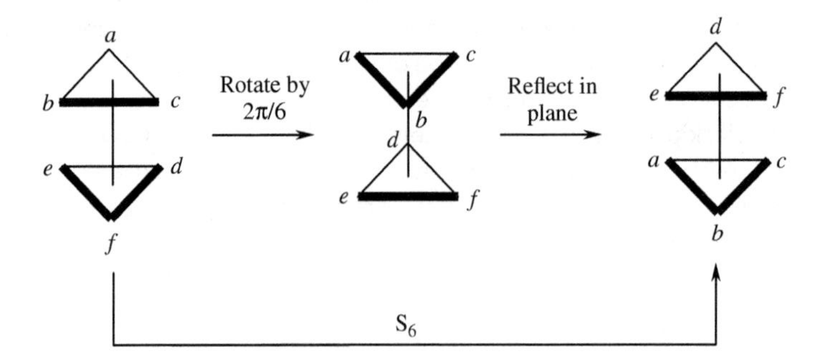

FIGURE 8.4
Stepwise illustration of an S_6 rotation, showing how an improper rotation consists of a rotation around the axis by $2\pi/6$ radians, followed by reflection through a perpendicular plane.

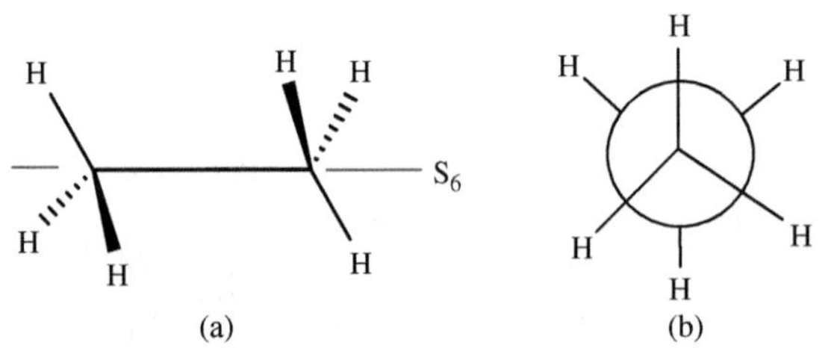

FIGURE 8.5
(a, b) The staggered conformation of ethane has an S_6 improper rotational axis that is collinear with its C_3 proper rotational axis. The Newman projection for staggered ethane shown at right is oriented so as to look directly down the S_6 axis.

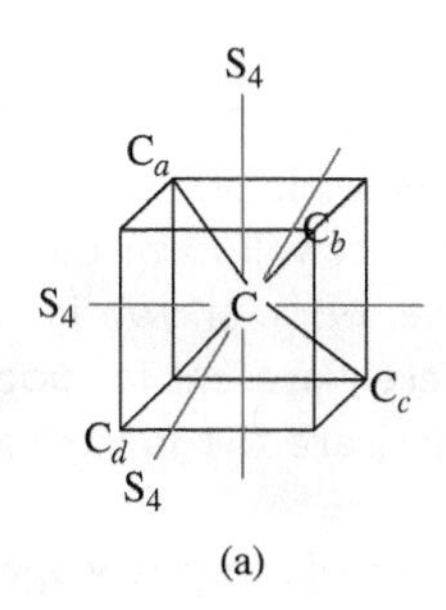

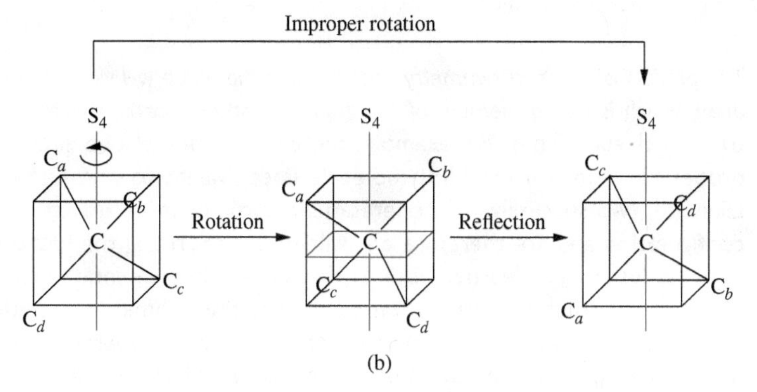

FIGURE 8.6
(a) Illustration of the three S_4 improper rotational axes in the tetrahedral CCl_4 molecule as collinear with the three Cartesian axes and (b) an illustration of how the S_4 improper rotation is equivalent to a combination of a rotation followed by a reflection.

TABLE 8.1 The different types of point symmetry elements and operations.

Symmetry Operation	Symmetry Element	What Happens?
Proper rotation, C_n	Proper rotational axis	Rotation around the axis by $2\pi/n$ radians
Inversion, *i*	Point (known as the *inversion center*)	Convert each point (*x*, *y*, *z*) into (−*x*, −*y*, −*z*)
Reflection, σ	Mirror plane	Reflection through a plane
Improper rotation, S_n	Improper rotational axis	Rotation by $2\pi/n$ radians, followed by reflection in a ⊥ plane
Identity, *E*	Entirety of space	Nothing (the entire object remains unchanged)

Example 8-2. Identify all of the symmetry elements present in each of the following ions or molecules: (a) $PtCl_4^{2-}$, (b) NH_3, and (c) $[Co(en)_3]^{3+}$.

Solution. (a) $PtCl_4^{2-}$ is square planar and has the following symmetry elements: *E*, C_4, C_2 (collinear with C_4), $2C_2'$, $2C_2''$, σ_h, $2\sigma_v$, $2\sigma_d$, *i*, and S_4 (collinear with C_4). (b) NH_3 is trigonal pyramidal and has the following symmetry elements: *E*, C_3, and $3\sigma_v$. (c) $[Co(en)_3]^{3+}$ is shown in Figure 7.1(b) and has the following symmetry elements: *E*, C_3, and $3C_2$.

8.2 SYMMETRY GROUPS

When a complete set of all the possible point symmetry operations for a molecule has been identified, the resulting list will form the basis for a mathematical group. Consider the ammonia molecule, NH_3, shown in Figure 8.7 (the labels on the H atoms are imaginary and are merely present for bookkeeping purposes). The symmetry operations present in NH_3 are the following: C_3, C_3^2, E ($=C_3^3$), σ_{v1}, σ_{v2}, and σ_{v3}.

A *symmetry group* is a collection of symmetry operations that can be interrelated according to the following rules:

1. *The product of any two symmetry operations in the group will yield another operation, which is also a member of the group.* In other words, the group needs to have closure. Thus, for example, the performance of two successive C_3 proper rotations on the NH_3 molecule is equivalent to one C_3^2 rotation. Likewise, two successive σ_{v1} operations return the molecule to its original configuration and are therefore equivalent to performing the identity operation, *E*. Just as you learned your "times tables" in elementary school, you can also construct a multiplication table using the symmetry operations for ammonia. By convention, the product of the symmetry operation in column *A* with the operation in row *B* is written as *AB*, where operation *B* is performed on the molecule first, followed by operation *A*. The order in which the operations are performed is critical because the symmetry operations do not necessarily commute (those symmetry groups for which all the operations do commute with one another are called *cyclic* or *Abelian* groups). The multiplication table for NH_3 is shown in Table 8.2.

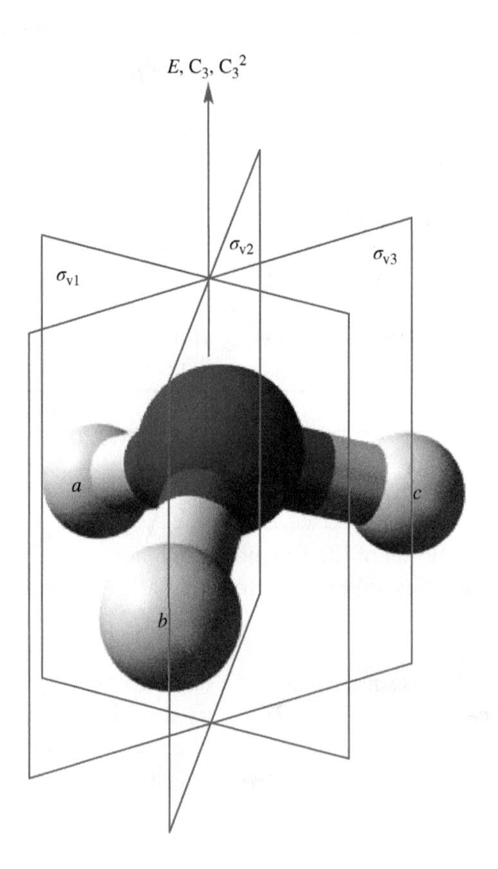

FIGURE 8.7
The complete list of symmetry operations present in the NH_3 molecule.

TABLE 8.2 Multiplication table for the symmetry operations in NH_3.

NH_3	E	C_3	C_3^2	σ_{v1}	σ_{v2}	σ_{v3}
E	E	C_3	C_3^2	σ_{v1}	σ_{v2}	σ_{v3}
C_3	C_3	C_3^2	E	σ_{v2}	σ_{v3}	σ_{v1}
C_3^2	C_3^2	E	C_3	σ_{v3}	σ_{v1}	σ_{v2}
σ_{v1}	σ_{v1}	σ_{v3}	σ_{v2}	E	C_3^2	C_3
σ_{v2}	σ_{v2}	σ_{v1}	σ_{v3}	C_3	E	C_3^2
σ_{v3}	σ_{v3}	σ_{v2}	σ_{v1}	C_3^2	C_3	E

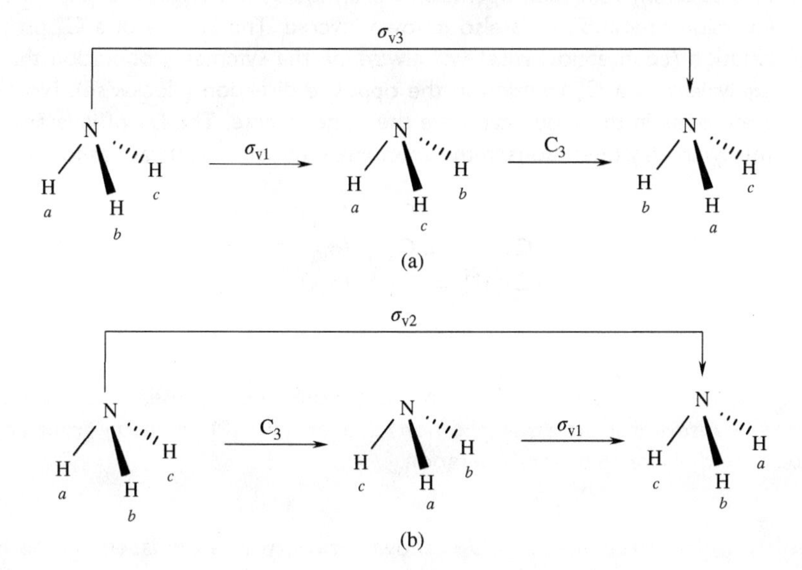

FIGURE 8.8
(a) The product $C_3\ \sigma_{v1}$ is identical to σ_{v3}, whereas (b) the product $\sigma_{v1}\ C_3$ is identical to σ_{v2}, showing how the product of these two operations does not commute in the NH_3 symmetry group. By convention, the operations are performed in the reverse order from the way that they are written.

There are several features of the multiplication table for NH_3 that are worth noting. First, this symmetry group is not Abelian. The product $C_3\ \sigma_{vl}$ (reflection followed by counterclockwise rotation) is identical to performing a σ_{v3} operation on the original molecule, as shown in Figure 8.8(a). However, the product $\sigma_{vl}\ C_3$ (counterclockwise rotation followed by reflection) is equivalent to doing a σ_{v2} operation, as shown in Figure 8.8(b). Thus, the symmetry operations in the NH_3 group do not commute. Second, a close inspection of the multiplication table reveals that each row and each column of the table are nothing more than unique rearrangements of the list of symmetry operations. This phenomenon is universal and is known as the *rearrangement theorem*. Third, the *order* of the group is defined as the total number of all the possible symmetry operations. Thus, the order of the NH_3 symmetry group is six and is given the symbol $h = 6$.

2. *One of the operations in the group (the identity operation) must commute with each of the others to leave them unchanged.* It was this requirement that forced us to introduce the identity operation (E) in the first place. Notice from the multiplication table that E commutes with every other operation to yield back the original operation. Thus, the first row and the first column in the multiplication table will be identical to the symmetry operations in the corresponding headings.
3. *The associate law of multiplication must apply to all the symmetry operations.* In other words, we should be able to group the operations according to

Equation (8.3). Thus, for example, the product σ_{v3} $(C_3\ \sigma_{v1}) = \sigma_{v3}\ \sigma_{v3} = E$ is identical to the product $(\sigma_{v3}\ C_3)\ \sigma_{v1} = \sigma_{v1}\ \sigma_{v1} = E$. Remember that it is always the last symmetry operation in each pairing that is actually performed first.

$$A(BC) = (AB)C \tag{8.3}$$

4. *Every symmetry operation in the group has an inverse operation that is also a member of the group*. In this context, the word "inverse" should not be confused with "inversion." The mathematical *inverse* of an operation is its reciprocal, such that $A\ A^{-1} = A^{-1}\ A = E$, where the symbol A^{-1} represents the inverse of operation A. The identity element will always be its own inverse. Likewise, the inverse of any reflection operation will always be the original reflection. The inversion operation (i) is also its own inverse. The inverse of a C_n proper rotation (counterclockwise) will always be the symmetry operation that is equivalent to a C_n rotation in the opposite direction (clockwise). No two operations in the group can have the same inverse. The list of inverses for the symmetry operations in the ammonia symmetry group are as follows:

$$\begin{array}{llll} E^{-1} & = E & (\sigma_{v1})^{-1} & = \sigma_{v1} \\ (C_3)^{-1} & = C_3^{\,2} & (\sigma_{v2})^{-1} & = \sigma_{v2} \\ (C_3^{\,2})^{-1} & = C_3 & (\sigma_{v3})^{-1} & = \sigma_{v3} \end{array}$$

Example 8-3. Make a list of all the symmetry operations present in the water molecule. Also, construct a multiplication table for the H_2O symmetry group and determine the inverse of each operation.

Solution. The structure of H_2O is shown below, where the labels on the H atoms are imaginary. VSEPR theory predicts tetrahedral electron geometry with a bent molecular geometry and bond angles less than 109.5°. The complete set of symmetry operations is E, C_2, σ_{v1}, and σ_{v2}. The multiplication table is shown below. In this symmetry group, each operation is its own inverse.

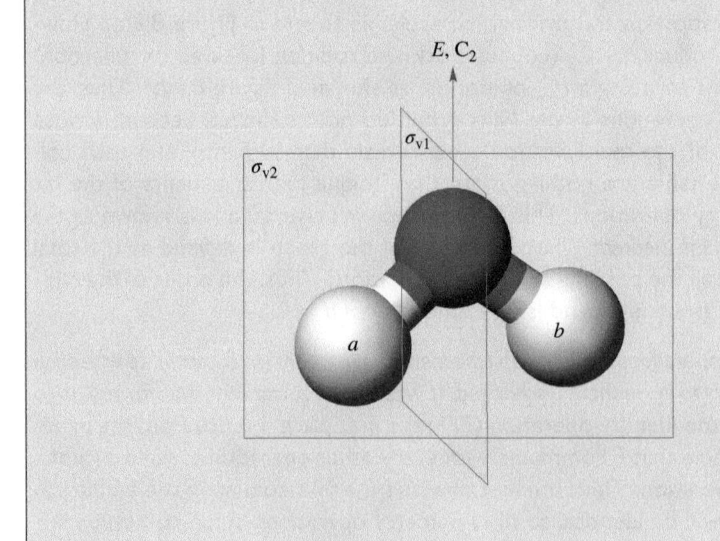

H_2O	E	C_2	σ_{v1}	σ_{v2}
E	E	C_2	σ_{v1}	σ_{v2}
C_2	C_2	E	σ_{v2}	σ_{v1}
σ_{v1}	σ_{v1}	σ_{v2}	E	C_2
σ_{v2}	σ_{v2}	σ_{v1}	C_2	E

Example 8-4. Make a list of all the symmetry operations present in the triphenylphosphine ligand shown below, construct a multiplication table for the $P(C_6H_6)_3$ symmetry group, and determine the inverse of each operation.

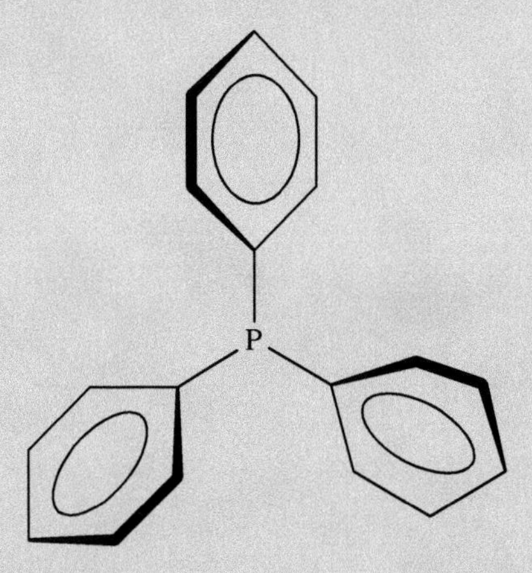

Solution. The structure of triphenylphosphine is shown above. The complete set of symmetry operations is E, C_3, and C_3^2. There are no mirror planes of symmetry because of the propeller nature of the phenyl groups. The multiplication table is shown below.

PPh_3	E	C_3	C_3^2
E	E	C_3	C_3^2
C_3	C_3	C_3^2	E
C_3^2	C_3^2	E	C_3

The inverses are $E^{-1} = E$, $(C_3)^{-1} = C_3^2$, and $(C_3^2)^{-1} = C_3$.

Example 8-5. Make a list of all the symmetry operations present in BH_3, construct a multiplication table for the BH_3 symmetry group, and determine the inverse of each operation.

Solution. The structure of BH_3 is shown below, where the labels on the H atoms are imaginary. VSEPR theory predicts a trigonal planar electron and molecular geometry with bond angles of 120°. The complete set of symmetry operations is E, C_3, C_3^2, C_2, C_2', C_2'', σ_h, S_3, S_3^2, σ_{v1}, σ_{v2}, and σ_{v3}. The multiplication table is shown below. Once about half of the products were determined, the rearrangement theorem was used to help complete the multiplication table.

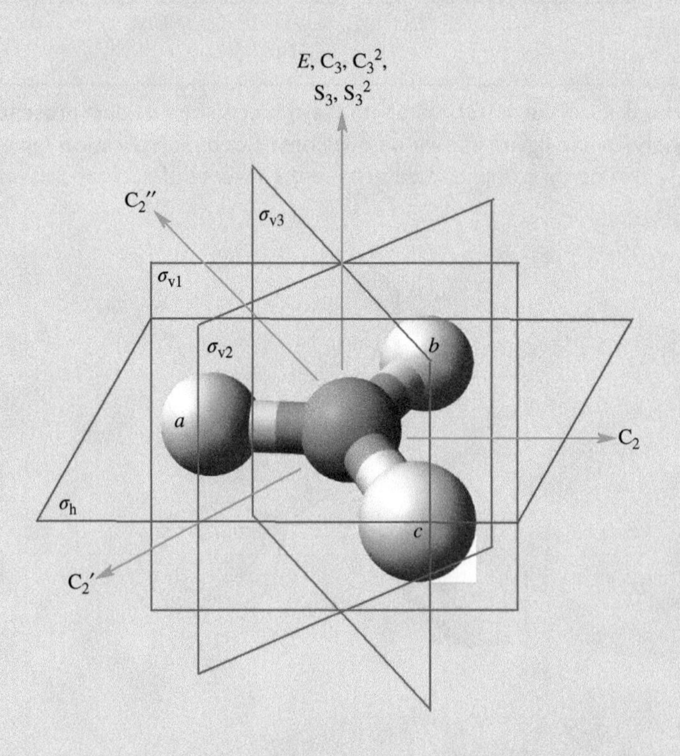

BH_3	E	C_3	C_3^2	C_2	C_2'	C_2''	σ_h	S_3	S_3^2	σ_{v1}	σ_{v2}	σ_{v3}
E	E	C_3	C_3^2	C_2	C_2'	C_2''	σ_h	S_3	S_3^2	σ_{v1}	σ_{v2}	σ_{v3}
C_3	C_3	C_3^2	E	σ_{v3}	σ_{v1}	σ_{v2}	S_3	S_3^2	σ_h	C_2''	C_2	C_2'
C_3^2	C_3^2	E	C_3	σ_{v2}	σ_{v3}	σ_{v1}	S_3^2	σ_h	S_3	C_2'	C_2''	C_2
C_2	C_2	C_2'	C_2''	E	C_3	C_3^2	σ_{v1}	σ_{v2}	σ_{v3}	σ_h	S_3	S_3^2
C_2'	C_2'	C_2''	C_2	C_3^2	E	C_3	σ_{v2}	σ_{v3}	σ_{v1}	S_3^2	σ_h	S_3
C_2''	C_2''	C_2	C_2'	C_3	C_3^2	E	σ_{v3}	σ_{v1}	σ_{v2}	S_3	S_3^2	σ_h
σ_h	σ_h	S_3	S_3^2	σ_{v1}	σ_{v2}	σ_{v3}	E	C_3	C_3^2	C_2	C_2'	C_2''
S_3	S_3	S_3^2	σ_h	σ_{v3}	σ_{v1}	σ_{v2}	C_3	C_3^2	E	C_2''	C_2	C_2'
S_3^2	S_3^2	σ_h	S_3	σ_{v2}	σ_{v3}	σ_{v1}	C_3^2	E	C_3	C_2'	C_2''	C_2
σ_{v1}	σ_{v1}	σ_{v2}	σ_{v3}	σ_h	S_3	S_3^2	C_2	C_2'	C_2''	E	C_3	C_3^2
σ_{v2}	σ_{v2}	σ_{v3}	σ_{v1}	S_3^2	σ_h	S_3	C_2'	C_2''	C_2	C_3^2	E	C_3
σ_{v3}	σ_{v3}	σ_{v1}	σ_{v2}	S_3	S_3^2	σ_h	C_2''	C_2	C_2'	C_3	C_3^2	E

The inverses are $E^{-1} = E$, $(C_3)^{-1} = C_3^2$, $(C_3^2)^{-1} = C_3$, $(C_2)^{-1} = C_2$, $(C_2')^{-1} = C_2'$, $(C_2'')^{-1} = C_2''$, $\sigma_h^{-1} = \sigma_h$, $(S_3)^{-1} = S_3^2$, $(S_3^2)^{-1} = S_3$, $\sigma_{v1}^{-1} = \sigma_{v1}$, $\sigma_{v2}^{-1} = \sigma_{v2}$, and $\sigma_{v3}^{-1} = \sigma_{v3}$.

8.3 MOLECULAR POINT GROUPS

A general observation is that the symmetry groups we have encountered in the examples to present are far from being unique to these individual molecules. Thus, for example, PCl_3, POF_3, and XeO_3 all share the same set of symmetry operations and identical multiplication tables as the one for NH_3. As a result, a more general notation known as a *molecular point group* is used to describe identical symmetry groups. *Molecular point groups* are given labels according to one or the other of two historical designations: the *Hermann–Mauguin notation*, which is used primarily by X-ray crystallographers, or the *Schoenflies notation*, which is used by most chemists and spectroscopists. Our focus in this text is on the more commonly used Schoenflies notation. The point group for any given molecule can be determined by listing all of the molecule's symmetry elements and then following the flowchart given in Figure 8.9. While initially you might need to refer to the flowchart frequently in order to determine the molecular point group, after some practice the identification of a molecule's point group should become so routine that consultation of the flowchart will no longer be necessary.

Using the flowchart in order to determine the molecular point group of NH_3, there are no C_∞ rotational axes (a C_∞ axis will only occur in a linear molecule). The ammonia molecule does, however, have a principal proper rotational axis, C_3. It does not have any C_5 or C_4 axes and there is only one (not four) C_3 axes. Thus, the molecule belongs to one of the point groups in the lower box of Figure 8.9. There are no C_2 axes in NH_3 that are perpendicular to the principal axis and there are no S_6 improper rotational axes. There are also no horizontal mirror planes, but there are $n = 3$ vertical mirror planes. Thus, the molecular point group for NH_3 is C_{3v}.

Example 8-6. Determine the molecular point group in the Schoenflies notation for each of the following molecules: (a) H_2O, (b) PPh_3, and (c) BH_3.

Solution. (a) C_{2v}, (b) C_3, and (c) D_{3h}.

Example 8-7. Determine the molecular point groups for each of the molecules pictured below.

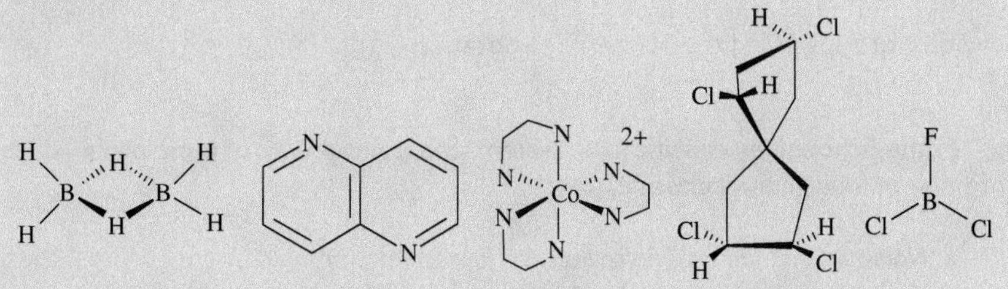

(a) (b) (c) (d) (e)

Solution. (a) D_{2d}, (b) C_{2h}, (c) D_3, (d) S_4, and (e) C_s.

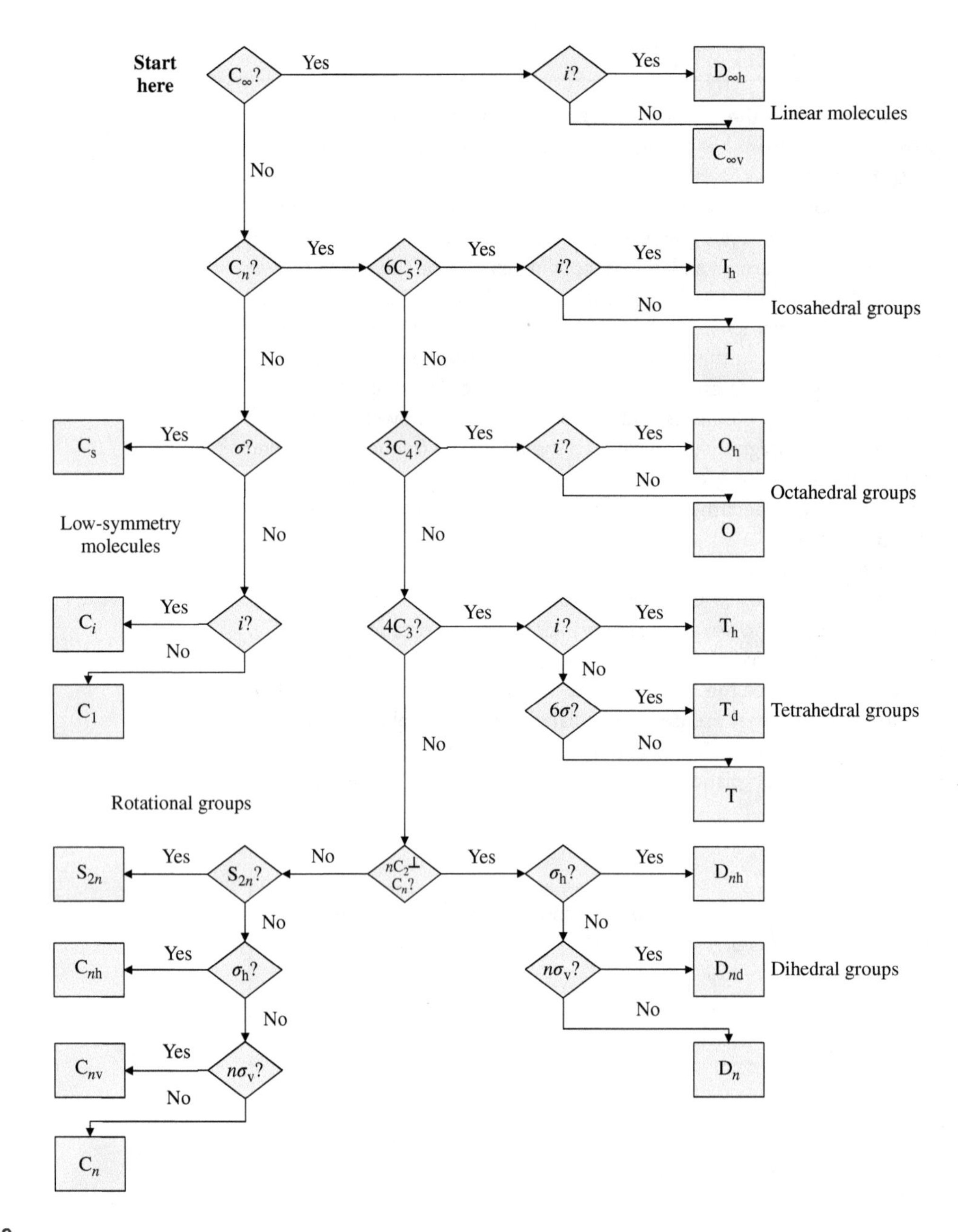

FIGURE 8.9
Flowchart for determining molecular point groups using the Schoenflies notation.

In the Schoenflies classification system, the point group of a molecule will fall into one of four general classifications:

- *Nonrotational groups (molecules with low symmetry)*:
  These include molecules that have no rotational axes (proper or improper) and include the point groups: C_1 (no symmetry elements other than the identity), C_i (only the identity and an inversion center), and C_s (only the identity and a mirror plane).
- *Single-axis rotational groups*:
  These point groups include molecules that have only a single rotational axis (there may, however, be more than one type of operation that can

occur around this axis—for instance, it can have a collinear C_3 and S_6 axis). Molecules that have only proper rotations (in addition to the identity) belong to the C_n cyclic point groups (where $n = 2, 3, \ldots, \infty$). The $C_{\infty v}$ point group is a special case that occurs in linear molecules that are noncentrosymmetric. If a molecule has only a single proper rotational axis C_n and it also has a horizontal mirror plane, it will belong to the class of C_{nh} point groups. If the molecule has a single C_n axis and *n* vertical mirror planes exist, the molecule will belong to the class of C_{nv} point groups. Lastly, if the molecule has a proper rotational axis and a collinear improper rotational axis, it will belong to the class of S_{2n} point groups.

- *Dihedral groups*:

  Molecules belonging to this class of point groups will have *n* C_2 axes that are perpendicular to the principal C_n axis. If these are the only symmetry elements that the molecule has (besides the identity), the molecular point group will be D_n. If a horizontal mirror plane is also present, the molecule will necessarily have a S_n improper rotational axis that is collinear with its principal axis and *n* vertical or dihedral mirror planes. When *n* is even, it will also have an inversion center. These molecules will belong to the class of D_{nh} point groups. The $D_{\infty h}$ point group is a special case for linear molecules that are also centrosymmetric. If no horizontal mirror plane is present, but there are *n* vertical or dihedral mirror planes, the molecule will belong to a D_{nd} point group.

- *Cubic groups (molecules with high symmetry)*:

  These point groups have many different rotational axes and are derived from one of the five Platonic solids. The three most common point groups are identical with the symmetry of their associated regular polyhedron: T_d (tetrahedral = 4 triangular faces), O_h (octahedral = 6 triangular faces or cubic = 6 square faces), and I_h (icosahedral = 20 triangular faces or dodecahedral = 12 pentagonal faces). The groups T and T_h are related to the T_d point group (in fact, they are subgroups), but they lack some of the higher symmetry features. The T point group consists of only the rotational operations in T_d. Likewise, the point groups O and I are subgroups of O_h and I_h, respectively, that include only rotational operations. Lastly, the T_h subgroup of T_d includes an inversion center.

The astute student of chemistry might have noticed that some portion of the multiplication tables for NH_3, PPh_3 and BH_3 are common to one another. For instance, the first nine entries in the upper left-hand corner of each of these multiplication tables are identical. From this observation alone, it is clear that these symmetry groups are somehow related to each other. In fact, the point group for PPh_3, C_3, is a subgroup of both the C_{3v} point group of NH_3 and the D_{3h} point group of BH_3. The symmetry operations in a *subgroup* meet all of the mathematical criteria to form a smaller group on their own accord. Thus, the order of a subgroup must be an integral divisor of the order of the larger group from which it is derived. In order to generate the C_{3v} point group from C_3, a mirror plane must be introduced. Because of the rotational symmetry of the C_3 group, once one mirror plane is introduced, a set of three (*n*) mirror planes is required. In order to generate the D_{3h} point group from C_3, a horizontal mirror plane must be introduced. The presence of the mirror plane implies that there will also be *n* C_2 axes perpendicular to the principal axis and an improper S_n axis collinear with the principal axis. For certain mathematical applications of molecular point groups having high symmetry, the appropriate subgroup is often used to simplify the number of calculations involved.

Table 8.3 lists the molecular point groups of the common molecular geometries derived from VSEPR theory in Chapter 7. Only those molecules that belong to a

TABLE 8.3 The molecular point groups corresponding to the common VSEPR molecular geometries (assuming that all of the ligands are composed of identical atoms).

Molecular Geometry	Molecular Formula	Molecular Point Group
Linear	AX_2	$D_{\infty h}$
Trigonal planar	AX_3	D_{3h}
Bent	AX_2E	C_{2v}
Tetrahedral	AX_4	T_d
Trigonal pyramidal	AX_3E	C_{3v}
Bent	AX_2E_2	C_{2v}
Trigonal bipyramidal	AX_5	D_{3h}
Seesaw	AX_4E	C_{2v}
T-shaped	AX_3E_2	C_{2v}
Linear	AX_2E_3	$D_{\infty h}$
Octahedral	AX_6	O_h
Square pyramidal	AX_5E	C_{4v}
Square planar	AX_4E_2	D_{4h}

particular set of point groups will have a dipole moment. In order for a molecule to have a dipole moment, the sum of the individual bond dipoles must be nonzero. Thus, any molecule containing an inversion center cannot have a nonzero dipole moment. Likewise, any molecule with more than one rotational axis cannot be polar because the dipole moment must lie along a single axis of the molecule. This restriction rules out any of the dihedral or cubic point groups. Lastly, any molecule having a horizontal mirror plane cannot be polar because the magnitudes of the individual bond dipoles will be equal on opposite sides of the plane. Therefore, only the C_1, C_s, C_n, and C_{nv} point groups are polar.

The symmetry of a molecule also places restrictions on whether or not it is possible for the molecule to be optically active. Optically active (or chiral) molecules can exist in one of two different isomeric forms known as *enantiomers*, each of which rotates plane-polarized light in a specific direction. In order for a molecule to be optically active, its optical isomers must consist of nonsuperimposable mirror images. This will occur if the molecule has no other symmetry besides the identity element or a proper rotation. As a result, any molecule having an improper rotational axis (S_n) cannot be optically active. This includes the nongenuine improper rotations, S_1 (mirror plane) and S_2 (inversion) operations. Thus, only molecules having the point groups C_1, C_n, D_n, T, O, and I can be optically active.

Example 8-8. Which of the following molecules are polar? (a) PF_5, (b) $Co(en)_3{}^{3+}$, (c) *cis*-dichloroethene, and (d) the boat form of cyclohexane.

Solution. (a) PF_5 is nonpolar (trigonal bipyramidal, D_{3h} point group). (b) $Co(en)_3{}^{3+}$ is nonpolar (D_3 point group). (c) *cis*-Dichloroethene is polar (C_{2v} point group). (d) The boat form of cyclohexane is polar (C_{2v} point group).

Example 8-9. Which of the molecules in Example 8-8 are optically active?

Solution. The only molecule lacking an improper rotational axis, mirror plane, or inversion center is $Co(en)_3{}^{3+}$, whose optical isomers are shown below. The two mirror images are clearly nonsuperimposable.

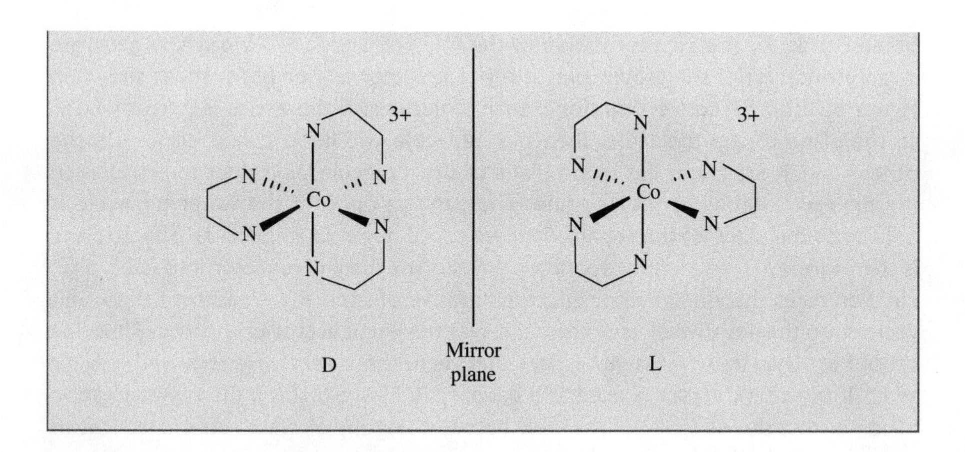

8.4 REPRESENTATIONS

Because the symmetry properties of molecules conform to mathematical groups, we can apply the mathematics of *group theory* to molecules in order to determine some of their physical and spectroscopic properties. Our general strategy here will be to use an internal set of coordinates of the molecule (such as vectors, molecular vibrations, or atomic orbitals) as a *basis set* for generating a representation of the different symmetry operations in the point group. A *representation* is a set of matrices, each of which corresponds to a symmetry operation and combine in the same way that the symmetry operators in the group combine. Thus, the multiplication table for the matrices that represent each symmetry operation must also multiply together in the same way that the symmetry operators themselves multiply.

Consider the water molecule, which has the C_{2v} point group. Suppose that we wanted to use a unit vector along the z-axis as an internal coordinate to generate a set of matrices for the different symmetry operations in the group. By convention, the principal axis of the molecule is always defined as the z-axis, as shown in Figure 8.10(a).

C_{2v}	E	C_2	σ_{v1}	σ_{v2}
Γ_z	[1]	[1]	[1]	[1]
Γ_y	[1]	[−1]	[−1]	[1]
Γ_x	[1]	[−1]	[1]	[−1]

The identity operation will transform the z-unit vector into itself. So, too, will the C_2 operator, σ_{v1}, and σ_{v2}. Thus, the set of 1 × 1 matrices that represent the four symmetry operations is given by the symbol Γ_z at the base of the figure. In

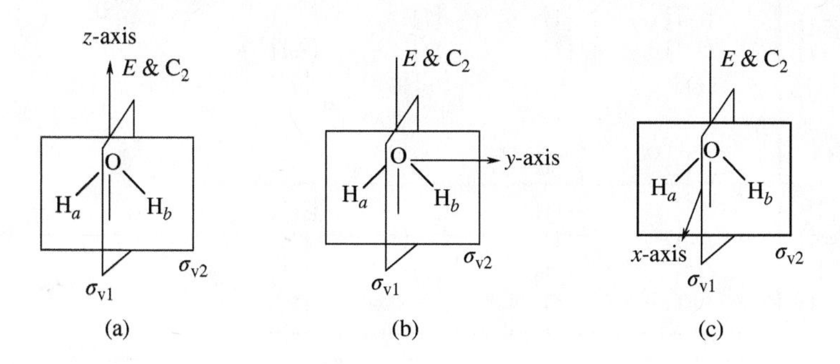

FIGURE 8.10
Generation of a set of 1 × 1 matrix representations of the symmetry operations in the C_{2v} point group using unit vectors along the (a) *z*-axis, (b) *y*-axis, and (c) *x*-axis as the basis set.

other words, Γ_z is a representation of the C_{2v} point group. We can also generate a representation of the group using the *y*-unit vector as our basis set, as shown in Figure 8.10(b). By convention, for planar molecules, if the *z*-axis is perpendicular to the plane of the molecule, then the molecule will lie in the *xz*-plane. On the other hand, if the *z*-axis lies in the plane of the molecule, as it does for H_2O, then the molecule will lie in the *yz*-plane. Thus, the *yz*-plane in the water molecule is defined as the plane of the paper (what we called σ_{v2} in Example 8-3). The *xz*-plane is therefore σ_{v1}. The identity operation transforms the *y*-unit vector into itself, a C_2 rotation takes the *y*-unit vector into the negative of itself, σ_{v1} transforms the *x*-unit vector into the negative of itself and σ_{v2} leaves the *y*-unit vector unchanged. Thus, the set of 1×1 matrices representing the *y*-unit vector basis set is given below the figure as Γ_y. If the *x*-unit vector is used as a basis, as in Figure 8.10(c), the representation becomes that shown for Γ_x in the table below. Using any of these sets of matrices to represent the four different symmetry operators, we obtain a result that is consistent with the multiplication table for the C_{2v} point group, as shown in Table 8.4.

If we had used the p_z, p_x, and p_y orbitals as basis sets in the C_{2v} point group, as shown in Figure 8.11, the resulting 1×1 matrix representations would have been the same as those obtained using the corresponding Cartesian axis' unit vector: Γ_z, Γ_x, and Γ_y, respectively.

Miniature rotational vectors can also be used as a basis set for the generation of representations of the symmetry operations in the C_{2v} point group, as shown in Figure 8.12. Rotational vectors can be visualized as collinear with the corresponding coordinate axis, but where the axis itself is rotating in a counterclockwise direction.

TABLE 8.4 Multiplication tables for the C_{2v} point group, showing how the 1 × 1 matrix representations in Figure 8.10 multiply together in the same way that the symmetry operations do.

C_{2v}	E	C_2	σ_{v1}	σ_{v2}
E	E	C_2	σ_{v1}	σ_{v2}
C_2	C_2	E	σ_{v2}	σ_{v1}
σ_{v1}	σ_{v1}	σ_{v2}	E	C_2
σ_{v2}	σ_{v2}	σ_{v1}	C_2	E

Γ_z	[1]	[1]	[1]	[1]
[1]	[1]	[1]	[1]	[1]
[1]	[1]	[1]	[1]	[1]
[1]	[1]	[1]	[1]	[1]
[1]	[1]	[1]	[1]	[1]

Γ_y	[1]	[−1]	[−1]	[1]
[1]	[1]	[−1]	[−1]	[1]
[−1]	[−1]	[1]	[1]	[−1]
[−1]	[−1]	[1]	[1]	[−1]
[1]	[1]	[−1]	[−1]	[1]

Γ_x	[1]	[−1]	[1]	[−1]
[1]	[1]	[−1]	[1]	[−1]
[−1]	[−1]	[1]	[−1]	[1]
[1]	[1]	[−1]	[1]	[−1]
[−1]	[−1]	[1]	[−1]	[1]

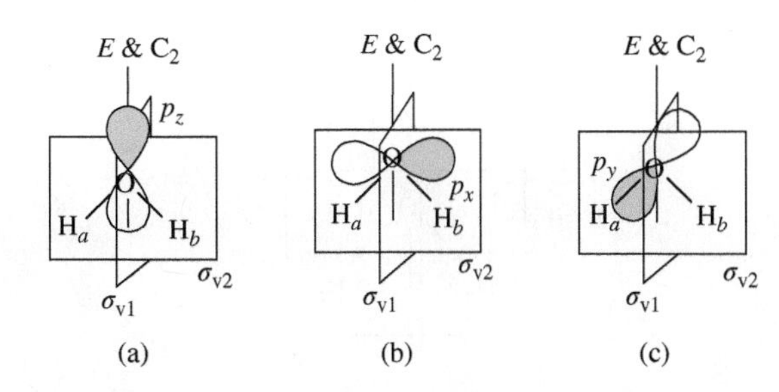

FIGURE 8.11
Generation of a set of 1 × 1 matrix representations of the symmetry operations in the C_{2v} point group using the (a) p_z orbital, (b) p_y orbital, and (c) p_x orbital as a basis set. The shaded lobe of the orbitals indicates a positive sign for the wave function, whereas the hollow lobe indicates a negative sign for the wave function. The three p orbitals transform with the same representations as the corresponding coordinate axes.

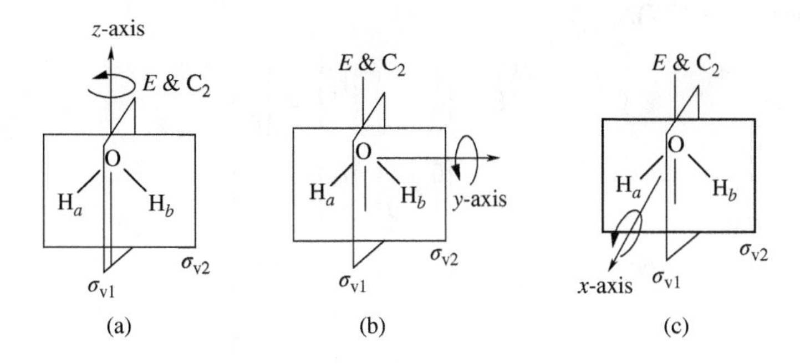

FIGURE 8.12
Rotational vector basis sets in the C_{2v} point group.

Reflections of a rotational vector in any mirror plane that contains that vector will therefore cause the vector to rotate in the opposite (or clockwise) direction, transforming it into the opposite of itself, or a matrix of [−1]. The representations for the three rotational vectors shown in Figure 8.12 are listed in the table.

C_{2v}	E	C_2	σ_{v1}	σ_{v2}
R_z	[1]	[1]	[-1]	[-1]
R_x	[1]	[-1]	[-1]	[1]
R_y	[1]	[-1]	[1]	[-1]

Notice that the R_x representation is identical with (i.e., has the same symmetry properties as) the Γ_y representation and that the R_y representation transforms the same way as Γ_x. The R_z representation, on the other hand, is a unique representation that we have not yet encountered. However, the matrices for the R_z representation still multiply in the same exact way that the symmetry operations in the C_{2v} point group multiply.

Virtually any internal coordinate that one can imagine will serve as a basis for generating representations of the symmetry operations in the point group. Suppose we were interested in the symmetry properties of the sigma bonds in the H_2O molecule. We could then use the two bond vectors shown in Figure 8.13 as our basis set. This would lead to a set of 2 × 2 matrix representations. The identity operation will act on r_1 and r_2 to convert them into themselves. A C_2 operation will transform r_1 into r_2 and r_2 into r_1. Reflection of the bond vectors in the yz-plane will yield them back unchanged. Finally, reflection in the xz-plane will interconvert the bond vectors with each other. It is our goal to determine this set of 2 × 2 transformation matrices, as shown in Equations (8.4)–(8.7).

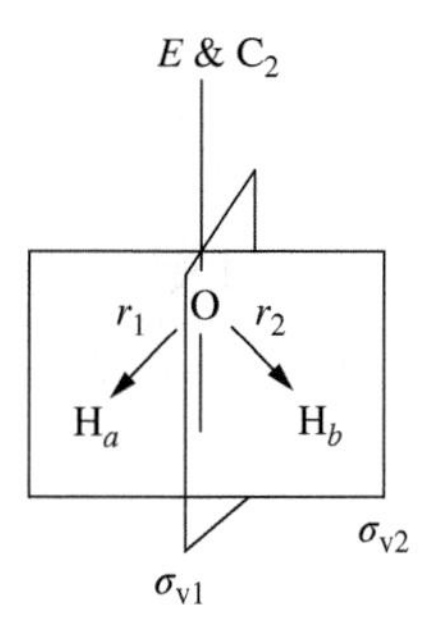

FIGURE 8.13
Bond vector basis set for the H_2O molecule in C_{2v}.

The matrix math showing each of these transformations can be seen in Equations (8.4)–(8.7):

$$E \quad \begin{bmatrix} 1 & 0 \\ 0 & 1 \end{bmatrix} \begin{bmatrix} r_1 \\ r_2 \end{bmatrix} = \begin{bmatrix} r_1 \\ r_2 \end{bmatrix} \tag{8.4}$$

$$C_2 \quad \begin{bmatrix} 0 & 1 \\ 1 & 0 \end{bmatrix} \begin{bmatrix} r_1 \\ r_2 \end{bmatrix} = \begin{bmatrix} r_2 \\ r_1 \end{bmatrix} \tag{8.5}$$

$$\sigma_{v1}(xz) \quad \begin{bmatrix} 0 & 1 \\ 1 & 0 \end{bmatrix} \begin{bmatrix} r_1 \\ r_2 \end{bmatrix} = \begin{bmatrix} r_2 \\ r_1 \end{bmatrix} \tag{8.6}$$

$$\sigma_{v2}(yz) \quad \begin{bmatrix} 1 & 0 \\ 0 & 1 \end{bmatrix} \begin{bmatrix} r_1 \\ r_2 \end{bmatrix} = \begin{bmatrix} r_1 \\ r_2 \end{bmatrix} \tag{8.7}$$

A brief review of matrix multiplication is provided by Equation (8.8), where h is the number of rows in the second matrix, c_{ij} is the matrix element in row i and column j of the product matrix, a_{ik} is the matrix element in row i and column k of the first matrix, and b_{kj} is the matrix element in row k and column j of the second matrix in the multiplication. The matrices that are to be multiplied with each other must first be conformable, which essentially means that the number of columns in the first matrix must be equivalent to the number of rows in the second matrix. The product matrix will therefore have the same number of rows as the first matrix and the same number of columns as the second one:

$$c_{ij} = \sum_{k=1}^{h} a_{ik} b_{kj} \tag{8.8}$$

Table 8.5 shows how the multiplication tables for the symmetry operations in the C_{2v} point group and the 2 × 2 bond vector (Γ_{bv}) representations of those symmetry operations multiply together in a homologous manner.

Another common basis set for molecules is one that represents the number of degrees of freedom that the molecule has by placing miniature vectors at 90° angles to each other on each of the atoms in the molecule. This type of basis set is shown in Figure 8.14 for the H_2O molecule. The identity operation, naturally, transforms each vector into itself and therefore consists of all one's along the diagonal from the upper left-hand corner of the matrix to the lower right-hand corner. A C_2 operation will transform a into a, b into $-b$, c into $-c$, d into g, e into $-h$, f into -i, g into d, h into −e, and i into $-f$. The σ_{v1} operation will convert a into a, b into $-b$, c into c, d into g, e into $-h$, f into i, g into d, h into −e, and i into f. Finally, the σ_{v2} operation transforms a into a, b into b, c into $-c$, d into d, e into e, f into $-f$, g into g, h into h, and i into $-i$. The resulting set of 9 × 9 transformation matrices representing the symmetry operations in the C_{2v} point group are shown at the bottom of the figure.

TABLE 8.5 Multiplication tables for the symmetry operations and 2 × 2 bond vector representations in the C_{2v} point group.

C_{2v}	E	C_2	σ_{v1}	σ_{v2}
E	E	C_2	σ_{v1}	σ_{v2}
C_2	C_2	E	σ_{v2}	σ_{v1}
σ_{v1}	σ_{v1}	σ_{v2}	E	C_2
σ_{v2}	σ_{v2}	σ_{v1}	C_2	E

Γ_{bv}	$\begin{bmatrix} 1 & 0 \\ 0 & 1 \end{bmatrix}$	$\begin{bmatrix} 0 & 1 \\ 1 & 0 \end{bmatrix}$	$\begin{bmatrix} 0 & 1 \\ 1 & 0 \end{bmatrix}$	$\begin{bmatrix} 1 & 0 \\ 0 & 1 \end{bmatrix}$
$\begin{bmatrix} 1 & 0 \\ 0 & 1 \end{bmatrix}$	$\begin{bmatrix} 1 & 0 \\ 0 & 1 \end{bmatrix}$	$\begin{bmatrix} 0 & 1 \\ 1 & 0 \end{bmatrix}$	$\begin{bmatrix} 0 & 1 \\ 1 & 0 \end{bmatrix}$	$\begin{bmatrix} 1 & 0 \\ 0 & 1 \end{bmatrix}$
$\begin{bmatrix} 0 & 1 \\ 1 & 0 \end{bmatrix}$	$\begin{bmatrix} 0 & 1 \\ 1 & 0 \end{bmatrix}$	$\begin{bmatrix} 1 & 0 \\ 0 & 1 \end{bmatrix}$	$\begin{bmatrix} 1 & 0 \\ 0 & 1 \end{bmatrix}$	$\begin{bmatrix} 0 & 1 \\ 1 & 0 \end{bmatrix}$
$\begin{bmatrix} 0 & 1 \\ 1 & 0 \end{bmatrix}$	$\begin{bmatrix} 0 & 1 \\ 1 & 0 \end{bmatrix}$	$\begin{bmatrix} 1 & 0 \\ 0 & 1 \end{bmatrix}$	$\begin{bmatrix} 1 & 0 \\ 0 & 1 \end{bmatrix}$	$\begin{bmatrix} 0 & 1 \\ 1 & 0 \end{bmatrix}$
$\begin{bmatrix} 1 & 0 \\ 0 & 1 \end{bmatrix}$	$\begin{bmatrix} 1 & 0 \\ 0 & 1 \end{bmatrix}$	$\begin{bmatrix} 0 & 1 \\ 1 & 0 \end{bmatrix}$	$\begin{bmatrix} 0 & 1 \\ 1 & 0 \end{bmatrix}$	$\begin{bmatrix} 1 & 0 \\ 0 & 1 \end{bmatrix}$

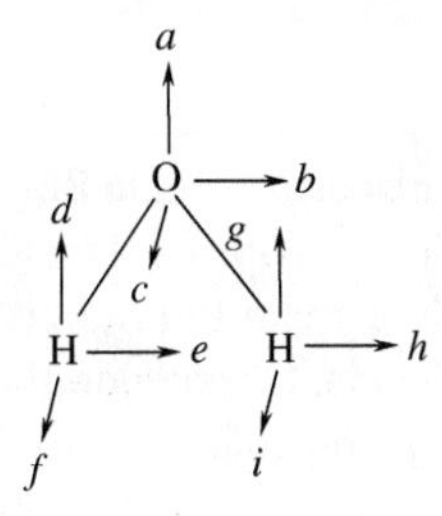

FIGURE 8.14
Degrees of freedom basis set for the H_2O molecule in C_{2v}.

Believe it or not, these 9 × 9 matrices also multiply together in the same manner as the symmetry operations do! One such example is shown in Equation (8.9): the product $C_2\,\sigma_{vI} = \sigma_{v2}$. You might want to try to prove this for yourself for at least one of the other nontrivial products in the C_{2v} multiplication table.

C_{2v}	E	C_2	σ_{vI}	σ_{v2}
Γ_{dof}	$\begin{bmatrix} 1&0&0&0&0&0&0&0&0 \\ 0&1&0&0&0&0&0&0&0 \\ 0&0&1&0&0&0&0&0&0 \\ 0&0&0&1&0&0&0&0&0 \\ 0&0&0&0&1&0&0&0&0 \\ 0&0&0&0&0&1&0&0&0 \\ 0&0&0&0&0&0&1&0&0 \\ 0&0&0&0&0&0&0&1&0 \\ 0&0&0&0&0&0&0&0&1 \end{bmatrix}$	$\begin{bmatrix} 1&0&0&0&0&0&0&0&0 \\ 0&-1&0&0&0&0&0&0&0 \\ 0&0&-1&0&0&0&0&0&0 \\ 0&0&0&0&0&0&1&0&0 \\ 0&0&0&0&0&0&0&-1&0 \\ 0&0&0&0&0&0&0&0&-1 \\ 0&0&0&1&0&0&0&0&0 \\ 0&0&0&0&-1&0&0&0&0 \\ 0&0&0&0&0&-1&0&0&0 \end{bmatrix}$	$\begin{bmatrix} 1&0&0&0&0&0&0&0&0 \\ 0&-1&0&0&0&0&0&0&0 \\ 0&0&1&0&0&0&0&0&0 \\ 0&0&0&0&0&0&1&0&0 \\ 0&0&0&0&0&0&0&-1&0 \\ 0&0&0&0&0&0&0&0&1 \\ 0&0&0&1&0&0&0&0&0 \\ 0&0&0&0&-1&0&0&0&0 \\ 0&0&0&0&0&1&0&0&0 \end{bmatrix}$	$\begin{bmatrix} 1&0&0&0&0&0&0&0&0 \\ 0&1&0&0&0&0&0&0&0 \\ 0&0&-1&0&0&0&0&0&0 \\ 0&0&0&1&0&0&0&0&0 \\ 0&0&0&0&1&0&0&0&0 \\ 0&0&0&0&0&-1&0&0&0 \\ 0&0&0&0&0&0&1&0&0 \\ 0&0&0&0&0&0&0&1&0 \\ 0&0&0&0&0&0&0&0&-1 \end{bmatrix}$

$$\underset{C_2}{\begin{bmatrix} 1 & 0 & 0 & 0 & 0 & 0 & 0 & 0 & 0 \\ 0 & -1 & 0 & 0 & 0 & 0 & 0 & 0 & 0 \\ 0 & 0 & -1 & 0 & 0 & 0 & 0 & 0 & 0 \\ 0 & 0 & 0 & 0 & 0 & 0 & 1 & 0 & 0 \\ 0 & 0 & 0 & 0 & 0 & 0 & 0 & -1 & 0 \\ 0 & 0 & 0 & 0 & 0 & 0 & 0 & 0 & -1 \\ 0 & 0 & 0 & 1 & 0 & 0 & 0 & 0 & 0 \\ 0 & 0 & 0 & 0 & -1 & 0 & 0 & 0 & 0 \\ 0 & 0 & 0 & 0 & 0 & -1 & 0 & 0 & 0 \end{bmatrix}} \underset{\sigma_{v1}}{\begin{bmatrix} 1 & 0 & 0 & 0 & 0 & 0 & 0 & 0 & 0 \\ 0 & -1 & 0 & 0 & 0 & 0 & 0 & 0 & 0 \\ 0 & 0 & 1 & 0 & 0 & 0 & 0 & 0 & 0 \\ 0 & 0 & 0 & 0 & 0 & 0 & 1 & 0 & 0 \\ 0 & 0 & 0 & 0 & 0 & 0 & 0 & -1 & 0 \\ 0 & 0 & 0 & 0 & 0 & 0 & 0 & 0 & 1 \\ 0 & 0 & 0 & 1 & 0 & 0 & 0 & 0 & 0 \\ 0 & 0 & 0 & 0 & -1 & 0 & 0 & 0 & 0 \\ 0 & 0 & 0 & 0 & 0 & 1 & 0 & 0 & 0 \end{bmatrix}} \underset{=}{=} \underset{\sigma_{v2}}{\begin{bmatrix} 1 & 0 & 0 & 0 & 0 & 0 & 0 & 0 & 0 \\ 0 & 1 & 0 & 0 & 0 & 0 & 0 & 0 & 0 \\ 0 & 0 & -1 & 0 & 0 & 0 & 0 & 0 & 0 \\ 0 & 0 & 0 & 1 & 0 & 0 & 0 & 0 & 0 \\ 0 & 0 & 0 & 0 & 1 & 0 & 0 & 0 & 0 \\ 0 & 0 & 0 & 0 & 0 & -1 & 0 & 0 & 0 \\ 0 & 0 & 0 & 0 & 0 & 0 & 1 & 0 & 0 \\ 0 & 0 & 0 & 0 & 0 & 0 & 0 & 1 & 0 \\ 0 & 0 & 0 & 0 & 0 & 0 & 0 & 0 & -1 \end{bmatrix}} \tag{8.9}$$

Example 8-10. Prove that the 9 × 9 matrices for $\sigma_{v1}\ \sigma_{v2} = C_2$.

Solution. The matrices multiply together as shown below.

$$\begin{bmatrix} 1 & 0 & 0 & 0 & 0 & 0 & 0 & 0 & 0 \\ 0 & -1 & 0 & 0 & 0 & 0 & 0 & 0 & 0 \\ 0 & 0 & 1 & 0 & 0 & 0 & 0 & 0 & 0 \\ 0 & 0 & 0 & 0 & 0 & 0 & 1 & 0 & 0 \\ 0 & 0 & 0 & 0 & 0 & 0 & 0 & -1 & 0 \\ 0 & 0 & 0 & 0 & 0 & 0 & 0 & 0 & 1 \\ 0 & 0 & 0 & 1 & 0 & 0 & 0 & 0 & 0 \\ 0 & 0 & 0 & 0 & -1 & 0 & 0 & 0 & 0 \\ 0 & 0 & 0 & 0 & 0 & 1 & 0 & 0 & 0 \end{bmatrix} \begin{bmatrix} 1 & 0 & 0 & 0 & 0 & 0 & 0 & 0 & 0 \\ 0 & 1 & 0 & 0 & 0 & 0 & 0 & 0 & 0 \\ 0 & 0 & -1 & 0 & 0 & 0 & 0 & 0 & 0 \\ 0 & 0 & 0 & 1 & 0 & 0 & 0 & 0 & 0 \\ 0 & 0 & 0 & 0 & 1 & 0 & 0 & 0 & 0 \\ 0 & 0 & 0 & 0 & 0 & -1 & 0 & 0 & 0 \\ 0 & 0 & 0 & 0 & 0 & 0 & 1 & 0 & 0 \\ 0 & 0 & 0 & 0 & 0 & 0 & 0 & 1 & 0 \\ 0 & 0 & 0 & 0 & 0 & 0 & 0 & 0 & -1 \end{bmatrix} = \begin{bmatrix} 1 & 0 & 0 & 0 & 0 & 0 & 0 & 0 & 0 \\ 0 & -1 & 0 & 0 & 0 & 0 & 0 & 0 & 0 \\ 0 & 0 & -1 & 0 & 0 & 0 & 0 & 0 & 0 \\ 0 & 0 & 0 & 0 & 0 & 0 & 1 & 0 & 0 \\ 0 & 0 & 0 & 0 & 0 & 0 & 0 & -1 & 0 \\ 0 & 0 & 0 & 0 & 0 & 0 & 0 & 0 & -1 \\ 0 & 0 & 0 & 1 & 0 & 0 & 0 & 0 & 0 \\ 0 & 0 & 0 & 0 & -1 & 0 & 0 & 0 & 0 \\ 0 & 0 & 0 & 0 & 0 & -1 & 0 & 0 & 0 \end{bmatrix}$$

Example 8-11. Consider the planar molecule BBrClF shown in Example 8-7(e). Assume that the molecule lies in the *xy*-plane. List all of the symmetry operations possible in this molecule and determine the molecular point group. Construct a multiplication table for the molecule. Determine the 1 × 1 matrices representing each symmetry operation using the following basis sets: (a) *x*-unit vector, (b) *y*-unit vector, (c) *z*-unit vector, (d) R_x rotational vector, (e) R_y rotational vector, and (f) R_z rotational vector. Also determine the 3 × 3 matrices representing each symmetry operation using the three bond vectors as a basis set. Finally, determine the 12 × 12 matrices representing each symmetry operation using the degrees of freedom basis set.

Solution. There are only two symmetry operators for this molecule: *E* and σ_h. The molecular point group is C_s.

The multiplication table is as follows:

C_s	E	σ_h
E	E	σ_h
σ_h	σ_h	E

(a, b) Both the *x*- and *y*-unit vectors lie in the plane of the molecule and are symmetric with respect to *E* and σ_h. Hence, $\Gamma_x = \Gamma_y = [1]$ for *E* and σ_h.

(c) The z-unit vector is perpendicular to the plane of the molecule and is symmetric with respect to E but is transformed into the negative of itself with σ_h. Thus, $\Gamma_z = [1]$ for E and $[-1]$ for σ_h. (d, e) Both the R_x and R_y rotational vectors lie in the xy-plane and are symmetric with respect to the E operation, but they are transformed into their mirror image opposites (rotating in the opposite direction) with the σ_h operation. Thus, $R_x = R_y = [1]$ for E and $[-1]$ for σ_h and these bases therefore transform in the same manner as Γ_z. (f) The R_z vector, which lies perpendicular to the mirror plane, on the other hand, is symmetric to both E and σ_h and transforms the same way as Γ_x and Γ_y with a matrix of [1] for E and [1] for σ_h. (g) Using the bond vectors shown below as a basis set, the vectors a, b, and c all transform into themselves with both the E and σ_h operations. Thus, the 3 × 3 matrix is the one shown in the product that is written below.

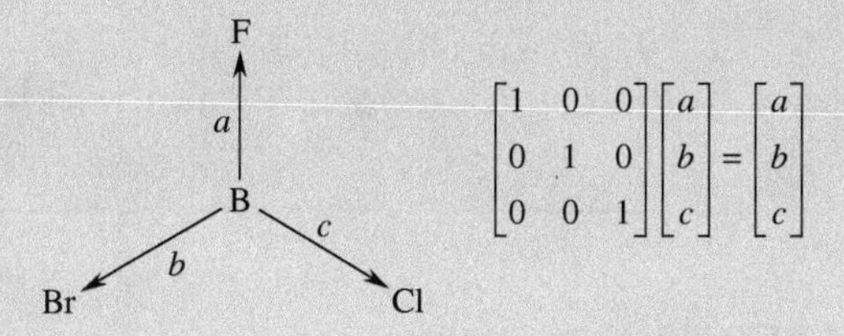

(h) The degrees of freedom basis set is shown below, along with its corresponding 12 × 12 matrix.

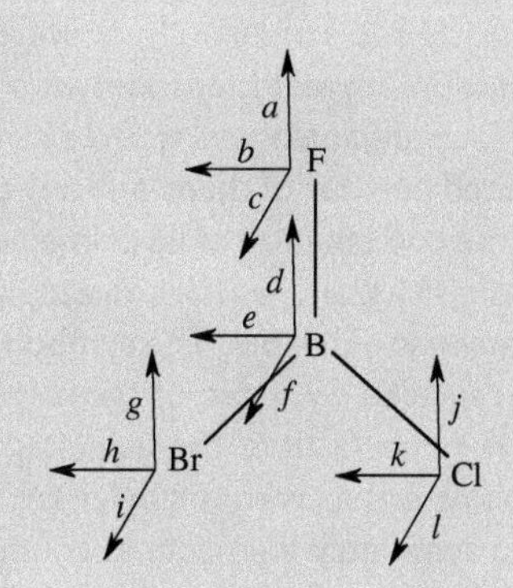

E

$$\begin{bmatrix} a \\ b \\ c \\ d \\ e \\ f \\ g \\ h \\ i \\ j \\ k \\ l \end{bmatrix} \begin{bmatrix} 1 & 0 & 0 & 0 & 0 & 0 & 0 & 0 & 0 & 0 & 0 & 0 \\ 0 & 1 & 0 & 0 & 0 & 0 & 0 & 0 & 0 & 0 & 0 & 0 \\ 0 & 0 & 1 & 0 & 0 & 0 & 0 & 0 & 0 & 0 & 0 & 0 \\ 0 & 0 & 0 & 1 & 0 & 0 & 0 & 0 & 0 & 0 & 0 & 0 \\ 0 & 0 & 0 & 0 & 1 & 0 & 0 & 0 & 0 & 0 & 0 & 0 \\ 0 & 0 & 0 & 0 & 0 & 1 & 0 & 0 & 0 & 0 & 0 & 0 \\ 0 & 0 & 0 & 0 & 0 & 0 & 1 & 0 & 0 & 0 & 0 & 0 \\ 0 & 0 & 0 & 0 & 0 & 0 & 0 & 1 & 0 & 0 & 0 & 0 \\ 0 & 0 & 0 & 0 & 0 & 0 & 0 & 0 & 1 & 0 & 0 & 0 \\ 0 & 0 & 0 & 0 & 0 & 0 & 0 & 0 & 0 & 1 & 0 & 0 \\ 0 & 0 & 0 & 0 & 0 & 0 & 0 & 0 & 0 & 0 & 1 & 0 \\ 0 & 0 & 0 & 0 & 0 & 0 & 0 & 0 & 0 & 0 & 0 & 1 \end{bmatrix} = \begin{bmatrix} a \\ b \\ c \\ d \\ e \\ f \\ g \\ h \\ i \\ j \\ k \\ l \end{bmatrix}$$

$$\sigma_h$$

$$\begin{bmatrix} a \\ b \\ c \\ d \\ e \\ f \\ g \\ h \\ i \\ j \\ k \\ l \end{bmatrix} \begin{bmatrix} 1 & 0 & 0 & 0 & 0 & 0 & 0 & 0 & 0 & 0 & 0 & 0 \\ 0 & 1 & 0 & 0 & 0 & 0 & 0 & 0 & 0 & 0 & 0 & 0 \\ 0 & 0 & -1 & 0 & 0 & 0 & 0 & 0 & 0 & 0 & 0 & 0 \\ 0 & 0 & 0 & 1 & 0 & 0 & 0 & 0 & 0 & 0 & 0 & 0 \\ 0 & 0 & 0 & 0 & 1 & 0 & 0 & 0 & 0 & 0 & 0 & 0 \\ 0 & 0 & 0 & 0 & 0 & -1 & 0 & 0 & 0 & 0 & 0 & 0 \\ 0 & 0 & 0 & 0 & 0 & 0 & 1 & 0 & 0 & 0 & 0 & 0 \\ 0 & 0 & 0 & 0 & 0 & 0 & 0 & 1 & 0 & 0 & 0 & 0 \\ 0 & 0 & 0 & 0 & 0 & 0 & 0 & 0 & -1 & 0 & 0 & 0 \\ 0 & 0 & 0 & 0 & 0 & 0 & 0 & 0 & 0 & 1 & 0 & 0 \\ 0 & 0 & 0 & 0 & 0 & 0 & 0 & 0 & 0 & 0 & 1 & 0 \\ 0 & 0 & 0 & 0 & 0 & 0 & 0 & 0 & 0 & 0 & 0 & -1 \end{bmatrix} = \begin{bmatrix} a \\ b \\ -c \\ d \\ e \\ -f \\ g \\ h \\ -i \\ j \\ k \\ -l \end{bmatrix}$$

8.5 CHARACTER TABLES

The number of representations for the symmetry operations of any point group is limited only by the imagination of the user and the way that he or she defines the internal coordinates of the basis set. Fortunately, however, each point group has a limited number of fundamental representations known as *irreducible representations* (*IRRs*). The IRRs of a point group are analogous to the primary number factors of any real number. There are a limited number of them, and any other factor of the number can always be reduced to some combination of its prime number factors. For the C_{2v} point group, there are only four IRRs; and we have already encountered each of them. The four IRRs for C_{2v} are shown in Table 8.6. The symbols for these representations (e.g., Γ_x or R_z) have been replaced by a set of more general descriptors known as *Mulliken symbols*. Mulliken symbols have the advantage of universality—that is, a Mulliken symbol in one point group with a similar or identical name in another point group will have certain symmetry elements in common. We will learn how to decipher the meanings of Mulliken symbols a bit later on.

You might have noticed that many of the representations for the different symmetry operations have all of their nonzero elements lying along the diagonal from the upper left-hand corner of the matrix to the lower right-hand corner (i.e., in elements c_{ii}). Matrices of this type are called *diagonal matrices*. The product of two diagonal matrices always commutes (in other words, $AB = BA$, as long as A and

TABLE 8.6 The character table containing the complete list of irreducible representations (IRRs) for the C_{2v} point group, along with their Mulliken symbol names.

C_{2v}	E	$C_2\ (z)$	$\sigma_{v1}\ (xz)$	$\sigma_{v2}\ (yz)$			
A_1	1	1	1	1	z	x^2, y^2, z^2	$z^3, z(x^2-y^2)$
A_2	1	1	−1	−1	R_z	xy	xyz
B_1	1	−1	1	−1	x, R_y	xz	$xz^2, x(x^2-3y^2)$
B_2	1	−1	−1	1	y, R_x	yz	$yz^2, y(3x^2-y^2)$

B are both diagonal matrices). The 9 × 9 matrix representing the identity operation in the Γ_{dof} representation for the H_2O molecule is an example of a diagonal matrix.

The trace or *character* of any matrix is the sum of its diagonal elements from upper left to lower right (Σc_{ii}). The characters of matrices are given the symbol χ. For the *E* operation in the Γ_{dof} representation of H_2O, $\chi = 9$. The characters for the C_2, σ_{v1}, and σ_{v2} matrices in the same representation are −1, 3, and 1, respectively. The characters of the IRRs of a point group are usually listed in *character tables*, such as those found in Appendix B. The character table for the C_{2v} point group is reproduced in Table 8.6. In any character table, the Schoenflies notation for the point group is shown in the upper left-hand corner. The top row of symbols is reserved for all of the different symmetry operations that are possible in that group, where the operations have been organized into classes (*vide infra*). The first column of the character table contains the Mulliken symbols for each of the IRRs. The characters of the IRR matrices are listed in the second section of the table just beneath the symmetry operations. The third section of the table describes how the common basis sets (unit vectors along the coordinate axes *x*, *y*, and *z* and the rotational vectors R_x, R_y, and R_z) transform in the point group. The p_x, p_y, and p_z orbitals will transform in the same way that the *x*, *y*, and *z* unit vectors do (respectively), as they have exactly the same symmetry properties. The fourth region of the character table lists the squares and binary products of the *x*, *y*, and *z* basis sets, particularly those corresponding with how the *d* orbitals transform in the point group. Finally, the right-most region of the character table is reserved for the cubes or ternary products of *x*, *y*, and *z*, which relate to how the *f* orbitals transform within the point group.

In some point groups, the *x*-, *y*-, and *z*-unit vectors will not always transform individually as 1 × 1 matrices. Consider the NH_3 molecule, for instance, which has the C_{3v} character table (Table 8.7). Using a unit vector along the *z*-axis as the basis, each of the six symmetry operations will transform the *z*-unit vector into itself, yielding a representation of 1 × 1 matrices having [1] for each of their matrix elements and corresponding to the characters for the A_1 IRR. Using a unit vector along the *x*-axis as the basis, the identity operation will transform the vector into itself. However, the C_3 and C_3^2 operations will transform the *x*-unit vector into some linear combination of the *x*- and *y*-unit vectors that lies in the *xy*-plane. The rotation of a generic unit vector lying in the *xy*-plane around a proper rotational axis lying along the *z*-axis is shown in Figure 8.15.

Using trigonometry, the following statements will be true (where r is the length of the basis vector):

$$x_1 = r\cos\beta \quad x_2 = r\cos(\alpha+\beta) \tag{8.10}$$

$$y_1 = r\sin\beta \quad y_2 = r\sin(\alpha+\beta) \tag{8.11}$$

Because of the trigonometric identities which state that $\cos(\alpha+\beta) = \cos\alpha\cos\beta - \sin\alpha\sin\beta$ and $\sin(\alpha+\beta) = \sin\alpha\cos\beta + \sin\beta\cos\alpha$, x_2 and y_2 can be rewritten as

TABLE 8.7 Character table for NH_3.

C_{3v}	E	$2C_3$	$3\sigma_v$			
A_1	1	1	1	z	x^2+y^2, z^2	z^3, $x(x^2-3y^2)$
A_2	1	1	−1	R_z		$y(3x^2-y^2)$
E	2	−1	0	(x, y), (R_x, R_y)	(x^2-y^2, xy), (xz, yz)	(xz^2, yz^2), $[xyz, z(x^2-y^2)]$

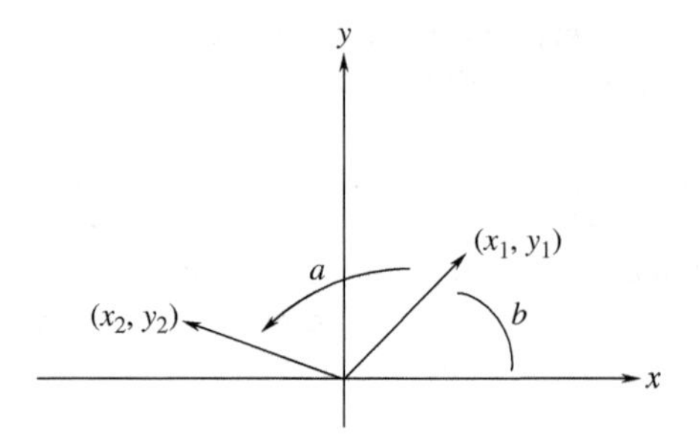

FIGURE 8.15
Rotation of a vector having length r around the z-axis by an angle α.

$$x_2 = r\cos\alpha\cos\beta - r\sin\alpha\sin\beta = x_1\cos\alpha - y_1\sin\alpha \quad (8.12)$$
$$y_2 = r\sin\alpha\cos\beta + r\sin\beta\cos\alpha = x_1\sin\alpha + y_1\cos\alpha \quad (8.13)$$

Thus, the 2 × 2 matrix that converts the vector (x_1, y_1) into the vector (x_2, y_2) is therefore given by Equation (8.14):

$$\begin{bmatrix} \cos\alpha & -\sin\alpha \\ \sin\alpha & \cos\alpha \end{bmatrix} \begin{bmatrix} x_1 \\ y_1 \end{bmatrix} = \begin{bmatrix} x_2 \\ y_2 \end{bmatrix} \quad (8.14)$$

Because the *x*-unit vector transforms into a linear combination of *x*- and *y*-unit vectors under any rotation about the *z*-axis, we cannot use the *x*-unit vector by itself as a basis. Instead, we must couple the *x*- and *y*-unit vectors together as a single basis. The resulting 2 × 2 matrices representing the symmetry operations will employ Equation (8.14), using the appropriate angle for either a C_3 or a $C_3{}^2$ rotation. The resulting IRR is said to be doubly degenerate. Whenever two bases transform together in a molecular point group, they will be listed in the third region of the character table in parentheses as a pair separated by a comma. Thus, the representation for the *x*- and *y*-unit vectors in the C_{3v} point group corresponds to the *E* IRR, which has them listed together as the pair (*x*, *y*) in the character table.

At this point, there are some theorems regarding the IRRs that are well worth noting. If we were ever forced to construct a character table for a point group from scratch instead of simply relying on the information provided in Appendix B, these rules would be very useful in helping us complete the character table.

Rules regarding IRRs:

1. There is always one IRR in every point group that has all of its characters equal to one (this is called the *totally symmetric IRR*).
2. The number of IRRs for a point group is always equal to the number of classes of symmetry operations.
3. The sum of the squares of the characters in any IRR must equal the order of the group (*h*), which in turn is equal to the total number of symmetry operations in the group.
4. The characters of the matrices belonging to the same class are identical in any given IRR.
5. The characters for the identity operation must always be positive.
6. The sum of the squares of the characters for the identity operation across all the IRRs in the group will also equal the order of the group.
7. The characters of any two IRRs in the group will be orthogonal, which means that they will satisfy Equation (8.15):

$$\sum_n \chi_i \chi_j = 0 \text{ whenever } i \neq j \quad (8.15)$$

In certain point groups, some of the symmetry operations belong to the same class and will therefore have identical characters for their matrices. Mathematically, a *class* is a set of operations that are conjugate to each other, or related by a similarity transform. In the world of matrix math, two matrices (A and B) will be *conjugate* with each other if there exists a third matrix Q, which is also a member of the group, such that B and A are related to each other by a *similarity transform*, given by Equation (8.16):

$$B = Q^{-1}AQ \tag{8.16}$$

At this point, there are also some rules for conjugate matrices that are worth noting:

1. Every matrix A is conjugate with itself.
2. If A is conjugate with B, then B will also be conjugate with A.
3. If A is conjugate with B and C, then B and C must be conjugate with each other.
4. The identity operation (E) is always in a class by itself.

The similarity transforms for the C_{2v} point group are given below:

- E is in a class by itself, as $A^{-1}\ E\ A = A^{-1}\ A = E$ irrespective of what the operation A is (rule 4 above).
- C_2 will also be in a class by itself because

$$\begin{aligned}
E^{-1}C_2E &= EC_2E &&= C_2E &&= C_2\\
(C_2)^{-1}C_2C_2 &= C_2C_2C_2 &&= EC_2 &&= C_2\\
(\sigma_{v1})^{-1}C_2\sigma_{v1} &= \sigma_{v1}C_2\sigma_{v1} &&= \sigma_{v2}\sigma_{v1} &&= C_2\\
(\sigma_{v2})^{-1}C_2\sigma_{v2} &= \sigma_{v2}C_2\sigma_{v2} &&= \sigma_{v1}\sigma_{v2} &&= C_2
\end{aligned}$$

- σ_{v1} is also in its own class because

$$\begin{aligned}
E^{-1}\sigma_{v1}E &= E\sigma_{v1}E &&= \sigma_{v1}E &&= \sigma_{v1}\\
(C_2)^{-1}\sigma_{v1}C_2 &= C_2\sigma_{v1}C_2 &&= \sigma_{v2}C_2 &&= \sigma_{v1}\\
(\sigma_{v1})^{-1}\sigma_{v1}\sigma_{v1} &= \sigma_{v1}\sigma_{v1}\sigma_{v1} &&= E\sigma_{v1} &&= \sigma_{v1}\\
(\sigma_{v2})^{-1}\sigma_{v1}\sigma_{v2} &= \sigma_{v2}\sigma_{v1}\sigma_{v2} &&= C_2\sigma_{v2} &&= \sigma_{v1}
\end{aligned}$$

- By default, σ_{v2} must therefore be in its own class, but we could also prove this as follows:

$$\begin{aligned}
E^{-1}\sigma_{v2}E &= E\sigma_{v2}E &&= \sigma_{v2}E &&= \sigma_{v2}\\
(C_2)^{-1}\sigma_{v2}C_2 &= C_2\sigma_{v2}C_2 &&= \sigma_{v1}C_2 &&= \sigma_{v2}\\
(\sigma_{v1})^{-1}\sigma_{v2}\sigma_{v1} &= \sigma_{v1}\sigma_{v2}\sigma_{v1} &&= C_2\sigma_{v1} &&= \sigma_{v2}\\
(\sigma_{v2})^{-1}\sigma_{v2}\sigma_{v2} &= \sigma_{v2}\sigma_{v2}\sigma_{v2} &&= E\sigma_{v2} &&= \sigma_{v2}
\end{aligned}$$

There are four classes of symmetry operations in the C_{2v} point group, so there must be four IRRs, as shown in Table 8.6. At least one of these IRRs must be totally symmetric and have all of its characters equal to one. This is the A_1 IRR. The sum of the squares of the characters for the identity operation across all the IRRs must equal the order of the group ($h = 4$) and each character for the identity operator must be a positive number. Thus, all four IRRs must have 1 as the character for their identity operation. Furthermore, the sum of the squares of the characters for each IRR must also equal the order of the group ($h = 4$). Thus, the characters for the other three operations in the group can only be 1 or -1. In order for the other three IRRs to be orthogonal to the first, they must have two characters that are 1 and two characters that are −1. For example, the A_1 and A_2 matrices are orthogonal because $(1)(1) + (1)(1) + (1)(-1) + (1)(-1) = 0$ and the A_2 and B_1 matrices are orthogonal because $(1)(1) + (1)(-1) + (-1)(1) + (-1)(-1) = 0$. It is easy to prove that all the other IRRs are also orthogonal with each other.

Example 8-12. Using the symmetry elements defined in Figure 8.7 for the NH_3 molecule and the multiplication table for the corresponding symmetry operations in the C_{3v} point group, organize the operations into classes.

Solution. Let us use the inverses in the NH_3 group to determine which operations belong to the same class by taking all the similarity transforms of each symmetry operation.

- The identity operation (E) is always in a class by itself, as the product of any operator A with E equals A and the product of $A^{-1}A = E$.
- C_3 and C_3^2 will belong to the same class because the following are true:

$$\begin{aligned}
E^{-1}C_3E &= EC_3E = C_3E = C_3 \\
(C_3)^{-1}C_3C_3 &= C_3^2C_3C_3 = EC_3 = C_3 \\
(C_3^2)^{-1}C_3C_3^2 &= C_3C_3C_3^2 = C_3^2C_3^2 = C_3 \\
(\sigma_{v1})^{-1}C_3\sigma_{v1} &= \sigma_{v1}C_3\sigma_{v1} = \sigma_{v2}\sigma_{v1} = C_3^2 \\
(\sigma_{v2})^{-1}C_3\sigma_{v2} &= \sigma_{v2}C_3\sigma_{v2} = \sigma_{v3}\sigma_{v2} = C_3^2 \\
(\sigma_{v3})^{-1}C_3\sigma_{v3} &= \sigma_{v3}C_3\sigma_{v3} = \sigma_{v1}\sigma_{v3} = C_3^2
\end{aligned}$$

- The three reflection operations form a class with each other because

$$\begin{aligned}
E^{-1}\sigma_{v1}E &= E\sigma_{v1}E = \sigma_{v1}E = \sigma_{v1} \\
(C_3)^{-1}\sigma_{v1}C_3 &= C_3^2\sigma_{v1}C_3 = \sigma_{v2}C_3 = \sigma_{v3} \\
(C_3^2)^{-1}\sigma_{v1}C_3^2 &= C_3\sigma_{v1}C_3^2 = \sigma_{v3}C_3^2 = \sigma_{v2} \\
(\sigma_{v1})^{-1}\sigma_{v1}\sigma_{v1} &= \sigma_{v1}\sigma_{v1}\sigma_{v1} = E\sigma_{v1} = \sigma_{v1} \\
(\sigma_{v2})^{-1}\sigma_{v1}\sigma_{v2} &= \sigma_{v2}\sigma_{v1}\sigma_{v2} = C_3^2\sigma_{v2} = \sigma_{v3} \\
(\sigma_{v3})^{-1}\sigma_{v1}\sigma_{v3} &= \sigma_{v3}\sigma_{v1}\sigma_{v3} = C_3\sigma_{v3} = \sigma_{v2}
\end{aligned}$$

Example 8-13. Using the rules for IRRs, determine the characters for each IRR in the C_{3v} point group.

Solution. The C_{3v} point group has three classes of symmetry operations (E, $2C_3$, and $3\sigma_v$); thus, there will be three IRRs. The order of the group is $h = 6$ because there are a total of six symmetry operations. Because the character for the identity operation must always be positive and the sum of the squares of the characters for the identity element must equal the order of the group and there are only three IRRs, these characters must be 1, 1, and 2. There must be a totally symmetric IRR with all the characters equal to 1, as shown by Γ_1 below. Note that although there are only three classes of symmetry operations, the sum of the squares of this IRR is still equal to six because we must take the sum over all the symmetry operations in the group. The other nondegenerate IRR must also have the sum of the squares of its characters equal to six. The only way to do this is to have characters of all 1 or −1. In order for the second IRR to be orthogonal to the first, it is essential to place the 1 in the class of $2C_3$ operations and the −1 in the class of $3\sigma_v$ operations. This way the sum of the products for each operation in the two IRRs will equal zero: $(1)(1) + (1)(1) + (1)(1) + (1)(-1) + (1)(-1) + (1)(-1) = 0$. The result is given by Γ_2 below. The final IRR must have characters of 1 or −1 for the $2C_3$ operations and 0 for the $3\sigma_v$ operations in order for the sum of the squares of the characters for this IRR to equal six (as the identity operation has a character of 2). In order to be orthogonal with the first IRR, the character for the $2C_3$ operations must be −1 and not 1, so that $(1)(2) + (1)(-1) + (1)(-1) + (1)(0) + (1)(0) + (1)(0) = 0$. The result is Γ_3. Note that Γ_2 and Γ_3 are also orthogonal to each other: $(1)(2) + (1)(-1) + (1)(-1) + (-1)(0) + (-1)(0) + (-1)(0) = 0$. A quick comparison with the character table for the C_{3v} point group in Table 8.7 shows the answer below is correct.

C_{3v}	E	$2C_3$	$3\sigma_v$
Γ_1	1	1	1
Γ_2	1	1	−1
Γ_3	2	−1	0

The Mulliken symbol for an IRR provides a short-hand way of determining some of the primary symmetry properties associated with that IRR. Here are some rules for determining the Mulliken symbol for an IRR:

1. For finite groups, the IRR will be either *A* or *B* if the character for its identity operation is 1 (nondegenerate), *E* if it is 2 (doubly degenerate), or *T* if it is 3 (triply degenerate).
2. If the point group has a principal axis, then the symbol *A* is used to describe any IRR that is symmetric with respect to rotation around C_n and *B* is used if the rotation around the principal axis is anti-symmetric.
3. Whenever there is more than one IRR with the same name, subscripts or superscripts are used to distinguish between them. If the molecule has a center of inversion, the *g* subscript (*gerade*) will be used if the IRR is symmetric with respect to inversion and the *u* subscript (*ungerade*) will be used if it is anti-symmetric to inversion.
4. The superscripts ′ and ″ are used (when necessary) to represent whether the IRR is symmetric or anti-symmetric (respectively) to a horizontal mirror plane.
5. Additional subscripts (when necessary) include 1 or 2 to represent whether the IRR is symmetric or anti-symmetric with respect to a C_2 perpendicular to the principal axis (or S_n or σ_v if there is no $\perp C_2$), respectively.
6. For infinite point groups, Greek letters are used, where Σ is used for IRRs having a character of 1 for their identity operation and Π, Δ, Φ, or Γ are used if the character for the identity operation is larger than 1.

If we now reexamine the IRRs in Table 8.6 for the C_{2v} point group, we see that all four IRRs will have either *A* or *B* Mulliken symbols because each of them has a character of 1 for the identity operation and the C_{2v} point group is a finite one. The first two IRRs are symmetric with respect to rotation around the principal axis and therefore belong to *A* Mulliken symbols, whereas the latter two IRRs are anti-symmetric to C_2 and are labeled as *B*. Because there are two *A* IRRs and two *B* IRRs, we need to further distinguish between them. There is no inversion center, so we cannot use *g* or *u* subscripts. There is also no other rotational axis. However, there is a σ_v. The first and third IRRs in Table 8.6 are symmetric with respect to reflection in the *xz*-plane and are therefore labeled as A_1 and B_1, respectively. The second and fourth IRRs are anti-symmetric with respect to σ_{v1} (*xz*) and belong to the A_2 and B_2 Mulliken symbols, respectively.

Example 8-14. Provide appropriate Mulliken symbols for the three IRRs in the C_{3v} point group of ammonia.

Solution. The characters for the three IRRs in NH_3 were given in the solution to Example 8-13. The characters for the identity operation for Γ_1 and Γ_2 are both 1, making these IRRs either *A* or *B*. Furthermore, these two IRRs are both symmetric with respect to rotation around the principal axis. Thus, they are both *A*-type Mulliken symbols. In order to distinguish between them, Γ_1 is symmetric with respect to a σ_v reflection, whereas Γ_2 is not. Thus, Γ_1 will be A_1 and Γ_2 will be A_2. The third IRR, Γ_3, has a character of 2 for the identity operation and is simply labeled as *E* (no subscript is necessary as this is the only *E* representation in the point group).

Example 8-15. Provide appropriate Mulliken symbols for the IRRs in the D_{3h} point group, given the characters in the table below.

D_{3h}	E	$2C_3$	$3C_2$	σ_h	$2S_3$	$3\sigma_v$
Γ_1	1	1	1	1	1	1
Γ_2	1	1	−1	1	1	−1
Γ_3	2	−1	0	2	−1	0
Γ_4	1	1	1	−1	−1	−1
Γ_5	1	1	−1	−1	−1	1
Γ_6	2	−1	0	−2	1	0

Solution. Both Γ_1 and Γ_2 have characters of 1 for the identity and are symmetric with respect to rotation about the C_3 axis. Thus, they will be *A*. The same is also true for Γ_4 and Γ_5. The other two IRRs, Γ_3 and Γ_6, will be *E* because their characters for the identity are each 2. To further distinguish between the four *A*'s and two *E*'s, we look next to the horizontal mirror plane. Because Γ_1, Γ_2, and Γ_3 are symmetric with respect to σ_h, they will have $'$ superscripts, whereas Γ_4, Γ_5, and Γ_6 are anti-symmetric to σ_h and will have $''$ superscripts. Because there are still two A' and two A'' IRRs, we need to further distinguish them using the subscript 1 if they are symmetric with respect to a $C_2 \perp C_3$ and 2 if they are anti-symmetric to this C_2 rotation. Thus, the resulting Mulliken symbols are as follows: $\Gamma_1 = A_1'$, $\Gamma_2 = A_2'$, $\Gamma_3 = E'$, $\Gamma_4 = A_1''$, $\Gamma_5 = A_2''$, and $\Gamma_6 = E''$.

Example 8-16. Provide appropriate Mulliken symbols for the IRRs in the O_h point group, given the characters in the table below.

O_h	E	$8C_3$	$6C_2$	$6C_4$	$3C_2'$	i	$6S_4$	$8S_6$	$3\sigma_h$	$6\sigma_d$
Γ_1	1	1	1	1	1	1	1	1	1	1
Γ_2	1	1	−1	−1	1	1	−1	1	1	−1
Γ_3	2	−1	0	0	2	2	0	−1	2	0
Γ_4	3	0	−1	1	−1	3	1	0	−1	−1
Γ_5	3	0	1	−1	−1	3	−1	0	−1	1
Γ_6	1	1	1	1	1	−1	−1	−1	−1	−1
Γ_7	1	1	−1	−1	1	−1	1	−1	−1	1
Γ_8	2	−1	0	0	2	−2	0	1	−2	0
Γ_9	3	0	−1	1	−1	−3	−1	0	1	1
Γ_{10}	3	0	1	−1	−1	−3	1	0	1	−1

Solution. The following will be *A* IRRs because they have a character of 1 for the identity and are symmetric with respect to C_3 (although C_4 has a higher value of *n* than C_3, the C_3 axes are considered to be the principal axes in the O_h point group because they are collinear with the S_6 axis, which has an even higher value of *n*): Γ_1, Γ_2, Γ_6, and Γ_7. The first two of these are *gerade* with respect to inversion, whereas the latter two are *ungerade*. To further distinguish between the two A_g and two A_u IRRs, we look to the $C_2 \perp C_3$. Γ_1 and Γ_6 are symmetric with respect to this C_2, whereas Γ_2 and Γ_7 are not. Thus, $\Gamma_1 = A_{1g}$, $\Gamma_2 = A_{2g}$, $\Gamma_6 = A_{1u}$, and $\Gamma_7 = A_{2u}$. There are two *E* IRRs: Γ_3 and Γ_8, which have characters of 2 for the identity. The former of these is symmetric with respect to inversion, whereas the latter is not. Thus, $\Gamma_3 = E_g$ and $\Gamma_8 = E_u$. The remaining four IRRs will have *T* symmetry because they have characters of 3 for the identity. The first two of these (Γ_4 and Γ_5) are *gerade*, whereas the last two (Γ_9 and Γ_{10}) are *ungerade*. Because Γ_4 and Γ_9 are both symmetric with respect to the $C_2 \perp C_3$ while Γ_5 and Γ_{10} are anti-symmetric to this operation, the remaining Mulliken symbols are $\Gamma_4 = T_{1g}$, $\Gamma_5 = T_{2g}$, $\Gamma_9 = T_{1u}$, and $\Gamma_{10} = T_{1u}$.

8.6 DIRECT PRODUCTS

The *direct product* of two (or more) IRRs is defined by Equation (8.17), where the characters for each class of operations (*R*) in the group are given by Equation (8.18). Direct products are useful in certain quantum mechanical calculations, especially those defining the spectroscopic selection rules for allowed transitions between quantum states. They are introduced here in order to illustrate how the binary products and squares in the fourth region of the character table were determined:

$$\Gamma_a \Gamma_b \ldots = \Gamma_{ab} \qquad (8.17)$$

$$\chi_a(R)\chi_b(R) \ldots = \chi_{ab}(R) \qquad (8.18)$$

Using the character table for the C_{2v} point group, the product of the IRRs for Γ_x (B_1) with Γ_z (A_1) are shown by Table 8.8. The product $\Gamma_x\Gamma_z = B_1A_1 = \Gamma_{xz} = B_1$.

TABLE 8.8 The direct product B_1A_1 in the C_{2v} point group.

C_{2v}	E	C_2	σ_{v1}	σ_{v2}		
B_1	1	−1	1	−1	x	
A_1	1	1	1	1	z	Same as:
B_1A_1	1	−1	1	−1	xz	B_1

Thus, the xz binary product belongs to the B_1 IRR, as shown in region 4 of the C_{2v} point group.

Example 8-17. Calculate the direct product of each of the following in the C_{2v} point group: (a) xy, (b) xyz, and (c) A_2B_2.

Solution. (a) The direct product of xy is the A_2 IRR as shown below:

C_{2v}	E	C_2	σ_{v1}	σ_{v2}		
B_1	1	−1	1	−1	x	
B_2	1	−1	−1	1	y	Same as:
B_1B_2	1	1	−1	−1	xy	A_2

(b) The direct product of xyz is the same as the direct product $(xy)z$ or $x(yz)$. A general property of direct products is that they commute. Therefore, the triple direct product xyz is the same as A_2A_1. The resulting IRR has the A_2 Mulliken symbol, as shown below:

C_{2v}	E	C_2	σ_{v1}	σ_{v2}		
A_2	1	1	−1	−1	xy	
A_1	1	1	1	1	z	Same as:
A_2A_1	1	1	−1	−1	xyz	A_2

(c) The direct product A_2B_2 is B_1, as shown in the table below:

C_{2v}	E	C_2	σ_{v1}	σ_{v2}	
A_2	1	1	−1	−1	
B_2	1	−1	−1	1	Same as:
A_2B_2	1	−1	1	−1	B_1

Some of the direct products listed in the fourth region of the character tables correspond with how the d orbital having the same name transforms in the point group. The d_{xz} orbital, for instance, is shown in Figure 8.16. Using this orbital as a basis to generate a representation of the symmetry operations in the C_{2v} point

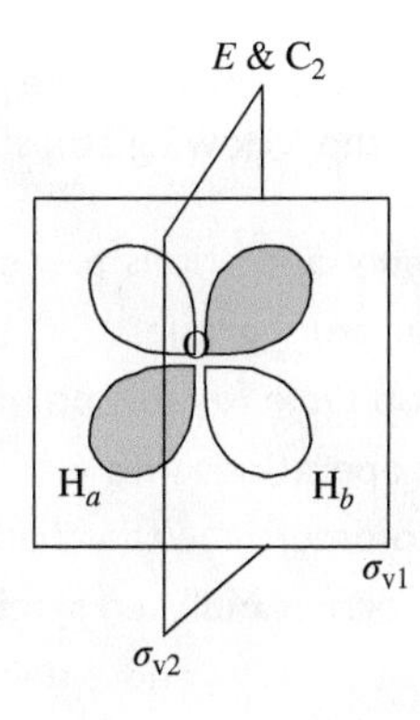

FIGURE 8.16
The d_{xz} orbital basis for the H_2O molecule.

group yields the B_1 IRR because the identity operation transforms d_{xz} into itself, C_2 converts it into $-d_{xz}$ (taking into account the sign of the wave function for each lobe), σ_{v1} (the *xz*-plane) converts it into itself, and σ_{v1} transforms d_{xz} into $-d_{xz}$. Similarly, the $d_z{}^2$, d_{xy}, and d_{yz} orbitals transform as the z^2, *xy*, and *yz* direct products transform. The $d_{x^2-y^2}$ orbital transforms as the difference of the x^2 and y^2 direct products. The *f* orbitals will also transform as linear combinations of the direct products corresponding to their names, as shown in the fifth region of the character table for C_{2v}.

Appendix C lists the binary direct products for the IRRs in the more common molecular point groups. The subscripts and superscripts for the direct products are omitted in these tables. For centrosymmetric molecules, the products of the *g* and *u* subscripts will yield the following subscript:

$$gg = g \quad uu = g \quad gu = ug = u$$

For molecules having $'$ or $''$ superscripts, the product superscript will transform as follows:

$$(')(') = (') \quad ('')('') = (') \quad (')('') = ('')(') = ('')$$

Some generalizations that can be made from the values in the direct product tables are

1. The direct product of any IRR with the totally symmetric IRR will always be equal to itself.
2. If each IRR in the product is nondegenerate (has a character of 1 for the identity operation), then the direct product IRR will also be nondegenerate.
3. The product of a nondegenerate IRR with a degenerate IRR will always yield a degenerate IRR having the same dimension as the original degenerate IRR.
4. The direct product of two degenerate IRRs will always be a reducible representation (RR). RRs can be factored into IRRs using the steps outlined in Section 8.6. Where there is more than one IRR in the direct product, the factor in brackets is the spatially anti-symmetric product.
5. The direct product of any IRR with itself will either be the totally symmetric IRR or contain the totally symmetric IRR as one of its factors after reduction. Furthermore, this will be the only direct product that contains the totally symmetric IRR as one of its factors.

Example 8-18. Construct a character table for the planar *trans*-dichloroethylene molecule by performing each of the following steps:

(a) Identify all of the symmetry operations present in this molecule.
(b) Determine the molecular point group.
(c) Construct a multiplication table for this point group.
(d) Organize the symmetry operations into classes.
(e) Use the rules for IRRs to determine the characters of each IRR.
(f) Label each IRR with its correct Mulliken symbol.
(g) Determine to which IRR the *x*-, *y*-, and *z*-unit vectors transform.
(h) Determine to which IRR the R_x, R_y, and R_z rotational vectors transform.
(i) Determine to which IRR each *d* orbital transforms.

Solution. The structure of *trans*-dichloroethylene molecule is

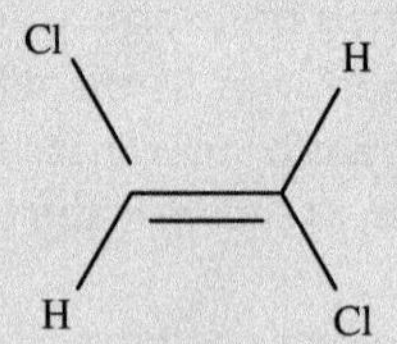

(a) There are four symmetry operations present in the molecule: E, C_2, i, and σ_h.

(b) The molecular point group is C_{2h}.

(c) The multiplication table is

C_{2h}	E	C_2	i	σ_h
E	E	C_2	i	σ_h
C_2	C_2	E	σ_h	i
i	i	σ_h	E	C_2
σ_h	σ_h	i	C_2	E

(d) Each operation is its own inverse. The identity operation is always in a class by itself. C_2 is also in its own class because

$$E^{-1}C_2E = EC_2E = C_2E = C_2$$
$$(C_2)^{-1}C_2C_2 = C_2C_2C_2 = EC_2 = C_2$$
$$i^{-1}C_2i = iC_2i = \sigma_h i = C_2$$
$$\sigma_h^{-1}C_2\sigma_h = \sigma_h C_2\sigma_h = i\sigma_h = C_2$$

The inversion operator is also in its own class, as shown below. By default, the horizontal reflection operation must also be in its own class:

$$E^{-1}iE = EiE = iE = i$$
$$(C_2)^{-1}iC_2 = C_2iC_2 = \sigma_h C_2 = i$$

$$i^{-1}ii \quad = iii \quad = Ei \quad = i$$
$$\sigma_h{}^{-1}i\sigma_h = \sigma_h i\sigma_h = C_2\sigma_h = i$$

(e) There are four classes, so there must be four IRRs. Because the sum of the squares of the characters for the identity operation must equal four (the order of the group) and each must always be positive, each IRR has a character of 1 for the identity. Furthermore, the sums of the squares of the characters for all the operators in each IRR must equal four, so each other character must be either 1 or −1. Finally, in order to make all four IRRs orthogonal with each other, there must always be two operations with characters of 1 and two with characters of −1. Thus, the four IRRs are shown below.

C_{2h}	E	C_2	i	σ_h
Γ_1	1	1	1	1
Γ_2	1	1	−1	−1
Γ_3	1	−1	1	−1
Γ_4	1	−1	−1	1

(f) Γ_1 is symmetric with respect to C_2 and i, so it is called A_g. Γ_2 is symmetric with respect to C_2, but it is anti-symmetric to i: A_u. Γ_3 is anti-symmetric to C_2, but it is symmetric to i, so it will be B_g. Finally, Γ_4 is anti-symmetric to both C_2 and i: B_u.

(g) By convention, the principal axis is always the z-axis and the horizontal mirror plane is therefore the xy-plane. A unit vector along the x-axis will transform into $+x$ with E, $-x$ with C_2, $-x$ with i, and $+x$ with σ_h. Thus, the x-unit vector transforms with characters 1, −1, −1, 1 as the B_u IRR. The y-unit vector will be converted into $+y$ by E, $-y$ by C_2, $-y$ by i, and $+y$ by σ_h, so that it will also transform as the B_u IRR. Finally, the z-unit vector will be transformed into $+z$ by E, $+z$ by C_2, $-z$ by i, and $-z$ by σ_h, which is the same as the A_u IRR.

(h) A rotational vector along the x-axis, R_x, will be converted into $+R_x$ by E, $-R_x$ by C_2, $+R_x$ by i, and $-R_x$ by σ_h, transforming as the B_g IRR. The same will also be true for R_y. R_z, on the other hand, will be carried into $+R_z$ by E, $+R_z$ by C_2, $+R_z$ by i, and $+R_z$ by σ_h, or the A_g IRR.

(i) The d_{xy} orbital will lie in the xy-plane (horizontal mirror plane) with alternating positive and negative signs of the wave functions in its four clover-leaf shaped lobes, each pointing between the x- and y-axes. Thus, it will be converted into itself by E, itself by C_2, itself by inversion, and itself by σ_h, transforming as the A_g IRR. The $d_{x^2-y^2}$ orbital also lies in the xy-plane, but with its lobes pointing directly on the x- and y-coordinate axes. It will also be symmetric with respect to E, C_2, i, and σ_h, transforming as A_g. The d_{z^2} orbital lies on the z-axis, with a positive lobe above and below the xy-plane and a negative ring in the xy-plane. It is therefore symmetric to E, C_2, i, and σ_h, also transforming as the A_g IRR. The d_{xz} orbital lies in the xz-plane perpendicular to the horizontal mirror plane. It will be symmetric to E, anti-symmetric to C_2, symmetric to i, and anti-symmetric to σ_h, making it belong to the B_g IRR. The d_{yz} orbital will lie in the yz-plane and transform in exactly the same manner as B_g.

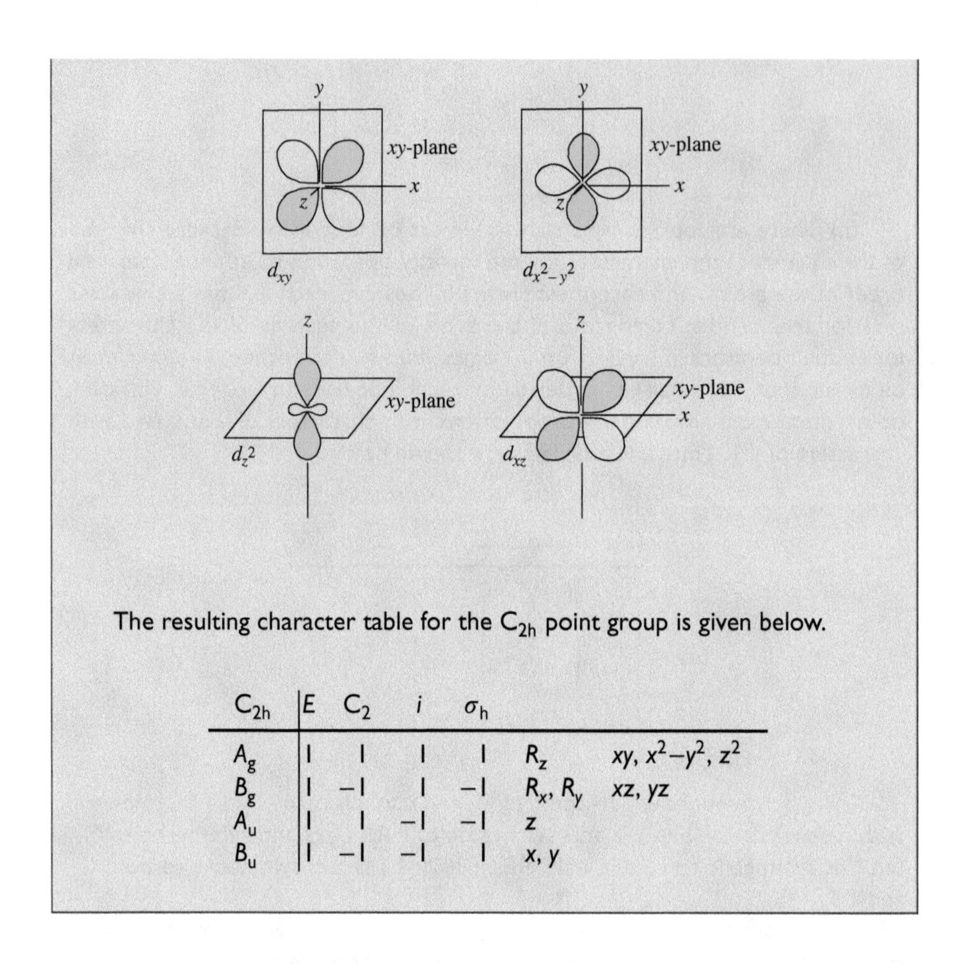

The resulting character table for the C_{2h} point group is given below.

C_{2h}	E	C_2	i	σ_h		
A_g	1	1	1	1	R_z	xy, x^2-y^2, z^2
B_g	1	−1	1	−1	R_x, R_y	xz, yz
A_u	1	1	−1	−1	z	
B_u	1	−1	−1	1	x, y	

8.7 REDUCIBLE REPRESENTATIONS

Any of the other representations for the symmetry operations in a point group that do not belong to the set of IRRs are called *RRs*. Using the H_2O molecule as an example, we have already shown (in Figure 8.14) how we can use a set of orthogonal vectors on each atom in the molecule to generate a set of 9 × 9 matrices representing the symmetry operations in the C_{2v} point group. These matrices are reproduced below.

C_{2v}	E	C_2	σ_{v1}	σ_{v2}
Γ_{dof}	$\begin{bmatrix} 1&0&0&0&0&0&0&0&0 \\ 0&1&0&0&0&0&0&0&0 \\ 0&0&1&0&0&0&0&0&0 \\ 0&0&0&1&0&0&0&0&0 \\ 0&0&0&0&1&0&0&0&0 \\ 0&0&0&0&0&1&0&0&0 \\ 0&0&0&0&0&0&1&0&0 \\ 0&0&0&0&0&0&0&1&0 \\ 0&0&0&0&0&0&0&0&1 \end{bmatrix}$	$\begin{bmatrix} 1&0&0&0&0&0&0&0&0 \\ 0&-1&0&0&0&0&0&0&0 \\ 0&0&-1&0&0&0&0&0&0 \\ 0&0&0&0&0&0&1&0&0 \\ 0&0&0&0&0&0&0&-1&0 \\ 0&0&0&0&0&0&0&0&-1 \\ 0&0&0&1&0&0&0&0&0 \\ 0&0&0&0&-1&0&0&0&0 \\ 0&0&0&0&0&-1&0&0&0 \end{bmatrix}$	$\begin{bmatrix} 1&0&0&0&0&0&0&0&0 \\ 0&-1&0&0&0&0&0&0&0 \\ 0&0&1&0&0&0&0&0&0 \\ 0&0&0&0&0&0&1&0&0 \\ 0&0&0&0&0&0&0&-1&0 \\ 0&0&0&0&0&0&0&0&1 \\ 0&0&0&1&0&0&0&0&0 \\ 0&0&0&0&-1&0&0&0&0 \\ 0&0&0&0&0&1&0&0&0 \end{bmatrix}$	$\begin{bmatrix} 1&0&0&0&0&0&0&0&0 \\ 0&1&0&0&0&0&0&0&0 \\ 0&0&-1&0&0&0&0&0&0 \\ 0&0&0&1&0&0&0&0&0 \\ 0&0&0&0&1&0&0&0&0 \\ 0&0&0&0&0&-1&0&0&0 \\ 0&0&0&0&0&0&1&0&0 \\ 0&0&0&0&0&0&0&1&0 \\ 0&0&0&0&0&0&0&0&-1 \end{bmatrix}$

At the same time, we have also shown how to generate the unique set of fundamental representations for the C_{2v} point group known as *IRRs*. The Γ_{dof} representation, shown above, is clearly not one of the four IRRs for this molecular point group. However, it can be factored into a linear combination of those fundamental representations. In fact, *any RR* (generated from an appropriate basis set in the molecule's point group) *can always be factored into some combination of that point group's IRRs*. Here is how it works. Just as every symmetry operation has an inverse, every matrix X also has an inverse matrix X^{-1}, such that the product $XX^{-1} = E$, the identity matrix (which has ones along its diagonal and zeroes in the off-diagonal positions). There will always be some matrix Q and its inverse Q^{-1} that can be used to perform a similarity transform on the RR A to convert it into some other matrix B with which it is conjugate. Many different similarity transforms of the matrix are possible. Our ultimate goal is to take the initial undiagonalized matrix A, perform one or more similarity transforms on it, and convert it into a *block-diagonalized matrix*, where the only nonzero matrix elements lie in a series of blocks along the diagonal from the upper left-hand side of the matrix to its lower right-hand side. An example of a block-diagonalized matrix is shown in Figure 8.17.

Once the matrix has been block-diagonalized, the different blocks of the larger matrix representation (the RR) will themselves form a set of smaller matrix representations (the IRRs) which follow the same multiplication table that the symmetry operations themselves follow. Using the degrees of freedom example for the H_2O molecule, we see that the 9×9 matrix representing the identity operation is already a diagonal matrix:

$$E = \begin{bmatrix} 1 & 0 & 0 & 0 & 0 & 0 & 0 & 0 & 0 \\ 0 & 1 & 0 & 0 & 0 & 0 & 0 & 0 & 0 \\ 0 & 0 & 1 & 0 & 0 & 0 & 0 & 0 & 0 \\ 0 & 0 & 0 & 1 & 0 & 0 & 0 & 0 & 0 \\ 0 & 0 & 0 & 0 & 1 & 0 & 0 & 0 & 0 \\ 0 & 0 & 0 & 0 & 0 & 1 & 0 & 0 & 0 \\ 0 & 0 & 0 & 0 & 0 & 0 & 1 & 0 & 0 \\ 0 & 0 & 0 & 0 & 0 & 0 & 0 & 1 & 0 \\ 0 & 0 & 0 & 0 & 0 & 0 & 0 & 0 & 1 \end{bmatrix} \qquad C_2 = \begin{bmatrix} 1 & 0 & 0 & 0 & 0 & 0 & 0 & 0 & 0 \\ 0 & -1 & 0 & 0 & 0 & 0 & 0 & 0 & 0 \\ 0 & 0 & -1 & 0 & 0 & 0 & 0 & 0 & 0 \\ 0 & 0 & 0 & 0 & 0 & 0 & 1 & 0 & 0 \\ 0 & 0 & 0 & 0 & 0 & 0 & 0 & -1 & 0 \\ 0 & 0 & 0 & 0 & 0 & 0 & 0 & 0 & -1 \\ 0 & 0 & 0 & 1 & 0 & 0 & 0 & 0 & 0 \\ 0 & 0 & 0 & 0 & -1 & 0 & 0 & 0 & 0 \\ 0 & 0 & 0 & 0 & 0 & -1 & 0 & 0 & 0 \end{bmatrix}$$

The C_2 matrix, however, is not a diagonal one. In order to convert the C_2 matrix into block-diagonal form, we will use the similarity transform shown below. For readers who might be familiar with linear algebra, the P matrix is formed by using the eigenvectors of the original C_2 matrix to form the different columns of Q. The inverse of Q is then obtained to yield P^{-1} (you should prove to yourself that the product $Q^{-1}Q= E$). We then take the similarity transform of the C_2 matrix, Q^{-1} A Q to obtain a conjugate matrix B that has been block-diagonalized, as shown by Equation (8.19). You should do the matrix multiplication yourself to prove that the similarity transform converts our original matrix into a diagonalized matrix. It is not important for you to understand where Q^{-1} and Q came from:

$$\begin{bmatrix} 1 & 2 & 1 & 0 & 0 & 0 \\ 3 & 1 & 1 & 0 & 0 & 0 \\ 1 & 2 & 1 & 0 & 0 & 0 \\ 0 & 0 & 0 & 3 & 1 & 0 \\ 0 & 0 & 0 & 1 & 2 & 0 \\ 0 & 0 & 0 & 0 & 0 & 3 \end{bmatrix}$$

FIGURE 8.17
A block-diagonalized matrix consisting of a 3×3 block, a 2×2 block, and a 1×1 block along the diagonal that runs from the upper left-hand corner to the lower right-hand corner of the matrix.

$$
\overset{Q^{-1}}{\begin{bmatrix} 1 & 0 & 0 & 0 & 0 & 0 & 0 & 0 & 0 \\ 0 & 1 & 0 & 0 & 0 & 0 & 0 & 0 & 0 \\ 0 & 0 & 1 & 0 & 0 & 0 & 0 & 0 & 0 \\ 0 & 0 & 0 & 0.5 & 0 & 0 & -0.5 & 0 & 0 \\ 0 & 0 & 0 & 0 & 0.5 & 0 & 0 & 0 & 0 \\ 0 & 0 & 0 & 0 & 0 & 0.5 & 0 & 0.5 & 0 \\ 0 & 0 & 0 & 0.5 & 0 & 0 & 0.5 & 0 & 0.5 \\ 0 & 0 & 0 & 0 & 0.5 & 0 & 0 & -0.5 & 0 \\ 0 & 0 & 0 & 0 & 0 & 0.5 & 0 & 0 & -0.5 \end{bmatrix}}
\overset{A}{\begin{bmatrix} 1 & 0 & 0 & 0 & 0 & 0 & 0 & 0 & 0 \\ 0 & -1 & 0 & 0 & 0 & 0 & 0 & 0 & 0 \\ 0 & 0 & -1 & 0 & 0 & 0 & 0 & 0 & 0 \\ 0 & 0 & 0 & 0 & 0 & 0 & 1 & 0 & 0 \\ 0 & 0 & 0 & 0 & 0 & 0 & 0 & -1 & 0 \\ 0 & 0 & 0 & 0 & 0 & 0 & 0 & 0 & -1 \\ 0 & 0 & 0 & 1 & 0 & 0 & 0 & 0 & 0 \\ 0 & 0 & 0 & 0 & -1 & 0 & 0 & 0 & 0 \\ 0 & 0 & 0 & 0 & 0 & -1 & 0 & 0 & 0 \end{bmatrix}}
$$

$$
\overset{Q}{\begin{bmatrix} 1 & 0 & 0 & 0 & 0 & 0 & 0 & 0 & 0 \\ 0 & 1 & 0 & 0 & 0 & 0 & 0 & 0 & 0 \\ 0 & 0 & 1 & 0 & 0 & 0 & 0 & 0 & 0 \\ 0 & 0 & 0 & 1 & 0 & 0 & 1 & 0 & 0 \\ 0 & 0 & 0 & 0 & 1 & 0 & 0 & 1 & 0 \\ 0 & 0 & 0 & 0 & 0 & 1 & 0 & 0 & 1 \\ 0 & 0 & 0 & -1 & 0 & 0 & 1 & 0 & 0 \\ 0 & 0 & 0 & 0 & 1 & 0 & 0 & -1 & 0 \\ 0 & 0 & 0 & 0 & 0 & 1 & 0 & 0 & 1 \end{bmatrix}}
\overset{=}{=}
\overset{B}{\begin{bmatrix} 1 & 0 & 0 & 0 & 0 & 0 & 0 & 0 & 0 \\ 0 & -1 & 0 & 0 & 0 & 0 & 0 & 0 & 0 \\ 0 & 0 & -1 & 0 & 0 & 0 & 0 & 0 & 0 \\ 0 & 0 & 0 & -1 & 0 & 0 & 0 & 0 & 0 \\ 0 & 0 & 0 & 0 & -1 & 0 & 0 & 0 & 0 \\ 0 & 0 & 0 & 0 & 0 & -1 & 0 & 0 & 0 \\ 0 & 0 & 0 & 0 & 0 & 0 & 1 & 0 & 0 \\ 0 & 0 & 0 & 0 & 0 & 0 & 0 & 1 & 0 \\ 0 & 0 & 0 & 0 & 0 & 0 & 0 & 0 & 1 \end{bmatrix}}
\tag{8.19}
$$

The 9×9 matrix representing the σ_{v2} operation is already in diagonal form, as shown below; however, the matrix for σ_{v1} is not. We can diagonalize the 9×9 matrix representing σ_{v1} using the similarity transform $Q^{-1}\ M\ Q$, where Q^{-1} and Q were the same matrices used to block-diagonalize C_2, as shown by Equation (8.20):

$$
\sigma_{v2} = \begin{bmatrix} 1 & 0 & 0 & 0 & 0 & 0 & 0 & 0 & 0 \\ 0 & 1 & 0 & 0 & 0 & 0 & 0 & 0 & 0 \\ 0 & 0 & -1 & 0 & 0 & 0 & 0 & 0 & 0 \\ 0 & 0 & 0 & 1 & 0 & 0 & 0 & 0 & 0 \\ 0 & 0 & 0 & 0 & 1 & 0 & 0 & 0 & 0 \\ 0 & 0 & 0 & 0 & 0 & -1 & 0 & 0 & 0 \\ 0 & 0 & 0 & 0 & 0 & 0 & 1 & 0 & 0 \\ 0 & 0 & 0 & 0 & 0 & 0 & 0 & 1 & 0 \\ 0 & 0 & 0 & 0 & 0 & 0 & 0 & 0 & -1 \end{bmatrix}
\qquad
\sigma_{v1} = \begin{bmatrix} 1 & 0 & 0 & 0 & 0 & 0 & 0 & 0 & 0 \\ 0 & -1 & 0 & 0 & 0 & 0 & 0 & 0 & 0 \\ 0 & 0 & 1 & 0 & 0 & 0 & 0 & 0 & 0 \\ 0 & 0 & 0 & 0 & 0 & 0 & 1 & 0 & 0 \\ 0 & 0 & 0 & 0 & 0 & 0 & 0 & -1 & 0 \\ 0 & 0 & 0 & 0 & 0 & 0 & 0 & 0 & 1 \\ 0 & 0 & 0 & 1 & 0 & 0 & 0 & 0 & 0 \\ 0 & 0 & 0 & 0 & -1 & 0 & 0 & 0 & 0 \\ 0 & 0 & 0 & 0 & 0 & 1 & 0 & 0 & 0 \end{bmatrix}
$$

$$
\overset{Q^{-1}}{\begin{bmatrix} 1 & 0 & 0 & 0 & 0 & 0 & 0 & 0 & 0 \\ 0 & 1 & 0 & 0 & 0 & 0 & 0 & 0 & 0 \\ 0 & 0 & 1 & 0 & 0 & 0 & 0 & 0 & 0 \\ 0 & 0 & 0 & 0.5 & 0 & 0 & -0.5 & 0 & 0 \\ 0 & 0 & 0 & 0 & 0.5 & 0 & 0 & 0 & 0 \\ 0 & 0 & 0 & 0 & 0 & 0.5 & 0 & 0.5 & 0 \\ 0 & 0 & 0 & 0.5 & 0 & 0 & 0.5 & 0 & 0.5 \\ 0 & 0 & 0 & 0 & 0.5 & 0 & 0 & -0.5 & 0 \\ 0 & 0 & 0 & 0 & 0 & 0.5 & 0 & 0 & -0.5 \end{bmatrix}}
\overset{M}{\begin{bmatrix} 1 & 0 & 0 & 0 & 0 & 0 & 0 & 0 & 0 \\ 0 & -1 & 0 & 0 & 0 & 0 & 0 & 0 & 0 \\ 0 & 0 & 1 & 0 & 0 & 0 & 0 & 0 & 0 \\ 0 & 0 & 0 & 0 & 0 & 0 & 1 & 0 & 0 \\ 0 & 0 & 0 & 0 & 0 & 0 & 0 & -1 & 0 \\ 0 & 0 & 0 & 0 & 0 & 0 & 0 & 0 & 1 \\ 0 & 0 & 0 & 1 & 0 & 0 & 0 & 0 & 0 \\ 0 & 0 & 0 & 0 & -1 & 0 & 0 & 0 & 0 \\ 0 & 0 & 0 & 0 & 0 & 1 & 0 & 0 & 0 \end{bmatrix}}
$$

$$
\begin{array}{ccc} Q & = & N \end{array}
$$

$$
\begin{bmatrix}
1 & 0 & 0 & 0 & 0 & 0 & 0 & 0 & 0\\
0 & 1 & 0 & 0 & 0 & 0 & 0 & 0 & 0\\
0 & 0 & 1 & 0 & 0 & 0 & 0 & 0 & 0\\
0 & 0 & 0 & 1 & 0 & 0 & 1 & 0 & 0\\
0 & 0 & 0 & 0 & 1 & 0 & 0 & 1 & 0\\
0 & 0 & 0 & 0 & 0 & 1 & 0 & 0 & 1\\
0 & 0 & 0 & -1 & 0 & 0 & 1 & 0 & 0\\
0 & 0 & 0 & 0 & 1 & 0 & 0 & -1 & 0\\
0 & 0 & 0 & 0 & 0 & 1 & 0 & 0 & 1
\end{bmatrix}
=
\begin{bmatrix}
1 & 0 & 0 & 0 & 0 & 0 & 0 & 0 & 0\\
0 & -1 & 0 & 0 & 0 & 0 & 0 & 0 & 0\\
0 & 0 & 1 & 0 & 0 & 0 & 0 & 0 & 0\\
0 & 0 & 0 & -1 & 0 & 0 & 0 & 0 & 0\\
0 & 0 & 0 & 0 & -1 & 0 & 0 & 0 & 0\\
0 & 0 & 0 & 0 & 0 & 1 & 0 & 0 & 0\\
0 & 0 & 0 & 0 & 0 & 0 & 1 & 0 & 0\\
0 & 0 & 0 & 0 & 0 & 0 & 0 & 1 & 0\\
0 & 0 & 0 & 0 & 0 & 0 & 0 & 0 & -1
\end{bmatrix}
\tag{8.20}
$$

The diagonalized 9 × 9 matrices representing the four symmetry operations in the C_{2v} point group are shown below.

C_{2v}	E	Diagonalized C_2 (B)	Diagonalized σ_{v1} (N)	σ_{v2}

$$
\Gamma_{dof}\quad
\begin{bmatrix}
1&0&0&0&0&0&0&0&0\\
0&1&0&0&0&0&0&0&0\\
0&0&1&0&0&0&0&0&0\\
0&0&0&1&0&0&0&0&0\\
0&0&0&0&1&0&0&0&0\\
0&0&0&0&0&1&0&0&0\\
0&0&0&0&0&0&1&0&0\\
0&0&0&0&0&0&0&1&0\\
0&0&0&0&0&0&0&0&1
\end{bmatrix}
\begin{bmatrix}
1&0&0&0&0&0&0&0&0\\
0&-1&0&0&0&0&0&0&0\\
0&0&-1&0&0&0&0&0&0\\
0&0&0&-1&0&0&0&0&0\\
0&0&0&0&-1&0&0&0&0\\
0&0&0&0&0&-1&0&0&0\\
0&0&0&0&0&0&1&0&0\\
0&0&0&0&0&0&0&1&0\\
0&0&0&0&0&0&0&0&1
\end{bmatrix}
\begin{bmatrix}
1&0&0&0&0&0&0&0&0\\
0&-1&0&0&0&0&0&0&0\\
0&0&1&0&0&0&0&0&0\\
0&0&0&-1&0&0&0&0&0\\
0&0&0&0&-1&0&0&0&0\\
0&0&0&0&0&1&0&0&0\\
0&0&0&0&0&0&1&0&0\\
0&0&0&0&0&0&0&1&0\\
0&0&0&0&0&0&0&0&-1
\end{bmatrix}
\begin{bmatrix}
1&0&0&0&0&0&0&0&0\\
0&1&0&0&0&0&0&0&0\\
0&0&-1&0&0&0&0&0&0\\
0&0&0&1&0&0&0&0&0\\
0&0&0&0&1&0&0&0&0\\
0&0&0&0&0&-1&0&0&0\\
0&0&0&0&0&0&1&0&0\\
0&0&0&0&0&0&0&1&0\\
0&0&0&0&0&0&0&0&-1
\end{bmatrix}
$$

We are now in a position to extract a set of nine 1 × 1 matrices for the symmetry operations from the elements lying in the corresponding positions along the diagonal from upper left to lower right in the block-diagonalized 9 × 9 matrices. These nine 1 × 1 matrices (sans brackets) are shown in Table 8.9.

Because they are unitary matrices, their elements are identical to their characters. You should then notice that each line of characters in Table 8.9 corresponds

TABLE 8.9 The characters lying along the positions a_{ii} on the diagonal of the above 9 × 9 diagonalized matrices representing the symmetry operations in the C_{2v} point group.

C_{2v}	E	C_2	σ_{v1}	σ_{v2}	IRR
i =1	1	1	1	1	A_1
2	1	−1	−1	1	B_2
3	1	−1	1	−1	B_1
4	1	−1	−1	1	B_2
5	1	−1	−1	1	B_2
6	1	−1	1	−1	B_1
7	1	1	1	1	A_1
8	1	1	1	1	A_1
9	1	1	−1	−1	A_2

with one of the IRRs in the point group (listed at right for your convenience). Thus, we have reduced the 9 × 9 representation given by Γ_{dof} into its IRR factors, as shown in Equation (8.21). Notice that the sum of the characters for each IRR for any given operation equals the character for that operation in the original 9 × 9 matrix.

$$\Gamma_{dof} = 3A_1 + A_2 + 2B_1 + 3B_2 \tag{8.21}$$

As this example has shown, the process of reducing an RR into its irreducible factors is not necessarily an easy one! The C_{2v} point group had only 4 symmetry operations and 4 IRRs, but the O_h point group has 48 symmetry operations and 10 IRRs. Imagine having to do matrix multiplication for a set of ten 48 × 48 matrices! Nonetheless, in application after application, we will be required to reduce one or more RRs. There is got to be an easier way! Thankfully, there is.

Because diagonalized matrices can be characterized by their traces, we will not have to write out all of the matrices in our calculations. Instead, we can just make use of their characters. Imagine having a number-crunching machine that can tell you exactly how many of each type of IRR will contribute to a given RR. In other words, it can take any large RR and literally "crunch" it into its irreducible factors. The magical machine that does this for us is called the *great orthogonality theorem*, and it is given by Equation (8.22). The abbreviations in the great orthogonality theorem are that n_i is the number of times that a given IRR will factor into the original RR, h is the order of the group, the summation is over all the classes in the point group, g_c is the number of operations in the class, χ_{IRR} is the character for that class of operations in the IRR, and χ_{RR} is the character for the same class in the RR:

$$n_i = \frac{1}{h}\sum_c g_c \chi_{IRR} \chi_{RR} \tag{8.22}$$

Using the H_2O example, the characters for the Γ_{dof} representation and for each of the IRRs in the C_{2v} point group are shown in Table 8.10. In order to determine how many times the A_1 IRR occurs in Γ_{dof} representation, we apply Equation (8.22) as follows: $n(A_1) = \frac{1}{4}[(1)(1)(9) + (1)(1)(-1) + (1)(1)(1) + (1)(1)(3)] = 3$. Similarly, $n(A_2) = \frac{1}{4}[(1)(1)(9) + (1)(1)(-1) + (1)(-1)(1) + (1)(-1)(3)] = 1$; $n(B_1) = \frac{1}{4}[(1)(1)(9) + (1)(-1)(-1) + (1)(1)(1) + (1)(-1)(3)] = 2$; and $n(B_2) = \frac{1}{4}[(1)(1)(9) + (1)(-1)(-1) + (1)(-1)(1) + (1)(1)(3)] = 3$. Notice that this is the exact same result as Equation (8.21), where we used similarity transforms on the 9 × 9 matrices to diagonalize them and then extracted the characters from the blocks along the diagonal. However, the great orthogonality theorem has saved us a great deal of unnecessary work.

In his book *Molecular Symmetry and Group Theory*, Robert Carter introduces a tabular method for applying the great orthogonality theorem, which is illustrated

TABLE 8.10 Tabular method for reducing representations using the great orthogonality theorem.

C_{2v}	E	C_2 (z)	σ_{v1} (xz)	σ_{v2} (yz)	$h = 4$	$n_i =$
Γ_{dof}	9	−1	1	3	Σ	Σ/h
A_1	9	−1	1	3	12	3
A_2	9	−1	−1	−3	4	1
B_1	9	1	1	−3	8	2
B_2	9	1	−1	3	12	3

in Table 8.10. Across the top of the table lie the symmetry operations organized into classes as they appear in the character table for C_{2v}. The next row lists the characters for the RR for Γ_{dof}. The rows beneath this list the triple product $g_c\ \chi_{IRR}\ \chi_{RR}$ in table entries. At right, there is a column for the summation of these triple products and for n_i, the number of times each IRR factors into the original RR. This shorthand method for application of the great orthogonality theorem might prove to be a useful tool for the average student. In addition, there are also several good websites available on the Internet that will allow you to input the RR for a given point group and factor it into its IRRs.

Example 8-19. Use the great orthogonality theorem and tabular method to reduce the Γ_{dof} representation for methane in the T_d point group.

T_d	E	$8C_3$	$3C_2$	$6S_4$	$6\sigma_d$
Γ_{dof}	15	0	−1	−1	3

Solution. The tabular method yields $\Gamma_{dof} = A_1 + E + T_1 + 3T_2$.

T_d	E	$8C_3$	$3C_2$	$6S_4$	$6\sigma_d$	$h = 24$	$n_i =$
Γ_{dof}	15	0	−1	−1	3	Σ	Σ/h
A_1	15	0	−3	−6	18	24	1
A_2	15	0	−3	6	−18	0	0
E	30	0	−6	0	0	24	1
T_1	45	0	3	−6	−18	24	1
T_2	45	0	3	6	18	72	3

Occasionally, we will encounter point groups such as C_3 or C_4, where some of the characters are complex numbers containing imaginary parts. These will always belong to doubly degenerate *E* IRR. Whenever this occurs, the character for the IRR containing the imaginary component will appear in brackets as two rows of symbols on the character table, making it impossible to determine the triple product using the great orthogonality theorem when reducing representations. In cases such as these, it is common to combine the two rows of characters to eliminate the imaginary components and yield a single real number, which can then be used in the great orthogonality theorem. An example follows for the C_3 point group. When using these substitute characters in the theorem, it is important to realize that the numerical result obtained for *n* for any *E* IRR will be twice that of reality as two rows of characters were combined into a set of single characters when constructing the surrogate characters for the *E* IRR. Thus, the result calculated for *n* must be divided by two, as shown.

Example 8-20. Reduce the $\Gamma_{x,y,z}$ representation for the C_3 point group by taking the sums of the components of the characters for *E* representation to make a single real character for each operation in the *E* IRR.

C_3	E	C_3	C_3^2
$\Gamma_{x,y,z}$	3	0	0

Solution. First, we will sum the components of E to make a single real set of characters for the E IRR.

Un-useable: contains imaginary components

C_3	E	C_3	C_3^2
A	1	1	1
E	$\left\{\begin{matrix} 1 \\ 1 \end{matrix}\right.$	$\begin{matrix} \varepsilon \\ \varepsilon* \end{matrix}$	$\left.\begin{matrix} \varepsilon* \\ \varepsilon \end{matrix}\right\}$

Useable form: made by taking the sum of the characters for E

C_3	E	C_3	C_3^2
A	1	1	1
E	{2	$2\cos(2\pi/3)$	$2\cos(2\pi/3)$}

$$\varepsilon = e^{2\pi i/3} = \cos(2\pi/3) + i\sin(2\pi/3) \text{ and } \varepsilon * = \cos(2\pi/3) - i\sin(2\pi/3)$$

Then, we will use the tabular method for the great orthogonality theorem to reduce $\Gamma_{x,y,z}$.

C_3	E	C_3	C_3^2	$h = 3$	$n_i =$
$\Gamma_{x,y,z}$	3	0	0	Σ	Σ/h
A	3	0	0	3	1
E	{6	0	0}	{6}	{2} = 1

Remembering that the characters for E are surrogates for two rows of characters, the real value of n for the E representation is half of the calculated one (left in brackets to remind us that it was derived from the surrogate characters). Thus, $\Gamma_{x,y,z} = A + E$ and not $A + 2E$. Indeed, this result is consistent with the overall dimensionality of the RR. The dimension of $\Gamma_{x,y,z}$ is 3 (because that is its character for the identity operation). Thus, the sum of the dimensions of the IRRs comprising $\Gamma_{x,y,z}$ must also equal 3. The result $A + E$ is consistent with this rule, because A representations are always unitary and (by definition) the E representation has a dimension of 2. Had we forgotten to divide the results for E by two in our table, we would have erroneously obtained the result $A + 2E$, which has a total dimension of 5 and is inconsistent with the dimensionality of our original RR.

Another complication that sometimes arises when using the great orthogonality theorem occurs for the infinite point groups. The order of these groups is infinity, making it impossible for us to arrive at a precise value for Σ/h. In cases such as these, our strategy will be to use the simplest subgroup of the molecular point group to do the reduction and then use a correlation table to yield back the appropriate IRRs in the infinite point group. A subgroup is a set of symmetry operations that is part of a larger point group but that itself satisfies the mathematical criteria to exist as an independent group. The characters for the operations in a subgroup will be identical to those for the same operations in the larger group. For the $C_{\infty v}$ infinite group, the simplest subgroup is C_{2v}. Some of the correlations between the

TABLE 8.11 Partial correlation table between $C_{\infty v}$ and C_{2v}.

$C_{\infty v}$	C_{2v}
$A_1 = \Sigma^+$	A_1
$A_2 = \Sigma^-$	A_2
$E_1 = \Pi$	$B_1 + B_2$
$E_2 = \Delta$	$A_1 + A_2$

TABLE 8.12 Partial correlation table between $D_{\infty h}$ and D_{2h}.

$D_{\infty h}$	D_{2h}
Σ_g^+	A_g
Σ_g^-	B_{1g}
Π_g	$B_{2g} + B_{3g}$
Δ_g	$A_g + B_{1g}$
Σ_u^+	B_{1u}
Σ_u^-	A_u
Π_u	$B_{2u} + B_{3u}$
Δ_u	$A_u + B_{1u}$

two groups are shown in Table 8.11. For the $D_{\infty h}$ point group, we will use D_{2h} as the subgroup. A partial list of the correlations between $D_{\infty h}$ and D_{2h} are shown in Table 8.12. Although they will not be used here, the correlation tables between the more common molecular point groups and their corresponding subgroups are listed in Appendix D. Notice that the total dimensionality of the IRRs in the subgroup and the original molecular point group must always be the same.

Example 8-21. Use a set of three orthogonal miniature vectors on each atom in the CO_2 molecule to generate the Γ_{dof} representation in the D_{2h} subgroup of $D_{\infty h}$. Then use the great orthogonality theorem and the correlation table in Table 8.12 to determine the IRRs that comprise the Γ_{dof} representation for the CO_2 molecule.

Solution. The basis set for generating the Γ_{dof} representation is given below:

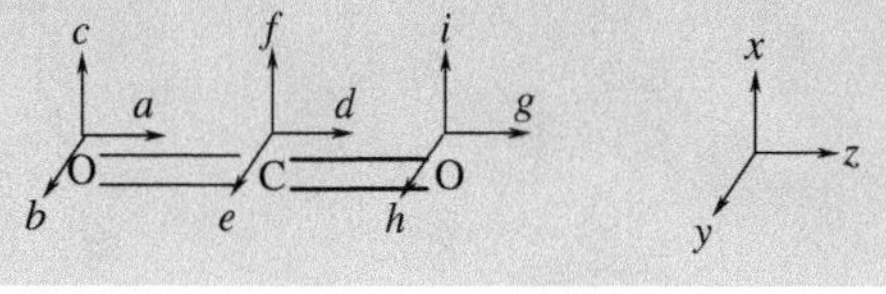

c f i x
a d g
O C O z
b e h y

The RR in the D_{2h} subgroup is shown in the second line of the table below. All nine of the vectors transform into themselves using the identity. A C_2 rotation around the *z*-axis leaves vectors *a*, *d*, and *g* unchanged, but it converts the other six vectors into their opposites, thus yielding an overall character of −3. Performing a C_2 operation around the *x*-axis, only vector *f* transforms as itself, whereas vectors *d* and e go into the opposites of themselves, yielding a character of −1, as all other vectors will be off the diagonal. Likewise, for a C_2 rotation around the *y*-axis, only vector e will transform into itself, whereas vectors *d* and *f* convert into their negatives and all other vectors yield off-diagonal elements, giving a character of −1. Inversion will transform vectors *d*, e, and *f* into their opposites, yielding a character of −3. Reflection in the *xy*-plane transforms e and *f* into themselves and *d* into −*d*, giving a character of 1. Reflection in the *xz*-plane retains vectors *a*, *c*, *d*, *f*, *g*, and *i*, while converting *b*, e, and *h* into their opposites, yielding a character of 3. Reflection in the *yz*-plane converts *a*, *b*, *d*, e, *g*, and *h* into themselves and *c*, *f*, and *i* into their opposites, so this operation also has a character of 3.

The remainder of the table below uses the tabular method of application of the great orthogonality theorem to reduce Γ_{dof} into its irreducible components.

D_{2h}	E	C_2 (z)	C_2 (y)	C_2 (x)	i	σ_{xy}	σ_{xz}	σ_{yz}	$h = 8$	$n_i =$
Γ_{dof}	9	−3	−1	−1	−3	1	3	3	Σ	Σ/h
A_g	9	−3	−1	−1	−3	1	3	3	8	1
B_{1g}	9	−3	1	1	−3	1	−3	−3	0	0
B_{2g}	9	3	−1	1	3	−1	3	−3	8	1
B_{3g}	9	3	1	−1	−3	−1	−3	3	8	1
A_u	9	−3	−1	−1	3	−1	−3	−3	0	0
B_{1u}	9	−3	1	1	3	−1	3	3	16	2
B_{2u}	9	3	−1	1	3	1	−3	3	16	2
B_{3u}	9	3	1	−1	3	1	3	−3	16	2

The result in the D_{2h} subgroup is $\Gamma_{dof} = A_g + B_{2g} + B_{3g} + 2B_{1u} + 2B_{2u} + 2B_{3u}$Using the correlation table in Table 8.12, this translates into the following IRRs in the $D_{\infty h}$ point group: $\Gamma_{dof} = \Sigma_g^+ + \Pi_g + 2\Sigma_u^+ + 2\Pi_u$.

EXERCISES

8.1. Determine the molecular point group for each of the following molecules or ions:

a. ClF_3
b. IF_5
c. I_3^-
d. XeO_4
e. $XeOF_4$
f. XeF_4
g. SF_2
h. SO_2
i. CO_2
j. SO_4^{2-}
k. NO_2^+
l. ClO_2F
m. CCl_4
n. SF_6

o. SNF_3
p. XeO_3
q. KrF_2
r. HOCl
s. CCl_3^+
t. NOF

8.2. Determine the molecular point group for each of the following molecules (look up their structures if necessary):

a. *trans*-Bromochloroethene
b. Dichloromethane
c. *cis*-Dichloroethene
d. Phosphorous pentafluoride
e. Cyclopropane
f. White phosphorous
g. Cubane
h. Cyclohexane (chair conformation)
i. Cyclohexane (boat conformation)
j. A tennis ball (including the seam)

8.3. Determine the molecular point group for each of the following boranes:

a. $B_6H_6^{2-}$ (*closo*)
b. B_5H_9 (*nido*)
c. B_4H_{10} (*arachno*)

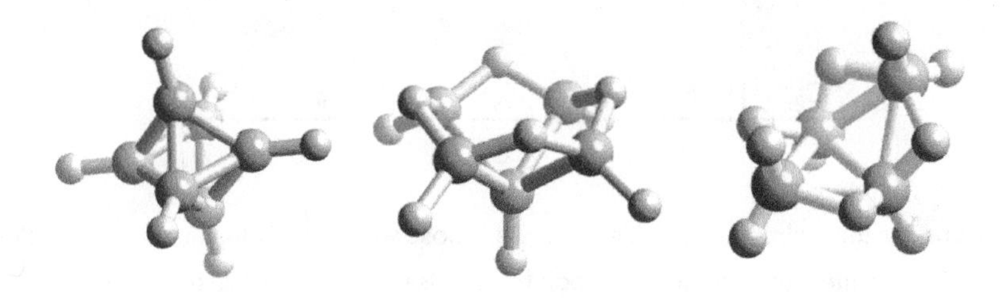

8.4. Which of the ions or molecules in Problems 8.1 and 8.2 are polar?

8.5. Which of the ions or molecules in Problems 8.1 and 8.2 are chiral?

8.6. Without the use of a character table, answer each of the following about the IF_5 molecule:

a. Make a complete list of all the possible symmetry operations in the molecule.
b. Determine the molecular point group.
c. Determine the inverse of each symmetry operation.
d. Determine which symmetry operations belong to the same class.
e. Determine the characters for each of the matrices representing the three *p* orbitals, p_x, p_y, and p_z.
f. Determine the characters for each of the matrices representing the three rotational axes, R_x, R_y, and R_z.
g. Determine the characters for each of the matrices representing the five *d* orbitals.
h. Determine all of the IRRs for the point group and assign the appropriate Mulliken symbol to each IRR.
i. Using the rules for IRRs, generate the complete character table for this point group.

8.7. Use miniature orthogonal vectors on each atom in the CO molecule to generate the 6 × 6 matrices corresponding to each symmetry operation representing the six degrees of freedom in this molecule. What are the characters for each of the symmetry operations in this RR? Use the great orthogonality theorem and the correlation table in Table 8.11 to determine which IRRs factor into this RR.

8.8. Use the character table in Appendix B and the great orthogonality theorem to reduce the following RR into its IRRs:

D_{6h}	E	$2C_6$	$2C_3$	C_2	$3C_2'$	$3C_2''$	i	$2S_3$	$2S_6$	σ_h	$3\sigma_d$	$3\sigma_v$
Γ	18	0	0	0	−2	0	0	0	0	6	0	2

8.9. Use the character tables in Appendix B and the great orthogonality theorem to determine each of the following direct products:

a. $B_{1g} \times B_{2u}$ in D_{4h}
b. $T_1 \times T_2$ in T_d
c. $B_g \times B_g$ in C_{2h}
d. $A_2 \times B_1 \times B_2$ in C_{4v}
e. $A_{1g} \times T_{1g} \times T_{2g}$ in O_h

8.10. Use the character table in Appendix B and the great orthogonality theorem to reduce the Γ_{dof} representation for boric acid, H_3BO_3, given below into its IRRs:

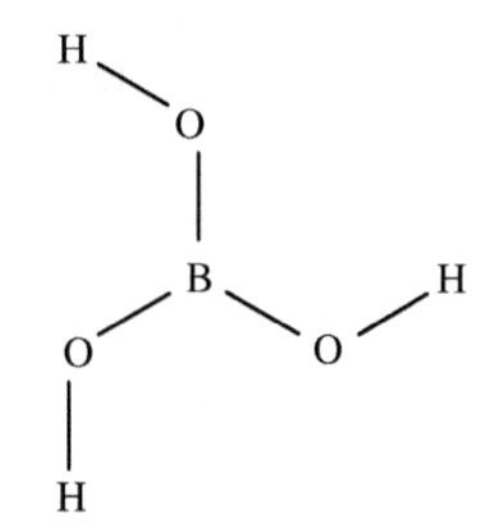

C_{3h}	E	C_3	C_3^2	σ_h	S_3	S_3^5
Γ_{dof}	21	0	0	7	−2	−2

8.11. Consider the hydrogen peroxide molecule, whose solid state structure is shown below.

a. Determine the molecular point group for H_2O_2.
b. Using a line segment to represent each of the three bonds, generate the RR using this basis set and use the great orthogonality theorem to reduce it into its IRRs.
c. Using miniature Cartesian axis vectors on each atom in the H_2O_2 molecule, generate the RR using this basis set, and use the great orthogonality theorem to reduce it into its IRRs.

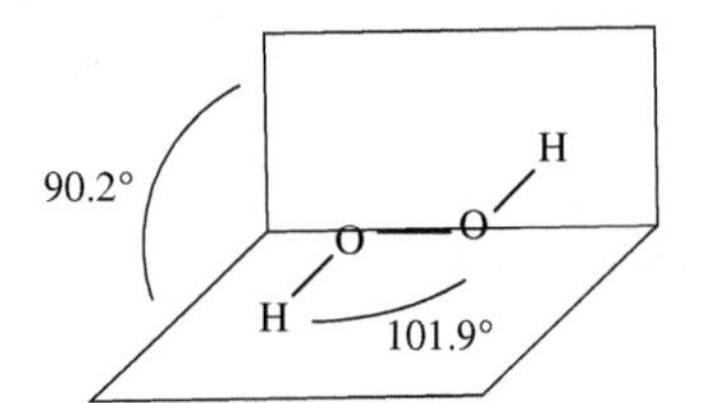

BIBLIOGRAPHY

1. Carter, R. L. *Molecular Symmetry and Group Theory*, John Wiley & Sons, Inc., New York, 1998.
2. Cotton, F. A. *Chemical Applications of Group Theory*, 2nd ed., John Wiley & Sons, Inc., New York, 1971.
3. Harris, D. C.; Bertolucci, M. D. *Symmetry and Spectroscopy*, Dover Publications, Inc., New York, 1978.
4. Housecroft, C. E.; Sharpe, A. G. *Inorganic Chemistry*, 3rd ed., Pearson Education Limited, Essex, England, 2008.

5. Huheey, J. E.; Keiter, E. A.; Keiter, R. L. *Inorganic Chemistry: Principles of Structure and Reactivity*, 4th ed., Harper Collins College Publishers, New York, 1993.
6. Kettle, S. F. A. *Symmetry and Structure*, 2nd ed., John Wiley & Sons, Inc., New York, 1995.
7. Miessler, G. L.; Tarr, D. A. *Inorganic Chemistry*, 4th ed., Pearson Education Inc., Upper Saddle River, NJ, 2011.
8. Tinkham, M. *Group Theory and Quantum Mechanics*, McGraw Hill, Inc., New York, 1964.

Vibrational Spectroscopy | 9

"Group theory is essential to the modern practice of spectroscopy."

—*Dan Harris*

9.1 OVERVIEW OF VIBRATIONAL SPECTROSCOPY

We have seen how the three-dimensional structure of a molecule determines its symmetry. The symmetry of a molecule, in turn, will determine its spectroscopy. *Spectroscopy* deals with the interaction of electromagnetic waves with the charged particles of matter. The process of the *absorption* of electromagnetic radiation occurs when a photon is *resonant* with (or has the same energy as) the difference between a ground state energy level and any of the quantized excited state energy levels, as was the case for the electronic transitions we observed using the Bohr model of the atom in Chapter 3. *Emission* is the opposite process, where a photon is emitted as a result of a transition from a high-lying energy level to some lower-lying energy level. While the emissions that gave rise to the line spectrum of hydrogen involved transitions between electronic energy levels, the spectra of molecules is complicated by the motions of the atoms in molecules. Each atom in a molecule can have three degrees of freedom (corresponding with motion along the three Cartesian axes), so that every molecule has a total of $3N$ degrees of freedom, where N is the number of atoms in the molecule. In general, there are three types of molecular motion: translation, rotation, and vibration.

The energy levels for each type of molecular motion are quantized. Translation occurs when the molecule is moving through space and it has components along the x, y, and z-axes, so that there are always three degrees of translational freedom in any given molecule. The translational energy levels are small and so closely spaced to each other that translational motion can be approximated as a continuum and will be treated as such. Rotational motion occurs when the molecule is rotating or spinning around one of the three Cartesian axes. Therefore, there are three degrees of rotational freedom for any nonlinear molecule (linear molecules have only two degrees of rotational freedom because rotation around the C_∞ axis does not change the state of the molecule). The spacing between rotational energy levels is typically on the order of 1 cm^{-1} and occurs in the microwave region of the electromagnetic

Vibrational modes of water. [Used with permission from Marivi Fernandez-Serra.]

Principles of Inorganic Chemistry, First Edition. Brian W. Pfennig.

spectrum. Lastly, vibrational motion involves changes in the position of the nuclei in molecules, which in turn affects the equilibrium bond lengths and bond angles. By subtraction, there are either $3N-5$ or $3N-6$ degrees of vibrational freedom depending on whether the molecule is linear or nonlinear. Vibrational absorptions occur in the infrared (IR) region of the electromagnetic spectrum. The different types of energy levels in molecules are superimposed on one another, as shown in Figure 9.1.

This section focuses exclusively on the vibrational spectra of molecules, although the principles developed here can also be applied to their rotational and electronic spectra. Vibrational spectroscopy confines itself to that narrow region of the electromagnetic spectrum where the absorption of a photon has sufficient energy to cause the nuclei of the excited molecule to vibrate. For all practical purposes, this implies a wavelength in the range 2.5–50 μm (or 200–4000 cm^{-1} in wavenumbers) with corresponding energies of 2.4–48 kJ/mol. Although it might seem natural to assume that the vibrations in molecules involve completely random gyrations of all the atoms in the molecule, the actual vibrational motions of the atoms are always a superposition (or linear combination) of certain fundamental vibrations known as the *normal modes* of vibration. As mentioned previously, there are a total of $3N-6$ (or $3N-5$ if the molecule is linear) normal modes of vibration, and it is the different types and symmetries of these normal modes with which we shall be concerned. Each normal mode of vibration occurs at one of the natural frequencies of the molecule in much the same way as a tuning fork vibrates at a natural frequency. A more rigorous definition is that a normal mode of vibration is one in which "each atom is displaced from its equilibrium position by an amount that corresponds to its maximum amplitude and that when they are allowed to relax they will all undergo a motion at the same frequency so that they will simultaneously pass through the equilibrium configuration."

A diatomic molecule will have only a single vibrational mode and it will be a stretching mode that lies along the internuclear axis, as shown in Figure 9.2. One can consider the two nuclei in the *AX* bond to be attached by an imaginary spring that allows them to simultaneously stretch to their maximum amplitude, pass back

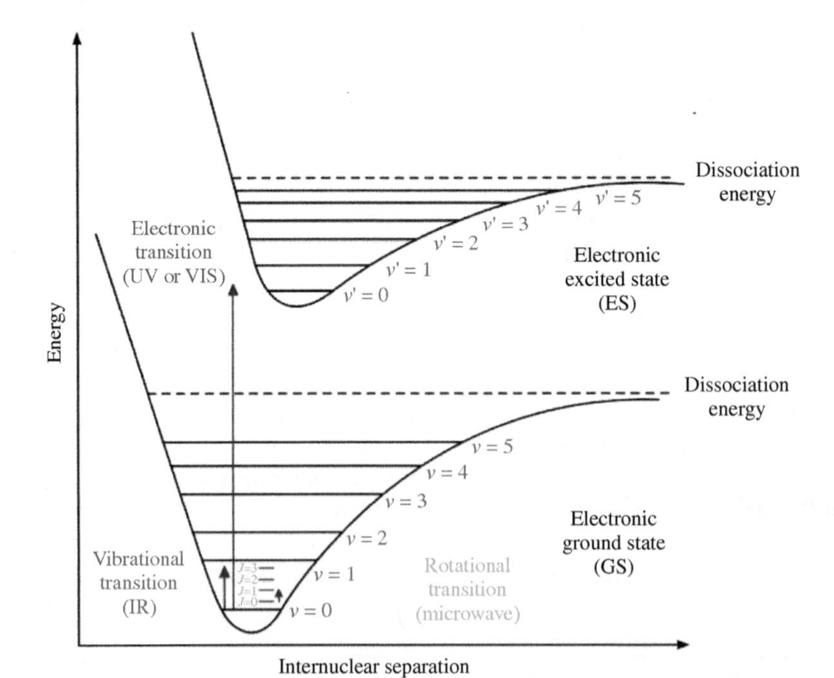

FIGURE 9.1
Illustration of the different types and relative spacings of the quantized energy levels in molecules. The vibrational levels are superimposed on the electronic energy levels and the rotational levels in turn are superimposed on the vibrational energy levels. Each vibrational energy level will have its own set of rotational energy levels, although they are only shown here for the lowest energy level where $v = 0$. The translational energy levels are not shown.

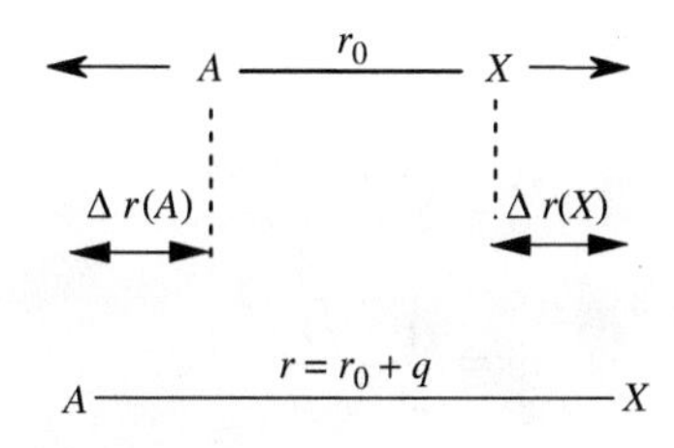

FIGURE 9.2
Definition of the normal coordinate q in the stretching vibration of a diatomic molecule AX.

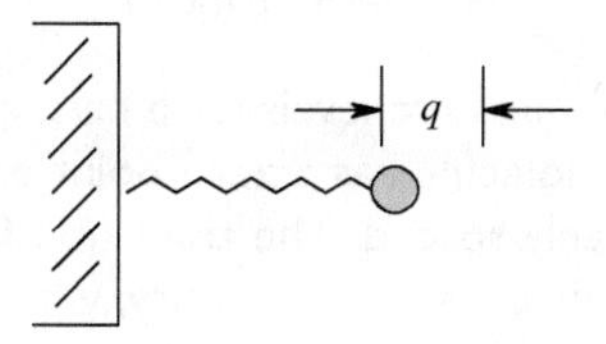

FIGURE 9.3
Harmonic oscillator (ball on a spring) model of molecular vibration.

through their equilibrium bond length, and then contract to some minimum distance in sync with one another. This stretching motion will occur along what we will define as the normal coordinate q. The *normal coordinate* is defined as some internal coordinate in the molecule along which the progress of the normal mode of vibration can be observed. In Figure 9.2, the normal coordinate q is defined along the molecular axis where its magnitude is given by Equation (9.1). Because the two atoms in the bond might have different atomic masses, the equality given by Equation (9.2) must be true in order to maintain the center of mass throughout the course of the molecular vibration:

$$q = r - r_0 = \Delta r(A) + \Delta r(X) \tag{9.1}$$

$$m(A)\Delta r(A) = m(X)\Delta r(X) \tag{9.2}$$

Using the harmonic oscillator model developed in Chapter 3, the vibrational modes of molecules can be modeled using Hooke's law, which describes the positional movement of a ball affixed to a spring as it is alternately stretched and compressed, as shown in Figure 9.3 and Equation (9.3). The potential energy of the vibrating spring can be calculated using Equation (9.4), assuming that the spring acts as a harmonic oscillator, where q is the displacement of the spring and k is a measure of the stiffness of the spring, known as the *force constant*:

$$F = ma = m\frac{d^2x}{dt^2} = -kq \tag{9.3}$$

$$V(x) = -\int F(x)\,dx = \int kq\,dx = \frac{1}{2}kq^2 \tag{9.4}$$

The mathematical solutions to the time-independent Schrödinger equation for the harmonic oscillator model were given in Chapter 3 and were given by Equations (3.74)–(3.77); they are also shown in Figure 3.32. The vibrational energy levels are quantized according to Equation (9.5) using the vibrational quantum number v, where ω is defined by Equation (9.6) and μ (or the reduced mass for the nuclei) is given by Equation (9.7). The vibrational quantum number can take only nonnegative, integral values: $v = 0, 1, 2, \ldots$

$$E_v = \left(v + \frac{1}{2}\right) h\omega \tag{9.5}$$

$$\omega = \frac{1}{2\pi}\sqrt{\frac{k}{\mu}} \tag{9.6}$$

$$\mu = \frac{m(A)\, m(X)}{m(A) + m(X)} \tag{9.7}$$

The resulting vibrational quantum levels for a harmonic oscillator are shown in Figure 9.4. Notice that the molecule has a zero-point energy even at 0 K and that the vibrational levels are evenly spaced. The transition from the $v = 0$ level to the $v = 1$ level is known as the *fundamental* frequency, whereas transitions from $v = 0$ to higher levels are known as the *overtones*.

In reality, most molecules, even diatomic ones, do not behave harmonically. At very large values of q, the restoring force decreases and the molecule is able to dissociate. Therefore, an asymmetrical potential energy curve given by a Morse potential is often used to calculate the quantum mechanical energy levels for molecular vibrations, which are shown in Figure 9.5. The Morse potential is largely a modification of the harmonic oscillator model that adds an empirical correction for the curvature at the bottom of the potential energy well, as given by Equation (9.8), where D_e is the well depth defined in Figure 9.5 and β is simply a measure of the curvature. The resulting energy levels are given by Equation (9.9), where ω_e has nearly the same value as ω and the second term is only a small correction to the first. The end result is that the vibrational energy levels in real molecules are no longer evenly spaced and they become closer together at larger values of v. Although the harmonic oscillator model is imperfect, it still serves as a reasonable first approximation, and we will continue to use it as our model for the remainder of this text:

$$V = D_e(1 - e^{-\beta q})^2 \tag{9.8}$$

$$E = \left(v + \frac{1}{2}\right) h\omega_0 - \frac{\left[\left(v + \frac{1}{2}\right) h\omega_0\right]^2}{4D_e} \tag{9.9}$$

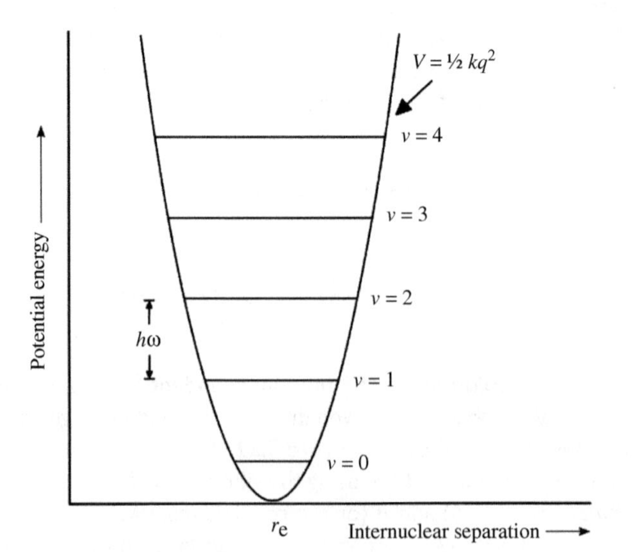

FIGURE 9.4
Harmonic oscillator model of molecular vibrations, showing the spacing of the different quantum mechanical energy levels.

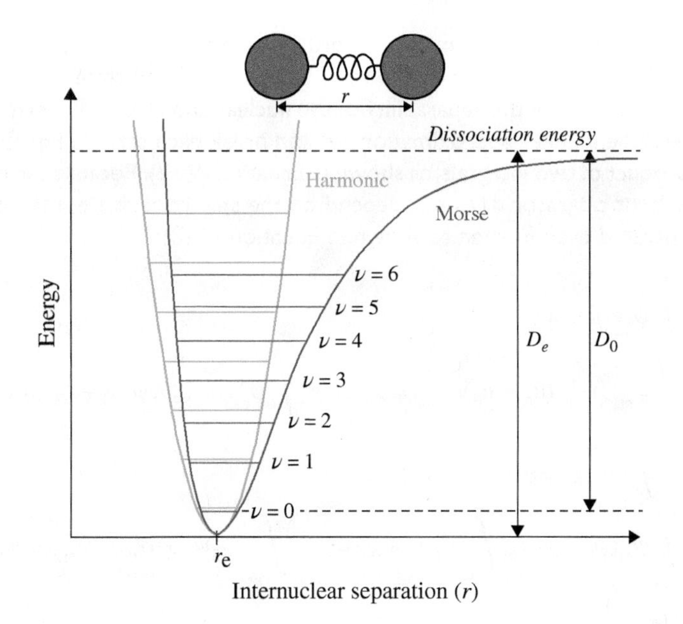

FIGURE 9.5 A comparison of the Morse potential (blue) and the harmonic oscillator potential (green), showing the effects of anharmonicity of the potential energy curve, where D_e is the depth of the well. [© Mark M Samoza/CCC-BY-SA_3.0/GFDL /Wikimedia Commons; reproduced from http://en.wikipedia.org/wiki /Morse_potential (accessed December 27, 2013).]

9.2 SELECTION RULES FOR IR AND RAMAN-ACTIVE VIBRATIONAL MODES

There are three commonly used experimental techniques for the determination of the vibrational frequencies of molecules: IR, Raman, and resonance Raman spectroscopies. The three techniques differ by the manner in which the quantum mechanical vibrational spacings in molecules are measured. Thus, each technique has its own set of *selection rules*, or rules for which transitions will be spectroscopically observed. The selection rules for electronic and vibrational transitions are derived from the time-dependent form of the Schrödinger equation by a consideration of the way in which the oscillating electric field of a light wave acts on a dipole moment component of the molecule. Just as the motion of a traveling wave across a body of water will cause a floating cork to oscillate vertically in harmonic motion as the wave crests and troughs pass by the cork, so too the electrons in molecules will oscillate as the electromagnetic field passes through the molecule. For the duration of this chapter, we focus only on the electric component of the electromagnetic wave.

The probability of a transition between two different stationary states (such as the quantized energy levels in Figure 9.4 or 9.5) will increase as the energy of the light wave approaches (or is in resonance with) that of the energy difference between two of the molecule's fixed energy levels. The absorption of light is therefore a resonance process that depends on the nuclear and electronic coordinates of the molecule. Because the electric field component of the electromagnetic wave is oscillating in time, we will need to use the time-dependent form of the Schrödinger equation developed in Chapter 3. For even the smallest of molecules, the resulting Hamiltonian becomes hopelessly complicated for most of us mere mortals, and so we will apply the Born–Oppenheimer approximation to help us simplify the math.

The *Born–Oppenheimer approximation* is the assumption that because the nuclei are several thousand times more massive than the electrons, the nuclei are essentially clamped in place during the absorption so that the wave function can be separated into its respective electronic and vibrational parts. The probability of a transition occurring between two different energy levels is proportional to the square of the transition moment integral, M_{01}, shown in Equation (9.10), where $\hat{u}$ is the dipole moment operator. Expansion of this equation yields Equation (9.11), where e is the electronic coordinate, n is the nuclear coordinate, and s is the spin coordinate. The subscripts ES and GS stand for the excited state and ground state wave functions, respectively. Because of the separability of the nuclear and electronic terms given by the Born–Oppenheimer approximation, we can break each term in Equation (9.11) into the product of two integrals, as shown in Equation (9.12). Because the electronic dipole moment operator does not depend on the spin coordinate, the latter term can be separated even further, as shown in Equation (9.13):

$$M_{01} = \int \psi * \hat{\mu} \psi \, d\tau \tag{9.10}$$

$$M_{01} = \int \psi^*_{esES} \psi^*_{vES} (\hat{u}_n + \hat{u}_e) \psi_{esGS} \psi_{vGS} d\tau = \int \psi^*_{esES} \psi^*_{vES} \hat{u}_n \psi_{esGS} \psi_{vGS} d\tau + \int \psi^*_{esES} \psi^*_{vES} \hat{u} e \psi_{esGS} \psi_{vGS} d\tau \tag{9.11}$$

$$M_{01} = \int \psi^*_{esES} \psi_{esGS} d\tau_{es} \int \psi^*_{vES} \hat{u}_n \psi_{vGS} d\tau_n + \int \psi^*_{vES} \psi_{vGS} d\tau_n \int \psi^*_{esES} \hat{u}_e \psi_{esGS} d\tau_{es} \tag{9.12}$$

$$M_{01} = \int \psi^*_{esES} \psi_{esGS} d\tau_{es} \int \psi^*_{vES} \hat{u}_n \psi_{vGS} d\tau_n + \int \psi^*_{vES} \psi_{vGS} d\tau_n \int \psi^*_{eES} \hat{u}_e \psi_{eGS} d\tau_e \int \psi^*_{sES} \psi_{sGS} d\tau_s \tag{9.13}$$

The first integral defines the overlap of the electronic wave functions for the ground and excited electronic states. The second term, which will become the focus of our discussion of vibrational spectroscopy, requires a change in the dipole moment of the molecule in order for a transition to occur from the vibrational (v) GS to the vibrational ES. The third term has to do with the amount of overlap between the different vibrational wave functions involved in the ground and excited vibrational states of the transition. This term is often called the *Franck–Condon factor*. The fourth integral governs the electronic orbital selection rules. It has to do with the symmetries of the electronic wave functions in the ground and excited electronic states. The fifth and final integral is the spin selection rule, which essentially states that the spin multiplicities of the ground and excited states must be the same in order for the transition to be observed.

Of the three experimental techniques for measuring the vibrational frequencies of molecules, IR spectroscopy is the most commonly used. As shown in Figure 9.6(a), a continuous source, such as a Globar lamp, generates IR radiation. In a typical scanning instrument, the wavelength is selected using a diffraction grating, or monochromator. The radiation is then split into two beams. One beam serves as the reference beam so that the intensity of the incident light (I_0) can be measured. The second beam passes through the sample (which is usually mounted between two salt plates or pressed into a transparent pellet). The intensity of the transmitted beam is then measured by the detector as I_t and compared with I_0. The percent transmittance is the ratio $(I_t/I_0) \times 100\%$.

IR spectroscopy is an absorption process between two vibrational levels, as shown in Figure 9.6(b). Therefore, it is the second integral in Equation (9.13), $\int \psi *_{vES} \hat{u}_n \psi_{vGS} d\tau_n$, that is of relevance. If this integral is nonzero, then the IR absorption will be allowed by the rules of quantum mechanics. The only circumstances under which this integral will be allowed are when the absorption of IR radiation causes a change in the molecule's permanent dipole moment. Because each of the wave functions and the dipole moment operator themselves have symmetry, the triple integral will be nonzero if and only if the triple direct product

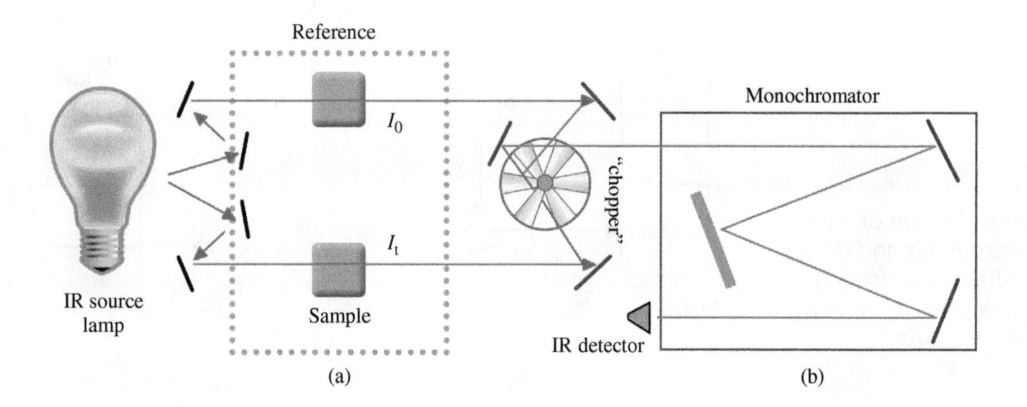

FIGURE 9.6
(a) Schematic diagram of a basic IR spectrometer, where the percent transmittance is equal to $(I_t/I_0) \times 100\%$, (b) IR vibrational transition between the $v = 0$ and $v = 1$ quantum levels. [Blatt Communications.]

contains the totally symmetric IRR (abbreviated here as the generic "A_1") for the point group of the molecule, as shown by Equation (9.14):

$$\Gamma_{vES}\Gamma_{\hat{u}}\Gamma_{vGS} \subset \text{“}A_1\text{”} \tag{9.14}$$

Typically, only the ground vibrational level ($v = 0$) is substantially populated at room temperature. Furthermore, we will choose our normal coordinate q such that every symmetry operation in the point group converts q into $\pm q$, so that q^2 will always be converted into itself. Thus, the symmetry of Γ_{vGS} will always be equal to the totally symmetric IRR. The symmetry of the dipole moment operator $\Gamma_{\hat{u}}$, on the other hand, will be the same as the symmetries of the x-, y-, and z-unit vectors, as the dipole moment operator has components along the x-, y-, and z-axes and transforms as a vector in three-dimensional space according to Equation (9.15):

$$\hat{u} = \begin{pmatrix} \hat{u}_x \\ \hat{u}_y \\ \hat{u}_z \end{pmatrix} \tag{9.15}$$

Finally, the symmetry of the excited vibrational level is given by Γ_{vES}. Because Γ_{vGS} is always the totally symmetric IRR, the product $\Gamma_{\hat{u}}\ \Gamma_{vGS}$ will have the same symmetry as $\Gamma_{\hat{u}}$. As we learned earlier, the only way for a direct product to contain the totally symmetric IRR is for both of the species to have identical symmetries. *Thus, in order for a vibrational mode to be IR-active, the symmetry of the vibrational mode, Γ_{vES}, must be identical to the symmetry of one of the components (x, y, or z) of the dipole moment operator $\Gamma_{\hat{u}}$.*

Raman spectroscopy, on the other hand, is a scattering process. The source is typically a LASER beam emitting coherent, monochromatic light in the visible region of the electromagnetic spectrum having some initial intensity, I_0. The LASER beam is directed onto the sample, as shown in Figure 9.7(a). Some of the light is scattered by the sample in all directions. Most of the scattered light has the same wavelength as the incident LASER beam (this is referred to as *Rayleigh scatter*). However, a small portion of the scattered light will differ in wavelength by one or more vibrational quantums, as shown in Figure 9.7(b). This type of scatter is known as *Raman scatter*. On average, only one in 10^6 photons of light will exhibit Raman scatter. Therefore, an intense LASER beam is typically used as the source. The scattered light is almost always measured at a 90° angle to the incident light in order to avoid saturating the detector

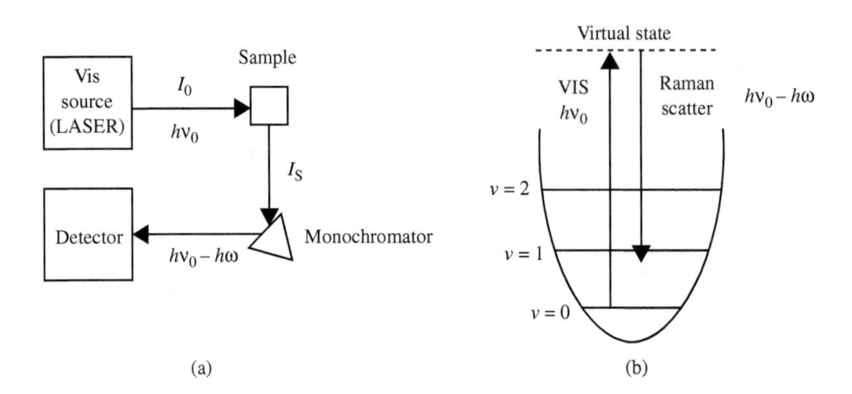

FIGURE 9.7
(a) Schematic diagram of a basic Raman spectrometer and (b) example of Raman scattering that differs by one vibrational quantum state. [Blatt Communications.]

with reflected light off the sample surface. The scattered light is passed through a monochromator to filter out the Rayleigh scatter and to determine the wavelength of the Raman-scattered light. This wavelength will fall in the visible region, but it will differ from the original LASER beam by a frequency corresponding to one of the normal modes of vibration for the molecule. The intensity of the scattered light I_s is recorded, typically using a CCD detector. From a phenomenological standpoint, the visible light is inelastically scattered from a virtual state. The Raman-scattered light will differ in frequency from the incident light, as shown in Figure 9.7(b).

The selection rules for Raman spectroscopy require a change in the induced dipole moment of the molecule, μ_{ind}, as the nuclei are attracted to the negative pole of the electromagnetic radiation and the electrons are attracted to its positive pole. The induced dipole moment is defined by Equation (9.16) as the product of the polarizability α with the magnitude of the electric field E component of the light wave. The polarizability changes as the molecule vibrates from its equilibrium position, and thus it is a tensor quantity, given by the components in Equation (9.17) where we have taken advantage of the fact that the tensor is symmetric so that $\alpha_{xy} = \alpha_{yx}$, for instance. In order for a vibrational mode to be Raman-active, the triple direct product shown in Equation (9.18) must contain the totally symmetric IRR for the molecular point group. Again, the symmetry of the ground vibrational level will be totally symmetric because of the way that we define our normal coordinate. Thus, the only way for the triple direct product to contain the totally symmetric IRR is if the excited vibrational level has the same symmetry as one of the components of the polarizability operator. Because the polarizability operator transforms the same way as the squares and binary products of x, y, and z do, *a vibrational mode will be Raman-active if it corresponds with one of the functions in the fourth part of the molecule's character table:*

$$\hat{u}_{\text{ind}} = \hat{\alpha} E \tag{9.16}$$

$$\hat{\alpha} = \begin{bmatrix} \alpha_{xx} & \alpha_{xy} & \alpha_{xz} \\ \alpha_{yx} & \alpha_{yy} & \alpha_{yz} \\ \alpha_{zx} & \alpha_{zy} & \alpha_{zz} \end{bmatrix} \tag{9.17}$$

$$\Gamma_{\text{vES}} \Gamma_{\hat{\alpha}} \Gamma_{\text{vGS}} \subset \text{“}A_1\text{”} \tag{9.18}$$

Some useful rules regarding the symmetries of the vibrational modes follow:

1. There are $3N-5$ vibrational modes for a linear molecule and $3N-6$ vibrational modes for a nonlinear molecule.
2. A vibrational mode will be IR-active if it spans the same IRR as the x-, y-, or z-unit vectors.

3. A vibrational mode will be Raman-active if it spans the same IRR as one of the squares or binary products of x, y, and z.
4. For a centrosymmetric molecule (one that contains an inversion center), the IR and Raman-active modes will be mutually exclusive (the IR-active modes will be *ungerade* and the Raman-active modes will be *gerade*).
5. It is possible that some of the vibrational modes are neither IR nor Raman-active and we shall call these *silent modes*.
6. The totally symmetric IRR for a point group will always be Raman-active because it always transforms as one of the squares or binary products of x, y, and z.
7. The totally symmetric vibrational modes can be identified in Raman spectroscopy by their depolarization ratio. When the incident light striking the same is plane-polarized using a polarizing lens, the depolarization ratio, given in Equation (9.19), is the ratio of the intensity of the plane-polarized light that is perpendicular to the original beam divided by the intensity of the plane-polarized light that is parallel to the original beam. In practice, the depolarization ratio is ~0 for polarized bands (the totally symmetric IRR) and ~0.75 for depolarized bands:

$$\rho = \frac{I_\perp}{I_\parallel} \tag{9.19}$$

9.3 DETERMINING THE SYMMETRIES OF THE NORMAL MODES OF VIBRATION

Our focus in this section is to determine the symmetries of the $3N-6$ or $3N-5$ normal modes of vibration for a molecule. We have already shown how group theory can be useful in the determination of the symmetry properties of such basis sets as unit vectors, rotational vectors, and atomic orbitals. As it turns out, each normal mode of vibration will form a basis for one of the IRRs of the point group of the molecule. In order to determine the symmetries of the vibrational modes, we will first generate a reducible representation using all $3N$ degrees of freedom by placing orthogonal vectors on every atom in the molecule. This will generate a set of $3N \times 3N$ matrices for each of the symmetry operations in the molecular point group. However, we can simplify the procedure by recognizing that only those elements that lie along the diagonal of the matrix from upper left to lower right will contribute to the character for the Γ_{dof} representation and only those atoms that are unmoved by the symmetry operation will have nonzero values along the diagonal. Any atom that is moved into a different position as a result of a symmetry operation will lie off the diagonal and will contribute zero to the character.

Each atom that is unmoved by the symmetry operation will contribute those values listed in Table 9.1 to the character in the Γ_{dof} representation. Each unmoved atom will contribute 3 to the identity operation because each of the three mini-Cartesian axes on the atoms will be unaffected by the symmetry operation. For a proper rotation, each unmoved atom will contribute $1 + 2\cos(2\pi/n)$ to the character for proper rotation because the 3×3 rotational matrix defined in Figure 8.15 is given by Equation (9.20). Any atom that lies in the mirror plane

$$\begin{bmatrix} \cos(2\pi/n) & -\sin(2\pi/n) & 0 \\ \sin(2\pi/n) & \cos(2\pi/n) & 0 \\ 0 & 0 & 1 \end{bmatrix} \begin{bmatrix} x_1 \\ y_1 \\ z_1 \end{bmatrix} = \begin{bmatrix} x_2 \\ y_2 \\ z_2 \end{bmatrix} \tag{9.20}$$

TABLE 9.1 The characters for the Γ_{dof} representation of any molecule can be determined by multiplying the number of atoms that are invariant under (not moved by) a given symmetry operation by the values listed in column 2.

Type of Symmetry Operation	Value
Identity, E	3
Proper rotation, C_n	$1+2\cos(2\pi/n)$
Reflection, σ	1
Inversion, i	−3
Improper rotation, S_n	$-1+2\cos(2\pi/n)$

will contribute a character of 1 to the reflection operation because two of the mini-Cartesian axes on the atom will lie in the plane and the other will be reflected into its opposite. For the inversion operation, if there is an atom at the inversion center of the molecule, it will contribute −3 to the character for inversion because each mini-Cartesian axis vector will be transformed into its exact opposite. Finally, for an improper rotation operation, any unmoved atom will contribute a character of $-1+2\cos(2\pi/n)$. The x- and y-components of the 3×3 improper rotational matrix will be the same as those in Equation (9.10), but the mini-Cartesian vector lying along the z-axis will be converted into the negative of itself by the reflection in the perpendicular plane. Thus, in order to determine the characters for the Γ_{dof} representation, all one needs to do is to determine how many atoms in the molecule are unmoved by each symmetry operation and then multiply this number times the value listed in Table 9.1 for the corresponding operation.

Once the characters for the Γ_{dof} representation have been determined, we will reduce this representation using the great orthogonality theorem in order to determine the symmetries of the $3N$ degrees of freedom. Subtraction of the three IRRs corresponding to x, y, and z (the translational degrees of freedom along the x-, y-, and z-axes, respectively) and the two or three IRRs corresponding to R_x, R_y, and R_z (if nonlinear, the rotational degrees of freedom) will leave the symmetries of the vibrational modes.

Consider the vibrational modes of the H_2O molecule. Water has $3N = 9$ degrees of freedom (as shown in Figure 8.14) and is a nonlinear molecule. Thus, H_2O should exhibit $3N–6=3$ vibrational degrees of freedom. The Γ_{dof} representation for H_2O was shown in Table 8.9 and is shown again in Table 9.2. Using the shorthand method to determine the characters for the different symmetry operations in the Γ_{dof} representation, we get a character of $3(3) = 9$ for E. Only

TABLE 9.2 The symmetries of the 3N degrees of freedom in the water molecule.

C_{2v}	E	C_2 (z)	σ_{v1} (xz)	σ_{v2} (yz)	$h=4$	$n_i=$
Γ_{dof}	9	−1	1	3	Σ	Σ/h
A_1	9	−1	1	3	12	3
A_2	9	−1	−1	−3	4	1
B_1	9	1	1	−3	8	2
B_2	9	1	−1	3	12	3

the O atom is unmoved with respect to a C_2 rotation, and it will therefore have a character of $1 + 2\cos(180°) = -1$. For the $\sigma_{v1}(xz)$ reflection, which is perpendicular to the plane of the molecule, only the O atom is unmoved and the character for this operation is therefore 1. Finally, for the $\sigma_{v2}(yz)$ reflection, all three atoms lie in the plane and are unmoved by the symmetry operation, so its character will be 3.

Using the tabular method of applying the great orthogonality theorem, the Γ_{dof} reduces to $3\ A_1 + A_2 + 2\ B_1 + 3\ B_2$. Subtraction of the three translational degrees of freedom (B_1, B_2, and A_1 which transform as x, y, and z, respectively) yields $2\ A_1 + A_2 + B_1 + 2\ B_2$. Subtraction of the three rotational degrees of freedom (B_2, B_1, and A_2, which transform as R_x, R_y, and R_z, respectively) yields $2\ A_1 + B_2$ for the symmetries of the three vibrational modes. As all three vibrational modes belong to the same IRR as x, y, or z, all three of these normal modes will be IR-active. The two A_1 vibrations are z-polarized, which means that if plane-polarized light is used for the excitation, only z-polarized light can effect this vibrational transition. The B_2 vibrational mode is y-polarized. Because all three vibrational modes correspond to IRRs representing the squares or binary products of x, y, and z, all three of the normal modes will also be Raman-active.

The symmetries of the three vibrational modes for the H_2O molecule are shown in Figure 9.8. The numbers of the vibrational modes are typically assigned from the highest to the lowest symmetry. The first vibrational mode is a symmetric stretch. Stretching modes are characterized by changes in the bond lengths but no changes in the bond angles. The symbol for a stretching vibration is ν, followed by the subscript s for symmetric, a for anti-symmetric, or the numerical assignment for the vibrational mode. The symmetry of the mode is then listed in parentheses. Notice that if we were to use the stretching motion of ν_1 as a basis set, the vibration would be symmetric to all four symmetry operations in the point group. It would therefore have characters of all ones and would transform as the A_1 IRR. The second vibrational mode for H_2O is a bending mode. Bending modes involve changes in the bond angles of the molecule and are given the symbol δ. The symmetry of this particular bending vibration is also A_1 because the motion is symmetric with respect to every symmetry operation in the C_{2v} point group. The third vibrational mode is another stretching mode. However, this mode is anti-symmetric with respect to the C_2 and σ_{v1} symmetry operations. Thus, its characters will be +1, −1, −1, +1 for the different symmetry operations in the group, which means that it will transform as the B_2 IRR.

Some commonly used symbols for other types of molecular vibrations are listed in Table 9.3. The subscript s implies that the vibrational mode is totally symmetric, *a* that it is anti-symmetric, and *d* that it is degenerate. a+ sign indicates that the atom is moving out of the plane of the paper and a− sign indicates that it is moving behind the plane of the paper. In addition to stretching and bending motions, there are also rocking, wagging, and twisting motions involving all three atoms defining a bond angle

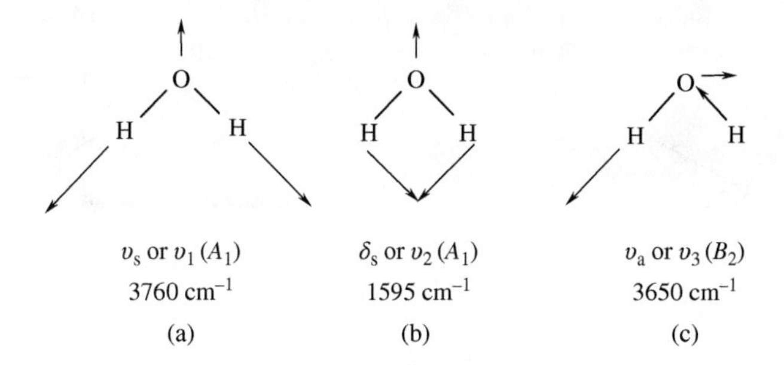

FIGURE 9.8
(a–c) The symmetries of the three normal modes of vibration for the water molecule.

TABLE 9.3 Symbols and names for selected types of molecular vibrations.

Type of Vibration	Symbol/Name	Type of Vibration	Symbol/Name
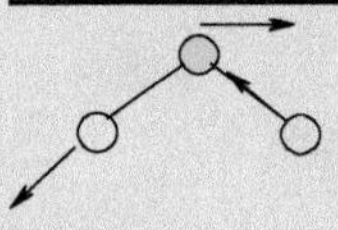	ν_a Asymmetric stretch	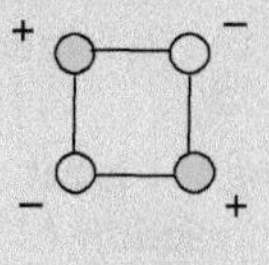 + − − +	π Out-of-plane bending
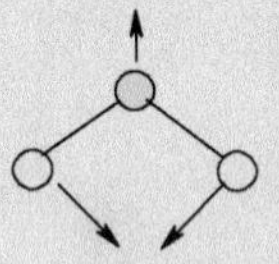	δ_s Symmetric bend	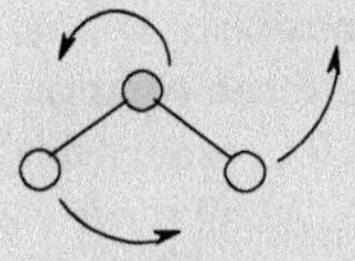	ρ_r Rocking
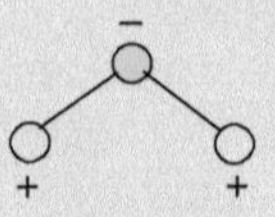 − + +	ρ_w Wagging	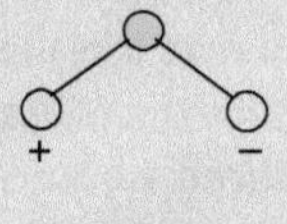 + −	ρ_t Twisting

The + and − signs indicate motion above and below the plane of the paper, respectively.

and these are given the symbol ρ, followed by the appropriate subscript (*r*, *w*, or *t*, respectively). When more than three atoms are involved in an out-of-plane bending motion, the symbol π is used.

The magnitudes of the arrows depend on the relative masses of the atoms involved in the vibration so that the center of mass is preserved during the vibration, as required by Equation (9.2). This is the reason why the arrows in Figure 9.8 for the O atom are shorter than those for the two H atoms. The H atoms are lighter and will have greater relative displacements. It is not uncommon in vibrational spectroscopy to isotopically label one or more of the atoms in a molecule. The different masses of the isotopes allow one to determine whether or not the radiolabeled atom is involved in the vibration, and they are therefore very useful in the assignment of the normal modes of vibration.

Example 9-1. Given the wavenumbers and vibrational modes shown below for the BCl_3 molecule in the two common isotopes of B, determine the symmetries of each vibrational mode; determine whether the mode is IR-active, Raman-active, or both; and rationalize the observed isotope effect. For the degenerate vibrational modes, only a single stretching frequency is observed.

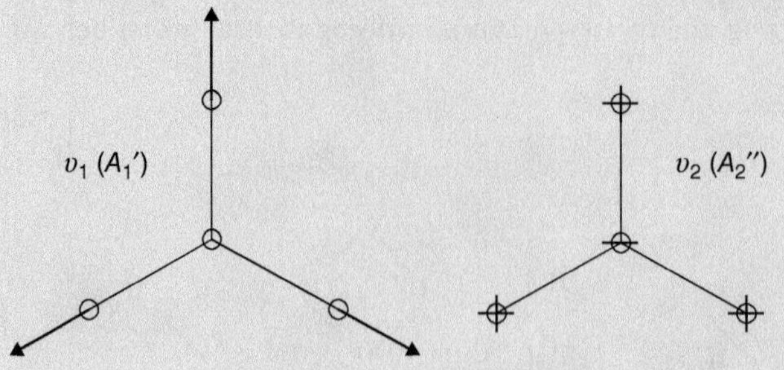

υ_1 (A_1') υ_2 (A_2'')

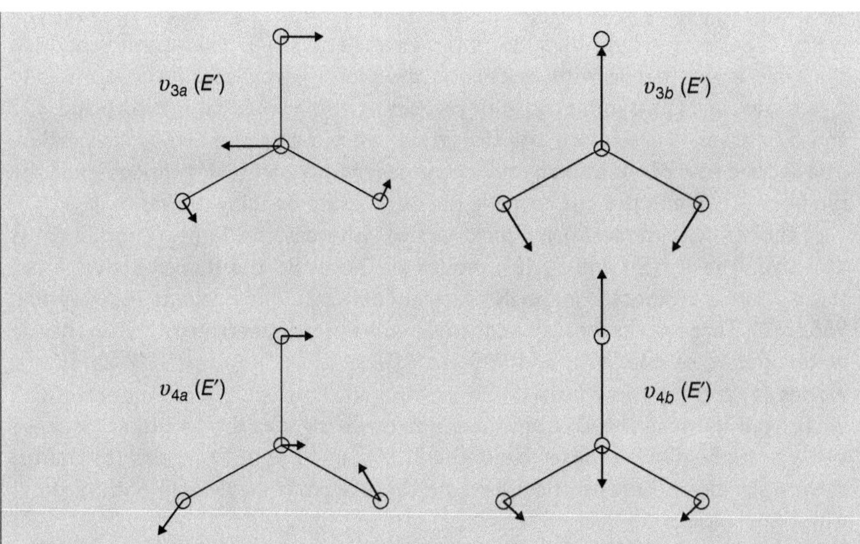

[Reproduced from Cotton, F. A. *Chemical Applications of Group Theory*, 2nd ed., Wiley-Interscience: New York, 1971. This material is reproduced with permission of John Wiley & Sons, Inc.]

Isotope	Mode a	Mode b	Mode c	Mode d
$^{10}BCl_3$	471 cm^{-1}	480 cm^{-1}	244 cm^{-1}	995 cm^{-1}
$^{11}BCl_3$	471 cm^{-1}	460 cm^{-1}	243 cm^{-1}	956 cm^{-1}

Solution. The molecule belongs to the D_{3h} molecular point group. Placing miniature orthogonal vectors on each atom in BCl_3 as a basis for generating the Γ_{dof} representation and using the rules in Table 9.1, we obtain the reducible representation shown in the following table. Using the tabular method to reduce this representation into its irreducible components, the symmetries for the 12 degrees of freedom in BCl_3 are A_1', A_2', 3 E', 2 A_2'', and E''. Subtraction of the three translational and three rotational degrees of freedom yields A_1', 2 E', and A_2'' for the six vibrational degrees of freedom. Two of the vibrational modes are doubly degenerate.

D_{3h}	E	$2C_3$	$3C_2$	σ_h	$2S_3$	$3\sigma_v$	$h = 12$	$n_i =$
Γ_{dof}	12	0	−2	4	−2	2	Σ	Σ/h
A_1'	12	0	−6	4	−4	6	12	1
A_2'	12	0	6	4	−4	−6	12	1
E'	24	0	0	8	4	0	36	3
A_1''	12	0	−6	−4	4	−6	0	0
A_2''	12	0	6	−4	4	6	24	2
E''	24	0	0	−8	−4	0	12	1

A quick inspection of the vibrational motion of the first mode in the table shows that it is totally symmetric with respect to every symmetry operation in the D_{3h} point group. This mode is therefore ν_s (A_1'). The second mode in the table is symmetric with respect to the identity, vertical mirror plane, and C_3 classes, but anti-symmetric with respect to the horizontal mirror plane, C_2, and S_3 classes. It therefore transforms as the out-of-plane Π (A_2'') vibration. The second row of vibrational modes are degenerate stretching modes given the symbol ν_d (E'') and the last row are the degenerate bending modes δ_d (E').

The ν_s (A_1') mode (a) is Raman-active, whereas the Π (A_2'') mode (b) is IR-active. The ν_d (E') and δ_d (E') modes are both IR- and Raman-active. Thus, three peaks are expected in the IR spectrum of $^{11}BCl_3$ and occur at 243, 460, and 956 cm^{-1}. Three peaks are also expected in the Raman spectrum of $^{11}BCl_3$, having wavenumbers of 243, 471, and 956 cm^{-1}. The ν_s (A_1') and ν_d (E') vibrational modes (a) and (c) show little or no isotope effect when the atomic mass of B is changed because the B atom does not move significantly in either of these two vibrations. On the other hand, the Π (A_2'') and δ_d (E'') modes (b) and (d) show a significant isotope effect because the B atom is involved in both of these vibrational modes.

Example 9-2. Determine the symmetries of the vibrational modes for the CO_2 molecule and determine which modes are IR-active and which modes are Raman-active.

Solution. The degrees of freedom basis set for the CO_2 molecule was shown in Example 8.21. Using the D_{2h} subgroup of the infinite point group $D_{\infty h}$ to generate the reducible representation Γ_{dof} in D_{2h} and then using the correlation table to determine the symmetries in $D_{\infty h}$, the degrees of freedom transform as $\Sigma_g^+ + \Pi_g + 2\Sigma_u^+ + 2\Pi_u$, as shown in Example 8.21. Subtraction of the three translational degrees of freedom yields $\Sigma_g^+ + \Pi_g + \Sigma_u^+ + \Pi_u$. Subtraction of the two rotational degrees of freedom leaves $\Sigma_g^+ + \Sigma_u^+ + \Pi_u$ as the symmetries of the four vibrational modes in CO_2. These modes and their corresponding symbols are shown later. There should be $3N-5=4$ vibrational modes. Although it appears as though we have only obtained three vibrational modes, the Π_u mode is doubly degenerate, with the two modes having the same vibrational frequency. Because the Σ_u^+ and Π_u modes transform the same way as x, y, or z, these three vibrational modes will be IR-active. The Σ_g^+ mode is the only one corresponding to a square or binary product of x, y, and z, and it is thus the only Raman-active mode for CO_2. As a rule, whenever the molecule is centrosymmetric, its vibrational modes will be mutually exclusive. That is, the IR and Raman spectra will not share any vibrational modes in common.

Group theory can be used to determine far more than just the symmetries of the vibrational modes and how many vibrations will be IR- or Raman-active. We have shown how the structure of a molecule determines its symmetry, which in turn determines its vibrational spectrum. The paradigm can also be used in reverse. If the structure is unknown, the number of peaks in the IR (or Raman) spectrum can often be used to determine the molecule's structure where the geometry is not known in advance. For instance, molecules of the type AX_4 can be either tetrahedral (T_d) or square planar (D_{4h}). Using the mini-Cartesian axis vectors on each atom in the T_d point group gives the reducible representation Γ_{dof} in Table 9.4, which reduces to $A_1 + E + 2T_1 + 3T_2$. Subtraction of the translations (T_2) and rotations (T_1) leaves the following vibrational modes: $A_1 + E + T_1 + 2T_2$. There are potentially two IR-active peaks (the $2T_2$ modes) and four Raman-active peaks ($A_1 + E + 2T_2$). I say

TABLE 9.4 The symmetries of the 3*N* degrees of freedom for a tetrahedral AX_4 molecule.

T_d	E	$8C_3$	$3C_2$	$6S_4$	$6\sigma_d$	$h=24$	$n_i =$
Γ_{dof}	15	0	−1	−1	3	Σ	Σ/h
A_1	15	0	−3	−6	18	24	1
A_2	15	0	−3	6	−18	0	0
E	30	0	−6	0	0	24	1
T_1	45	0	3	−6	−18	24	2
T_2	45	0	3	6	18	72	3

TABLE 9.5 The symmetries of the 3*N* degrees of freedom for a square planar AX_4 molecule.

D_{4h}	E	$2C_4$	C_2	$2C_2'$	$2C_2''$	i	$2S_4$	σ_h	$2\sigma_v$	$2\sigma_d$	$h=16$	$n_i =$
Γ_{dof}	15	1	−1	−3	−1	−3	−1	5	3	1	Σ	Σ/h
A_{1g}	15	2	−1	−6	−2	−3	−2	5	6	2	16	1
A_{2g}	15	2	−1	6	2	−3	−2	5	−6	−2	16	1
B_{1g}	15	−2	−1	−6	2	−3	2	5	6	−2	16	1
B_{2g}	15	−2	−1	6	−2	−3	2	5	−6	2	16	1
E_g	30	0	2	0	0	−6	0	10	0	0	16	1
A_{1u}	15	2	−1	−6	−2	3	2	−5	−6	−2	0	0
A_{2u}	15	2	−1	6	2	3	2	−5	6	2	32	2
B_{1u}	15	−2	−1	−6	2	3	−2	−5	−6	2	0	0
B_{2u}	15	−2	−1	6	−2	3	−2	−5	6	−2	16	1
E_u	30	0	2	0	0	6	0	10	0	0	48	3

potentially because it is always possible that two or more of the vibrational modes are accidentally degenerate.

The Γ_{dof} representation for the D_{4h} molecular point group is shown above in Table 9.5. Using the tabular method, this reduces to $A_{1g} + A_{2g} + B_{1g} + B_{2g} + E_g + 2A_{2u} + B_{2u} + 3E_u$. After subtraction of the translations ($A_{2u} + E_u$) and rotations ($A_{2g} + E_g$), the following vibrational modes remain: $A_{1g} + B_{1g} + B_{2g} + A_{2u} + B_{2u} + 2E_u$. There are potentially three IR-active peaks (A_{2u} $+2E_u$ modes) and three Raman-active peaks ($A_{1g} + B_{1g} + B_{2g}$). The B_{2u} vibration is silent. On the basis of the differing numbers of IR- and Raman-active modes expected on the basis of the group theoretical analysis, it should be possible to distinguish between the two different geometries.

As a second example, consider the organometallic species $Fe(CO)_5$, which is five-coordinate. The two most common coordination geometries for five-coordinate molecules are trigonal bipyramidal (D_{3h}) and square pyramidal (C_{4v}). Given that $Fe(CO)_5$ exhibits two ν(CO) peaks in its IR spectrum, we might be able to use this information to determine its probable molecular geometry. Because we are only interested in the CO stretching vibrations, it would be a complete waste of our time to determine the symmetries of all 27 vibrational modes, decide which correspond to ν(CO) stretches, and then determine how many of those stretches will be IR-active. Instead, we will use the CO stretches themselves as a basis set in the two limiting geometries to determine the symmetries of the ν(CO) stretches. We will then assess how many of these ν(CO) stretches will be observable in the IR spectrum. The basis sets for the five ν(CO) stretches in $Fe(CO)_5$ in each of the two probable geometries are shown in Figure 9.9.

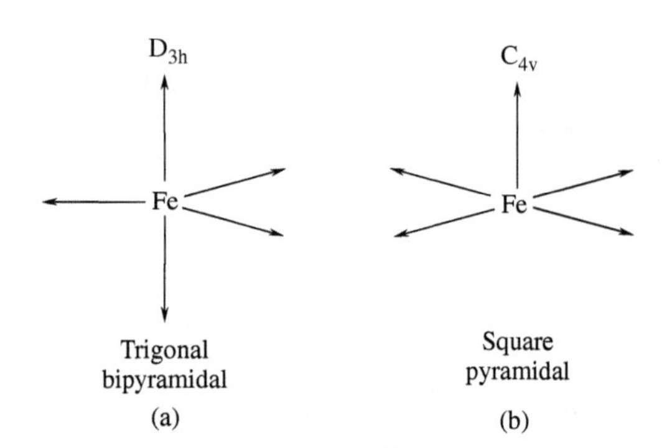

FIGURE 9.9
Vectors are used as a basis to represent the five ν(CO) stretches in $Fe(CO)_5$ in the trigonal bipyramidal (a) and square pyramidal (b) molecular geometries.

In generating the Γ_{CO} representation in each of the two point groups, only those vectors that are transformed into themselves or into the opposite of themselves will lie along the diagonal of the resulting 5×5 matrices representing the symmetry operations. Therefore, only these vectors will contribute to the characters in the Γ_{CO} representation. If the vector is transformed into itself, it will contribute +1 to the character; if it is transformed into its opposite, it will contribute −1 to the character. Table 9.6 shows how the Γ_{CO} representation factors into the different IRRs for the D_{3h} point group. Thus, Γ_{CO} factors into $2\,A_1' + A_2'' + E'$ in the D_{3h} point group. The A_2'' and E' vibrational modes are IR-active, predicting two peaks in the IR spectrum. Table 9.7 shows how the Γ_{CO} representation factors into the different IRRs for the C_{4v} point group. For the C_{4v} point group, Γ_{CO} factors into $2\,A_1 + B_1 + E$. The A_1 and E vibrational modes are IR-active. As there are $2\,A_1$ modes in this point group, three peaks are expected in the IR spectrum of the square pyramidal geometry.

Thus, the trigonal bipyramidal geometry is more consistent with the experimental observations. The $Fe(CO)_5$ molecule is, in fact, trigonal bipyramidal, as predicted on the basis of its symmetry.

TABLE 9.6 Factoring the Γ_{CO} representation for $Fe(CO)_5$ in D_{3h}.

D_{3h}	E	$2C_3$	$3C_2$	σ_h	$2S_3$	$3\sigma_v$	$h=12$	$n_i =$
Γ_{CO}	5	2	1	3	0	3	Σ	Σ/h
A_1'	5	4	3	3	0	9	24	2
A_2'	5	4	−3	3	0	−9	0	0
E'	10	−4	0	6	0	0	12	1
A_1''	5	4	3	−3	0	−9	0	0
A_2''	5	4	−3	−3	0	9	12	1
E''	10	−4	0	−6	0	0	0	0

TABLE 9.7 Factoring the Γ_{CO} representation for $Fe(CO)_5$ in C_{4v}.

C_{4v}	E	$2C_4$	C_2	$2\sigma_v$	$2\sigma_d$	$h=8$	$n_i =$
Γ_{CO}	5	1	1	3	1	Σ	Σ/h
A_1	5	2	1	6	2	16	2
A_2	5	2	1	−6	−2	0	0
B_1	5	−2	1	6	−2	8	1
B_2	5	−2	1	−6	2	0	0
E	10	0	−2	0	0	8	1

Example 9-3. Considering the two possible structures for boric acid shown below, determine the symmetry of each normal mode of vibration and how many IR- or Raman-active O–H peaks are expected to be present in the vibrational spectra of each species.

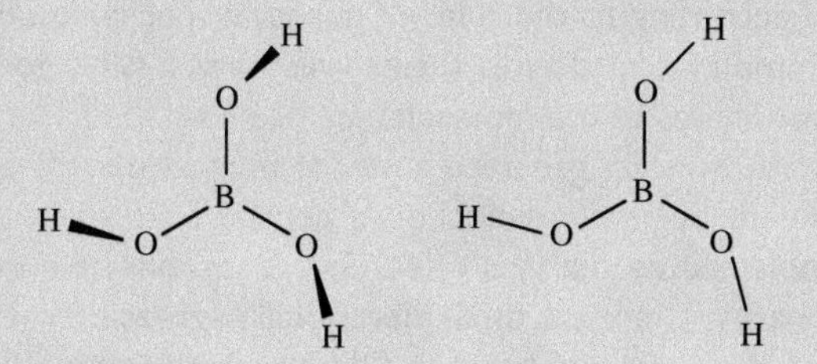

Solution. The structure on the left has C_3 symmetry. Using a bond vector to represent the O–H stretches, the characters of the 3 × 3 matrices representing the symmetry operations in this point group are shown below. There can be two IR-active peaks (*A* and *E*) and two Raman-active peaks (*A* and *E*).

C_3	E	C_3	C_3^2	$h = 3$	$n_i =$
Γ_{OH}	3	0	0	Σ	Σ/h
A	3	0	0	3	1
E	{6}	{0}	{0}	{6}	1

The structure on the right has C_{3h} symmetry. Using a bond vector to represent the O–H stretches, the characters of the 3 × 3 matrices representing the symmetry operations in this point group are shown below. Here, there can only be one IR-active peak (E'') and two Raman-active peaks (A_1' and E').

C_{3h}	E	C_3	C_3^2	σ_h	S_3	S_3^2	$h = 6$	$n_i =$
Γ_{OH}	3	0	0	3	0	0	Σ	Σ/h
A'	3	0	0	3	0	0	6	1
E'	{6}	{0}	{0}	{6}	{0}	{0}	{12}	1
A''	3	0	0	−3	0	0	0	0
E''	{6}	{0}	{0}	{-6}	{0}	{0}	{0}	0

9.4 GENERATING SYMMETRY COORDINATES USING THE PROJECTION OPERATOR METHOD

To this point in our discussion of molecular vibrations, we have shown how to generate the symmetries of the vibrational modes by reduction of the Γ_{dof} representation and subtraction of translational and rotational modes. When pictures of the normal modes of vibration are available, we have also been able to demonstrate how to assign the different symmetries to the correct normal modes of vibration. However, we have not yet attempted to show the origins of the normal modes themselves. The reason for this is because a full-blown *normal mode analysis* (*NMA*) is a complex endeavor that involves the force constants of the vibrations, the masses of the atoms involved, and the vibrational selection rules. While the generation of the normal coordinates is therefore beyond the scope of this book, it is, however, possible for us to generate the *symmetry coordinates* of molecules. In practice, the normal coordinates are composed of linear combinations of the symmetry coordinates, weighted according to the rules of quantum mechanics. However, in many instances, the symmetry coordinates themselves form a fairly good approximation of the normal coordinates of the molecule.

Let us illustrate how to generate a set of normal coordinates for the H_2O molecule. In order to begin, we will choose certain changes in the internal coordinates of the molecule as our basis sets for generating the symmetries of the symmetry coordinates. There are three vibrational degrees of freedom in the water molecule. Because the molecule has two OH bonds and one HOH bond angle, it would make sense to utilize these as our internal basis sets, as shown in Figure 9.10. Defining the *yz*-plane as the plane of the molecule, the reducible representations for the Γ_{OH} and Γ_{HOH} basis sets are shown in Table 9.8. By this point, it is a trivial exercise to show how the Γ_{OH} bond vector basis transforms as the A_1 and B_2 IRRs, whereas the Γ_{HOH} representation has A_1 symmetry. These are exactly the same results obtained previously for the symmetries of the three vibrational modes in the H_2O molecule: $2A_1 + B_2$. Furthermore, we previously saw that the symmetries of the two OH stretching modes for H_2O were A_1 and B_2, whereas the symmetry of the HOH bending mode was A_1, as we also obtain here.

Once the IRRs for each basis set have been determined, it is possible to sketch the symmetry coordinates for the molecules by application of the projection operator. The *projection operator* method, which is used to generate *symmetry-adapted linear combinations* (*SALCs*), is an important tool in group theory and will be widely used in the following chapter, so it is essential that it be mastered right away. The projection operator, which is given by Equation (9.21), will generate linear combinations of functions having a particular symmetry from a single basis function. In this equation, the operator P is the projection operator acting on basis function ϕ_t for the *i*th IRR, d_i is the dimension of the *i*th IRR, h is the order of the group, the summation is over all the operations in the group, χ_i^R is the character for the symmetry

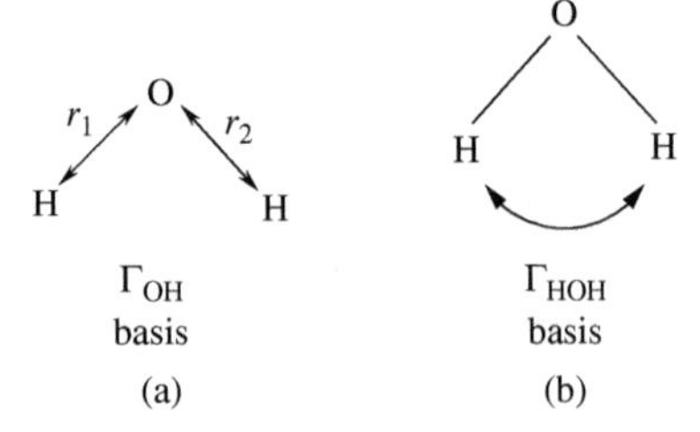

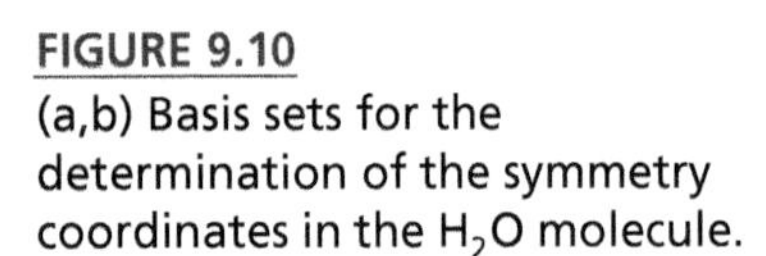

FIGURE 9.10
(a,b) Basis sets for the determination of the symmetry coordinates in the H_2O molecule.

TABLE 9.8 Representations for the basis sets shown in Figure 9.10 for the symmetry coordinates of the H_2O molecule.

C_{2v}	E	C_2	σ_{v1}	σ_{v2}	IRRs
Γ_{OH}	2	0	0	2	$A_1 + B_2$
Γ_{HOH}	1	1	1	1	A_1
Γ_{vib}	3	1	1	3	$2A_1 + B_2$

operation R in the ith IRR, and R_j is the result of the jth symmetry operator acting on the basis function:

$$\hat{P}_i(\phi_t) = \frac{d_i}{h}\sum_R \chi_i^R \hat{R}_j(\phi_t) \tag{9.21}$$

For example, application of the projection operator for the A_1 IRR on the r_1 basis function yields the following: $\frac{1}{4}[(1)E(r_1)+(1)C_2(r_1)+(1)\sigma_{v1}(r_1)+(1)\sigma_{v2}(r_1)] = \frac{1}{4}[r_1+r_2+r_2+r_1] = \frac{1}{2}[r_1+r_2]$. Likewise, application of the projection operator for the A_1 IRR on the r_2 basis function will yield $\frac{1}{2}[r_1+r_2]$. The projection operator, in essence, has taken the basis function r_1 (or r_2) and projected it out over all of the other functions to make a symmetry coordinate having A_1 symmetry. This is the first SALC. Its corresponding symmetry coordinate consists of an equally weighted linear combination of the r_1 and r_2 basis functions, as shown in Figure 9.11(a). This A_1 symmetry coordinate is remarkably similar to the A_1 normal mode of vibration for the H_2O molecule.

Application of the projection operator for the B_2 IRR on the r_1 basis function yields the following: $\frac{1}{4}[(1)E(r_1)+(-1)C_2(r_1)+(-1)\sigma_{v1}(r_1)+(1)\sigma_{v2}(r_1)] = \frac{1}{4}[r_1 - r_2 - r_2+r_1] = \frac{1}{2}[r_1-r_2]$. This is the second SALC. Thus, the symmetry coordinate for the B_2 IRR is a negative linear combination, making it anti-symmetric, as shown in Figure 9.11(b).

The final symmetry coordinate is the A_1 bending vibration. Because it is already a unique function, we do not need to apply the projection operator to it in order to expand it over all the bond angle coordinates. This symmetry coordinate is shown in Figure 9.11(c). Notice that all three symmetry coordinates form fairly good approximations to the actual normal modes of vibration for the H_2O molecule illustrated in Figure 9.8.

Application of the projection operator to degenerate IRRs, such as E or T, can become a little tricky. Consider the BCl_3 example, which we began to investigate in Example 9-1. This molecule has six vibrational degrees of freedom, which transformed as the $A_1' + A_2'' + 2E''$ IRRs in the previous example. Using internal coordinates to define the vibrations, there are three B–Cl bonds, shown in Figure 9.12(a), which can form a triply degenerate Γ_{BCl} representation. Table 9.9 illustrates how the reduction of Γ_{BCl} yields $A_1' + E'$ for the symmetries of the three stretching vibrational modes. There are also three Cl–B–Cl bond angles in the molecule. Using changes in the three bond angles as a second basis set, as shown in Figure 9.12(b), we generate the triply degenerate Γ_θ representation. This also reduces to $A_1' + E'$. However, we should recognize that it is impossible for us to expand all three bond angles

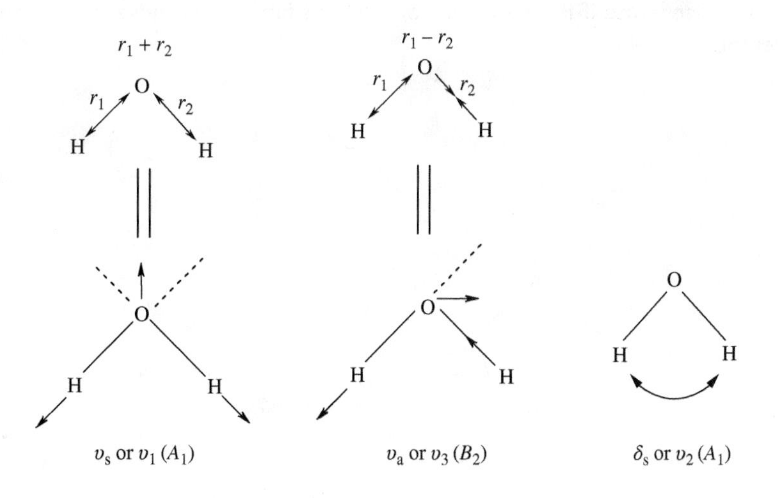

FIGURE 9.11
(a) The A_1 stretching symmetry coordinate, (b) the B_2 stretching symmetry coordinate, and (c) the A_1 bending symmetry coordinate. Compare these with the normal modes of vibration shown in Figure 9.8.

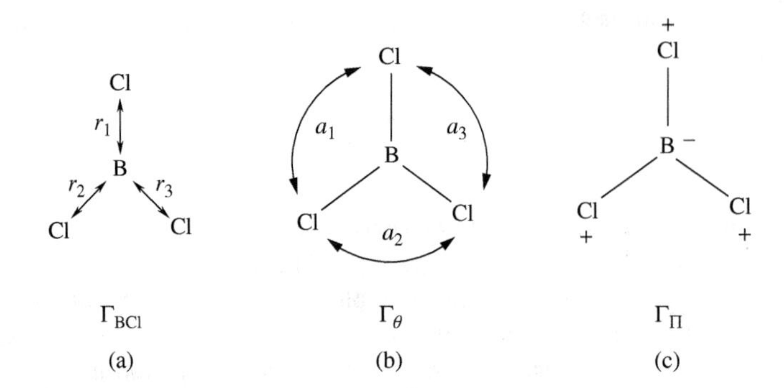

FIGURE 9.12
Basis sets for the symmetry coordinates of BCl_3.

TABLE 9.9 Representations for the basis sets shown in Figure 9.12 for the symmetry coordinates of the BCl_3 molecule.

D_{3h}	E	$2C_3$	$3C_2$	σ_h	$2S_3$	$3\sigma_v$	IRRs
Γ_{BCl}	3	0	1	3	0	1	$A_1'+E'$
Γ_θ	3	0	1	3	0	1	$(A_1')+E'$
Γ_π	1	1	−1	−1	−1	1	A_2''
Γ_{vib}							$A_1'+A_2''+2E'$

simultaneously because expansion of any two bond angles must cause a subsequent reduction in the third bond angle. Furthermore, our sixth vibrational degree of freedom must have A_2'', not A_1', symmetry. Thus, the totally symmetric bending A_1' IRR must be discarded as an unsuitable symmetry coordinate.

One vibrational degree of freedom therefore remains. It can be shown that the out-of-plane bending motion shown in Figure 9.12(c) forms the final basis. According to Table 9.9, the out-of-plane vibration transforms as the A_2'' IRR. The A_2'' IRR is a unique one and the symmetry coordinate for out-of-plane bending motion will be identical to the Γ_π basis set in Figure 9.12(c), as shown in Figure 9.13(a).

In order to determine the symmetry coordinates for the three stretching vibrations, we must apply the projection operator to one of the three basis functions in the Γ_{BCl} basis set. Application of the projection operator for the A_1' IRR on the r_1 basis function yields the SALC shown in Table 9.10. This SALC is shown in Figure 9.13(b). If the A_1' projection operator had been applied to the r_2 or r_3 basis function instead, the exact same SALC would have resulted. This will always be true for a nondegenerate IRR. The choice of the basis function in these cases is entirely irrelevant.

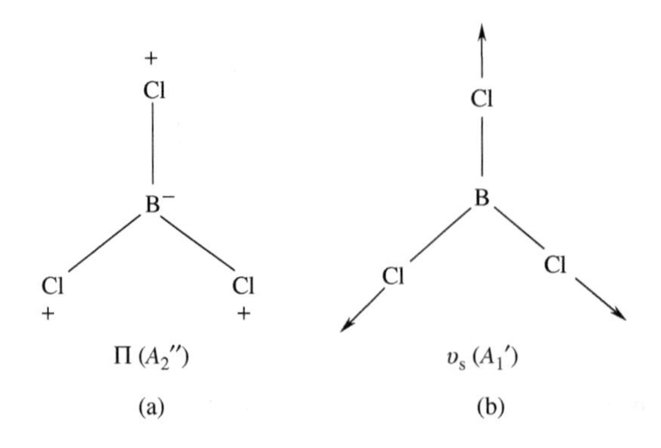

FIGURE 9.13
Symmetry coordinates for (a) the π (A_2'') and (b) ν_s (A_1') vibrational modes of BCl_3.

TABLE 9.10 Application of the projection operator method using the r_1 bond stretch as the basis.

D_{3h}	E	C_3	C_3^2	C_2	C_2'	C_2''	σ_h	S_3	S_3^2	σ_v	σ_v'	σ_v''	$1/h = 1/12$
r_1	r_1	r_2	r_3	r_1	r_3	r_2	r_1	r_2	r_3	r_1	r_3	r_2	
$\phi_1(A_1')$	r_1	r_2	r_3	r_1	r_3	r_2	r_1	r_2	r_3	r_1	r_3	r_2	$1/4[r_1+r_2+r_3]$
$\phi_1(E')$	$2r_1$	$-r_2$	$-r_3$	0	0	0	$-2r_1$	r_2	r_3	0	0	0	$1/6[2r_1-r_2-r_3]$

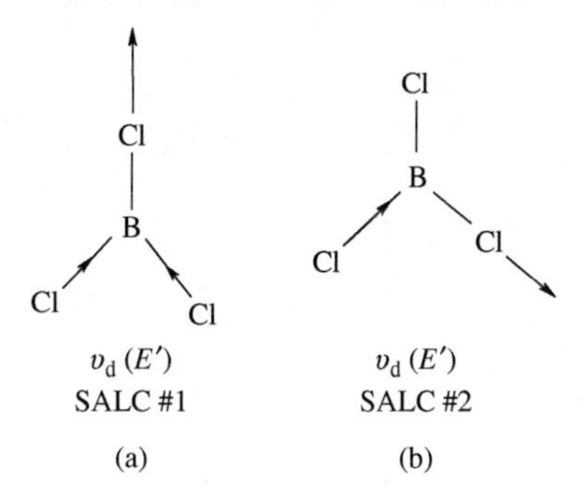

FIGURE 9.14
(a,b) Symmetry coordinates for the two degenerate E' stretching vibrational modes of BCl_3.

Application of the projection operator for the E' IRR on the r_1 basis function yields the result shown in Table 9.10. The shape of this SALC is shown in Figure 9.14(a). Application of the E' projection operator on r_2 in turn yields $\phi_2 = (1/6)[2r_2 - r_3 - r_1]$ and on r_3 yields $\phi_3 = (1/6)[2r_3 - r_1 - r_2]$. Unlike the A_1' projection operator, where the choice of basis function did not matter, the result for degenerate IRRs is dependent on which basis function is employed. We know that the E' IRR is supposed to be doubly degenerate, but we currently have three different mathematical results. This presents us with somewhat of a dilemma.

Our solution to this problem will be to arbitrarily accept one of the three solutions as one of the two acceptable SALCs and to make the other SALC a linear combination of the remaining two solutions. To illustrate, let us arbitrarily accept the ϕ_1 solution as SALC#1: $(1/6)[2r_1 - r_2 - r_3]$. The shape of this E' stretching SALC is shown in Figure 9.14(a). Once we have chosen this solution to form one of our two SALCs, the other degenerate SALC must consist of a linear combination of the other two solutions and it must also be orthogonal to SALC#1. To determine the appropriate linear combination of ϕ_2 and ϕ_3 to use, we will orient our molecule such that r_1 (which was used to generate ϕ_1 for SALC#1) lies on one of the coordinate axes; let us say on the y-axis, as shown in Figure 9.15. Then, from the shape of the basis set in Figure 9.12(a), we can see that r_2 (from which we derived ϕ_2) has a negative component along the x-axis and r_3 (from which we derived ϕ_3) has a positive component lying along the x-axis, so that an appropriate linear combination orthogonal to the first would consist of $-\phi_2 + \phi_3$. Substituting, we obtain $(1/6)[-2r_2 + r_3 + r_1 + 2r_3 - r_1 - r_2] = (1/2)[-r_2 + r_3]$ as SALC#2.

Both of these SALCs ultimately need to be normalized, so the coefficients in front of the brackets do not really matter at this point. In order to normalize a wave function, the sum of the square of the coefficients in front of the wave functions making up the linear combination must be equal to one. The final wave functions must also be orthogonal to each other. It is the coefficients in front of each basis function that are relevant. For SALC#1, these coefficients are 2, −1, and

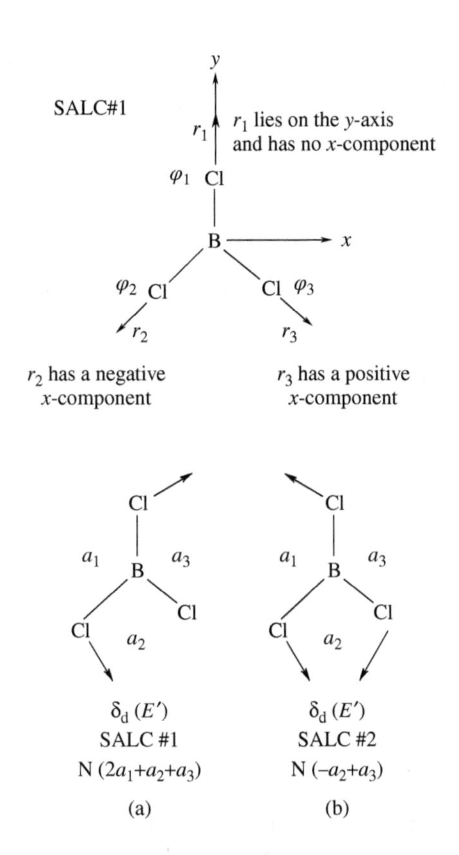

FIGURE 9.15
Vector diagram for the Γ_{BCl} basis set of BCl_3, showing how orientation of SALC#1 along the y-axis results in SALC#2 lying along the x-axis in order for the two SALCs to be orthogonal. SALC#2 is a linear combination of $-\phi_2+\phi_3$ because r_2 has a negative x-component and r_3 has a positive x-component.

FIGURE 9.16
(a,b) The E' bending symmetry coordinates for BCl_3.

−1. For SALC#2, they are 0, −1, and +1. A quick check will show that these two SALCs are in fact orthogonal: $(2)(0)+(-1)(-1)+(1)(-1) = 0$. It is an easy matter to prove that both of the E' SALCs are also orthogonal to the A_1' stretching mode, which had characters of 1, 1, and 1. The shape of the E' SALC#2 is shown in Figure 9.14(b).

Besides the four symmetry coordinates shown in Figures 9.13 and 9.14, there are still two degenerate bending modes having E' symmetry. We have already applied the projection operator to the E' IRR and found that two orthogonal results are $1/6[2r_1+r_2+r_3]$ and $\frac{1}{2}[-r_2+r_3]$. Thus, the shapes of the two E' bending symmetry coordinates are shown in Figure 9.16. Notice that in SALC#1, bond angle a_1 increases twice as much as angles a_2 and a_3 decrease, and in SALC#2, bond angle a_2 decreases while angle a_3 increases, consistent with the mathematical forms of these two SALCs. A comparison of the six symmetry coordinates in Figures 9.13, 9.14, and 9.16 with the six normal modes of vibration for a D_{3h} molecule shown in Example 9-1 illustrates how the symmetry coordinates and normal modes are in fairly good agreement.

Example 9-4. Using miniature orthogonal vectors on each atom in the N_2F_2 molecule (shown below), determine the symmetries of the normal modes of vibration for the molecule and indicate which modes are IR-active and which are Raman-active. Then, use internal coordinates such as bond length stretches, bond angle changes, and out-of-plane bending motions, along with application of the projector operator, to generate a set of symmetry coordinates for the molecule. Show that there is a one-to-one correspondence between the IRRs of the six normal modes of vibration determined from the Γ_{dof} representation and the IRRs of the six symmetry coordinates.

Solution. The molecule lies in the C_{2h} point group. The basis set for the Γ_{dof} representation and its reduction using the great orthogonality theorem are shown in the figure and table below.

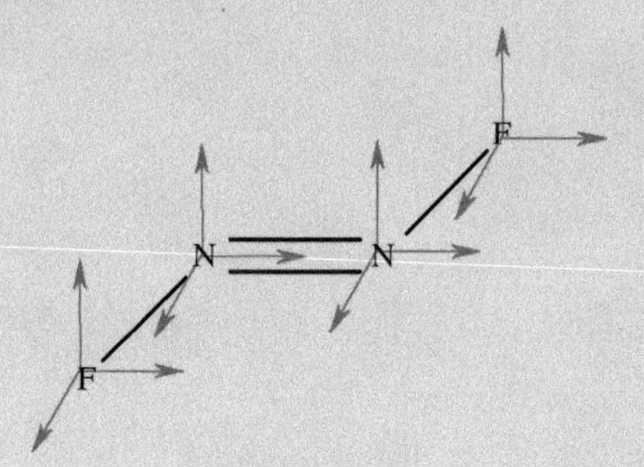

C_{2h}	E	C_2	i	σ_h	$h = 4$	$n_i =$
Γ_{dof}	12	0	0	4	Σ	Σ/h
A_g	12	0	0	4	16	4
B_g	12	0	0	−4	8	2
A_u	12	0	0	−4	8	2
B_u	12	0	0	4	16	4

The Γ_{dof} reduces to $4A_g + 2B_g + 2A_u + 4B_u$. After subtraction of the three translational and three rotational degrees of freedom, the symmetries of the six vibrational modes are $3A_g + A_u + 2B_u$. There can be one A_u and two B_u peaks in the IR spectrum and three A_g peaks in the Raman spectrum of the molecule. Because this is a centrosymmetric molecule, the IR- and Raman-active modes are mutually exclusive. The six internal coordinates for the N_2F_2 molecule are one N=N bond stretch, two N−F bond stretches, two FNN bond angles, and one out-of-plane bend. The characters for each of these four basis sets and their reductions into their IRR components are shown in the table below.

C_{2h}	E	C_2	i	σ_h	IRRs
Γ_{NN}	1	1	1	1	A_g
Γ_{NF}	2	0	0	2	$A_g + B_u$
Γ_{FNN}	2	0	0	2	$A_g + B_u$
Γ_{π}	1	1	−1	−1	A_u
Γ_{vib}					$3A_g + A_u + 2B_u$

The N=N stretching symmetry coordinate is identical to its basis set vibration and has A_g symmetry, as shown below. The out-of-plane bending symmetry coordinate is also identical to its basis set, but it has A_u symmetry. Application of the projection operator for the A_g IRR on the r_1 basis function for a

NF bond stretch yields the following result: $\frac{1}{4}[(1)E(r_1)+(1)C_2(r_1)+(1)i(r_1)+(1)\sigma_h(r_1)] = \frac{1}{4}[r_1+r_2+r_2+r_1] = \frac{1}{2}[r_1+r_2]$. Application of the projection operator for the B_u IRR on the r_1 basis function for a NF bond stretch yields the following result: $\frac{1}{4}[(1)E(r_1)+(-1)C_2(r_1)+(-1)i(r_1)+(1)\sigma_h(r_1)] = \frac{1}{4}[r_1-r_2-r_2+r_1] = \frac{1}{2}[r_1-r_2]$. The corresponding shapes of the A_g and B_u symmetry coordinates for the NF bond stretches are shown below.

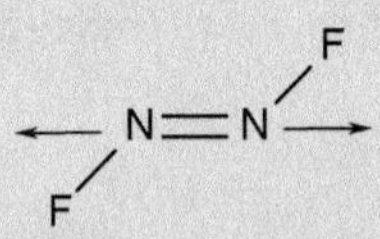

υ_s (A_g) 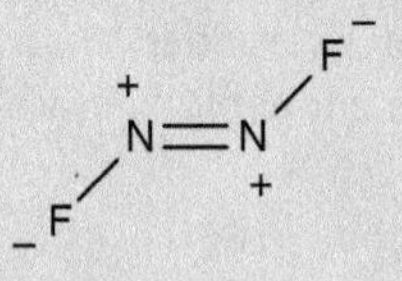 Π (A_u)

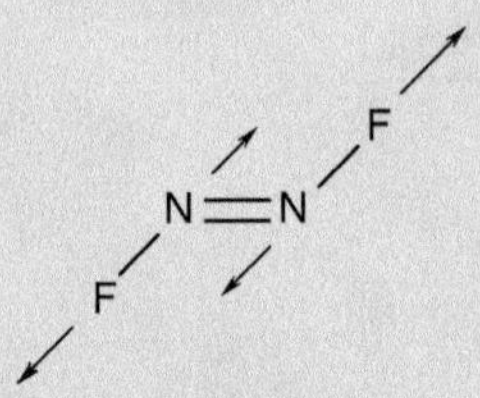

υ_s (A_g) 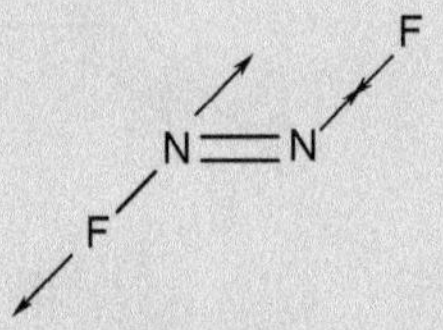 υ_a (B_u)

Application of the projection operator for the A_g IRR on the a_1 basis function for an NNF bond angle change yields the following result: $\frac{1}{2}[a_1+a_2]$, while application of the projection operator for the B_u IRR on the a_1 basis function for a NNF bond angle change yields the following result: $\frac{1}{2}[a_1-a_2]$. The corresponding NNF bending symmetry coordinates are shown below.

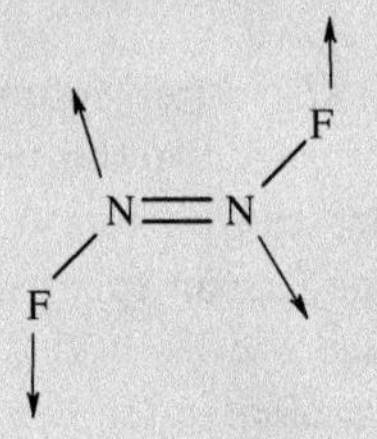

δ_s (A_g) 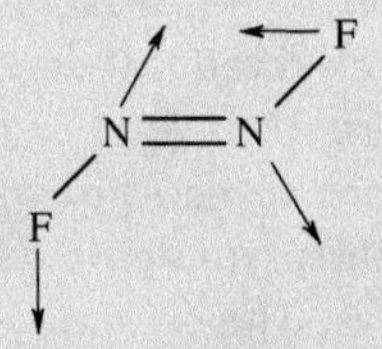 δ_a (B_u)

For molecular point groups with very large character tables, it is sometimes more useful to simplify the problem using a smaller subgroup of the molecular point group. If the molecule contains a principal axis (other than one of the high-symmetry point groups such as T_d, O_h, or I_h), the pure rotational subgroup having the same value of *n* as the principal axis can often be used. This can greatly simplify the amount of work necessary when applying the projection operator method. Consider the $PtCl_4^{2-}$ ion, which has a square planar geometry and belongs to the D_{4h} point group. We can use miniature Cartesian vectors on each atom to generate the Γ_{dof} representation, reduce it into its IRRs, and subtract the translational and rotational modes to determine the symmetries of the nine vibrational modes. The result, which is shown in Table 9.11, is that the vibrational modes are $A_{1g}+B_{1g}+B_{2g}+A_{2u}+B_{2u}+2E_u$. Next, we can use the Pt–Cl bond stretches shown in Figure 9.17(a) as a basis set to determine which of those vibrational modes are stretching vibrations. In this case, the RR for the stretching vibrations is given by Γ_{PtCl} in Table 9.11. Using the great orthogonality theorem, this RR reduces to $A_{1g}+B_{1g}+E_u$. By difference, the

TABLE 9.11 Representations for the basis sets shown in Figure 5.31 for the $PtCl_4^{2-}$ anion and their irreducible components.

D_{4h}	E	$2C_4$	C_2	$2C_2'$	$2C_2''$	i	$2S_4$	σ_h	$2\sigma_v$	$2\sigma_d$	IRRs
Γ_{dof}	5	1	1	−3	−1	−3	−1	5	3	1	$A_{1g}+(A_{2g})+B_{1g}+B_{2g}+(E_g)+(A_{2u})+A_{2u}+B_{2u}+(E_u)+2E_u$
Γ_{PtCl}	4	0	0	2	0	0	0	4	2	0	$A_{1g}+B_{1g}+E_u$
Γ_{oop}	4	0	0	−2	0	0	0	4	2	0	$A_{2u}+B_{2u}+(E_g)$
Γ_{ip}	4	0	0	−2	0	0	0	4	−2	0	$(A_{2g})+B_{2g}+E_u$

IRRs corresponding to translational or rotational degrees of freedom are enclosed in parentheses.

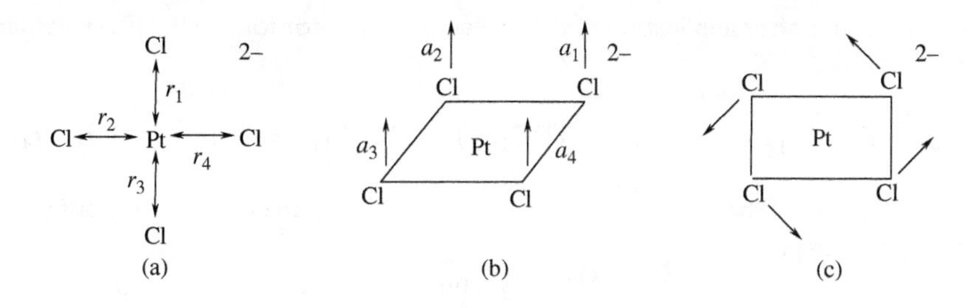

FIGURE 9.17 Basis sets for the (a) PtCl stretching vibrational modes, (b) out-of-plane bending modes, and (c) in-plane bending modes for the $PtCl_4^{2-}$ anion.

bending modes must have $B_{2g}+A_{2u}+B_{2u}+E_u$ symmetry. We can further categorize the bending modes into out-of-plane bends using the basis set in Figure 9.17(b) to generate the Γ_{oop} representation in Table 9.11, which reduces to $A_{2u}+B_{2u}+E_g$. As E_g is not included on the list of vibrational modes, it is excluded (in fact, it corresponds to rotation around the *x*- and *y*-axes). By default, the in-plane bends must have $B_{2g}+E_u$ symmetry. Using the basis set shown in Figure 9.17(c), we generate the Γ_{ip} representation in Table 9.11, which reduces to $A_{2g}+B_{2g}+E_u$. As the A_{2g} IRR is not on our list of vibrational modes, it too is excluded (in fact, it corresponds with rotation around the *z*-axis).

There are a total of 16 symmetry operations in the D_{4h} point group. Having to apply the projection operator to a basis function using all 16 symmetry operations would be a very tedious exercise. If we could substitute the C_4 subrotational group in its place, we could reduce the number of symmetry operations down to only four, which would save us considerable time. However, there is one small catch: the *E* IRR in the C_4 subgroup has some imaginary characters, as shown below:

$$\begin{array}{cccccc} & E & C_4 & C_2 & C_4^3 \\ E = & \Big\{ \begin{matrix} 1 \\ 1 \end{matrix} & \begin{matrix} i \\ -i \end{matrix} & \begin{matrix} -1 \\ -1 \end{matrix} & \begin{matrix} -i \\ i \end{matrix} \Big\} \end{array}$$

We can convert these imaginary characters into real ones by taking linear combinations of the two rows of characters. We will replace the first row of characters by taking the positive linear combination $(a+b)$ and the second row by taking the negative linear combination $(a-b)$ to obtain the following:

$$\begin{array}{cccccc} & E & C_4 & C_2 & C_4^3 & \\ E = & \Big\{ \begin{matrix} 2 \\ 0 \end{matrix} & \begin{matrix} 0 \\ 2i \end{matrix} & \begin{matrix} -2 \\ 0 \end{matrix} & \begin{matrix} 0 \\ -2i \end{matrix} \Big\} & \begin{matrix} (a+b) \\ (a-b) \end{matrix} \end{array}$$

We then divide each row by the greatest common multiple in that row to obtain characters having only real components, as shown below:

$$E = \begin{array}{c} E \quad C_4 \quad C_2 \quad C_4^3 \\ \left\{ \begin{array}{cccc} 1 & 0 & -1 & 0 \\ 0 & 1 & 0 & -1 \end{array} \right\} \end{array} \quad \begin{array}{l} \text{(divided by 2)} \\ \text{(divided by } 2i\text{)} \end{array}$$

With these preliminaries out of the way, we are ready to apply the projection operator in order to determine the nine symmetry coordinates for the $PtCl_4^{2-}$ ion. Beginning with the A_{1g} stretching mode (which correlates with the A IRR in the C_4 subgroup), application of the projection operator to the r_1 basis function yields

$$A_{1g} \quad 1/4[(1)E(r_1) + (1)C_4(r_1) + (1)C_2(r_1) + (1)C_4^3(r_1)] = 1/4[r_1 + r_2 + r_3 + r_4]$$

The results after application of the projection operator for the B_{1g} (B) stretching mode are

$$B_{1g} \quad 1/4[(1)E(r_1) + (-1)C_4(r_1) + (1)C_2(r_1) + (1)C_4^3(r_1)] = 1/4[r_1 - r_2 + r_3 - r_4]$$

Application of the projection operator for the E_u (E) stretching vibrations yields the degenerate pair:

$$E_u \quad 1/2[r_1 - r_3] \text{ and } 1/2[r_2 - r_4]$$

Once the numerical results for the projection operator for all of the IRRs in the C_4 subgroup have been determined (as they now have), it is a simple matter to list the mathematical forms for the remaining vibrational modes. The out-of-plane bending modes are

$$A_{2u} \quad 1/4[a_1 + a_2 + a_3 + a_4]$$
$$B_{2u} \quad 1/4[a_1 - a_2 + a_3 - a_4]$$

Finally, the in-plane bending modes are

$$B_{2u} \quad 1/4[b_1 - b_2 + b_3 - b_4]$$
$$E_u \quad 1/2[b_1 - b_3] \text{ and } 1/2[b_2 - b_4]$$

Each of the symmetry coordinates for $PtCl_4^{2-}$ is shown in Figure 9.18.

9.5 RESONANCE RAMAN SPECTROSCOPY

The third experimental technique for the measurement of vibrational frequencies is resonance Raman spectroscopy. The resonance Raman technique is similar to Raman spectroscopy in that a LASER beam (typically having a wavelength in the UV or visible range) is used to excite the sample. However, instead of a random excitation to a virtual state of the molecule, the LASER beam is tuned to a wavelength inside or very near to one of the allowed electronic transitions of the molecule. A comparison of the two techniques is shown in Figure 9.19. When the energy of the LASER beam is in resonance with the electronic transition, the intensities of those vibrational modes associated with nuclear changes from the ground to the excited electronic state are selectively enhanced. One of the main advantages of the resonance Raman technique is the greater intensity of the Raman-scattered frequencies. Resonant enhancements $10^3 - 10^6$ times more intense than traditional Raman

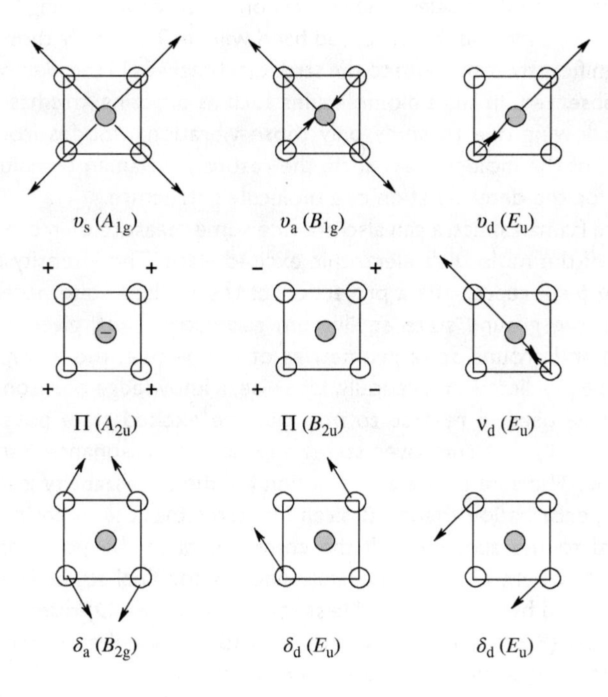

FIGURE 9.18
Symmetry coordinates for the vibrational modes of $PtCl_4^{2-}$.

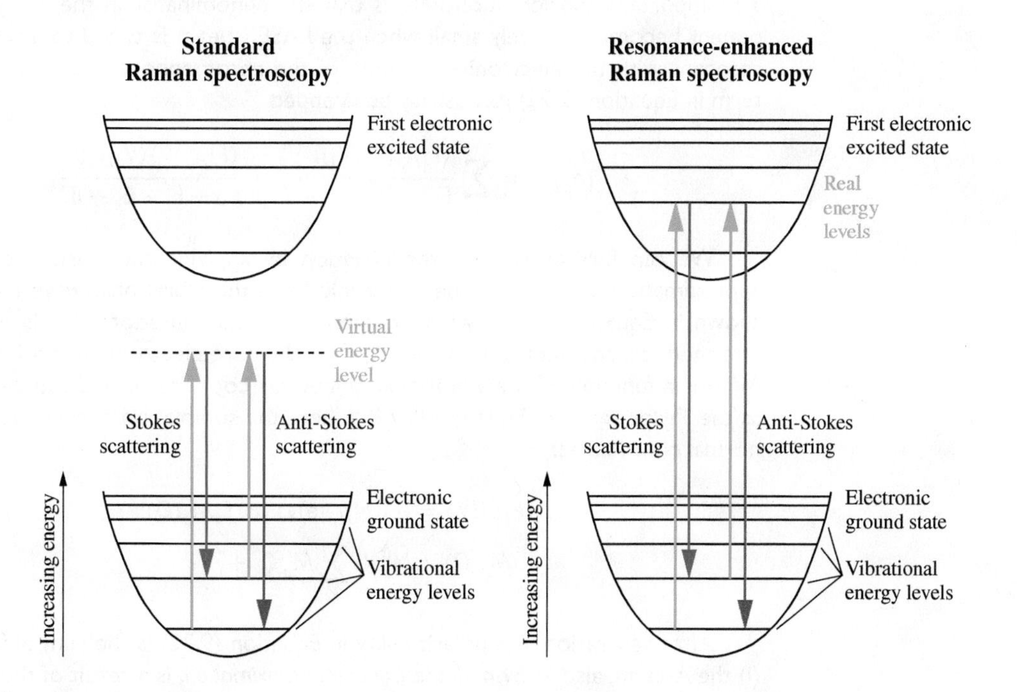

FIGURE 9.19
A comparison of standard Raman versus resonance Raman spectroscopic techniques. [Copyright Semrock, Inc., reproduced by permission from http://www.semrock.com/ultraviolet-uv-raman-spectroscopy.aspx (accessed December 21, 2013).]

spectroscopy are not unusual. Furthermore, the larger the contribution of a given molecular vibration to the nuclear reorganization term accompanying the electronic transition, the more intense the observed band will be. Thus, only those vibrational modes that significantly contribute to the structural changes in the electronic excited state will be observed. In large biomolecules such as proteins, this has the decided advantage of allowing one to study only those vibrational modes from the chromophoric part of the molecule. As such, the resonance Raman technique is a very powerful one for the determination of a molecule's structure.

Resonance Raman spectra can also provide some measure of information about the structure of the molecule's electronic excited state. The intensity of each resonance Raman peak represents a projection of the excited state potential energy surface against the ground state equilibrium geometry. Thus, given a description of the excited and ground state geometries of a molecule, the resonance Raman spectrum can be predicted theoretically. Likewise, a knowledge of resonance Raman intensities can be used in reverse to map out the excited state potential energy surface. Traditionally, the sum over states approach to resonance Raman intensities is employed. The sum over states solution for the polarizability is derived from second-order perturbation theory. Basically, it states that the intensity of the band is proportional to the sum over all the components of the polarizability tensor $\Sigma\ [\alpha_{\rho\lambda}]_{IF}[\alpha_{\rho\lambda}]_{IF}{}^*$, where i is the initial state and f is the final state. The polarizability tensor is defined by the Kramer–Heisenberg–Dirac (KHD) dispersion formula, given by Equation (9.22). The details of this equation are unimportant to the average reader. The summation occurs over all of the vibronic states (vibrational levels superimposed on the electronic states), μ is the transition dipole moment (related to what we previously defined as M_{0I}), F is the final state, I is the initial state, V is the intermediate state, E_L is the incident photon energy, and $i\Gamma$ is a damping factor. The important consideration here is that the denominator in the first time component becomes infinitely small when the LASER beam is tuned to a wavenumber resonant with the electronic transition of the chromophore and thus the second term in Equation (9.22) can usually be dropped:

$$[\alpha_{\rho\lambda}]_{IF} = \sum_V \frac{\langle F|\mu_\rho|V\rangle\langle V|\mu_\lambda|I\rangle}{E_V - E_I - E_L - i\Gamma} + \frac{\langle F|\mu_\lambda|V\rangle\langle V|\mu_\rho|I\rangle}{E_V - E_I + E_L - i\Gamma} \tag{9.22}$$

We can further simplify the equation by applying the Born–Oppenheimer approximation to separate the electronic from the vibrational wave functions, as shown in Equation (9.23), where f, v, and i are the vibrational levels of the final, intermediate, and initial states and $M_\rho(Q)$ is the pure electronic transition moment. $M(Q)$ is a function of the vibrational or nuclear coordinates and can be expanded as the Taylor series in Equation (9.24), where the summation occurs over all of the normal coordinates:

$$\langle F|\mu_\rho|V\rangle = \langle f|\langle g|\mu_\rho|e\rangle v\rangle = \langle f|M_\rho(Q)|v\rangle \tag{9.23}$$

$$M_\rho(Q) = M_\rho^0 + \sum_k M'_\rho Q_k + \cdots \tag{9.24}$$

After separation, the polarizability in Equation (9.22) is the sum of four terms: (i) the A-term, also known as the *Condon approximation*, is a result of the pure electronic transition moment and vibrational overlap integrals; (ii) the B-term results from vibronic coupling of the resonant excited state with other excited states; (iii) the C-term has to do with the vibronic coupling of the ground state with one or more excited states; and (iv) the D-term results from vibronic coupling of the resonant excited state to other excited states coupling in both the electronic transition moments.

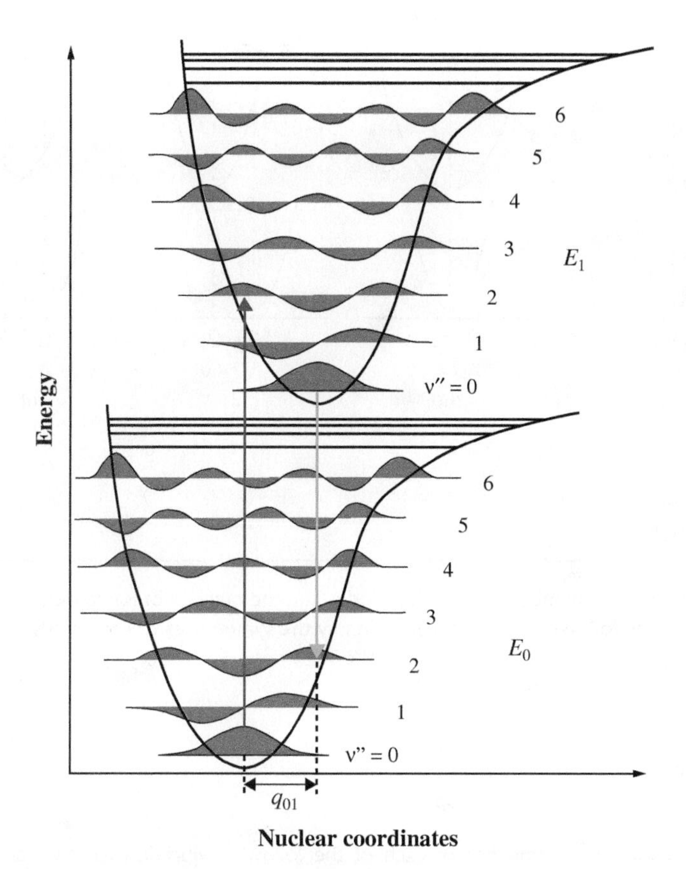

FIGURE 9.20
The Franck–Condon factors, shown here for the process of absorption (blue line) followed by emission (green line). In a resonance Raman experiment, the LASER beam is tuned to be in resonance with an allowed electronic transition (indicated here by the blue arrow), which takes the electron from some initial state I to some intermediate state V. Unlike fluorescence, there is no vibrational relaxation of the excited state. The electron is scattered from the intermediate V state down to the $v''=1$ vibrational level of the ground electronic state, which we will call the final state F. The Franck–Condon overlap integrals involve the amount of overlap between the wave functions, which are shown on this diagram in brown. [© Mark M Samoza/CC-BY-SA_2.5/GFDL /Wimimedia Commons; reproduced from http://en.wikipedia.org/wiki /Franck%E2%80%93Condon_ principle (accessed December 21, 2013).]

We will focus our attention here only on the A-term scattering. The A-term scattering is given by Equation (9.25), where $<f|v>$ and $<v|i>$ are known as the *Franck–Condon factors*. The *Franck–Condon factors* are the overlap integrals between the vibrational wave functions shown in Figure 9.20. In order for the band to be observed, these integrals must be nonzero. The integrals will be nonzero if the electronic transition is symmetry allowed and if the vibrational wave functions are nonorthogonal (if their overlap is nonzero):

$$[\alpha_{\mathrm{IF}}](E_{\mathrm{L}}) = M^2 \sum_{V} \frac{\langle f|v\rangle\langle v|i\rangle}{\varepsilon_{\mathrm{V}} - \varepsilon_{\mathrm{i}} + E_0 - E_{\mathrm{L}} - i\Gamma} \tag{9.25}$$

In Figure 9.21(a), the overlap integral is zero. There are two ways in which the Franck–Condon overlap integrals can be non zero: (i) if the ground and excited states have different vibrational wavenumbers (the force constants are different, for instance), as shown in Figure 9.21(b), or (ii) if the displacement of the excited state potential minimum versus the ground state minimum occurs along the nuclear coordinate Q_k, as shown in Figure 9.21(c); or both, as in Figure 9.21(d). Condition (ii) can only occur for the totally symmetric vibrational modes as long as the point group of the molecule remains the same in the ground and excited electronic states. Thus, the totally symmetric normal modes of vibration will be resonantly enhanced by A-term scattering. Normal modes with other symmetries can still be enhanced by B-, C-, or D-term scattering, but these vibrational modes are usually less intense than the totally symmetric ones.

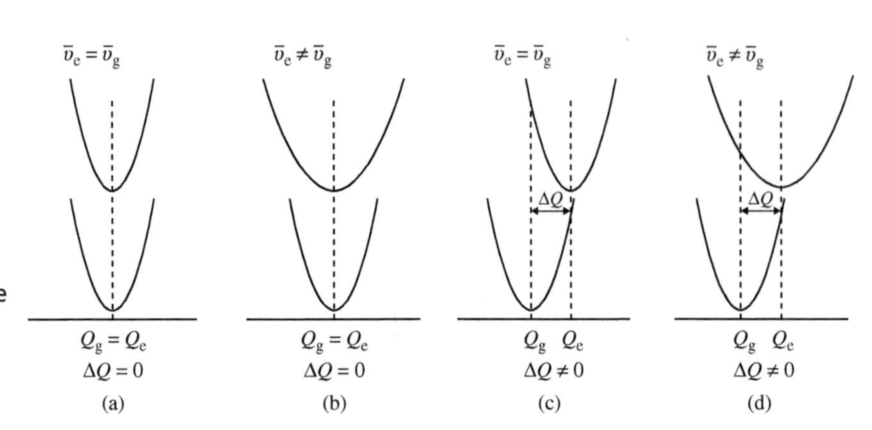

FIGURE 9.21
The Franck–Condon factors under different circumstances. (a) When the ground and excited states have the same wavenumber and zero displacement along the nuclear coordinate Q, the overlap integral for the transition will be zero because the shapes of the wave function in the ground and excited states will be identical. (b) When the excited state has a different wavenumber for the vibration than the ground state (if their force constants are different), then the Franck–Condon integrals will be nonzero. (c) The integrals will also be nonzero if the ground state and excited state potential energy minima are displaced along the nuclear coordinate, even if their wavenumbers are identical. (d) If both the wavenumbers of the ground and excited states differ and there is a nuclear displacement, the Franck–Condon integrals will also be nonzero. [Blatt Communications.]

EXERCISES

9.1. Determine the number of IR-active modes and the number of Raman-active modes for each of the following molecules and identify the symmetries of each mode:

a. NH_3
b. PF_5
c. H_2O_2
d. BrF_5
e. Staggered ethane
f. $[PtCl_6]^{2-}$
g. $[NiCl_4]^{2-}$
h. N_2O
i. SF_4
j. BrF_3

9.2. Determine the symmetries of each of the following vibrational modes and indicate the appropriate symbol for each mode. The X_2Y_2 molecule is planar. The symbol + represents motion above the plane of the paper and – represents motion below this plane.

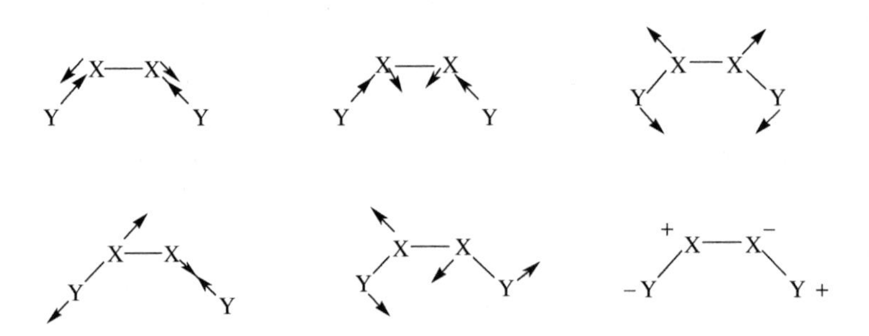

9.3. Consider the square pyramidal $XeOCl_4$ molecule.

a. Determine the point group.
b. Determine the symmetries of all the vibrational modes.
c. Determine the symmetry of the Xe = O stretching vibration.
d. Determine the symmetries of the four Xe–Cl stretching vibrations.
e. Determine the symmetries of the three Cl–Xe–Cl bending modes.
f. Determine the symmetries of the four O–Xe–Cl bending modes.
g. Use the projection operator method to determine the shapes of all 12 symmetry coordinates.
h. Indicate which modes are IR-active and which are Raman-active.

9.4. The IR spectrum of BF_3 has three peaks at 480, 691, and 1449 cm^{-1}. Determine whether BF_3 belongs to the C_{3v} or D_{3h} point group.

9.5. There are two possible geometrical isomers of $M(CO)_4X_2$ compounds. Use group theory to predict the number of IR-active ν(CO) modes for the *trans*-$M(CO)_4X_2$ and *cis*-$M(CO)_4X_2$ isomers.

9.6. Use group theory to determine the symmetries of all the stretching vibrational modes in Al_2Cl_6 (shown below). Determine which are IR-active and which are Raman-active.

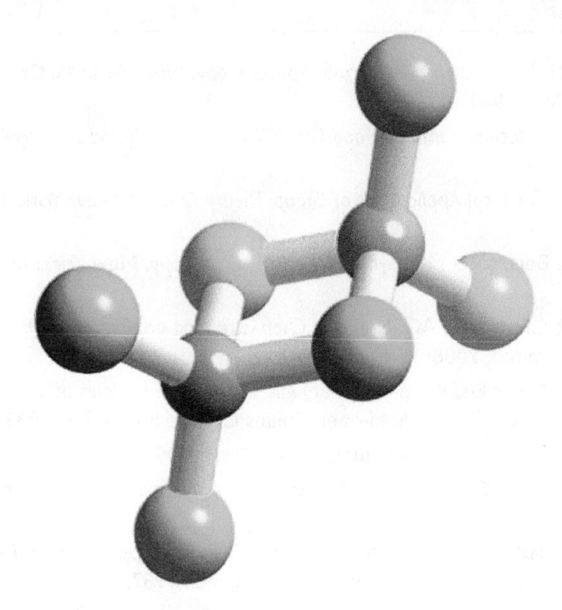

9.7. Given the molecule shown below, answer the following questions:

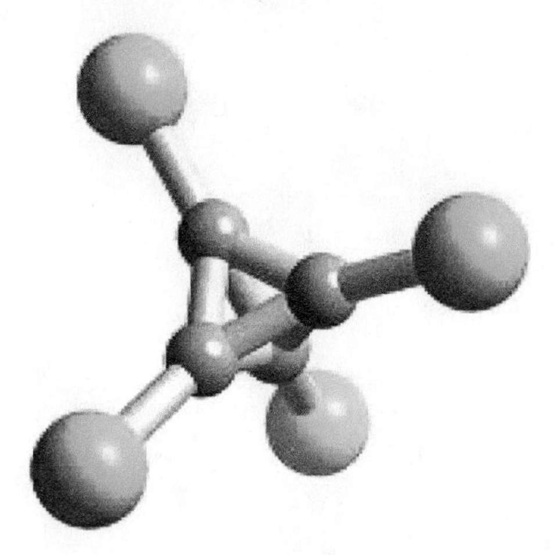

a. Using an appropriate basis, determine the symmetries of all the normal modes of vibration inB_4Cl_4.

b. How many and which peaks are expected in the infrared (IR) spectrum of the compound?

c. How many and which peaks are expected in the Raman spectrum of the compound? Using an appropriate basis, determine the symmetries of the B–Cl stretching vibrational modes.

d. Using an appropriate basis, determine the symmetries of the B–B stretching vibrational modes.

e. Using the process of elimination, determine the symmetries of the bending modes in B_4Cl_4.

9.8. Use the projection operator method to determine the symmetries and shapes of all the Pt–Cl stretching vibrational modes in the octahedral $[PtCl_6]^{2-}$ ion.

BIBLIOGRAPHY

1. Barrow GM. *Introduction to Molecular Spectroscopy*. New York: McGraw-Hill Book Company, Inc.; 1962.
2. Carter RL. *Molecular Symmetry and Group Theory*. New York: John Wiley & Sons, Inc.; 1998.
3. Cotton FA. *Chemical Applications of Group Theory*. 2nd ed. New York: John Wiley & Sons, Inc.; 1971.
4. Harris DC, Bertolucci MD. *Symmetry and Spectroscopy*. New York: Dover Publications, Inc.; 1978.
5. Housecroft CE, Sharpe AG. *Inorganic Chemistry*. 3rd ed. Essex, England: Pearson Education Limited; 2008.
6. Huheey JE, Keiter EA, Keiter RL. *Inorganic Chemistry: Principles of Structure and Reactivity*. 4th ed. New York: Harper Collins College Publishers; 1993.
7. Kettle SFA. *Symmetry and Structure*. 2nd ed. New York: John Wiley & Sons, Inc.; 1995.
8. Miessler GL, Tarr DA. *Inorganic Chemistry*. 4th ed. Upper Saddle River, NJ: Pearson Education Inc.; 2011.
9. Myers AB, Mathies RA. In: Spiro TG, editor. *Biological Applications of Resonance Raman Spectroscopy*. Vol. II. New York: Wiley & Sons, Inc.; 1987.
10. Wilson EB Jr, Decius JC, Cross PC. *Molecular Vibrations: The Theory of Infrared and Raman Vibrational Spectroscopy*. New York: Dover Publications, Inc.; 1955.

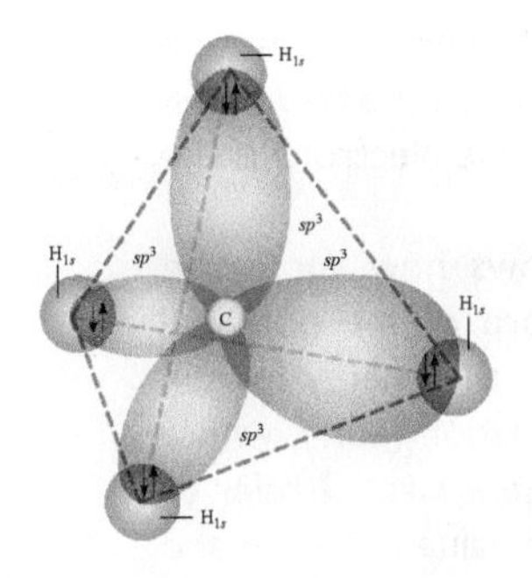

Covalent Bonding | 10

"Attempts to regard a molecule as consisting of specific atoms or ionic units held together by discrete numbers of bonding electrons or electron pairs are considered as more or less meaningless."

—*Robert Mulliken*

10.1 VALENCE BOND THEORY

The quantum mechanical model of atomic structure, developed by Schrödinger, Heisenberg, and Dirac in the mid-to-late 1920s, states that everything that can be known about the electrons in atoms is contained within the wavefunction solutions to the Schrödinger equation. Owing to the Heisenberg uncertainty principle, however, the precise position of the electrons cannot be known at any given time. In order to make a connection between the mathematical wavefunction solutions to the Schrödinger equation and physical reality, Max Born suggested that the square of the wavefunction is related to the probability density and that the electrons therefore exist in specific regions of space known as the *atomic orbitals (AOs)*. At the same time, the Pauli principle restricted electrons having the same spin quantum number to spatially different AOs. As a result, no single orbital can hold more than a pair of electrons, explaining Lewis's original "rule of two."

The first significant quantum mechanical model of chemical bonding, known as *valence bond theory (VBT)*, which was first developed by the German duo Walter Heitler and Fritz London and later popularized and expanded upon by Linus Pauling, was introduced in 1928. Pauling's famous book, *The Nature of the Chemical Bond*, is still one of the seminal books in the history of chemistry and at least part of the reason for which he won the 1954 Nobel Prize in chemistry. VBT is a localized bonding model that expands on the Lewis concept of electron pairs serving as the "glue" that holds the nuclei together in chemical bonds. Building on the work of their mentor Erwin Schrödinger, Heitler and London advocated the concept of electron exchange between the H atoms in the H_2 molecule. As the two H nuclei approach one another, their electron clouds (or orbitals) begin to overlap, so that the electron which originally belonged to the first H atom is now also attracted to the second H nucleus. The same thing occurs for the electron which originally belonged to the

sp^3 hybrid orbitals in CH_4. [Blatt Communications.]

Principles of Inorganic Chemistry, First Edition. Brian W. Pfennig.

second H atom as it becomes attracted to the first nucleus. At a certain point, the two electrons can rapidly exchange, jumping back and forth between the two nuclei at a rate of billions of times per second. This mutual sharing of the electrons is the hallmark of a covalent bond.

The potential energy diagram shown in Figure 10.1 shows how successive amendments to the VB model lead to a more precise record of the potential energy curve for the hydrogen molecule. Curve (*a*) represents the energy for the isolated H atoms, given by the total wavefunction shown in Equation (10.1), where the subscripts *A* and *B* refer to the two H nuclei and the numbers 1 and 2 refer to imaginary labels placed on the two electrons. It has a minimum value at 90 pm and only 25 kJ/mol.

$$\psi(H_2) = \psi_{A(1)}\psi_{B(2)} \tag{10.1}$$

Once we bring the nuclei together and allow them to exchange electrons, the potential energy decreases significantly, leading to curve (*b*) in Figure 10.1. The total wavefunction describing this second curve is given by Equation (10.2). The minimum now occurs at 87 pm and 304 kJ/mol, which is closer to the experimental values.

$$\psi(H_2) = \psi_{A(1)}\psi_{B(2)} + \psi_{A(2)}\psi_{B(1)} \tag{10.2}$$

A correction for the shielding of one electron from experiencing the full strength of either nuclear attraction by the presence of the other electron improves the curve, leading to a minimum at 74 pm and 365 kJ/mol. Finally, curve (*c*) allows for the possibility of an ionic contribution to the bonding, where both of the electrons reside either on nucleus *A* or on nucleus *B*. The probability of an ionic contribution in H_2, where there is no difference in electronegativity between the two H atoms is very small, and so a weighting factor λ is included as a coefficient in Equation (10.3) in front of each of the ionic terms (where $\lambda \ll 1$). The minimum now occurs at 74.6 pm and 397 kJ/mol.

$$\psi(H_2) = \psi_{A(1)}\psi_{B(2)} + \psi_{A(2)}\psi_{B(1)} + \lambda\psi_{A(1)}\psi_{A(2)} + \lambda\psi_{B(1)}\psi_{B(2)} \tag{10.3}$$

After application of these corrections to the VB model, the theoretical curve (*c*) closely matches that of the experimental curve for the H_2 molecule when the two electrons have opposite spin, as shown in Figure 10.1 by curve (*d*), the attractive

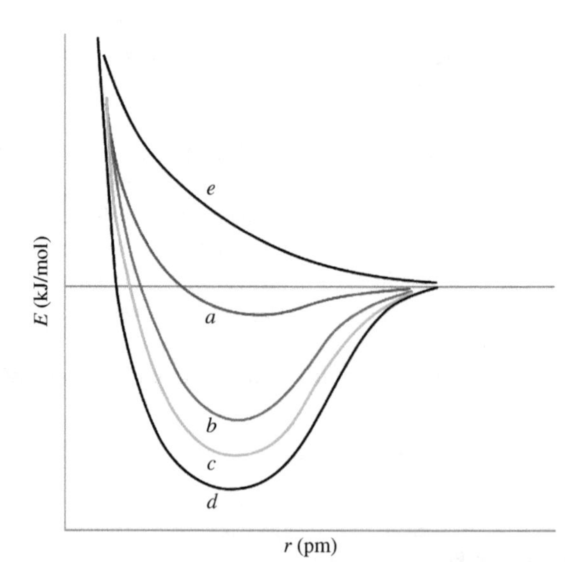

FIGURE 10.1
Artist's rendering of the theoretical potential energy curves (*a*)–(*c*) constructed using the data given by Equations (10.1)–(10.3), along with the experimentally observed attractive potential (*d*) for the H_2 molecule. Curve (*e*) is the experimental repulsive term. [Blatt Communications.]

term. This minimum occurs at 74 pm and 432 kJ/mol. Curve (e) at the top of the diagram corresponds with the experimental values for two electrons in the H_2 molecule having identical spins, or the repulsive term.

At its core, the valence bond model states that chemical bonds form as a result of the overlap of the valence AOs on two different nuclei such that both nuclei can share a pair of electrons. The H–F single bond in hydrogen fluoride, for instance, results from the overlap of the half-filled 1*s* orbital on H with the 2*p* orbital on F that contains only a single electron. By sharing the pair of electrons, the H atom assumes a more stable configuration with two electrons surrounding it (so that it has an electron configuration analogous to the noble gas He), while the F atom achieves a full octet, as shown in Figure 10.2.

For polyatomic molecules, such as CH_4, this relatively simplistic picture of the bonding needs to be modified in order to be consistent with the actual experimental results. Although chemists knew that the typical C atom has a valence of four, physicists argued that it should only have a valence of two, because there are just two unpaired electrons in its outermost 2*p* AO. This is where Pauling's contribution to the VB model proved utterly invaluable to its ultimate success. In order to merge these seemingly opposing views, Pauling had the genius to invent the concept of hybridization of AOs. Hybridization is nothing more than a mathematical trick made possible only because of the fact that the Hamiltonian is a linear, Hermitian operator. As a result, any linear combination of the eigenfunctions is also an acceptable solution to the Schrödinger equation. It is similar to the way in which the same musical note can have a different timbre depending on the musical instrument of origin. The timbre of a specific note is simply a linear combination of the fundamental frequency and its overtones, where the weighting factors in front of each overtone are variable from one instrument to the next. Thus, in a sense, there is a certain musical quality to the formation of hybridized orbitals!

Because the electronic ground state of the C atom contains only two orbitals having unpaired electrons, while methane has four equivalent C–H single bonds, it is necessary to first promote an electron from the 2*s* orbital on C to the remaining empty 2*p* orbital, as shown in Figure 10.3. For C, this process of electron promotion costs 406 kJ/mol to achieve; nonetheless, the resulting excited state is required by the necessity for C to form four equivalent C–H bonds.

But this only solves one part of the problem. If we were to just take the one 2*s* and three 2*p* AOs on the C atom and form overlaps between these and the four 1*s* AOs on the H atoms, the resulting molecule would have at least three of its C–H bonds at right angles to each other because the three 2*p* orbitals on the C atom are orthogonal to one another. And yet all of the experimental evidence says that methane is tetrahedral, having identical bond angles of 109.5°. Furthermore, because of the differing amount of overlap between the 2*s*(C)–1*s*(H) and 2*p*(C)–1*s*(H) orbitals, three of the C–H bonds in methane would be expected to have a different bond length from the fourth. In reality, all four bond lengths in CH_4 are identical at 114 pm. In order to address this second problem, Pauling suggested that linear combinations of the s and *p* orbitals on the C atom should be constructed

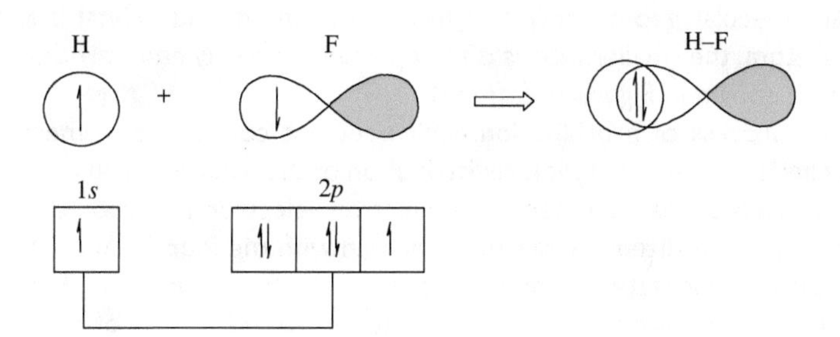

FIGURE 10.2
According to VBT, the H–F bond forms as a result of the overlap of the 1*s* orbital on H with the 2*p* orbital on F so that the two nuclei can share a pair of electrons. In this and all subsequent figures, the hollow orbital lobes indicate positive sign of the wavefunction while the shaded lobes indicate negative sign of the wavefunction.

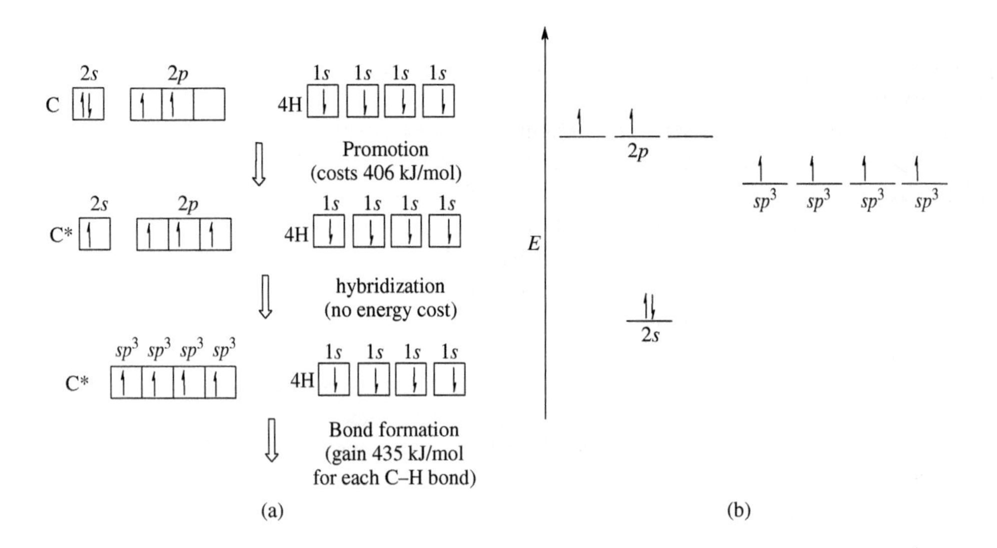

FIGURE 10.3
(a,b) Valence bond diagram for the formation of the hybrid orbitals in methane.

in order to make a set of four hybridized C atoms that all have identical shapes and energies. In the case of CH_4, the appropriate linear combinations to take in order to achieve a tetrahedral configuration are those given by Equations (10.4)–(10.7).

$$\psi_1 = \frac{1}{2}[s + p_x + p_y + p_z] \tag{10.4}$$

$$\psi_2 = \frac{1}{2}[s + p_x - p_y - p_z] \tag{10.5}$$

$$\psi_3 = \frac{1}{2}[s - p_x + p_y - p_z] \tag{10.6}$$

$$\psi_4 = \frac{1}{2}[s - p_x - p_y + p_z] \tag{10.7}$$

In taking linear combinations of the AOs, conservation of orbitals must always apply. In other words, if one 2*s* orbital and three 2*p* orbitals are used in construction of the linear combinations, a total of four hybridized orbitals must necessarily result. Furthermore, the total energy of the four hybridized orbitals must be identical to the sum of the energies of the unhybridized AOs, as shown in Figure 10.3(b). The electron contour diagram for one of the sp^3 hybrids is shown in Figure 10.4. Notice that the nodal surface is slightly offset from the nucleus (although many other textbooks have a tendency to sketch the hybrid orbitals inaccurately).

If all four hybridized orbitals were plotted with the same origin, as shown in Figure 10.5, their major lobes would each be oriented at exactly 109.5° from each other, in perfect agreement with the experimental bond angles in CH_4 and a tetrahedral molecular geometry. When more than one hybrid orbital is shown on the central atom, the smaller lobe is often omitted for clarity and only the larger lobe is drawn, as shown in Figure 10.5 (right).

The process of hybridization itself does not consume any energy, because it simply involves a mathematical redistribution of the total electron density. However, the increased overlap and decreased electron–electron repulsion that result when the four sp^3-hybridized orbitals on C overlap with the four 1*s* AOs on the H atoms more than compensate for the initial cost of electron promotion. Each C–H single bond in methane gains 435 kJ/mol of energy. The centrality of hybrid orbitals to

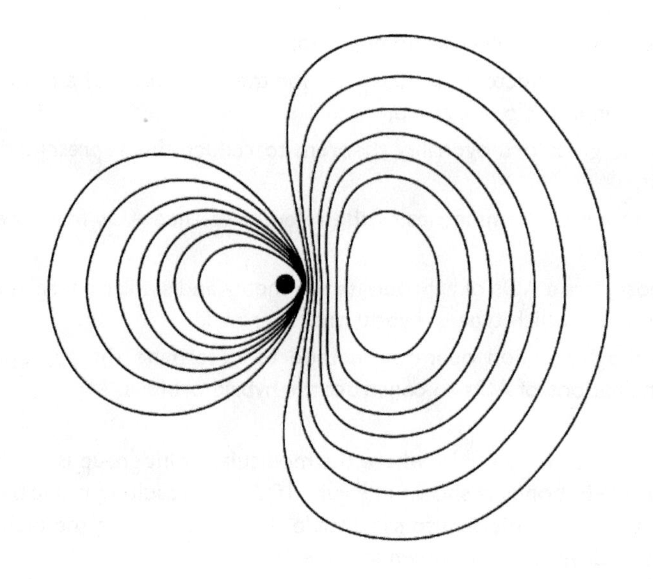

FIGURE 10.4
Contour diagram for an sp^3 hybrid orbital, where the dashed line represents a nodal surface.

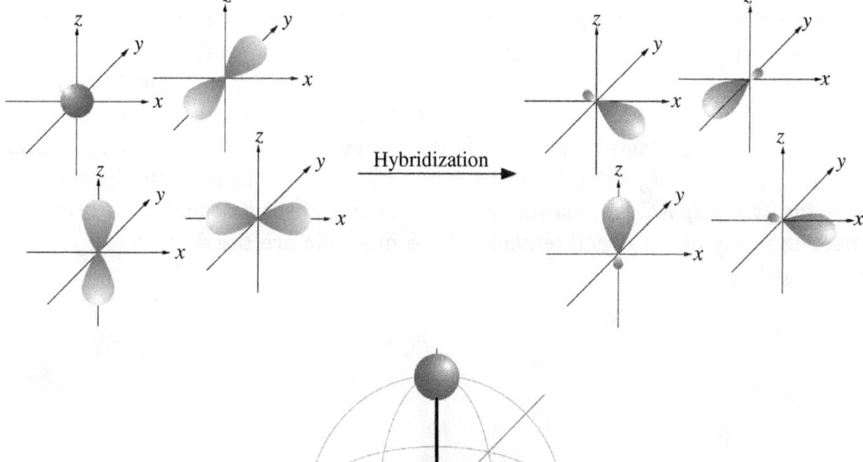

FIGURE 10.5
Formation of the four sp^3 hybrid orbitals on the C atom in the methane molecule when plotted in three dimensions, demonstrating how hybridization leads to a tetrahedral geometry.
[E. Generalic, http://www.periodni.com, used with permission.]

the valence bond model is so important that Linus Pauling once quipped that "if quantum theory had been developed by the chemist rather than the spectroscopist, it is possible that the tetrahedral orbitals ... would play the fundamental role in atomic theory, in place of *s* and *p* orbitals."

To this point, we have not been specific about why we chose to make sp^3-hybridized orbitals in methane and how we decided on which linear combinations of the 2*s* and 2*p* AOs to use in order to construct the hybrid orbitals. In order to address these unanswered questions, we make use of the group theoretical methods developed in Chapter 8. Our general strategy here is to

1. use VSEPR (valence-shell electron-pair) theory to predict the molecular geometry;

2. determine the molecular point group;
3. use the bond vectors as a basis set for the generation of a reducible representation in the point group;
4. use the great orthogonality theorem to reduce this representation into its IRRs;
5. determine the symmetries of the relevant valence AOs from the character table;
6. choose those AOs of appropriate symmetry and similar energies in order to determine which type of hybrid orbital to make;
7. use the projection operator method to determine the appropriate linear combinations of AOs to construct the hybrid orbitals.

Taking CH_4 as an example, where the molecular point group is T_d, the basis set for the four C–H bonds is shown in Figure 10.6. The resulting reducible representation and its decomposition into irreducible components using the tabular method introduced in Chapter 8 are shown in Table 10.1.

The characters for the Γ_{CH} representation are determined by the number of basis vectors that remain unmoved by the symmetry operation, as any transformation of one vector into another will lie off the diagonal in the 4×4 matrix representing the symmetry operation and will contribute nothing to the character. Following reduction, the C–H single bonds in CH_4 transform as the $A_1 + T_2$ IRRs. The A_1 IRR corresponds with either an *s*-orbital (as *s*-orbitals are always totally symmetric) or the d_{z^2} orbital, while the T_2 IRR corresponds with the three *p*-orbitals or the d_{xy}, d_{xz}, and d_{yz} orbitals. Given that the 3*d*-orbitals lie at much higher energy than the 2*s* and 2*p* AOs for carbon, the lowest energy hybrid orbitals that form while meeting the symmetry requirements of the molecule are the A_1 (*s*) and T_2 (p_x, p_y, p_z), or sp^3 hybrids.

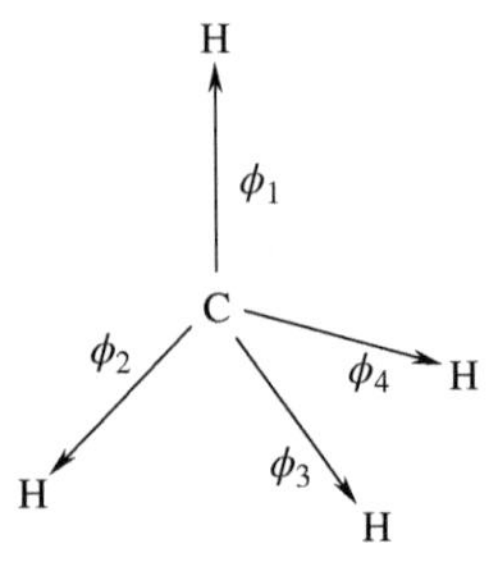

FIGURE 10.6
Basis set for the four C–H bonds in the methane molecule.

TABLE 10.1 Reduction of the C–H bond vector basis representation in methane into its irreducible components.

T_d	E	$8C_3$	$3C_2$	$6S_4$	$6\sigma_d$	$h=24$	$n_i =$
Γ_{CH}	4	1	0	0	2	Σ	Σ/h
A_1	4	8	0	0	12	24	1
A_2	4	8	0	0	−12	0	0
E	8	−8	0	0	0	0	0
T_1	12	0	0	0	−12	0	0
T_2	12	0	0	0	12	24	1

In order to determine the mathematical form of the four sp^3 hybrid orbitals, we can apply the projection operator method. Application of the projection operator for the A_1 IRR on the ϕ_1 basis function yields $\frac{1}{4}[\phi_1 + \phi_2 + \phi_3 + \phi_4]$, which is identical to the result obtained in Equation (10.4) after the resulting wavefunction has been normalized and the appropriate s and p AOs have been substituted for the basis functions $\phi_1 - \phi_4$. Likewise, application of the projection operator for the T_2 IRR yields (after normalization) the three wavefunctions given by Equations (10.5)–(10.7). Thus, application of group theoretical methods to the methane molecule can determine not only which type of hybridization will occur but also the mathematical forms of the resulting hybrid orbitals.

Example 10-1. Use group theory to determine the hybrid orbitals for the B atom in BH_3. List the mathematical expressions for the linear combinations of AOs that comprise these hybrids. Finally, sketch the valence bond diagram for the BH_3 molecule.

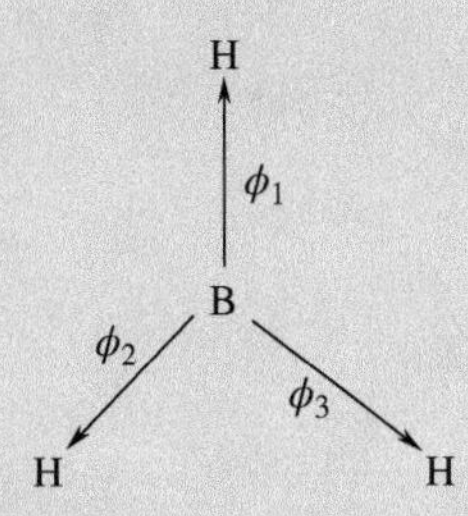

Solution. Using the B–H bond vectors as basis sets for the generation of a Γ_{BH} representation in the D_{3h} point group, we obtain the results shown in the table.

D_{3h}	E	$2C_3$	$3C_2$	σ_h	$2S_3$	$3\sigma_v$	$h = 12$	$n_i =$
Γ_{BH}	3	0	1	3	0	1	Σ	Σ/h
A_1'	3	0	3	3	0	3	12	1
A_2'	3	0	−3	3	0	−3	0	0
E'	6	0	0	6	0	0	12	1
A_1''	3	0	3	−3	0	−3	0	0
A_2''	3	0	−3	−3	0	3	0	0
E''	6	0	0	−6	0	0	0	0

This representation reduces to $A_1' + E'$, which best corresponds with the $2s$, $2p_x$ and $2p_y$ orbitals. Thus, the hybridization of the B atom in BH_3 is sp^2, with the molecule lying in the xy-plane. The empty $2p_z$ orbital is not used in the hybridization. Application of the projection operator yields the same results as those obtained in Chapter 9 for the symmetry coordinates of the B–H stretching vibrations in the isomorphous BCL_3 molecule. These hybrid orbitals take the following mathematical forms (after normalization and substitution of AOs for the basis functions):

$$\psi_1(A_1') = 1/\sqrt{3}(s + p_x + p_y)$$

$$\psi_2(E') = 1/\sqrt{6}(s - p_x - p_y)$$

$$\psi_3(E') = 1/\sqrt{2}(-p_x + p_y)$$

The valence bond picture for the bonding in BH_3 is given below.

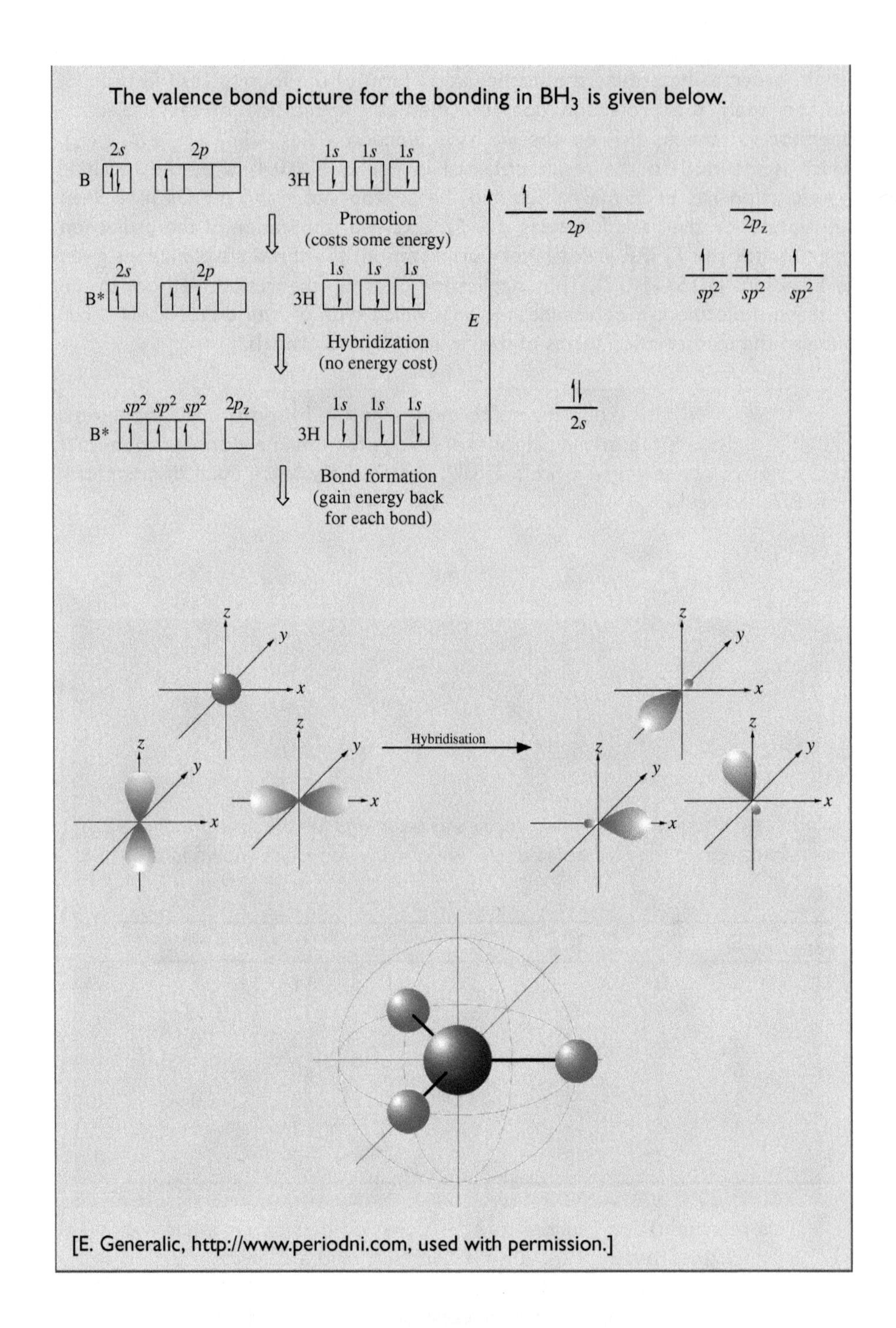

[E. Generalic, http://www.periodni.com, used with permission.]

The type of hybrid orbital that forms depends on the electron geometry of the molecule. A summary of this correlation for the most common electron geometries is provided in Table 10.2.

The process of forming the *sp* hybrid orbitals for a linear electron geometry is shown in Figure 10.7. Taking the z-axis as the molecular axis, the p_z orbital is the one that is used to make the hybrids, while the p_x and p_y orbitals are unaffected.

Next, let us consider an example where the electron and molecular geometries are different, such as that which occurs for H_2O. The hybridization of the O atom in the water molecule is sp^3 because the electron geometry around the central atom

TABLE 10.2 The type of hybrid orbital that forms as a function of the electron geometry.

Electron Geometry	Type of Hybrid Orbital
Linear	sp (sp_z)
Trigonal planar	sp^2 (sp_xp_y)
Tetrahedral	sp^3
Trigonal bipyramidal	dsp^3($d_{z^2}sp^3$)
Octahedral	d^2sp^3 ($d_{z^2}d_{x^2-y^2}sp^3$)

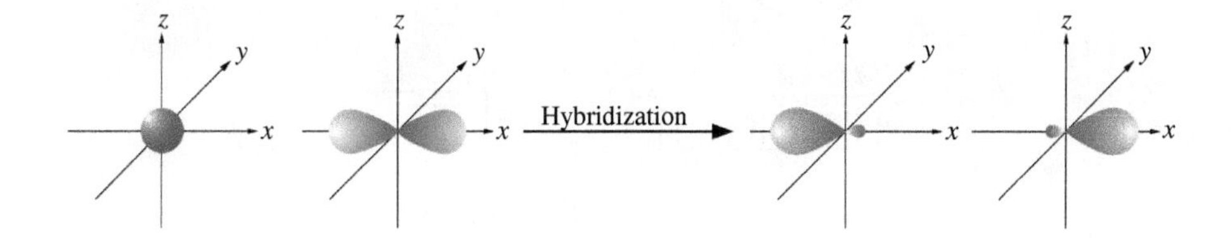

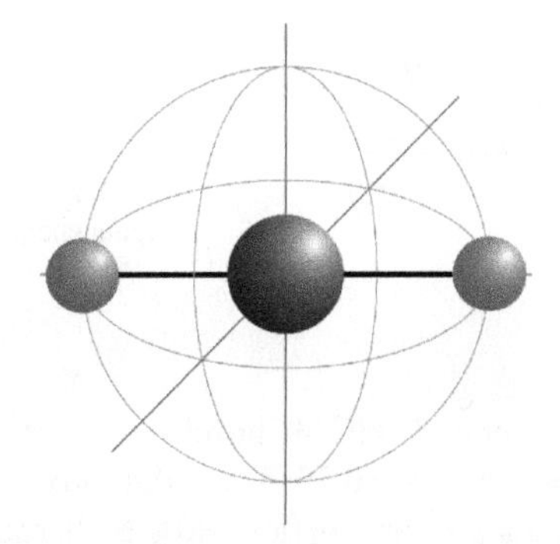

FIGURE 10.7
Formation of the *sp* hybrid orbitals in the linear electron geometry. [E. Generalic, http://www.periodni.com, used with permission.]

is tetrahedral. To a first approximation, the valence bond model for H_2O is shown in Figure 10.8.

Because the central atom already has two unpaired electrons that it can use to form O–H single bonds with the 1*s* AOs on each H atom, the promotion step is unnecessary in this molecule. However, the hybridization step is still required in order to achieve approximately the right molecular geometry (bond angles close to 109.5°). In all of the examples we have encountered thus far, each of the hybrid orbitals was equivalent with one another. That is to say, all four sp^3 hybrid orbitals in CH_4 contain exactly 25% *s*-character and 75% *p*-character. Each sp^3 hybrid had the same energy and shape, differing only in spatial orientation. There is no reason that this need always be the case.

In molecules such as H_2O, the composition of the four hybrid orbitals (two bonding and two lone pairs) on the O atom will be slightly different. In such cases, the actual percent composition of each hybrid orbital will be dictated by the molecule's geometry. For sp^n hybridized orbitals ($n = 1-3$), the relationship between the bond angle (θ) and the decimal percent *s*-character (S) is given by Equation (10.8). For hybrids that also contain some *d*-orbital character, the relationship between the bond angle and directionality of the hybrid orbitals is somewhat more complex and is not discussed in this textbook.

$$\cos\theta = \frac{S}{S-1} \tag{10.8}$$

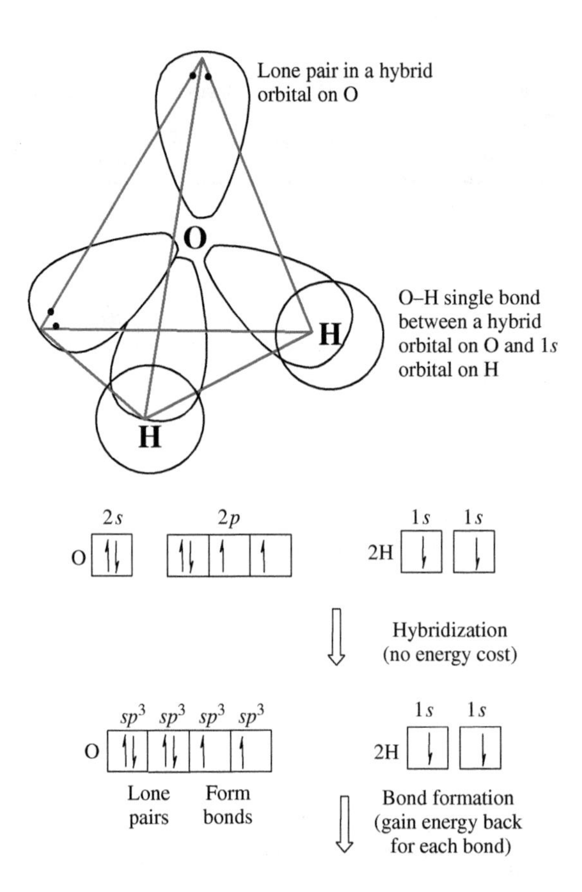

FIGURE 10.8
Valence bond model applied to the water molecule. The sp^3 hybridization of the O atom is dictated by the tetrahedral electron geometry of the molecule. Two of the sp^3 hybrid orbitals are used to form bonds with the H atoms, while the other two sp^3 hybrids contain the two lone pairs.

In H_2O, the experimental HOH bond angle is 104.5°. Application of Equation (10.8) yields a value of $S = 0.20$. Thus, the two O–H bonding orbitals can be considered to have 20% s-character and 80% p-character. Because the overall hybridization in water was sp^3, this result implies that the two lone pairs in H_2O have 30% s-character (because $2 * 20\% + 2 * 30\% = 100\%$) and 70% p-character (because $2 * 80\%$ and $2 * 70\% = 300\%$) so that the overall hybridization still averages to sp^3.

All sp^n-hybridized orbitals are more directional than s or p AOs are by themselves. Increased directionality implies increased orbital overlap. This is at least part of the reason that it is worth paying the price of electron promotion in the first place—the increased overlap will gain more energy back when the hybrid orbitals on the central atom overlap with the ligand orbitals to form bonds. The amount of overlap in sp^n hybrid orbitals increases with the percent s-character. The more s-character in the hybrid orbital, the lower the overall energy of that orbital will be and the larger the size of the "big" lobe of the sp^n hybrid. Thus, sp orbitals have greater overlap with the ligand orbitals than do sp^2 or sp^3 orbitals.

The experimental bond angle in NH_3 is 107.2°. Application of Equation (10.8) yields 23% s-character and 77% p-character in each of the three N–H single bonds. Because $3(23.3\%) = 69\%$, the lone pair in ammonia has approximately 31% s-character. As a result, the lone pair in NH_3 will have slightly greater s-character than will the lone pair in H_2O. The greater basicity of NH_3 relative to H_2O can at least in part be attributed to the fact that the lone pair in ammonia will have a greater amount of overlap with the 1s-orbital on a proton than will a lone pair on water.

This analysis can also be applied to explain an empirical relationship known as *Bent's rule*, which states that more electronegative ligands prefer to occupy hybrid orbitals having a lower percentage of s-character. On the other hand, the more electropositive ligands will prefer hybrid orbitals having a greater percentage of s-character.

Consider the PCl_2F_3 molecule, for instance. VSEPR theory would predict a trigonal bipyramidal molecular geometry with the two Cl ligands in the equatorial sites, where there is more room for them. Bent's rule yields exactly the same result. The hybridization for the D_{3h} point group is $d_{z^2}sp^3$. Upon closer inspection, the trigonal bipyramidal geometry can be considered the sum of a trigonal planar geometry (lying in the equatorial plane) superimposed on a linear geometry (lying on the C_3 axis). Using group theory, we can take the three equatorial bonds as one basis set in order to determine their symmetries and the two axial ligands as a separate basis. The trigonal planar geometry then results in sp_xp_y hybridization, while the linear component transforms as $d_{z^2}p_z$. Thus, the site with the higher percentage s-character and the greater degree of orbital overlap will be the equatorial position. Because each Cl is more electropositive than F, the Cl ligands will prefer to occupy the equatorial positions in the trigonal bipyramid. The argument here is that the P–Cl bonds are more covalent than the highly polarized P–F bonds, and their greater covalency requires more orbital overlap between the two atoms. The increased amount of overlap with the higher percentage s-character hybrid orbitals therefore favors this particular configuration of PCl_2F_3. The greater percent s-character in the equatorial positions can also be used to rationalize why the equatorial P–F bond length (154 pm) is shorter (stronger) than that of the axial P–F bond length (159 pm).

Example 10-2. Rationalize the following experimental bond angles using Bent's rule.

Species	X–S–X angle
H_2S	92.1°
SF_2	98.0°
SCl_2	102.7°

Solution. All of the molecules have tetrahedral electron geometries and bent molecular geometries. The percent s-character calculated for the S–X bonding hybrid orbitals using Equation (10.8) are listed here. The percent s-character in each lone pair was calculated by using the formula: $100 = 2S + 2\Sigma$, where S is the percent s-character in the two S–X bonds and Σ is the percent s-character in the two lone pairs.

Species	X–S–X angle	S (%)	Σ (%)
H_2S	92.1°	3.5	46.5
SF_2	98.0°	12.2	37.8
SCl_2	102.7°	18.0	32.0

As you can see from these examples, it is possible to obtain nonintegral values for the hybridization. Thus, for example, the O–H single bonds in H_2O are formally sp^4, while those for the N–H single bonds in NH_3 are $sp^{3.35}$. This concept is known as *variable hybridization*. It was necessary to tweak the character of the hybrid orbitals in order to explain the observed geometry and the deviations from the ideal bond angles. A similar tweak is required when there are two different types of ligands on the central atom, as is the case for CH_3Cl. Whenever there are two different ligands, the hybridization parameters (λ_i and λ_j) for ligands *i* and *j*, respectively can be calculated from Equations (10.9)–(10.11), where *S* is the decimal percent *s*-character in the sp^n hybrid and *P* is the decimal percent *p*-character in the hybridized orbital.

$$S = \frac{1}{1 + \lambda^2} \quad (10.9)$$

$$P = \frac{\lambda^2}{1 + \lambda^2} \quad (10.10)$$

$$\cos(\theta_{ij}) = \frac{-1}{\sqrt{\lambda_i \lambda_j}} \quad (10.11)$$

Equation (10.11) is also known as *Coulson's theorem*. Referring to CH_3Cl as our example, let us assign ligand *i* to be the H atoms and ligand *j* to be the Cl. The experimental HCH and HCCl bond angles in this molecule are 110.5° and 108.0°, respectively. Using Equation (10.11) with $\theta = 110.5°$, we obtain the result that $\lambda_i = 1.7$. Application of Equations (10.9) and (10.10) yields $S = 0.26$ and $P = 0.74$. Thus, the C–H bonds are said to be $sp^{2.9}$-hybridized. Using Equation (10.11) with $\theta = 108.0°$ and $\lambda_i = 1.7$, we obtain $\lambda_j = 1.9$. Application of Equations (10.9) and (10.10) yields $S = 0.22$ and $P = 0.78$. The C–Cl bond is therefore $sp^{3.5}$-hybridized. These findings are in support of Bent's rule, which states that the more electronegative Cl ligand will prefer a hybrid orbital on the C having less *s*-character. Another way of rationalizing Bent's rule in this case is to introduce some ionic character into the polar C–Cl bond. In the extreme case, we can add a resonance structure having the form $CH_3^+Cl^-$. In this canonical structure, there is less electron density on C pointed in the direction of the Cl ligand. Therefore, there will be an increase in the amount of electron density on the C atom pointed at each of the three H atoms. Because the *s*-orbital used in hybridization is lower in energy than the *p*-orbitals, there will be a greater degree of *s*-character in the C–H bonds, as the results of our calculations indicate.

We now turn our attention to the application of VBT to double- and triple-bonded molecules. Taking molecular oxygen as an example, the Lewis structure for O_2 contains an O=O double bond with two lone pairs of electrons on each of the O atoms. Thus, the electron geometry around each O is trigonal planar and the expected hybridization is sp^2 based on the correlation listed in Table 10.2.

As shown in Figure 10.9 (top), each O will have three sp^2 hybrid orbitals lying in the *xy*-plane, where two of these orbitals will contain a pair of electrons (representing the two lone pairs) and the other will have one unpaired electron with which to form a single bond to the other O atom. When this occurs, a *sigma* bond is formed because the overlap of the two sp^2 hybrid orbitals is a head-on type overlap. In addition, each O atom will have one remaining unhybridized p_z orbital lying above and below the *xy*-plane, which also contains a single unpaired electron. The p_z orbitals on each O atom can overlap in a sideways type overlap to form a *pi* bond between the two O atoms. Thus, an O=O double bond consists of one *sigma* and one *pi* bond, as shown in Figure 10.9 (bottom).

In a similar manner, VBT can be used to explain why ethylene is a planar molecule. As in O_2, the two C atoms are sp^2 hybridized and lie in the *xy*-plane—only in $H_2C{=}CH_2$, each sp^2 hybrid contains a single electron. One of these hybrids is used to form the C–C *sigma* bond, while the other two hybrids are used to form the C–H *sigma* bonds. The C–C *pi* bond forms as a result of the sideways overlap between the two p_z orbitals. In order for the two p_z orbitals to overlap with each other, there can be no free rotation around the C=C double bond. Therefore, both CH_2 halves of the ethylene molecule must lie in the same plane, as shown in Figure 10.10.

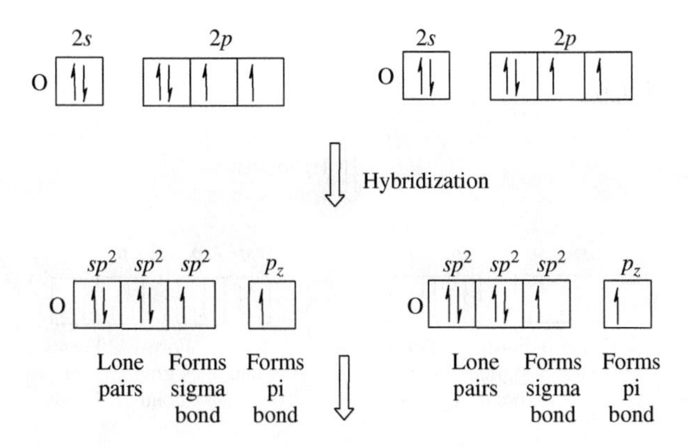

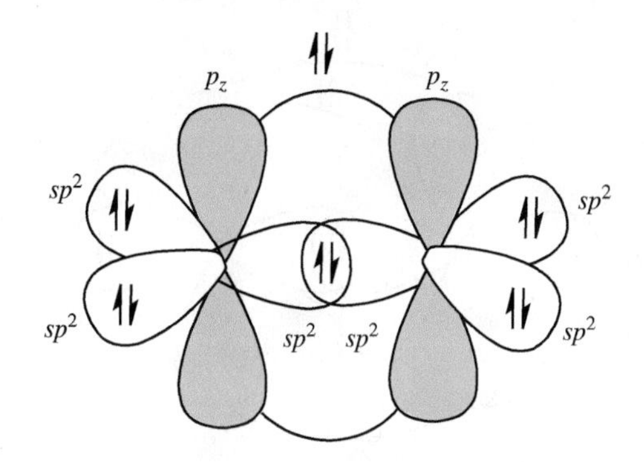

FIGURE 10.9
Valence bond model for molecular oxygen, showing both the *sigma* and *pi* bonds.

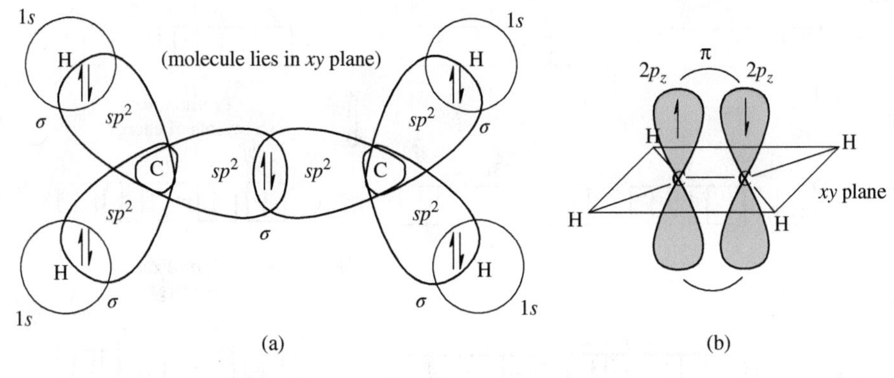

FIGURE 10.10
Valence bond model for ethylene, showing both the *sigma* bonding network (a) and the *pi* bond (b). The presence of *pi* bonding precludes the molecule from rotating around the C–C inter-nuclear axis and therefore restricts the molecule to the *xy*-plane.

The valence bond model for the nitrogen molecule is shown in Figure 10.11. The best Lewis structure for N_2 has a N≡N triple bond with a lone pair on each N atom, or a linear electron geometry. Therefore, the expected hybridization from Table 10.2 is *sp*. One of the *sp* hybrids will be used to form a *sigma* bond between the two N atoms, while the other *sp* hybrid will contain a lone pair. In addition, each N atom will have two unhybridized *p*-orbitals. These two orbitals can overlap in a sideways manner to form two *pi* bonds between the N atoms.

For molecules where the valence of the central atom violates the octet rule (the so-called hypervalent molecules), the valence bond model is forced to resort to the inclusion of *d*-orbitals in the hybridization. Thus, for instance, the hybridization of phosphorous in PF_5 is formally dsp^3, as shown in Figure 10.12, even though it is doubtful that the high-lying 3*d* orbitals actually participate in the bonding to any significant extent.

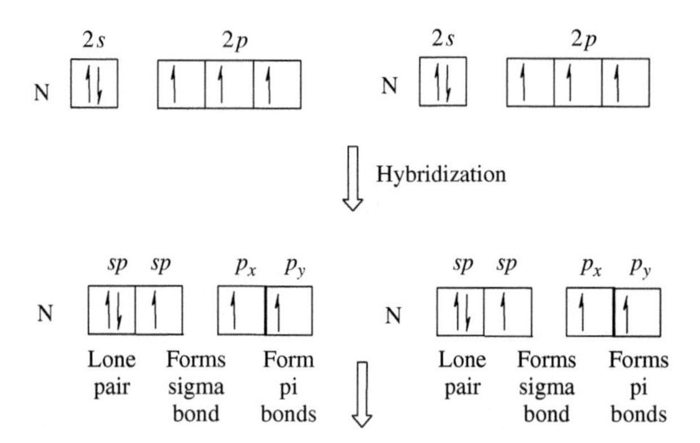

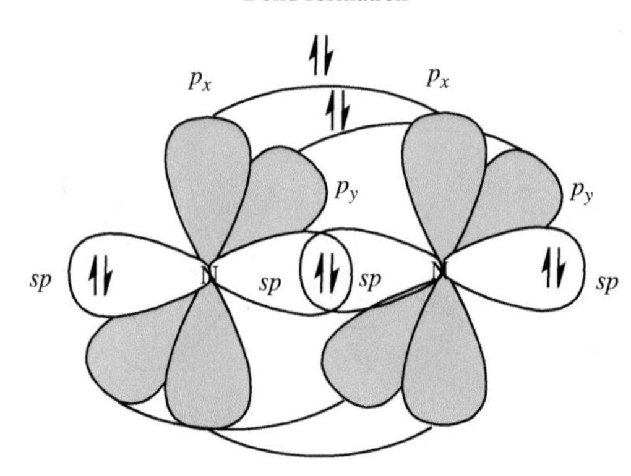

FIGURE 10.11
Valence bond model for N_2, showing one *sigma* and two *pi* bonds.

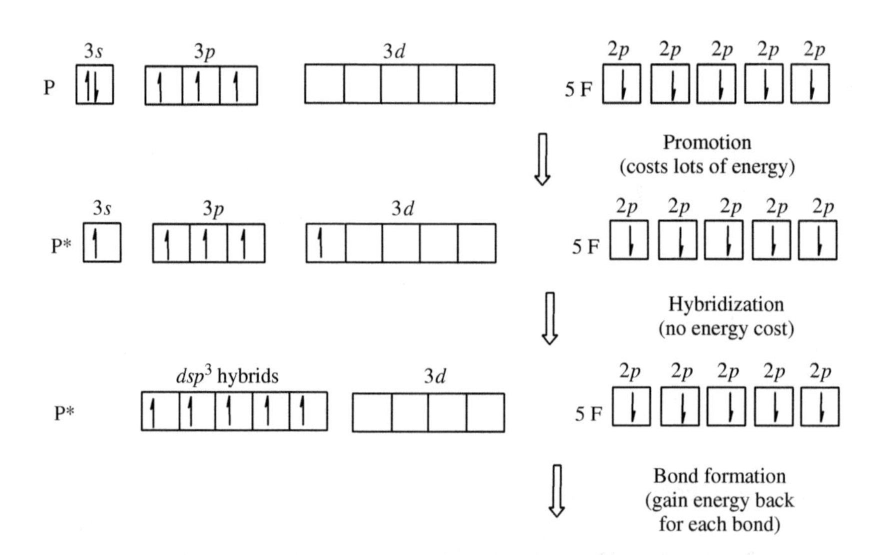

FIGURE 10.12
Valence bond model for PF_5, showing the supposed participation of the 3*d* orbitals.

It should be noted that hypervalent molecules are only stable when strongly polar covalent bonds are present; that is, when the electronegativities of the ligands are large. In fact, computations have shown that the electronegativity differences in PF_5 are large enough that the F atoms withdraw enough electron density from the P central atom that it effectively obeys Lewis' "rule of eight." One can therefore argue that ionic canonical structures such as those shown in Figure 10.13 contribute to the resonance hybrid for PF_5. This conclusion is supported by the fact that hypervalent molecules lacking electronegative ligands, such as PH_5, simply do not exist—even though there would certainly be plenty of room for five H atoms to coordinate to

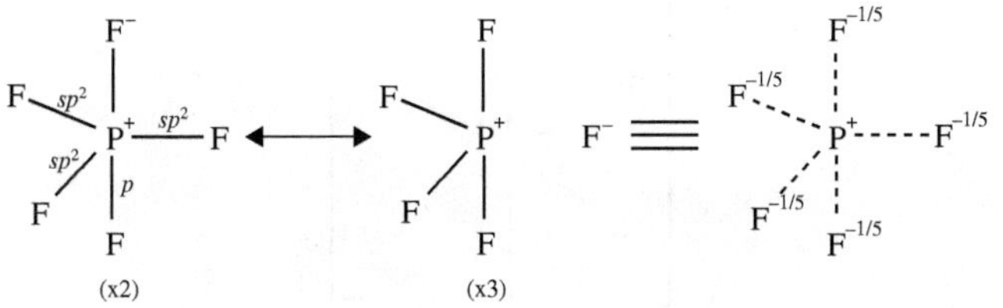

FIGURE 10.13
Inclusion of five ionic canonical structures in the resonance hybrid for PF_5 so as to allow the total valence on the P central atom to obey the "octet rule." [Reproduced from http://en.wikipedia.org/wiki/Hypervalent_molecule (accessed December 28, 2013).]

the central P atom. Because the term *hypervalent* is actually misleading, other authors have used the term *hypercoordination* because its meaning is less ambiguous.

The main problem that VBT has with trying to explain the bonding in molecules having hypercoordination is the fact that it is built on the premise that chemical bonds are localized in two-center, two-electron (2c–2e) bonds (or in bonding pairs) in order to remain consistent with simple Lewis structures. In reality, as we shall see in our forthcoming discussion of molecular orbital theory (MOT), the bonding in most polyatomic molecules is better treated from a delocalized perspective. Foreshadowing this approach is the concept of a three-centered, four-electron (3c–4e) bond in molecules such as PF_5. If we consider the three F atoms in the equatorial trigonal plane of PF_5 to form bonds with three equivalently shaped sp^2 hybridized orbitals, this would leave one unhybridized $3p_z$ orbital on the P atom to be used in the bonding along the z-axis. Because there are two F atoms lying on the z-axis in the two axial sites, the $3p_z$ orbital must bond in some manner with both of the axial F atoms. One potential solution to this apparent dilemma is a 3c–4e bond, as shown in Figure 10.14. The two axial F atoms use their valence $2p_z$ orbitals to overlap with the unhybridized $3p_z$ orbital on the P atom. Depending on the sign of the wavefunction, three possible linear combinations result between the three centers involved in the 3c–4e bond: (i) a bonding orbital occurs when the overlap of each *p*-orbital occurs with constructive interference, as shown at the bottom of Figure 10.14; (ii) an antibonding orbital occurs when the overlap of each *p*-orbital occurs with destructive interference, as shown at the top of the figure; and (iii) a nonbonding orbital results in the linear combination of just the two *p*-orbitals on the F ligands, as shown in the middle of the figure. The energies of the corresponding orbitals are shown in the diagram.

There are a total of 40 valence electrons in the PF_5 molecule. Thirty of these electrons can be considered to be the 3 lone pairs of electrons that exist on each

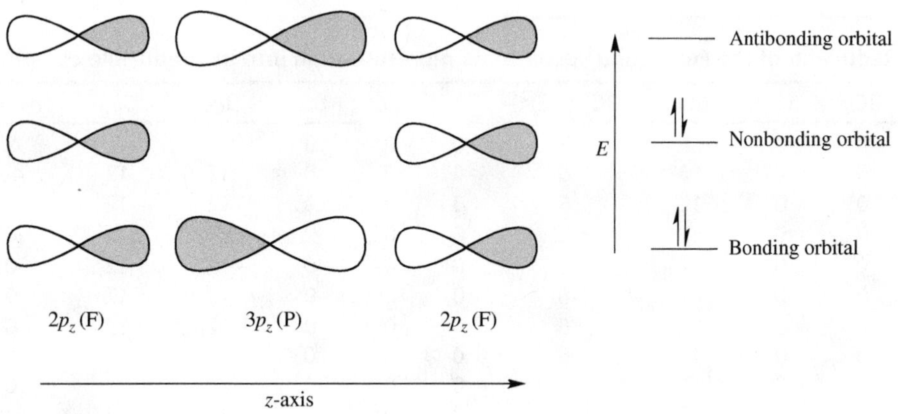

FIGURE 10.14
Three-centered, four-electron (3c–4e) bonding along the *z*-axis in the PF_5 molecule.

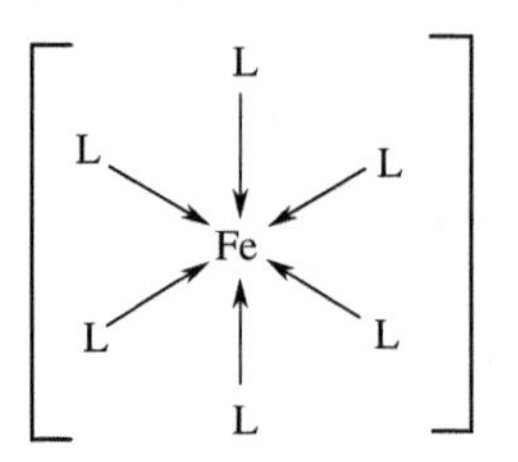

FIGURE 10.15
Basis set for the determination of the hybridization of Fe^{3+} in FeL_6.

of the 5 F ligands in the Lewis structure of the molecule. This leaves 10 remaining valence electrons to be accounted for in the VB model. Six of these electrons are involved in the formation of the three $P{-}F_{eq}$ single bonds, which leaves four valence electrons to be placed in the 3c–4e axial energy levels. Two of these valence electrons occupy the lower lying bonding orbital in Figure 10.14, while the other two occupy a nonbonding orbital. Thus, the net amount of bonding on the z-axis is a single bond, so that each $P{-}F_{ax}$ bond would have a bond order of only one-half. This model of the bonding in PF_5 is consistent with the fact that the $P{-}F_{ax}$ bond lengths (157.7 pm) are longer than the $P{-}F_{eq}$ bond lengths (153.4 pm). At the same time, it provides a bonding argument where it is unnecessary for the valence on the central P atom to violate the "octet rule."

The valence bond model works rather well for main group compounds, but it is much less useful for transition metal compounds. This is because the nonbonding electrons in the transition metals typically lie in the outer shell of the core, rather than in the valence shell, as they do in the main group elements. Unlike the main group compounds, therefore, the geometries of transition metal compounds are typically dictated by a combination of the ligand close-packing model and crystal field stabilization energies (Chapter 16). Thus, the valence bond model for transition metal compounds only involves the metal–ligand bonds as the basis set. Consider, for example, the octahedral FeL_6 coordination compound shown in Figure 10.15.

Application of group theory to the Fe–L single-bond basis set shown in Figure 10.15 yields the reducible representation shown in Table 10.3. One difference between the covalent bonds between main group elements and the bonding in transition metal compounds is that the ligand donates both pairs of electrons to an empty orbital on the metal in the latter case. This type of bond is known as a *coordinate covalent bond.* The ligand is said to coordinate to the metal. Hence, the direction of the bond vector arrow in Figure 10.15 is intentional. The

TABLE 10.3 Reduction of the Fe–F bond vector basis representation into its irreducible components.

O_h	E	$8C_3$	$6C_2$	$6C_4$	$3C_2\ (=C_4^2)$	i	$6S_4$	$8S_6$	$3\sigma_h$	$6\sigma_d$	$h=48$	$n_i =$
Γ_{FeL}	6	0	0	2	2	0	0	0	4	2	Σ	Σ/h
A_{1g}	6	0	0	12	6	0	0	0	12	12	48	1
A_{2g}	6	0	0	12	6	0	0	0	12	−12	0	0
E_g	12	0	0	0	12	0	0	0	24	0	48	1
T_{1g}	18	0	0	12	−6	0	0	0	−12	−12	0	0
T_{2g}	18	0	0	12	−6	0	0	0	−12	12	0	0
A_{1u}	6	0	0	12	6	0	0	0	−12	−12	0	0
A_{2u}	6	0	0	−12	6	0	0	0	−12	12	0	0
E_u	12	0	0	0	12	0	0	0	−24	0	0	0
T_{1u}	18	0	0	12	−6	0	0	0	12	12	48	1
T_{2u}	18	0	0	−12	−6	0	0	0	12	−12	0	0

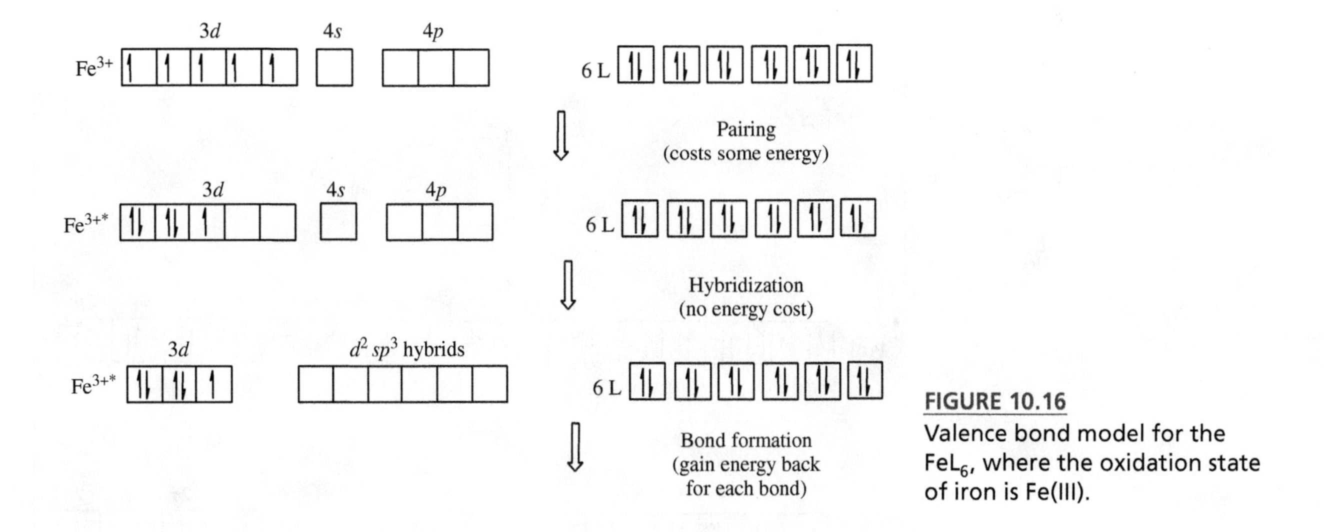

FIGURE 10.16
Valence bond model for the FeL_6, where the oxidation state of iron is Fe(III).

Γ_{FeF} representation in the O_h point group reduces to $A_{1g} + E_g + T_{1u}$, which best corresponds to d^2sp^3 hybridization.

The resulting valence bond model for FeL_6 is shown in Figure 10.16. As a result of the coordinate covalent bonding, each ligand (L) will donate two electrons to an empty d^2sp^3 hybrid orbital on Fe^{3+} to form the six-coordinate covalent Fe–L bonds. Furthermore, it makes perfect sense that the d_{z^2} and the $d_{x^2-y^2}$ orbitals are the ones involved in the formation of the d^2sp^3 hybrids, as these two orbitals have their lobes along the coordinate axes, where they will be pointing directly at the ligands, which approach the metal ion from the positive and negative poles of the three Cartesian axes.

Example 10-3. Use group theory to determine the hybrid orbitals on the Pt^{2+} ion in the square planar tetrachloroplatinate(II) ion. Sketch the valence bond diagram for the $PtCl_4^-$ ion.

Solution. Using the Pt–Cl bond vectors as basis sets for the generation of a Γ_{PtCl} representation in the D_{4h} point group, we obtain the results shown in the table. In this case, because a transition metal acts as the central atom, there is a high likelihood that one or more of the *d*-orbitals will be involved in the hybridization.

D_{4h}	E	$2C_4$	C_2	$2C_2'$	$2C_2''$	i	$2S_4$	σ_h	$2\sigma_v$	$2\sigma_d$	$h = 16$	$n_i =$
Γ_{PtCl}	4	0	0	2	0	0	0	4	2	0	Σ	Σ/h
A_{1g}	4	0	0	4	0	0	0	4	4	0	16	1
A_{2g}	4	0	0	−4	0	0	0	4	−4	0	0	0
B_{1g}	4	0	0	4	0	0	0	4	4	0	16	1
B_{2g}	4	0	0	−4	0	0	0	4	−4	0	0	0
E_g	8	0	0	0	0	0	0	−8	0	0	0	0
A_{1u}	4	0	0	4	0	0	0	−4	−4	0	0	0
A_{2u}	4	0	0	−4	0	0	0	−4	4	0	0	0
B_{1u}	4	0	0	4	0	0	0	−4	−4	0	0	0
B_{2u}	4	0	0	−4	0	0	0	−4	4	0	0	0
E_u	8	0	0	0	0	0	0	8	0	0	16	1

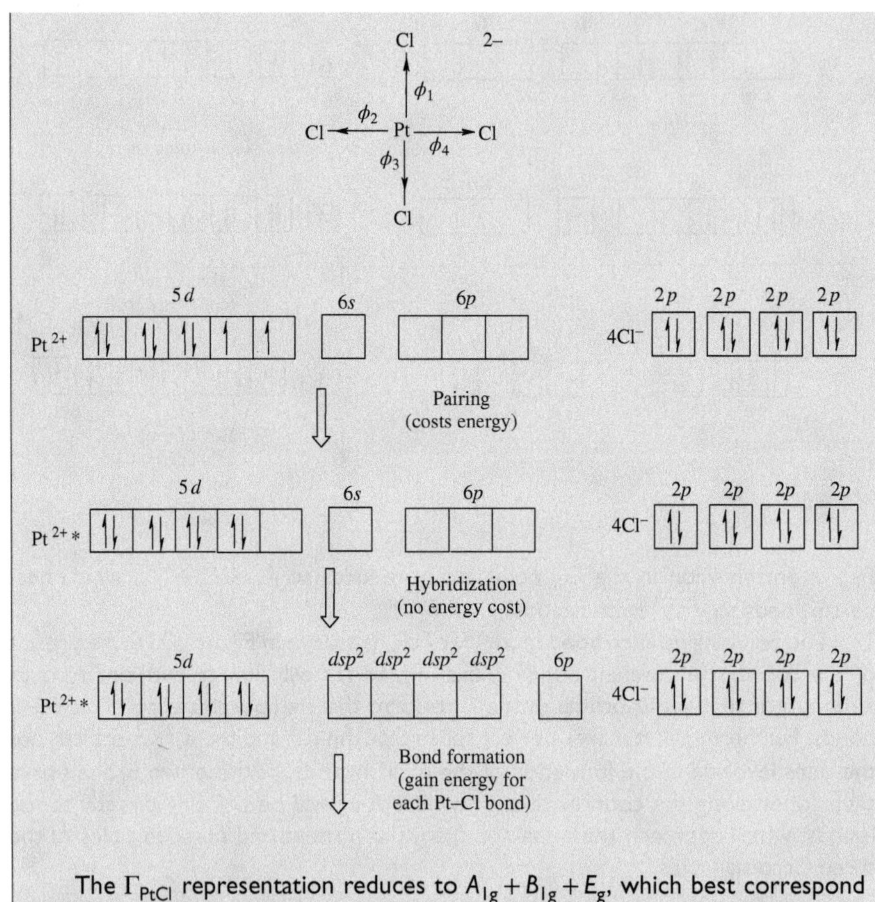

The Γ_{PtCl} representation reduces to $A_{1g} + B_{1g} + E_g$, which best correspond with the 6s, $5d_{x^2-y^2}$, and $6p_x$, $6p_y$ orbitals, respectively. Thus, the hybridization of the Pt atom in $PtCl_4^{2-}$ is dsp^2. Furthermore, the theory tells us which *p*-orbitals are involved. Those Pt AOs whose highest electron densities (lobes) lie in the same plane as the molecule will naturally be the most conducive for overlap with the valence AOs on the Cl^- ligands. These are the p_x and p_y orbitals. From the set of *d*-orbitals, the $d_{x^2-y^2}$ orbital not only lies in the *xy*-plane, but it also has its lobes pointed directly at the four Cl^- ligands lying on the *x*- and *y*-coordinate axes.

While the valence bond model is intuitively simple and does an excellent job rationalizing molecular structure, it is not without its shortcomings. One of those, as we have already seen, is the restriction that the bonding electrons be localized between two and only two nuclei in the molecule (2c–2e bonding). In the ozone molecule, for instance, VBT predicts that there will be one O–O single bond and one O=O double bond. However, the two experimental bond lengths in O_3 are identical and somewhat intermediate between that of a typical O–O single bond and a typical O=O double bond. Thus, it was necessary to "invent" the concept of resonance in order to explain the molecular structure of ozone in valence bond terms. Second, the theory utterly fails at the prediction of the magnetic properties of transition metal complexes, such as why $NiCl_4^{2-}$ is tetrahedral and paramagnetic (contains unpaired electrons and is attracted to an external magnetic field), while the isoelectronic ion $PtCl_4^{2-}$ is square planar and diamagnetic (has all of its electrons paired and is weakly repelled by an external magnetic field). Nor can it explain why $NiCl_4^{2-}$ is paramagnetic, while $Ni(CN)_4^{2-}$ is diamagnetic. In fact, for certain transition metal complexes, the species is paramagnetic at one temperature and diamagnetic at another temperature.

Perhaps the most egregious failure of VBT, however, is its inability to explain the photoelectron spectroscopy (PES) of even the simplest of molecules. This technique employs a high-energy photon (typically having a wavelength of 58.4 nm from a He discharge lamp) to bombard a gaseous sample of the molecule. Using the photoelectric effect, given by Equation (3.8), as long as the energy of the photon ($h\nu$) is greater than the ionization energy (IE) of a given electron, the electron will be ejected from the molecule having a kinetic energy T. The detector counts the relative number of ejected electrons and their kinetic energies. If we assume that the energies of the electrons in the orbitals have the same magnitude as that of their IEs (an assumption known as *Koopman's theorem*), the electronic distribution of the molecule's valence electrons can be determined. Typically, the radiation is sufficiently energetic to eject only the outermost electrons from the molecule. However, a different technique, known as *ESCA* (*electron spectroscopy for chemical analysis*), which uses X-ray photons, can be used to ionize some of the inner electrons. Often, the two techniques are used in conjunction with one another.

The photoelectron spectrum for CH_4 is shown in Figure 10.17. Without providing any explanation about their band shapes, it is clear from the figure that there are essentially two different energy levels: one centered around 13.5 eV and another at 23.0 eV. Furthermore, integration of the areas under these peaks indicates that six electrons correspond to the former band and two electrons to the latter band. The two peaks therefore represent the eight valence electrons in the CH_4 molecule. When this technique is coupled with ESCA, a third peak having a much higher IE (~291 eV) and a relative area of two electrons is observed. The third peak corresponds to the nonbonding 1s electrons on the C atom. The valence bond model predicts that all eight valence electrons will have the same IE (because all four C–H bonds are equivalent and have the same amount of orbital overlap), but the PES data suggest two different sets of energy levels in the valence shell of the molecule: one which is triply degenerate and another which is nondegenerate. On hindsight, this result should not be surprising, given that the highest dimensional IRR in the T_d point group has only a threefold degeneracy.

It is important to recognize at this point that VBT is only a model. As such, it is useful only to the extent that it helps us visualize and better understand the nature of the chemical bonding in molecules. Given the simplicity of Lewis structures, the intricacies of the VSEPR model, and the relatively straightforward concept of hybridization, it is remarkable how far the valence bond model can be pushed in the first place. In the next section, MOT will be introduced as an alternative and complementary model to VBT. In many ways, MOT will prove itself to be the more powerful of the two theories in that it can explain most of the shortcomings of the

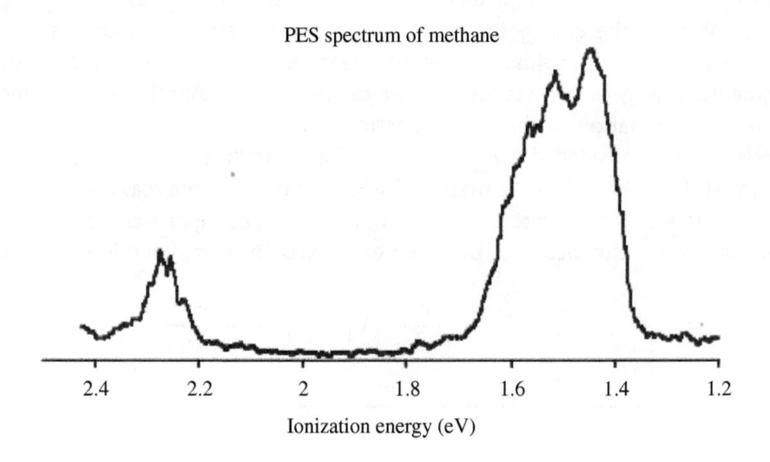

FIGURE 10.17
Photoelectron spectrum of methane. [Reproduced by permission, Mike Prushan.]

valence bond model. However, the construction of molecular orbitals (MOs) is considerably more challenging from a mathematical standpoint and much less intuitive than the formation of hybrid orbitals. Thus, while MOT is arguably more comprehensive than VBT, it is significantly more difficult to understand. Given the advent of inexpensive molecular modeling software in recent years, however, the energies of MOs can be calculated much more efficiently than in the past and the shapes of the resulting MOs can be visualized on the computer screen in three dimensions.

10.2 MOLECULAR ORBITAL THEORY: DIATOMICS

Shortly after the development of VBT, an alternative model, known as *MOT*, was introduced by the American physicist Robert Mulliken (and others) around 1932. MOT is a delocalized bonding model, where the nuclei in the molecule are held in fixed positions at their equilibrium geometries and the Schrödinger equation is solved for the entire molecule to yield a set of *MOs*. In practice, it is possible to solve the Schrödinger equation exactly only for one-electron species, such as H_2^+. Whenever more than one electron is involved, the wave equation can only yield approximate solutions because of the *electron correlation problem* that results from Heisenberg's principle of indeterminacy. If one cannot know precisely the position and momentum of an electron, it is impossible to calculate the force field that this one electron exerts on every other electron in the molecule. As a result of this mathematical limitation, an approximation method must be used to calculate the energies of the MOs.

The approximation method of choice is the *LCAO-MO method*, which is an acronym that stands for linear combinations of atomic orbitals to make molecular orbitals. Unlike VBT, however, the linear combinations used to construct MOs derive from the AOs on two or more *different* nuclei, whereas the linear combinations used to make hybrid orbitals in VBT involved only the valence orbitals on the central atom. Using the *variation theorem*, the energy of a particle can be determined from the integral in Equation (10.12), which is also written in its Dirac (or bra-ket) notation.

$$E = \frac{\int_{-\infty}^{\infty} \psi * \hat{H}\psi d\tau}{\int_{-\infty}^{\infty} \psi * \psi d\tau} = \frac{\langle \psi|\hat{H}|\psi\rangle}{\langle \psi|\psi\rangle} \tag{10.12}$$

If there is more than one electron in the molecule, the Schrödinger equation cannot be solved to give an exact solution. However, it might be possible to guess a function ψ_g that is a reasonable approximation to the solution. Application of Equation (10.12) then yields an energy E_g, which will always be greater than the actual energy of the system if the Schrödinger equation could be solved exactly. If we strategically build some adjustable parameters into our original function ψ_g, we can then minimize the energy by varying those parameters in an iterative manner. When the energy of the adjustable function reaches a minimum or approaches the experimental energy of the system, then we can be satisfied that the original function is a close approximation to the actual function.

Now let us consider the H_2 molecule. The Hamiltonian for H_2 is given by Equation (10.13), where M is the mass of the H nucleus, m_e is the mass of an electron, the subscripts A and B refer to the two H nuclei, the subscripts 1 and 2 refer to the two electrons, and the distances between each particle are defined in Figure 10.18.

$$\hat{H} = -\frac{h^2}{8\pi^2 M}(\nabla_A^2 - \nabla_B^2) - \frac{h^2}{8\pi^2 m_e}(\nabla_1^2 - \nabla_2^2) - \frac{e^2}{4\pi\varepsilon_0 r_{1A}} - \frac{e^2}{4\pi\varepsilon_0 r_{1B}} - \frac{e^2}{4\pi\varepsilon_0 r_{2A}} - \frac{e^2}{4\pi\varepsilon_0 r_{2B}} + \frac{e^2}{4\pi\varepsilon_0 r_{12}} + \frac{e^2}{4\pi\varepsilon_0 R} \tag{10.13}$$

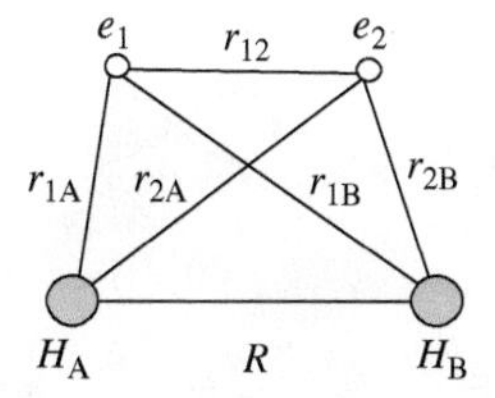

FIGURE 10.18
Definitions of the interparticle distances in the H_2 molecule.

Because the nuclei are several orders of magnitude more massive than the electrons and will therefore move more slowly than the electrons, the *Born–Oppenheimer approximation* states that the nuclear and electronic motions of the molecule can be treated separately. Essentially, we can assume that the nuclei in the molecule are stationary and solve the equation solely for the electronic motion. This causes the nuclear or first term in Equation (10.13) to drop out.

Because there is more than one electron in the H_2 molecule, we cannot solve Equations (10.12) and (10.13) exactly. As a first approximation, let us assume that the wavefunction for H_2 is some linear combination of the 1s AOs of the two isolated H atoms, as given by Equation (10.14), where ψ is the wavefunction of our molecular orbital and ϕ is the wavefunction for a 1s AO. The constants c_1 and c_2 are simply weighting factors. These are the adjustable parameters of our trial wavefunction. Because the energies of the two 1s AOs for each H atom are identical, $c_1 = c_2$ in this example.

$$\psi = c_1\phi_A \pm c_2\phi_B \tag{10.14}$$

At this point, let us return to the one-electron example H_2^+, whose exact solution *can* be determined; and then we will extrapolate the results for H_2^+ to the hydrogen molecule. The functions ϕ_A and ϕ_B are the solutions to the Schrödinger equation for H_A and H_B, respectively. Because both of them are also normalized, the denominator in Equation (10.12) will be equal to one and $\hat{H}_A\phi_A$ will equal $E_A\phi_A$. Substitution of Equation (10.14) using the positive linear combination into Equation (10.12) yields Equation (10.15), where N is the normalization coefficient for the MO wavefunction:

$$\begin{aligned} E &= \int_{-\infty}^{\infty} (N\psi) * \hat{H}(N\psi)d\tau = N^2\int_{-\infty}^{\infty} (c_1\phi_A + c_2\phi_B) * \hat{H}(c_1\phi_A + c_2\phi_B)d\tau \\ &= N^2[c_1^2\int_{-\infty}^{\infty} \phi_A * \hat{H}\phi_A d\tau + c_1c_2\int_{-\infty}^{\infty} \phi_A * \hat{H}\phi_B\, d\tau + c_2c_1\int_{-\infty}^{\infty} \phi_B * \hat{H}\phi_A\, d\tau \\ &+ c_2^2\int_{-\infty}^{\infty} \phi_B * \hat{H}\phi_B\, d\tau] = N^2[c_1^2\hat{H}_{AA} + 2c_1c_2\hat{H}_{AB} + c_2^2\hat{H}_{BB}] \end{aligned} \tag{10.15}$$

The integrals H_{AA} and H_{BB} are the *Coulomb integrals*, sometimes given by the symbol J, while the integrals $H_{AB} = H_{BA}$ (because they are Hermitian operators) are the *exchange integrals*, sometimes given by the symbol K. In order to determine the normalization coefficient, N, we must use Equations (10.16) and (10.17).

$$\begin{aligned} 1 &= \int_{-\infty}^{\infty} (N\psi) * (N\psi)\, d\tau = N^2\int_{-\infty}^{\infty} (c_1\phi_A * + c_2\phi_B *)(c_1\phi_A + c_2\phi_B)d\tau \\ &= N^2\int_{-\infty}^{\infty} c_1^2\phi_A^2 d\tau + 2c_1c_2\phi_A\phi_B d\tau + c_2^2\phi_B^2 d\tau) = N^2(c_1^2 + 2c_1c_2\int_{-\infty}^{\infty} \phi_A\phi_B d\tau + c_2^2) \\ &= N^2(c_1^2 + 2c_1c_2S_{AB} + c_2^2) \end{aligned} \tag{10.16}$$

$$N = (c_1^2 + 2c_1c_2S_{AB} + c_2^2)^{-1} \tag{10.17}$$

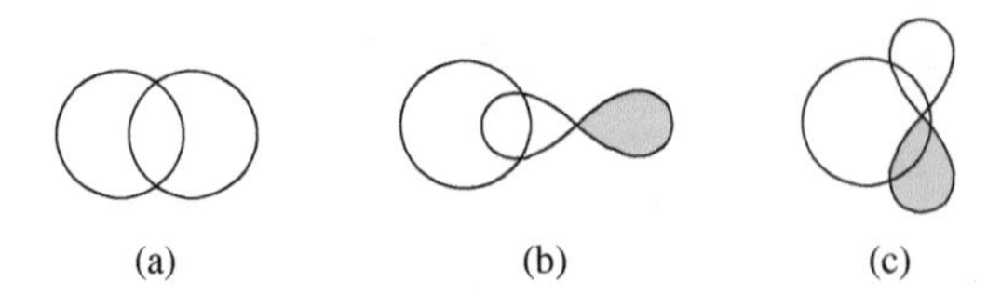

FIGURE 10.19
(a) Nonzero overlap between s and s-orbitals, (b) nonzero overlap between s and p-orbitals, and (c) net zero overlap between s and p-orbitals.

The resulting equation for the energy therefore reduces to Equation (10.18):

$$E(c_1^2 + 2c_1c_2S_{AB} + c_2^2) = c_1^2\hat{H}_{AA} + 2c_1c_2\hat{H}_{AB} + c_2^2\hat{H}_{BB} \tag{10.18}$$

The symbol S_{AB} is the *overlap integral* and it relates to the amount of orbital overlap between the two AOs that are part of the linear combination. As shown in Figure 10.19(a), the overlap integral is nonzero whenever two s-orbitals are combined. If an s-orbital and a p-orbital are combined, the overlap integral will only be nonzero if there is a net constructive interference between the wavefunctions. This will only occur when the positive lobe of a p-orbital overlaps with the s-orbital in a head-on manner, as shown in Figure 10.19(b). If the p-orbital overlaps in a sideways manner with the s-orbital, as shown in Figure 10.19(c), the amount of +/+overlap and +/− overlap will cancel and S_{AB} will equal zero. Thus, only certain combinations of AOs will result in the formation of molecular orbitals, and we can use principles of group theory to determine the appropriate combinations. Furthermore, the magnitude of the overlap integral depends on the degree of orbital overlap. Thus, a linear combination of two 2s AOs will result in a larger S_{AB} than the overlap of two 1s AOs at the same internuclear distance. Likewise, the overlap of any two 1s AOs will increase as the orbitals are brought closer together.

In order to minimize the energy of the trial wavefunction, we must differentiate Equation (10.18) with respect to both c_1 and c_2, yielding Equations (10.19) and (10.20):

$$\frac{\partial E}{\partial c_1}(c_1^2 + 2c_1c_2S_{AB} + c_2^2) + E(2c_1 + 2c_2S_{AB}) = 2c_1\hat{H}_{AA} + 2c_2\hat{H}_{AB} \tag{10.19}$$

$$\frac{\partial E}{\partial c_2}(c_1^2 + 2c_1c_2S_{AB} + c_2^2) + E(2c_2 + 2c_1S_{AB}) = 2c_2\hat{H}_{BB} + 2c_1\hat{H}_{AB} \tag{10.20}$$

Setting both of these differentials equal to zero, we obtain Equations (10.21) and (10.22), which lead to the *secular determinant* given by Equation (10.23):

$$c_1(H_{AA} - E) + c_2(H_{AB} - ES) = 0 \tag{10.21}$$

$$c_1(H_{AB} - ES) + c_2(H_{BB} - E) = 0 \tag{10.22}$$

$$\begin{vmatrix} H_{AA} - E & H_{AB} - ES \\ H_{AB} - ES & H_{BB} - E \end{vmatrix} = 0 \tag{10.23}$$

For H_2, where the energies of the two 1s AO wavefunctions are identical, $H_{AA} = H_{BB}$. The solutions to Equation (10.23) are given by Equations (10.24) and (10.25).

$$E_+ = \frac{H_{AA} - H_{AB}}{1 - S} \tag{10.24}$$

$$E_- = \frac{H_{AA} + H_{AB}}{1 + S} \tag{10.25}$$

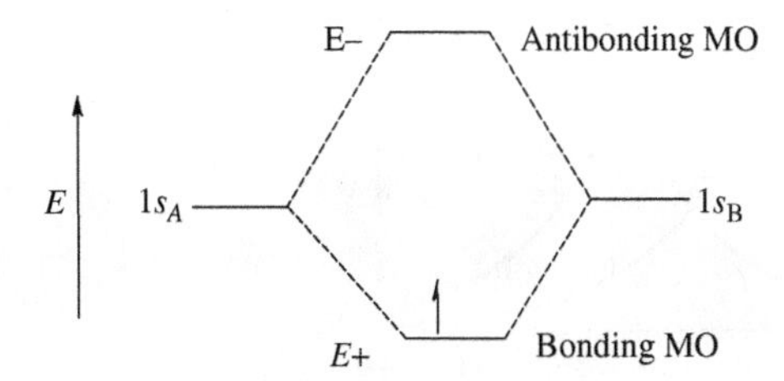

FIGURE 10.20 Molecular orbital diagram for H_2^+.

The energy given by the positive linear combination in Equation (10.24) is stabilized with respect to the energies of the 1s AOs on the isolated H atoms. This MO is defined as a *bonding MO*. On the other hand, the energy of the negative linear combination in Equation (10.25) is destabilized with respect to the isolated atoms and it is therefore an *antibonding MO*. Notice that because of the different denominators, the antibonding MO is destabilized somewhat more than the bonding MO is stabilized, as shown in Figure 10.20. The lone electron in H_2^+ fills in the lowest energy MO, the bonding MO, in accordance with the Aufbau principle.

After substitution of E into Equations (10.21) and (10.22), we find that for the bonding MO (E_+), $c_2 = c_1$ and for the antibonding MO, $c_2 = -c_1$. Thus, the mathematical equations for the bonding and antibonding wavefunctions are given by Equations (10.26) and (10.27), respectively. A graph of the two wavefunctions and their squares is shown in Figure 10.21. The bonding MO has an increase in electron density between the nuclei, as compared with two isolated H atoms, while the antibonding MO has a node between the two nuclei.

$$\psi_+ = \frac{1}{\sqrt{2+2S}}(\phi_A + \phi_B) \tag{10.26}$$

$$\psi_+ = \frac{1}{\sqrt{2-2S}}(\phi_A - \phi_B) \tag{10.27}$$

When the Schrödinger equation was solved for the H atom, we obtained an infinite number of solutions: 1s, 2s, 2p, 3s, 3p, 3d, and so on. Although it would appear from this example that there are only two MOs for the H_2^+ ion, there are in fact an infinite number of MOs. When applying the LCAO-MO method, we could have included higher order terms in our trial wavefunction, as shown in Equation (10.28).

$$\psi = c_1 1s_A + c_2 1s_B + c_3 2s_A + c_4 2s_B + c_5 2p_{zA} + c_6 2p_{zB} + \cdots \tag{10.28}$$

For simplicity, only the positive linear combination was shown in Equation (10.28). However, the total number of MOs we could form would simply depend on the number of AOs included in our trial wavefunction. The lone electron will always fill in the lowest energy MO. Because the energies of the 2s and 2p AOs lie at much higher energies than the 1s orbitals do, their weighting coefficients in Equation (10.28) for the lowest energy MO will be small compared to the magnitude of c_1 and c_2. Thus, our strategy for forming MOs is limited to taking linear combinations of only those AOs that meet the following set of criteria:

1. Conservation of orbitals must apply (no. of AOs = no. of MOs).
2. The energies of the combining AOs must be similar.
3. The combining orbitals must have a nonzero overlap ($S_{AB} \neq 0$).
4. The symmetries of the combining orbitals must be the same.

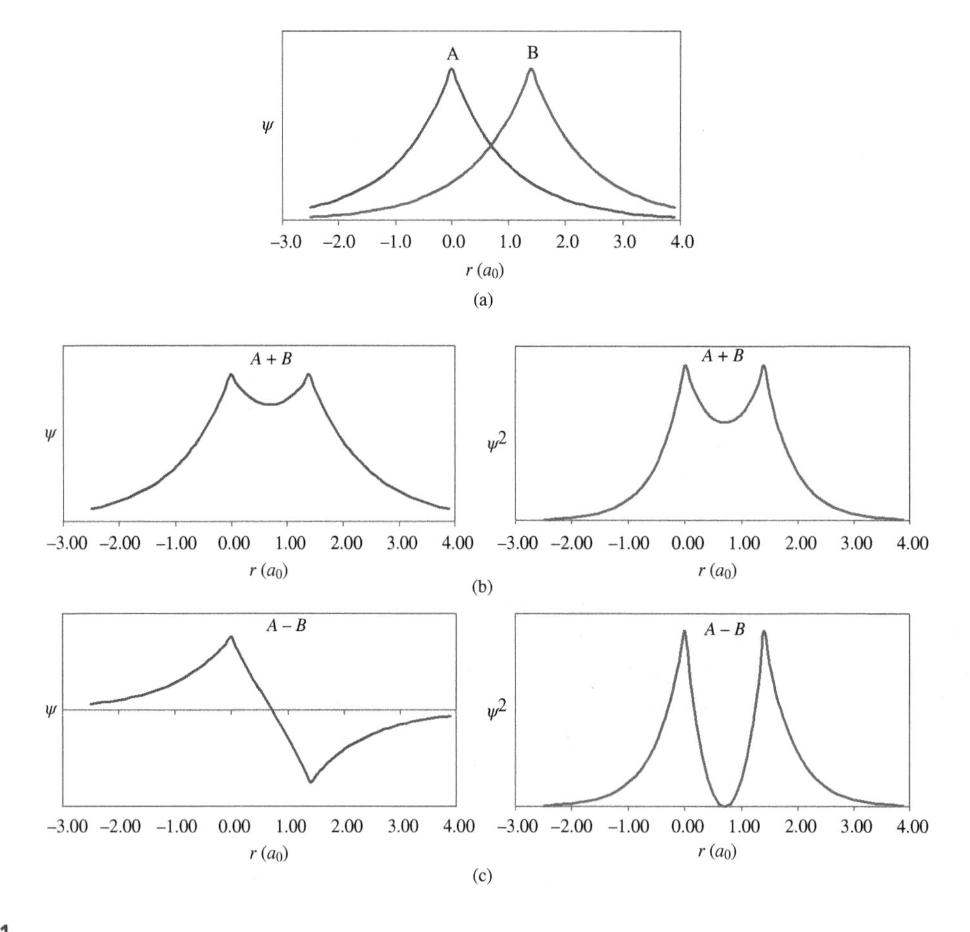

FIGURE 10.21
(a) Plot of ψ as a function of internuclear separation in units of a_0 for two isolated H atoms A and B which are 74 pm apart from each other, (b) plots of ψ and ψ^2 for the bonding (+) molecular orbital of H_2^+, (c) and plots of ψ and ψ^2 for the antibonding (−)molecular orbital of H_2^+.

TABLE 10.4 Reducible representation for the 1*s* atomic orbitals basis set in the H_2^+ ion.

$D_{\infty h}$	E	$2C_\infty$	$\infty\sigma_v$	i	$2S_\infty$	∞C_2
Γ_{1s}	2	2	2	0	0	0

In order to determine the symmetries of the combining orbitals, we can use the valence AOs on the separate atoms to form a basis set for the IRRs of the point group. Using the 1*s* AOs as a basis set for the H_2^+ ion in the $D_{\infty h}$ point group, we obtain the reducible representation Γ_{1s} shown in Table 10.4.

Reduction of Γ_{1s} using the great orthogonality theorem yields two IRRs: Σ_g^+ and Σ_u^+. By convention, one-electron MO diagrams use lower case symbols to represent the symmetries of the MOs. In addition, the subscripts and superscripts are usually dropped and replaced with the subscript *b* for the bonding MOs and an asterisk superscript for the antibonding MOs. Application of the projection operator yields the following linear combinations after normalization:

$$\sigma_b(\Sigma_g^+) = \frac{1}{\sqrt{2}}(\phi_A + \phi_B) \quad (10.29)$$

$$\sigma * (\Sigma_u^+) = \frac{1}{\sqrt{2}}(\phi_A - \phi_B) \quad (10.30)$$

With the exception of the overlap integral in the denominator of the normalization coefficients, Equations (10.29) and (10.30) are identical to Equations (10.26) and (10.27).

Using the MO diagram for $H_2{}^+$ in Figure 10.22 as a template, the "one-electron" MO diagram for H_2 will also consist of a *sigma* bonding and *sigma* antibonding MO. The only significant difference between the two MO diagrams will be the addition of a second electron in the hydrogen molecule. According to the Aufbau principle, this electron will also occupy the bonding MO, but the Pauli exclusion principle states that two electrons occupying the same orbital must have opposite electron spins, as shown by the MO diagram for H_2 in Figure 10.22. The "one-electron" MO diagram ignores the contribution of electron–electron repulsions, which are treated more precisely using the term symbol methods which were first introduced in Chapter 3.

The *bond order* (B.O.) is defined by Equation (10.31), where B.E. is the number of electrons occupying bonding MOs and A.E. is the number of electrons occupying antibonding MOs. The bond order for H_2 is therefore one, which is identical to the single bond predicted by its Lewis structure using VBT.

$$\text{B.O.} = \frac{1}{2}(\text{B.E.} - \text{A.E.}) \quad (10.31)$$

Furthermore, the total wavefunction for the H_2 molecule using MOT is given by Equation (10.32). With the exception of the lack of weighting factors for the ionic terms, Equation (10.32) is identical to the total wavefunction derived using VBT in Equation (10.3).

$$\begin{aligned}\psi(H_2) &= \psi_{\sigma b(1)}\psi_{\sigma b(2)} = [\psi_{A(1)} + \psi_{B(1)}][\psi_{B(1)} + \psi_{B(2)}] \\ &= \psi_{A(1)}\psi_{B(2)} + \psi_{A(2)}\psi_{B(1)} + \psi_{A(1)}\psi_{A(2)} + \psi_{B(1)}\psi_{B(2)} \quad (10.32)\end{aligned}$$

As long as MOT is corrected to deemphasize the ionic terms, then the theoretical potential energy curve will nicely match the experimental potential energy given by curve (e) in Figure 10.1.

The one-electron MO diagram for the fictional molecule He_2, shown in Figure 10.23, readily shows why the dihelium molecule does not exist. The bond order for He_2 is zero. Furthermore, because the antibonding MO is destabilized

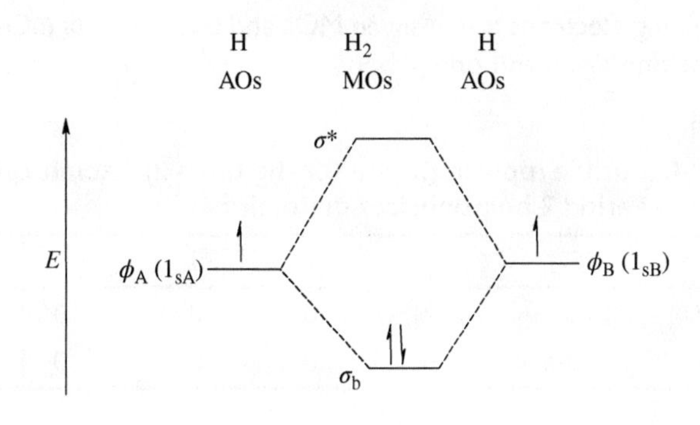

FIGURE 10.22
One-electron molecular orbital diagram for H_2.

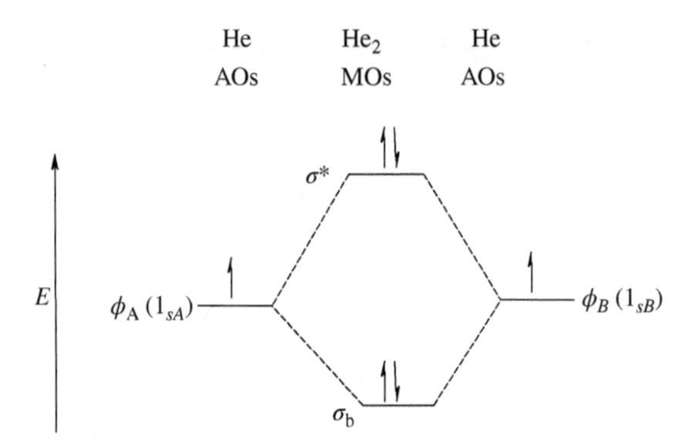

FIGURE 10.23
One-electron MO diagram showing why the He_2 molecule does not exist.

more than the bonding MO is stabilized, it would actually cost energy to make He_2 from two He atoms. While He_2 is unstable, the molecular ion He_2^+ does exist, having a bond order of ½. Nonintegral bond orders arise naturally in the MO model, whereas they only existed in resonance hybrids using VBT.

For the second period homonuclear diatomics, both the 2*s* and 2*p* AOs are used in the formation of the MOs. Using the *p*-orbitals as a basis set in the $D_{\infty h}$ point group, we obtain the IRRs shown in Table 10.5. Because the p_x and p_y orbitals transform as a pair, they are taken together as a single basis set. Recalling that the character for any C_n rotation is given by $2\cos\theta$, we obtain a value of $4\cos\theta$ for the C_∞ operation in the $\Gamma_{px,py}$ representation and a value of 0 for the C_2 operation. The rest of the characters in Table 10.5 are easy to discern considering how many of the basis AOs remain unchanged by each symmetry operation.

Application of the projection operator to determine the mathematical forms of the linear combinations, followed by normalization of the wavefunctions and then taking the square of the wavefunctions leads to the shapes of the molecular orbitals depicted in Figure 10.24. Thus, the set of six 2*p* AOs form six MOs: a σ_b and σ^* pair and two doubly degenerate π_b and π^* MOs. In general, *sigma* MOs are cylindrical and are characterized by the absence of a node along the internuclear axis, while *pi* MOs have a nodal plane along this axis. Likewise, bonding MOs are characterized by the absence of a nodal plane perpendicular to the internuclear axis and an increase in electron density between the nuclei, while the antibonding MOs have a perpendicular nodal plane between the two nuclei.

Because the magnitude of the overlap integral is larger when the orbitals combine "head-on" than when they combine "sideways," the σ_b MO will be more stabilized with respect to the isolated 2*p* AOs than will be the degenerate pair of π_b MOs. The one-electron MO diagram for O_2, which is shown in Figure 10.25, exemplifies this feature. The bond order for O_2, calculated using Equation (10.31), is two, which is consistent with that predicted using VBT. Furthermore, the Lewis structure for O_2 has a total of four lone pairs of electrons in its valence shell. These four pairs of nonbonding electrons can be identified in the MO diagram for O_2 as the pairs of σ_b and σ^* or π_b and π^* which essentially cancel each other out. In addition, two of the uncancelled bonding electrons are in *sigma* MOs and two are in *pi* MOs, so that the O_2 molecule has one *sigma* and one *pi* bond.

TABLE 10.5 Reducible representation for the three 2*p* atomic orbitals basis set in the Period 2 homonuclear diatomics.

$D_{\infty h}$	E	$2C_\infty$	$\infty\sigma_v$	i	$2S_\infty$	∞C_2	IRRs
Γ_{pz}	2	2	2	0	0	0	$\Sigma_g^+ + \Sigma_u^+$
$\Gamma_{px,py}$	4	$4\cos\theta$	0	0	0	0	$\pi_u + \pi_g$

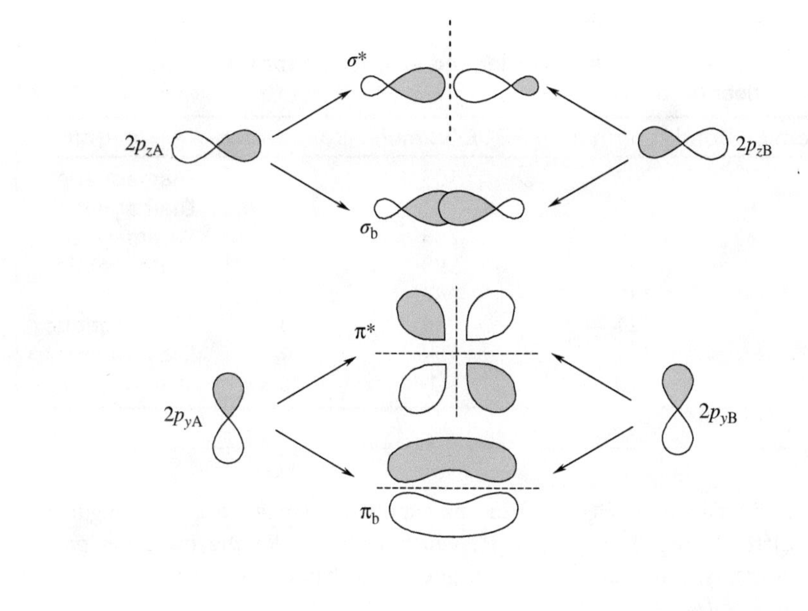

FIGURE 10.24
Sigma and *pi* molecular orbitals made by taking linear combinations of the $2p$ AOs in a homonuclear diatomic molecule. The $2p_x$ AOs will also generate a set of π_b and π^* MOs having the same shapes and energies as those derived from the $2p_y$ AOs, but lying perpendicular to the plane of the paper.

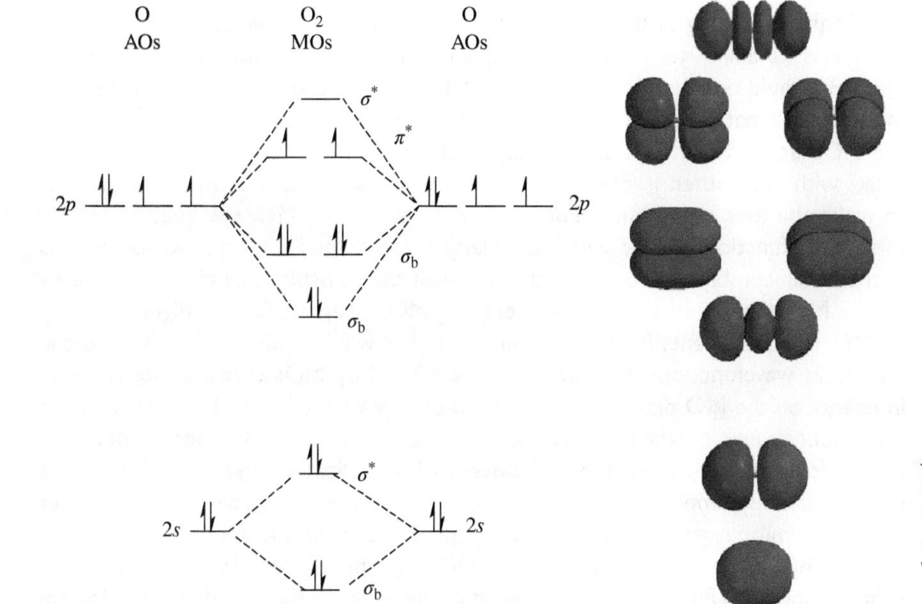

FIGURE 10.25
One-electron molecular orbital diagram for O_2. Molecular orbitals were generated using Wavefunction's Spartan Student Edition version 5.0.

At this point, one of the strengths of MOT with respect to VBT becomes evident. Because of the degeneracy of the π_b MOs, coupled with Hund's rule of maximum multiplicity, the MO diagram for O_2 correctly predicts that oxygen molecules are paramagnetic—something which VBT could not explain. The electron configuration for the O_2 molecule is written as follows: KK $(\sigma_{2s})^2(\sigma^*_{2s})^2(\sigma_{2p})^2(\pi_{2p})^4(\pi^*_{2px})^1(\pi^*_{2py})^1$ to show that the molecule is indeed paramagnetic. The inner shell of 1s core electrons is abbreviated as KK because no net bonding occurs between them. Closed shells have the following symbols using the old-school convention that $n = 1$ is K, $n = 2$ is L, $n = 3$ is M, and so on.

The experimentally determined bond lengths, bond dissociation energies, bond orders, and magnetic properties of the Period 2 homonuclear diatomics are listed in Table 10.6. Changing only the total number of valence electrons, the MO diagram shown in Figure 10.25 correctly predicts the bond order and magnetic properties of Li_2, Be_2, N_2, O_2, F_2, and Ne_2; however, it fails to predict the correct magnetism of B_2 and C_2. For instance, C_2 has a total of eight valence electrons.

TABLE 10.6 Observed bond order and magnetism for the Period 2 homonuclear diatomics.

Diatomic	Bond Length (pm)	BDE (kJ/mol)	Bond Order	Magnetism
Li_2	267	105	1	Diamagnetic
Be_2	245	9	0	Diamagnetic
B_2	159	289	1	Paramagnetic
C_2	124	599	2	Diamagnetic
N_2	110	942	3	Diamagnetic
O_2	121	494	2	Paramagnetic
F_2	141	154	1	Diamagnetic
Ne_2	310	<1	0	Diamagnetic

Using the MO diagram in Figure 10.25, the expected valence electron configuration is $KK(\sigma_{2s})^2(\sigma^*_{2s})^2(\sigma_{2pz})^2(\pi_{2px})^1(\pi_{2py})^1$, which would make the molecule paramagnetic. In reality, however, C_2 is diamagnetic and has the electron configuration $KK(\sigma_{2s})^2(\sigma^*_{2s})^2\ (\pi_{2px})^2(\pi_{2py})^2$.

At the beginning of this discussion, it was stated that one of the strengths of MOT was its ability to explain the magnetic properties of molecules, something that VBT could not do. Now it appears as though there are suddenly holes in this model, too. What is wrong? In truth, nothing at all is wrong with the theory. Because the MOs derived from the $2s$ and $2p_z$ AOs both have *sigma* symmetry, they can "mix" with each other. In other words, all four AOs will form linear combinations to make the four *sigma* MOs. For the lowest energy σ_b MO, the coefficients for the $2s$ wavefunctions will be significantly larger than those for the $2p_z$ wavefunctions in the linear combination trial function because the $2s$ orbitals lie closest in energy on the MO diagram to the lowest energy σ_b MO. Likewise, for the highest energy σ^* MO, the coefficients for the $2p_z$ wavefunctions will be much greater than those for the $2s$ wavefunctions. However, for the σ^* and σ_b MOs that are intermediate in energy on the MO diagram (and close in energy to each other), the coefficients for all four atomic orbital basis functions will be similar. This MO concept of *mixing* is therefore very similar to the VB concept of hybridization. Whenever two MOs have the same symmetry and similar energies, the electrons occupying those orbitals will lie in similar regions of space and will therefore repel one another. This causes the two intermediate *sigma* MOs on the MO diagram in Figure 10.25 to repel each other, as shown in Figure 10.26. The result of the mixing is the modified MO diagram shown at the right.

When the "mixing" MO diagram shown at right in Figure 10.26 is used for B_2 and C_2, then the predicted and observed magnetic properties will be the same. For instance, C_2 will have eight total valence electrons, giving it the electron configuration $KK(\sigma_b)^2(\sigma^*)^2(\pi_{2px})^2(\pi_{2py})^2$, which would be diamagnetic because all the electrons are paired. The question then arises as to why the mixing MO diagram is necessary in some cases, but not in others. As one proceeds from the left to right across the second period of the periodic table, the energies of the $2s$ and $2p$ AOs both decrease as a result of the increased effective nuclear charge across the row. Because the $2p$ orbitals are shielded to a certain extent by the $2s$ orbitals, the energies of the $2s$ orbitals decrease more precipitously than those of the $2p$ orbitals, as shown in Table 10.7.

As a result, those elements on the left-hand side of the periodic table have their $2s$ and $2p$ AOs closer in energy to one another than elements from the right-hand side. Thus, there will be considerably more mixing in elements from the left-hand side because the energies of the interacting MOs derived from these AOs will be

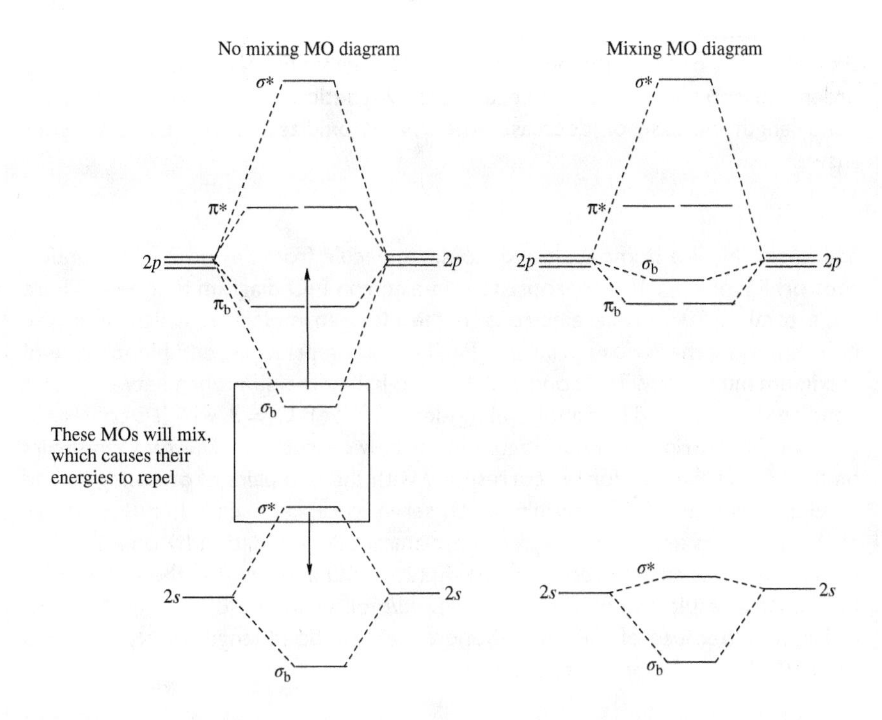

FIGURE 10.26 Whenever two MOs have the same symmetry and similar energies, they will mix with each other; in extreme cases, this can lead to the MO diagram shown at right.

TABLE 10.7 The experimentally determined energy gap between the 2*s* and 2*p* orbitals for the second period elements.

Element	Li	Be	B	C	N	O	F
-*E*(2*s*), eV	5.39	9.32	12.9	16.6	20.3	28.5	37.8
-*E*(2*p*), eV	3.54	6.59	8.3	11.3	14.5	13.6	17.4
E(2*p*)-*E*(2*s*), eV	1.85	2.73	4.6	5.3	5.8	14.9	20.4

closer together. Notice how the energy gap between the 2*s* and the 2*p* orbitals is much larger for O and F than it is for Li, Be, B, C, and N. As a result, mixing will occur in Li_2, Be_2, B_2, C_2, and N_2 but not in O_2 and F_2, as shown in Figure 10.27. Further evidence that mixing occurs in N_2, which lies close to the mixing/nonmixing border, is supported by the PES spectrum of N_2, which shows that the four π_b electrons are ionized at slightly higher energies than the two σ_b ($2p_z$) electrons.

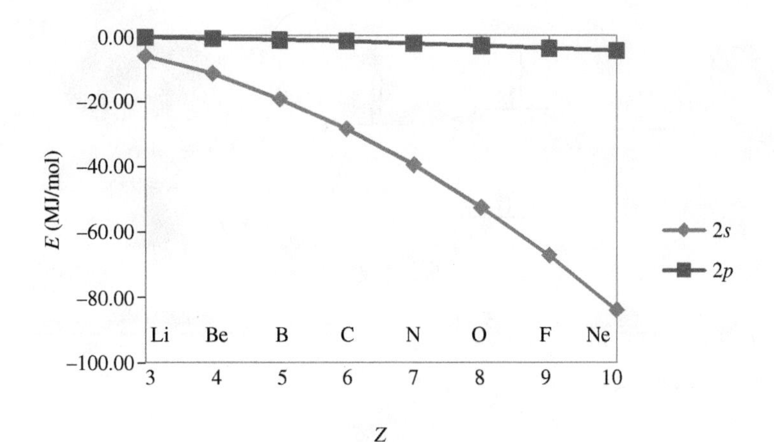

FIGURE 10.27 Relative energies of the valence MOs for the Period 2 homonuclear diatomics. Only the unpaired electrons in the paramagnetic B_2 and O_2 are shown in this diagram.

Example 10-4. Sketch the one-electron MO diagram for N_2. Calculate the bond order and determine if the molecule is diamagnetic or paramagnetic. Will the bond length increase or decrease when N_2 is oxidized to N_2^+? Explain your answer.

Solution. N_2 is a homonuclear diatomic molecule from Period 2 which undergoes orbital mixing. The appropriate one-electron MO diagram is shown. There are a total of 10 valence electrons in the nitrogen molecule, which fill in the MOs following the Aufbau principle, Pauli exclusion principle, and Hund's rule of maximum multiplicity. The bond order is one-half of the difference between eight bonding electrons and two antibonding electrons, or B.O. = 3, which is consistent with the N≡N triple bond predicted by its Lewis structure. The two lone pairs on the Lewis structure for N_2 correspond with the two pairs of opposing σ_b and σ^* electrons in the MO diagram, which serve to cancel each other. Because all of the electrons are paired, N_2 will be diamagnetic. Oxidation by one electron to N_2^+ removes an electron from the σ_b $(2p_z)$ MO and reduces the bond order to 2.5. As a result, the BDE of N_2 is 942 kJ/mol, while the BDE of N_2^+ is only 842 kJ/mol. Because of the lower bond order, the bond length of N_2 increases from 109.8 to 111.6 pm in its oxidized form.

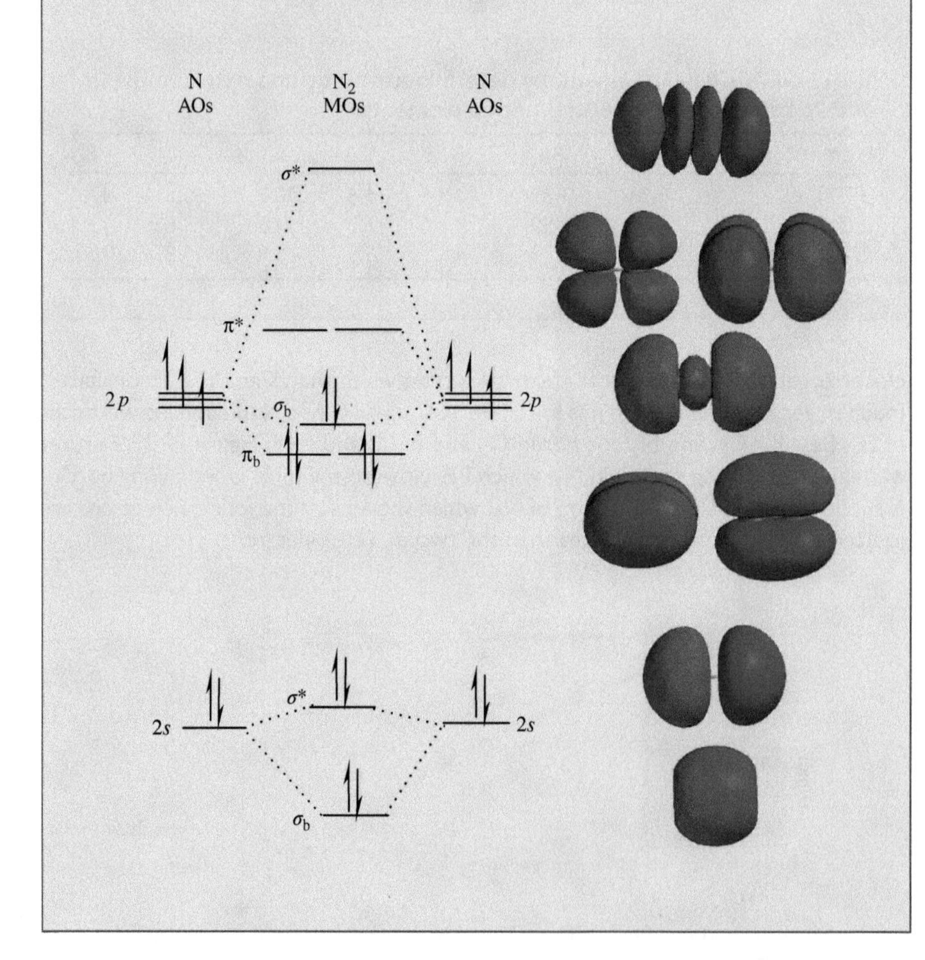

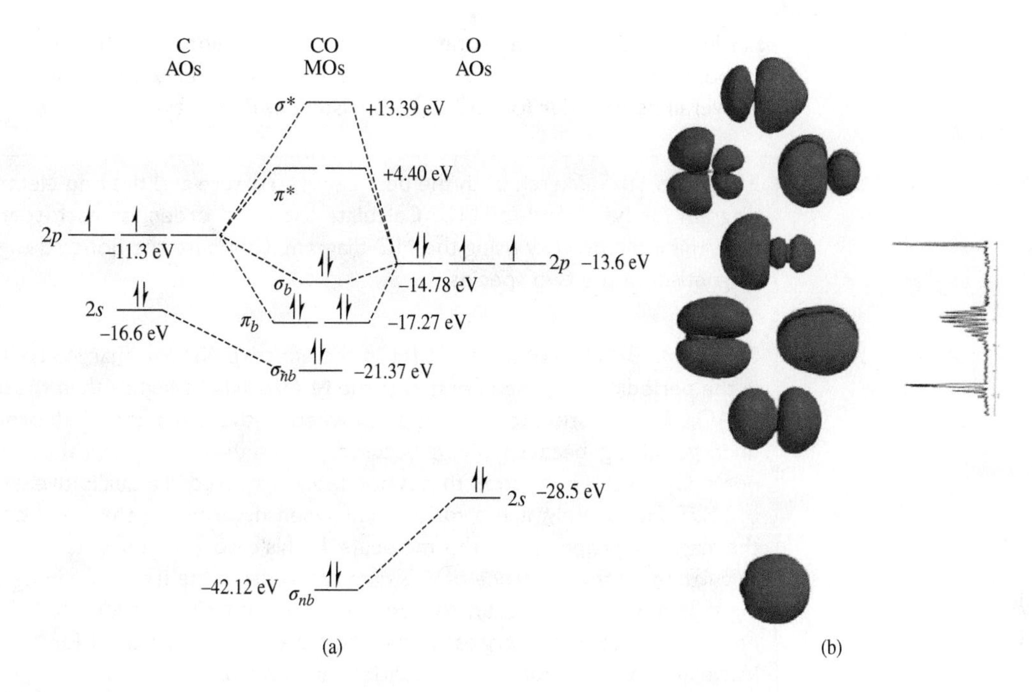

FIGURE 10.28
(a) One-electron molecular orbital diagram for CO, showing the calculated shapes of the MOs and the photoelectron spectrum of CO (b). The energies and the shapes of the MOs were calculated using Wavefunction's Spartan Student Edition version 5.0. The energies of the AOs are from Table 10.7.

In order to sketch the one-electron MO diagrams for heteronuclear diatomics, some understanding of the relative energies of the combining AOs must be known in advance in order to determine which AOs are close enough in energy to form linear combinations where the MO wavefunction is not simply that of a single AO. The MO diagram for the CO molecule is shown in Figure 10.28. Using the periodic trend that the energies of the 2*s* and 2*p* orbitals decrease across the second period, the relative energies of the C and O AOs can be placed on the extreme left and right of the one-electron MO diagram. The C AOs lie at higher energies than those of the O AOs. Because CO is isoelectronic with N_2, which undergoes mixing, it is likely that the MO diagram for CO will also exhibit mixing. This result is borne out by the PES data for CO shown at right in the figure.

In the heteronuclear molecule, the percentage contribution of AO character in the σ_b ($2p_z$) and π_b MOs is larger for the O atom than it is for the C atom because the energies of these MOs are closer to AOs derived from O than from C. This is reflected in the shapes of these MOs, which show more electron density on the O atom than on the C atom. It is also consistent with the expected polarity of the bond, where the bonding electrons are drawn closer to the O atom as a result of its larger electronegativity. Thus, MOT can also explain the relative polarity of bonds in heteronuclear diatomics. Furthermore, the Lewis structure of CO has a lone pair of electrons on C and a second lone pair on O. The lowest energy MO on the diagram is a *sigma* MO that is much closer in energy to the 2*s* AO on oxygen than it is to the 2*s* AO on C. Therefore, the weighting coefficient in the LCAO-MO method will consist predominantly of O character. Because the composition of this MO is almost identical to its corresponding AO, the character of the resulting MO is nonbonding (nb). The same is true for the second lowest MO, which consists largely of C character (recall that the 2*s* and 2*p* AOs mix in C, a fact reflected in the shape of this MO). These nonbonding MOs can be used to indicate the two lone pairs in the CO molecule—one lies on the O atom and the other on the C atom. Nonbonding MOs are not counted in the determination of the bond order. Thus, the overall bond order for CO is 3, consistent with its Lewis structure.

Example 10-5. Sketch both the best Lewis structure and the one-electron MO diagram for NO^+ and for NO. Calculate the bond order for each species and determine the polarity using the MO diagram. Compare the bond strengths and magnetism of the two species.

Solution. Because of the increase in the effective nuclear charge across a row in the periodic table, the energies of the N AOs will lie higher than those of the O AOs. The question then arises as to whether the *sigma* 2*s* and 2*p* orbitals will undergo mixing, because mixing occurs between these atomic orbitals in N_2 but not in O_2. This is a question that is not easily answered in a qualitative discussion of MOT. Fortunately, it is a moot point when determining the bond order and the magnetic properties of this molecule. In this case, only a quantum mechanical calculation or the presence of PES data can address the issue of mixing. Assuming that mixing does occur, the one-electron MO diagram for NO^+ is shown here. The reported energies of the MOs are those calculated for NO^+ using Wavefunction's Spartan Student Edition version 5.0.

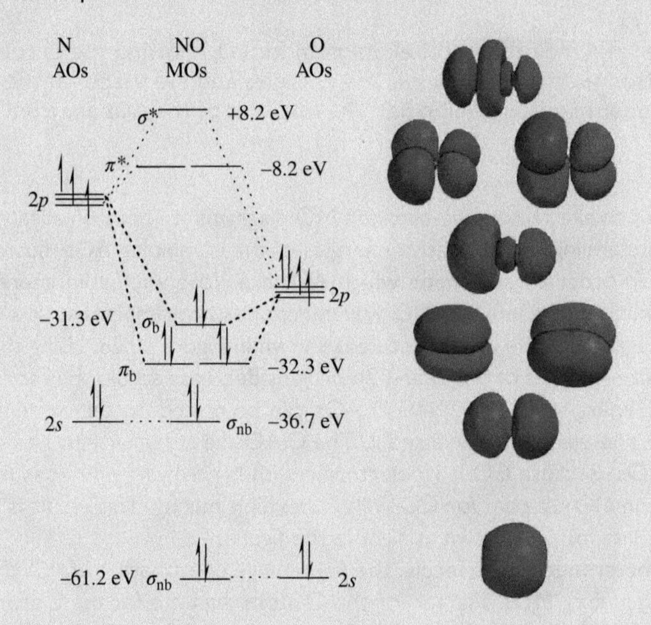

There are a total of 10 valence electrons. The bond order is ½ (6−0) = 3. Therefore, the bond dissociation energies of NO^+ and NO are both expected to be larger than that for a typical N=O double bond, which is 607 kJ/mol. The experimental BDE for the NO molecule is 626 kJ/mol. NO^+ is diamagnetic, while NO is paramagnetic and a free radical.

As was the case for CO, the 2*s* AOs of N and O are too far apart in energy to form bonding and antibonding MOs. They therefore come directly across in the MO diagram as nonbonding MOs and are not included when determining the bond order. Because the bonding MOs lie closer in energy to AOs derived from O than they do from the AOs on N, the NO bond will be polarized toward the more electronegative O atom. The NO molecule represents a good example of the limitations of VBT in accurately describing the bonding. The best Lewis structure for NO places the odd electron on the N atom and contains an N=O double bond. Any canonical form attempting to place a triple bond between the two nuclei would violate the octet rule. The observed BDE for NO is more consistent with the bond order of 2.5 derived from MOT than it is from the bond order of 2 derived from its Lewis structure.

Example 10-6. Given the PES data in the table (where the number in parentheses is the relative number of electrons ejected from that energy level), sketch the one-electron MO diagram for HF, determine the bond order, rationalize the polarity on the basis of the MO diagram, explain the lone pairs in the Lewis structure, and predict the magnetism of the molecule.

H	F	HF
13.6 eV (1)	17.4 eV (5)	16.1 eV (4)
	37.8 eV (2)	18.6 eV (2)
		39.0 eV (2)

Solution. The one-electron MO diagram for HF using the PES data from the table is shown here.

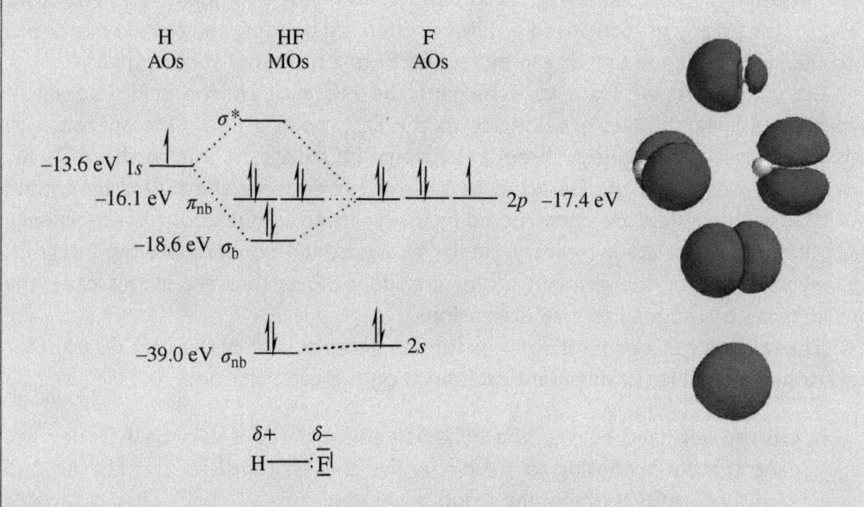

The energy of the 1s AO on H is similar to that of the 2*p* AOs on F, so that a *sigma* bonding and antibonding combination is formed by the overlap of these two orbitals. The remaining AOs will be nonbonding in nature. The slight differences in energy between the nonbonding MOs and the corresponding AOs from which they are derived have to do with the small degree of mixing with higher lying AOs. The bond order is ½ (2−0) = 1, consistent with the Lewis structure. Because the bonding MO has a higher percentage F character to it, the H−F bond will be polarized toward the F atom. In the Lewis structure, there are three lone pairs on the F atom. The corresponding electron pairs in the MO diagram are all nonbonding MOs derived from AOs on the F atom. HF is diamagnetic.

The MO diagram for HF can also predict the approximate bond-dissociation energy for HF. The BDE will be approximately equal to the energy difference between the σ_b and σ^* MOs. Because the σ^* MO is unpopulated, its exact energy is unknown as there are no electrons in this orbital to be ejected by the PES technique. However, we can estimate the energy of the σ^* MO because we know that the σ_b MO will be stabilized approximately as much as the σ^* MO will be destabilized. The σ_b MO is stabilized with respect to the $2p$ AOs on F by 1.2 eV. Thus, the σ^* MO should be destabilized with respect to the 1s AO on H by approximately 1.2 eV, placing its energy at about −12.4 eV. The energy difference between the σ^* and σ_b MOs is therefore 6.2 eV. Given that 1 eV = 96.5 kJ/mol, this corresponds to an energy difference of 6.0×10^2 kJ/mol. The actual BDE of HF is 5.7×10^2 kJ/mol, a difference of only 6%!

10.3 MOLECULAR ORBITAL THEORY: POLYATOMICS

In all of the examples thus far, we have applied a localized version of the LCAO-MO method, where the MOs were derived by taking linear combinations of the AO wavefunctions of only two of the atoms in the molecule. In other words, the electrons that occupied these MOs were localized between two nuclei, just as they were in VBT. However, MOT does not restrict us to this arrangement. The alternative is a delocalized approach, where the electrons are not forced *a priori* to be localized between specific nuclei in the molecule. While the delocalized approach is more consistent with the observed electronic spectra and PES data of molecules, it is also inherently less intuitive because our minds have been trained to think of the electrons in chemical bonds as always occurring between just two nuclei. Therefore, we need to discard any preconceived notions of chemical bonding and retrain our brains into thinking about the bonding in molecules from a more holistic perspective.

Let us say that we want to determine the MO diagram for BeH_2. Beryllium hydride is a linear molecule belonging to the $D_{\infty h}$ point group. One approach to this problem is to imagine a three-dimensional MO diagram, where the AOs for the two H atoms and the Be atom lie on the extremes of the MO diagram and the MOs in the middle are constructed by taking linear combinations of the valence AOs having appropriate symmetry, similar energies, and nonzero overlap integrals. However, whenever the pendant atoms are identical, as they are in this case, the problem can be reduced to two dimensions.

The following is a general strategy for the development of the MO diagram for polyatomic molecules having identical ligands on the central atom:

1. On the left-hand side of the MO diagram, order the valence AOs for the central atom according to their energies and list the IRRs that correspond with those AOs in the point group of the molecule.
2. On the right-hand side of the MO diagram, order the AOs of the ligands.
3. Use the different sets of AOs on the ligands as basis sets to generate reducible representations known as *symmetry-adapted linear combinations* (*SALCs*). Any basis function that is unchanged by a given symmetry operation will contribute 1 to the character, any basis function that transforms into the opposite of itself will contribute −1, and any basis function that is transformed into a basis function on a different ligand will be an off-diagonal element and contribute 0 to the character.
4. Use the great orthogonality theorem to determine the symmetries of the SALCs. The projection operator method can be used to determine the mathematical forms of the SALCs and their corresponding shapes.
5. For any AO on the central atom and SALC on the ligands that has the same symmetry and similar energy, bonding and antibonding MOs can be made by taking the positive and negative linear combinations, respectively.

TABLE 10.8 Symmetries of the 2H SALCs in BeH_2.

$D_{\infty h}$	E	$2C_\infty$	$\infty\sigma_v$	i	$2S_\infty$	∞C_2	IRRs
Γ_{2H}	2	2	2	0	0	0	$\sigma_g^+ + \sigma_u^+$

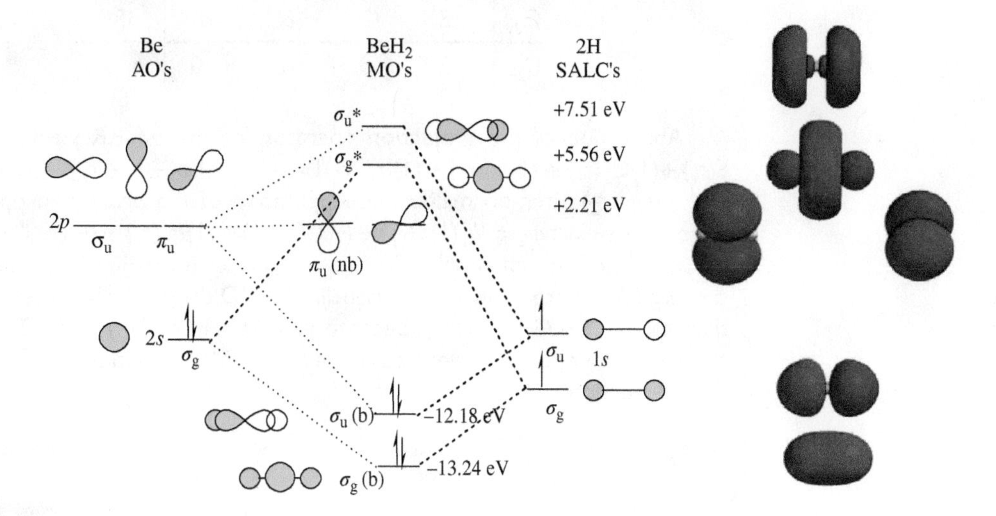

FIGURE 10.29
Qualitative one-electron MO diagram for BeH_2. The energies and shapes of the MOs were calculated using Wavefunction's Spartan Student Edition version 5.0.

6. Any remaining AOs or SALCs that do not have a corresponding partner with the same symmetry and similar energy will become nonbonding MOs.

Taking BeH_2 as an example, the symmetry of the 2*s* AO in the $D_{\infty h}$ point group is σ_g, while the symmetry of the $2p_z$ AO is σ_u, and the symmetries of the $2p_x$ and $2p_y$ AOs are π_u (recall that lower case symbols are used for the IRRs in one-electron MO diagrams and that the subscripts and superscripts are often omitted). Only the 1*s* AOs on the two H atoms can be used in the bonding. Taking the set of 1*s* AOs having positive sign of the wavefunction as the basis set, the reducible representation for the two SALCs is shown in Table 10.8.

After reduction, $\Gamma_{2H} = \sigma_g + \sigma_u$. The projection operator for the σ_g IRR (after normalization) yields the wavefunction in Equation (10.33), while that for the σ_u IRR yields the wavefunction given by Equation (10.34). The corresponding shapes of these two SALCs are shown at the right in Figure 10.29.

$$\phi_1 = \frac{1}{\sqrt{2}}(1s_A + 1s_B) \tag{10.33}$$

$$\phi_2 = \frac{1}{\sqrt{2}}(1s_A - 1s_B) \tag{10.34}$$

Example 10-7. Use the delocalized MO model to qualitatively sketch the MO diagram for the H_2O molecule and to determine the bond order for each O–H bond.

Solution. The H_2O molecule is bent and belongs to the C_{2v} point group. The 2*s* AO on oxygen transforms as the a_1 IRR, the $2p_z$ as a_1, the $2p_y$ as b_2, and the $2p_x$ as b_1. We will assume that the molecule lies in the *xz*-plane. Using the 1*s* AOs on the pendant H atoms as a basis set, the corresponding reducible representation is given in the table and reduces to $a_1 + b_1$.

C_{2v}	E	C_2	σ_v	σ_v'	IRRs
Γ_{2H}	2	0	2	0	$a_1 + b_1$

Application of the projection operator for the a_1 IRR yields: $\phi_1 = \frac{1}{4}[(1)E(r_1) + (1)C_2(r_1) + (1)\sigma_v(r_1) + (1)\sigma_v'(r_1)] = \frac{1}{4}[\ r_1 + r_2 + r_1 + r_2]$, which equals $1/2^{1/2}\ (r_1 + r_2)$ after normalization. Application of the projection operator for the b_1 IRR yields: $\phi_2 = \frac{1}{4}[(1)E(r_1) + (-1)C_2(r_1) + (1)\sigma_v(r_1) + (-1)\sigma_v'(r_1)] = \frac{1}{4}[\ r_1 - r_2 + r_1 - r_2]$, which equals $1/2^{1/2}\ (r_1 - r_2)$ after normalization. The shapes of the two SALCs are shown at the right in the MO diagram. The a_1 SALC is lower than the b_1 SALC in energy because it lacks any nodal planes. The energies and shapes of the MOs were calculated using Wavefunction's Spartan Student Edition, version 5.0.

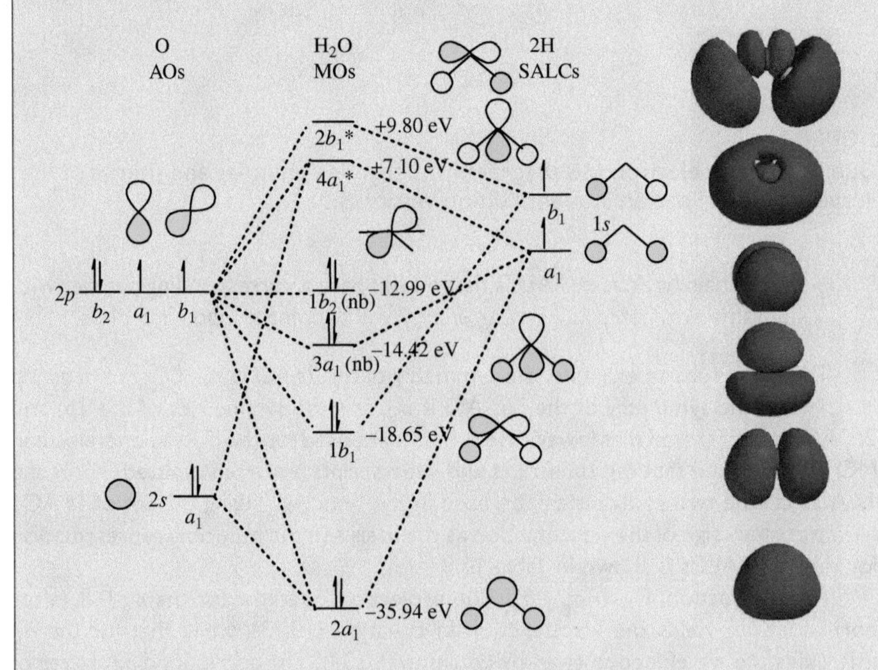

The three a_1 orbitals combine to make bonding, nonbonding, and antibonding a_1 MOs (the common practice of assigning MOs as *sigma* or *pi* rigorously applies only for linear molecules). The two b_1 orbitals combine to make bonding and antibonding b_1 MOs. The b_2 AO on oxygen will become a nonbonding b_2 MO. As with the MOs in BeH_2, the relative energies of the MOs will depend on the energies of the orbitals from which they are derived and on the number of nodal planes between the nuclei.

The MO diagram for H_2O is shown in Example 10-7. It is common practice to label MOs having the same symmetry numerically, beginning with the lowest energy MO. The lowest energy MO shown in the MO diagram is labeled as $2a_1$

because the lower lying $1a_1$ (nb) MO deriving from the 1s AO on the O atom is not shown in the diagram. The $3a_1$ MO is intermediate between a bonding and nonbonding MO. Some authors refer to this type of orbital as "slightly bonding." For purposes of the bond order calculation, we will consider it to be essentially nonbonding. The overall bond order in H_2O is therefore 2, which is consistent with the two O–H single bonds predicted by its Lewis structure. The two lone pairs in the Lewis structure on the O atom can be rationalized in the MO diagram from the two electron pairs in the $3a_1$ and $1b_2$ nonbonding MOs, both of which derive from the O atom. Furthermore, the electrons in the $2a_1$ and $1b_1$ bonding MOs all lie closer in energy to orbitals derived from the O atom than they do to orbitals derived from the H atoms. In other words, the weighting coefficients in these two wavefunctions are skewed toward the O atom, consistent with the polarity of the bond predicted on the basis of electronegativity arguments. The calculated shapes of the MOs are shown at the right in the diagram. Notice how the calculated shapes of the MOs can be obtained in a qualitative sense by making linear combinations of the appropriate AO on the O atom with SALCs from the 2H atoms having the same symmetry.

One of the strengths of VSEPR theory was that it could correctly predict the geometries of most small molecules. We now show how MOT is also capable of predicting molecular geometry. Let us begin this process by considering the class of molecules having the formula AH_2, where A can represent any element. There are two limiting geometries having this molecular formula: linear ($D_{\infty h}$) and bent (C_{2v}). The one-electron MO diagrams for these two limiting cases were shown in Figure 10.29 and Example 10-7.

In 1953, A. D. Walsh made a plot of the "orbital binding energies" as a function of the bond angle. The total energy for a molecule will then be the sum of its orbital binding energies. Whichever configuration has the lowest overall energy will dictate the molecule's geometry. Today, the energies of the MOs are used in place of orbital binding energies. The *Walsh diagram* for AH_2 is nothing more than a correlation diagram, where each MO in the $D_{\infty h}$ point group has a corresponding MO in the C_{2v} point group. For each pair of corresponding MOs, a consideration of the degree of orbital overlap can be used to determine if the energy of that orbital is lower in the linear or in the bent geometry. Thus, a plot of the relative orbital energies can be made as a function of the bond angle.

The Walsh diagram for AH_2 molecules is shown in Figure 10.30. The $2a_1$ MO in the bent configuration is lower in energy than the corresponding $2\sigma_g$ MO in the linear geometry because of the increased orbital overlap between the ligands as the bond angle decreases from 180°. Thus, any AH_2 molecule having 1–2 valence electrons will prefer to populate the $2a_1$ MO in the bent molecular geometry. The H_3^+ ion, whose structure has been determined to form an equilateral triangle, is an example of a two-electron AH_2 molecule.

A consideration of the next highest level of MOs in each geometry shows that there is considerably more overlap in the $1\sigma_u$ MO of the linear geometry (where the $2p_z$ orbital interacts directly with the 2H SALC) than in the $1b_1$ MO of the bent geometry (where the $2p_z$ orbital and the SALC are at an angle to one another). Thus, any AH_2 molecule having 3–4 valence electrons will prefer the linear molecular geometry. We have already encountered one such example in BeH_2, which has four valence electrons. Lithium dihydride (which has three valence electrons) is another example.

Moving up the MO diagram to the third set of corresponding MOs, the $1\pi_{ux}$ MO in $D_{\infty h}$ is nonbonding, while the corresponding $3a_1$ MO in C_{2v} is bonding. Thus, any AH_2 molecule having 5–6 valence electrons will favor the bent molecular

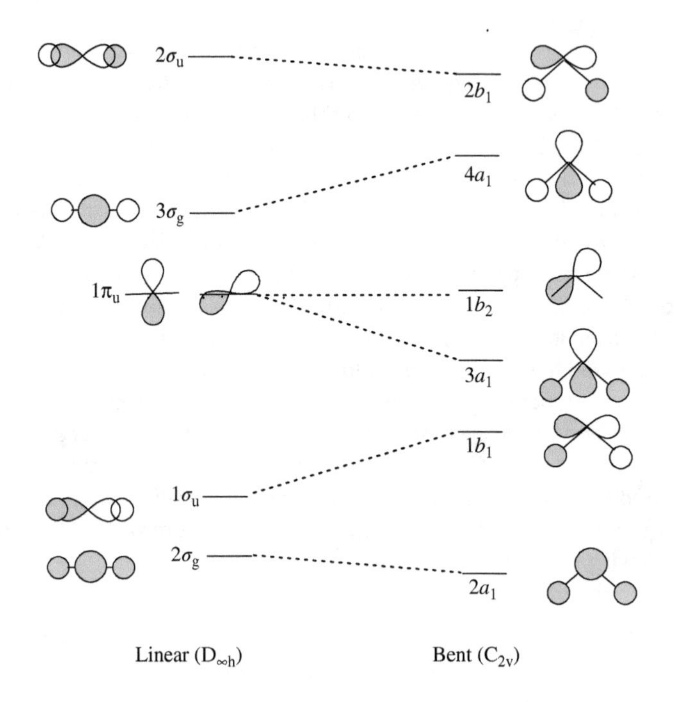

FIGURE 10.30
Walsh diagram for AH_2 molecules.

geometry, as is the case for BH_2, which has five valence electrons and an equilibrium bond angle of 131°. Moving up to the fourth energy level, the $1\pi_{uy}$ MO in $D_{\infty h}$ is nonbonding, while the corresponding $1b_2$ MO in C_{2v} is also nonbonding. Thus, these two orbitals have the same energy. Any AH_2 molecule having 7–8 total valence electrons will prefer the bent molecular geometry, because it is the *overall* orbital energy that matters. The water molecule, which has eight valence electrons, has a bent molecular geometry (104.5°).

Continuing up the MO diagram to the fifth energy level, the $3\sigma_g$ MO in $D_{\infty h}$ is less antibonding than the $4a_1$ MO in C_{2v}. Thus, any AH_2 species containing 9–10 valence electrons will prefer a linear geometry. Finally, the $2\sigma_u$ MO in $D_{\infty h}$ is more antibonding than the $2b_2$ MO in C_{2v}, so that 11- to 12-electron AH_2 molecules would prefer to be bent. Because of the population of antibonding MOs at these higher levels, very few such species actually exist. Walsh diagrams are extremely accurate in the prediction of molecular geometries. A summary of the results for AH_2 molecules is provided in Table 10.9.

TABLE 10.9 Predicted molecular geometries of AH_2 molecules using Walsh diagrams as a function of the total number of valence electrons in the molecule.

Number of Valence e^-	Predicted Geometry
1–2	Bent
3–4	Linear
5–8	Bent
9–10	Linear
11–12	Bent

Example 10-8. Sketch the one-electron MO diagram for NH_3.

Solution. The ammonia molecule belongs to the C_{3v} point group. The 2*s* and $2p_z$ AOs on both transform (separately) as the a_1 IRR, while the $2p_x$ and $2p_y$ AOs transform together as the e IRR. Using the 1*s* AOs on the three H atoms as a basis set for the SALCs, the reducible representation Γ_{3H} is given in the table and reduces to: $a_1 + e$.

Basis functions for the three H 1*s* AOs in NH_3 (the N is above the plane of the paper and the three H atoms are below the plane):

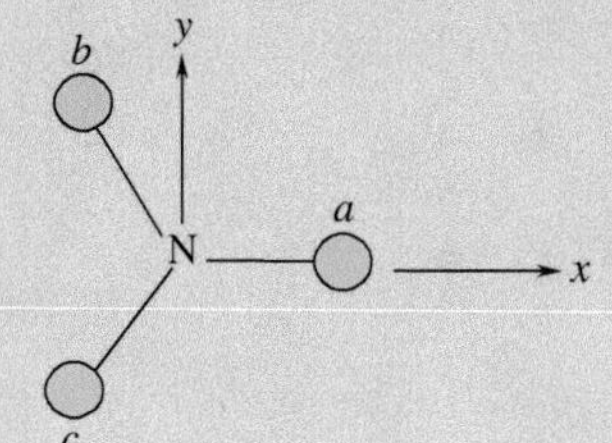

C_{3v}	E	$2C_3$	$3\sigma_v$	IRRs
Γ_{4H}	3	0	1	$a_1 + e$

Using the C_{3v} point group, the projection operator for a_1 yields: $\phi_1 = 1/3^{1/2}\,[a + b + c]$. Application of the projection operator for the e IRR on the *a* basis function gives: $\phi_a = N\,[2a - b - c]$. If we had applied the projection operator for e to the *b* basis function instead, we would have obtained the result: $\phi_b = N\,[2b - a - c]$. Likewise, had the same projection operator acted on the basis function *c*, the result would have been: $\phi_c = N\,[2c - a - b]$. Because the e IRR is doubly degenerate and we currently have three results for it, we will arbitrarily choose ϕ_a to be an acceptable solution that lies along the *x*-axis. After normalization, $\phi_2 = 1/6^{1/2}\,[2a - b - c]$. In order to ensure orthogonality, the final wavefunction ϕ_3 must have components along the *y*-axis. It is therefore a linear combination of $\phi_b - \phi_c$, which (after normalization) yields: $\phi_3 = 1/2^{1/2}\,[b - c]$. The shapes of the three SALCs are shown (where again the *N* is above the plane of the paper and the Hs are below the plane).

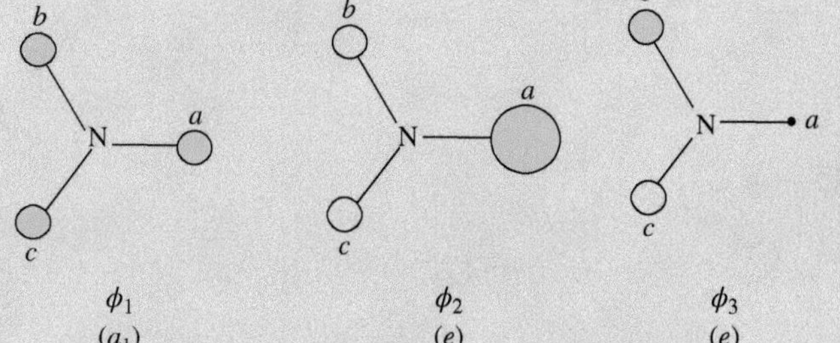

The three a_1 orbitals combine to form bonding, nonbonding, and antibonding MOs; and the two sets of e orbitals combine to form two sets of bonding and antibonding MOs. The qualitative one-electron MO diagram for NH_3 is shown here. Because the $3a_1$ MO is nonbonding, the overall bond order for NH_3 is three, which is consistent with the three N–H single bonds in its Lewis structure.

The energies and shapes of the MOs were calculated using Wavefunction's Spartan Student Edition, version 5.0.

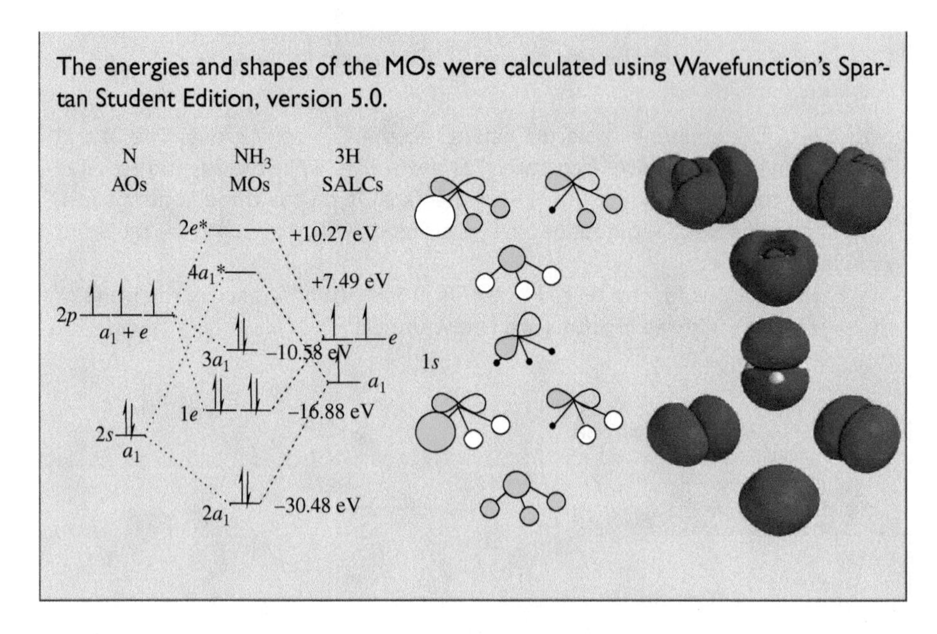

The Walsh diagram for AH_3 molecules is shown in Figure 10.31. The two limiting geometries in this case are trigonal planar (D_{3h}) and trigonal pyramidal (C_{3v}). The MO diagram for NH_3 is shown in Example 10-8 as a reference. The MO diagram for the trigonal planar BH_3 molecule is very similar, except that the Mulliken symbols for the MOs have the ′ superscripts. The relative energies of the MOs in Figure 10.31 were determined on the basis of whether the amount of orbital overlap increases or decreases as the bond angle changes. For instance, the $1a_1$ MO in the trigonal pyramidal C_{3v} geometry has more overlap (between the ligands) than the corresponding $1a_1'$ MO in D_{3h}. A summary of the results of the predicted molecular geometries for AH_3 molecules is provided in Table 10.10.

The PES of methane was shown in Figure 10.17. The presence of two peaks at the valence level (one integrating to six electrons and the other integrating to two electrons) could not be explained by VBT. We now generate the delocalized MO diagram for CH_4. The symmetry of the 2*s* AO on C in the T_d point group is a_1, while the three 2*p* AOs transform together as the t_2 IRR. Using the 1*s* AOs on the

2e′ — 2e
3a1
2a′
1a2″
2a1
1e′ — 1e
1a1′ — 1a1
Trigonal planar (D_{3h}) Trigonal pyramidal (C_{3v})

FIGURE 10.31
Walsh diagram for AH_3 molecules.

TABLE 10.10 Predicted molecular geometries of AH_3 molecules using Walsh diagrams as a function of the total number of valence electrons in the molecule.

Number of Valence e^-	Predicted Geometry
1–2	Trigonal pyramidal
3–6	Trigonal planar
7–8	Trigonal pyramidal
9–10	Trigonal planar
11–14	Trigonal pyramidal

TABLE 10.11 Symmetries of the 4H SALCs in CH_4.

T_d	E	$8C_3$	$3C_2$	$6S_4$	$6\sigma_d$	IRRs
Γ_{4H}	4	1	0	0	2	$a_1 + t_2$

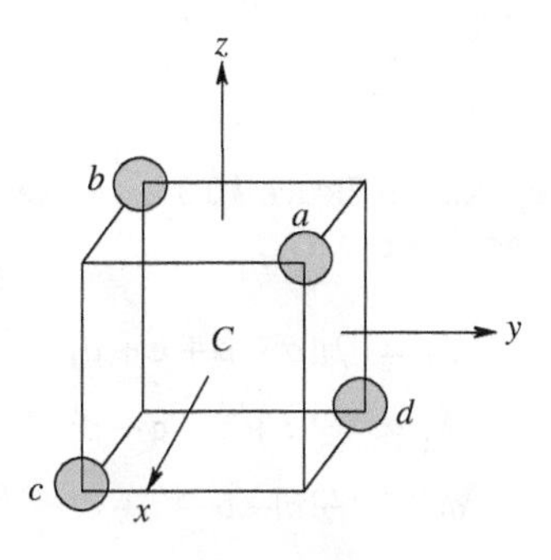

FIGURE 10.32
Basis set showing the 4H 1s AOs in CH_4.

four H atoms as a basis set, the reducible representation Γ_{4H} has the characters shown in Table 10.11 and reduces to $a_1 + t_2$.

In order to simplify the formation of SALCs, we use the T subgroup to generate the mathematical functions for the SALCs using the projection operator method. The basis set of hydrogen 1s AOs is shown in Figure 10.32.

The a_1 SALC will be totally symmetric as given by Equation (10.36) and will have the same shape and sign of the wavefunction as the basis set shown in Figure 10.32. When the projection operator for the t_2 IRR acts on the basis function a, the result is $P_a = \frac{1}{2}[3a-b-c-d]$. When the same projection operator for t_2 acts on b, it yields $P_b = \frac{1}{2}[3b-a-c-d]$. Similarly, we obtain the functions $P_c = \frac{1}{2}[3c-a-b-d]$ and $P_d = \frac{1}{2}[3d-a-b-c]$ when the projection operator for t_2 acts on functions c and d, respectively. There can only be three forms of the wavefunction for the t_2 SALCs and we have at present four solutions. In order to ensure that the three t_2 SALCs are orthogonal to one another, we will choose to take components along the three coordinate axes. Along the z-axis, we will take the linear combination: $P_a + P_b - P_c - P_d$, which yields (after normalization) the SALC given by Equation (10.36). Along the y-axis: $P_a + P_d - P_b - P_c$, which gives Equation (10.37). Finally, along the x-axis, we have: $P_a + P_c - P_b - P_d$, which is given by Equation (10.38). The four SALCs for the 4H ligands in CH_4 therefore have the forms shown in Figure 10.33. The 2s AO on C overlaps with the ϕ_1 SALC, the $2p_z$ AO with ϕ_2, the $2p_y$ AO with ϕ_3, and the $2p_x$ AO with ϕ_4 to form bonding and antibonding linear combinations. The one-electron

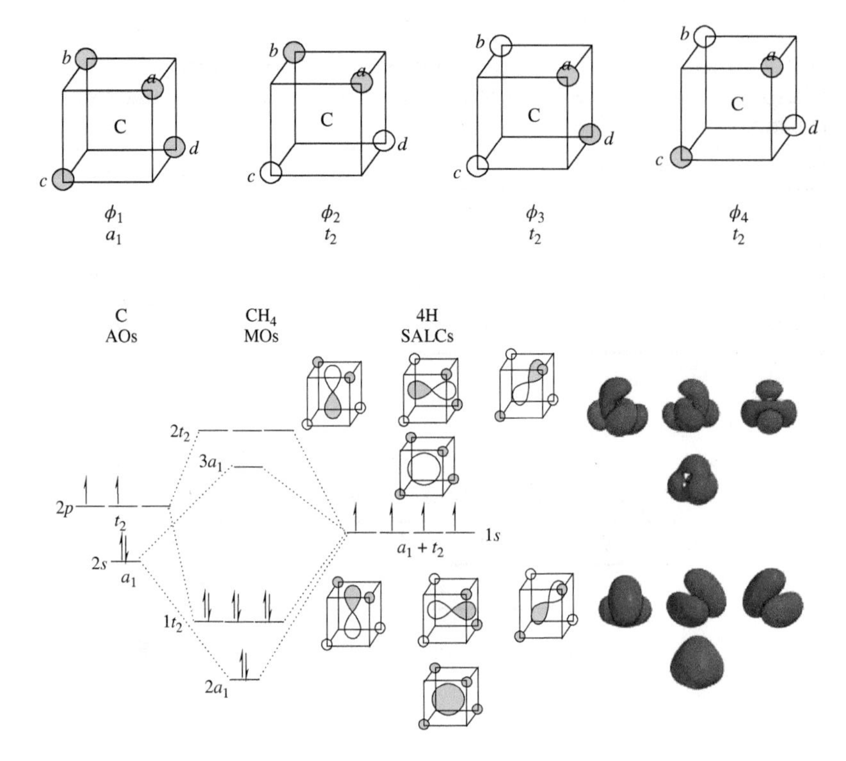

FIGURE 10.33
Shapes of the four SALCs for the 4H atoms in CH_4.

FIGURE 10.34
One-electron MO diagram for CH_4. The energies and shapes of the MOs were calculated using Wavefunction's Spartan Student Edition, version 5.0.

MO diagram for CH_4 is depicted in Figure 10.34. The two sets of bonding MOs are consistent with the observed PES of CH_4.

$$\phi_1 = \tfrac{1}{2}[a + b + c + d] \quad (10.35)$$

$$\phi_2 = \tfrac{1}{2}[a + b - c - d] \quad (10.36)$$

$$\phi_3 = \tfrac{1}{2}[a - b - c + d] \quad (10.37)$$

$$\phi_4 = \tfrac{1}{2}[a - b + c - d] \quad (10.38)$$

Example 10-9. Sketch the one-electron MO diagram for BeH_4^{2-}, assuming square planar geometry.

Solution. Square planar molecules belong to the D_{4h} point group. In D_{4h}, the 2s AO transforms as a_{1g}, the $2p_z$ AO transforms as a_{2u}, and the $2p_x$ and $2p_y$ AOs transform together as the e_u IRR. Using the 1s AOs on the four H atoms as a basis set for the SALCs, the reducible representation Γ_{4H} is given in the table and reduces to: $a_{1g} + b_{1g} + e_u$.

D_{4h}	E	$2C_4$	C_2	$2C_2'$	$2C_2''$	i	$2S_4$	σ_h	$2\sigma_v$	$2\sigma_d$	IRRs
Γ_{4H}	4	0	0	2	0	0	0	4	2	0	$a_{1g} + b_{1g} + e_u$

The totally symmetric a_{1g} SALC naturally is the positive linear combination of all four basis functions:

$$\phi_1 = \tfrac{1}{2}[a + b + c + d]$$

The b_{1g} SALC is symmetric with respect to E, C_2, C_2'',i, σ_h, and σ_v but anti-symmetric to C_4, C_2'', S_4, and σ_d, which yields the following linear combination:

$$\phi_2 = {}^1\!/\!_2[a - b + c - d]$$

Finally, application of the e_u projection operator to the a basis function and the b basis function yields the following orthogonal linear combinations:

$$\phi_3 = (1/2^{1/2})[a - c]$$

$$\phi_4 = (1/2^{1/2})[b - d]$$

The shapes of the four H SALCs are shown here. The a_{1g} SALC will form bonding and antibonding linear combinations with the 2s AO on Be. The e_u SALCs will likewise form bonding and antibonding linear combinations with the $2p_x$ and $2p_y$ AOs on Be. The a_{2u} ($2p_z$) AO on Be will become a nonbonding MO, as will the b_{1g} SALC.

a_{1g} b_{1g} e_u

The one-electron MO diagram for square planar BeH_4^{2-} is shown, along with the qualitative shapes of the MOs predicted using group theory. The energies and shapes of the MOs to the right of the MO diagram were calculated using Wavefunction's Spartan Student Edition, version 5.0.

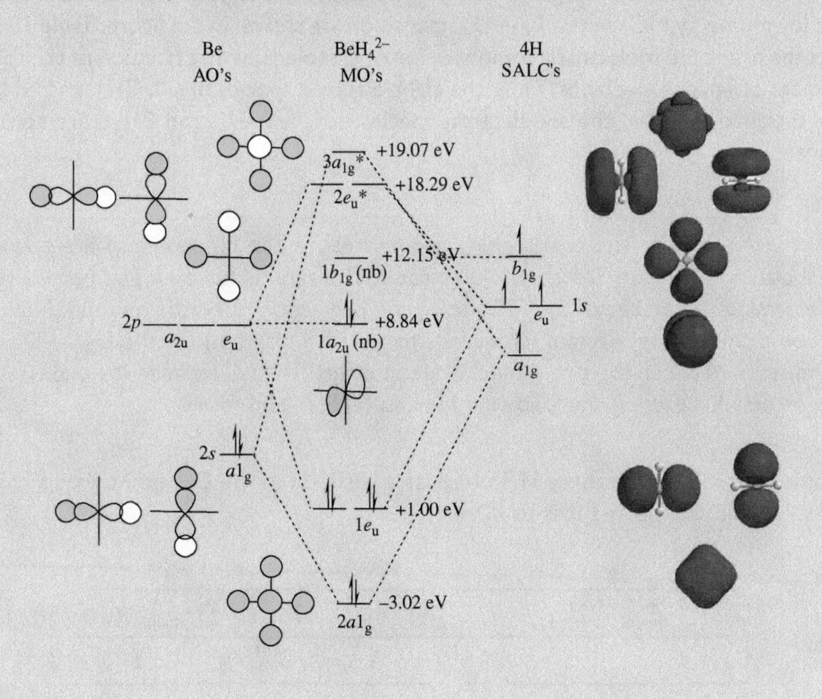

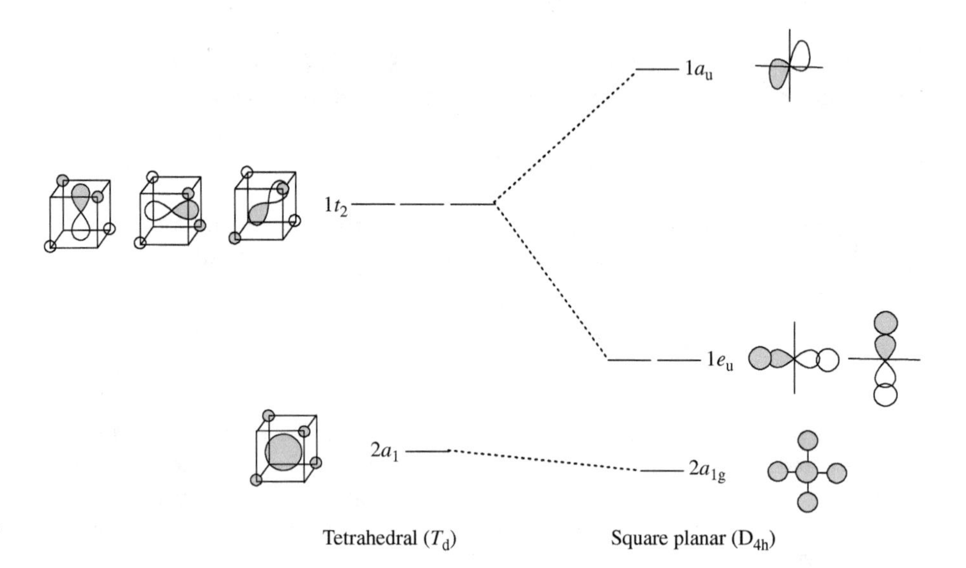

FIGURE 10.35
Partial Walsh diagram for AH_4 molecules.

TABLE 10.12 Predicted molecular geometries of AH_4 molecules as a function of the total number of valence electrons in the molecule using the partial Walsh diagram shown in Figure 10.34.

Number of Valence e^-	Predicted Geometry
1–6	Square planar
7–8	Tetrahedral

A qualitative Walsh diagram for AH_4 molecules is shown in Figure 10.35. Only the lower energy MOs in the $T_d \rightarrow D_{4h}$ transition are shown in the figure. Table 10.12 lists the predicted molecular geometries for AH_4 molecules as a function of the total number of valence electrons. Thus, the eight-electron species BH_4^-, CH_4, and NH_4^+ are tetrahedral, while the six-electron species LiH_4^-, BeH_4, and BH_4^+ are square planar.

Example 10-10. The two limiting geometries for the H_3 molecule are trigonal planar and linear. Sketch one-electron MO diagrams for each geometry and determine the shapes of the MOs using the projection operator method. Next, draw a correlation diagram (analogous to a Walsh diagram) for the $D_{3h} \rightarrow D_{\infty h}$ transition between the two molecular geometries. Then determine the expected geometry for each of the following species: (a) H_3^+ and (b) H_3^-.

Solution. Using the three H 1s AOs as a basis set in the D_{3h} point group, the representation Γ_{3H} reduces to $a_1' + e'$.

D_{3h}	E	$2C_3$	$3C_2$	σ_h	$2S_3$	$3\sigma_v$	IRRs
Γ_{3H}	3	0	1	3	0	1	$a_1' + e'$

In this case, we will form the MOs by making symmetry-adapted linear combinations of all three H 1s AOs directly. The a_1' MO is totally symmetric and has the wavefunction: $\phi_1 = 1/3^{1/2}\ [a+b+c]$. The wavefunctions for the doubly degenerate e′ SALCs are: $\phi_2 = 1/6^{1/2}\ [2 - b - c]$ and $\phi_3 = 1/2^{1/2}\ [b - c]$. The shapes of the MOs and the structure of the MO diagram for trigonal planar H_3 are shown. The a_1' MO will be bonding and the e′ MOs will be antibonding. In order to maintain a *barycenter* of energy when the AOs mix to form MOs, the a_1' MO will be stabilized approximately twice as much as the e′ MOs will be destabilized with respect to the 3H 1s AOs.

In the $D_{\infty h}$ point group, the symmetry of the 1s AO on the central H atom is σ_g^+, while the Γ_{2H} representation shown here reduces to $\sigma_g^+ + \sigma_u^+$.

$D_{\infty h}$	E	$2C_\infty$	$\infty\sigma_v$	i	$2S_\infty$	∞C_2	IRRs
Γ_{2H}	2	2	2	0	0	0	$\sigma_g^+ + \sigma_u^+$

The σ_g^+ AO and σ_g^+ SALC form bonding and antibonding MOs, while the σ_u^+ MO will be nonbonding. The shapes of the MOs and the structure of the MO diagram for linear H_3 are shown.

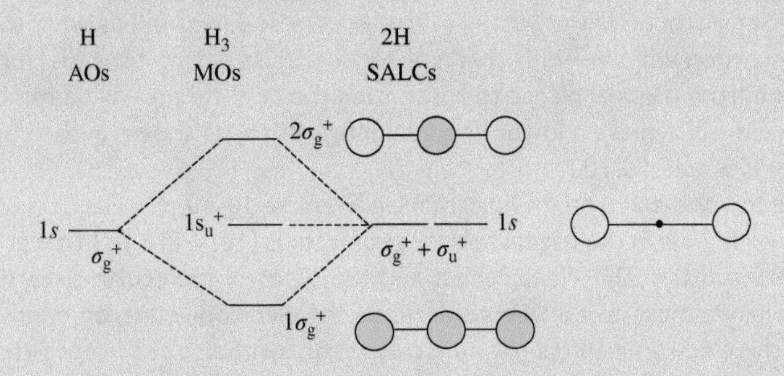

The correlation diagram for the $D_{3h} \rightarrow D_{\infty h}$ transition in the H_3^- ion is shown.

The lowest energy MO has more overlap in the trigonal planar a_1' MO than when it is in the linear $1\sigma_g^+$ MO. Moving up to the second level, the nonbonding $1\sigma_u^+$ MO in the linear geometry will be more stable than the e′ antibonding MO in D_{3h}. In the third level, the e′ MO in the trigonal planar geometry has only one node, while the $2\sigma_g^+$ MO in the linear configuration has two nodes. Thus, the e′ MO will have the lower energy of the pair.

The correlation diagram predicts the following geometries based on the number of valence electrons:

Number of Valence e^-	Predicted Geometry
1–2	Trigonal planar
3–4	Linear
5–6	Trigonal planar

Thus, two-electron H_3^+ is expected to be trigonal planar, while four-electron H_3^- is predicted to be linear.

The H_3^- example shown in Example 10-10 provides a natural introduction to the concept of a Jahn–Teller distortion. The *Jahn–Teller effect*, which was first published in 1937, states that any nonlinear molecule in an orbitally-degenerate state will undergo a distortion of its geometry in order to lower the symmetry, remove the degeneracy, and lower the energy. Using the correlation diagram in Example 10-10 for the H_3^- anion, we find that if the species had been forced into a trigonal planar molecular geometry, the MO diagram for D_{3h} symmetry would place one electron in each of the degenerate e′ MOs. This being a nonlinear molecule in an orbitally degnerate state, the molecule will be forced to undergo a change in its geometry, or a Jahn–Teller distortion. The Jahn–Teller theorem does not tell us what the nature of that distortion will be—only that a distortion must occur. If the H_3^- undergoes a distortion from trigonal planar to linear, then the degeneracy will be removed and the molecule will achieve a lower energy configuration, as shown in the correlation diagram in Example 10-10.

Using the one-electron MO diagrams in Example 10-10, the electron configuration for the H_3^- ion in the trigonal planar geometry is $(1a_1)^2\ (1e')^2$. In order to more fully understand the Jahn–Teller effect and the types of molecular distortions that are possible, we must take into consideration the electron–electron repulsions and consider the electronic states (or molecular term symbols) of the species. The $1a_1$ MO is fully occupied and therefore has spherical symmetry overall, and it is therefore assigned the totally symmetric term symbol $^1A_1'$. It is a singlet state because the electrons are paired.

On the other hand, there are six possible ways to arrange two electrons in the 1e′ degenerate set of MOs. These six microstates give rise to three distinct electronic states having the term symbols $^1A_1'$, $^1E'$, and $^3A_2'$. For two electrons in a doubly degenerate e′ MO, the orbital component of the six microstates can be determined by taking the direct product of $e' \times e'$, because the valence electron configuration was $(1e')^2$. This direct product is shown in Table 10.13. Using the great orthogonality theorem or the direct product tables in Appendix C to reduce this direct product into its irreducible components, the result is $A_1' + E_1' + A_2'$. By convention, term symbols are represented by upper case Mulliken symbols. Because there are only two electrons in the e′ MOs, the spin multiplicity for the three terms can only be 1 (if the electrons have opposite spins) or 3 (if the electrons have parallel

TABLE 10.13 Symmetric and antisymmetric components of the direct product $e' \times e'$ in the D_{3h} point group.

D_{3h}	E	$2C_3$	$3C_2$	σ_h	$2S_3$	$3\sigma_v$	IRRs
$\chi(R)$	2	−1	0	2	−1	0	
$E' \times E'$	4	1	0	4	1	0	$A_1' + E' + A_2'$
R^2	E	C_3^2	E	E	C_3^2	E	
$\chi(R^2)$	2	−1	2	2	−1	2	
$[\chi(R)]^2$	4	1	0	4	1	0	
χ^+	3	0	1	3	0	1	$A_1' + E'$
χ^-	1	1	−1	1	1	−1	A_2'

spins). We also know that there are a total of six possible microstates, which means that either the nondegenerate A_1 or the A_2 term symbol will be a triplet and the other two terms must be singlets. The Pauli exclusion principle states that the total wavefunction must be antisymmetric with respect to exchange of electrons. Because triplet spin wavefunctions are symmetric with respect to the exchange of two electrons and singlet spin wavefunctions are antisymmetric, the orbital component of the triplet term symbol must be antisymmetric. The characters for the symmetric component of the direct product are given by Equation (10.39), while the characters for the antisymmetric direct product are given by Equation (10.40). In both equations, $\chi(R)$ is the character for the operation R and $\chi(R^2)$ is the character for the square of operation R. In this case, the A_2 term symbol is the antisymmetric one. Alternatively, using the direct product tables in Appendix C, the antisymmetric components of the direct product are the ones listed in brackets.

$$\chi^+ = \frac{1}{2}\{[\chi(R)]^2 + \chi(R^2)\} \quad (10.39)$$

$$\chi^- = \frac{1}{2}\{[\chi(R)]^2 - \chi(R^2)\} \quad (10.40)$$

If the ground state of H_3^- is either $^3A_2'$ or $^1A_1'$, then the ion should be stable in the trigonal planar geometry. However, if the ground state of H_3^- is the orbitally degenerate $^1E'$ term, then the species must undergo a Jahn–Teller distortion in order to lower the symmetry, lower the energy, and remove the degeneracy. The molecule can only lower its symmetry by undergoing a molecular vibration having the same symmetry as one of the components of its direct product. The A_2' IRR does not correspond with any of the vibrational modes in the D_{3h} point group. The A_1' vibration is totally symmetric and would not cause a change in the point symmetry of the molecule. Therefore, the only remaining vibrational modes that could lead to the Jahn–Teller distortion are the E' modes, whose shapes are shown in Figure 10.36. One of these vibrational modes leads to an isosceles C_{2v} geometry, while the other leads to the linear $D_{\infty h}$ geometry we already encountered in the correlation diagram given in Example 10-10. In either case, the symmetry has been lowered, which removes the degeneracy and lowers the total energy of the molecule. This example illustrates the essence of the Jahn–Teller effect.

10.4 MOLECULAR ORBITAL THEORY: PI ORBITALS

In all of the polyatomic MO diagrams that we have examined up until this point, the ligands have always been H atoms. Because the SALCs derived from 1s AOs always

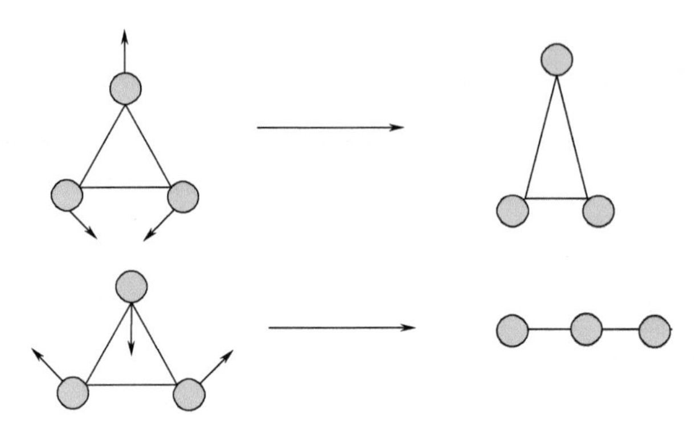

FIGURE 10.36
The effects of the two E' molecular vibrations on the trigonal planar geometry. These are the only possible Jahn–Teller distortions for the D_{3h} H_3^- ion.

have a direct overlap with the AOs on the central atom, these types of bonds are more generally referred to as *sigma* bonds. In this section, we learn how to handle the one-electron MO diagrams for molecules that also contain *pi* bonds. One of the simplest such examples occurs for the O_3 molecule because the energies of the corresponding AOs for all three O atoms are identical. Ozone has a bent molecular geometry and belongs to the C_{2v} point group. Assuming that the molecule lies in the *xz*-plane, the symmetries of the AOs on the central O atom are $2s = a_1$, $2p_z = a_1$, $2p_x = b_1$, and $2p_y = b_2$. Using the four sets of basis functions for the 2*s* and 2*p* AOs on the pendant O atoms, the symmetries of the SALCs are given in Table 10.14. The second basis function in Table 10.14 (Γ_1) was selected in order to maximize overlap along the bonding axes. The remaining two basis functions (Γ_2 and Γ_3) were derived from 2*p* orbitals on the O ligands and must be orthogonal to Γ_1 and to each other.

Using the projection operator to determine the wavefunctions for the SALCs, we obtain the shapes observed in Figure 10.37.

By matching AOs on the central atom with SALCs on the ligands having the same symmetry and similar energies, the one-electron MO diagram in Figure 10.38 can be constructed. The relative energies of the bonding MOs is based on the degree of overlap between the AOs on the central atom and the SALCs on the ligands. The greater the overlap, the more bonding the MO will be. The ordering also takes into account the number of nodal planes between the nuclei.

TABLE 10.14 The symmetries of the SALCs for the two O ligands in the ozone molecule.

Basis Set	C_{2v}	E	C_2	$\sigma_v\ (xz)$	$\sigma_v'\ (yz)$	IRR's
	Γ_s	2	0	2	0	$a_1 + b_1$
	Γ_1	2	0	2	0	$a_1 + b_1$
	Γ_2	2	0	2	0	$a_1 + b_1$
	Γ_3	2	0	−2	0	$a_2 + b_2$

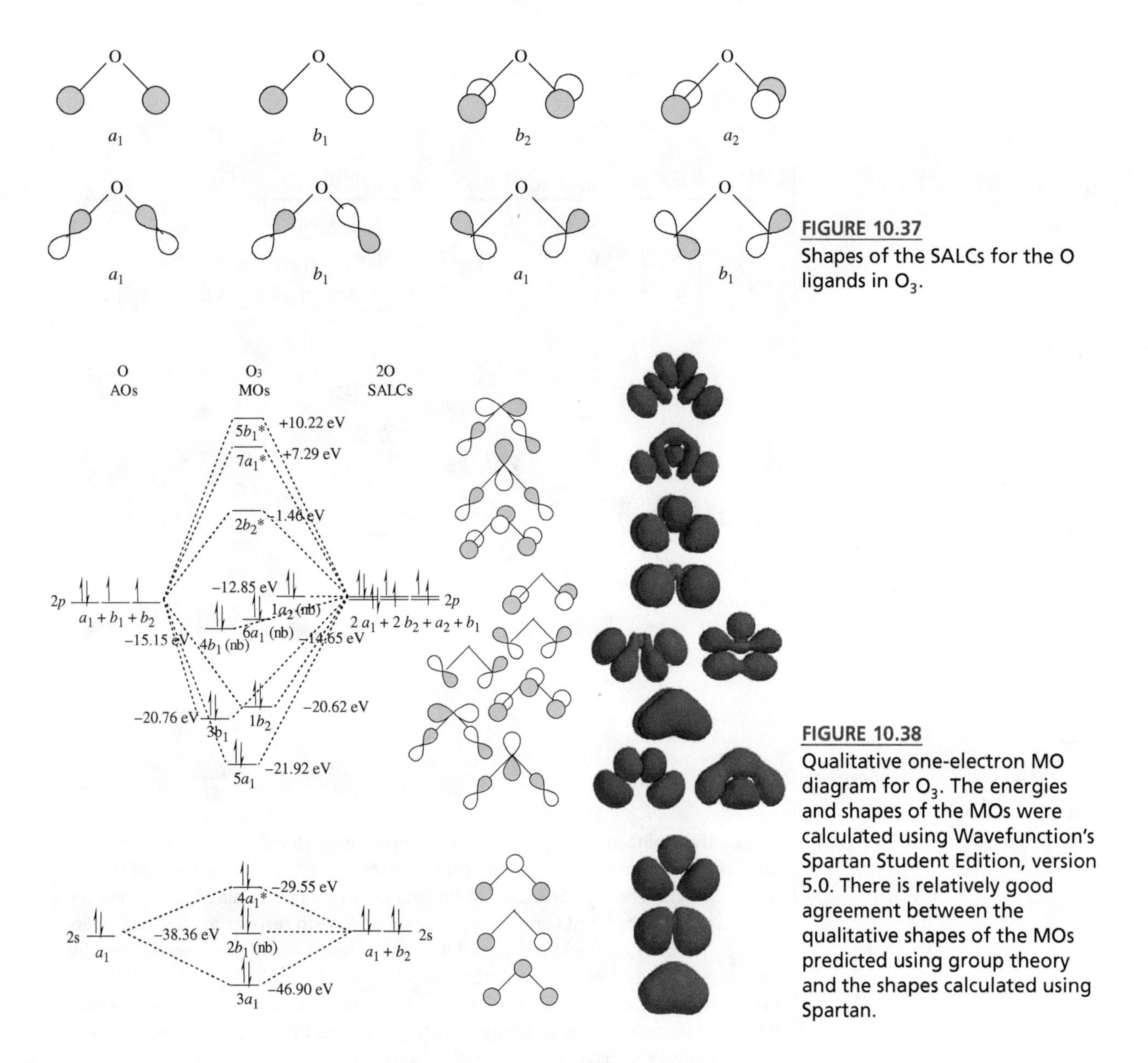

FIGURE 10.37
Shapes of the SALCs for the O ligands in O_3.

FIGURE 10.38
Qualitative one-electron MO diagram for O_3. The energies and shapes of the MOs were calculated using Wavefunction's Spartan Student Edition, version 5.0. There is relatively good agreement between the qualitative shapes of the MOs predicted using group theory and the shapes calculated using Spartan.

The overall bond order for the ozone molecule is 3, indicating an average O–O bond order of 1.5. This is the identical result that we obtained using VBT when we considered the average bond strength of the two equivalent resonance structures. However, in the case of MOT, the 1.5 bond order arose naturally. Furthermore, an examination of the shapes of the bonding MOs reveals that the electron density is delocalized over all three O atoms, negating the necessity in VBT for us to think about the bonding in terms of two equivalent canonical structures.

Example 10-11. Sketch the qualitative one-electron MO diagram for CO_2.

Solution. The AOs for C will lie higher in energy than those for O because of the smaller effective nuclear charge on the carbon nucleus. The symmetries of the C AOs in the $D_{\infty h}$ point group are as follows: $2s = \sigma_g^+$, $2p_z = \sigma_u^+$, and $2p_x$, $2p_y = \pi_u$. Using the D_{2h} subgroup, the symmetries of the 2s and 2p SALCs on the two O atoms are shown in the table. The corresponding IRR's in $D_{\infty h}$ are also shown.

D_{2h}	E	$C_{2(z)}$	$C_{2(y)}$	$C_{2(x)}$	I	σ_{xy}	σ_{xz}	σ_{yz}	IRR's: ($D_{2h} \rightarrow D_{\infty h}$)
Γ_s	2	2	0	0	0	0	2	2	$a_g + b_{1u} \rightarrow \sigma_g^+ + \sigma_u^+$
$\Gamma_{p(z)}$	2	2	0	0	0	0	2	2	$a_g + b_{1u} \rightarrow \sigma_g^+ + \sigma_u^+$
$\Gamma_{p(x,y)}$	4	−4	0	0	0	0	0	0	$b_{2g} + b_{3g} + b_{2u} + b_{3u} \rightarrow \pi_g + \pi_u$

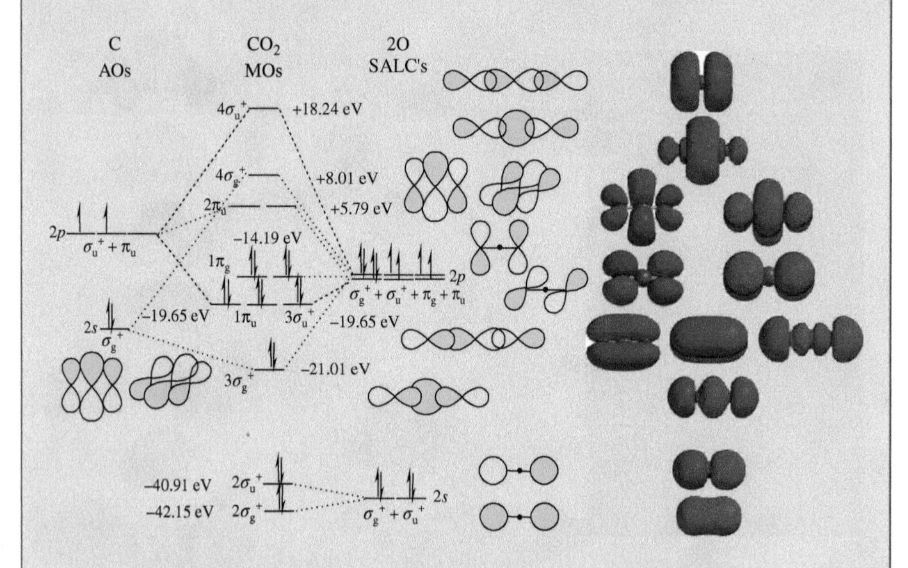

By this point in time, you should not even need to follow the formal procedure of the projection operator method to determine the symmetries and shapes of the SALCs. The σ_g^+ SALC must be totally symmetric to all of the symmetry operations. The σ_u^+ SALC must be antisymmetric with respect to all of the inversion, S_∞, and C_2 operations. The π SALCs will have a nodal plane containing the internuclear axis, with π_g symmetric with respect to inversion and π_u antisymmetric to inversion. The shapes of the SALCs can be seen in the one-electron MO diagram, where the energies and shapes of the MOs were calculated using Wavefunction's Spartan Student Edition, version 5.0.

Next, let us consider a slightly more complicated example: BF_3. According to VSEPR theory, boron trifluoride is trigonal planar and belongs to the D_{3h} molecular point group. The energies of the valence orbitals on B will lie higher in energy than those on F because B lies more to the left in the periodic table. The symmetries of the AOs on the central B atom are as follows: $2s = a_1'$, $2p_z = a_2''$, and $2p_x$, $2p_y = e'$. Using the basis functions shown in Table 10.15, the symmetries of the pendant F atoms are $2s = a_1' + e'$, $2p_z = a_2'' + e''$, $2p$ (pointed toward the central atom) $= a_1' + e'$, and $2p$ (perpendicular to the bond axis but still in the plane of the molecule) $= a_2' + e'$. Using the projection operator, the wavefunctions for the SALCs are given by Equations (10.41)–(10.43).

$$\phi_a = \frac{1}{\sqrt{3}}(a + b + c) \tag{10.41}$$

TABLE 10.15 The symmetries of the SALCs for the three pendant F atoms in the BF_3 molecule.

Basis Set	D_{3h}	E	$2C_3$	$3C_2$	σ_h	$2S_3$	$3\sigma_v$	IRRs
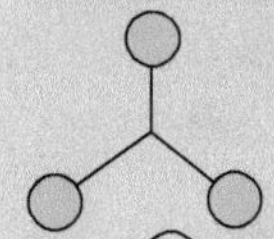	Γ_s	3	0	1	3	0	1	$a_1' + e'$
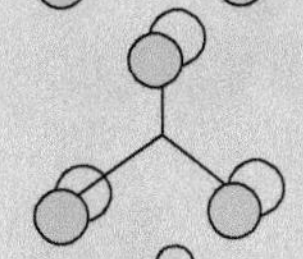	Γ_{pz}	3	0	−1	−3	0	1	$a_2'' + e''$
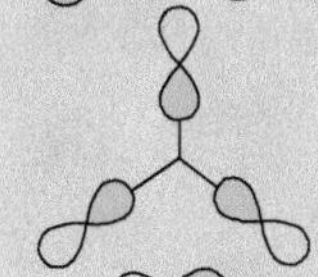	Γ_1	3	0	1	3	0	1	$a_1' + e'$
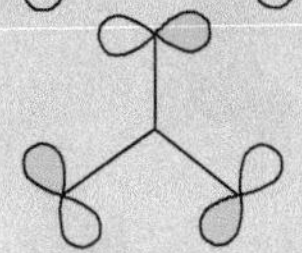	Γ_2	3	0	−1	3	0	−1	$a_2' + e'$

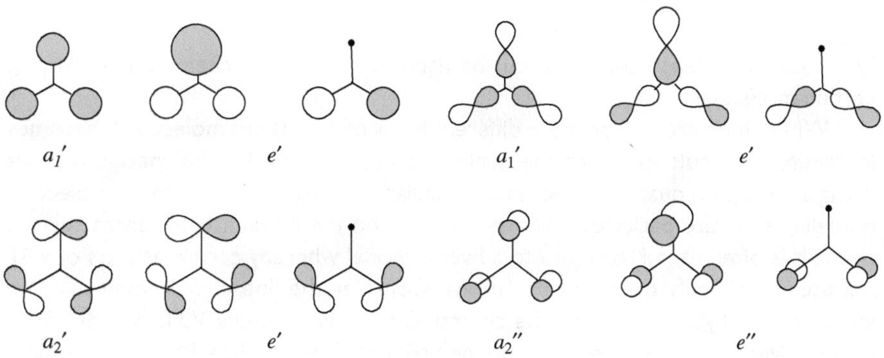

a_1' e' a_1' e'

a_2' e' a_2'' e''

FIGURE 10.39 Shapes of the SALCs for the three fluoride ligands in BF_3.

$$\phi_{e1} = \frac{1}{\sqrt{6}}(2a - b - c) \tag{10.42}$$

$$\phi_{e2} = \frac{1}{\sqrt{2}}(b - c) \tag{10.43}$$

The shapes of the SALCs are shown in Figure 10.39.

The qualitative one-electron MO diagram for BF_3 is shown in Figure 10.40. There are a total of 24 valence electrons in the molecule, which fill the MO diagram to the level of the nonbonding F $2p$ MOs. The overall bond order is 4, indicating an average B–F bond order of 1.3. The nonintegral bond order can only be explained in VBT by invoking three equivalent canonical forms of BF_3, each of which has two B-F single bonds and one B=F double bond. In fact, there is such strong bonding in the BF_3 molecule that the $2a_2''$ LUMO (lowest unoccupied molecular orbital) can accept a pair of electrons from another atom without breaking a B-F bond (because the average B.O. is only reduced to one after accepting an additional electron pair). Hence, BF_3 can act as a strong Lewis acid (Lewis acids are electron pair acceptors).

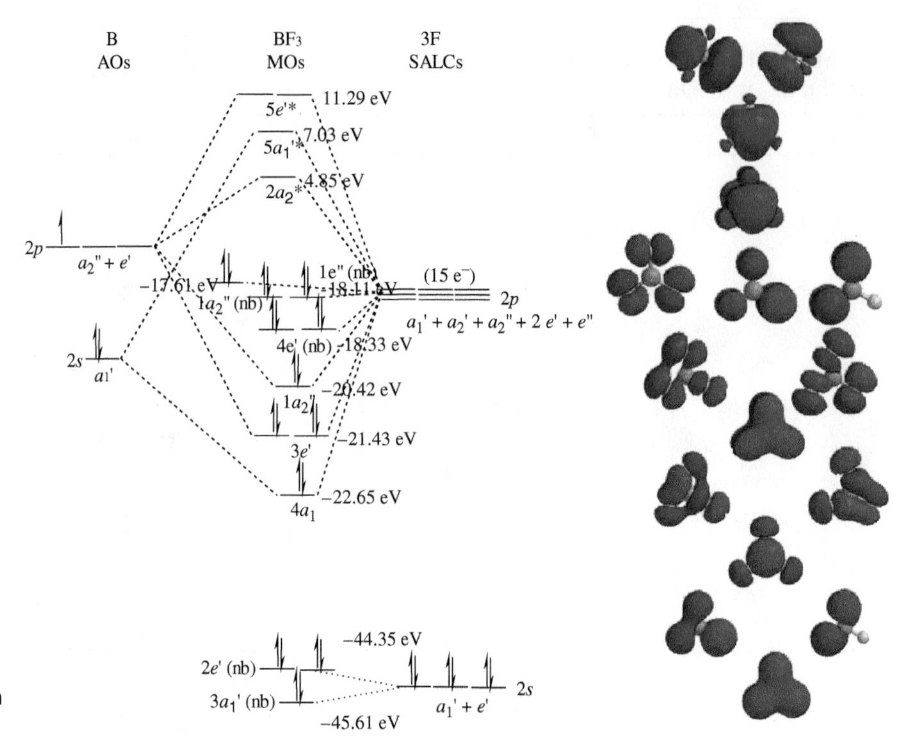

FIGURE 10.40
Qualitative one-electron MO diagram for BF_3. The energies and shapes of the MOs were calculated using Wavefunction's Spartan Student Edition, version 5.0.

This example shows how MOT can be used to predict the chemical reactivity of certain molecules.

When there are two or more different kinds of ligands in a molecule, it becomes increasingly difficult to sketch the qualitative MO diagram for the species without doing a full-blown quantum mechanical calculation using one of the many molecular modeling software packages that are currently on the market. In instances such as these, it is often useful to return to a hybrid model whereby certain aspects of VBT are used to simplify the MO calculations. Consider the linear HCN molecule, for instance. The hybridization on the central C atom is *sp*. Using VBT, one *sp* hybrid forms a *sigma* bond with the 1*s* AO on the pendant H atom, while the other *sp* orbital forms a *sigma* bond with an *sp* hybrid orbital on the N atom. The two unhybridized 2*p* AOs on the C atom are then free to form *pi* bonds with the two unhybridized 2*p* orbitals on the N atom, as shown in Figure 10.41. In the $C_{\infty v}$ point group, the a_1 IRR is also called σ^+ and the e_1 IRR is also called π. We employ these alternative Mulliken symbols in the MO diagram.

We begin by forming *sp* hybrids on the C atom on the left-hand side of the MO diagram by taking the valence bond approach and combining the 2*s* and $2p_z$ AOs to make two *sp* hybrid orbitals. The $2p_x$ and $2p_y$ AOs on the C atom remain unhybridized. One of the *sp* hybrid orbitals on the C atom forms bonding and antibonding σ^+ SALCs with the 1*s* AO on the H atom and these are localized between the C and H atoms. On the right-hand side of the MO diagram, the 2*s* and $2p_z$ AOs on the N atom combine to form two *sp* hybrid orbitals, while the $2p_x$ and $2p_y$ AOs remain unhybridized. The remaining *sp* hybrid on the CH fragment interacts with one of the *sp* hybrids on the N atom to form *sigma* bonding and antibonding MOs localized between the C and N atoms. The two unhybridized 2*p* orbitals on C and on N then interact to form two *pi* bonding and antibonding MOs localized between the C and N atoms. The overall bond order in HCN is four: a C–H single bond and a C≡N triple bond, as the MO diagram and the Lewis structure for HCN both predict. The bonding MOs are polarized toward the N atom, as a consideration of

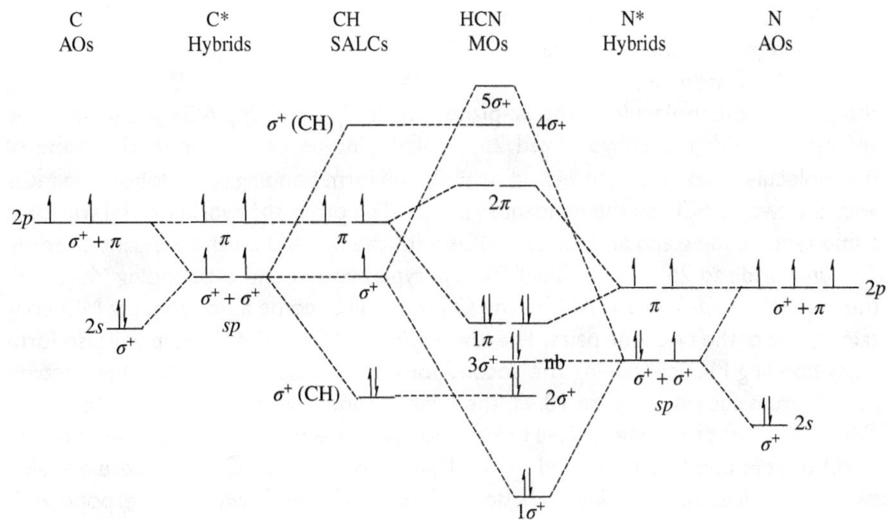

FIGURE 10.41
Qualitative one-electron MO diagram for the HCN molecule. The MOs are numbered beginning with the valence level.

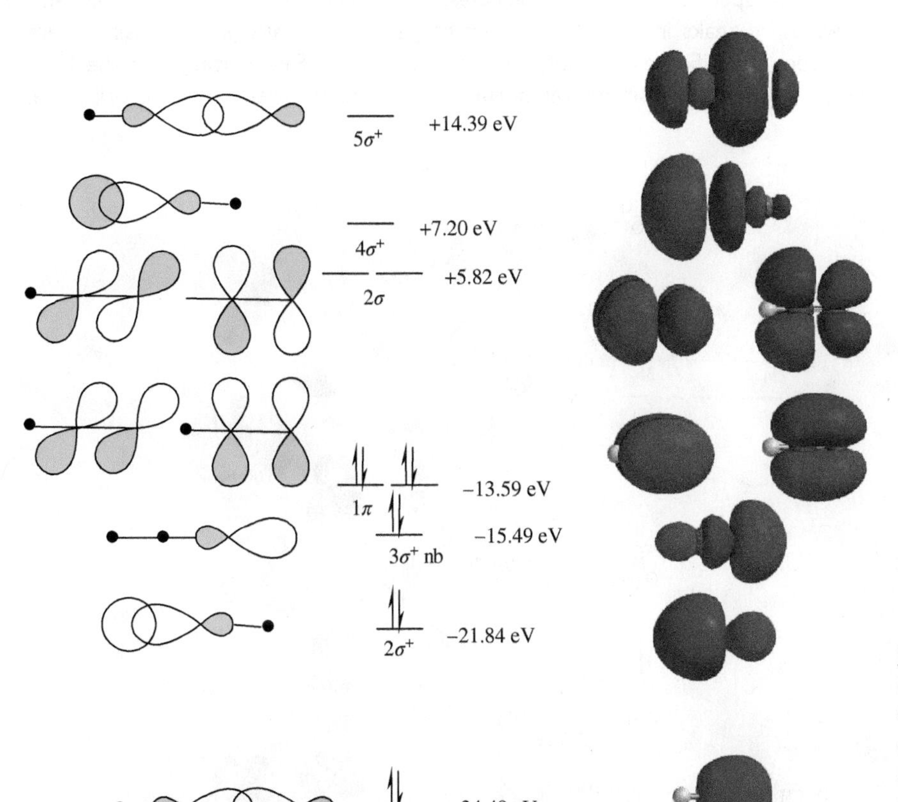

FIGURE 10.42
MO diagram and contour shapes of the MOs for HCN. The energies and shapes of the MOs at the right were calculated using Wavefunction's Spartan Student Edition, version 5.0.

the differences in the electronegativities of the three atoms also predicts. Finally, the MO diagram shows one lone pair of electrons derived from the N atom (the $3\sigma^+$ nonbonding MO), which is also consistent with the Lewis structure of HCN. Naturally, the localized MO approach is an imperfect solution to a very difficult problem. Nonetheless, it does provide a reasonable approximation to the shapes of the actual molecular orbitals (shown in Figure 10.42).

We now turn our attention to the formaldehyde molecule, $H_2C{=}O$. Formaldehyde is a trigonal planar molecule having C_{2v} point symmetry. The symmetries of the AOs on the C atom are as follows: $2s = a_1$, $2p_z = a_1$, $2p_y = b_2$, and $2p_x = b_1$. Defining the plane of the molecule as the *xz*-plane, the 2*s*, $2p_x$, and $2p_z$ AOs will form three sp^2 hybrids with the unhybridized $2p_y$ orbital lying perpendicular to the plane of the molecule. Two of the sp^2 hybrid orbitals will form bonding and antibonding MOs with the two 1*s* AOs on the pendant H atoms. The other sp^2 hybrid orbital will form *sigma*-type bonding and antibonding MOs with the $2p_z$ AO on the pendant O atom. The unhybridized $2p_y$ AO on C will form *pi*-type bonding and antibonding MOs with the $2p_y$ AO on O. The $2p_x$ AO on the O atom will become a nonbonding MO containing one of the two lone pairs. The lower lying 2*s* AO on the O atom will also form a nonbonding MO containing the second lone pair. According to the one-electron MO diagram shown in Figure 10.43, the overall bond order in formaldehyde is four. Two of the bonds belong to C–H single bonds, while the other two belong to the C=O double bond, as predicted by the Lewis structure for CH_2O. There are also two sets of filled nonbonding MOs derived from AOs on O which correspond with the two lone pairs in the Lewis structure.

It is sometimes possible to use qualitative MO theory to explain the electronic absorption spectra of simple molecules. The UV/VIS spectrum of formaldehyde shows three peaks in the 300–360 nm range. These peaks (in decreasing energy) have been identified as $\pi \rightarrow \pi^*$, $n \rightarrow \sigma^*$, and $n \rightarrow \pi^*$. Recognizing that the lowest energy electronic transitions will occur between the frontier molecular orbitals and

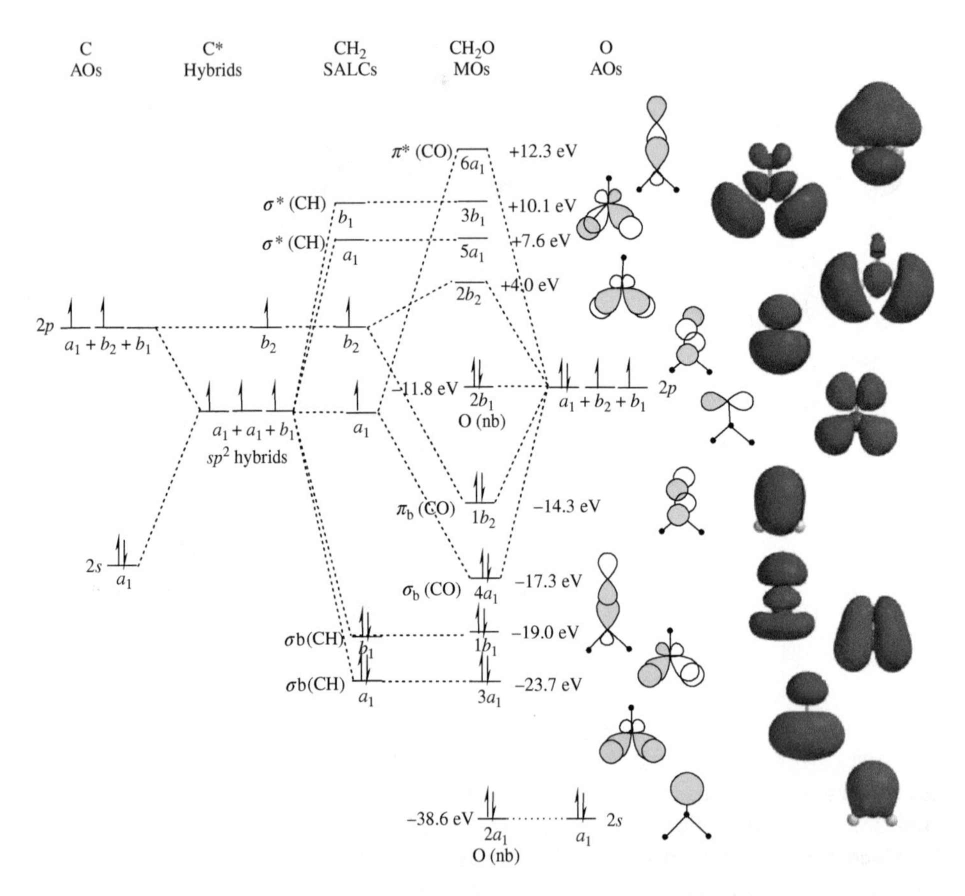

FIGURE 10.43
One-electron MO diagram for formaldehyde, starting with a localized VB approach.

using the one-electron MO diagram in Figure 10.43 as a guide, the $\pi \rightarrow \pi^*$ transition corresponds with the $2b_2 \leftarrow 1b_2$ absorption because the b_2 MOs have pi-type symmetry (overlap above and below the plane of the molecule). By convention, absorptions are written using an arrow connecting the ground state on the right with the excited state on the left. The $n \rightarrow \sigma^*$ transition can be defined as the $5a_1/6a_1 \leftarrow 2b_1$ absorption and the $n \rightarrow \pi^*$ peak as the $2b_2 \leftarrow 2b_1$ absorption.

To be totally rigorous, we should take into account the electron–electron repulsions when assigning electronic transitions and substitute the energies of the different states (or terms) for the one-electron MO orbital energies. In the case of formaldehyde, the electrons are paired in all of the occupied MOs. Thus, the term symbol corresponding with each filled level of the MO diagram will belong to the totally symmetric IRR. Because the electrons are all paired, these terms will all have the $^1A_1$ term symbol. Thus, we can make a one-to-one correspondence between transitions involving orbitals on the one-electron MO diagram for CH_2O and their corresponding $^1A_1$ terms. For example, the $\pi \rightarrow \pi^*$ peak corresponding to the $2b_2 \leftarrow 1b_2$ transition is actually a transition between these two states: $^1A_1(2b_2) \leftarrow {}^1A_1(1b_2)$. The ability of MO theory to accurately explain the electronic spectroscopy of molecules is one of its principal assets.

Sometimes, we are not interested in knowing the shape of the entire MO diagram. In aromatic systems, we are often only interested in studying the shape of the *pi* molecular orbitals. Let us take the case of the cyclopentadienyl ion (abbreviated Cp^-), a fairly common ligand in organometallic chemistry. The Cp^- ligand has six *pi* electrons delocalized over a five-carbon ring structure, as shown in Figure 10.44(a). The molecular point group is D_{5h}. Because the *pi* MOs derive from the $2p_z$ AOs on the five carbon atoms that lie perpendicular to the plane, we can use these (or a set of vectors representing them) as a basis set to generate the symmetries of the five *pi* MOs, as shown in Figure 10.44(b). The resulting representation is shown in Table 10.16 and reduces to the $e_1'' + a_2'' + e_2''$ IRRs.

Using the C_5 subgroup to determine the shapes of the SALCs with the projection operator method, the a SALC is totally symmetric, having the mathematical form shown in Equation (10.44) and the shape shown in Figure 10.45.

$$\phi_a = \frac{1}{\sqrt{5}}(a + b + c + d + e) \tag{10.44}$$

For the e_1 and e_2 SALCs, we must first take linear combinations of the two rows of characters for each IRR and divide by the greatest common multiple to obtain the

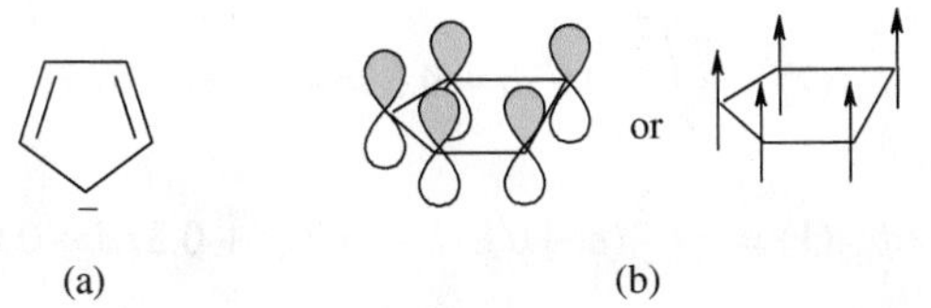

(a) (b)

FIGURE 10.44
(a) The cyclopentadienyl ion; (b) the p_z AOs or vectors perpendicular to the plane can be used as a basis set to generate the symmetry-adapted linear combinations used to formulate the *pi* MOs.

TABLE 10.16 Representation for the p_z-orbitals in the cyclopentadienyl ion.

D_{5h}	E	$2C_5$	$2C_5^2$	$5C_2$	σ_h	$2S_5$	$2S_5^3$	$5\sigma_v$	IRRs:
Γ_{pi}	5	0	0	−1	−5	0	0	1	$e_1''+a_2''+e_2''$

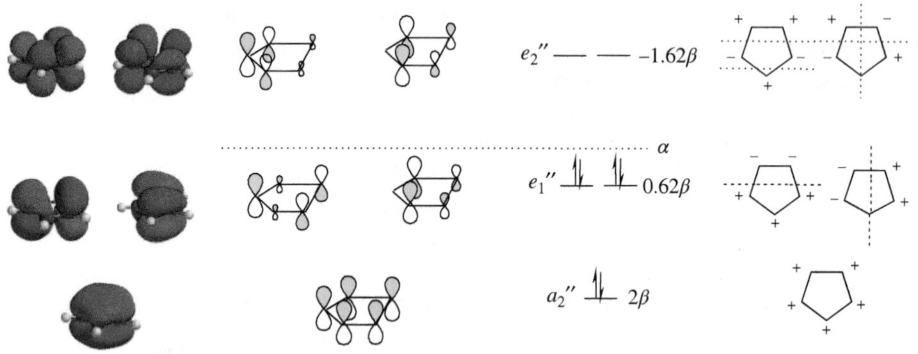

FIGURE 10.45
One-electron molecular orbital diagram for the *pi* MOs in the cyclopentadienyl anion. The shapes of the MOs shown at the left were calculated using Wavefunction's Spartan Student Edition, version 5.0. A cartoon representation of the MOs is shown at the right, where + means the sign of the wavefunction is positive above the plane of the molecule and – means it is negative above the plane. The Hückel energies for the MOs are listed in the center of the MO diagram.

real components of the characters given:

$$e_1: \begin{Bmatrix} 1 & \cos 72^\circ & \cos 144^\circ & \cos 144^\circ & \cos 72^\circ \\ 0 & \sin 72^\circ & \sin 144^\circ & -\sin 144^\circ & -\sin 72^\circ \end{Bmatrix}$$

$$= \begin{Bmatrix} 1 & 0.31 & -0.81 & -0.81 & 0.31 \\ 0 & 0.95 & 0.59 & -0.59 & -0.95 \end{Bmatrix}$$

$$e_2: \begin{Bmatrix} 1 & \cos 144^\circ & \cos 72^\circ & \cos 72^\circ & \cos 144^\circ \\ 0 & \sin 144^\circ & -\sin 72^\circ & \sin 72^\circ & -\sin 144^\circ \end{Bmatrix}$$

$$= \begin{Bmatrix} 1 & -0.81 & 0.31 & 0.31 & -0.81 \\ 0 & 0.95 & -0.59 & 0.59 & -0.95 \end{Bmatrix}$$

Thus, the wavefunctions for the two e_1 SALCs are given by Equation (10.45), while the equations for the e_2 SALC are given by Equation (10.46). The shapes of the MOs are shown on the MO diagram in Figure 10.45.

$$\phi_{e1}(1) = \sqrt{\frac{2}{5}}(a + 0.31b - 0.81c - 0.81d + 0.31e) \qquad (10.45a)$$

$$\phi_{e1}(2) = \sqrt{\frac{2}{5}}(0.95b + 0.59c - 0.59d - 0.95e) \qquad (10.45b)$$

$$\phi_{e2}(1) = \sqrt{\frac{2}{5}}(a - 0.81b + 0.31c + 0.31d - 0.81e) \qquad (10.46a)$$

$$\phi_{e2}(2) = \sqrt{\frac{2}{5}}(0.95b - 0.59c + 0.59d - 0.95e) \qquad (10.46b)$$

Qualitatively, the energies of the MOs for Cp^- can be determined on the basis of the number of nodal planes they possess. The a_2'' MO has no nodal planes and will be the lowest in energy. The e_1'' MOs have one nodal plane and will lie somewhere near the middle of the MO diagram, while the e_2'' MOs have two nodal planes and will have the highest energy. Notice that if you connected the energies of the five *pi* MOs with lines, the shape will match that of the cyclopentadienyl ion itself if it

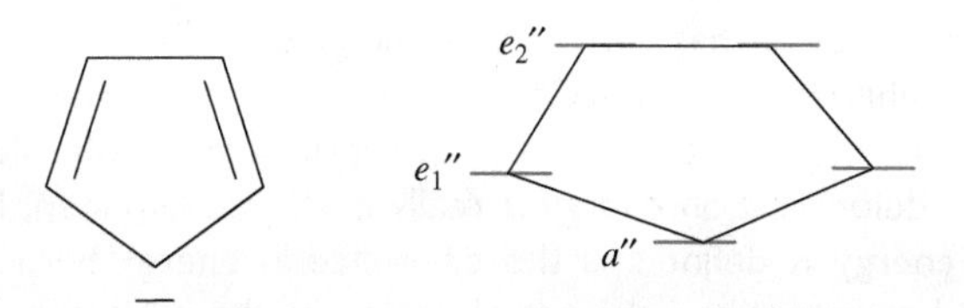

FIGURE 10.46
The shape formed by connecting the different energy levels of the *pi* MOs for any conjugated ring system will always be identical to the shape of the ring system standing on one of its points (the polygon rule).

were standing on one of its points, as shown in Figure 10.46. In fact, this is a general property of any conjugated ring system and is referred to as the *polygon rule.*

If we wanted to know the energies of the orbitals more precisely than this, we could use the *Hückel approximation*. The Hückel method assumes that the H_{ii} Coulomb integrals in the secular determinant for the conjugated *pi* system have the same energy as the 2*p* AOs on the corresponding C atom, which we will call α. The S_{ii} overlap integrals are assumed to be unity. The exchange integrals H_{ij} will be zero unless the two C atoms are adjacent to one another, in which case their energies will depend on the degree of interaction between the C atoms (abbreviated as β). The S_{ij} overlap integral between these orbitals is assumed to be zero (which is perhaps a gross simplification). Thus, the secular determinant for the Cp^- ion is given by Equation (10.47). Solving the determinant for the energy ($E = \alpha - x\beta$) yields the values given in Equation (10.48). More generally, the energies of the *pi* MOs in any conjugated ring system are always relative to the original $2p_z$ AOs on carbon (α), where *x* is given by Equation (10.49).

$$\begin{vmatrix} H_{11}-E & H_{12}-ES_{12} & H_{13}-ES_{13} & H_{14}-ES_{14} & H_{15}-ES_{15} \\ H_{21}-ES_{21} & H_{22}-E & H_{23}-ES_{23} & H_{24}-ES_{24} & H_{25}-ES_{25} \\ H_{31}-ES_{31} & H_{32}-ES_{32} & H_{33}-E & H_{34}-ES_{34} & H_{35}-ES_{35} \\ H_{41}-ES_{41} & H_{42}-ES_{42} & H_{43}-ES_{43} & H_{44}-E & H_{45}-ES_{45} \\ H_{51}-ES_{51} & H_{52}-ES_{52} & H_{53}-ES_{53} & H_{54}-ES_{54} & H_{55}-E \end{vmatrix}$$

$$= \begin{vmatrix} \alpha-E & \beta & 0 & 0 & \beta \\ \beta & \alpha-E & \beta & 0 & 0 \\ 0 & \beta & \alpha-E & \beta & 0 \\ 0 & 0 & \beta & \alpha-E & \beta \\ \beta & 0 & 0 & \beta & \alpha-E \end{vmatrix} = 0 \qquad (10.47)$$

$$\begin{aligned} E_1 &= \alpha - 2\beta \\ E_2 = E_3 &= \alpha - 0.62\beta \\ E_4 = E_5 &= \alpha + 1.62\beta \end{aligned} \qquad (10.48)$$

$$x_k = -2\cos\frac{2k\pi}{n}\left(k = 0, \pm 1, \pm 2, \ldots \begin{Bmatrix} \pm\frac{n-1}{2} \text{ for odd } n \\ \pm\frac{n}{2} \text{ for even } n \end{Bmatrix} \right) \qquad (10.49)$$

Notice that the MOs are evenly distributed in energy with respect to the *barycenter*. If the orbitals were completely filled, the total energy would be the same (within the bounds of the Hückel approximation anyway) as the energy of five filled $2p_z$ atomic orbitals. The four electrons in the e_2'' MOs would be destabilized by $4(1.62\beta) = 6.48\beta$, while the six electrons in the a_2'' and e_1'' MOs would be stabilized by the same amount: $2(2\beta) + 4(0.62\beta) = 6.48\beta$. The barycenter energy is indicated by the horizontal dashed line in Figure 10.45. Because there are only six *pi* electrons in the Cp^- ion, the total energy of these electrons is $6\alpha + 6.48\beta$. The energy of the six *pi* electrons in three isolated ethylene molecules is $6\alpha + 6\beta$.

The Cp^- ligand therefore has a *resonance energy* of 0.48β, or about 12 kJ/mol. The additional stabilization of the Cp^- ligand results from the delocalization of electron density throughout the entire *pi* network of the molecule. The concept of resonance or delocalization energy is really a VB phenomenon. In this context, the resonance energy is defined as the difference in energy between the lowest energy canonical structure and the actual energy of the molecule. The resonance energy can often be estimated experimentally from the electronic spectrum of the molecule.

Example 10-12. Using the polygon rule and Equation (10.49), sketch the one-electron MO diagram for the *pi* MOs in benzene. Use group theory to determine the symmetries and shapes of these MOs.

Solution. Using the polygon rule, the shape of the MO diagram should look like a benzene ring standing on one of its corners. There will be one very antibonding MO, two degenerate less antibonding MOs, two degenerate slightly bonding MOs, and one very bonding MO. Using the $2p_z$ AOs on the six C atoms as a basis yields the following reducible representation:

D_{6h}	E	$2C_6$	$2C_3$	C_2	$3C_2'$	$3C_2''$	i	$2S_3$	$2S_6$	σ_h	$3\sigma_v$	$3\sigma_d$
Γ_{pi}	6	0	0	0	−2	0	0	0	0	−6	2	0

The Γ_{pi} representation reduces to $a_{2u} + b_{2g} + e_{1g} + e_{2u}$. Using the C_6 subgroup and the projection operator method, the following wavefunctions for the MOs are obtained (after normalization):

$$\phi_a = \frac{1}{\sqrt{6}}(a + b + c + d + e + f)$$

$$\phi_b = \frac{1}{\sqrt{6}}(a - b + c - d + e - f)$$

$$\phi_{e1}(1) = \frac{1}{\sqrt{12}}(2a + b - c - 2d - e + f)$$

$$\phi_{e1}(2) = \frac{1}{2}(b + c - e - f)$$

$$\phi_{e2}(1) = \frac{1}{\sqrt{12}}(2a - b - c + 2d - e - f)$$

$$\phi_{e2}(2) = \frac{1}{2}(b - c + e - f)$$

Thus, the a_{2u} MO will have zero nodes and lie lowest in energy, the e_{1g} set of MOs will each have one nodal plane and lie just beneath the barycenter, the e_{2u} MOs will have two nodal planes and lie right above the barycenter, and the b_{2g} MO that has three nodal planes will lie highest in energy. Using Equation (10.49), where $n = 6$ (even), the following energies are obtained:

$$(a_{2u}) : E_0 = \alpha + 2\beta$$

$$(e_{1g}) : E_{\pm 1} = \alpha + \beta$$

$$(e_{2u}) : E_{\pm 2} = \alpha - \beta$$

$$(b_{2g}) : E_{\pm 3} = \alpha - 2\beta$$

Thus, the one-electron MO diagram for the *pi* MOs in benzene is shown here.

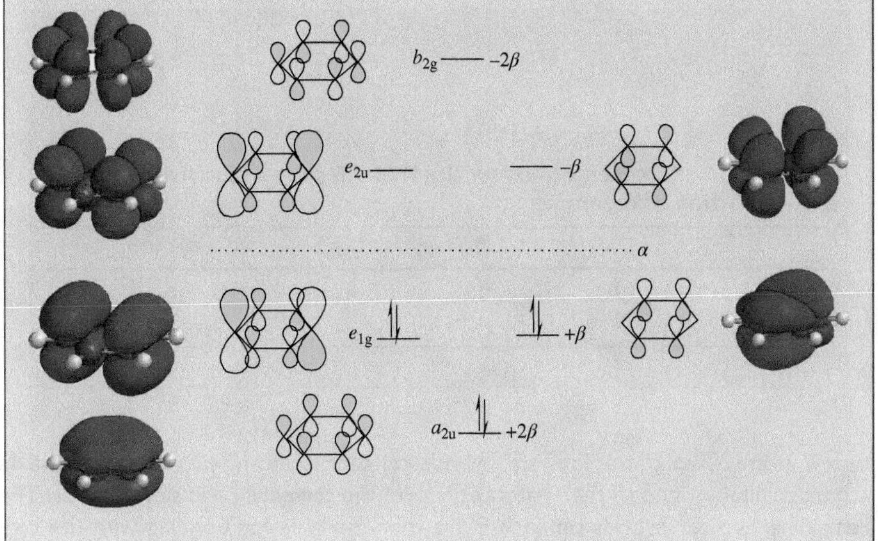

b_{2g} —— -2β

e_{2u} —— —— $-\beta$

α

e_{1g} —⇅— —⇅— $+\beta$

a_{2u} —⇅— $+2\beta$

10.5 MOLECULAR ORBITAL THEORY: MORE COMPLEX EXAMPLES

One of the more interesting molecules that is rather difficult to rationalize with the valence bond model is diborane, whose molecular structure is shown in Figure 10.47. In order for the bridging H atoms to bond to both of the B atoms, the valence electron in each bridging H must be delocalized over all three atoms. These "three-centered, two-electron bonds" are problematic for the VB model, which likes to pigeon-hole the valence electrons into regions of space that are localized between two and only two nuclei. Furthermore, the molecular geometry with respect to the bridging H's simply cannot be explained using VSEPR theory. Because there are two domains around each bridging H the predicted electron geometry would be linear with a B–H–B bond angle of 180°. On the other hand, the concept of a three-centered, two-electron bond is not a problem for MOT, where delocalized bonding is a common phenomenon.

Because we are really only interested in the bonds involving the B–H–B bridges in diborane, we can therefore focus on just that portion of the molecule in our MO analysis. We begin the problem by forming four equivalent sp^3 hybrid orbitals on the

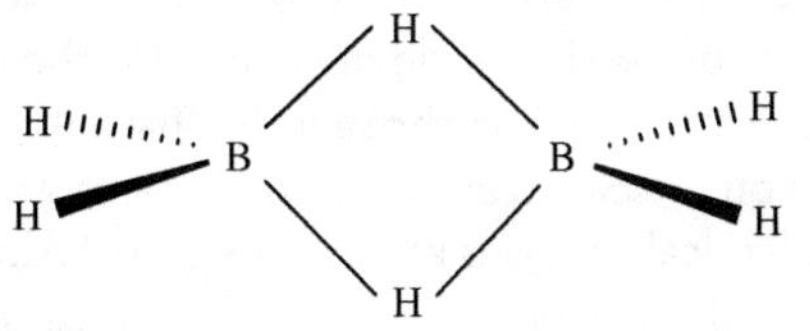

H H B B H H H H

FIGURE 10.47
The structure of diborane, B_2H_6.

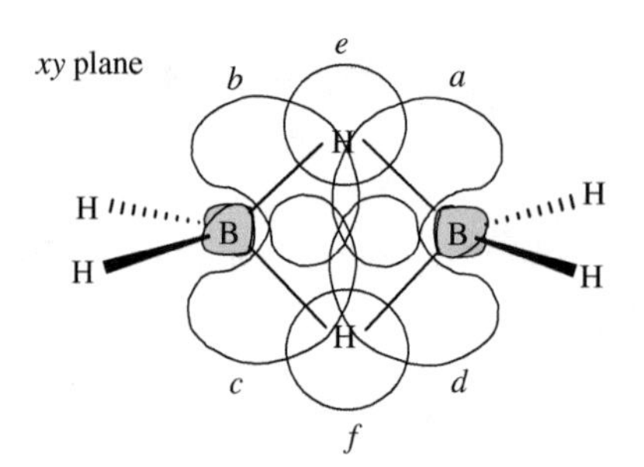

FIGURE 10.48
Basis orbitals for the B–H–B portion in B_2H_6.

TABLE 10.17 Representations for the H and B orbitals involved in the B–H–B portion of diborane.

D_{2h}	E	C_2^z	C_2^y	C_2^x	i	σ_{xy}	σ_{xz}	σ_{yz}	IRRs:
Γ_H	2	0	0	2	0	2	2	0	$a_g + b_{3u}$
Γ_B	4	0	0	0	0	4	0	0	$a_g + b_{1g} + b_{2u} + b_{3u}$

two B atoms. Two of the four sp^3 hybrids on each B atom will overlap with a 1s atomic orbital on two of the H atoms to form the terminal B–H single bonds. The remaining two sp^3 hybrids on each B are then available for bonding with the two bridging H atoms. Using the basis orbitals shown in Figure 10.48, the SALCs for the two bridging H atoms will have $a_g + b_{3u}$ symmetry, while the SALCs for the two B atoms will have $a_g + b_{1g} + b_{2u} + b_{3u}$ symmetry, as shown in Table 10.17.

Using the projection operator, the mathematical forms of the six SALCs (after normalization) are given by Equations (10.50)–(10.55):

$$a_{1g}(H) : \phi_1 = \frac{1}{\sqrt{2}}(e + f) \tag{10.50}$$

$$b_{3u}(H) : \phi_2 = \frac{1}{\sqrt{2}}(e - f) \tag{10.51}$$

$$a_{1g}(B) : \phi_3 = \frac{1}{2}(a + b + c + d) \tag{10.52}$$

$$b_{1g}(B) : \phi_4 = \frac{1}{2}(a - b + c - d) \tag{10.53}$$

$$b_{2u}(B) : \phi_5 = \frac{1}{2}(a - b - c + d) \tag{10.54}$$

$$b_{3u}(B) : \phi_6 = \frac{1}{2}(a + b - c - d) \tag{10.55}$$

The one-electron MO diagram for diborane and the shapes of the MOs are shown in Figure 10.49. In sketching the shapes of the MOs, only the largest lobes of the four sp^3 hybrid orbitals were shown; however, the smaller lobes on the opposite side can also overlap. Because of this fact, the two nonbonding MOs will not have identical energies. The b_{1g} MO will be slightly lower in energy than the b_{2u} MO because the smaller lobes of the sp^3 hybrid orbitals in b_{1g} show positive overlap above and below the internuclear axis while those in b_{2u} have negative overlap here. The four sp^3 hybrid orbitals on the two B atoms have the same energy after hybridization, but they do not have identical energies after making the SALCs. Thus, their IRRs are shown in parentheses at the extreme left of the MO diagram. Likewise, the two 1s

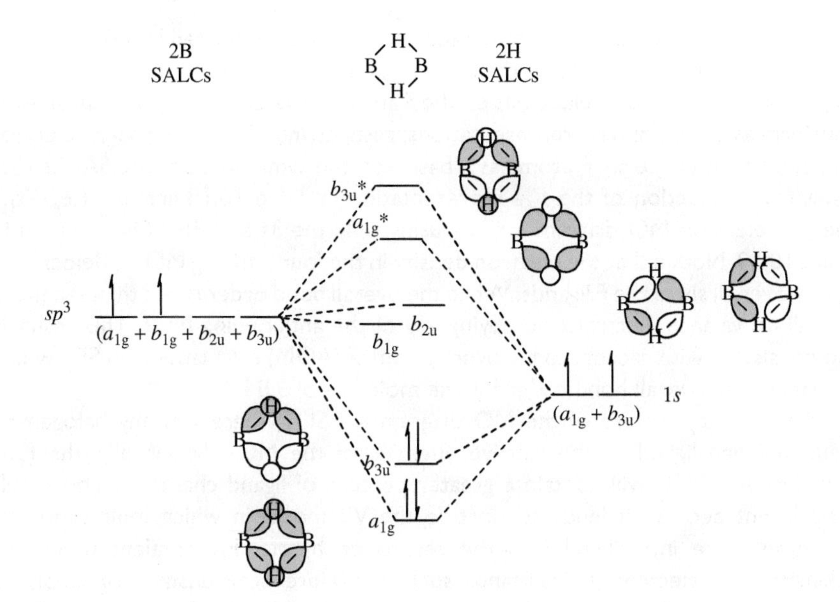

FIGURE 10.49
Partial one-electron MO diagram for the B–H–B portion of B_2H_6.

AOs on the two H atoms have the same energy before we combine them to make SALCs, but they have different energies after taking the linear combinations. Their Mulliken symbols are also shown in parentheses on the right-hand side of the MO diagram.

The shapes of the bonding a_{1g} and b_{3u} MOs clearly show that the electron pairs occupying these MOs are delocalized over all three nuclei. Because of the small size of the H atoms, there also exists a certain amount of direct B–B bonding in diborane that helps to stabilize the molecule. The MO diagram is also consistent with the PES spectrum of diborane, which shows six peaks of comparable intensity at 11.8, 13.3, 13.9, 14.7, 16.1, and 21.4 eV. Four of these peaks are due to ionization from the four terminal B–H bonding MOs (not shown in the figure), while the two peaks at 13.3 and 14.7 eV correspond with the b_{3u} and a_{1g} MOs, respectively, on the B–H–B bridges.

One of the other shortcomings of VBT was in the way that it dealt with so-called hypervalent molecules, such as SF_6. According to the definition proposed by Musher, a hypervalent molecule is one that contains a Group 15–18 central atom having a formal oxidation number higher than its lowest oxidation state. In this specific example, it would appear that the central S atom has 12 valence electrons around it in a traditional Lewis structure illustrating localized covalent bonding. Because this exceeds the valence of S, the molecule was said to be hypervalent or hypercoordinated. In order to explain how S might be able to "expand its octet," some authors argued that S could use its low-lying 3*d*-orbitals in order to make the d^2sp^3 hybrid orbitals on the central S atom that were necessary to explain its octahedral coordination. An alternative model argued that the electronegativity difference between S and F is sufficient enough to include ionic terms in the resonance hybrid, having the formula $F^- \cdots SF_4^{2+} \cdots F^-$. Thus, the S atom could maintain a noble gas electronic configuration. A third approach developed by Pimental in the 1950s involved three-centered, four-electron (3c–4e) bonding, as was discussed in Section 10.1 for PF_5.

According to VBT, the hybridization on the central S atom is d^2sp^3, where the S atom is forced to use two of its low-lying 3*d* orbitals in the formation of the hybrids. The percent *d*-character in a d^2sp^3 hybrid is 33%. However, modern quantum mechanical calculations on the SF_6 molecule indicate that there is only about 8% *d*-character in the S-F bonds, which is considerably less than the 33% expected for a d^2sp^3 hybrid. Clearly, while the *d*-orbitals seem to participate to a limited extent in the bonding, their influence is considerably less than that predicted by VBT. In fact, it is entirely possible to rationalize the bonding in SF_6 using a MO treatment involving only the valence 3*s* and 3*p* electrons on the S atom. In the O_h point group, these AOs transform as the a_{1g} and t_{1u} representations, respectively. Taking a *sigma*-type donor orbital on each of the six F atoms as a basis set, the symmetries of the SALCs that result from reduction of the Γ_{6F} representation in Table 10.18 are: $a_{1g}+e_g+t_{1u}$. The one-electron MO diagram for SF_6 using only the 3*s* and 3*p* AOs is shown in Figure 10.50. Notice that the electron density in the four bonding MOs is delocalized equally over all six of the F ligands. While the overall bond order is less than six, there are still no valence electrons occupying any of the antibonding MOs. This result is also consistent with recent atomic overlap matrix (AOM) calculations on SF_6, which determined an overall bond order for the molecule of 3.84.

Another key feature of the MO diagram for SL_6, where L is any halogen or pseudohalogen ligand, is the relative energies of the MOs. Specifically, the four *sigma*-bonding MOs will contain a greater percent of ligand character. This result is significant because it lends support to the VB model in which ionic canonical structures were introduced into the resonance hybrid. Hypervalent molecules lacking strongly electronegative ligands, such as SH_6, are highly unstable or simply do not exist. Furthermore, the ionic character of the bonding in SF_6 can be stabilized by what is known as *negative hyperconjugation*. One of the filled 2*p*-orbitals on the F ligands can interact with the unoccupied σ^* (SF) MO, which has the potential to break one of the S–F bonds to form an ionic canonical structure, as shown in Figure 10.51. Because the energy of the σ^* (3*s*) LUMO will lie lower in the MO diagram as the electronegativity of the ligand increases, this hyperconjugation interaction is stronger for more electronegative ligands. In conclusion, the best model of chemical bonding for hypervalent molecules is one which favors ionic bonding and negative hyperconjugation over *d*-orbital participation. This is yet another example of how MOT is in many ways superior to its VB counterpart.

TABLE 10.18 Symmetries of the SALCs formed from σ-donor orbitals on the six F atoms in the octahedral molecule SF_6.

O_h	E	$8C_3$	$6C_2$	$6C_4$	$3C_2'$	i	$6S_4$	$8S_6$	$3\sigma_h$	$6\sigma_d$	IRRs:
Γ_{6F}	6	0	0	2	2	0	0	0	4	2	$a_{1g}+e_g+t_{1u}$

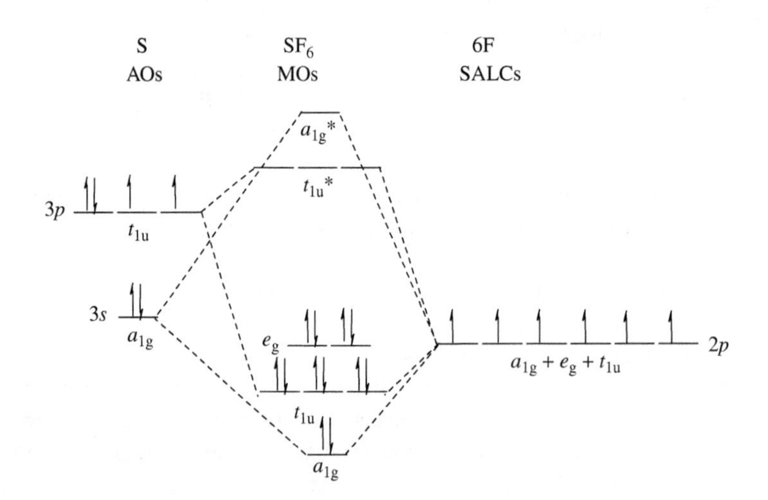

FIGURE 10.50
One-electron MO diagram for the SF_6 molecule using only the 3*s* and 3*p* atomic orbitals on the S atom.

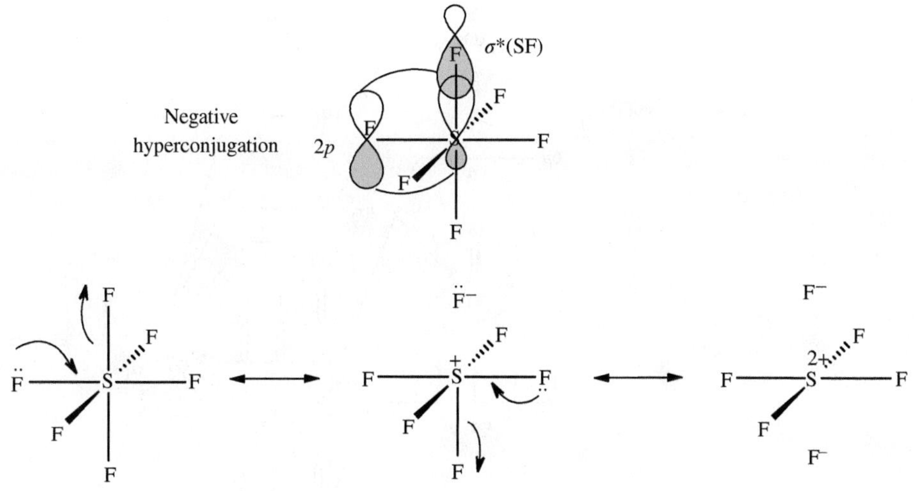

FIGURE 10.51
Negative hyperconjugation stabilizing the ionic canonical structures in the resonance hybrid for SF_6.

In the previous example, we showed how it was possible to construct an MO diagram for SF_6 without using any *d*-orbitals on the central atom. Because Os is a transition metal, in the case of OsO_4, the *d*-orbitals *must* contribute to the bonding. The electron configuration for Os is [Xe] $6s^2$ $5d^6$. We therefore use the 5*d*, 6*s*, and the empty 6*p* AOs on Os to form MOs with SALCs from the O ligands. Recall that when the (*n*−1)*d* orbitals are filled with electrons, they are lower in energy than the ns orbitals. The 5*d* AOs transform as $e+t_2$, the 6*s* AO as a_1, and the 6*p* AOs as t_2. For the four O atoms, the SALCs are constructed using the valence 2*s* and 2*p* AOs. Using the four 2*s* AOs on the O atoms as a basis, the symmetries of the SALCs (after reduction) are a_1+t_2, as shown in Table 10.19. The four 2*p* AOs on the O atoms that directly lie along the Os–O bonds (the *sigma*-type orbitals) also transform as a_1+t_2. The remaining eight (*pi*-type) 2*p* SALCs reduce to $e+t_1+t_2$. A qualitative MO diagram for OsO_4 is shown in Figure 10.52. There are a total of 32 valence electrons that fill the MO diagram to the $1t_1$ (nb) level.

The overall bond order in the molecule is nine, which is slightly larger than the four Os=O double bonds predicted by VBT. One interesting feature of this MO diagram, which is supported by PES data, is that some of the *pi*-type MOs ($2t_2$ and 1e) are more tightly bound than the *sigma*-type MOs ($2a_1$). This contrasts markedly with the usual prediction in VBT that the *sigma* orbitals are more tightly held than the *pi* orbitals in double bonds. One of the chemical properties of osmium tetroxide is that it is an extremely powerful oxidizing agent. Because of its large overall bond order,

TABLE 10.19 Symmetries of the different basis sets for the four O atoms in OsO_4.

T_d	E	$8C_3$	$3C_2$	$6S_4$	$6\sigma_d$	IRRs:
Γ_s	4	1	0	0	2	a_1+t_2
Γ_σ	4	1	0	0	2	a_1+t_2
Γ_π	8	−1	0	0	0	$e+t_1+t_2$

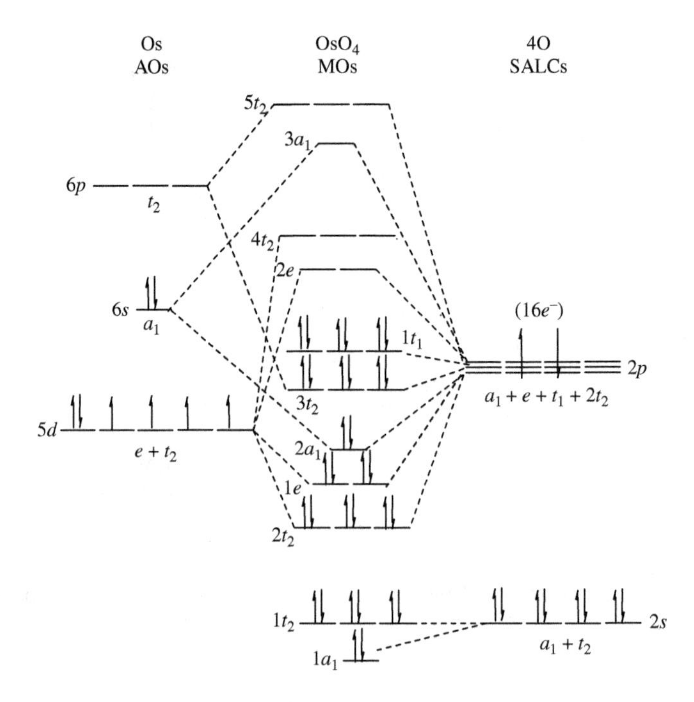

FIGURE 10.52
Qualitative MO diagram for OsO_4.

the molecule can easily accept electrons into the relatively low-lying 2e antibonding MO without significantly weakening the Os–O bonding.

Example 10-13. Sketch a qualitative MO diagram for the octahedral coordination compound, $[Fe(H_2O)_6]^{2+}$. For each coordinate covalent bond, the H_2O ligands act as Lewis bases (σ-donors) and each donate a pair of electrons to the metal–ligand bond. Using only the σ-donating orbitals on the H_2O ligands as a basis set for the ligands (as we did for the F atoms in SF_6), determine the symmetries of the six H_2O SALCs. For the Fe^{2+} ion, include the valence 3*d*, 4*s*, and 4*p* AOs. Sketch the shapes of all the filled MOs.

Solution. In the O_h point group, the symmetries of the 3*d* AOs are $e_g + t_{2g}$, the 4*s* orbital is a_{1g}, and the 4*p* AOs transform as t_{1u}. The basis set for the six *sigma*-donating ligands is shown at the top of the next page. The SALCs from the H_2O ligands are (as shown here): $a_{1g} + e_g + t_{1u}$.

O_h	E	$8C_3$	$6C_2$	$6C_4$	$3C_2'$	i	$6S_4$	$8S_6$	$3\sigma_h$	$6\sigma_d$	IRRs:
Γ_{6F}	6	0	0	2	2	0	0	0	4	2	$a_{1g} + e_g + t_{1u}$

σ_z σ_{-x} σ_y Fe σ_{-y} σ_x σ_{-z}

Combining orbitals having the same symmetry and similar energies from the AOs on Fe^{2+} and SALCs on the ligands, we obtain the following MO diagram:

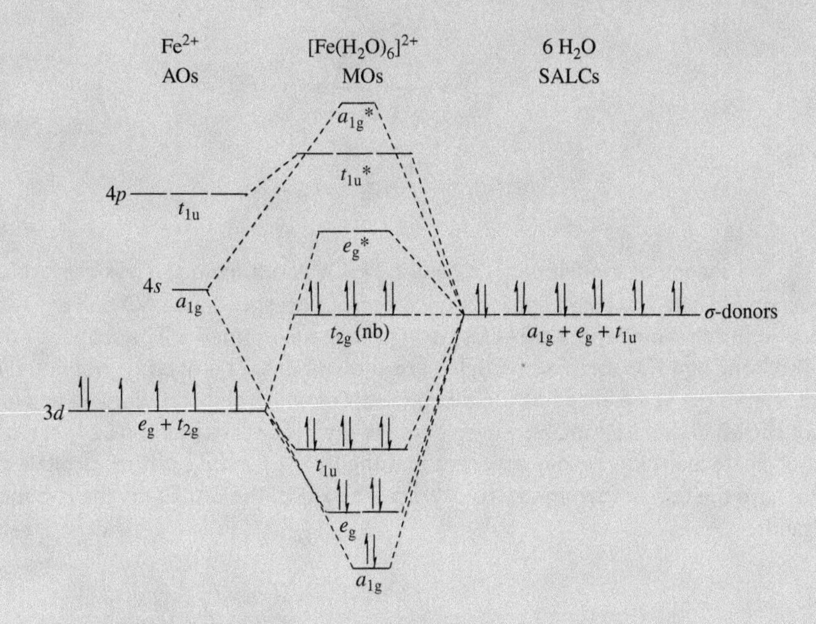

In order to determine the shapes of the filled MOs, we must first determine the shapes of the six SALCs for the ligands. Using the O sub-rotational group and the projection operator method, the mathematical forms of the six SALCs are derived as follows. Using the σ-donor orbitals lying along the x, $-x$, y, $-y$, z, and $-z$ axes as our basis functions, the a_1 SALC will be totally symmetric:

$$\phi_{a1} = \frac{1}{\sqrt{6}}(\sigma_x + \sigma_{-x} + \sigma_y + \sigma_{-y} + \sigma_z + \sigma_{-z})$$

In order to obtain the two e SALCs, we first apply the projection operator to the basis function σ_z to obtain the result: $\phi_z = N(2\sigma_z + 2\sigma_{-z} - \sigma_x - \sigma_{-x} - \sigma_y - \sigma_{-y})$. If the projection operator acts on σ_x, we obtain $\phi_x = N(2\sigma_x + 2\sigma_{-x} - \sigma_y - \sigma_{-y} - \sigma_z - \sigma_{-z})$. Likewise, $\phi_y = N(2\sigma_y + 2\sigma_{-y} - \sigma_x - \sigma_{-x} - \sigma_z - \sigma_{-z})$. We need only two orthogonal solutions for the doubly degenerate e representation. Accepting ϕ_z to be the first of those solutions, the other must be a linear combination of ϕ_x and ϕ_y which is also orthogonal to ϕ_z. Taking the negative linear combination of $\phi_x - \phi_y$, we obtain the second representation for e shown here. It can be readily shown that the coefficients in front of each basis function (in the order x, $-x$, y, $-y$, z, and $-z$) for the two e SALCs satisfy the criterion for orthogonality: $(-1)(1) + (-1)(1) + (-1)(-1) + (-1(-1) + (2)(0) + (2)(0) = 0$.

$$\phi_e(1) = \frac{1}{2\sqrt{3}}(-\sigma_x - \sigma_{-x} - \sigma_y - \sigma_{-y} + 2\sigma_z + 2\sigma_{-z})$$

$$\phi_e(2) = \frac{1}{2}(\sigma_x + \sigma_{-x} - \sigma_y - \sigma_{-y})$$

When the projection operator for the t_1 IRR acts on σ_z, we obtain $\phi_z = N(\sigma_z + \sigma_{-z})$. When the same operator acts on σ_x, we obtain $\phi_x = N(\sigma_x + \sigma_{-x})$; and when it acts on σ_y, we get $\phi_y = N(\sigma_y + \sigma_{-y})$. These three solutions are already orthogonal to one another. Following normalization, the three t_1 SALCs have the forms:

$$\phi_{t1}(1) = \frac{1}{\sqrt{2}}(\sigma_x - \sigma_{-x})$$

$$\phi_{t1}(2) = \frac{1}{\sqrt{2}}(\sigma_y - \sigma_{-y})$$

$$\phi_{t1}(3) = \frac{1}{\sqrt{2}}(\sigma_z - \sigma_{-z})$$

The shapes of the filled a_{1g}, e_g, and t_{1g} MOs result from the positive orbital overlap formed by taking a linear combination of the appropriate AO on Fe^{2+} with one of the abovementioned SALCs having the same symmetry. The shapes of the filled bonding MOs for $[Fe(H_2O)_6]^{2+}$ are shown here. The filled t_{2g} nonbonding MOs have the same shape as the *d*-orbitals from which they are derived and are not shown in the diagram. However, because their lobes are pointed between the coordinate axes, it is readily apparent that the d_{xy}, d_{xz}, and d_{yz} atomic orbitals do not have the correct symmetry to overlap with any of the SALCs on the σ-donor ligands.

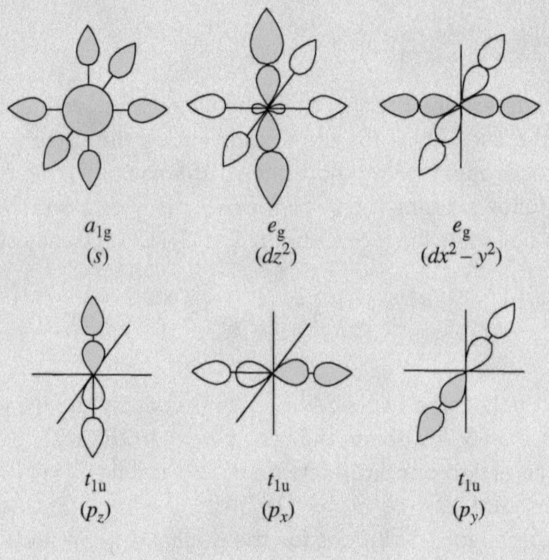

a_{1g} (s) e_g (dz^2) e_g ($dx^2 - y^2$)

t_{1u} (p_z) t_{1u} (p_x) t_{1u} (p_y)

One of the earliest known complexes to contain a metal–metal quadruple bond was the $[Re_2Cl_8]^{2-}$ ion, which was first reported in 1965. The molecule consists of two square planar $ReCl_4^-$ fragments that are bonded to each other through the Re ions. The Re–Re bond distance in $[Re_2Cl_8]^{2-}$ (2.24 Å) is considerably shorter than the average Re–Re bond length (2.75 Å). Despite the extremely short bond length, however, the two square planar $ReCl_4^-$ fragments have an eclipsed configuration! The point group of the molecule is therefore D_{4h}. Given the significant electron–electron repulsions between the bonding pairs on any two eclipsed chloride ligands (especially in such close proximity), VSEPR theory is unable to explain the observed geometry. Let us see if the MO diagram for $[Re_2Cl_8]^{2-}$ can shed any light on this apparent dilemma.

Ignoring the Cl^- ligands for the moment and concentrating only on the Re–Re bonding, the symmetries of the valence Re atomic orbitals in the D_{4h} point group are $s = a_{1g}$; $p_z = a_{2u}$; p_x, $p_y = e_u$; $d_{z^2} = a_{1g}$; $d_{x^2-y^2} = b_{1g}$; $d_{xy} = b_{2g}$; and d_{xz}, $d_{yz} = e_g$.

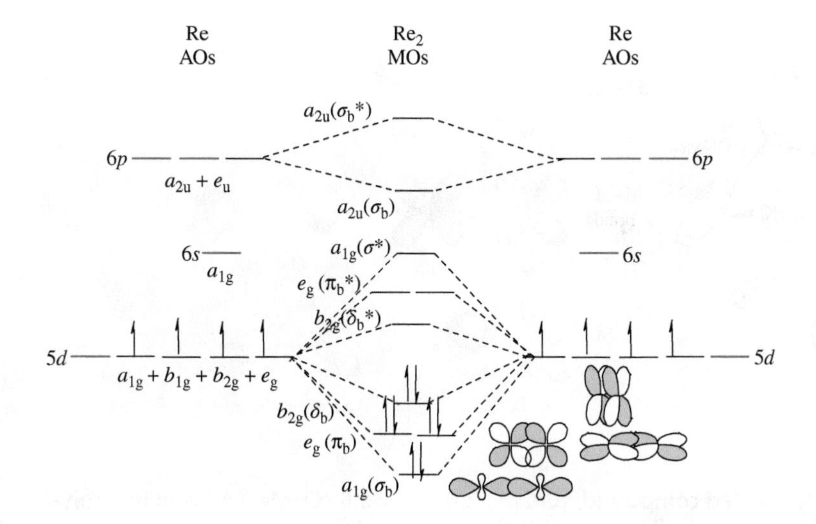

FIGURE 10.53
Partial one-electron MO diagram for $[Re_2Cl_8]^{2-}$ focusing on the Re–Re bonding. Those atomic orbitals in red are used to make the square planar dsp^2 hybrid orbitals that form the eight Re–Cl single bonds (which are omitted here for clarity).

The one-electron MO diagram for the Re–Re bond in $[Re_2Cl_8]^{2-}$ ion is shown in Figure 10.53, where the shapes of the bonding MOs have also been included. The hybridization within each square planar $ReCl_4^-$ fragment is $d_{x^2-y^2}sp_xp_y$. The AOs having these symmetries are therefore used in making the *sigma* bonds to the Cl^- ligands. The p_z AOs on each Re atom overlap to form *sigma*-type MOs. However, these MOs are both higher in energy than those formed by the four remaining *d* AOs.

According to the MO diagram, the overall Re–Re bond order in $[Re_2Cl_8]^{2-}$ is four, which explains the unusually short Re–Re bond distance in the species. The a_{1g} MO is a *sigma*-type bonding orbital, the e_g MOs form two *pi*-type bonds, and the b_{2g} bonding MO is defined as a *delta bond*. Notice from the relative energy of the b_{2g} MO that the *delta* bond in $[Re_2Cl_8]^{2-}$ is weaker than its two *pi* bonds, which themselves are weaker than its *sigma* bond. The two $ReCl_4^-$ fragments are forced to be eclipsed because the same *d*-orbital must be used on each Re in order to have the four lobes on the two d_{xy} orbitals overlap with one another to form the *delta* bond. Metal–metal quadruple bonds are not unique to the $[Re_2Cl_8]^{2-}$ ion. Indeed, a large variety of compounds having the general formula $[M_2X_8]^{n-}$, where M = Mo, W, Os, Re, Tc and X = Cl^-, Br^-, exhibit molecular structures consistent with an M-M bond order of four. Other quadrupally bonded complexes, where a ligand such as acetate bridges the two metal centers, are also consistent the presence of a *delta* bond.

In 2005, the first metal–metal quintuple bond was proposed for the dichromium compound shown in Figure 10.54, which shows how a second *delta* bond can form between the *d*-orbitals of the two metals.

10.6 BORANE AND CARBORANE CLUSTER COMPOUNDS

The boranes represent an interesting class of cluster compounds that are based on *deltahedra*. Deltahedra are simply polyhedrons whose faces are all composed of equilateral triangles. The boranes consist of a class of compounds that contain only B and H atoms. Boron trihydride (BH_3) is the parent compound. Similar to many boranes, it is highly reactive and readily dimerizes into diborane, B_2H_6, a compound whose three-centered B–H–B, two-electron bonding was discussed previously. Although they are too reactive to occur naturally, a wide variety of boranes have been synthesized in the laboratory, and the complexity of their interesting structures can be rationalized on the basis of *Wade's rules*. Wade's rules are themselves a subset of the more general *polyhedral skeletal electron pair (SEP) theory*. They were first postulated in 1971 by Kenneth Wade and further developed by D. M. P. Mingos. Also referred

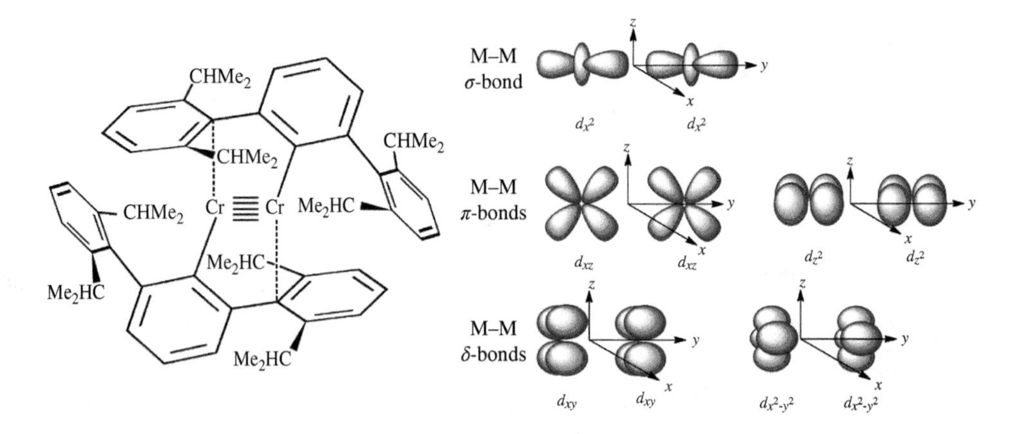

FIGURE 10.54
Proposed structure of the first pentuply bonded compound, $[CrC_6H_3\text{-}2,6\text{-}(C_6H_3\text{-}2,6\text{-}(CHMe_2)_2)_2]$ and its orbital interactions. [(a) Reproduced from http://en.wikipedia.org/wiki/Quintuple_bond (accessed January 2, 2014) and (b) http://commons.wikimedia.org/wiki/File:Quintuple_bond_orbital_diagram.png (accessed January 2, 2014).]

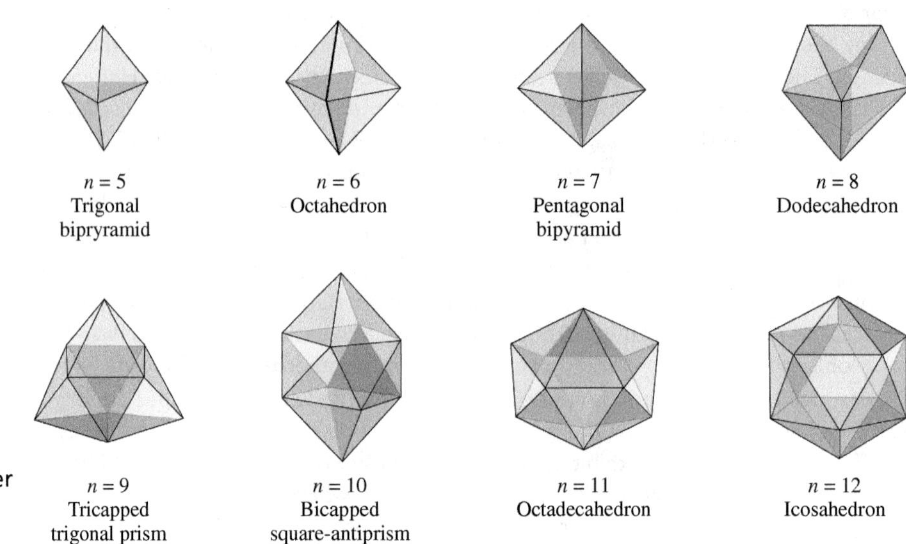

FIGURE 10.55
Deltahedral cages having $n = 5–12$, where n is the number of vertices. [Blatt Communications.]

to as the *4n rules*, Wade's rules apply to those classes of cluster compounds having approximately four electrons per vertex in their deltahedra structures. The names and structures of the deltahedra with 5–12 vertices are shown in Figure 10.55. These closed deltahedral cages are appropriately referred to as *closo* (from the Greek word for cage).

Application of Wade's rules to the boranes is as follows:

1. Determine the valence electron count by adding three electrons for each B atom and one electron for each H atom. If the borane is negatively charged, then additional electrons will need to be added for the charge on the anion.
2. Determine the type of borane cluster from Table 10.20, where n is the number of B atoms. *Closo* structures have n vertices. *Nido* structures (from the Greek word for nest) are based on the $n + 1$ deltahedron with a single vertex missing. Similarly, the *arachno* (spider), *hypho* (net), and *klado* (branch) boranes

TABLE 10.20 Nomenclature and structural formulas for the different types of borane clusters.

Type of Cluster	General Formula	Number of Valence e⁻	Number of Vertices	Number of Vacancies	Number of SEPs or Bonding MOs
closo-	$B_nH_n^{2-}$	$4n+2$	n	0	$n+1$
nido-	$B_nH_n^{4-}$	$4n+4$	$n+1$	1	$n+2$
arachno-	$B_nH_n^{6-}$	$4n+6$	$n+2$	2	$n+3$
hypho-	$B_nH_n^{8-}$	$4n+8$	$n+3$	3	$n+4$
klado-	$B_nH_n^{10-}$	$4n+10$	$n+4$	4	$n+5$

are based on the deltahedra listed in Table 10.20 with the corresponding number of vertices removed from the parent structure.

3. Use the correlation diagram illustrated in Figure 10.56 to determine which vertices will be vacant from the parent deltahedral structure.
4. Those B atoms having three of more connected vertices in the resulting structure will each have one terminal H atom, while those B atoms having only two connected vertices in the cluster will each have two terminal H atoms. All of the remaining H atoms in the molecular formula will exist as part of the structural framework as bridging H atoms.
5. Subtracting two electrons for each of the terminal B–H bonds from the total valence electron count and then dividing this result by two yields the number of structural electron pairs (SEPs) in the framework.

Let us consider $B_6H_6^{2-}$ as an example. The total valence electron count is $6(3)+6(1)+2=26$ valence electrons. There are six B atoms in the structure, so $n=6$. Because $4n+2=26$ e⁻, this structure is a *closo* cluster based on the deltahedron with *n* vertices, or an octahedron. Each of the B atoms in the octahedron has four connecting vertices. Therefore, each B will have one terminal H atom connected to it, leading to the structure shown in Figure 10.57, where all of the B–B bonds are 172 pm in length. Subtracting two electrons for each of the six terminal B–H bonds from the total valence electron count yields $26-6(2)=14$ valence electrons available for bonding within the framework, or 7 SEPs.

The underlying basis of Wade's rules and the list of 4*n* rules in Table 10.20 derive from the molecular orbital diagrams of the cluster compounds. Again using the $B_6H_6^{2-}$ ion as an example, two sets of basis functions can be used to generate the molecular orbitals involved in the bonding. First, we will arbitrarily assign the z-axis for each B atom in the cluster to point toward the center of the parent deltahedron. Then we will form a set of *sp*-hybridized orbitals using the p_z orbital on each, as shown in the example provided in Figure 10.58. One of the *sp* hybrids on each B atom will be used to form the terminal B–H bonds, while the other *sp* hybrid will participate in a delocalized set of molecular orbitals in the center of the cage. We will call this latter set of *sp* hybrid orbitals the radial orbitals. Next, we will define a set of tangential basis functions on each B atom from the unhybridized p_x and p_y orbitals. Application of the group theoretical principles developed earlier in this chapter yields the symmetries of the framework MOs listed in Table 10.21.

Ignoring those MOs which are solely involved in the terminal B–H bonding and focusing only on the framework MOs affords the bonding MOs shown in Figure 10.59. Notice that the number of bonding molecular orbitals is identical to the number of SEPs. A more complete MO diagram for the framework is shown in Figure 10.60, where the energies of the MOs are shown more explicitly. The seven SEPs will completely fill the lower lying sets of a_{1g}, t_{1u}, and t_{1g} bonding molecular

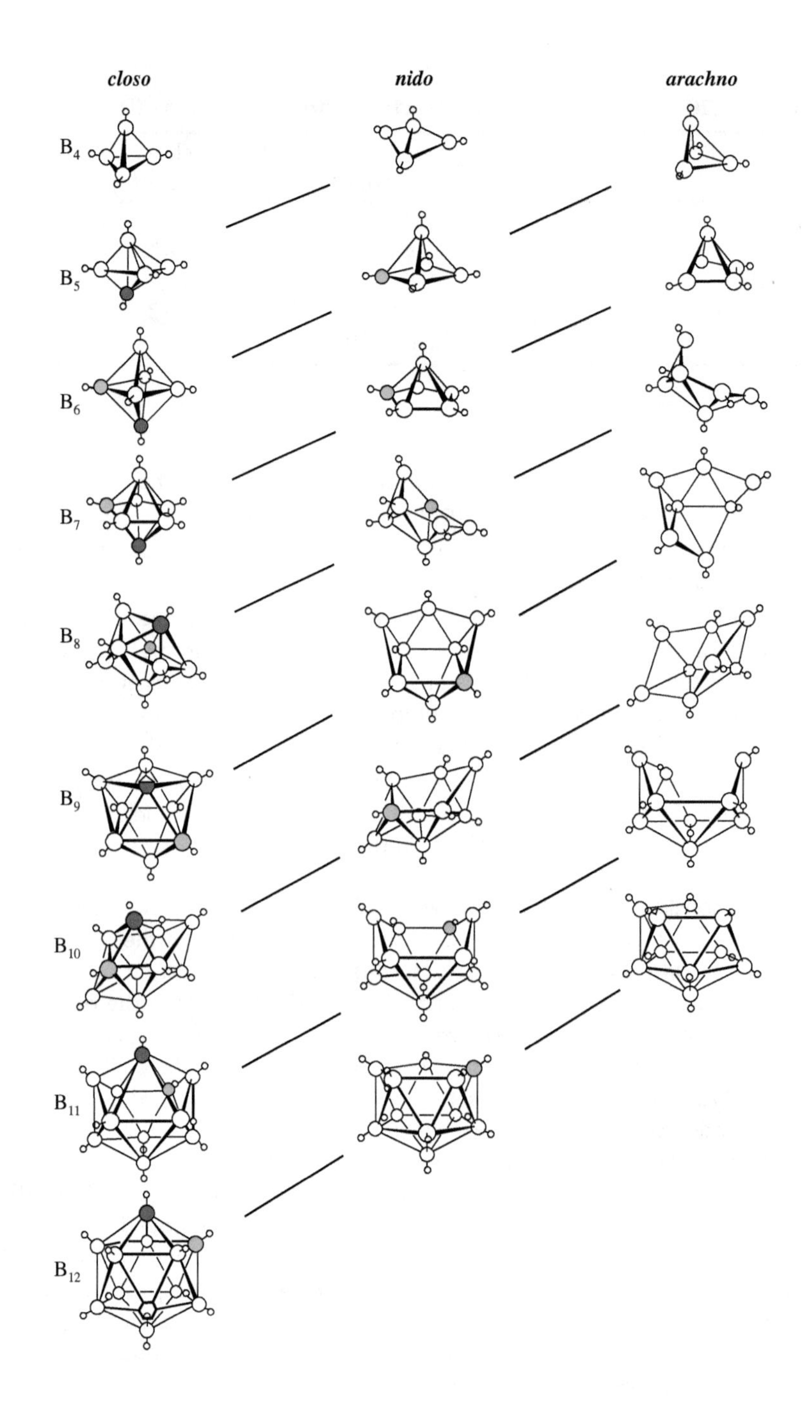

FIGURE 10.56
Structural relationship between the *closo*, *nido*, and *arachno* boranes for differing values of *n* ranging from 4 through 12, where the diagonal lines indicate structures that have the same number of skeletal electron pairs (SEPs). The red vertex is removed first, followed by the green vertex. [Reproduced from Rudolph, R. W. *Accts Chem Res* **1976**, *9*, 447. Used with permission.]

orbitals. The large HOMO-LUMO gap on the MO diagram can be used to explain the stability of the compound predicted on the basis of Wade's rules.

As a second example, let us consider the case of B_4H_{10}. The total valence electron count is $4(3) + 10(1) = 22$ valence electrons. There are four B atoms in the structure, so $n = 4$. Because $4n + 6 = 22\ e^-$, according to Table 10.20, the resulting structure must be the *arachno* cluster derived from the deltahedron with $n + 2$ vertices, or an octahedron. Using the structural correlation shown in Figure 10.56, the framework that results when two of the octahedral vertices are removed is the one shown in Figure 10.61, where the two remaining triangular faces share an edge in common and the four B atoms lie at the remaining four vertices.

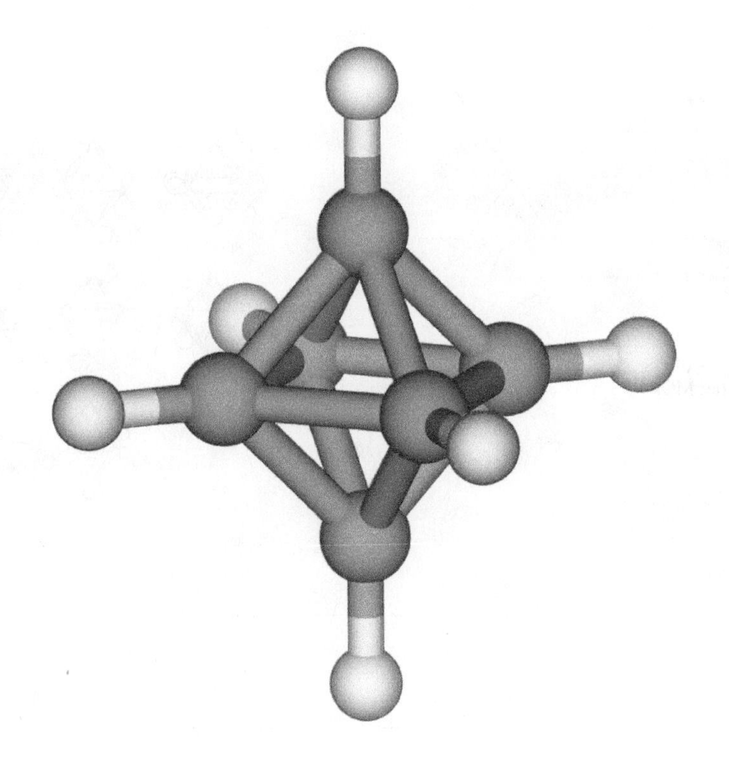

FIGURE 10.57
Structure of $B_6H_6^{2-}$ or *closo*-hexahydrohexaborate(2−). [Reproduced from http://commons.wikimedia.org/wiki/File:B6H6_otahedron_view1.png (accessed February 28, 2014), attributed to Episodesn under the Creative Commons Attribution 3.0 Unported license (accessed February 28, 2014).]

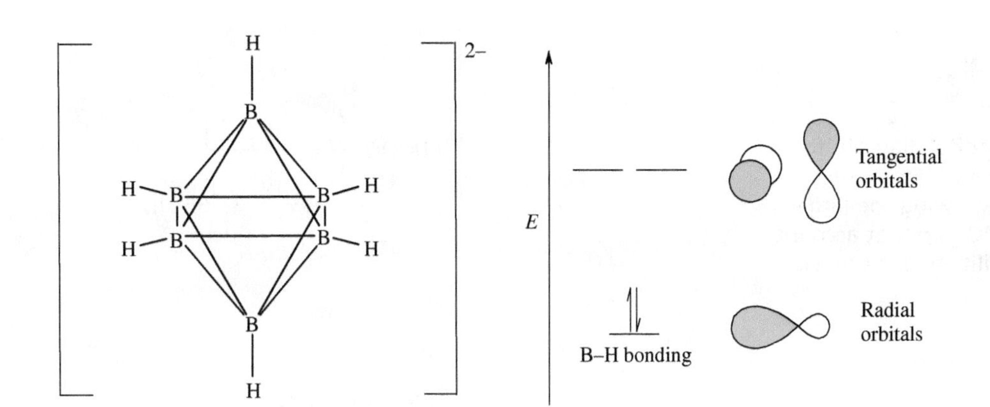

FIGURE 10.58
Basis orbitals on each B atom in the $B_6H_6^{2-}$ cluster for the formation of the framework MOs.

TABLE 10.21 Generation of the framework MOs in the $B_6H_6^{2-}$ ion using the basis set pictured in Figure 10.58.

O_h	E	$8C_3$	$6C_2$	$6C_4$	$3C_2'$	i	$6S_4$	$8S_6$	$3\sigma_h$	$6\sigma_d$	IRRs:
Γ_{radial}	6	0	0	2	2	0	0	0	4	2	$a_{1g}+e_g+t_{1u}$
$\Gamma_{tangential}$	12	0	0	0	−4	0	0	0	0	0	$t_{1g}+t_{2g}+t_{1u}+t_{2u}$

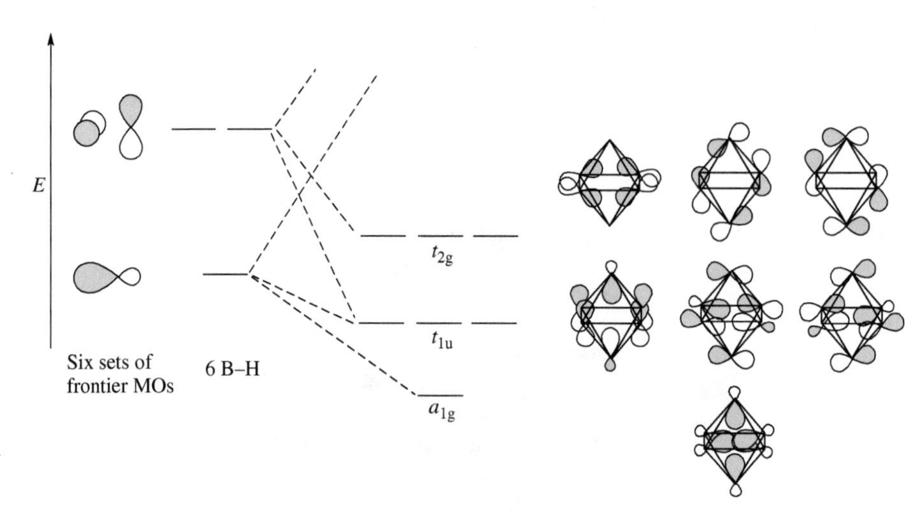

FIGURE 10.59
Shapes of the bonding molecular orbitals in the $B_6H_6^{2-}$ framework.

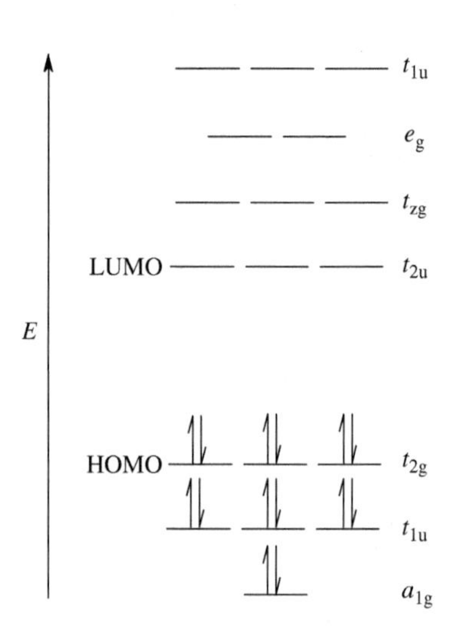

FIGURE 10.60
Molecular orbital diagram for the framework MOs in the $B_6H_6^{2-}$ ion, showing the large HOMO-LUMO gap that accounts for the stability of the cluster.

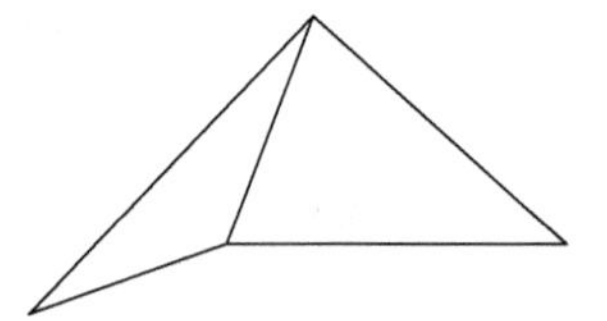

FIGURE 10.61
The structure that results when two vertices are removed from an octahedron in the order shown in Figure 10.58.

The two B atoms in the center of Figure 10.61 are each connected to three other vertices and will therefore have only a single terminal H attached to them, while the other two B atoms (at the extreme left and extreme right in Figure 10.61) are each connected to only two other vertices and will therefore have two terminal H atoms attached to them. Thus, there are a total of six terminal B–H bonds. The four remaining H atoms form part of the framework and lie between neighboring B atoms to form four B–H–B bridges, as shown in Figure 10.62. Subtracting two electrons for each of the six terminal B–H bonds from the total valence electron count yields $22 - 6(2) = 10$ valence electrons available for bonding within the framework, or 5 SEPs.

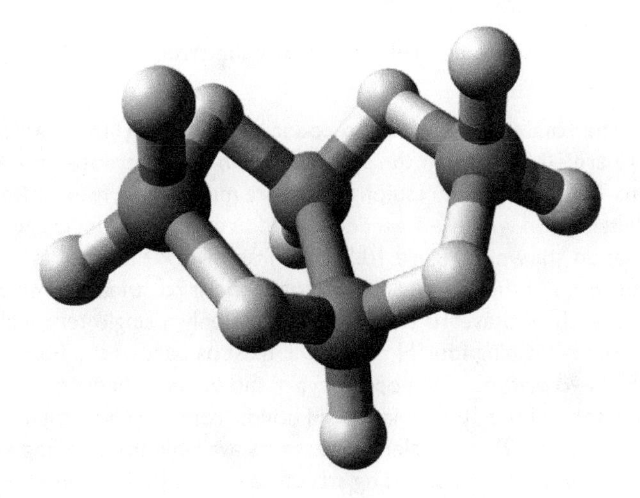

FIGURE 10.62
The structure of B_4H_{10}, or *arachno*-tetraborane. [Reproduced from http://en.wikipedia.org/wiki/Tetraborane (accessed February 28, 2014).]

Example 10-14. Use Wade's rules to predict the structure of B_5H_9.

Solution. The total valence electron count is $5(3)+9(1)=24$ valence electrons. There are five B atoms in the structure, so $n=5$. Because $4n+4=24\ e^-$, according to Table 10.20, the resulting structure must be the *nido* cluster derived from the deltahedron with $n+1$ vertices, or an octahedron. Using the structural correlation shown in Figure 10.56, the framework that results is a square pyramid. The B atom at the apex of this pyramid will be connected to four other vertices and will therefore have a single terminal H. The four B atoms that comprise the square base will have three connections: two to the neighboring B atoms in the square and one to the apical B atom. Thus, these four B atoms will also have only one terminal H atom attached to them. As a result, the remaining four H atoms will exist as part of the framework and form four B–H–B bridges in the base of the square pyramid. Subtracting two electrons for each of the five terminal B–H bonds from the total valence electron count yields $24-5(2)=14$ valence electrons available for bonding within the framework, or 7 SEPs. The resulting structure of B_5H_9 is shown here. The apical B-basal B bond length is 166 pm, while the four basal–basal bond distances are 172 pm because of the intervening B–H–B bridge.

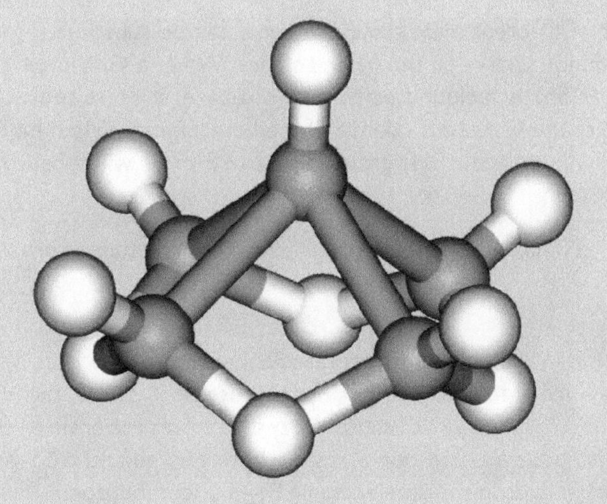

[Reproduced from http://en.wikipedia.org/wiki/Pentaborane (accessed March 24, 2014).]

Example 10-15. Use Wade's rules to predict the structure of $B_{10}H_{14}$.

Solution. The total valence electron count is $10(3) + 14(1) = 44$ valence electrons. There are 10 B atoms in the structure, so $n = 10$. Because $4n + 4 = 44$ e^-, according to Table 10.20, the resulting structure must be the *nido* cluster derived from the deltahedron with $n + 1$ vertices, or an octadecahedron. Using the structural correlation shown in Figure 10.56, the framework that results is missing a single vertex. Each of the 10 B atoms has connections to at least three other vertices, so that each of these 10 B atoms will have only a single terminal H atom. As a result, the remaining four H atoms will exist as part of the framework and form four B–H–B bridges in the open part of the octadecahedron. Subtracting 2 electrons for each of the 10 terminal B–H bonds from the total valence electron count yields $44 - 10(2) = 24$ valence electrons available for bonding within the framework, or 12 SEPs. The resulting structure of $B_{10}H_{14}$ is shown here.

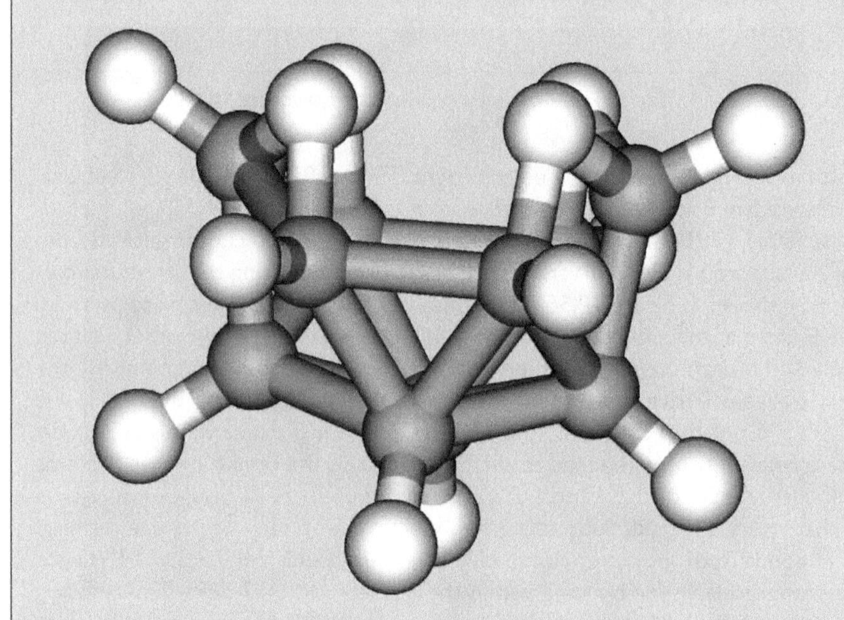

[Reproduced from http://en.wikipedia.org/wiki/Decaborane (accessed March 24, 2014).]

A number of different main group elements can be substituted for B to yield a class of compounds known as the *heteroboranes*. When a Group 14 element (such as C, Si, Ge, or Sn) is included as part of the cage, it must substitute for a BH group because of the increased valence of the heteroatom. When N, P, or As is the heteroatom, it will replace a BH_2 group. Similarly, S or Se will substitute for BH_3.

Example 10-16. Use Wade's rules to predict the structure of *ortho*-carborane, $C_2B_{10}H_{12}$.

Solution. Using Wade's rules, the total valence electron count is $2(4) + 10(3) + 12(1) = 50$ valence electrons and $n = 12$. Because $4n + 2 = 50$, this carborane is based on a *closo* icosahedron. Because the two C atoms are *ortho* to one another, the resulting structure is the one shown, where two adjacent C atoms replace two of the BH groups in the isoelectronic $B_{12}H_{14}$ icosahedron.

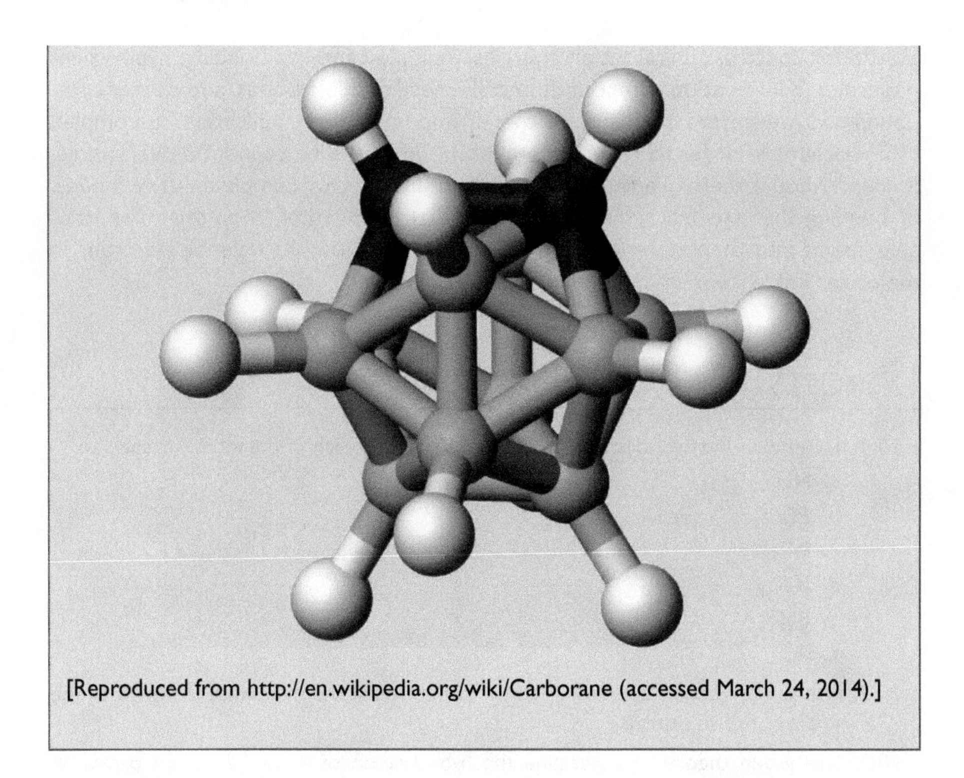

[Reproduced from http://en.wikipedia.org/wiki/Carborane (accessed March 24, 2014).]

Example 10-17. Use Wade's rules to predict the structure of $N_2B_8H_{12}$.

Solution. Using Wade's rules, the total valence electron count is $2(5) + 8(3) + 12(1) = 46$ valence electrons and $n = 10$. Because $4n + 4 = 46$, this carborane is based on the *arachno* $n + 2$ icosahedron where each N replaces a BH_2 group. The resulting structure of $N_2B_8H_{12}$ is shown here.

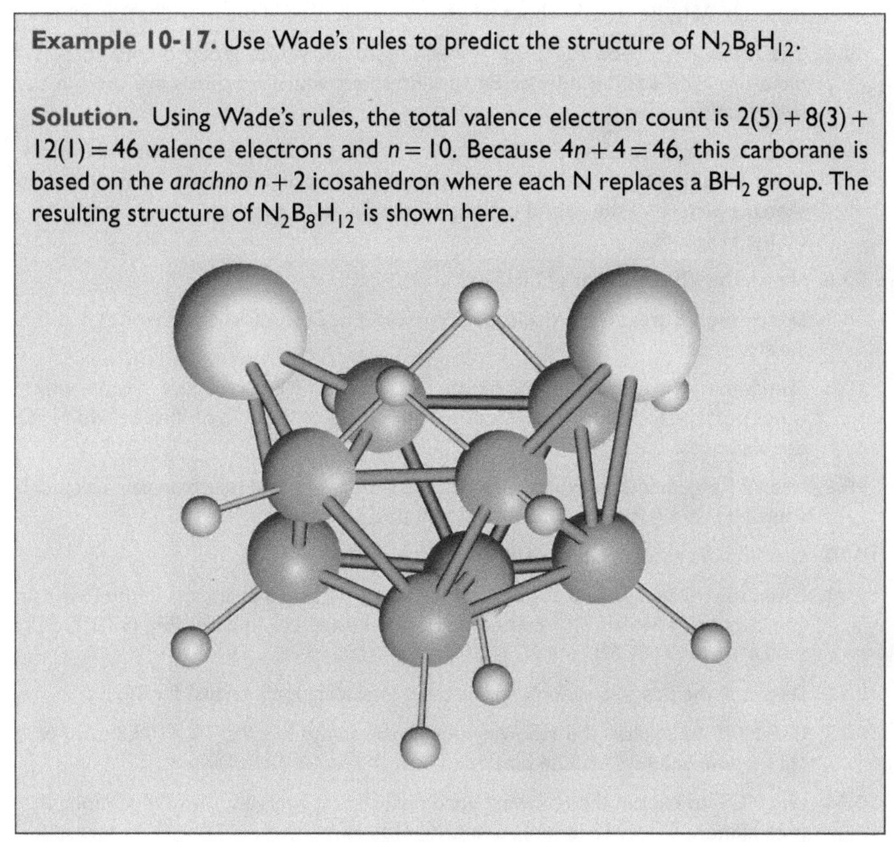

As these examples have shown, the MO model of bonding is an extremely powerful one. It can be used to rationalize molecular shapes, nonintegral bond orders, bond polarities, the magnetic properties of molecules, PES data, electronic absorption spectra, delocalization energies, multiple-centered bonding, hypervalent molecules, coordination compounds, metal–metal multiple bonds, and cluster compounds. However, the simplicity of the VB model and its application to complex MO problems in order to reduce them into a simpler form cannot be overlooked. Students should therefore think of VBT and MOT as two complementary models of bonding that are mathematically related. Taken together, they can offer valuable insight into the nature of chemical bonding and how the valence electrons in molecules influence their chemical properties.

EXERCISES

10.1. Determine the hybridization of the central atom in each of the following species:

a. NH_3
b. PF_5
c. BO_3^{3-}
d. XeF_2
e. $BeCl_2$
f. SF_6

10.2. Use the projection operator method to determine the mathematical forms of the five dsp^3 hybrid orbitals.

10.3. Use group theory to determine the hybridization of A in the square pyramidal molecule AH_5. Be specific about which *d*-orbital is used in the hybridization scheme.

10.4. The $[Mo(CN)_8]^{3-}$ ion has a dodecahedral symmetry. Use group theory to determine the hybridization on Mo. Be specific about which *d*-orbitals are used in the hybridization scheme.

10.5. Sketch the valence bond treatment of BeH_2, using boxes to represent the orbitals. Show the process of electron promotion, hybridization, and bond formation. Also sketch a picture of the hybrid orbitals on the Be atom overlapping with the 1*s* AOs on the H ligands.

10.6. Sketch the VB treatment of NH_3.

10.7. Sketch the VB treatment of allene, $CH_2{=}C{=}CH_2$. Discuss whether or not allene is a planar molecule.

10.8. Sketch the VB treatment of the nitrate ion. From a VBT perspective, explain why it is not possible to include any resonance structures of NO_3^- that include two N=O double bonds.

10.9. Use VBT arguments to explain why BH_3 is a good Lewis acid (electron pair acceptor), while NH_3 is a good Lewis base (electron pair donor).

10.10. Use VB concepts to explain why PF_5 exists, but NF_5 cannot.

10.11. Given the following bond angles, determine the percent *s*-character in the lone pair and determine which PX_3 compound is the better Lewis base: PF_3 (97.8°), PCl_3 (100.4°), PBr_3 (101.0°), and PI_3 (102°).

10.12. Sketch all the possible canonical structures (including ionic terms) for SF_6.

10.13. Use MOT to explain the following equilibrium bond lengths: N_2 (109.8 pm), N_2^+ (111.6 pm), and N_2^{2-} (122.4 pm).

10.14. Use MOT to explain the following bond dissociation energies: O_2 (494 kJ/mol), O_2^+ (626 kJ/mol), O_2^- (393 kJ/mol), and O_2^{2-} (138 kJ/mol).

10.15. Which of the following species are paramagnetic: B_2, N_2, N_2^{2-}, O_2, O_2^{2-}, F_2?

10.16. Sketch the one-electron MO diagram for BN. Label each MO as *sigma* or *pi* and as bonding, nonbonding, or antibonding; fill in the electrons; and determine the bond order and magnetism.

10.17. Sketch the one-electron MO diagram for NO. Label each MO as *sigma* or *pi* and as bonding, nonbonding, or antibonding; fill in the electrons; and determine the bond order and magnetism.

10.18. The gas-phase ionization potentials for some diatomic molecules are given in this table. Use MOT to explain the trend in the ionization potentials.

Molecule	IE (eV)
Li_2	5.11
C_2	11.4
N_2	15.6
O_2	12.1
F_2	15.7

10.19. Use the MO diagram of the hydroxide ion OH^- to explain why it is a good Lewis base (electron pair donor).

10.20. Compare and contrast the one-electron MO diagrams for the isoelectronic ligands CO and CN^-. Using MOT, determine which of these ligands is the better *sigma*-donor. The better *sigma*-donating ligand is the one which has the greatest percent C character in its highest occupied molecular orbital (HOMO).

10.21. Use a Walsh diagram to predict whether the shape of the CH_3^+ carbocation intermediate in an S_N1 reaction will be trigonal planar or trigonal pyramidal. How does the molecular geometry of the CH_3^+ intermediate explain the stereochemical product distribution (retention, inversion, or racemic) in an S_N1 reaction?

10.22. Sketch the one-electron MO diagram for the H_4 molecule having a linear geometry. Hint: in constructing your MO diagram, you might wish to use the symmetry-adapted linear combinations for H_2 as a starting point on either side of your MO diagram and then combine these SALCs together to form the MOs in the middle of the diagram. Label each MO with its group theoretical symbol; state which orbitals are bonding, nonbonding, and antibonding; fill in the electrons; and determine the overall bond order and magnetism of the molecule.

10.23. Sketch the one-electron MO diagram for the H_4 molecule having square planar geometry (with the H atoms lying in a plane at the corners of a square). Label each MO with its group theoretical symbol; state which orbitals are bonding, nonbonding, and antibonding; fill in the electrons; and determine the overall bond order and magnetism of the molecule.

10.24. Using the MO diagrams for linear H_4 and for square planar H_4, sketch a correlation diagram (in the same manner as a Walsh diagram) in order to determine the preferred molecular geometry for this species.

10.25. Explain why square planar H_4^+ cannot exist.

10.26. Using each H–H bond in the fictional square planar H_4^+ ion as a basis set, determine the symmetries of the four stretching vibrational modes. Which of these vibrational modes will cause the molecule to Jahn–Teller distort?

10.27. Sketch the one-electron MO diagram for the azide ion, N_3^-. Label each MO with its group theoretical symbol; state which orbitals are bonding, nonbonding, and antibonding; fill in the electrons; and determine the overall bond order and magnetism of the molecule. Hint: use the D_{2h} subgroup of $D_{\infty h}$ and a correlation table to determine the symmetries of the resulting MOs.

10.28. Sketch the one-electron MO diagram for the linear molecule H–A–A–H, where A is a Period 2 main group element. In this case, you will need to determine the SALCs for the two H atoms on one side of your MO diagram and the SALCs for the two A atoms on the other side of the MO diagram. Then, combine SALCs having appropriate symmetries and energies to form MOs in the center of the MO diagram.

Label each MO with its group theoretical symbol; state which orbitals are bonding, nonbonding, and antibonding; fill in the electrons; and determine the overall bond order and magnetism of the molecule. Also, sketch the predicted shape of each of the MOs.

10.29. Sketch the one-electron MO diagram for the linear FHF^- ion. Label each MO with its group theoretical symbol; state which orbitals are bonding, nonbonding, and antibonding; fill in the electrons; and determine the overall bond order and magnetism of the molecule. Also, sketch the predicted shape of each of the MOs.

10.30. Sketch the one-electron MO diagram for the nitrite ion, NO_2^-. Label each MO with its group theoretical symbol; state which orbitals are bonding, nonbonding, and antibonding; fill in the electrons; and determine the overall bond order and magnetism of the molecule. Also, sketch the predicted shape of each of the MOs.

10.31. Sketch the one-electron MO diagram for the *pi* MOs in square planar cyclobutadiene. Label each MO with its group theoretical symbol; state which orbitals are bonding, nonbonding, and antibonding; fill in the electrons; and determine the overall bond order and magnetism of the molecule. Also, sketch the predicted shape of each of the MOs. Then use the MO diagram to explain why cyclobutadiene is an extremely unstable molecule.

10.32. Sketch the one-electron MO diagram in a MM quintuple bond using only the set of *d*-orbitals. If M=Cr, what must be the oxidation state of each chromium ion?

10.33. Use Wade's rules to predict the structure of $B_5H_5^{4-}$. What category of borane is it? What deltahedron is the structure based on? How many vertices are missing from this deltahedron (if any) and how many skeletal electron pairs does it possess?

10.34. Use Wade's rules to predict the structure of B_5H_{11}. What category of borane is it? What deltahedron is the structure based on? How many vertices are missing from this deltahedron (if any) and how many skeletal electron pairs does it possess?

10.35. Use Wade's rules to predict the structure of CB_5H_7. What category of carborane is CB_5H_7? What deltahedron is the structure based on? How many vertices are missing from this deltahedron (if any) and how many skeletal electron pairs does it possess?

10.36. Classify each of the following species as *closo*, *nido*, *arachno*, or *hypho*:

a. B_9H_{15}
b. B_4H_8
c. $B_9H_{12}^-$
d. $B_6H_{11}^+$
e. CB_5H_9
f. $C_2B_6H_{10}$
g. B_9H_9NH
h. $B_9H_{11}S$

BIBLIOGRAPHY

1. Albright, T. A.; Burdett, J. K. *Problems in Molecular Orbital Theory*, Oxford University Press, New York, 1992.
2. Albright, T. A.; Burdett, J. K.; Whangbo, M.-H. *Orbital Interactions in Chemistry*, John Wiley & Sons, Inc., New York, 1985.
3. Carter, R. L. *Molecular Symmetry and Group Theory*, John Wiley & Sons, Inc., New York, 1998.
4. Cotton, F. A. *Chemical Applications of Group Theory*, 2nd ed., John Wiley & Sons, Inc., New York, 1971.
5. Douglas, B.; McDaniel, D.; Alexander, J. *Concepts and Models of Inorganic Chemistry*, 3rd ed., John Wiley & Sons, Inc., New York, 1994.

6. Gillespie, R. J.; Popelier, P. L. A. *Chemical Bonding and Molecular Geometry*, Oxford University Press, New York, 2001.
7. Harris, D. C.; Bertolucci, M. D. *Symmetry and Spectroscopy*, Dover Publications, Inc., New York, 1978.
8. Housecroft, C. E.; Sharpe, A. G. *Inorganic Chemistry*, 3rd ed., Pearson Education Limited, Essex, England, 2008.
9. Huheey, J. E.; Keiter, E. A.; Keiter, R. L. *Inorganic Chemistry: Principles of Structure and Reactivity*, 4th ed., Harper Collins College Publishers, New York, 1993.
10. Kettle, S. F. A. *Symmetry and Structure*, 2nd ed., John Wiley & Sons, Inc., New York, 1995.
11. Khaniana, Y.; Badiei, A. E-J. Chem., 2009, 6, 169–176.
12. Li, W.-K.; Zhou, G.-D.; Wai Mak, T. C., *Advanced Structural Inorganic Chemistry*, Oxford University Press, New York, 2008.
13. McQuarrie, D. A.; Simon, J. D. *Physical Chemistry: A Molecular Approach*, University Science Books, Sausalito, CA, 1997.
14. McWeeny, R. *Coulson's Valence*, 3rd ed., Oxford University Press, New York, 1979.
15. Miessler, G. L.; Tarr, D. A. *Inorganic Chemistry*, 4th ed., Pearson Education Inc., Upper Saddle River, NJ, 2011.
16. Mingos, D. M. P. Acc. Chem. Res., 1984, 17, 311–319.

Metallic Bonding | 11

"The important thing in science is not so much to obtain new facts as to discover new ways of thinking about them."

—*William Bragg*

11.1 CRYSTALLINE LATTICES

In the previous chapter, we have discussed different ways of thinking about the nature of covalent bonding in molecules. In most of the examples we have encountered so far, the molecules existed in the gas phase, so we only needed to focus our attention on the interactions of atoms in molecules. On the other hand, most metals exist in the solid state at room temperature. In the ensuing discussion of metallic bonding, it is therefore essential that we consider the arrangement of the metal atoms with respect to one another within the crystalline lattice.

Crystalline solids consist of atoms, ions, or molecules that are arrayed into a long-range, regularly ordered structure known as a *crystalline lattice*. A crystal consists of a pattern of objects that repeats itself periodically in three-dimensional space, so that it has the property of translational symmetry. A *lattice* is simply a three-dimensional array of lattice points, where the atoms, ions, or molecules are held together in the solid state by a balance of attractive and repulsive forces. Lattice points are geometrical constructs; it is not a necessary condition for a physical entity, such as an atom or ion, to actually occupy the lattice point. Indeed, many lattice points are simply empty space, around which a basis, or *motif*, of particles is centered. Two examples of two-dimensional lattices are shown in Figure 11.1.

The lattice points in a crystalline solid can be connected by lines to form parallelepipeds in such a way that the parallelepipeds fill all space without any gaps between them. The *unit cell* of a crystalline lattice is defined as the simplest array of lattice points (or parallelepipeds) from which a crystal can be constructed using translational symmetry and which best represents the overall symmetry of the crystal.

In theory, there are a variety of ways in which a unit cell can be defined, but only one of these definitions will represent the actual internal symmetry of the crystal. In the example shown in Figure 11.2, structure (a) is not a unit cell because it lacks

Mockup of a crystalline lattice. [Photo credit: B. Pfennig, taken at the Terra Mineralia Museum in Freiberg, Germany.]

Principles of Inorganic Chemistry, First Edition. Brian W. Pfennig.

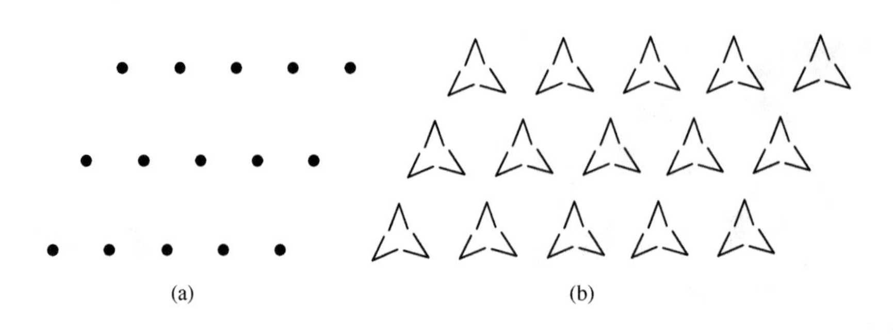

FIGURE 11.1
Examples of crystalline lattices containing (a) atoms at the lattice points and (b) a motif (or basis) at the lattice points.

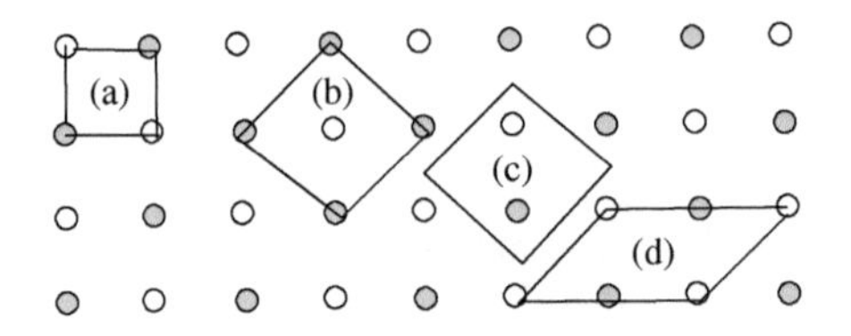

FIGURE 11.2
The choice of a unit cell requires the smallest parallelepiped that connects identical lattice points to fill all space without any gaps or overlaps and that also has the highest degree of symmetry common to the external crystal. Structure (a) does not contain identical lattice points at all the corners and cannot be a unit cell. Structures (b)–(d) are all suitable; however, structure (b) has the highest rotational symmetry and is therefore the better choice.

translational symmetry. It is impossible to build a two-dimensional structure from squares of (a) that will fill all space without any gaps or overlaps because the four corners of this building block consist of different types of points. On the other hand, each of the parallelograms (b), (c), and (d) can be used to form a suitable lattice. Furthermore, each of these three species occupies an identical area and contains the same number of each type of lattice point when fractional lattice points—those that are shared between different cells—are taken into consideration. The cell defined by structure (d) has only a twofold rotational axis, whereas that in structure (c) has none at all. Lastly, the fourfold symmetry of structure (b) is the most representative of the entire two-dimensional array and we therefore choose this to be the most appropriate definition of the unit cell for our two-dimensional lattice.

Crystalline lattices are commonly described in terms of their symmetry properties. Crystals can have two fundamentally different kinds of symmetry: translational symmetry and point symmetry. Translational symmetry is a repeating type of symmetry that describes the method of propagation between lattice points lying on a screw axis or a glide plane, both of which are illustrated in Figure 11.3. Point symmetry, on the other hand, is the type where at least one point within the object remains unchanged with respect to every kind of symmetry operation. Point symmetry operations include proper rotations, reflections, improper rotations (rotation-reflection), and inversion; these were already discussed in Chapter 8.

In order to be consistent with an extended crystalline lattice, however, only certain types of rotational operations are allowed at the lattice points. These include C_1, C_2, C_3, C_4, and C_6 rotations. Any other rotational symmetry operation would be unable to fill all space without any gaps or overlaps. This can be proved geometrically by considering the two-dimensional lattice shown in Figure 11.4. Starting with lattice point A, a second lattice point exists at distance a along the x-axis at point B. Rotation of the line segment AB by $2\pi/n$ radians generates the line segment AB',

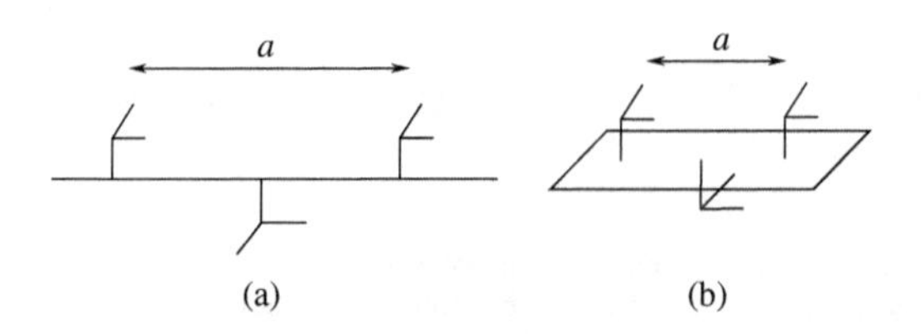

FIGURE 11.3
Examples of two translational symmetry elements: (a) screw axis = translation, rotation and (b) glide plane = translation, reflection.

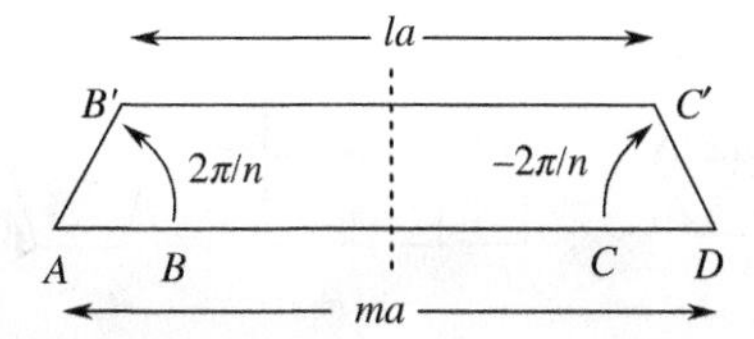

FIGURE 11.4
Two-dimensional lattice representation showing how the only values of n consistent with the generation of a second row of lattice points are 1, 2, 3, 4, and 6.

where B' is a lattice point in the next row. Likewise, rotation of the line segment DC by $-2\pi/n$ radians generates a new line segment DC', where C' is also in the second row of lattice points. Geometrically, the distance between points $B'C'$ must equal some integral value (l) of the x-axis lattice spacing, which we will call a. Likewise, the distance between points A and D in the first row of lattice points must also equal some integral value (m) of a, where m is a positive integer. Thus, the relationship given by Equation (11.1) must hold true. Dividing both sides by a and solving for the cosine of the angle, we obtain Equation (11.2). Because the cosine can only range between -1 and $+1$, the only integral values of n for which Equation (11.2) can exist are 1, 2, 3, 4, and 6, as shown in Table 11.1, proving that the only rotational axes consistent with the two-dimensional lattice are the C_1, C_2, C_3, C_4, and C_6 axes.

$$la = ma - 2a \cos(2\pi/n) \tag{11.1}$$

$$\cos(2\pi/n) = (m - l)/2 \tag{11.2}$$

Another way of thinking about the geometrical constraints on the types of rotational axes consistent with a crystalline lattice is to consider the two-dimensional example of a set of bathroom tiles shown in Figure 11.5. The entire floor of the bathroom could be covered with tiles composed of points (C_1), line segments (C_2), equilateral triangles (C_3), squares (C_4), or hexagons (C_6). Pentagons (C_5), heptagons (C_7), and octagons (C_8), on the other hand, would always leave a gap. Given these restrictions, it can be shown that there are only 14 unique ways of arranging lattice points in three-dimensional space; these are the 14 Bravais lattices shown in Figure 11.6. Every crystalline lattice will belong to one of these 14 fundamental lattice types.

The 14 Bravais lattices can be further classified into seven crystal systems on the basis of their external dimensions: the edge lengths a, b, and c; and the angles between the edges α, β, and γ. For reference, α is the angle formed between edges b and c, β between a and c, and γ between a and b. The seven crystal systems and their corresponding lattice types are listed in Table 11.2. The point groups that are consistent with each crystal system are also listed in the table as a reference. As it turns out, only 32 crystallographic point groups (those involving the symmetry of the motif around each lattice point) are allowed.

TABLE 11.1 The only allowed values of n which satisfy Equation (11.2) in such a way that l, m, and n will be integers.

n	$2\pi/n$	$\cos(2\pi/n)$
1	180°	−1
2	120°	−1/2
3	90°	0
4	60°	+1/2
6	0°, 360°	+1

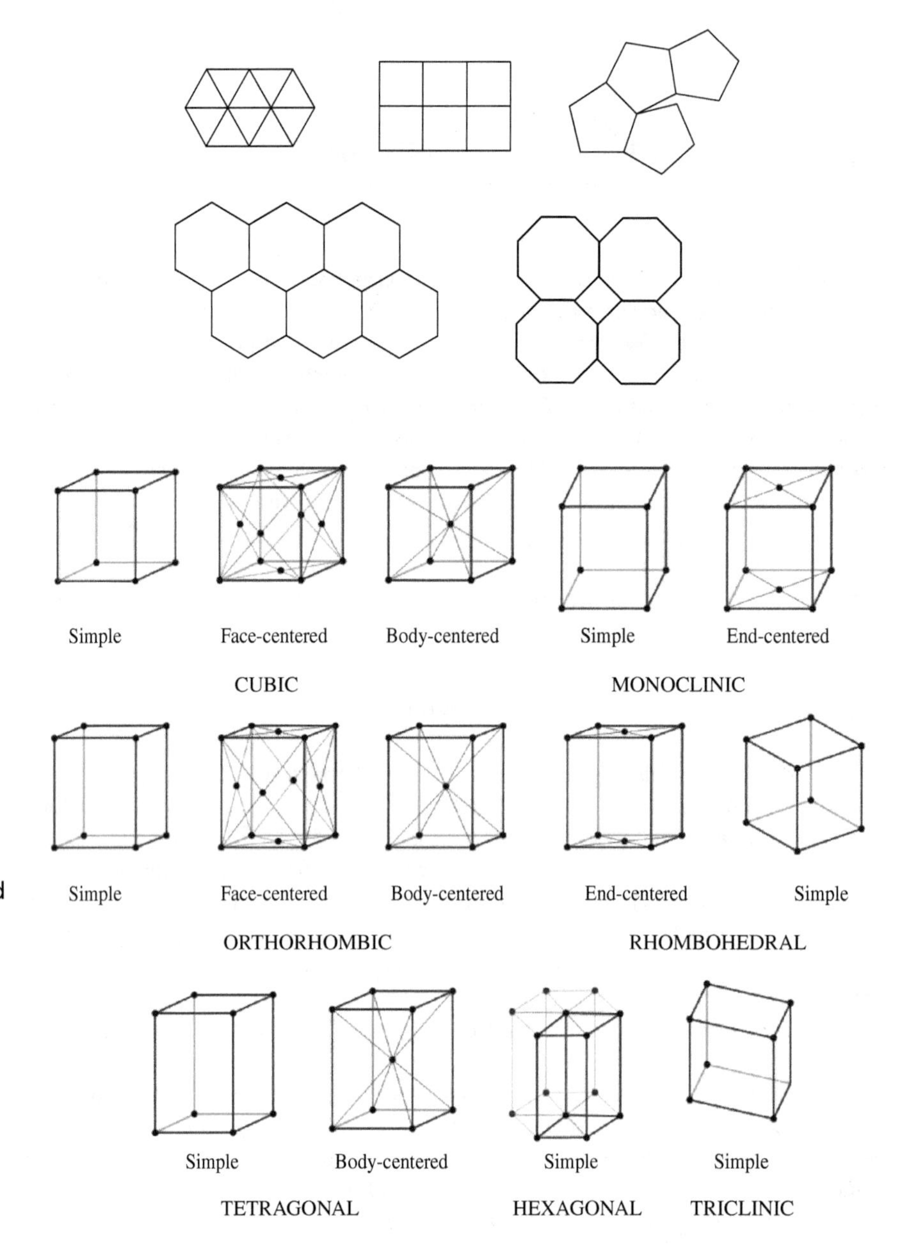

FIGURE 11.5
Two-dimensional illustration of why only certain rotational axes can fill all space without any gaps or overlaps. Threefold, fourfold, and sixfold axes are allowed, for instance, but fivefold and eightfold axes are not.

FIGURE 11.6
The 14 Bravais lattices organized into 7 crystal systems (simple or primitive = P, body-centered = I, face-centered = F, end-centered = C, rhombohedral = R). The hexagonal unit cell is inscribed in a larger geometric entity in order to emphasize the bond angles. [E. Generalic, http://www.periodni.com, modified with permission.]

Crystallographers do not typically use the Schoenflies system of notation for the point groups. Instead, they use an alternative system known as the *International* or *Hermann-Mauguin system*. The Hermann-Mauguin notation for the molecular point groups uses numbers to represent the proper rotation axes. For example, a C_2 proper rotation is simply 2, C_4 is 4, and so on. In the Schoenflies system, an improper rotation, such as S_6, is a rotation-reflection, whereas in the Hermann-Mauguin system an improper rotation is a rotation-inversion and its symbol is a number with a bar over it. Thus, $\bar{1}$ is the same as *i* in the Schoenflies notation, $\bar{2}$ is the same as σ_h, $\bar{3}$ is the same as S_6, $\bar{4}$ is the same as S_4, and $\bar{6}$ is the same as S_3. Mirror planes are represented by the symbol m. If both a mirror plane and a principal axis exist, the plane that contains the axis is given by its numeral followed by the symbol m. For example, a vertical mirror plane containing a C_3 rotational axis would be written as 3m. If the mirror plane is perpendicular to the principal axis, then the symbol is written with a slash, such as 3/m.

TABLE 11.2 The seven crystal systems organized according to their edge lengths and bond angles.

Crystal System	Bravais Lattices	Edge Lengths	Defined Angles	Allowed Point Groups
Cubic	P, I, F	$a=b=c$	$\alpha=\beta=\gamma=90°$	T, T_h, T_d, O, O_h
Tetragonal	P, I	$a=b\neq c$	$\alpha=\beta=\gamma=90°$	C_4, S_4, C_{4h}, D_{2d}, C_{4v}, D_4, D_{4h},
Orthorhombic	P, F, I, C	$a\neq b\neq c$	$\alpha=\beta=\gamma=90°$	C_{2v}, D_2, D_{2h}
Hexagonal	P	$a=b\neq c$	$\alpha=\beta=90°$ $\gamma=120°$	C_{3h}, C_6, C_{6h}, D_{3h}, C_{6v}, D_6, D_{6h}
Rhombohedral (trigonal)	P (or R)	$a=b=c$	$\alpha=\beta=\gamma\neq 90°$	C_3, S_6, C_{3v}, D_3, D_{3d}
Monoclinic	P, C	$a\neq b\neq c$	$\alpha=\gamma=90°$ $\beta\neq 90°$	C_s, C_2, C_{2h}
Triclinic	P	$a\neq b\neq c$	$\alpha\neq\beta\neq\gamma$	C_1, C_i

Those point symmetry groups that are consistent with each crystal system are also listed.

The Hermann-Mauguin notation is written to be as succinct as possible in the description of the symmetry elements present in the point group. Consider, for example, the C_{3v} point group in the Schoenflies notation, which has a threefold proper rotational axis and three vertical mirror planes. In the Hermann-Mauguin system, this point group is simply given the symbol 3m, as the presence of the other two vertical mirror planes is implied by the existence of the first. Likewise, C_{4v} is simply given the symbol 4m because the presence of one vertical mirror plane implies the presence of a second vertical mirror plane as well as the two dihedral mirror planes. For the D_{2h} point group, you might expect the symbol to be 2/m 2/m 2/m because of the presence of three mutually perpendicular twofold rotational axes and their corresponding orthogonal mirror planes. However, the symbol is simplified to the more compact mmm, which conveys all of the same information. The Schoenflies and corresponding Hermann-Mauguin symbols for the 32 crystallographic point groups are listed in Table 11.3. This table also includes the accepted crystallographic point group number in the first column of the table.

The *cubic* crystal systems have all three edges equivalent ($a=b=c$) and all of their internal angles are orthogonal ($\alpha=\beta=\gamma=90°$). The tetragonal system has two equal sides ($a=b\neq c$) and all three angles at 90°. The *orthorhombic* system also has all three angles at 90°, but none of the edges are equal in length. The *hexagonal* Bravais lattice has two 90° angles and a 120° angle, with two sides of equal length ($a=b\neq c$). The *rhombohedral* or *trigonal* system has all three edges the same length and all three angles identical, but none of them are 90°. The monoclinic system has two 90° angles, but no identical edge lengths. Lastly, the triclinic system has $a\neq b\neq c$ and $\alpha\neq\beta\neq\gamma$.

Every crystal system contains a primitive (P) Bravais lattice type as one of its examples. The other lattice types can be derived from the primitive cell by performing some sort of centering operation. Centering means that the crystal structure will be exactly the same if you picked up all the atoms around the origin and moved them to the centered position. Three different types of centering are possible, depending on the crystal system: body-centered (I), face-centered (F), and end-centered (C or A or B). Not every crystal system will exhibit all or any of these different centering types. For example, a face-centered tetragonal Bravais lattice does not exist. If one were to pick up all the atoms around the origin of a primitive tetragonal unit cell and move them onto the face-centers, the result would be equivalent to a smaller body-centered tetragonal, as shown in Figure 11.7. Looking down the top face of a fictional face-centered tetragonal unit cell, the hollow circles represent atoms that are in the plane of the paper, while the solid circles are located exactly halfway down the unit cell. The dark solid lines indicate the smaller body-centered tetragonal unit cell that results.

TABLE 11.3 The 32 crystallographic point groups with their Schoenflies and Hermann-Mauguin point group symbols.

Number	Schoenflies Symbol	Hermann-Mauguin Symbol	Crystal System
1	C_1	1	Triclinic
2	C_i	$\bar{1}$	Triclinic
3	C_s	M	Monoclinic
4	C_2	2	Monoclinic
5	C_{2h}	2/m	Monoclinic
6	C_{2v}	mm2	Orthorhombic
7	D_2	222	orthorhombic
8	D_{2h}	mmm	Orthorhombic
9	C_4	4	Tetragonal
10	S_4	$\bar{4}$	Tetragonal
11	C_{4h}	4/m	Tetragonal
12	C_{4v}	4 mm	Tetragonal
13	D_{2d}	$\bar{4}$2m	Tetragonal
14	D_4	422	Tetragonal
15	D_{4h}	4/mmm	Tetragonal
16	C_3	3	Rhombohedral
17	S_6	$\bar{3}$	Rhombohedral
18	C_{3v}	3 m	Rhombohedral
19	D_3	32	Rhombohedral
20	D_{3d}	$\bar{3}$m	Rhombohedral
21	C_{3h}	$\bar{6}$	Hexagonal
22	C_6	6	Hexagonal
23	C_{6h}	6/m	Hexagonal
24	D_{3h}	$\bar{6}$m2	Hexagonal
25	C_{6v}	6 mm	Hexagonal
26	D_6	622	Hexagonal
27	D_{6h}	6/mmm	Hexagonal
28	T	23	Cubic
29	T_h	m3	Cubic
30	T_d	$\bar{4}$3m	Cubic
31	O	432	Cubic
32	O_h	m3m	Cubic

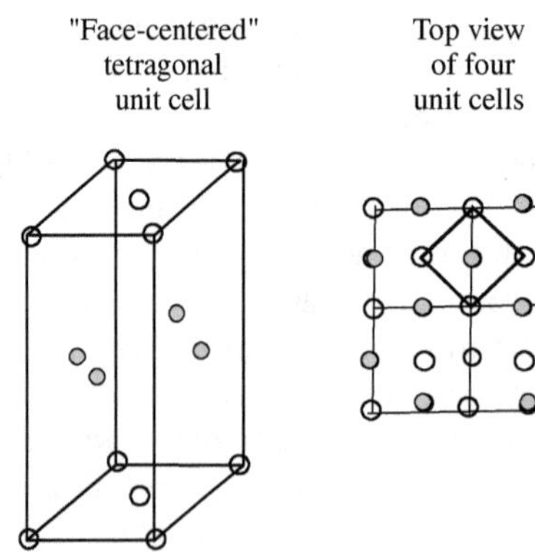

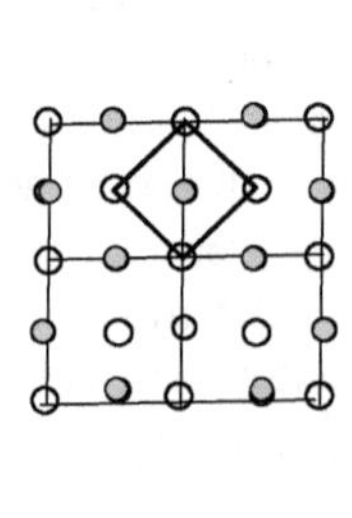

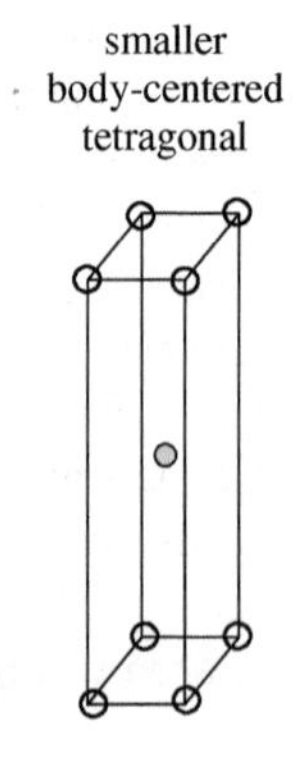

FIGURE 11.7
Diagram showing that a "face-centered" tetragonal unit cell would be equivalent to a smaller body-centered tetragonal unit cell.

When the 32 crystallographic point groups are combined with the different lattice translations (screw axes and glide planes), ignoring redundant results, there are a total of 230 possible combinations known as the *space groups*. The descriptions, representations, and symmetry properties of all 230 space groups are tabulated in the *International Tables of Crystallography*. Each entry lists the space group number (unique identifier), point group, crystal system, lattice dimensions, positions of the bases (coordinates), and the conditions required for certain X-ray diffraction planes to be observed. As long as a suitable single crystal can be grown, the space group can usually be determined experimentally using X-ray crystallography. The space group numbers and their point group symmetry elements are also listed in Appendix E.

11.2 X-RAY DIFFRACTION

The technique of X-ray diffraction takes advantage of the fact that the wavelengths of the X-ray region of the electromagnetic spectrum are similar in length to the sizes of atoms. As a result, X-rays can be scattered (or diffracted) by the electron cloud of the atoms, a phenomenon that was first observed by Max von Laue in 1912. The highly ordered alignment of the atoms in a crystalline lattice permits constructive interference of the diffracted X-ray wavelengths. The foundations of X-ray diffraction were laid by a unique father–son team during the years 1913–1914. Working at the University of Leeds, William Henry Bragg was a great experimentalist and designed the first X-ray spectrometer. His son, William Lawrence Bragg, was independently working in the field while a doctoral student at Cambridge under the tutelage of J. J. Thomson. Together, the pair shared the 1915 Nobel Prize in physics for "their services in the analysis of crystal structure by means of X-rays." At the time, William Lawrence was the youngest person ever to have won the Nobel Prize. He was also the first winner to lecture on his research at the awards ceremony.

The derivation of the Bragg diffraction law is shown in Figure 11.8 and given by Equation (11.3). Two parallel wavefronts of X-rays having a monochromatic wavelength (λ) impinge upon the face of a crystal. The first X-ray beam is scattered by the electron cloud of an atom in the surface layer, while the second beam passes between the atoms in the first layer and is scattered by an atom from the second layer. The two planes of lattice points are separated by a distance (d). In order for constructive interference of the two diffracted X-ray beams to occur, the extra distance traveled by the second beam must be an integral multiple (n) of the wavelength. Using the properties of similar triangles, it can be shown that the second wave travels an extra distance (shown in red) of $2d \sin \theta$. Hence:

$$n\lambda = 2d \sin \theta \tag{11.3}$$

Depending on the nature of the crystalline lattice and the orientation of the incident X-ray beam with respect to the different crystal faces, the d-spacings between the planes of lattice points will differ. Rotation of the crystal with respect to the

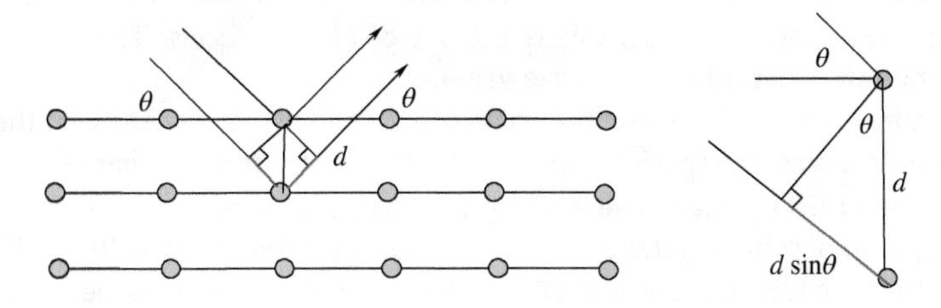

FIGURE 11.8
Derivation of the Bragg diffraction law.

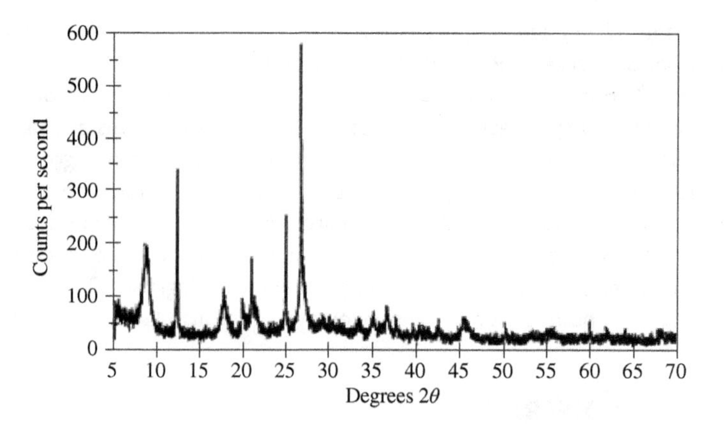

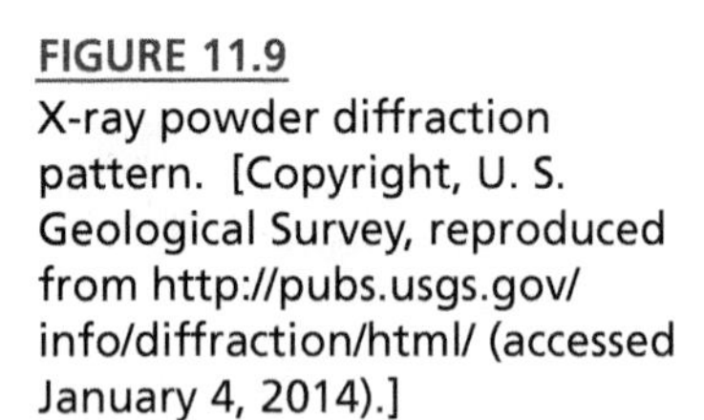
FIGURE 11.9
X-ray powder diffraction pattern. [Copyright, U. S. Geological Survey, reproduced from http://pubs.usgs.gov/info/diffraction/html/ (accessed January 4, 2014).]

incident beam will result in constructive interference at several different values of θ. In actual practice, the X-ray beam is fixed and the crystal is mounted on a rotating goniometer. The detector is then rotated at 2θ in order to observe the diffracted X-ray beams. The final result is a graph or table of 2θ values, each corresponding to a set of planes in the crystal where constructive interference is observed. The relative intensities of each of the diffracted lines are also recorded, as shown by the example in Figure 11.9. Collectively, this information can often be used to identify the crystal type.

Although the space group is determined by the internal symmetry of the crystal, it is the external face of the crystal with which the X-rays in a diffraction experiment interacts. A parallel set of X-ray beams will see these faces as a set of planes containing the lattice points. The different planes of lattice points within a crystal are known as *Miller planes* and are given a label, which is known as the *Miller index* in the format (*hkl*). The values of *h*, *k*, and *l* derive from the small whole number ratio of the reciprocals of the crystalline axes, which we will call *a*, *b*, and *c* and which are not necessarily orthogonal to one another. Examples of several different Miller planes in a primitive cubic lattice are shown in Figure 11.10. Using *a*, *b*, and *c* as the coordinate axes, an arbitrary point lying on any of the planes is chosen as the origin. Next, we examine the next closest parallel plane of points to the original plane. The intercepts (or *Weiss parameters*) of this second plane with the coordinate axes are then recorded. The reciprocals of these intercepts, after conversion to the lowest whole number ratio, are defined as the Miller indices. Using the Miller planes in Figure 11.10(a) as an example and choosing the indicated point as the origin, the closest neighboring plane has the intercepts $a = 1$, $b = 1$, and $c = \infty$ (never). Thus, these are the (1 1 0) planes. Whenever there is more than one option for one of the neighboring planes to have a negative number for one of its intercepts, by convention, the negative intercept is always chosen in the order *a* before *b* before *c*. Hence, in Figure 11.10(b), there are two neighboring planes to the original. The first of these is shown with intercepts $a = -1$, $b = 1$, and $c = \infty$. The other (not shown) would lie to the left of these two unit cells and have intercepts at $a = 1$, $b = -1$, and $c = \infty$. According to convention, the plane with the a-intercept is the one which is chosen to be negative. Hence, this is the set of $(\bar{1}\ 1\ 0)$ planes. The bar over the number indicates that it has a negative value.

You might have noticed that the spacing between planes depends on the Miller indices. For instance, the spacing between the (0 1 2) planes is half that of the spacing between the $(\bar{1}\ 0\ 1)$ planes. The spacing between planes will also depend on the crystal type. In a cubic crystal system, the spacings between (1 0 0), (0 1 0), and (0 0 1) planes are identical because all of the edge lengths are the same. In this case,

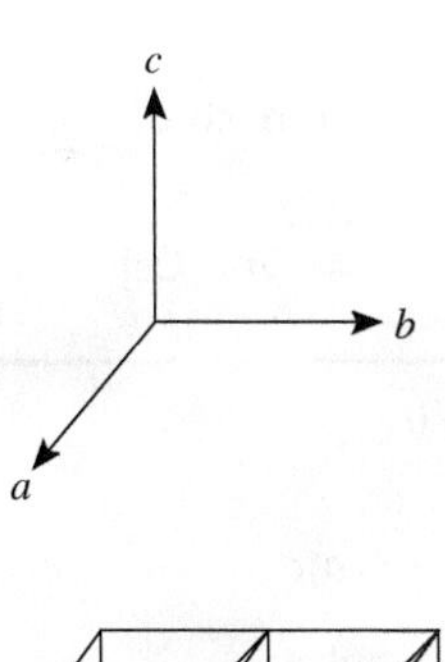

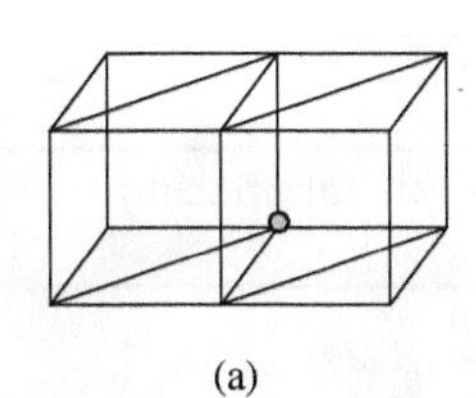

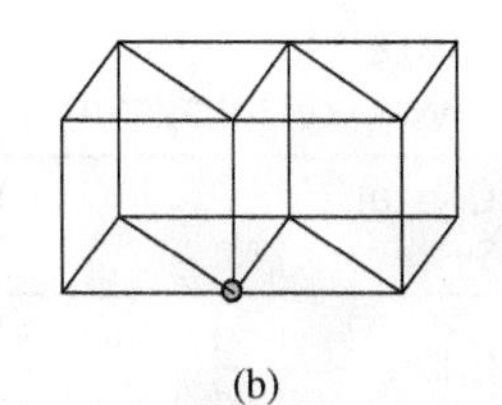

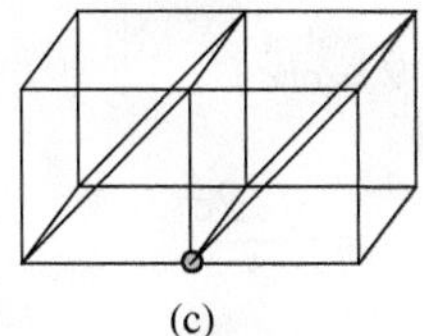

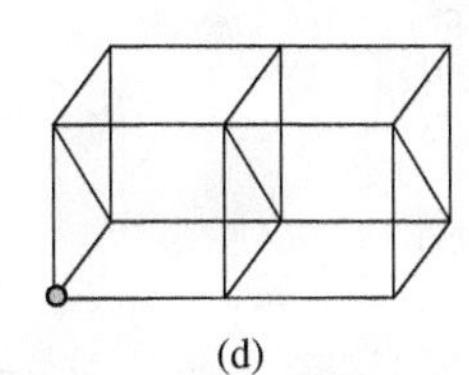

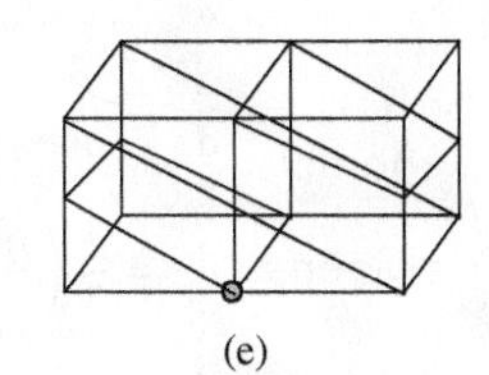

FIGURE 11.10
Sets of Miller planes, having the indices: (a) (1 1 0), (b) ($\bar{1}$ 1 0), (c) (1 $\bar{1}$ 0), (d) ($\bar{1}$ 0 1), and (e) (0 1 2). An arbitrary origin is indicated by the blue dot.

the three indicated planes are said to lie in the same family of Miller planes, as they all have the same magnitude of d_{hkl}. By convention, this particular family of Miller planes is referred to as the $\langle 1\ 0\ 0 \rangle$ set of planes (where the non-zero planes are listed in the order $a > b > c$. In an orthorhombic system, however, where all of the edge lengths of the Bravais lattice have unequal lengths, these same three Miller planes would have different values of d_{hkl} and would therefore belong to three different families of Miller planes.

The process of assigning each of the 2θ diffraction angles to each of the different d_{hkl} lattice spacings is known as *indexing*. Using the Bragg diffraction law, the value of d can be calculated using Equation (11.3). The corresponding values of $\langle h\ k\ l \rangle$ can be calculated using the geometrical formula in Table 11.4 appropriate to the lattice. Thus, the first step in indexing the X-ray diffraction powder pattern is to predict the lattice type. The next step is to make an assumption about the Miller index for the diffraction line that has the smallest value of θ. For example, we might assume that the first line corresponds with the $\langle 1\ 0\ 0 \rangle$ family of planes. Using the formula from Table 11.4, the lattice parameters can then be calculated. Consider the set of experimental data in Table 11.5, where only the 2θ values in column one are initially known. The Bragg diffraction law allows us to determine the interplanar spacings in column two. First, we will make the assumption of first-order diffraction ($n = 1$) because $n = 2$ for the (1 0 0) planes is the same as $n = 1$ for the (2 0 0) planes, and so on. Assuming a cubic lattice and that the first diffraction line (where $2\theta = 22.0°$) is the family of $\langle 1\ 0\ 0 \rangle$ planes, the value of a can be calculated using the formula in Table 11.4. Using this value of a to calculate the $h^2 + k^2 + l^2$ values in column three, if the results are all integers, then we can be fairly confident that all of our initial assumptions were correct. If, on the other hand, this value of a results in a non-integral result for any of the lines in column three, then our initial assumption that the first diffraction peak corresponded to the $\langle 1\ 0\ 0 \rangle$ family of planes must have been incorrect. We must then guess a different $\langle h\ k\ l \rangle$ for the first line, recalculate a, and see if the revised value of a will yield only integers in column three. Eventually, a value of a will be discovered that is suitable for all the other diffraction lines. If this is not the case, then perhaps our initial assumption of a cubic lattice was wrong and we might have to try a different lattice type.

Whenever the lattice type is something other than primitive, certain sets of planes will yield destructive interference in the centered cell where there normally would have been constructive interference in the primitive cell. The example shown in Figure 11.11 illustrates how this can happen in the case of a body-centered cell. The (1 0 0) reflection is missing because the body center yields a set of equivalent

TABLE 11.4 Interplanar spacings and unit cell volumes for each of the seven crystal systems.

Crystal System	Interplanar Spacing	Unit Cell Volume
Cubic	$\frac{1}{d^2} = \frac{h^2+k^2+l^2}{a^2}$	$V = a^3$
Tetragonal	$\frac{1}{d^2} = \frac{h^2+k^2}{a^2} + \frac{l^2}{c^2}$	$V = a^2c$
Ortho-rhombic	$\frac{1}{d^2} = \frac{h^2}{a^2} + \frac{k^2}{b^2} + \frac{l^2}{c^2}$	$V = abc$
Hexagonal	$\frac{1}{d^2} = \frac{4}{3}\left(\frac{h^2+hk+k^2}{a^2} + \frac{l^2}{c^2}\right)$	$V = \frac{\sqrt{3}a^2c}{2}$
Monoclinic	$\frac{1}{d^2} = \frac{1}{\sin^2\beta}\left(\frac{h^2}{a^2} + \frac{k^2\sin^2\beta}{b^2} + \frac{l^2}{c^2} - \frac{2hl\cos\beta}{ac}\right)$	$V = abc\sin\beta$
Triclinic	$\frac{1}{d^2} = \frac{1}{V^2}\left[h^2b^2c^2\sin^2\alpha + k^2a^2c^2\sin^2\beta + l^2a^2b^2\sin^2\gamma + 2hkabc^2(\cos\alpha\cos\beta - \cos\gamma) + 2kla^2bc(\cos\beta\cos\gamma - \cos\alpha) + 2hlab^2c(\cos\alpha\cos\gamma - \cos\beta)\right]$	$V = abc\left(1 - \cos^2\alpha - \cos^2\beta - \cos^2\gamma + 2\cos\alpha\cos\beta\cos\gamma\right)^{1/2}$

TABLE 11.5 Sample data for X-ray diffraction pattern indexing, using $\lambda = 1.54$ Å.

2θ	d_{hkl}, Å	$h^2+k^2+l^2$	<h k l>	Comments
22.0°	4.035	1	<1 0 0>	Our initial assumption yields $a = 4.035$ Å
31.0°	2.881	2	<1 1 0>	
38.0°	2.365	3	<1 1 1>	
45.0°	2.012	4	<2 0 0>	
50.0°	1.822	5	<2 1 0>	
56.0°	1.640	6	<2 1 1>	
65.0°	1.433	8	<2 2 0>	
70.0°	1.342	9	<2 2 1> or <3 0 0>	No unique family of planes can be identified here
74.0°	1.279	10	<3 1 0>	
78.0°	1.224	11	<3 1 1>	

planes that are exactly halfway in between the (1 0 0) set of planes in the corresponding primitive cell and which are exactly out of phase with the originals. Hence, destructive interference occurs and the (1 0 0) planes are not observed. The "missing" planes are known as *systematic absences*. Different lattice types have different rules for which families of planes will not be observed; these rules are known as the *diffraction conditions*. The diffraction conditions for each of the different types of centering are listed in Table 11.6. Certain types of translational symmetry elements

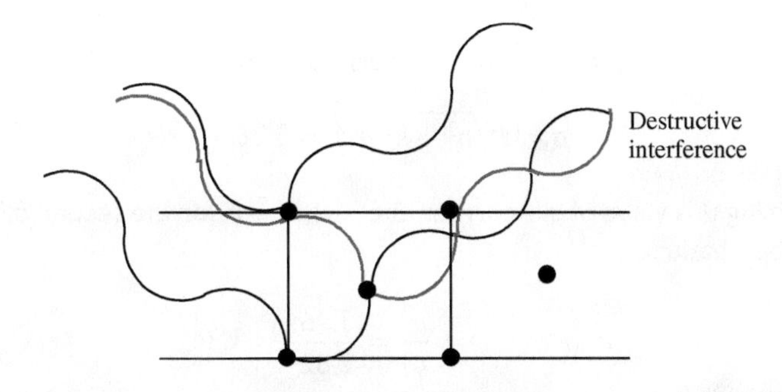

FIGURE 11.11
Illustration of how body-centering leads to destructive interference for parallel wave fronts that would have constructively interfered in a primitive unit cell.

TABLE 11.6 Diffraction conditions for different types of centering.

Type of Centering	Diffraction Conditions
Body-centered (I)	$(h+k+l)$ even
End-centered (A)	$(k+l)$ even
End-centered (B)	$(h+l)$ even
End-centered (C)	$(h+k)$ even
Face-centered (F)	$(h+k)$ even $(h+l)$ even $(k+l)$ even

can also lead to systematic absences, but these are not discussed here. As a result, the families of Miller planes that are absent (or missing) from an X-ray diffraction powder pattern can provide important clues about the lattice type and the space group.

The exact space group can often be identified on the basis of the intensities of each peak in the diffraction pattern. These intensities are proportional to a structure factor that takes into consideration the position and scattering strength of each atom in the crystal. For complex structures, a model must be used and refined against the diffraction data. Because of the limitations of the modeling process, thermal fluctuations in the motions of the nuclei, and uncertainties involved in the assignment of certain space groups, X-ray crystal structures should not be construed as 100% accurate representations of the actual molecular structure. There are sometimes occasions where the exact space group of the crystalline lattice cannot be unambiguously assigned.

Example 11-1. Using an X-ray beam with $\lambda = 2.291$ Å, the following values of 2θ were obtained for a cubic crystal system: 41.2°, 47.9°, 70.1°, 84.7°, 89.4°, 108.7°, 124.6°, 130.5°, 168.5°. Index each of the X-ray diffraction peaks to the appropriate family of Miller planes, determine the Bravais lattice, and indicate the unit cell parameters.

Solution. Using the Bragg diffraction law in Equation (11.3), the d_{hkl} spacings are calculated for each 2θ reflection. The first example is shown here:

$$d = \frac{\lambda}{2 \sin \theta} = \frac{2.291}{2 \sin(20.6)} = 3.26$$

The complete set of d_{hkl} values are 3.26, 2.82, 1.99, 1.70, 1.63, 1.41, 1.29, 1.26, and 1.15 Å. Next, we will guess that the first peak corresponds with the <1 0 0> family of planes. Using the formula from Table 11.4 for a cubic crystalline lattice, the lattice parameter a can be calculated from the first diffraction peak as follows:

$$a = d\sqrt{h^2 + k^2 + l^2} = 3.26$$

Assuming this value of a is correct, the $h^2 + k^2 + l^2$ for the second diffraction peak can be obtained:

$$h^2 + k^2 + l^2 = \frac{a^2}{d^2} = \frac{3.26^2}{2.82^2} = 1.34$$

Because this value is not an integer, our assumption that the first line can be assigned to the <1 0 0> family of planes must be incorrect.

We therefore go back and make the assumption that the first line corresponds to the <1 1 0> family of planes. In this case, $a = 2^{1/2}$ (3.26 Å) = 4.60 Å. Using this value of a to calculate $h^2 + k^2 + l^2$ yields the answer 2.66, which is still not an integer. Thus, our second guess was also incorrect. We next guess that the first diffraction peak corresponds with the <1 1 1> family of planes. If this is correct, then $a = 3^{1/2}$ (3.26 Å) = 5.64 Å. Using this value of a, $h^2 + k^2 + l^2 = 4.00$ for the second peak. Finally, we have an integer. Using a = 5.64 Å and d = 1.99 Å for the third line affords $h^2 + k^2 + l^2 = 8.03$, which is close enough to being an integer to be acceptable. Using a = 5.64 Å, the values listed in the following table are obtained for the remaining diffraction peaks. Because each of these sums is roughly integral, we can be confident that a = 5.64 Å is the correct value.

We are told that this is a cubic lattice; thus, $a = b = c$ = 5.64 Å and $\alpha = \beta = \gamma = 90°$ are the lattice parameters. We can index each of the diffraction peaks using the values of $h^2 + k^2 + l^2$ and the fact that h, k, and l must each be integers. The results are listed in this table. Because each value of $h + k$, $k + l$, and $h + l$ is even, the unit cell is a face-centered cubic.

2θ	d_{hkl}	$h^2 + k^2 + l^2$	<h k l>
41.2	3.26	3	<1 1 1>
47.9	2.82	4	<2 0 0>
70.1	1.99	8	<2 2 0>
84.7	1.70	11	<3 1 1>
89.4	1.63	12	<2 2 2>
108.7	1.41	16	<4 0 0>
124.6	1.29	19	<3 3 1>
130.5	1.26	20	<4 2 0>
168.5	1.15	24	<4 2 2>

11.3 CLOSEST-PACKED STRUCTURES

One of the driving forces for crystallization is the maximum occupancy of space. This is particularly true for metallic solids, whose crystalline structures can often be considered as the packing of identically sized spheres. Metallic bonding is considered to be nondirectional. Unlike the nonmetals, which can fill their valence shells by sharing only a few pairs of electrons, the metals require much larger coordination numbers in order to satisfy their outermost valence shells. As a result, the metals

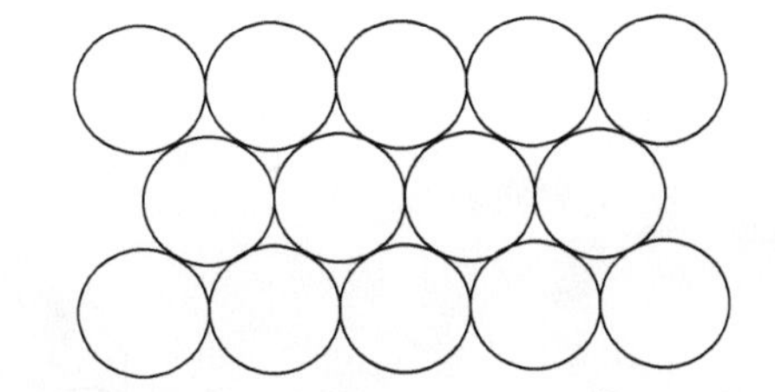

FIGURE 11.12
Closest-packing of spheres in two-dimensional space.

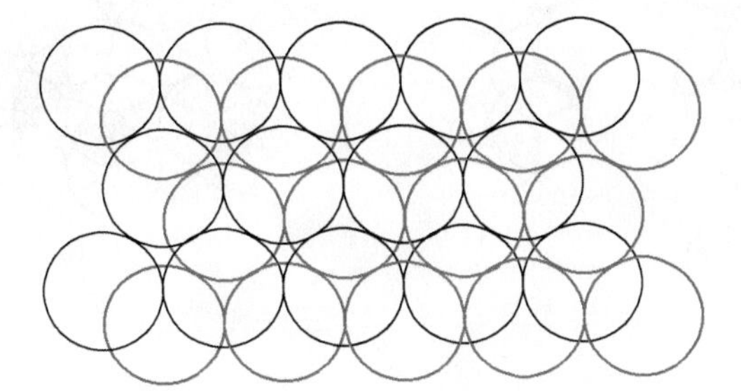

FIGURE 11.13
The bottoms of the closest-packed spheres in the second (red) layer will occupy indentations in the first (black) layer. In this case, all of the upward-pointing indentations are occupied.

will want to have as many nearest neighbors as possible in order to maximize the attractive forces between them.

The most efficient way of packing metal atoms into two-dimensional space is to stagger them in rows where each atom is surrounded by six nearest neighbors, as shown in Figure 11.12. This is the result that you might obtain if you poured a bag of equally sized marbles into the bottom of a rectangular glass aquarium.

Once the bottom of the aquarium has been filled; if we were to introduce a second layer of marbles on top of the first layer, there are two ways that we could do this and the marbles could arrange. We could place the marbles in the triangular-shaped indentations of the first layer that are all pointing downward in the figure or we could place them in all the indentations that are pointing upward. In either case, our second layer would look identical to the first, but it would be somewhat displaced so that the bottom of all the marbles in the second layer would occupy one of the types of holes or indentations in the first layer, as shown in Figure 11.13.

Once the second layer of marbles in the aquarium has been filled, we can now introduce a third layer of marbles that would sit in either all of the upward-pointing indentations or all of the downward-pointing indentations of the second layer, as shown in Figure 11.14. If the third layer lies in all the downward-pointing indentations of the second layer, then the first and third layers will be coincident and the lattice will repeat in an *ABABAB* alternating manner. If, on the other hand, the third layer lies in all of the upward-pointing indentations of the second layer, it will take a fourth layer to be coincident with the first and the lattice will repeat in an *ABCABC* pattern. Both of these three-dimensional structures are equally efficient at filling all space, with the marbles occupying 74% of the overall volume. However, the two different scenarios will belong to different crystalline lattices—the former *(ABABAB)* is the hexagonal closest-packed (hcp), while the latter *(ABCABC)* is the cubic closest-packed (ccp), both of which are shown in Figure 11.15.

It is easily shown that the hcp structure is the same as two interpenetrating hexagonal crystalline lattices, as shown in Figure 11.16(a). The hexagonal relationship is clearly shown in the diagram by connecting four atoms each from two adjacent *A* layers to form the corners of a primitive hexagonal unit cell. Connecting the nearest-neighbor atoms from two adjacent *B* layers will form the corners of a second hexagonal unit cell. Together, the two unit cells form an interpenetrating lattice, where the *B* atoms are displaced by one-half a unit cell length along the *c*-axis.

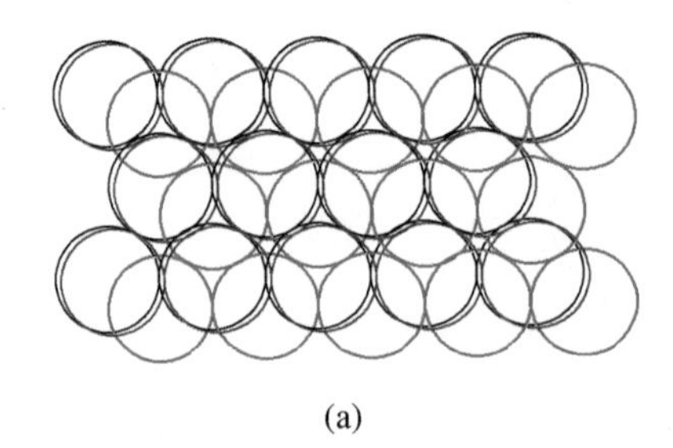

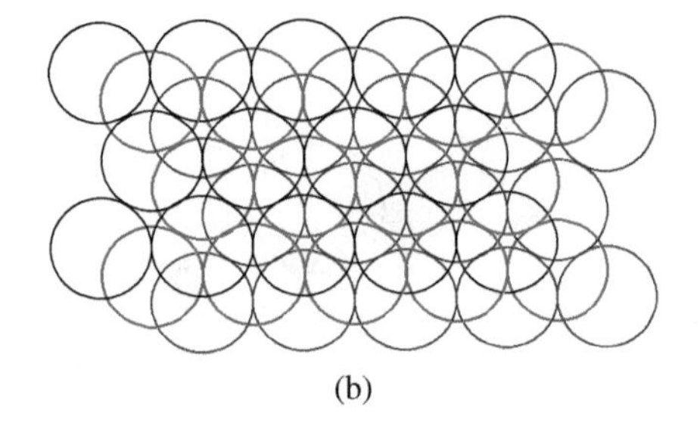

FIGURE 11.14
Addition of a third layer to the closest-packing of spheres in the first two layers. (a) If the third (blue) layer occupies the downward-pointing indentations of the second (red) layer in Figure 11.16, then the first (black) and third layers will become coincident, forming an *ABABAB* repeat pattern. (b) If the third layer occupies the upward-facing indentations, then an *ABCABC* repeating lattice will result.

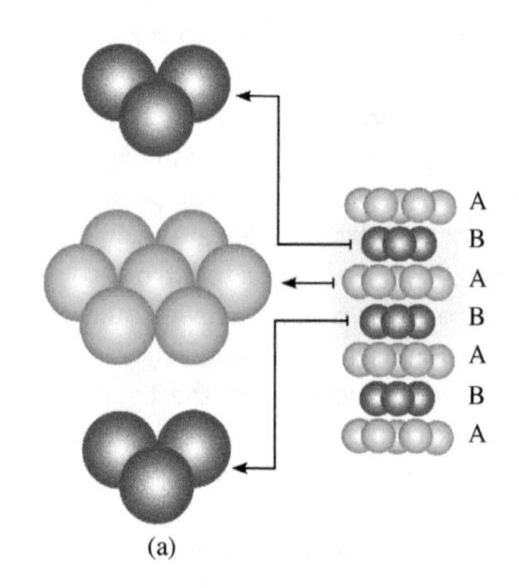

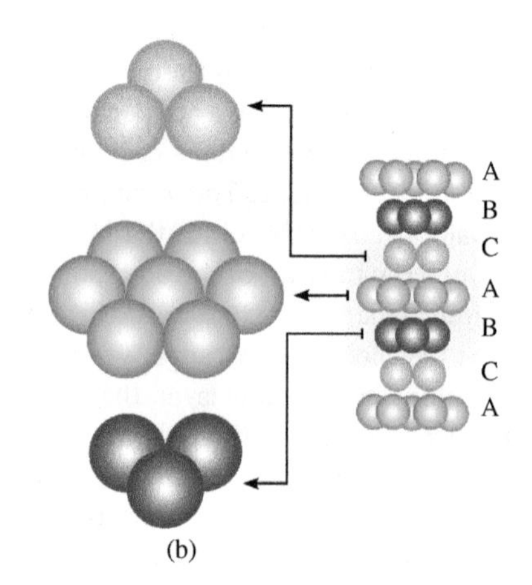

FIGURE 11.15
(a) The hexagonal closest-packed (hcp) lattice is the same as two interpenetrating hexagonal Bravais lattices that are displaced by one-half unit cell along the c-coordinate axis, while (b) the cubic closest-packed (ccp) lattice is identical to the face-centered cubic Bravais lattice. Those atoms marked by the designation ½ lie exactly one-half of a unit cell above or below the plane of atoms lacking this designation. [Blatt Communications.]

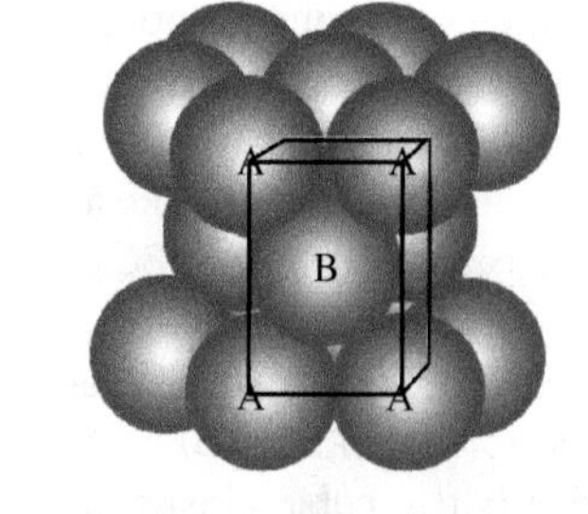

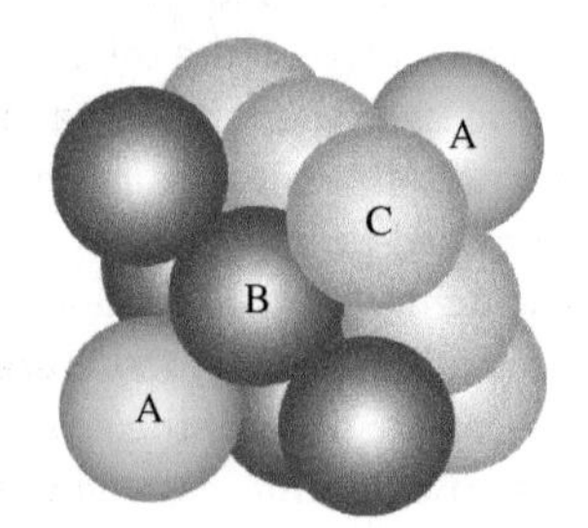

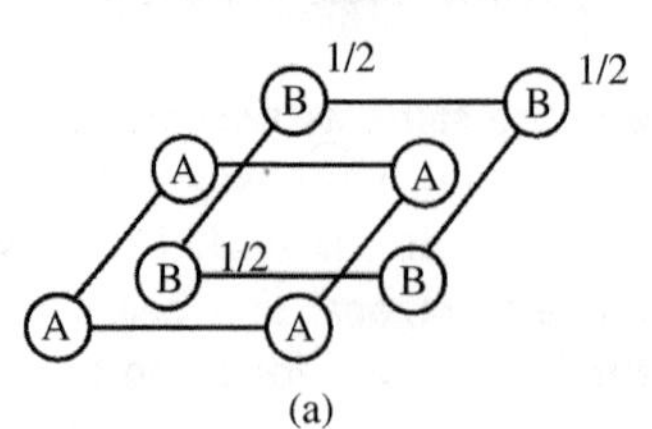

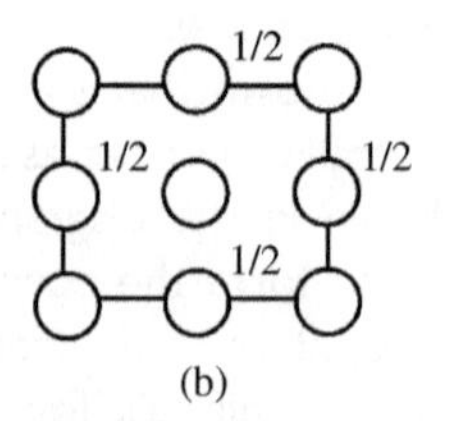

FIGURE 11.16
(a) The hexagonal closest-packed (hcp) structure *(ABABAB)* and (b) the cubic closest-packed (ccp) structure *(ABCABC)*. [Blatt Communications.]

Because each of the corner *A* atoms of a hcp unit cell are shared between eight different unit cells, while the intervening *B* atom is contained within a single unit cell, the hcp contains a total of two atoms (one *A* and one *B*) per unit cell.

Figure 11.16(b) shows how the ccp structure is identical to the face-centered cubic Bravais lattice. In this case, each of the eight corner atoms is shared by eight different unit cells and the six face-centers are each shared by two unit cells. Hence, there are a total of four atoms per unit cell within the ccp structure. While the ccp contains more atoms per unit cell than the hcp, some of its edges are also longer, so that the two closest-packed structures have identical densities and packing efficiencies. Many—but not all—metallic solids will also assume one of the two closest-packed structures.

Example 11-2. Prove that the atoms of both the cubic and hcp lattices occupy the same amount of space and that they both pack with 74% efficiency.

Solution. Let us first consider the case of the ccp or face-centered cubic unit cell. The closest contact between atoms in this crystalline lattice occurs along the face-diagonals of a cube, as shown in the given diagram.

$$a^2 + a^2 = (4r)^2$$

$$2a^2 = 16r^2$$

$$a = 8^{1/2}r$$

Using the Pythagorean theorem, the relationship between the edge length (a) and the radius (r) is given by the following equation:

$$a = 2\sqrt{2}r = 2.83r$$

As mentioned in the text, there are four atoms per unit cell (eight corners times one-eighth occupancy plus six face-centers times one-half occupancy). The volume of four spheres in terms of the radius is given by

$$V\ (\text{atoms}) = 4\left(\frac{4}{3}\pi r^3\right) = 16.76r^3$$

The volume of a single unit cell is a^3. In terms of r, this is

$$V\ (\text{unit cell}) = a^3 = (2.83r)^3 = 22.63r^3$$

Thus, the percent of the unit cell volume that is occupied by atoms is

$$\%\,\text{occupancy} = \frac{16.76r^3}{22.63r^3} \times 100\% = 74\%.$$

We now consider the case of the hcp unit cell. This unit cell contains two atoms per cell (eight corners times one-eighth occupancy plus one internal atom). Thus, the volume of the atoms is

$$V\ (\text{atoms}) = 2\left(\frac{4}{3}\pi r^3\right) = 8.38r^3$$

The hcp unit cell can be considered as two trigonal prisms having a single face in common. The volume of the unit cell is $V = ahc$. Because the atoms touch one another along the short edge of the hcp, $a = 2r$. It is the relationship between c and r that is more difficult to ascertain. The process can be made simpler if one were to realize that three of the A atoms in the base of Figure 11.16(a) and the enclosed B atom form a tetrahedron with edges of length a. The height of the tetrahedron is equal to exactly one-half the c-axis of the hcp unit cell, or $c/2$. If one were to drop a perpendicular from the atom at the peak of the tetrahedron, it would fall precisely in the center of the equilateral triangle that comprises its base. The distance from any of the corners to this point is represented by m in the given diagram. The value of m can be calculated from the smaller right triangle shown at right.

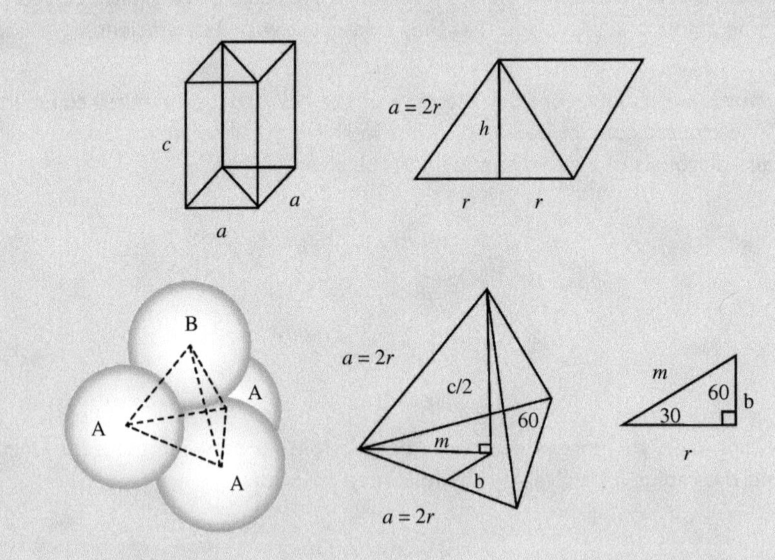

[Blatt Communications.]

$$m = \frac{r}{\sin 60°} = 1.155r$$

$$m^2 + \left(\frac{c}{2}\right)^2 = a^2$$

$$c^2 = 4(a^2 - m^2) = 4[(2r)^2 - (1.155r)^2] = 10.66r^2$$

$$c = \sqrt{10.66r^2} = 3.266r$$

The relationship between h and r can be determined using b, as follows:

$$b = m \sin(30°) = 0.5m$$

$$h = m + b = 1.5m = 1.733r$$

Finally, the volume of the unit cell can be calculated:

$$V(\text{unit cell}) = ahc = (2r)(1.733r)(3.266r) = 11.32r^3$$

The efficiency of packing in the hcp structure is

$$\%\,\text{occupancy} = \frac{8.38r^3}{11.32r^3} \times 100\% = 74\%$$

Example 11-3. Calculate the density of silver in g/cm^3 if the Ag atoms crystallize in a face-centered cubic and have a metallic radius of 144 pm.

Solution. Density is mass divided by volume. There are four Ag atoms per unit cell in a face-centered cubic (ccp) because each of the eight corner atoms is shared between eight unit cells ($8 \times {}^1/_8$) and each of the six face-centers is shared between two unit cells ($6 \times {}^1/_2$). According to the periodic table, one mole of Ag atoms has a mass of 107.9 g. The mass of one unit cell is

$$\frac{6.022 \times 10^{23}\ \text{atoms}}{107.9\,\text{g}} = \frac{4\,\text{atoms}}{m}$$

$$m = 7.167 \times 10^{-22}\,\text{g}$$

In the ccp lattice, the atoms touch along the diagonal of a face-center, such that $a = 8^{1/2}r$. The volume of a cube is $V = a^3$. Hence, the volume of a unit cell is

$$V = a^3 = (\sqrt{8}r)^3 = [\sqrt{8}(1.44 \times 10^{-8}\,\text{cm})]^3 = 6.76 \times 10^{-23}\,\text{cm}^3$$

Finally, the density of silver is calculated as

$$d = \frac{m}{V} = \frac{7.167 \times 10^{-22}\,\text{g}}{6.76 \times 10^{-23}\,\text{cm}^3} = 10.6\,\text{g/cm}^3$$

11.4 THE FREE ELECTRON MODEL OF METALLIC BONDING

In the previous chapters, we discussed various models of bonding for covalent and polar covalent molecules (the VSEPR and LCP models, valence bond theory, and molecular orbital theory). We shall now turn our focus to a discussion of models describing metallic bonding. We begin with the *free electron model*, which assumes that the ionized electrons in a metallic solid have been completely removed from the influence of the atoms in the crystal and exist essentially as an "electron gas." Freshman chemistry books typically describe this simplified version of metallic bonding as a "sea of electrons" that is delocalized over all the metal atoms in the crystalline solid. We shall then progress to the band theory of solids, which results from introducing the periodic potential of the crystalline lattice.

The metallic elements lie on the left-hand side of the periodic table, where the number of valence electrons is considerably less than the number of valence orbitals available to accommodate them. As a result, the metals maximize their bonding interactions by having as many nearest-neighbor interactions as possible—often forming either a ccp or hcp crystalline lattice. Because the metals have a smaller effective nuclear charge than their nonmetal counterparts, the valence electrons of the metals are ionized fairly easily; it is this resulting "sea" of delocalized electrons that serves to hold the metal atoms together in the crystalline state. Thus, as a first approximation, the ionized electrons in a metallic solid can be considered a sort of "electron gas," where the electrons are free to behave independently from one another and where they do not really feel the potential of any individual atomic nucleus. In this model, the atomic orbitals essentially lose their identity and the electrons are considered to be moving in a nearly free potential. As a result, the electrons can be thought of as a sort of ideal gas that obeys the classical kinetic molecular theory. This rather crude approximation is known as the *Drude-Lorentz theory* of metallic bonding. Some obvious shortcomings of the model are that the negatively charged electrons will repel each other and that electrons are also subject to the Pauli exclusion principle. Another significant difference is that the electrons are much lighter than gas particles and this will have an important influence on the number of conducting electrons in a given volume of the metal.

To begin, we assume that the electrons exist in a cube of metal atoms, where all the edges of the cube are of length L. As the potential of the electrons inside the cube is a constant, we can set $V = 0$ and solve the wave equation for the three-dimensional particle in a box model, which yields the result given by Equation (11.4), where the quantum numbers n_x, n_y, and n_z are positive integers taking the values 1, 2, 3, ...

$$E = \frac{(n_x^2 + n_y^2 + n_z^2)h^2}{8mL^2} \tag{11.4}$$

In order to proceed with the derivation, we now consider each of the quantum numbers as points in a three-dimensional lattice forming their own cube. A two-dimensional representation is shown in Figure 11.17, where every point represents a unique combination of n_x, n_y, and n_z. The distance R from the origin (shown as a vector in Figure 11.17) can be calculated according to Equation (11.5).

$$R^2 = n_x^2 + n_y^2 + n_z^2 = \frac{8mL^2E}{h^2} \tag{11.5}$$

The number of allowed orbitals within a given radius R is given by one-eighth of the total points inside the sphere containing the lattice of quantum numbers. The factor of one-eighth arises from the fact that each of the three quantum numbers has to be a positive integer and this will only apply in the octant where n_x, n_y, and n_z are ≥ 1. Because each orbital can hold at most two electrons, the number (N)

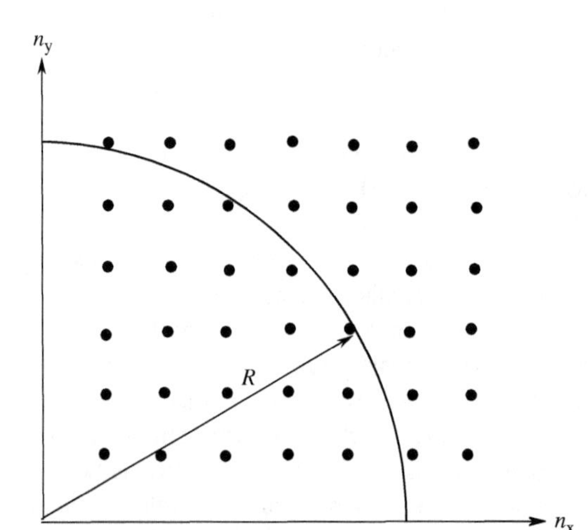

FIGURE 11.17
Two-dimensional graph of the allowed quantum numbers represented as points in a lattice in the free electron model of metallic bonding. The distance R from the origin can be calculated using Equation (11.5).

of electrons that can be accommodated in all of the allowed energy levels up to a certain maximum energy, E_{max}, will be given by Equation (11.6).

$$N = 2\left(\frac{1}{8}\right)\left(\frac{4}{3}\right)\pi R^3 = \frac{\pi}{3}\left(\frac{8mL^2E_{max}}{h^2}\right)^{3/2} = \frac{8\pi}{3}\left(\frac{2mE_{max}}{h^2}\right)^{3/2} L^3 \tag{11.6}$$

If we substitute the electron density $\rho = N/L^3$ and solve Equation (11.6) for E_{max}, we obtain the energy of the highest occupied energy level, or the *Fermi level*, at $T = 0$ K, as shown in Equation (11.7).

$$E_F = E_{max} = \frac{h^2}{2m}\left(\frac{3\rho}{8\pi}\right)^{2/3} \tag{11.7}$$

Example 11-4. Given that Al metal crystallizes in a face-centered cubic lattice with $a = 405.0$ pm, calculate E_{max} in units of eV using Equation (11.7).

Solution. First, we need to determine the electron density using the unit cell dimensions. A body-centered cubic lattice contains four Al atoms per unit cell ($8 \times 1/8$ from the corner atoms and $6 \times 1/2$ from the face-centers). The volume of the unit cell is given by $a^3 = (4.050 \times 10^{-10}\ \text{m})^3 = 6.643 \times 10^{-29}\ \text{m}^3$. Furthermore, each Al atom can ionize three electrons. Thus, $\rho = (4 \times 3)/6.643 \times 10^{-29}\ \text{m}^3 = 1.806 \times 10^{29}\ \text{m}^{-3}$. Next, we use Equation (11.7) to calculate E_{max}, where $m_e = 9.109 \times 10^{-31}$ kg and $h = 6.626 \times 10^{-34}$ J·s. Solving, we obtain $E_{max} = 1.867 \times 10^{-19}$ J. Given that 1 eV $= 1.602 \times 10^{-19}$ C, $E_{max} = 11.65$ eV. The experimental value for the width of the conduction band (CB) of Al, as determined from PES, is 11.8 eV. Thus, the calculated value is in reasonable agreement with the experimental value.

Another major difference between ideal gases and the electrons in the free electron model is that the density of conduction electrons is much larger than the density of an ideal gas, which causes E_{max} to be greater for the ionized electrons. Thus, while at room temperature a majority of the particles in an ideal gas will exist in thermally excited translational energy levels, only a very small fraction of electrons in the top-filled energy level can become thermally excited. The *density of states* (DOS), $D(E)$, is defined as the number of states available to the electrons within a given energy range per unit volume. The DOS within the CB can also be calculated using the free electron model by differentiating Equation (11.6) with respect to the energy and then dividing by the volume, as shown in Equation (11.8) with the result being given by Equation (11.9).

$$\frac{1}{L^3}\frac{dN}{dE} = \frac{3}{2}\frac{8\pi}{3}\left(\frac{2m}{h^2}\right)^{3/2} E^{1/2} \tag{11.8}$$

$$D(E) = 4\pi\left(\frac{2m}{h^2}\right)^{3/2} E^{1/2} \tag{11.9}$$

The DOS at 0 K is shown in Figure 11.18 using the free electron model. However, the Drude–Lorentz model employs the classical equipartition of energy and does not take into account the fact that quantum mechanics places restrictions on the placement of the electrons (as a result of the Pauli exclusion principle). A revised theory, known as the *Sommerfield model*, allows for this modification. At temperatures above 0 K, the fraction, $f(E)$, of allowed energy levels with energy E follows the

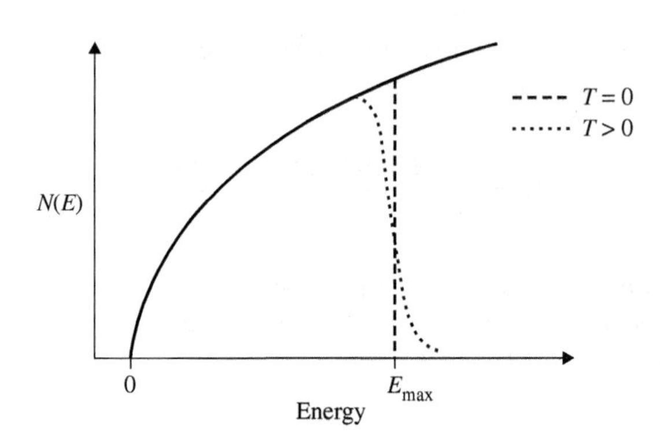

FIGURE 11.18
Density of states for an arbitrary metallic solid using the free electron model. The dashed curve is for $T=0\,K$, while the dotted curve is for $T>0\,K$ and employs the Fermi–Dirac distribution given by Equation (11.10). [© P.A. Cox, *The Electronic Structure and Chemistry of Solids*, 1987, by permission of Oxford University Press.]

Fermi–Dirac distribution given by Equation (11.10), which also takes into account the fact that electrons are indistinguishable to interchange. Thus, the actual population of the different energy levels in Figure 11.18 is given by the dotted curve at the top edge of the CB. The symbol E_F represents the *Fermi level*, which is just the top-filled energy level in the band.

$$f(E) = \frac{1}{1 + \exp[(E - E_F)/kT]]} \tag{11.10}$$

We now shift our focus in the free electron model to the momentum operator. According to Table 3.2, the quantum mechanical momentum operator in one-dimension is given by Equation (11.11).

$$\hat{p}\phi = -i\hbar\frac{\partial\phi_x}{\partial x} = p_x\phi_x \tag{11.11}$$

One possible solution to this differential equation is given by Equation (11.12), where k_x is an arbitrary constant.

$$\phi_x = e^{ik_x x} \tag{11.12}$$

Example 11-5. Prove that Equation (11.12) is a solution to Equation (11.11).

Solution. Taking the first partial derivative, we obtain:

$$-i\hbar\frac{\partial}{\partial x}(e^{ik_x x}) = -i^2\hbar k_x e^{ik_x x} = \hbar k_x \phi_x$$

This is an eigenfunction, where the eigenvalue is equal to the momentum in the *x*-direction.

$$p_x = \hbar k_x$$

As the kinetic energy T is equal to $\frac{1}{2}mv^2$ and we are already assuming that $V=0$, we can solve for the energy in terms of the arbitrary constant k, as shown in Equation (11.13).

$$E = V + T = \frac{1}{2}mv^2 = \frac{p^2}{2m} = \frac{\hbar^2k^2}{2m} \tag{11.13}$$

The constant k is also related to the wavelength through the deBroglie relationship, as shown in Equation (11.14).

$$\lambda = \frac{h}{p} = \frac{h}{\hbar k} = \frac{2\pi}{k} \tag{11.14}$$

Extrapolating to three dimensions, Equation (11.12) becomes Equation (11.15):

$$\phi(r) = e^{ik_x x} e^{ik_y y} e^{ik_z z} = e^{i\vec{k}\cdot\vec{r}} \tag{11.15}$$

Equation (11.15) is the equation for a plane wave, and the meaning of the k-vector is now apparent as the propagation vector. Similarly, in three dimensions, Equation (11.13) becomes Equation (11.16):

$$E = \frac{\hbar^2 \vec{k}^2}{2m} = \frac{\hbar^2}{2m}(k_x^2 + k_y^2 + k_z^2) \tag{11.16}$$

Applying the boundary conditions that $\phi(L, 0, 0) = \phi(0, L, 0) = \phi(0, 0, L) = 0$, we obtain the solutions given by Equation (11.17):

$$e^0 = e^{ik_x L} = e^{ik_y L} = e^{ik_z L} = 1 \tag{11.17}$$

Using Euler's formula, the boundary conditions reduce to Equation (11.18):

$$\cos(k_x x) + i\sin(k_x x) = 1, \text{etc.} \tag{11.18}$$

Thus, the only acceptable values for k_x, k_y, and k_z are given by Equation (11.19):

$$k_x = 0, \pm\frac{2\pi}{L}, \pm\frac{4\pi}{L}, \ldots \tag{11.19}$$

Each of the energy levels, $E_k > 0$, given by Equation (11.16) will be doubly degenerate as k can take on either a positive or a negative value. The results of the free electron model are shown in Figure 11.19.

The free electron model is successful in explaining many of the experimental properties of solids, including the range of energy levels, the DOS, the temperature dependence of the heat capacity, the Pauli magnetic susceptibility for most metals, and the relationship between the electrical conductivity and thermal conductivity. The model works best for early row metals, such as the alkali and alkaline earth metals. It does not, however, work well for the transition metals, where the *d*-orbitals become involved. For one, the *d*-orbitals are more contracted than the *s*-orbitals and *p*-orbitals, leading to a narrower band of energy levels. Secondly, because the *d*-orbitals can accommodate as many as 10 electrons, they also have a much larger DOS.

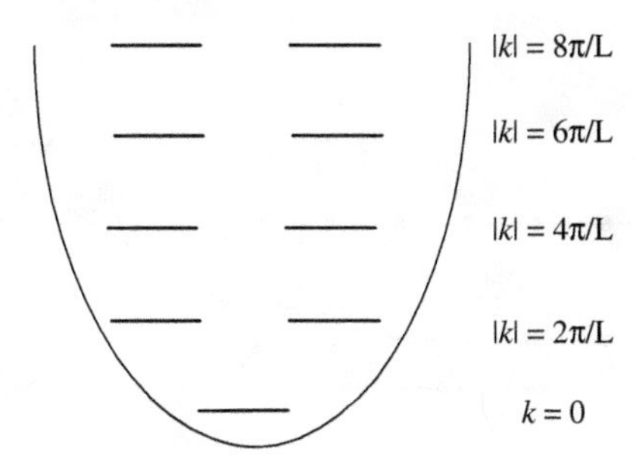

FIGURE 11.19
Energy levels in the free electron model as a function of *k*.

11.5 BAND THEORY OF SOLIDS

Up until this point, we have been treating the electrons in a metal as totally free, experiencing no attractions whatsoever to the positively charged nuclei. Clearly, this is only a crude approximation. In real life, the electrons will exist in a periodic potential $V(r)$ formed by the underlying translational symmetry of the crystalline lattice. We begin with a discussion of the wavefunction for an electron moving along a one-dimensional chain of atoms, such as the one pictured in Figure 11.20, where a is the length of the unit cell, or a unitary translation of an arbitrary lattice point along the row of atoms. Because the translational group T_N is isomorphous with the cyclic point group C_N when $N=\infty$, we can approximate the linear chain of atoms as resulting from an arc segment of atoms linked together in a very large ring, such that the curvature of the ring is nearly imperceptible, as shown in Figure 11.20.

This approximation has the benefit of allowing us to introduce the Born-von Karman boundary conditions given by Equation (11.20), where N is the number of atoms in the ring:

$$\psi(x + Na) = \psi(x) \tag{11.20}$$

In 1928, the Swedish physicist Felix Bloch proved that the solutions to the Schrödinger equation for a periodic potential take the form shown in Equation (11.21), where e^{ikx} is the equation for a plane wave and $u(x)$ is simply a function having the same periodicity as that of the lattice.

$$\psi(x) = e^{ikx}\,u(x) \tag{11.21}$$

Setting Equation (11.21) equal to Equation (3.35), as shown in Equation (11.22), we obtain the result given in Equation (11.23), which is equivalent to that obtained in Equation (11.14). Thus, the quantum number k is, in one sense, a wavenumber.

$$\psi(x) = Ae^{2\pi i\left(\frac{x}{\lambda}\right)} = u(x)e^{ikx} \tag{11.22}$$

$$\frac{2\pi xi}{\lambda} = ikx \tag{11.23}$$

As chemists, we are already familiar with the types of functions u(x) that might have translational symmetry in a metallic solid—these are the atomic orbitals located on the atoms at the lattice points. In a homogeneous metallic solid, the valence AOs can combine with one another to form MOs. For a homonuclear diatomic, such as Li_2, there are two MOs: one bonding and one antibonding, as shown in Figure 11.21. For a metallic solid containing only three metal atoms, three MOs will result: one bonding MO (with no nodes), one nonbonding MO (having one node), and one antibonding MO (having two nodes). For n metal atoms, n MOs will form, each pair of which has a slightly different energy from its neighboring pair on the MO diagram. The relative energies of each MO can be determined on the basis of the number of nodal planes that exist. The lowest energy MO will have zero

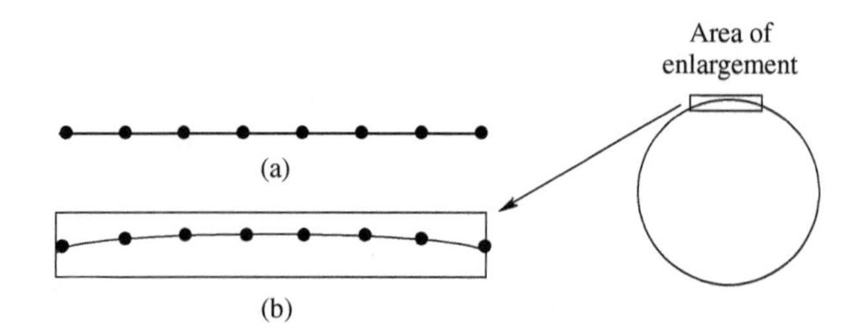

FIGURE 11.20
Representation of (a) a linear chain of atoms and (b) the approximation of a linear chain of atoms as the enlarged area from a ring of atoms having a very large radius.

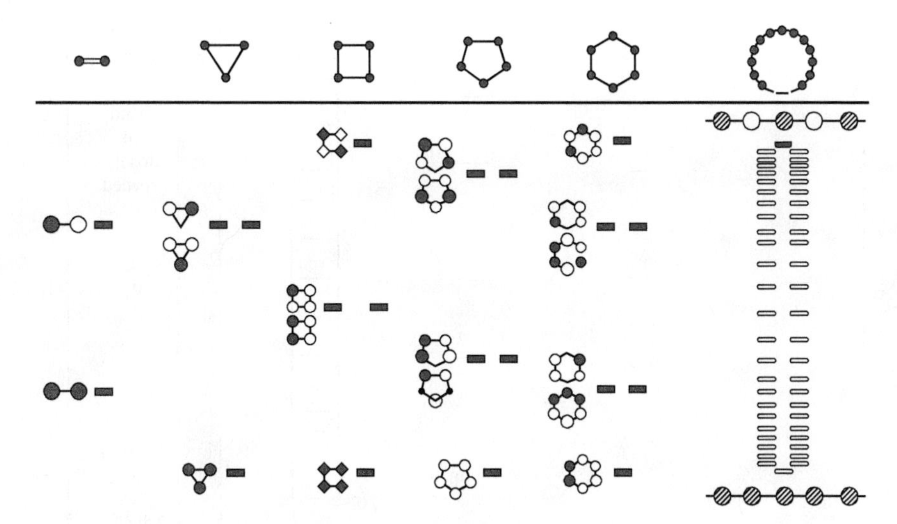

FIGURE 11.21
Molecular orbital diagram for the fictional ring structures of Li metal having n atoms. [Reproduced from Hoffmann, R. *Solids and Surfaces: A Chemist's View of Bonding in Extended Structures*, Wiley-VCH: New York, 1989. This material is reproduced with permission of John Wiley & Sons, Inc.]

nodes and the highest will have $n-1$ nodes. For an infinite number of metal atoms, the MOs become so closely spaced that they blur together into a "band" of MOs, as shown at the right in Figure 11.21. It is the energies and shapes of these bands that will determine the physical and chemical properties of metals. This improvement over the Drude–Sommerfield model of metallic bonding is known as *band theory*.

Our approach to understanding the band theory of solids is to make symmetry-adapted linear combinations of the atomic orbitals on each of the atoms at the lattice points using the translational symmetry of the crystal. In practice, this approach is directly analogous to the way we made molecular orbitals in covalently bonded compounds by taking linear combinations of atomic orbitals having similar energies and the appropriate symmetry to yield a net nonzero overlap. Thus, if there are N atomic orbitals in the lattice, there will be N molecular orbitals (known as the *Bloch orbitals*) in the crystalline solid, as shown in Figure 11.22. In fact, the use of linear combinations of atomic orbitals to make Bloch orbitals actually predates the LCAO-MO method. The wavefunctions for the MOs in the crystal are given by the Bloch functions shown in Equation (11.24), where N is the number of atomic orbitals, ψ_k is the kth MO, ϕ_n is the wavefunction for the nth AO basis function, and a is the lattice spacing. In the C_N cyclic group, k is therefore an index labeling the irreducible representation. As shown in Figure 11.21, k is also the number of nodes in the resulting Bloch function. Finally, as shown in Example 11-5 and Equation (11.14), k is a wavevector and its magnitude is directly related to the momentum of the electron.

$$\psi_k = N^{1/2}\sum_n e^{ikna}\phi_n \qquad (11.24)$$

The resulting Bloch orbitals are subject to the cyclic boundary conditions imposed by the ring structure (also known as *Born-von Karman boundary conditions*). Every Bloch function belongs to a different IRR of the C_N cyclic group. Because the e^{ikna} coefficient in Equation (11.24) is a periodic function of k, there is a limited range of values that k can assume: $-\pi/a \leq k \leq +\pi/a$. This range of k values is also known as the first *Brillouin zone*, the zone of unique values of k. The energies for each of the Bloch orbitals can be calculated using Equation (11.25), where α and β have the same meanings as they did in Hückel theory.

$$E(k) = \alpha + \beta\cos(ka) \qquad (11.25)$$

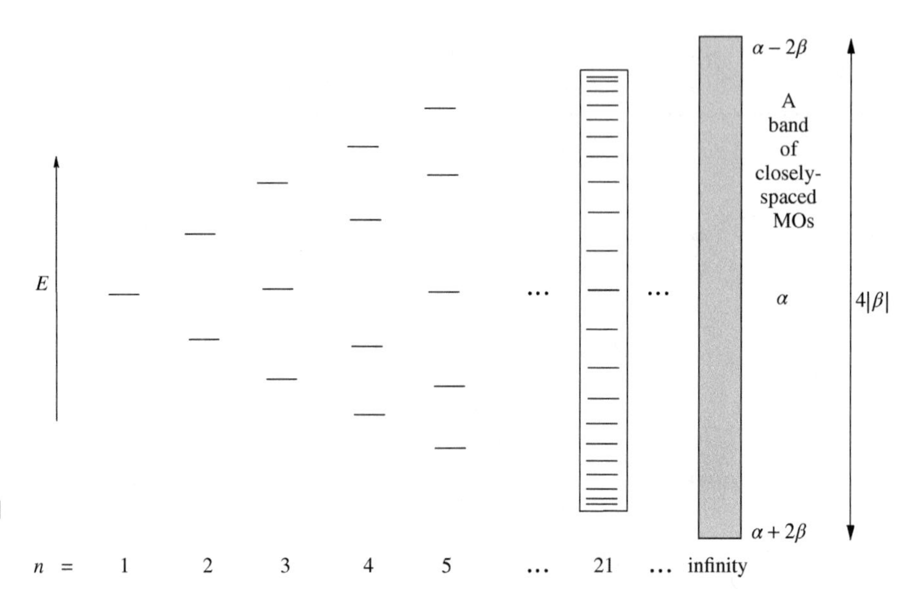

FIGURE 11.22 Molecular orbital diagram for a linear chain of n metal atoms, showing how the closely-spaced MOs blur into a band as n approaches infinity.

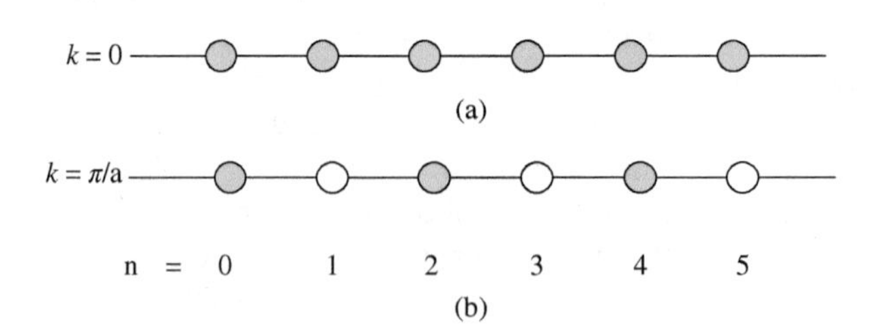

FIGURE 11.23 Bloch orbitals for the two extremes of the first Brillouin zone for a chain of Li atoms (2s AO basis functions): (a) $k=0$, (b) $k=\pi/a$. The shaded spheres represent the positive sign of the wavefunction, while the hollow spheres represent the negative sign of the wavefunction.

Consider a chain of H atoms, where the 1s atomic orbital on each atom is used as a basis set to derive the Bloch orbitals according to Equation (11.25). When $k=0$, the coefficients in Equation (11.24) are $e^0=\cos(0)+i\sin(0)=1$. Thus, the shape of the $k=0$ Bloch orbital results from the entirely positive linear combination of AO basis functions shown in Figure 11.23(a). Because there are no nodes in the resulting Bloch orbital, this orbital will be the most bonding one and it will lie lowest in energy on the Bloch diagram, having an energy of $\alpha+2\beta$. At the other extreme, when $k=\pi/a$, the coefficients in Equation (11.24) are equal to $e^{\pi in}=\cos(n\pi)+i\sin(n\pi)=\cos(n\pi)=(-1)^n$, because $\cos(n\pi)=+1$ whenever n is even and $\cos(n\pi)=-1$ whenever n is odd. The shape of the resulting linear combination of atomic orbitals is shown in Figure 11.23(b). This Bloch orbital contains the maximum number of nodes and will therefore lie at the highest energy on the Bloch diagram, having an energy of $\alpha-2\beta$. The shapes of the real component of the resulting Bloch functions at $k=0$ and $k=\pi/a$, as well as at some intermediate value of k, are shown in Figure 11.24. Notice that the wavelength of the plane wave is inversely proportional to k, as demonstrated by Equation (11.14).

The energies of the Bloch orbitals curve gradually upward between these two extremes in the first Brillouin zone. The net result is that the energy of the original basis AO (α) is spread out by the overlap with its neighbors to form a band of MOs having a bandwidth equal to $4|\beta|$. The resulting Bloch diagram is shown in Figure 11.25.

Because the distance between the neighboring H atoms on the ring affects the magnitude of the exchange integral β, the bandwidth will also depend on the size of the spacing a. The closer the H atoms are to one another, the larger the value

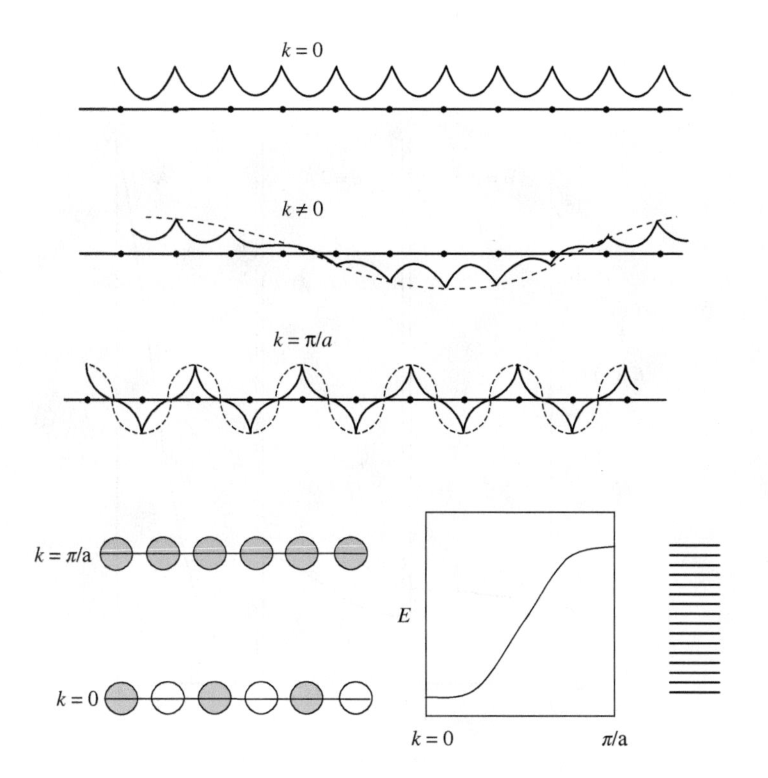

FIGURE 11.24
The real (solid) and imaginary (dashed) components of the Bloch waves. [© P.A. Cox, *The Electronic Structure and Chemistry of Solids*, 1987, by permission of Oxford University Press.]

FIGURE 11.25
Bloch diagram for the first Brillouin zone resulting from the overlap of 1*s* AO basis functions for a ring of H atoms. The bandwidth is given by $4|\beta|$.

of β and the greater the dispersion (or bandwidth). This can readily be seen by the results of calculations for $a = 3$, 2, and 1 Å shown in Figure 11.26.

The amount of overlap will also depend on the nature of the AO basis function. Suppose that our basis function in the one-dimensional chain consists of the *p* atomic orbitals. If the *p*-orbitals lie perpendicular to the internuclear axis, the magnitude of their overlap will be significantly less than that for the *s* atomic orbitals (just as a π-bond has less overlap than a σ-bond). Thus, the bandwidth for the *pi* MOs in a metallic solid will be narrower than that for the *sigma* MOs.

Another consideration is the direction that the band "runs." Assuming that the one-dimensional chain lies along the z-axis, the basis functions for the p_x and p_y atomic orbitals will lie perpendicular to the linear chain. In this case, the band will run "up" in the first Brillouin zone, as shown in Figure 11.27(a), because the most bonding Bloch orbital will occur at $k = 0$ and the most antibonding one at $k = \pi/a$. On the other hand, if the set of p_z atomic orbitals is used as the basis, the Bloch orbital at $k = 0$ will have the maximum number of nodes and will be the most antibonding, while the Bloch orbital at $k = \pi/a$ will have the least number of nodes and will lie lowest in energy. This band will run "down" in energy, as shown in Figure 11.27(b). Although the first Brillouin zone also extends to $k = -\pi/a$, the Bloch diagram usually only shows the range $k = 0$ to $k = \pi/a$ as $E(k) = E(-k)$. Therefore, the Bloch diagram between $k = 0$ and $k = -\pi/a$ will be the mirror image of the one that is already shown.

The DOS is defined as the number of energy levels between E and $E + dE$ in the band, where dE is some arbitrarily small energy interval. Thus, the DOS is inversely proportional to the slope of E versus k. As a result, the flatter the band is, the greater the DOS. Because the energy levels for the Bloch orbitals are more closely spaced at the extreme bottom and extreme top of the band, as shown in Figure 11.22, the DOS is greatest at the band edges and smallest in the middle. The DOS diagrams for the two different types of *p*-orbital bands in Figure 11.27 are shown in Figure 11.28.

One of the more interesting examples of one-dimensional linear chains occurs for the tetracyanoplatinates: $[Pt(CN)_4]^{2-}$. These square planar anions stack in the

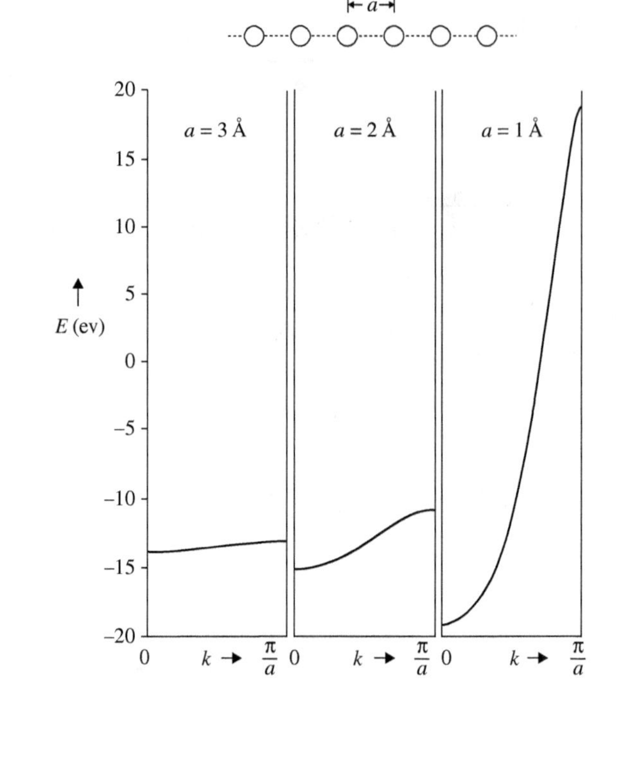

FIGURE 11.26
Bloch diagram for the overlap of H 1*s* AOs having spacings of 3, 2, and 1 Å. [Reproduced from Hoffmann, R. *Solids and Surfaces: A Chemist's View of Bonding in Extended Structures*, Wiley-VCH: New York, 1989. This material is reproduced with permission of John Wiley & Sons, Inc.]

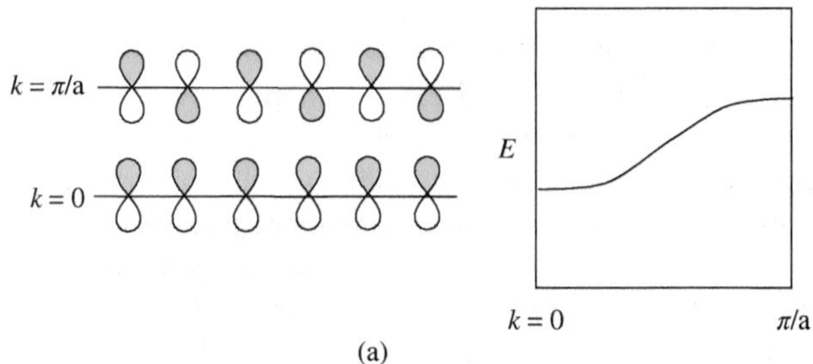

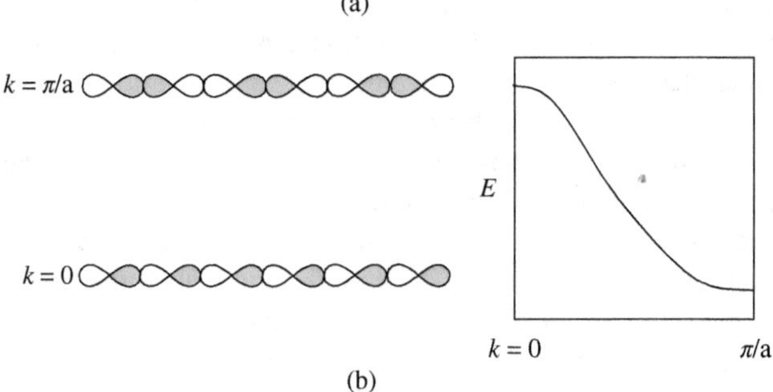

FIGURE 11.27
Bloch functions and Bloch diagrams for the *p* AO basis sets, where the *p*-orbitals lie (a) perpendicular to and (b) parallel with the linear chain connecting the atoms in the crystal. The greater overlap in (b) implies a larger bandwidth for (b) than for (a).

solid state just like the pieces in a sliced loaf of bread, as shown in Figure 11.29. For example, the species $K_2[Pt(CN)_4]$ stacks along the z-axis of the molecule with a Pt–Pt distance of ~3.3 Å. While the electrical properties of this molecule are fairly uninteresting (it is an electrical insulator), if the material is partially oxidized with Cl_2, the resulting product $K_2[Pt(CN)_4Cl_{0.3}]$, which has a shorter Pt–Pt distance of ~2.7 Å, is an electrical conductor. We will now attempt to rationalize these results using our knowledge of one-dimensional band theory.

To begin the problem, we need to sketch the one-electron MO diagram for the $[Pt(CN)_4]^{2-}$ monomer. In order to simplify matters, we will model the compound

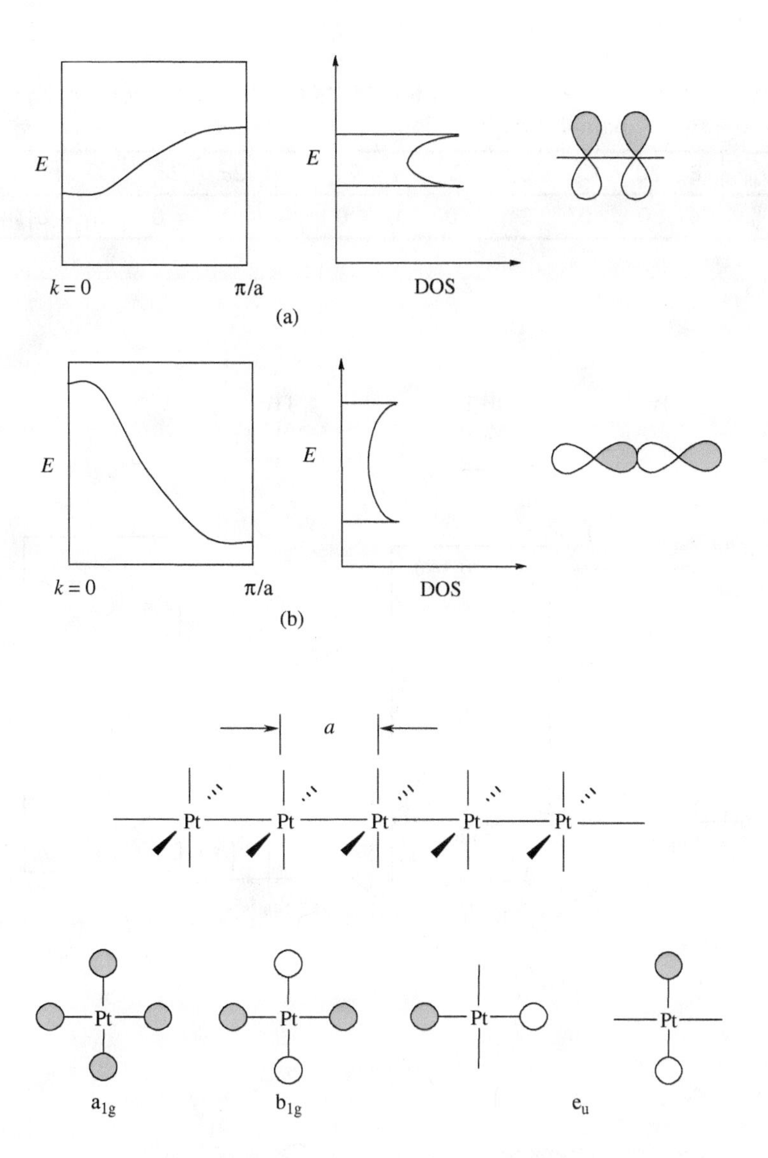

FIGURE 11.28
(a, b) Bloch diagrams and density of states (DOS) plots for two different bands formed by overlapping *p*-orbitals in a one-dimensional chain. The narrower the *E* versus *k* band, the larger the DOS will be.

FIGURE 11.29
A one-dimensional linear chain of $[Pt(CN)_4]^{2-}$ or $[PtH_4]^{2-}$ anions, having a repeat distance in the unit cell of length *a*.

FIGURE 11.30
Symmetries of the SALCs for the four H^- ligands in $[PtH_4]^{2-}$.

with the simpler $[PtH_4]^{2-}$ anion, which has a square planar geometry. Using the D_{4h} point group, the symmetries of the valence AOs on Pt are: $5d_z{}^2 = a_{1g}$, $5d_{x^2-y^2} = b_{1g}$, $5d_{xy} = b_{2g}$, $5d_{xz,yz} = e_g$, $6s = a_{1g}$, $6p_z = a_{2u}$, and $6p_{x,y} = e_u$. Using the 1s AOs on each H^- ligand as a basis, the reducible representation for 4H is given in Table 11.7 and it reduces to $a_{1g} + b_{1g} + e_u$. Using the projection operator method, the shapes of the four SALCs are shown in Figure 11.30. By taking linear combinations of the AOs on Pt with the SALCs on the ligands having the same symmetry and similar energies, the one-electron MO diagram can be constructed, as shown in Figure 11.31.

Focusing only on the frontier MOs (also known as *Bloch orbitals*), we use Equation (11.24) to consider what the shapes of the Bloch orbitals will look like at the two extreme edges of the first Brillouin zone (at $k = 0$ and $k = \pi/a$), as shown in Figure 11.32. Our goal is to determine two things: (i) the direction in which the band will run (up or down?) and (ii) how wide or narrow the band will be. For example, the a_{2u} Bloch orbital at $k = 0$ will lie higher in energy than at $k = \pi/a$; this band will therefore run "down" on the Bloch diagram. Because the a_{2u} overlap between monomers has *sigma*-type bonds, the bandwidth is expected to be fairly large. Using this type of analysis for each of the basis sets, a qualitative Bloch diagram and DOS plot for $[PtH_4]^{2-}$ can be sketched, as shown in Figure 11.33.

TABLE 11.7 Representation for the SALCs formed by the four H^- ligands in the square planar $[PtH_4]^{2-}$ anion.

D_{4h}	E	$2C_4$	C_2	$2C_2'$	$2C_2''$	i	$2S_4$	σ_h	$2\sigma_v$	$2\sigma_d$	IRRs:
Γ_{4H}	4	0	0	2	0	0	0	4	2	0	$a_{1g}+b_{1g}+e_u$

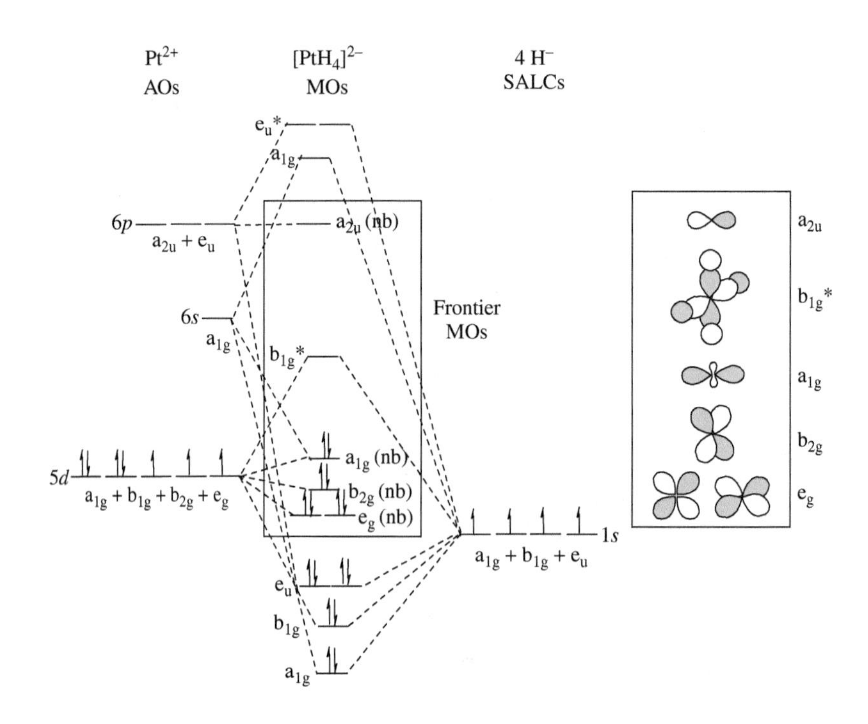

FIGURE 11.31
Qualitative one-electron MO diagram for $[PtH_4]^{2-}$, including the shapes of the frontier MOs.

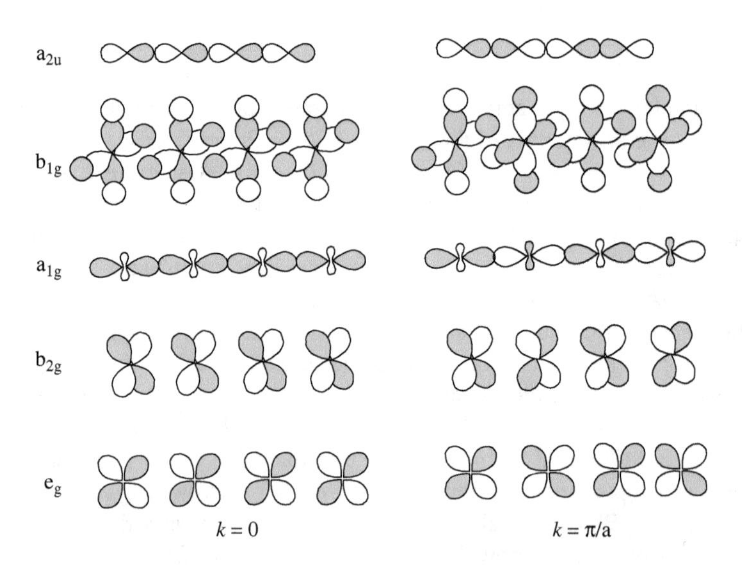

FIGURE 11.32
Shapes of the frontier Bloch orbitals for a linear chain of $[PtH_4]^{2-}$ anions at $k=0$ and $k=\pi/a$.

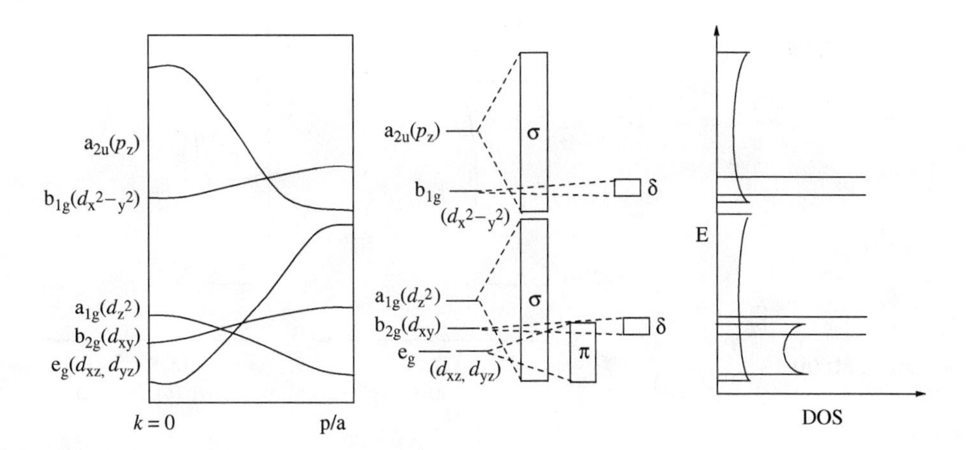

FIGURE 11.33
Qualitative Bloch diagram and DOS plot for $[PtH_4]^{2-}$.

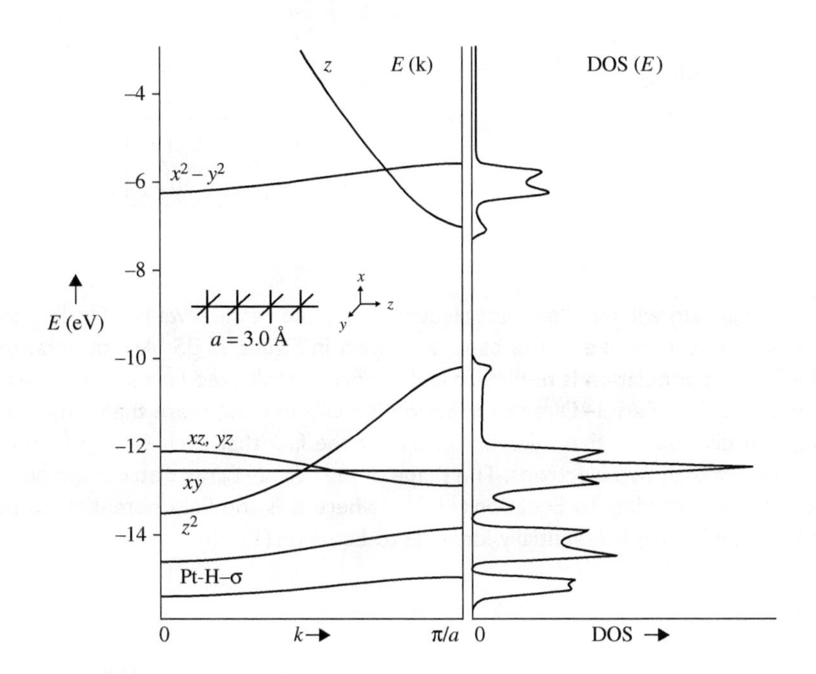

FIGURE 11.34
Calculated Bloch diagram and DOS plot for $[PtH_4]^{2-}$. [Reproduced from Hoffmann, R. *Solids and Surfaces: A Chemist's View of Bonding in Extended Structures*, Wiley-VCH: New York, 1989. This material is reproduced with permission of John Wiley & Sons, Inc.]

The calculated Bloch diagram for $[PtH_4]^{2-}$ and its corresponding DOS plot are shown in Figure 11.34. It is clear from a comparison of the two diagrams that the qualitative predictions made in Figure 11.33 bear a remarkably accurate resemblance to the calculated results shown in Figure 11.34.

One main difference between molecular orbital theory and band theory is that in the latter model the bands can overlap with one another. This is because the electrons have no way of distinguishing between the different bands. They will naturally fill in all of the available MOs from bottom to top in the DOS profile, according to the Aufbau principle and the Pauli exclusion principle. Thus, we need some way of indicating on the DOS plot which bands (or parts of bands) are filled and which are empty. The *Fermi level* (E_F) is defined as the energy of the highest occupied orbital at a temperature of absolute zero. For a simple polymer, such as the fictional linear chain of H atoms, the Fermi level will lie halfway up the band of MOs formed by overlap of the 1s AOs on the H atoms comprising the solid. Thus, exactly half of

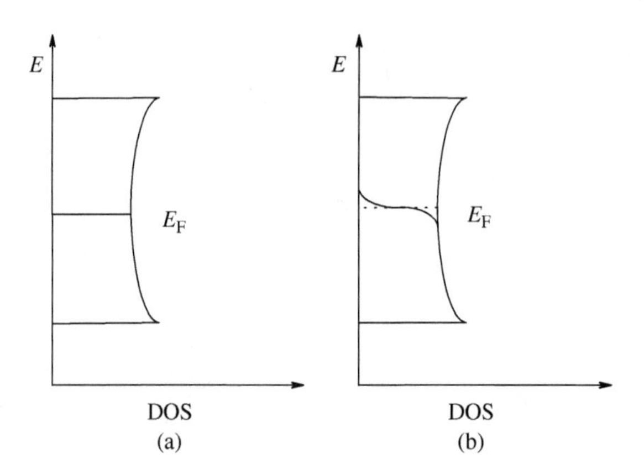

FIGURE 11.35
Fermi level distribution at: (a) $T = 0$ K, (b) $T > 0$ K.

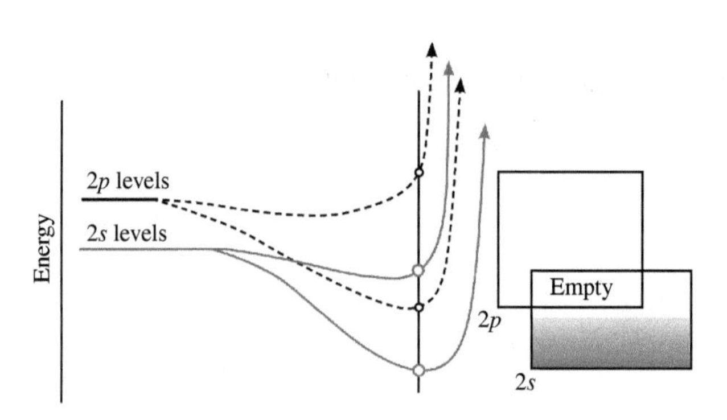

FIGURE 11.36
Energy diagram for a chain of Li atoms as a function of the internuclear separation r. The band diagrams at right occur when $r = a$ (the lattice spacing). [Adapted from Hoffmann, R. *Solids and Surfaces: A Chemist's View of Bonding in Extended Structures*, Wiley-VCH: New York, 1989. This material is reproduced with permission of John Wiley & Sons, Inc.]

the DOS diagram will be filled with electrons, and the Fermi level is a flat line that lies directly in the middle of the band, as shown in Figure 11.35. At temperatures above 0 K, the population is nonlinear and a more complicated Fermi–Dirac distribution exists. The Fermi–Dirac distribution is really nothing more than a modified Boltzmann distribution that takes into account the fact that each energy level can hold no more than two electrons. The shape of the Fermi–Dirac distribution decays exponentially according to Equation (11.26), where μ is the field potential. Notice that Equation (11.26) is essentially identical to Equation (11.10).

$$f(E) = \frac{1}{e^{(E-\mu)/kT} + 1} \tag{11.26}$$

We are now in a position to discuss the conductivity of solids. Whenever the Fermi level falls in the middle of a band, as it does for a fictional linear chain of Li atoms (whose band structure is shown in Figure 11.36), the *valence band* (VB) (the band of MOs containing electrons) and the CB (an empty band of MOs) overlap with one another. Thus, it takes a minimal amount of energy to excite an electron from a filled MO in the top of the VB into an empty MO in the CB, where that electron is delocalized over the entire linear chain. Likewise, the hole (h^+) or electron vacancy that is left behind in the VB is also free to migrate throughout the entire lattice. Electrons "sink" to the bottom of a band (or fill in the lowest energy MOs), while the holes "rise" to the top of a band. When an external potential is applied, the energy of the band next to the positive bias is bent downward and the solid becomes conductive.

For purposes of this textbook, we define a conductor as any substance that has a conductivity (σ) greater than 100 S/m (S = Siemens, the SI unit for conductance). Silver metal has one of the highest known conductivities at 6.3×10^7 S/m when measured at 20 °C. The conductivity of an electrical conductor decreases with increasing temperature because the atoms can vibrate more in their lattice sites at higher T, disrupting the overlap of the valence AOs. This is a phenomenon known as *carrier scattering* and it has to do with the presence of an impurity in the crystalline solid interfering with its normal conductivity. On the other hand, substances having conductivities less than 10^{-5} S/m are electrical insulators and do not conduct electricity. In an electrical insulator, the VB and CB are separated from each other by a large *band gap*. The energy difference between the highest occupied MO of the VB and the lowest unoccupied MO of the CB is too great to be achieved under ambient conditions, and the electrons are therefore trapped in the immediate vicinity around a single nucleus. Substances where the band gap is moderately small are known as *semiconductors*. Semiconductors have conductivities ranging from 10^{-5}–10^{2} S/m and their conductivities increase with temperature because the higher thermal energy can be used to excite electrons from the VB into the CB.

Returning to the one-dimensional chain of $[Pt(CN)_4]^{2-}$ ions having $a = 3.3$ Å, the MO diagram in Figure 11.31 is filled up to the a_{1g} (*nb*) molecular orbital. When the frontier orbitals form bands, some of the bands from different Bloch orbitals overlap with each other. In the DOS plots shown in Figures 11.33 and 11.34, there are essentially two sets of overlapping bands. The electrons will completely fill the bottom of these two sets. Therefore, there will be a band gap between the top of the a_{1g} band in the lower set of bands and the bottom of the a_{2u} band in the upper set of bands. The presence of this band gap explains why a linear chain of $K_2[Pt(CN)_4]$ molecules is an electrical insulator.

When the material is partially oxidized to form $K_2[Pt(CN)_4Cl_{0.3}]$, however, some of the electrons are removed from the top of the a_{1g} band. The tops of bands are antibonding because they have the greatest number of nodes in their Bloch orbitals. Thus, the partial oxidation of the $[Pt(CN)_4]^{2-}$ ion removes some of the antibonding electrons, and the overall Pt–Pt bonding is strengthened. This is precisely the result observed by the decrease in Pt–Pt bond length from 3.3 Å in $K_2[Pt(CN)_4]$ to 2.7 Å in $K_2[Pt(CN)_4Cl_{0.3}]$. In fact, one study has shown that (within a certain range) the Pt–Pt distance in the linear chain is inversely proportional to the degree of oxidation, which is completely consistent with the band theory model. Partial oxidation also causes an electron deficiency in the top of the a_{1g} band, so that the valence and CBs overlap with one another, and the species becomes a conductor.

You might also recall that as the linear tetracyanoplatinate chain becomes oxidized, it changes its geometry from eclipsed to staggered, as shown in Figure 11.37. This probably occurs because of the shorter Pt–Pt distance and the resulting steric interactions between the ligands. The net effect of the geometric change is to double the length of the unit cell. Because the number of bands in a crystal depends on the

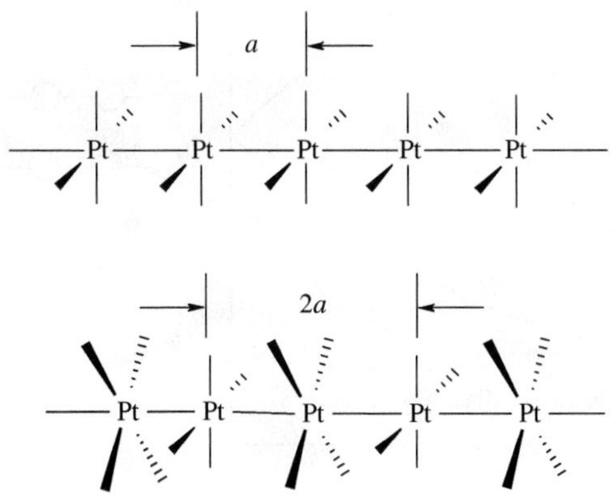

FIGURE 11.37 Partial oxidation of a linear chain of $[Pt(CN)_4]^{2-}$ ions causes a change in the geometry from eclipsed to staggered, which doubles the unit cell.

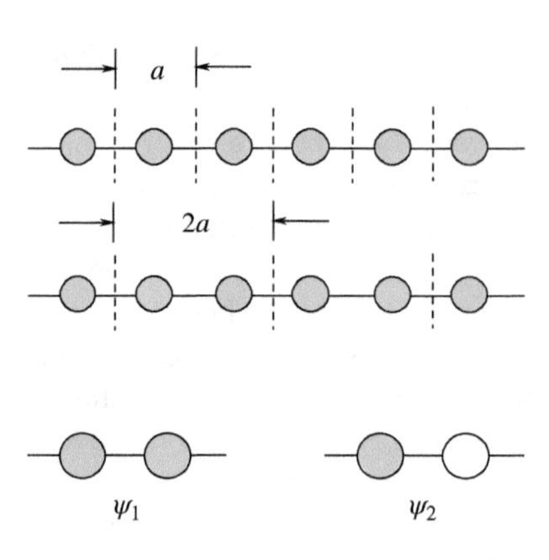

FIGURE 11.38
Doubling the unit cell for a linear chain of H atoms. The 1*s* orbital basis set is shown.

FIGURE 11.39
Shapes of the two MOs for the doubled unit cell shown in Figure 11.38.

number of MOs in the unit cell, if the unit cell is doubled, there will be twice as many bands.

When the unit cell is doubled, the Brillouin zone will be half as long, because k exists in reciprocal space. Hence, the first Brillouin zone will range from $-\pi/2a$ to $+\pi/2a$. Let us again consider the example of a linear chain of H atoms. If the unit cell is doubled, as shown in Figure 11.38, then we must first take linear combinations of the two 1s AOs in the unit cell to make MOs and then take linear combinations of the MOs at the extreme edges of the first Brillouin zone to construct the Bloch orbitals.

The first (bonding) MO results from the positive linear combination: $\psi_1 = 1/2^{0.5}$ $[\phi_{1sA} + \phi_{1sB}]$, while the second (antibonding) MO results from the negative linear combination: $\psi_2 = 1/2^{0.5}$ $[\phi_{1sA} - \phi_{1sB}]$, as shown in Figure 11.39. The ψ_1 band will run up in the first Brillouin zone, while the ψ_2 band will run down. At $k = \pi/2a$, the two Bloch orbitals have exactly the same degree of bonding and antibonding character (displaced by one-half of a unit cell translation), and they are therefore degenerate. The shape of the Bloch diagram for the doubled unit cell is shown in Figure 11.40.

Notice that the Bloch diagram for the doubled unit cell is identical to the Bloch diagram for the original unit cell doubled back over itself, as shown in Figure 11.41. The calculated band structure for a staggered linear chain of $[PtH_4]^{2-}$ ions is shown

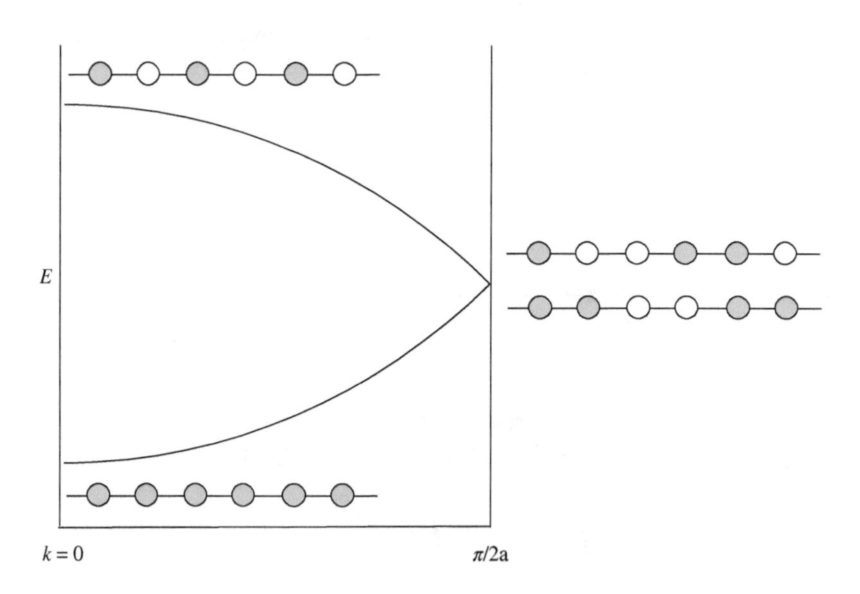

FIGURE 11.40
Bloch diagram for the linear chain in Figure 11.38 for the doubled unit cell.

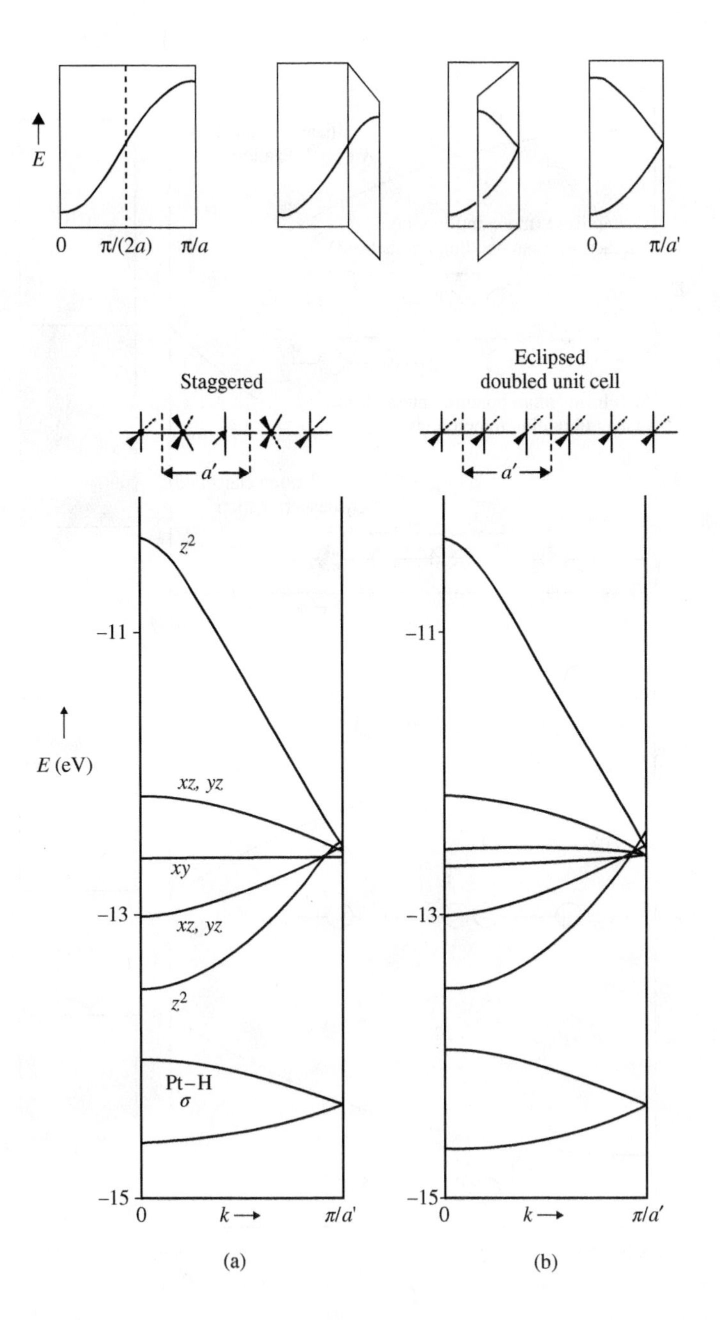

FIGURE 11.41
Doubling the unit cell has the effect of halving the first Brillouin zone, such that the new Bloch diagram can be obtained by halving the original Bloch diagram and folding it back over itself. [Reproduced from Hoffmann, R. *Solids and Surfaces: A Chemist's View of Bonding in Extended Structures*, Wiley-VCH: New York, 1989. This material is reproduced with permission of John Wiley & Sons, Inc.]

FIGURE 11.42
Bloch diagram for: (a) a staggered linear chain of $[PtH_4]^{2-}$ ions where the unit cell includes two Pt ions; (b) an eclipsed linear chain of $[PtH_4]^{2-}$ ions having only one Pt per unit cell where the bands have been folded in half and then doubled back over. [Reproduced from Hoffmann, R. *Solids and Surfaces: A Chemist's View of Bonding in Extended Structures*, Wiley-VCH: New York, 1989. This material is reproduced with permission of John Wiley & Sons, Inc.]

in Figure 11.42(a) and compared to the folded-back band structure of the eclipsed linear chain in Figure 11.42(b). The two Bloch diagrams are virtually identical.

The band structure shown in Figure 11.40 for the doubled unit cell of H atoms is not entirely accurate, as our chemical intuition tells us. If we could somehow line up all the H atoms into a one-dimensional chain, we might expect that they would spontaneously begin to pair up to make H_2 molecules. This is a result of a *Peierls distortion*, which is the solid-state equivalent to the Jahn–Teller effect. Because $E(k) = E(-k)$, each of the bands in the Bloch diagram is degenerate. Thus, whenever the Fermi level lies precisely in the middle of a band, there will always be a phonon (or lattice vibration) that lowers the symmetry, removes the degeneracy, and lowers the energy. A typical Peierls distortion for a one-dimensional linear chain involves a stretching vibration along the chain direction, as shown in Figure 11.43(a). At the bottom of the band, the Peierls distortion has little effect. Any increase in overlap

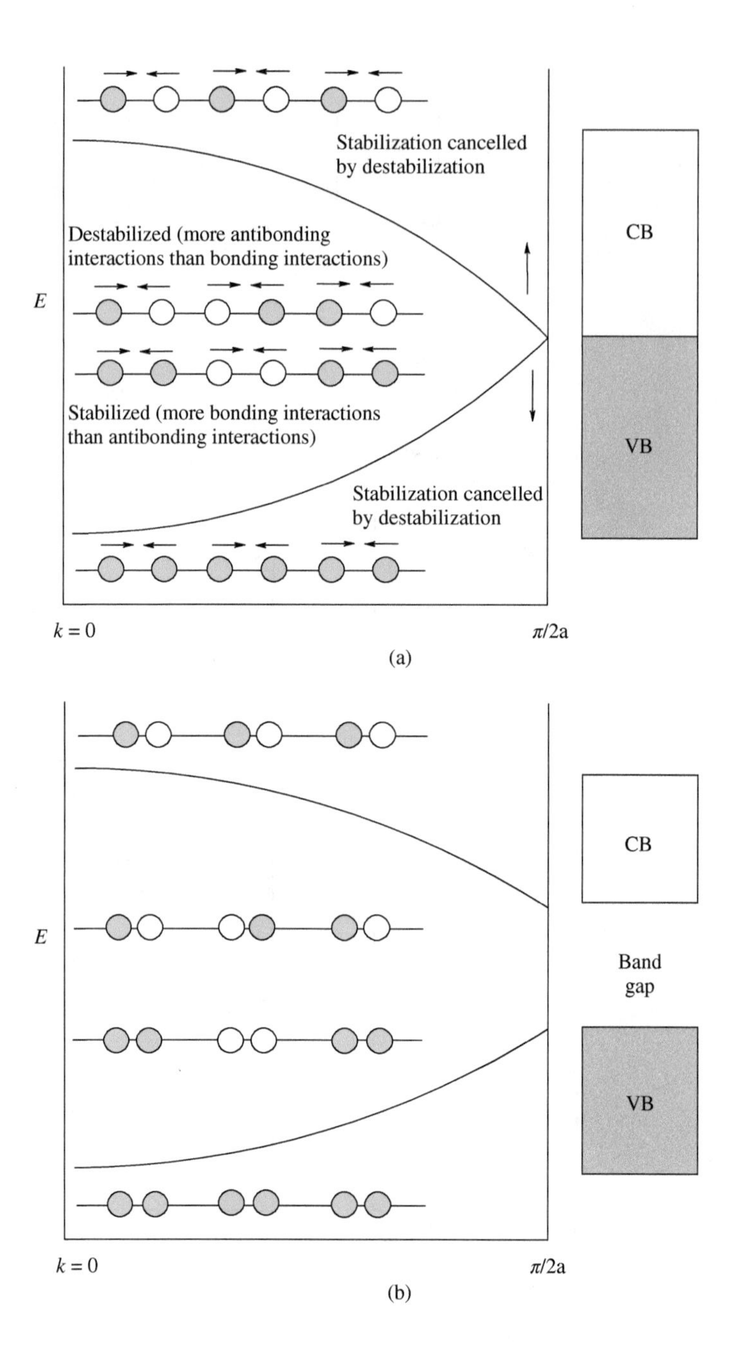

FIGURE 11.43
(a, b) A lattice vibration known as a *Peierls distortion* lowers the symmetry of the lattice, removes the degeneracy at $k = \pi/2a$, lowers the overall energy, and opens up a band gap at the Fermi level.

between two neighboring Li atoms is exactly compensated for by a decrease in overlap with its other neighbor. The same thing is true at the top of the band. However, in the middle of the band, at the Fermi level, the effect is to stabilize one of the degenerate Bloch orbitals at the edge of the first Brillouin zone and to destabilize the other, as shown in Figure 11.43(b). This opens a band gap at the Fermi level, which decreases the conductivity.

In the linear chain tetracyanoplatinates, the unoxidized $K_2[Pt(CN)_4]$ is an electrical insulator. Partial oxidation of the material removes electrons from the top of the $d_z{}^2$ band, which causes the material to become an electrical conductor. Because the tops of bands are antibonding, the partial oxidation also increases the overall bond order and decreases the Pt–Pt distance. The shorter Pt–Pt bond distance will broaden all of the bands because of the increased orbital overlap at shorter distances.

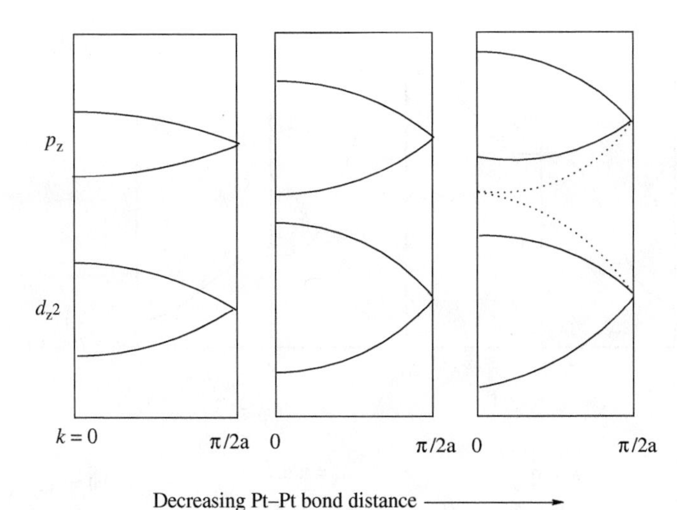

FIGURE 11.44
The band structure for the p_z and d_{z^2} sigma-bonding MOs in the staggered linear chain tetracyanoplatinates as the degree of partial oxidation increases. As the Pt–Pt bond distance decreases, the bands approach one another at $k=0$. As the two bands become degenerate at $k=0$ (dashed curves at right), a Peierls distortion opens up a band gap at the band edge.

This broadening will be especially significant for the two σ-bonding Bloch orbitals (the MOs derived the p_z and d_{z^2} AOs). As the Pt–Pt bond distance shortens, the top of the d_{z^2} band begins to approach the same energy as the bottom of the p_z band, as shown in Figure 11.44. If the two bands ever became degenerate, a Peierls distortion would open up a band gap between them. Thus, our model is entirely consistent with the observation that the partially oxidized tetracyanoplatinates are metallic at room temperature and that their conductivity drops off markedly at lower temperatures, where they are only semiconducting. Recall that for typical metals, the conductivity increases with a decrease in temperature. Here, the conductivity decreases as the temperature is lowered because there is less thermal energy to overcome the band gap created by the Peierls distortion.

In summary, the band theory of solids can be used to explain the structure, spectroscopy, and electrical properties of one-dimensional solids, such as the linear chain tetracyanoplatinates. However, most crystalline solids have translational symmetry in more than one dimension. Can band theory be extended to two and even three dimensions? The answer is, of course, yes. However, the shapes of the bands become significantly more complicated as the number of dimensions increases. Let us just consider two dimensions for the moment and an imaginary lattice composed of only H atoms. In two dimensions, the wavefunction for the Bloch orbitals is given by Equation (11.27):

$$\psi_{k_a,k_b} = \frac{1}{\sqrt{N}}\sum_m \sum_n e^{ik_a ma} e^{ik_b nb}\,\phi_{mn} \tag{11.27}$$

The Bloch diagram is divided into four regions, as shown in Figure 11.45 for a two-dimensional lattice of H atoms. At edge Γ, which is always considered to be the origin in reciprocal space, the coefficients in Equation (11.27) reduce to $(1)^m(1)^n$; at X they are $(-1)^m(1)^n$; at Y they are $(1)^m(-1)^n$; and at M they are $(-1)^m(-1)^m$. The shapes of the Bloch orbitals at each of these band edges are shown in Figure 11.46. The Bloch orbital at Γ is the lowest in energy, because it lacks any nodal planes. The Bloch orbital at M has the highest energy, because the Bloch orbital will be most antibonding when the signs of the wavefunction alternate along both of the connecting axes. The s-orbitals therefore run up from Γ to either of the band edges X, Y, or M. In this fictional example, however, where the separation between H atoms along the *a*-axis is shorter than along the *b*-axis, the Bloch orbital at Y (where the H atoms have a positive overlap along the shorter internuclear axis) will be lower in energy than the Bloch orbital at X; hence, the more gradual slope for the transition

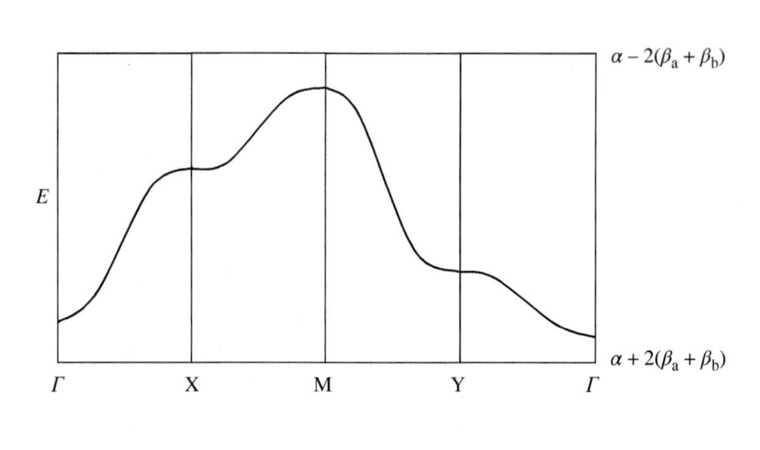

FIGURE 11.45
Bloch diagram for a two-dimensional lattice of H atoms, where the H atoms lie closer to each other along the *a*-axis than along the *b*-axis.

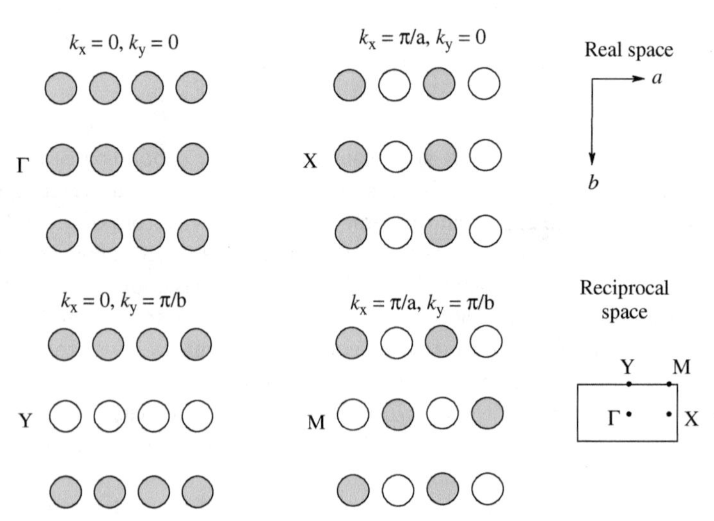

FIGURE 11.46
Bloch orbitals for a two-dimensional lattice of H atoms at each of the band edges, where the H–H bond distance is shorter along the *a*-axis than along the *b*-axis.

along Γ→Y than along Γ→X. The situation is similar (albeit somewhat harder to visualize) in three dimensions.

11.6 CONDUCTIVITY IN SOLIDS

Band theory is a quantum-mechanical treatment of the bonding in metals. When the number of atoms *N* in a crystal is very large, the spacing between the energy levels in the band is infinitesimally small so that promotion to the next highest energy level within the band can occur quite readily. Whenever a band is fully occupied with valence electrons, it is known as the VB. If a band is unoccupied or empty, it is known as the CB. In a metal such as Li, as shown in Figure 11.36, the band constructed from a linear combination of the valence 2*s* AOs is exactly half-filled, so that the VB and the CB derive from the same set of orbitals and are only distinguished from each other by the presence or absence of electrons. The energy of the highest filled MO at 0 K is known as the *Fermi level.* Because the Fermi level lies inside a band in Li, this metal is a conductor. For a metal such as Mg, the band formed by the 2*s* AO basis set (in other words, the VB) is completely filled. The only reason that Mg is a conductor is because 2*s* band overlaps with the empty band formed by the 2*p* basis set (or the CB), as shown in Figure 11.47 (at left).

A simplified version of band theory can therefore be used to make a rather sweeping generalization about the electrical properties of the elements, as shown in Figure 11.48. One primary difference between the metals and nonmetals is related

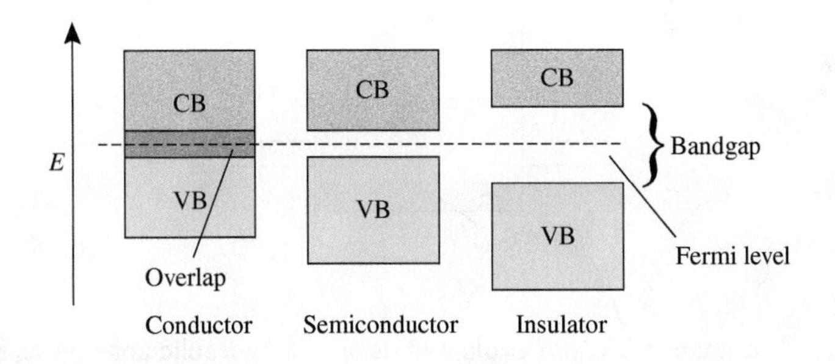

FIGURE 11.47
Simplified band diagram showing the differences between a conductor, a semiconductor, and an insulator. The main difference has to do with the band gap energy, or separation between the top of the valence band (VB) and the bottom of the conduction band (CB). The Fermi level is indicated by a dashed line on the diagram.

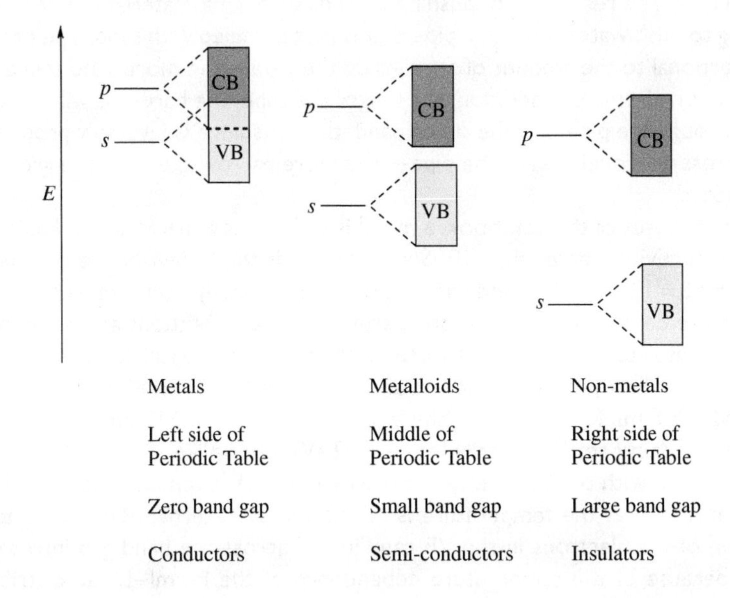

FIGURE 11.48
A simplistic rationalization for the electrical conductivity of the elements in the periodic table based on their band structures.

to the energies of their valence s and *p* AOs. This energy difference increases across a row of the periodic table because of increased shielding of the *p*-electrons by the more penetrating *s*-electrons. The metals, which lie to the left in the periodic table, are excellent electrical conductors because their valence and CBs overlap and there is essentially no energy gap at all between the bands. On the other end of the spectrum, the nonmetals, which lie to the right in the periodic table, are typically electrical insulators because of the sizable band gap that exists between their valence and CBs. In the middle of the periodic table, there is a gray area that is intermediate in its electrical properties between those of a metal and those of a nonmetal. These are the metalloids, which are usually delineated by a bold diagonal "staircase" on the periodic table. The metalloids tend to be *semiconductors* because they have only a small band gap.

The conductivity (σ) of a material is given by Equation (11.28), where ρ is the resistivity of the metal. The resistivity has units of $\Omega{\cdot}$m and is given by Equation (11.29), which is a simple algebraic rearrangement of Pouillet's law, given by Equation (11.30), where R is the electrical resistance (measured in ohms, Ω), A is the cross-sectional area of the material, and l is the length of the material.

$$\sigma = \frac{1}{\rho} \quad (11.28)$$

$$\rho = R\frac{A}{l} \quad (11.29)$$

$$R = \frac{\rho l}{A} \quad (11.30)$$

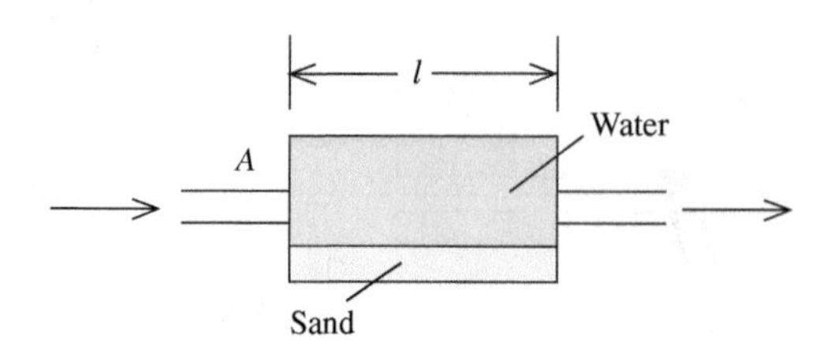

FIGURE 11.49
The hydraulic analogy between water in a pipe filled with sand and the electrical resistance in a material.

Electrical resistance is often explained using the hydraulic analogy, as shown in Figure 11.49. The resistance of pushing current through a material is a little similar to trying to push water through a pipe that is partially filled with sand. The resistivity is proportional to the amount of sand inside the pipe—the more sand there is, the greater the resistance. In addition, the longer the tube, the harder it will be to push water through the pipe. On the other hand, the resistance is inversely proportional to the cross-sectional area of the pipe—the more narrow the pipe, the greater the resistance.

For purposes of this textbook, a metal is defined as a material that has an electrical conductivity greater than 10^2 S/m (where S is the SI unit Siemens, which has the value $1\ S = 1\ \Omega^{-1}$). The conductivities of metals typically decrease with increasing temperature because there are more disruptive lattice vibrations at higher temperatures. An insulator is defined as a material that has an electrical conductivity of less than 10^{-5} S/m. Lastly, semiconductors are defined as having conductivities between 10^{-5} and 10^2 S/m. The *band gap* (the energy between the VB and CB) in a semiconductor is fairly small (typically less than 2 eV); but unlike the metals, the bands do not overlap with one another. At low temperatures, a semiconductor will act as an insulator; but as the temperature is raised and an external potential is applied, a portion of the electrons in the VB can "jump" across the band gap into the CB. This is because of the temperature dependence of the Fermi–Dirac distribution, which spreads out at higher temperatures over a greater range of energy levels, as shown in Figure 11.50. Unlike the metals, whose conductivities generally decrease with increasing temperature, the conductivity of a semiconductor increases as the temperature is raised because a greater fraction of the electrons possess enough thermal energy to jump across the barrier. The electrons can also cross the band gap in a semiconductor by the absorption of visible light.

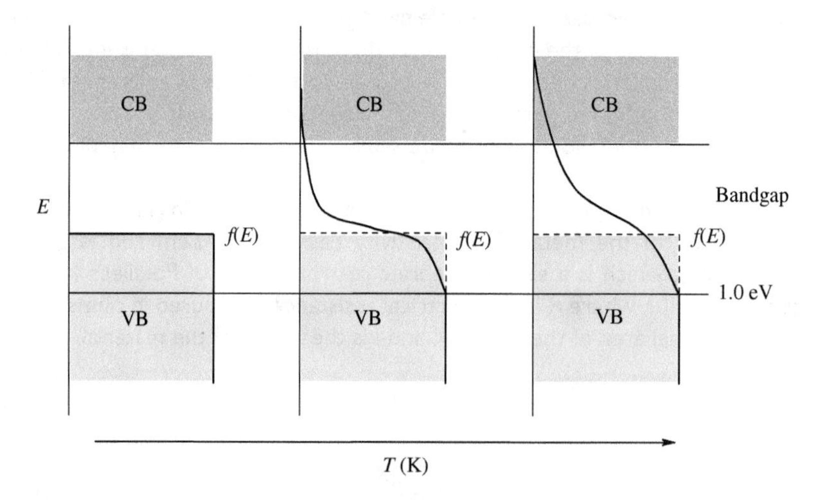

FIGURE 11.50
The Fermi–Dirac distribution as a function of temperature in a semiconductor.

TABLE 11.8 The electrical resistivities and conductivities of a variety of different materials at 20 °C.

Material	ρ, Ω·m	σ, S/m
Silver	1.6×10^{-8}	6.3×10^{7}
Copper	1.7×10^{-8}	6.0×10^{7}
Gold	2.4×10^{-8}	4.1×10^{7}
Aluminum	2.8×10^{-8}	3.5×10^{7}
Tungsten	5.6×10^{-8}	1.8×10^{7}
Nickel	7.0×10^{-8}	1.4×10^{7}
Iron	1.0×10^{-7}	1.0×10^{7}
Platinum	1.1×10^{-7}	9.4×10^{6}
Lead	2.2×10^{-7}	4.6×10^{6}
Mercury	9.8×10^{-7}	1.0×10^{6}
Graphite (in basal plane)	$2.5–5.0\times10^{-6}$	$2–3\times10^{5}$
Graphite (⊥ to plane)	3.0×10^{-3}	3.3×10^{2}
Germanium	0.46	2.2
Silicon	6.4×10^{2}	1.6×10^{-3}
Diamond	1×10^{12}	$\sim10^{-13}$

The flow of current through a material is directly proportional to the conductivity through Ohm's law. Thus, the larger the value of σ, the greater the flow of current and the better the conductor will be. Table 11.8 lists the conductivities for a variety of common materials. In order to better understand how current flows through a material, imagine the charge carriers (electrons in this case) in a metal as the individual water molecules inside the pipe shown in Figure 11.49. In the absence of an applied electric field, the pipe is resting on a horizontal surface and the water molecules move randomly through the pipe in all directions owing to their ambient molecular motion. At low temperatures, they will be moving fairly slowly, but their average kinetic energy will increase with temperature.

If an electric field is applied to the material, this has the effect of tipping the pipe so that it is no longer horizontal. If the electrons were in a wire, they would begin to flow preferentially in the direction of the more positive potential just as the water molecules inside the pipe would begin to flow downhill. The rate of flow depends on the angle at which the pipe has been tipped. Thus, the drift velocity v_d of the electrons depends on the magnitude of the applied potential. The current i is defined by Equation (11.31), where ΔQ depends on the product of the number of charge carriers (electrons in this example) times the charge q (in this case, $q = e = 1.602\times10^{-19}$ C). The number of charge carriers, in turn, depends on ΔV (the larger the pipe, the greater the number of water molecules in motion). Thus, the numerator term in Equation (11.31) is given by Equation (11.32), where n is defined as the charge carrier density. Because Δt is the same as $\Delta x/v_d$, we then arrive at Equation (11.33) for the current, where A is the cross-sectional area of the wire.

$$i = \frac{\Delta Q}{\Delta t} \tag{11.31}$$

$$\Delta Q = nq\Delta V \tag{11.32}$$

$$i = nqAv_d \tag{11.33}$$

In a metal, the Fermi level lies inside the CB, so that the electrons will flow whenever an external electric field is applied to the material. However, in a semiconductor at 0 K, the Fermi level lies at the top of the VB and there is a barrier to the flow of electrons known as the *band gap*. As the temperature increases, according to the Fermi–Dirac distribution given by Equation (11.10), some of the electrons can achieve enough energy to cross the band gap into the CB. Because the current is proportional to the charge carrier density *n*, we would like to be able to determine this value. The conduction electron population $N(E)\,dE$ can be calculated as the product of the DOS given by Equation (11.9) times the Fermi–Dirac distribution given by Equation (11.10), as shown in Equation (11.33). The charge carrier density *n* will simply be the integral of $N(E)\,dE$ evaluated from E_{gap} to infinity, having the result shown in Equation (11.34).

$$N(E)\,dE = D(E)f(E)\,dE = \frac{8\sqrt{2}\pi m^{3/2}}{h^3}\sqrt{(E-E_{gap})}\exp\left[\frac{1}{(E-E_F)/kT+1}\right]dE \qquad (11.34)$$

$$n = \int_{E_{gap}}^{\infty} N(E)\,dE = AT^{3/2}e^{-E_{gap}/2kT},\quad \text{where}A = \frac{2^{5/2}(m\pi k)^{3/2}}{h^3} \qquad (11.35)$$

Example 11-6. Use Equation (11.34) to calculate the charge carrier density for Si at 25 °C and at 50 °C, given that the band gap in Si is 1.1 eV.

Solution. First, we will convert the energy of the band gap into SI units:
$E_{gap} = 1.1\text{ eV} \times 1.602 \times 10^{-19}\text{ J/eV} = 1.8 \times 10^{-19}\text{ J}$.
Next, we will calculate *A* in SI units:

$$A = \frac{2^{5/2}[(9.109\times10^{-31}\text{ kg})(3.14)(1.38\times10^{-23}\text{ J/K})]^{3/2}}{(6.626\times10^{-34}\text{Js})^3}$$
$$= 4.83\times10^{21}\text{ e}^-/\text{m}^3\text{K}^{3/2}$$

Finally, we will plug in the temperatures to determine the charge carrier density:

$$T = 298\text{ K}: \quad n = 4.83\times10^{21}\text{ e}^-/\text{m}^3\text{K}^{3/2}(298\text{K})^{3/2}$$
$$\exp\left[\frac{-1.8\times10^{-19}\text{J}}{2\,(1.38\times10^{-23}\text{J/K})\,(298\text{K})}\right] = 1.2\times10^{16}\text{ e}^-/\text{m}^3$$
$$T = 323\text{ K}: \quad n = 4.83\times10^{21}\text{ e}^-/\text{m}^3\text{K}^{3/2}(323\text{ K})^{3/2}$$
$$\exp\left[\frac{-1.8\times10^{-19}\text{J}}{2\,(1.38\times10^{-23}\text{J/K})\,(323\text{K})}\right] = 7.5\times10^{16}\text{ e}^-/\text{m}^3$$

Some of the metalloids, such as Si (band gap = 1.1 eV) and Ge (band gap = 0.74 eV), are *intrinsic semiconductors*, which means that the pure substance can exhibit semiconducting properties. This is one of the reasons that the entire modern electronics era is based on Si microchips. The conductivity of Si can be enhanced, however, with the addition of small amounts of dopants that increase the number of charge carriers. For every so many Si atoms, for example, if we were to replace one of these Si atoms with an As atom, the conductivity would increase because the lattice acquires an extra electron (As has five valence electrons, while Si only has four). There is no room for the extra electrons from the As atoms to occupy MOs in the VB of Si, so they occupy a very narrow set of

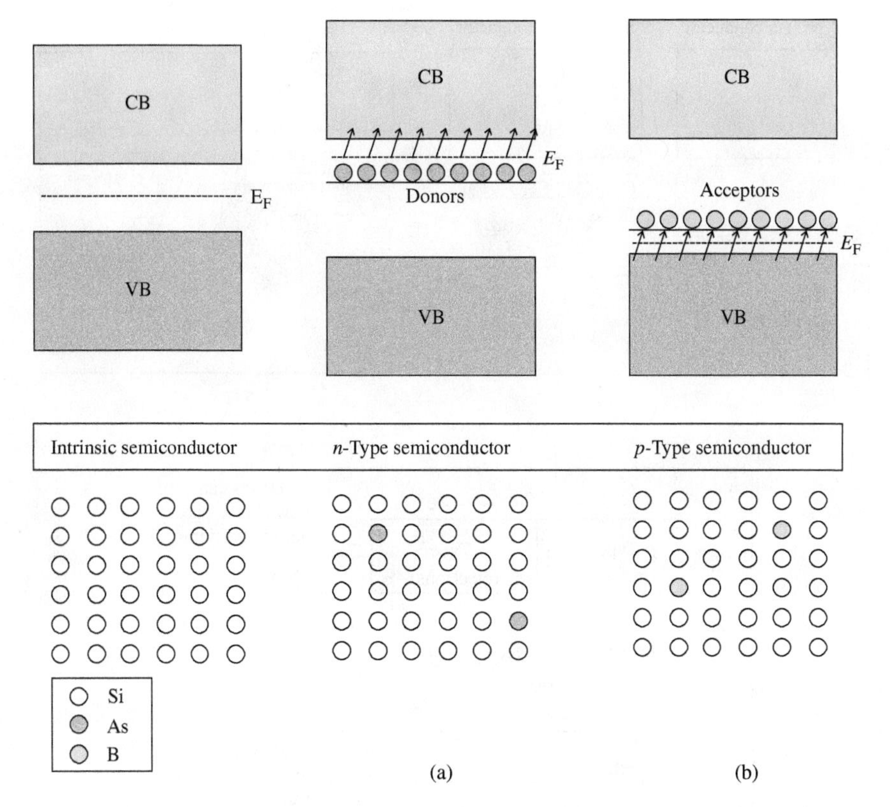

FIGURE 11.51
(a, b) The difference between an intrinsic semiconductor, an *n*-type semiconductor, and a *p*-type semiconductor.

orbitals of their own (narrow because they are few and far between so that their valence orbitals have very little overlap). The As orbital energies are intermediate between those of the VB and CB of Si, as shown in Figure 11.51(a).

Whenever the dopant adds electrons to the lattice, the material is called an *n-type semiconductor* because the dopant adds negative charge carriers. The conductivity of an n-type semiconductor is enhanced because the electrons in the As orbitals lie higher in energy than those in the VB of Si. The energy difference (band gap) between the highest occupied MOs and the lowest unoccupied MOs is now significantly smaller. This makes it easier for electrons to jump across the barrier into the CB.

An alternative way to improve the conductive properties of a semiconductor (or to convert an insulator into an *extrinsic semiconductor*) is to dope it with an element that contains less electrons. One example is to dope a Si wafer such that B atoms occupy some of the lattice sites of the material. Because B has only three valence electrons, the B dopant provides a set of low-lying empty MOs, as shown in Figure 11.51(b). The band gap between the top of the VB for Si and these empty MOs is very small, so that some of the electrons in the VB can be thermally excited into the empty B MOs. When this occurs, the electron deficiency creates holes in the VB, which act as positive charge carriers. Because the dopant increases the number of positive charge carriers, the material is called a *p-type semiconductor*.

When a p-type and an n-type semiconductor are connected to each other, as shown in Figure 11.52, the result is a *p–n junction*. Almost every modern electronic device (from laptops to cell phones to mp3 players) contains a p–n junction as part of its integrated circuitry. In the absence of an applied potential, there is no net flow of charge carriers in the p–n junction. The surplus of electrons in the n-type semiconductor is attracted to the surplus of holes in the p-type semiconductor. When the electrons and holes meet in the center, they annihilate each other in a process known as *recombination* and a nonconductive *depletion zone* (or barrier) is formed.

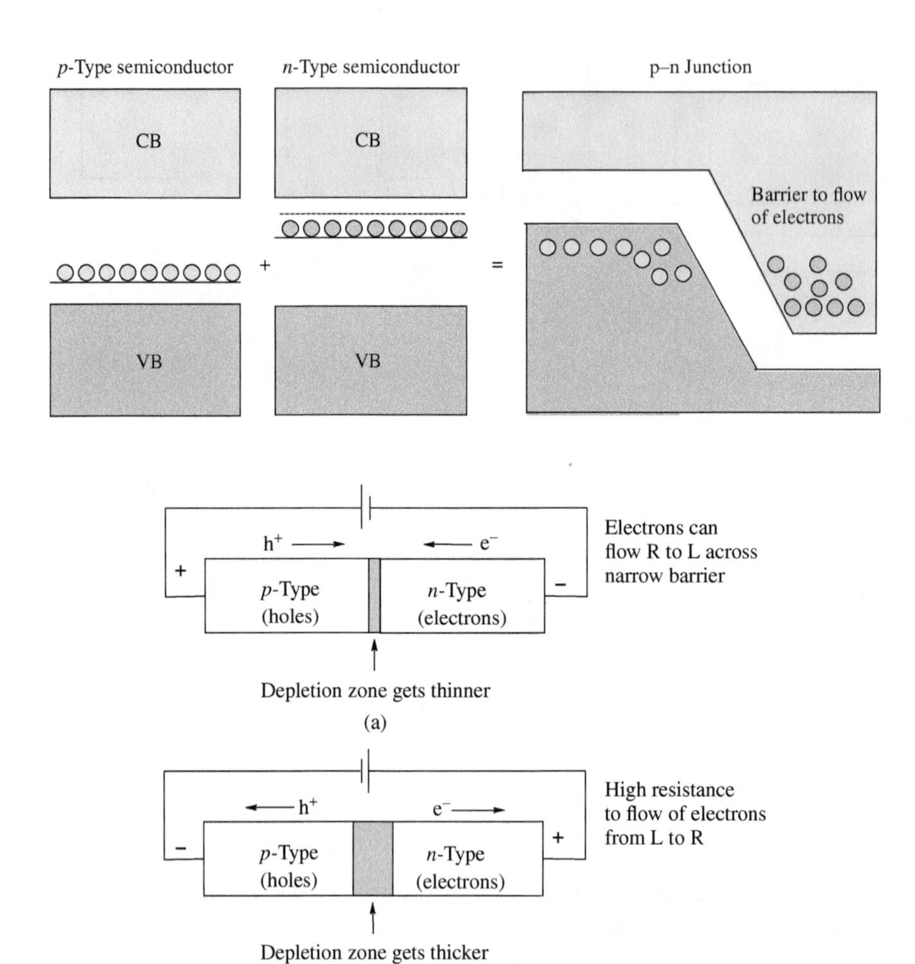

FIGURE 11.52
A *p–n* junction.

FIGURE 11.53
A *p–n* junction under: (a) forward bias allows the flow of electrons from right to left and (b) reverse bias prevents the flow of electrons from left to right.

When the p–n junction is connected to an external power source, such as a battery, with the positive electrode attached to the p-type semiconductor and the negative electrode attached to the n-type semiconductor, the material operates under a *forward bias*, as shown in Figure 11.53(a). The holes in the p-type semiconductor are repelled away from the positive terminal and toward the depletion zone, while the electrons in the n-type semiconductor are repelled away from the negative terminal toward the depletion zone. As a result of this bias, the barrier at the p–n junction gets thinner and, ultimately, electrons are able to tunnel across the barrier so that current flows through the material from right to left in the diagram. On the other hand, if the p-type semiconductor is connected to the negative terminal of a power source and the n-type semiconductor connected to the positive terminal, the material acts under a *reverse bias*, as shown in Figure 11.53(b). The holes in the p-type semiconductor are attracted to the negative terminal and the electrons in the n-type semiconductor are attracted to the positive terminal. The net result is that the depletion zone at the p–n junction gets larger and there is a high resistance to electron flow across the circuit. Thus, the p–n junction acts as a *diode*, an electrical switch that allows current to flow in only one direction (from right to left in Figure 11.53) and not in the opposite direction.

Diodes lie at the heart of all sorts of electronic devices, from photovoltaic cells (where light energy is used to produce the charge carriers) to transistors (which consist of two p–n junctions in series, such as the common n–p–n bipolar junction transistor, or BJT). The transistor is arguably the most important invention of the

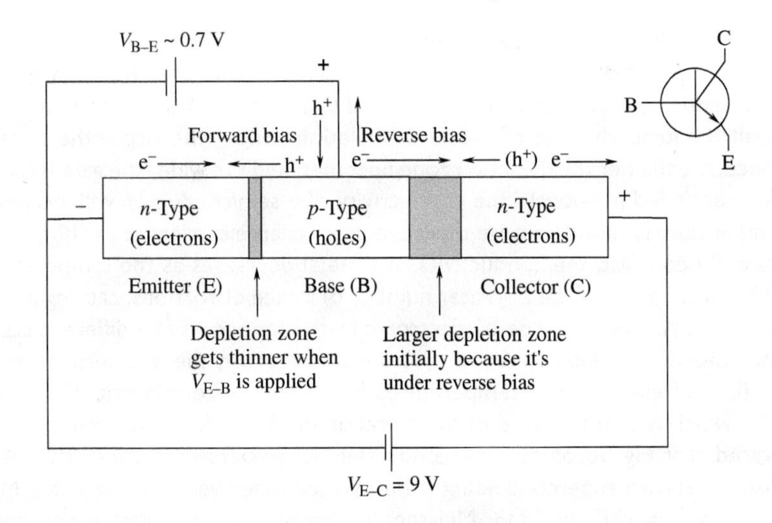

FIGURE 11.54
Schematic diagram of an *n–p–n* bipolar junction transistor (BJT).

twentieth century. Without it, the modern age of communications could never have occurred. Just try to imagine a world without transistor radios, televisions, computers, electronic guitars, and cell phones! The transistor was invented in 1947 at Bell Labs by John Bardeen, Walter Brattain, and William Shockley. A typical BJT, such as the n–p–n type transistor, is shown in Figure 11.54.

The transistor consists of three parts: an n-type emitter (E), a p-type base (B), and an n-type collector (C). The essential purpose of the transistor is to amplify current. The way that this is done is that a small voltage is applied between the emitter and the base, with the positive terminal attached to the base and the negative terminal attached to the emitter. This causes some of the holes in the base (the majority charge carriers in the p-type semiconductor) to migrate toward the p–n junction with the emitter. At the same time, some of the electrons in the emitter (the majority charge carriers in an n-type semiconductor are the electrons) to migrate toward the p–n junction with the base. As a result of this forward bias between the emitter and base, a swarm of electrons will flow across the barrier and into the base. Essentially, the E–B potential acts to open (or partially open) the gate (depletion zone) that is blocking the flow of electrons between the emitter and base. For Si, the gate will begin to open when a potential of ~0.5 V is applied. The gate is fully open at a potential of 0.7 V.

On the other side of the BJT, there is also a depletion zone between the base and the collector. In fact, because this second p–n junction is under reverse bias, the size of the B–C barrier is larger than that of the E–B barrier. Thus, at this point, the second gate remains closed. That is, it remains closed unless we apply a second potential difference across the entire circuit (perhaps from a 9 V battery). The electrons (which are the majority charge carriers) in the collector will then be attracted to the positive terminal of the power source and the holes (minority charge carriers) in the collector will migrate toward the p–n junction with the base.

Now, as long as the E–B gate is open, there is still that swarm of electrons flowing into the base to consider. Some of these electrons are attracted to the positive terminal attached to the base, where they collide with holes heading toward the negative terminal on the emitter to undergo charge recombination. However, there are many other electrons flooding into the base and these can migrate their way across the entire base to the p–n junction with the collector. Here, they are attracted to the holes that have been collecting in the collector on the opposite side of the p–n junction, the depletion zone gets thinner, and electrons flow across the barrier. Because of the larger potential difference across the entire E–B–C circuit than just across E–B, the net result is that a small amount of current applied to E–B is "converted" into a large flow of current across the entire circuit. Thus, the BJT acts as an amplifier of current, where the gain can be controlled by the magnitude of the potential applied between the base and the emitter. This potential difference essentially controls the size of the E–B depletion zone—the larger the potential difference, the thinner the depletion zone becomes, and the wider the gate will open.

As mentioned previously, the conductivity of a semiconductor will increase as the temperature is raised because there are more charge carriers at the higher temperature. By contrast, the conductivity of a metal decreases as the temperature is raised because there will be a greater number of lattice distortions, causing an electronic instability. Another type of electronic instability occurs in a different class of material known as a *superconductor*. In a superconductor, there is zero resistance to the flow of electricity at temperatures lower than a certain critical temperature, T_c, which is characteristic of the material. In 1911, Heike Kamerlingh Onnes discovered that Hg becomes superconducting at temperatures lower than 4.2 K. The highest known superconducting transition for many years was that for MgB_2, which has a $T_c = 39$ K. In 1933, Meissner Ochsenfeld noticed that superconductors exclude an internal magnetic field, a phenomenon now known as the *Meissner effect*. Thus, a superconducting solid cooled below its T_c will appear to float above a bar magnet. As a result of the Meissner effect, superconducting materials have found practical application in the magnets of nuclear magnetic resonance (NMR), magnetic resonance imaging (MRI), and particle accelerators. In 1986, Bednorz and Müller discovered that if La_2CuO_4 is doped with Ba and partially oxidized, it will become superconducting at $T_c = 35$ K. By changing the cations to Y, the familiar 1–2–3 superconductor $YBa_2Cu_3O_7$ was discovered to have $T_c > 90$ K, a temperature higher than the boiling point of liquid nitrogen (77 K), and the modern field of high-temperature superconductors was born. Even in the absence of an applied potential, a small current can still flow with zero resistance through the superconductor. Because these materials can be cooled relatively inexpensively using liquid nitrogen, high-temperature superconductors have been envisioned in a variety of practical applications, including high-performance transformers, electric motors, and magnetically levitated high-speed trains. They are already commercially employed in superconducting quantum interference devices (SQUIDs), which are the most sensitive magnetometers on the market. Currently, the highest temperature superconductor has a $T_c = 150$ K.

The principle model of superconductivity is the *BCS theory*, which was proposed by Bardeen, Cooper, and Schrieffer in 1957, for which they won the 1972 Nobel Prize. The theory was also independently discovered by Bogoliubov. According to the BCS theory, the conducting electrons in a superconducting solid cause a transitory deformation in the lattice, which creates a region of positive charge density. A second electron having opposite spin is then attracted to the positive charge and the two electrons wind up as a pair (known as a *Cooper pair*) having a certain binding energy between them. As long as the binding energy of the Cooper pair is greater than the energy of the lattice vibrations that tend to disrupt the pair, the two electrons will travel together throughout the lattice having zero resistance. This will be true at lower temperatures, where the disrupting forces from the oscillating atoms in the lattice have insufficient energy to overcome the binding energy of the Cooper pair. In one analogy, the Cooper pair moves through the lattice similar to fans in a stadium doing the wave.

From the quantum mechanical standpoint, the charge carriers in a superconductor are Cooper pairs with opposite momenta that result from an electron–phonon interaction that mixes the wavefunction product $\phi(k)\ \phi(-k)$, an occupied orbital, with $\phi(k')\ \phi(-k')$, which is unoccupied. This mixing opens a band gap at the Fermi level between $\phi(k)$ and $\phi(k')$ so that the Cooper pairs cannot dissociate at low temperatures (below T_c). High-temperature superconductors typically have a larger degree of mixing and low-frequency (or soft) phonon vibrational modes. The superconducting properties of the high-temperature copper oxides, such as $YBa_2Cu_3O_7$, are believed to result from systematic vacancies in the oxide layers of the tetragonal perovskite-type structure shown in Figure 11.55.

11.7 CONNECTIONS BETWEEN SOLIDS AND DISCRETE MOLECULES

One of the underlying themes of this textbook is to draw connections between seemingly different topics. The comparisons between metallic and covalent bonding are shown in Table 11.9 in terms of the differing terminology between the band theory of the atoms in metals and the discrete MOs of atoms in molecules. As mentioned in Chapter 6, the three different types of chemical bonding can be considered as lying at the apexes of a triangle of bonding. We have already discussed the MO theory of covalent bonding and shown how this is related to the band theory of metallic bonds. In the following chapter, we shall introduce ionic bonding to the mix. As we examine each of the three most common types of bonding in detail, it is important to bear in mind that there is a gray area in between each of the extremes; and that it is essential to be able to examine the bonding in any given substance from a variety of different perspectives.

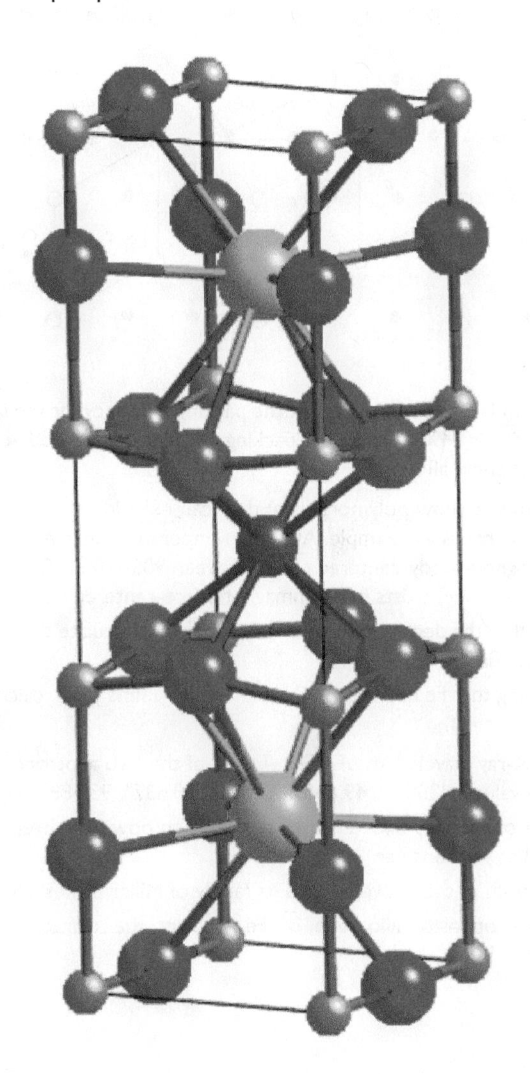

FIGURE 11.55
The crystalline structure of the high-temperature superconductor $YBa_2Cu_3O_7$, showing the different planes of copper oxide layers. Systematic oxide vacancies along the *z*-axis occur in the Y planes of the material. (Color legend: Y = purple, Ba = green, Cu = blue, and O = red).

TABLE 11.9 A comparison between terms used in the band theory of solids and those in MOT of discrete molecules.

Band Theory	Molecular Orbital Theory
Bloch orbitals	Molecular orbitals
VB	HOMO
CB	LUMO
E_F	$E°$
E_{gap}	HOMO-LUMO gap
Phonon (lattice vibration)	Molecular vibration
Peierls distortion	Jahn-Teller distortion
n-Doping	Reduction, pH (base)
p-Doping	Oxidation, pH (acid)

EXERCISES

11.1. Which of the following is the best choice for a unit cell? Explain your answer.

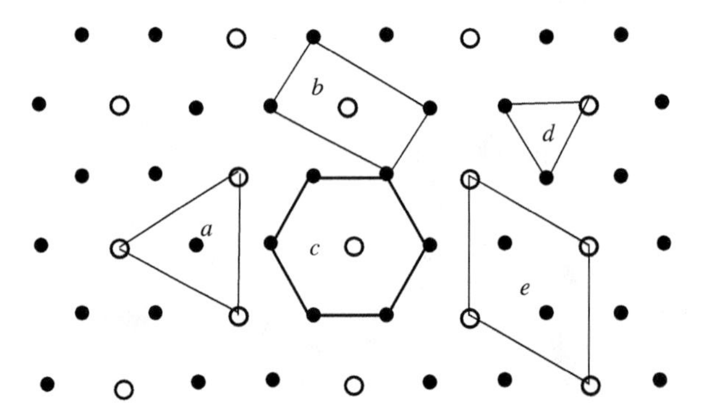

11.2. In Example 11-2, it was proved that the packing efficiency for the face-centered cubic Bravais lattice is 74%. What is the packing efficiency of the body-centered cubic Bravais lattice? Show all work.

11.3. Some elements show polymorphism: they can exist in more than one type of crystalline form. Iron is an example. At room temperature, iron exists as ferrite, or α-Fe, and is a magnetic body-centered cube. Between 906–1403 °C, however, an allotrope of iron (called γ-Fe) exists in a nonmagnetic face-centered cube.

a. Given that the density of ferrite is 7.874 g/cm^3, calculate the metallic radius of Fe in ferrite (in Å).

b. Assuming that Fe has the same radius in its gamma form, calculate the density of γ-Fe (in g/cm^3).

11.4. Using an X-ray wavelength of 1.541 Å, one of the two allotropes of Fe gave the following 2θ values: 42.778°, 49.645°, 73.081°, 88.637°, 93.688°, and 114.700°.

a. Which of the two allotropes gives the given powder pattern? Explain how you arrived at your answer.

b. Index each of the above lines to its family of Miller planes <*h k l*>.

c. For the opposite allotrope of Fe, predict the diffraction pattern between $0° < 2\theta < 120°$.

11.5. Nickel crystallizes in a body-centered cubic unit cell. Given that the density of Ni is 8.90 g/cm^3, calculate the metallic radius of Ni.

11.6. Using a wavelength of 154 pm, the following 2θ values were obtained for a metallic solid: 22, 31, 38, 45, 50, 56, 65, 70, 74, and 78°.

a. Index each of the powder pattern lines to its family of Miller planes.

b. Is this a face-centered cube, body-centered cube, or a primitive cube? Explain.

c. Sketch at least two parallel (2 2 0) planes through several unit cells of the metal. Be sure to label your axes correctly as a, b, and c.

11.7. The element tin exists in two different allotropes. White tin (the β-form) crystallizes with unit cell dimensions of $a=b=5.831$ Å and $c=3.182$ Å. This more common allotrope is a malleable form of metal that is stable at temperatures above 13.2 °C. Gray tin (the α-form) crystallizes with unit cell dimensions of $a=b=c=6.489$ Å and is a brittle solid that falls apart easily. Urban legend has it that the white tin buttons on Napoleon's soldiers converted to the gray form during his winter invasion of Russia, causing them to fall apart in the cold. Using the crystal structures of each allotrope shown here, determine the density of both allotropes of tin.

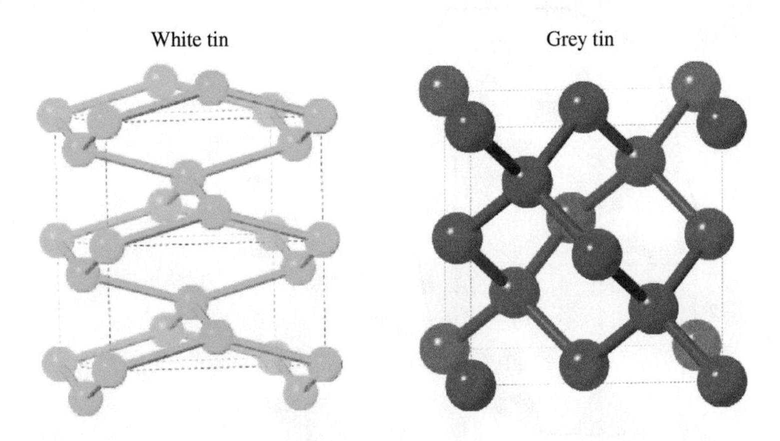

11.8. Using the information given in Exercise 7 and an X-ray wavelength of 1.541 Å, predict the X-ray powder diffraction pattern (2θ values) for the first eight lines of each allotrope.

11.9. Determine the crystallographic point group for each of the following crystals, where the rotational axes and mirror planes are indicated. Use both the Schoenflies and Hermann–Mauguin notations.

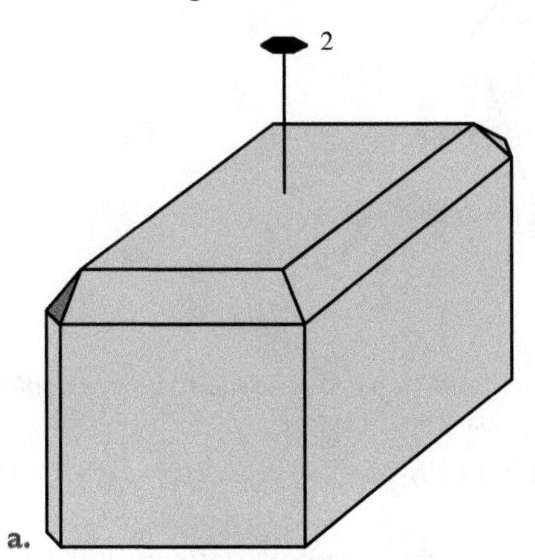

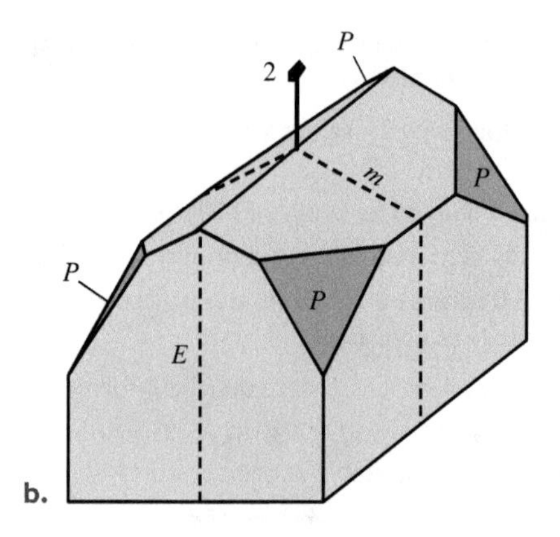

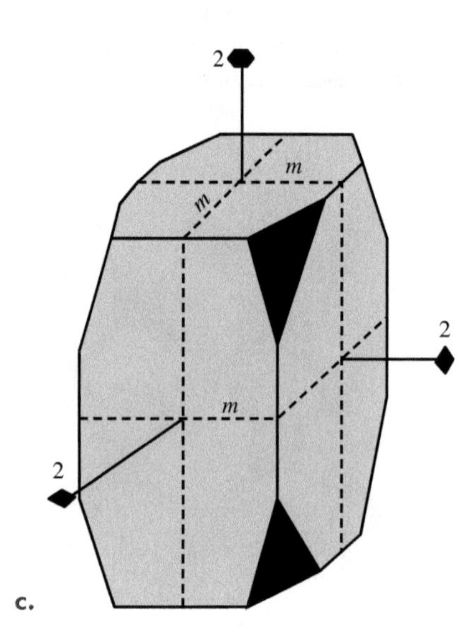

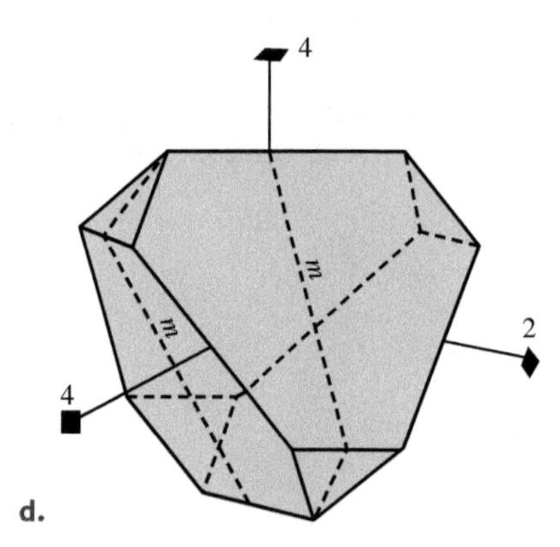

[Reproduced with permission from http://www.tulane.edu/~sanelson/eens211/32crystalclass.htm (accessed January 4, 2013).]

11.10. Given that Na metal crystallizes in a body-centered cubic lattice with $a = 429.1$ pm, calculate E_{max} in eV.

11.11. Given that Mg metal crystallizes in the hexagonal closest-packed lattice with $a = b = 320.9$ pm, and $c = 521.1$ pm, calculate E_{max} in eV.

11.12. Assume there exists a planar lattice (in the yz plane) of boron which has the repetition shown here.

a. Choosing the above-mentioned dashed lines as our linear unit cell, the symmetry of each unit is C_{2v}. Looking at just the monomer, use the p_x orbitals (perpendicular to the plane of the paper) as a basis set for generating the four *pi* molecular orbitals. Identify which is which in the given diagram by indicating the appropriate Mulliken symbol next to each MO.

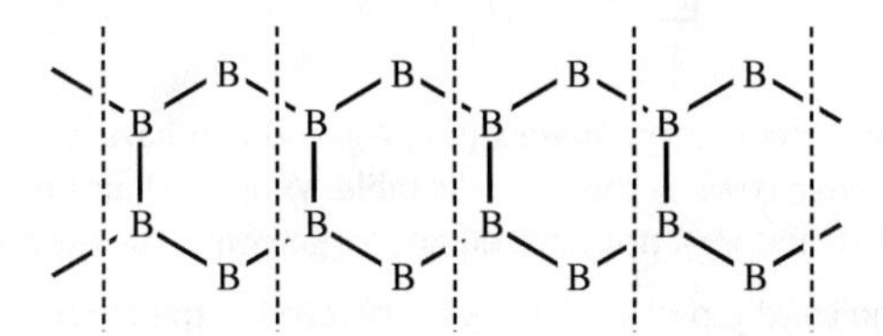

b. Now allow these four pi MOs to extend throughout the lattice. Sketch the Bloch orbitals for each MO at $k = 0$ and at $k = \pi/a$.

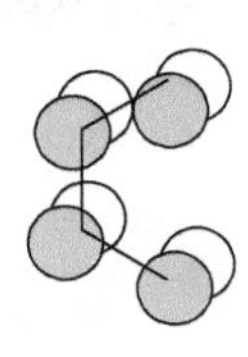
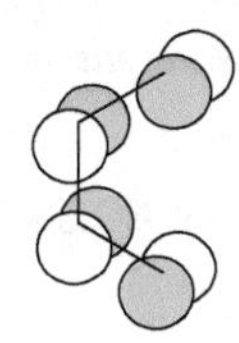
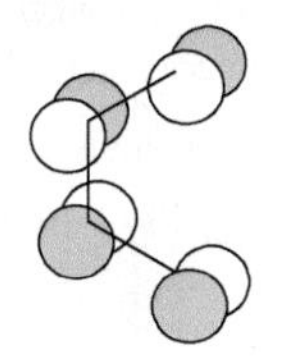
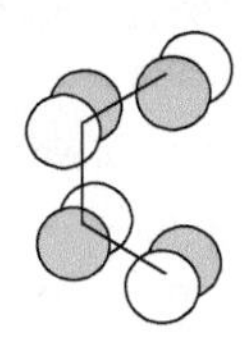

c. Sketch the band structure for these four MOs on an E versus k band diagram.

11.13. Assume a square lattice ($a = b$) composed of Li atoms and using the 2s AOs as the basis set. Sketch the shapes of the Bloch orbitals at each of the following positions: Γ, X, M. Then sketch the Bloch diagram for this lattice by analogy to the one shown in Figure 11.45. [The designation Y is redundant in this case as the Bloch orbital at Y will have the same shape as the one at X, so it is omitted from the Bloch diagram and the progression just goes from $\Gamma \rightarrow X \rightarrow M \rightarrow \Gamma$.]

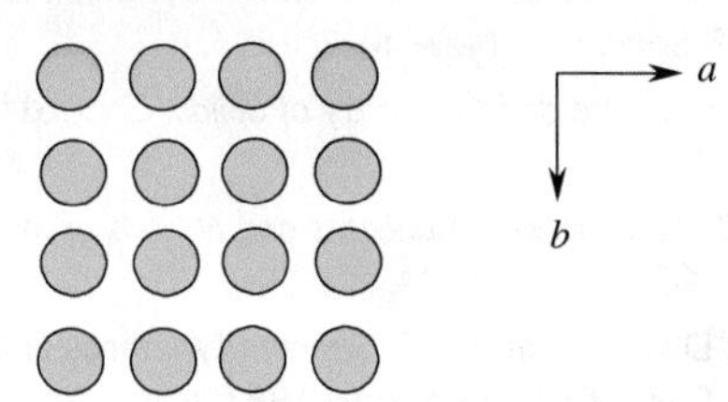

11.14. Assume a square lattice composed of Li atoms and using the $2p_x$ AOs as the basis set (the p_x orbitals have their lobes lying along a). Sketch the shapes of the Bloch orbitals at each of the following positions: Γ, X, M. Then sketch the Bloch diagram for this lattice from $\Gamma \rightarrow X \rightarrow M \rightarrow \Gamma$. Do the same thing for a basis set using the $2p_y$ orbitals on the Li atoms (the p_y orbitals have their lobes lying along b). Will the band structures for the p_x and p_y AO basis sets be the same or different? Explain. Finally, sketch the shapes of the Bloch orbitals and the Bloch diagram for a basis set using the $2p_z$ orbitals on the Li atoms.

11.15. The structure of NbI_4 at high temperatures is shown here using the coordinate axis defined in the diagram. The local symmetry around each Nb(IV) ion is octahedral, so that the hybridization is d^2sp^3. Given the definition of the coordinate system, it would appear that the d_{xy} and d_{z^2} orbitals are the ones that would have the greatest overlap with the I^- ligands and would be involved in the formation of Nb-I σ-bonds. Sketch the expected band diagram for the remaining d-orbitals ($d_{x^2-y^2}$, d_{xz}, and d_{yz}), which are involved in Nb–Nb bonding along the one-dimensional chain. In your analysis for each of the three bands, indicate whether the band runs up or down and its relative dispersion. For each basis, sketch the shape of the Bloch orbitals at $k = 0$ and $k = \pi/a$.

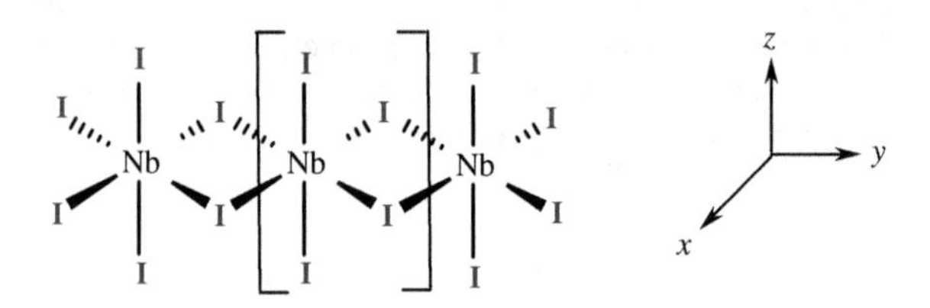

11.16. Rationalize why the coinage metals (Cu, Ag, and Au) have the largest conductivities in their respective rows of the periodic table. Why do their neighbors, the Group 12 metals (Zn, Cd, and Hg), not have equally large conductivities?

11.17. Given that the band gap of CdS is 2.4 eV, determine the color of this semiconductor. Explain how you arrived at your answer.

11.18. Calculate the charge carrier density for Ge at 25 °C, given that the band gap in Ge is 0.74 eV at 0 K.

11.19. What would be the most logical atom to dope into a crystal of In_2O_3 in order to make an *n*-doped semiconductor? Explain your answer.

11.20. Predict whether WO_3 is likely to be an *n*-type or a *p*-type semiconductor.

11.21. If the charge carrier density for GaAs at 302 K is 2.5×10^{25} e^-/m^3, calculate the band gap energy.

BIBLIOGRAPHY

1. Altman SL. *Band Theory of Solids: An Introduction from the Point of View of Symmetry*. Clarendon Press: Oxford; 1991.
2. Ball P. *Made To Measure*. Princeton, NJ: Princeton University Press; 1997.
3. Burdett JK. *Chemical Bonding in Solids*. Oxford University Press: New York; 1995.
4. Canadell E. Electronic Structure of Solids. In: King RB, editor. *Encyclopedia of Inorganic Chemistry*. John Wiley & Sons, Inc.: New York; 1994.
5. Cox PA. *The Electronic Structure and Chemistry of Solids*. Oxford University Press: New York; 1987.
6. Douglas B, McDaniel D, Alexander J. *Concepts and Models of Inorganic Chemistry*. 3rd ed. New York: John Wiley & Sons, Inc.; 1994.
7. Ebsworth EAV, Rankin DWH, Cradock S. *Structural Methods in Inorganic Chemistry*. Boston, MA: Blackwell Scientific Publications; 1987.
8. Hoffmann R. Angew. *How Chemistry and Physics Meet in the Solid State*. Chem. Int. Ed. Engl. 1987;26:846–878.
9. Hoffmann R. *Solids and Surfaces: A Chemist's View of Bonding in Extended Structures*. New York: VCH Publishers; 1988.
10. Housecroft CE, Sharpe AG. *Inorganic Chemistry*. 3rd ed. Essex, England: Pearson Education Limited; 2008.
11. Huheey JE, Keiter EA, Keiter RL. *Inorganic Chemistry: Principles of Structure and Reactivity*. 4th ed. New York: Harper Collins College Publishers; 1993.
12. Kittel C. *Introduction to Solid State Physics*. 6th ed. New York: John Wiley & Sons, Inc.; 1986.
13. Krebs H, Walter PHL. *Fundamentals of Inorganic Crystal Chemistry*. New York: McGraw-Hill; 1968.
14. Miessler GL, Tarr DA. *Inorganic Chemistry*. 4th ed. Upper Saddle River, NJ: Pearson Education Inc.; 2011.

15. Müller U. *Inorganic Structural Chemistry*. West Sussex, England: John Wiley & Sons, Ltd.; 1993.
16. Rodgers GE. *Descriptive Inorganic, Coordination, and Solid-State Chemistry*. 3rd ed. Belmont, CA: Brooks/Cole Cengage Learning; 2012.
17. Sands DE. *Introduction to Crystallography*. New York: W. A. Benjamin, Inc.; 1969.
18. Smith JV. *Geometrical and Structural Crystallography*. New York: John Wiley & Sons, Inc.; 1982.

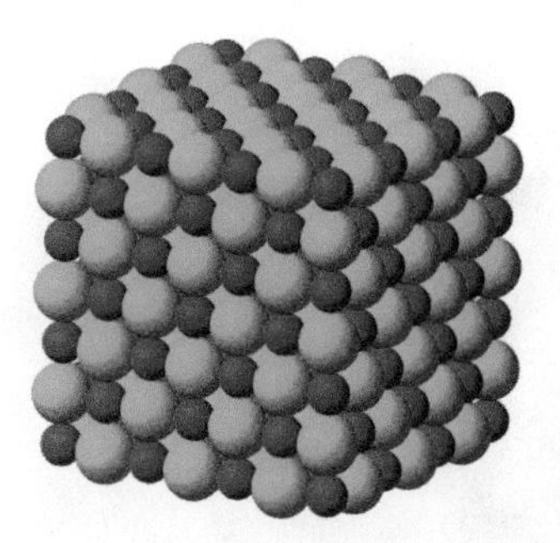

Ionic Bonding | 12

"My attitude was, why shouldn't I use the understanding that I have developed of the nature of crystals in inorganic substances to proceed to predict their structures?"

—*Linus Pauling*

12.1 COMMON TYPES OF IONIC SOLIDS

Just as most metals are solids at room temperature because of their omnidirectional bonding, most ionic compounds also exist in the solid state at standard temperature and pressure. Because of their omnidirectional bonding, most metallic solids prefer to crystallize in either the hexagonal or cubic closest-packed lattices in order to maximize their bonding interactions. Because every anion is surrounded by a number of cations and every cation is surrounded by a number of anions, the bonding in an ionic solid is likewise omnidirectional and is exceptionally strong. As a result, ionic solids are typically hard, brittle, have high melting points, and dissolve in polar solvents having large dielectric constants. Most ionic compounds are electrical insulators as solids because the ions are fixed in place in the solid state, but they become conductors in the molten state. The exception is for ionic solids that contain certain types of defects in their crystalline lattices, for these can also conduct in the solid state.

Unlike metallic solids, however, ionic solids are unable to closest pack in the crystalline state for two different reasons. First, the sizes of the anions and the cations in an ionic solid are rarely the same. Second, and more importantly, because the ions have opposite charges, the lowest energy configuration is one that minimizes the repulsion of like charges and maximizes the attraction between ions of opposite charge.

Although ionic solids themselves cannot form closest-packed structures, that does not prevent them from trying. Many ionic solids can be represented as closest-packed arrays of one type of ion with the other type of ion occupying vacancies (or holes) in this lattice. Thus, for example, the halite structure, shown in Figure 12.1, consists of a face-centered cube of chloride ions with the smaller sodium ions distributed in one type of vacancy that exists within the lattice.

NaCl lattice. [Reproduced from http://en.wikipedia.org/wiki/Sodium_chloride (accessed January 11, 2014).]

Principles of Inorganic Chemistry, First Edition. Brian W. Pfennig.

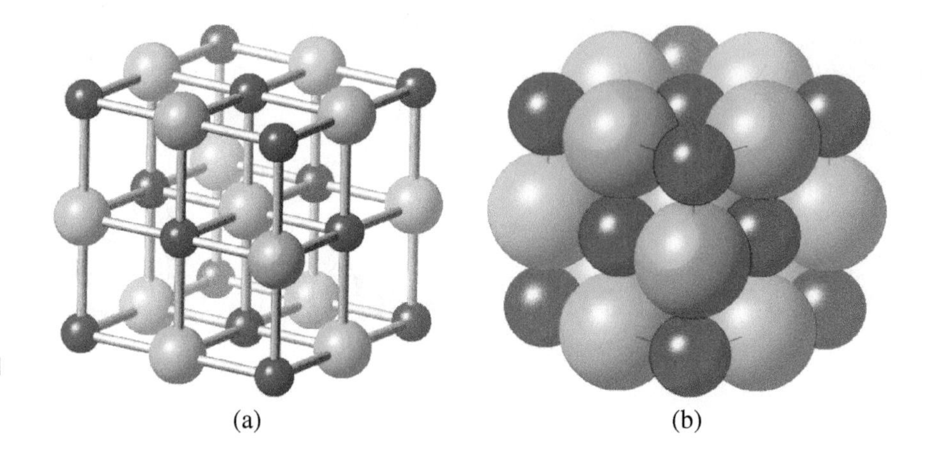

FIGURE 12.1
Halite (NaCl) unit cell in (a) ball and stick and (b) space-filling representations.

Because the coordination geometry of each sodium ion is based on an octahedron, this particular type of vacancy (also known as an *interstitial site*) is known as an *octahedral hole*. Without the intervening sodium ions, the cubic closest-packed array of chloride ions is inherently unstable as a result of the many close anion–anion repulsions. By their occupation of the octahedral holes, however, the sodium ions help to stabilize the lattice by preventing the anions from touching each other directly and by maximizing cation–anion attractions. In order to remove a single ion from the lattice, the ion would need to be removed from the direct influence of six octahedrally situated ions of the opposite charge. As a result, a lot of energy would need to be provided in order to break the electrostatic attractions in the lattice and to make the molten salt. The melting point of NaCl is therefore fairly large (801 °C).

Figure 12.2 shows two types of holes, or vacancies, in a face-centered cubic unit cell where the smaller types of ions in an ionic solid might fit. As shown in Figure 12.2(a), the vacancy in the exact body center of a face-centered cubic unit cell has an octahedral coordination geometry with six equidistant nearest neighbors. Therefore, this type of hole is called an *octahedral hole* or *o-hole*. The diagram in the

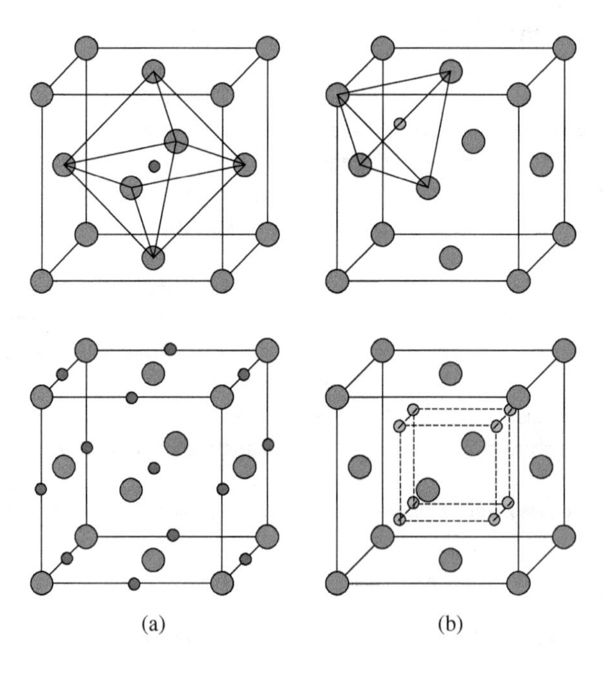

FIGURE 12.2
(a) Diagram of an octahedral hole (red) in the body center of a face-centered cubic unit cell (top). At the bottom, the other octahedral holes in the unit cell are also shown. (b) Diagram of a tetrahedral hole (green) in a face-centered cubic unit cell (top). At the bottom, the other tetrahedral holes in the unit cell are shown to inscribe a smaller cube inside the main cube.

bottom of the figure shows other octahedral holes in the face-centered cubic unit cell that lie along the edge centers of the cubic unit cell. Because each edge is shared with a total of four different unit cells, these 12 edge center o-holes contribute in the aggregate three ions to each unit cell and the body-centered o-hole contributes one additional ion. Thus, there are four o-holes per face-centered cubic unit cell, the same number of o-holes as there are sites in the face-centered cube. Hence, a structure such as NaCl can maintain its 1 Na^+ : 1 Cl^- ratio by forming a face-centered cube of the anions with the cations occupying each of the o-holes in the unit cell.

Other types of holes or interstitial sites can also exist in a face-centered cubic unit cell, as demonstrated by the smaller vacancies shown in Figure 12.2(b). By connecting any corner ion with its three nearest neighbor face-centers, a tetrahedral shape will result. In the center of each tetrahedron, there exists a vacancy which has a tetrahedral coordination geometry, known as a *tetrahedral hole* or *t-hole*. Because there are eight corner ions in a face-centered cubic, there will be eight t-holes per unit cell, which is double the number of lattice sites in the unit cell. Thus, by filling every t-hole, a 2 : 1 stoichiometry of anions : cations can be formed, as is case in molecules such as CaF_2. Other ionic solids with 1 : 1 ratios, such as BeO, consist of a face-centered cubic unit cell of one type of ion, with the other type of ion occupying every other t-hole.

Some common ionic lattices based on closest-packed structures are shown in Table 12.1. As we have already mentioned, the halite structure shown in the table consists of a face-centered cube of one type of ion with the opposite ion occupying every octahedral hole. The face-centered cube contains four anions per unit cell (8 corners × 1/8 each, plus 6 face-centers × 1/2 each). There are also four octahedral holes per unit cell (1 body-center × 1 each, plus 12 edge-centers × 1/4 each). Thus, the 1:1 stoichiometry is satisfied when every octahedral hole is filled with cations. This type of structure is adopted by many of the alkali metal halides, including all those of lithium, sodium, potassium, and rubidium, in addition to CsF, AgCl, AgBr, LiH, and PbS. A large number of oxides will also take this structure, including MgO, CaO, SrO, BaO, CdO, FeO, MnO, and TiO.

Each of the crystal structures shown in Table 12.1 can be illustrated in the familiar ball-and-stick representation where the ostensible bonding is represented by a hollow spoke, as shown at the left in the diagram. This provides a convenient way of determining the coordination geometry around each type of ion in the crystalline lattice. However, as was mentioned previously, ionic bonds are not localized similar to the directional bonds in covalent Lewis structures, and therefore this type of model should not be taken too literally. An alternative representation using polyhedrons can also be employed, where each polyhedron contains a cation at its center and the anions are located at the vertices. The polyhedral model has the advantage of simplicity, especially in more complex crystalline lattices, such as in layered solids or chain compounds. For the halite lattice, each Na^+ ion sits at the center of an octahedron of chloride ions, and these octahedrons all share their edges in common with their neighbors in the extended lattice.

The octahedrons surrounding the Ni^{3+} ions in NiAs, on the other hand, are edge-shared in one direction while face-shared in another, leading to a layered structure with an *ABABAB*-type repeating motif. The following molecules all crystallize in the nickel arsenide structure: TiS, TiSe, TiTe, CrS, CrSe, CrTe, CrSb, NiS, NiSe, NiTe, NiSb, and NiSn.

The zinc blende or sphalerite structure in Table 12.1 consists of a face-centered cubic lattice of the anions, but with the cations occupying only every other tetrahedral hole. Compounds that assume this structure therefore have a 1 : 1 ratio and include the following species: BeO, BeS, ZnO, ZnS, ZnSe, MnS, CdS, HgS, SiC, GaP, AlP, InAs, CuF, and CuCl.

TABLE 12.1 Common ionic solids based on closest-packed structures. In all of the examples, the cation is shown in red and the anion is shown in blue.

Formula	Type of Holes	Cubic Closest-Packed Arrangement		Hexagonal Closest-Packed Arrangement	
AB	All o holes filled	Halite or rock salt (NaCl) 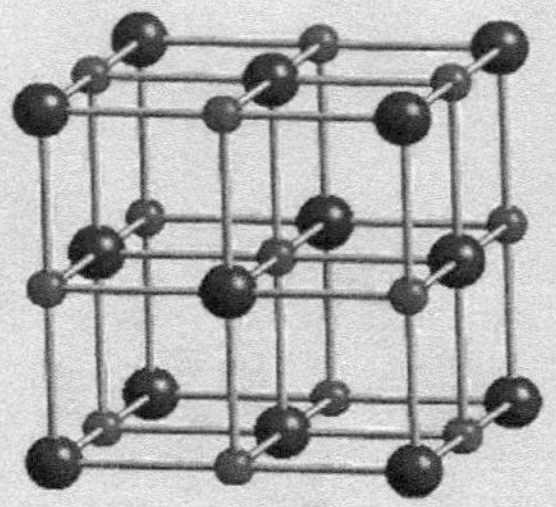	Edge-shared octahedrons 	Nickel arsenide (NiAs) 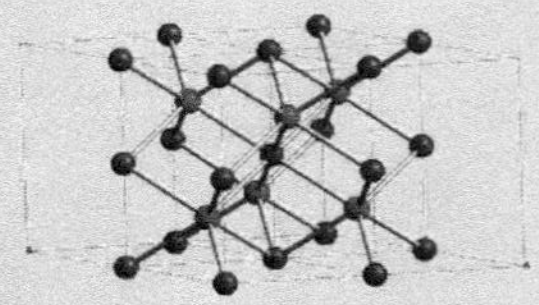	Edge- and face-shared octahedrons
AB	Half t holes filled	Zinc blende or sphalerite (ZnS) 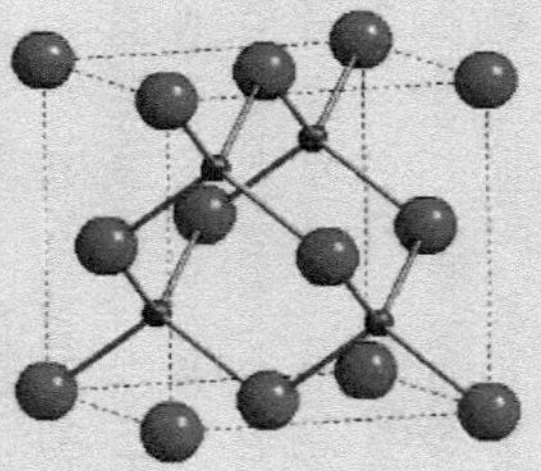	Corner-shared tetrahedrons 	Wurtzite (ZnS) 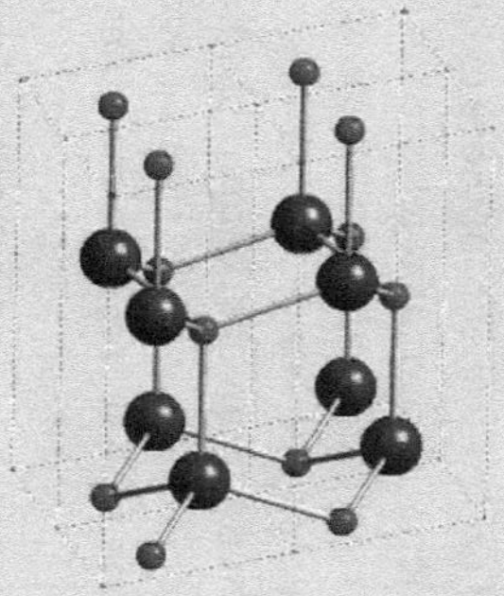	Canted corner-shared tetrahedrons 

AB_2	All t holes filled	Fluorite (CaF_2)	Face-shared tetrahedrons	No examples	No examples
A_2B	Half o holes filled (layers)	Cadmium chloride ($CdCl_2$)	Edge-shared octahedrons (layers)	Cadmium iodide (CdI_2)	Edge-shared octahedral (layers)

(continued)

TABLE 12.1 ***(Continued)***

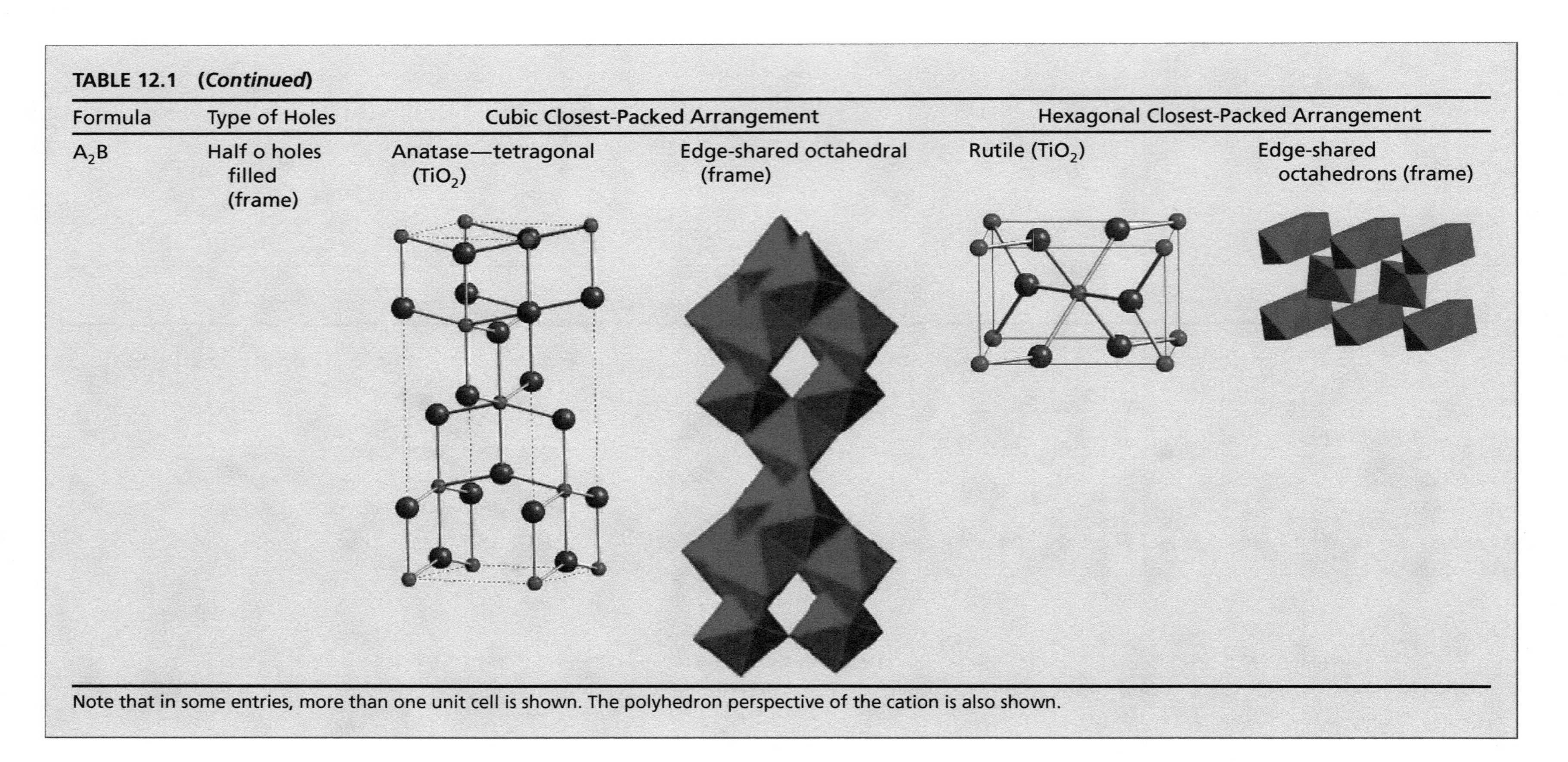

Formula	Type of Holes	Cubic Closest-Packed Arrangement		Hexagonal Closest-Packed Arrangement	
A_2B	Half o holes filled (frame)	Anatase—tetragonal (TiO_2)	Edge-shared octahedral (frame)	Rutile (TiO_2)	Edge-shared octahedrons (frame)

Note that in some entries, more than one unit cell is shown. The polyhedron perspective of the cation is also shown.

The zinc blende lattice is named after its parent compound, ZnS. Zinc sulfide also exists in a different structure known as the *wurtzite lattice*. Molecules that can exist in more than one type of crystalline form exhibit *polymorphism*. The wurtzite lattice is comprised of one type of ion forming a hexagonal closest-packed unit cell, with the other type of ion occupying half of the tetrahedral holes. The following molecules can assume the wurtzite lattice: ZnO, ZnS, ZnSe, ZnTe, BeO, AgI, CdS, MnS, SiC, AlN, and NH_4F. Both types of lattices consist of corner-shared tetrahedrons, but the tetrahedrons in wurtzite are canted in alternating layers.

The fluorite structure in Table 12.1 consists of a face-centered cubic of the cations with the anions occupying the tetrahedral holes in the lattice. In this example, it is the cation that is larger than the anion and therefore the anions occupy the interstitial sites in the lattice. Each tetrahedral hole is oriented toward one of the corners of the cube and is coordinated by the corner ion and the three closest face centers. Because there are eight corners in a cube, there are eight tetrahedral holes per unit cell. This satisfies the 1 : 2 stoichiometry of species such as the fluoride salts of the alkaline earth metals, Cd, Hg(II), and Pb(II), as well as the oxides of Zr and Hf. The packing of the ions is tighter in the fluorite structure than in the zinc blende lattice, as evidenced by the face-shared tetrahedrons in the fluorite structure. The anti-fluorite structure (not shown) simply has the positions of the ions reversed, so that the anions assume the cubic closest-packed positions and the cations occupy the tetrahedral holes. Examples include the lattices of Li_2O, Na_2O, K_2O, Na_2S, Na_2Se, and K_2S. There is no structural analog to the fluorite unit cell using a hexagonal closest-packed lattice of anions because the tetrahedral holes in this lattice are too close to each other to accommodate the cations without an excessive amount of cation–cation repulsion.

The cadmium chloride lattice is based on the cubic closest-packed structure of anions where the cations occupy every other octahedral hole. Recall that there is the same number of o-holes in a face-centered cube as there are Cl^- lattice sites. Thus, we fill only half of the octahedral interstitial sites with Cd^{2+} ions in order to maintain the 1 : 2 stoichiometry of the compound. This also results in a more open structure, leading to a layered compound. Each of the following molecules crystallizes in the cadmium chloride structure: $MgCl_2$, $MnCl_2$, TaS_2, and NbS_2. The cadmium iodide analog is based on a hexagonal closest-packed structure of the iodide ions with every other octahedral hole filled by a cadmium ion, and it too leads to a layered lattice.

Titanium(IV) oxide is another example of a molecule that exhibits polymorphism. The anatase structure of TiO_2 belongs to a tetragonal unit cell. It might help if you visualize the tetragonal unit cell as a face-centered cube that has been stretched along its *c* coordinate axis, with the Ti^{4+} ions occupying half of the octahedral holes in such a way that the edge-shared octahedrons form chains in the extended crystalline lattice. Other examples of the anatase structure include $CaSi_2$, $EuSi_2$, USi_2, $CaGe_2$, HgI_2, and $Eu(NH_2)_2$.

Lastly, the rutile structure of TiO_2 is based on a distorted hexagonal unit cell comprised of the O^{2-} ions, with the Ti^{4+} ions occupying every other octahedral hole. The octahedral coordination of the central Ti^{4+} ion is readily apparent in the unit cell shown in Table 12.1. On closer inspection, the trigonal planar coordination geometry of each O^{2-} ion can also be observed. Examples of this type of unit cell include a large number of M(IV) oxides, such as MnF_2, NiF_2, CuF_2, CoF_2, FeF_2, ZnF_2, PdF_2, CrO_2, MnO_2, MoO_2, NbO_2, GeO_2, PtO_2, RuO_2, RhO_2, SnO_2, OsO_2, WO_2, IrO_2, and PbO_2.

An example of a common ionic solid that is not based on a closest-packed structure is the cesium chloride lattice shown in Figure 12.3. The cesium chloride structure consists of a body-centered cube of one type of ion with the opposite type sitting in the cubic hole at the center of the unit cell. The cubic hole is larger than

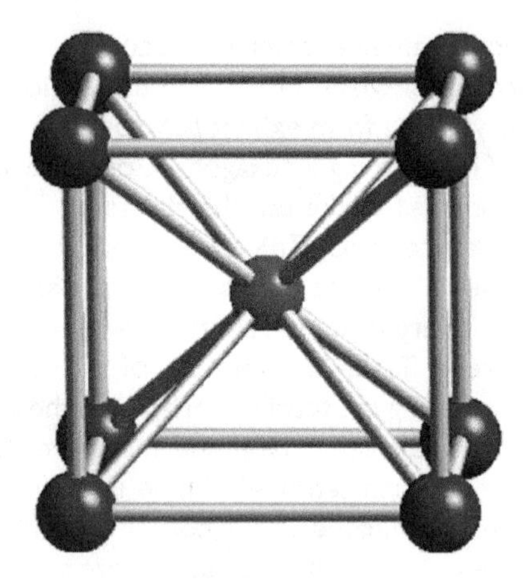

FIGURE 12.3
Crystal structure of cesium chloride.

either the octahedral or the tetrahedral holes in the previous examples. This type of ionic solid is therefore favored for 1 : 1 stoichiometric compounds where the anion and cation are of comparable sizes. Other examples of compounds crystallizing in the CsCl lattice include CsBr, CsI, CsCN, CsSH, TlCl, IBr, TlSb, TlCN, CuZn, CuPd, AlNi, BeCu, and LiHg.

Example 12-1. If the ionic radii of Cs^+ and Cl^- are 181 and 167 pm, respectively, calculate the density of the unit cell for CsCl in units of g/cm^3.

Solution. First, we assume that the oppositely charged ions touch along the body diagonal of the cube in the space-filling model of CsCl. If the unit cell length is given by a, the body diagonal will therefore be $3^{1/2}a$. The body diagonal will also be equal to the sum of two Cs^+ and two Cl^- radii, or 696 pm. Setting $3^{1/2}a$ equal to 696 and solving for the edge length yields $a = 402$ pm. Thus, the volume of the unit cell (a^3) is 6.49×10^{-23} cm^3. There is one Cs^+ ion and one Cl^- per unit cell, as each of the eight corner ions are shared between eight different unit cells and the body diagonal belongs exclusively to one unit cell. Thus, the mass of one unit cell is $(132.91 + 35.45$ g/mol$)/(6.022 \times 10^{23}$ mol$^{-1}) = 2.80 \times 10^{-22}$ g. Because density is mass divided by volume, $d = 4.31$ g/cm^3.

12.2 LATTICE ENTHALPIES AND THE BORN–HABER CYCLE

To a first approximation, the lattice energy of an ionic solid can be calculated from a purely electrostatic model based on the attractive and repulsive interactions of the oppositely charged ions with their neighbors, as shown by Equation (12.1). The attractive forces of oppositely charged ions are given by Coulomb's law and go as the square of the electronic charge, 1.602×10^{-19} C, times the charges on each ion, divided by the permittivity of space in a vacuum ($4\pi\varepsilon_0$) times the distance between them (r). The attractive term is modified by multiplication by (i) Avogadro's constant ($N_A = 6.022 \times 10^{23}$ mol^{-1}) to yield the lattice energy per mole of substance and (ii) by the Madelung constant (A), which is a measure of how the ions interact with one another in the crystalline lattice.

$$U = U_{attr} + U_{rep} = \frac{AN_A Z^+ Z^- e^2}{4\pi\varepsilon_0 r} + \frac{N_A B}{r^n} \qquad (12.1)$$

The Madelung constant can be determined geometrically based on the lattice type. For example, a Na^+ ion in NaCl will have six nearest-neighbor Cl^- ions (an attractive interaction) that lie on the face-centers at a distance of arbitrary length l because each Na^+ ion lies in an octahedral hole of the face-centered cube formed by the Cl^- ions. The original Na^+ ion's second-nearest neighbors are the 12 Na^+ ions (a repulsive interaction) that lie on the edge-centers of the face-centered cube as shown in Figure 12.3(a). If the Cl^- face-centers are 1 unit of distance from the central Na^+ ion, then the 12 edge-center ions will lie at a distance of $2^{1/2}$. The eight third-nearest neighbor Cl^- ions to the original Na^+ body-center lie at the corners of the face-centered cubic unit cell at a distance of $3^{1/3}$. The attractive and repulsive interactions with this one Na^+ ion extend throughout the entire crystalline lattice, forming a convergent series such as the one given by Equation (12.2). For NaCl, the series converges to 1.74756. The geometrical Madelung constants for other types of ionic lattices are listed in Table 12.2.

The repulsive term in Equation (12.1) is equal to Avogadro's constant times the Born constant (B), divided by r^n, where n is the compressibility factor, which depends on the electron configuration of the ions. The compressibility factors for different electron configurations are given in Table 12.3. Where the two different kinds of ions do not have the same electron configuration, the average value is used for n. For instance, the value of n for CaF_2 would be 8, as Ca^{2+} = [Ar] (n = 9) and F^- = [Ne] (n = 7). In general, the magnitude of the compressibility factor increases with the size

TABLE 12.2 Geometrical Madelung constants for several ionic lattice types.

Lattice Type	Madelung Constant, A
NaCl, rock salt	1.74756
CsCl, cesium chloride	1.76267
ZnS, wurtzite	1.64132
ZnS, zinc blende	1.63806
CaF_2, fluorite	2.51939
TiO_2, rutile	2.4080
TiO_2, anatase	2.400
SiO_2, β-quartz	2.2197
Al_2O_3, corundum	4.17187

TABLE 12.3 Born compressibility factors (based on the electron configurations of the ions).

Electronic Configuration	Compressibility Factor, n
[He]	5
[Ne]	7
[Ar], [Cu^+]	9
[Kr], [Ag^+]	10
[Xe], [Au^+]	12

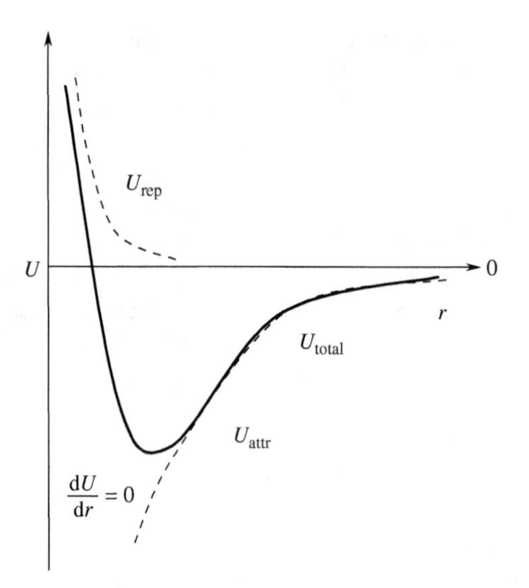

FIGURE 12.4
Potential energy diagram for the lattice energy of an ionic solid, showing the attractive (coul) and repulsive (rep) components as dotted and dashed lines, respectively.

of the ion because the electron clouds of larger ions are more polarizable and diffuse.

$$A\,(\text{NaCl}) = +\frac{6}{1} - \frac{12}{\sqrt{2}} + \frac{8}{\sqrt{3}} - \frac{6}{2} + \frac{24}{\sqrt{5}} - \cdots = 1.74756 \tag{12.2}$$

A plot of the lattice energy (u) versus the interionic separation (r) yields the familiar Lennard-Jones potential energy diagram shown in Figure 12.4. Recognizing that the equilibrium bond length (r_0) occurs at the minimum (U_0) in Figure 12.4 and that this minimum has $dU/dr = 0$, one can derive the Born–Landé equation, given by Equation (12.6), which allows for an estimate of the lattice energy for any ionic solid with a known lattice type based on a purely electrostatic model.

$$\frac{dU}{dr} = 0 = -\frac{AN_AZ^+Z^-e^2}{4\pi\varepsilon_0 r_0^2} - \frac{nN_AB}{r_0^{n+1}} \tag{12.3}$$

Solving for the Born constant, B, we get:

$$B = -\frac{AN_AZ^+Z^-e^2r_0^{n+1}}{4\pi\varepsilon_0 nN_Ar_0^2} = -\frac{AZ^+Z^-e^2r^{n-1}}{4\pi\varepsilon_0 n} \tag{12.4}$$

Substitution of B into Equation (12.1) yields:

$$U_0 = \frac{AN_AZ^+Z^-e^2}{4\pi\varepsilon_0 r_0} - \frac{AN_AZ^+Z^-e^2r_0^{n-1}}{4\pi\varepsilon_0 nr_0^n} \tag{12.5}$$

Combining terms, we obtain the Born–Landé equation:

$$U_0 = \frac{AN_AZ^+Z^-e^2}{4\pi\varepsilon_0 r_0}\left(1 - \frac{1}{n}\right) \tag{12.6}$$

Example 12-2. Calculate the lattice energy for NaCl using Equation (12.6) if the length of a unit cell edge is 566 pm. Assume that the Na^+ and Cl^- ions touch each other along each edge in a hard-spheres-type model.

Solution. The Madelung constant (A) for NaCl is 1.74756, $Z^+ = 1$, $Z^- = -1$, and $n = 8$. The interionic separation, $r_0 = 283$ pm (because one edge is equal to two Cl^- radii and two Na^+ radii). [In actuality, the hard-spheres model is only an approximation and the electron clouds of the two ions do overlap to a certain extent, so that the interionic separation in NaCl, as determined from X-ray diffraction data, is 281.4 pm.]

$$U_0 = \frac{1.75746(6.022 \times 10^{23}\ \text{mol}^{-1})(1)(-1)(1.602 \times 10^{-19}\ \text{C})^2}{1.113 \times 10^{-10}\ \text{C}^2/\text{Jm}\,(2.83 \times 10^{-10}\ \text{m})}\left(1 - \frac{1}{8}\right)$$

$$U_0 = -755\ \text{kJ/mol}$$

The experimental value for the lattice energy of NaCl is −787 kJ/mol, making the percent error only 4%.

If the lattice type and Madelung constant for an ionic solid are not yet known, the lattice energy can still be approximated using an empirical equation developed by Kapustinskii, shown in Equation (12.7). Only the charges on each ion, the stoichiometric total number of ions (υ), and the interionic separation need to be known in order to make the calculation. If r_0 is entered into Equation (12.7) in units of pm, the calculated lattice energy will be in units of kJ/mol. Using the values in Example 12-1 for NaCl and $\upsilon = 2$, $U_0 = -746$ kJ/mol, a difference of only 5% from the experimental value is obtained.

$$U_0 = \frac{120,200\ \nu Z^+ Z^-}{r_0}\left(1 - \frac{34.5}{r_0}\right) \quad (12.7)$$

The strength of an ionic bond is proportional to its lattice energy. As shown by both Equations (12.6) and (12.7), the lattice energy becomes more negative as (i) the charges on the ions increase, (ii) the number of ions increases, and (iii) the interionic distance decreases. The experimental lattice energies for a number of ionic solids are listed in Table 12.4.

Note that the lattice energy for $AlCl_3$ is considerably larger than for $MgCl_2$ or NaCl, even though each cation has the same electron configuration. All three of the given factors support this observation (the ionic radii of six-coordinate Na^+, Mg^{2+}, and Al^{3+} are 102, 72, and 54 pm). When comparisons are made between the sodium halides, the decreasing lattice energy as one descends the halogens is a direct result of the periodic trend for the ionic radius of the halide, which increases down the

TABLE 12.4 Lattice energies for selected ionic solids.

Ionic Solid	U_0, kJ/mol	Ionic Solid	U_0, kJ/mol
NaF	−914	LiCl	−840
NaCl	−787	NaCl	−787
NaBr	−728	KCl	−701
NaI	−681	RbCl	−682
LiF	−1,036	CsCl	−630
BeO	−4,443	Al_2O_3	−15,916
$MgCl_2$	−2,526	Fe_2O_3	−14,774
$AlCl_3$	−5,492		

group. Likewise, a comparison of the lattice energies of the alkali metal chlorides increases as the ionic radius of the cation decreases. The main differences between the lattice energies of the isoelectronic LiF and BeO are the increased charge and smaller ionic radius of the ions in the latter compound.

The more negative lattice energy for Al_2O_3 compared to Fe_2O_3 results from the smaller ionic radius of the Al^{3+} cation. The Al^{3+} ion has an ionic radius of 54 pm, as compared with the 65 pm ionic radius of Fe^{3+}. This example serves to illustrate the importance of the interionic separation term (r_0) in Equations (12.6) and (12.7). A difference of less than 10 pm in the ionic radius causes Al_2O_3 to have a lattice energy that is > 1100 kJ/mol more negative than that for Fe_2O_3. In turn, the more negative lattice energy of Al_2O_3 is the main thermodynamic driving force for the thermite reaction, given by Equation (12.8). The thermite reaction, which is a common chemical demonstration, is so exothermic (−851.5 kJ/mol) that it generates temperatures up to 3000 °C (hot enough to produce molten iron). This reaction was used during the Civil War to repair torn-up railroad tracks. Rumor has it that some mischievous students at MIT once used the reaction to weld trolley cars to their tracks and to weld closed the gates to Harvard Yard. It is amazing what only a few picometers difference in the ionic radius can do!

$$2\,Al(s) + Fe_2O_3(s) \rightarrow 2\,Fe(s) + Al_2O_3(s) \tag{12.8}$$

Experimentally, lattice enthalpies are measured as part of a thermodynamic cycle known as the *Born–Haber cycle*. Using the additive nature of Hess's law, the lattice enthalpy can be derived from a series of known enthalpies for other physical and chemical processes, as shown in Figure 12.5 for NaCl.

$Na(s) + \frac{1}{2}\,Cl_2(g) \rightarrow NaCl(s)$	$\Delta H_f^\circ = -411$ kJ/mol
$Na(g) \rightarrow Na(s)$	$-\Delta H_a^\circ = -107$ kJ/mol
$Na^+(g) + e^- \rightarrow Na(g)$	$-\Delta H_{IE}^\circ = -496$ kJ/mol
$Cl(g) \rightarrow \frac{1}{2}\,Cl_2(g)$	$-\frac{1}{2}$BDE $= -121$ kJ/mol
$Cl^-(g) \rightarrow Cl(g) + e^-$	EA = 348 kJ/mol
$Na^+(g) + Cl^-(g) \rightarrow NaCl(s)$	$U_0 = -787$ kJ/mol

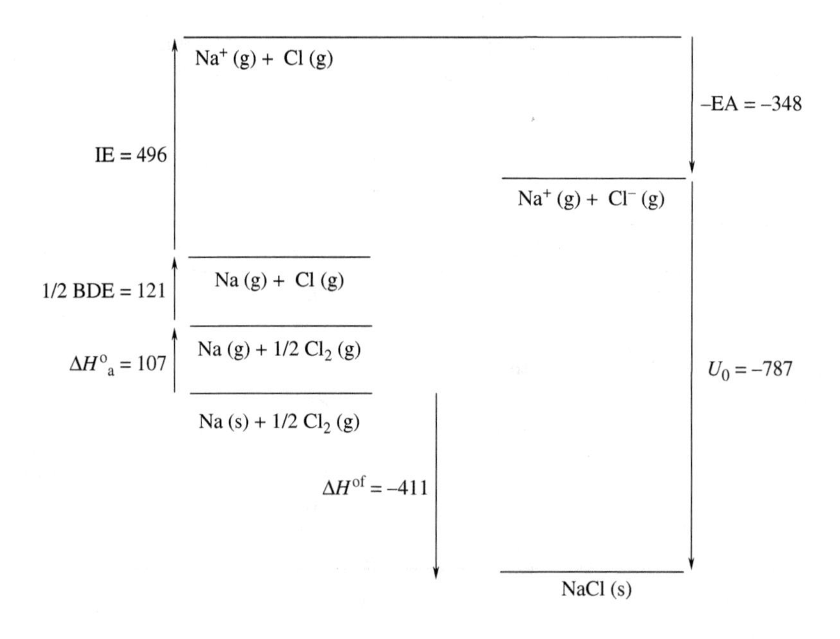

FIGURE 12.5
The Born–Haber cycle for NaCl (s) is simply a manifestation of Hess's law that the enthalpies are additive.

By summing the enthalpy of formation of NaCl (s), the reverse of the atomization energy for Na (s), the reverse of the first ionization energy for Na (g), the reverse of one-half of the bond dissociation energy for Cl_2 (g), and the electron affinity for Cl (g), one obtains the same chemical equation as that for the lattice enthalpy of NaCl (s). Because the thermodynamic values for each of these different processes are known, the lattice energy can be calculated using Hess's law. Lattice energies that are calculated in this way are referred to as *theoretical lattice energies*. In practice, however, the lattice enthalpy is usually experimentally determined, and the Born–Haber cycle is used to calculate the electron affinity of the more electronegative element or the thermochemical radii of polyatomic ions.

Sometimes the Born–Haber cycle is used in reverse to predict whether the enthalpy of formation of an unknown ionic compound is favorable or not. For example, the theoretical enthalpy of formation of NaF_2 can be predicted using the Born–Haber cycle as shown here. The fictional NaF_2 molecule would dissociate into three ions: one Na^{2+} and two F^- ions. The lattice energy of NaF_2 was calculated using the Kapustinskii equation, where $r(F^-) = 133$ pm and assuming that $r(Na^{2+}) \approx r(Mg^{2+}) = 72$ pm.

$$U_0 = \frac{120,200(3)(+2)(-1)}{205}\left(1 - \frac{34.5}{205}\right) = -2926 \text{ kJ/mol}$$

$$\begin{array}{ll}
\text{Na(s)} \rightarrow \text{Na(g)} & \Delta H_a^\circ = 107 \text{ kJ/mol} \\
\text{Na(g)} \rightarrow \text{Na}^+\text{(g)} + \text{e}^- & \Delta H_{IE1}^\circ = 496 \text{ kJ/mol} \\
\text{Na}^+\text{(g)} \rightarrow \text{Na}^{2+}\text{(g)} + \text{e}^- & \Delta H_{IE2}^\circ = 4562 \text{ kJ/mol} \\
\text{F}_2\text{(g)} \rightarrow 2\text{ F(g)} & \text{BDE} = 158 \text{ kJ/mol} \\
2\text{F(g)} + 2\text{ e}^- \rightarrow 2\text{ F}^-\text{(g)} & -2\text{EA} = -656 \text{ kJ/mol} \\
\text{Na}^{2+}\text{(g)} + 2\text{F}^-\text{(g)} \rightarrow \text{NaF}_2\text{(s)} & U_0 = -2926 \text{ kJ/mol} \\
\hline
\text{Na(s)} + \text{F}_2\text{(g)} \rightarrow \text{NaF}_2\text{(s)} & \Delta H_f^\circ = 1741 \text{ kJ/mol}
\end{array}$$

This is a very large, positive enthalpy of formation. Because the entropy of formation of NaF_2 will be negative, the free energy of formation for NaF_2 (s) is assuredly positive. Thus, NaF_2 would be thermodynamically unstable.

Example 12-3. Use the Born–Haber cycle to predict whether the ionic solid CaF (s) is likely to form. Assume that the ionic radius of Ca^+ is identical to that for K^+.

Solution. The ionic radius of K^+ is 138 pm, while that of F^- is 133 pm. Thus, $r_0 = 271$ pm in the Kapustinskii equation:

$$U_0 = \frac{120,200(2)(+1)(-1)}{271}\left(1 - \frac{34.5}{271}\right) = -774 \text{ kJ/mol}$$

$$\begin{array}{ll}
\text{Ca(s)} \rightarrow \text{Ca(g)} & \Delta H_a^\circ = 178 \text{ kJ/mol} \\
\text{Ca(g)} \rightarrow \text{Ca}^+\text{(g)} + \text{e}^- & H_{IE1}^\circ = 590 \text{ kJ/mol}
\end{array}$$

$\frac{1}{2} F_2(g) \rightarrow F(g)$	$\frac{1}{2}$ BDE = 79 kJ/mol
$F(g) + e^- \rightarrow F^-(g)$	− EA = −328 kJ/mol
$Ca^+(g) + F^-(g) \rightarrow CaF(s)$	$U_0 = -774$ kJ/mol
$Ca(s) + \frac{1}{2} F_2(g) \rightarrow CaF(s)$	$\Delta H_f^\circ = -255$ kJ/mol

Thus, it would appear that it is possible to form CaF (s) if the enthalpy term dominates over the negative entropy of formation. However, even if CaF (s) forms, it would rapidly disproportionate into Ca (s) and CaF_2 (s):

$$2\,CaF(s) \rightarrow Ca(s) + CaF_2(s) \qquad \Delta H_{rxn}^\circ = -1220 - 2(-255) = -710 \text{ kJ/mol}$$

12.3 IONIC RADII AND PAULING'S RULES

The radius of an ion will not be the same as for that of the neutral atom. Anions are larger than their respective atoms because of the additional electron–electron repulsion, while cations are smaller than their corresponding atoms because the effective nuclear charge is larger. Pauling used a theoretical model to calculate ionic radii. Recognizing that there is an inverse relationship between the radius and the effective nuclear charge, Pauling set up the ratio in Equation (12.9), where r_0 is given by Equation (12.10).

$$\frac{r^+}{r^-} = \frac{Z^-}{Z^+} \tag{12.9}$$

$$r_0 = r^+ + r^- \tag{12.10}$$

Solving Equations (12.9) and (12.10) simultaneously for r^+ and r^-, where the interionic distance in the crystal is known, yields the *univalent radii*. Univalent radii assume that the ions interact using a hard-spheres model such that their electron clouds do not interpenetrate with one another. For ions that are not univalent, such as Mg^{2+}, a correction must be made for the compressibility because the larger the nuclear charge, the more likely the ion will be able to polarize the electron cloud of its neighbor. The correction factor, shown in Equation (12.14) can be derived from the Born constant, as demonstrated below in Equations (12.11)–(12.13). The corrected Pauling ionic radii for selected ions are listed in Table 12.5.

$$B = -\frac{AZ^+Z^-e^2r_0^{n-1}}{4\pi\varepsilon_0 n} \tag{12.11}$$

$$r_0^{n-1} = \frac{4\pi\varepsilon_0 nB}{Ae^2}\left(-\frac{1}{Z^+Z^-}\right) \tag{12.12}$$

$$r_0 = \left(\frac{4\pi\varepsilon_0 nB}{Ae^2}\right)^{1/n-1}\left(-\frac{1}{Z^+Z^-}\right)^{1/n-1} = k\left(-\frac{1}{Z^+Z^-}\right)^{1/n-1} \tag{12.13}$$

$$\frac{r_{corr}}{r_{univ}} = \left(-\frac{1}{Z^+Z^-}\right)^{1/n-1} \tag{12.14}$$

TABLE 12.5 Pauling's ionic radii (pm).

Period 2		Period 3		Period 4		Period 5		Period 6	
Li^{+}	60	Na^{+}	95	K^{+}	133	Rb^{+}	148	Cs^{+}	169
Be^{2+}	31	Mg^{2+}	65	Ca^{2+}	99	Sr^{2+}	113	Ba^{2+}	135
B^{3+}	20	Al^{3+}	50	Ga^{3+}	62	In^{3+}	81	Tl^{3+}	95
C^{4+}	15	Si^{4+}	41	Ge^{4+}	53	Sn^{4+}	71	Pb^{4+}	84
		P^{5+}	34	As^{5+}	47	Sb^{5+}	62	Bi^{5+}	74
		S^{6+}	29	Se^{6+}	42	Te^{6+}	56		
N^{3-}	171	P^{3-}	212	As^{3-}	222	Sb^{3-}	245		
O^{2-}	140	S^{2-}	184	Se^{2-}	198	Te^{2-}	221		
F^{-}	136	Cl^{-}	181	Br^{-}	195	I^{-}	216		

Example 12-4. Use Equations (12.9) and (12.10) to calculate the univalent radii of Na^{+} and F^{-} if the interionic distance in NaF is 231 pm.

Solution. Using Slater's rules to calculate Z^* for the isoelectronic ions:

$$[Na^+] = [F^-] = [Ne] \quad (1s)^2(2s,\ 2p)^8$$

$$Na^+ \quad Z^* = 11-[7(0.35) + 2(0.85)] = 6.85$$

$$F^- \quad Z* = 9-[7(0.35) + 2\ (0.85)] = 4.85$$

$$r^+/r^- = 4.85/6.85 = 0.708$$

$$r^+ + r^- = 231 \text{ pm} = 0.708r^- + r^- = 1.708r^-$$

$$r^- = 135 \text{ pm and } r^+ = 96 \text{ pm}$$

Example 12-5. Use Pauling's equations and the results from Example 12-4 to calculate the univalent and corrected radii of Mg^{2+} and O^{2-}. Then, use the corrected ionic radii to predict the interionic distance in MgO.

Solution. Using Slater's rules to calculate Z^* for the isoelectronic ions:

$$[Mg^{2+}] = [O^{2-}] = [Na^+] = [F^-] = [Ne] \quad (1s)^2(2s,\ 2p)^8$$

$$Mg^{2+} \quad Z^* = 12-[7(0.35) + 2(0.85)] = 7.85$$

$$O^{2-} \quad Z^* = 8-[7(0.35) + 2\ (0.85)] = 3.85$$

Because of the inverse relationship between r and Z^*, the proportionality constant can be calculated from the univalent radius of either Na^{+} or F^{-}, both of which are isoelectronic with Mg^{2+} and O^{2-}.

$$r = k/Z^*$$

$$k = Z^*r = 6.85(96 \text{ pm}) \text{ for } Na^+, \text{ or } 4.85(135 \text{ pm}) \text{ for } F^-$$

In either case, $k = 658$ pm. Using this value of k to calculate the univalent radii of Mg^{2+} and O^{2-}, we obtain:

$$r^{+} = 658\ \text{pm}/7.85 = 84\ \text{pm for } Mg^{2+}(\text{univalent})$$

$$r^{-} = 658\ \text{pm}/3.85 = 171\ \text{pm for } O^{2-}(\text{univalent})$$

Correcting for the compressibility using Equation (12.14) with $n = 7$, we obtain:

$$r_{corr} = 84\ \text{pm}\ (1/4)^{1/6} = 66\ \text{pm for } Mg^{2+}$$

$$r_{corr} = 171\ \text{pm}\ (1/4)^{1/6} = 136\ \text{pm for } O^{2-}$$

$$r_0 = r_+ + r_- = 66 + 136 = 202\ \text{pm (predicted)}$$

The experimentally determined interionic distance in MgO is 210 pm. Note that if the univalent radii had not been corrected for the compressibility, the predicted interionic distance would have been 255 pm, which is considerably different from the experimental value.

These corrected ionic radii are a much better approximation to the experimental values derived from X-ray diffraction data and electron density contour maps, such as the one shown in Figure 12.6 for TiC. The experimentally determined values, defined by where the electron density of one ion ends in the contour map,

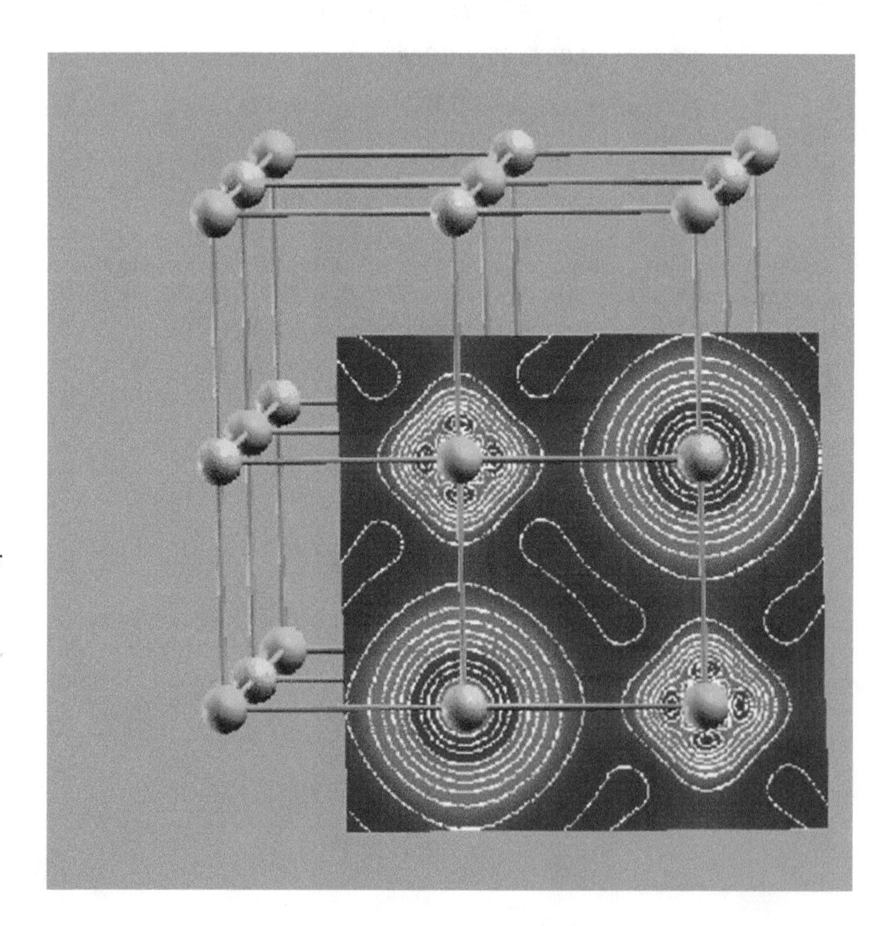

FIGURE 12.6
Electron density contour map for TiC, which assumes the halite lattice structure. [Reproduced by permission. Blaha, P.; Schwarz, K., Madsen, G., Kvasnicka, D.; Luitz, J. (2001): "WIEN2k, An Augmented Plane Wave + Local Orbitals Program for Calculating Crystal Properties". K. Schwarz (ed.), Techn. Universität Wien, Austria. ISBN 3-9501031-1-2.]

are known as the *crystal radii.* The modern (revised Shannon and Prewitt) crystal radii for representative ions are listed in Table 5.2. These values are the average of experimentally derived radii for a large number of ionic compounds.

In general, the crystal radii of cations will be influenced by one or more of the following properties: (i) the charge of the ion, (ii) the coordination number that the ion exhibits in the crystalline lattice, and (iii) the configuration of the *d*-electrons in transition metals. As the charge on the cation increases, the electrons are held tighter by the positively charged nucleus, and the crystal radius decreases. For example, for four-coordinate Mn^{n+}, the crystal radii for the 4+, 5+, 6+, and 7+ oxidation states are 39, 33, 25.5, and 25 pm. Likewise, the greater the coordination number of the cation, the more anionic neighbors it will have in the crystalline solid. The larger the coordination number, the greater the degree of anion–anion repulsion, and the larger the crystal radius. Consider, for example, the crystal radii of La^{3+} in 6-, 7-, 8-, 9-, 10-, and 12-coordinate environments, which are 103.2, 110, 116.0, 121.6, 127, and 136 pm. For transition metal cations, the configuration of the *d*-electrons will also play a role. As we will learn in a later chapter, the five *d*-orbitals in a six-coordinate, octahedral environment will split into two different energy levels: a higher lying e_g (antibonding) set and a lower lying t_{2g} set, as shown in Figure 12.7.

As early as 1929, Linus Pauling, who is arguably the most influential chemist of the twentieth century, developed a set of five principles that can be used to rationalize the structures of many ionic solids. *Pauling's rules,* which follow, are based on a completely ionic, hard-spheres electrostatic model.

Rule 1. You can draw a polyhedron of anions around every cation in a crystalline lattice, such that: (i) the interionic separation can be determined as the sum of the ionic radii, according to Equation (12.10), and (ii) the coordination number of the cation can be determined using the radius ratio rule (*vide infra*). The radius ratio rule sets the minimum r^+/r^- ratio that can exist for a cation with a given coordination number. This ratio can be determined from a geometrical consideration of the minimum cationic radius necessary to keep the anions in a particular coordination geometry from just touching each other.

A derivation of the minimum radius ratio for octahedral coordination (CN = 6) is shown in Figure 12.8 and Equations (12.15)–(12.18).

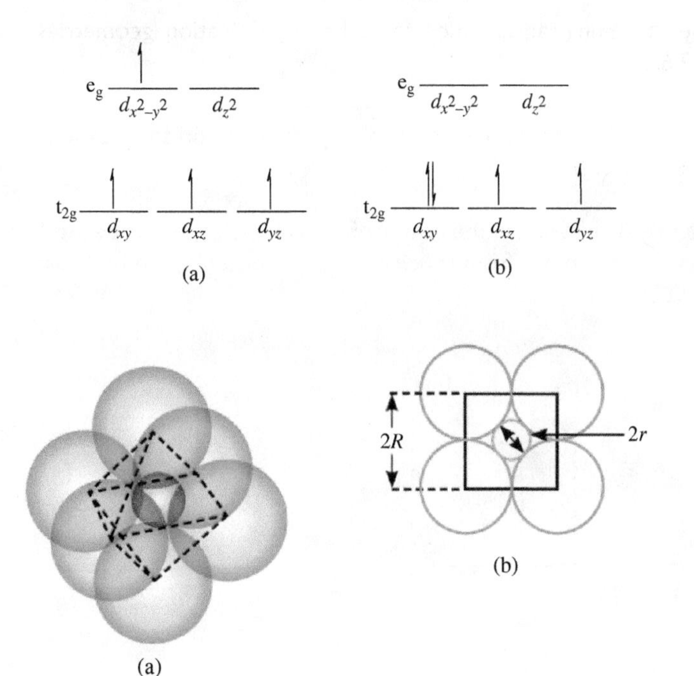

FIGURE 12.7
Splitting of the *d*-orbitals in an octahedral coordination environment for Mn^{3+}: (a) the high-spin (HS) configuration, and (b) the low-spin (LS) configuration.

FIGURE 12.8
(a) Illustration of a cation surrounded by six anions in an octahedral geometry. Figure (b) shows a cross-section of the octahedron, where *R* represents the radius of the anion and *r* is the radius of the cation. [Blatt Communications.]

TABLE 12.6 The radius ratio (r/R) ranges for two ions of different sizes, according to the coordination number of the ion having radius r.

Radius ratio range (r/R)	Coordination Number	Type of Hole in Which the Smaller Ion Resides
0.155–0.224	3	Trigonal
0.225–0.413	4	Tetrahedral
0.414–0.731	6	Octahedral
0.732–0.999	8	Cubic
≥1.000	12	Cuboctahedral

Using the Pythagorean theorem for the right triangle formed by connecting the centers of three adjacent anions, one obtains:

$$(2R)^2 + (2R)^2 = (2R + 2r)^2 \tag{12.15}$$

Taking the square root of both sides, Equation (12.15) reduces to

$$\sqrt{8}R = 2R + 2r \tag{12.16}$$

Collecting all the R terms on one side of the equation yields:

$$(\sqrt{8} - 2)R = (2\sqrt{2} - 2)R = 2r \tag{12.17}$$

Solving for the radius ratio, r/R, one obtains the minimum radius ratio necessary to keep the anions from touching one another:

$$\frac{r}{R} = \frac{2\sqrt{2} - 2}{2} = \sqrt{2} - 1 = 0.414 \tag{12.18}$$

The minimum radius ratios for other coordination geometries are listed in Table 12.6.

Example 12-6. Prove that the minimum radius ratio for a cation having cubic coordination is 0.732.

Solution. The cation in the center of the cube will need to be just large enough to prevent the anions from touching each other along the body diagonal of the cube, as shown.

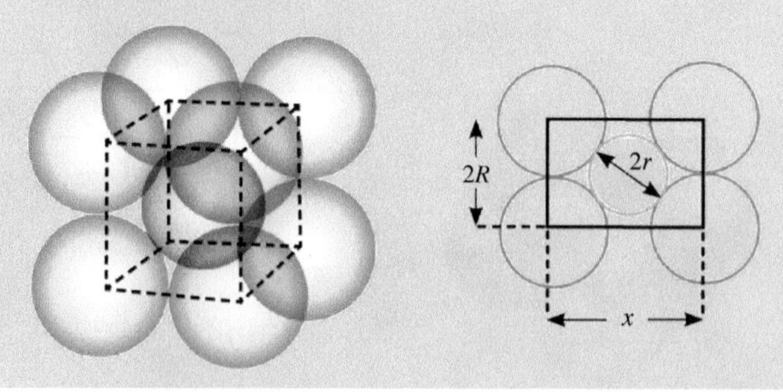

Because the distance x represents a face diagonal of the cube, it will be equal to the square root of the sum of the squares of the edges:

$$x = \sqrt{(2R)^2 + (2R)^2} = 2\sqrt{2}R$$

Because the body diagonal is equal to $2R + 2r$, the Pythagorean theorem yields the equality:

$$(2R)^2 + (2\sqrt{2}R)^2 = (2R + 2r)^2 = 12R^2$$

Taking the square root of both sides and collecting all the Rs on the left, one obtains:

$$(2\sqrt{3} - 2)R = 2r$$

The minimum radius ratio is therefore:

$$\frac{r}{R} = \frac{2\sqrt{3} - 2}{2} = \sqrt{3} - 1 = 0.732$$

Figure 12.9 demonstrates how the lattice energy becomes more negative as r^+ decreases but then flattens out as anion–anion contact begins to occur when r^+ gets too small to keep the anions from touching anymore. For each coordination number, the energy reaches a minimum at the minimum radius ratio listed in Table 12.6. The end result is that it is easier for a small cation to settle into the holes of a lattice having a smaller coordination number than it is for the lattice to expand its coordination number.

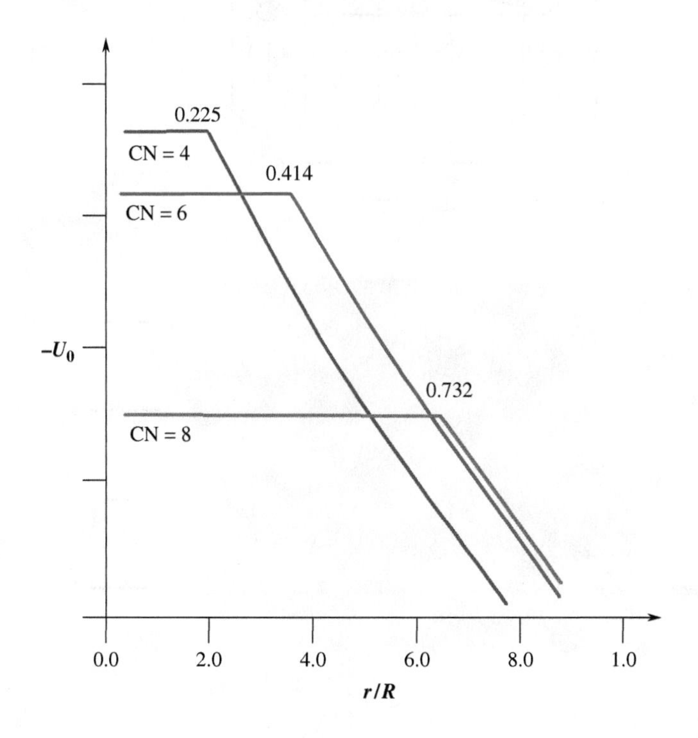

FIGURE 12.9
Lattice stabilization energy for 4-, 6-, and 8-coordinate geometries as a function of the radius ratio r/R. [Blatt Communications.]

Using the Pauling ionic radii of Na^+ (95 pm) and Cl^- (181 pm) from Table 12.5, the predicted geometry of a Na^+ ion in NaCl is octahedral ($r/R = 0.53$). The Pauling ionic radii should be used instead of the revised Shannon and Prewitt crystal radii because the latter already assume that the coordination number is known in advance. Thus, using the Shannon and Prewitt crystal radii in the radius ratio rule is essentially a circular argument. As shown in Figure 12.1, the NaCl lattice can be visualized as a face-centered cube of chloride ions, with the smaller sodium ions fitting into every octahedral hole. An alternate polyhedral representation of the NaCl lattice is the one shown in Figure 12.9, where the lattice is comprised of edge-sharing octahedrons.

The radius ratio rule successfully explains the structure of crystalline solids only about 50% of the time. In particular, it predicts CN = 8 more times than is actually observed and CN = 6 at a lower frequency. This is because there is only a 1% difference in the stability of these two coordination geometries (as shown in Figure 12.10). Other factors, such as the covalency of the bonding, also play a role in the choice of the crystalline lattice. For example, the radius ratio rule predicts a coordination number of 6 for the ions in HgS ($r/R = 110/184 = 0.597$), but the actual coordination geometry is 4 (the black β-form of HgS crystallizes in the sphalerite lattice). The small difference in electronegativity between the two elements introduces a significant amount of covalent character to the bonding, so that covalent factors (such as directional bonding) play a greater role than simply the size effects of hard-sphere ions assumed by Pauling's rules.

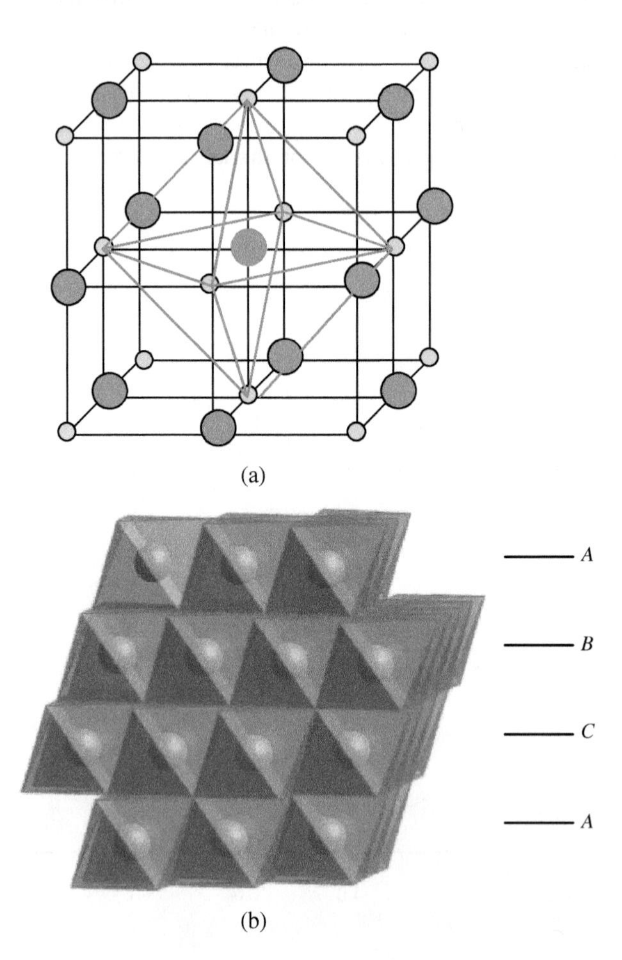

FIGURE 12.10
Representation of the NaCl crystal as a lattice composed of edge-shared octahedrons (b). (a) illustrates how an octahedron of anions can form around every cation in the structure. (b) shows the *ABCABC* repeating pattern of the face-centered cubic lattice.

Example 12-7. Employ the radius ratio rule using the Pauling ionic radii listed in Table 12.5 to predict the lattice types of (a) CsCl, (b) SrF_2, and (c) KBr.

Solution. The ionic radii of Cs^+, Sr^{2+}, K^+, Cl^-, Br^-, and F^-, according to the data in Table 12.5 are 169, 113, 133, 181, 195, and 136 pm, respectively.

(a) CsCl: $169/181 = 0.934 \rightarrow$ predicts CN = 8. As shown in Figure 12.3, CsCl assumes a primitive cubic lattice of the larger Cs^+ ions, with the Cl^- ions sitting in the cubic (body-center) holes.
(b) SrF_2: $113/136 = 0.831 \rightarrow$ predicts CN = 8 for Sr^{2+}. The 2 : 1 ratio in this compound, implies that the coordination number for F^- must be half that for Sr^{2+} (CN = 4). SrF_2 assumes a fluorite structure such as the one shown in Table 12.1. This type of lattice is composed of a face-centered cube of Sr^{2+} ions with the smaller F^- ions sitting in every tetrahedral hole.
(c) KBr: $133/195 = 0.682 \rightarrow$ predicts CN = 6. KBr assumes a halite structure such as the type shown in Figure 12.1. The halite lattice consists of a face-centered cube of Br^- ions with a K^+ ion occupying each of the octahedral holes.

Rule 2. An ionic solid will be stable if the sum of the electrostatic bond strengths (ebs's) that reach an ion is identical to the charge on that ion. In other words, its bond valence should equal its oxidation number. The electrostatic bond strength of an ion in a solid is given by Equation (12.19). This rule ensures that there is electrical neutrality in the immediate vicinity of each ion, and hence is also known as the *principle of charge balance.*

$$\text{ebs} = \frac{m}{n} = \frac{\text{charge on the ion}}{\text{no. of nearest neighbors}} \tag{12.19}$$

The exact lattice type of an ionic solid can often be uniquely defined using a combination of Rules 1 and 2. Consider the two ionic salts SrF_2 and SnO_2. Both compounds exhibit a 1 : 2 cation : anion ratio. Using the radius ratio rule for SrF_2 ($113/136 = 0.831$), one predicts a coordination number of eight for Sr^{2+}. Thus, the electrostatic bond strength of Sr^{2+} would be 2/8, or 1/4. In order for the principle of charge balance to hold true, the sum of the electrostatic bond strengths that reach each F^- ion must be equal to its oxidation state (−1). Because the Sr^{2+} ion contributes an electrostatic bond strength of 1/4 to each Sr–F bond, the F^- ion must contribute an electrostatic bond strength of −1/4. In order for the sum of the ebs's on F^- to equal its −1 charge, there must be four bonds to each F^-, or a coordination number of four. Hence, SrF_2 assumes the fluorite structure, shown in Figure 12.11(a). In terms of Pauling's polyhedrons, SrF_2 consists of alternating edge-shared cubes, with the Sr^{2+} ions located in the center of each cube, as shown in Figure 12.11(a). This alternating arrangement is necessary for the F^- ions to be four-coordinate. Note that every corner (or each fluoride ion) is shared between exactly four cubes.

Let us compare this result with that for stannic oxide. For SnO_2, the radius ratio rule ($71/140 = 0.51$) predicts the Sn^{4+} ion to be six-coordinate, with an electrostatic bond strength of 4/6, or 2/3. Thus, each O^{2-} ion would need an electrostatic bond strength of −2/3. Because the total charge on each oxide ion is −2, there would need to be three bonds to each O^{2-} ion in order to satisfy the principle of charge balance. As a result, SnO_2 crystallizes in the rutile structure shown in Figure 12.11(b). In terms of polyhedrons, the SnO_2 lattice can be constructed from staggered rows

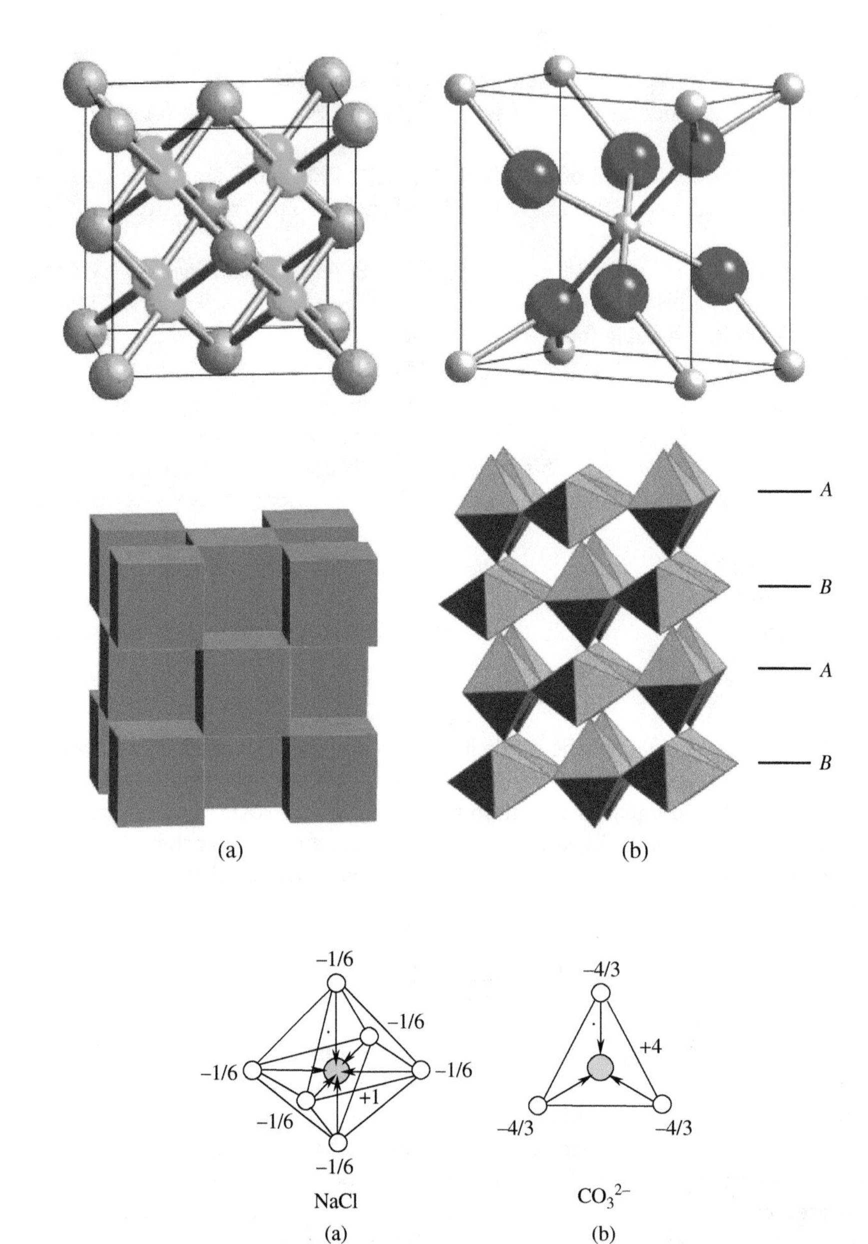

FIGURE 12.11
(a) The structure of SrF_2 (fluorite lattice), showing an alternating arrangement of the cubic Sr^{2+} ions surrounded by four-coordinate F^- ions. (b) The structure of SnO_2 (rutile lattice), composed of staggered rows of Sn^{4+} octahedrons, with each O^{2-} ion being shared between three edge-shared octahedrons.

FIGURE 12.12
(a) Sodium chloride is isodesmic, while (b) the carbonate ion is anisodesmic.

of edge-shared octahedrons, forming the familiar *ABABAB* pattern of a hexagonal lattice, with the Sn^{4+} ions occupying the centers of each octahedron. The staggered arrangement allows for each corner ion (each oxide) to be shared between three different octahedrons.

Crystals in which all of the bonds are of equal strength are called *isodesmic*. Sodium chloride is an isodesmic crystal because the sum of the ebs's from the six Cl^- ions exactly balances the charge on the Na^+ ion, as shown in Figure 12.12(a). Each Na–Cl bond has the exact same strength. This is not the case, however, for the C^{4+} ion in the carbonate ion. Using the radius ratio rule, the C^{4+} ion in CO_3^{2-} is predicted to be three-coordinate. Thus, it has an electrostatic bond strength of 4/3 and the contribution from each O^{2-} ion to the C–O bonding is −4/3, as shown in Figure 12.12(b). If the O^{2-} ions are bonded to only one C^{4+} ion, as they are in carbonate, there is still a −2/3 charge on each O^{2-} ion that's unaccounted for. This −2/3 charge allows the oxide ions to coordinate to some other cation in the crystal. In Na_2CO_3, the remaining −2/3 charge on each O^{2-} ion would need to be balanced by coordination to the Na^+ ions. Thus, the bonding in sodium carbonate is *anisodesmic*. The O^{2-} ions are more strongly coordinated to the C^{4+} ion than they are to the Na^+ ions. This unequal bonding is precisely the reason that we usually consider the bonding in sodium carbonate as a combination of covalent bonding between C and O within the carbonate ion and ionic bonding between the Na^+ and CO_3^{2-} ions. However, Pauling's initial assumption of an entirely hard-spheres model precludes any formal treatment of covalent bonding in the Na_2CO_3 lattice.

Rule 3. The sharing of edges and faces by two polyhedrons decreases the stability of an ionic structure because of the increase in cation–cation repulsions as the cations get closer together. For any given coordination geometry, the degree of cation–cation repulsion increases in the order: corner-shared < edge-shared < face-shared, as shown in Figure 12.13. The cation–cation repulsions also increase with a decrease in the coordination number. For the face-shared polyhedra in Figure 12.13, note that the relative distances between neighboring cations in the four-coordinate structure (at top) is much shorter than between the six-coordinate structure (at bottom). The net result of Rule 3 is that most ionic solids will prefer to be corner-shared rather than edge- or face-shared. This is particularly true for polyhedra having small coordination numbers and large oxidation numbers. The smaller the coordination number, the closer the cations can come to each other, leading to cation–cation repulsion. The larger the oxidation number of the cation, the greater the magnitude of the Coulombic repulsion between cations will become.

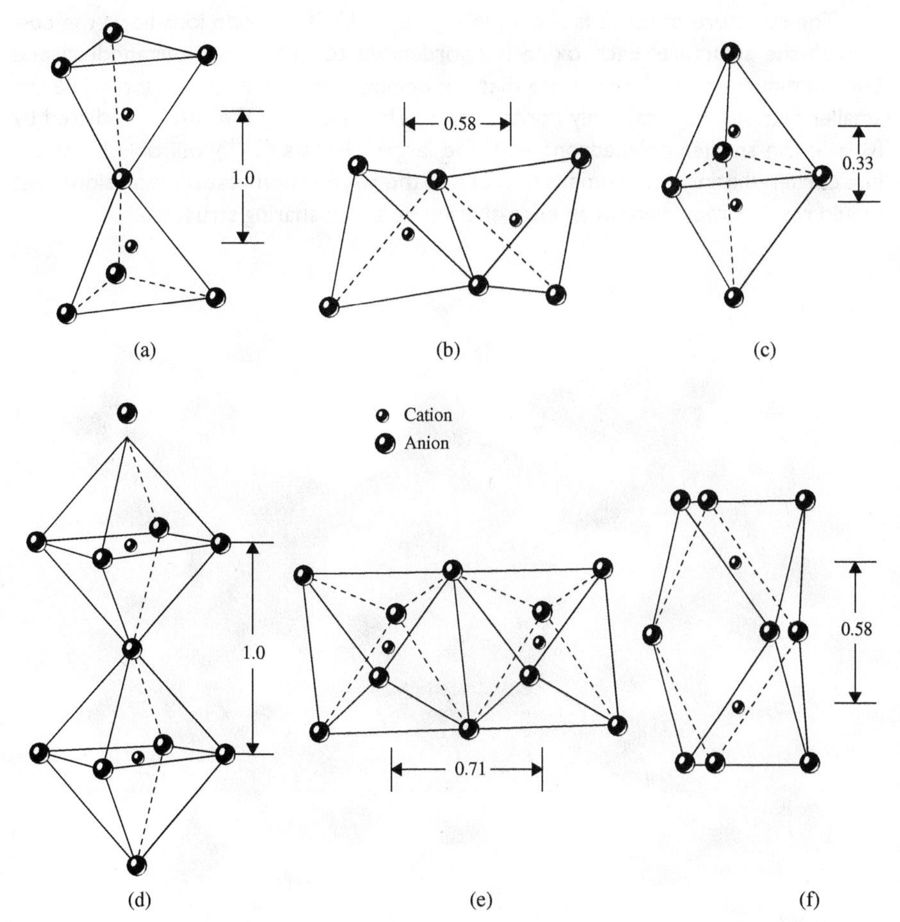

FIGURE 12.13
Cation–cation repulsion in (a) corner-shared tetrahedrons, (b) edge-shared tetrahedrons, (c) face-shared tetrahedrons, (d) corner-shared octahedrons, (e) edge-shared octahedrons, and (f) face-shared octahedrons. The degree of cation–cation repulsion increases with the degree of sharing and the oxidation number of the cation and decreases with an increase in the coordination number. [Reproduced from Klein, C. *Manual of Mineral Science*, 22nd ed., Wiley: New York, 1977. This material is reproduced with permission of John Wiley & Sons, Inc.]

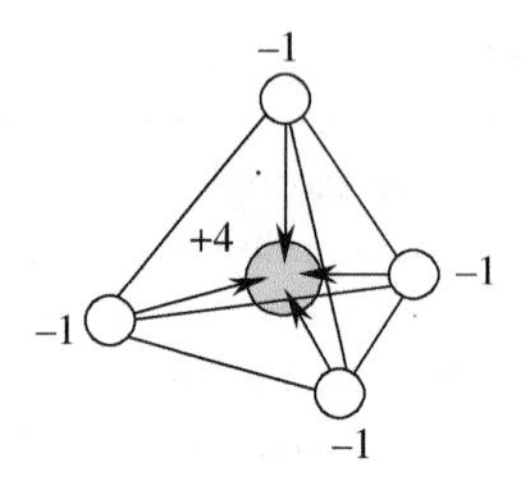

FIGURE 12.14
The SiO_4^{4-} tetrahedron is the basic building block of the silicate class of minerals. The bonding in this structure is mesodesmic because the charge remaining on each O atom following coordination to the Si^{4+} ion at the center of the tetrahedron is exactly one-half the charge on the oxide ion.

The silicate class of minerals is a perfect example of Pauling's first three rules at work. The radius ratio rule (Rule 1) predicts that each Si^{4+} ion will be four-coordinate (because $41/140 = 0.29$). Because the electrostatic bond strength (Rule 2) of Si^{4+} is 4/4, or +1, each O^{2-} will contribute −1 to the Si–O bonding, as shown in Figure 12.14.

The bonding in each SiO_4^{4-} tetrahedron is *mesodesmic* because the −2 charge on oxide is exactly half-balanced by coordination to the silicon. As a result, each oxygen carries an additional −1 charge that can be used to coordinate with other cations in the crystalline lattice. Consider the mineral topaz, which happens to be the author's birthstone. The structure of topaz is based on the molecular formula Al_2SiO_4, where one oxide ion is replaced by two hydroxide or two fluoride ions. Al^{3+} is six-coordinate in the family of aluminosilicates. Given that the ebs for Si^{4+} is 4/4, or +1, while the ebs of Al^{3+} is 3/6, or ½, the only way to satisfy the principle of charge neutrality (Rule 2) in this structure is for each O^{2-} ion to coordinate with one Si^{4+} and two Al^{3+}: $-2 = 1(-1) + 2(-1/2)$.

The structure of topaz is shown in Figure 12.15. The oxide ions lie at the corners in the structure. Each oxide is coordinated to one silicon tetrahedron and two aluminum octahedrons. Note that the octahedrons are edge-sharing, while the smaller tetrahedrons can only corner-share. This is exactly the result predicted by Rule 3: the smaller polyhedrons with the larger charges (Si^{4+}) will only be stable in a corner-sharing environment because of the large cation–cation repulsions that would result if they were in an edge-sharing or a face-sharing structure.

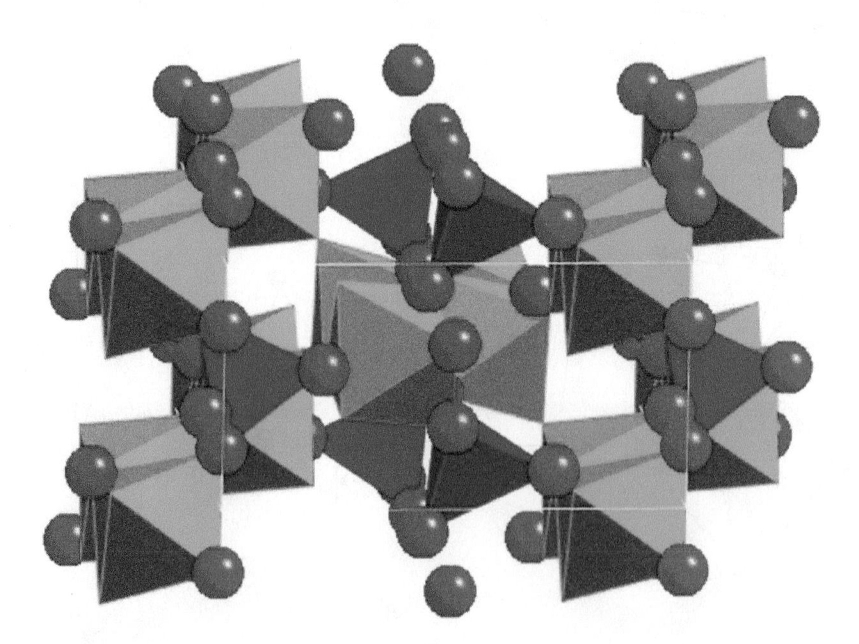

FIGURE 12.15
A polyhedron representation of the mineral topaz, where each oxygen (red) is coordinated to one Si^{4+} tetrahedron (blue) and to two Al^{3+} octahedrons (green). The F^- ions are omitted for clarity.

Example 12-8. The *perovskites* are an important family of compounds having the general formula ABO_3. $BaTiO_3$ is an example. Use Pauling's first three rules to predict the structure of $BaTiO_3$. The ionic radii of Ba^{2+} and Ti^{4+} with O^{2-} as the anion are 142 and 61 pm, respectively.

Solution. Using the ionic radii for Ba^{2+} and Ti^{4+} provided in the problem and the ionic radius of O^{2-} from Table 12.5, the radius ratio rule predicts that each Ba^{2+} will be 12-coordinate, or cuboctahedral (142/140 = 1.01). For Ti^{4+}, the expected coordination is octahedral (61/140 = 0.44). Accordingly, the ebs's for Ba^{2+} and Ti^{4+} calculated using Rule 2 are 1/6 and 2/3, respectively. The only way to balance these charges using O^{2-} is for each oxide to coordinate to four Ba^{2+} and two Ti^{4+} ions: $-2 = 4(-1/6) + 2(-2/3)$. A polyhedral view of $BaTiO_3$ is shown, where the Ti^{4+} ions sit in the center of a layer of corner-shared octahedrons, known as the *BO_2 layer*. The Ba^{2+} ions lie in a different layer between the planes of octahedrons, known as the *AO layer*. Each Ba^{2+} is 12-coordinate, as it is equidistant from four O^{2-} ions in the top BO_2 layer, four Os in the AO layer, and four O^{2-} ions in the next BO_2 layer.

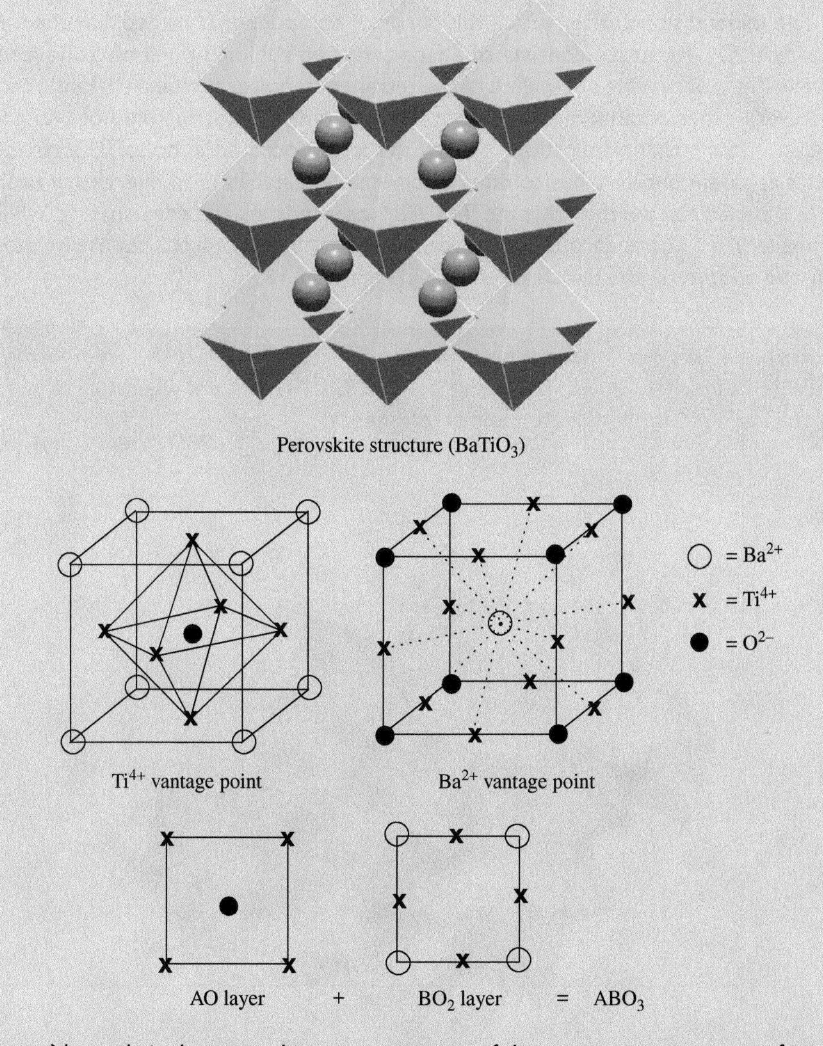

Not only is the perovskite structure one of the most common types of mineral lattices on earth but the perovskites also have a variety of interesting physical properties. Several kinds of perovskites, including $BaTiO_3$, are *ferroelectric*, which means that they have a permanent electric dipole moment. Ferroelectric materials are often used in the read-access memory (RAM) of computers because of their high dielectric constants. Other types of perovskites have interesting magnetic properties. A number of high-temperature superconductors (compounds that have zero resistance to the flow of electricity) also consist of structures that are based on the perovskite unit cell.

Rule 4. In a crystal structure that contains two or more types of cations, those with the higher valence and the smaller coordination number do not typically share polyhedron elements in common with each other. This rule again results from the large cation–cation repulsions that result if these polyhedrons were forced to have ions in common. The *spinels* are a prevalent class of minerals that have the general formula AB_2O_4, where A is typically Mg^{2+}, Fe^{2+}, Mn^{2+}, or Zn^{2+} and b is Al^{3+}, Cr^{3+}, or Fe^{3+}. Similar to the perovskites, the spinels are often magnetic. Magnetite, for example, has the formula $Fe^{III}(Fe^{II},Fe^{II})O_4$. This mineral is present in small quantities in the brains of pigeons and is believed to be the primary agent responsible for their uncanny sense of direction.

The mineral spinel, after which this family of compounds is named, has the formula $MgAl_2O_4$. Its lattice consists of a face-centered cubic unit cell of oxide ions, with the Mg^{2+} occupying one-eighth of the tetrahedral holes and the Al^{3+} ions occupying every other octahedral hole. Note that the radius ratio rule did not work in this case—the smaller 3+ cations occupy the larger octahedral holes. The reason for this apparent anomaly has to do with crystal field stabilization energies, a topic that will be left for another chapter. The Al^{3+} octahedrons are edge-sharing, while the smaller Mg^{2+} tetrahedrons shared corners with the octahedrons but are isolated from one another in the lattice, as shown in Figure 12.16.

Example 12-9. Rationalize the scheelite structure of $CaWO_4$ shown in terms of Pauling's first four rules. The ionic radius of Ca^{2+} is 114 pm when O^{2-} is the anion. The W^{6+} ion in this example is tetrahedral.

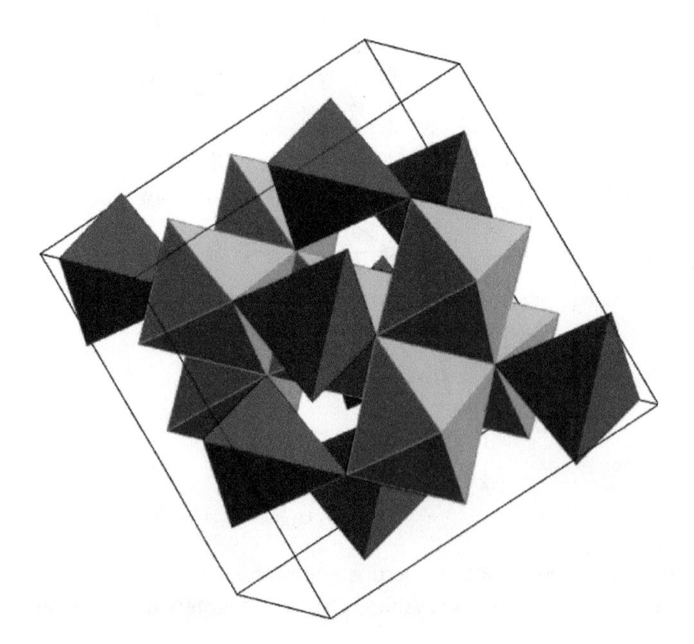

FIGURE 12.16
The structure of spinel, $MgAl_2O_4$, showing how the aluminum octahedrons (shown in blue) are edge-sharing and the magnesium tetrahedrons (shown in red) don't touch each other in the crystalline lattice.

Solution. Using Rule 1, the radius ratio for Ca^{2+} predicts CN = 8 (114/140 = 0.814), which is exactly what we observe in the diagram (note that the polyhedron is hexagonal bipyramidal instead of cubic, however). Rule 2 can be used to calculate the electrostatic bond strengths of each cation: ebs(Ca^{2+}) = 2/8 or 1/4, ebs(W^{6+}) is 6/4 or 3/2. If two W^{6+} tetrahedrons shared an O^{2-} ion, then the bond valence of oxygen would exceed −2, as shown: 2(−3/2) = −3. Thus, the W^{6+} tetrahedrons, which have a very large oxidation state and a small coordination number, must be isolated from one another in the crystal, according to Rule 4 (as observed). The Ca^{2+} hexagonal bipyramids are edge-sharing. Rule 3 states that, in general, edge-shared polyhedrons are less stable than corner-shared ones. This is especially true for polyhedrons with small coordination numbers and high oxidation numbers because of increased cation–cation repulsions. Here, the coordination number is large (eight) and the oxidation number is fairly small (two), allowing the observed edge-sharing of hexagonal bipyramids to occur.

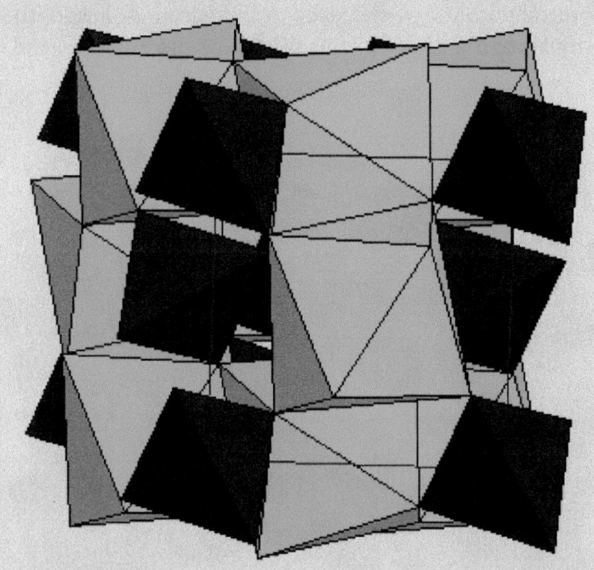

[Joesph Smyth, University of Colorado; reproduced from http://ruby.colorado.edu/~smyth/min/scheelite.html (accessed January 18, 2014), used with permission.]

Rule 5. There are a limited number of different kinds of coordination environments for an ion in a crystal. This rule, also known as the *rule of parsimony*, suggests that there is "beauty in simplicity." It would be highly unlikely, for instance, to observe a crystalline structure where some Al^{3+} ions are tetrahedral and others are octahedral, even though the energy difference between CN = 4 and CN = 6 is relatively small.

12.4 THE SILICATES

The silicates represent the single most common family of minerals found in the earth's crust. They owe their pervasiveness to the fact the elements Si and O comprise more than 70% of the earth's crust by mass (46.6% O and 26.7% Si). As mentioned in the previous section, the basic building block of the silicates is the mesodesmic SiO_4^{4-} tetrahedron. Because Pauling's rules (specifically rules 3 and 4) predict that polyhedrons with low coordination numbers and high oxidation states tend to avoid one another in the crystalline lattice, the SiO_4^{4-} tetrahedrons will either be isolated from one another or exist as corner-shared tetrahedrons. Edge- and face-sharing SiO_4^{4-} tetrahedrons are never observed in the silicates. The silicates can be classified into six different types, according to the manner in which their tetrahedral building blocks are arranged within the crystalline lattice.

The *neosilicates* (orthosilicates) are composed of isolated SiO_4^{4-} tetrahedrons. The mineral olivine, Mg_2SiO_4, is an example. The radius ratio rule for Mg^{2+} and O^{2-} predicts an octahedral coordination for Mg^{2+} (65/140 = 0.46). Thus, the ebs for Mg^{2+} is 1/3, while that of Si^{4+} is 1. In order to satisfy the principle of charge neutrality, each O^{2-} must coordinate to three Mg^{2+} ions and one Si^{4+} ion: $-2 = 3(-1/3) + 1(-1)$. The structure of olivine is shown in Figure 12.17. There are two types of distorted edge-sharing octahedrons for Mg^{2+} (*M1* and *M2*), shown in dark green and red, in addition to the isolated tetrahedrons containing Si^{4+}, shown in blue.

The *sorosilicates* (pyrosilicates) contain two tetrahedrons that are corner-shared as double-isolated islands of $Si_2O_7^{6-}$. This class of silicates is fairly rare. One example, thortveitite, is shown in Figure 12.18.

The *cyclosilicates* (ring silicates) consist of 3- to 6-membered rings of corner-shared tetrahedrons, where each tetrahedron is linked to exactly two others. The formulas of the basic building blocks for the 3-, 4-, 5-, and 6-membered

FIGURE 12.17
The crystal structure of olivine (fosterite), showing the isolated tetrahedrons of the Si atoms (in blue).

FIGURE 12.18
Crystal structure of thortveitite, whose formula is $Sc_2Si_2O_7$. The corner-linked $Si_2O_7^{6-}$ silica tetrahedrons (red) are isolated from each other by the edge-sharing scandium octahedrons (violet).

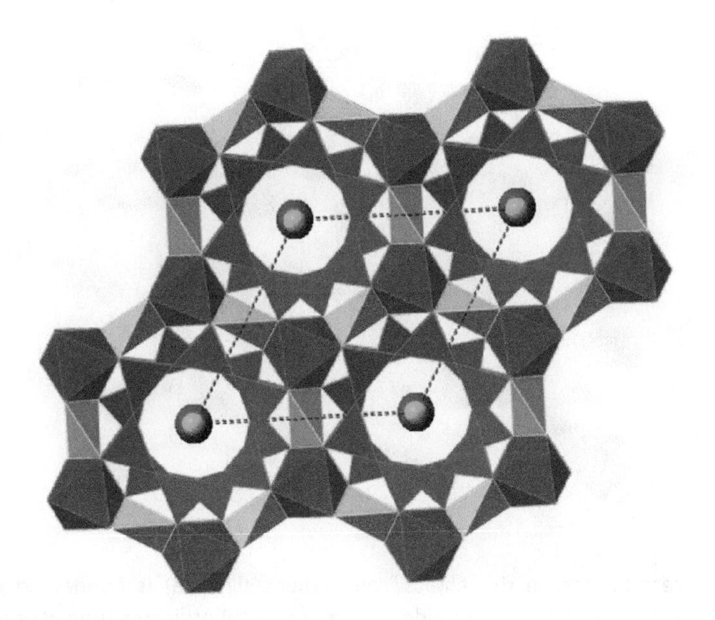

FIGURE 12.19
Crystal structure of beryl, whose idealized molecular formula is $Be_3Al_2Si_6O_{18}$. In this rendition, some of the cations are replaced by Na^+ and Cs^+ ions (shown in green and blue, respectively). The corner-linked Si tetrahedrons are in red forming a ring structure. The Al octahedrons (shown in magenta) and the distorted Be tetrahedrons (shown in light blue) are edge-sharing with one another. It should be noted that the Be^{2+} tetrahedrons can edge-share, while the Si^{4+} tetrahedrons can only corner-share because the larger positive charge on the latter cations results in greater cation–cation repulsion.

rings are $Si_3O_9{}^{6-}$, $Si_4O_{12}{}^{8-}$, $Si_5O_{15}{}^{10-}$, and $Si_6O_{18}{}^{12-}$, respectively. The mineral beryl, $Be_3Al_2Si_6O_8$, shown in Figure 12.19, is but one of many examples. The Al^{3+} ion is six-coordinate, while Be^{2+} is four-coordinate. Thus, the electrostatic bond strengths of Al^{3+}, Be^{2+}, and Si^{4+} are 1/2, 1/2, and 1, respectively. These coordination geometries uniquely define the coordination of each O^{2-} ion to one Al^{3+}, one Be^{2+}, and one Si^{4+} ion using Pauling's second rule.

The *ionsilicates* (chain silicates) are comprised of either single or double chains of linked silica tetrahedrons. The single chains (pyroxene group) have the formula $Si_2O_6{}^{4-}$ and consist of a chain of tetrahedrons that are linked together by two of their corner oxygens. The double chains (amphibole group) have the formula $Si_4O_{11}{}^{6-}$ and consist of two interconnected chains of linked tetrahedrons, where the tetrahedrons are linked together by three of their corner oxygens. Both types of inosilicates are illustrated in Figure 12.20.

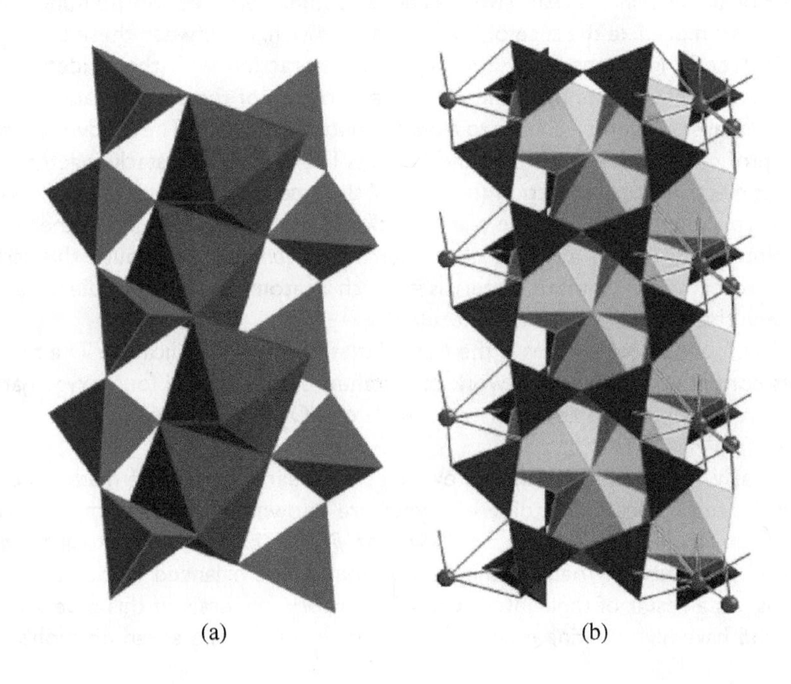

FIGURE 12.20
Single-chain and double-chain inosilicates. (a) Diopside is an example of a pyroxene or single-chain inosilicate, having the molecular formula $CaMgSi_2O_6$. (b) Cummingtonite is an amphibole or double-chain inosilicate, having the formula $(Fe,Mg)_7Si_8O_{22}(OH)_2$. The chains of Si tetrahedrons are shown in blue.

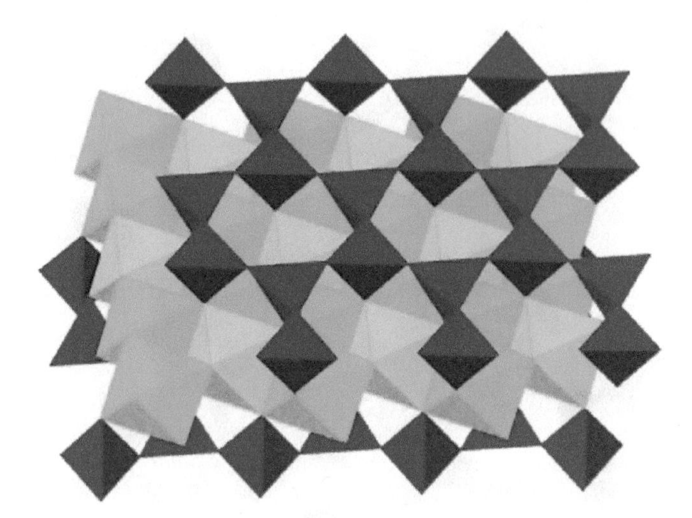

FIGURE 12.21
The structure of the clay talc, whose molecular formula is $Mg_3Si_4O_{10}(OH)_2$, showing the sheets of Si tetrahedrons (blue) separated from each other by sheets of Mg octahedrons (green).

Each tetrahedron in the *phyllosilicates* (sheet silicates) is connected to three other tetrahedrons through an oxide linkage. The phyllosilicates have essentially the same structure as a double chain inosilicate, except that the chains extend into a two-dimensional, sheet-like network, as shown in Figure 12.21. Several important examples of sheet silicates include micas and clays, such as talc and kaolinite. In the clays, such as talc, water molecules are intercalated between the sheets. The water molecules are held in the host by hydrogen bonding and dipole–dipole forces. There are a large number of sites where water can adhere within the sheets, which explains why these materials are so absorbent. While hydrated, clay can be molded into virtually any shape that one desires. When the molecule is heated in a furnace, the intercalated water can often be removed to leave behind a hard and rigid structure.

Biotite, which is also known as *iron mica*, is an iron-containing phyllosilicate that is found in many different types of igneous and metamorphic rocks. The chemical formula of biotite is $K(Mg,Fe)_3(AlSi_3)O_{10}(OH)_2$. Biotite, similar to most micas, is a layered material that is easily cleaved into very thin sheets. The side view of biotite, shown in Figure 12.22, illustrates why this is the case. The mica is composed of two sheets of aluminosilicate that have the apexes of their tetrahedrons pointing toward each other, much like the slice of bread on a sandwich. In between these two layers, the Fe^{3+} and Mg^{2+} ions form a strong ionic interaction with the oxides to hold the layers together. The hydroxide ions are also part of the electrostatic glue that hold two layers together, similar to a peanut butter sandwich. The sandwiches stack on top of one another with only the K^+ ions left to hold the stack together. This weaker electrostatic interaction (because of the smaller positive charge) allows the mineral to be cleaved along the planes between the sandwiches. In another sense, therefore, biotite is an example of an *intercalation compound*, although this term is usually reserved for synthetic materials in which an atom, ion, or molecule is inserted between the layers of the host molecule.

The final class of silicates is the *tectosilicates* (framework silicates). The tectosilicates consist of an infinite network of tetrahedrons, where all four oxygen atoms are shared, giving them an empirical formula of SiO_2. An example of a tectosilicate is shown in Figure 12.23.

Nearly 75% of all the minerals existing in the earth's crust are made up of the tectosilicates. At least nine different types are known to occur. Examples include both high (α) and low (β) quartz, as well as β-cristobalite. The tectosilicates are themselves electrically neutral and do not need to be balanced by other types of cations. As a result of their interlocking framework, minerals of this type are very hard and have high melting points. Quartz, for example, is a seven on Moh's scale

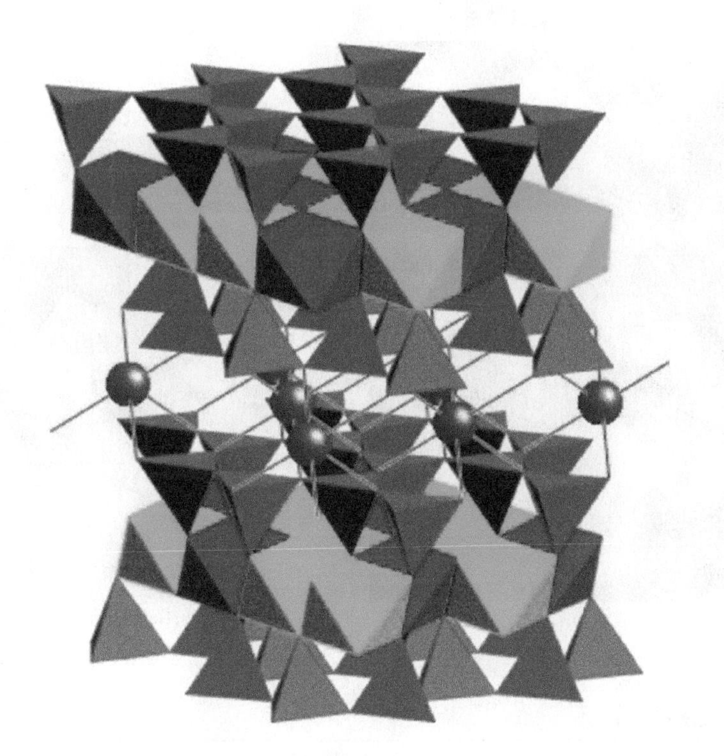

FIGURE 12.22
The structure of biotite, an iron-containing mica. This phyllosilicate is composed of two sheets of silicate tetrahedrons (shown in blue) interspersed with alumina octahedrons that point toward one another and are linked together by Fe^{3+}, Mg^{2+}, and OH^- ions. These double layers are in turn linked together by a weaker interaction with K^+ ions (shown in magenta). The micas cleave fairly easily along these latter planes.

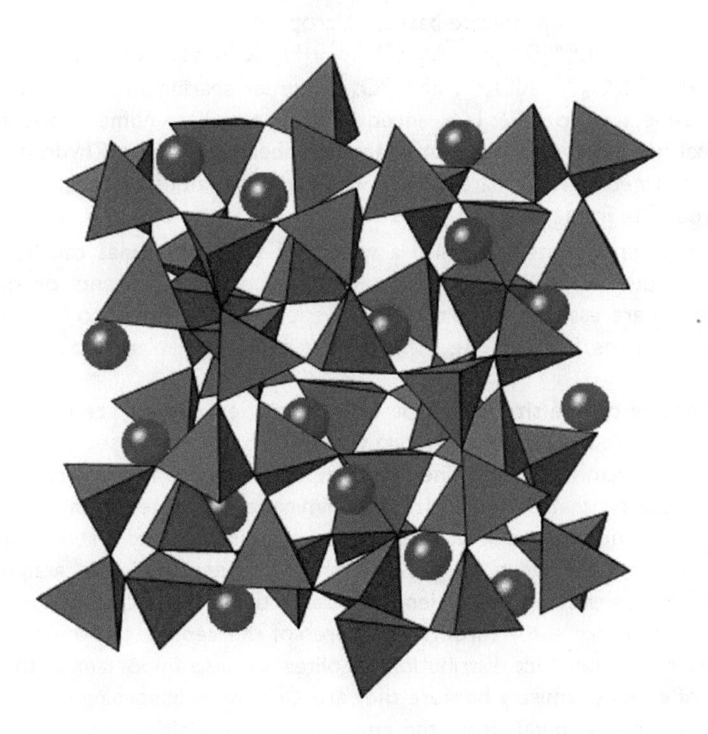

FIGURE 12.23
The basic structure of a tectosilicate is composed of an infinite network of interlocking silicate tetrahedrons. The example shown here is leucite, whose molecular formula is $KAlSi_2O_6$.

of hardness and melts at a temperature of about 1600 °C. Its structure consists of interlocking chains of silica tetrahedrons on a threefold screw axis, as shown in Figure 12.24.

Feldspar minerals are a special case of the tectosilicates, where Al^{3+} octahedrons replace some of the Si^{4+} tetrahedrons. Other cations are therefore necessary for charge balance. The mineral sanidine, which has the formula $KAlSi_3O_8$, is a common example.

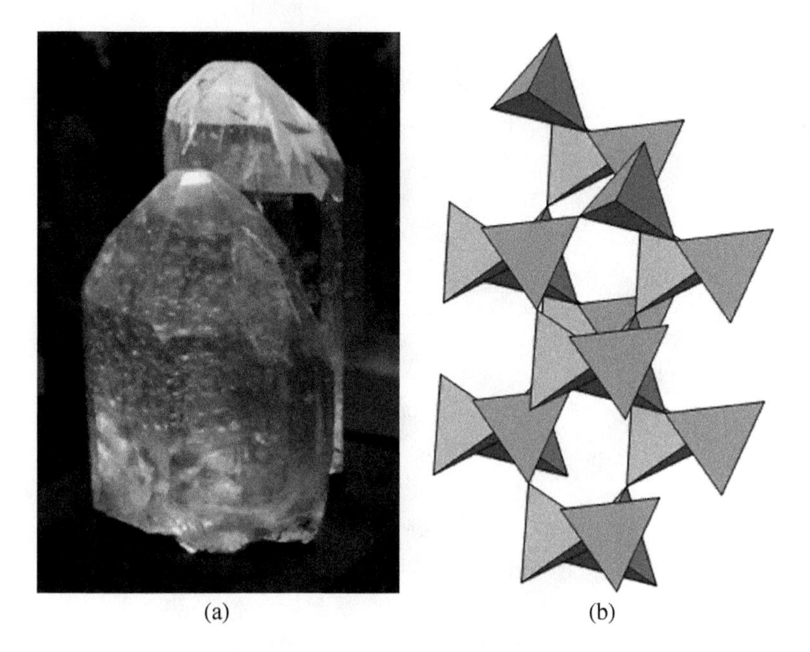

FIGURE 12.24
(a) A large crystal of quartz in the Museum of Natural History in Washington, DC. (b) Crystalline structure of quartz, showing the threefold axis of Si tetrahedrons.

12.5 ZEOLITES

Zeolites are important silicate-based, microporous cage structures having the general formula $M_q^{m+}[Si_{x-y}P_{1-x}Al_{1-y}O_4] \cdot n\ H_2O$. In essence, they consist of a framework of SiO_4^{4-}, AlO_4^{5-}, and PO_4^{3-} corner-sharing tetrahedrons with an oxygen-to-metal ratio of 2 : 1, balanced by an appropriate number of cations for charge balance. They also contain a large number of waters of hydration within their porous network. To date, over 130 different varieties of zeolites have been discovered. The framework structure contains cavities or channels, usually ranging from 3–10 Å in diameter, in which a variety of small molecules can fit. Because of their porous nature, zeolites are often used as drying agents or molecular sieves, which are essentially molecule-sized sponges. They can also be used in ion exchange columns because the M^{m+} cations can freely migrate throughout the structure.

Depending on the size and shape of their cavities, different zeolites can serve as shape-selective or size-selective catalysts. ZSM-5 has been used by ExxonMobil to convert methanol into gasoline. This particular zeolite has a very acidic interior. It is believed that the acid catalyzes elimination of water from the methanol to form a carbene intermediate. This carbene then inserts into a second methanol molecule to form the hydrocarbon. ZSM-5 has also been used as a catalyst for the alkylation of toluene to form xylenes. Because the linear shape of the *p*-xylene isomer can fit more easily through the pores of the zeolite, there is a degree of selectivity to the product distribution. Zeolites are also important in the emerging field of green chemistry because they are efficient at absorbing harmful organics and toxic heavy metals from the environment. In addition, zeolites have been used as less toxic alternatives to phosphates in commercial detergents and water softeners.

The basic building block of zeolites is the β-cage structure formed by interlocking tetrahedrons and shown in Figure 12.25. The different sizes of zeolitic cavities are determined by the ways in which these cages are connected to one another. Sodalite, which is shown in Figure 12.26(a), is a gemstone containing chlorine that is commonly used to make necklaces. Zeolite-A, which is shown in Figure 12.26(b), is used in detergents and as an air pollutant adsorbent.

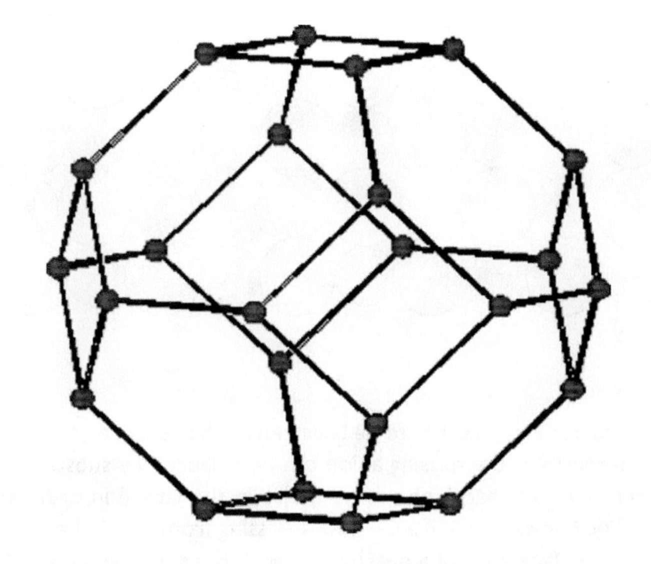

FIGURE 12.25
The β-cage building block of zeolites consists of interlocking 24 interlocking SiO_4^{4-} and AlO_4^{5-} tetrahedrons. Each line represents an oxygen bridge between tetrahedrons. [Reproduced with permission from P. McArdle, NUI, Galway, Ireland from http://www.nuigalway.ie/cryst/oscail_tutorial/zeolite/zeolites.htm (accessed January 14, 2014).]

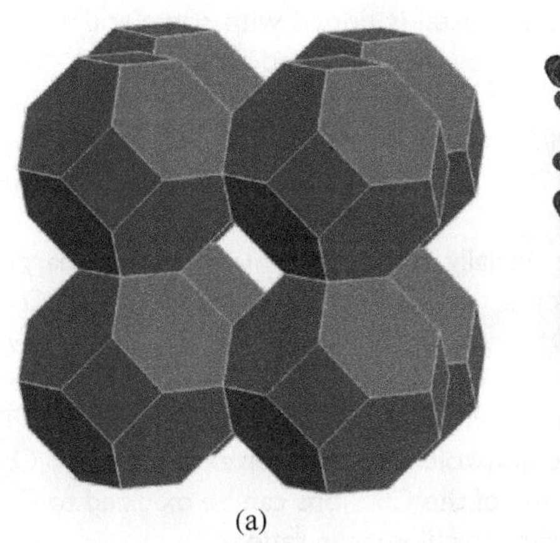

FIGURE 12.26
Structures of (a) sodalite and (b) ZSM-5.

12.6 DEFECTS IN CRYSTALS

The discussion of crystalline lattices has proceeded with the implicit assumption that every crystal is a perfect one. In reality, however, most crystals contain defects. As we have seen previously, whenever two or more gaseous ions combine to make a crystalline solid, the process is favored by enthalpy and a large amount of energy is released as the lattice enthalpy. At the same time, the process of crystallization is entropically unfavorable. If the enthalpy term is greater than the entropy term, the resulting lattice will approach that of a perfect crystal. However, whenever the entropy term is comparable in magnitude with the enthalpy of formation, the resulting solid will necessarily contain defects in its crystalline lattice. Because of the temperature dependence of the entropy term, the number of defects typically increases with temperature.

Several different types of crystal defects can be present; often, more than one type is present at the same time. A *Schottky defect* occurs when one or more atoms in the crystalline lattice are missing. For example, NaCl contains approximately 1 missing ion per 430,000 ions at 1000 K and the crystalline lattice of CrO consists of ~8% vacancies. For an ionic solid, the charges need to be balanced. Thus, a number

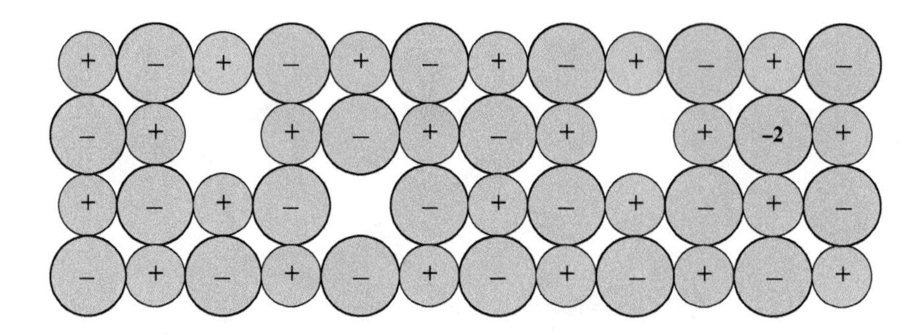

FIGURE 12.27
Two different examples of Schottky defects in a two-dimensional crystalline lattice; a missing cation is balanced by a missing nearby anion, or a missing anion is balanced by another anion of similar size but more negative charge nearby.

of missing anions, for instance, has to be balanced by the absence of an equal number of cations. Alternatively, the missing anion can be balanced by substituting an anion having a larger negative charge somewhere else in the lattice in order to maintain charge neutrality. For example, if a Cl^- ion is missing from AgCl, the charges can be balanced by the replacement of a nearby Ag^+ with the similarly-sized S^{2-} ion. Both examples of Schottky defects are illustrated in Figure 12.27.

Sometimes an electron can become trapped in one of the vacant sites. This might happen in sodium chloride if the crystal is doped with some sodium metal. Some of the sodium metal will get oxidized into Na^+ and an electron. When the electron fills a vacancy where there should have been a Cl^- ion, the result is that the compound can absorb light in the visible region. It is for this reason that this type of crystal defect is known as an *F center* (from the German word "Farbe" which means "color").

In certain types of ionic solids, especially in transition metal oxides, charge balance can be accomplished by forming a nonstoichiometric compound. FeO is an ionic solid of alternating Fe^{2+} and O^{2-} ions. If a Schottky defect is present where one of the Fe^{2+} cations is missing from the lattice, it is possible to compensate for the missing charges by oxidizing two other irons in the lattice from Fe^{2+} to Fe^{3+}. Thus, the resulting empirical formula will be nonstoichiometric; for example, $Fe_{0.95}O$. The same thing happens in CuS, where some of the Cu^+ ions can be oxidized to Cu^{2+} if there is a deficiency of cations from the stoichiometric ratio.

A second common type of defect in crystals is known as a *Frenkel defect.* Frenkel defects occur when one of the ions (usually the smaller ion) becomes displaced from its normal position and occupies an interstitial site in the crystalline lattice. This occurs more frequently when there is a large difference in size between the cations and the anions. For example, the Ag^+ ions in AgBr usually sit in the octahedral holes formed by a face-centered cubic lattice of Br^- ions. However, every so often, one of these Ag^+ ions might find itself displaced to one of the smaller tetrahedral holes in the lattice. In the zinc blende ionic lattice, where every other tetrahedral hole is

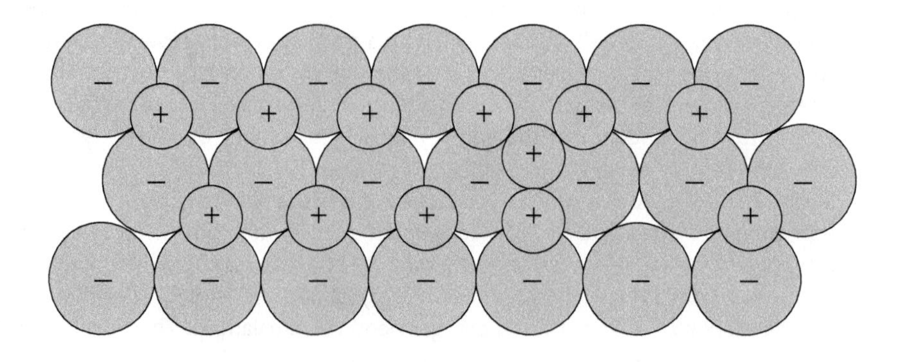

FIGURE 12.28
Example of a Frenkel defect where one of the smaller ions has been displaced to a nearby interstitial site in the crystalline lattice.

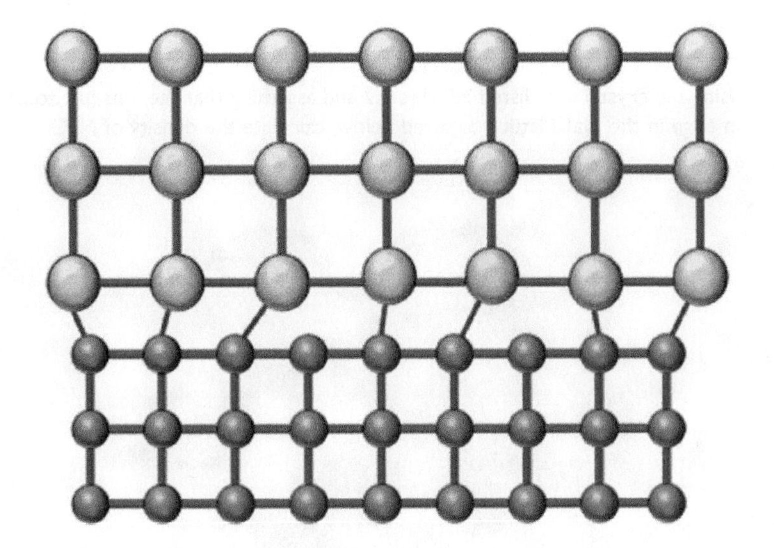

FIGURE 12.29
Example of an edge dislocation in a metallic solid. [Image created by Greg Sun, University of Massachusetts Boston. Used with permission.]

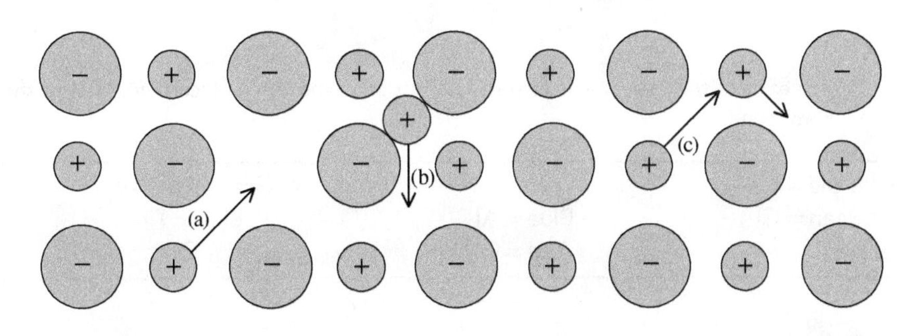

FIGURE 12.30
Illustration of the different mechanisms of ionic conduction in crystalline solids having defects: (a) vacancy migration, (b) interstitial migration, (c) concerted intersticialcy mechanism.

occupied, it is possible that one of the smaller ions in a tetrahedral interstitial site will be displaced so that two neighboring tetrahedral holes are occupied. An example of a Frenkel defect is illustrated in Figure 12.28.

Another type of defect, known as an *edge dislocation*, occurs most often in metallic solids where an extra half plane of atoms inserts itself into the lattice, as shown in Figure 12.29. The point of termination of this half plane is known as the *dislocation*. The presence of a dislocation in a metallic solid makes it more susceptible to deformation. Metals having a large number of edge dislocations, such as lead or white tin, are very malleable. In other substances, such as copper or iron, it is even possible to hammer out the dislocation by mechanical force.

The presence of lattice defects can often lead to some interesting properties in crystalline solids. For example, perfect ionic solids are insulators in the solid state. In order for the solid to become an electrical conductor, there needs to be some mechanism whereby charges can move freely throughout the crystalline lattice. Defects in the crystal can lead to conductivity through one of several mechanisms. Through the vacancy mechanism, ions are allowed to hop from one vacant site to another throughout the lattice, with the conductivity increasing with the number of vacancies. A second pathway for electrical conduction can occur through an interstitial mechanism, where the ions hop from one interstitial site to the next. Often, a combination of the two mechanisms known as the *interstitialcy mechanism* can also take place. The three different mechanisms of ionic conduction are shown in Figure 12.30.

EXERCISES

12.1. Using the crystal radii listed in Table 5.2 and assuming that the ions just touch along an edge in the NaCl lattice pictured below, calculate the density of NaCl in units of g/cm^3.

12.2. For each of the ionic solids pictured here, determine the molecular formula of the compound.

a. Purple = Cd	b. Yellow = Mg	c. Gray = Pb
Green = Cl	Blue = Al	L. blue = Ti
	Red = O	Red = O

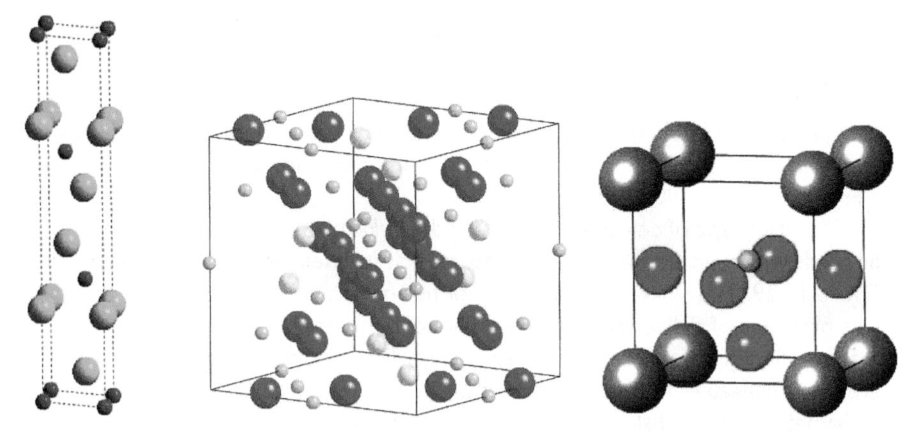

12.3. Rank the following from the most to the least negative lattice enthalpy: NaCl, AlF_3, LiF, $CaCl_2$.

12.4. Knowing that CaF_2 crystallizes in the fluorite lattice, use the appropriate ionic radii from Table 5.2, the proper Madelung constant, the correct compressibility factor, and the Born–Landé equation to calculate the lattice enthalpy of CaF_2.

12.5. Given that the lattice energy of $CaSO_4$ is −2630 kJ/mol and the crystal radius of six-coordinate Ca^{2+} is 114 pm, use the Kapustinskii equation to calculate the thermochemical radius of the sulfate ion.

12.6. Use data from Table 5.2 to determine whether RbI will crystallize in the NaCl lattice or CsCl lattice. Then use the appropriate Madelung constant and compressibility factor in the Born–Landé equation to calculate the lattice energy. Given that the sublimation enthalpy of Rb is 82 kJ/mol and its first ionization energy is 403 kJ/mol, the bond dissociation energy of I_2 (s) is 107 kJ/mol and its electron affinity is 295 kJ/mol, calculate the standard enthalpy of formation of RbI using the Born–Haber cycle.

12.7. Use the Kapustinskii equation to calculate the lattice enthalpies of TlCl and $TlCl_3$ given that the ionic radii of Cl^-, Tl^+, and Tl^{3+} are 167, 173, and 103 pm. Given the following thermochemical data, calculate the enthalpy of formation for TlCl and $TlCl_3$: atomization energy of Tl = 182 kJ/mol, bond dissociation energy of Cl_2 = 242 kJ/mol, electron affinity of Cl = 349 kJ/mol, first ionization energy of Tl = 589 kJ/mol, second ionization energy of Tl = 1971 kJ/mol, and third ionization energy of Tl = 2878 kJ/mol. Notice that TlCl is more stable than $TlCl_3$ (an example of the inert pair effect). In fact, $TlCl_3$ will disproportionate at 40 °C according to the following chemical reaction: $TlCl_3 \rightarrow TlCl + Cl_2$.

12.8. Geometrically prove that the minimum radius ratio for a tetrahedral hole (CN = 4) is 0.225.

12.9. Use the radius ratio rule to predict the coordination geometries of each ion in the following compounds:

a. MgO

b. CaF_2

c. AlN

d. TiO_2 (the Pauling radius for Ti^{4+} is 68 pm)

12.10. In the crystalline lattice of $BaTiO_3$, the ionic radii of the Ba^{2+}, Ti^{4+}, and O^{2-} ions are 149, 75, and 121 pm. Use the radius ratio rule to predict the coordination number of Ba^{2+} and Ti^{4+} in $BaTiO_3$. Then calculate the electrostatic bond strengths of each cation and use this information to determine how many Ba^{2+} ions and how many Ti^{4+} ions each O^{2-} will be coordinated to in the crystalline lattice.

12.11. The crystal structure of andradite (a mineral in the garnet class) is shown here. The chemical formula for andradite is $Ca_3Fe_2Si_3O_{12}$ with ionic radii as follows: Ca^{2+} (114 pm), Fe^{3+} (69 pm), Si^{4+} (40 pm), and O^{2-} (121 pm).

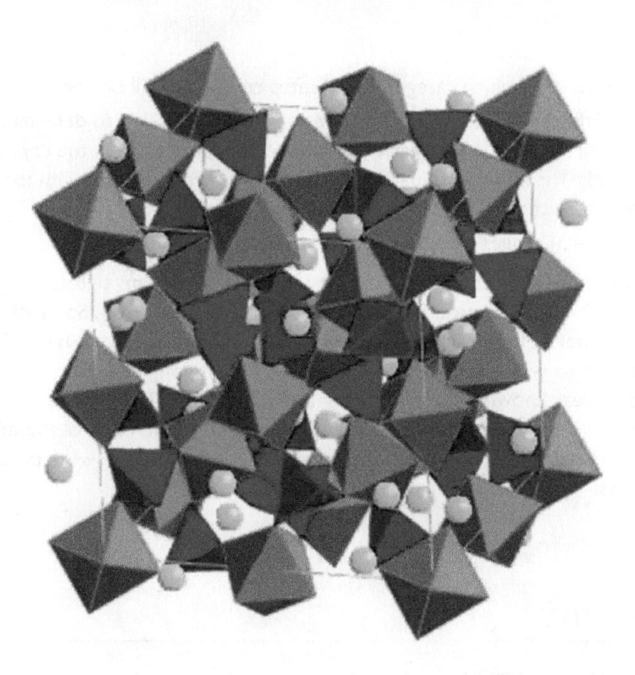

a. Determine to which category of silicates andradite will belong.

b. Calculate the ebs of each cation and use this information to determine how many Ca^{2+}, Fe^{3+}, and Si^{4+} ions each O^{2-} will be coordinated to in the crystalline lattice.

c. Use Pauling's rules to rationalize the structure.

12.12. The crystal structure of cordierite is shown here. The chemical formula for cordierite is $Ca_3Fe_2Si_3O_{12}$ with ionic radii as follows: Mg^{2+} (53 pm), Al^{3+} (86 pm), Si^{4+} (40 pm), and O^{2-} (121 pm).

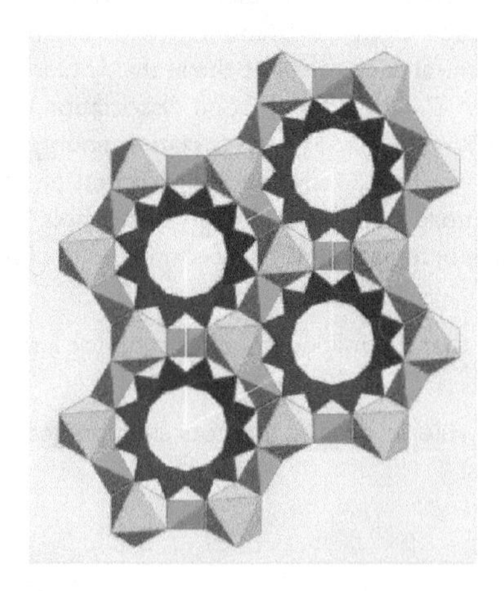

a. Determine to which category of silicates cordierite will belong.

b. Calculate the ebs of each cation and use this information to determine how many Mg^{2+}, Al^{3+}, and Si^{4+} ions each O^{2-} will be coordinated to in the crystalline lattice. Hint: Mg^{2+} is four-coordinate (one of those borderline cases with the radius ratio rule where the lower coordination number is favored).

c. Use Pauling's rules to rationalize the structure.

12.13. The structure of talc, which is a hydrated magnesium silicate, was shown in Figure 12.21. The chemical formula for talc is $Mg_3Si_4O_{10}(OH)_2$. Some of the O atoms at the corners of the silica tetrahedra between the layers contain the OH^- groups. Determine to which category of silicates talc will belong and explain why talc is one of the softest known minerals (having a hardness of only 1 on the Moh scale). It is so soft that it can be scratched with a fingernail. A loose form of the mineral called *talcum powder* is used to prevent diaper rash or to adsorb sweat on the hands of gymnasts.

BIBLIOGRAPHY

1. Douglas B, McDaniel D, Alexander J. *Concepts and Models of Inorganic Chemistry*. 3rd ed. Inc., New York: John Wiley & Sons; 1994.
2. Gillespie RJ, Popelier PLA. *Chemical Bonding and Molecular Geometry*. New York: Oxford University Press; 2001.
3. Housecroft CE, Sharpe AG. *Inorganic Chemistry*. 3rd ed. Essex, England: Pearson Education Limited; 2008.

4. Huheey JE, Keiter EA, Keiter RL. *Inorganic Chemistry: Principles of Structure and Reactivity*. 4th ed. New York: Harper Collins College Publishers; 1993.
5. Krebs H, Walter PHL. *Fundamentals of Inorganic Crystal Chemistry*. New York: McGraw-Hill; 1968.
6. Klein C. *Manual of Mineral Science*. 22nd ed. New York: John Wiley &Sons, Inc.; 2002.
7. Miessler GL, Tarr DA. *Inorganic Chemistry*. 4th ed. Upper Saddle River, NJ: Pearson Education Inc.; 2011.
8. Müller U. *Inorganic Structural Chemistry*. West Sussex, England: John Wiley & Sons, Ltd.; 1993.
9. Rodgers GE. *Descriptive Inorganic, Coordination, and Solid-State Chemistry*. 3rd ed. Belmont, CA: Brooks/Cole Cengage Learning; 2012.

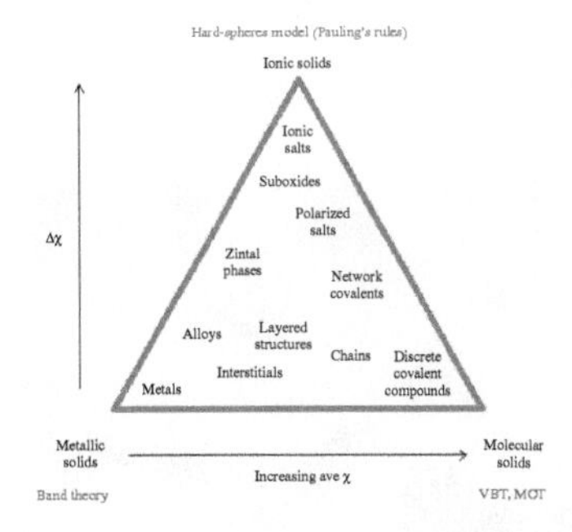

Structure and Bonding 13

"We might say that the description of a bond is essentially the description of the pattern of the charge cloud."

—*Charles Coulson*

13.1 A REEXAMINATION OF CRYSTALLINE SOLIDS

In our previous discussion of the solid state, we showed how crystalline solids can be classified into different crystal systems based on the external symmetries of their crystalline lattices. More traditionally, crystalline solids have been classified into different types depending on the motif that occupies or surrounds the lattice points in the crystal structure. Atomic solids have atoms at these lattice points, ionic solids have ions, and molecular solids have molecules. Atomic solids can be further classified into three categories depending on the type of bonding or intermolecular forces that hold the atoms together. Metallic solids are held together by metallic bonds, network covalent solids by covalent bonds, and Group 18 (noble gas) solids by London dispersion forces. Examples of each type are illustrated in Figure 13.1.

Thus, in a sense, the type of chemical bonding or intermolecular forces holding the particles together in the solid phase can be used as a sorting tool in the classification of crystalline lattices. The physical properties of each type of crystalline lattice are ultimately connected to the different kinds of forces that exist within the crystalline structure, their directionalities, and their relative strengths. In many solids, more than one chemical force will affect its properties. The major difference between the different types of forces is the distance over which the forces can act. Ionic, covalent, and metallic bonding occur over relatively long distances and are thus very difficult to break. The typical strength of a chemical bond is between 200 and 600 kJ/mol. As a result, ionic, network covalents, and most metallic solids have comparatively large melting points. The normal melting points of corundum (Al_2O_3), diamond (C), and tungsten (W) are 1950, 3527, and 3422 °C.

Intermolecular forces, which include hydrogen bonding, ion–dipole, dipole–dipole, ion–induced dipole, dipole–induced dipole, and London dispersion forces, are all short-range in character. These types of chemical forces occur between molecules or between chains and layers in the solid state. As a result, they are typically several orders of magnitude weaker than a chemical bond. As indicated by the data in Table 13.1, the relative strengths of intermolecular forces relate to the distances over which they can act (r). *Ion–dipole forces* occur either when a cation attracts the negative end of a dipolar molecule or an anion attracts the positive

Principles of Inorganic Chemistry, First Edition. Brian W. Pfennig.

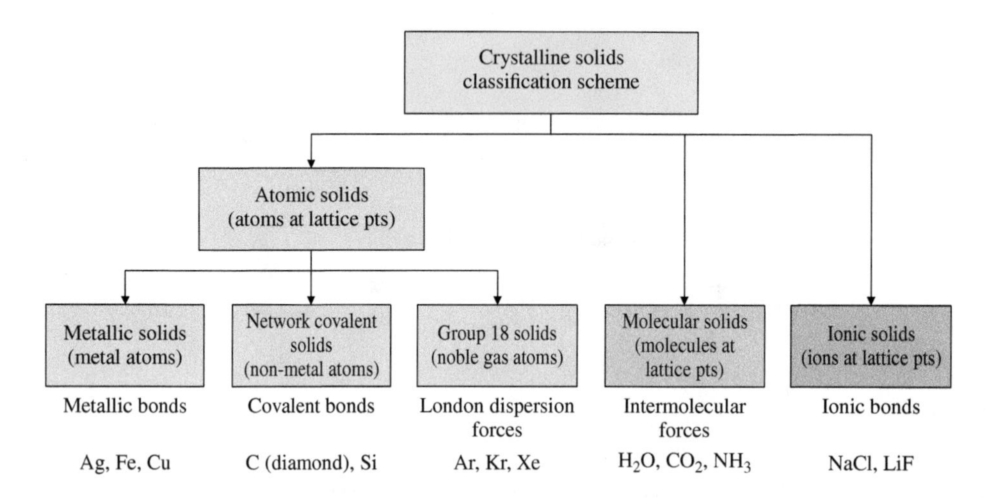

FIGURE 13.1
Classification scheme and properties of crystalline solids.

end. Ion–dipole forces are therefore directional in nature. The potential energy of the interaction is inversely proportional to the square of the distance between the ion and the center of the dipole. *Dipole–dipole forces* are similar, except these occur between the positive end of one polar molecule and the negative end of a second dipole. The energy of interaction is somewhat less than that of an ion–dipole force, both as a result of the diminished partial charge of the atoms in a dipolar molecule as compared with that of a full-fledged ion and because the interaction occurs as the inverse of the cube of the distance between their centers. Hence, while the strength of a typical ion–dipole force is 10–20 kJ/mol, that of a dipole–dipole force is only 1–5 kJ/mol.

Collectively known as *van der Waals forces*, ion–induced dipole, dipole–induced dipole, and London dispersion forces occur over even shorter distances. All of these intermolecular forces involve a temporary dipole moment induced on a nonpolar species by an ion, a polar molecule, or another nonpolar substance. *Ion–induced* and *dipole–induced intermolecular forces* occur when an ion or polar molecule (respectively) causes a temporary polarization of the electron cloud in a neighboring nonpolar molecule. Ion–induced dipoles (1–3 kJ/mol) are stronger than dipole–induced dipoles (0.05–2 kJ/mol) because of the stronger ionic charge and the longer distance-dependence shown in the table. Ion–induced dipole forces occur over slightly longer ranges ($1/r^4$) than do dipole–induced dipole forces ($1/r^6$).

The final category of van der Waals forces is *London dispersion forces*. Dispersion forces occur in atoms or in nonpolar molecules as a result of the transient dipoles that form from random electronic motion. Once a temporary dipole moment is achieved in one of the molecules, it can polarize the electron cloud of its neighbor, which can then polarize the electron cloud of yet another neighboring molecule. The net result is analogous to people doing the "stadium wave." At random, a number of people stand up and raise their arms in the air. This, in turn, causes their neighbors to all stand up. The ripple effect rapidly works its way around the entire stadium. In both cases, the effect is transitory; the people will only stand for a limited period of time just as the polarization of electron clouds is a temporary phenomenon. The only difference between the two is that London dispersion forces operate on a much shorter timescale, typically lasting only 10^{-14}–10^{-15} s! Because of their tenuous and short-lived nature, London dispersion forces are extremely weak (0.05–2 kJ/mol). They are also very short-ranged ($1/r^6$).

TABLE 13.1 Types of chemical forces, their distance dependence, and their average strengths.

Type of Chemical Force	Cartoon Representation	Distance Dependence (r)	Ave Strength (kJ/mol)
Ionic bond	r; (+) (−)	$E \cong \frac{Z^+Z^-e^2}{4\pi\varepsilon_0 r}$	300–600
Hydrogen bond	δ^+ δ^-; (H) . . . (X); X = F, O, N, Cl	Complex	20–40
Ion–dipole	r; (+) (δ^- δ^+)	$E = \frac{\lvert Z\rvert \mu e}{4\pi\varepsilon_0 r^2}$	10–20
Dipole–dipole	r; (δ^- δ^+) (δ^- δ^+)	$E = \frac{-2\mu_1\mu_2}{4\pi\varepsilon_0 r^3}$	1–5
Ion–induced dipole	r; (+) (δ^- δ^+)	$E = \frac{-Z^2\alpha e^2}{2r^4}$	1–3
Dipole–induced dipole	r; (δ^- δ^+) (δ^- δ^+)	$E = \frac{-\mu^2\alpha}{r^6}$	0.05–2
London dispersion force	r; (δ^- δ^+) (δ^- δ^+)	$E = \frac{-2\overline{\mu\alpha}}{r^6}$	0.05–2

The strength of any of the induced dipole forces, including London dispersion forces, will increase with the *polarizability* of the atom or nonpolar molecule's electron cloud. The less tightly the outer electrons are held by the atom's nucleus, the more easily the electron cloud will be influenced by its neighbors. To extend the stadium analogy, the more inebriated the fans, the lower their inhibitions, and the more likely they will participate in doing the wave. Despite their individual weakness, however, the collective strength of intermolecular forces in solids should not be underestimated. While a single tipsy fan might be little more than a nuisance, a whole crowd of them can turn into a raging mob!

Hydrogen bonding, the strongest of all the intermolecular forces, can occur between an acidic hydrogen (one that has a partial positive charge by virtue of its attachment to a highly electronegative element) and the atoms F, O, and N. Sulfur is sometimes also included in this list, although the strength of a hydrogen bond to sulfur is considerably less than that of the others. Not only are N, O, and F atoms among the most electronegative in the periodic table, but they are also among the smallest. This ensures that their charge densities will be very large and that the short-range interaction between them will be fairly strong. Typical lengths of hydrogen bonds are between 270 and 310 pm. The ideal hydrogen bond has a bond angle Y–H···X of 180°, although most examples are actually somewhat less than this. Hydrogen bonding is in many ways a hybrid between ion–dipole or dipole–dipole forces and covalent bonding. A molecular orbital treatment shows that a symmetrical hydrogen bond, such as the one that occurs in FHF^-, can be treated as a pair of electrons occupying a three-centered, delocalized molecular orbital.

The properties of crystalline lattices depend on the strengths of the chemical bonds or intermolecular forces that hold the atoms, ions, or molecules together in the lattice. Metallic solids exhibit a wide range of hardness values, melting points, and densities, depending on their lattice type and the strength of the metallic bond, but they are all electrical conductors. Network covalent solids are generally hard materials with very high melting points and densities because of the strongly directional covalent bonds that hold them together. These solids are typically semiconductors or insulators. Group 18 atomic solids are extremely soft and have low melting points and low densities because they are only held together by weak London dispersion forces. Molecular solids also tend to be soft, with low melting points and densities, as they, too, are bound to each other in the solid only by intermolecular forces. Group 18 and molecular solids are insulators in the solid state. Ionic solids, on the other hand, are hard, with extremely high melting points and large densities as a result of the strong electrostatic attraction and the omnidirectional character of ionic bonding. Most ionic solids are electrical insulators as solids, unless there are a considerable number of defects or imperfections in their crystals. When molten, these materials become conductive and are often used as electrolytes.

Such a classification scheme, however, is overly simplistic and does not accurately represent the many types of intermediate bonding interactions in solids. In many cases, more than one of the traditional types of classification can be applied. Consider, for example, the bonding in graphite, which is shown in Figure 13.2(a). Within the layers, the average C–C bond distance is 142 pm and can best be described as directional covalent bonding (formed by the overlap of sp^2 hybridized orbitals). Between the layers, the C–C separation is considerably longer at 335 pm. Here, the interaction is predominantly van der Waals forces (between the electrons in the unhybridized *p*-orbitals that lie perpendicular to the plane). Graphite is therefore an intermediate case between a network covalent and a molecular solid. Similarly, mercuric bromide, which has the structure shown in Figure 13.2(b) is a hybrid of ionic and network covalent bonding. At first glance, $HgBr_2$ approximates an ionic lattice having the fluorite (CaF_2) structure. On closer inspection, however, the lattice is really composed of a layered structure with a short 248 pm Hg–Br bond distance within the layers and a longer 323 pm Hg–Br interaction between the layers.

13.2 INTERMEDIATE TYPES OF BONDING IN SOLIDS

Ketalaar's triangle of bonding was introduced in Chapter 6 in order to explain the transition between the three primary types of chemical bonding discussed in detail in

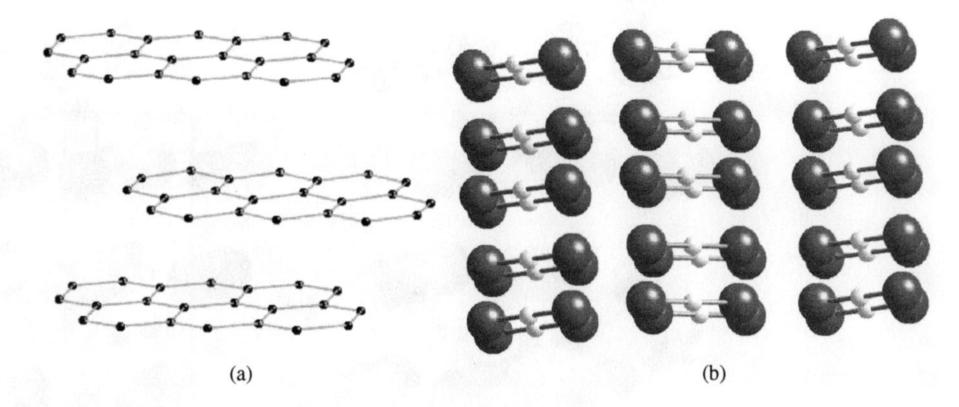

FIGURE 13.2
Crystal structures of (a) graphite and (b) $HgBr_2$.

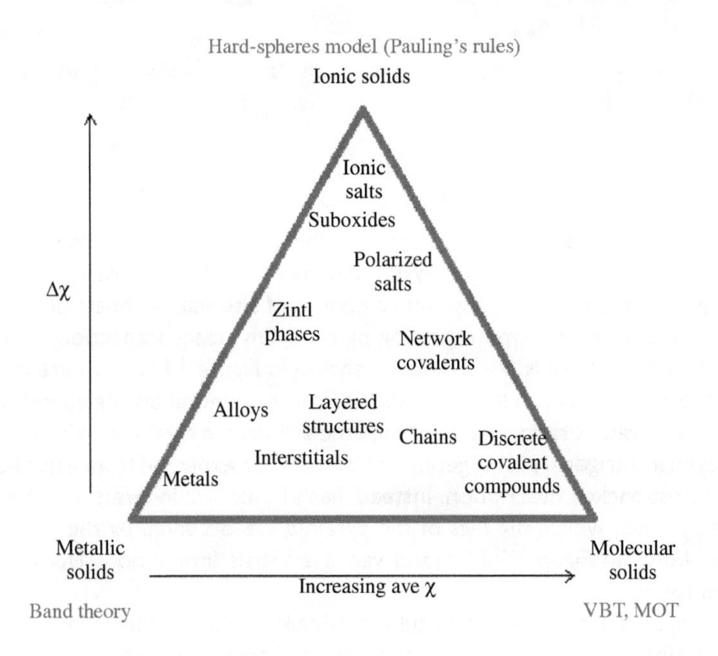

FIGURE 13.3
Bonding triangle applied as a classification scheme for the different types of crystalline solids.

Chapters 10–12. We have previously shown how the introduction of ionic character into the covalent model of bonding can explain many of the unique properties of polar covalent bonds. Alternatively, we can begin with a purely ionic, electrostatic model of bonding and then introduce covalent character using Fajans' rules. The van Arkel triangle of bonding shown in Figure 13.3 extends these concepts into the solid state. As before, metallic, ionic, and covalent (molecular) solids lie at the apexes of the triangle, with a vast gray area of intermediate structures in between them. Others have expanded van Arkel's triangle into a tetrahedron by including intermolecular forces at a fourth apex. Whatever the classification scheme, however, it is readily apparent that a wide array of crystalline structures exists, ranging from ionic salts to interstitials to layered compounds, and that a multitude of chemical forces (both close-range and long-range) are necessarily involved. Thus, employing a completely hard-spheres, electrostatic model to the description of an ionic solid, for example, would be a gross oversimplification and this type of model should not be expected to apply to all or even to a majority of ionic solids. In this section, we examine some of the more nuanced and intermediate types of bonding in crystalline solids.

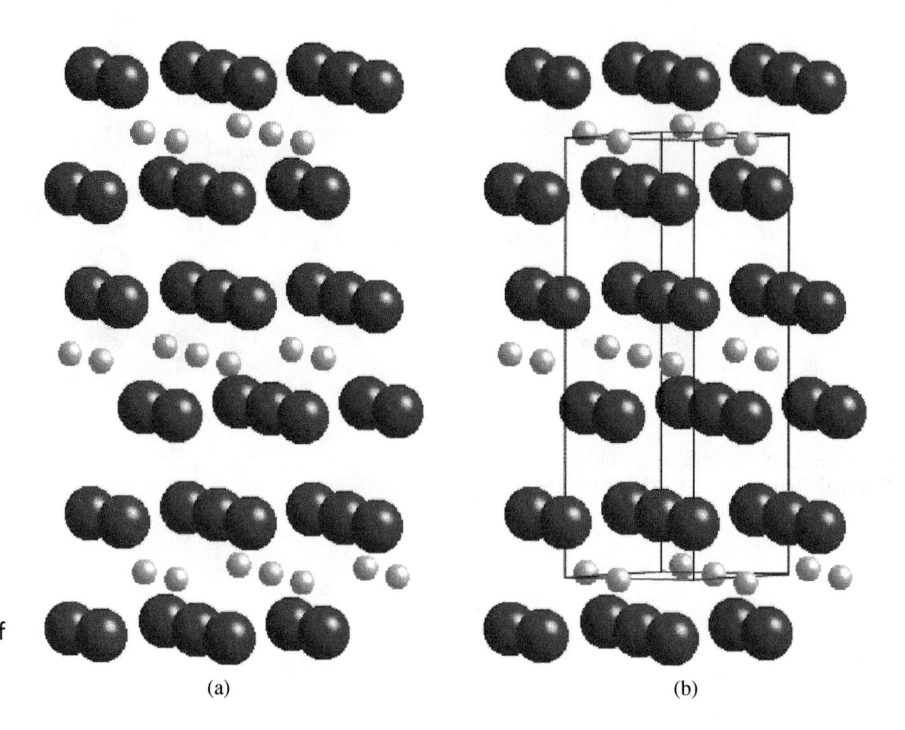

FIGURE 13.4
Single crystal structure of CdI_2 (a) showing only the positions of the ions, and (b) showing the hexagonal unit cell.

First, we consider some structures which line along the ionic-covalent edge of the van Arkel triangle. The single crystal structure of CdI_2 is shown in Figure 13.4(a). At first glance, the structure appears to consist of alternating sheets of I^-, Cd^{2+}, I^- ions arranged in an *ABC* type repeating pattern. On closer inspection, a hexagonal closest-packed unit cell is discernible, as shown in Figure 13.4(b), where each Cd^{2+} has both a coordination number of six and the octahedral arrangement predicted using Pauling's radius ratio rule. But while the I^- ions are three-coordinate, they are not arrayed in a trigonal planar geometry, as might be expected from a hard-spheres, ligand closest-packed description. Instead, each I^- ion sits in layers at the apex of a trigonal pyramid, where the legs of the pyramid are occupied by the smaller Cd^{2+} ions, as shown in Figure 13.5(a), and van der Waals interactions link the I^- ions between layers.

The layered structure is particularly evident in the polyhedral view shown in Figure 13.5(b), where the octahedrons are edge-shared within each layer. The nature of the bonding in CdI_2 is decisively different than that which is expected of a purely ionic solid. Comparing the cadmium iodide lattice to that of rutile, for instance, where the Ti^{4+} ions sit in the centers of corner-sharing octahedrons, the smaller difference in electronegativity in the former structure adds a significant degree of covalent character to the bonding. This mirrors the result predicted by Fajans' rules, where the completely filled *d*-subshell of the Cd^{2+} ion is fairly ineffective in screening the positive nuclear charge on the cation and the large, diffuse nature of the 5*p* valence AOs on the I^- ion result in a significant degree of polarization between the two ions. As the bonding becomes more covalent along the right edge of the van Arkel triangle, it also becomes more directional in nature as it transitions from an ionic solid, to a layered structure, and ultimately to isolated molecular solids. The cadmium iodide structure represents one of those intermediate cases between an ionic solid and a molecular solid. Indeed, the melting point of CdI_2 (387 °C) is significantly lower than its more ionic group congeners (CdF_2 melts at 1110 °C and crystallizes in the antifluorite lattice, while $CdCl_2$ has a melting point of 564 °C and forms a layered lattice similar to CdI_2). The cadmium iodide crystal structure

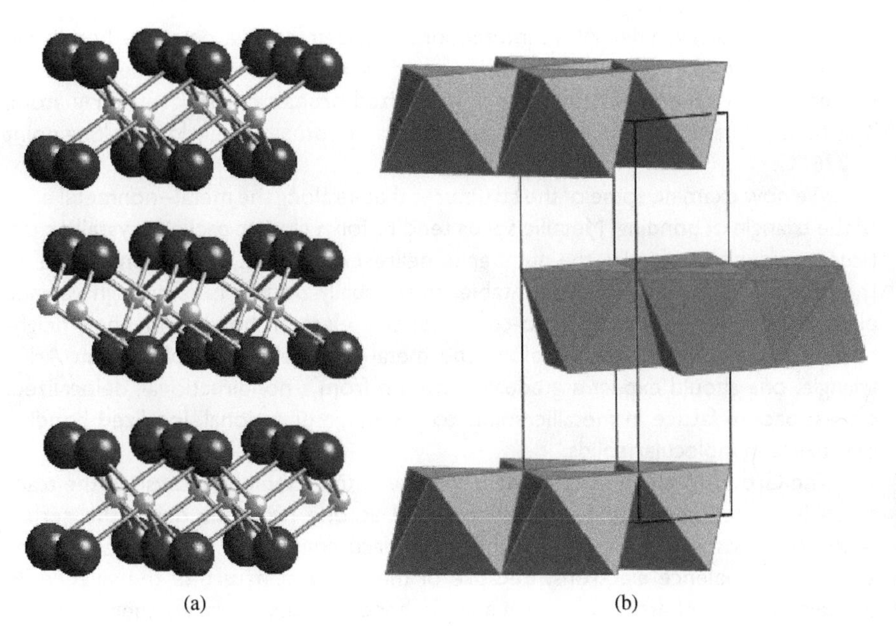

FIGURE 13.5
Single crystal structure of CdI_2 (a) showing its layered structure and (b) in polyhedral view as edge-shared octahedrons.

actually becomes unstable when the bonding is more ionic because of the strong anion–anion repulsions between the layers.

Ionic bonding is characterized by crystal structures having large coordination numbers so as to maximize the attractive interactions between the cations and anions. Covalent bonding, on the other hand, yields highly directional bonds with smaller coordination numbers as a result of the increased orbital overlap caused by sharing electrons. Let us consider the differences between $CaCl_2$, $CdCl_2$, and $HgCl_2$, for example. Because the difference in electronegativity between the atoms decreases across the series, we can expect a gradual transition from more ionic character in $CaCl_2$ to more covalent character in $HgCl_2$. As one proceeds further along the ionic-covalent edge of the van Arkel triangle pictured in Figure 13.3, the increasing covalent nature of the bonding leads to even more directional bonding. Predictably, $CaCl_2$ crystallizes in an ionic lattice, as pictured in Figure 13.6(a), and melts at 772 °C. Cadmium chloride, on the other hand, forms a layered/network covalent structure, as shown in Figure 13.6(b), and melts at 564 °C. Finally, the most covalent congener in the series, $HgCl_2$, crystallizes as a molecular solid, as shown in Figure 13.6(c), with discrete linear Cl–Hg–Cl linked to each other in a chain-like

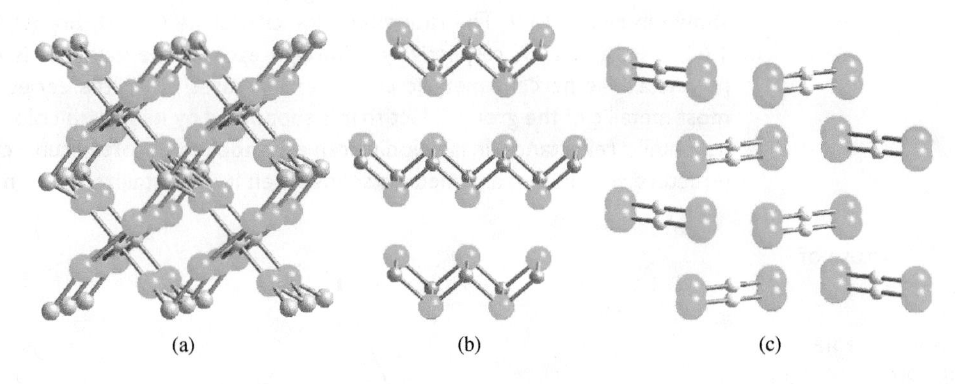

FIGURE 13.6
Crystal structures of (a) $CaCl_2$, (b) $CdCl_2$, and (c) $HgCl_2$, showing the transition from ionic solid to molecular solid.

structure through van der Waals interactions between the Cl atoms. The bonding in this species is more covalent than it is ionic and its linear geometry can best be rationalized using a VB model employing *sp*-hybridized orbitals on the central Hg atom. The covalent nature of $HgCl_2$ is also reflected in its comparatively low melting point of 276 °C.

We now examine some of the structures that lie along the metal–nonmetal edge of the triangle of bonding. Metallic solids tend to form closest-packed crystalline lattices in order to maximize the number of nearest neighbors. Because metals lie on the left-hand side of the periodic table, they usually do not have enough valence electrons to form traditional two-centered, two-electron bonds with their neighbors. Thus, as one progresses along the metal–nonmetal edge of the van Arkel triangle, one should expect a gradual transition from a nondirectional, delocalized, closest-packed lattice in metallic solids to the more directional, localized bonding that exists in molecular solids.

The Group 15 elements provide a representative example because of the transition from nonmetal to semimetal to metal as one proceeds down the series. Nitrogen exists as a diatomic gas under standard conditions. Dinitrogen contains a total of 10 valence electrons. Because of the compact nature of the valence 2*p* orbitals, the two N atoms can form a triple bond in order to satisfy their octets in the diatomic molecule. This is an example of the uniqueness principle, or the second period anomaly, where the chemical properties of elements from the second period differ significantly from elements in the same group that lie lower in the periodic table. The more diffuse 3*p* valence orbitals of P, on the other hand, cannot overlap to form multiple bonds (except under very extreme conditions). Phosphorus exists in several different allotropes, all of which contain single P–P bonds. White phosphorus crystallizes as the molecular solid P_4, where the P atoms form a tetrahedron, as shown in Figure 13.7. Assuming that each P atom is bonded to each of the other three P atoms using conventional two-centered, two-electron bonds, the six P–P bonds in the tetrahedron would utilize 12 of the 20 valence electrons in P_4. The remaining eight valence electrons could be placed as lone pairs on each P atom in order to satisfy their octets and to explain their tetrahedral electron geometries.

The rest of the pnictogens (Group 15 elements) crystallize as layered structures, as shown in Figure 13.8, where there are two different atom–atom distances. The distance to the nearest three neighboring atoms within each layer is defined as r_1 in Table 13.2, while the distance to the three closest atoms in an adjacent layer is defined as r_2. The ratio r_2/r_1 has also been calculated in Table 13.2. The smaller this ratio, the less layering and more metallic the structure will be. It is clear from the data in the table that there is a gradual transition down the series from As to Sb to Bi as the bonding in the elements becomes more metallic and less directional.

A similar trend exists for the halogens. The structure of crystalline iodine is shown in Figure 13.9. The ratios r_2/r_1 for crystalline Cl_2 (s), Br_2 (s), and I_2 (s) are 1.68, 1.46, and 1.29, respectively. Although each of the halogens is classified as a nonmetal, the percent metallic character increases down the series. Iodine is the most metallic of the group, a fact that is supported by its grayish color as a solid and its metallic reflectance. In fact, iodine can even adopt a distorted cubic closest-packed structure similar to many metallic solids when it is crystallized at high pressure.

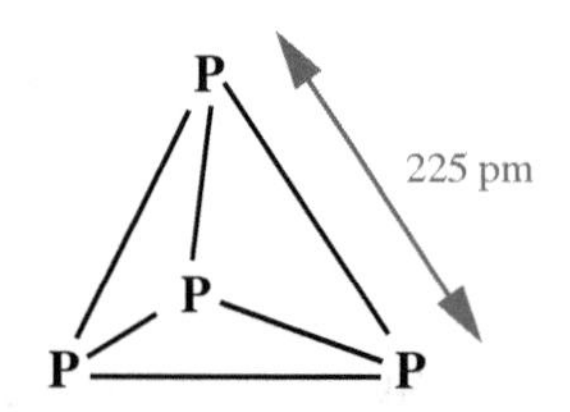

FIGURE 13.7
The tetrahedral structure of white phosphorus, P_4. [Reproduced from http://commons.wikimedia.org/wiki/File:Tetraphosphorus-liquid-2D-dimensions.png (accessed March 13, 2014).]

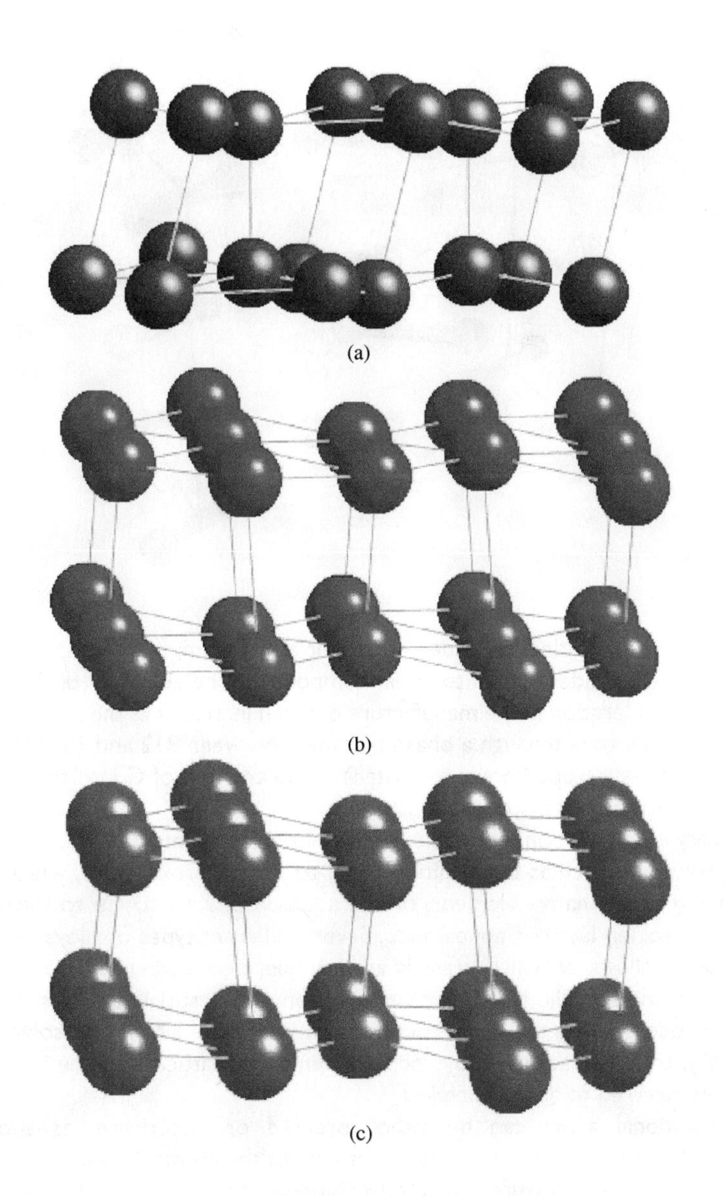

FIGURE 13.8
The crystal structures of (a) As, (b) Sb, and (c) Bi, showing how the bonding becomes less directional down the series.

TABLE 13.2 The different atom–atom distances in Group 15 solids.

Atom–Atom Distance (pm)	α-As	Sb	Bi
Nearest neighbors within a layer (r_1)	252	288	306
Nearest neighbors in adjacent layer (r_2)	312	335	351
Ratio of r_2/r_1	1.24	1.16	1.15

Lastly, we turn our attention to the gray area between metallic and ionic solids. An *interstitial* is a compound that is formed when another atom (typically having about the same electronegativity as the metal) is small enough to occupy some of the interstitial sites in a metallic solid. The elements H, B, C, and N are common examples. For instance, H atoms can occupy the interstitial sites in Pd metal to form compounds having the general formula PdH_n. Interstitials can also be considered as solid solutions, where the smaller atom serves as the solute and the larger

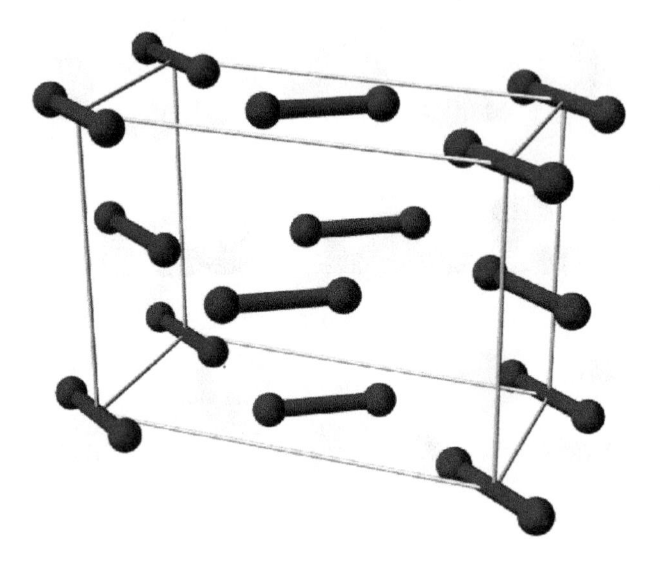

FIGURE 13.9
The crystal structure of I_2 (*s*). [Reproduced from http://en.wikipedia.org/wiki/File:Iodine-unit-cell-3D-balls-B.png (accessed March 13, 2014).]

metal as the solvent. Metal carbides represent another important class of materials that can be considered as interstitial compounds. The solubility of C in Fe (an important consideration in the manufacture of steel) increases as the body-centered α-allotrope of Fe goes through a phase transition between 912 and 1394 °C to the face-centered γ-allotrope known as *austenite*. The addition of C into the Fe lattice helps to strengthen the material.

An *alloy* is a phase consisting of one or more components. Alloys can also exist as solid solutions, such as the addition of Sn to Cu to form bronze, which makes it tougher than the native element, or the addition of Zn into Cu to form brass, which has a golden lustrous appearance. Several different types of alloys can occur. Substitutional alloys, of which brass is an example, occur when the solute atoms substitute in the metallic lattice for solvent atoms. Interstitial alloys, as discussed previously, occur when the solute occupies an interstitial site in the solvent's lattice. Finally, transformational alloys occur when a new lattice is formed; these are sometimes referred to as *intermetallics*.

Substitutional alloys can be either ordered or disordered, as shown in Figure 13.10. Often, their properties are related to the average number of conduction electrons in the mixture. Most substitutional alloys obey the *Hume–Rothery rules*, which state that (a) the atomic radii of the two metals should not differ by more than 15%, (b) the two elements should form the same crystal structure, (c) they should also have similar electronegativities, and (d) the solute atoms will dissolve easier when the solvent has either the same or a higher valence than the solute. Thus, mixtures of Zn and Cu, which have a radius ratio of 1.04 and an electronegativity difference of 0.25 tend to form solid solutions over a wide range of stoichiometries, while the more electronegative As can only form a solid solution with Cu at concentrations up to 6% (atom : atom). Certain element combinations, such as Au and Cu, form ordered lattices having specific ratios. For instance, when the Cu : Au ratios are 1 : 1 and 3 : 1, the resulting structures form a well-defined crystalline phases. The Hume–Rothery rules can be used to determine the minimum radius ratio for the different phase boundaries: 1.36 for fcc, 1.48 for bcc, and 1.69 for hcp. These rules are derived from the band theory of solids and can be successfully applied to a remarkably large number of different substitutional alloys.

The structures of ordered alloys such as AuCu or $AuCu_3$ fill space more effectively than the disordered (noncrystalline) solid solutions that form at higher temperatures. In many ordered alloys, the metal atoms will pack together into dense

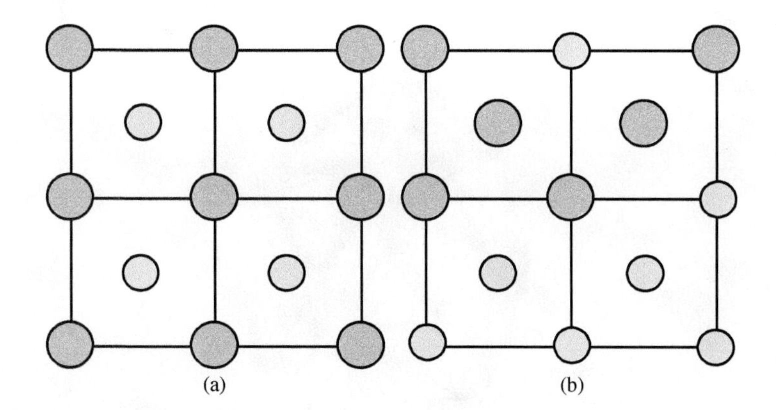

FIGURE 13.10
(a) Ordered substitutional alloy of *AB* and (b) disordered solid solution of *AB*.

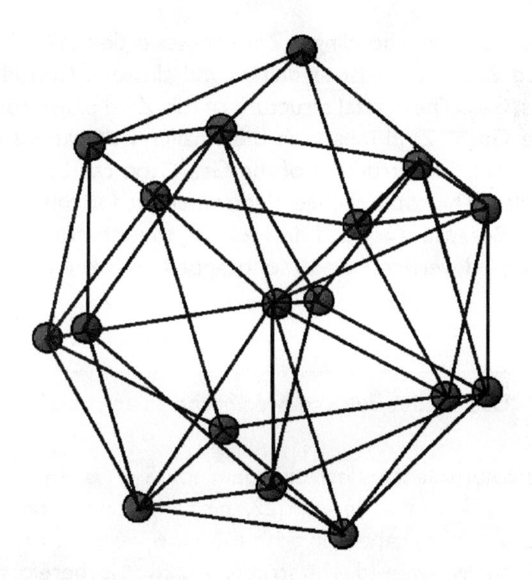

FIGURE 13.11
Friauf polyhedron for $MgCu_2$.

deltahedra known as the *Friauf polyhedra.* These densely packed polyhedra can fit together to fill space in the crystalline solid. The resulting solid is typically denser than the average of the two materials alone. For example, the packing in $MgCu_2$, whose structure is shown in Figure 13.11, leads to a density of 5.75 g/cm^3, which is larger than the average density of the two individual metals (5.37 g/cm^3). This compound is but one example of a more general class of intermetallics known as the *Laves phases.* The Laves phases have an AB_2 stoichiometry and they crystallize in one of three structure types, of which the diamond-type cubic $MgCu_2$ is one example (the Mg atoms form a diamond lattice). The other two types of Laves phases are the hexagonal $MgZn_2$ and hexagonal $MgNi_2$ structures. Over 1400 intermetallic compounds having one of these three structure types are known!

The *Zintl phases* are a class of compounds named after Edward Zintl, who pioneered their exploration in the early 1930s. A Zintl phase is a compound that contains a mixture of Group 1 or Group 2 metals and a cluster of *p*-elements (containing either a metal or semimetal) that are bonded to each other in such a way that they are electronically balanced. Zintl phases lie intermediate in the triangle of bonding between purely metallic bonding found in alloys, where the difference in electronegativity between the metal atoms is small, and ionic solids, which have a larger difference in electronegativity. Zintl phases are typically semiconducting (or weakly conducting), diamagnetic (or weakly paramagnetic), brittle compounds having high melting points.

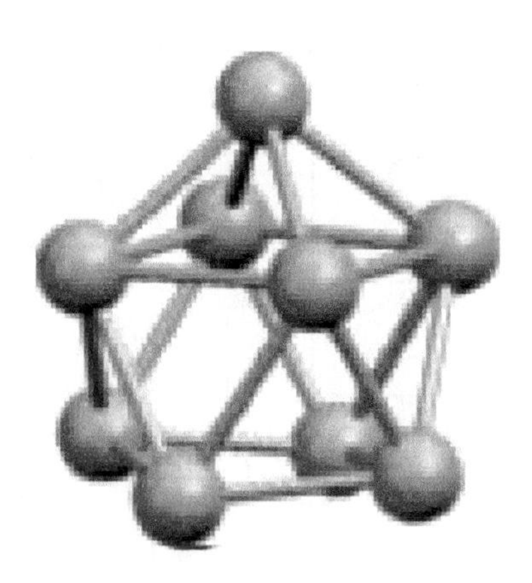

FIGURE 13.12
The structure of the *nido*-Ge_9^{4-} Zintl ion. [Sevov group, University of Notre Dame; reproduced from http://www3.nd.edu/~sevovlab/research.html (accessed March 2, 2014), used with permission.]

As with the boranes, the bonding in Zintl phases is delocalized, and many of the negatively charged Zintl ions exist as deltahedral clusters. Consider, for example, the bonding in Cs_4Ge_9. The crystal structure of this Zintl phase consists of isolated deltahedra of the Ge_9^{4-} Zintl ions with the smaller Cs^+ cations occupying some of the interstitial sites. The structure of the Ge_9^{4-} ion can be predicted using the Wade–Mingos rules. The total valence electron count for this ion is $9(4)+4=40$ valence electrons. Because $n=9$ and $4n+4=40$, the structure is based on a *nido* deltahedron with $n+1$ vertices, or a monocapped square antiprism, as shown in Figure 13.12.

Example 13-1. Use Wade's rules to predict the structure of the S_4^{2+} cluster.

Solution. The total valence electron count for S_4^{2+} is $4(6)-2=22$ valence electrons and $n=4$. Given that $4n+6=22$, this compound belongs to the *arachno* structure based on the deltahedron with $n+2=6$ vertices, or the octahedron with two of its vertices removed. The structure of S_4^{2+} is therefore square planar, as shown here.

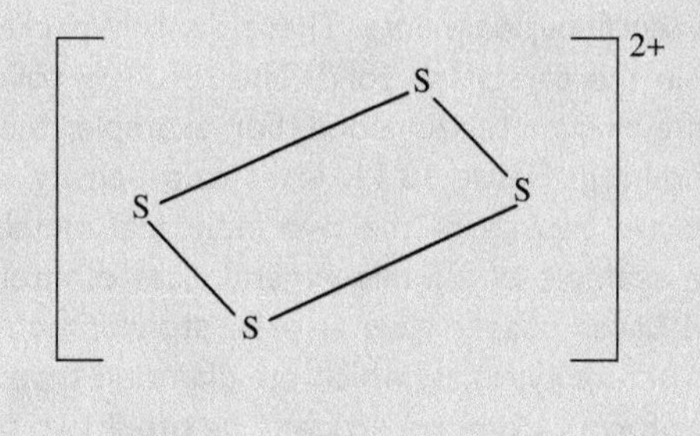

By now, you should have some appreciation for the different types of crystalline lattices and have some idea of the difficulty in predicting the exact structure that a compound will form. There are a variety of factors involved in the process of crystallization, including (but not limited to) closest packing of spheres, closest packing of polyhedra, electrostatic interactions, hydrogen bonding, hybridization, crystal field effects, and the degree of polarization. When it comes to the solid state, simple stoichiometries do not necessarily imply simple structures.

13.3 QUANTUM THEORY OF ATOMS IN MOLECULES (QTAIM)

Because chemical reactions simply involve rearrangements of the ways that atoms are bonded together with each other, an understanding of molecular structure is fundamental to the study of chemical reactivity. While the structure of molecules can be determined by any number of spectroscopic techniques, the most useful method in the crystalline state is X-ray diffraction. As a result of their orderly structure, a monochromatic beam of electrons will be reflected from the surface of a crystal to yield a diffraction pattern. In a single crystal X-ray diffraction experiment, the crystal is usually mounted on the tip of a goniometer, which is rotated with respect to the incident X-ray beam. In order to determine molecular structure, the diffraction pattern is then fit to a three-dimensional map of the electron density in the unit cell. Because the X-rays are diffracted by the electrons, the strongest diffraction will occur from those regions in the crystal having the highest electron density. Typically, the electron density will be highest near the nuclei of the atoms because of the filled quantum shells in the inner core surrounding the nucleus.

Thus, while X-ray diffraction can tell us where the positions of the nuclei are in a crystalline solid, it does not tell us how to connect them. Once the positions of the atoms in the unit cell have been determined, however, we still have to determine how the atoms are actually bonded to each other. In many cases, our chemical intuition and the bonding models developed in the previous chapters can be used to determine where the pairs of valence electrons will lie between the different atoms. On the basis of our knowledge of Lewis structures and the preferred valences of atoms, we can easily "connect" the atoms together to form a molecule using a simple ball-and-sticks approach. Thus, for example, the tetrahedral shape of the methane molecule can be rationalized by drawing a single bond between the central C atom and each of the four H atoms at the apexes of the tetrahedron. The eight valence electrons in the molecule can be apportioned in this manner into four pairs of bonding electrons that are localized between the C and H atoms, providing the "glue" that holds the molecule together. This analysis provides a satisfying result because each H atom will now have two bonding electrons in its sphere of influence (analogous to the noble gas He) and the C atom will have a completely filled octet. We can achieve the tetrahedral arrangement of the H atoms by hybridizing the atomic orbitals on carbon to form four equivalent sp^3 hybrid orbitals using the VB model in order to maximize the orbital overlap with the 1*s* atomic orbitals on the H atoms.

Now let us consider a slightly more challenging example—that of the H_2O molecule. Water also has a total of eight valence electrons. Because of our knowledge of atomic radii, our chemical intuition tells us that we should probably draw a single bond between the central O atom and each of the two H atoms because they are certainly close enough to each other for orbital overlap. However, how do we know that the two H atoms are not bonded to each other? The distance between these atoms is only 151.5 pm. Is this really too far apart for some finite overlap of the two 1*s* atomic orbitals on the two H atoms? It is only our preconceived knowledge of valence which prevents us from drawing a single bond between the two H atoms. Instead, we use VBT to form sp^3 hybrid orbitals on the central O atom. Two of these hybrids are used to form the O–H single bonds, while the other two contain the remaining four valence electrons as lone pairs. This is also a satisfying result because it explains the bent molecular geometry of the molecule. In fact, the H–O–H bond angle can even be rationalized as less than the ideal 109.5° of a perfect tetrahedron because of repulsion between the two lone pairs.

FIGURE 13.13
Polychromic electron density distribution in the H_2O molecule.

But is this necessarily a true picture of the bonding in H_2O? In actuality, *ab initio* calculations on the water molecule are inconsistent with the implied directional nature of the lone pairs. Instead, the negative charge on the O atom is more evenly spread out over the entire area where the lone pairs were expected to be, as illustrated by the polychromic electron density distribution shown in Figure 13.13. This result is supported by the observation of the nearly trigonal planar hydrogen bonding found in the restricted sites in the hydration of proteins, suggesting that sp^2 hybridization with an unhybridized $2p_z$ AO might provide a better description than the traditional sp^3 hybrid orbitals shown in most general chemistry textbooks.

The question of where to draw the bonds becomes especially complicated in molecules such as diborane, which do not conform to the conventional two-centered, two-electron bonding usually implied by a Lewis structure. Diborane has a total of 12 valence electrons. Molecular orbital theory indicates that the B–H terminal bonds are of the traditional type and account for 8 of the 12 valence electrons. Each B atom has one remaining valence electron that it can use in its bonding and each of the bridging H atoms also contributes one valence electron. Molecular orbital calculations are consistent with a three-centered, two-electron B–H–B bridge. The electron deficient three-centered bonds are weaker than two-centered, two-electron bonds, as evident by the bond lengths (119 pm for the terminal B–H bonds vs 131 pm for the B–H bridges) and ν(BH) stretching frequencies (2500 and 2100 cm^{-1} for the terminal and bridging B–H vibrations, respectively).

One interesting consequence of the bonding in diborane is the less-than-expected H_{br}–B–H_{br} bond angle. If all four of the bonds to each B atom were of the conventional type, one would expect this bond angle to be approximately 109.5°, consistent with VSEPR theory. However, the observed H_{br}–B–H_{br} bond angle is significantly less than this at only 97°. One plausible explanation for this is that the two bridging H atoms are close enough to form a weak H···H bonding interaction of their own.

As the given example illustrates, the concept of a chemical bond, which is so fundamental to our understanding of chemistry, is in fact a nebulous one. How, then, should we define a chemical bond? One modern approach derives from the quantum theory of atoms in molecules (QTAIM or simply AIM), which was developed by Bader and his colleagues in the 1970s and paralleled the increasingly sophisticated development of computational quantum mechanical methods. The AIM approach is a topological method that examines the electron density distribution of atoms in molecules as determined from X-ray crystallography. While the wavefunction for a polyatomic molecule cannot be solved exactly, the electron density distribution is in fact an observable quantity. Furthermore, the Hellman–Feynman theorem tells us that the effective force acting on a nucleus in a molecule can be calculated from classical physics as the sum of all the Coulombic interactions with the other nuclei and the electron density distribution $\rho(r)$.

The topology of $\rho(r)$ for the diborane molecule in the plane containing the terminal H atoms is shown in Figure 13.14(a). The greatest electron density lies near the two B nuclei because of the inner core electrons and has been truncated in the figure. There are smaller peaks around the four H nuclei. As a result, the topology of $\rho(r)$ can be used to determine the relative positions of each atom in the molecule, which reflects the conventional application of X-ray crystallography. A two-dimensional contour map for the diborane molecule is shown in Figure 13.14(b) in the plane of the terminal H atoms. The different rings reflect the magnitude of the electron density in the same way that the contours on a topographical map of geological terrain represent the altitude above sea level. The electron density is a constant along any of the wavy solid lines in the two-dimensional contour map. The steepest regions, similar to those on a map, occur where the contour lines are more closely spaced.

Now imagine that you were at sea level and you began hiking to higher and higher altitudes. Being the intrepid college student that you are, at every turn you decide to take the steepest possible climb. In the field of topology, this path is known as the *gradient path*. The gradient of the electron density ($\nabla\rho(r)$) is simply a vector that points in the direction of the steepest ascent. The gradient path is therefore always perpendicular to the contour line. Notice that there are two types of gradient paths in Figure 13.14(b). The first of these connects the two lowest points around the nucleus and defines what is known as the *atomic basin*. The atomic basin essentially defines the outer limits of the nuclear attractor. Notice that the atomic surface boundary of a B atom in the diborane molecule is shared with the surface boundary of the two terminal H atoms, but that it does not share a surface boundary with the other B atom. This is an indication that the two B atoms are not directly bonded to one another. The second type of gradient path begins and ends at the nuclei and links pairs of neighboring nuclear attractors.

Every so often along one of the gradient paths, you will encounter what is known as a *critical point*. A critical point is an extremum in a function where the curvature is such that $\nabla\rho(r) = 0$. A two-dimensional function has three types of critical points: maxima, minima, and saddle points. Clearly, a critical point exists at the position of each of the nuclei, because these are the regions of maximum electron density. However, it is the saddle points in between the nuclei with which we shall be most interested. It is the partitioning of the electron density in these regions to the individual nuclei that helps us to determine where the chemical bonds might lie. The line of maximum charge density connecting any two nuclei is known as the *bond path*. The presence of a bond path is a necessary, but not a sufficient, condition indicating that a chemical bond between the two nuclei is likely. If a critical point lies along the molecular axis between any two nuclei, it is known as a *bond critical point*, ρ_b. This represents the point at which the shared electron density between the atoms reaches a minimum. The bond critical points for the diborane molecule in the plane of the terminal H atoms are indicated in Figure 13.14(b) as dots.

The relief and contour electron density maps for diborane in the plane of the bridging H atoms are shown in Figures 13.15(a) and (b), respectively. One notable feature here is that the atomic surface boundaries for the two bridging H atoms actually touch one another. Although there is no direct bond path between the two bridging H atoms, the larger value of the electron density in this plane can be used to explain the fact that DFT calculations indicate that the H$\cdots$H interaction

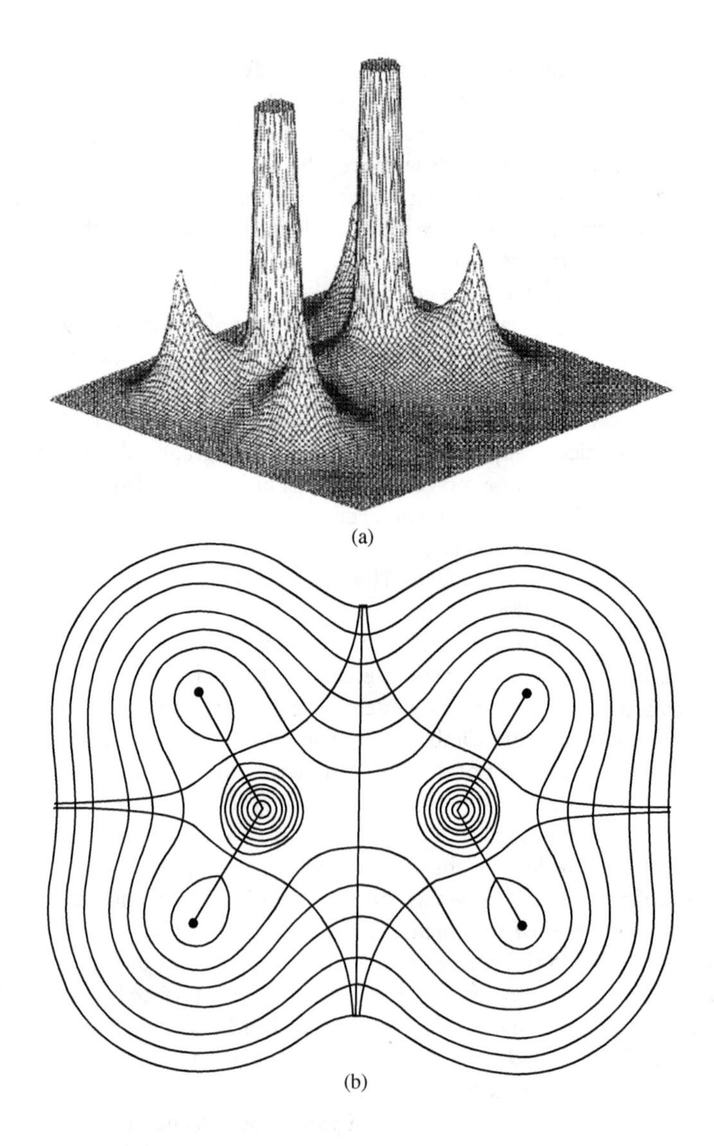

FIGURE 13.14a and b
Electron contour diagram of diborane in the plane of the terminal H atoms. [Reproduced from *J Amer Chem Soc* **1979**, *101*, 1389–1395. Used with permission of the American Chemical Society.]

between the two bridging H atoms is approximately three times stronger than the B $\cdots$ B interaction. This helps to explain the smaller-than-expected H_{br}–B–H_{br} bond angle. Furthermore, an examination of the bond paths in Figure 13.15(b) illustrates an inward curvature, which also supports the assertion of a secondary H $\cdots$ H interaction. The observation of bond paths in a wide range of unusual molecules using the AIM method has helped to address a number of unresolved issues about chemical bonding and even hydrogen bonding which were unable to be explained by the more conventional bonding models discussed in the previous three chapters.

In the case of a homonuclear diatomic, the bond critical point will lie exactly halfway in between the two nuclei, such that the distance between the nucleus and the bond critical point can be defined as the covalent radius of the atom. In a heteronuclear diatomic, the position of the bond critical point is an indicator of the degree of polarity in the bonding. In fact, the magnitude of ρ_b is reflective of the degree of electron density built up in the bonding region. Covalent bonds will therefore have the highest values of ρ_b because of the increase in electron density between the two nuclei, while a pure ionic bond will have $\rho_b = 0$. Thus, the magnitude of ρ_b can be used as a measure of the degree of ionic character in a chemical bond.

Another indicator of the percent ionic character in a chemical bond has to do with the atomic charge in a molecule. We have already shown how the different

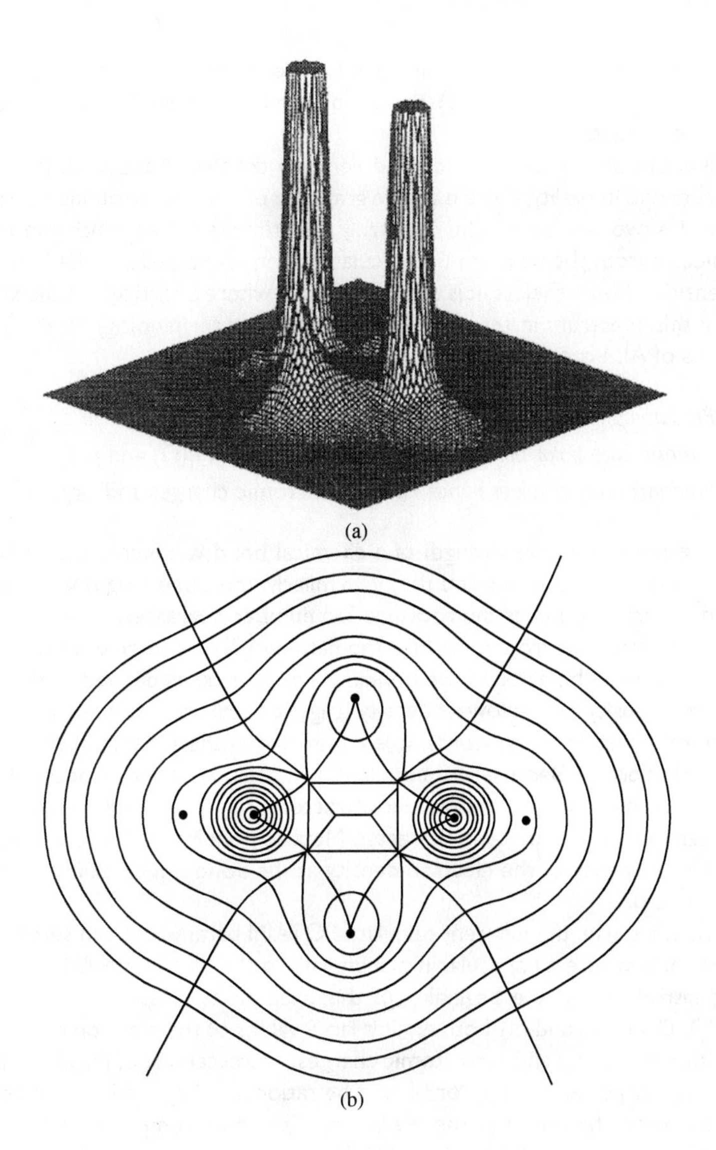

FIGURE 13.15a and b
Electron contour diagram of diborane in the plane of the bridging H atoms. [Reproduced from *J Amer Chem Soc* **1979**, *101*, 1389–1395. Used with permission of the American Chemical Society.]

gradient paths can be used to map out the atomic basin for an atom in a molecule. Although these shapes can often be rather irregular, in principle, it is possible to calculate the atomic volume as the integral of all the volume elements over the atomic basin. The electron population $N(\Omega)$ can then be obtained by integrating the density of a volume element $d\tau$ over the atomic basin Ω, as shown in Equation (13.1):

$$N(\Omega) = \int_{\Omega} \rho \, d\tau \qquad (13.1)$$

The atomic charge $q(\Omega)$ can then be calculated by subtracting $N(\Omega)$ from the atomic number Z, according to Equation (13.2):

$$q(\Omega) = Z - N(\Omega) \qquad (13.2)$$

The atomic charge will always be intermediate between the oxidation number of the atom (which is based on a purely ionic bond) and its formal charge (which is based on a purely covalent bond). Thus, for example, the $q(\Omega)$ calculated using the AIM model for the F atom in HF is −0.7073 (which is intermediate between its oxidation number of −1 and its formal charge of 0). In contrast, $q(\Omega)$ for the Cl atom in HCl is only −0.2420. Because this value lies closer to the formal charge of Cl (0) than to its oxidation number (−1), the bonding in HCl is significantly more covalent than the bonding in HF.

The entire notion of an ionic bond versus a covalent bond is, in fact, an antiquated concept. In reality, there is a huge gray area of polar compounds intermediate between the two extremes. The only truly quantitative values which can delineate one molecule from the next are the calculated atomic charges and the bond critical point densities. Nevertheless, it is useful to discuss where a particular molecule might fall along this spectrum in terms of the more familiar terminology. In this context, the results of AIM can be summarized as follows:

1. *Predominantly ionic bonds* have large atomic charges and small ρ_b.
2. *Intermediate polar bonds* have moderate values of $q(\Omega)$ and ρ_b.
3. *Predominantly covalent bonds* have small atomic charges and large ρ_b.

As a general rule, the strength of a chemical bond will increase with both the charge on the bonded atoms and the ρ_b. Similarly, the bond length will shorten as $q(\Omega)$ and ρ_b increase and as the coordination number decreases.

A third parameter from QTAIM—the flatness (f%)—can be used to gauge the degree of metallic character in the bonding. The flatness is defined as the ratio of the electron density on the lowest density (cage) critical point to the highest density bond critical point. In other words, it is a measure of the flatness of the terrain of the electron density. Because the bonding in metallic solids is delocalized—a sea of electrons punctuated by the nuclear attractors, the flatter the electron density, the greater the percent metallic character. Metallic bonding is also characterized by relatively small values of the electron density at the bond critical points and atomic charges of zero.

Thus, in a sense, the different calculated QTAIM parameters can serve as a way of determining where a particular molecule falls in the classic van Arkel triangle of bonding introduced at the beginning of this chapter, as illustrated by the data in Table 13.3. Diamond and N_2 both exhibit large values of the electron density at the bond critical point (ρ_b) and zero atomic charges, characteristic of primarily covalent bonding. The larger value of ρ_b for N_2 can be rationalized by the increased electron density between the nuclei in the N≡N triple bond as compared with the singly bonded C atoms in the diamond crystal. The slight difference in the flatness (f%) between the two reflects the network covalent nature of the bonding in diamond as compared with the molecular solid N_2. On the other hand, CaF_2 and LiO_2 have small values for ρ_b, indicating that the electron density reaches a minimum at the bond critical point and the two atoms can be considered as approximately closed-shell ions of opposite charge. Furthermore, the atomic charges on the ions approach those values representative of their oxidation states. Notice also that the bonding in Li_2O is slightly more metallic ($f\% = 7.3$) than it is in CaF_2 ($f\% = 1.3$). This result is also consistent with the smaller average electronegativity of the atoms in Li_2O than in CaF_2. The bonding in the Al and Li examples is predominantly metallic in character, having relatively small values for ρ_b and $q(\Omega)$, but notably larger values for the flatness (f%). As expected on the basis of their electronegativities and relative positions in the periodic table, Li has a greater percent metallic character ($f\% = 89.2$) than Al ($f\% = 56.8$). The nonzero atomic charge in Li is a relatively unusual feature in metallic bonding and is reflective of the nonnuclear maxima (NNM) of the electron density in the Li crystal.

In conclusion, many of the bonding models that we examined in previous chapters rely on circular arguments, to a given extent, because we define the radius of an atom depending on its molecular context. Thus, the radius ratio rule for ionic solids was self-consistent only in the sense that it relied on the ionic radii

TABLE 13.3 Topological parameters calculated using QTAIM for a variety of crystalline solids.

Crystal	AB	ρ_b	q(A)	q(B)	f, %
Diamond	C–C	0.2659	0.000	NA	4.8
N_2	N–N	0.7380	0.000	NA	0.0
CaF_2	Ca–F	0.0297	1.821	−0.911	1.3
Li_2O	Li–O	0.0278	0.897	−1.792	7.3
Al	Al–Al	0.0308	0.000	NA	56.8
Li	Li–NNM*	0.0071	0.825	NA	89.2

*NNM, nonnuclear maxima (see text for definition).

determined by X-ray crystallography of a large number of ionic crystals. Likewise, in the LCP model of molecular geometry, the ligand radii depended on the nature of the central atom because of the extent of sharing of the electron density in the bonding. The percent ionic character in polar covalent bonds depended on the difference in electronegativity between the two atoms, but the electronegativity itself was determined from the ionic resonance energy in polar covalent bonds. Thus, in a certain sense, our understanding of chemical bonding is almost entirely empirical, lending credence to the chapter's opening quotation that "We might say that the description of a bond is essentially the description of the pattern of the charge cloud." The quantum theory of atoms in molecules is simply a computational way of extracting useful quantitative parameters from the experimental pattern or topology of the electron density. In the next chapter, we shall see how the structure of molecules (and specifically the making and breaking of chemical bonds) is integrally related to a molecule's reactivity.

EXERCISES

13.1. For each of the crystal structures listed here, indicate whether it is a metallic, Group 18, network covalent, ionic, or molecular solid and identify the Bravais lattice.

a. Perovskite ($CaTiO_3$)

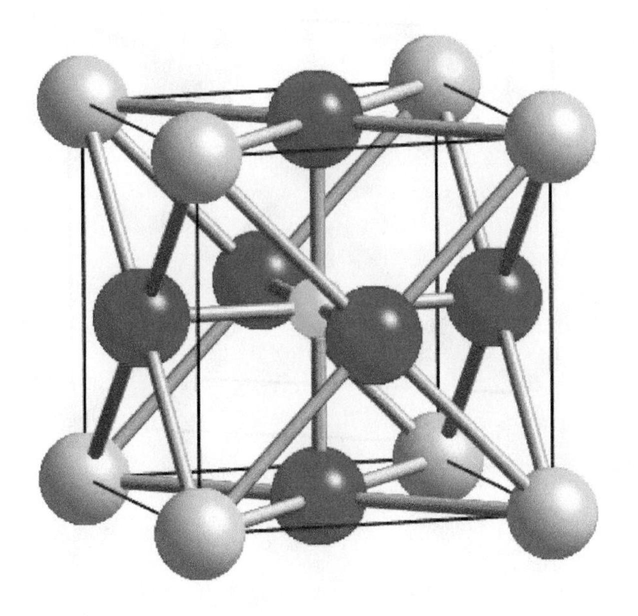

b. Buckminster fullerene (C_{60})

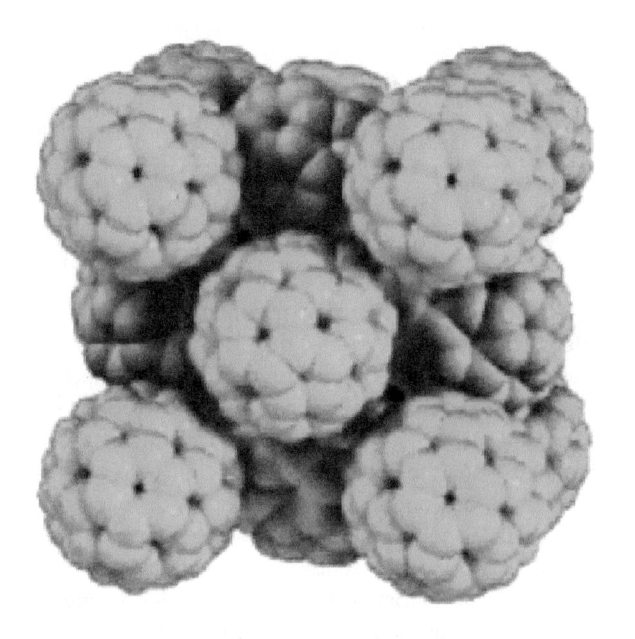

[Reproduced with permission of Jianwen Jiang.]

c. Ruthenium (IV) oxide (RuO_2)

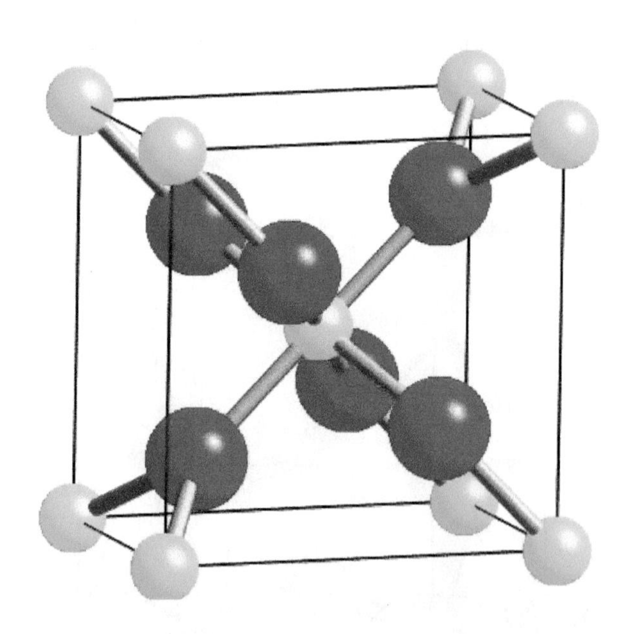

d. YCo_5

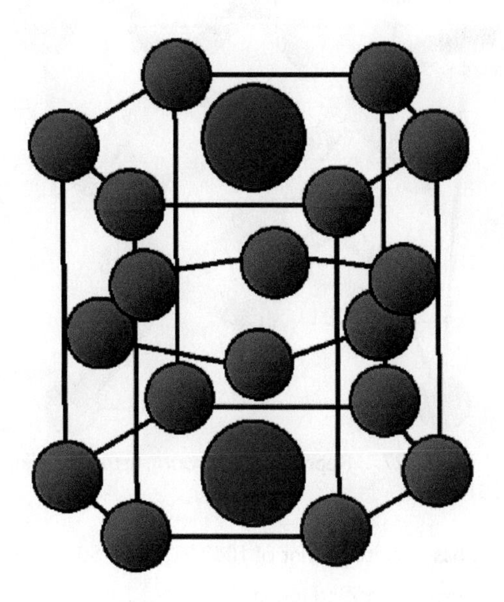

e. $YBa_2Cu_3O_7$ superconductor

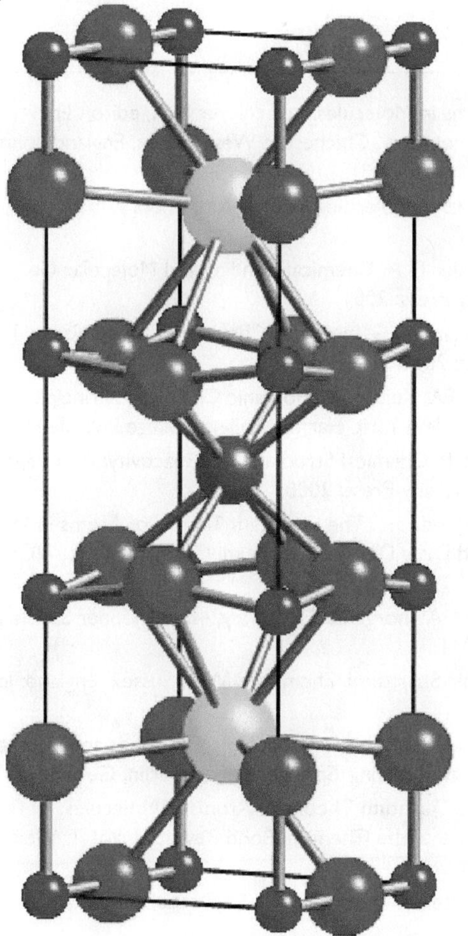

f. Silicon (Si)

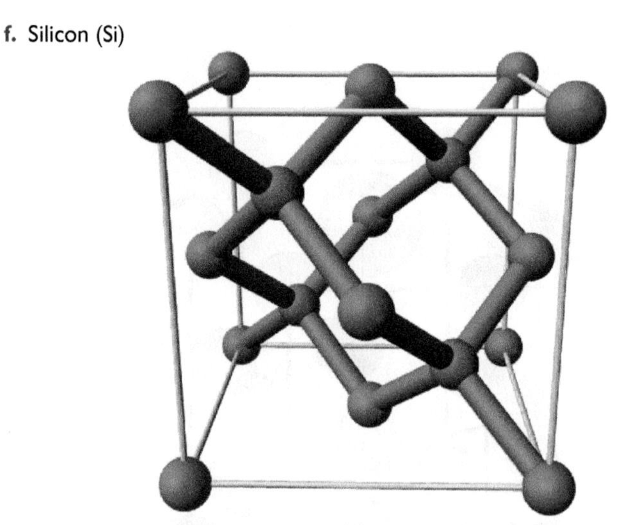

[Attributed to benjah-bmm27. Reproduced from http://en.wikipedia.org/wiki/Silicon (accessed March 31, 2014).]

13.2. Explain why AlF_3 has a melting point of 1040 °C, while SiF_4 sublimes at −77 °C.

13.3. Predict the structure of the Zintl ion In_{11}^{8-}.

13.4. Predict the structure of the Zintl ion Tl_6^{8-}.

BIBLIOGRAPHY

1. Bader RFW. Atoms in Molecules. In: Schleyer PVR, editor. Encyclopedia of Computational Chemistry. Chichester, West Sussex, England: John Wiley & Sons, Ltd.; 1998.
2. Bader RFW. Atoms in Molecules: A Quantum Theory. New York: Oxford University Press; 1990.
3. Gillespie RJ, Popelier PLA. Chemical Bonding and Molecular Geometry. New York: Oxford University Press; 2001.
4. Housecroft CE, Sharpe AG. Inorganic Chemistry. 3rd ed. Essex, England: Pearson Education Limited; 2008.
5. Huheey JE, Keiter EA, Keiter RL. Inorganic Chemistry: Principles of Structure and Reactivity. 4th ed. New York: Harper Collins College Publishers; 1993.
6. Keeler J, Wothers P. Chemical Structure and Reactivity: An Integrated Approach. New York: Oxford University Press; 2008.
7. Matta CF, Boyd RJ, editors. The Quantum Theory of Atoms in Molecules: From Solid State to DNA and Drug Design. Weinheim, Germany: Wiley-VCH Verlag GmbH & Co.; 2007.
8. Miessler GL, Tarr DA. Inorganic Chemistry. 4th ed. Upper Saddle River, NJ: Pearson Education Inc.; 2011.
9. Müller U. Inorganic Structural Chemistry. West Sussex, England: John Wiley & Sons, Ltd.; 1993.
10. Popelier PLA. Quantum Chemical Topology: On Bonds and Potentials. In: Wales DJ, editor. Structure and Bonding. Springer-Verlag: Berlin, Germany; 2005.
11. Popelier PLA. The Quantum Theory of Atoms in Molecules. In: Frenking G, Shaik S, editors. The Nature of the Chemical Bond Revisited. Vol. 1. Weinheim, Germany: Wiley-VCH; 2005.
12. West AR. Solid State Chemistry and Its Applications. John Wiley & Sons, Inc.: New York; 1991.

Structure and Reactivity | 14

"During the course of chemical reactions, the interaction of the highest filled (HOMO) and lowest unfilled (antibonding) molecular orbital (LUMO) in reacting species is very important to the stabilization of the transition structure."

—*Kenichi Fukui*

14.1 AN OVERVIEW OF CHEMICAL REACTIVITY

The majority of textbooks in the field of inorganic chemistry contain whole chapters of information on the myriad of chemical reactions that occur in main group and transition metal chemistry. The sheer variety of types of reactions can be overwhelming; thus, the familiar approach has been to attack the problem by examining the chemical reactivity of each family of the periodic table one group at a time. For example, a textbook might focus on the similarities of the different alkali metals and their reactions in one chapter and examine the chemistry of the halogens in another. The transition metals and the lanthanides and actinides are often treated as separate blocks. While this piecemeal approach to the study of inorganic chemistry has its advantages, it also tends to overlook similarities between the chemistry of molecules from very different parts of the periodic table. Rather than present yet another survey of the descriptive chemistry of the elements, the focus of this text is to examine the connections between chemical structure and reactivity.

Chemical reactions involve rearrangements of the ways in which atoms and electrons are organized in molecules. During the course of a chemical reaction, typically, some of the bonds in the reactants are broken and new bonds are formed in the process of making the products. As such, every chemical reaction requires either the transfer of an atom(s) and/or electron(s) between the reacting species. The redistribution of electron density requires that the valence molecular orbitals change during the course of the reaction. In fact, the core electrons are not involved in the chemical reaction to any significant extent. Although it is possible, and even very likely, that *all* of the MOs change to a certain extent, it is often the case that the initial changes are largest between the frontier molecular orbitals (FMOs) of the reacting species. The interactions between a filled MO and a vacant MO having

[Reproduced from http://en.wikipedia.org/wiki/File:Nucleophile-HOMO-carbonyl-LUMO-overlap-3D-balls.png (accessed March 13, 2014)].

Principles of Inorganic Chemistry, First Edition. Brian W. Pfennig.

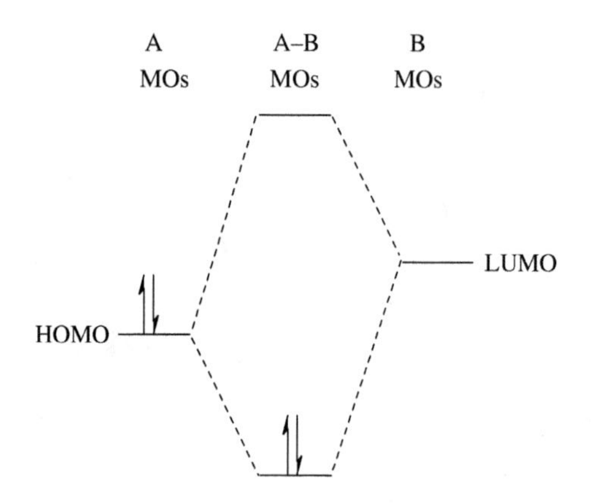

FIGURE 14.1
The interaction of the HOMO on species A with the LUMO on species B can lead to a lower energy configuration in A–B.

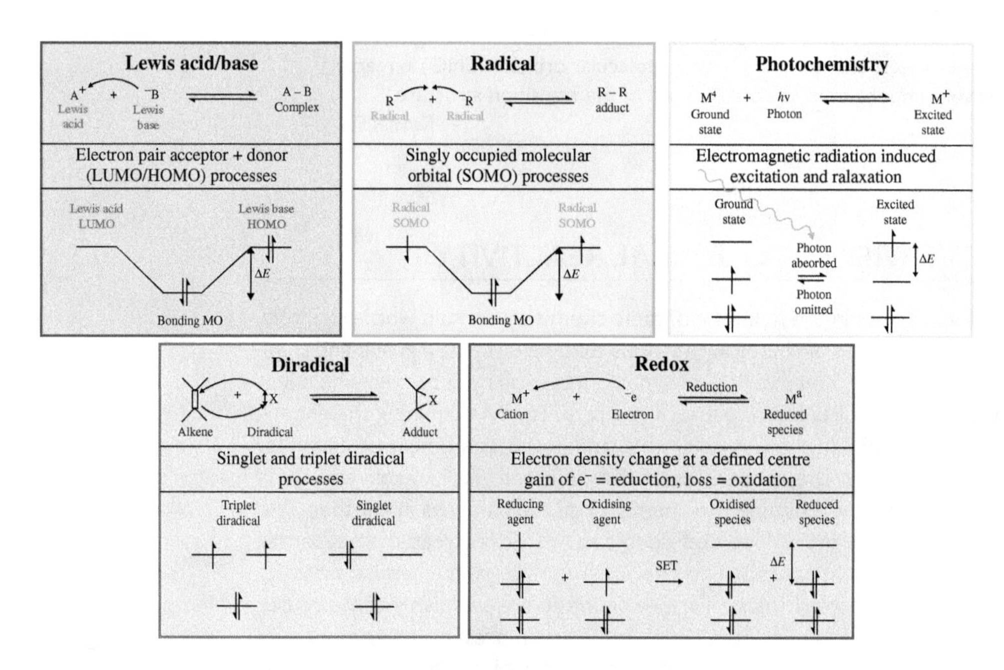

FIGURE 14.2
Five basic types of reaction chemistries based on the interaction of their FMOs. [© Mark R Leach, meta-synthesis (accessed October 22, 2013).]

similar energies and the appropriate symmetries for maximal electronic overlap often results in a lowering of the overall energy, as shown in Figure 14.1, and can serve as the thermodynamic driving force for the reaction.

We have spent a great deal of time in the preceding chapters discussing various types of chemical bonding (metallic, ionic, and covalent) and their relationships with one another. In a similar manner, we can categorize the thousands of different chemical equations on the basis of the topology and energy of their interacting FMOs, as shown by the examples in Figure 14.2. The FMOs for a molecule are defined as the HOMO (highest-occupied molecular orbital), LUMO (lowest-unoccupied molecular orbital), and SOMO (singly occupied molecular orbital). The two most important types of chemical reactions are Lewis acid–base reactions and oxidation–reduction (redox) reactions. We shall examine each in turn and then attempt to show the underlying structural relationship between the two.

14.2 ACID–BASE REACTIONS

One of the most important types of chemical reactions is the acid–base reaction. However, the definition of which species constitute acids or bases has evolved over the years as the breadth of known chemical reactions has continued to proliferate. For this reason, it is necessary to first introduce the more common historical definitions of acids and bases so that we may better understand how they each fit into the lexicon of chemical reactivity. Just as there were several complimentary models to facilitate our understanding of chemical bonding, so too there are numerous definitions of what it means to be an acid or a base. Which of these definitions we choose will depend on the complexity of the specific acid–base interaction at hand. Ultimately, however, every acid–base reaction entails a change in the way that the valence electrons are arranged in the atomic or molecular orbitals of the participating species. Therefore, the most modern definition of acid–base chemistry builds upon the MO concepts developed in previous chapters and provides the context for a natural continuation of that discussion.

The traditional Arrhenius definition of acids and bases derives from the early experiments of Arrhenius and Ostwald on the theory of electrolytic dissociation, specifically as it applies to the autoionization of water given by Equation (14.1). At 298 K, $K_w = 1.0 \times 10^{-14}$ and the equilibrium concentrations of H_3O^+ and OH^- are identical.

$$2\,H_2O\,(l) \Longleftrightarrow H_3O^+(aq) + OH^-(aq) \qquad (14.1)$$

According to the Arrhenius definition, an acid is any species that increases the concentration of H^+ or removes OH^- and a base is any species that increases the concentration of OH^- or removes H^+. In reality, however, H^+ cannot exist in isolation in aqueous solution, as the free energy for the solvation reaction in Equation (14.2) is extremely exergonic:

$$H^+(aq) + H_2O\,(l) \rightarrow H_3O^+(aq) \qquad (14.2)$$

With this caveat, we will use the terms H^+ and H_3O^+ (the hydronium ion) interchangeably. Using the Arrhenius definition, HCl (g) is an acid because it generates H_3O^+ (aq) ions when it is dissolved in aqueous solution, according to Equation (14.3). Boric acid is also an Arrhenius acid, as it removes OH^- ions, as shown in Equation (14.4). On the other hand, NaOH is an Arrhenius base, as it increases the concentration of the hydroxide ion in aqueous solution.

$$HCl\,(g) + H_2O\,(l) \rightarrow H_3O^+(aq) + Cl^-(aq) \qquad (14.3)$$

$$B(OH)_3(s) + OH^-(aq) \rightarrow B(OH)_4{}^-(aq) \qquad (14.4)$$

Arrhenius acids and bases represent a subset of a more generalized model of acid–base theory known as *solvent theory*. In solvent theory, an acid is defined as any substance that increases the concentration of the cationic species that results from autoionization of the solvent, whereas a base increases the concentration of the anionic species from autoionization. Table 14.1 lists the ion products for some of the more common solvents that undergo autoionization.

The autoionization of liquid ammonia is given by Equation (14.5). Thus, any species that is capable of increasing the concentration of ammonium ion in liquid ammonia is considered to be an acid and any species that generates $NH_2{}^-$ is a base:

$$2\,NH_3(l) \Longleftrightarrow NH_4{}^+ + NH_2{}^- \qquad (14.5)$$

TABLE 14.1 Acid and base forms and the ion products for selected solvents that undergo autoionization.

Solvent	Acid Form	Base Form	K_{ion} (298 K)
H_2O	H_3O^+	OH^-	10^{-14}
NH_3	NH_4^+	NH_2^-	10^{-27}
H_2SO_4	$H_3SO_4^+$	HSO_4^-	10^{-4}
CH_3COOH	$CH_3COOH_2^+$	CH_3COO^-	10^{-15}
CH_3OH	$CH_3OH_2^+$	CH_3O^-	10^{-19}
CH_3CH_2OH	$CH_3CH_2OH_2^+$	$CH_3CH_2O^-$	10^{-20}
CH_3CN	CH_3CNH^+	CH_2CN^-	10^{-29}

As G. N. Lewis said, "We frequently define an acid or a base as a substance whose aqueous solution gives, respectively, a higher concentration of hydrogen ion or hydroxide ion than that furnished by pure water. This is a very one-sided definition." In 1923, Brønsted and Lowry expanded the definitions of acids and bases to include species that do not involve solvent participation. According to the Brønsted–Lowry definition, an acid is any proton donor, whereas a base is any proton acceptor. This broader definition expanded acid–base theory to include gaseous species, such as HCl (g) and NH_3 (g). It also allowed for the inclusion of acid–base reactions occurring in nonionizing solvents, such as benzene, as shown by Equation (14.6):

$$\text{HCl (benzene)} + \text{NH}_3\text{(benzene)} \rightarrow \text{NH}_4\text{Cl (s)} \qquad (14.6)$$

The relative strength of a Brønsted–Lowry acid can be quantified by its acid-dissociation constant, K_a, which is defined in Equation (14.7):

$$HA + H_2O \Longleftrightarrow H_3O^+ + A^- \qquad K_a = [H_3O^+][A^-]/[HA] \qquad (14.7)$$

Strong acids, such as HCl, HBr, HI, HNO_3, H_2SO_4 (first proton only), $HClO_4$, and $HClO_3$, have $K_a > 1$. The actual strength of these acids cannot be measured in aqueous solution because of the *leveling effect*, which states that the strongest acid in any given solution will be the cation that results from autoionization of the solvent. In aqueous solution, any acid that is stronger than the hydronium ion will be "leveled" to that of H_3O^+ because it generates H_3O^+ in aqueous solution, as shown for HCl in Equation (14.3). For *weak acids* ($K_a < 1$), the majority of the molecules remain in their protonated form. The larger the value of K_a, the greater the concentration of H_3O^+ ions in solution, and the stronger the acid will be. A summary of the K_a values for some common acids is given in Table 14.2.

Using the thermodynamic cycle shown in Figure 14.3, it is possible to calculate ΔH for the acid dissociation equilibrium given by Equation (14.7). Even if the entropy term is disregarded, it is clear that there are a number of competing factors that will determine the acidity of a molecule. For the binary acids within a family, the dominant factor is the relative strengths of the bond dissociation energies. For reasons that were discussed in Chapter 5, the BDE for HF (565 kJ/mol) > HCl (431 kJ/mol) > HBr (366 kJ/mol) > HI (299 kJ/mol). Hence, it will be less difficult to break the H–I bond than it will be to break the H–F bond, and HI can dissociate its proton more easily (HI is a strong acid, while HF has a $K_a = 6.3 \times 10^{-4}$). Across a series in the periodic table, the BDEs are less important and it is the relative magnitude of the electron affinity that plays the greater role. Thus, while the BDEs for N–H (453 kJ/mol), O–H (492 kJ/mol), and H–F (565 kJ/mol) are increasing across the series (primarily due to the increased percent ionic character in the bonding), HF is by far the strongest acid of the three. This is because the EA of F is so much greater than that for O or N.

TABLE 14.2 Acid dissociation constants at 298 K for some common inorganic Brønsted–Lowry acids.

Acid	K_a	Acid	K_a
Monoprotic acids			
HF	6.3×10^{-4}	H_3AsO_4	5.6×10^{-3}
HN_3	1.9×10^{-5}	H_3AsO_3	5.1×10^{-10}
HCN	6.2×10^{-10}	$HClO_2$	1.1×10^{-2}
CH_3COOH	1.8×10^{-5}	HClO	3.0×10^{-8}
C_6H_5COOH	6.5×10^{-5}	HBrO	2.5×10^{-9}
$C_5H_5NH^+$	5.6×10^{-6}	HIO	2.3×10^{-11}
HNO_2	4.3×10^{-4}	HIO_3	1.7×10^{-1}
HCOOH	1.8×10^{-4}	NH_4^+	5.6×10^{-10}
CH_3COOH	1.8×10^{-5}	$CH_3NH_3^+$	2.8×10^{-11}
CCl_3COOH	3.0×10^{-1}	$(CH_3)_3NH^+$	1.5×10^{-10}
Polyprotic acids	K_{a1}	K_{a2}	K_{a3}
H_2SO_4	Strong acid	1.2×10^{-2}	
H_2SO_3	1.4×10^{-2}	6.3×10^{-8}	
H_2CO_3	4.3×10^{-7}	5.6×10^{-11}	
H_2S	1.3×10^{-7}	7.1×10^{-15}	
H_3BO_3	5.4×10^{-10}	1.8×10^{-13}	1.6×10^{-14}
H_3PO_4	7.6×10^{-3}	6.2×10^{-8}	2.1×10^{-13}

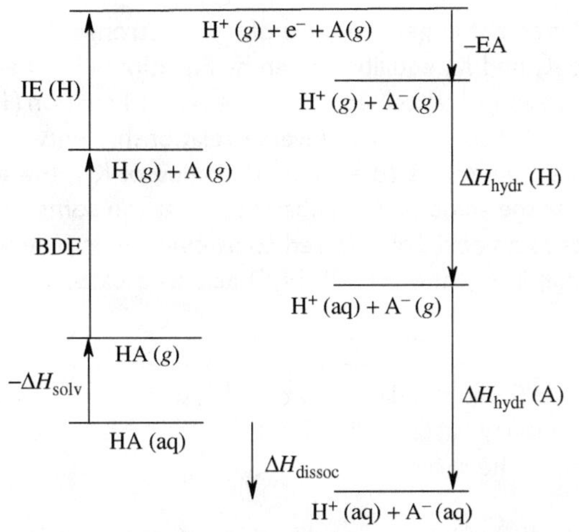

$H^+(g) + e^- + A(g)$ | −EA | $H^+(g) + A^-(g)$ | IE (H) | $H(g) + A(g)$ | ΔH_{hydr} (H) | BDE | $H^+(aq) + A^-(g)$ | $HA(g)$ | $-\Delta H_{solv}$ | ΔH_{hydr} (A) | $HA(aq)$ | ΔH_{dissoc} | $H^+(aq) + A^-(aq)$

FIGURE 14.3 Factors affecting the acid dissociation constants of binary acids.

The relative strengths of the oxyacids also depend on a variety of factors, including the size and electronegativity of the heteroatom, the degree of resonance stabilization of the corresponding anion, and inductive effects. Within a structurally homologous series of oxyacids, such as HClO, HBrO, and HIO, the largest factor determining the relative acidity is the electronegativity of the heteroatom. The dissociable proton is attached to an O atom that is bonded to the heteroatom. The more electronegative this atom is, the greater the electron withdrawal away from the O atom and the easier it will be to lose the proton. Inductive effects can also play a role. Hence, the more withdrawing Cl_3CCOOH is a stronger acid than CH_3COOH. It is for similar reasons that the superacid CF_3SO_3H (trifluoromethanesulfonic acid) is over 1000 times stronger than H_2SO_4. The term *superacid* was first coined by James Conant in 1927 to mean that these acids are stronger than the mineral acids. The strongest known superacid is $HSbF_6$, which is 2×10^{16} times more powerful than sulfuric acid.

For a homologous series of oxyacids where the heteroatom is the same but the number of O atoms differs, such as is the case for HClO, $HClO_2$, $HClO_3$, and $HClO_4$, the acidity will increase with the number of O atoms. There are several underlying reasons for this trend. First, the oxidation number of the heteroatom increases with the number of O atoms. The larger the oxidation number of the heteroatom, the more withdrawing it will be. Secondly, and perhaps more importantly, the resonance stabilization of the anion that results from the dissociation will increase with the number of O atoms. For organic acids, the hybridization of the atom to which the dissociable proton is attached will also play a role. It is well known that the strength of a chemical bond increases with the percent s-character in the hybrid orbital.

Each of the acids listed in Table 14.2 has a corresponding *conjugate base*, which differs from the acid only in the absence of a proton in its chemical structure. For instance, the conjugate base of H_2S in Equation (14.8) is the HS^- ion. Likewise, the *conjugate acid* of NH_3 (a base) is NH_4^+:

$$\underset{\text{acid}}{H_2S\,(aq)} + \underset{\text{base}}{NH_3(aq)} \rightarrow \underset{\text{conj. base}}{HS^-(aq)} + \underset{\text{conj. acid}}{NH_4^+(aq)} \qquad (14.8)$$

The relative strengths of Brønsted–Lowry bases can be measured based on their K_b base dissociation constants. A strong base, such as NaOH or $Mg(OH)_2$, has a $K_b > 1$ and will dissociate 100% in aqueous solution. The definition of K_b for the generic weak base A^- is given by Equation (14.9):

$$A^- + H_2O \Longleftrightarrow OH^- + HA \qquad K_b = [OH^-][HA]/[A^-] \qquad (14.9)$$

For *weak bases*, the larger the value of K_b, the stronger the base will be. Because the sum of the K_a and K_b equilibria given by Equations (14.7) and (14.9) is identical to the autoionization of water equilibrium given in Equation (14.1), $K_w = K_a\ K_b$ for any conjugate acid–base pair. The inverse relationship between K_a and K_b implies that the stronger the acid is (the larger the value of K_a), the weaker its conjugate base (the smaller the value of K_b). Substances that can sometimes act as a base and other times act as an acid are referred to as *amphoteric*. The water molecule is an obvious example; in Equation (14.7), H_2O acts as a base, and in Equation (14.9), it acts as an acid.

Example 14-1. Given the data in Table 14.2, calculate K_b for each of the following Brønsted–Lowry bases at 298 K: (a) CN^-, (b) N_3^-, and (c) F^-. Which is the strongest base of the three?

Solution. The K_b can be calculated from the K_a of the conjugate base, as follows: $K_b = K_w/K_a$. (a) $K_b = 1.6 \times 10^{-5}$, (b) 5.3×10^{-10}, and (c) 1.6×10^{-11}. Cyanide is the strongest base of the three because it has the largest value of K_b.

A more universal measure of the basicity of a species is its gas-phase *proton affinity* (PA), which is defined in Equation (14.10). Like electron affinities, proton affinities are reported as positive values even though their enthalpies of reaction are always negative:

$$B(g) + H^+(g) \rightarrow BH^+(g) \qquad PA = -\Delta H_{rxn} \qquad (14.10)$$

The larger the PA, the harder it will be to remove a proton from the base, and the stronger the base will be. Proton affinities are typically measured in thermodynamic cycles, such as the one shown in Figure 14.4. However, once the PA of a particular base is known, ion cyclotron resonance spectroscopy can be used

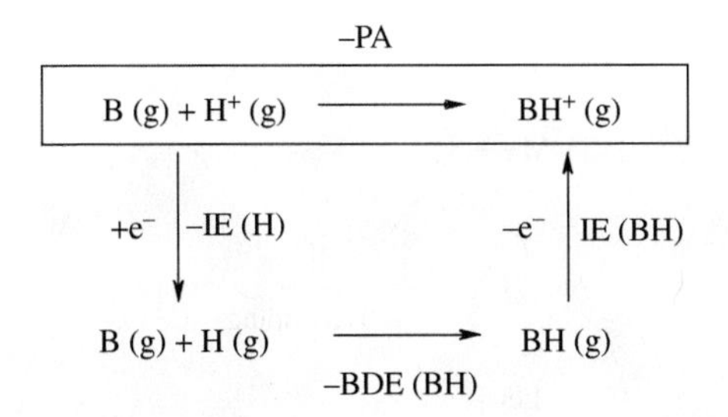

FIGURE 14.4 Thermodynamic cycle for the calculation of proton affinities (PAs).

TABLE 14.3 Proton affinities (PAs) for selected nonmetal hydrides.

Group 14	PA, kJ/mol	Group 15	PA, kJ/mol	Group 16	PA, kJ/mol	Group 17	PA, kJ/mol
CH_4	552	NH_3	854	H_2O	697	HF	399
SiH_4	~648	PH_3	789	H_2S	712	HCl	564
		AsH_3	750	H_2Se	717	HBr	589
						HI	628

indirectly to measure the equilibrium constant for proton exchange between that base and another base having an unknown PA.

The proton affinities of the nonmetal hydrides are listed in Table 14.3. Across a series, the changes in the BDEs are usually quite small compared with those for the BH ionization energies. The magnitude of the PA increases as the size of the IE (BH) decreases. For example, the proton affinities decrease in the order $NH_3 > H_2O > HF$ (which also matches the trend in their solution basicity) because of the increase in IE across the series. Within some families, such as the Group 17 hydrides, the PA increases down the column as the IE decreases. For the Group 15 hydrides, on the other hand, the changes in the BDE down a column are larger than the changes in the ionization energy. Thus, the PAs in this family track with the B–H bond dissociation energies and decrease down the column.

Example 14-2. The gas-phase PA for CH_3^- is 1743 kJ/mol, whereas that for NH_2^- is 1689 kJ/mol. Explain why the PA of $CH_4 < NH_3$, but the PA of $CH_3^- > NH_2^-$.

Solution. Proton affinities tend to decrease across a row of the periodic table because the changes in the IE (BH) term are larger than those of the BDE term in Figure 14.2. Because IEs are positive, the larger the IE (BH) term, the more negative the PA will be. This explains why CH_3^- is a better base than NH_2^- in the gas phase. However, unlike NH_3, CH_4 has no lone pairs of electrons with which to bind a proton. Therefore, NH_3 is a much better base than CH_4.

In 1923, Gilbert Newton Lewis defined an acid as an electron pair acceptor and a base as an electron pair donor. This definition is even more inclusive than the previous one because it includes all Brønsted–Lowry acids and bases as a subset and provides the foundation for the field of coordination chemistry. A *coordination compound* is the product of a Lewis acid–base reaction, such as the one shown in Equation (14.11) and Figure 14.5, in which the metal ion (Lewis acid) and *ligand* (Lewis base) are held together by a coordinate covalent bond.

$$Cu^{2+}(aq) + 4\,NH_3(aq) \rightarrow [Cu(NH_3)_4]^{2+}(aq) \quad (14.11)$$

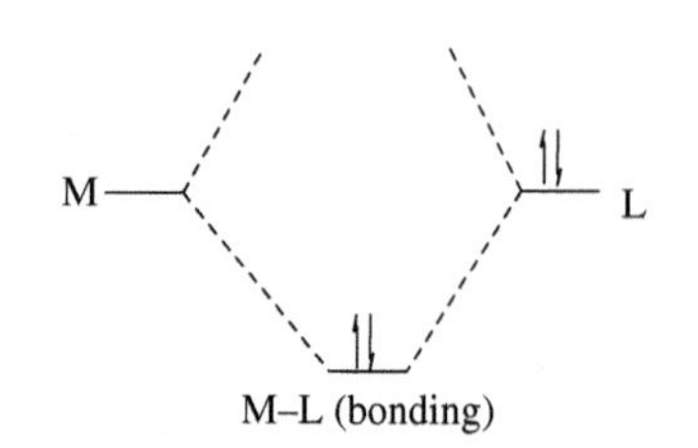

Lewis acid Lewis base

M + L ⟶ M–L bond

Cu^{2+} + NH_3 ⟶ $Cu(NH_3)^{2+}$

FIGURE 14.5
Molecular orbital interaction between a metal (Lewis acid) and a ligand (Lewis base) to form a coordinate covalent bond.

As its name implies, a *coordinate covalent bond* (also known as a *dative bond* in the older literature) is a covalent bond where the valence electrons are shared between the metal and the ligand, but where the ligands donate both of the shared electrons to the bond. Thus, the ligand (L) "coordinates" (or adds on to) the metal (M), as shown in Figure 14.5. Coordination compounds are discussed in more detail in Chapter 15.

When metal ions are dissolved in aqueous solution, the metal ion can act as a Lewis acid and the H_2O molecules as Lewis bases to form a hydrated metal cation. The thermodynamic driving force for the hydration reaction (ΔH_{hyd}, measured in kJ/mol) depends on the charge (Z) on the ion and its ionic radius (r in pm), according to Equation (14.12). This is because the strength of the ion–dipole interaction between the metal cation and the water molecule increases with the charge on the ion and decreases with the distance between them. In some cases, the interaction is sufficiently strong that the hydrated metal ion can undergo a hydrolysis equilibrium, such as the one given by Equation (14.13), where n is the number of coordinated H_2O ligands and z is the charge on the complex cation.

Most coordination compounds are octahedral ($n = 6$), although other geometries can also exist depending on the size and charge of the ion. The pK_a for the hydrolysis reaction shown in Equation (14.13) for a large number of metal ions is known to follow the empirical relationship given by Equation (14.14), where χ is the Pauling electronegativity of the cation. If the electronegativity of the metal is less than 1.50, the dependence on the electronegativity in Equation (14.14) is simply ignored in the calculation. The pK_a values for a variety of representative metal ions are listed in Table 14.4. The experimental pK_a values for the different metal ions are plotted in Figure 14.6 [the graph at left is for metals having electronegativities less than 1.50, whereas the graph at right includes the correction for electronegativity given in Equation (14.14)]:

$$\Delta H_{hyd} = -\frac{60,900\,Z^2}{r + 50} \tag{14.12}$$

$$[M(H_2O)_n]^{z+} + H_2O \Longleftrightarrow [M(H_2O)_{n-1}(OH)]^{(z-1)+} + H_3O^+ \tag{14.13}$$

$$pK_a = 15.14 - 88.16\left[\frac{Z^2}{r} + 0.096\,(\chi - 1.50)\right] \tag{14.14}$$

TABLE 14.4 Experimental values of pK_a for the hydrolysis of selected metal ions.

+1 ions	pK_a	+2 ions	pK_a	+3 ions	pK_a	+4 ions	pK_a
K	14.5	Ba	13.5	La	8.5	Th	3.2
Na	14.2	Sr	13.3	Y	7.7	Np	1.5
Li	13.6	Ca	12.8	Lu	7.6	U	0.6
Tl	13.2	Mn	10.6	Pu	7.0	Pu	0.5
Ag	12.0	Cd	10.1	Al	5.0	Hf	0.2
		Cr	10.0	Sc	4.3	Zr	−0.3
		Ni	9.9	In	4.0	Pa	−0.8
		Co	9.6	Cr	4.0	Ce	−1.1
		Fe	9.5	Ga	2.6		
		Zn	9.0	Ti	2.2		
		Pb	7.7	Fe	2.2		
		Be	6.2	Bi	1.1		
		Sn	3.4	Tl	0.6		
		Hg	3.4	Au	−1.5		

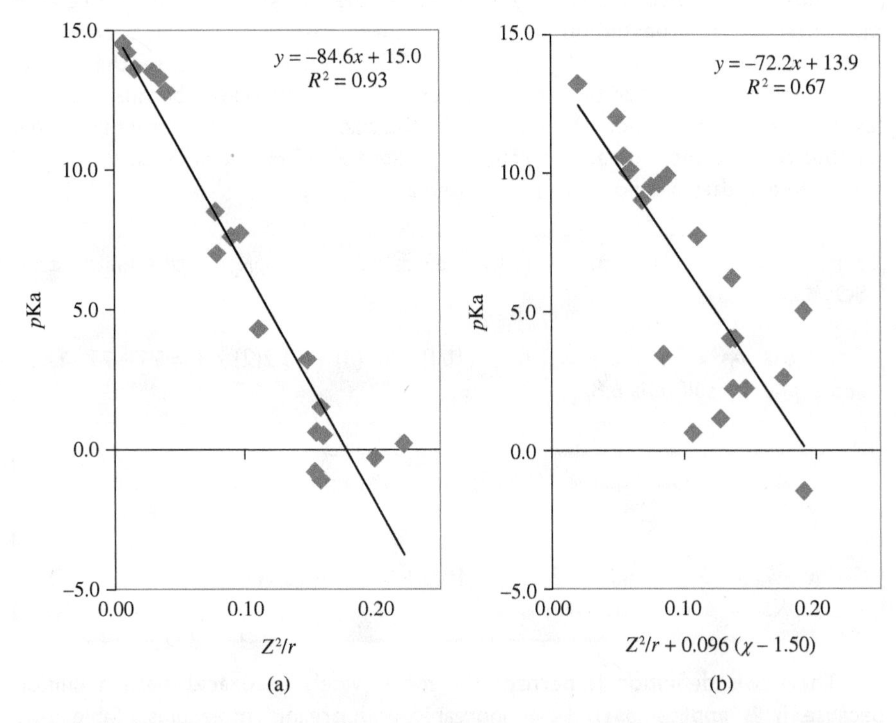

FIGURE 14.6 Plot of the pK_a of representative metal ions from Table 14.4 versus Z/r^2 for metals having $\chi < 1.50$ (a) and versus $Z/r^2 + 0.096(\chi - 1.50)$ for metals having $\chi > 1.50$ (b). Electronegativity and ionic radius values are from www.webelements.com. Whenever these values were known, Pauling radii were used; otherwise, the Shannon crystal radii were used for the octahedral (HS) metal ion. The pK_a values are from Wulfsberg.

Example 14-3. Use Equation (14.14) to calculate the expected pK_a of Fe^{3+}, given that $r = 78$ pm and $\chi = 1.83$.

Solution. Using Equation (14.14), we obtain the following:

$$pK_a = 15.14 - 88.16\left[\frac{3^2}{78} + 0.096(1.83 - 1.50)\right] = 2.17$$

The experimental value of pK_a for Fe^{3+} is 2.2, which agrees with the calculated value.

Example 14-4. Use Equation (14.14) to calculate the expected pK_a of Li^+, given that $r = 90$ pm and $\chi = 0.98$.

Solution. Using Equation (14.14) and omitting the correction for electronegativity yields the following result. The experimental value of pK_a for Li is 13.6 for only a 4% error.

$$pK_a = 15.14 - 88.16\left[\frac{1^2}{90}\right] = 14.16$$

According to Wulfsberg, a cation is nonacidic if its $pK_a > 14$. As a rule, the metal ions remain largely hydrated at pH values up to the pK_a of the cation. When the pH exceeds the pK_a, the metal hydroxides or metal oxides will typically form, many of which are insoluble and will precipitate out of solution.

In a similar manner, the pK_b values of the oxo anions MO_x^{y-} can be calculated using the empirical relationship given by Equation (14.15), where x is the number of O atoms and y is the charge on the complex anion:

$$pK_b = 10.0 + 5.7x - 10.2y \pm 1 \tag{14.15}$$

Thus, for instance, the basicity increases as the charge on the anion becomes more negative, as expected on the basis of the electrostatic attraction between the anion and a proton. Hence, pK_b for $PO_4^{3-} > SO_4^{2-} > ClO_4^-$. Likewise, the basicity decreases as the number of oxo groups on the anion increases because there is less resonance stabilization of the anion and the size of the anion increases with the number of O atoms. The anions remain hydrated at pOH values up to the pK_b of the anion and they are normally protonated as $pOH > pK_b$.

Example 14-5. Use Equation (14.15) to calculate the pK_b of the sulfite ion, SO_3^{2-}.

Solution. As $x = 3$ and $y = 2$, $pK_b = 10.0 + 5.7(3) - 10.2(2) \pm 1 = 5.7–6.7$. The actual pK_b for sulfite is 6.8.

Example 14-6. Use Equation (14.15) to calculate the pK_b of the phosphate ion, PO_4^{3-}.

Solution. As $x = 4$ and $y = 3$, $pK_b = 10.0 + 5.7(4) - 10.2(3) \pm 1 = 1.2–3.2$. The actual pK_b for phosphate is 1.6.

The Lewis definition is perhaps the most widely used acid–base definition because it is applicable to both inorganic and organic molecules. Table 14.5 summarizes some of the different types of molecular species that exhibit Lewis acid–base behavior.

As a general rule, the relative strengths of Lewis bases roughly parallel their ability to accept a proton and the strengths of Lewis acids increase as the size of the metal ion decreases and their charges increase. It has been known for some time that there are trends in the thermodynamic stability of coordination compounds and the Lewis acidity of the metal ion. For example, the Irving–Williams series (published in 1953) states that the stability of a divalent transition metal complex with a given ligand increases in the order:

$$Ba^{2+} < Sr^{2+} < Ca^{2+} < Mg^{2+} < Mn^{2+} < Fe^{2+} < Co^{2+} < Ni^{2+} < Cu^{2+} > Zn^{2+}$$

TABLE 14.5 The different types of compounds exhibiting Lewis acid–base behavior.

Lewis Acid Behavior	Lewis Base Behavior
Metal ions (Cu^{2+}, Ag^+, Fe^{3+})	Ligands (NH_3, CN^-, pyridine)
Electrophiles (attacking L. acids)	Nucleophiles (attacking L. bases)
Electrofuges (leaving L. acids)	Nucleofuges (leaving L. bases)
Spectator cations	Spectator anions
Electron-deficient *pi* systems	Electron-rich *pi* systems
Electron-deficient species (BF_3, $AlCl_3$)	H^+ abstractors

TABLE 14.6 Ahrland's classification of metal ions and ligands.

Type a Species	Type b Species
Li^+, Na^+, K^+, Rb^+, Cs^+,	Cu^+, Ag^+, Au^+, Tl^+, Hg_2^{2+},
Be^{2+}, Mg^{2+}, Ca^{2+}, Sr^{2+}, Ba^{2+}	Cu^{2+}, Pb^{2+}, Ni^{2+}, Pd^{2+}, Pt^{2+}, Hg^{2+}
Ti^{4+}, Cr^{3+}, Fe^{3+}, Co^{3+}	Tl^{3+}
$N >> P > As > Sb$	$N << P > As > Sb$
$O >> S > Se > Te$	$O << S < Se \sim Te$
$F > Cl > Br > I$	$F < Cl < Br < I$
H^+	

This order is largely due to the decrease in ionic radius across a period; however, ligand field effects (see Chapter 16) also play a role. Another observation was that certain ligands prefer (i.e., form more stable complexes with) certain kinds of metal ions, whereas other ligands prefer a different type of metal ion. In 1956, Ahrland, Chatt, and Davies classified metals and ligands into two categories: types a and b, as shown in Table 14.6.

Perhaps the most useful classification scheme, and certainly the most widely known, is the hard–soft acid–base (HSAB) theory developed by Pearson in 1963. HSAB theory is essentially an extension of Ahrland's type a and b classification scheme. As in the previous example, hard acids prefer hard bases and soft acids prefer soft bases. A classification of compounds into hard, soft, or borderline categories is shown in Table 14.7. In general, a *hard* acid has a small ionic radius, high positive charge, no electron pairs in its valence shell, low EA, is strongly solvated, and/or has a high-energy LUMO. Hard bases have small radii, are hard to oxidize, are weakly polarizable, have very electronegative centers, are strongly solvated, and/or have high-energy HOMOs. *Soft* acids, on the other hand, have large radii, have low or partial positive charge, have electron pairs in their valence shells, are easy to polarize and oxidize, and have low-energy LUMOs with large magnitude coefficients. Soft bases tend to have large radii, have intermediate values for the electronegativity, are easily polarized and oxidized, and have low-energy HOMOs with large magnitude coefficients.

A wide variety of data support the empirical observations of HSAB theory. For example, the overall formation constants (β_2) for the reaction shown in Equation (14.16) are given as

$$Ag^+(aq) + 2\,L\,(aq) \rightarrow AgL_2^{\ +}(aq)$$

$$\begin{array}{lll} Ag^+(\text{soft}) & L = Cl^-(\text{hard}) & \beta_2 = 10^5 \\ & L = NH_3(\text{hard}) & \beta_2 = 10^7 \\ & L = S_2O_3^{\,2-}(\text{soft}) & \beta_2 = 10^{13} \\ & L = CN^-(\text{soft}) & \beta_2 = 10^{21} \end{array} \qquad (14.16)$$

Because Ag^+ is a soft Lewis acid, the stabilities of the silver complex cations increase as the ligands become softer. A similar trend is observed for the hard base NH_3 in Equation (14.17). The harder metal ions form more stable complexes:

$$M^{n+}(aq) + 6\,NH_3(aq) \rightarrow M(NH_3)_6^{\,n+}(aq)$$

$$\begin{array}{lll} NH_3(\text{hard}) & M = Co^{2+}(\text{intermediate}) & \beta_6 = 10^5 \\ & M = Co^{3+}(\text{hard}) & \beta_6 = 10^{33} \\ & M = Ni^{2+}(\text{intermediate}) & \beta_6 = 10^8 \end{array} \qquad (14.17)$$

Further evidence supporting HSAB theory occurs for the dimethylsulfoxide (dmso) complexes of different transition metals. DMSO is an ambidentate ligand, where the ligand can coordinate to a metal either through the hard O atom or the soft S atom. When $M = Cu^{2+}$ (hard), IR evidence shows the presence of two O-bound dmso ligands. When $M = Pd^{2+}$ (soft), two S-bound dmso ligands appear in the IR. For $M = Ru^{2+}$ (intermediate), a mixture of O- and S-bound dmso ligands results.

TABLE 14.7 Pearson's hard–soft classification scheme for Lewis acids and bases.

HSAB	Hard	Borderline	Soft
Acids	H^+, HX, Li^+, Na^+, K^+, Be^{2+}, Mg^{2+}, Ca^{2+}, Sr^{2+}, Sn^{2+}, Sc^{3+}, Ga^{3+}, In^{3+}, La^{3+}, Ce^{4+}, Gd^{3+}, As^{3+}, Ir^{3+}, Zr^{4+}, Th^{4+}, Pu^{4+}, Cr^{3+}, Co^{3+}, Fe^{3+}, Mn^{2+}, Al^{3+}, Si^{4+}, Ti^{4+}, Sn^{4+}, UO_2, I^{7+}, I^{5+}, Cl^{7+}, $(CH_3)_2Sn^{2+}$, $Be(CH_3)_2$, BF_3, BCl_3, $B(OR)_3$, $AlCl_3$, $Al(CH_3)_3$, AlH_3, $Ga(CH_3)_3$, $In(CH_3)_3$, SO_3, RSO_2, $ROSO_2$, N^{3+}, $RPO_2^{\,+}$, $ROPO_2^{\,+}$, CO_2, NC^+	Fe^{2+}, Co^{2+}, Ni^{2+}, Cu^{2+}, Zn^{2+}, Os^{2+}, Rh^{3+}, Ir^{3+}, Ru^{3+}, $B(CH_3)_3$, R_3C^+, Sn^{2+}, Pb^{2+}, Sb^{3+}, NO^+, SO_2	Pd^{2+}, Pt^{2+}, Pt^{4+}, Cu^+, Ag^+, Au^+, Tl^+, Cd^{2+}, $Hg_2^{\,2+}$, Hg^{2+}, Cs^+, Tl^{3+}, $Tl(CH_3)_3$, RH_3, RS^+, RSe^+, RTe^+, CH_3Hg^+, GaX_3, BH_3, CH_2, O, HO^+, RO^+, R_3C, carbenes, π-acceptors, Cl, Br, I, Br_2, Br^+, I_2, I^+, N, M^0, bulk M's
Bases	NH_3, RNH_2, N_2H_4, H_2O, OH^-, O^{2-}, ROH, OR^-, R_2O, CH_3COO^-, $CO_3^{\,2-}$, $NO_3^{\,-}$, $PO_4^{\,3-}$, $SO_4^{\,2-}$, $ClO_4^{\,-}$, F^-, Cl^-	Br^-, $N_3^{\,-}$, N_2, pyridine, benzyl amine, $NO_2^{\,-}$, $SO_3^{\,2-}$	H^-, R^-, C_2H_4, CN^-, benzene, CO, RNC, SCN^-, R_3P, $(RO)_3P$, R_3As, R_2S, RSH, RS^-, $S_2O_3^{\,2-}$, I^-

TABLE 14.8 Proton affinities (PAs) of substituted amines toward B and $B(CH_3)_3$, showing the effects of steric interactions on the basicity of the amines.

Amine Base	PA, kJ/molw/ B (g)	PA, kJ/molw/ $B(CH_3)_3$
$(CH_3)_3N$	929	73.7
$(CH_3)_2NH$	912	80.6
CH_3NH_2	884	73.8
NH_3	846	57.5

HSAB theory is important in bioinorganic chemistry, as well. For instance, mercuric reductase (MerA) is a bacterial enzyme that can reduce toxic Hg^{2+} ions to Hg^0. The active site of MerA contains two cysteine residues (Cys-207 and Cys-628). The Cys amino acid residues bind to the soft Hg^{2+} ions through their soft HS side chains. Only a few other metals can bind in the active site of MerA and inhibit this reduction. These are the soft metals Ag^+, Au^{3+}, and Cd^{2+}, as well as the intermediate metals Cu^{2+} and Co^{2+}. No hard metals are capable of binding to the soft Cys amino acid residues in the MerA enzyme.

The acidity or basicity of a species, however, does not just depend on its electronic properties. Steric effects also play an important role. The proton affinities for a series of substituted amines with respect to B (g) and $B(CH_3)_3$ are listed in Table 14.8. As expected, the PA for B (g) increases with the number of methyl groups on the amine because of the inductive effect of the methyl groups donating electron density to the N lone pair. However, the PA for the more sterically crowded $B(CH_3)_3$ is more complex. First of all, the PA for $B(CH_3)_3$ of all four amines is significantly decreased compared to that for B. Secondly, the inductive effect of the methyl groups increases the PA across the series of amines to a certain degree, but then the steric interaction between methyl groups on the B and on the N take precedence and the PA of $(CH_3)_3N$ for $B(CH_3)_3$ decreases from the expected trend.

An alternative set of factors to HSAB theory that takes into account both electrostatic and covalent contributions is the Drago–Wayland parameters. The enthalpy of formation for the generic acid–base reaction in Equation (14.18) can be calculated from Equation (14.19) using the empirical *E* and *C* factors listed in Table 14.9. The subscripts A and B refer to acid and base, respectively:

$$A(g) + B(g) \rightarrow A - B(g) \quad \Delta H^\circ_{AB} \tag{14.18}$$

$$-\Delta H^\circ{}_{AB}\,(kcal/mol) = E_A E_B + C_A C_B \tag{14.19}$$

In the Drago–Wayland equation, the *E* parameter is a measure of the ionic interactions (roughly parallel to hardness), whereas the *C* parameter is a measure of the covalent interaction (roughly parallel to softness). Equation (14.19) can be used to predict the enthalpies of reaction for hundreds of acid–base reactions with a remarkable degree of accuracy. Large values of both E_A and E_B imply a hard–hard interaction, leading to a large ionic term, which is favorable, as shown by Equation (14.20). Likewise, large values of C_A and C_B correspond with a soft–soft interaction, leading to a large covalent term, which is also thermodynamically favorable, as shown by Equation (14.21). A mixture of the two (large C_A with large E_B or vice versa) represents a much less favorable soft–hard interaction, as shown by Equation (14.22):

$$\begin{array}{lll} HCN(g) + & NH_3(g) \rightarrow NH_4CN\ (s) & \\ \text{hard} & \text{hard} & \Delta H^\circ_{calc} = -5.1\ kcal/mol \end{array} \tag{14.20}$$

TABLE 14.9 Drago–Wayland parameters using Equation (14.19) for selected acids (A) and bases (B).

Acids	E_A	C_A	Bases	E_B	C_B
I_2	0.50	2.00	NH_3	2.31	2.04
H_2O	1.54	0.13	CH_3NH_2	2.16	3.12
SO_2	0.56	1.52	$(CH_3)_2NH$	1.80	4.21
CH_3OH	1.25	0.75	$(CH_3)_3N$	1.21	5.61
CH_3CH_2OH	1.34	0.69	$(CH_3CH_2)NH_2$	2.35	3.30
t-Butyl alcohol	1.36	0.51	$(CH_3CH_2)_3N$	1.32	5.73
Phenol	2.27	1.07	$(CH_3CH_2)_3N$	0.99	11.0
H^+	45.0	13.0	Pyridine	1.78	3.54
CH_3^+	19.7	12.6	CH_3CN	1.64	0.71
Li^+	11.7	1.45	$P(CH_3)_3$	1.46	3.44
K^+	3.78	0.10	Benzene, C_6H_6	0.70	0.45
NH_4^+	4.31	4.31	Methyl acetate	1.63	0.95
$CH_3NH_3^+$	2.18	2.38	Ethyl acetate	1.62	0.98
$(CH_3)_2NH_2^+$	3.21	0.70	Acetone	1.74	1.26
$(CH_3)_3NH^+$	2.60	1.33	CH_3OH	1.80	0.65
$(CH_3)_4N^+$	1.96	2.36	CH_3CH_2OH	1.85	1.09
H_3O^+	13.3	7.89	Dimethyl ether	1.68	1.50
$B(OCH_3)_3$	0.54	1.22	Diethyl ether	1.80	1.63
$B(CH_3CH_2)_3$	1.70	2.71	CH_3OH	1.80	0.65
HF	2.03	0.30	CH_3CH_2OH	1.85	1.09
HCl	3.69	0.74	H_2O	2.28	0.10
HCN	1.77	0.50	F^-	9.73	4.28
H_2S	0.77	1.46	Cl^-	7.50	3.76
PF_3	0.61	0.36	Br^-	6.74	3.21
AsF_3	1.48	1.14	I^-	5.48	2.97
$(CH_3)_3Sn^+$	7.05	3.15	CN^-	7.23	6.52
$(C_5H_5)Ni^+$	11.9	3.49	OH^-	10.4	4.60
$Fe(CO)_5$	0.10	0.27	CH_3O^-	10.0	4.42
CH_3COOH	1.72	0.86	$(CH_3)_2SO$ (dmso)	2.40	1.47
CF_3COOH	2.07	1.06	$(CH_3)_2S$	0.25	3.75

$$SO_2 + (CH_3)_3N \rightarrow (CH_3)_3NSO_2$$

$$\text{soft} \quad \text{soft} \qquad \Delta H^\circ_{calc} = -9.7\,\text{kcal/mol} \tag{14.21}$$

$$SO_2 + H_2O \rightarrow H_2SO_3$$

$$\text{soft} \quad \text{hard} \qquad \Delta H^\circ_{calc} = -1.1\,\text{kcal/mol} \tag{14.22}$$

Additionally, the solubilities of many inorganic salts are directly related to HSAB concepts. Consider the series of silver halides listed in Table 14.10. Because the Ag^+ ion is a soft Lewis acid, it will form a stronger interaction with the soft ligand I^- than it will with the harder Cl^- ion. Thus, the equilibrium constant for the formation of AgI is larger than that for the formation of AgCl. Because the formation constant is the inverse of the solubility product, AgI will therefore have the smallest K_{sp}. This makes sense intuitively because the more covalent the compound is, the weaker will be its ion–dipole interaction with the polar water molecule. Furthermore, because of the relationship between the formation constant and the standard reduction potential given by Equation (14.23), HSAB theory can also be used to rationalize the E° values listed in the table. The larger the formation constant, the larger ΔE° will be, and the more negative will be the standard reduction potential for

TABLE 14.10 Solubility products, calculated formation constants, and calculated and experimental $E°$ values (V vs NHE) for the silver halides at 298 K.

AgX Reaction	K_{sp}	K_f (calc)	$E°$ (calc), V	$E°$ (exptl), V
$AgCl \Leftrightarrow Ag^+ + Cl^-$ (Cl^- is hard)	1.8×10^{-10}	5.6×10^9	0.22	0.22
$AgBr \Leftrightarrow Ag^+ + Br^-$ (Br^- is intermediate)	7.7×10^{-13}	1.3×10^{12}	0.083	0.073
$AgI \Leftrightarrow Ag^+ + I^-$ (I^- is soft)	7.7×10^{-17}	1.3×10^{16}	−0.15	−0.15

the silver halide. The calculated value for $E°$ in Column 4 is based on the equation: $\Delta E° = E°(Ag^+/Ag) - E°(AgX/Ag, X^-)$.

$$K_f = \frac{1}{K_{sp}} = e^{-\frac{\Delta G°}{RT}} = e^{\frac{nF\Delta E°}{RT}} \tag{14.23}$$

The dependence of the solubility on HSAB theory is intimately related to the electronegativity. Hard acids have electronegativities between 0.7 and 1.6, whereas hard bases have electronegativities between 3.4 and 4.0. Thus, the hard–hard interaction has a large amount of ionic character to it. On the other hand, soft acids have electronegativities in the range 1.9–2.5, whereas soft bases have electronegativities of 2.1–3.0. As a result, soft–soft interactions will be largely covalent in nature. The second factor affecting the solubilities of inorganic salts is the relative acid–base strength as it relates to Z^2/r. Strongly acidic cations with strongly basic anions almost always form insoluble salts. Examples include $MgSO_4$ or TiO_2. On the other hand, acidic cations with nonbasic anions or nonacidic cations with basic anions tend to form soluble salts. Hence, $Fe(NO_3)_3$ and Na_3PO_4 are both soluble in aqueous solution.

14.3 FRONTIER MOLECULAR ORBITAL THEORY

With the advent of inexpensive molecular modeling software, the FMO definition of acids and bases has become increasingly popular. According to this definition, an acid reacts via its LUMO, whereas a base reacts using its HOMO. The MO diagram for H_2O was derived in Chapter 10 and is reproduced in Figure 14.7. The molecule has C_{2v} symmetry. The HOMO is the $1b_2$ nonbonding MO derived from a $2p$ AO on the oxygen atom.

When H_2O reacts with a proton to form the hydronium ion, H_3O^+, the $1b_2$ HOMO on H_2O will form a bonding and antibonding combination with the 1s LUMO on H^+, as shown in Figure 14.8. The new bonding MO (shown at right in the diagram) is isoenergetic with the $1b_1$ MO in H_2O (they each have a single nodal plane); thus, the two form a degenerate e MO in H_3O^+. The overall effect of the interaction diagram is to lower the energy of the $1b_2$ (nonbonding) orbital as it becomes an e bonding MO in H_3O^+. Thus, the reaction $H_2O + H^+ \rightarrow H_3O^+$ is favored by enthalpy, even though it is entropically unfavorable. Note that the energies and symmetries of the MOs for H_3O^+ in the interaction diagram in Figure 14.8 are identical to those in the MO diagram for H_3O^+ derived by taking linear combinations of the AOs on O with the SALCs of the three H atoms in the C_{3v} point group.

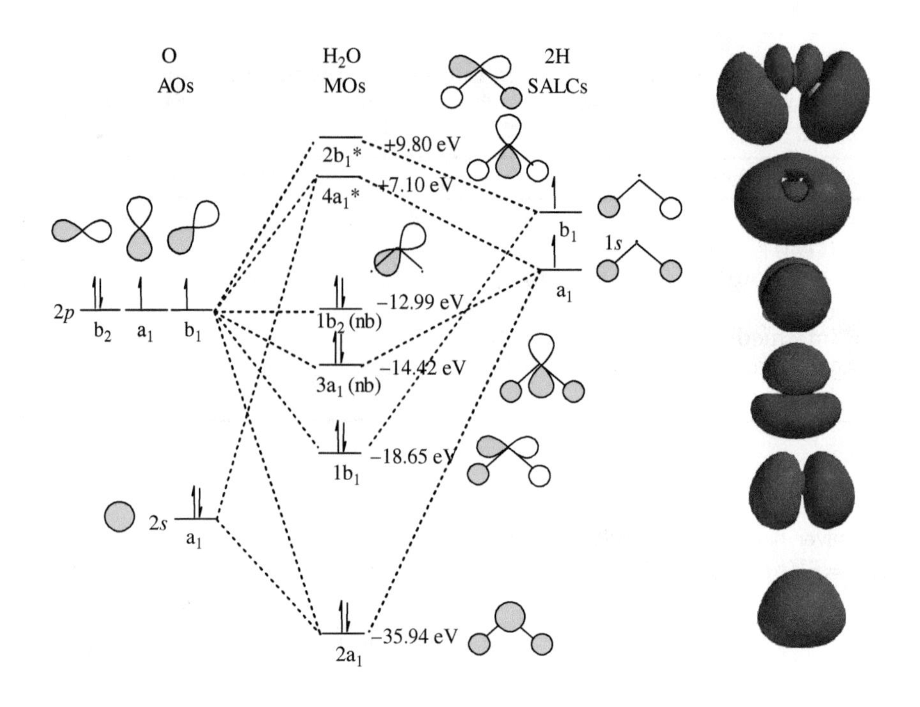

FIGURE 14.7
One-electron MO diagram for H_2O.

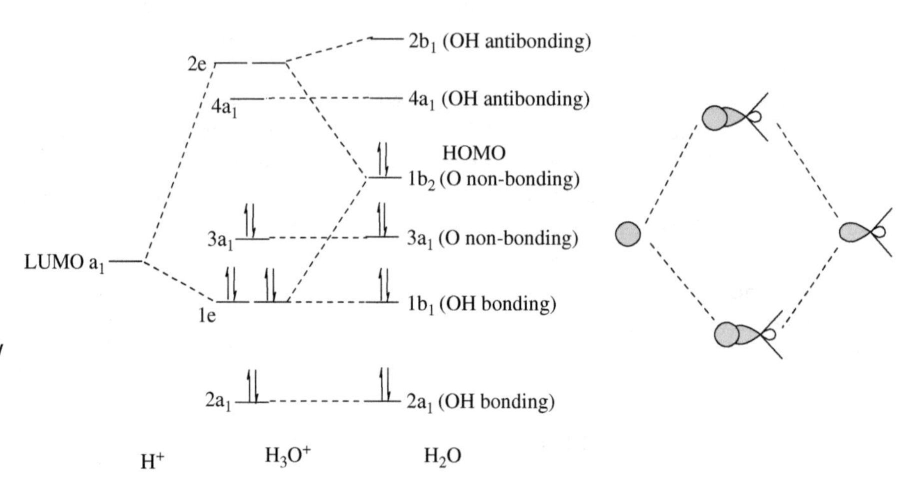

FIGURE 14.8
Acid–base interaction MO diagram for H_3O^+, showing how the LUMO on H^+ and HOMO on H_2O interact to form a bonding and antibonding pair of MOs in H_3O^+.

Example 14-7. (a) Construct the one-electron MO diagram for the NH_4^+ ion by taking linear combinations of the valence AOs on the central N atom with the four tetrahedral SALCs for the pendant H atoms. Sketch the shapes of the $1t_2$ bonding MOs. (b) Construct the interaction MO diagram for NH_4^+ by taking the MO diagram for NH_3 and adding a proton to it. Sketch the shapes of the frontier MOs and show how they are analogous to the ones generated in part (a).

Solution. (a) Using the T_d point group, the symmetries of the AOs for N are $2s = a_1$ and $2p_{x,y,z} = t_2$. The symmetries of the four SALCs are also $a_1 + t_2$ after reduction of the Γ_{NH} representation given below.

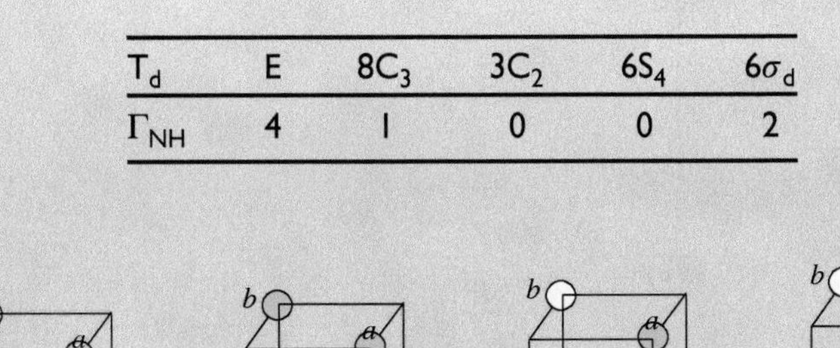

T_d	E	$8C_3$	$3C_2$	$6S_4$	$6\sigma_d$
Γ_{NH}	4	1	0	0	2

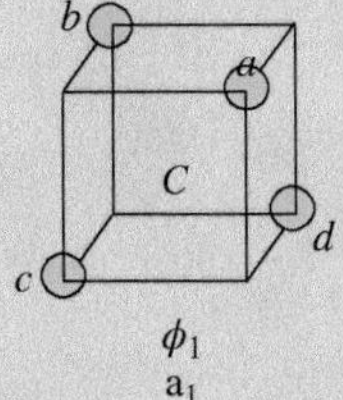

ϕ_1
a_1

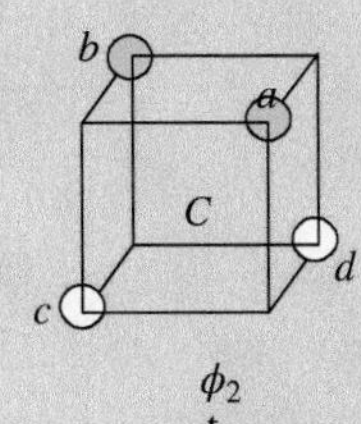

ϕ_2
t_2

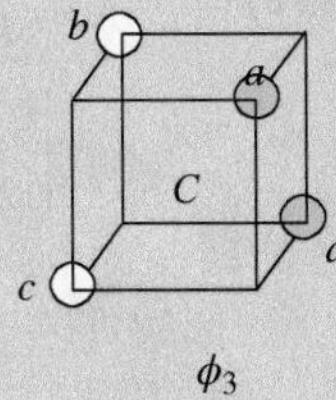

ϕ_3
t_2

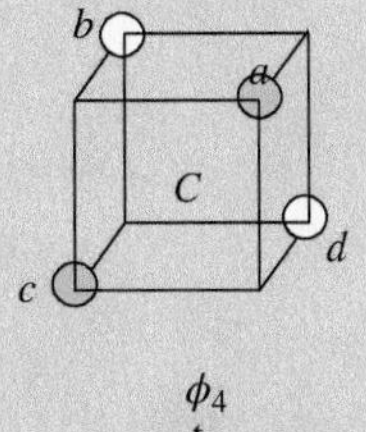

ϕ_4
t_2

The one-electron MO diagram for NH_4^+ is shown later.

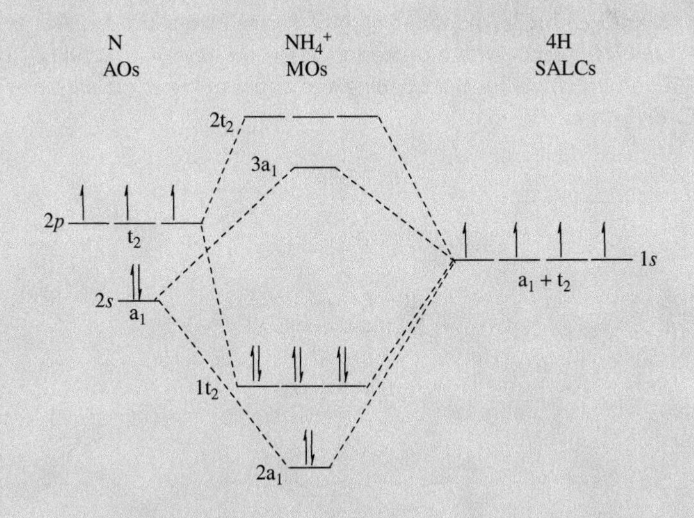

The shapes of the t_2 MOs for NH_4^+ can be determined by taking linear combinations of the three $2p$ AOs on N with the three t_2 SALCs in such a way as to maximize the orbital overlap. A qualitative depiction of the degenerate set of t_2 MOs is shown below.

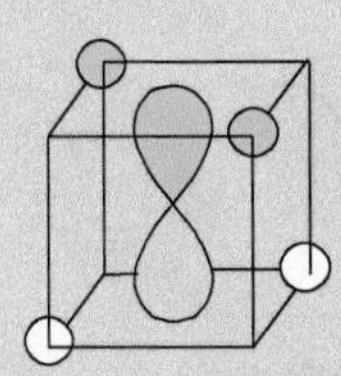

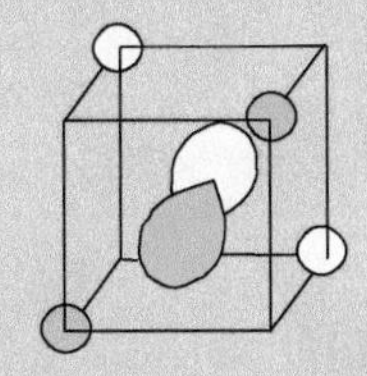

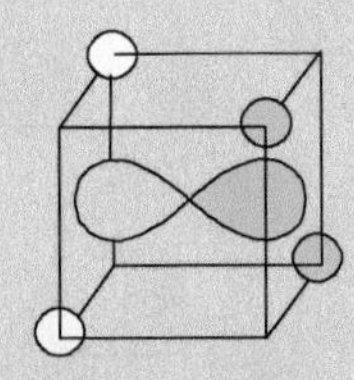

(b) The one-electron MO diagram for NH_3 was developed in Chapter 10 and is reproduced below. The molecule belongs to the C_{3v} point group.

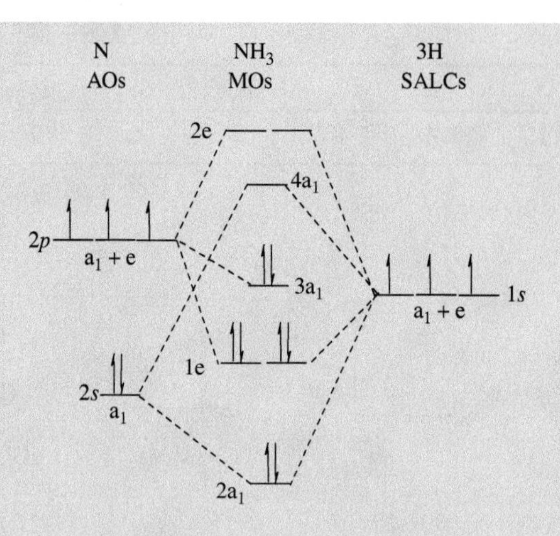

The HOMO is the $3a_1$ nonbonding MO derived from the $2p_z$ AO of the N atom. When NH_3 reacts with a proton to form the tetrahedral NH_4^+ ion, the $3a_1$ HOMO on NH_3 will form a bonding and antibonding combination with the 1s LUMO on H^+, as shown below.

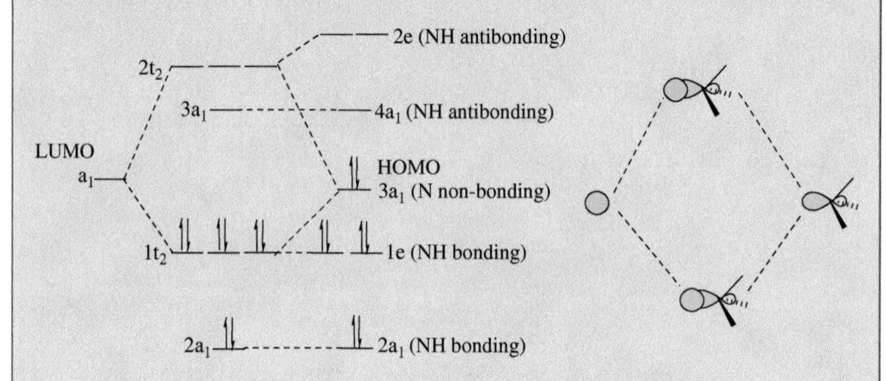

The shapes of the two e MOs for NH_3 are based on the overlap of two of the 2*p* AOs on N with the two e SALCs depicted below. The a_a nonbonding lone pair is also shown (at right).

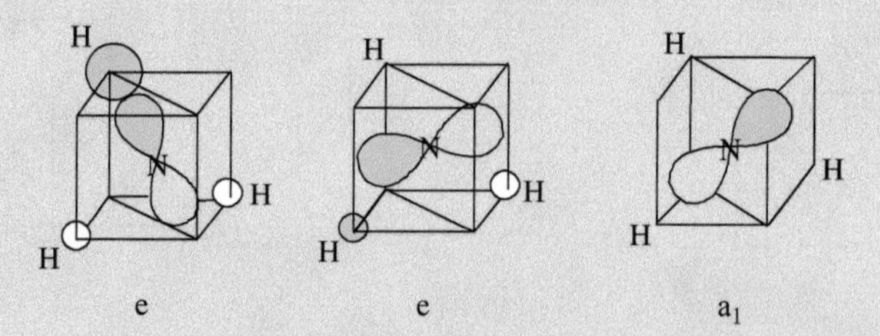

When the C_{3v} ammonia molecule reacts with H^+ to become the T_d ammonium ion and the NH_3 MOs form linear combinations with the H^+ 1s AO, a slight frame of reference shift occurs to yield the three degenerate t_2 MOs shown below. These MOs are analogous to those derived previously by combining the N AOs directly with the four H SALCs to make NH_4^+ MOs.

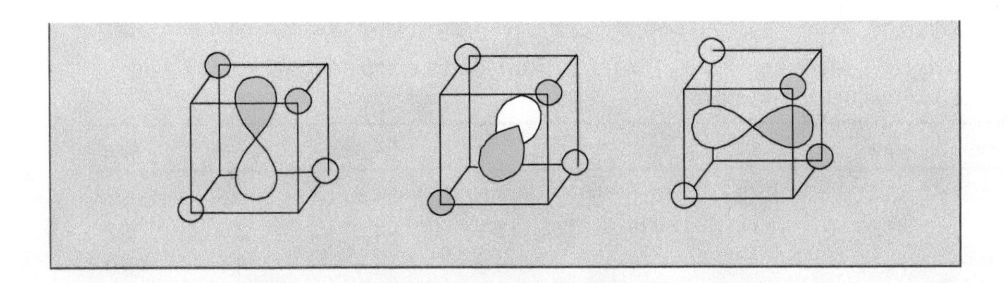

Example 14-8. The proton affinities for NH_3, H_2O, and HF are 854, 697, and 399 kJ/mol, respectively, as reported in Table 14.3. Explain this trend in terms of the energies of the bonding MOs in the acid–base interaction diagrams of each species with H^+.

Solution. The energy of the HOMO for each species decreases in the order: NH_3, H_2O, HF because the energies of the 2*p* AOs decrease in this same order across a row of the periodic table. Given that the energy of the 1*s* AO for H^+ is higher in energy than any of the 2*p* AOs on the heteroatoms, as shown in the diagram below, the HOMO for NH_3 will lie closest in energy to the LUMO for H^+. Thus, the bonding MO formed in the $NH_3 + H^+ \rightarrow NH_4^+$ interaction diagram will be lower in energy than the analogous bonding MOs formed in H_3O^+ and HFH^+. The lower in energy the bonding MO is, the greater the magnitude of the PA.

HOMO of the base | LUMO of the acid

E

NH_3 | H^+

H_2O

HF

In 1988, Ralph Pearson, building on the work of Klopman, published a paper defining *absolute hardness*, η, as half the difference between the ionization energy and the electron affinity (measured in units of eV), as shown by Equation (14.24). Using this definition, a hard compound will be one where the IE and EA are dissimilar and a soft compound will be one where the IE and EA have approximately the same magnitude. Because the ionization energy of a species is the amount of energy necessary to remove an electron from its HOMO and the electron affinity is a measure of how much energy is gained by the addition of an electron to its LUMO, the absolute hardness could be defined alternatively as half the difference in energy between a compound's HOMO and its LUMO:

$$\eta = \frac{\text{IE} - \text{EA}}{2} \tag{14.24}$$

$$\chi_{\text{MJ}} = \frac{\text{IE} + \text{EA}}{2} \tag{14.25}$$

The absolute hardness is not to be confused with the Mulliken–Jaffe definition of electronegativity, given by Equation (14.25), where the electronegativity χ is defined as the *average* of the IE and EA (for the valence electrons). The absolute hardness and

TABLE 14.11 The ionization energy (IE), electron affinity (EA), Mulliken–Jaffe electronegativity (χ), and absolute hardness (η) for selected Lewis acids and bases.

Species	IE, eV	EA, eV	χ, eV	η, eV	Species	IE, eV	EA, eV	χ, eV	η, eV
Li^+	75.64	5.39	40.52	35.12	$(CH_3)_3N$	7.8	−4.8	1.5	6.3
Mg^{2+}	80.14	15.04	47.59	32.55	PF_3	12.3	−1.0	5.7	6.7
Al^{3+}	119.99	28.45	74.22	45.77	PH_3	10.0	−1.9	4.1	6.0
Cu^+	20.29	7.73	14.01	6.28	SO_2	12.3	1.1	6.7	5.6
Zn^{2+}	39.72	17.96	28.84	10.88	C_6H_6	9.3	−1.2	4.1	5.3
Ru^{2+}	28.47	16.76	22.62	5.86	C_5H_5N	9.3	−0.6	4.4	5.0
Rh^{2+}	31.06	18.08	24.57	6.49	F^-	17.42	3.40	10.4	7.01
Pd^{2+}	32.93	19.43	26.18	6.75	Cl^-	13.01	3.62	8.31	4.70
Ag^+	21.49	7.58	14.53	6.96	Br^-	11.84	3.36	7.60	4.24
Au^+	20.5	9.23	14.9	5.6	I^-	10.45	3.06	6.76	3.70
Hg^{2+}	34.2	18.76	26.5	7.7	OH^-	13.17	1.83	7.50	5.67
BF_3	15.81	−3.5	6.2	9.7	NO_2^-	> 10.1	2.30	> 6.2	> 3.9
H_2O	12.6	−6.4	3.1	9.5	CN^-	14.02	3.82	8.92	5.10
NH_3	10.7	−5.6	2.6	8.2					

The electronegativities are not adjusted to the Pauling scale.

related parameters for a variety of acids and bases are listed in Table 14.11. While the electronegativity represents the average valence electron energy, absolute hardness is a measure of the breadth of the LUMO–HOMO energy difference, as shown in Figure 14.9 for the halide ions. The energy difference is greatest for the F^- ion, which falls under the "hard" category in Pearson's HSAB model. Iodide, which has

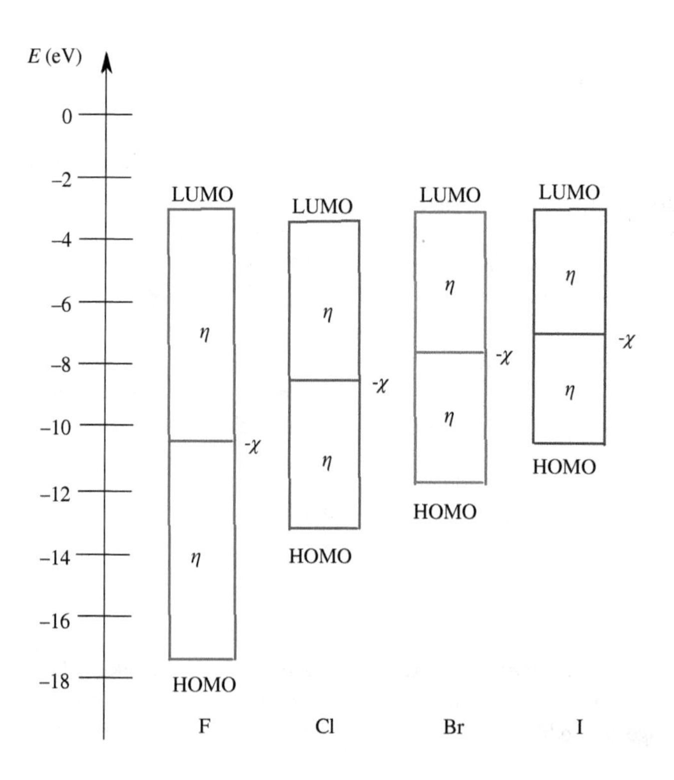

FIGURE 14.9
Mulliken–Jaffe electronegativity (χ) and absolute hardness (η) for the halide ions, defined using the frontier molecular orbital model.

the smallest energy difference, falls under the "soft" classification. This quantitative scale allows for direct comparisons between two different species belonging to the same classification. For instance, Table 14.7 indicates that Li^+ and Al^{3+} are both hard Lewis acids, but it did not specify exactly where along the hardness spectrum each of these two ions fell. Using the data in Table 14.11, it is clear that the Al^{3+} ion is the stronger Lewis acid of the pair. This also makes sense intuitively, as Al^{3+} has a higher charge density than Li^+.

The FMO definition also helps explain *why* Pearson's hard–hard and soft–soft interactions form stable complexes. Hard compounds have a large HOMO–LUMO gap, as shown in Figure 14.9 for F. Therefore, hard Lewis acid–base complexes tend to form strongly ionic compounds, such as LiF, where the interaction is dominated by electrostatic attractions. Soft compounds, on the other hand, have a small HOMO–LUMO gap, as shown in Figure 14.9 for I, so that these types of interactions form covalently bonded acid–base adducts, where the strength of the interaction is controlled primarily by the energies of the FMOs that participate in the bonding.

14.4 OXIDATION–REDUCTION REACTIONS

Perhaps not surprisingly, there is also a connection between the HSAB properties of ions and their redox properties. For example, the *noble* metals, those that are resistant to corrosion or oxidation in air, are all derived from soft-acid cations (Ru, Rh, Pd, Ag, Os, Ir, Pt, Au, and Hg). In fact, Pd, Pt, Au, and Hg can only be dissolved in *aqua regia* (or "king's water"), which is a 1 : 3 (v : v) mixture of concentrated HNO_3 with HCl. The $E°$ values for selected noble metals are listed in Table 14.12. As expected, these metals have very positive standard reduction potentials. For comparison, the standard reduction potentials for a variety of other materials were already presented in Table 5.11.

Because, in general, the soft-acid cations tend to have large electronegativities, it follows that there might be a trend in the standard reduction potentials of the metal cations and their Pauling electronegativities. In fact, there is quite an excellent correlation between the two, as demonstrated by the data in Figure 14.10.

Wulfsberg has classified the redox properties of the elements into five broad categories on the basis of their electronegativities. The hard-acid metal cations having $\chi < 1.4$ tend to have very negative standard reduction potentials ($E° < -1.6$ V) and comprise the *very active metals*. These include all of the alkali metals and most of the alkaline earth metals (with the exception of Be). The heavier alkali metals are extremely reactive and can reduce water to hydrogen gas. The alkali metals can also dissolve in liquid ammonia to yield blue solutions containing a solvated electron. The second category consists of the borderline-acid cations having $1.4 < \chi < 1.9$. These metal ions typically have standard reduction potentials in the range $-1.6\,V < E° < 0\,V$ and are known as the *moderately active metals*. Examples include most of the first-row transition metals and the early period second- and third-row metals, as well as Be, Al, and Ga. The third category contains the soft-acid metals having $1.9 < \chi < 2.55$. These include the later period second- and third-row transition metals, as well as In, Tl, Sn, and Pb. These metals typically have $E° > 0$ and are known collectively as the *inactive metals* because they cannot reduce hydrogen. As stated earlier, the noble metals are particularly unreactive. The fourth category are the soft-basic anions formed from the less electronegative nonmetals ($1.9 < \chi < 2.8$) such as B, C, P, S, As, Se, Te, and I. These are referred to as the *relatively inactive nonmetals*. Lastly, the *inactive nonmetals* are those derived from borderline or hard-basic anions and having electronegativities greater than 2.8. These include the most electronegative nonmetals N, O, F, Cl, and Br. With the exception of N_2, the elemental forms of these compounds are potent

TABLE 14.12 Standard reduction potentials for selected ions of the noble metals.

Reduction Half-Reaction	$E°$ (V vs NHE)
$Au^{3+} + 3\ e^- \rightarrow Au$	1.50
$Pt^{2+} + 2\ e^- \rightarrow Pt$	1.18
$Ir^{3+} + 3\ e^- \rightarrow Ir$	1.16
$Pd^{2+} + 2\ e^- \rightarrow Pd$	0.99
$OsO_4 + 8\ H^+ + 8\ e^- \rightarrow Os + 4\ H_2O$	0.84
$Ag^+ + e^- \rightarrow Ag$	0.80
$Hg_2{}^{2+} + 2\ e^- \rightarrow 2\ Hg$	0.80

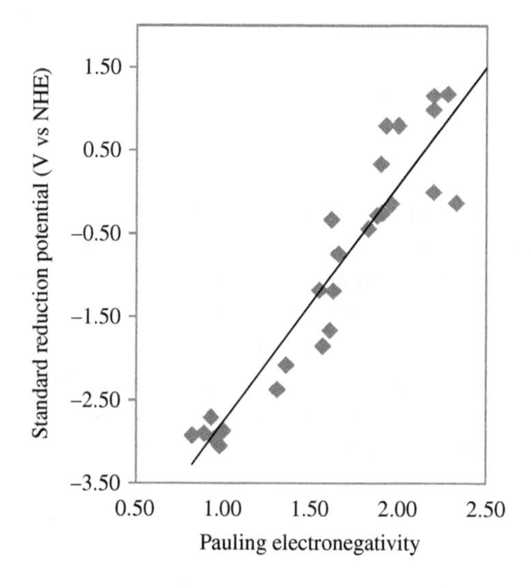

FIGURE 14.10
Plot of the standard reduction potential $E°$ (V vs NHE) for the most common oxidation state of representative metal cations versus their corresponding Pauling electronegativities. Standard reduction potentials are from Tables 5.11 and 14.12; Pauling electronegativities are from Table 5.12 or http://www.webelements.com.

oxidizing agents. Taken collectively, Wulfseberg's five rules can be summarized into a single general statement: "soft-acid cations make good oxidizing agents and soft-base anions make good reducing agents."

In theory, the standard reduction potential for a metal ion can be calculated using a Born–Haber-type thermochemical cycle. The reduction half-reaction is the sum of the negative of the atomization energy, the negative of the ionization energy, and the negative of the hydration enthalpy, as shown in Equations (14.26)–(14.28):

$$M(g) \rightarrow M(s) \qquad -\Delta H_a \qquad (14.26)$$

$$M^+(g) + e^- \rightarrow M(g) \qquad -IE \qquad (14.27)$$

$$M^+(aq) \rightarrow M^+(g) \qquad -\Delta H_{hydr} \qquad (14.28)$$

The equation that results from this type of analysis is given by Equation (14.29), where z is the charge on the ion, $E°$ is in V versus NHE, and the units for enthalpy are in kJ/mol:

$$E° = \frac{1}{96.5}\left[\frac{\Delta H_a^\circ}{z} + \frac{\sum IE}{z} + \frac{\Delta H_{hyd}^\circ}{z} - 439\,\text{kJ/mol}\right] \qquad (14.29)$$

Example 14-9. Calculate $E°$ for the Na^+ ion using Equation (14.29).

Solution. Using the sublimation energy for Na given in Chapter 12 (107 kJ/mol), the ionization energy from Table 5.7 (496 kJ/mol) and the enthalpy of hydration for Na^+ given in Table 5.5 (−405 kJ/mol), after substitution:

$$E° = \frac{1}{96.5}\left[\frac{107}{1} + \frac{496}{1} - \frac{405}{1} - 439\right] = -2.50\,\text{V}$$

The actual value of $E°$ for the half-reaction $Na^+ + e^- \rightarrow Na$ is −2.71 V versus NHE.

14.5 A GENERALIZED VIEW OF MOLECULAR REACTIVITY

In his *Chemogenesis WebBook*, Dr. Mark Leach argues that "the Lewis acid and Lewis base concept organizes and 'explains' the majority of reaction chemistry that school and university students are expected to be familiar with." Employing FMO concepts, Leach categorizes Lewis acids into six different groups on the basis of the topology of their interacting orbitals: the proton, s-LUMO ions, onium ions, lobe-LUMO acids, π-LUMO acids, and the heavy metal acids. Similarly, there are four types of Lewis bases: s-HOMO bases, complex anions, lobe-HOMO bases, and π-HOMO bases. As Lewis acids combine with Lewis bases, they form the interaction matrix shown in Figure 14.11. Thus, from the innumerable different types of chemical reactions in the literature that fall into the Lewis acid–base category (including much of organic, main group, and organometallic chemistry), there are only 24 basic types of molecular interactions.

The first type of Lewis acid is also the most important—the proton. Because of its small size and its large charge : size density, the proton can interact with any of the four different types of Lewis base FMOs. The transfer of a proton from one Lewis base to another encompasses all of the chemical reactivity relating to Brønsted–Lowry acid–base chemistry. The second type of Lewis acid is the s-LUMO acids, which are composed of the alkali and alkaline earth metal cations that react via their *ns* LUMOs. The s-LUMO acids fall under Pearson's "hard" classification. The smaller s-LUMO acids are especially polarizing and tend to form covalently bonded Lewis acid–base adducts. The third class of Lewis acid is the onium ions, such as NH_4+ and H_3O^+. The defining characteristic of the onium ions is the presence of an electronegative central atom surrounded by H^+ or R^+ ligands to form a complex cation. Onium ions are typically hard ions and they participate in charge-controlled or ionic reactions. If they transfer an H^+ ligand, they act as a Brønsted–Lowry acid; if they transfer R^+, they behave as an alkylating agent. The fourth class of Lewis acid is the electron-deficient lobe-LUMO acids such as BF_3 or carbocations, which react using an empty *p*-orbital. A number of organic functional groups that are susceptible to nucleophilic attack, such as alkyl halides and $R_2C{=}O$ carbonyls, belong to the lobe-LUMO classification. The fifth type of Lewis acid is the electron poor π-LUMO acids, such as the allyl cation or TCE (tetracyanoethylene), and these often contain delocalized cationic π-systems that act as "soft" Lewis acids by accepting electron density using their low-lying π^* MOs. The final category of Lewis acid is the heavy metal acids, which include the elemental metals, as well as their metal cations. These are typically "hard" Lewis acids such as the third-row transition metals, post-transition metals, lanthanides, and actinides, particularly those ions having

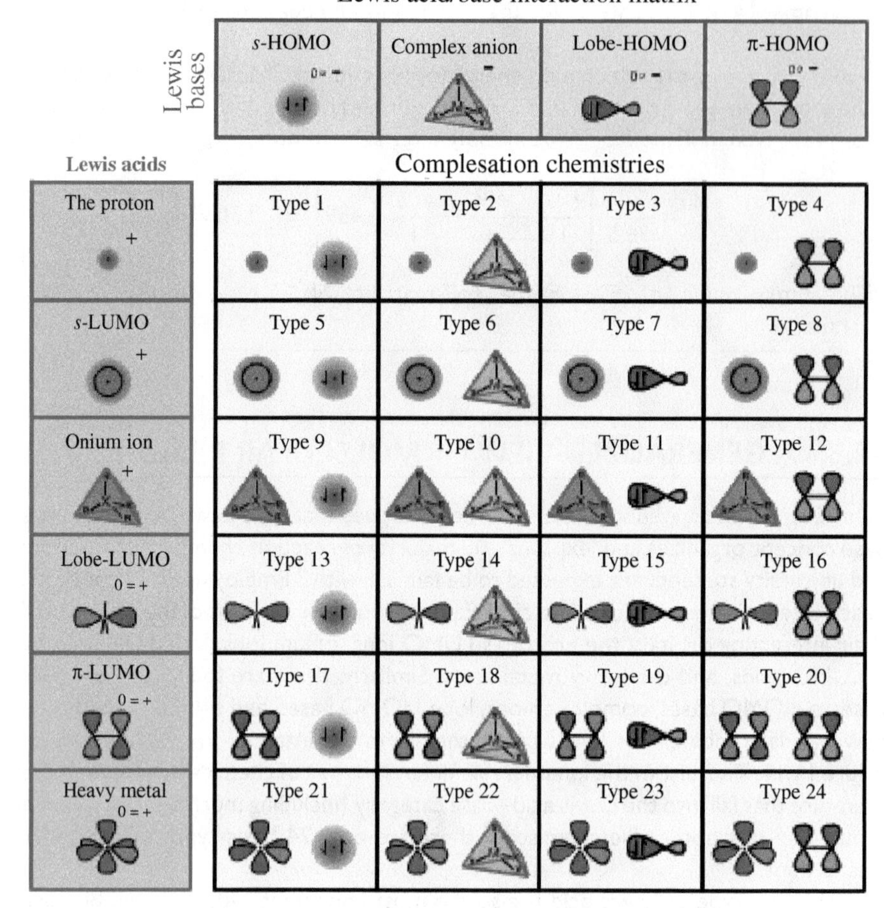

FIGURE 14.11
Interaction matrix for the 24 different types of chemical reactions involving Lewis acids and bases, organized according to the topology of their interacting frontier molecular orbitals. [© Mark R Leach, meta-synthesis (accessed October 22, 2013).]

high oxidation numbers. The heavy metals have a variety of empty orbitals that they can use in their bonding, and they frequently participate in π-backbonding.

On the basis of their FMO topologies, there are four general types of Lewis bases. The first class of Lewis base is the s-HOMO bases, H^- and H_2. They react using their closed-shell 1s AO or σ-bonding MO, respectively, and behave as "soft" bases that either undergo proton abstractions or act as reducing agents. The second class of Lewis base is the complex anion bases, which include AlH_4^-, SbF_6^-, and $Cr_2O_7^{2-}$. The complex anion Lewis bases belong to the "hard" base classification and often participate as nucleophiles in ligand substitution reactions. The third, and most common, type of Lewis base is the lobe-HOMO bases, such as CH_3^-, H_2O, OH^-, NH_3, and CN^-. This important class of Lewis bases reacts using a filled sp^n-hybrid orbital or unhybridized p-AO to form a σ-bond with a Lewis acid. They often participate as Brønsted–Lowry bases, ligands, nucleophiles, leaving groups, and spectator ions. The fourth and final category of Lewis base is the electron-rich π-HOMO bases, which consist of molecules containing delocalized π-electron systems. This class of Lewis bases, which includes the allyl anion, benzene, and cyclopentadienyl ion, is especially important in the field of organometallic chemistry.

The six types of Lewis acids and four types of Lewis bases can interact in 24 distinct ways, as shown in Figure 14.11. This type of FMO analysis has been successfully employed to explain a plethora of different reaction mechanisms: from

Lewis acids \ Lewis bases	s-HOMO	Complex anion	Lobe-HOMO	π-HOMO
The proton	1 1σ MO complexes	2 George olah type super acids	3 Common brønsted acids	4 Protonation of π-systems
s-LUMO	5 Saline hydrides	6 Ionic salts or hydride donor reagents (If ligand = H)	7 Common proton abstracting bases and ionic salts	8 Stabilised π-anion complexes
Onium ion	9 Reduction of onium ion	10 Ionic salts	11 Ionic salt, proton transfer or alkylation	12 Proton transfer or alkylation
Lobe-LUMO	13 Reduction of the lewis acid	14 Friedel-crafts enium ion $[X]^+$ reagents	15 Mechanistic pathways S_N1 S_N2 S_E1 S_E2	16 Electrophilic addition and SEAr substitution
π-LUMO	17 Reduction of the organic π-system	18 Stabilised salts of cationic π-system	19 Nucl. attack on the organic π-system	20 Diels-Alder cycloaddition & charge transfer cmplx.
Heavy metal	21 Hydrides	22 Ionic salts	23 Co-ordination chemistry	24 Organometallic chemistry

FIGURE 14.12 The different types of Lewis acid–base interactions sorted according to their more traditional classifications in inorganic, main group, organic, coordination, and organometallic chemistry. [© Mark R Leach, meta-synthesis (accessed October 22, 2013).]

classical Brønsted–Lowry acid–base chemistry to redox chemistry, from organic chemistry to organometallics, and from main group chemistry to transition metal complexes. A summary of the different types of interaction chemistry is shown in Figure 14.12. The distinctions between the traditional subdisciplines of chemistry can be loosely mapped to the different types of FMO acid–base interactions. For instance, almost all Olah-type superacids (such as H_2SO_4, HSO_3CF_3, and $HClO_4$) are Type 2 complexes; the common Brønsted–Lowry acids (HCl, HBr, and HI) are Type 3 complexes; organic nucleophilic substitution (S_N1 and S_n2) reactions belong to Type 15; Diels–Alder cycloaddition reactions fall under Type 20; coordination chemistry (the subject of Chapters 15–17) belongs to Type 23; and most of organometallic chemistry (Chapters 18 and 19) fits into the Type 24 classification.

The principal objectives of this particular classification scheme are to demystify the myriad of reaction mechanisms in the chemical literature and to illustrate the parallels among organic, inorganic, and organometallic reaction chemistry. In this context, the acid–base reaction is reduced to its most basic form: a systematic analysis of the fundamental types of interactions between the LUMO of the Lewis acid and the HOMO of the Lewis base. Although the rest of the molecule cannot be wholly ignored because electronegativity, inductive effects, and steric interactions will also affect the MO diagram, the aim of keeping the focus on the topology of the interacting FMOs is intended to help the average student of chemistry "to see the forest through the trees."

On a final note, we shall attempt to show how FMO theory can be used to connect the seemingly diverse Lewis acid–base reactions and oxidation–reduction reactions. When the LUMO of the acid and HOMO of the base have similar energies and appropriate symmetries for nonzero overlap, they can combine to form a

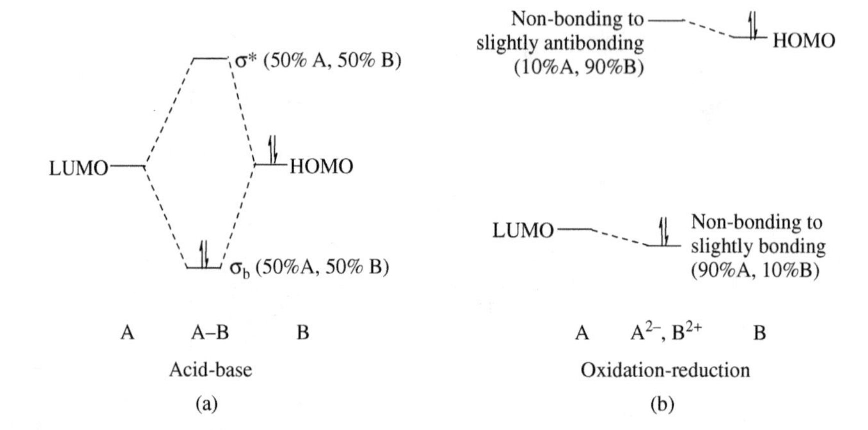

FIGURE 14.13
Frontier MO diagram illustrating the similarities and differences between (a) an acid–base reaction and (b) an oxidation–reduction reaction.

bonding (and antibonding) MO, as shown in Figure 14.13(a). Because the bonding MO is lower in energy than either of the two reactants, the products will be more stable than either of the reactants. The thermodynamic driving force for the reaction is related to the degree of orbital overlap and how closely the energy match is between the LUMO of the acid and the HOMO of the base. A useful pedagogical example would be the reaction of $H^+ + H^- \rightarrow H_2$. In this case, the empty 1s AO on H^+ (its LUMO) and the filled 1s AO on H^- (its HOMO) have identical energies. Taking linear combinations of the two AOs results in the formation of a sigma bonding MO and a sigma antibonding MO, as shown in Figure 14.13(a). On the other hand, when the energies of the LUMO on A and HOMO on B are dissimilar (where one is significantly higher in energy than the other), then the complete transfer of electrons is possible and an oxidation–reduction reaction results, as shown in Figure 14.13(b).

From this perspective, acid–base and oxidation–reduction reactions lie at the two extremes of a continuous spectrum of chemical reactivity. An acid–base reaction is analogous to the formation of a coordinate covalent bond (or Lewis acid–base adduct) because the electrons in the product are shared approximately equally in a bonding MO that spans both atoms A and B. These types of reactions are referred to as *FMO-controlled.* The bonding MO will have equal weighting coefficients from each of the AOs that were used to construct the MOs. On the other hand, an oxidation–reduction reaction is analogous to the formation of an ionic bond, where one or more electrons are transferred from one species to another to form ions. These types of reactions are referred to as *charge-controlled.* In this case, the weighting coefficients for the AOs used to construct the lower-lying MO are dominated by those of a single AO. Thus, the resulting MO is essentially of nonbonding character.

Taking the nonbonding $1b_2$ HOMO and $4a_1$* LUMO in H_2O (Figure 14.7) as the reference molecule A, H_2O can undergo four basic types of chemical reactions, as shown in Figure 14.14. If the energy of the HOMO of molecule B is significantly higher in energy than the FMOs in H_2O, as is the case when Na metal reacts with water according to Equation (14.30), this is an example of a Type 13 interaction. Because of the energy mismatch between the FMOs, an oxidation–reduction reaction results, where water is acting as an oxidant:

$$2\ Na(s) + 2\ H_2O\ (l) \rightarrow 2\ NaOH\ (aq) + H_2(g) \qquad (14.30)$$

If, on the other hand, the energy of the HOMO for molecule B is comparable in energy to the FMOs of water, then an acid–base reaction will occur. If the

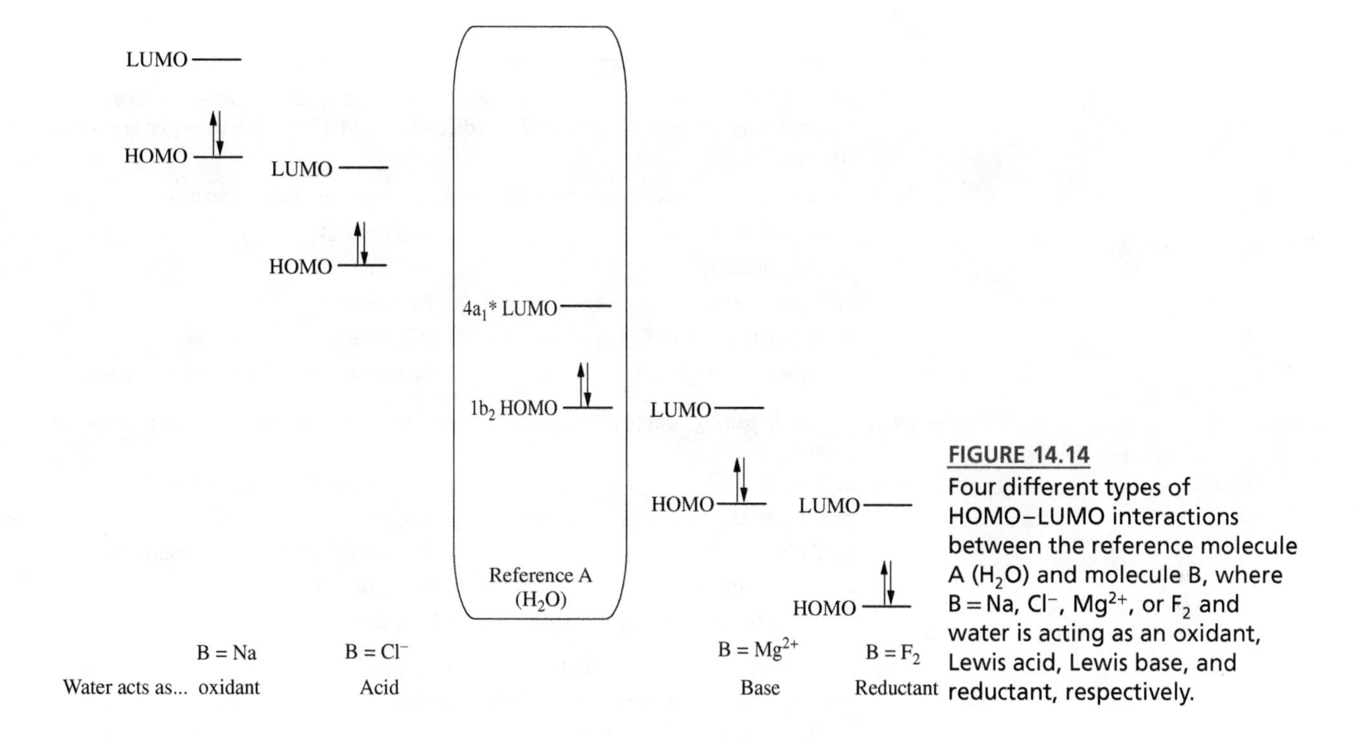

FIGURE 14.14 Four different types of HOMO–LUMO interactions between the reference molecule A (H_2O) and molecule B, where B = Na, Cl^-, Mg^{2+}, or F_2 and water is acting as an oxidant, Lewis acid, Lewis base, and reductant, respectively.

energy of the HOMO on B is higher than the LUMO on A, such as the case where B = Cl^-, a Type 15 interaction occurs, where H_2O acts as a Lewis acid, as shown by Equation (14.31). If the energy of the HOMO on B is lower than the LUMO on A, then H_2O can act as a Lewis base. This is the case for the Type 23 interaction given by Equation (14.32), when B = Mg^{2+}:

$$Cl^-(aq) + 6\ H_2O\ (l) \rightarrow [Cl(H_2O)_6]^-(aq) \qquad (14.31)$$

$$Mg^{2+}(aq) + 6\ H_2O\ (l) \rightarrow [Mg(H_2O)_6]^{2+}(aq) \qquad (14.32)$$

Finally, if the FMOs on B are significantly lower in energy than the FMOs on A, then a redox reaction will result, where H_2O is acting as a reducing agent, such as the reaction with F_2 shown in Equation (14.33). Each H_2O molecule donates a pair of electrons to fluorine to generate fluoride ions as the product species:

$$2\ F_2(g) + 2\ H_2O\ (l) \rightarrow 4\ F^-(aq) + 4\ H^+(aq) + O_2(g) \qquad (14.33)$$

In this chapter, we have shown how the majority of chemical reactions can fall into one of two categories: acid–base reactions or oxidation–reduction reactions. The FMO definition of acid–base chemistry is by far the most detailed and inclusive of the theories introduced in this chapter. If the energy match of the interacting FMOs is similar, the reaction will be FMO-controlled and appear as a classical Lewis acid–base reaction, whereas if there is an energy mismatch between the FMOs, the reaction will be charge-controlled and an oxidation–reduction reaction will occur. Furthermore, we have shown how virtually every type of Lewis acid–base reaction mechanism from classical inorganic chemistry to organic chemistry and everything in between can be rationalized on the basis of the topology of the interacting FMOs. Thus, the topological FMO model perfectly illustrates a common theme of this text: the symmetry and structure of molecules are intimately connected to their chemical reactivities.

EXERCISES

14.1. The bicarbonate ion is amphoteric, which means that it can act either as an acid or a base. Using the data in Table 14.2, predict whether HCO_3^- is a stronger acid or a stronger base.

14.2. For each pairing, determine which will be the strongest acid and explain your answer:

a. PH_3 or H_2S
b. PH_3 or AsH_3
c. H_2Se or H_2S
d. F_3COSO_3 or CH_3OSO_3
e. H_2SeO_3 or H_2SeO4
f. HIO or HIO_4
g. Al^{3+} or Fe^{3+}
h. Fe^{2+} or Fe^{3+}
i. BCl_3 or BF_3
j. Benzoic acid or 3,5-dinitrobenzoic acid

14.3. For each pairing, determine which will be the strongest base and explain your answer:

a. NH_3 or NCl_3
b. NF_3 or BF_3
c. CaO or CO_2
d. CaO or MgO
e. $Be(OH)_2$ or $Ca(OH)_2$
f. $Ca(OH)_2$ or $Sr(OH)_2$
g. PH_3 or AsH_3
h. PCl_3 or PPh_3 (triphenylphosphine)
i. SO_3^{2-} or SO_4^{2-}
j. NH_3 or $N(CH_3)_3$

14.4. For each of the following chemical reactions, identify the acid and base. Furthermore, indicate whether the acid is a Brønsted–Lowry acid.

a. $SO_3 + H_2O \rightarrow HSO_4^- + H^+$
b. $Al_2O_3 + 2\ OH^- + 3\ H_2O \rightarrow 2\ [Al(OH)_4]^-$
c. $CH_3[B_{12}] + Hg^{2+} \rightarrow [B_{12}]^+ + CH_3Hg^+$ (where $[B_{12}]$ represents Vitamin B_{12})
d. $SiCl_4 + 4\ H_2O \rightarrow Si(OH)_4 + 4\ HCl$
e. $AsF_3 + SbF_5 \rightarrow [AsF_2] + [SbF_6]$
f. $Na_2O + H_2O \rightarrow 2\,Na^+ + OH^-$
g. $NaHCO_3 + HCl \rightarrow H_2O + CO_2 + NaCl$
h. $KCl + SnCl_2 \rightarrow K^+ + [SnCl_3]^-$
i. $FeCl_3 + 6\ KCN \rightarrow K_3[Fe(CN)_6] + 3\ KCl$
j. $2\ HSO_3^- \rightarrow H_2SO_3 + SO_3^{2-}$

14.5. What is the strongest possible acid in a solution of acetic acid?

14.6. Calculate the expected pK_a of Mg^{2+}, given that $r = 86$ pm and $\chi = 1.31$.

14.7. Calculate the expected pK_a of HClO.

14.8. Calculate the expected pK_b of the nitrite ion, NO_2^-.

14.9. Will Cu^{2+} react more strongly with OH^- or NH_3? Explain your answer.

14.10. Will Fe^{3+} react more strongly with OH^- or NH_3? Explain your answer.

14.11. Which of the following coordination compounds will be more stable: $NiCl_4^{2-}$ or $CuCl_4^{2-}$? Explain your answer.

14.12. Which of the following coordination compounds will have the larger overall formation constant, β_4: $[Cr(OH)_4]^-$ or $[Zn(OH)_4]^{2-}$? Explain your answer.

14.13. The SCN^- ligand is an ambidentate ligand. When it is S-bound, it has the name thiocyanate and the M–S–C bond angle is bent, but when it is N-bound, it has the name isothiocyanate and the M–N–C bond angle is ~180°. Predict the structures of the metal–ligand bonding when the SCN^- ligand reacts with Fe^{2+} to form the bright red $Fe(SCN)^{2+}$ ion. Predict the method of bonding when Cr^{3+} reacts with the SCN^- ligand. Explain why the S-bound species is bent.

14.14. Will the equilibrium constant for $CH_3HgI + HCl \Longleftrightarrow CH_3HgCl + HI$ be greater or less than 1? Explain.

14.15. Use the Drago–Wayland parameters in Table 14.9 to calculate ΔH for the following reaction in units of kcal/mol: $I_2 + C_6H_6 \rightarrow I_2{\cdot}C_6H_6$.

14.16. Use the Drago–Wayland parameters in Table 14.9 to determine which of the following Lewis acid–base adducts will be more thermodynamically stable: $BF_3{\cdot}NH_3$ or $BF_3{\cdot}N(CH_3)_3$.

14.17. Predict the water solubility of each of the following salts:

a. $MgSO_4$
b. $FeBr_3$
c. K_2CO_3
d. $Cu(OH)_2$

14.18. Which of the following coordination compounds is likely to have the more positive standard reduction potential, $[Cu(H_2O)_4]^{2+}$ or $[Cu(NH_3)_4]^{2+}$? Explain your answer.

14.19. Which of the following will have the smallest value for the solubility product, K_{sp}, ZnS or HgS? Explain your answer.

14.20. Which of the following silver complexes will be most easily reduced: $[Ag(NH_3)_2]^+$ or $[Ag(S_2O_3)_2]^{3-}$? Explain your answer.

14.21. The preferred method of extracting gold from its ore is the MacArthur–Forrest process, in which gold is first converted into a soluble coordination complex using the following equation:

$$Au + 8\ NaCN + O_2 + 2H_2O \rightarrow 4Na[Au(CN)_2] + 4\ NaOH$$

a. Use Equation (14.29) to calculate $E°$ for the following half-reaction, given that $\Delta H_a = 368$ kJ/mol, IE = 890 kJ/mol, and ΔH_{hyd} for $Au^+ = -615$ kJ/mol: $Au^+ (aq) + e^- \rightarrow Au\ (s)$.

b. Given that $E° = +0.60$ V versus NHE for the complex anion $[Au(CN)_2]^-$, calculate the stability constant for the following reaction:

$$Au^+(aq) + 2\ CN^-(aq) \Longleftrightarrow [Au(CN)_2]^-(aq) \qquad K = [Au(CN)_2{}^-]/[Au^+][CN^-]^2$$

c. Use Equation (14.29) to predict ΔH_{hyd} for Au^{3+} (aq) given that $\Delta H_a = 368$ kJ/mol, $IE_1 = 890$ kJ/mol, $IE_2 = 1980$ kJ/mol, $IE_3 = 2900$ kJ/mol, and $E° = 1.50$ V versus NHE for the following half-reaction: $Au^{3+} (aq) + 3\ e^- \rightarrow Au\ (s)$.

d. Use the above data to rationalize why the MacArthur–Forrest method is the preferred way of extracting gold from its ore.

14.22. Predict the product when each of the following is reacted with H_2O (write the balanced chemical equation):

a. Ca (s)
b. HCl (g)
c. Fe^{3+} (aq)

14.23. Classify each of the following reactions into one of the categories in Figure 14.11:

a. $Zr^{4+} + SiO_4{}^{4-} \rightarrow ZrSiO_4$
b. $Zn^{2+} + 4\ H^- \rightarrow ZnH_4{}^{2-}$
c. $FeCl_2 + 2\ LiC_5H_5 \rightarrow (C_5H_5)_2Fe$ (ferrocene) + 2 LiCl
d. $XeF_4 + 2\ H_2 \rightarrow Xe + 4\ HF$

BIBLIOGRAPHY

1. Douglas B, McDaniel D, Alexander J. *Concepts and Models of Inorganic Chemistry*. 3rd ed. New York: John Wiley & Sons, Inc.; 1994.
2. Housecroft CE, Sharpe AG. *Inorganic Chemistry*. 3rd ed. Essex, England: Pearson Education Limited; 2008.

3. Huheey JE, Keiter EA, Keiter RL. *Inorganic Chemistry: Principles of Structure and Reactivity*. 4th ed. New York: Harper Collins College Publishers; 1993.
4. Leach, M. R. *The Chemogenesis WebBook*, www.meta-synthesis.com, 1999–2014.
5. Miessler GL, Tarr DA. *Inorganic Chemistry*. 4th ed. Upper Saddle River, NJ: Pearson Education Inc.; 2011.
6. Mingos DMP. *Essential Trends in Inorganic Chemistry*, Oxford University Press: New York; 1998.
7. Wulfsberg GP. Periodic Table: Trends in the Properties of the Elements. In: King RB, editor. *Encyclopedia of Inorganic Chemistry*. Vol. 1. John Wiley & Sons, Inc.: New York; 1994.

An Introduction to Coordination Compounds 15

"Alfred Werner awoke at 2 a.m. one night in late 1892. In a dream he had solved the riddle of the constitution of 'molecular compounds'"

—*George Kauffman*

15.1 A HISTORICAL OVERVIEW OF COORDINATION CHEMISTRY

One of the most distinguishing characteristics of transition metals is the wide array of colors that they form in their compounds. It should therefore come as no surprise that the history of coordination chemistry has its foundations in the field of art. Many of the early pigments and dyes used by artists were, in fact, composed of coordination compounds, although an understanding of what that term meant was not developed until the late nineteenth century. Prussian blue is a coordination compound containing both iron(II) and iron(III) linked together by cyanide ligands ($KFe^{III}[Fe^{II}(CN)_6]$). It was discovered in 1704 by Diesbach and gets its name from the fact that it was used to dye the uniforms of the Prussian army a deep blue color. It has been used by artists as a pigment in paintings from the mid-eighteenth century until the early 1900s. Prussian blue has also been used in cyanotype photography, blueprints, and blue jeans. The bright yellow dye aureolin, or cobalt yellow, has been in use as a pigment since 1831. Its chemical formula is $K_3[Co(NO_2)_6]{\cdot}6H_2O$. Alizarin red is an aluminum lake dye derived from the roots of the madder plant. It was used in both ancient Persia and Egypt, as well as by the Greeks and the Romans. Alexander the Great might have once used the dye to make bloody splotches on his soldiers' uniforms in order to encourage the Persians to attack what they believed was a demoralized army. The French and British military have also used alizarin on their uniforms. When Paul Revere said that "the redcoats are coming," he was referring to the bright red color of the uniforms of the colonial British soldiers. Today, the alizarin red dye is used as a stain in histology. The structures of these three coordination compounds are shown in Figure 15.1. For Prussian blue, the K^+ ions are not shown and for alizarin, the waters of hydration have been omitted.

Structure of cisplatin. [Reproduced from http://en.wikipedia.org/wiki/Coordination_complex (accessed January 17, 2014).]

Principles of Inorganic Chemistry, First Edition. Brian W. Pfennig.

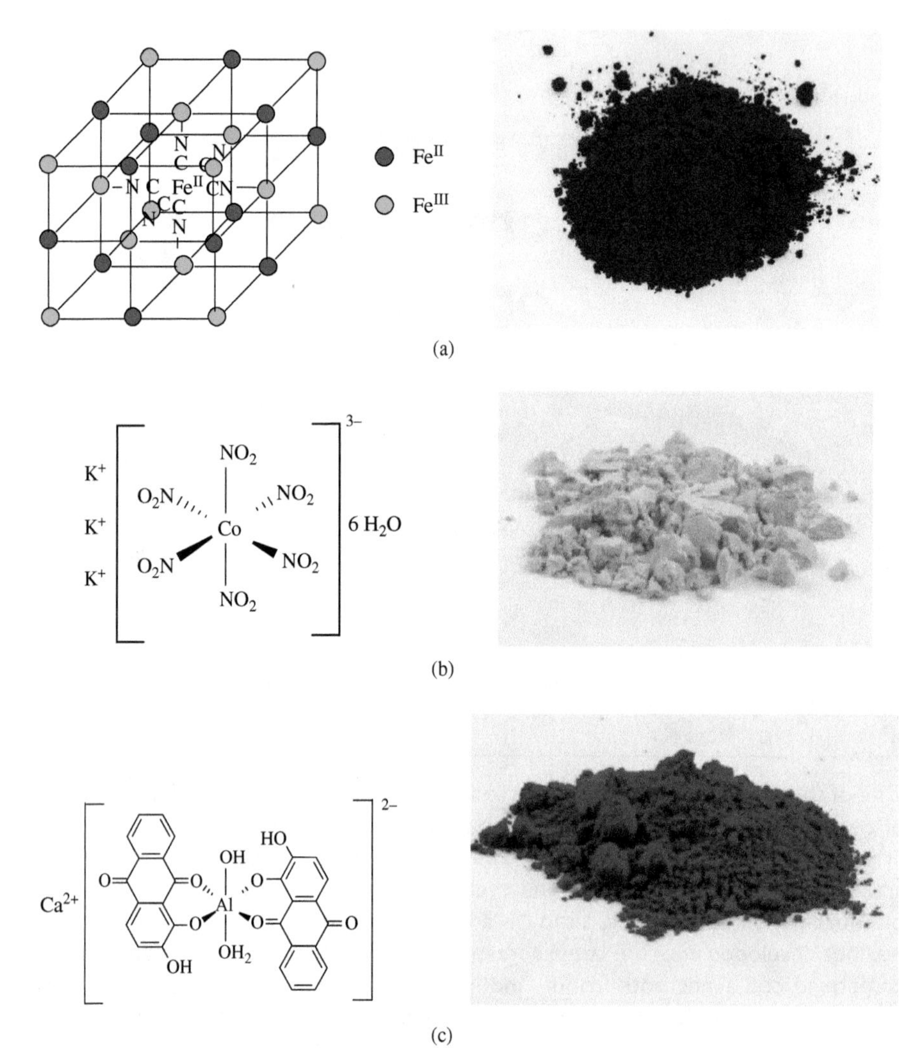

FIGURE 15.1
Structures of the (a) Prussian blue, (b) aureolin, and (c) alizarin dyes. [Pigment photos courtesy of Kremer Pigments, Inc.]

By definition, a *coordination compound* is any compound where the coordination number of the central atom is larger than its oxidation number. Thus, for instance, the anion $[Al(OH)_4]^-$ is considered a coordination compound, while the SO_4^{2-} ion is not. An alternative definition is that a coordination compound is formed by joining independent molecules or ions (known collectively as the *ligands*) to a central atom using coordinate covalent bonds. A *coordinate covalent* (or *dative*) *bond*, similar to a covalent bond, is one in which the metal and the ligand both share a pair of electrons, but where both the electrons in the bonding pair originate from the ligand. Thus, in a sense, a coordination compound is really a Lewis acid–base adduct. The requirement of a metal excludes main group compounds having a nonmetallic central atom, such as PCl_5 or SF_6.

The first real theory of coordination chemistry was developed by Graham in 1837. He argued that metal ions bonded to the ammonium ion by the displacement of one of the H atoms by the metal. This early model, however, could not explain the coordination of tertiary amines to transition metals, nor could it account for molecules such as $CoCl_3{\cdot}6NH_3$, in which the number of NH_3 molecules exceeds the valence of the metal. In 1871, Blomstrand adapted Graham's ammonium model by stating that because N is pentavalent, the ammonia molecules could form

TABLE 15.1 Cobalt(III) coordination compounds, their proposed structures using the Blomstrand–Jørgensen and Werner models, and the predicted number of AgCl equivalents upon reaction with excess $AgNO_3$.

Name (Color)	Jørgensen's Chain Formulation	Number of AgCl	Werner's Coordination Formulation	Number of AgCl
Luteo (yellow)	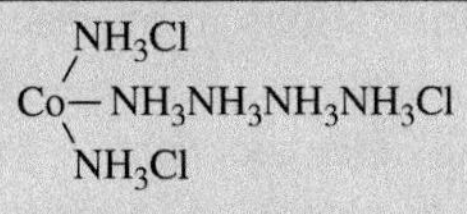 $Co(NH_3Cl)(NH_3NH_3NH_3NH_3Cl)(NH_3Cl)$	3	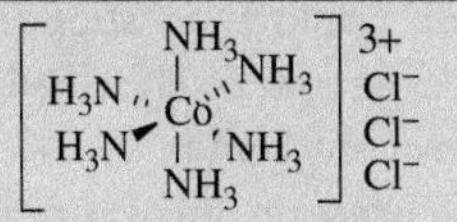 $[Co(NH_3)_6]^{3+}$ Cl^- Cl^- Cl^-	3
Ourpureo (purple)	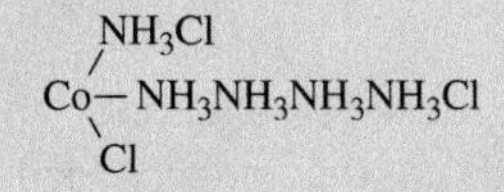 $Co(NH_3Cl)(NH_3NH_3NH_3NH_3Cl)(Cl)$	2	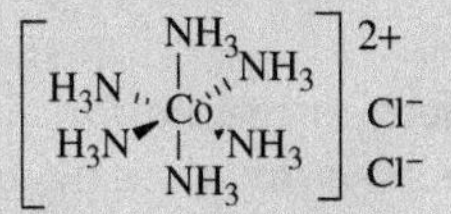 $[Co(NH_3)_5Cl]^{2+}$ Cl^- Cl^-	2
Praseo (green)	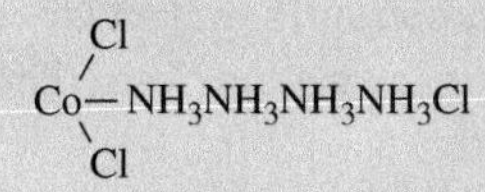 $Co(Cl)(NH_3NH_3NH_3NH_3Cl)(Cl)$	1	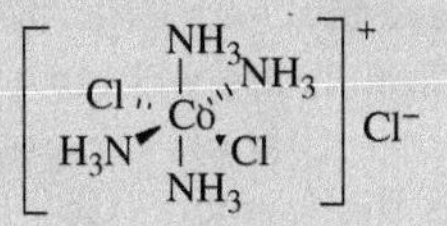 $[Co(NH_3)_4Cl_2]^{+}$ Cl^-	1
Violeo (violet)	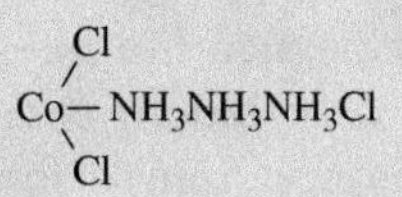 $Co(Cl)(NH_3NH_3NH_3Cl)(Cl)$	1	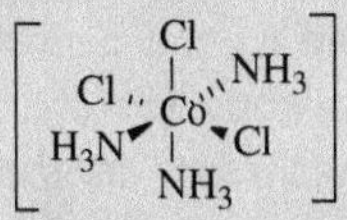 $[Co(NH_3)_3Cl_3]$	0

long chains of pentavalent N atoms with the halide atoms at the terminus, e.g., $-NH_3-NH_3-NH_3-X$, by analogy to the way that tetravalent C forms chains of $-CH_2-CH_2-CH_2-X$.

The synthetic chemist Sophus Mads Jørgensen expanded upon Blomstrand's chain theory and postulated the chemical structures shown in Table 15.1 for a series of colorful Co(III) coordination compounds. In order to explain the number of chloride ions that could be precipitated using $AgNO_3$, Jørgensen postulated that those halides that were "nearer" (or directly bonded) to the central atom could not be easily precipitated, while those attached to the weaker N–X bonds at the ends of the ammonia chains could readily dissociate and precipitate with $AgNO_3$. In order to support his theory, he synthesized a wide variety of Co(III) coordination compounds, thereby adding to the accumulating volume of data in this burgeoning field. The Blomstrand–Jørgensen chain theory was not only successful at explaining the number of chloride ligands that would precipitate upon the addition of $AgNO_3$ but it could also account for molecules where the number of ammonia ligands exceeded the valence of the metal. This model held up for nearly a quarter of a century until it was finally challenged by Werner's coordination theory in 1893.

One night in late 1892, at the age of 26, Alfred Werner was awakened from his sleep at two in the morning. He had just had a dream about the nature of bonding in coordination compounds. For the next 15 h, he recorded his insights in what eventually became his third scientific paper. Unlike the chain theory, Werner postulated that all of the cobalt compounds in Table 15.1 had two types of valences: primary (*Hauptvalenz*) and secondary valence. Furthermore, every metal had a fixed *coordination number* equal to the number of ligands that were directly bonded to the central atom. The primary valence of the central metal ion was the same as its oxidation number. In the series of Co compounds listed in Table 15.1, the primary valence is three and the coordination number is six because these species adopt octahedral *coordination geometries*. The secondary valence was the charge on the complex cation formed by the metal and its contingent of ligands. Thus, for the *luteo* compound, which had only the neutral NH_3 molecules as ligands, the secondary valence on the complex cation $[Co(NH_3)_6]^{3+}$ was also three. This trivalent cation could then support three chloride ions in an electrostatic (ionic) attraction. Werner later classified those ligands attached to the central metal by coordinate covalent bonds as belonging to the *inner coordination sphere* and those ions that were present only for charge balance as belonging to the *outer coordination sphere*.

The modern formalism for distinguishing between the inner and outer coordination spheres is to place all those ligands in the inner coordination sphere inside the square brackets when writing the compound's molecular formula and the outer-sphere ligands are written outside of the square brackets. Thus, the formula for the *luteo* coordination compound is written as $[Co(NH_3)_6]Cl_3$. The *purpureo* compound has the formula $[Co(NH_3)_5Cl]Cl_2$. In this case, one of the chlorides is covalently bonded to the central Co(III) ion (alongside five NH_3 ligands), making the secondary valence (or charge on the complex cation) two. Thus, there are two Cl^- ions necessary in the outer coordination sphere for charge balance. When the *purpureo* complex is reacted with an excess of $AgNO_3$ under ambient conditions, only the two chloride ions in the outer coordination sphere (the ionically bound Cl^-) will precipitate as AgCl (s). Thus, Werner made a distinction between the inner sphere (bound) chlorides as "nonionogenic" and the outer sphere (electrostatically attracted) chlorides as "ionogenic." In fact, the commonly used term coordination *complex* is a direct result of the fact that many of Werner's inner sphere coordination compounds were also complex cations.

It is important to note that Werner's coordination theory took place *before* the work of G.N. Lewis. Indeed, the electron had not yet even been discovered in 1893 when his paper was published. Thus, it was truly a revolutionary concept for the time, and one which would eventually earn the young Alfred Werner the 1913 Nobel Prize in chemistry and the rightful title of "father of coordination chemistry." Because the Blomstrand–Jørgensen model had already been in place for nearly a quarter of a century, however, it was not so easily relinquished by the scientific community. Jørgensen continued to challenge Werner's theory for nearly a decade until he finally admitted defeat in 1907. The proverbial nail in the coffin for the chain theory of coordination compounds was primarily supplied by the synthesis and isolation of two isomers of the tetrammine cobalt (III) compounds listed in Table 15.1. The first of these was synthesized by Gibbs in 1857 and was green in color (*praseo*). Werner himself synthesized the second isomer in 1907, a violet-colored species given the name *violeo*. Although the latter compound was not very stable, Werner was able to show that it contained only a single ionogenic chloride ion. Werner and his colleague Miolati had recently begun investigations on the electrical conductances of coordination compounds, which lent a great deal of credence to his theory of inner and outer coordination spheres. The presence of two geometrical isomers of Werner's $[Co(NH_3)_4Cl_2]Cl$ formulation was a natural consequence of its octahedral geometry, but it could not be rationalized using Jørgensen's chain model. The *trans*-isomer (where the two nonionogenic chlorides are directly opposite each other at a Cl–Co–Cl bond angle of 180°) was the green isomer, while the violet form of the compound had the two chlorides in the *cis*-configuration with a Cl–Co–Cl bond angle of 90°. While the absence of a third isomeric form of the tetraammine complex did not conclusively prove the octahedral coordination geometry, it was enough for Jørgensen, who was a tremendous synthetic chemist, to finally abandon his chain theory formulation.

15.2 TYPES OF LIGANDS AND NOMENCLATURE

In a coordination compound, ligands "coordinate" to the central metal using coordinate covalent bonds. In other words, the ligand acts as a Lewis base or an electron donor, while the metal acts as a Lewis acid or an electron acceptor. Monodentate ligands (literally meaning "one-toothed") bind to the metal in a single coordination site. The range of compounds capable of ligation include most anions having a lone pair of electrons, inorganic Lewis bases, and organic compounds having functional groups containing lone pairs of electrons or basic heteroatoms in their ring structures. Examples of monodentate ligands include Cl^-, H_2O, R-NH_2, and pyridine. There are also a few ligands that are ambidentate, which means that they can act as a bridge between two metal centers and can coordinate from both ends of the molecule. Such species include CN^-, SCN^-, N_2, and 4,4′-dipyridyl. Sometimes the ligand even has different names depending on which end of the molecule is coordinating. Thus, for example, when the SCN^- ligand is S-bound to a metal, it is referred to as *thiocyanate*, but when it is N-bound its name changes to *isothiocyanate*. The names of some common monodentate ligands are listed in Table 15.2.

In addition to the halides and univalent ions such as CN^-, SCN^-, and OH^-, a number of neutral molecules can also act as ligands. Usually, these molecules will contain one of the following donor groups as part of their molecular formulas: R_3N, R_3P, R_3As, R_3Sb, R_2O, R_2S, R_2Se, or R_2Te. When NH_3 coordinates to a metal, it is known as the *ammine ligand*, which is not to be confused with the amines (such as RNH_2, R_2NH, and R_3N), which can also act as ligands. The amines differ from the ammine ligand by their greater size and their electronic effects. The most common heteroatoms in organic molecules that can act as ligands are the more electronegative atoms N and O; and, less commonly, P and S.

Ligands that can coordinate to the metal at two or more of its coordination sites are known as *chelates*, from the Greek word for "claw." In order for a molecule to act as a chelating ligand, there must be sufficient space between the coordinating parts of the molecule to avoid steric strain when the ligand is bound. Typically, *bidentate* ("two-toothed") ligands bind the metal ion using two different sets of lone pair electrons and form four- or five-membered rings upon coordination to the metal. Some ligands, such as the amino acid glycine, can coordinate through the carboxylate only, the amine only, or as a chelate using both ends of the molecule. Tridentate ("three-toothed"), tetradentate ("four-toothed"), and hexadentate ("six-toothed") ligands also exist. Most chelating ligands will attach to the metal using adjacent coordination sites. The names, abbreviations, and chemical structures of some common chelates are listed in Table 15.3.

The recommended IUPAC nomenclature rules for coordination compounds are as follows:

1. The cation is listed first, then a space, followed by the anion.
2. Any ligands that are part of the inner coordination sphere are listed first, followed immediately by the name of the metal with its oxidation state in parentheses.
3. The following prefixes are used to indicate the number of each type of ligand in the inner sphere:
   a. If the ligand is monodentate: di (2), tri (3), tetra (4), penta (5), hexa (6), hepta (7), octa (8), nona (9), and deca (10).

TABLE 15.2 The names, abbreviations, and chemical structures of selected monodentate ligands.

Common Name of Ligand	Chemical Formula
Fluoro	F^-
Chloro	Cl^-
Bromo	Br^-
Iodo	I^-
Cyano	CN^-
Thiocyanato	SCN^-
Isothiocyanato	NCS^-
Aqua	H_2O
Ammine	NH_3
Carbonyl	CO
Thiocarbonyl	CS
Phosphine	R_3P
Triphenylphosphine	PPh_3
Pyridine	C_5H_5N (py)
Amido	NH_2^-
Azido	N_3^-
Piperidine	$C_5H_{11}N$ (pip)
Methylamine	CH_3NH_2
Dimethylamine	$(CH_3)_2NH$
Trimethylamine	$(CH_3)_3N$
Nitrosyl	NO
Nitrito	NO_2^-
Nitryl	NO_2
Hydrido	H^-
Hydroxo	OH^-
Sulfo	S^{2-}
Dimethylsulfoxo	$(CH_3)_2SO$ (dmso)
sulfinyl or thionyl	SO
Sulfonyl	SO_2
Carbonato	CO_3^{2-}
Oxo	O^{2-}
Sulfato	SO_4^{2-}
Thiourea	$NH_2(CS)NH_2$ (tu)
Thiosulfato	$S_2O_3^{2-}$

Common names are used instead of IUPAC names.

   b. If the ligand is a chelate, the prefix is followed by the name of the ligand in parentheses: bis (2), tris (3), tetrakis (4), pentakis (5), hexakis (6), heptakis (7), octakis (8), nonakis (9), and decakis (10).

4. Other prefixes, such as *cis*- and *trans*-, are used to designate geometrical or optical isomers, whenever this is necessary.
5. If the ligand is a bridge between two metals, the symbol μ- is used as a prefix before the bridging ligand.
6. The ligands in the inner coordination sphere are listed in alphabetical order of the ligand's root name (ignoring any prefixes).
7. If the anion is a coordination compound, the suffix -ate is used; some metals have special names in this case: ferrate (Fe), stannate (Sn), plumbate (Pb), argentate (Ag), and aurate (Au)

TABLE 15.3 The common names, abbreviations, and chemical structures of selected chelating ligands.

Common Name of Ligand	Abbreviation	Chemical Structure
Ethylenediamine	en	$NH_2CH_2CH_2NH_2$
Diethylenetriamine	dien	$H_2NCH_2CH_2NHCH_2CH_2NH_2$
Triethylenetetraamine	trien	$H_2NCH_2CH_2NHCH_2CH_2NHCH_2CH_2NH_2$
2,2′-Dipyridyl	bpy	
1,10-Phenanthroline	phen	
Oxalato	ox	$^{-}O(O)C–C(O)O^{-}$
1,2-*bis*(Diphenylphosphino)-ethane	dppe	$Ph_2PCH_2CH_2PPh_2$
Acetylacetonato	acac	
Ethylenediamine-tetraacetato	edta	
Porphyrin (a variety of porphyrins exist having different side chains)		

Example 15-1. Name each of the following coordination compounds: (a) $K_4[Fe(CN)_6]$, (b) $[Co(NH_3)_5Cl]Cl_2$, (c) $[Fe(acac)_3]$, (d) $[Pt(en)_2Cl_2]Cl_2$, (e) $Na[PtBrCl(NO_2)NH_3]$, (f) $[Ru(bpy)_2Cl_2]$.

Solution. (a) potassium hexacyanoferrate(II), (b) pentaamminechlorocobalt(III) chloride, (c) tris(acetylacetato)iron(III), (d) dichlorobis(ethylenediamine) -platinum(IV) chloride, (e) sodium amminebromochloronitrito-platinate(II), (f) dichlorobis(2,2′-bipyridyl)ruthenium(II).

15.3 STABILITY CONSTANTS

The thermodynamic stabilities of coordination compounds are typically measured using stability or formation constants, as shown in Equations (15.1)–(15.4) for $Cu(NH_3)_4{}^+$. The tetraaquacopper(II) cation is used as the starting material in Equation (15.1) because the hydration enthalpy is so negative that most metal ions cannot exist as "naked" cations in aqueous solution. It is not always possible for the stepwise constants to be measured individually, so typically only the *overall formation constant* β_n is reported, where n is the number of ligands attached to the metal ion. If the stepwise stability constants do happen to be known, then the overall constant can be determined from the product of each individual formation constant. The stepwise formation constants for coordination compounds usually decrease in magnitude as the value of n increases. This is an entropic effect that has to do with the number of available substitutions. Thus, for example, addition of NH_3 to $[Cu(H_2O)_4]^{2+}$ in Equation (15.1) has four possible positions available for substitution, whereas addition of NH_3 to $[Cu(NH_3)_3(H_2O)]^{2+}$ in Equation (15.4) has only one possible position available for substitution.

$$[Cu(H_2O)_4]^{2+} + NH_3 \Longleftrightarrow [Cu(NH_3)(H_2O)_3]^{2+} + H_2O \qquad K_1 = 1.9 \times 10^4 \tag{15.1}$$

$$[Cu(NH_3)(H_2O)_3]^{2+} + NH_3 \Longleftrightarrow [Cu(NH_3)_2(H_2O)_2]^{2+} + H_2O \qquad K_2 = 3.9 \times 10^3 \tag{15.2}$$

$$[Cu(NH_3)_2(H_2O)_2]^{2+} + NH_3 \Longleftrightarrow [Cu(NH_3)_3(H_2O)]^{2+} + H_2O \qquad K_3 = 1.0 \times 10^3 \tag{15.3}$$

$$[Cu(NH_3)_3(H_2O)]^{2+} + NH_3 \Longleftrightarrow [Cu(NH_3)_4]^{2+} + H_2O \qquad K_4 = 1.5 \times 10^2 \tag{15.4}$$

$$[Cu(H_2O)_4]^{2+} + 4\,NH_3 \Longleftrightarrow [Cu(NH_3)_4]^{2+} + 4\,H_2O \qquad \beta_2 = K_1K_2K_3K_4 = 1.1 \times 10^{13}$$

Formation constants can be measured using a variety of instrumental techniques, including cyclic voltammetry, polarography, pH measurements, IR, UV/VIS, and NMR, to name but a few. Because the values of stability constants can range over many orders of magnitude, they are frequently reported as logarithms. Table 15.4 lists the stability constants for selected coordination compounds. As with any equilibrium constant, the magnitude of the formation constant is dependent on the temperature, as well as on the overall ionic strength. As expected on the basis of hard–soft acid–base (HSAB) theory, the stability constant for the soft-acid Ag^+ complexes increases with the softness of the ligand, as shown in Equations (15.5) and (15.6). The stability constants for the hydroxo complexes of Cr^{3+} and Zn^{2+} also follow the HSAB prediction. Hydroxide is a hard ligand and Cr^{3+} is a hard acid, while Zn^{2+} is

TABLE 15.4 Selected stability constants for transition metal compounds, reported as the logarithm of the overall stability constant at 298 K.

Complex	log K_f	Complex	log K_f
$Ag(CN)_2^-$	8.75	$Cr(OH)_4^-$	29.9
$Ag(NH_3)_2^+$	7.2	$Co(SCN)_4^{2-}$	3.0
$Ag(S_2O_3)_2^{3-}$	13.5	$Cu(CN)_4^{2-}$	25.0
$AgCl_2^-$	5.0	$Fe(CN)_6^{4-}$	35.0
$Cu(NH_3)_2^{2+}$	7.7	$Fe(CN)_6^{3-}$	42.0
$Cu(en)^{2+}$	10.6	$Ni(en)^{2+}$	7.5
$Cu(NH_3)_4^{2+}$	13.1	$Ni(en)_2^{2+}$	13.9
$CdBr_4^{2-}$	2.5	$Ni(en)_3^{2+}$	18.2
$Ni(NH_3)_2^{2+}$	5.1	$Au(CN)_2^-$	38.3
$Ni(NH_3)_6^{2+}$	9.1	$Zn(OH)_4^-$	17.7

only a borderline acid. Thus, the overall stability constant for $Cr(OH)_4^-$ is about 12 orders of magnitude larger than that for $Zn(OH)_4^-$.

The larger stability constant for $Cu(en)^{2+}$, which is actually for the reaction given in Equation (15.5), compared to β_2 for $Cu(NH_3)_2^{2+}$, given by Equation (15.6), has to do with entropic considerations relating to the chelating ligand. The standard enthalpies of formation for the two reactions, both of which result in the substitution of two H_2O ligands on Cu^{2+} with two σ-donating N atoms, are roughly identical. However, the standard entropy of formation is much more positive for the ethylenediamine complex than it is for the ammine complex. This phenomenon is known as the *chelate effect*, which states that the stability of a coordination compound containing a chelating ligand will always be greater than the overall stability constant for their corresponding non-chelating analogs. The entropy term is greater with the chelate because there are less molecular entities on the reactant side of the equation for the chelating ligand than for the non-chelating ligand.

$$[Cu(H_2O)_6]^{2+} + en \Longleftrightarrow [Cu(en)(H_2O)_4]^{2+} + 2\,H_2O$$
$$\Delta H^\circ = -54\,kJ/mol,\ \Delta S^\circ = 23\,J/K\cdot mol \qquad K_1 = 4.0\times 10^{10} \tag{15.5}$$
$$[Cu(H_2O)_6]^{2+} + 2\,NH_3 \Longleftrightarrow [Cu(NH_3)_2(H_2O)_4]^{2+} + 2\,H_2O$$
$$\Delta H^\circ = -46\,kJ/mol,\ \Delta S^\circ = -8.4\,J/K\cdot mol \qquad \beta_2 = 5.0\times 10^{7} \tag{15.6}$$

Example 15-2. Given that $K_1 = 2.1\times 10^3$ and $K_2 = 8.2\times 10^3$, calculate the overall stability constant β_2 for the formation of $Ag(NH_3)_2^+$.

Solution. The two formation constants are simply multiplied together: $\beta_2 = K_1K_2 = 1.7\times 10^7$. The reason why $K_2 > K_1$ in this case has to do with the different geometries of the two coordination compounds. The actual stepwise chemical equations are as follows:

$$[Ag(H_2O)_4]^+ + NH_3 \Longleftrightarrow [Ag(NH_3)(H_2O)_3]^+ + H_2O \quad K_1 = 2.1\times 10^3$$
$$[Ag(NH_3)(H_2O)_3]^+ + NH_3 \Longleftrightarrow [Ag(NH_3)_2]^+ + 3\,H_2O \quad K_2 = 8.2\times 10^3$$

Because of the greater number of products, the second equilibrium is strongly favored by entropy.

Example 15-3. Which of the following would be expected to have the largest overall formation constant: $[Ni(en)_3]^{2+}$, $[Ni(edta)]^{2-}$, or $[Ni(NH_3)_6]^{2+}$? Explain your answer.

Solution. The hexadentate complex $[Ni(edta)]^{2-}$ would have the largest overall formation constant because of the chelate effect.

Example 15-4. For each pair, determine which coordination compound will have the larger overall stability constant: (a) $Hg(NH_3)_4{}^{2+}$ or $Hg(CN)_4{}^{2-}$, (b) $Fe(SCN)_5{}^{2-}$ or $FeF_5{}^{2-}$, and (c) $[Co(NH_3)_6]^{3+}$ or $[Co(NH_3)_6]^{2+}$.

Solution. Typically, the relative magnitude of the overall stability constants can be determined from HSAB theory.

(a) The Hg^{2+} ion is a soft-acid cation, so it will prefer the soft-basic CN^- ion over the hard base NH_3. The value of $\log(\beta_4)$ for $Hg(NH_3)_4{}^{2+}$ is 19.3, while that for $Hg(CN)_4{}^{2-}$ is 41.5, supporting this analysis.

(b) The Fe^{3+} ion is a hard-acid cation, so it will prefer the hard-basic anion F^- over the soft-basic anion SCN^-. The actual values of $\log(\beta_5)$ in this case are 7.1 for $Fe(SCN)_5{}^{2-}$ and 15.3 for $FeF_5{}^{2-}$.

(c) The NH_3 ligand is a hard base, so it will prefer the hard-acid cation Co^{3+} over the borderline acid cation Co^{2+}. The experimental values for $\log(\beta_6)$ in this case are 33.7 for $[Co(NH_3)_6]^{3+}$ and 4.7 for $[Co(NH_3)_6]^{2+}$.

Example 15-5. Given the standard reduction potential for the following half-reactions, calculate the stability constant for $Ag(CN)_2{}^-$:

$$Ag^+ + e^- \rightarrow Ag \qquad E^\circ = 0.80\,V$$

$$Ag(CN)_2{}^- + e^- \rightarrow Ag + 2\,CN^- \qquad E^\circ = -0.31\,V$$

Solution. Rearrangement of these half-reactions yields the formation of the complex anion $Ag(CN)_2{}^-$:

$$Ag^+ + e^- \rightarrow Ag \qquad G^\circ = -nFE^\circ = -77\,kJ/mol$$

$$Ag + 2\,CN^- \rightarrow Ag(CN)_2{}^- + e^- \qquad G^\circ = -nFE^\circ = -30\,kJ/mol$$

$$Ag^+ + 2\,CN^- \Longleftrightarrow Ag(CN)_2{}^- \qquad \Delta G^\circ = -107\,kJ/mol$$

Using the expression that $\Delta G^\circ = -RT\ \ln(K)$ and solving for K, we get $K = 5.7 \times 10^{18}$.

15.4 COORDINATION NUMBERS AND GEOMETRIES

The *coordination number* of a compound is defined as the number of attachment sites of the various ligands to the metal center. The valence-shell electron-pair repulsion (VSEPR) model does not work well for transition compounds having partially filled *d*-subshells. The *Kepert model* is sometimes used instead. As with the VSEPR model, the metal ion is assumed to be spherical with the ligands lying along the surface of the sphere. The ligands will repel one another for either electronic or steric reasons and will tend to distribute themselves around the sphere so as to avoid each other. In the Kepert model, the lone pair electrons (which are the low-lying *d*-electrons in the transition metal ions) are ignored and will not affect the molecular geometry in any way. Thus, the coordination geometry of transition metal compounds is dictated by the number of ligands (or sites of attachment) around the metal ion. As in the VSEPR model, a coordination number of 2 leads to a linear geometry, 3 = trigonal planar, 4 = tetrahedral, 5 = trigonal bipyramidal, and 6 = octahedral. Similar to the VSEPR model, the Kepert model also has a number of exceptions and should be taken only loosely as a guide. As a general rule, metals having low oxidation states and small ionic radii, especially with large ligands, will tend to have small coordination numbers, while metals having higher oxidation numbers and larger ionic radii, especially with small ligands, will have large coordination numbers.

Very low coordination numbers are extremely rare among transition metal compounds. Typically, coordination numbers of 2 will only occur for the +1 cations of the Group 11 metals or for the Hg^{2+} ion. These metals all have a spherically symmetric, filled d^{10}-subshell and a low oxidation number. Where substances such as $Ag(NH_3)_2^+$ have been characterized, they typically have linear geometries and can easily react to form more stable species having a higher coordination number, such as $Ag(NH_3)_4^+$. Low-coordinate geometries can also exist when the ligands are quite bulky, such as $[N(SiMe_3)_2]^-$ or $[P(\textit{cyclo}\text{-}C_6H_{11})_3]$.

Even three-coordinate transition metal complexes are uncommon. Again, these species are only favored by sterically demanding ligands and low oxidation numbers. Some examples include the zero-valent compounds $Pt(PPh_3)_3$ and $Cr\{N(SiMe_3)_2\}_3$. Most of the three-coordinate species have trigonal planar molecular geometries, predominantly as the result of steric repulsions between the bulky ligands.

On the other hand, four-coordinate transition metal complexes are quite common, ranking second only to six-coordinate species in their importance in the field of coordination chemistry. The two limiting four-coordinate geometries are tetrahedral (T_d) and square planar (D_{4h}), with a whole spectrum of intermediate and distorted structures in between these two extremes. Tetrahedral species are favored by the Kepert model on the basis of steric interactions. Thus, coordination compounds with large ligands, such as Cl^-, Br^-, and I^-, especially tend to form tetrahedral compounds. Among the metals, tetrahedral coordination is favored by small metals in lower oxidation numbers, metals having noble gas electron configurations (Be^{2+}, Al^{3+}), and transition metals with closed *d*-subshells (such as d^0 or d^{10}), for example, MnO_4^- or $Ni(CO)_4$. A large number of Cu(II) compounds are also tetrahedral. Tetrahedral compounds do not exhibit geometric isomers, but optical isomers (as is the case for tetrahedral carbon) are possible whenever all four ligands are different.

The square planar geometry is less prevalent because whenever the ligands are small enough to assume this more sterically hindered molecular shape, the molecule is usually able to add two more ligands to increase its coordination geometry to octahedral. The d^8 electron geometry favors square planar coordination on the basis of electronic effects having to do with crystal field theory (see Section 16.2). Coordination compounds containing Ni^{2+}, Pd^{2+}, Pt^{2+}, Au^{3+}, and Rh^+ are commonly square planar, especially if they contain π-accepting ligands. Examples include $PtCl_4^{2-}$, $Au(CN)_4^-$, and $[RhCl(PPh)_3]$. Square planar compounds can exhibit *cis-trans* isomerization, but optical isomers are precluded by the presence of a horizontal mirror plane.

Five-coordinate species represent a rare class of coordination compounds. This rather unusual coordination number is a delicate balance between electronic factors which favor disproportion into four- and six-coordinate species and directional covalent bonds which help to stabilize them. The two limiting geometries (trigonal bipyramidal and square pyramidal) are therefore similar in energy and easily interconverted. As a result, five-coordinate species can exist as *fluxional* molecules, where their ligands can exchange places rapidly on the NMR time scale.

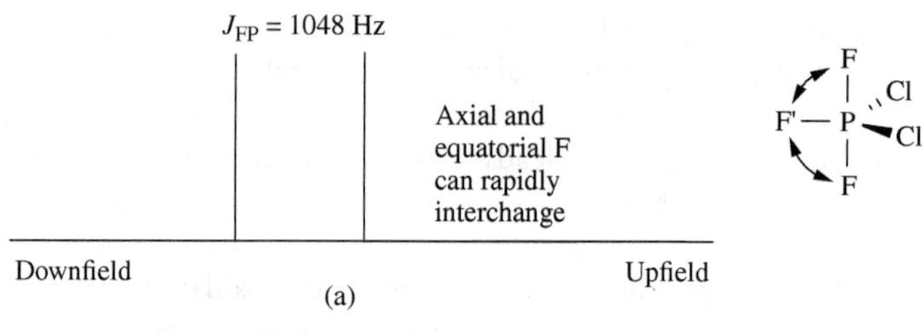

FIGURE 15.2
Cartoon representation of the ^{19}F-NMR spectrum of PCl_2F_3 at: (a) −22 °C and (b) −143 °C. [After Holmes, R. R.; Carter, R. P., Jr.; Peterson, G. E. *Inorg. Chem.* **1964**, *3*, 1748–1754].

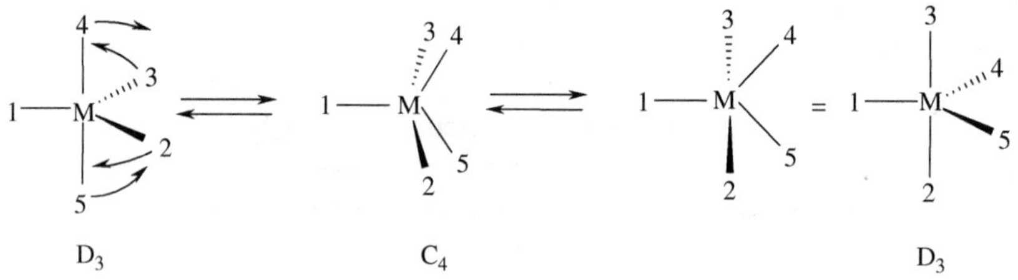

FIGURE 15.3
Mechanism of a Berry pseudorotation, by which the axial and equatorial sites of a trigonal bipyramid can interchange via a square pyramidal intermediate.

For example, the ^{19}F-NMR spectrum of PCl_2F_3 at −22 °C shown in Figure 15.2(a) illustrates how all three F atoms are interchangeable even though the $P{-}F_{ax}$ and $P{-}F_{eq}$ bond lengths are not identical. Thus, a lone doublet appears in the ^{19}F-NMR spectrum with a J_{FP} coupling constant of 1048 Hz. At this temperature, the F atoms can interchange with one another in a process known as a *Berry pseudorotation*, which is shown in Figure 15.3. At −143 °C, two distinct F resonances are observed in the ^{19}F-NMR, as shown in Figure 15.2(b), with a doublet of doublets having J_{FP} = 1048 Hz and $J_{FF'}$ = 124 Hz occurring downfield and a doublet of triplets having $J_{F'P}$ = 1048 Hz and $J_{F'F}$ = 124 Hz in the upfield region. At this lower temperature, the F atoms in the axial and equatorial positions are "frozen" on the NMR time scale and cannot interconvert.

Figure 15.4 shows a correlation diagram between the trigonal bipyramidal and square pyramidal molecular geometries for the frontier molecular orbitals. The square pyramidal geometry is only favored for low-spin d^5 and d^6 electronic configurations with small ligands. In every other case, there is both an electronic and steric preference for the trigonal bipyramidal molecular geometry.

At the trigonal bipyramidal extreme, the lower lying e″ MOs are nonbonding, the e′ degnerate set is weakly antibonding in the equatorial plane, and the high-lying a_1' MO is strongly antibonding along the axial direction. Thus, for low-spin d^0–d^4 electron configurations, the overall configuration is nonbonding, and the axial sites will be more antibonding than the equatorial sites. Acceptor ligands will prefer to be where there is the largest excess of electrons, which will be in the positions of the weakest bonds. Thus, there is an electronic preference for the acceptor ligands to occupy the weaker axial positions and for donor ligands to occupy the equatorial sites, as shown in Figure 15.5(a). For the d^8 electronic configuration (and less so for d^6 and d^7), the weakly antibonding e′ MOs will be occupied while the strongly antibonding a_1' MO will be empty. Thus, the weaker bonds will lie equatorial in these

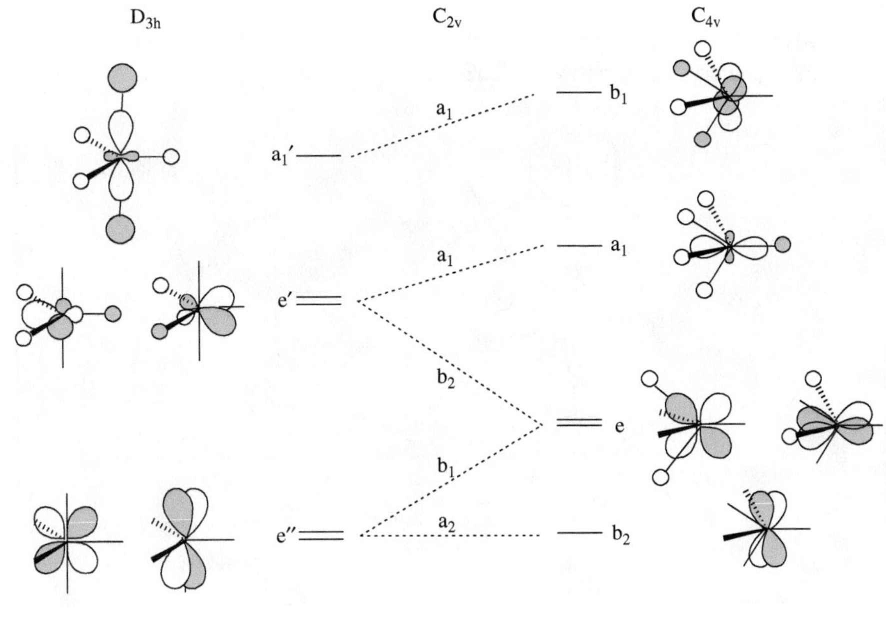

FIGURE 15.4
Correlation diagram along the Berry pseudorotation coordinate as one progresses from a trigonal bipyramidal to a square pyramidal electron configuration. [After Rossi, A. R.; Hoffmann, R. *Inorg. Chem.* **1975**, *14*, 365–374.]

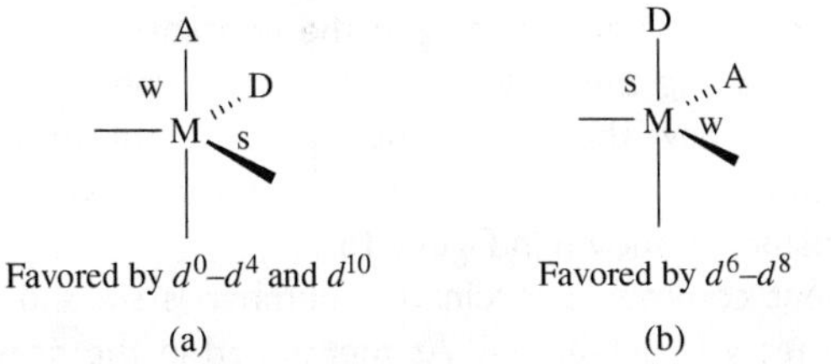

Favored by d^0–d^4 and d^{10}

(a)

Favored by d^6–d^8

(b)

FIGURE 15.5
In the trigonal bipyramidal geometry, there is an electronic preference for: (a) the acceptor (A) ligands to occupy the weaker (w) axial sites and donor (D) ligands to occupy the stronger (s) equatorial sites for the d^0–d^4 and d^{10} electron configurations; (b) the acceptor ligands to occupy the weaker equatorial sites and donor ligands the stronger axial sites for the d^6–d^8 electron configurations.

electronic configurations and the stronger bonds will lie along the axial direction. As a result, any acceptor ligands will have an electronic preference for the weaker bonds in the equatorial plane and donor ligands will occupy the axial sites, as shown in Figure 15.5(b).

Example 15-6. Predict the preferred positions for each ligand in the following trigonal bipyramidal molecules: (a) $Ta(CH_3)_3L_2$ (where L has the structure shown here) and (b) $CoBr_2(PPh_3)_3$.

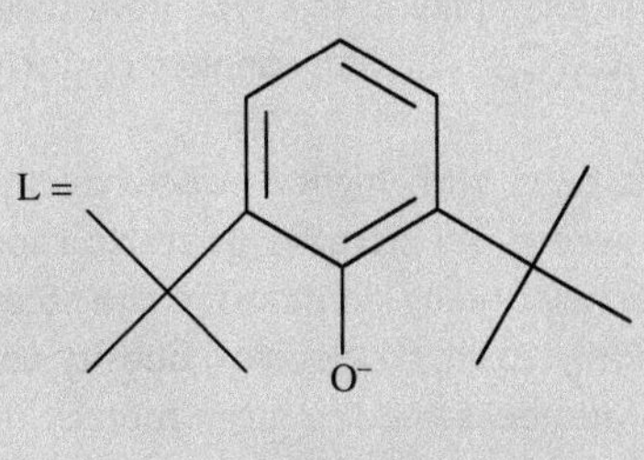

L =

O^-

Solution. (a) Ta(V) has a d^0 electron configuration. According to Figure 15.5(a), the acceptor ligands will therefore prefer the axial sites and the donor ligands will prefer the equatorial sites. The CH_3^- groups are better electron donors and the OR^- (or L) groups are the better acceptors. Thus, $Ta(CH_3)_3L_2$ will have the structure shown here. (b) Co(II) is d^7 and will prefer to have its donor ligands in the axial sites and its acceptor ligands in the equatorial sites, as shown in

Figure 15.5(b). For this molecule, PPh_3 is the donor and Br^- is the acceptor. Thus, the preferred geometry for $CoBr_2(PPh_3)_3$ is shown.

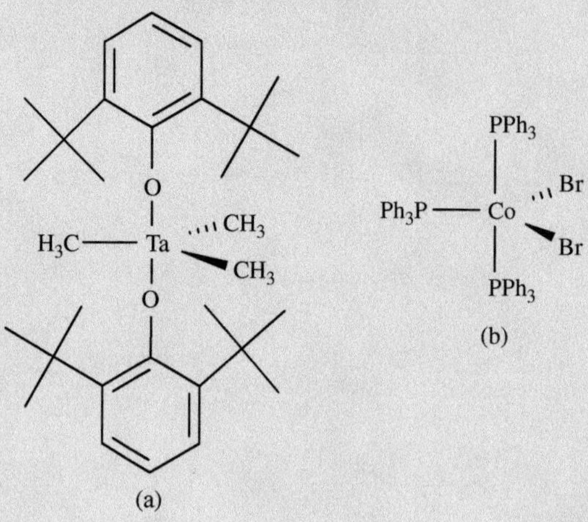

(a)

(b)

For the square pyramidal geometry, the b_2 MO is essentially nonbonding. The degenerate set of e MOs is antibonding in the equatorial plane and the a_1 MO is strongly antibonding along the axial direction. Thus, the low-spin d^0–d^6 and d^{10} electron configurations have the stronger bonding along the axial direction, as shown in Figure 15.6(a) and the d^7–d^8 electron configurations have the stronger bonding in the equatorial plane, as shown in Figure 15.6(b).

By far, the most common coordination number is six and the most common coordination geometry is octahedral. As mentioned in the previous section, both geometric and optical isomers are possible in the octahedral geometry. The most common distortion from octahedral geometry is a tetragonal distortion, where the ligands lying along one axis are either compressed ("z-in") or elongated ("z-out"), as shown in Figure 15.7, reducing the symmetry to D_{4h}. Trigonal distortions are also possible, where opposite triangular faces of the octahedron are either compressed or elongated, leading to a trigonal antiprism and a symmetry reduction from O_h to D_{3d}, as shown in Figure 15.8.

A few hexa-coordinate compounds also crystallize in the trigonal prismatic molecular geometry, shown in Figure 15.9, although this molecular shape is rare and requires some extra steric or electronic benefit in order to be stabilized over a regular octahedron. Most examples of this type have three bidentate ligands (usually dithiolates, such as $S_2C_2(C_6H_5)_2$ that connect opposite triangular faces in the trigonal prism.

Although compounds exist with higher coordination numbers than six, they are exceptionally rare. There is an increase in steric repulsion as the number of ligands increases and a corresponding decrease in the strength of the metal–ligand bonding because there is only so much electron density to go around. As a general rule, higher coordination numbers require large metals in high oxidation numbers

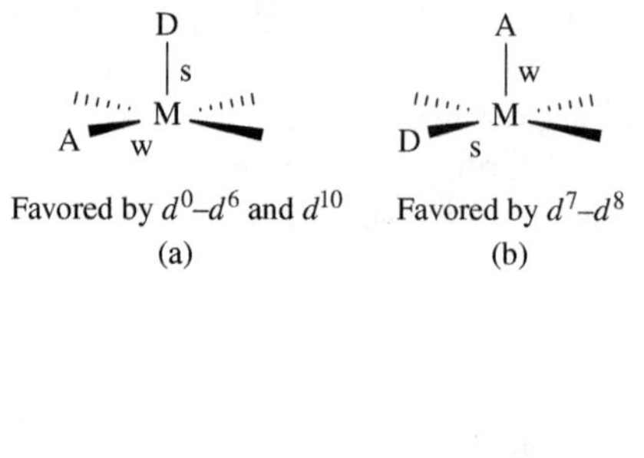

Favored by d^0–d^6 and d^{10}
(a)

Favored by d^7–d^8
(b)

FIGURE 15.6
(a, b) Electronic preferences in the square pyramidal geometry.

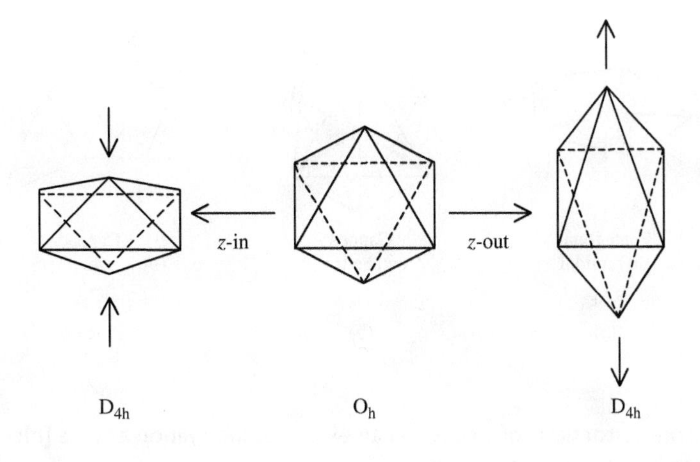

FIGURE 15.7
Tetragonal distortions from octahedral geometry.

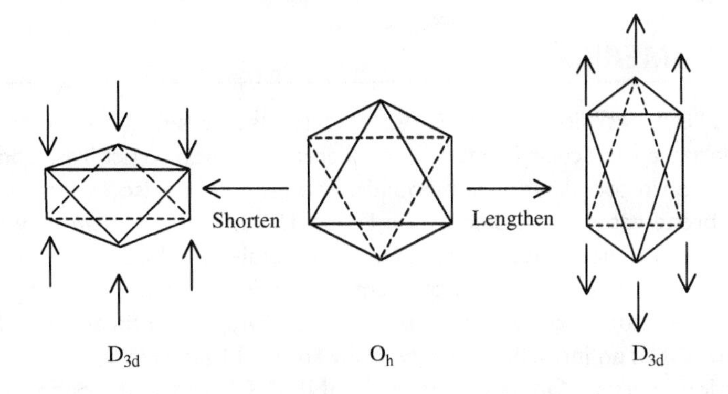

FIGURE 15.8
Trigonal distortions from octahedral geometry.

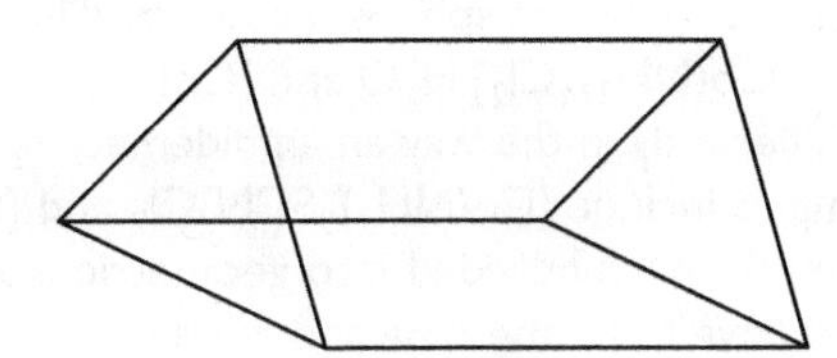

FIGURE 15.9
The trigonal prismatic molecular geometry.

and small, hard ligands. A high oxidation number is necessary in order to offset some of the electron density that accumulates on the central atom as more and more ligands coordinate to the metal center. Because of the high oxidation state, the metal will be a hard Lewis acid, which will prefer to bind only with hard ligands. A large metal with small ligands is required in order to avoid steric interactions between the ligands crowding around the metal center. Among seven-coordinate compounds, three different geometries are common (Figure 15.10): the pentagonal bipyramid (which has a pentagonal-shaped equatorial plane with axial ligands above and below the plane), a capped octahedron (which has an extra atom above one of the faces), and a capped trigonal prism (which has an extra atom above one of the faces). Seven-coordinate compounds most often occur with polydentate ligands, which force the molecule into an unusual geometry.

Among the higher coordination numbers, eight-coordinate compounds are the most common. They most often occur when the central metal ion belongs to the lanthanide or actinide series, although complexes with Zr, Hf, Nb, Ta, Mo, and W are also well-known. The two most common coordination geometries having $CN = 8$ are the trigonal dodecahedron and the square antiprism (Figure 15.11), both of which are

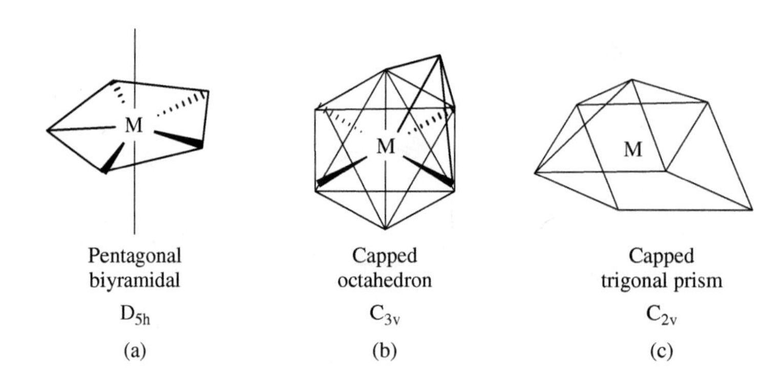

FIGURE 15.10
The three most common geometries for seven-coordinate compounds: (a) the pentagonal bipyramid, (b) the capped octahedron, and (c) the capped trigonal prism.

derived from distortions of a cube. As an example, the cyanometalate $[Mo(CN)_8]^{4-}$ crystallizes in the square antiprism.

15.5 ISOMERISM

The word "isomers" literally means "equal units." Isomers occur when compounds having the same molecular formula have different chemical structures. Isomerism is quite common in coordination compounds. As a general rule, isomers can be divided into two broad categories, as shown in Figure 15.12: *structural isomers*, which have the same stoichiometry but different kinds of metal–ligand bonding; and *stereoisomers*, which have the same types of chemical bonds but different arrangements of the atoms. Structural isomers, in turn, can be subdivided into three types. *Ionization isomers* consist of an interchange of (typically anionic) ligands in the inner and outer coordination spheres. One example is $[Co(NH_3)_5Cl]Br$ and its ionization isomer $[Co(NH_3)_5Br]Cl$. Similarly, *hydration isomers* result from the interchange of a water of hydration in the outer coordination sphere with a ligand in the inner coordination sphere, such as the pair $[Co(NH_3)_4Cl_2]{\cdot}H_2O$ and $[Co(NH_3)_4H_2OCl]Cl$. *Linkage isomers* are species that differ only in the way an ambidentate ligand is coordinated to the metal center. Examples include $[Co(NH_3)_5SCN]Cl_2$ and $[Co(NH_3)_5NCS]Cl_2$.

Stereoisomers can also be subdivided into geometric isomers and optical isomers. *Geometric isomers* have the same type and number of each kind of ligand, but differ in the geometrical arrangement of those ligands around the central metal ion. The most common set of geometrical isomers are *cis* and *trans*, which occur in both octahedral and square planar molecular geometries. The *cis* isomer has two identical ligands at 90° angles with respect to each other, while the *trans* isomer has them

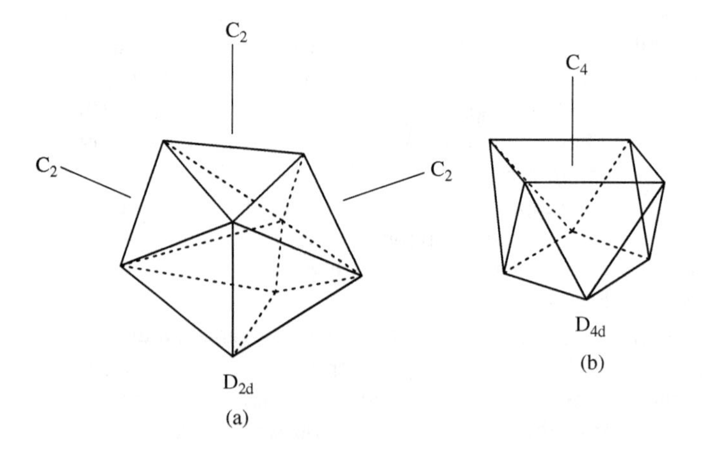

FIGURE 15.11
The two most common geometries for eight-coordinate compounds: (a) the trigonal dodecahedron and (b) the square antiprism.

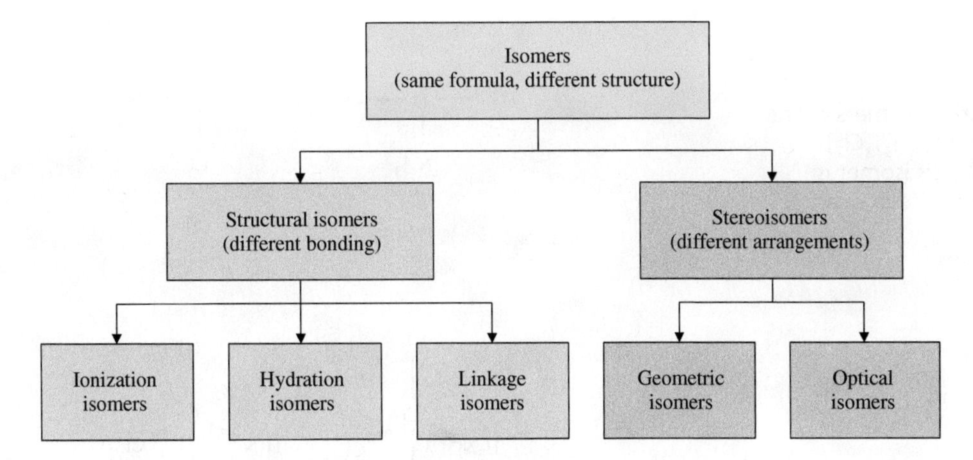

FIGURE 15.12
The different types of isomerism possible in coordination compounds.

opposite each other at 180°. Werner and his students synthesized over 50 different geometrical isomers of octahedral cobalt and chromium compounds. The most famous of these were the isomers of $[Co(NH_3)_4Cl_2]Cl$, which caused Jørgensen to finally admit that his chain model of coordination compounds was invalid. The *cis*-$[Co(NH_3)_4Cl_2]Cl$ isomer (*violeo*) is violet, while the *trans* isomer (*praseo*) is green. Both species are shown in Figure 15.13.

Figure 15.14 shows another pair of important geometrical isomers. In 1964, Barnett Rosenberg was studying the effects of electric fields on the cell division of *Escherichia coli* when he serendipitously discovered that a specific Pt(II) compound that was formed as a side product in his electrochemical cell caused the bacteria to stop dividing, inducing filamentous growth. Ultimately, it was determined that the chemical causing this unusual behavior was the *cis* isomer of $[Pt(NH_3)_2Cl_2]$, now known as *cisplatin*. Cisplatin is one of the most important anticancer medications known to date; it (along with several derivatives) has been effectively used to treat testicular, head, neck, and ovarian cancers. The *cis* geometry is important in the way that the drug binds to the DNA base pairs of cancer cells. While the *trans* isomer can also bind to DNA, its effectiveness as a chemotherapeutic agent is minimal.

For octahedral compounds having the general formula MA_3B_3, two other geometrical isomers exist, as shown in Figure 15.15. The facial (*fac*) isomer of $[Co(NH_3)_3Cl_3]$ is shown at the left, where the three chloro ligands all lie along one of the six triangular faces of the octahedron, giving this isomer its name. The meridional (*mer*) isomer, on the other hand, has all three chloro ligands lying in the same plane, along the meridian of the molecule.

Optical isomers are molecules that have the same number and type of each kind of ligand, but they are non-superimposable mirror images of each other. As such, optical isomers are *chiral* and will rotate plane-polarized light in opposite directions.

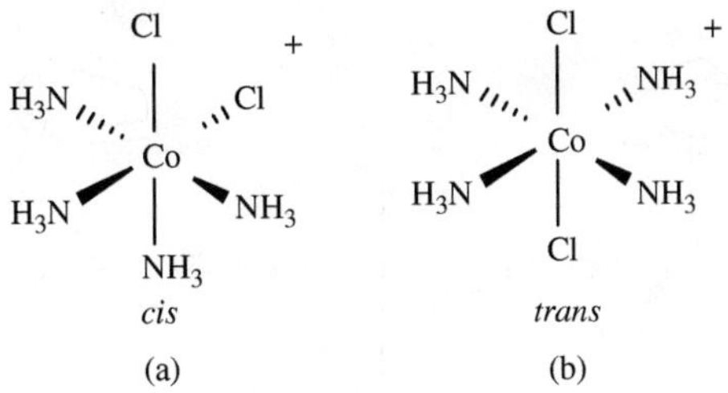

cis (a) *trans* (b)

FIGURE 15.13
(a, b) The geometrical isomers of *violeo* and *praseo* $[Co(NH_3)_4Cl_2]^+$ which caused Werner's coordination theory to triumph over the Blomstrand and Jørgensen's chain model.

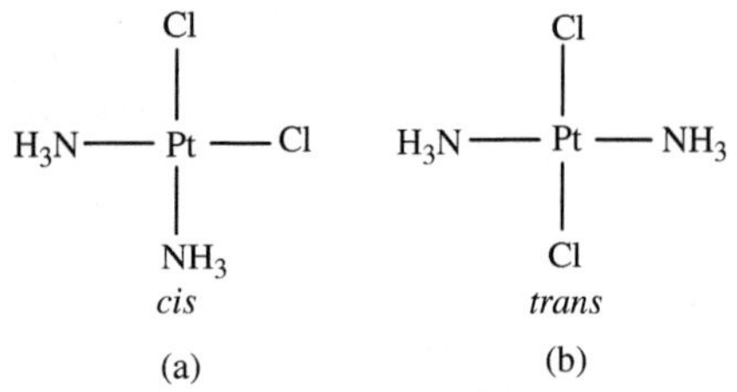

FIGURE 15.14
(a, b) Geometrical isomers of the square planar $[Pt(NH_3)_2Cl_2]$ compound. The *cis* isomer is better known as the anticancer drug *cisplatin*.

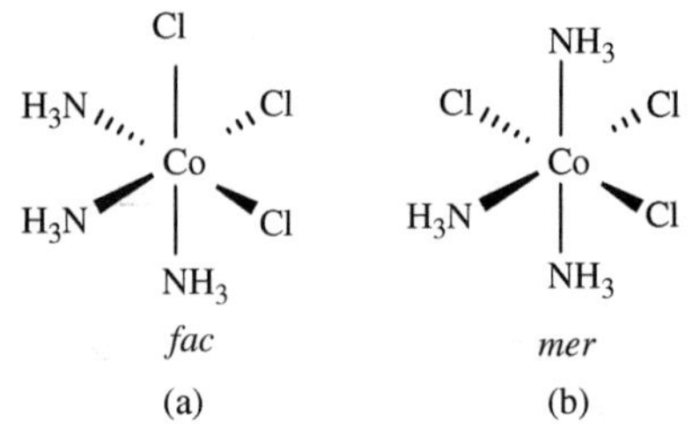

FIGURE 15.15
The facial (*fac*; a) and meridional (*mer*; b) isomers of $[Co(NH_3)_3Cl_3]$.

The angle of rotation can be measured using an experimental technique known as *circular dichroism* (CD). This method plots the difference in the molar absorptivity ($\Delta\varepsilon$) of the optical isomer that rotates left-circular polarized (lcp) light and that of right-circular polarized (rcp) light as a function of the wavelength (λ), as given by Equation (15.7).

$$\Delta\varepsilon = \varepsilon_{lcp} - \varepsilon_{rcp} \text{ versus } \lambda \tag{15.7}$$

As discussed in Chapter 8, in order for a molecule to have optical isomers, it must lack an improper rotational axis. Thus, only those compounds belonging to the following molecular point groups can be optically active: C_1, C_n, D_n, T, O, and I. Certain molecules based on a tetrahedral geometry, such as the iron compounds shown in Figure 15.16, are chiral.

Octahedral coordination compounds having the general formula ML_2X_2, where L is a bidentate ligand and X is a halide, will also exhibit optical isomerism. Werner and his students were among the first chemists to synthesize an optically-active coordination compound. Specifically, Werner was able to resolve the optical isomers of *cis*-$[CoCl(NH_3)(en)_2]^{2+}$, which are shown in Figure 15.17, by precipitation with silver *d*-α-bromocamphor-π-sulfonate.

Octahedral compounds having the general formula ML_3, where L is a bidentate ligand, will also be optically active. Thus, for example, the strongly luminescent $[Ru(bpy)_3]^{2+}$ compound, which phosphoresces a bright orange color in fluid solution, has the optical isomers shown in Figure 15.18. The Δ-isomer has the appropriate geometry to fit into the groove of right-handed B-DNA, while the Λ-isomer cannot.

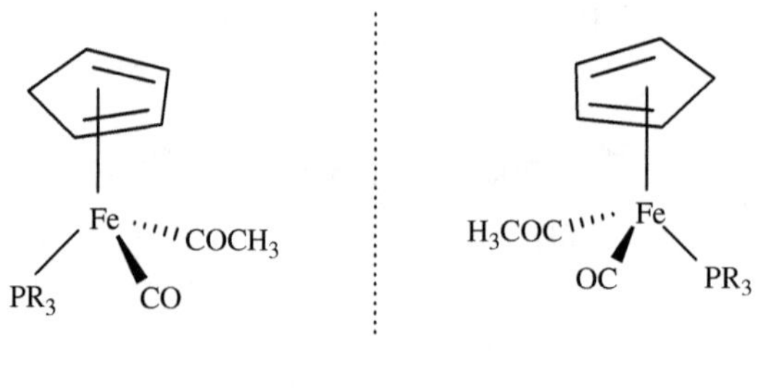

FIGURE 15.16
Optical isomers are non-superimposable mirror images. The dashed line represents the mirror plane in which the two isomers are reflected.

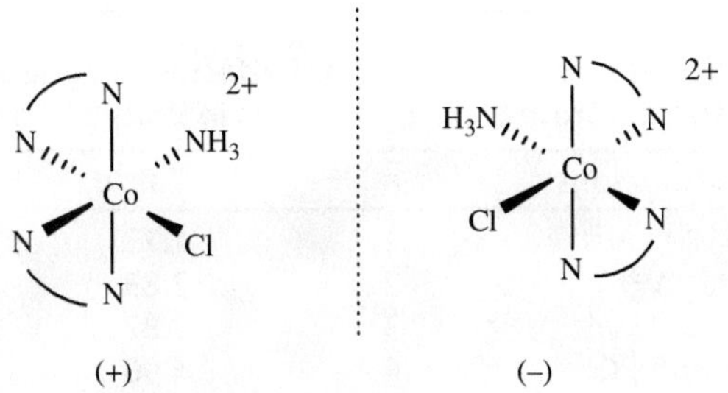

FIGURE 15.17
The optical isomers of *cis*-$[CoCl(NH_3)(en)_2]^{2+}$ first isolated by Werner and King. The two N atoms connected by a curved line are cartoon representations for the ethylenediamine ligand.

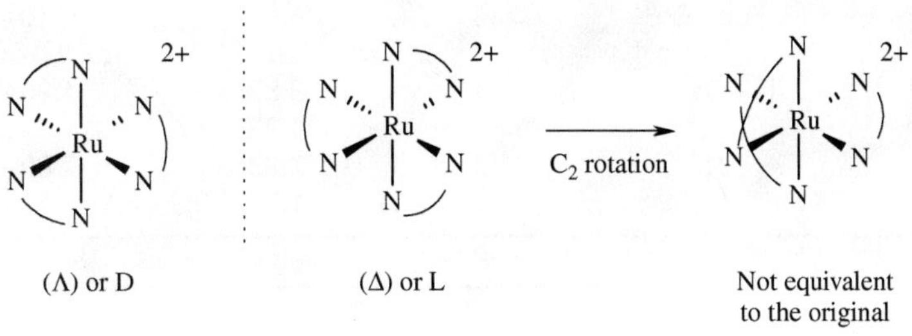

FIGURE 15.18
Optical isomers of $[Ru(bpy)_3]^{2+}$, showing how rotation of the second isomer by 180° does not lead to a superimposable mirror image to the original. The two N atoms connected by a curved line are cartoon representations for the 2,2′-dipyridyl ligand.

15.6 THE MAGNETIC PROPERTIES OF COORDINATION COMPOUNDS

The range of colors that coordination compounds can exhibit is but one of many reasons why they are interesting to study from a theoretical point of view. Their magnetic properties are a second reason. The macroscopic magnetic properties of the transition metals are due to the magnetic moments associated with individual electrons. Each electron in an atom has a magnetic moment that arises from its orbital motion around the nucleus and its spin, which generates a magnetic moment along its axis of rotation. The effective magnetic moment (μ) of an atom in a vacuum is related to its total angular momentum J, according to Equation (15.8), where g is the gyromagnetic ratio, S is the total spin quantum number (the sum of the m_s values), L is the total angular quantum number (the sum of the m_l quantum numbers), and μ_B is a unit known as the *Bohr magneton*, which is equal to 9.27×10^{-24} A m^2.

$$\mu = g\sqrt{S(S+1) + [0.25L(L+1)]}\,\mu_B \tag{15.8}$$

For the lanthanides, both the spin and angular momenta contribute to the effective magnetic moment; but for the transition metals, the orbital component is largely quenched, leaving only the spin component given by Equation (15.9). This quenching of the spin-orbit coupling is actually due to the strong effects of the crystal field on the *d*-electrons. The value of g for the spin-only magnetic moment is 2.00023, or approximately 2. Because S will always be equal to one-half the number of unpaired electrons (n), the value of the effective magnetic moment, p, can be calculated from the number of unpaired electrons that the molecule has using Equation (15.9).

$$\mu = 2\sqrt{S(S+1)}\,\mu_B = \sqrt{n(n+2)}\,\mu_B \tag{15.9}$$

The number of unpaired electrons, in turn, can be predicted on the basis of crystal field theory (see Chapter 16) and whether or not the molecule has a high-spin (HS) or low-spin (LS) electronic configuration. The experimental and calculated values of the effective magnetic moments for the first-row transition metals are listed in Table 15.5.

TABLE 15.5 Calculated and experimental effective magnetic moments for the octahedral first-row transition metals in units of Bohr magnetons.

d^n	Metal Ions	n	μ (Calc'd)	μ (Expt'l)
d^1	Ti^{3+}, V^{4+}	1	1.73	1.68–1.78
d^2	V^{3+}	2	2.83	2.75–2.85
d^3	Cr^{3+}, V^{2+}	3	3.87	3.70–3.90
HS d^4	Mn^{3+}, Cr^{2+}	4	4.90	4.75–5.00
LS d^4	Mn^{3+}, Cr^{2+}	2	2.83	3.18–3.30
HS d^5	Fe^{3+}, Mn^{2+}	5	5.92	5.65–6.1
LS d^5	Fe^{3+}, Mn^{2+}	1	1.73	1.80–2.5
HS d^6	Fe^{2+}, Co^{3+}	4	4.90	5.10–5.70
LS d^6	Fe^{2+}, Co^{3+}	0	0	
HS d^7	Co^{2+}	3	3.87	4.30–5.20
LS d^7	Co^{2+}	1	1.73	1.8
d^8	Ni^{2+}	2	2.83	2.80–3.50
d^9	Cu^{2+}	1	1.73	1.70–2.20

The most common way to ascertain the magnetic properties of materials is to measure their molar magnetic susceptibility, χ_m. The magnetic susceptibility is related to the effective magnetic moment according to Equation (15.10) and is measured using either a Gouy balance or an Evans balance.

$$\mu = 2.828\,(\chi_m T)^{1/2} \tag{15.10}$$

Using the Guoy method, as shown in Figure 15.19, the solid is packed into a glass capillary tube (similar to an NMR tube) and weighed using an analytical balance. The material is then reweighed in the presence of a U-shaped magnet. Those compounds having a nonzero magnetic moment will either be attracted to or repelled by one of the poles of the magnet. If the substance is *paramagnetic* (if it has unpaired electrons), it will be attracted to the magnetic field and the mass will appear to increase. If the compound is *diamagnetic* (if all the electrons are paired), on the other hand, it will be weakly repelled by the magnetic field and the mass will appear to decrease. By using a substance having a known molar magnetic susceptibility as a calibration standard, χ_m for the unknown material can be determined.

An Evans balance, on the other hand, works by placing two pairs of magnets on opposite ends of a beam. When a magnetic sample that is fixed in place in the sample compartment is introduced next to one pair of the magnets, it will either attract or repel that pair. This deflection is detected optically, and a current is placed through the other pair of magnets in order to compensate for the deflection. The strength of the magnetic moment in the sample is proportional to the amount of current necessary to keep the beam in a fixed position. The direction of the deflection (whether it is repulsion or attraction) is indicated on the digital readout using a plus or minus sign. Nowadays, extremely precise magnetic susceptibilities can be measured using an instrument known as a *SQUID magnetometer* (SQUID stands for superconducting *qu*antum *i*nterference *d*evice).

There are essentially five basic types of magnetic behavior based on the temperature dependence of the magnetic susceptibility and bulk magnetic properties, as shown in Figure 15.20: diamagnetism, paramagnetism, ferromagnetism, antiferromagnetism, and ferrimagnetism.

Diamagnetic compounds have molar magnetic susceptibilities that are small, negative, and temperature independent, with typical values of χ_m ranging from 10^{-5} to

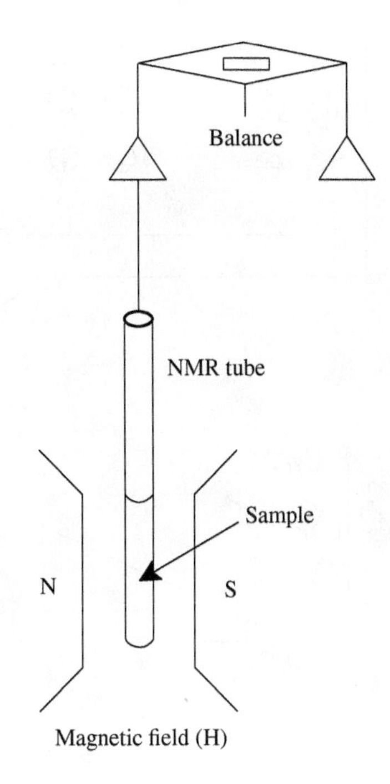

FIGURE 15.19
Schematic diagram of a Guoy balance.

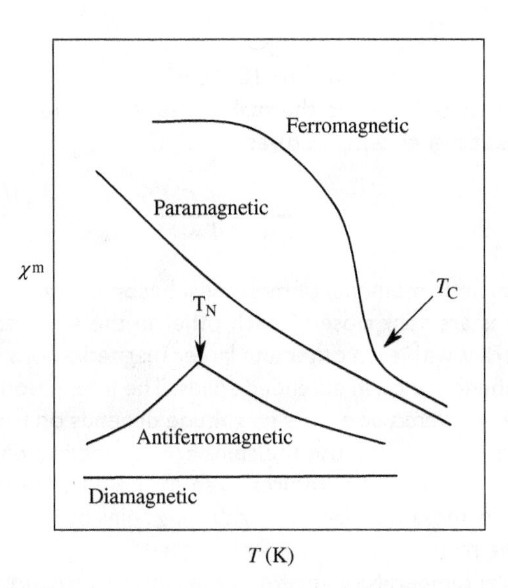

FIGURE 15.20
The temperature dependence of magnetic susceptibility for the different kinds of bulk magnetism.

10^{-6} cm^3/mol. In a diamagnetic substance, all of the electrons are paired. In the absence of a magnetic field, the spins are all paired and randomly oriented, as shown in Figure 15.21. In the presence of an external magnetic field, a weak magnetic moment is induced, and the individual moments will align themselves in opposition to the external magnetic field.

For a *paramagnetic* substance, the magnetic susceptibility is inversely proportional to the temperature and can often be fit to the *Curie-Weiss law*, given by Equation (15.11), where χ_0 is the Pauli paramagnetic component (which is temperature independent), C is the Curie constant (which depends on the substance) and Θ is a constant having units of K. In the absence of a magnetic field, the unpaired electrons in a paramagnetic material are randomly oriented, but when an external magnetic field is introduced, they will align their spins along the same direction as the field, as shown in Figure 15.22. The molar magnetic susceptibilities for paramagnetic

FIGURE 15.21
The magnetic susceptibility of a diamagnetic material is temperature-independent.

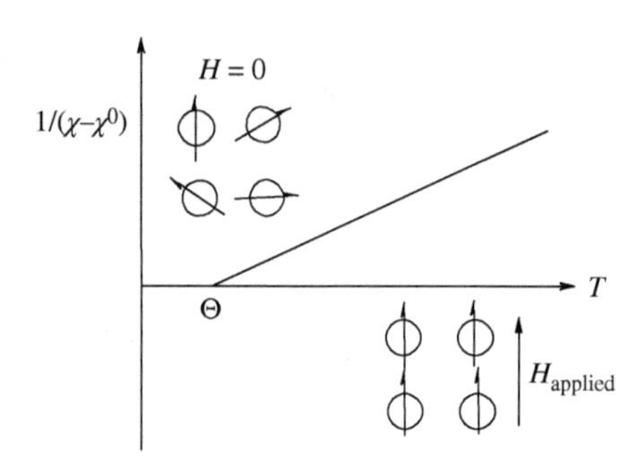

FIGURE 15.22
The temperature dependence of the magnetic susceptibility of a paramagnetic material follows the Curie–Weiss law.

materials generally range from 10^{-3} to 10^{-6} cm^3/mol. The inverse dependence on the temperature results from the thermal vibrations of the solid, which serve to disorder the spins at higher temperatures.

$$\frac{1}{\chi - \chi_0} = \frac{C}{T - \Theta} \tag{15.11}$$

When the magnetic moments of molecules become large enough and/or when the magnetic atoms are very close to each other in the solid state, their magnetic moments can interact with each other and larger magnetizations will occur. This is a rather common phenomenon in extended solids. The interaction between the spins is called the *exchange interaction* and its magnitude depends on the degree of overlap of the charge distributions. Because the spins are interacting with each other, they will naturally order at low temperatures even in the absence of a magnetic field. When the exchange interaction favors neighboring spins being antiparallel, as shown in Figure 15.23, the material will be *antiferromagnetic*.

Above a critical temperature known as the *Neel temperature* (T_N), the spins are disordered and the magnetic susceptibility will follow the Curie–Weiss law, showing an inverse dependence on T. Below T_N, however, the spins will order themselves into

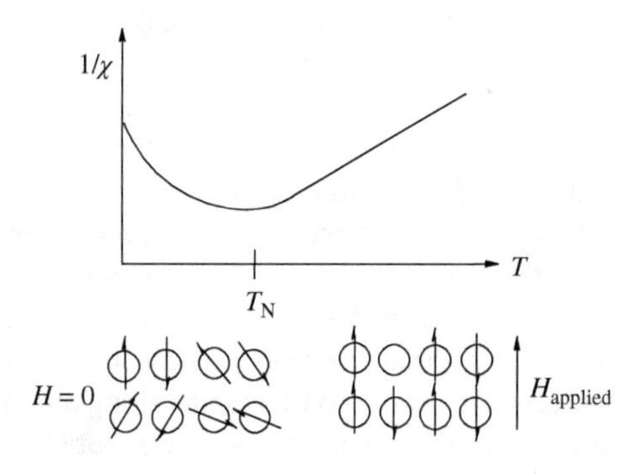

FIGURE 15.23
The temperature dependence of the magnetic susceptibility of an antiferromagnetic material.

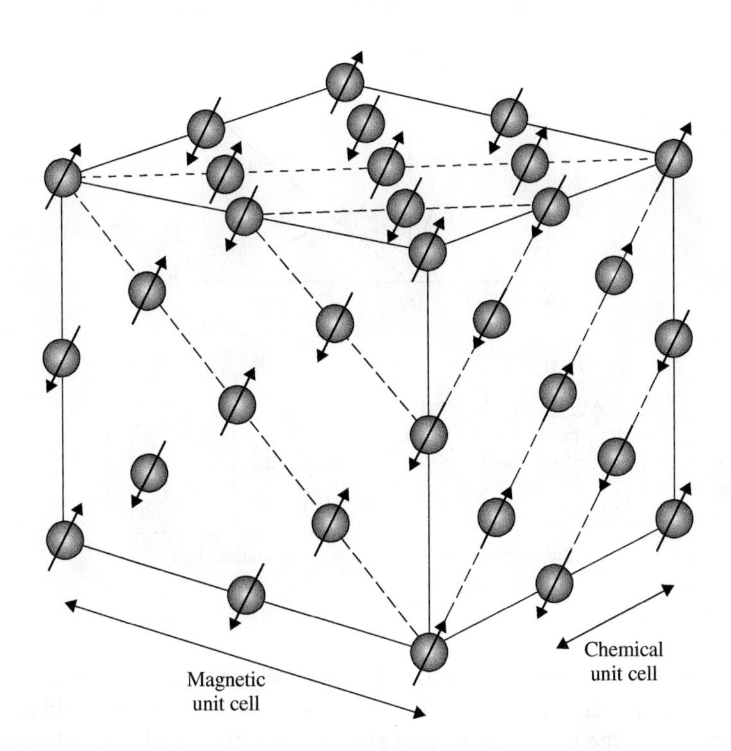

FIGURE 15.24
Magnetic unit cell for the MnO antiferromagnet. [Reproduced from Kittel, C. *Introduction to Solid State Physics*, 6th ed., John Wiley & Sons, Inc: New York, 1986. This material is reproduced with permission of John Wiley & Sons, Inc.]

antiparallel arrangements with each neighbor having opposite spin. As a result, the magnetic susceptibility will decrease below the Neel temperature. Many metal oxides are antiferromagnetic because the short O^{2-} bridges between the metal ions allow for a strong magnetic interaction, while the orbitals on the O atoms can interchange their spins by a superexchange mechanism. This antiferromagnetic ordering is shown in Figure 15.24 for MnO, which crystallizes in the rock salt-type lattice.

Less commonly, the exchange interaction between the spins on different metal ions will be a positive one, where the magnetic moments align with each other at low temperatures even in the absence of an external magnetic field, as shown in Figure 15.25. When this occurs, the material is *ferromagnetic.*

Above a critical temperature, known as the *Curie temperature*, T_C, thermal disorder will cause the magnetic susceptibility to follow the inverse dependence of the

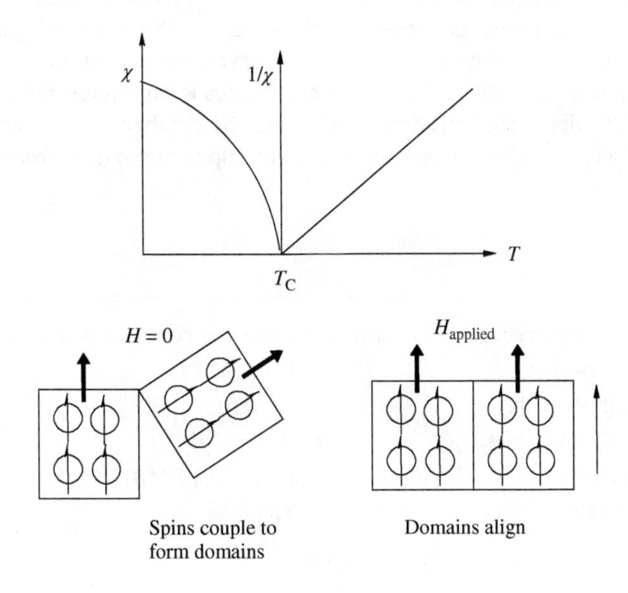

FIGURE 15.25
Below the Curie temperature (T_C), the magnetic moments can interact and couple with one another to form domains where all the spins are parallel. The net magnetic moments of the different domains are randomly oriented in the absence of an applied field. When an external magnetic field is introduced, the domain walls reorient so that the net magnetic moments of each domain align with the applied field. Note that below T_C, χ is plotted versus T and above T_C, $1/\chi$ is plotted versus *T*.

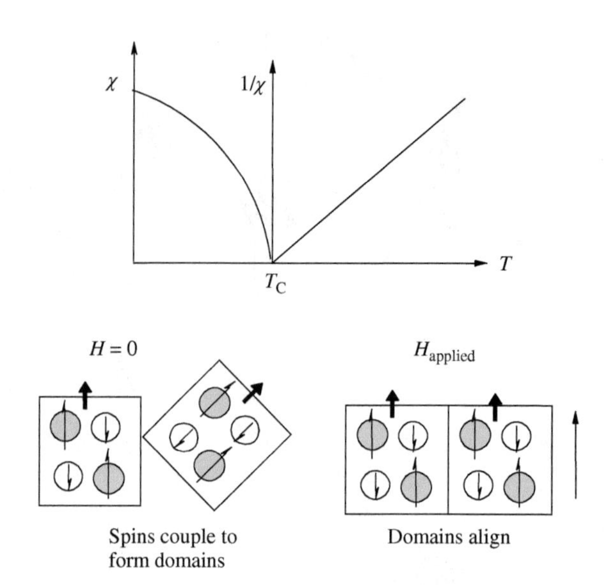

FIGURE 15.26
Below the Curie temperature, the magnetic moments on two different atoms can interact and couple with one another to form domains where all the spins are antiparallel. The net magnetic moments of the different domains are randomly oriented in the absence of an applied field. When an external magnetic field is introduced, the domain walls reorient so that the net magnetic moments of each domain align with the applied field. Note that below T_C, χ is plotted versus T and above T_C, $1/\chi$ is plotted versus T.

Curie–Weiss law. However, below T_C, the spins will order into magnetic domains with their spins parallel to one another. When an external magnetic field is applied to a ferromagnet, the parallel spins that are aligned within randomly oriented magnetic domains begin to align themselves with the magnetic field. As a result, magnetic susceptibilities as high as 10^6 cm^3/mol are possible! If the external magnetic field is suddenly removed, a ferromagnet will retain most of its long-range ordering until the domains are once again randomized by the thermal motions of the solid. This is exactly the process that occurs when an iron nail is "magnetized" by placing it in a strong magnetic field. Once the field is removed, the nail can retain some of its magnetic properties. Ferromagnetic materials are used in magnetic hard drives for the long-term storage of digital information.

The final type of bulk magnetism is *ferrimagnetism*. A ferrimagnet is really just a special case of antiferromagnetism, in which the interacting atoms have different magnetic moments. At temperatures lower than the Curie temperature, the interacting spins will align their opposing magnetic moments, but because of their different magnitudes, they fail to completely cancel each other, as shown in Figure 15.26. The net magnetic moment for each pairing is small even if all of the atoms themselves have large magnetic moments. However, the aggregate effect of the highly ordered, uncanceled magnetic moments leads to the observed magnetism. The mineral magnetite, Fe_3O_4, is a ferrimagnet. Magnetite crystallizes in an inverse spinel structure, where one Fe^{3+} sits in the tetrahedral holes and the octahedral holes are occupied by both Fe^{2+} and Fe^{3+}. Ferrimagnetic spinels are important in microwave frequency inductors.

EXERCISES

15.1. Provide the correct name for each of the following coordination compounds:

a. $[Co(NH_3)_6]Cl_3$
b. $[Cr(NH_3)_5Cl]Cl_2$
c. $[CoCl(NO_2)(NH_3)_4]Cl$
d. $[PtCl(NH_2CH_3)(NH_3)_2]Cl$
e. $K_2[PdBr_4]$
f. $[Co(en)_3]Cl_3$
g. $[Cr(H_2O)_6](NO_3)_3$
h. $K_3[Fe(C_2O_4)_3]$
i. $[Cr(NCS)_4(NH_3)_2]Cl$
j. $K[AuCl_4]$

15.2. Write the molecular formula for each of the following coordination compounds:

a. triamminetriaquachromium(III) chloride
b. pentaamminechloroplatinum(IV) bromide
c. dichlorobis(ethylenediamine) platinum(IV) chloride
d. tris(ethylenediamine)cobalt(III) sulfate
e. potassium hexacyanoferrate(II)
f. sodium tetrachloronickelate(II)
g. diamminetetrachloroplatinum(IV)
h. pentacarbonyliron(0)
i. ammonium diaquabis(oxalato) nickelate(II)
j. diamminesilver(I) dicyanoargentate(I)

15.3. For each pair, determine which coordination compound will have the larger overall stability constant:

a. $[Cu(CN)_4]^{2-}$ or $[Cu(CN)_4]^{3-}$
b. $[Co(en)_3]^{3+}$ or $[Ni(en)_3]^{2+}$
c. $[Co(C_2O_4)_3]^{4-}$ or $[Fe(C_2O_4)_3]^{3-}$

15.4. Given the standard reduction potentials listed here, determine the overall formation constant for $Zn(NH_3)_4{}^{2+}$:

$$Zn^{2+} + 2\,e^- \rightarrow Zn \qquad E^\circ = -0.762\ V$$

$$Zn(NH_3)_4{}^{2+} + 2\,e^- \rightarrow Zn + 4\,NH_3 \qquad E^\circ = -1.015\ V$$

15.5. Explain why $\log(\beta)$ is 9.8 for $[Ni(py)_6]^{2+}$ and nearly twice that value for $[Ni(bpy)_3]^{2+}$.

15.6. If $\log(K_1) = 5.84$ for $[Ni(en)(H_2O)_2]^{2+}$ and $\log(\beta_2) = 10.62$ for $[Ni(en)_2]^{2+}$, determine the stepwise formation constant for the following equilibrium:

$$[Ni(en)(H_2O)_2]^{2+} + en \rightarrow [Ni(en)_2]^{2+} + 2\,H_2O \qquad K_2 = ?$$

15.7. Given the following information, determine the standard entropy change for the formation of each complex ion and comment on the relative magnitudes of each.

a. $[Ni(NH_3)_4]^{2+}$ $\log(\beta_4) = 7.44$ $\Delta H^\circ = -53.1\ kJ/mol$
b. $[Ni(en)_2]^{2+}$ $\log(\beta_2) = 10.62$ $\Delta H^\circ = -56.5\ kJ/mol$

15.8. Sketch the three-dimensional structure of $[Pd(PPh_3)_4Br]^+$, paying particular attention to which ligands lie in each site.

15.9. Sketch the three-dimensional structure of $[Cr(PR_3)_3Cl_2]$, paying particular attention to which ligands lie in each site.

15.10. Sketch all the possible isomers for each of the following coordination compounds:

a. $[Ru(bpy)_2Cl_2]$
b. $[Pt(NH_3)_2(OH)(H_2O)]Cl$
c. $K_3[Mn(C_2O_4)_3]$

15.11. Calculate the predicted spin-only effective magnetic moment for each of the following gas-phase ions in units of Bohr magnetons:

a. Mn^{4+}
b. V^{3+}
c. Cu^+

15.12. Write the ground-state term symbols (in the form ${}^{2S+1}L_J$) for each of the following gas-phase ions and then calculate the predicted effective magnetic moment in units of Bohr magnetons:

a. Ce^{3+}
b. Sm^{3+}
c. Yb^{3+}

15.13. Which of the following is expected to have the largest effective magnetic moment: $[Cr(H_2O)_6]^{3+}$, $[Fe(H_2O)_6]^{2+}$, $[Cu(H_2O)_6]^{2+}$, or $[Zn(H_2O)_6]^{2+}$?

15.14. Which of the following compounds cannot be paramagnetic: $Cr(ClO_4)_3$, $KMnO_4$, $TiCl_3$, or $VOBr_2$?

15.15. If the effective magnetic moment of a first-row transition metal is 6.92 Bohr magnetons, what is its electron configuration?

15.16. Which of the following ions is paramagnetic: Cu^+, Zn^{2+}, Fe^{2+}, or Sc^{3+}?

BIBLIOGRAPHY

1. Atkins P, Overton T, Rourke J, Weller M, Armstrong F, Hagerman M. *Shriver& Atkins' Inorganic Chemistry*. 5th ed. New York: W. H. Freeman and Company; 2010.
2. Carter RL. *Molecular Symmetry and Group Theory*. New York: John Wiley & Sons, Inc.; 1998.
3. Clare BW, Kepert DL. Coordination Numbers and Geometries. In: King RB, editor. *Encyclopedia of Inorganic Chemistry*. John Wiley & Sons, Inc.: New York; 1994.
4. Cotton FA. *Chemical Applications of Group Theory*. 2nd ed. John Wiley & Sons, Inc.: New York; 1971.
5. Douglas B, McDaniel D, Alexander J. *Concepts and Models of Inorganic Chemistry*. 3rd ed. New York: John Wiley & Sons, Inc.; 1994.
6. Harris DC, Bertolucci MD. *Symmetry and Spectroscopy*. New York: Dover Publications, Inc.; 1978.
7. Housecroft CE, Sharpe AG. *Inorganic Chemistry*. 3rd ed. Essex, England: Pearson Education Limited; 2008.
8. Huheey JE, Keiter EA, Keiter RL. *Inorganic Chemistry: Principles of Structure and Reactivity*. 4th ed. New York: Harper Collins College Publishers; 1993.
9. Kauffman GB. Coordination Chemistry: History. In: King RB, editor. *Encyclopedia of Inorganic Chemistry*. John Wiley & Sons, Inc.: New York; 1994.
10. Lawrence GA. *Introduction to Coordination Chemistry*. Chichester, West Sussex, England: John Wiley & Sons Ltd; 2010.
11. Miessler GL, Tarr DA. *Inorganic Chemistry*. 4th ed. Upper Saddle River, NJ: Pearson Education Inc.; 2011.
12. Rodgers GE. *Descriptive Inorganic, Coordination, and Solid-State Chemistry*. 3rd ed. Belmont, CA: Brooks/Cole Cengage Learning; 2012.

Structure, Bonding, and Spectroscopy of Coordination Compounds 16

"Thus the best features of both the valence-bond picture and the crystal field theory are incorporated in the ligand field theory, and it is this theory with which we shall be mostly concerned."

—*Carl Ballhausen*

16.1 VALENCE BOND MODEL

The goal of this chapter is to gain an understanding of how the structure and bonding in coordination compounds can be used to rationalize their electronic spectra and magnetic properties. We begin our discussion with a review of atomic structure. The solutions to the three-dimensional wave equation for the hydrogen atom were given in Chapter 4 and the mathematical forms of the wave function in spherical polar coordinates were given in Table 4.1. Unsöld's theorem states that the total probability for any filled subshell or orbitals (e.g., the five *d*-orbitals) must be spherically symmetric. The angular wave function solutions to the time-independent Schrödinger equation for the set of *d*-orbitals take the general forms listed in Equations (16.1)–(16.5), where the subscript on the *d*-orbital indicates the value of the magnetic quantum number m_l and where the symbols *x*, *y*, and *z* are abbreviations for the angular wave functions of the p_x, p_y, and p_z orbitals.

$$d_2 = (x + iy)^2 \tag{16.1}$$

$$d_1 = z(x + iy) \tag{16.2}$$

$$d_0 = \frac{1}{2}(2z^2 - x^2 - y^2) \tag{16.3}$$

$$d_{-1} = z(x - iy) \tag{16.4}$$

$$d_{-2} = (x - iy)^2 \tag{16.5}$$

Recall also that the wavelike properties of electrons allow for superposition; in other words, any linear combination of these wavefunctions will also be an acceptable solution to the Schrödinger equation. Thus, we obtain the more familiar "cubic"

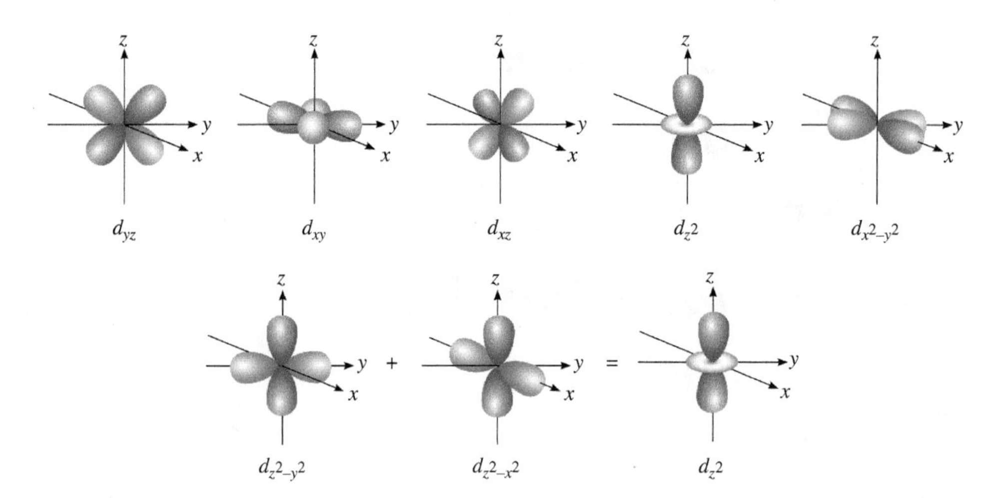

FIGURE 16.1
Shapes of the five *d*-orbitals. [Blatt Communications.]

forms of the *d*-orbitals given by Equations (16.6)–(16.10):

$$d_{x^2-y^2} = \frac{1}{2}(d_2 + d_{-2}) = x^2 - y^2 \tag{16.6}$$

$$d_{xy} = \frac{1}{2}(d_2 - d_{-2}) = 2xy \tag{16.7}$$

$$d_{xz} = \frac{1}{2}(d_1 + d_{-1}) = 2xz \tag{16.8}$$

$$d_{yz} = \frac{1}{2}(d_1 - d_{-1}) = 2yz \tag{16.9}$$

$$d_{z^2} = d_0 = \frac{1}{2}(2z^2 - x^2 - y^2) \tag{16.10}$$

Equations (16.6)–(16.10) perhaps help clarify where the familiar names of the five *d*-orbitals come from. The shapes of the "cubic" *d*-orbitals are shown in Figure 16.1. The unusual shape of the $d_z{}^2$ orbital is explained by the graphical linear combination shown in the figure. Students should readily familiarize themselves with the names and shapes of these five orbitals, which are so central to the discussion that follows.

In Chapter 10, the valence bond (VB) model was introduced as a localized bonding model for covalent bonds. Because the coordinate covalent bond is simply a subset of covalent bonding, it might be reasonable to begin a discussion of the bonding in coordination compounds by employing a VB approach. Because most coordination compounds have the octahedral geometry, we begin with the group theoretical analysis of the σ-bonding by determining the symmetries of the basis function formed using the six bond vectors depicted in Figure 16.2. The reducible representation for the six σ-bonds is shown in Table 16.1 and reduces to $a_{1g} + t_{1u} + e_g$. The a_{1g} IRR corresponds with the *s*-orbital, the three *p*-orbitals transform as t_{1u}, and the e_g orbital represents the $d_x{}^2{}_{-y}{}^2$ and $d_z{}^2$ atomic orbitals (AOs). Therefore, the hybridization of an octahedral coordination compound is d^2sp^3. The results match our chemical intuition because the ligands in an octahedral coordination geometry lie along the *x*, *y*, and *z* coordinate axes. Thus, the six σ-donating ligands will have maximum overlap with the three *p*-orbitals, which also lie along the coordinate axes; the $d_x{}^2{}_{-y}{}^2$ orbital, which has its lobes along the *x* and *y* axes; the $d_z{}^2$ orbital, which has its larger

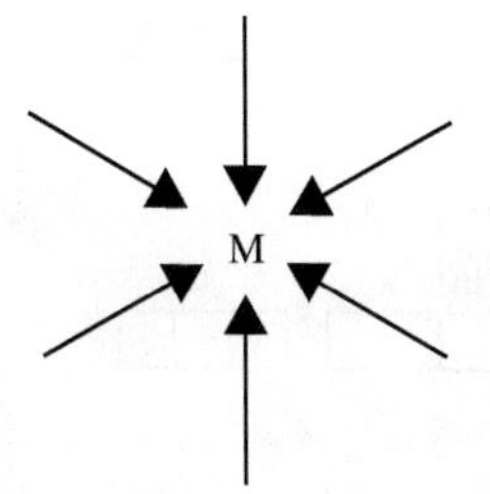

FIGURE 16.2
Bond vector basis set for the hybridization of a metal ion (M) in an octahedral coordination geometry.

TABLE 16.1 Reducible representation for the σ-bonding basis set shown in Figure 16.2.

O_h	E	$8C_3$	$6C_2$	$6C_4$	$3C_2$	i	$6S_4$	$8S_6$	$3\sigma_h$	$6\sigma_d$	IRRs
Γ_{FeF}	6	0	0	2	2	0	0	0	4	2	$a_{1g} + t_{1u} + e_g$

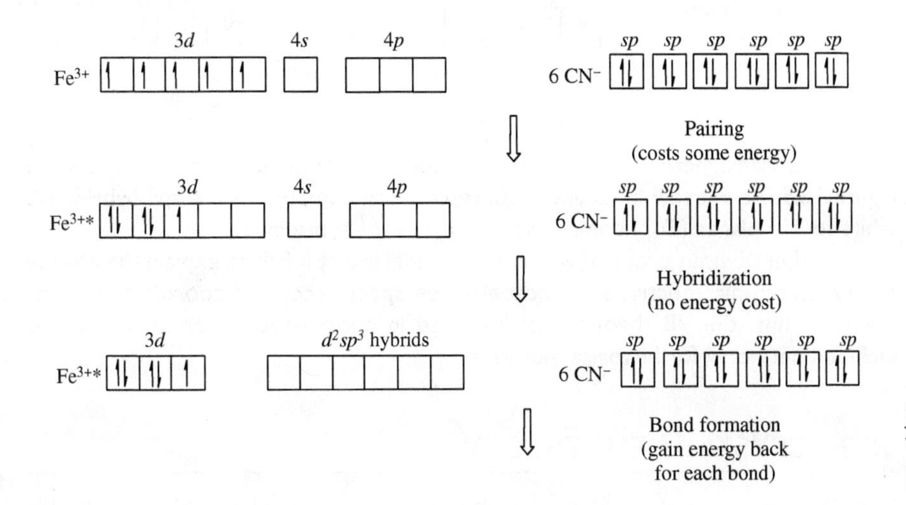

FIGURE 16.3
Valence bond treatment for the $[Fe(CN)_6]^{3-}$ ion.

lobe lying along the z axis and the donut region in the *xy* plane; and, of course, the spherically symmetric *s*-orbital.

Using the techniques developed in Chapter 10, the purported VB development for $[Fe(CN)_6]^{3-}$ is shown in Figure 16.3. Each of the empty d^2sp^3 hybrid orbitals on the Fe^{3+} ion can accept a pair of electrons from one of the *sp* hybrid orbitals on the CN^- ligands to form the six Fe–CN coordinate covalent bonds. The experimental magnetic moment of the ferricyanide ion is 2.20 BM, which is in reasonable agreement with the value of 1.73 BM predicted by Equation (15.5).

One problem with the VB model is that not every Fe^{3+} coordination compound has the same magnetic properties. For example, the experimental magnetic moment of $[FeF_6]^{3-}$(5.85 BM) is more consistent with five unpaired electrons than it is with one electron. The only way to rationalize the magnetic behavior of $[FeF_6]^{3-}$ is to use the VB development shown in Figure 16.4, where the 3*d* electrons remain unpaired and the much higher lying 4*d* orbitals are used in the construction of the six d^2sp^3 hybrids. Clearly, this is not a very satisfying result.

Another shortcoming of the VB model is its inability to predict the geometry of four-coordinate compounds. For example, the $[NiCl_4]^{2-}$ ion is tetrahedral and paramagnetic, while the isoelectronic $[Ni(CN)_4]^{2-}$ and $[PtCl_4]^{2-}$ ions are square

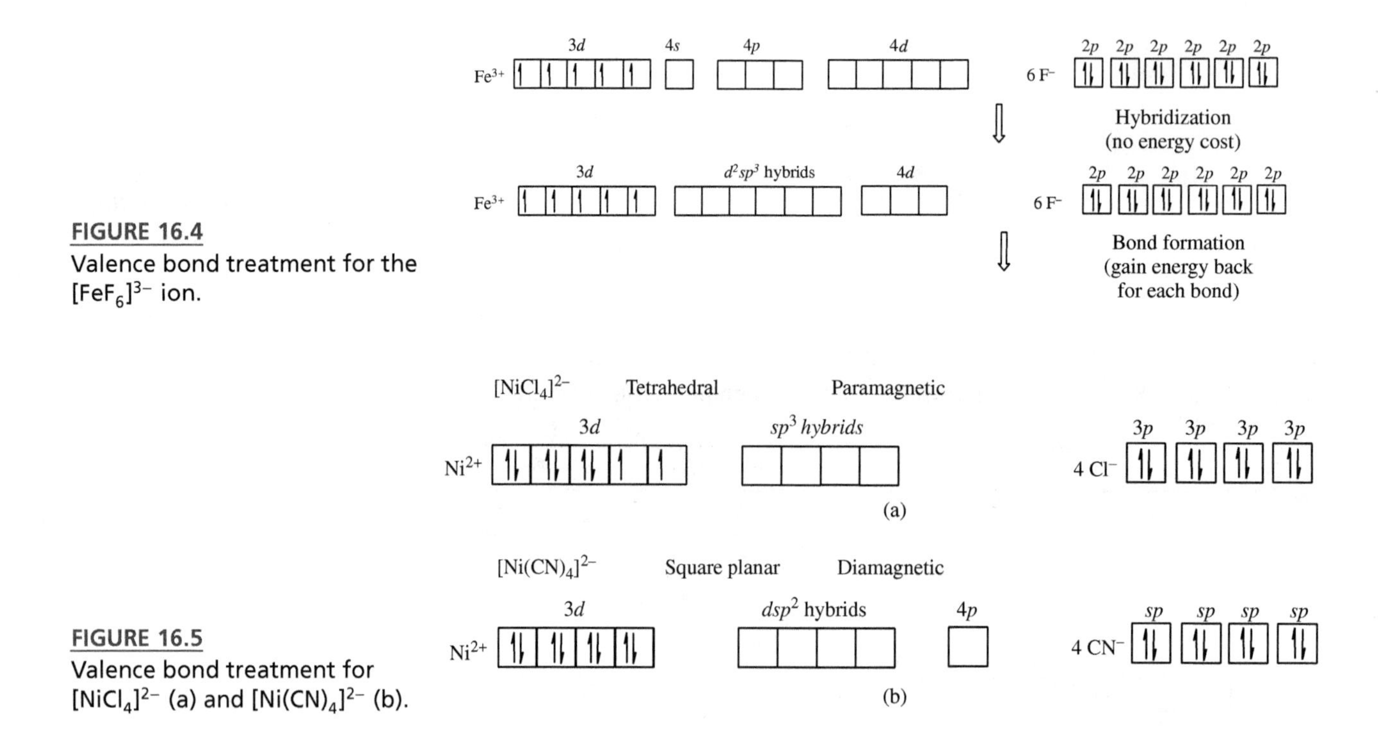

FIGURE 16.4
Valence bond treatment for the $[FeF_6]^{3-}$ ion.

FIGURE 16.5
Valence bond treatment for $[NiCl_4]^{2-}$ (a) and $[Ni(CN)_4]^{2-}$ (b).

planar and diamagnetic. The VB treatment for the two different cases is shown in Figure 16.5. As we saw in Chapter 10, tetrahedral compounds are sp^3 hybridized, while square planar (D_{4h}) compounds are $d_{x^2-y^2}sp^2$ hybridized.

Another obvious problem with the VB model is that it fails to explain the diverse colors, electronic spectra, and photoelectron spectroscopy of coordination compounds. Thus, the VB theory is of little use in coordination chemistry and it is included here only for historical purposes.

16.2 CRYSTAL FIELD THEORY

The VB model of chemical bonding developed in Chapter 10 is incapable of explaining the magnetic properties of coordination compounds, the preference of some metals for square planar over tetrahedral coordination, and the rich electronic spectra of coordination compounds. As a result of its many shortcomings, an alternative model called *crystal field theory* (CFT) was proposed by the physicist Hans Bethe in 1929. Originally developed to explain the electronic transitions in colored minerals, CFT takes a completely electrostatic (ionic) approach. According to this model, the metal ion can be represented as a sphere of positive charge at the center of the coordination compound (or occupying one of the cationic sites in a crystalline lattice), while the ligands behave as negative point charges distributed around the metal ion with the appropriate coordination geometry.

Let us first consider the case for the most common coordination geometry of octahedral. The crystal field (CF) treatment of the problem is developed in Figure 16.6, which shows what happens to the energies of the degenerate set of *d*-orbitals when an imaginary sphere of negative charge is introduced to a positively charged transition metal cation in the gas phase. In the first step, the energies of all the *d*-orbitals are raised as the distance separating the charges (*r*) in the denominator of Coulomb's law decreases. In the next step, the sphere of negative charge is divided by six and placed into specific sites around the metal ion in the positions where the ligands lie. For octahedral coordination, these negative point

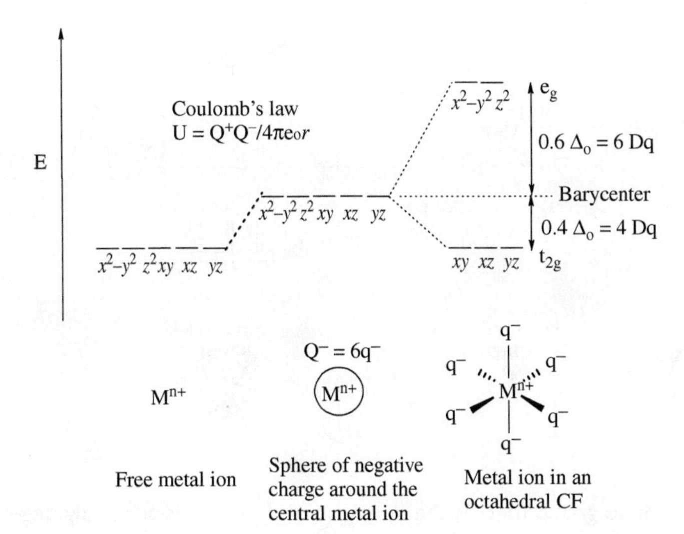

FIGURE 16.6
The interaction of a free metal ion in the gas phase with a sphere of negative charge causes the energy of the *d*-orbitals to increase as a result of the smaller value of *r*. Redistribution of the negative charge in an octahedral CF causes some of the orbitals to be raised with respect to the barycenter, while others are stabilized. The splitting between the two energy levels is defined as Δ_o or 10 Dq.

charges lie along the positive and negative directions of the three Cartesian axes. Because the value of *r* has not changed in this second step (it simply involved a redistribution of the negative charge), the overall energy of the *d*-orbitals remains unchanged by this second perturbation. However, the degeneracy of the orbitals is removed, as some of the orbitals lie closer to the negative point charges than others. When the orbitals are filled with electrons, the repulsion between electrons in the *d*-orbitals and the negative point charges representing the ligands in the crystal will cause some of the orbitals to be raised in energy.

The interaction of a free metal ion in the gas phase with a sphere of negative charge causes the energy of the *d*-orbitals to increase as a result of the smaller value of r. Redistribution of the negative charge in an octahedral CF causes some of the orbitals to be raised with respect to the barycenter, while others are stabilized. The splitting between the two energy levels is defined as Δ_o or 10 Dq. Because the $d_{x^2-y^2}$ and d_{z^2} orbitals have lobes that point directly at the negative point charges as shown in Figure 16.7, they will be raised in energy as a result of electron-electron repulsions. Because the total energy of the *d*-orbitals must remain unchanged by this second step, the energies of the remaining three *d*-orbitals must be lowered relative to the *barycenter*.

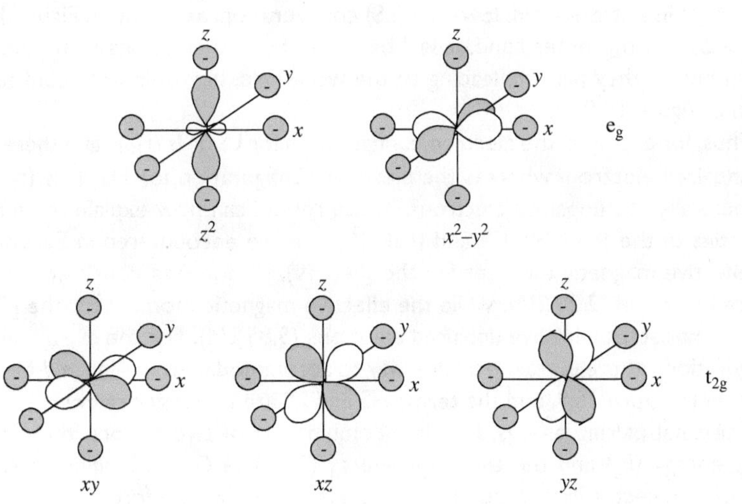

FIGURE 16.7
Interaction of the five *d*-orbitals with the negative point charges in an octahedral crystal field. The e_g orbitals have a more direct interaction with the negative point charges and are raised in energy, while the t_{2g} orbitals point between the negative point charges and are lowered in energy.

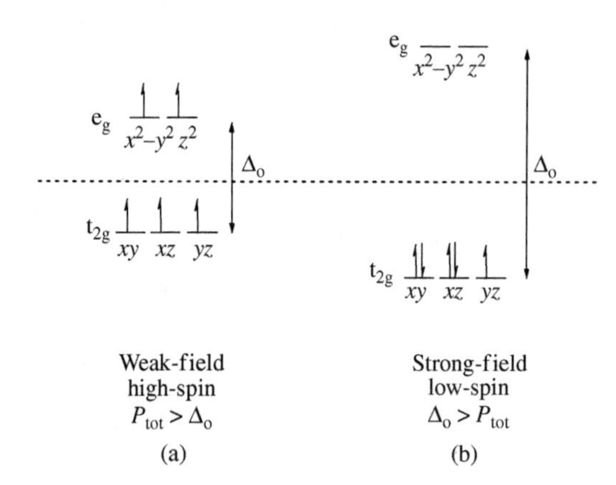

FIGURE 16.8
The relative energies of the five *d*-orbitals (a) in a weak CF, as is the case for the high-spin $[FeF_6]^{3-}$ complex ion and (b) in a strong CF, as is the case for the low-spin $[Fe(CN)_6]^{3-}$ ion. The dashed line represents the barycenter.

According to group theory, the d_{xy}, d_{xz}, and d_{yz} orbitals are triply degenerate in an octahedral CF and have their lobes pointed between the coordinate axes and away from the negative point charges. Thus, the *d*-orbitals will split into two different sets in an octahedral CF. Group theory tells us that the higher lying, doubly degenerate set of orbitals has e_g symmetry, while the lower lying, triply degenerate set is t_{2g}. If the energy difference between the two levels is defined as Δ_o or 10 Dq, then the e_g set will be raised by $0.6\Delta_o$ (6 Dq) and the t_{2g} set will be lowered by $0.4\Delta_o$ (4 Dq) relative to the barycenter. Thus, six of the electrons in the d^{10} electron configuration would be stabilized by 4 Dq each (for a total stabilization of 24 Dq) and the other four electrons would be destabilized by 6 Dq each (for a total destabilization of 24 Dq). Overall, the energy of the 10 electrons occupying the *d*-orbitals in an octahedral CF would be identical to that of the barycenter. It is this splitting of the *d*-orbitals by the CF that causes the coordination compounds to be colored. If light of the appropriate wavelength is shone on the molecule, an electron from the lower lying t_{2g} set can be photochemically excited to an empty orbital in the e_g set, causing an electronic absorption in the visible region.

As the valence *d*-electrons fill the empty orbitals in Figure 16.6 according to the Aufbau principle, they will first populate the lower lying t_{2g} set, going in with their spins paired according to Hund's rule. Because the degenerate orbitals are indistinguishable, there is only one way to write the electron configurations for d^1-d^3. For the d^4-d^7 electron configurations, however, there are two possible ways for the valence electrons in a metal to occupy the t_{2g} and e_g sets of orbitals. If the magnitude of $\Delta_o > P_{tot}$ (where P_{tot} is the total pairing energy), then the electrons will pair up first in a strong-field, low-spin (LS) configuration, as shown in Figure 16.8(a). If $P_{tot} > \Delta_o$, on the other hand, it will be easier for the electrons to fill the upper e_g level before they pair up, leading to the weak-field, high-spin (HS) configuration shown in Figure 16.8(b).

Thus, for example, the electron configuration for LS d^5 is $(t_{2g})^5$ and there is only one unpaired electron, whereas the electron configuration for HS d^5 is $(t_{2g})^3(e_g)^2$ and there are five unpaired electrons. These results can now explain the magnetic properties of the $[Fe(CN)_6]^{3-}$ and $[FeF_6]^{3-}$ ions we encountered in Section 16.1. The effective magnetic moment for the $[Fe(CN)_6]^{3-}$ ion was consistent with one unpaired electron (2.20 BM), while the effective magnetic moment for the $[FeF_6]^{3-}$ ion was consistent with five unpaired electrons (5.85 BM). For the d^8-d^{10} electron configurations, there is again only one way to accommodate the valence *d*-electrons, as shown in Figure 16.9, and the terms HS and LS are no longer relevant.

The total pairing energy, P_{tot}, is a combination of two factors: the Coulomb pairing energy (P_c) and the exchange energy (P_e). The *Coulomb pairing energy* is a result of the electrostatic repulsion of two electrons that are forced to occupy the

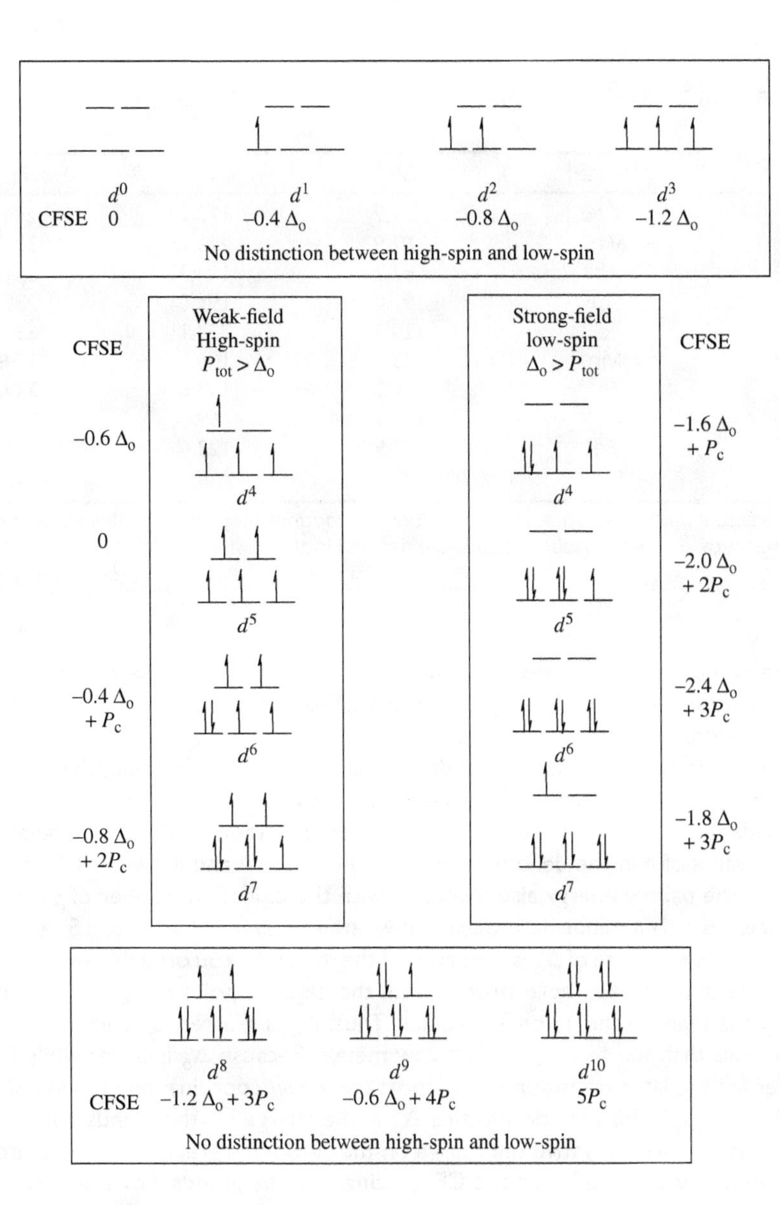

FIGURE 16.9
Crystal field splitting diagram for the different d^n configurations in an octahedral CF. If the magnitude of Δ_o is greater than P_{tot}, then the electrons will prefer to pair up in the lower-lying t_{2g} set before they begin to fill the higher-lying e_g orbitals. In this *strong-field* case, the resulting electron configuration will be *LS*, as shown at right. On the other hand, if P_{tot} is greater than Δ_o, once the t_{2g} level is half-filled, the next electron will prefer to occupy an e_g orbital rather than paying the price to pair up with another electron in the t_{2g} level. In this *weak-field* case, the resulting electron configuration will be *high-spin*, as shown at the left.

same region of space when the electrons are paired up in the same orbital. The Coulomb pairing energy is destabilizing. For all of the d^4–d^7 electron configurations, there is a difference in the number of Coulomb pairing energies between the HS and the LS cases. For example, HS d^6 will have only one P_c, while LS d^6 has three P_c. The magnitude of the Coulomb pairing energy depends on the metal ion. P_c is greater for first-row transition metals than it is for metals from the later transition series because the 3*d* orbitals are smaller than the 4*d* and 5*d* orbitals. As a result of their larger P_c values, the first-row transition metals tend to form HS complexes ($P_{tot} > \Delta_o$) more often than their second- or third-row counterparts. Because the ionic radius of a metal decreases as the charge of the metal increases, metals in high oxidation states have larger values of P_c than those having lower oxidation numbers. Thus, the values of P_c in Table 16.2 for the iron series are Fe^{3+} (120.2 kJ/mol) > Fe^{2+} (89.2 kJ/mol) > Fe^{+} (87.9 kJ/mol). For isoelectronic metals, the magnitude of P_c increases across the series: Cr^{+} (67.3 kJ/mol) < Mn^{2+} (91.0 kJ/mol) < Fe^{3+} (120.2 kJ/mol).

The P_e term is related to the loss of the exchange energy when electrons having parallel spins are forced to be antiparallel. Statistically, the exchange energy is related

TABLE 16.2 Pairing energies (kJ/mol) for selected metal ions in the gas phase.

d^n	Metal	ΣP_c	P_e	P_{tot}
d^4	Cr^{2+}	71.2	173.1	244.3
	Mn^{3+}	87.9	213.7	301.6
d^5	Cr^{+}	67.3	144.3	211.6
	Mn^{2+}	91.0	194.0	285.0
	Fe^{3+}	120.2	237.1	357.4
d^6	Mn^{+}	73.5	100.6	174.2
	Fe^{2+}	89.2	139.8	229.1
	Co^{3+}	113.0	169.6	282.6
d^7	Fe^{+}	87.9	123.6	211.5
	Co^{2+}	100	150	250

The actual pairing energies in coordination compounds may be up to 30% smaller as a result of covalency. Abbreviations are defined in the text.

to the number of exchanges between electrons having the same energy and the same spin (parallel spins). The greatest loss of exchange energy occurs for the d^5 configuration.

The magnitude of Δ_o also depends on a variety of factors, including the oxidation state of the metal, the series of the metal, and the charge on the ligand. The greater the oxidation state of the metal, the smaller its ionic radius will be. This leads to a smaller value of r in the denominator of Coulomb's law and a greater CF splitting. Because the pairing energy also increases with the oxidation number of the metal, this factor is only a minor determinant in whether a complex is HS or LS. A second factor affecting the size of Δ_o is the series of the metal. The 5*d* orbitals extend farther into space and interact more strongly with the negative point charges representing the ligands than do the 4*d* or 3*d* orbitals. Thus, Δ_o is greater for third-row transition metals than for second- or first-row metals. Because Δ_o is larger while P_{tot} is smaller for the later transition series, most third-row coordination compounds are LS ($\Delta_o > P_{tot}$). A third factor affecting Δ_o is the charge on the ligands. One might think that the more negative the charge on the ligand, the greater the numerator in Coulomb's law and the larger the CF splitting. For the ligands, however, there are more important factors at work that affect the size of Δ_o, and CFT needs to be modified in order to account for the effects of covalent bonding. A revised model, known as *ligand field theory* (LFT), embodies concepts from both CFT and molecular orbital theory (MOT) and is discussed in a later section.

The *crystal field stabilization energy* (CFSE) is defined as the energy of the d^n electron configuration in a CF relative to that of the barycenter. The CFSE is the sum of the energies of all the *d*-electrons as a result of the CF splitting and takes into consideration their Coulomb and exchange pairing energies. The CFSEs for the different d^n configurations in an octahedral CF are shown in Figure 16.9. For example, the CFSE of HS d^6, which has the electron configuration $(t_{2g})^4(e_g)^2$, is $-0.4\Delta_o + P_c$. The four electrons in the t_{2g} set are stabilized (have negative energy with respect to the barycenter) by $0.4\Delta_o$ each, while the two electrons in the e_g set are destabilized by $0.6\Delta_o$ each; this results in an overall stabilization imposed by the octahedral geometry of $-0.4\Delta_o * 4 + 0.6\Delta_o * 2 = -0.4\Delta_o$. In addition, there is one Coulomb pairing energy as a result of two electrons that are forced to pair up in the t_{2g} orbital. For LS d^6, which has the electron configuration $(t_{2g})^6$, the CFSE is $-2.4\Delta_o + 3P_c$. Table 16.3 summarizes the *difference* in the CFSE for the strong-field (LS) minus the weak-field (HS) cases.

TABLE 16.3 The difference in the CFSE for the strong-field (LS) minus the weak-field (HS) case for d^4–d^7 in an octahedral CF.

d^n	ΔCFSE (LS–HS)
d^4	$-1.0\Delta_o + P_c$
d^5	$-2.0\Delta_o + 2P_c$
d^6	$-2.0\Delta_o + 2P_c$
d^7	$-1.0\Delta_o + P_c$

Hans Bethe's classic 1929 paper on CFT, entitled "The Splitting of Terms in Crystals," had its origins in solid-state physics and was largely ignored by most chemists. Although the theoretician John van Vleck adopted CFT to explain the magnetic and spectroscopic properties of transition metal complexes outside of a crystalline environment, the Pauling VB model was more widely applicable, and interest in CFT waned until it was revived again by Leslie Orgel in the 1950s. One of the earliest thermodynamic studies in support of CFT was published by Don McClure in 1959. McClure showed that the absolute magnitudes of the lattice enthalpies of the divalent metal halides from the first transition series depended on the CFSE of the metal ion, as shown in Figure 16.10.

In general, the lattice enthalpy increases across the series as a result of the decreasing size of the ionic radius that accompanies the increase in the effective nuclear charge. However, the actual lattice enthalpies do not fit to a smooth, linear progression across the row. Instead, there are distinct peaks and valleys in the experimental data. Those d^n configurations having the smallest deviation from the expected linear trend occur for the d^0, d^5, and d^{10} electron configurations, each of which has a CFSE = 0 (ignoring the pairing energy). Some of the larger deviations from the linear progression (which correspond to more stable lattice enthalpies) occur for the d^3 and d^8 electron configurations, which (not coincidentally) happen to have the largest CFSEs ($-1.2\Delta_o$).

Other evidence for CFT comes from the enthalpies of hydration for the divalent metal ions of the first transition series. Again, a linear trend is expected, with the enthalpy of hydration decreasing across the series. The experimental data are shown by the solid blue line in Figure 16.11. The largest deviations from the expected trend occur for those d^n electron configurations having the greatest CFSE, namely d^3 and d^8. If the CFSE is subtracted from the experimental enthalpy of hydration, the expected linear trend is observed, as shown by the solid red line.

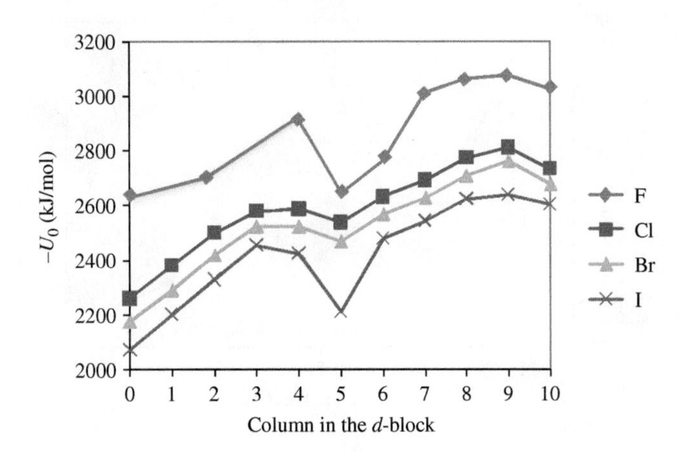

FIGURE 16.10
The absolute value of the magnitude of the lattice enthalpies (kJ/mol) for the divalent metal halides of the first transition series. [Data were obtained from http://www.webelements.com (accessed March 29, 2014). Thermochemical data were chosen over calculated values whenever they were known.]

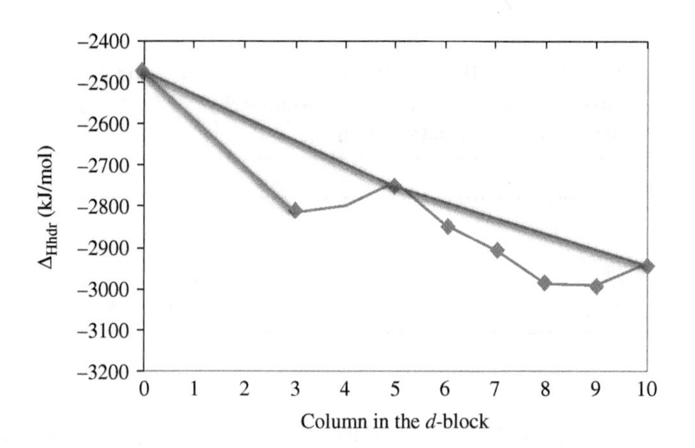

FIGURE 16.11
Hydration enthalpies for the first row divalent transition metals as a function of d^n electron count. The blue line represents the experimental data, while the red line approximates the experimental values minus the CFSE. [Data are from Shriver and Atkins, *Inorganic Chemistry*, 4th edition.]

With the advent of reliable X-ray crystallographic data, the ionic radii of the divalent and trivalent metal ions of the first transition series were reported by Shannon and Prewitt. In the case of coordination compounds having LS electron configurations, the ionic radius decreases across the series as the t_{2g} set is filled until the d^6 electron configuration is reached, as shown in Figure 16.12. The decrease across the row occurs (in part) because the low-lying t_{2g} orbitals are somewhat bonding in nature. Then, from d^7 to d^{10}, there is a gradual increase in the ionic radius as the higher lying, antibonding e_g set of orbitals is occupied. For the corresponding HS compounds, the radius decreases across the series until it reaches the d^3

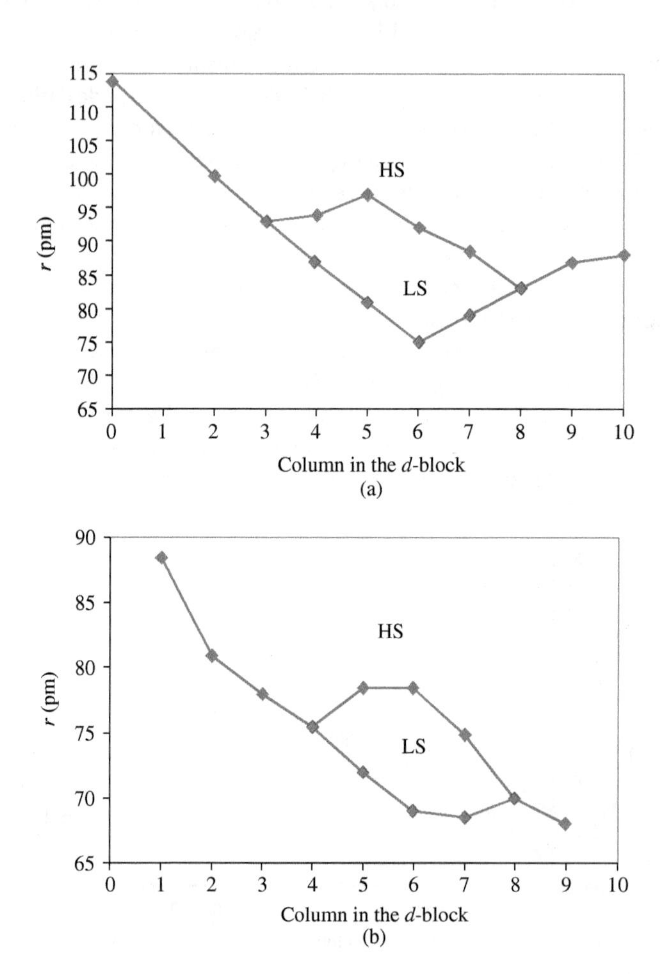

FIGURE 16.12
Ionic radii for divalent (a) and trivalent (b) first-row transition metal ions. [Data were obtained from http://www.webelements.com (accessed March 29, 2014).]

configuration, where the t_{2g} orbitals are half-filled. The fourth and fifth electrons then occupy the antibonding e_g set and the ionic radius increases. It gradually decreases for d^6-d^8 as the electrons again fill the more bonding t_{2g} level before finally increasing for d^9 and d^{10} as the e_g level is further occupied.

Group theory can be used to determine the symmetries of the different energy levels for the *d*-orbital splitting patterns in geometries other than octahedral. The relative energies of each level can then be determined based on how strongly the negative point charges—representing the ligands—interact with the *d*-orbitals. For example, the *d*-orbitals in the tetrahedral CF split as follows: t_2 (d_{xy}, d_{xz}, d_{yz}) and e ($d_{x^2-y^2}$, d_{z^2}). The positions of the negative point charges in a tetrahedral molecule lie at opposite corners of a cube, as shown in Figure 16.13.

Because the t_2 orbitals lie in orthogonal planes with their lobes pointing at the centers of four edges of the cube, these orbitals have a moderately strong interaction with the four negative point charges. The lobes of the e orbitals, on the other hand, point at the centers of the faces and have little interaction with the negative point charges. Thus, the t_2 level will lie higher in energy than the e level in the CF splitting diagram for a tetrahedral compound, as shown in Figure 16.14.

Because there are less negative point charges and none of the interactions are direct, as they are in the octahedral ligand field (LF) for the e_g set, the energy separation between the t_2 and e levels in the tetrahedral geometry (Δ_t) will be smaller than that in the octahedral geometry (Δ_o). As a general rule, $\Delta_t \cong 4/9\Delta_o$. Because Δ_t is so small, the pairing energy is almost always larger than Δ_t. Thus, almost every tetrahedral compound will be HS.

The third most common coordination geometry is square planar (D_{4h}). In this geometry, the *d*-orbitals split into four groups, as follows: a_{1g} (d_{z^2}), b_{1g} ($d_{x^2-y^2}$), b_g (d_{xy}), and e_g (d_{xz}, d_{yz}). Taking the principal C_4 axis of rotation as the z-axis, the ligands will approach the metal ion in the *xy* plane along the coordinate axes at $+x$, $-x$, $+y$, and $-y$. Thus, the orbital that interacts most with the negative point charges is the $d_{x^2-y^2}$ orbital, because this orbital also lies in the *xy*-plane with its lobes pointed directly at those charges. Therefore, the b_{1g} molecular orbital (MO) will lie highest in energy. The second highest MO will be the b_{2g} level because the d_{xy} orbital also lies in the *xy*-plane even though the lobes are pointed between the coordinate axes.

M

(a)

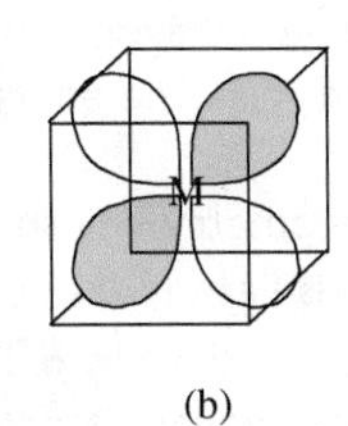

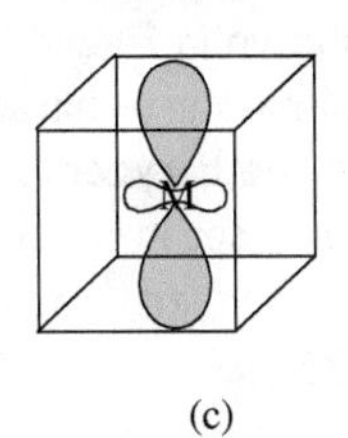

FIGURE 16.13
(a) Positions of the ligands in a tetrahedral CF, where the ligands lie at every other corner of the cube; (b) a representative t_2 orbital has its lobes pointing at the centers of the edges of the cube; and (c) a representative e orbital has its lobes pointing at the centers of the faces. The t_2 orbitals interact more strongly with the ligand positions than do the e orbitals.

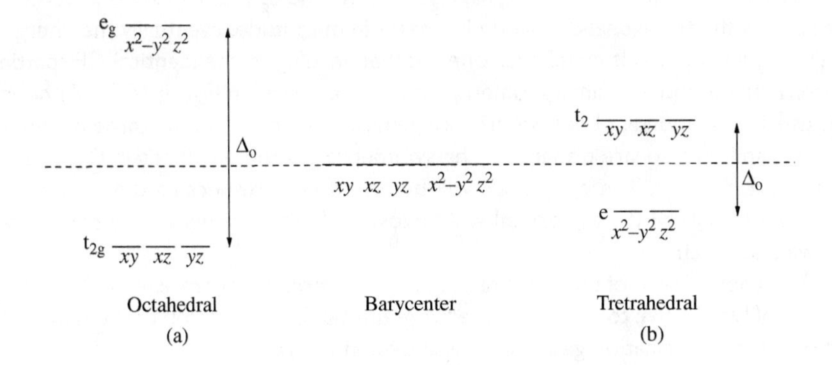

FIGURE 16.14
Crystal field splitting patterns comparing octahedral (a) and tetrahedral (b) coordination geometries.

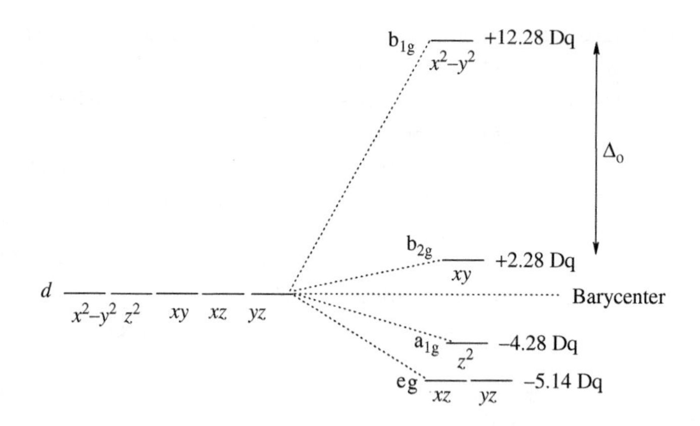

FIGURE 16.15
Crystal field splitting pattern for the square planar geometry.

Next in energy will be the a_{1g} level because the d_{z^2} orbital has a ring of electron density in the *xy*-plane even though the other half of the orbital lies along the *z*-axis. Lastly, the e_g orbitals will lie lowest in energy because the d_{xz} and d_{yz} orbitals have a nodal plane that is coincident with the *xy*-plane, which is where the negative point charges lie. Thus, the CF splitting pattern for the square planar geometry is the one shown in Figure 16.15.

The difference in energy between the $d_{x^2-y^2}$ and the d_{xy} orbitals is defined as Δ_o and it has the same value as it would have in an octahedral geometry. The relative energies of each orbital in units of Δ_o are based on calculations of the amount of orbital overlap between each *d*-orbital and a sigma-donating ligand. Steric interactions favor a tetrahedral geometry for four-coordinate compounds (as per valence-shell electron-pair repulsion (VESPR) theory), but square planar geometry can occur for electronic reasons, which happens most often with a LS d^8 electron configuration. The reason for this is because there is such a large separation between the b_{1g} and b_{2g} MOs. If the complex is LS d^8, it will have an extremely favorable CFSE of $-24.56\Delta_o$. Thus, square planar compounds are always LS.

An alternative method for deriving the square planar diagram is to start with an octahedral CF and then imagine what would happen if a tetragonal distortion occurred along the *z*-axis ("z-out") until the *z*-axis ligands were completely pulled off, as shown in Figure 16.16. The tetragonal distortion will lower the energy of any *d*-orbital having *z*-character and raise the energies of the other orbitals in order to maintain the barycenter.

The e_g set in octahedral geometry will split into a higher lying b_{1g} ($d_{x^2-y^2}$) and lower lying a_{1g} (d_{z^2}) as the *z*-axis ligands are removed. Note that the d_{z^2} is lowered by $10.28\Delta_o$ while the $d_{x^2-y^2}$ is only raised by $6.28\Delta_o$. In other words, the two orbitals do not split evenly when the tetragonal distortion is introduced. At the same time, the t_{2g} set in octahedral geometry will split into the b_{2g} (d_{xy}) set, which will be raised by $6.28\Delta_o$ with respect to the original t_{2g} level, and the e_g (d_{xz}, d_{yz}) set is lowered by $1.14\Delta_o$. As the tetragonal distortion increases in magnitude, eventually the energies of the b_{2g} and a_{1g} levels cross over one another, leading to the identical CF splitting pattern for the square planar geometry that was observed in Figure 16.15. Although it is usually unimportant because they are generally unoccupied, the three *p*-orbitals will also split in a square planar CF. The symmetries of the *p*-orbitals in D_{4h} are a_{2u} (p_z) and e_u (p_x, p_y). The e_u orbitals will be raised with respect to the barycenter, while the energy of the a_{2u} orbital will be lowered with respect to the barycenter by twice as much.

The energy levels of the *d*-orbitals in other CF geometries are listed in Table 16.4 in units of Dq relative to the barycenter. A graphical representation of the same data for selected coordination geometries is shown in Figure 16.17.

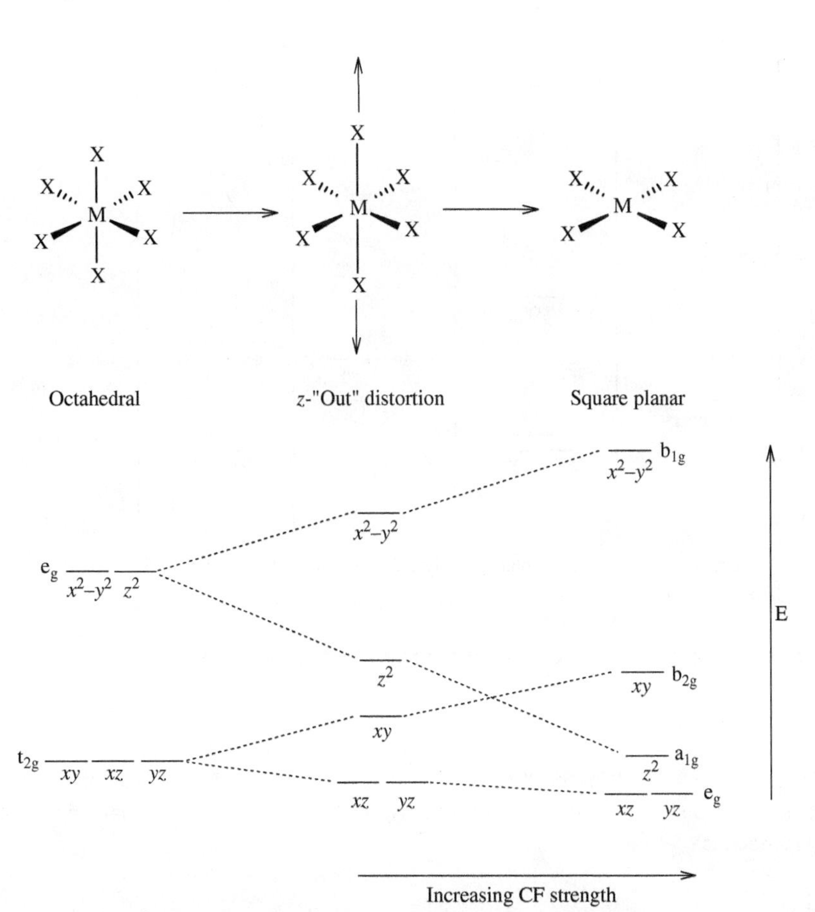

FIGURE 16.16
Generation of the square planar crystal field splitting pattern by starting with an octahedral crystal field and removing the ligands along the z-axis by way of a tetragonal distortion.

TABLE 16.4 Crystal field splitting patterns of the *d*-orbitals for selected geometries in units of Dq relative to the barycenter.

CN	Geometry	d_{z^2}	$d_{x^2-y^2}$	d_{xy}	d_{xz}	d_{yz}
1	Linear ($D_{\infty h}$)	$5.14(\sigma_g^+)$	$-3.14(\delta_g)$	$-3.14(\delta_g)$	$0.57(\pi_g)$	$0.57(\pi_g)$
2	Linear ($D_{\infty h}$)	$10.28(\sigma_g^+)$	$-6.28(\delta_g)$	$-6.28(\delta_g)$	$1.14(\pi_g)$	$1.14(\pi_g)$
3	Trigonal planar (D_{3h})	$-3.21(a_1')$	$5.46(e')$	$5.46(e')$	$-3.86(e'')$	$-3.86(e'')$
4	Tetrahedral (T_d)	$-2.67(e)$	$-2.67(e)$	$1.78(t_2)$	$1.78(t_2)$	$1.78(t_2)$
4	Square planar (D_{4h})	$-4.28(a_g)$	$12.28(b_{1g})$	$2.28(b_{2g})$	$-5.14(e_g)$	$-5.14(e_g)$
5	Trigonal bipyramidal (D_{3h})	$7.07(a_1')$	$-0.82(e')$	$-0.82(e')$	$-2.72(e'')$	$-2.72(e'')$
5	Square pyramidal (C_{4v})	$0.86(a_1)$	$9.14(b_1)$	$-0.86(b_2)$	$-4.57(e)$	$-4.57(e)$
6	Octahedral (O_h)	$6.00(e_g)$	$6.00(e_g)$	$-4.00(t_{2g})$	$-4.00(t_{2g})$	$-4.00(t_g)$
6	Trigonal prismatic (D_{3h})	$0.96(a_1')$	$-5.84(e')$	$-5.84(e')$	$5.36(e'')$	$5.36(e'')$
7	Pentagonal bipyramidal (D_{5h})	$4.93(a_1')$	$2.82(e_2')$	$2.82(e_2')$	$-5.28(e_1'')$	$-5.28(e_1'')$
8	Cubic (T_d)	$-5.34(e)$	$-5.34(e)$	$3.56(t_2)$	$3.56(t_2)$	$3.56(t_2)$
8	Square antiprismatic (D_{4d})	$-5.34(a_1)$	$-0.89(e_2)$	$-0.8(e_2)$	$3.56(e_3)$	$3.56(e_3)$
12	Icosahedral (I_h)	$0.00(h_g)$	$0.00(h_g)$	$0.00(h_g)$	$0.00(h_g)$	$0.00(h_g)$

The point group and appropriate Mulliken symbol for each geometry are listed in parentheses.

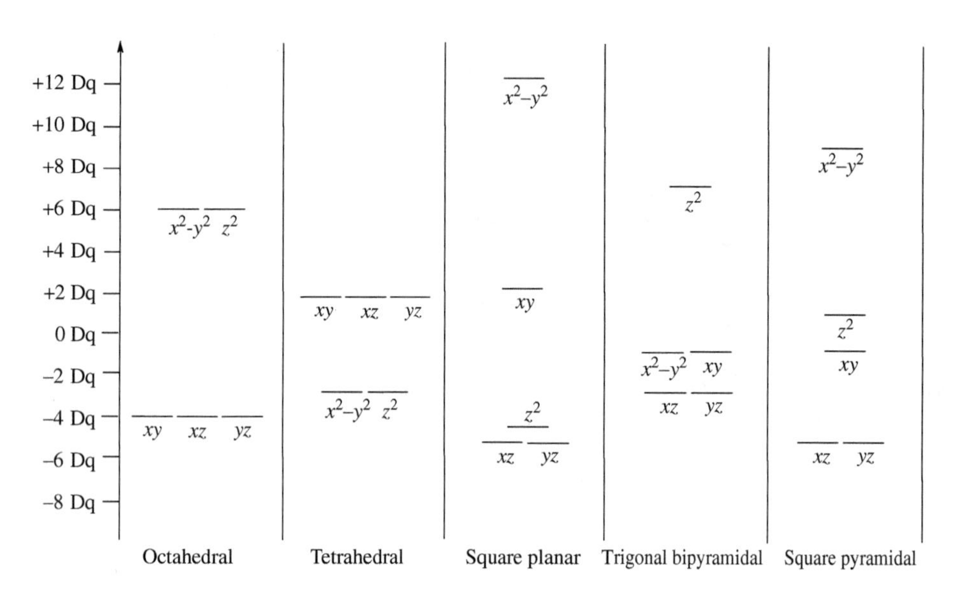

FIGURE 16.17
The crystal field splitting diagrams for the five *d*-orbitals in different coordination geometries.

Example 16-1. Using data from Table 16.4, calculate the CFSEs for every d^n configuration (HS and LS) in both the trigonal bipyramidal and square pyramidal geometries.

Solution. The electron configurations and CFSEs for each d^n electron count in both the trigonal bipyramidal and square pyramidal geometries are listed in the table. CFSEs are in Dq units.

Trigonal Bipyramidal		CFSE	Square Pyramidal		CFSE
d^1	$(e'')^1$	−2.72	d^1	$(e)^1$	−4.57
d^2	$(e'')^2$	−5.44	d^2	$(e)^2$	−9.14
d^3 HS	$(e'')^2(e')^1$	−6.26	d^3 HS	$(e)^2(b_2)^1$	−10.00
d^3 LS	$(e'')^3$	−8.16	d^3 LS	$(e)^3$	−13.71
d^4 HS	$(e'')^2(e')^2$	−7.08	d^4 HS	$(e)^2(b_2)^2$	−10.86
d^4 LS	$(e'')^4$	−10.88	d^4 LS	$(e)^4$	−18.28
d^5 HS	$(e'')^3(e')^2$	−9.80	d^5 HS	$(e)^3(b_2)^2$	−15.43
d^5 LS	$(e'')^4(e')^1$	−11.70	d^5 LS	$(e)^4(b_2)^1$	−19.14
d^6	$(e'')^4(e')^2$	−12.52	d^6	$(e)^4(b_2)^2$	−20.00
d^7 HS	$(e'')^4(e')^2(a_1')^1$	−5.45	d^7	$(e)^4(b_2)^2(a_1)^1$	−19.14
d^7 LS	$(e'')^4(e')^3$	−13.34			
d^8 HS	$(e'')^4(e')^3(a_1')^1$	−15.01	d^8	$(e)^4(b_2)^2(a_1)^2$	−18.28
d^8 LS	$(e'')^4(e')^4$	−14.16			
d^9	$(e'')^4(e')^4(a_1')^1$	−7.09	d^9	$(e)^4(b_2)^2(a_1)^2(b_1)^1$	−9.14

Although CFT can explain the magnetism of coordination compounds and predict that they are colored, it is not a quantitative theory. It cannot predict, for instance, the exact color of a transition metal complex because there is no way of rationalizing the relative magnitude of Δ_o *a priori*. Even for a metal ion having

TABLE 16.5 Crystal field splitting parameters (Δ_o) as a function of d^n.

Coordination Compound	Δ_o, kK	Coordination Compound	Δ_o, kK
$[Ti(H_2O)_6]^{3+}$	20.1	$[FeCl_6]^{3-}$	11.6
$[TiF_6]^{3-}$	18.9	$[Fe(C_2O_4)_3]^{3-}$	14.1
$[TiCl_6]^{3-}$	13.0	$[Fe(CN)_6]^{3-}$	35.0
$[TiCl_6]^{4-}$	8.4	$[Ru(H_2O)_6]^{3+}$	28.6
$[VF_6]^{2-}$	16.1	$[Ru(C_2O_4)_3]^{3-}$	28.7
$[VCl_6]^{2-}$	15.4	$[IrCl_6]^{4-}$	27.0
$[V(H_2O)_6]^{3+}$	19.0	$[Fe(H_2O)_6]^{2+}$	9.4
$[VF_6]^{3-}$	16.1	$[Fe(phen)_3]^{2+}$	13.1
$[VCl_6]^{3-}$	12.0	$[Fe(CN)_6]^{4-}$	32.2
$[V(CN)_6]^{3-}$	23.9	$[Ru(H_2O)_6]^{2+}$	19.8
$[V(H_2O)_6]^{2+}$	12.3	$[Ru(en)_3]^{2+}$	28.1
$[VCl_6]^{4-}$	7.2	$[Ru(CN)_6]^{4-}$	33.8
$[V(bpy)_3]^{2+}$	15.9	$[Co(H_2O_6]^{3+}$	20.8
$[Cr(H_2O_6]^{3+}$	17.4	$[Co(NH_3)_6]^{3+}$	22.9
$[Cr(NH_3)_6]^{3+}$	21.6	$[Co(en)_3]^{3+}$	23.2
$[Cr(en)_3]^{3+}$	21.9	$[CoF_6]^{3-}$	13.1
$[CrF_6]^{3-}$	14.9	$[Co(CN)_6]^{3-}$	32.2
$[CrCl_6]^{3-}$	13.2	$[Rh(H_2O)_6]^{3+}$	27.2
$[CrBr_6]^{3-}$	12.7	$[Rh(NH_3)_6]^{3+}$	34.0
$[Cr(CN)_6]^{3-}$	26.7	$[RhF_6]^{3-}$	23.3
$[Mo(H_2O)_6]^{3+}$	26.0	$[RhCl_6]^{3-}$	20.4
$[MoCl_6]^{3-}$	19.2	$[RhBr_6]^{3-}$	19.0
$[MoBr_6]^{3-}$	18.3	$[Rh(CN)_6]^{3-}$	45.5
$[MoI_6]^{3-}$	16.6	$[Ir(NH_3)_6]^{3+}$	41.2
$[MnF_6]^{2-}$	21.8	$[Ir(en)_3]^{3+}$	41.4
$[MnCl_6]^{2-}$	17.9	$[IrCl_6]^{3-}$	25.0
$[TcF_6]^{2-}$	28.4	$[PtF_6]^{4-}$	33.0
$[ReF_6]^{2-}$	32.8	$[PtCl_6]^{4-}$	29.0
$[Cr(H_2O)_6]^{2+}$	14.0	$[PtBr_6]^{4-}$	25.0
$[CrF_6]^{2-}$	22.0	$[Co(H_2O)_6]^{2+}$	9.2
$[CrCl_6]^{2-}$	10.2	$[Co(NH_3)_6]^{2+}$	10.2
$[Mn(H_2O)_6]^{3+}$	21.0	$[Ni(H_2O)_6]^{2+}$	8.5
$[MnF_6]^{3-}$	21.7	$[Ni(NH_3)_6]^{2+}$	10.8
$[MnCl_6]^{3-}$	17.5	$[Ni(en)_3]^{2+}$	11.5
$[Mn(CN)_6]^{3-}$	31.0	$[Ni(bpy)_3]^{2+}$	12.7
$[Mn(H_2O_6]^{2+}$	7.3	$[NiF_6]^{4-}$	7.3
$[Mn(en)_3]^{2+}$	10.0	$[NiCl_6]^{4-}$	7.0
$[MnCl_6]^{4-}$	6.7	$[NiBr_6]^{4-}$	6.8
$[Fe(H_2O)_6]^{3+}$	13.7		

the same oxidation number and the same coordination geometry, the size of the CF splitting parameter can vary over an enormous range, as shown by the empirical data listed in Table 16.5. For instance, $[CrBr_6]^{3-}$ is HS with $\Delta_o = 12{,}700\ cm^{-1}$ (or 12.7 kK) but $[Cr(CN)_6]^{3-}$ is LS with $\Delta_o = 26{,}700\ cm^{-1}$ (or 26.7 kK), more than twice as large!

The empirical data in Table 16.5 and other sources has led to what is known as the *spectrochemical series*, which is a listing of the ligands in decreasing magnitude of their relative CF splitting parameters in coordination compounds:

CO, $CN^- > PR_3 > R^- > NO_2^- >$ phen > bpy > en > NH_3, py > $NCS^- > H_2O$, $C_2O_4^{2-} > OH^- > F^- > Cl^- > SCN^- > Br^- > I^-$

To summarize, the magnitude of Δ_o in a coordination compound will increase with:

1. the oxidation number of the metal ion (larger Q^+ in Coulomb's law); for example, Δ_o for $[VCl_6]^{4-}$ (7.2 kK) < $[VCl_6]^{3-}$ (12.0 kK) < $[VCl_6]^{2-}$ (15.4 kK)
2. the series of the metal (1st row < 2nd row < 3rd row), due to the greater reach of the *d*-orbitals; for example, Δ_o for $[Co(NH_3)_6]^{3+}$ (22.9 kK) < $[Rh(NH_3)_6]^{3+}$ (34.0 kK) < $[Ir(NH_3)_6]^{3+}$ (41.2 kK)
3. the number of ligands and the coordination geometry (size of Q^- and directness of the overlap of *d*-orbitals); for example, Δ_t for $[FeCl_4]^-$ (5.2 kK) $\ll \Delta_o$ for $[FeCl_6]^{3-}$ (11.6 kK) [the rule of thumb here is that $\Delta_t \cong (4/9)\Delta_o$]
4. the nature of the ligands (where the ligand falls in the spectrochemical series); for example, Δ_o for $[CrCl_6]^{3-}$ (13.2 kK) < $[Cr(H_2O)_6]^{3+}$ (17.4 kK) < $[Cr(en)_3]^{3+}$ (21.9 kK).

Example 16-2. For each of the following coordination compounds, sketch the CF splitting diagram, label the *d*-orbitals, label each energy level with its group theoretical symbol, fill in the electrons, and predict the magnetism: (a) $[FeCl_6]^{3-}$, (b) $[Co(CN)_6]^{3-}$, (c) $[Co(H_2O)_6]^{2+}$, (d) $[IrCl_6]^{3-}$, and (e) $[NiCl_4]^{2-}$.

Solution

(a) Fe(III) is a first-row metal in a moderately high oxidation state with a fairly weak-field ligand (Cl^-). Two of the three factors favor a HS complex. From Table 16.5, Δ_o = 11.6 kK and from Table 16.2, P_{tot} = 357.4 kJ/mol (or 29.9 kK). Because $\Delta_o \ll P_{tot}$, this will be an HS d^5 complex with the CF splitting diagram given here. With five unpaired electrons, $[FeF_6]^{3-}$ will be strongly paramagnetic.

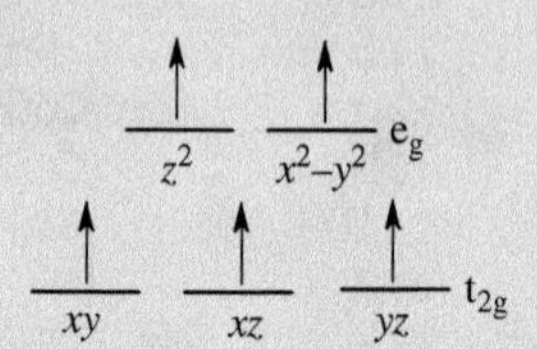

(b) Co(III) is a first-row metal in a moderately high oxidation state and with an extremely strong-field ligand (CN^-). Two of the three factors favor a LS complex. From Table 16.5, Δ_o = 32.2 kK and from Table 16.2, P_{tot} = 282.6 kJ/mol (or 23.6 kK). Because $\Delta_o > P_{tot}$, this will be an LS d^6 complex with the given CF splitting diagram. With no unpaired electrons, $[Co(CN)_6]^{3-}$ will be diamagnetic.

e_g: z^2, x^2-y^2
t_{2g}: xy, xz, yz

(c) Co(II) is a first-row metal in a moderately low oxidation state with a fairly weak-field ligand (H_2O). All signs point to a HS d^7 complex. From Table 16.5, Δ_o = 9.2 kK and from Table 16.2, P_{tot} = 250 kJ/mol (or 20.9 kK). Because $\Delta_o \ll P_{tot}$, this will be an HS d^7 complex with the given CF splitting diagram. With three unpaired electrons, $[Co(H_2O)_6]^{2+}$ will be paramagnetic.

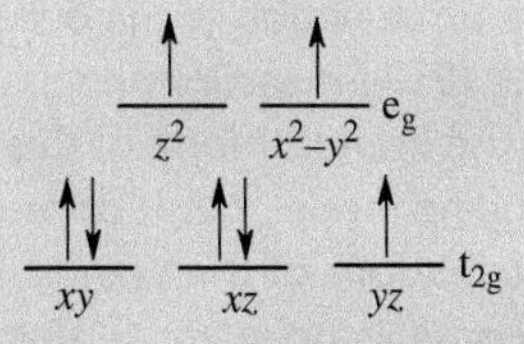

e_g: z^2 ↑, x^2-y^2 ↑

t_{2g}: xy ↑↓, xz ↑↓, yz ↑

(d) Ir(III) is a third-row metal in a moderately high oxidation state with a moderately weak-field ligand (Cl^-). Two of the three factors favor a LS d^6 complex. From Table 16.5, $\Delta_o = 25$ kK and from Table 16.2, $P_{tot} = 250$ kJ/mol (or 20.9 kK). Because $\Delta_o > P_{tot}$, this will be a LS d^6 complex with the given CF splitting diagram. With no unpaired electrons, $[IrCl_6]^{3-}$ will be diamagnetic.

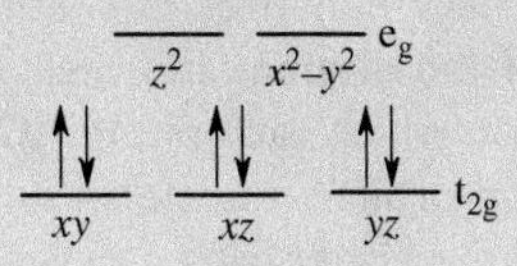

e_g: z^2, x^2-y^2

t_{2g}: xy ↑↓, xz ↑↓, yz ↑↓

(e) Ni(II) is a first-row metal in a moderately low oxidation state with a weak-field ligand (Cl^-) and a tetrahedral coordination geometry. This is most definitely a HS d^8 complex, although the designation HS and LS is irrelevant for the d^8 electron configuration. The experimental value of Δ_t for $[NiCl_4]^{2-}$ is 3.7 kK, which is much smaller than the CF splitting parameters we have seen for any octahedral coordination compound. Thus, this ion will have the CF splitting diagram shown. With two unpaired electrons, $[NiCl_4]^{2-}$ will be paramagnetic.

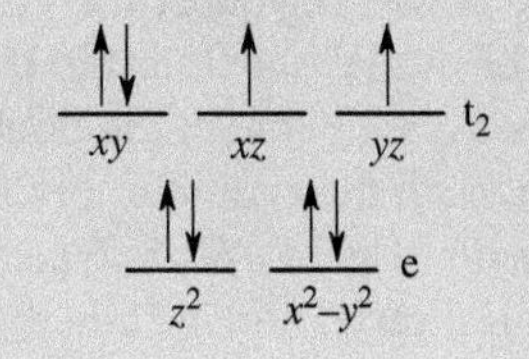

t_2: xy ↑↓, xz ↑, yz ↑

e: z^2 ↑↓, x^2-y^2 ↑↓

Because it is a purely electrostatic (ionic) model, CFT cannot explain some of the trends in the spectrochemical series. Why, for instance, would the negatively charged OH^- ligand lie lower in the spectrochemical series than the neutral H_2O ligand, both of which bind through the O atom? And why is not the smaller (more concentrated negative point charge) and better Lewis base NH_3 higher in the spectrochemical series than the larger, soft PR_3 ligand? The answer, of course, has to do with the fact that the ligands are not merely negative point charges. They are *ligands* and they form coordinate *covalent* bonds with the metal ions, not ionic bonds. Hence, we need to modify our theory to deal with the realities at hand.

16.3 LIGAND FIELD THEORY

There is currently a lot of evidence in the chemical literature for covalency in coordination compounds. Nuclear magnetic resonance (NMR) studies have shown that there is a great deal of delocalization of the unpaired electrons over the ligands. This is especially true for some of the stronger field ligands, such as CO and CN^-. Evidence that these ligands can accept π-electron density back from the metal comes from the frequencies of ν(CO) as a function of the buildup of electron density on the metal. Electron spin resonance (ESR) measurements also suggest that the unpaired electrons can delocalize onto the ligands. None of this means that we need to completely disregard CFT as a bonding model, but it does mean that we need to modify it to include the covalent contributions of the ligands. Remembering that CFT actually predates MOT, it should come as no surprise that as the latter theory became increasingly popular in explaining the chemistry of covalent bonds, it was also examined for its applicability in coordinate covalent bonding.

This modified version of CFT, which includes a MOT component that allows for the covalent nature of the metal–ligand bond, is known as *ligand field theory* (LFT). To begin, we consider the MO diagram for an octahedral complex where the metal–ligand bonding is strictly of the σ-bonding type. In this case, the metal acts as a Lewis acid and the ligands act as σ-donors, or Lewis bases. Employing the usual group theoretical approach, the symmetries of the metal ion valence orbitals are listed on the left-hand side of the diagram and their Mulliken symbol designations are determined from the way in which each of the AOs transforms in the O_h point group. Thus, the lower lying *d*-orbitals transform as $e_g + t_{2g}$, the *s*-orbital as the spherically symmetric a_{1g}, and the higher lying *p*-orbitals as the triply degenerate t_{1u} set. Using the σ-donor basis set shown in Figure 16.18 to generate the symmetries of the symmetry-adapted linear combinations (SALCs), we obtain the following IRRs from the reduction of the representation in Table 16.1: $a_{1g} + e_g + t_{1u}$.

The resulting MO diagram is shown in Figure 16.19 and the shapes of the six bonding MOs will be derived from the overlap of the appropriate AOs on the metal ion with the SALCs on the ligand group having the same symmetry properties, as shown in Figure 16.20. The 12 ligand electrons will fill the MO diagram up to the top of the e_g bonding MO, yielding an overall bond order of 6, as expected. Any valence electrons from the metal ion will start to fill the MO diagram beginning at the nonbonding t_{2g} MO in the boxed region of the MO diagram. Thus, the middle part of the MO diagram in the LF model is identical to the splitting of the *d*-orbitals in CFT, with the triply-degenerate t_{2g} set beneath the doubly-degenerate e_g orbitals. In the σ-donor-only MO diagram, both the t_{2g} and e_g orbitals are non-bonding, yielding the moderately-sized Δ_o shown in Figure 16.19. Typical σ-donor-only ligands lie in the middle of the spectrochemical series and include the sp^n hybridized ligands H_2O, NH_3, and en.

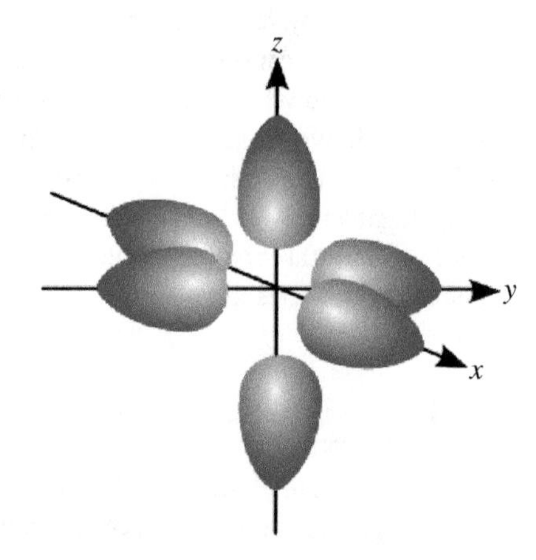

FIGURE 16.18
Basis set for the generation of the symmetries of the six SALCs in an octahedral ligand field where the ligands are acting solely as σ-donors. [Blatt Communications.]

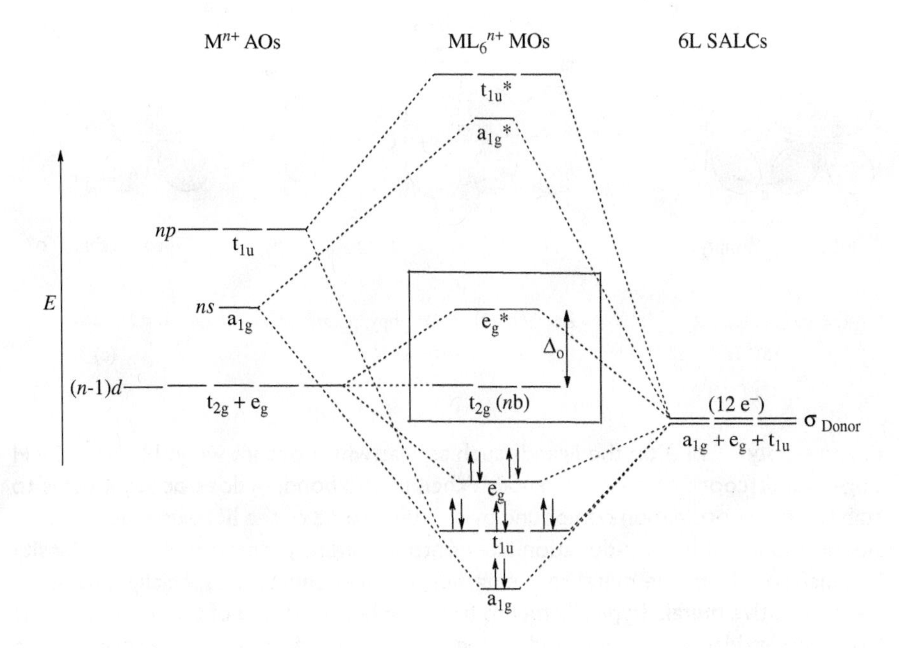

FIGURE 16.19
MO diagram for the formation of $[ML_6]^{n+}$, where the metal–ligand bonding involves only σ-interactions.

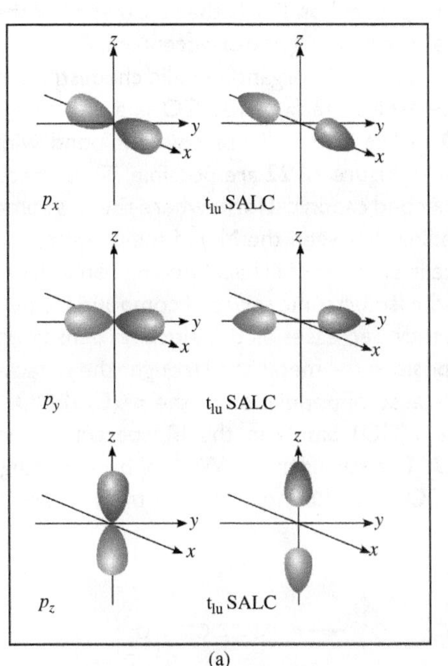

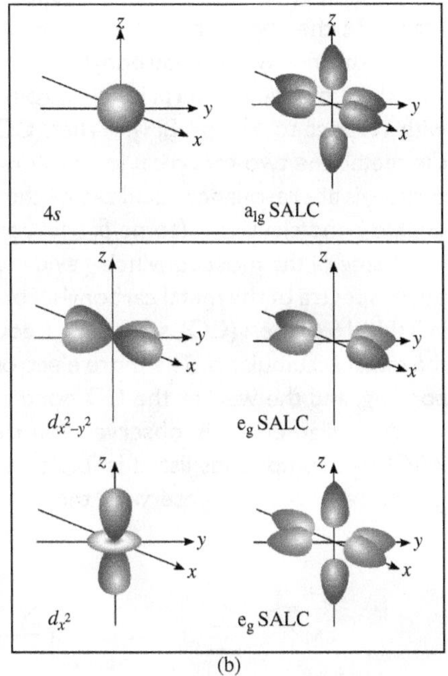

FIGURE 16.20
(a, b) Shapes of the AOs on the metal and the SALCs on the six σ-donor ligands in the octahedral $[ML_6]^{n+}$ coordination compound, whose MO diagram is shown in Figure 16.19. [Blatt Communications.]

While it is satisfying that the LF (MO) approach yields much the same result as CFT, we have yet to explain why ligands such as CO, CN^-, PR_3, R^-, NO_2^-, phen, and bpy (particularly those that are neutral) lie so high in the spectrochemical series. The answer involves the fact that in addition to acting as σ-donor ligands, this subset of ligands can also undergo what is known as *pi backbonding*. Figure 16.21 illustrates three different types of *pi* backbonding ligands. *Pi* backbonding can occur from a filled t_{2g} orbital on the metal to (i) an empty, low-lying *d*-orbital on the ligands, such as that which occurs for PR_3, AsR_3, or SR_2; (ii) an empty π^* MO on a ligand containing multiple bonds, such as that which occurs in CO, CN^-, NO_2^-, bpy, or phen; or

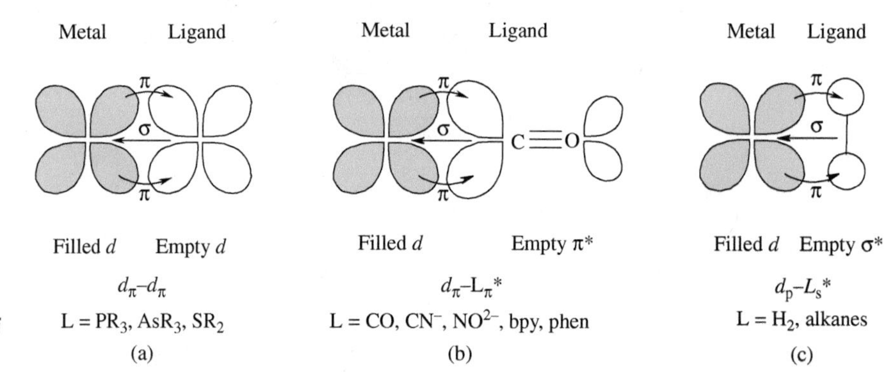

FIGURE 16.21
(a–c) Three different types of *pi* backbonding interactions.

(iii) an empty σ^* MO on the ligands such as that which occurs when H–H or C–H single bonds coordinate to the metal. When *pi* backbonding does occur, it helps to stabilize the coordination compound by shuttling some of the built-up electron density on the metal from σ-donation back onto the ligands. Because electron density is transferred from the metal to the ligands, *pi* backbonding is especially favored by electropositive metals (typically metals from the left-hand side of the periodic table) having low oxidation numbers and strong σ-donor ligands in the other coordination sites.

The most obvious evidence supporting *pi* backbonding is the fact that all of the ligands at the top of the spectrochemical series are good π-acceptors. This also helps explain why metal carbonyls are so prevalent in organometallic chemistry and why their bonding is surprisingly strong, despite the fact that CO is a weak base with respect to H^+ or BH_3. When CO makes a coordinate covalent bond with the metal, the two canonical forms shown in Figure 16.22 are possible. Thus, metal carbonyls are resonance stabilized by the second canonical form where there is some double-bond character (some *pi* backbonding) between the M and the C atom.

Some of the most convincing evidence in support of *pi* backbonding comes from the IR spectra of the metal carbonyls. For the isolectronic series of compounds listed in Table 16.6, the ν(CO) stretching frequency decreases as the electron density on the metal accumulates. The more electropositive the metal, the stronger the *pi* backbonding, and the weaker the CO bond because of population of the π^*(CO) MO.

A similar effect is observed for the ν(CO) bands in the IR spectra of the $W(CO)_5L$ compounds listed in Table 16.7. The stronger the W–L *pi* backbonding, the more it will compete with the W–CO *pi* backbonding of the *trans* carbonyl

$$\text{M} \quad :\text{C}{\equiv}\text{O}: \longrightarrow \left[\overset{-}{\text{M}}{-}\text{C}{\equiv}\text{O}^{+} \longleftrightarrow \text{M}{=}\text{C}{=}\text{O}:\right]$$

FIGURE 16.22
Two canonical forms for metal-carbonyl bonding, showing the *pi* character in the resonance hybrid.

TABLE 16.6 IR frequencies for the ν(CO) stretch in a series of isoelectronic metal carbonyls.

Compound	M^{n+}	ν(CO), cm^{-1}
$Mn(CO)_6^+$	Mn^+	2090
$Cr(CO)_6$	Cr^0	2000
$V(CO)_6^-$	V^-	1860
$Ti(CO)_6^{2-}$	Ti^{2-}	1748

TABLE 16.7 IR frequencies for the *trans* ν(CO) stretch in a series of $W(CO)_5L$ compounds.

Compound	π_{bb} L?	ν(CO), cm^{-1}
$W(CO)_5PF_3$	Yes	2007
$W(CO)_5PCl_3$	Yes	1990
$W(CO)_5PPh_3$	Yes	1942
$W(CO)_5PH_3$	Yes	1921
$W(CO)_5OEt_2$	No	1908
$W(CO)_5py$	No	1895

ligand, hence weakening the latter bond. For the series of ligands (L), the strongest *pi* backbonding will occur for PF_3, which can *pi* backbond through its empty *d*-orbitals and which has the very electron-withdrawing F groups attached to it. Thus, as the W–L bond is strengthened, the W–CO bond weakens and there is less *pi* backbonding to the CO ligand. Less *pi* backbonding means less population of the π^*(CO) MO and a higher energy ν(CO) vibrational mode.

There is also crystallographic evidence in support of *pi* backbonding. For example, the Cr–CO bond lengths in $Cr(CO)_5PPh_3$ pictured in Figure 16.23 are longer in the *cis* positions than in the position *trans* to the weaker *pi* backbonding PPh_3 group. This indicates that there is greater *pi* backbonding to the carbonyl when it is opposite the triphenylphosphine group than when it has to compete with *pi* backbonding to the second carbonyl. As expected, the CO bond length is shorter in the *cis* positions where the CO *pi* backbonding is weaker than in the *trans* position.

Figure 16.24 shows the MO diagram for an octahedral coordination compound that can undergo *pi* backbonding. The symmetries of the metal orbitals were already determined in the O_h point group as: $(n-1)d = e_g + t_{2g}$, $ns = a_{1g}$, and $np = t_{1u}$. The symmetries of the SALCs for the six ligands, which each coordinate to the metal ion through σ-donation, were also determined: $a_{1g} + e_g + t_{1u}$. As in Figure 16.19, these six SALCs will form the six σ-coordinate covalent bonding MOs by overlapping with the appropriate AOs on the metal ion (although only the e_g interaction is shown in Figure 16.24). We shall now derive the symmetries of the empty, high-lying π^* SALCs on the ligands using the basis set shown in Figure 16.25. Using Table 16.8 to reduce the Γ_π representation yields the following IRRs: $t_{1g} + t_{2g} + t_{1u} + t_{2u}$. The t_{1g}, t_{1u}, and t_{2u} SALCs will form π_{nb} MOs because they do not have any other AOs on the metal ion with which to overlap. On the other hand, the t_{2g} SALC can now overlap with the t_{2g} AOs (the d_{xy}, d_{xz}, and d_{yz}) to form bonding and antibonding MOs. The 12 σ-donating ligands will fill the lowest six bonding MOs in the MO diagram. Any valence electrons from the metal will begin to fill the t_{2g} bonding MO. Because the

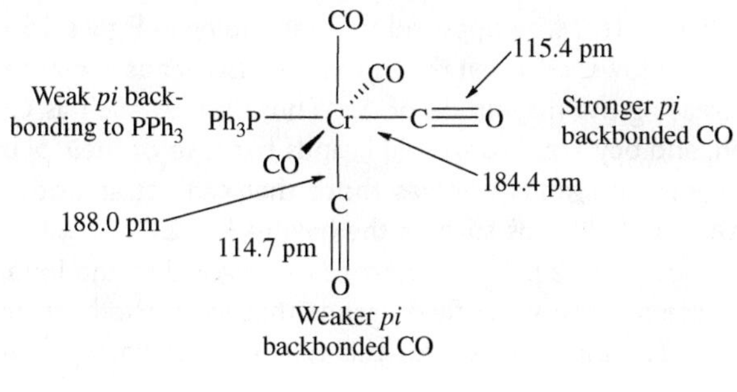

FIGURE 16.23
X-ray crystallographic data for $Cr(CO)_5PPh_3$ in support of *pi* backbonding.

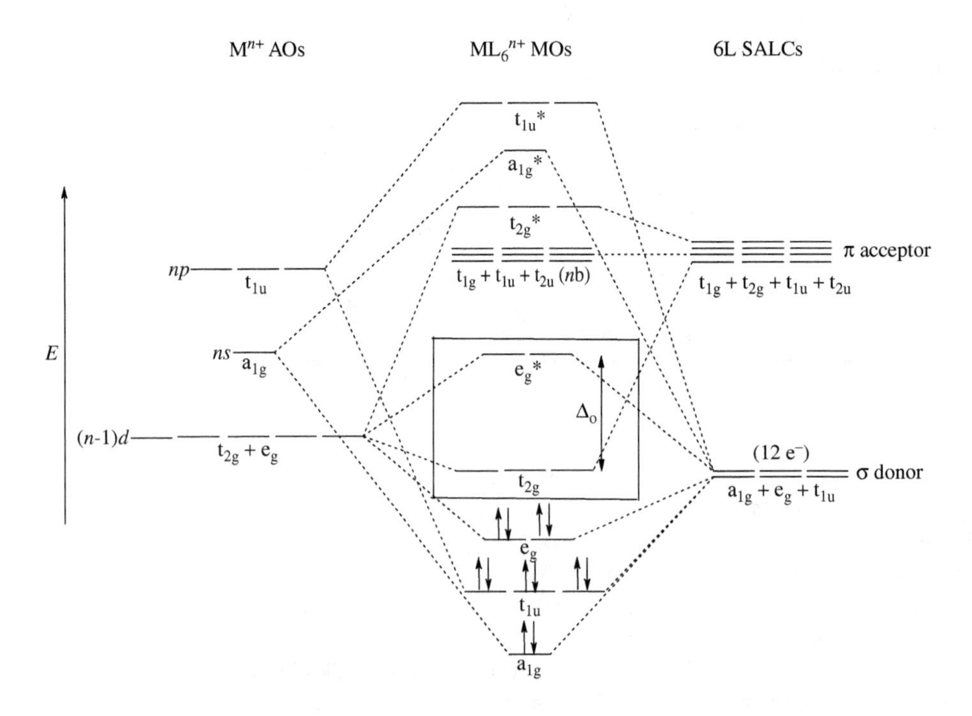

FIGURE 16.24
MO diagram for an octahedral $[ML_6]^{n+}$ compound, where the ligands can act as both σ-donors and π-acceptors.

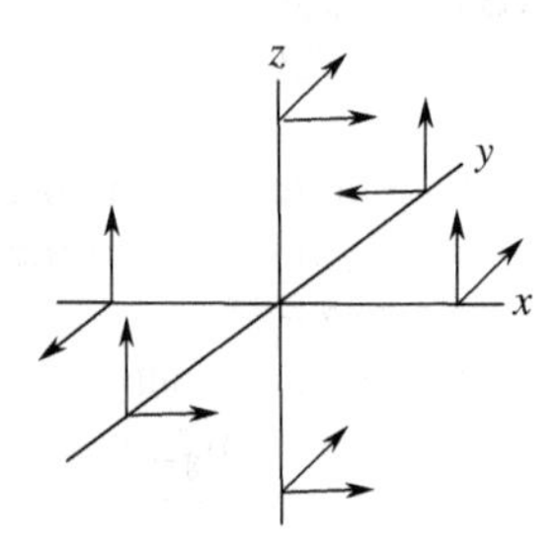

FIGURE 16.25
Pi basis set in an octahedral coordination geometry.

TABLE 16.8 Reducible representation for the π-bonding basis set shown in Figure 16.25.

O_h	E	$8C_3$	$6C_2$	$6C_4$	$3C_2$	i	$6S_4$	$8S_6$	$3\sigma_h$	$6\sigma_d$	IRRs
Γ_π	12	0	0	0	−4	0	0	0	0	0	$t_{1g}+t_{2g}+t_{1u}+t_{2u}$

t_{2g} is bonding in Figure 16.24, as opposed to nonbonding in Figure 16.19, the energy of the t_{2g} MO will be lowered when the ligand can also act as a π-acceptor. This has the effect of increasing the magnitude of Δ_o. Thus, ligands such as CO, CN^-, PR_3, NO_2^-, R^-, phen, and bpy are strong-field ligands because of their *pi* backbonding.

A third category of ligands involves those that can act as both σ-donors and π-donors. These include ligands such as the halides F^-, Cl^-, Br^-, I^-, and other ligands that have a second lone pair of electrons orthogonal to the lone pair that acts as a σ-donor. These are the weak-field ligands that lie at the bottom of the spectrochemical series. The MO diagram for this type of coordination compound in an

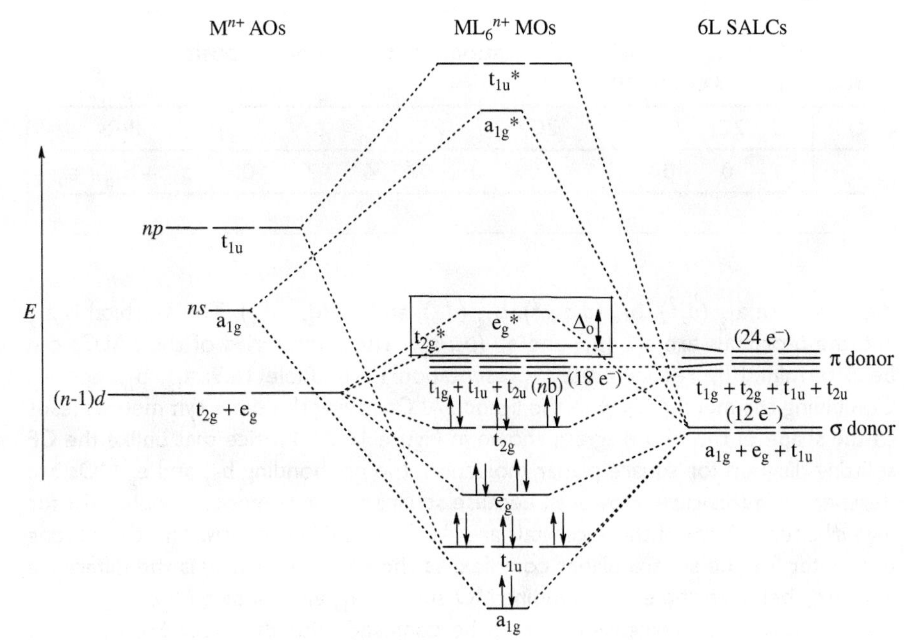

FIGURE 16.26
MO diagram for an octahedral $[ML_6]^{n+}$ compound, where the ligands can act as both σ-donors and π-donors. The t_{1g}, t_{1u}, and t_{2u} SALCs on the ligands π^*-orbitals are omitted for clarity.

octahedral LF is shown in Figure 16.26. In this case, the ligand π-orbitals are filled, so the 18 ligand electrons will fill the a_{1g} and t_{1u} bonding MOs, as well as the e_g and t_{2g} bonding MOs. Thus, any valence electrons on the metal will begin to fill the t_{2g}* MO, which is antibonding. This weakens the M–L bonding and decreases the magnitude of Δ_o because the t_{2g}* MO is raised with respect to the corresponding t_{2g} nonbonding MO in Figure 16.19. Thus, those ligands that are both good σ-donors and π-donors will lie at the bottom of the spectrochemical series.

The MO diagram for the square planar $[ML_4]^{n+}$ ion having σ-donor-only ligands is shown in Figure 16.27. Using the D_{4h} point group, the symmetries of the

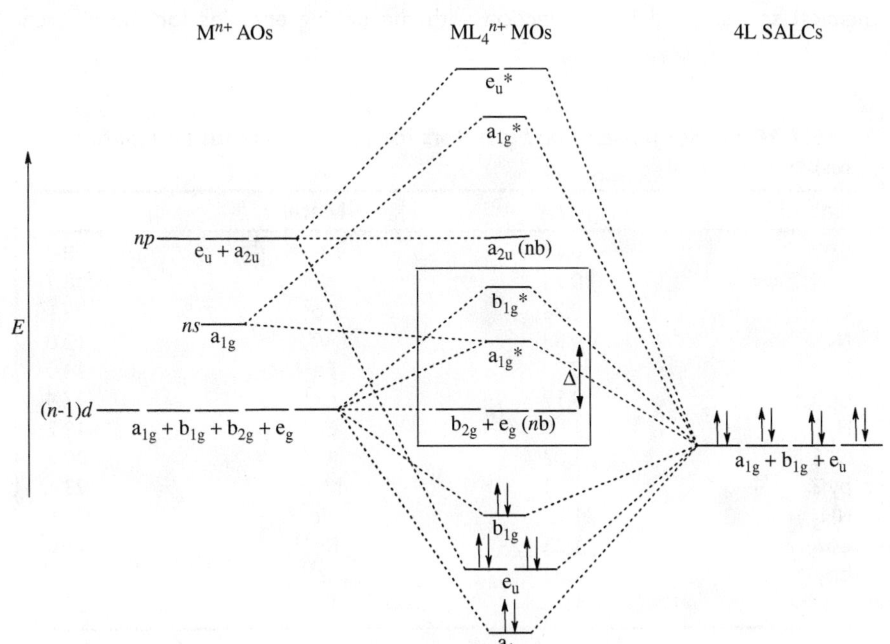

FIGURE 16.27
MO diagram for the $[ML_4]^{n+}$ ion in a square planar geometry having only σ-donor ligands.

TABLE 16.9 Reducible representation for the σ-bonding basis set in a square planar geometry.

D4h	E	$2C_4$	C_2	$2C_2'$	$2C_2''$	i	$2S_4$	σ_h	$2\sigma_v$	$2\sigma_d$	IRRs
Γ_σ	4	0	0	2	0	0	0	4	2	0	$a_{1g} + b_{1g} + e_u$

d-orbitals are a_{1g} (d_{z^2}), b_{1g} ($d_{x^2-y^2}$), b_{2g} (d_{xy}), and e_g (d_{xz}, d_{yz}). The s-orbital is a_{1g} and the *p*-orbitals are a_{2u} (p_z) and e_u (p_x, p_y). The symmetries of the SALCs can be determined by reducing the representation Γ_σ in Table 16.9: a_{1g}, b_{1g}, and e_u. Combining the metal AOs with the ligand SALCs having the same symmetries leads to the shape of the MO diagram shown in Figure 16.27. Notice that unlike the CF splitting diagram for square planar molecules, the nonbonding b_{2g} and e_g MOs are degenerate in this case. However, because square planar compounds only exist for high d^n counts, both of these orbitals are likely to be filled anyway. The CF splitting parameter for the square planar complex, as shown in the figure, is the difference in energy between the e_g nonbonding MO and the a_{1g} antibonding MO.

In 1971, C. K. Jørgensen (not to be confused with the elder Sophus Mads Jørgensen) established a table of empirical *f* and *g* factors that could be used to estimate the value of Δ_o for an unknown coordination compound. The *f* factors listed in Table 16.10 apply to the ligand (with H_2O assigned a value of 1.00), while the *g* factors refer to the metal ion. For an octahedral complex containing only one type of ligand, the product *fg* is approximately equal to the CF splitting parameter Δ_o in units of kK (1 kK = 1000 cm^{-1}), as shown in Equation (16.11).

$$\Delta_o(\text{kK}) \cong f \times g \tag{16.11}$$

If there is more than one type of ligand, then the *f* factors must be weighted according to the fraction of sites that each ligand occupies in the octahedral coordination geometry. The table of *f* values provides a more thorough and semi-quantitative measure of the strengths of the different ligands in the spectrochemical series. Used in conjunction with the pairing energies for the different

TABLE 16.10 Jørgensen *f* and *g* factors for selected ligands and metals, respectively.

Ligand	*f*	Metal	*g*
Br^-	0.72	Mn^{2+}	8.0
SCN^-	0.73	Ni^{2+}	8.7
Cl^-	0.78	Co^{2+}	9
N_3^-	0.83	V^{2+}	12.0
F^-	0.9	Fe^{3+}	14.0
$C_2O_4^{2-}$	0.99	Cr^{3+}	17.4
H_2O	1.00	Co^{3+}	18.2
NCS^-	1.02	Ru^{2+}	20
py	1.23	Mn^{4+}	23
NH_3	1.25	Mo^{3+}	24.6
en	1.28	Rh^{3+}	27.0
bpy	1.33	Ir^{3+}	32
CN^-	1.7	Pt^{4+}	36

metals, the data in Table 16.10 can be used (in principle) to predict whether a given coordination compound will be HS or LS.

As expected, ligands having the largest values of *f* are those that can *pi* backbond and therefore lie at the top of the spectrochemical series (CN^-, bpy); those in the middle (en, NH_3, NCS^-, H_2O, $C_2O_4^{2-}$) are σ-donor ligands; and those having the smallest values of *f* (F^-, Cl^-, Br^-) are both σ-donor and π-donor ligands.

Example 16-3. For each of the following octahedral coordination compounds, estimate the value of Δ_o and predict the valence electron configuration: (a) $[Co(en)_3]^{3+}$, (b) $[Rh(NH_3)_6]^{3+}$, (c) $[Fe(H_2O)_6]^{3+}$, (d) $[Ru(CN)_6]^{4-}$, and (e) $[MnF_6]^{2-}$.

Solution. Using Equation (16.11) to calculate Δ_o:

(a) $\Delta_o = (1.28)(18.2) = 23.3$ kK (expt'l $\Delta_o = 23.2$ kK). Converting the data in Table 16.2 from kJ/mol into kK (1 kJ/mol = 0.0836 kK), P_{tot} for the free ion is Co^{3+} is 23.6 kK. Because the pairing energy for coordination compounds is typically 15–30% smaller than for the free ion because of the *nephelauxetic effect* (delocalization of electron density onto the ligands) in this case, $\Delta_o > P_{tot}$ and the complex will be LS: $(t_{2g}{}^6e_g{}^0)$.

(b) $\Delta_o = (1.25)(27.0) = 33.8$ kK (expt'l $\Delta_o = 34.1$ kK). Notice that *g* for Rh^{3+} is larger than for the isoelectronic Co^{3+} in (a) because Rh^{3+} is a second-row transition metal. The pairing energy for Rh^{3+} is not listed in Table 16.2, but it should be smaller than for Co^{3+} because Rh^{3+} is a second-row transition metal. Thus, $[Rh(NH_3)_6]^{3+}$ should also be LS: $(t_{2g}{}^6e_g{}^0)$.

(c) $\Delta_o = (1.00)(14.0) = 14.0$ kK (expt'l $\Delta_o = 14.0$ kK). P_{tot} for the free ion Fe^{3+} is 29.9 kK. Even at 15–30% lower than this value, $P_{tot} > \Delta_o$, so $[Fe(H_2O)_6]^{3+}$ should be HS: $(t_{2g}{}^3e_g{}^2)$.

(d) $\Delta_o = (1.7)(20) = 34$ kK (expt'l $\Delta_o = 33.8$ kK). The total pairing energy for isoelectronic Fe^{2+} is 19.2 kK, and Ru^{2+} will be smaller than this, so the hexacyanoruthenate(II) ion will be LS: $(t_{2g}{}^6e_g{}^0)$.

(e) $\Delta_o = (0.9)(23) = 21$ kK (expt'l $\Delta_o = 21.8$ kK). For Mn^{4+} as a free ion, P_{tot} will be larger than the $P_{tot} = 25.2$ kK reported for Mn^{3+} in Table 16.2. Thus, it is very likely that $P_{tot} > \Delta_o$ for $[MnF_6]^{2-}$. Either way, the d^3 complex will have the configuration $(t_{2g}{}^3e_g{}^0)$.

Example 16-4. The standard reduction potentials for three octahedral Co(III) coordination compounds are: −1.83, −0.11, and +0.83 V versus NHE, respectively. Using LFT, determine which of the following ligands corresponds with which standard reduction potential: NH_3, CN^-, and H_2O.

For example, $[CoL_6]^{3+} + e^- \rightarrow [CoL_6]^{2+}$ $E°$ (V vs NHE)

Solution. As a result of the reduction process, the coordination compound will gain an extra electron in its valence t_{2g} or e_g MO. The lower in energy this orbital lies, the more favorable the reduction will be, and the more positive the value of $E°$. Thus, those compounds having larger values of Δ_o will have more positive standard reduction potentials. Using the *f* factors in Table 16.10, the magnitude of Δ_o will increase in the order: $H_2O < NH_3 < CN^-$. The standard reduction potentials are as follows:

$[Co(H_2O)_6]^{3+/2+}$	$E° = -1.83$ V
$[Co(NH_3)_6]^{3+/2+}$	$E° = -0.11$ V
$[Co(CN)_6]^{3-/4-}$	$E° = +0.83$ V

To summarize, the LF model is an extension of CFT where the nature of the metal–ligand bonding is taken into consideration instead of just treating the ligands as negative point charges. Three general scenarios ensue. (i) The ligand acts as both a σ-donor and a π-acceptor, in which case the t_{2g} MO will be a bonding MO. This will strengthen the bonding through *pi* backbonding and lead to the strong-field case where Δ_o is large and the ligand lies near the top of the spectrochemical series. (ii) The ligand acts as a σ-donor only, in which case the t_{2g} MO is nonbonding and a medium-field having an average-sized Δ_o results. (iii) The ligand acts as both a σ-donor and π-donor, in which case the t_{2g} MO is antibonding and the weak-field case results where Δ_o is small and the ligand lies near the bottom of the spectrochemical series. Thus, LFT retains all of the advantages that CFT had (the same general splitting patterns result), except that LFT can also explain the spectrochemical series. The apparent anomalies of ligand placement in the spectrochemical series using CFT can now be rationalized as follows. (i) $H_2O > OH^-$ because the sp^3 hybridization on the aqua ligand makes it impossible to donate pi-electron density to the metal, whereas the hydroxo ligand can act as a sigma and pi-donor. (ii) $PR_3 > NH_3$ because the phosphine ligands can serve as π-acceptors using their low-lying empty *d*-orbitals to *pi* backbond to the metal, while NH_3 can only serve as a σ-donor. While we have used the labels "strong-field," "medium-field," and "weak-field" in a loose classification of the ligands in the spectrochemical series, it is important to remember that there are still other factors affecting the magnitude of Δ_o and that not every coordination compound containing a "weak-field" ligand will be HS; for example, $PtCl_6^{2-}$ is LS because it has a third-row metal with a 4+ charge. Taken together, the CFT and LFT models of bonding can rationalize much of the structure, bonding, and spectroscopy of coordination compounds.

16.4 THE ANGULAR OVERLAP METHOD

The *angular overlap model* (AOM) attempts to parameterize the results of LFT on the basis of the amount of overlap between the ligand group orbitals and the metal *d*-orbitals. Because the results are based on empirical observations (from spectroscopic data), the method is not very different from using the Jørgensen *f* and *g* factors in Table 16.10, although it does have the distinct advantage of splitting out the ligand contribution into a σ-bonding component (e_σ) and a π-bonding component (e_π). In the AOM, the energy of a *d*-orbital in the MO diagram can be determined by summing the bonding parameters of each ligand.

We begin by considering the σ-bonding interactions. Taking the overlap of a ligand p_z orbital with the d_{z^2} orbital on the metal as our reference point, as shown in Figure 16.28, the bonding MO will be stabilized relative to the energy of the σ-donor ligand by an amount of e_σ. At the same time, the antibonding MO will be raised relative to the metal d_{z^2} AO by e_σ (actually, the antibonding MO is destabilized slightly more in energy than the bonding orbital is stabilized, but we ignore this minor detail in our discussion). The d_{z^2} AO will interact with all the other ligands in an octahedral LF, but not necessarily to the same degree. Using the ligand numbering system

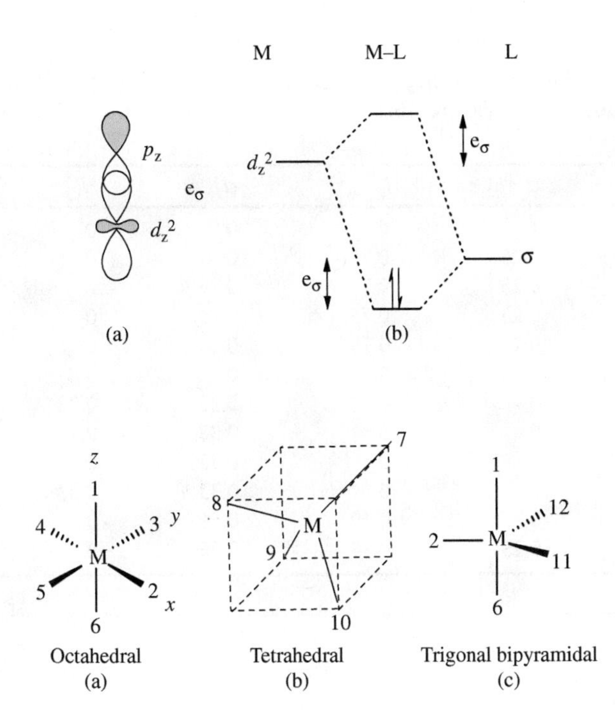

FIGURE 16.28
(a, b) Definition of the reference point e_σ as the overlap of a d_{z^2} AO on the metal ion with a p_z orbital on the ligand.

FIGURE 16.29
(a–c) The numbering system used by the AOM in different coordination geometries.

TABLE 16.11 The positions of the ligands (as defined in Figure 16.29) in the AOM as a function of the coordination geometry of the metal.

Coordination Geometry	Ligand Positions
Linear	1, 6
Trigonal planar	2, 11, 12
T-shaped	1, 3, 5
Tetrahedral	7, 8, 9, 10
Square planar	2, 3, 4, 5
Trigonal bipyramidal	1, 2, 6, 11, 12
Square pyramidal	1, 2, 3, 4, 5
Octahedral	1, 2, 3, 4, 5, 6

defined in Figure 16.29, the d_{z^2} AO will interact equally with ligands in positions 1 and 6, which lie along the z-axis, but it will interact much less so with ligands 2–5, which lie in the xy-plane and overlap only with the donut part of the d_{z^2} AO. The analysis can be carried through to other geometries, as well, where the ligands are in the positions defined by Table 16.11. The relative contributions of each of the d-orbitals to the σ-interaction with the ligands in positions 1–12 are listed in Table 16.12.

The relevant portion of the MO diagram for an octahedral coordination compound containing only σ-donor ligands is shown in Figure 16.30. When calculating the relative energies of the MOs, the change in energy for a given d-orbital is the sum of the values for that orbital in Table 16.12 for all of the ligand positions appropriate to the coordination geometry. Thus, for an octahedral compound, the d_{z^2} and $d_{x^2-y^2}$ interactions will be $3e_\sigma$ and the d_{xy}, d_{xz}, and d_{yz} interactions will be zero, because these are the sums for ligands in positions 1–6. Likewise, the change in energy for a given ligand will be the sum of the numbers for all of the d-orbitals in the appropriate row of Table 16.12. Thus, each of the ligands in positions 1–6 in an octahedral LF

TABLE 16.12 Sigma bonding interactions between the ligands in the positions defined by Figure 16.29 and each of the *d*-orbitals (in units of e_σ).

Position	d_{z^2}	$d_{x^2-y^2}$	d_{xy}	d_{xz}	d_{yz}
1	1	0	0	0	0
2	0.25	0.75	0	0	0
3	0.25	0.75	0	0	0
4	0.25	0.75	0	0	0
5	0.25	0.75	0	0	0
6	1	0	0	0	0
7	0	0	0.33	0.33	0.33
8	0	0	0.33	0.33	0.33
9	0	0	0.33	0.33	0.33
10	0	0	0.33	0.33	0.33
11	0.25	0.19	0.56	0	0
12	0.25	0.19	0.56	0	0

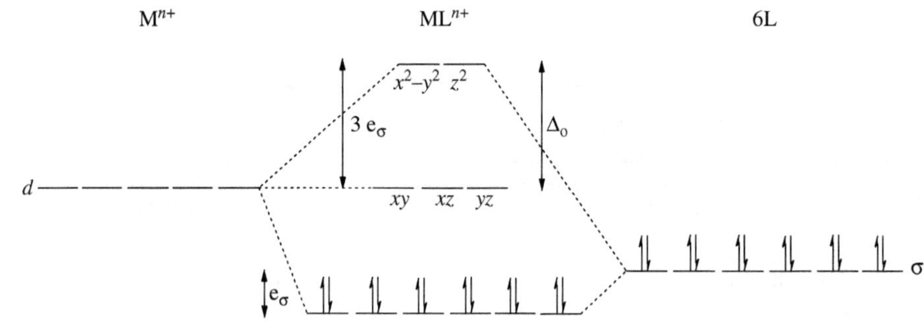

FIGURE 16.30
Relative energies of the metal *d*-orbitals and ligand orbitals using the AOM in an octahedral coordination compound, where the ligands act solely as σ-donors.

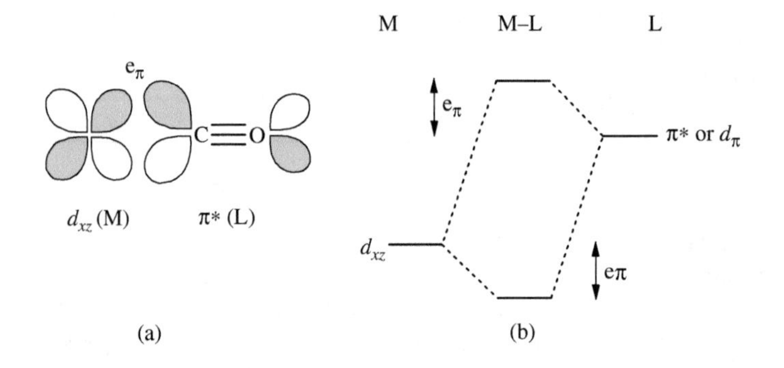

FIGURE 16.31
(a, b) Definition of the reference point e_π as the overlap of a *d*-orbital on the metal ion having π-symmetry with either a π* MO on the ligand or a high-lying empty *d*-orbital on the ligand.

will change by 1e_σ. These are exactly the results indicated on the abbreviated MO diagram shown in Figure 16.30. Thus, the value of Δ_o in this case is equal to 3e_σ.

We next consider the π-bonding interactions. When the ligand acts as a π-acceptor, the reference interaction e_π is defined by the interaction diagram shown in Figure 16.31. In this case, the π* antibonding MOs for the ligand (e.g., CO or CN^-) or empty *d*-orbitals on the ligand (e.g., PR_3) lie higher in energy than the *d*-orbitals on the metal. Thus, the metal ion's *d*-orbitals will be lowered by e_π and the ligand's π-orbitals will be raised by e_π, as indicated in the diagram. Table 16.13 lists the relative contributions of each *d*-orbital to the π-bonding overlap with the

TABLE 16.13 Pi bonding interactions between the ligands in the positions defined by Figure 16.29 and each of the *d*-orbitals (in units of e_π).

Position	d_{z^2}	$d_{x^2-y^2}$	d_{xy}	d_{xz}	d_{yz}
1	0	0	0	1	1
2	0	0	1	1	0
3	0	0	1	0	1
4	0	0	1	1	0
5	0	0	1	0	1
6	0	0	0	1	1
7	0.67	0.67	0.22	0.22	0.22
8	0.67	0.67	0.22	0.22	0.22
9	0.67	0.67	0.22	0.22	0.22
10	0.67	0.67	0.22	0.22	0.22
11	0	0.75	0.25	0.25	0.75
12	0	0.75	0.25	0.25	0.75

ligands as a function of the ligand position number. As with the previous example, when calculating the relative energies of the MOs, the change in energy for a given *d*-orbital is the sum of the values for that orbital in Table 16.13 for all of the ligand positions appropriate to the coordination geometry. Likewise, the change in energy for a given ligand will be the sum of the numbers for all of the *d*-orbitals in the appropriate row of Table 16.13.

Using the values in Tables 16.12 and 16.13 as a guide, the abbreviated MO diagram for an octahedral compound having both σ-donor and π-acceptor ligands is shown in Figure 16.32. Each ligand σ-SALC is stabilized by $1e_\sigma$ and each ligand π^*-SALC is destabilized by $2e_\pi$. The σ-interacting d_{z^2} and $d_{x^2-y^2}$ AOs are destabilized by $3e_\sigma$ and the π-interacting d_{xy}, d_{xz}, and d_{yz} AOs are stabilized by $4e_\pi$. Thus, the value of Δ_o for this strong-field case is $3e_\sigma + 4e_\pi$.

Lastly, we consider the case of ligands that can act as both σ-donors and π-donors. In this case, the π-interacting ligand orbitals will lie lower in energy than the metal's *d*-orbitals, so the ligand orbitals will be stabilized by e_π and the σ-interacting d_{z^2} and $d_{x^2-y^2}$ AOs will be destabilized by $3e_\sigma$, as shown in Figure 16.33.

The resulting abbreviated MO diagram using the AOM for an octahedral compound having both σ-donor and π-donor ligands is shown in Figure 16.34. The ligand orbitals are stabilized by $1e_\sigma + 2e_\pi$, the σ-interacting d_{z^2} and $d_{x^2-y^2}$ AOs are

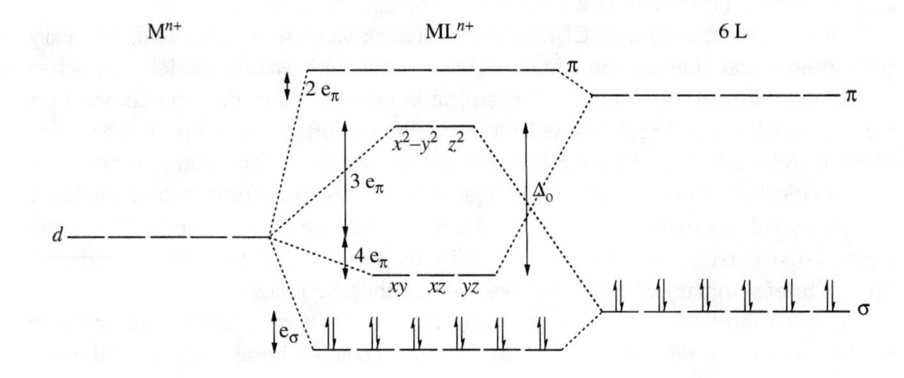

FIGURE 16.32
Relative energies of the metal *d*-orbitals and ligand orbitals using the AOM in an octahedral coordination compound, where the ligands act as σ-donors and π-acceptors.

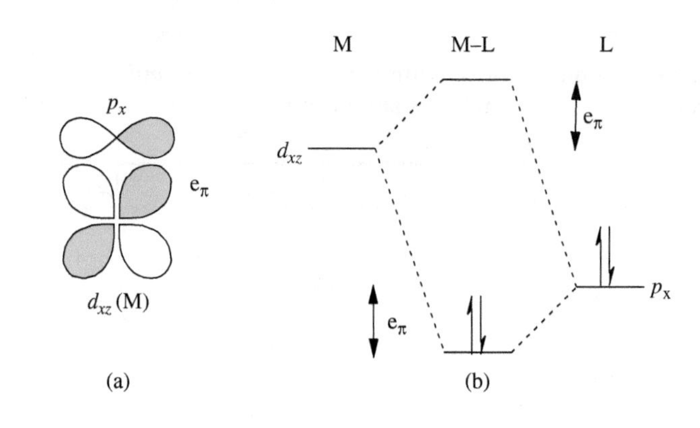

FIGURE 16.33
(a, b) Definition of the reference point e_π as the overlap of a *d*-orbital on the metal ion having π-symmetry with a *p* AO on the ligand.

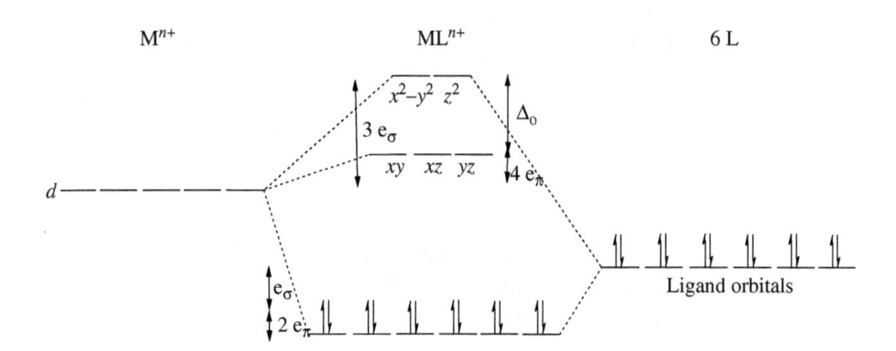

FIGURE 16.34
Relative energies of the metal *d*-orbitals and ligand orbitals using the AOM in an octahedral coordination compound, where the ligands act as σ-donors and π-donors.

destabilized by $3e_\sigma$, and the π-interacting d_{xy}, d_{xz}, and d_{yz} AOs are destabilized by $4e_\pi$. Thus, the magnitude of Δ_o in the weak-field case is $3e_\sigma - 4e_\pi$.

The magnitudes of the e_σ and e_π values listed in Table 16.14 were derived from spectroscopic data. First of all, the exact sizes of e_σ and e_π depend on the metal ion and ligands. Second, because σ-bonds are always stronger than their corresponding π-bonds (as a result of the increased overlap), e_σ is always greater than e_π. Third, the magnitude of both e_σ and e_π decreases as the size of the ligand or the size of the metal increases. This is because the longer M–L bond length results in less orbital overlap. Last, the magnitudes of both e_σ and e_π decrease as the electronegativity of the ligand decreases. This is because the less electronegative ligands exert less pull on the metal's *d*-electrons in the coordinate covalent bond. The combined effect of the smaller size of F^- and its large electronegativity provide an explanation for why $F^- > Cl^- > Br^- > I^-$ in the spectrochemical series.

In the square planar geometry (not shown), the AOM has an additional parameter e_{ds} which extends the theory to include the overlap of mixing of the $(n-1)d_z{}^2$ AO and the ns AO. This has the effect of decreasing the energy of the $d_z{}^2$ orbital in a square planar LF so that its energy is given by $e_\sigma - 4e_{ds}$.

One of the founders of CFT, John Hasbrouck van Vleck, once said, "A really good theoretical chemist can obtain right answers with wrong models." In other words, even though our models of chemical bonding for coordination compounds are oversimplified and each has its limitations, we can often use them "as a basis for correct understanding of rather complicated phenomena." The complexity of the mathematical solutions to the Schrödinger equation for transition metal complexes is well beyond the realm of this text and it is easy to lose sight of the forest for the trees. What is truly important is our ability to grasp the essence of the material. Thus, a brief summary of what we have learned thus far follows.

The VB model is a localized bonding theory that treats coordinate covalent bonds in much the same way as it does a normal covalent bond, except for the fact

TABLE 16.14 Selected values of the AOM parameters for metal ions in different coordination geometries as a function of the ligand.

Metal Ion	Ligand	e_σ, cm^{-1}	e_π, cm^{-1}	e_{ds}, cm^{-1}
Octahedral				
Ni^{2+}	NH_3	3600		
	en	4000		
	py	4500	900	
Co^{2+}	py	3860	110	
Fe^{2+}	py	3700	100	
Co^{3+}	py	6100	750	
Cr^{3+}	NH_3	7180		
	en	7260		
	py	6150	−330	
	F^-	8200	2000	
	Cl^-	5700	980	
	Br^-	5380	950	
	I^-	4100	670	
	OH^-	8600	2150	
	H_2O	7550	1850	
	Glycine (NH_2)	6700		
	Glycine (O)	8800	2000	
	CN^-	7530	930	
Tetrahedral				
Ni^{2+}	PPh_3	5000	1750	
	Cl^-	3900	1500	
	Br^-	3600	1000	
	I^-	2000	600	
	Quinoline	4000	500	
Co^{2+}	PPh_3	3800	1000	
	Cl^-	3600	1400	
	Br^-	3300	1000	
	Quinoline	3500	500	
Square planar				
Cu^{2+}	Cl^-	5030	900	1320
Pd^{2+}	Cl^-	10,150	2000	2540
	Br^-	9500	1800	2380
Pt^{2+}	Cl^-	12,400	2800	3100
	Br^-	10,900	2200	2725

that the transition metals have to use their *d*-orbitals in the hybridization scheme. For an octahedral coordination compound, the hybridization is formally d^2sp^3, where the d_{z^2} and $d_{x^2-y^2}$ orbitals (the ones that have lobes that point directly at the ligands) are included in the hybridization. Two of the main shortcomings of the VB model are that it cannot explain the magnetism or the electronic spectroscopy of coordination complexes. CFT was developed by theoretical physicists before the invention of MOT as a way to describe the bonding in coordination compounds. In CFT, the metal ion and its ligands are treated using an electrostatic model where they act as point charges. Those *d*-orbitals on the metal ion that have the most direct overlap with the negative point charges are raised in energy, while those that have little or no interaction are lowered in energy, so as to maintain an overall barycenter. In an octahedral CF, it is the e_g orbitals that are raised in energy and the t_{2g} set is lowered.

The energy gap between the two sets of *d*-orbitals is defined as Δ_o. Depending on the relative magnitude of Δ_o and the total pairing energy, the coordination compound could be HS or LS for the d^4–d^7 electron configurations. The magnitude of Δ_o increases with the oxidation state of the metal, the row of the periodic table, the charges on the ligands, and where the ligand falls in the spectrochemical series. Because CFT could not explain the positions of the ligands in the spectrochemical series, a modified theory known as *ligand field theory* was developed; it allowed for the effects of metal–ligand bonding covalency and was basically nothing more than an MO approach to the problem. Finally, the AOM was devised as a way of parameterizing LFT so that it could be used in a quantitative sense.

Some key aspects to emerge from the different models of chemical bonding follow. Unlike the very electropositive alkali or alkaline earth metals, which lie to the left in the periodic table and tend to form ionic bonds with the nonmetals, the transition metals lie closer to the metal–nonmetal line. As a result, the difference in electronegativity between a transition metal ion and its ligands is not as great, and there is a significant amount of covalent character in the metal–ligand bonding. Because of this increased covalency, the actual positive charge on the metal is significantly smaller than its oxidation number, and the *d*-orbitals effectively expand in size (this is referred to as the *nephelauxetic effect* or the *expanding cloud effect*). In an octahedral coordination compound with σ-donor ligands, the 12 valence electrons originating from the ligands occupy MOs that have largely *s* and *p* character, so that the *d*-orbitals on the metal actually play little role in determination of the metal–ligand bond strength. In this case, any valence electrons from the metal occupy a nonbonding t_{2g} MO, creating a moderate-sized Δ_o. If the ligands act as π-acceptors in addition to σ-donation, then the t_{2g} orbital is lowered and becomes a weakly bonding MO, which strengthens the metal–ligand bonding and increases the size of Δ_o. If, on the other hand, the ligands act as both σ-donors and π-donors, the t_{2g} orbital is destabilized as an antibonding MO and the metal–ligand bonding is weakened, leading to a smaller value for Δ_o. Table 16.15 neatly summarizes the results of LFT and the AOM.

Example 16-5. Use the AOM parameters in Table 16.14 to calculate Δ_o for each of the following coordination compounds and compare Δ_o to the values listed in Table 16.5: (a) $[Cr(en)_3]^{3+}$, (b) $[Cr(CN)_6]^{3-}$, and (c) $[Ni(NH_3)_6]^{2+}$.

Solution. (a) For an octahedral compound having only a σ-donor ligand, such as $[Cr(en)_3]^{3+}$, $\Delta_o = 3e_\sigma = 3(7260) = 21{,}800\ cm^{-1}$ or 21.8 kK (Table 16.5 has 21.9 kK). (b) For an octahedral compound having a ligand that is a σ-donor and π-acceptor, such as $[Cr(CN)_6]^{3-}$, $\Delta_o = 3e_\sigma + 4e_\pi = 3(7530) + 4(930) = 26{,}300\ cm^{-1}$ or 26.3 kK (Table 16.5 has 26.7 kK). (c) For an octahedral compound having a ligand that is only a σ-donor, $\Delta_o = 3e_\sigma = 3(3600) = 10{,}800\ cm^{-1}$ or 10.8 kK (Table 16.5 has 10.8 kK).

TABLE 16.15 Summary of the results of ligand field theory and the angular overlap model for octahedral coordination compounds.

Type of LF	Type of Ligand	Examples	t_{2g}	e_g	Δ_o
Strong-field	σ-donor, π-acceptor	CO, CN^-, NO_2^-, PR_3	Bonding	Antibonding	$3e_\sigma + 4e_\pi$
Medium-field	σ-donor only	NH_3, H_2O, en	Non-bonding	Antibonding	$3e_\sigma$
Weak-field	σ-donor, π-donor	F^-, Cl^-, Br^-, OH^-	Antibonding	Antibonding	$3e_\sigma - 4e_\pi$

16.5 MOLECULAR TERM SYMBOLS

One of the most striking features of coordination compounds is reflected in the variety of different colors that they exhibit, both in the solid state and in solution. Irrespective of whether Bethe's original electrostatic CFT or the more sophisticated LFT is used, the net result of metal–ligand bonding is to remove the degeneracy of the *d*-orbitals. For a d^1 coordination compound, there is only a single electronic absorption band in the molecule's UV/VIS spectrum. For $[Ti(H_2O)_6]^{3+}$, this peak occurs at ~20 kK (500 nm), giving the complex its characteristic pinkish-purple color. Because electronic transitions occur between the frontier MOs, this electronic transition corresponds to the excitation of the lone t_{2g} (nb) MO into an empty e_g* MO. Thus, the magnitude of Δ_o for $[Ti(H_2O)_6]^{3+}$ can be measured directly from its UV/VIS spectrum. However, whenever there is more than one valence electron, electron–electron repulsions must be taken into consideration, and the simple "one-electron" MO diagram developed in the previous section is unable to accurately predict the number and energy of the electronic transitions. For example, the d^2 coordination compound $[V(H_2O)_6]^{3+}$ has two absorptions in its UV/VIS spectrum at 17.8 kK (562 nm) and 25.7 kK (389 nm).

In order to explain the electronic spectra of coordination compounds having more than one electron, the *molecular term symbols* for the complex must first be determined. As we learned in Chapter 4, term symbols relate to the different energy levels that arise from a particular electron configuration. In the case of transition metal complexes, the electron configuration can be determined from the one-electron MO diagram for the coordination compound. For instance, the V^{3+} ion in $[V(H_2O)_6]^{3+}$ has the d^2 electron configuration. As stated elsewhere, the electron configuration itself is an incomplete description of the way in which the electrons actually occupy the valence orbitals. For the d^2 electron configuration, there are 45 different ways that the two *d*-electrons can occupy the set of five *d*-orbitals, and some of these different sets of microstates will obviously have different energies. The 45 possible microstates can be organized into states (or terms), with each term having its own distinct energy. A general method for the determination of the term symbols for any given electron configuration was presented in Section 4.5. The term symbols for V^{3+} as a free ion are ^3F (21), ^3P (9), ^1G (9), ^1D (5), and ^1S (1), where the number in parentheses is the degeneracy, or the number of microstates having that particular energy. The term symbols for the different d^n electron configurations are listed in Table 16.16, where the first term in the list is always the ground-state (GS) term symbol. The degeneracy of each term can be determined by taking the product $(2L + 1)(2S + 1)$, where *L* is the total angular momentum quantum number (0 for *S*, 1 for *P*, 2 for *D*, etc.) and *S* is the total spin quantum number. The spin multiplicity for the term is equal to $2S + 1$. When $2S + 1 = 1$, it is a singlet state: when $2S + 1 = 2$, it is

TABLE 16.16 Term symbols for the different d^n configurations.

d^n	d^{10-n} (hole)	Term Symbols
0	10	^1S
1	9	^2D
2	8	^3F, ^3P, ^1G, ^1D, ^1S
3	7	^4F, ^4P, ^2H, ^2G, ^2F, ^2D, ^2D, ^2P
4	6	^5D, ^3H, ^3G, ^3F, ^3F, ^3D, ^3P, ^3P, ^1I, ^1G, ^1G, ^1F, ^1D, ^1D, ^1S, ^1S
5		^6S, ^4G, ^4F, ^4D, ^4P, ^2I, ^2H, ^2G, ^2G, ^2F, ^2F, ^2D, ^2D, ^2D, ^2P, ^2S

The ground-state term symbol is always listed first.

a doublet, and so on. The terms for d^{10-n} are the same as those for d^n because d^{10-n} has the same number of "holes" as d^n has electrons. This simplifying correlation is known as the *hole formalism*.

The relative energies of the different terms depend on the degree of electron–electron repulsions, and they therefore involve the evaluation of the complete Hamiltonian. By the complete Hamiltonian, we mean one that includes quite a large number of Coulomb integrals having to do with the attraction of each nucleus with each electron and the repulsion between each pair of electrons, as well as all the exchange integrals. The methods for approximating these integrals are numerous and varied, but the bottom line is that the energies of each term depend on essentially three parameters, known as the *Racah parameters* given the symbols *A*, *B*, and *C*. The Racah parameters depend on the metal and its oxidation state, as shown in Table 16.17. For a given d^n configuration, each term will have the same value for the Racah parameter *A*, so this parameter is generally ignored. For first-row transition metals, the value of *B* is approximately equal to 1 kK. The relative energies of the terms for the d^2–d^5 electron configurations are given in Figure 16.35. For d^6–d^8, the hole formalism can be used; and for d^0, d^1, d^9, and d^{10}, there is only a single term. This figure assumes that the ratio $C/B = 4.7$. The actual ratio will vary depending on which row in the periodic table houses the transition metal ion and on its oxidation number.

The symmetry of the GS term can always be determined using Hund's rules, as described in Section 4.5, where the spin multiplicity is maximized first, and then the total orbital angular momentum *L* is maximized. Thus, the GS term symbol for the d^2 electron configuration is the ^{3}F term. The energies of the other terms can be determined on the basis of their Racah parameters, according to Table 16.18.

TABLE 16.17 The Racah parameters *B* and *C/B* in units of cm^{-1} for selected transition metals as free ions as a function of their oxidation states.

Charge:	0	0	1+	1+	2+	2+	3+	3+	4+	4+
Racah:	*B*	*C/B*	*B*	*C/B*	*B*	*C/B*	*B*	*C/B*	*B*	*C/B*
Ti	380	2.4	583	3.4	714	3.7				
Zr	250	7.9	450	3.9	540	3.0				
Hf	280		440	3.4						
V	436	2.4	585	4.2	760	3.8	886	4.0		
Nb	300	8.0	260	7.7	530	3.8	604	4.1		
Ta	350	3.7	480	3.8			562	4.3		
Cr	790	3.2	655	4.1	796	4.2	933	4.0	1038	4.1
Mo	460	3.9	440	4.5					680	
W	370	3.1								
Mn	720	4.3	680	4.6	859	4.1	950	4.3	1088	4.1
Re	850	1.4	470	4.0						
Fe	805	4.4	764	4.5	897	4.3	1029	4.1	1122	4.2
Ru	600	5.4	670	3.5	620	6.5				
Co	780	5.3	798	5.5	989	4.3	1080	4.2	1185	4.2
Rh					620	6.5	720			
Ni	1025	4.1	1040	4.2	1042	4.4	1149	4.2	1238	4.2
Pd					830	3.2				
Pt					600				720	
Cu			1220	4.0	1240	3.8				

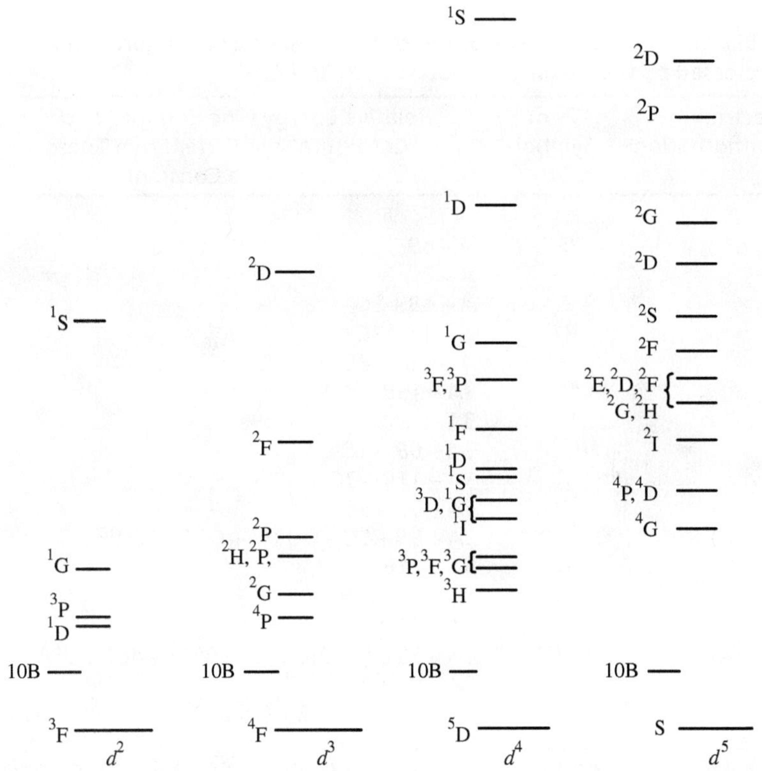

FIGURE 16.35
The energies of the terms for the d^2–d^5 electron configurations of the transition metals as free ions, assuming $C/B = 4.7$. [Reproduced from Figgis, B. N.; Hitchman, M. A. *Ligand Field Theory and Its Applications*, Wiley-VCH: New York, 2000. This material is reproduced with permission of John Wiley & Sons, Inc.]

All of the preceding text described a way to determine the term symbols and their relative energies for the different transition metals as free ions in the gas phase. However, it cannot be as simple as all of that! Application of a LF to the free ion causes a reduction in the symmetry and leads to splitting of some of the free ion terms. There are two possible scenarios: (i) the weak-field limit and (ii) the strong-field limit. The weak-field limit applies when there is only a small LF. In this event, the free ion term will be split by LS coupling (coupling of the orbital and spin angular momenta) and the LF will then be added as a weak perturbation. In the strong-field limit, on the other hand, the LF is so strong that the free ion will first be split by the LF and the LS coupling will then be introduced as a weaker perturbation. In the end analysis, there is a one-to-one correspondence between terms in the weak-field limit with those in the strong-field limit.

Beginning with the weak-field limit, the terms for the free ion will first be split by LS coupling. Because terms have the same total orbital angular momentum as the orbitals with the same names, they will split in an octahedral LF in exactly the same way that the orbitals split. Thus, an S term will transform in the O_h point group as an A_{1g} state, a P term as the T_{1g} state, and a D term will split into E_g and T_{2g} states (just as an *s*-orbital transforms as a_{1g}, the *p*-orbitals transform as t_{1u}, and the *d*-orbitals split into $e_g + t_{2g}$). The subscripts for the terms for the metal ion will always be *g*, however, because they refer to the *d*-orbitals and the *d*-orbitals are always *gerade*. In order to determine how the other term symbols will transform, one can use the formula in Equation (16.12) in the sub-rotational group of the molecule, where α is the angle of the proper rotation for each class of rotational operators. We will then generate a reducible representation in the sub-rotational group and use the great orthogonality theorem to determine its IRRs.

$$\chi(\alpha) = \frac{\sin[(L + 1/2)\alpha]}{\sin[\alpha/2]} \tag{16.12}$$

TABLE 16.18 The energies of the terms for each d^n configuration as expressed by the Racah parameters *A*, *B*, and *C*.

Electron Configuration	Term Symbol	Relative Energy (The Energies of d^{10-n} Configurations Differ From Those of d^n by a Constant)
d^1, d^9	^{2}D	
d^2, d^8	^{3}F	$A-8B$
	^{3}P	$A+7B$
	^{1}G	$A+4B+2C$
	^{1}D	$A-3B+2C$
	^{1}S	$A+14B+7C$
d^3, d^7	^{4}F	$3A-15B$
	^{4}P	$3A$
	^{2}H, ^{2}P	$3A-6B+3C$
	^{2}G	$3A-11B+3C$
	^{2}F	$3A+9B+3C$
	^{2}D′, ^{2}D″	$3A+5B+5C\pm(193B^2+8BC+4C^2)^{1/2}$
d^4, d^6	^{5}D	$6A-21B$
	^{3}H	$6A-17B+4C$
	^{3}G	$6A-12B+4C$
	^{3}F′, ^{3}F″	$6A-5B+(11/2)C\pm(3/2)(68B^2+4BC+C^2)^{1/2}$
	^{3}D	$6A-5B+4C$
	^{3}P′, ^{3}P″	$6A-5B+(11/2)C\pm(1/2)(912B^2-24BC+9C^2)^{1/2}$
	^{1}I	$6A-15B+6C$
	^{1}G′, ^{1}G″	$6A-5B+(15/2)\,C\pm(1/2)(708B^2-12BC+9C^2)^{1/2}$
	^{1}F	$6A+6C$
	^{1}D′, ^{1}D″	$6A+9B+(15/2)\,C\pm(3/2)(144B^2+8BC+C^2)^{1/2}$
	^{1}S′, ^{1}S″	$6A+10B+10C\pm2(193B^2+8BC+4C^2)^{1/2}$
d^5	^{6}S	$10A-35B$
	^{4}G	$10A-25B+5C$
	^{4}F	$10A-13B+7C$
	^{4}D	$10A-18B+5C$
	^{4}P	$10A-28B+7C$
	^{2}I	$10A-24B+8C$
	^{2}H	$10A-22B+10C$
	^{2}G′	$10A-13B+8C$
	^{2}G″	$10A+3B+10C$
	^{2}F′	$10A-9B+8C$
	^{2}F″	$10A-25B+10C$
	^{2}D′, ^{2}D″	$10A-3B+11C\pm3(57B^2+2BC+C^2)^{1/2}$
	^{2}D″	$10A-4B+10C$
	^{2}P	$10A+20B+10C$
	^{2}S	$10A-3B+8C$

For octahedral molecules, the O subgroup is used. L'Hôpital's rule is used to determine the character for the identity operation: $\chi(E)=2L+1$. Thus, for an F term ($L=3$) in an octahedral LF, the characters for the reducible representation are given in Table 16.19, which reduces to $A_{2g}+T_{1g}+T_{2g}$. By convention, lower case Mulliken symbols are used to describe the symmetries of the orbitals in the "one-electron" MO diagrams (where the electron–electron repulsions have been ignored), whereas upper case Mulliken symbols are used to describe the term symbols. The splitting patterns for some of the other terms in an octahedral LF are listed in Table 16.20.

TABLE 16.19 Generation of the splitting pattern of an F term in an octahedral LF, using the O sub-rotational group and Equation (16.12).

O	E	$6C_4$	$3C_2$	$8C_3$	$6C_2'$	IRR's:
Γ_F	7*	−1	−1	1	−1	$A_{2g}+T_{1g}+T_{2g}$

*From L'Hôpital's rule: 2(3) + 1 = 7.

TABLE 16.20 Splitting of the d^n terms in an octahedral LF.

Term	Components in an Octahedral LF
S	A_{1g}
P	T_{1g}
D	E_g+T_{2g}
F	$A_{2g}+T_{1g}+T_{2g}$
G	$A_{1g}+E_g+T_{1g}+T_{2g}$
H	$E_g+T_{1g}+T_{1g}+T_{2g}$
I	$A_{1g}+A_{2g}+E_g+T_{1g}+T_{2g}+T_{2g}$

The weak-field limit for the d^2 electron configuration in an octahedral LF is shown at the left in Figure 16.36. The 45 microstates are split into 11 different energy levels by LS coupling and the octahedral LF. The degeneracies of each term are listed in parentheses on the diagram (summing to 45 in total). Notice that the spin multiplicity of a term is unaffected by the LF.

In the strong-field limit, the strength of the LF dominates and LS coupling is treated as a secondary perturbation. Thus, we begin with the one-electron MO diagram for an octahedral LF, where the t_{2g} set lies beneath the e_g set. There are three possible electron configurations for d^2 in a strong octahedral LF: $t_{2g}{}^2e_g{}^0$ (which will be the GS), $t_{2g}{}^1e_g{}^1$ (intermediate in energy), and $t_{2g}{}^0e_g{}^2$ (the highest energy state), as shown at the right in Figure 16.36. Each of these energy levels (having the degeneracies listed in parentheses) is then split by the weaker LS coupling. This is where things get a little complicated because there are different ways to determine the effects of the LS coupling depending on the electron configuration. As a general rule, the higher the spin multiplicity, the lower in energy the respective term symbol will lie.

With the exception of the icosahedral geometry, the LS coupling perturbation to the strong-field limit essentially boils down to six different scenarios, depending on the exact nature of the electron configuration. After the LS coupling is determined in the strong-field limit, perturbation theory requires a one-to-one correspondence with terms from the weak-field limit to terms from the strong-field limit. By connecting those terms having like symmetries and spin multiplicities from both sides of the diagram and following the *no crossing rule* (lines connecting terms having identical term symbols cannot cross over one another), the *correlation diagram* shown in Figure 16.36 can be determined. The correlation diagrams for other d^n electron configurations can be derived in exactly the same way, although they are far more complicated.

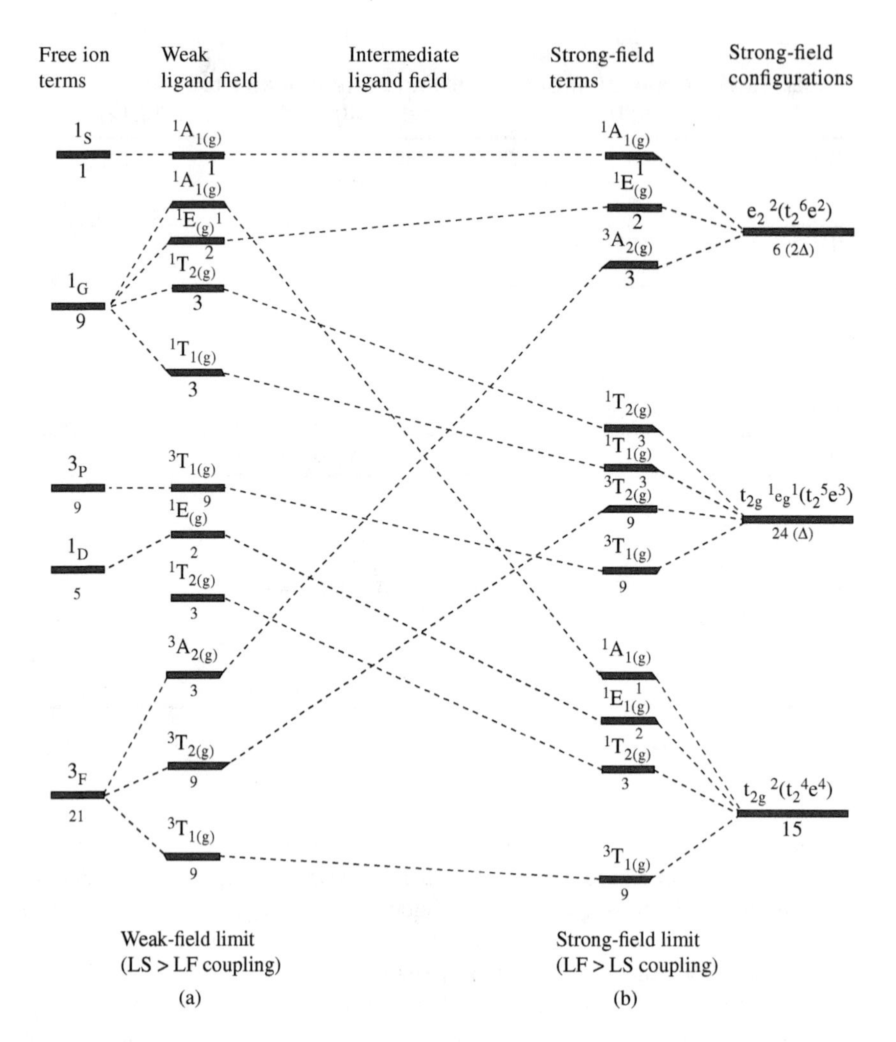

FIGURE 16.36
Correlation diagram between the weak-field limit (a) and strong-field limit (b) for a d^2 ion in an octahedral LF. [Reproduced from Figgis, B. N.; Hitchman, M. A. *Ligand Field Theory and Its Applications*, Wiley-VCH: New York, 2000. This material is reproduced with permission of John Wiley & Sons, Inc.]

16.5.1 Scenario 1—All the Orbitals are Completely Occupied

In this scenario, the overall electron distribution will be spherical and the term symbol will be the totally symmetric IRR for the point group. Because all of the electrons will be paired, this will necessarily be a singlet term symbol. In an octahedral LF, this will be the $^1A_{1g}$ term.

16.5.2 Scenario 2—There is a Single Unpaired Electron in One of the Orbitals

In this case, the term symbol will have the same designation as the MO containing the unpaired electron (except that it will now have the upper case term symbol) and it will be a doublet state. Thus, if the electron occupied a σ_g^+ orbital, the term symbol will be $^2\Sigma_g^+$.

16.5.3 Scenario 3—There are Two Unpaired Electrons in Two Different Orbitals

The symmetries of the resulting term symbols will be identical to the direct product of the symmetries of the two occupied orbitals. The spin multiplicity will either be singlet (when the electrons have opposite spins) or triplet (if the electrons have parallel spins). For example, the electron configuration $b_{1g}{}^{1}b_{2g}{}^{1}$ in the D_{4h} point group will give rise to the terms: ${}^{1}A_{2g} + {}^{3}A_{2g}$.

16.5.4 Scenario 4—A Degenerate Orbital is Lacking a Single Electron

This scenario is equivalent to the case of a single electron occupying the degenerate orbital because of the hole formalism. This will always give rise to a doublet state. Thus, the electron configuration $t_{2g}{}^{6}e_{g}{}^{3}$ in an O_h LF will have the ${}^{2}E_g$ term.

16.5.5 Scenario 5—There are Two Electrons in a Degenerate Orbital

The symmetries of the resulting terms will be identical to the direct product of the symmetries of the occupied orbitals. The spin multiplicity will either be singlet or triplet. However, some of the terms will violate the Pauli exclusion principle, which states that the exchange of any two electrons must be antisymmetric. Using the direct product tables in Appendix D, the term in brackets is orbitally antisymmetric. Because the triplet spin wave function is symmetric with respect to the exchange of electrons, the orbitally antisymmetric terms must be the triplet states. It follows that the orbitally symmetric terms must therefore be the singlet states. For example, the electron configuration $t_{2g}{}^{6}e_{g}{}^{2}$ will have the following terms (because $e_g \times e_g = a_{1g} + [a_{2g}] + e_g$ in the O_h point group): ${}^{1}A_{1g} + {}^{3}A_{2g} + {}^{1}E_g$.

16.5.6 Scenario 6—There are Three Electrons in a Triply Degenerate Orbital

The symmetries of the resulting terms will once again be identical to the overall direct product of the occupied orbitals. The spin multiplicity will either be 4 (if all three electrons have parallel spins) or 2. The characters for the doublet state under each operation in the sub-rotational point group are given by Equation (16.13), while those for the quartet state are given by Equation (16.14).

$$\chi(\text{doublet}) = (1/3)\{[\chi(R)]^3 - \chi(R^3)\} \quad (16.13)$$

$$\chi(\text{quartet}) = (1/6)\{[\chi(R)]^3 - 3\chi(R)\chi(R^2) + 2\chi(R^3)\} \quad (16.14)$$

Thus, for example, the electron configuration $t_{2g}{}^{3}e_{g}{}^{0}$ in an octahedral LF has the following terms, as shown in Table 16.21: ${}^{4}A_{2g} + {}^{2}E_g + {}^{2}T_{1g} + {}^{2}T_{2g}$.

TABLE 16.21 Term symbols for the $t_{2g}{}^{3}e_{g}{}^{0}$ electron configuration in the strong-field limit for an octahedral LF.

O	E	$6C_4$	$3C_2$	$8C_3$	$6C_2'$	
$\chi(R)$	3	−1	−1	0	1	
R^2	E	C_2	E	$C_3{}^2$	E	
$\chi(R^2)$	3	−1	3	0	3	
R^3	E	$C_4{}^3$	C_2	E	C_2'	
$\chi(R^3)$	3	−1	−1	3	1	
$[\chi(R)]^3$	27	−1	−1	0	1	IRRs:
$\chi(\text{doublet})$	8	0	0	−1	0	$E + T_1 + T_2$
$\chi(\text{quartet})$	1	−1	1	1	−1	A_2

Example 16-6. Using these scenarios, determine the terms for the strong-field limit for the d^2 electron configuration and compare your answer to the terms shown in Figure 16.36.

Solution. The three strong-field electron configurations are $t_{2g}{}^2e_g{}^0$, $t_{2g}{}^1e_g{}^1$, and $t_{2g}{}^0e_g{}^2$. The $t_{2g}{}^2e_g{}^0$ configuration will have a degeneracy of $6!/(2!4!) = 15$, based on the formula given in Chapter 4. Using Scenario 5, the direct product $t_{2g} \times t_{2g} = a_{1g} + e_g + [t_{1g}] + t_{2g}$. Thus, this energy level will split into ${}^1A_{1g}$ (1) + ${}^1E_g$ (2) + ${}^3T_{1g}$ (9) + ${}^1T_{2g}$ (3), where the degeneracies are listed in parentheses. The degeneracy of each term can be determined by taking the dimensions of the IRR (1 for A or B, 2 for E, and 3 for T) times the spin multiplicity. For the $t_{2g}{}^1e_g{}^1$ configuration, Scenario 3 applies. Both singlet and triplet terms will be observed. The direct product of $t_{2g} \times e_g$ is $t_{1g} + t_{2g}$, so the terms will be ${}^1T_{1g}$ (3) + ${}^3T_{1g}$ (9) + ${}^1T_{2g}$ (3) + ${}^3T_{2g}$ (9). Finally, the $t_{2g}{}^0e_g{}^2$ was already covered in the Scenario 5 example and yielded the following terms: ${}^1A_{1g}$ (1) + ${}^1E_g$ (2) + ${}^3A_{2g}$ (3). The terms match those shown in Figure 16.36 and the sum of all the degeneracies totals 45 microstates.

The correlation diagram for the d^8 electron configuration is the same as that for the d^2 (hole) configuration. Because electrons "sink" and holes "float" the only difference between the d^8 and d^2 correlation diagrams is that the terms for the secondary perturbations in both the weak and strong-field limits are inverted, as shown in Figure 16.37. The GS free ion terms for d^2 and d^8 are identical, but

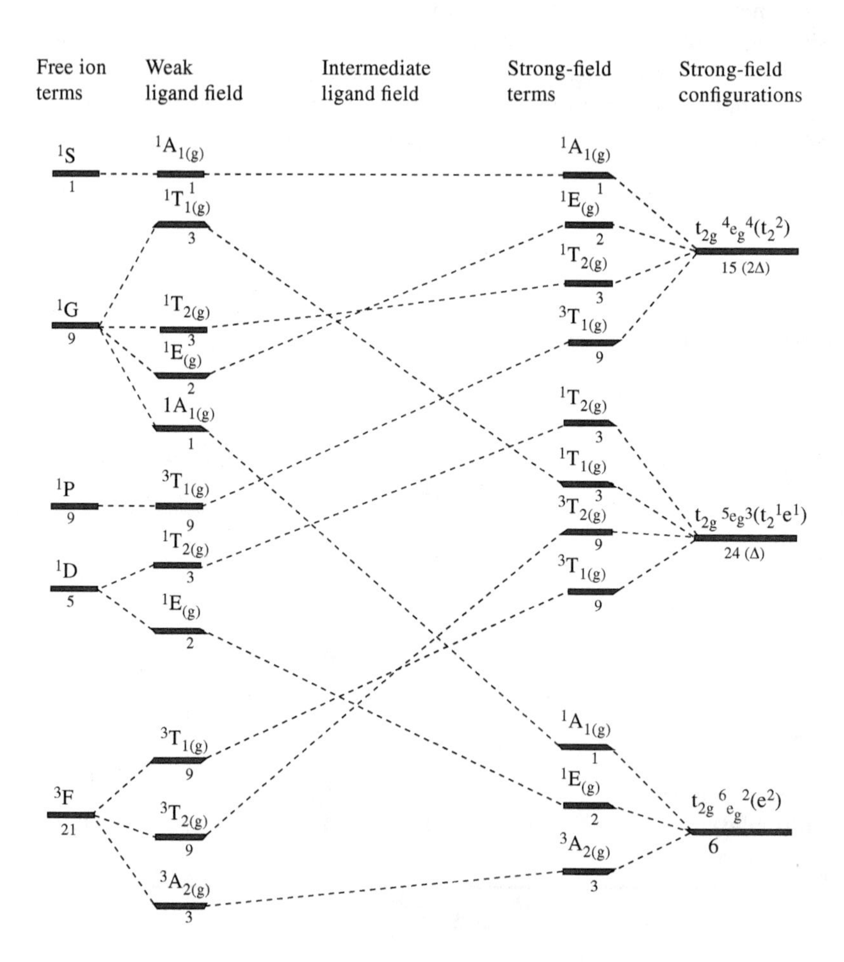

FIGURE 16.37
Correlation diagram for the d^8 electron configuration in an octahedral LF. [Reproduced from Figgis, B. N.; Hitchman, M. A. *Ligand Field Theory and Its Applications*, Wiley-VCH: New York, 2000. This material is reproduced with permission of John Wiley & Sons, Inc.]

TABLE 16.22 Ground-state term symbols for the different d^n electron configurations in an octahedral LF and in a tetrahedral LF.

d^n (O_h)	d^n (T_d)	GS Term
0, 10	10, 0	$^1A_{1g}$
1	9	$^2T_{2g}$
2	8	$^3T_{1g}$ (F)
3	7	$^4A_{2g}$
4	6	$^5E_{2g}$
5	5	$^6A_{1g}$
6	4	^5T_g
7	3	$^4T_{1g}$ (F)
8	2	$^3A_{2g}$
9	1	$^2E_{2g}$

the LF split terms are inverted. For d^2, the GS term in an O_h LF is $^3T_{1g}$; for d^8, it is $^3A_{2g}$.

Because the e and t_2 MOs in a tetrahedral LF are inverted from those in an octahedral LF, the correlation diagrams for the d^n electron configurations in a tetrahedral LF are related to those in an octahedral LF as follows: d^n (T_d) = d^{10-n} (O_h). Thus, d^8 (T_d) and d^2 (O_h) share the same correlation diagram. Table 16.22 summarizes the GS term symbols for the different d^n electron configurations in both the tetrahedral and octahedral LFs.

16.6 TANABE–SUGANO DIAGRAMS

Tanabe–Sugano diagrams are often used by spectroscopists to determine the number of allowed electronic transitions and their relative energies. The Tanabe–Sugano diagram for d^2 (O_h) is shown in Figure 16.38. A *Tanabe–Sugano diagram* is just a special kind of correlation diagram where the *y*-axis is E/B (energy), the *x*-axis is Δ/B, and the axes have been normalized so that the GS term symbol in both the weak-field and strong-field limits has the same energy. The energies of the terms as a function of the LF (Δ) are carefully plotted so that the Tanabe–Sugano diagram can be used to obtain quantitative information about the energies of the electronic transitions. There are, however, some minor differences between the Tanabe–Sugano diagrams and regular correlation diagrams. First, some of the lines are curved. This results when there are two states that have identical term symbols. Whenever two terms having the same symmetry approach each other along the energy coordinate, the states will repel one another, leading to the observed curvatures (this is known as the *no-crossing rule*). Second, for the d^4–d^7 electron configurations, there is a vertical line in the middle of the Tanabe–Sugano diagram where abrupt changes in energy occur between the terms and the GS term symbols change. This vertical line represents the transition from HS to LS in these electron configurations. The Tanabe–Sugano diagrams for the different d^n electron configurations in an octahedral LF are shown in Figures 16.38–16.44 on the ensuing pages. The d^0 and d^{10} configurations are closed-shell configurations having a single term: $^1A_{1g}$ and no *d–d* (or LF) electronic transitions. For d^1 and d^9, there is a single LF transition between the $^2T_{2g}$ and ^2Eg terms, where the energy difference between these two terms is Δ_o.

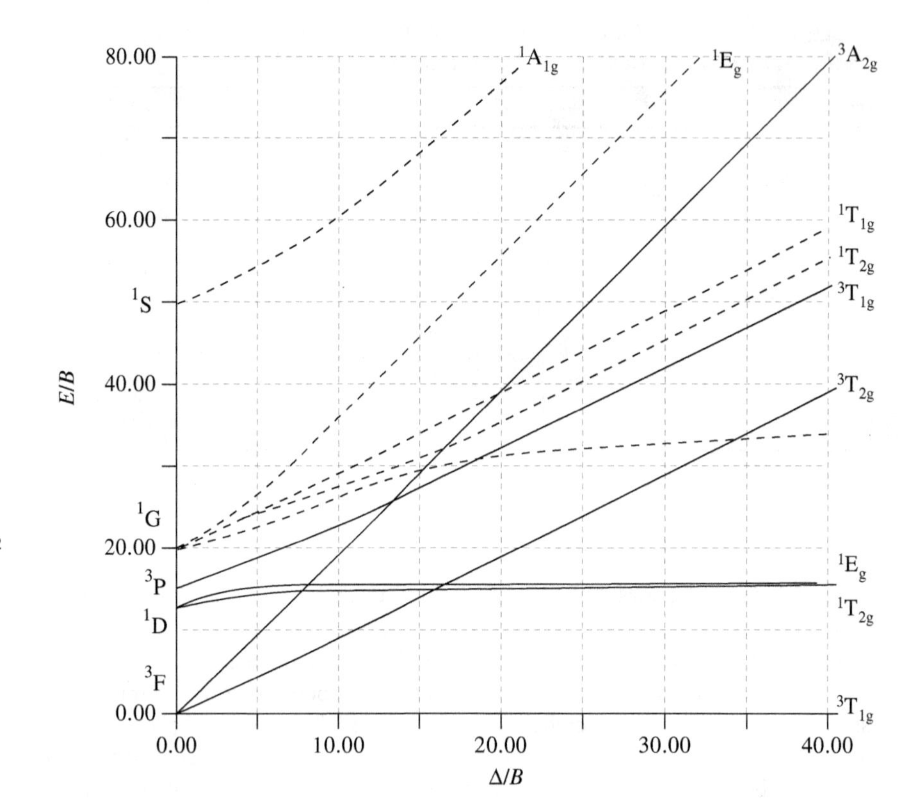

FIGURE 16.38
Tanabe–Sugano diagram for d^2 (O_h), where $B = 886\ cm^{-1}$ and $C/B = 4.42$. [Reproduced from Figgis, B. N.; Hitchman, M. A. *Ligand Field Theory and Its Applications*, Wiley-VCH: New York, 2000. This material is reproduced with permission of John Wiley & Sons, Inc.]

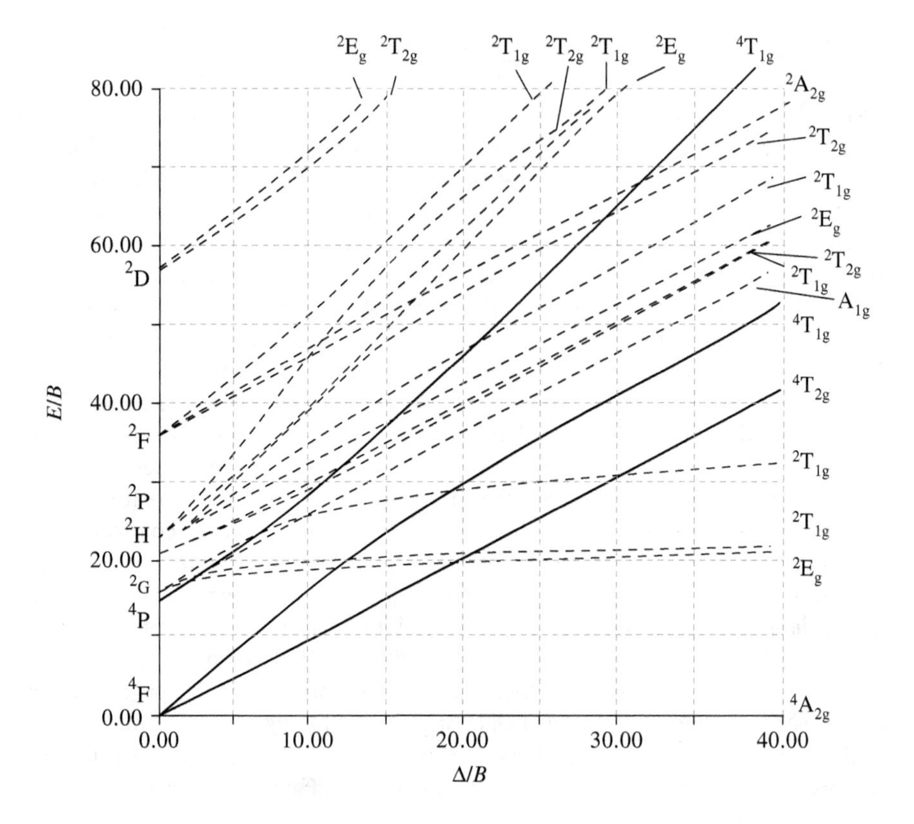

FIGURE 16.39
Tanabe–Sugano diagram for d^3 (O_h), where $B = 933\ cm^{-1}$ and $C/B = 4.5$. [Reproduced from Figgis, B. N.; Hitchman, M. A. *Ligand Field Theory and Its Applications*, Wiley-VCH: New York, 2000. This material is reproduced with permission of John Wiley & Sons, Inc.]

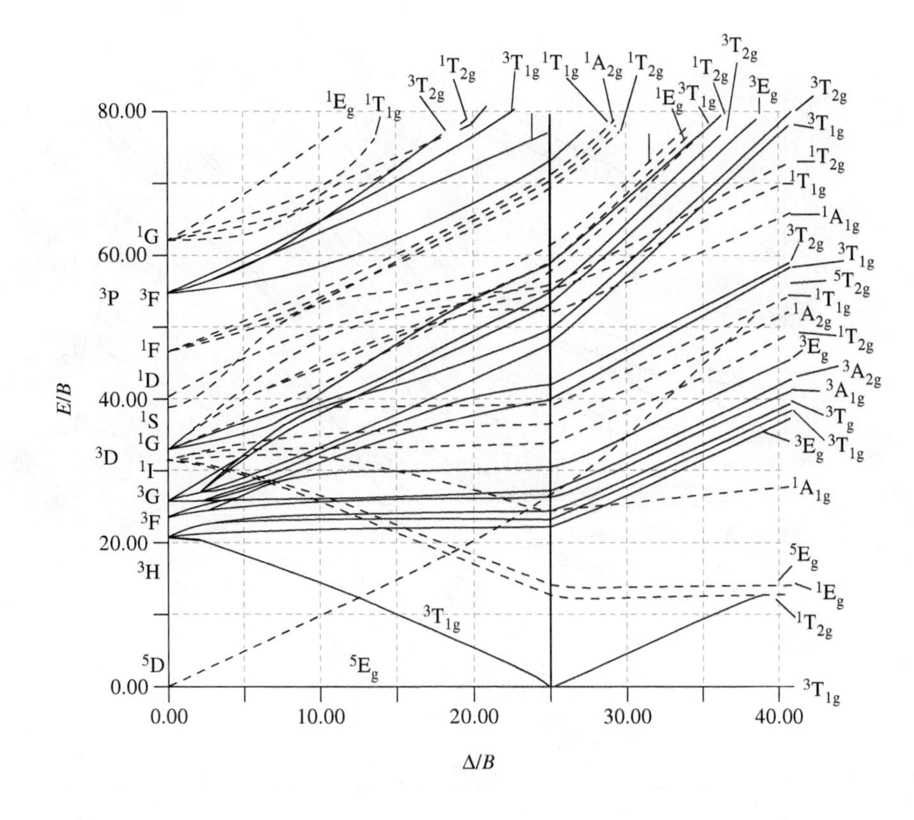

FIGURE 16.40

Tanabe–Sugano diagram for d^4 (O_h), where $B = 796\ cm^{-1}$ and $C/B = 4.6$. [Reproduced from Figgis, B. N.; Hitchman, M. A. *Ligand Field Theory and Its Applications*, Wiley-VCH: New York, 2000. This material is reproduced with permission of John Wiley & Sons, Inc.]

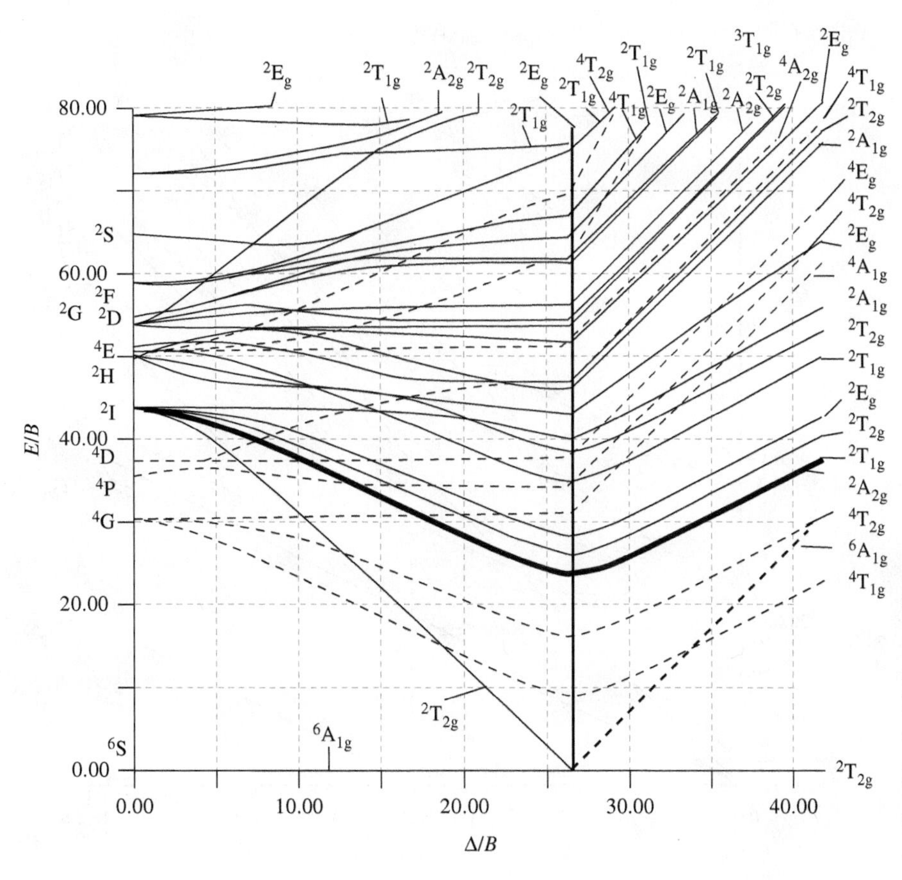

FIGURE 16.41

Tanabe–Sugano diagram for d^5 (O_h), where $B = 859\ cm^{-1}$ and $C/B = 4.48$. [Reproduced from Figgis, B. N.; Hitchman, M. A. *Ligand Field Theory and Its Applications*, Wiley-VCH: New York, 2000. This material is reproduced with permission of John Wiley & Sons, Inc.]

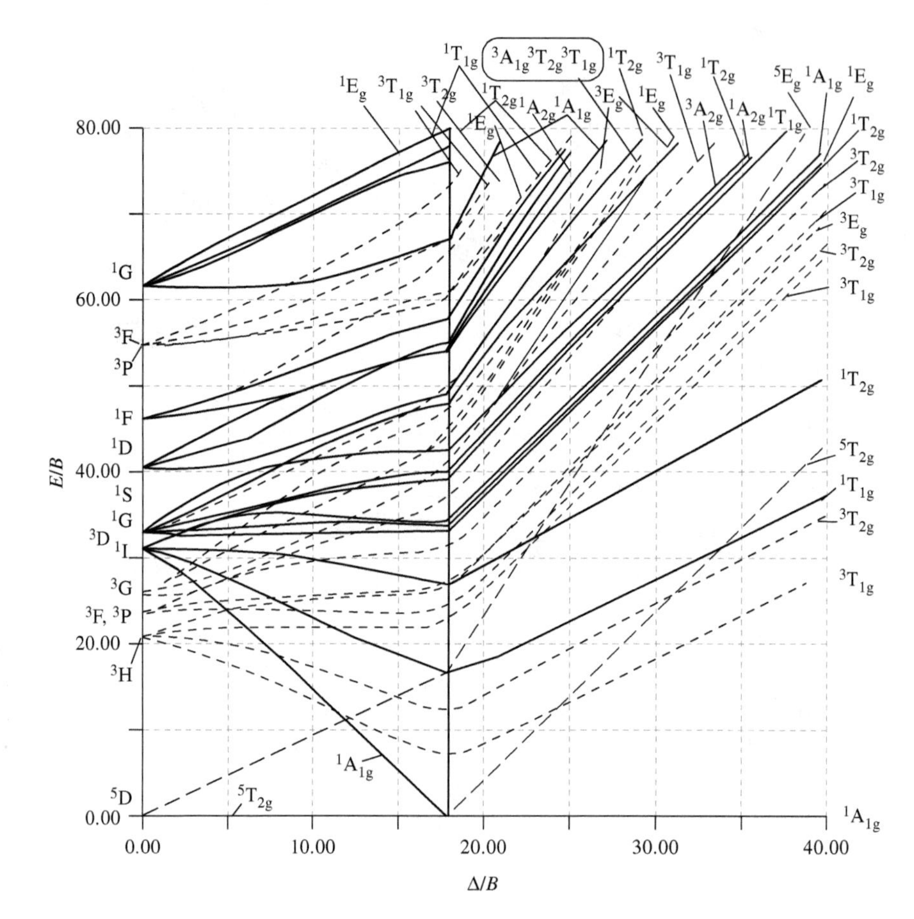

FIGURE 16.42

Tanabe–Sugano diagram for d^6 (O_h), where $B = 1080\ cm^{-1}$ and $C/B = 4.42$. [Reproduced from Figgis, B. N.; Hitchman, M. A. *Ligand Field Theory and Its Applications*, Wiley-VCH: New York, 2000. This material is reproduced with permission of John Wiley & Sons, Inc.]

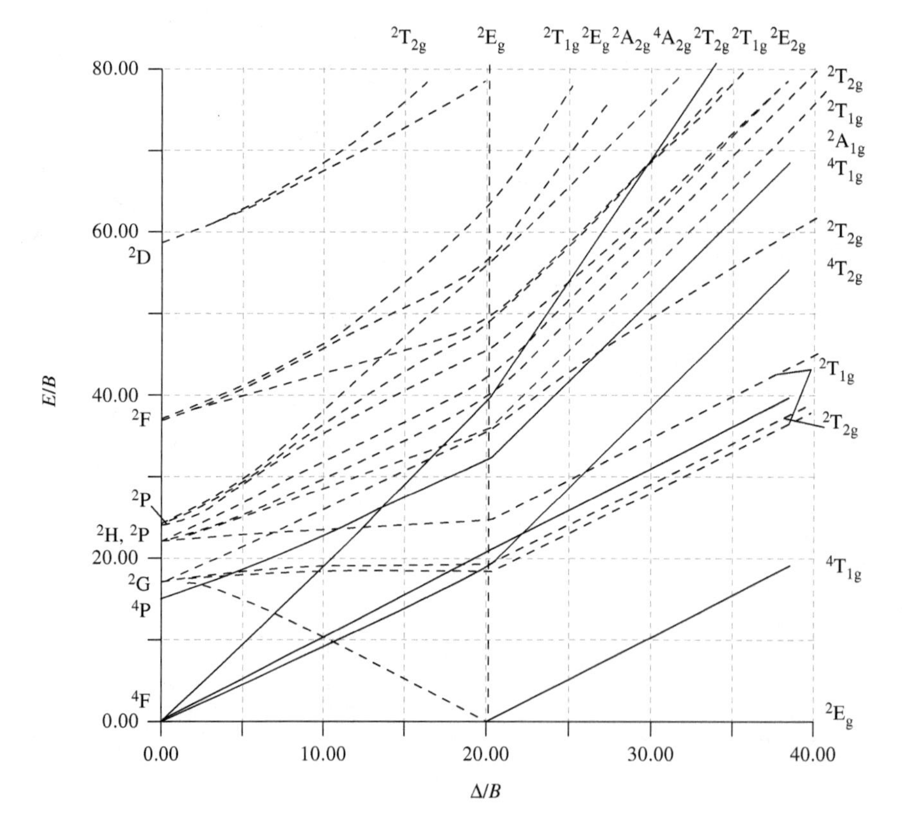

FIGURE 16.43

Tanabe–Sugano diagram for d^7 (O_h), where $B = 986\ cm^{-1}$ and $C/B = 4.63$. [Reproduced from Figgis, B. N.; Hitchman, M. A. *Ligand Field Theory and Its Applications*, Wiley-VCH: New York, 2000. This material is reproduced with permission of John Wiley & Sons, Inc.]

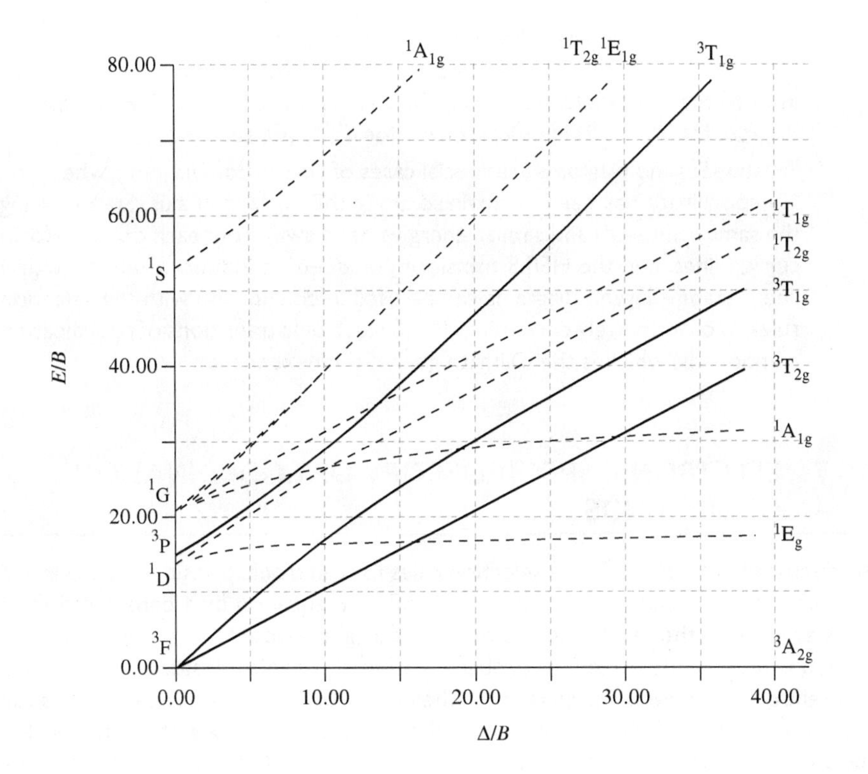

FIGURE 16.44
Tanabe–Sugano diagram for d^8 (O_h), where $B = 1042\ cm^{-1}$ and $C/B = 4.71$. [Reproduced from Figgis, B. N.; Hitchman, M. A. *Ligand Field Theory and Its Applications*, Wiley-VCH: New York, 2000. This material is reproduced with permission of John Wiley & Sons, Inc.]

Here is a brief summary of what we learned so far:

1. Whether CFT or LFT was used, the one-electron MO diagrams for coordination compounds ignored the interelectron repulsions present whenever there was more than one valence electron. We cannot quantitatively predict the electronic spectra of coordination compounds until we include the electron–electron repulsions.
2. When considering the polyelectronic picture, states (or terms) must be used to describe the actual arrangements of the electrons in the valence orbitals. Each term will have a different energy, although it may be composed of a degenerate set of microstates. The terms for the free ion can be determined using methods developed in Chapter 4.
3. In the weak-field limit, the LS coupling is treated first and the LF splitting is introduced as a secondary perturbation. Because the term symbols have the same total angular momentum as the orbitals after which they have been named, they will split in a weak LF in the same manner that the corresponding orbitals split. Whenever the terms split in energy, the overall degeneracy must remain the same.
4. The relative energies of the terms depend on the magnitudes of the Racah parameters *A*, *B*, and *C*. As a general rule, states having large spin multiplicities are usually (but not always) lower in energy than those with smaller spin multiplicities.
5. In the strong-field limit, the LF splitting is treated first and the LS coupling is introduced as a secondary perturbation. There are six general scenarios that allow for the prediction of the terms generated by the LS coupling.
6. When the states in the weak-field limit are connected with like states in the strong-field limit using the no-crossing rule, the result is a correlation diagram, where there is a one-to-one correspondence between the term symbols from

the weak- and strong-field limits. Only five correlation diagrams are necessary because $d^n = d^{10-n}$ (hole), where the terms for the secondary perturbations in both the weak-field and strong-field limits are inverted. The correlation diagram for d^n (T_d) is identical to that for d^{10-n} (O_h).

7. Tanabe–Sugano diagrams are special cases of correlation diagrams where the GS coordinate has been normalized along the horizontal axis. States having the same symmetry and similar energies bend away from each other to form curved lines, and the HS/LS transition for d^4–d^7 is indicated with a vertical line. Tanabe–Sugano diagrams can be used in conjunction with the selection rules to quantitatively determine the LF electronic transitions of coordination compounds, which is the ultimate goal of the next section.

16.7 ELECTRONIC SPECTROSCOPY OF COORDINATION COMPOUNDS

As discussed in Section 9.2, the selection rules for electronic transitions are derived from the time-dependent form of the Schrödinger equation by a consideration of the way in which the oscillating electric field of a light wave acts on a dipole moment component of the molecule. The probability of an electronic transition between two different quantum levels will increase as the energy of the light wave approaches (or is in resonance with) that of the energy difference between two of the molecule's fixed energy levels. In the context of this chapter, this will be the energy difference between the GS and excited-state (ES) term symbols for a given d^n electron configuration in either an octahedral or tetrahedral LF. When light having an appropriate frequency for absorption impinges upon the sample, some of the total energy flux (or incident light intensity, I_0) is absorbed (I_a) by the molecule, some of it is transmitted (I_t) through the sample, some of it is reflected (I_r), and some of it is scattered (I_s) by Rayleigh or Raman processes, as shown in Figure 16.45. The total energy flux is therefore given by Equation (16.15).

$$I_0(\text{photons/cm}^3) = I_a + I_t + I_r + I_s \tag{16.15}$$

By orienting the sample at 90° to the incident light beams and by diluting the substance in transparent solvent, the amounts of reflected and scattered light can be diminished until they are essentially negligible. Thus, Equation (16.15) reduces to Equation (16.16).

$$I_0 \cong I_a + I_t \tag{16.16}$$

The change in energy flux as it passes through the sample is given by Equation (16.17), where the Einstein coefficient of absorption B_{ij} is defined

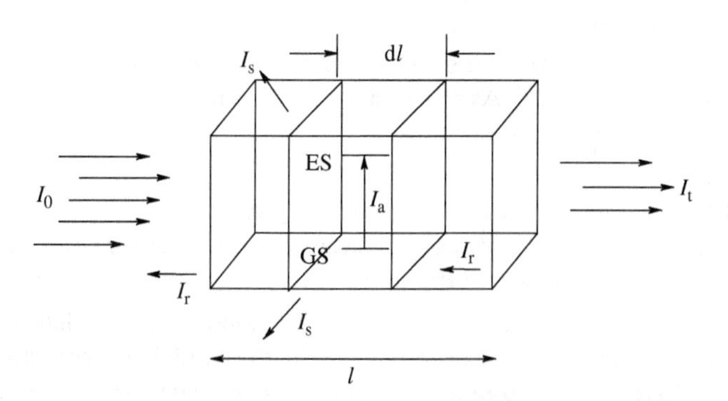

FIGURE 16.45
Incident light (I_0) impinges on a sample of length l. The incident light is absorbed (I_a), transmitted (I_t), reflected (I_r), or scattered (I_s).

by Equation (16.18). M_{01} is the transition moment integral, which was defined in Section 9.2, h is Planck's constant, and c is the speed of light in centimeter per second.

$$-dI = B_{ij} h\nu N_A (C/1000)\, dl \tag{16.17}$$

$$B_{ij} = \frac{8\pi^3}{3h^2} M_{01}^2 \frac{I}{c} \tag{16.18}$$

After combination of all the constants into α, Equation (16.18) reduces to Equation (16.19), and it shows that the change in the energy flux is directly proportional to the concentration of the colored compound. By integration of the left-hand side of Equation (16.19) from I_0 to I_t and the right-hand side from 0 to l, we obtain Equation (16.20), whose solution is given by Equation (16.21). Converting from natural log to base 10, we define the molar absorptivity, $\varepsilon = \alpha/2.303$, so that Equation (16.21) can be rearranged after substitution to yield the familiar Beer–Lambert law in Equation (16.22). The important thing to note here is that the molar absorptivity depends on the magnitude of the square of the transition moment integral.

$$-dI/I = \alpha C\, dl \tag{16.19}$$

$$\int_{I_0}^{I_t} \frac{dI}{I} = -\int_0^l \alpha C\, dl \tag{16.20}$$

$$I_t = I_0 e^{-\alpha C l} \tag{16.21}$$

$$A = \log\left(\frac{I_0}{I_t}\right) = \varepsilon l C \tag{16.22}$$

As discussed in Section 9.2, the transition moment integral has five components to it, as shown in Equation (16.23).

$$M_{01} = \int \psi_{esES}^* \psi_{esGS} d\tau_{es} \int \psi_{vES}^* \hat{u}_n \psi_{vGS} d\tau_n + \int \psi_{vES}^* \psi_{vGS} d\tau_n \int \psi_{eES}^* \hat{u}_e \psi_{eGS} d\tau_e \int \psi_{sES}^* \psi_{sGS} d\tau_s \tag{16.23}$$

The first term defines the overlap of the electronic GS and ES, whereas the second integral has to do with vibrational transitions. The third term (or Franck–Condon factor) has to do with the vibrational overlap. The final two terms form the basis for the orbital and spin selection rules, respectively.

The fourth integral governs the electronic *orbital selection rule*. It has to do with the symmetries of the electronic wave functions in the electronic GS and ES. In order for the electronic transition to be "orbitally allowed," the triple direct product $\Gamma_{GS}\Gamma_{\mu}\Gamma_{ES}$ must contain the totally symmetric IRR for the point group of the molecule. This will be true if and only if $\Gamma_{GS}\Gamma_{ES}$ contains the IRR for one of the Cartesian unit vectors (x, y, or z). A special case of the orbital selection rule is known as the *Laporte selection rule*. The *Laporte selection rule* states that for any centrosymmetric molecule, the only orbitally allowed electronic transitions are those that involve a change in the parity of the GS and ES wavefunctions. In other words, $g \rightarrow u$ or $u \rightarrow g$ in order for the electronic transition to be Laporte-allowed.

In octahedral symmetry, all LF transitions between d-orbitals are Laporte-forbidden because they always involve terms with the *gerade* subscripts, while (x, y, and z) transform as the *ungerade* term t_{1u}. Because the direct product g · u · g will always yield an *ungerade* term and the totally symmetric IRR is *gerade*, all d–d transitions in an O_h LF are orbitally forbidden. Nonetheless, as evidenced by the plethora of colors exhibited by coordination compounds, LF transitions are, in fact, experimentally observed. The reason for this is that *vibronic coupling* (the coupling of a vibrational mode to the electronic transition) can do an end-around to the orbital selection rule. Any coupled vibration that temporarily removes the center of inversion in the molecule will lower the symmetry of the molecule so that the Laporte selection rule is no longer applicable. Thus, *d–d* transitions are, in fact, observed for the transition metals, although their molar absorptivities are weak.

The fifth and final integral is the *spin selection rule*, which essentially states that the spin multiplicities of the GS and ES must be the same in order for the transition to be "spin-allowed." The spin selection rule essentially states that an electronic transition will be spin-allowed if and only if the spin multiplicity of the GS and the ES terms are identical. This is because the two spin functions are orthogonal whenever the spin multiplicities differ, so that the fifth integral in Equation (16.23) is zero. However, spin-forbidden transitions are also frequently observed in the experimental spectra of coordination compounds. In this case, spin-orbit coupling (which is especially significant when $Z > 30$) can transfer some "allowed-ness" to the otherwise spin-forbidden transition.

The spin angular momentum of an electron generates a small magnetic dipole. Treating the electron from a relativistic approach where the electron is fixed and the nucleus appears to orbit the electron (much the same way as the sun appears to orbit the earth), the "orbiting motion" of the nucleus (which is a charged particle) creates an orbital magnetic moment. The electron spin and nuclear orbital magnetic moments can couple together, mixing the spin character of those states having the same total angular momentum. Thus, a singlet GS might acquire a small percentage of triplet character as a result of spin-orbit coupling. Because the coupling involves the nucleus, it can be shown that the magnitude of the spin-orbit coupling goes as Z^4, where Z is the atomic number. Thus, spin-orbit coupling is negligible for the lighter transition metals ($Z \leq 30$), but it increases rather rapidly for $Z > 30$. When spin-orbit coupling is especially large, the LS coupling scheme is no longer valid and jj coupling should be used instead. The energy of a term in the jj coupling scheme is given by Equation (16.24), where λ is the spin-orbit coupling constant (selected values are given in Table 16.23). Some typical molar absorptivities for the different types of LF transitions are listed in Table 16.24.

Finally, we get to the heart of the matter at hand: the quantitative analysis of the electronic spectra of coordination compounds. Consider the electronic spectrum of $[Co(en)_3]^{3+}$, which has two peaks of moderate intensity at 338 and 464 nm. Converting from wavelength (which is inversely proportional to the energy) to frequency (or wavenumbers, which are directly proportional to the energy), we find that the two peaks occur at 29.6 and 21.6 kK, respectively. While the complex actually has D_{3d} symmetry, we will consider only the *N* atoms occupying the six octahedral sites around a central Co^{3+} ion. Cobalt(III) is d^6. Here, we have a first-row transition metal in a moderately high oxidation state with strong-field ligands. As shown in Example 16-3(a), we predict $\Delta_o = (1.25)(18.2) = 22.8$ kK. The total pairing energy (from Table 16.2) for the Co^{3+} free ion is 23.6 kK. Because the pairing energy for coordination compounds is 15–30% smaller than for the free ion, Δ_o will be slightly larger than P_{tot} in this case and the complex will be LS.

The GS term symbol for d^6 LS in an O_h LF is $^1A_{1g}$, as shown on the right-hand (strong-field) side of the d^6 Tanabe–Sugano diagram in Figure 16.42. There are five spin-allowed ESs on the LS side of the diagram. By contrast, had our analysis been wrong and the complex was HS, the GS term symbol would be $^5T_{2g}$ and there would only be one spin-allowed ES (^5E_g). Because we observe more than one spin-allowed transition in the UV/VIS spectrum of the compound, we can be certain that the

TABLE 16.23 Spin-orbit coupling constants (λ) for selected transition metal ions in their ground states (in units of cm^{-1}).

Metal	0	1+	2+	3+	4+	5+	6+
Ti	18	30	62	155			
Zr		(100)	(200)	(500)			
V	19	34	57	105	250		
Nb		(105)	(203)	(400)			
Ta				(700)			
Cr	−34	37	58	92	168	380	
Mo			(168)	(267)	(425)	(900)	
W			(375)	(600)	(1150)	(2700)	
Mn	−63	−64	60	89	138	238	540
Tc			(190)	(300)	(433)	(850)	(1700)
Re			(420)	(625)	(1100)	(1850)	(4200)
Fe	−138	−112	−100	92	130	197	333
Ru				(250)	(350)	(500)	(850)
Os				(600)	(1000)	(1500)	(2100)
Co	−390	−228	−172	−145	130	179	263
Rh					(340)	(463)	(700)
Ir					(1000)	(1380)	(2000)
Ni		−565	−315	−235	−198	173	238
Pd		(−1300)	(−800)				
Pt		(−3400)					
Cu			−830	−445	−320	−258	226
Ag			(−1800)				
Au			(−5000)				

Values in parentheses are estimated.

TABLE 16.24 Typical values of ε for LF transitions.

Spin-allowed	Orbitally allowed*	$\varepsilon = 10^3 - 10^6$ M^{-1} cm^{-1}
Spin-allowed	Orbitally forbidden	$\varepsilon = 10^0 - 10^3$ M^{-1} cm^{-1}
Spin-forbidden	Orbitally allowed*	$\varepsilon = 10^0 - 10^2$ M^{-1} cm^{-1}
Spin-forbidden	Orbitally forbidden	$\varepsilon = 10^{-3} - 10^0$ M^{-1} cm^{-1}

*For comparative purposes only (LF transitions in O_h are orbitally forbidden).

complex is, in fact, LS. Only two of the five possible spin-allowed transitions are observed because the other three peaks lie at much higher energies in the UV.

Now that we are certain of an LS complex, the two spin-allowed electronic transitions can be assigned as $^1T_{1g} \leftarrow {}^1A_{1g}$ (21.6 kK) and $^1T_{2g} \leftarrow {}^1A_{1g}$ (29.6 kK). By convention, absorptions are written with the ES on the left and the GS on the right of an arrow indicating the direction of the electronic transition. The ratio of the energy of the second peak to that of the first is (29.6)/(21.6) = 1.37. We will now locate the two lines for the $^1T_{2g}$ and $^1T_{1g}$ terms on the Tanabe–Sugano diagram. Using a ruler oriented with respect to Figure 16.42, we will slide the ruler left to right across the horizontal axis looking for the value of Δ/B where the E/B value for $^1T_{2g}$ is 1.37 times the E/B value for $^1T_{1g}$. This ratio occurs at a value of Δ/B of 39.5 on the horizontal axis, as shown in Figure 16.46. The E/B values for the two transitions at this point occur at 51 for the $^1T_{2g} \leftarrow {}^1A_{1g}$ transition and 37 for the $^1T_{1g} \leftarrow {}^1A_{1g}$ transition. Because the energies of the two bands are known,

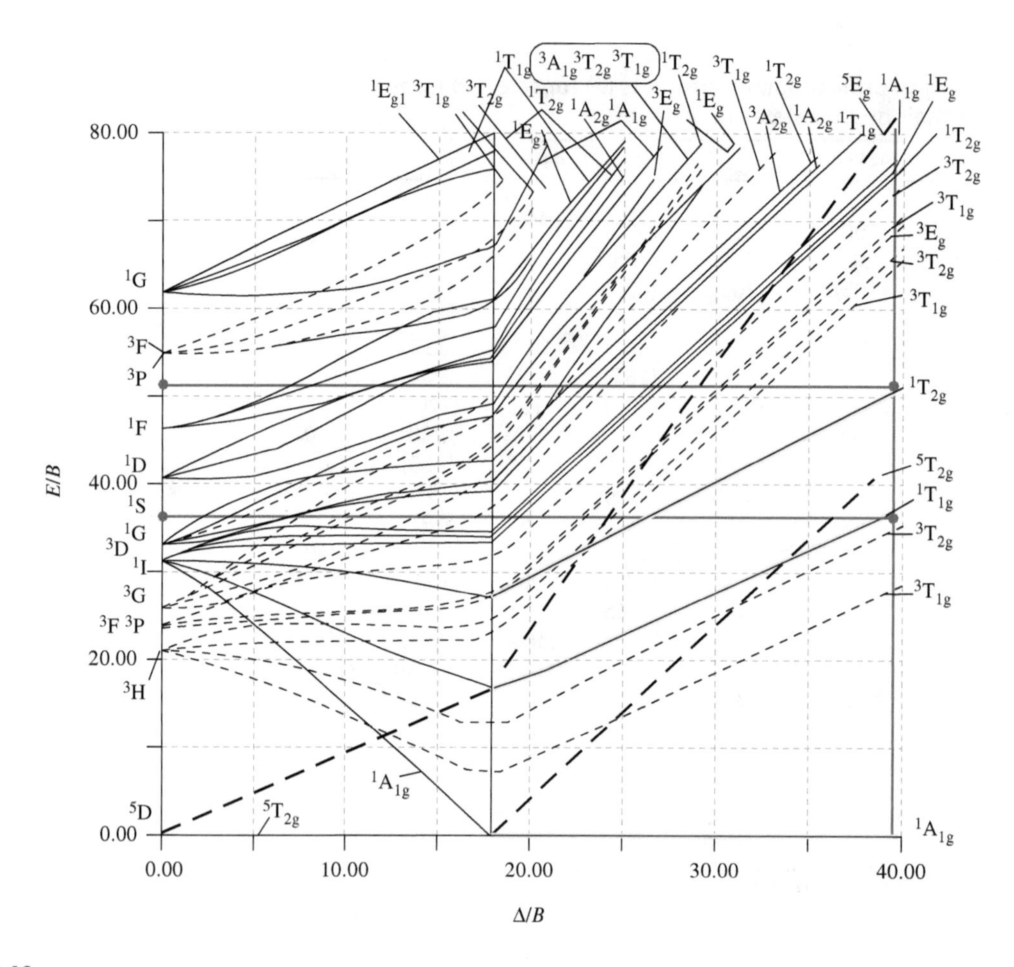

FIGURE 16.46

Tanabe–Sugano diagram for the d^6 coordination compound, $[Co(en)_3]^{3+}$, with the spin-allowed transitions highlighted and the correct values of E/B' and Δ/B' shown in red. [Adapted from Figgis, B. N.; Hitchman, M. A. *Ligand Field Theory and Its Applications*, Wiley-VCH: New York, 2000. This material is reproduced with permission of John Wiley & Sons, Inc.]

we can solve both equations for B and then average the two values. $E = 29.6$ kK for $^1T_{2g} \leftarrow {}^1A_{1g}$, so $B = E/(E/B) = 29.6/51 = 0.58$ kK. For the $^1T_{1g} \leftarrow {}^1A_{1g}$ transition, $B = 21.6/37 = 0.58$ kK. Averaging the two values, $B = 580\ \text{cm}^{-1}$ for the Co^{3+} ion in the complex. Using this value of B and the fact that $\Delta/B = 39.5$, the LF splitting parameter Δ_o is calculated as 23 kK, which is roughly the same value predicted on the basis of Jørgensen's *f* and *g* factors.

The value of B for metal ions in coordination compounds is always lower than the value of B that the metal has as a free ion, a fact known as the *nephelauxetic effect*, which derives from the German word for "expanding cloud." Because metal–ligand bonds have a covalent character, the effective size of the *d*-orbitals is larger in a transition metal complex than it is for the free ion. The metal electrons are somewhat delocalized over the metal–ligand bond. As a result, the value of B for the metal, which is a measure of the electron–electron repulsions, will always be smaller in a coordination compound than in the free ion. The nephelauxetic factor, β, is defined as the ratio of the Racah parameter in the coordination compound (B') versus that for the free ion (B), as shown in Equation (16.24)

$$\beta = B'/B = (1 - hk) \qquad (16.24)$$

Using Table 16.17, $B = 1080\ \text{cm}^{-1}$ for Co^{3+} as a free ion and therefore $\beta = 580/1080 = 0.54$ for $[Co(en)_3]^{3+}$. The axes on the Tanabe–Sugano diagrams would more correctly be labeled E/B' versus Δ/B'. Using experimental data for the nephelauxetic parameter from a large series of coordination compounds, Jørgensen derived a list of *h* and *k* factors for the ligands and metals, respectively. The *h* and *k* factors are somewhat analogous to his *f* and *g* factors. Using the *h* and *k* factors listed in Table 16.25, the nephelauxetic factor for an unknown coordination compound can be calculated using Equation (16.24). For $[Co(en)_3]^{3+}$, $h = 1.4$ and $k = 0.33$, yielding a predicted β of 0.54, which is identical to the value obtained experimentally. The nephelauxetic parameters should run parallel with the degree of covalent character in the metal–ligand bonding. For the ligands, F^- ($h = 0.8$) is highly ionic because of its large charge/size ratio, while I^- ($h = 2.7$) is polarizable and covalent. For the metal ions, the first-row transition metals have smaller radii and tend to form more ionic bonds, while the larger second- and third-row metals have greater covalent character. These general trends are also observed among the *h* and *k* factors in Table 16.25.

Tanabe–Sugano diagrams can also be used in reverse. Starting with the *f*, *g*, *h*, and *k* factors, one can predict the magnitude of Δ_o and β. Then, using the value of *B* for the free ion from Table 16.2 and the calculated value of β, Equation (16.24) can be used to predict the Racah parameter B' in the complex. Using these values for the ratio Δ/B', the horizontal position on the Tanabe–Sugano diagram can be determined, as shown in Figure 16.47. The E/B' ratios for the spin-allowed transitions that occur at this value of Δ/B' can then be read off the *y*-axis of the diagram. Because B' is known, the energies of the corresponding electronic transitions can be predicted. For example, $[Ni(H_2O)_6]^{2+}$ has $f = 1.00$, $g = 8.7$, $h = 1.0$, and $k = 0.12$. Using Equations (16.11) and (16.24), $\Delta = 8.7$ kK and $\beta = 0.88$. From Table 16.17, the value of *B* for the Ni^{2+} free ion is $1042\ \text{cm}^{-1}$, so B' is calculated as $920\ \text{cm}^{-1}$ or 0.92 kK. Using the ratio $\Delta/B' = 9.5$, a vertical line is drawn on the d^8 Tanabe–Sugano diagram. There are three spin-allowed transitions from the ^3A_g GS: $^3T_{2g} \leftarrow {}^3A_{2g}$ ($E/B' = 9.2$), $^3T_{1g}$ (F) $\leftarrow {}^3A_{2g}$ ($E/B' = 15$), and $^3T_{1g}$ (P) $\leftarrow {}^3A_{2g}$ ($E/B' = 28$). Because $B' = 0.92$ kK, the energies of these three transitions are given in Table 16.26. The theoretical and experimental values match very closely. Thus, the Tanabe–Sugano diagrams can not only predict which electronic transitions will be observed, but they can also be used to quantitatively predict the energies of each absorption band. This is quite an impressive feat!

TABLE 16.25 Jørgensen's *h* and *k* factors for the nephelauxetic series of ligands and metal ions.

Ligand	*h*	Metal Ion	*K*
F^-	0.8	Mn^{2+}	0.07
H_2O	1.0	V^{2+}	0.1
Urea	1.2	Ni^{2+}	0.12
NH_3	1.4	Mo^{3+}	0.15
En	1.5	Cr^{3+}	0.20
$C_2O_4^{2-}$	1.5	Fe^{3+}	0.24
Cl^-	2.0	Rh^{3+}	0.28
CN^-	2.1	Ir^{3+}	0.28
Br^-	2.3	Co^{3+}	0.33
N_3^-	2.4	Pt^{4+}	0.6
I^-	2.7	Pd^{4+}	0.7

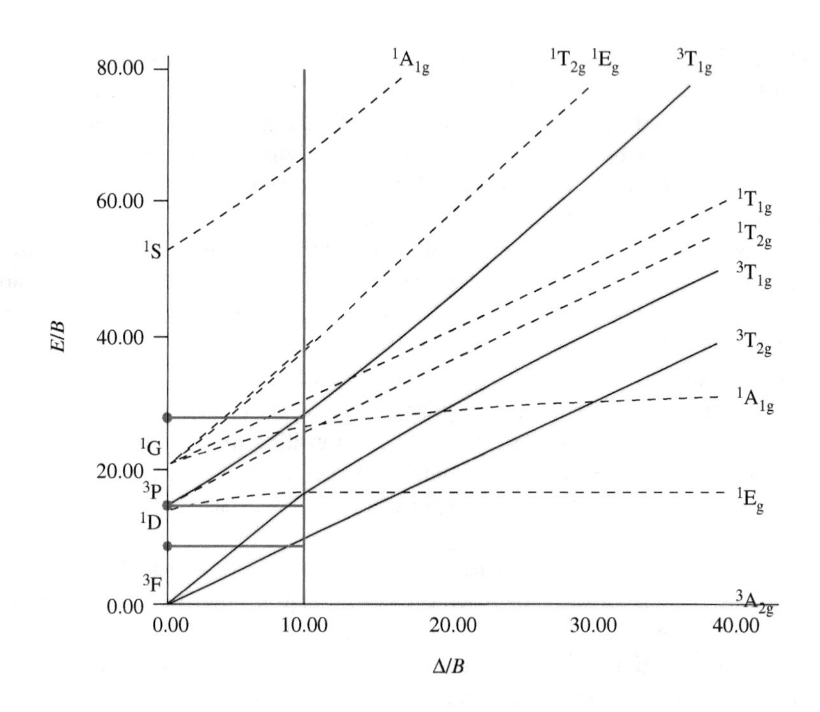

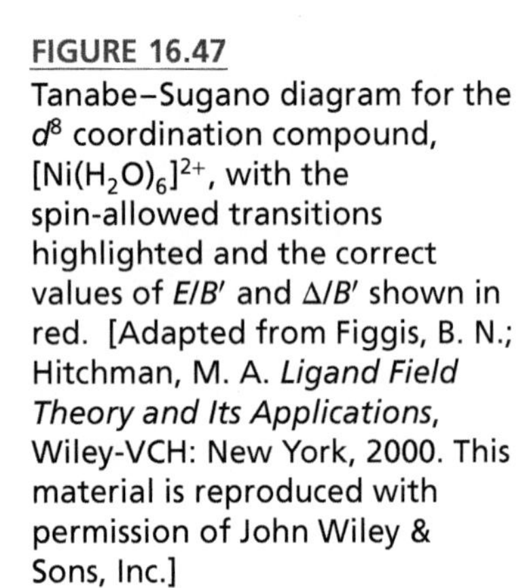

FIGURE 16.47
Tanabe–Sugano diagram for the d^8 coordination compound, $[Ni(H_2O)_6]^{2+}$, with the spin-allowed transitions highlighted and the correct values of E/B' and Δ/B' shown in red. [Adapted from Figgis, B. N.; Hitchman, M. A. *Ligand Field Theory and Its Applications*, Wiley-VCH: New York, 2000. This material is reproduced with permission of John Wiley & Sons, Inc.]

TABLE 16.26 Electronic transitions for $[Ni(H_2O)_6]^{2+}$ and their predicted and experimental energies.

Transition	E/B'	E_{calc}, kK	$E_{expt'l}$, kK
$^3T_{2g} \leftarrow {}^3A_{2g}$	9.2	8.5	8.5
$^3T_{1g}$ (F) $\leftarrow {}^3A_{2g}$	15	14	15.4
$^3T_{1g}$ (P) $\leftarrow {}^3A_{2g}$	28	26	26.0

Example 16-7. The electronic spectrum for $[V(H_2O)_6]^{3+}$ is shown here. Calculate the values of Δ_o, B', and β.

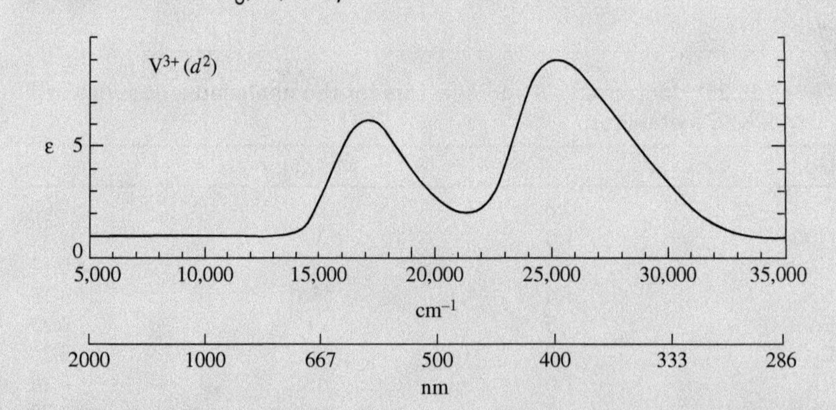

Solution. The two peaks occur at 17.2 and 25.6 kK. Vanadium(III) is a first-row metal in a moderately high oxidation state with a medium-field ligand. As an initial starting point, we will therefore use the Tanabe–Sugano diagram for a d^2

metal ion with our ruler placed somewhere near the middle of the diagram. At this energy, the two lowest energy spin-allowed electronic transitions are $^3T_{2g} \leftarrow {}^3T_{1g}$ and $^3T_{1g}$ (P) $\leftarrow {}^3T_{1g}$ (at values of $\Delta/B' < 13$, the $^3A_g \leftarrow {}^3T_{1g}$ band is lower in energy than the $^3T_{1g}$ (P) $\leftarrow {}^3T_{1g}$ band). The ratio of the two transition energies is (25.6)/(17.2) = 1.49. Sliding the ruler across the horizontal axis of the Tanabe–Sugano diagram, we find this ratio at $\Delta/B' = 25$. The corresponding values of E/B' are 24 and 35. Solving for B', we get 17.6/24 = 0.73 and 25.6/35 = 0.73. Thus, $B' = 730\ \text{cm}^{-1}$, $\Delta_o = 25(0.73) = 18$, and $\beta = (730/886) = 0.82$. We can also use these data to predict the values of g (18) and k (0.18) for V^{3+}.

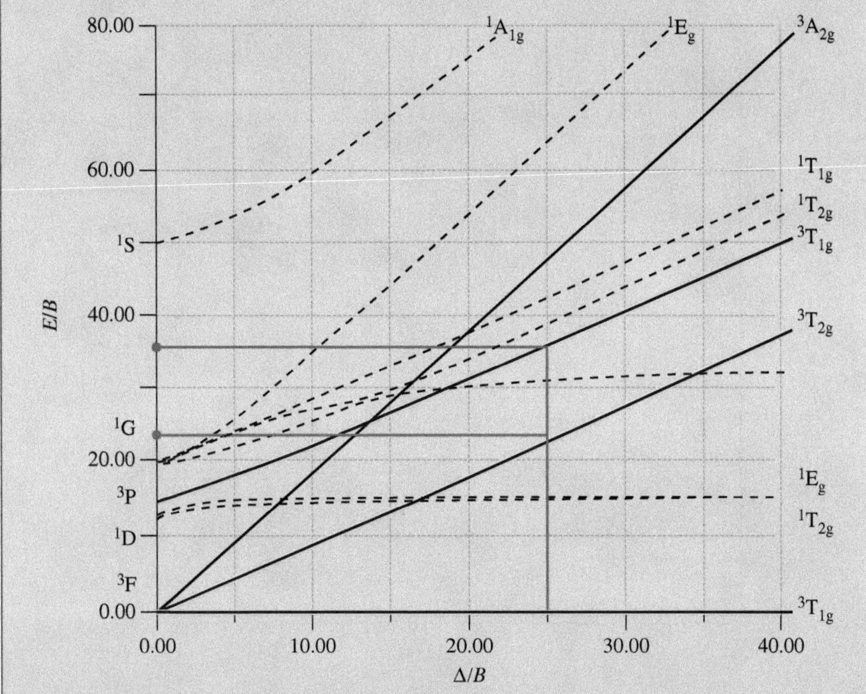

[Adapted from Figgis, B. N.; Hitchman, M. A. *Ligand Field Theory and Its Applications*, Wiley-VCH: New York, 2000. This material is reproduced with permission of John Wiley & Sons, Inc.]

Example 16-8. The $[CoCl_4]^{2-}$ ion exhibits an intense blue color, with a strong electronic transition at 15,000 cm^{-1} and a weaker one at 23,000 cm^{-1}. Two other strong bands occur in the near-IR and the IR at 5800 and 3200 cm^{-1}, respectively. Calculate Δ_t and β from these data.

Solution. Cobalt(II) is d^7. Because $[CoCl_4]^{2-}$ has tetrahedral symmetry, we will therefore use the $d^{10-7} = d^3$ Tanabe–Sugano diagram. There are three spin-allowed transitions, presumably corresponding to the three stronger absorptions, as follows:

$^4T_{1g}(P) \leftarrow {}^4A_{2g}$ 15.0 kK

$^4T_{1g}(F) \leftarrow {}^4A_{2g}$ 5.8 kK

$^4T_{2g} \leftarrow {}^4A_{2g}$ 3.2 kK

The ratios are: $E_3/E_1 = 4.7$ and $E_2/E_1 = 1.8$. Sliding our ruler across the horizontal axis of the d^3 Tanabe–Sugano diagram, we find these ratios occur at $\Delta/B' = 5.0$ with $E_3/B' = 21.4$, $E_2/B' = 8.2$, and $E_1/B' = 4.5$. The corresponding values of B' are 0.70, 0.71, and 0.71, so that the average value is 710 cm^{-1}. The value of B for Co^{2+} in Table 16.17 is 989 cm^{-1}, making $\beta = 0.72$. Thus, there is a fairly large amount of covalent character in the Co–Cl bonds. Using $\Delta/B' = 5.0$, the calculated value of Δ_t is 3.5 kK. Given the prediction that $\Delta_o = fg = (0.78)(9) = 7.0$ and $\Delta_t \cong (4/9)\Delta_o$, or 3.1 kK, the observed and theoretical values of Δ_t show reasonable agreement.

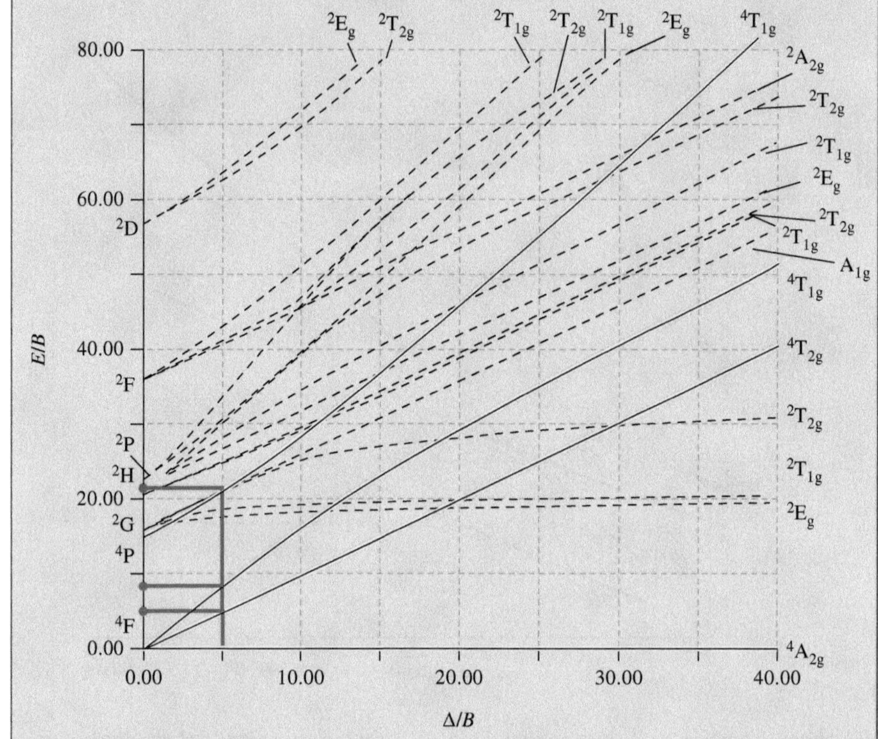

[Adapted from Figgis, B. N.; Hitchman, M. A. *Ligand Field Theory and Its Applications*, Wiley-VCH: New York, 2000. This material is reproduced with permission of John Wiley & Sons, Inc.]

In addition to LF electronic transitions, coordination compounds can also exhibit intraligand (IL), ligand-to-metal charge transfer (LMCT), metal-to-ligand charge transfer (MLCT), and metal-to-metal charge transfer (MMCT) transitions. IL transitions occur between MOs (usually of *pi* symmetry) within the ligand itself. Their energies are therefore similar to those of the corresponding transition in the uncoordinated ligand. LMCT transitions occur between a filled, low-lying MO on the ligand and an empty MO on the metal, as shown in Figure 16.48. The metal must be in a high oxidation state because the net effect of an LMCT absorption is to at least partially oxidize the ligand and partially reduce the metal. For example, LMCT bands occur from the halide ion to the Ti(IV) center in Cp_2TiX_2 organometallic compounds. The energy of an LMCT depends on the electrochemical potentials of the L and M. Thus, the more electronegative Cl^- ligand in Cp_2TiCl_2 has a higher energy LMCT band than that for the more electropositive I^- ion in Cp_2TiI_2.

MLCT transitions follow the opposite trend. MLCT transitions involve the transfer (or partial transfer) of an electron from a metal-centered orbital to a ligand-centered orbital (typically a π^* MO). Consequently, the metal must be oxidizable, and the energy of the MLCT transition decreases as the ligand's reduction

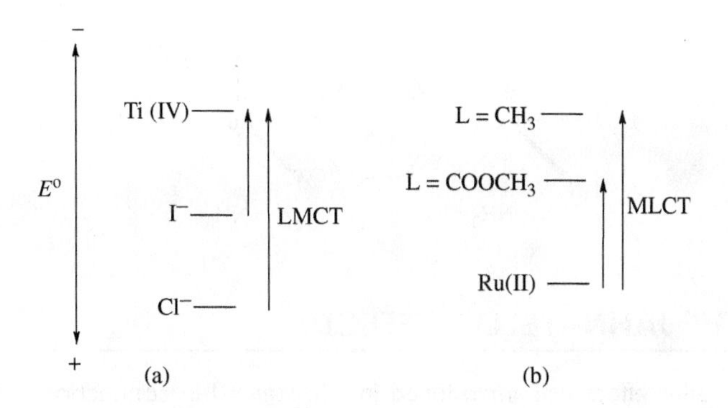

FIGURE 16.48
LMCT transitions for Cp_2TiX_2 organometallics (a) and MLCT transitions for $[Ru(NH_3)_5(py\text{-}L)]^{2+}$ compounds (b). The energies of the transitions are a function of the redox potentials of the metal and ligand.

TABLE 16.27 Hammett σ parameters and the λ_{max} of the MLCT bands for selected $[Ru(NH_3)_5(py\text{-}L)]^{2+}$ compounds.

L	σ Parameter	λ_{max} (nm) for MLCT
CH_3	−0.17	398
H	0.00	407
$CONH_2$	0.27	479
$COOCH_3$	0.39	497
CHO	0.50	545

potential becomes more negative. The λ_{max} values for the MLCT transitions in a series of $[Ru(NH_3)_5(py\text{-}L)]^{2+}$ compounds are listed in Table 16.27. The energies of the bands track with the redox potential of the ligand, as determined by its Hammett σ parameter. One must be careful with this type of analysis, however, because the redox potential of the metal also shifts with the electron-donating abilities of the ligand. As further evidence that these are MLCT in character, the transitions disappear altogether when the metal is oxidized from Ru(II) to Ru(III). A second example of an MLCT transition is the d[Ru(II)] $\rightarrow \pi^*$(bpy) MLCT band in $[Ru(bpy)_3]^{2+}$ and its related congeners. This compound has been widely studied because of its long-lived luminescence and the fact that its ES can undergo electron transfer reactions with other species in solution, leading to its potential use as a photocatalyst.

Lastly, MMCT bands occur between metal-centered orbitals on two different metals within the same complex, such as those in Prussian blue or in the Creutz–Taube ion, $[(NH_3)_5Ru\text{-}\mu(pz)\text{-}Ru(NH_3)_5]^{5+}$ shown in Figure 16.49. In the case of *mixed-valence compounds*, one of the metals must be in a low oxidation state so that it can act as an electron donor, while the other metal exists in a high oxidation state and acts as the electron acceptor. Within the realm of mixed-valence compounds, the degree of localization or delocalization of the charges on the metal ions in the GS can vary significantly with the nature of the metals and the bridging ligand. For example, the formal oxidation states of the two Ru ions in the Creutz-Taube ion are 2.5+, 2.5+ so that the GS electron is delocalized over both metal centers through *pi* backbonding to the short bridging ligand. If the pyrazine bridge in the Creutz–Taube ion is exchanged for 4,4′-dipyridyl, the greater distance between the metal centers localizes the electron on just one of the metals, so that the formal oxidation states of the two Ru ions are 2+ and 3+.

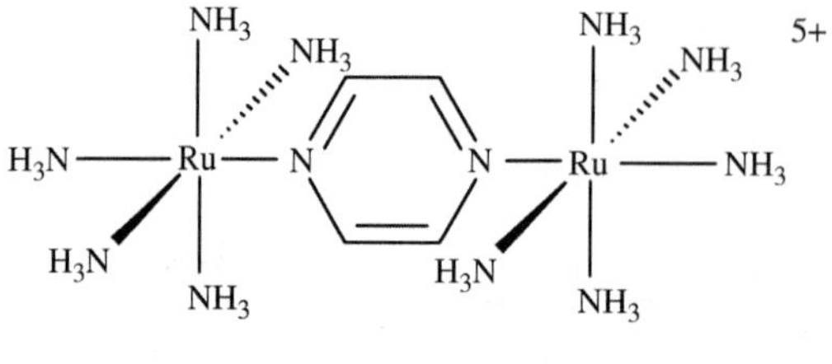

FIGURE 16.49
Structure of the Creutz–Taube ion, which exhibits an MMCT absorption.

16.8 THE JAHN–TELLER EFFECT

The Jahn–Teller effect was introduced in Chapter 10 in connection with the preferred geometries of small molecules using MOT. We revisit it here in the context of how it influences the electronic spectra of coordination compounds. The *Jahn–Teller theorem* states that whenever there are an odd number of electrons occupying an orbitally degenerate GS, the molecule will undergo a distortion along one of its vibrational coordinates in order to lower the symmetry, remove the degeneracy, and lower the energy. Any GS having an E or T term is therefore an orbitally degnerate GS and is expected to undergo the distortion. Using the GS term symbols in Table 16.22, Jahn–Teller distortions are expected for the following d^n configurations: d^1, d^2, d^4 (HS and LS), LS d^5, HS d^6, d^7 (HS and LS), and d^9. In other words, the Jahn–Teller effect applies to almost half of the octahedral GSs!

The effect is particularly pronounced, however, for those electron configurations where the odd electron occupies the e_g* (antibonding) MO. This is the case for HS d^4, HS d^6, LS d^7, and d^9. For these electron configurations, the distortion is exaggerated because the metal–ligand bonds are already weakened when the e_g* MO is populated. Jahn–Teller distortions in Cu(II) compounds (d^9) are particularly prevalent. While the Jahn–Teller theorem does not tell us *a priori* exactly what type of distortion will occur, tetragonal distortions are fairly common. If an "z-out" tetragonal distortion occurs, those orbitals having z-character in them will be stabilized by the distortion, as shown at the left in Figure 16.50. The magnitude of the distortion for each set of orbitals must remain symmetric with respect to a barycenter. For example, the d_{xz} and d_{yz} orbitals are stabilized by 1/3 δ_2, while the d_{xy} orbital is destabilized by twice that amount (2/3 δ_2). For an "z-in" tetragonal distortion, any orbital containing z-character will be destabilized by the distortion, as shown at the right in Figure 16.50. Because the d_{xy} orbital is stabilized by 2/3 δ_2 in this scenario versus only a 1/3 δ_2 stabilization of the d_{xz} and d_{yz} orbitals for the "z-out" configuration, d^1 coordination compounds will prefer to undergo the "z-in" tetragonal distortion because this lowers their overall energy to a greater extent. This is indeed the case observed for the $[TiCl_6]^{3-}$ ion.

The electronic spectrum of this d^1 coordination compound is expected to have a single LF transition in the absence of Jahn–Teller splitting. In point of fact, the spectrum of the $[TiCl_6]^{3-}$ ion consists of two closely spaced peaks separated by approximately 1400 cm^{-1}. As a result of the tetragonal distortion, the symmetry of the complex is lowered from O_h to D_{4h}. Using the correlation table in Appendix D, the $^2T_{2g}$ GS term in O_h symmetry splits into $^2B_{2g}$ (which becomes the new GS term in D_{4h}) and ^2E_g, as shown in Figure 16.51. Likewise, the ^2E_g term in O_h splits into $^2A_{1g}$ and $^2B_{1g}$. The two lowest energy spin-allowed transitions are $^2B_{1g} \leftarrow {}^2B_{2g}$ and $^2A_{1g} \leftarrow {}^2B_{2g}$. The separation of these peaks by 1400 cm^{-1} implies that the magnitude of δ_2 is also 1400 cm^{-1}.

Besides the splitting of terms observed in the electronic transitions of coordination compounds, other forms of experimental verification of the Jahn–Teller effect include the ESR spectra of coordination compounds, X-ray crystallographic data showing different M–L bond lengths, and thermodynamic data. As an example, consider the stepwise formation constants in Table 16.28 for $[Cu(NH_3)_6]^{2+}$, which

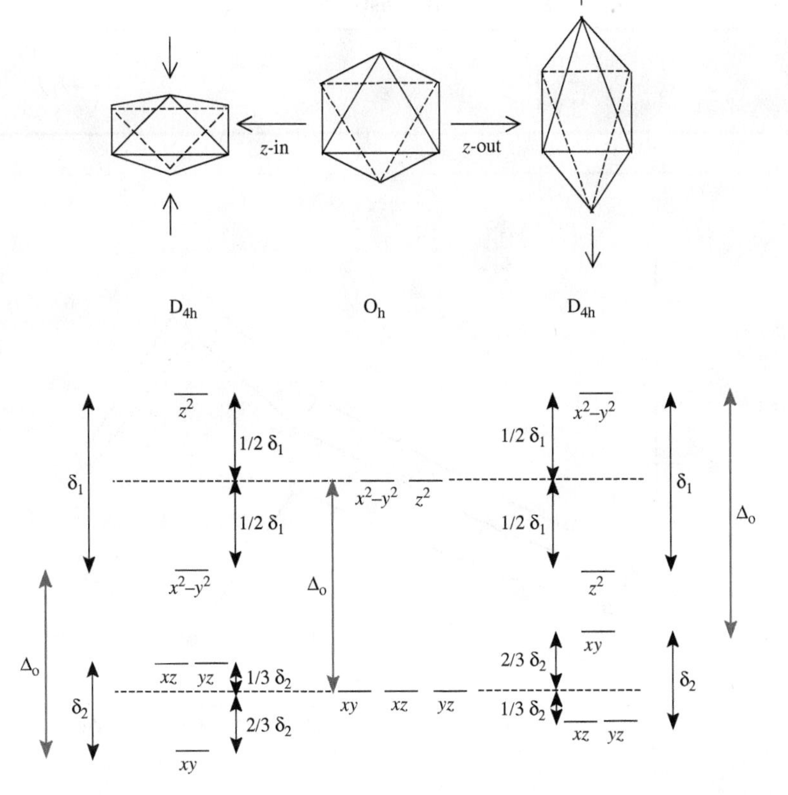

FIGURE 16.50
Tetragonal distortions from O_h symmetry and their effects on the energies of the *d*-orbitals. The "z-out" tetragonal distortion is shown at left and "z-in" tetragonal distortion is shown at right.

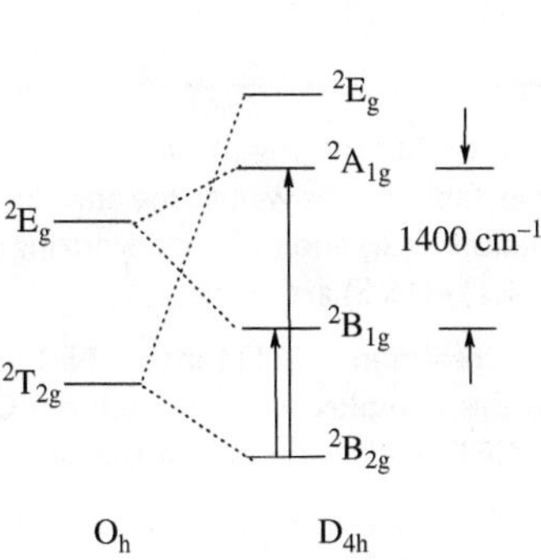

FIGURE 16.51
Splitting of the terms for a d^1 $[TiCl_6]^{3-}$ ion as it undergoes a tetragonal distortion from octahedral symmetry, illustrating the two observed spin-allowed transitions for the complex. Note: not drawn to scale.

has the d^9 electron configuration. As expected, the stability constants decrease with increasing n because there is less positive charge on the metal as it becomes more saturated with ligands. However, there is an abrupt drop in the stability of the Cu(II) compounds between $n=4$ and $n=5$. The reason that it is so difficult to add the fifth and sixth NH_3 ligands to Cu^{2+} is because the octahedral complex undergoes an "z-out" tetragonal distortion, which significantly weakens the two bonds along the z-axis.

The stepwise formation constants for the $[M(en)_3]^{2+}$ ions from the first row of transition metals are shown in Figure 16.52. In general, the stability constants increase across the row until they reach Cu^{2+} in accordance with the Irving–Williams series. The lone exception to this trend occurs for $[Cu(en)_3]^{2+}$, which is highly unstable. The strong instability of this coordination compound has to do with the strain imposed on the chelating ligand by the Jahn–Teller effect. The *mono* and *bis* complexes can relieve this strain by letting the H_2O ligands undergo the Jahn–Teller distortion, but the rigidity of the en ligand makes this much more challenging for the *tris* compound.

TABLE 16.28 Stepwise formation constants for $[Cu(NH_3)_n]^{2+}$.

K_1	2×10^4	K_4	2×10^2
K_2	4×10^3	K_5	3×10^{-1}
K_3	1×10^3	K_6	Very small

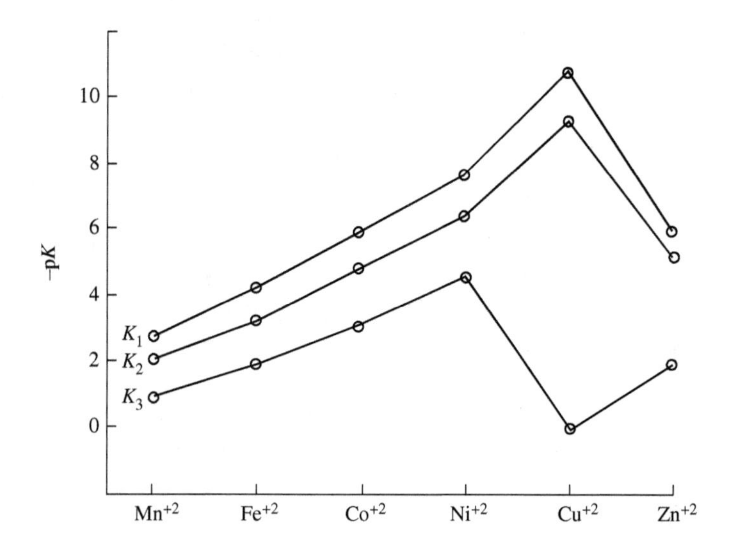

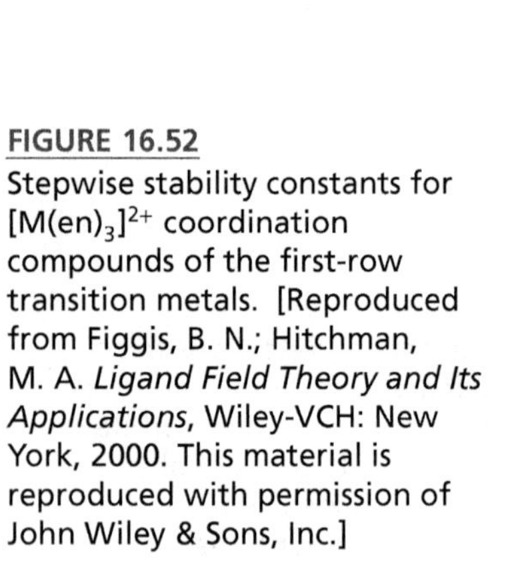

FIGURE 16.52
Stepwise stability constants for $[M(en)_3]^{2+}$ coordination compounds of the first-row transition metals. [Reproduced from Figgis, B. N.; Hitchman, M. A. *Ligand Field Theory and Its Applications*, Wiley-VCH: New York, 2000. This material is reproduced with permission of John Wiley & Sons, Inc.]

EXERCISES

16.1. Using the mathematical forms of the angular wavefunctions for the five *d*-orbitals and for the p_z orbital given in Table 4.1, as well as the angular wavefunctions for the p_x and p_y orbitals given in Equations (4.6) and (4.7) and ignoring the normalization constants, prove that Equations (16.1)–(16.5) are true.

16.2. Sketch the valence bond treatment (VBT) for the $[Ni(H_2O)_6]^{2+}$ ion. Can VBT predict what the magnetism of this complex ion is? Sketch the CF *d*-orbital splitting diagram for $[Ni(H_2O)_6]^{2+}$. Can CFT predict the magnetism of the complex?

16.3. Sketch the expected CF splitting diagram for the *d*-orbitals in ferrocene, Cp_2Fe, whose structure is shown here. Label the *d*-orbitals, label each energy level with its group theoretical symbol, fill in the electrons, and predict the magnetism.

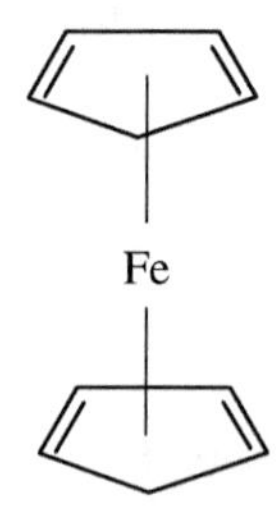

[Adapted from Figgis, B. N.; Hitchman, M. A. *Ligand Field Theory and Its Applications*, Wiley-VCH: New York, 2000. This material is reproduced with permission of John Wiley & Sons, Inc.]

16.4. Calculate the CFSE for each of the following coordination compounds in units of Dq:

a. $[Co(H_2O)_6]^{2+}$

b. $[Cu(NH_3)_4]^{2+}$

c. $[Cr(NH_3)_6]^{3+}$
d. $[CuCl_5]^{3-}$
e. $[Pt(CN)_4]^{2-}$

16.5. Using data from Table 16.4, calculate the CFSEs for every d^n configuration in both the tetrahedral (always HS) and square pyramidal geometries. Explain why LS d^8 coordination compounds will always prefer the square planar geometry.

16.6. For each of the following coordination compounds, sketch the CF splitting diagram, label the *d*-orbitals, label each energy level with its group theoretical symbol, fill in the electrons, and predict the magnetism.
a. $[CoCl_4]^{2-}$
b. $[Os(CN)_6]^{3-}$
c. $[FeF_6]^{3-}$
d. $[AuCl_4]^{-}$
e. $[Cu(NH_3)_6]^{2+}$

16.7. Use LFT and Equation (15.5) to predict the magnitude of the effective magnetic moment for each of the following coordination compounds:
a. $[Cr(en)_3]^{3+}$
b. $[Fe(H_2O)_6]^{2+}$
c. $[Mn(H_2O)_6]^{2+}$
d. $[Co(NH_3)_6]^{3+}$
e. $[Co(CN)_6]^{3+}$

16.8. Order the following coordination compounds from the most negative to the most positive standard reduction potential: $[Cr(NH_3)_5Cl]^{2+}$, $[Cr(NH_3)_5F]^{2+}$, $[Cr(NH_3)_5Br]^{2+}$, $[Cr(NH_3)_6]^{3+}$.

16.9. Sketch the MO diagram for the square planar $[Ni(CN)_4]^{2-}$ ion:
a. Determine the symmetries of the 3*d*, 4*s*, and 4*p* AOs on the Ni^{2+} ion.
b. Using appropriate basis sets, determine the symmetries of the σ-donor and π-acceptor SALCs.
c. Combine the AOs and the SALCs to make MOs. Label each MO with its Mulliken symbol.
d. Fill in the ligand valence electrons in red.
e. Fill in the metal valence electrons in blue.

16.10. Using the basis set shown in the diagram, sketch the MO diagram for $Ni(CO)_4$.
a. Determine the symmetries of the 3*d*, 4*s*, and 4*p* AOs on the Ni atom.
b. Determine the symmetries of the σ-donor and π-acceptor SALCs.
c. Combine the AOs and the SALCs to make MOs. Label each MO with its Mulliken symbol.
d. Fill in the ligand valence electrons in red.
e. Fill in the metal valence electrons in blue.

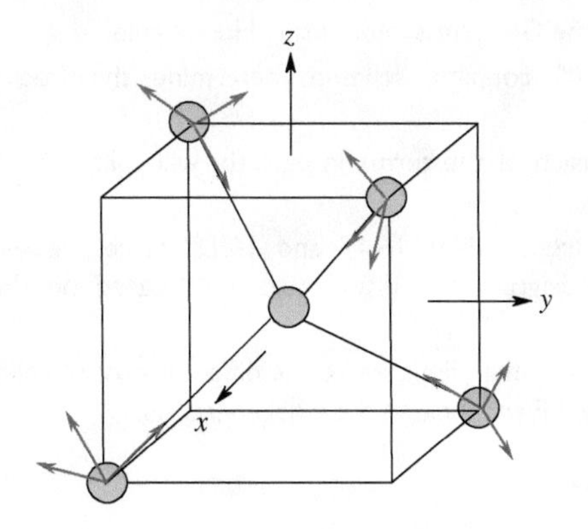

16.11. Using the Jørgensen *f* and *g* factors in Table 16.10, calculate the predicted energy of Δ_o in units of cm^{-1} for each of the following coordination compounds and then compare your answer with the literature values listed in Table 16.5:

a. $[CrCl_6]^{3-}$

b. $[V(H_2O)_6]^{2+}$

c. $[Ru(en)_3]^{2+}$

d. $[NiF_6]^{4-}$

e. $[Rh(CN)_6]^{3-}$

16.12. Use the AOM parameters in Table 16.14 to calculate Δ_o for each of the following coordination compounds:

a. $[Ni(en)_3]^{2+}$

b. $[Co(py)_6]^{2+}$

c. $[Cr(H_2O)_6]^{3+}$

16.13. Use the AOM to sketch the abbreviated MO diagram for the tetrahedral complex $[CoCl_4]^{2-}$. Show the stabilization and destabilization of the metal and ligand orbitals with their appropriate values of e_σ and e_π. Then use the AOM parameters in Table 16.14 to calculate Δ_t for this ion.

16.14. Explain why the Ag^+ ion is larger in a square planar geometry than it is in a tetrahedral geometry, but the reverse is true for Ni^{2+}.

16.15. For each pair of complexes, predict which will have the larger CF splitting parameter:

a. $[NiCl_4]^{2-}$ or $[NiCl_6]^{4-}$

b. $[NiS_4]^{6-}$ or $[Ni(H_2O)_4]^{2+}$

c. $[PtF_6]^{2-}$ or $[Pt(CN)_6]^{2-}$

d. $[CoCl_6]^{4-}$ or $[Co(PMe_3)_6]^{2+}$

e. $[Ru(NH_3)_6]^{2+}$ or $[Ru(NH_3)_6]^{3+}$

16.16. Explain why $[Co(H_2O)_6]^{3+}$ is a strong enough oxidizing agent to oxidize H_2O, but $[Co(NH_3)_6]^{3+}$ is stable in aqueous solution.

16.17. Calculate all the possible effective magnetic moments of Co(II) in tetrahedral, square planar, and octahedral complexes.

16.18. Most first-row transition metals in the 2+ oxidation state are octahedral. That being said, the preference for tetrahedral complexes being formed increases in the order: Ni < Fe < Co. Calculate the CFSEs for each complex in the tetrahedral and octahedral geometries. Do these values help explain the order? Does the AOM offer any help in addressing this problem? Explain.

16.19. An undergraduate taking IR spectra of $[Ni(CO)_4]$, $[Fe(CO)_4]^{2-}$, and $[Co(CO)_4]^-$ obtained three different spectra having $\nu(CO) = 1790$, 1890, and 2060 cm^{-1}, but she forgot to label which spectrum belonged to which compound. Use your knowledge about *pi* backbonding to help assign the compounds to the correct spectra. Explain your answer.

16.20. Consider the $[Cr(H_2O)_6]^{3+}$ ion. Using Figure 16.37 as a template for comparison:

a. Determine the GS term symbol using Hund's rule.

b. Using the LS coupling scheme, determine the free ion terms and their degeneracies.

c. Show how each of the terms in part (b) will split in an O_h LF and include the degeneracies.

d. Using data from Tables 16.17 and 16.18, place the weak-field terms on an energy-level diagram and draw it to scale based on the calculated values of each term.

e. Using the strong-field limit, write the different possible electron configurations in an octahedral LF and include their degeneracies.

f. Using the scenarios for molecular term symbols in the strong-field limit, determine how each of the terms in part (e) will split as a result of LS coupling and include their degeneracies.

g. Complete the correlation diagram for the Cr^{3+} ion by using the no-crossing rule to connect the terms.

16.21. Use Equation (16.12) to show how the G term will split in an octahedral LF. Show your work.

16.22. Consider the V^{3+} ion and show how the free ion terms will split in a trigonal bipyramidal ligand field in the weak-field limit.

16.23. Consider the complex $[MoCl_6]^{3-}$, which has two spin-allowed transitions at 418 and 522 nm. Using a Tanabe–Sugano diagram and $B = 610\ cm^{-1}$ for the free ion, calculate Δ_o, B', and β for the complex.

16.24. The orbital selection rule states that the triple direct product $\Gamma_{ES}\Gamma_{\mu}\Gamma_{GS} \subset A_{1g}$ for an octahedral compound. Because all *d*-orbitals are *gerade* and the dipole moment operator is *ungerade*, all LF transitions are therefore forbidden by this selection rule (the Laporte rule). The aforementioned selection rule is defeated by vibronic coupling. Vibronic coupling occurs when there is a vibrational mode that is antisymmetric to inversion, temporarily removing the inversion center in the ground state. This coupling of vibrational and electronic coordinates allows the transition to be orbitally allowed. In other words, the transition will be allowed if the quintuple direct product $\Gamma_{ES}\Gamma_{v'}\Gamma_{\mu}\Gamma_{v}\Gamma_{GS} \subset A_{1g}$. The ground vibrational level (Γ_v) is always A_{1g}. The electronic ground (Γ_{GS}) and excited states (Γ_{ES}) can be determined from the appropriate Tanabe–Sugano diagram. The dipole moment operator (Γ_{μ}) spans the same irreducible representation as *x*, *y*, and *z*. Using mini-Cartesian axis vectors on each atom in $[MoCl_6]^{3-}$ to represent the 3N degrees of freedom, generate the reducible representation for 3N, reduce it, determine the symmetries of the 3N-6 vibrational modes, and then determine which of these vibrational modes ($\Gamma_{v'}$) will yield a quintuple direct product that contains the A_{1g} IRR. These are the vibrational modes that can couple with the electronic transition to make it allowed.

16.25. Predict the values of Δ_o, B′, and β from Jørgensen's *f*, *g*, *h* and *k* factors for $[MoF_6]^{3-}$. Then, use these values to predict the energies (in kK) of the two spin-allowed transitions in this complex ion.

16.26. Use the appropriate Tanabe–Sugano diagram to explain why $[Mn(H_2O)_6]^{2+}$ is essentially colorless in aqueous solution. What is the term symbol for the ground state of this complex cation?

16.27. The compound $[M(H_2O)_6]^{2+}$ has a single electronic transition in its d^4 UV/VIS spectrum.

a. Is this an HS or an LS complex? Explain.

b. Given that the energy of this electronic transition is 709 nm, calculate the Jørgensen *g* factor for M.

16.28. Use Jørgensen's *f*, *g*, *h*, and *k* factors, along with the Racah parameter from Table 16.17 and the appropriate Tanabe–Sugano diagram to predict the energy of the first two spin-allowed absorption bands in the spectrum of $[Cr(H_2O)_6]^{3+}$.

16.29. The complex ion $[Ni(NH_3)_6]^{2+}$ has three spin-allowed electronic transitions at 355, 571, and 930 nm. Assign the transitions and determine Δ_o and B' for this compound.

16.30. Which complex has more covalency in its metal–ligand bonding, $[Pt(CN)_6]^{2-}$ or $[Cr(en)_3]^{3+}$? Explain.

16.31. $[Ti(SCN)_6]^{3-}$ has an asymmetric, slightly split band at 18.4 kK. Calculate Δ_o for this compound and predict the Jørgensen *g* factor for the Ti^{3+} ion. Also explain why the band is slightly split.

16.32. Given the series of compounds with the general formula $[Cr(NH_3)_5X]^{2+}$, where $X = Cl^-$, Br^-, and I^-, predict the relative energies of the LMCT bands in these three complexes.

16.33. When Co(II) is dissolved in aqueous solution, it makes the extremely pale pink-colored complex $[Co(H_2O)_6]^{2+}$. When concentrated HCl is added to the same ion, it forms the intensely blue-colored complex $[CoCl_4]^{2-}$. Explain the relative colors and intensities of the two coordination compounds.

16.34. When $Ni(NO_3)_2$ is dissolved in water, it forms a green-colored solution. Addition of concentrated NH_4OH changes the color of the solution to blue while addition of ethylenediamine changes it to violet. Predict formulas for each of the coordination compounds and explain the trends in the colors of the compounds.

16.35. The permanganate ion, MnO_4^-, has a deep violet color. Do you think the color of this compound is due to an LF, LMCT, MLCT, or MMCT transition? Explain your reasoning.

16.36. Will the energy of the MLCT band in $[Ru(bpy)_3]^{2+}$ increase or decrease when the bpy ligands are substituted with nitro groups? Explain your answer.

16.37. Explain why the ferrocyanide ion is colorless in aqueous solution, while the ferricyanide ion is yellow. What type of electronic transition causes the ferricyanide ion to be colored?

16.38. Of the first-row transition metal complexes having the formula $[M(NH_3)_6]^{2+}$, which metals are predicted to have a Jahn–Teller distortion? Which of these is expected to have the largest distortion and why?

16.39. Which is expected to have the higher energy MLCT band, $[V(CO)_6]^-$ or $[Cr(CO)_6]$? Explain your answer.

16.40. Which d^n electron configurations are expected to undergo a Jahn–Teller distortion in a tetrahedral coordination compound? Explain your answer.

BIBLIOGRAPHY

1. Atkins P, Overton T, Rourke J, Weller M, Armstrong F, Hagerman M. *Shriver & Atkins' Inorganic Chemistry*. 5th ed. New York: W. H. Freeman and Company; 2010.
2. Carter RL. *Molecular Symmetry and Group Theory*. John Wiley & Sons, Inc.: New York; 1998.
3. Clare BW, Kepert DL. Coordination Numbers and Geometries. In: King RB, editor. *Encyclopedia of Inorganic Chemistry*. John Wiley & Sons, Inc.: New York; 1994.
4. Cotton FA. *Chemical Applications of Group Theory*. 2nd ed. John Wiley & Sons, Inc.: New York; 1971.
5. Douglas B, McDaniel D, Alexander J. *Concepts and Models of Inorganic Chemistry*. 3rd ed. New York: John Wiley & Sons, Inc.; 1994.
6. Figgis BN, Hitchman MA. *Ligand Field Theory and Its Applications*. New York: Wiley-VCH; 2000.
7. Harris DC, Bertolucci MD. *Symmetry and Spectroscopy*. New York: Dover Publications, Inc.; 1978.
8. Housecroft CE, Sharpe AG. *Inorganic Chemistry*. 3rd ed. Essex, England: Pearson Education Limited; 2008.
9. Huheey JE, Keiter EA, Keiter RL. *Inorganic Chemistry: Principles of Structure and Reactivity*. 4th ed. New York: Harper Collins College Publishers; 1993.
10. Kauffman GB. Coordination Chemistry: History. In: King RB, editor. *Encyclopedia of Inorganic Chemistry*. John Wiley & Sons, Inc.: New York; 1994.
11. Lawrence GA. *Introduction to Coordination Chemistry*. Chichester, West Sussex, England: John Wiley & Sons Ltd; 2010.
12. Miessler GL, Tarr DA. *Inorganic Chemistry*. 4th ed. Upper Saddle River, NJ: Pearson Education Inc.; 2011.

13. Rodgers GE. *Descriptive Inorganic, Coordination, and Solid-State Chemistry*. 3rd ed. Belmont, CA: Brooks/Cole Cengage Learning; 2012.
14. Smith DW. Ligand Field Theory and Spectra. In: King RB, editor. *Encyclopedia of Inorganic Chemistry*. John Wiley & Sons, Inc.: New York; 1994.

Reactions of Coordination Compounds 17

"Almost every student of organic chemistry knows that most substitution reactions do not occur in a random manner. In a similar manner, substitution reactions among coordination compounds are not random."

—*George B. Kauffman*

17.1 KINETICS OVERVIEW

The opening artwork, entitled *Trail Riders*, by Thomas Hart Benton, depicts some early explorers on horseback criss-crossing over the rugged terrain of the American frontier. From the first time that I saw this painting in the National Gallery of Art, I had the distinct impression that it reminded me of a reaction coordinate diagram—with the trail of riders traversing in the low valleys and saddle points between the looming mountain peaks. Each peak reminded me of an activation barrier and each valley an intermediate as the stream of molecules, similar to the riders, sought out the lowest energy pathway between the surrounding hills.

Reaction coordinate diagrams, so often written as a single pathway between reactants and products on a two-dimensional canvas, are in fact much more complex and multidimensional. In order to truly understand how a chemical reaction occurs, all of the available kinetic and thermodynamic data should be known: (i) What is the driving force for the reaction? (ii) What are the elementary steps that lead from reactant to product? (iii) What factors govern the heights of the activation barriers? (iv) What are the structures of the intermediates? (v) How and where are the bonds broken or made? (vi) What is the stereochemistry of the reaction? In order to answer these questions, a variety of experimental techniques must be used. The dependence of the rate and product distribution on pH, temperature, pressure, and solvent must also be examined.

The kinetics of chemical reactions can provide many clues as to the nature of the bond-breaking and bond-making processes on the molecular level that lie at the core of any reaction mechanism. The reaction rate can be defined as the differential rate of loss of a reactant or the differential rate of formation of a product as a function of time. The rates of chemical reactions depend on a variety of factors, including the concentrations of the reactants, ionic strength, temperature, surface area, and

Trail Riders painting. [Thomas Hart Benton, *Trail Riders*, National Gallery of Art, Washington, DC.]

Principles of Inorganic Chemistry, First Edition. Brian W. Pfennig.

presence of a catalyst. The explicit dependence of the rate on the concentrations of the reactants is known as the *differential rate law*. For the generic reaction given in Equation (17.1), the differential rate law is given by Equation (17.2), where the sum of the exponents, $a + b$ in this case, is defined as the overall order of the reaction.

While it is often possible to postulate more than one mechanism for a reaction, any proposed mechanism must be consistent with the experimentally observed rate law. The rate constant (k) is a proportionality constant that relates the concentrations of the reactants to the rate of reaction. It shows an exponential dependence on the temperature, according to the *Arrhenius equation* given by Equation (17.3), and it includes all of the other factors influencing the rate in its pre-exponential (A) and activation barrier (E_a) terms. For example, the presence of a catalyst lowers the activation barrier by providing an alternative pathway for the reaction to occur, while the proper stereochemistry or orientation necessary for the combination of reactants leading to products is included in the Arrhenius pre-exponential constant.

$$xA + yB \rightarrow zC \tag{17.1}$$

$$-\frac{1}{x}\frac{d[A]}{dt} = -\frac{1}{y}\frac{d[B]}{dt} = +\frac{1}{z}\frac{d[C]}{dt} = k[A]^a[B]^b \tag{17.2}$$

$$k = Ae^{-\frac{E_a}{RT}} \tag{17.3}$$

The experimental rate law can be determined by monitoring the concentration of one of the reactants or products as a function of time using spectroscopic means. For instance, the Beer–Lambert law states that the absorbance of a colored compound is directly proportional to its concentration (for optically dilute solutions anyway), so that the absorbance can be measured as the course of the reaction proceeds. The data are then fit to a model, such as the function that results when integrating one of the differential rate law equations. The *integrated rate laws* for some commonly occurring kinetics are listed in Table 17.1. Half-life equations are also included for some of the reactions in this table, where the *half-life* ($\tau_{1/2}$) is defined as the length of time that it takes for half of the initial reactant concentration to disappear.

The *reaction coordinate diagram* is a two-dimensional slice of the multidimensional terrain of the chemical reaction that focuses on a single reaction coordinate. The progress of the reaction (the making and/or breaking of chemical bonds) often proceeds primarily along one of the vibrational coordinates of the molecule and is plotted on the x-axis, while the energy (or, more accurately, the Gibbs free energy) is plotted on the y-axis. If the elementary steps of the mechanism are known, the rate law for each step in the mechanism can be determined from the stoichiometry of the elementary step.

The overall rate of a chemical reaction can only occur as fast as its slowest step, or its *rate-determining step* (RDS), along this coordinate. The RDS is the component of the reaction mechanism that has the highest activation barrier in the reaction coordinate diagram. If this happens to be the first step in the mechanism, as shown in Figure 17.1, then determination of the theoretical rate law is trivial because the rate law for the first step will necessarily depend on only the reactant concentrations. However, if the RDS is a later step in the mechanism, the rate law might depend on the concentration of the intermediate, which is something that is not easily measured because it exists in only a very small amount and for a finite period of time. Thus, in order to get the rate law in terms of only reactant concentrations, we assume that the intermediate is in a steady state; in other words, it is present only in small amounts and the rate at which it is formed in one step and the rate at which it is consumed in another step are equal. This assumption is known as the

TABLE 17.1 Differential and integrated rate laws and half-lives for some commonly occurring chemical kinetics.

Reaction Type	Differential Rate Law	Integrated Rate Law	Half-Life ($\tau_{1/2}$)
$A \rightarrow P$ zero-order	$-\frac{d[A]}{dt} = k$	$[A] - [A]_0 = -kt$	$\tau_{1/2} = \frac{1}{2k}$
$A \rightarrow P$ 1st-order (irrev.)	$-\frac{d[A]}{dt} = k[A]$	$\ln[A] - \ln[A]_0 = -kt$	$\tau_{1/2} = \frac{\ln 2}{k}$
$A \Longleftrightarrow P$ 1st-order (equil.)	$-\frac{d[A]}{dt} = k_1[A] - k_{-1}[P]$	$\ln([A] - [A]_{eq}) - \ln([A]_0 - [A]_{eq}) = (-k_1 + k_{-1})t$	$\tau_{1/2} = \frac{\ln 2}{k_1 + k_{-1}}$
$2A \rightarrow P$ 2nd-order (irrev.)	$-\frac{d[A]}{dt} = k[A]^2$	$\frac{1}{[A]} - \frac{1}{[A]_0} = kt$	$\tau_{1/2} = \frac{1}{k[A]_0}$
$A + B \rightarrow P$ 2nd-order (complex)	$-\frac{d[A]}{dt} = k[A][B]$	$\ln\frac{[B]_0 - [C]}{[A]_0 - [C]} = kt([B]_0 - [A]_0)$ if $[B]_0 > [A]_0$	$\tau_{1/2} = \frac{1}{k([B]_0 - [A]_0)} \ln\left(\frac{2[B]_0 - [A]_0}{[B]_0}\right)$
$A + B \rightarrow P$ pseudo-1st-order ($[B] \gg [A]$)	$-\frac{d[A]}{dt} = k_{obs}[A]$, where $k_{obs} = k[B]$	$\ln[A] - \ln[A]_0 = -k_{obs}t$	$\tau_{1/2} = \frac{\ln 2}{k_{obs}}$
$A + B \Longleftrightarrow P$ 2nd-order (equil.)	$-\frac{d[A]}{dt} = k_1[A][B] - k_{-1}[P]$	$\ln\left(\frac{[A]_0[B]_0 - [P]_{eq}[P]}{[P]_{eq}([P]_{eq} - [P])}\right) + \ln\left(\frac{[P]^2_{eq}}{[A]_0[B]_0}\right) = \left(\frac{k([A]_0[B]_0 - [P]^2_{eq})}{[P]_{eq}}\right)t$	

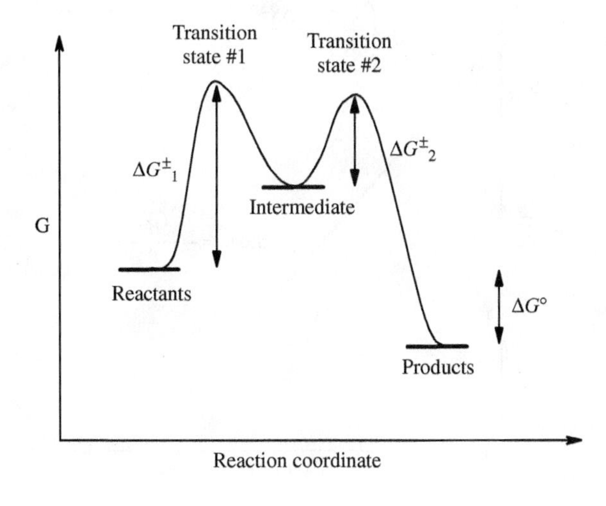

FIGURE 17.1
A typical reaction coordinate diagram illustrating the difference between an intermediate, which can often be isolated, and a transition state, which is usually too short-lived for direct observation.

steady-state approximation, and it can be used in order to obtain the overall rate law in terms of only the reactant concentrations. Of course, the results of the steady-state approximation are not always rigorously true, but most of the time it can be used to simplify the overall kinetic expression for the rate law.

The rate constants for a chemical reaction lead to the gross path of the reaction, but it is the activation parameters that truly provide clues about the mechanism. According to the standard *transition state theory*, at the top of each hill in the reaction coordinate diagram lies a transition state (TS), or an activated complex. The TS is extremely short-lived. Unlike an intermediate, it cannot be isolated. TS theory assumes that the activated complex is always in rapid equilibrium with the reactants, as shown in Equation (17.4), where $K^{\pm}$ is the equilibrium constant for the formation of the TS. It also states that the TS will lead to products only if the thermal energy of its vibrational frequency ($\nu = k_B T/h$) is equal to k_2. Thus, the rate of formation of the product is expressed as Equation (17.5), where $k_B = 1.38 \times 10^{-23}$ J/K and $h = 6.63 \times 10^{-34}$ J s. Given the relationship between the Gibbs free energy of activation ($\Delta G^{\circ\pm}$) and $K^{\pm}$ in Equation (17.6), the *Eyring equation*, given by Equation (17.7), relates the rate constant to the Gibbs free energy of activation.

$$A + B \overset{K^{\pm}}{\Longleftrightarrow} [A-B]^{\pm} \xrightarrow{k_2} P \tag{17.4}$$

$$\frac{d[P]}{dt} = k_2[A-B]^{\pm} = k_2 K^{\pm}[A][B] = \frac{k_B T}{h} K^{\pm}[A][B] \tag{17.5}$$

$$\Delta G^{\circ\pm} = -RT \ln K^{\pm} = \Delta H^{\circ\pm} - T\Delta S^{\circ\pm} \tag{17.6}$$

$$k_{\text{expt}} = \frac{k_B T}{h} e^{-\frac{\Delta G^{\circ\pm}}{RT}} \tag{17.7}$$

The standard state notation is generally dropped. Following rearrangement, substitution, and taking the natural logarithm of both sides, Equation (17.7) becomes Equation (17.8). By plotting the $\ln(k/T)$ versus $(1/T)$, as shown in Figure 17.2, the *enthalpy of activation* ($\Delta H^{\pm}$) can be determined from the slope of the graph and the *entropy of activation* ($\Delta S^{\pm}$) can be extracted from the *y*-intercept.

$$\ln\left(\frac{k}{T}\right) = \ln\left(\frac{k_B}{h}\right) - \frac{\Delta H^{\pm}}{RT} + \frac{\Delta S^{\pm}}{R} \tag{17.8}$$

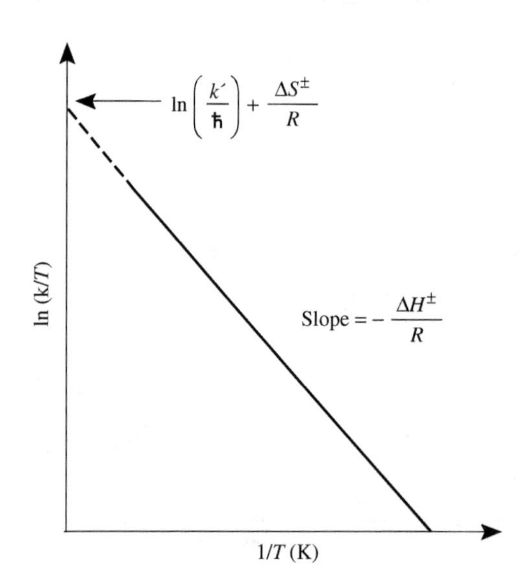

FIGURE 17.2
Eyring plot of $\ln(k/T)$ versus $(1/T)$.

The $\Delta H^{\pm}$ term is generally not very informative because the enthalpies of activation (which deal with the energy of bond breaking) for associative and dissociative mechanisms are similar. However, the $\Delta S^{\pm}$ term is very useful. When $\Delta S^{\pm}$ is large and negative, the implication is that there are less particles present in the TS than in the reactants. This would indicate an associative TS, where the transition metal increases its coordination number in the activated complex. On the other hand, if the $\Delta S^{\pm}$ term is large and positive, the TS will have more particles than the reactants, implying a dissociative mechanism.

The *volume of activation* ($\Delta V^{\pm}$) can also be measured by changing the overall pressure of the reaction vessel. This can be accomplished using a technique known as *pressure-tuning spectroscopy* (PTS) or by other methods. The relationship between the rate constant as a function of pressure and $\Delta V^{\pm}$ is shown in Equation (17.9), which is simply a modification of the van't Hoff equation. If $\Delta V^{\pm}$ is independent of the pressure, Equation (17.9) can be integrated at constant T to yield Equation (17.10). The volume of activation can then be determined from the slope of the line when $\ln(k)$ is plotted versus the pressure.

$$\left(\frac{\delta \ln k_1}{\delta P}\right)_T = -\frac{\Delta V^{\pm}}{RT} \tag{17.9}$$

$$\ln(k_1) = \ln(k_1)_0 - \frac{\Delta V^{\pm} P}{RT} \tag{17.10}$$

A positive volume of activation implies that the TS is occupying more volume than the reactants, indicating a dissociative mechanism. On the other hand, a negative volume of activation implies that the TS is occupying less volume than the reactants, indicating an associative mechanism. The determination of the activation parameters is therefore an invaluable tool in the study of inorganic reaction mechanisms.

17.2 OCTAHEDRAL SUBSTITUTION REACTIONS

One of the most common types of reactions of coordination compounds is a simple substitution reaction in which one of the ligands is replaced by another, as shown by the example given in Equation (17.11):

$$[Co(NH_3)_5X]^{2+} + Y^- \rightarrow [Co(NH_3)_5Y]^{2+} + X^- \tag{17.11}$$

Langford and Gray have classified octahedral substitution reactions into three general categories: *associative* (A), where the metal increases its coordination number in the intermediate; *dissociative* (D), where the metal decreases its coordination number in the intermediate; or *interchange* (I), where there is no discernible intermediate and in which the incoming ligand approaches the metal simultaneously with the loss of the original ligand. The basic reaction coordinate diagrams for the three different reactions are shown in Figure 17.3.

The associative mechanism is the inorganic equivalent to an S_N2 reaction in organic chemistry, although the associative intermediate can be isolated in this case. On the other end of the spectrum, the dissociative mechanism is analogous to an S_N1 reaction in organic chemistry. There is a large gray area in between the two extremes, with some interchange mechanisms where the incoming ligand has a greater influence than the leaving group (I_a) and others where the opposite is true (I_d). We will now consider the kinetics of each in its turn.

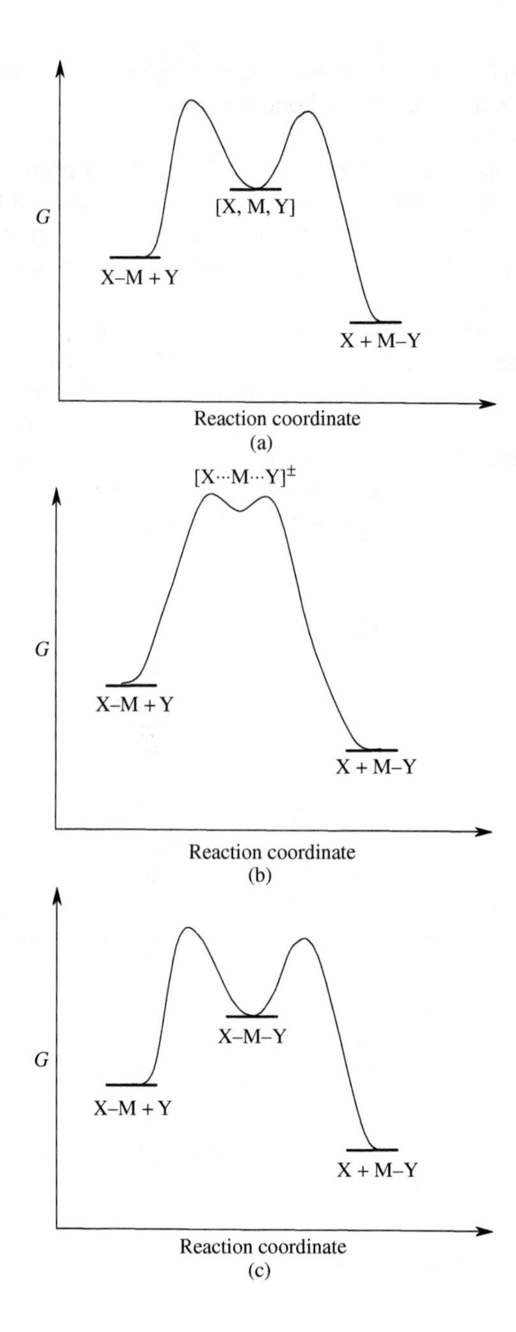

FIGURE 17.3
(a–c) Reaction coordinate diagrams for the three different types of octahedral substitution reactions, where M is the metal ion, X is the leaving group, and Y is the entering ligand. The other ligands are omitted for clarity. From top to bottom, the mechanisms are associative, interchange, and dissociative.

17.2.1 Associative (A) Mechanism

$$ML_5X + Y \Longleftrightarrow ML_5X \cdot Y \qquad K_1 = k_1/k_{-1}(\text{RDS}) \qquad (17.12)$$

$$ML_5X \cdot Y \rightarrow ML_5Y + X \qquad k_2 \qquad (17.13)$$

In the associative mechanism, which is given by Equations (17.12) and (17.13), the influence of the entering ligand on the rate constant is immediately evident in the equation for the rate law. Assuming that the RDS in the mechanism involves an equilibrium with association of Y to form a seven-coordinate intermediate and that the ensuing dissociation of X to form product is rapid, the rate law for an associative mechanism can be determined as follows:

$$\frac{d[ML_5Y]}{dt} = k_1[ML_5X][Y] \quad (17.14)$$

As a rule, the rates of associative reaction mechanisms: (i) increase with the concentration of Y, (ii) increase with the nucleophilicity of Y, (iii) decrease with the sterics of the overall coordination environment, (iv) show a negative $\Delta S^{\pm}$ (as there are less species in the intermediate than in the reactants), and (v) exhibit a negative $\Delta V^{\pm}$. Several types of octahedral substitution reactions seem to favor an associative (A) or, at the very least an associative interchange (I_a) mechanism. For example, the rate constants for ligand substitution in $[Cr(H_2O)_6]^{3+}$ vary by a factor of 2000 depending on the nature of the incoming ligand. The rate constants for substitution of the aqua ligand in $[Ru(edta)H_2O]^-$ also depend on the nature of the ligand and they all exhibit a negative entropy of activation.

17.2.2 Interchange (I) Mechanism

$$ML_5X + Y \Leftrightarrow ML_5X \cdot Y \rightarrow K_1 = k_1/k_{-1} \quad (17.15)$$

$$ML_5X \cdot Y \rightarrow ML_5Y + X \qquad k_2(\text{RDS}) \quad (17.16)$$

The generic interchange mechanism is given by Equations (17.15) and (17.16), where the second step is rate-determining. In the I mechanism, a rapid equilibrium forms between the reactant coordination complex and the incoming ligand, where the two are loosely paired together either as a tight ion pair or through covalent interactions. The distinction between I_a and I_d has to do with the degree of bond formation in this first step of the mechanism. The rate law for the I mechanism is given by Equation (17.17) and it involves the $ML_5X{\cdot}Y$ complex as an intermediate. We shall therefore employ the steady-state approximation for $ML_5X{\cdot}Y$, as shown in Equation (17.18).

$$\frac{d[ML_5Y]}{dt} = k_2[ML_5X \cdot Y] \quad (17.17)$$

$$\frac{d[ML_5X \cdot Y]}{dt} = k_1[ML_5X][Y] - k_{-1}[ML_5X \cdot Y] - k_2[ML_5X \cdot Y] = 0 \quad (17.18)$$

If $[Y] \gg [ML_5X]$, which we can adjust by the experimental conditions, then the following equations will be true, where the naught subscript indicates the total initial reactant concentrations of each species:

$$[M]_0 = [ML_5X] + [ML_5X \cdot Y] \quad (17.19)$$

$$[Y]_0 \cong [Y] \quad (17.20)$$

In this case, the steady-state approximation given by Equation (17.18) reduces to Equation (17.21) and the overall rate law becomes Equation (17.22):

$$k_1([M]_0 - [ML_5X \cdot Y])[Y]_0 - k_{-1}[ML_5X \cdot Y] - k_2[ML_5X \cdot Y] = 0 \quad (17.21)$$

$$\frac{d[ML_5Y]}{dt} = k_2[ML_5X \cdot Y] = \frac{k_2K_1[M]_0[Y]_0}{1 + K_1[Y]_0 + (k_2/k_{-1})} \cong \frac{k_2K_1[M]_0[Y]_0}{1 + K_1[Y]_0} \quad (17.22)$$

17.2.3 Dissociative (D) Mechanism

$$ML_5X \Longleftrightarrow ML_5 + X \qquad K_1 = k_1/k_{-1} \qquad (17.23)$$

$$ML_5 + Y \rightarrow ML_5Y \qquad k_2(\text{RDS}) \qquad (17.24)$$

In a dissociative mechanism, the metal ion dissociates X in the first step to form a five-coordinate intermediate and then adds Y in the slow step to form the substituted product. The rate law for the D mechanism is given by Equation (17.25) and depends on the concentration of the intermediate $[ML_5]$. Applying the steady-state approximation to $[ML_5]$, as shown in Equation (17.26), we obtain the overall rate law in terms of only the reactant concentrations given by Equation (17.27).

$$\frac{d[ML_5Y]}{dt} = k_2[ML_5] \qquad (17.25)$$

$$\frac{d[ML_5]}{dt} = k_1[ML_5X] - k_{-1}[ML_5][X] - k_2[ML_5][Y] = 0 \qquad (17.26)$$

$$\frac{d[ML_5Y]}{dt} = k_2[ML_5X \cdot Y] = \frac{k_2k_1[ML_5X][Y]}{k_{-1}[Y]_0 + k_2[Y]} \qquad (17.27)$$

The majority of octahedral substitution reactions are believed to occur by a D or I_d type mechanism, although it can be difficult to distinguish between these two mechanisms based on the similarity of their rate laws in the absence of the observation of any intermediates. In general, the rates of dissociative reaction mechanisms are largely independent of the nature of Y, increase with steric bulk around the metal ion of the overall coordination, show a positive $\Delta S^{\pm}$ (as there are more species in the intermediate than in the reactants), and exhibit a positive sign for $\Delta V^{\pm}$.

The classic studies of octahedral substitution reactions involved Co(III) ammine compounds, such as the example already shown previously in Equation (17.11). For this particular class of compounds, the experimental data yielded the following results:

1. The rate of substitution varies only a little with the nature of the incoming ligand Y^-, as shown by the data in Table 17.2. This would support a dissociative or I_d mechanism.
2. The rate of substitution varies over five orders of magnitude with the nature of the leaving group X, as shown in Table 17.3. The weaker the Co–X bond, the faster the reaction. These results support a dissociative intermediate, as the loss of X^- occurs in the RDS of the D mechanism.

TABLE 17.2 Rate constants for the reaction of $[Co(NH_3)_5(H_2O)]^{3+}$ with Y^- in aqueous solution at 45 °C.

Y^-	k, $M^{-1}\ s^{-1}$
NCS^-	1.3×10^{-6}
$H_2PO_4^-$	2.0×10^{-6}
Cl^-	2.1×10^{-6}
NO_3^-	2.3×10^{-6}
SO_4^{2-}	1.5×10^{-5}

TABLE 17.3 Rate constants for the reaction of $[Co(NH_3)_5X]^{2+}$ with H_2O in aqueous solution.

X^-	k, s^{-1}
NCS^-	5.0×10^{-10}
$H_2PO_4^-$	2.6×10^{-7}
Cl^-	1.7×10^{-6}
NO_3^-	2.7×10^{-5}
SO_4^{2-}	1.2×10^{-6}

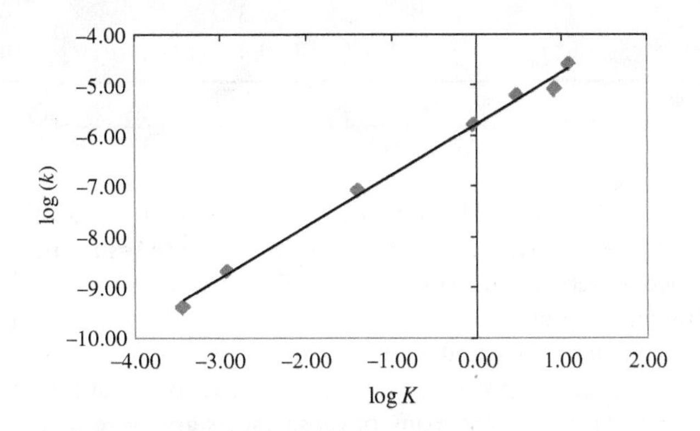

FIGURE 17.4
Linear free energy relationship for $[Co(NH_3)_5X]^{2+}$ hydrolysis reactions, where $X^- = NCS^-$, N_3^-, F^-, Cl^-, Br^-, I^-, and NO_3^-.

3. Further support for a D mechanism derives from the *linear free energy relationship* between the logarithms of the rate constant for hydrolysis of $[Co(NH_3)_5X]^{2+}$ and the equilibrium constant for the Co–X bond formation. This relationship occurs because the natural logarithm of the Arrhenius equation has a relationship similar to the natural logarithm of the equilibrium constant, as shown in Equations (17.28) and (17.29). Assuming that the Arrhenius pre-exponential term and the entropy are roughly constant and that the activation barrier depends on the enthalpy of reaction, log(k) will track linearly with log(K) as shown in Figure 17.4 for the reaction of $[Co(NH_3)_5X]^{2+}$ with H_2O.

$$\ln(k) = \ln(A) - \frac{E_a}{RT} \qquad (17.28)$$

$$\ln(K) = -\frac{\Delta H^\circ}{RT} + \frac{\Delta S^\circ}{R} \qquad (17.29)$$

4. Steric crowding on the Co(III) ion increases the rate of substitution. For example, the rate constant for the reaction of $[CoL_5Cl]^{2+}$ with H_2O in aqueous solution is $1.7 \times 10^{-6}\ s^{-1}$ when $L = NH_3$, but it increases to $3.7 \times 10^{-4}\ s^{-1}$ when $L = NMeH_2$. If the reaction mechanism were A or I_a, the rate of reaction would decrease with steric bulk around the metal because it would be harder for the incoming ligand to penetrate into the inner-coordination sphere.
5. Positive values for the entropy of activation and volume of activation suggest a dissociative mechanism. For example, $\Delta S^{\pm} = 28$ J/K·mol and $\Delta V^{\pm} = 1.2\ cm^3/mol$ for the octahedral substitution reaction of $[Co(NH_3)_5H_2O]^{3+}$.

TABLE 17.4 Activation parameters for octahedral substitution reactions (water exchange) of selected transition metal complexes.

Transition Metal Complex	$\Delta V^{\pm}$, cm^3/mol	$\Delta S^{\pm}$, J/K·mol
$[Co(NH_3)_5H_2O]^{3+}$	+1.2	28
$[Rh(NH_3)_5H_2O]^{3+}$	−4.1	3
$[Rh(H_2O)_6]^{3+}$	−4.1	29.3
$[Ir(NH_3)_5H_2O]^{3+}$	−3.2	11
$[Ir(H_2O)_6]^{3+}$	−5.7	2.1
$[Cr(NH_3)_5H_2O]^{3+}$	−5.8	0
$[Cr(H_2O)_6]^{3+}$	−9.6	11.6
$[Ru(NH_3)_5H_2O]^{3+}$	−4.0	−7.7
$[Ru(H_2O)_6]^{3+}$	−8.3	−48.2
$[Ru(H_2O)_6]^{2+}$	−0.4	16.1

The evidence listed overwhelmingly supports a D or I_d mechanism for substitution reactions involving Co(III) ammines. However, these results necessarily imply that *all* octahedral substitution reactions are dissociative in nature. While it does seem that the majority of octahedral substitution reactions are D or I_d, the data in Table 17.4 point to a more mixed interpretation. Most of the transition metal complexes in the table exhibit positive entropies of activation, indicating a TS having a dissociative character; however, many of these values are close to zero, making an unequivocal assignment of the reaction mechanism difficult. To make matters worse, most of the volumes of activation are negative, predicting an A or I_a mechanism. One of the main problems with relying too heavily on the activation parameters is that it is challenging to separate the solvent from $\Delta V^{\pm}$. The higher the charge density of the metal complex, the more strongly the solvent shell will adhere to the complex cation. For example, substitution reactions of Ti(III) appear to be strongly associative in nature. To further complicate matters, the mechanism often changes even for the same metal as the ligands or oxidation state of the coordination compounds change. For example, substitution reactions of $[Cr(NH_3)_5H_2O]^{3+}$ appear to be I_d, while those of $[Cr(H_2O)_6]^{3+}$ appear to be I_a. Similarly, most Ru(III) compounds have associative mechanisms, while Ru(II) compounds are more likely to be dissociative.

The rates of many octahedral substitution reactions are particularly sensitive to the pH of the solution. *Acid catalysis* occurs in reactions where the X^- leaving group can be protonated while it is still attached to the metal, thereby weakening the metal–ligand bond and facilitating its dissociation. For example, consider the rates of the reactions given by Equations (17.30) and (17.31). The rate of substitution is accelerated in the presence of acid for any ligands that can be protonated, but there is no change in the rate when the ligand is NH_3, which cannot be protonated while it is still attached to the metal.

$$[Cr(H_2O)_5X]^{2+} + H_3O^+ \rightarrow [Cr(H_2O)_6]^{3+} + HX \qquad k_1 \qquad (17.30)$$

$$[Cr(H_2O)_5X]^{2+} + H_2O \rightarrow [Cr(H_2O)_6]^{3+} + X^- \qquad k_0 \qquad (17.31)$$

For $X^- = F^-$, $k_1 = 1.4 \times 10^{-8}\ s^{-1}$ and $k_0 = 6.2 \times 10^{-10}\ s^{-1}$, a factor of over 20 times faster. For $X^- = CN^-$, the effect of acid catalysis is also observed, with $k_1 = 5.9 \times 10^{-4}\ s^{-1}$ and $k_0 = 1.1 \times 10^{-5}\ s^{-1}$. For NH_3, there is no noticeable difference between the two rate constants.

Octahedral substitution is also affected by *base catalysis* according to the *conjugate base mechanism* (S_N1CB). The rate constant for substitution of Cl^- in $[Co(NH_3)_5Cl]^{2+}$ is over a million times faster for OH^- than it is for H_2O. In fact, the rate law for the base catalysis reaction is complex second order: first order in $[Co(NH_3)_5Cl]^{2+}$ and first order in OH^-. In reality, however, the reaction takes place by proton abstraction, as shown by Equations (17.32)–(17.34):

$$[Co(NH_3)_5X]^{2+} + OH^- \Longleftrightarrow [Co(NH_3)_4(NH_2)X]^+ + H_2O \qquad (17.32)$$

$$[Co(NH_3)_4(NH_2)X]^+ \rightarrow [Co(NH_3)_4(NH_2)]^{2+} + X^- \quad \text{(RDS)} \qquad (17.33)$$

$$[Co(NH_3)_4(NH_2)]^{2+} + H_2O \rightarrow [Co(NH_3)_5OH]^{2+} \quad \text{(fast)} \qquad (17.34)$$

When the reaction is run in isotopically labeled H_2O, the ratio of $^{18}O/^{16}O$ in the ligand is the same as that in the solvent irrespective of the leaving group. If a truly associative mechanism was occurring, there should be a higher percentage of ^{18}O in the product than in the water. The electropositive amido group stabilizes the TS, making it easier to lose X^-. When the ammine group is sterically hindered (e.g., $NMeH_2$), the rate of reaction is accelerated because the steric crowding around the metal favors dissociation. Thus, the base catalysis reaction only *appears* to be associative because OH^- shows up in the experimentally observed rate law. In reality, the reaction has a dissociative character, as shown by the RDS in Equation (17.33). Tertiary amines, where there are no abstractable protons, react very slowly, if at all.

Octahedral substitution reactions also exhibit a *kinetic chelate effect*. Coordination compounds containing a chelating ligand react more slowly than their counterparts containing two monodentate ligands with similar M–L bond strengths. For example, the rate constant for substitution of $Ni[(bpy)]^{2+}$ is 3.3×10^{-4} s^{-1}, almost 10^5 times slower than for $[Ni(py)]^{2+}$, which has a rate constant of 38.5 s^{-1}. The process can be viewed as taking place in two steps, as shown in Equations (17.35) and (17.36). Because a chelating ligand such as en is structurally more rigid than NH_3, the chelate has to bend internal bond angles in order to be released from the transition metal. This increases the activation barrier for ligand loss and decreases the rate of reaction. Secondly, once the first end of the chelate has dissociated, because it is still attached to the metal by the other end, it has a tendency to dangle there long enough for the first end to reattach to the metal.

$$\text{M}\langle\text{L}\frown\text{L}\rangle \underset{k_{-1}}{\overset{k_1}{\rightleftharpoons}} \text{M–L}\frown\text{L} \qquad (17.35)$$

$$\text{M–L}\frown\text{L} \underset{k_{-2}}{\overset{k_2}{\rightleftharpoons}} \text{M} + \text{L}\frown\text{L} \qquad (17.36)$$

There are definite trends in the substitution rates of octahedral coordination compounds corresponding to the nature of the metal. Traditionally, metal ions have been grouped into four classes on the basis of their rates of water exchange. Class I metals exchange water extremely quickly, with rate constants on the order of 10^8 s^{-1}, only slightly slower than that for a diffusion-controlled reaction. This class of metal ions includes the alkali and heavier alkaline earth metals, as well as some of the lanthanides. Some first-row transition metals in low oxidation numbers also belong to Class I, such as Cu^{2+} and Cr^{2+}. The bonding is primarily ionic in these transition metals, with the metals having low charges and larger sizes. Class II metals exchange water with rate constants ranging from 10^5-10^8 s^{-1}. They include metal ions with larger charge densities than the Class I metals but with small ligand field

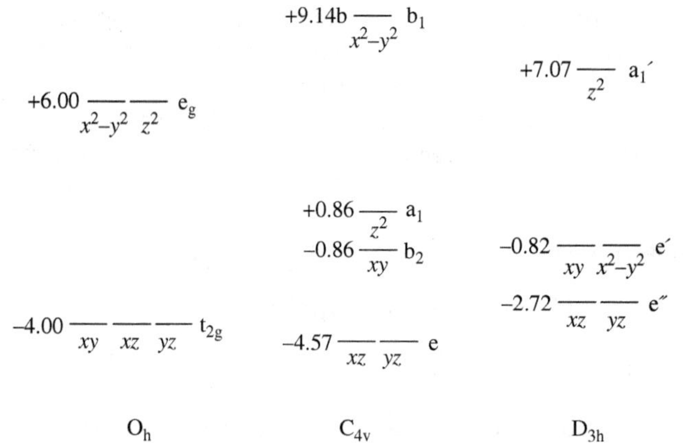

+6.00 — — e_g (x^2-y^2, z^2)
−4.00 — — — t_{2g} (xy, xz, yz)
O_h

+9.14b — b_1 (x^2-y^2)
+0.86 — a_1 (z^2)
−0.86 — b_2 (xy)
−4.57 — — e (xz, yz)
C_{4v}

+7.07 — a_1' (z^2)
−0.82 — — e' (xy, x^2-y^2)
−2.72 — — e'' (xz, yz)
D_{3h}

FIGURE 17.5
Ligand field energies for the *d*-orbitals in units of Dq for the octahedral, square pyramidal, and trigonal bipyramidal geometries.

stabilization energies (LFSEs). Examples include the lighter alkaline earth metals, In^{3+}, Ni^{2+}, Co^{2+}, Fe^{2+}, Mn^{2+}, Zn^{2+}, and Cd^{2+}. Class III metal ions exchange water more slowly, with rate constants on the order of $1-10^4$ s^{-1}. Most metal ions in this group have 3+ charges, large charge densities, and moderate LFSEs. Examples include Be^{2+}, Al^{3+}, V^{2+}, V^{3+}, Ga^{3+}, Fe^{3+}, and Pd^{2+}. Finally, Class IV metal ions exchange water very slowly, having rate constants ranging from $10^{-1}-10^{-9}$ s^{-1}. These ions are characterized by large LFSEs and include Cr^{3+}, Ru^{3+}, Pt^{2+}, and Co^{3+}. Coordination compounds that are kinetically sluggish are called *inert*, while those that react rapidly are called *labile*.

For those octahedral substitution reactions that occur by a dissociative mechanism, there are two likely geometries for the five-coordinate intermediate: square pyramidal and trigonal bipyramidal. The trigonal bipyramidal geometry is favored on the basis of steric considerations (VSEPR theory) and will likely result when the ligands on the coordination compound are bulky. The square pyramidal geometry is favored, however, for electronic reasons, as shown by the energies of the different *d*-orbitals in Figure 17.5, which was made using data from Table 16.4. The CFSE in the square pyramidal geometry is superior to that for trigonal bipyramidal geometry for most d^n configurations.

The *ligand field activation energy* (LFAE) is defined as the difference in the LFSE for the TS minus that of the reactants. Assuming a square pyramidal dissociative intermediate, the LFAEs for the different d^n electron configurations are listed in Table 17.5.

TABLE 17.5 Ligand field activation energies (LFAE) for an octahedral reactant and a square pyramidal transition state.

d^n	CFSE (C_{4v})		CFSE (O_h)		LFAE	
1	−4.57		−4.00		−0.57	
2	−9.14		−8.00		−1.14	
3	−10.00		−12.00		+2.00	
4	−9.14	−14.57	−6.00	−16.00	−3.14	+1.43
5	0.00	−19.14	0.00	−20.00	0.00	+0.86
6	−4.57	−20.00	−4.00	−24.00	−0.57	+4.00
7	−9.14	−19.14	−8.00	−18.00	−1.14	−1.14
8	−10.00		−12.00		+2.00	
9	−9.14		−6.00		−3.14	

The units are in Dq. Whenever a column splits, HS configurations are listed on the left and LS configurations are listed on the right.

Notice that the most positive LFAEs occur for the d^3, LS d^6, and d^8 electron configurations, which helps explain why Cr^{3+}, Co^{3+}, and Pt^{2+} are substitutionally inert.

Example 17-1. Some octahedral substitution reactions are believed to occur by an I_a mechanism, where the transition state is seven-coordinate. Assuming a pentagonal bipyramidal transition state, calculate the LFAEs for d^1–d^9. Which electron configurations will lead to substitutionally inert metal ions, assuming an I_a mechanism?

Solution. The energies of the *d*-orbitals in a pentagonal bipyramid were given in Table 16.4. The d_{z^2} orbital lies highest in energy at +4.93 Dq, the $d_{x^2-y^2}$ and d_{xy} orbitals form a degenerate pair at +2.82 Dq, and the d_{xz} and d_{yz} orbitals lie lowest in energy at −5.28 Dq. The LFSEs in D_{5h} and O_h are shown in the following table. Whenever a column splits, the HS case is on the left and the LS case is on the right. The LFAE is calculated by taking the difference LFSE (D_{5h}) − LFSE (O_h) and is listed in Column 4 of the table.

d^n	CFSE (D_{5h})		CFSE (O_h)		LFAE	
1	−5.28		−4.00		−1.28	
2	−10.56		−8.00		−2.56	
3	−7.74		−12.00		+4.26	
4	−4.92	−13.02	−6.00	−16.00	+1.08	+2.98
5	0.00	−18.30	0.00	−20.00	0.00	+1.70
6	−5.28	−15.48	−4.00	−24.00	−1.28	+8.52
7	−10.55	−12.66	−8.00	−18.00	−2.55	+5.34
8	−7.73		−12.00		+4.27	
9	−4.91		−6.00		+1.09	

In general, the LFAEs are more positive for the associative mechanism than for the dissociative mechanism, adding further support to the theory that most octahedral substitution reactions are dissociative. The largest LFAEs for the associative mechanism largely mirror those for the dissociative mechanism: d^3, LS d^6, and d^8, with the main exception being that LS d^7 is also now included as inert.

17.3 SQUARE PLANAR SUBSTITUTION REACTIONS

The compound $[Pt(NH_3)_2Cl_2]$ was first prepared in 1845 by two different synthetic routes. The α form of the compound (known as *Peyrone's salt*) was made by the reaction of $[PtCl_4]^{2-}$ with NH_3, while the β form (also called *Resiet's second chloride*) resulted when $[Pt(NH_3)_4]Cl_2$ was heated to 250 °C. Many years later, Alfred Werner showed that the two compounds were isomers of each other and suggested that they had square planar molecular geometries. Currently, we know the α form as the anticancer drug cisplatin, *cis*-$[Pt(NH_3)_2Cl_2]$, and the β form as its geometrical isomer *trans*-$[Pt(NH_3)_2Cl_2]$.

From his further studies of Pt(II) chemistry, Werner postulated that the addition and elimination steps in square planar substitution reactions were both stereospecific. Consider, for example, Peyrone's reaction, which is shown in Equation (17.37). Addition of only one equivalent of NH_3 or py to the tetrachloroplatinate(II) ion leads to the monosubstituted $[PtACl_3]$, where A = a *N*-donor ligand. The addition

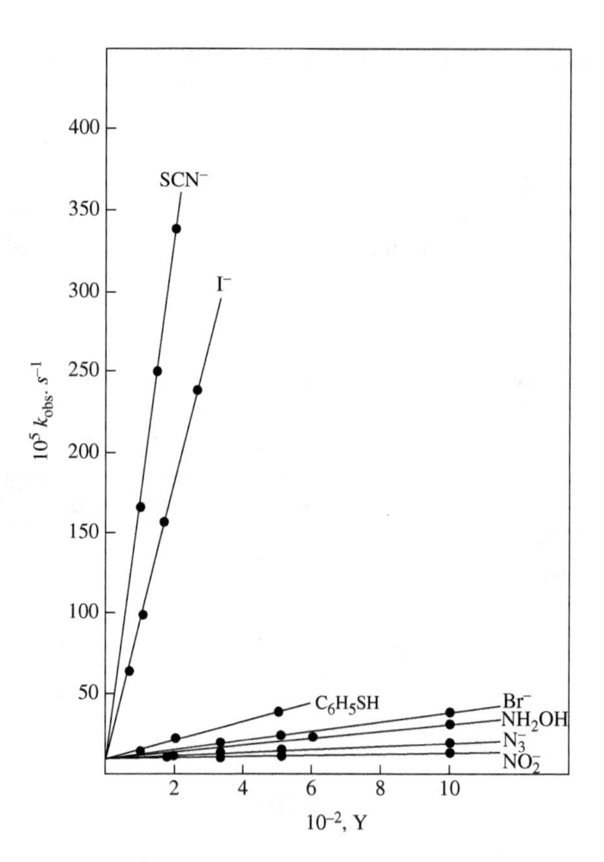

FIGURE 17.6
Rate constants for the square planar substitution reactions of $[Pt(py)_2Cl_2]$ as a function of the nucleophile Y in CH_3OH at 30 °C. [Reproduced from *J Amer Chem Soc* 1965 ,*87*, 241–246. Used with permission of the American Chemical Society.]

of the second ammine ligand always occurs *trans* to one of the Cl^- ligands, resulting 100% of the time in the *cis* isomer. Similarly, in Jørgensen's reaction, which is shown in Equation (17.38), when $[PtA_4]^{2+}$ is reacted with two HCl, the second Cl^- always attaches in the position *trans* to the first Cl^-, yielding 100% of the *trans* product. Although Werner recognized this as the concept of "trans elimination," it was not until 1926 when the Russian chemist Il'ya Chernyaev published his landmark paper on the *trans effect* that the significance of the stereoselectivity of square planar substitution compounds was fully recognized.

$$\left[\begin{array}{c} Cl \\ | \\ Cl - Pt - Cl \\ | \\ Cl \end{array}\right]^{2-} \xrightarrow[-Cl^-]{+A} \begin{array}{c} Cl \\ | \\ A - Pt - Cl^- \\ | \\ Cl \end{array} \xrightarrow[-Cl^-]{+A} \begin{array}{c} A \\ | \\ A - Pt - Cl \\ | \\ Cl \end{array} \qquad (17.37)$$

$$\left[\begin{array}{c} A \\ | \\ A - Pt - A \\ | \\ A \end{array}\right]^{2+} \xrightarrow[-A]{+Cl^-} \begin{array}{c} A \\ | \\ A - Pt - Cl^+ \\ | \\ A \end{array} \xrightarrow[-A]{+Cl^-} \begin{array}{c} A \\ | \\ Cl - Pt - Cl \\ | \\ A \end{array} \qquad (17.38)$$

The kinetics of square planar substitution reactions are complex, as shown in Figure 17.6. The experimentally determined rate law has two terms, as shown in Equation (17.39).

$$\text{rate} = k_1[ML_2TX] + k_2[ML_2TX][Y] \qquad (17.39)$$

The rate law can be rewritten in terms of the transition metal compound according to Equation (17.40), where X is the leaving group, Y is the nucleophile, and T is the ligand *trans* to the leaving group.

$$\text{rate} = k_{obs}[ML_2TX], \text{ where } k_{obs} = k_1 + k_2[Y] \qquad (17.40)$$

For very small values of [Y], the rate will be independent of the incoming ligand and k_{obs} will approach the value k_1, which is given by the *y*-intercept in Figure 17.6.

What follows is a summary of the kinetic data for square planar substitution reactions:

1. The rate of substitution decreases with the steric bulk of the ligands. For the substitution reaction of *trans*-$[Pt(PEt_3)_2LCl]$, the rate constants are listed in Table 17.6, where $L = PR_3$. This is suggestive of an associative TS.
2. The $\Delta V^{\pm}$ and $\Delta S^{\pm}$ activation parameters are both negative, again supporting an associative mechanism because there are more species present as the reactants than in the TS. Both the k_1 and k_2 terms are affected. For the reaction of *trans*-$[Pt(PEt_3)_2RBr]$ with I^-, the activation parameters are listed in Table 17.7.
3. Five-coordinate intermediates having trigonal bipyramidal molecular geometries have been isolated in several cases, including, for example, $[Ni(CN)_5]^{3-}$ and $[Pt(SnCl_3)_5]^{3-}$.

   All of these results support an A or I_a mechanism for square planar substitution. But why are there two terms?
4. The rate of substitution strongly depends on the coordinating ability of the solvent. The rate constants for square planar substitution of

TABLE 17.6 Rate constants for square planar substitution of Cl^- in *trans*-$[Pt(PEt_3)_2LCl]$ as a function of the bulkiness of the phosphine ligand, $L = PR_3$.

R	k_{obs}, s^{-1}
	8×10^{-2}
	2×10^{-4}
	1×10^{-6}

TABLE 17.7 Activation parameters ($\Delta V^{\pm}$ and $\Delta S^{\pm}$) for the reaction: *trans*-$[Pt(PEt_3)_2RBr] + I^- \rightarrow$ *trans*-$[Pt(PEt_3)_2RBI] + Br^-$.

Parameter	k_1 term	k_2 term
$\Delta V^{\pm}$ (cm^3/mol)	−16	−16
$\Delta S^{\pm}$ (J/K·mol)	−52	−115

TABLE 17.8 Rate constants for the square planar substitution reaction of *trans*-[Pt(py)$_2$Cl*Cl] with Cl$^-$ as a function of the coordinating ability of the solvent.

Solvent	$10^5\ k_{obs}$, s^{-1}
DMSO	380
H_2O	3.5
CH_3NO_2	3.2
CH_3CH_2OH	1.4
$CH_3CH_2CH_2OH$	0.42

trans-[Pt(py)$_2$Cl*Cl] with Cl$^-$ as a function of the solvent are reported in Table 17.8. The coordinating abilities of the solvent decrease down the column. The reaction occurs about 1000 times faster in DMSO, which is a good coordinating solvent, than it does in *n*-propyl alcohol.

The combined effects of the dependence of the substitution rate on the ability of the solvent to coordinate to the metal and the fact that both the k_1 and k_2 terms have negative activation parameters suggest that the first term in the experimental rate law is associative with the solvent. This term only appears to be first order because the concentration of the solvent is so much greater than that of the metal complex. In actuality, it is pseudo first order. The second term is associative with the nucleophile Y, showing complex second-order dependence. The overall mechanism is therefore a competition between the two terms, as shown in Figure 17.7.

5. As Chernyaev articulated in his 1926 paper, the rates of square planar substitution reactions strongly depend on the ligand that is *trans* to the leaving group. The observed *trans effect series* is as follows:

   CO, CN$^-$, C_2H_4 > PR_3, AsR_3, SR_2, H$^-$ > CH_3^- > $C_6H_5^-$, NO_2^- > I$^-$, SCN$^-$ > Br$^-$ > Cl$^-$ > py > NH_3 > OH$^-$ > H_2O, F$^-$

Some representative data in support of the trans effect are listed in Table 17.9.

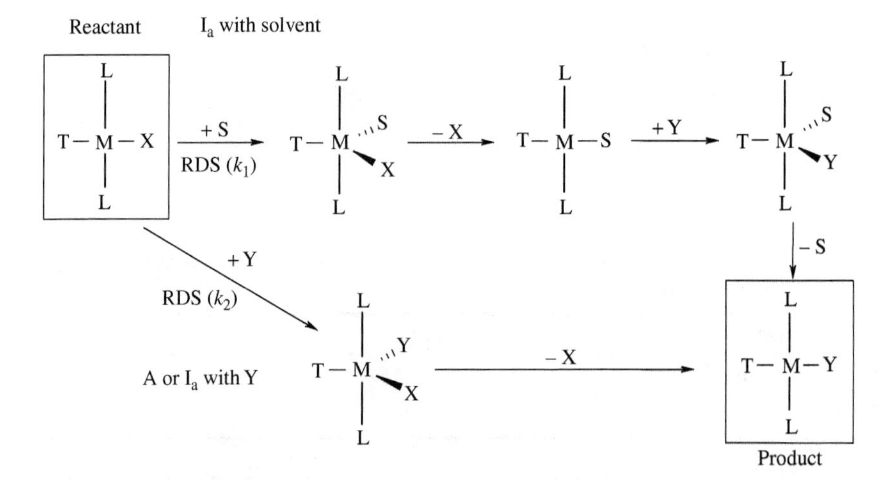

FIGURE 17.7
Proposed competing mechanisms for square planar substitution reactions to explain the observed rate law shown in Figure 17.6 and given by Equation (17.40).

TABLE 17.9 Effect of the *trans* ligand on the rate of substitution of *trans*-$[Pt(PEt_3)_2LCl]$.

L	k_1, s^{-1}	k_2, s^{-1}
PEt_3	1.7×10^{-2}	3.8
H^-	1.8×10^{-2}	4.2
CH_3^-	1.7×10^{-4}	6.7×10^{-2}
$C_6H_5^-$	3.3×10^{-5}	1.6×10^{-2}
p-$CH_3OC_6H_4^-$	2.8×10^{-5}	1.3×10^{-2}
Cl^-	1.0×10^{-6}	4.0×10^{-4}

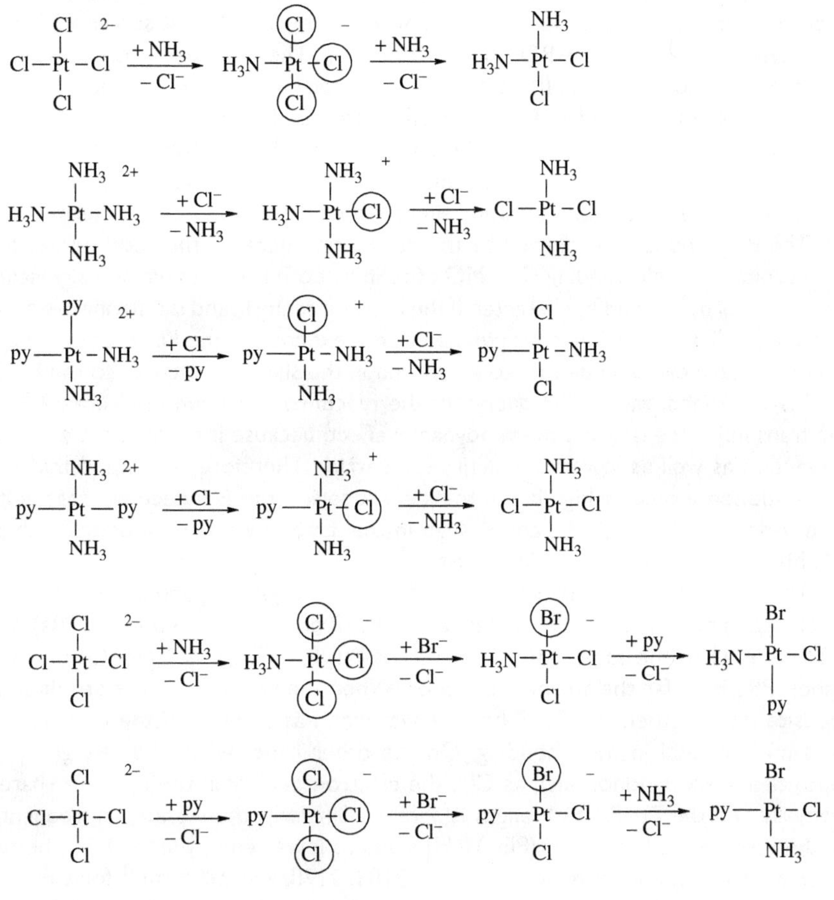

FIGURE 17.8 Selected stereochemically controlled syntheses using the trans effect. The stronger *trans*-directing ligand in each case is circled.

The trans effect is of great synthetic utility when a product with specific stereochemistry is desired. Several examples are shown in Figure 17.8, where the greater *trans*-directing ligand is circled. The preferred site of substitution is a trade-off between the trans effect and the strength of the Pt–X bond. The strength of the bonding decreases in the order: Pt–NH_3 > Pt–py > Pt–Cl.

Example 17-2. Predict the products of each of the following square planar substitution reactions:

(a) $[Pt(NO_2)Cl_3]^{2-} + NH_3 \rightarrow$
(b) $[PtCl_4]^{2-} + 2\ PR_3 \rightarrow$
(c) *cis*-$[Pt(py)_2(NH_3)_2]^{2+} + 2\ Cl^- \rightarrow$

Solution. (a) The stronger *trans*-directing ligand is NO_2^-, so the product will be *trans*-$[Pt(NO_2)(NH_3)Cl_2]^-$. (b) The first PR_3 ligand replaces any Cl^- ligand at random. Because PR_3 is a stronger *trans*-directing ligand than Cl^-, the product will be *trans*-$[Pt(PR_3)_2Cl_2]$. (c) Because the Pt–NH_3 bond is weaker than the Pt–py bond, the first Cl^- will replace one of the ammine ligands. Because Cl^- is then the strongest *trans*-directing ligand, the product will be *trans*-$[Pt(py)(NH_3)Cl_2]$.

The trans effect showed that there is a strong dependence of the reaction rate on the nature of the ligand in a position that is *trans* to the leaving group. In order to enhance the rate of substitution, the *trans*-directing ligand must somehow lower the activation barrier for the RDS in the mechanism. There are two possible ways to lower the activation barrier: (i) destabilize the reactants as shown in Figure 17.9(b) or (ii) stabilize the TS leading to the intermediates as shown in Figure 17.9(c). The term *trans influence* is used to describe only the first of these two possibilities, the destabilizing effect of the *trans* ligand on the ground-state structure or electronic properties of the reactants. The trans effect is the combination of both terms.

The Pt–X bond is weakened by the Pt–T bond because they both utilize the same orbitals in their bonding. The MO describing both bonds contains a significant percentage of $d_{x^2-y^2}$ and p_x character. If the *trans*-directing ligand is a strong σ-donor, then the Pt–T bond will be strengthened at the expense of the Pt–X bond. There is only so much electron density to go around in the shared orbital(s), so the Pt–X bond is weakened, raising the energy of the reactants as shown in Figure 17.9(b). The trans influence is also a thermodynamic effect, because it changes the enthalpy of reaction as well as lowers the activation barrier. Therefore, any ligand that is a good σ-donor should lie high in the trans effect series; this is indeed the case, with good σ-donors like H^-, PR_3, and I^- high in the list and weaker σ-donors such as Cl^-, NH_3, H_2O, and F^- lower in the list.

The trans influence is supported by X-ray crystallographic data on the metal–ligand bond lengths. For example, the Pt–P bond length in *trans*-$Pt(PR_3)_2Cl_2$ is 230 pm, while the same bond length in *cis*-$Pt(PR_3)_2Cl_2$ is 224 pm. Of the two ligands, PR_3 is by far the stronger σ-donor. When the two PR_3 ligands are directly opposite one another, the Pt–P bond is weakened as a result of the two sharing the same *d*-orbital in their bonding. On the other hand, when the PR_3 group is opposite a weak σ-donor such as Cl^-, the electron density in the $d_{x^2-y^2}$ is shared unequally and the Pt–P bond length is shorter. Similarly, the vibrational frequency for the Pt–H bond in *trans*-$[Pt(PEt_3)_2LH]$ shifts to lower energy when L is a better σ-donating ligand. For example, υ(Pt–H) = 2183, 2178, and 2156 cm^{-1} for L = Cl^-, Br^-, and I^-, respectively. For the same series of compounds, the chemical shift for the hydride ligand in the ^1H NMR of the compounds is shifted upfield: $\delta = 16.9$, 15.6, and 12.7 ppm versus TMS for L = Cl^-, Br^-, and I^-, respectively.

However, the trans influence is insufficient to explain the entire order of ligands in the trans effect series. For instance, the two highest ligands in the series are CO and CN^-, which, in fact, are fairly poor σ-donors. As we have seen before in the spectrochemical series, CO and CN^- are strong π-acceptors. The π-orbitals on the metal are the d_{xz} and d_{yz} orbitals. In a square planar LF, these two orbitals lie

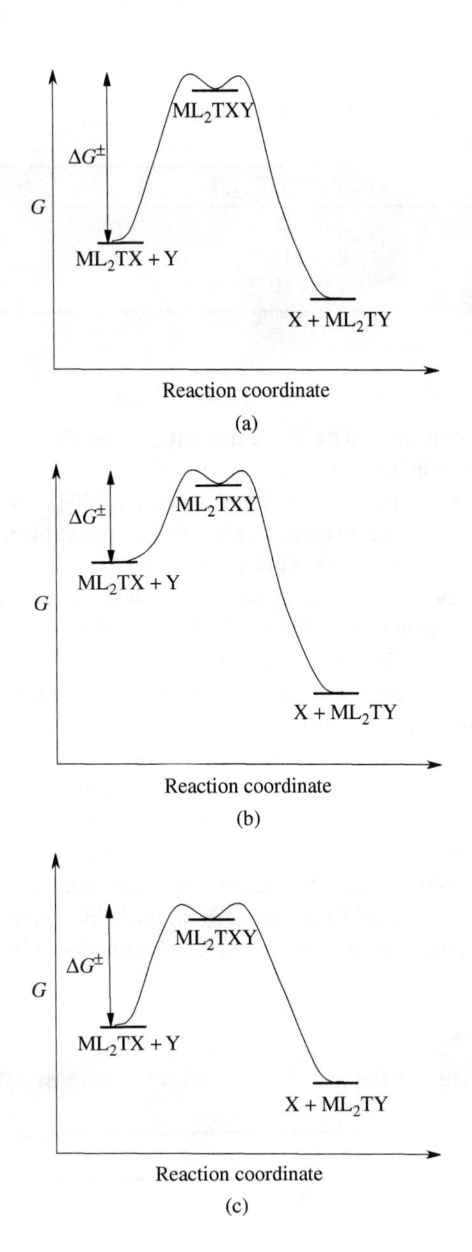

FIGURE 17.9
There are two ways that the trans effect can lower the activation barrier (a) of a square planar substitution reaction: (b) by weakening the Pt–X bonds in the reactants (the trans influence) or (c) by stabilizing the transition state leading to the intermediates.

at −5.14 Dq. Assuming a trigonal bipyramidal intermediate in the associative mechanism, the energies of the d_{xz} and d_{yz} orbitals are raised to −2.72 Dq. Thus, any ligand that can stabilize the π-orbitals in the TS will help lower the activation barrier for the reaction. Strong π-acceptor ligands, such as CO, CN^-, C_2H_4, and PPh_3, can therefore lower the energy of the TS by removing electron density from the metal as the coordination number increases from four to five. The trans effect is the combination of both GS destabilization and TS stabilization. Thus, those ligands that are either good σ-donors or strong π-acceptors (or both) are high in the trans effect series.

There is also evidence supporting a *cis* effect on the rate of square planar substitution reactions, although it is not quite as dramatic. However, steric factors can be very important in the *cis* position. Steric congestion enhances the rate of a dissociative substitution reaction and slows the rate of an associative one. Because the bulky ligands are closer to each other in the *cis* position than in the *trans* position,

TABLE 17.10 Effect of steric bulk on the rates (k_{obs}, s^{-1}) of square planar substitution reactions at 25 °C.

L	*cis*-$[Pt(PEt_3)_2LCl]$	*trans*-$[Pt(PEt_3)_2LCl]$
Phenyl	8.0×10^{-2}	1.2×10^{-4}
o-tolyl	2.0×10^{-4}	1.7×10^{-5}
Mesityl	1.0×10^{-6}	3.4×10^{-6}

the effects of steric bulk should be more dramatic when the larger groups are *cis* to one another, as shown in Table 17.10.

The rates of square planar substitution also vary significantly with the strength of the Pt–X bond. Dissociative mechanisms are expected to show a strong dependence on the leaving group because the RDS is the loss of X. However, an associative mechanism can also exhibit a similar dependence on the leaving group based on the degree of bond breaking in the TS. Table 17.11 shows the effect of the leaving group on the rate of substitution in a series of $[Pt(dien)X]^+$ compounds. As a general rule, the rates of substitution parallel that of the trans effect—the stronger the Pt–X bond, the slower the rate of substitution.

For an associative mechanism, one would expect that the rate of substitution would show a strong dependence on the nucleophile, as the nucleophile directly participates in the RDS. This is indeed the case for square planar substitution reactions, as illustrated by the data in Table 17.12, which show that the rates of reaction vary by five orders of magnitude with the nature of the incoming ligand Y.

The effect of the nucleophile is primarily an enthalpic effect. While it might be tempting to make comparisons of the rates of substitution with the basicity of the

TABLE 17.11 Effect of the leaving group on the rate of substitution of $[Pt(dien)X]^+$.

X	k_{obs}, s^{-1}	X	k_{obs}, s^{-1}
H_2O	1900	N_3^-	0.8
Cl^-	35	SCN^-	0.3
Br^-	23	NO_2^-	0.05
I^-	10	CN^-	0.02

TABLE 17.12 Effect of the nucleophile Y on the rate of substitution in *trans*-$[Pt(py)_2Cl_2]$ at 25 °C.

Y	k_2, M^{-1} s^{-1}	Y	k_2, M^{-1} s^{-1}
Cl^-	0.45	Br^-	3.7
NH_3	0.47	I^-	107
NO_2^-	0.68	SCN^-	180
N_3^-	1.55	PPh_3	250,000

incoming ligand toward H^+, the enhancement of the rate has more to do with HSAB theory. The Pt(II) ion is a soft Lewis acid, and it will therefore prefer soft bases such as PPh_3 and SCN^-.

17.4 ELECTRON TRANSFER REACTIONS

An electron transfer (ET) reaction is defined here as an oxidation–reduction reaction that occurs between two coordination compounds. The compounds may be the same species, but where the metals have different oxidation states (a *self-exchange reaction*), or are completely different species (a *cross-reaction*). An example of a self-exchange reaction is given in Equation (17.41), where *Co is an isotopically labeled cobalt; a cross-reaction is given by Equation (17.42).

$$[Co(NH_3)_6]^{3+} + [Co * (NH_3)_6]^{2+} \rightarrow [Co(NH_3)_6]^{2+} + [Co * (NH_3)_6]^{3+} \quad (17.41)$$

$$[Co(NH_3)_6]^{3+} + [Cr(bpy)_3]^{2+} \rightarrow [Co(NH_3)_6]^{2+} + [Cr(bpy)_3]^{3+} \quad (17.42)$$

Because the coordination compounds in a self-exchange reaction are identical, $\Delta G^\circ = 0$. For a cross-reaction, on the other hand, there is a nonzero thermodynamic driving force for the reaction. ET reactions can be further classified into outer-sphere and inner-sphere reaction mechanisms. An *outer-sphere* ET occurs without the breaking or making of bonds. During an *inner-sphere* ET, on the other hand, bonds are made and/or broken during the ET, which occurs through a bridging ligand. Both of the ET reactions shown are examples of outer-sphere ET.

ET reactions are typically bimolecular in the RDS and therefore display second-order kinetics. The reaction approximates a simple collision model, where the free energy of activation ($\Delta G^\pm$) involves three terms, as shown in Equation (17.43).

$$\Delta G^\pm{}_{tot} = \Delta G^\pm{}_t + \Delta G^\pm{}_i + \Delta G^\pm{}_o \quad (17.43)$$

The reactants first have to be brought to reaction distance, as shown in Figure 17.10; this requires a work term associated with the lost translational and rotational free energy involved in bringing the two reactants together. It also involves a change in the electrostatics as the distance between the two species moves from infinity to the reaction distance. This first term is given the abbreviation $\Delta G^\pm{}_t$ in Equation (17.43). Once the two coordination compounds are at reaction distance, there is still an activation barrier associated with the formation of the *precursor complex*, as shown in Figure 17.10. This barrier involves both inner-sphere ($\Delta G^\pm{}_i$) and outer-sphere ($\Delta G^\pm{}_o$) components.

Even self-exchange reactions (where $\Delta G^\circ = 0$) have a barrier to ET. This barrier has to do with the *Franck–Condon principle*, which states that because the nuclei in a coordination compound are much heavier than the electrons, nuclear motion will be slow on the electronic timescale and the energies of the participating orbitals must therefore be the same before the ET can take place. For a self-exchange reaction, such as the one illustrated in Figure 17.11, the higher oxidation state (hollow circles) typically has the smaller radius (the ionic radii of octahedral LS Co^{3+} and HS Co^{2+}, for example, are 68.5 and 88.5 pm, respectively). Thus, in order for the ET in Equation (17.41) to occur, the Co(III)-N bond lengths must increase and the Co(II)-N bond lengths must decrease until both of them have the same internuclear separation. This can occur if sufficient vibrational energy is supplied to those vibrational modes that lie along the reaction coordinate for ET.

The degree of the inner-sphere *reorganization energy* (λ_i) associated with the adjustment of the metal–ligand bond lengths is given by Equation (17.44), where n

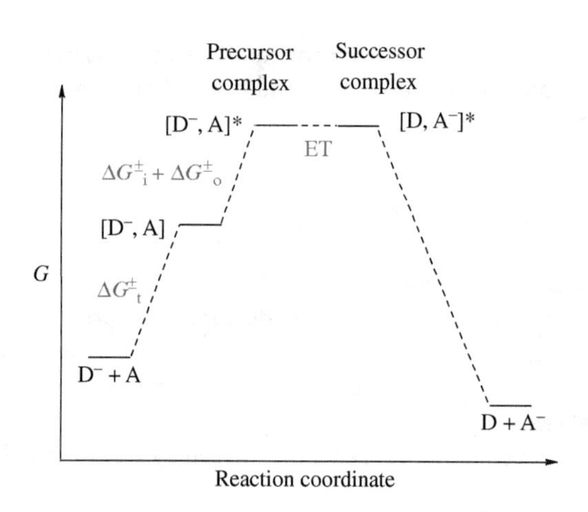

FIGURE 17.10
The mechanistic steps involved in the activation of the reactants (D = donor metal, A = acceptor metal) to reach the precursor complex necessary for ET. Following ET, the successor complex undergoes dissociation to form the products.

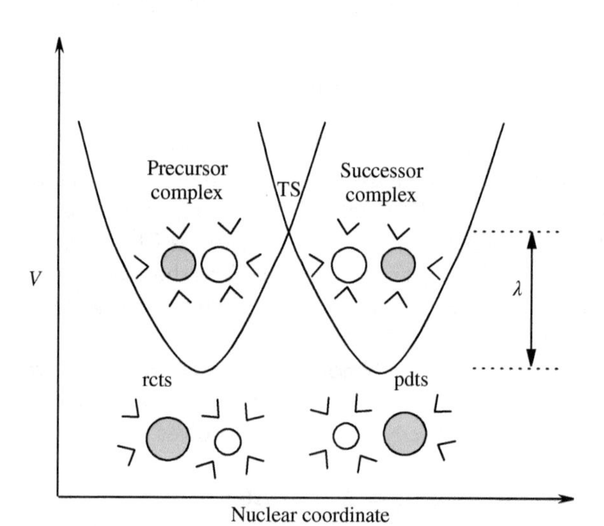

FIGURE 17.11
Potential energy diagram for a self-exchange electron transfer reaction.

is the nth normal mode of vibration along the reaction coordinate, k is the force constant for the vibration, and ΔQ_e is the displacement of the vibration along the nuclear reaction coordinate. At the same time, the solvent shell surrounding the two different ions is also different between the 3+ and 2+ oxidation states. Thus, not only must the metal–ligand bond lengths change before ET can occur but the solvent dipoles must also redistribute themselves around the two coordination compounds. This second term is called the *outer-sphere reorganization energy* (λ_o) and is given by Equation (17.45), where e is the electronic charge, a_2 and a_3 are the molecular radii of the reactants in the corresponding oxidation states, r is the distance that the electron must transfer, D_{op} is the optical dielectric constant (which is equal to the square of the refractive index), and D_s is the static dielectric constant of the solvent. If one makes the assumption (which is not always valid) that there is no entropic contribution to the ET, then $\Delta G^{\pm}{}_i \cong \lambda_i$ and $\Delta G^{\pm}{}_o \cong \lambda_o$. For a self-exchange reaction, the reorganization energy (λ) is therefore equal to the free energy of activation ($\Delta G^{\pm}$).

$$\lambda_i = \sum_n \frac{1}{2} k_n (\Delta Q_e)^2 \tag{17.44}$$

$$\lambda_o = e^2 \left(\frac{1}{2a_2} + \frac{1}{2a_3} - \frac{1}{r} \right) \left(\frac{1}{D_{op}} - \frac{1}{D_s} \right) \tag{17.45}$$

TABLE 17.13 Rate constants and activation parameters for selected self-exchange ET reactions at 25 °C.

Coordination Complex	k, $M^{-1}\ s^{-1}$	$\Delta H^{\pm}$, kJ/mol	$\Delta S^{\pm}$, J/K·mol
$[V(H_2O)_6]^{3+/2+}$	1.0×10^{-2}	53	−105
$[Cr(H_2O)_6]^{3+/2+}$	$<2\times10^{-5}$	88	−33
$[Fe(H_2O)_6]^{3+/2+}$	3.3	39	−105
$[Co(H_2O)_6]^{3+/2+}$	2.5	53	−59
$[Co(NH_3)_6]^{3+/2+}$	8×10^{-6}		
$[Co(phen)_3]^{3+/2+}$	4.0×10^{-2}	21	−142
$[IrCl_6]^{2-/3-}$	2.3×10^{5}		
$MnO_4^{-/2-}$	7.1×10^{2}	42	−38
$[Mo(CN)_8]^{3-/4-}$	3×10^{4}		
$[W(CN)_8]^{3-/4-}$	$<4\times10^{8}$		
$[Fe(CN)_6]^{3-/4-}$	2.26×10^{2}	23	−117

The rate constants for some self-exchange ET reactions are listed in Table 17.13. Notice that the entropies of activation are strongly negative, implying an associative mechanism.

Example 17-3. Using arguments from CFT, explain the differences between the rate constants for self-exchange listed in Table 17.13 for the following pairs: (a) $[V(H_2O)_6]^{3+/2+}$ versus $[Fe(H_2O)_6]^{3+/2+}$ and (b) $[Co(H_2O)_6]^{3+/2+}$ versus $[Co(NH_3)_6]^{3+/2+}$.

Solution

(a) The $V^{3+/2+}$ couple transfers an electron between two nonbonding t_{2g} orbitals as it goes from $t_{2g}^{\ 2}e_g^{\ 0}$ to $t_{2g}^{\ 3}e_g^{\ 0}$. The same is true for the $Fe^{3+/2+}$ couple as it goes from $t_{2g}^{\ 3}e_g^{\ 2}$ to $t_{2g}^{\ 4}e_g^{\ 2}$. However, the V–O bonds are stronger than the Fe–O bonds because the e_g antibonding MOs are empty in V. The stronger metal–ligand bonding and lighter metal in $[V(H_2O)_6]^{3+/2+}$ imply that the υ(M–O) stretching frequency will be higher with V than with Fe. Thus, there is a larger vibrational trapping in the vanadium complex than in the iron complex, and the rate of ET will be slower for $[V(H_2O)_6]^{3+/2+}$ self-exchange than for $[Fe(H_2O)_6]^{3+/2+}$ self-exchange (ignoring any contributions from the solvent reorganization energy).

(b) In the HS complex $[Co(H_2O)_6]^{3+/2+}$, the ET is between two nonbonding t_{2g} orbitals as it goes from $t_{2g}^{\ 4}e_g^{\ 2}$ to $t_{2g}^{\ 5}e_g^{\ 2}$. In $[Co(NH_3)_6]^{3+/2+}$, the complex is LS d^6 for Co(III) and HS d^7 for Co(II). Thus, the transferring electron populates an antibonding e_g MO as it goes from $t_{2g}^{\ 6}e_g^{\ 0}$ to $t_{2g}^{\ 5}e_g^{\ 2}$. This significantly weakens the Co–N bonding and causes a large structural reorganization energy. The molecule also undergoes a spin flip, which adds another 103 kJ/mol to the activation barrier. Thus, the activation barrier for self-exchange in $[Co(NH_3)_6]^{3+/2+}$ is considerably higher than that for $[Co(H_2O)_6]^{3+/2+}$, and the latter coordination compound undergoes self-exchange more than 10^5 times faster than does the former complex.

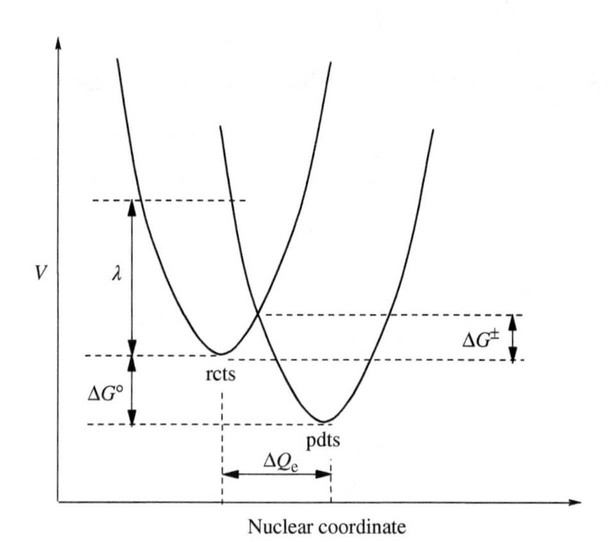

FIGURE 17.12
Potential energy diagram for a cross-electron transfer reaction, showing the relationships between λ, ΔG°, and ΔG^*.

For a cross-reaction, the free energy of activation depends not only on the reorganization energy but also on the ground state free energy difference between the reactant and product states, as shown in Figure 17.12. If the potential energies of the two states can be given by the harmonic oscillator approximation that their potential functions are parabolic with the displacement, then the free energy of activation can be calculated from Equation (17.46).

$$\Delta G^{\pm} = \frac{(\lambda + \Delta G^\circ)}{4\lambda} \tag{17.46}$$

Many different approaches to the treatment of ET theory have been developed over the years, from the fully classical model of Marcus and Hush to the Jortner's fully quantum mechanical model. According to ET theory, the rate constant for ET is given by the *Fermi golden rule*, according to Equation (17.47), where H_{ab} is the electronic coupling matrix element between the donor and acceptor metals and FC is the Franck–Condon factor, which has to do with the amount of vibrational overlap along the reaction coordinate. The main difference between the theories is the manner in which they treat the Franck–Condon factor. In the classical treatment, in order for ET to occur, the reactants must be activated until they have enough energy to reach the intersection region of the two parabolas in Figures 17.11 or 17.12. Using a quantum mechanical approach, ET can occur at lower energies than the intersection region because the electron can tunnel through the barrier.

$$k = \frac{4\pi^2}{h}|H_{ab}|^2(\text{FC}) \tag{17.47}$$

We can also use TS theory to calculate the rate constant for ET from the free energy of activation using the Eyring equation given by Equation (17.7). Taking the natural logarithm of this equation and substituting Equation (17.46) for $\Delta G^{\pm}$, we obtain the result given by Equation (17.48).

$$\ln(k) = \ln\left(\frac{k_B T}{h}\right) - \frac{(\lambda + \Delta G^\circ)^2}{4\lambda RT} \tag{17.48}$$

A plot of ln(k) versus ΔG°, such as the one shown in Figure 17.13, yields an inverted bell-shaped curve with a maximum rate constant (in the absence of diffusion-limited behavior) when $|\Delta G^\circ| = \lambda$. To the left of the maximum is the

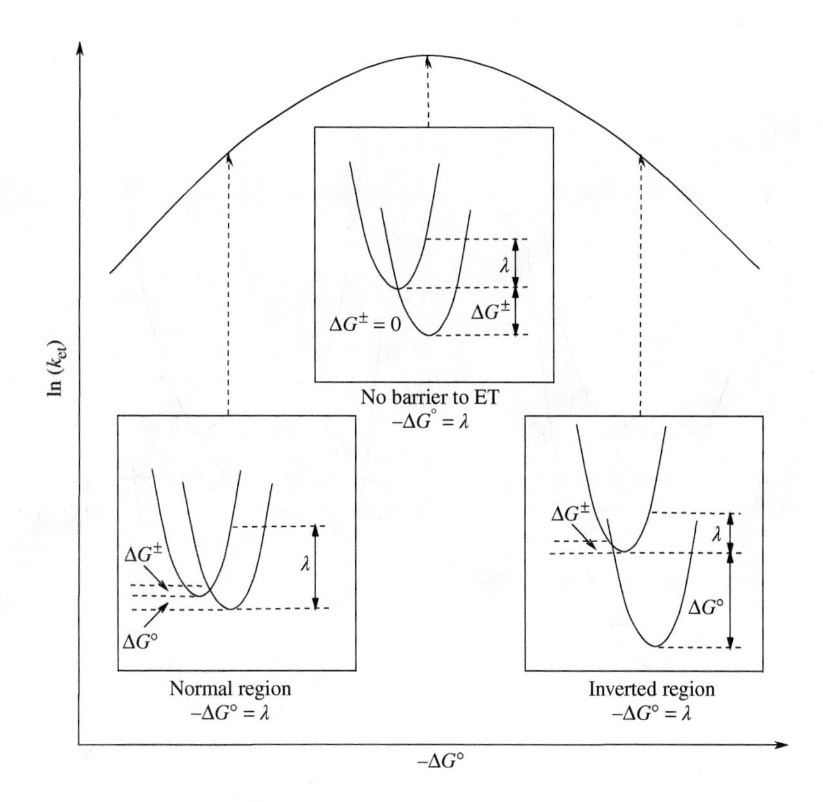

FIGURE 17.13
The dependence of the natural logarithm of the rate constant for ET on the thermodynamic driving force for the reaction.

so-called *normal region*, which agrees with how we intuitively expect the rate of ET to vary with the thermodynamic driving force. As ΔG° becomes more negative, the activation barrier decreases and the rate of ET increases with the magnitude of $|\Delta G^{\circ}|$. Eventually, when $-\lambda = \Delta G^{\circ}$, there is no vibrational barrier to ET and ln(k) reaches a maximum (often the ceiling for the ET rate constant reaches the diffusion-controlled limit, however). To the right of the maximum is the *Marcus inverted region*.

In the inverted region, the rate of ET decreases as ΔG° becomes more negative, as shown in Figure 17.13. At first glance, this might seem counterintuitive. Classically, it would appear as if there is no vibrational barrier to overcome because the two potential energy curves are imbedded with each other. However, the two wavefunctions are also orthogonal to each other so that a direct electronic transition between the two is not allowed. As with forbidden electronic transitions in the previous chapter, this obstacle can be overcome if there is a promoting vibration that can couple the two wavefunctions together through vibronic coupling. The inverse dependence on ΔG° occurs because there is a greater amount of vibrational overlap between the vibrational levels of the two electronic states when ΔG° is smaller, as shown in Figure 17.14. Typically, theories are used to justify huge volumes of experimental data; and, in fact, this was also the case for Marcus's theory of ET in the normal region. It is rare for a theory, however, to supersede the presence of experimental data, as is the case here for the Marcus inverted region. For several decades after Marcus postulated the existence of the inverted region, experimentalists failed to find any examples of ET reactions in support of the theory. It was not until 1984 that scientists were able to verify Marcus' prediction. Marcus won the Nobel Prize in chemistry several years later (in 1992) for his body of work on ET theory.

The majority of compounds that we will encounter as inorganic chemists, however, will fall in the normal region. In the late 1950s, Marcus developed his famous cross-relation, given by Equation (17.49), that relates the rate constant

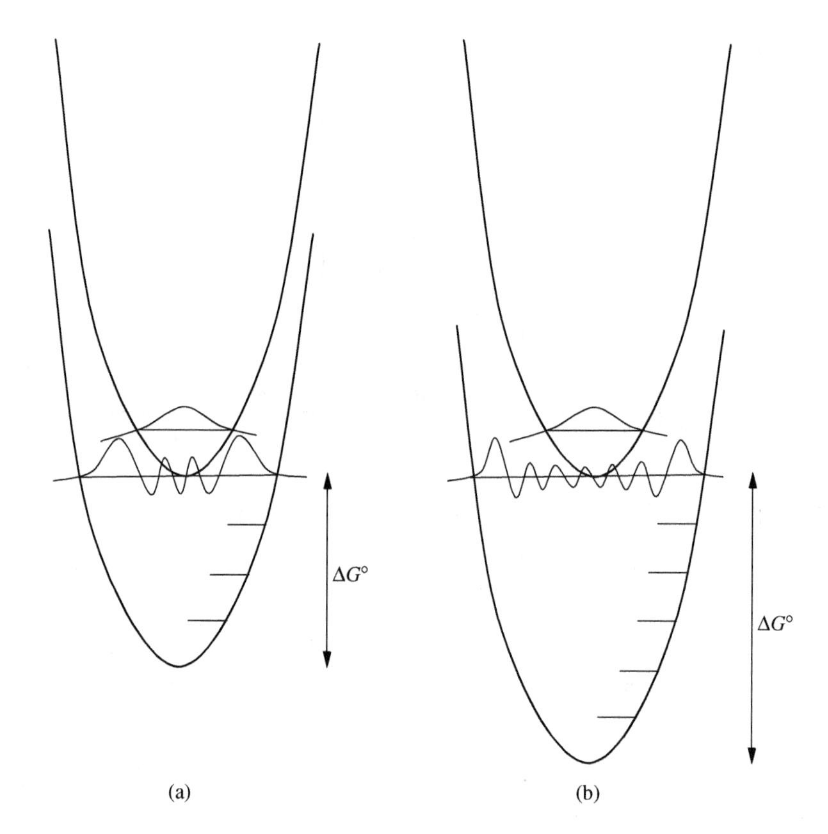

FIGURE 17.14
Electron transfer is faster in the Marcus inverted region when the magnitude of $\Delta G°$ is smaller because there is greater overlap between the ES and the GS, thus promoting vibrational wavefunctions in (a) than there is in (b).

for a cross-reaction (k_{12}) to the individual self-exchange rate constants (k_{11} and k_{22}) and to the thermodynamic driving force for the ET reaction (here measured as the equilibrium constant for the ET reaction, K_{12}). The values of K_{12} can be calculated from the difference in electrochemical potentials of the two coordination compounds from Equation (17.50). The f_{12} term is the frequency factor, which is typically close to unity for most coordination compounds. The experimental rate constants for some cross-exchange ET reactions and the predicted rate constants from Equation (17.49) are listed in Table 17.14. As a rule, there has been such good agreement between the experimental and calculated values of k_{12} for hundreds of

TABLE 17.14 Observed and calculated rate constants ($M^{-1}\ s^{-1}$) for selected cross-exchange ET reactions at 25 °C.

Cross-Reaction	$\log(K_{12})$	k_{12} (calc)	k_{12} (expt)
$[Ru(NH_3)_6]^{2+} + [Ru(NH_3)_5py]^{3+}$	4.40	4×10^6	1.4×10^6
$[Ru(NH_3)_6]^{2+} + [Co(phen)_3]^{3+}$	5.42	1×10^5	1.5×10^4
$[Mo(CN)_8]^{4-} + [IrCl_6]^{2-}$	2.18	8×10^5	1.9×10^6
$[Mo(CN)_8]^{4-} + MnO_4^-$	−4.07	6×10^1	2.7×10^2
$[Fe(CN)_6]^{4-} + [IrCl_6]^{2-}$	4.08	1×10^6	3.8×10^5
$[Fe(CN)_6]^{4-} + MnO_4^-$	3.40	6×10^4	1.7×10^5
$[Ru(en)_3]^{2+} + [Fe(H_2O)_6]^{3+}$		4×10^5	8.4×10^4
$[Fe(CN)_6]^{4-} + [Co(phen)_3]^{3+}$		2×10^3	6×10^6

compounds that when an exception occurs, there is likely to be a plausible justification for why it does not follow Marcus theory. Thus, for example, the apparent anomaly in the cross-reaction of $[Fe(CN)_6]^{4-} + [Co(phen)_3]^{3+}$ can be justified by the fact that the ions have different charges, so that the work term necessary to bring them to reaction distance requires less energy than the corresponding work terms for the $[Fe(CN)_6]^{4-/3-}$ and $[Co(phen)_3]^{3+/2+}$ self-exchange reactions, where the ions repel each other in each self-exchange.

$$k_{12} = (k_{11} k_{22} K_{12} f_{12})^{1/2} \tag{17.49}$$

$$K_{12} = e^{-\frac{\Delta G^\circ}{RT}} = e^{\frac{nF\Delta E^\circ}{RT}} \tag{17.50}$$

Example 17-4. Given the following information, calculate the cross-exchange rate constant for the reaction shown here. E° for the $[Fe(CN)_6]^{3-/4-}$ couple is 0.19 V versus SCE, while E° for the $[Mo(CN)_8]^{3-/4-}$ couple is 0.31 V versus SCE. The self-exchange rate constants for the two species are listed in Table 17.13 and the frequency factor f_{12} is 0.85.

$$[Fe(CN)_6]^{4-} + [Mo(CN)_8]^{3-} \rightarrow [Fe(CN)_6]^{3-} + [Mo(CN)_8]^{4-}$$

Solution. For the ET reaction given, $\Delta E^\circ = 0.12\,V$. Because of the relation: $\Delta G^\circ = -nF\Delta E^\circ$, the driving force for the reaction is $\Delta G^\circ = -12$ kJ/mol. Using $K_{12} = \exp(-\Delta G^\circ/RT)$, K_{12} is calculated as 1.1×10^2 at 298 K. Plugging the data into Equation (17.49), the calculated rate constant for cross-exchange is $2 \times 10^4\ M^{-1}\ s^{-1}$. The experimental value for the rate constant at 298 K is $3.0 \times 10^4\ M^{-1}\ s^{-1}$.

As we discovered in Example 17-3, the self-exchange rate constant for the $[Co(NH_3)_6]^{3+/2+}$ couple is very small as a result of the spin flip that accompanies the ET and the large inner-sphere reorganization energy that results when the antibonding e_g* MO is populated. The self-exchange rate constant for the $[Cr(H_2O)_6]^{3+/2+}$ couple is also small because the electron configuration changes from $t_{2g}{}^3e_g{}^0$ to $t_{2g}{}^3e_g{}^1$. Thus, it should come as no surprise that the cross-reaction shown in Equation (17.51) is extremely slow, with a second-order rate constant of $8.0 \times 10^{-5}\ M^{-1}\ s^{-1}$.

$$[Co(NH_3)_6]^{3+} + [Cr(H_2O)_6]^{2+} \rightarrow [Co(NH_3)_6]^{2+} + [Cr(H_2O)_6]^{3+} \tag{17.51}$$

It is therefore all the more surprising that the rate constant for the reaction shown in Equation (17.52) is $6.0 \times 10^5\ M^{-1}\ s^{-1}$, or approximately 10 orders of magnitude faster!

$$[Co(NH_3)_5Cl]^{2+} + [Cr(H_2O)_6]^{2+} \rightarrow [Cr(H_2O)_5Cl]^{2+} + Co^{2+} + H_2O + 5\,NH_3 \tag{17.52}$$

Clearly, there must be something fishy going on for there to be such a large discrepancy in the rate constant from that predicted by Marcus theory. The solution to this dilemma was first articulated by Henry Taube, who introduced the notion of inner-sphere ET reactions in the mid-1960s. Taube ran the abovementioned reaction in isotopically labeled *Cl^- solution. Because the radiolabeled Cl^- had a much higher concentration than that of the $[Co(NH_3)_5Cl]^{2+}$ coordination compound, one might expect there to be a reasonable amount of exchange of Cl^- ligands with *Cl^-

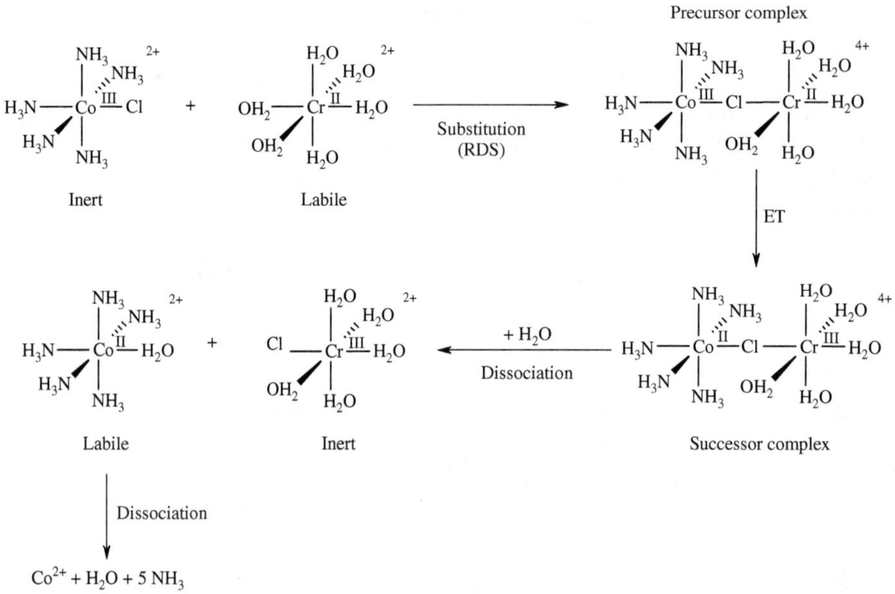

FIGURE 17.15
Proposed mechanism of the inner-sphere ET reaction in Equation (17.52).

from the solvent if the reaction occurred by dissociation of chloride from the cobalt, followed by addition of chloride to the chromium. However, hardly any of the radiolabeled chloride was found in the $[Cr(H_2O)_5Cl]^{2+}$ product! Taube's results supported the notion that the Cl^- ligand serves as a bridging ligand during the course of the reaction and that atom transfer accompanies ET.

The proposed mechanism in Figure 17.15 is consistent with the experimentally observed data. The first step of the mechanism involves substitution of a H_2O ligand on $[Cr(H_2O)_6]^{2+}$ to make the chloride-bridged precursor complex, $[(H_3N)_5Co^{III}\text{-}\mu(Cl)\text{-}Cr^{II}(H_2O)_5]^{4+}$. The substitution occurs readily because Co(III) is inert, while Cr(II) is labile. Once the bridging ligand has brought the two coordination compounds to within reaction distance, the electron can transfer from Co to Cr to form the successor complex, $[(H_3N)_5Co^{II}\text{-}\mu(Cl)\text{-}Cr^{III}(H_2O)_5]^{4+}$. The successor complex is unstable because Co(II) is labile, while Cr(III) is inert. Thus, the successor complex dissociates into $[Co(NH_3)_5H_2O]^{2+}$ and $[Cr(H_2O)_5Cl]^{2+}$, with the Cl^- staying with the more inert Cr(III). Because the former product is substitutionally labile, it rapidly dissociates its ligands to form Co^{2+} *(aq)* and 5 NH_3. In reality, Co^{2+} *(aq)* is the same as $[Co(H_2O)_6]^{2+}$, where the aqua ligands can exchange rapidly with the solvent. The overall rate of reaction for Equation (17.52) is so much faster than that for Equation (17.51) because the bridging Cl^- ligand lowers the energy of the activation barrier for ET by bringing the Co and Cr metal ions closer together than they could otherwise get in an outer-sphere ET reaction. In 1983, Henry Taube won the Nobel Prize in chemistry for his work on "the mechanisms of ET reactions."

In the abovementioned example, atom transfer was a consequence of the ET. The bridging chloride ligand in the precursor complex stayed with the more inert metal center throughout the entire reaction mechanism. Because the RDS is substitution to form the bridged complex, the rate of any inner-sphere substitution reaction is limited by the rate of substitution. Thus, if the rate of an ET reaction is ever faster than the rate of substitution, the ET reaction must occur by an outer-sphere pathway. As one might expect, the nature of the bridging ligand is very

TABLE 17.15 Rate constants for the inner-sphere ET reaction of $Co(NH_3)_5L]^{3+}$ and $[Cr(H_2O)_5]^{2+}$ as a function of the bridging ligand.

L	k, M^{-1} s^{-1}
NH_3	8.0×10^{-5}
F^-	2.5×10^{5}
Cl^-	6.0×10^{5}
Br^-	1.4×10^{6}
I^-	3.0×10^{6}
N_3^-	3.0×10^{5}
OH^-	1.5×10^{6}
NCS^-	19
SCN^-	1.9×10^{5}
H_2O	~0.1

Note that NH_3 cannot serve as a bridging ligand, but it is included here solely as a comparison.

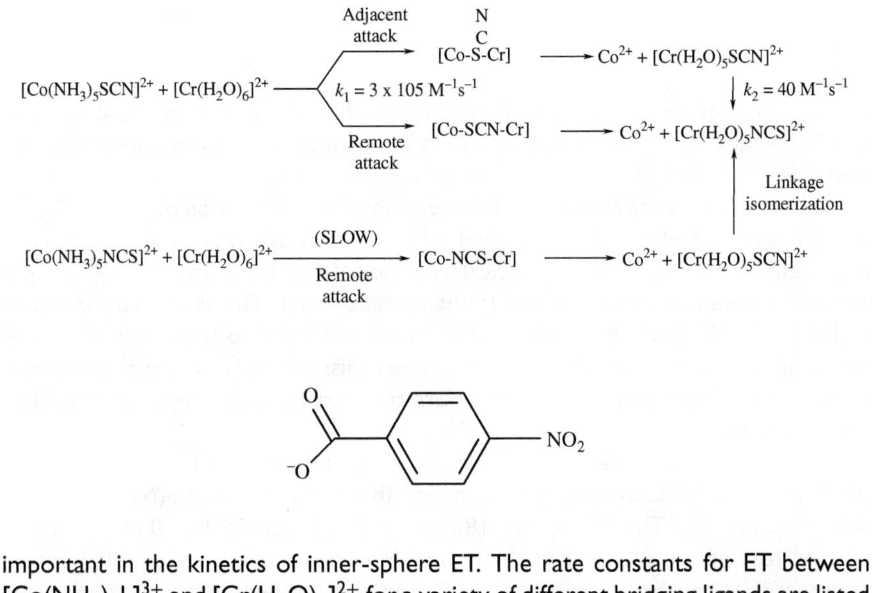

$[Co(NH_3)_5SCN]^{2+} + [Cr(H_2O)_6]^{2+}$ — Adjacent attack → [Co-S(CN)-Cr] → $Co^{2+} + [Cr(H_2O)_5SCN]^{2+}$

$k_1 = 3 \times 105\ M^{-1}s^{-1}$; $k_2 = 40\ M^{-1}s^{-1}$

Remote attack → [Co-SCN-Cr] → $Co^{2+} + [Cr(H_2O)_5NCS]^{2+}$

Linkage isomerization

$[Co(NH_3)_5NCS]^{2+} + [Cr(H_2O)_6]^{2+}$ — (SLOW) Remote attack → [Co-NCS-Cr] → $Co^{2+} + [Cr(H_2O)_5SCN]^{2+}$

FIGURE 17.16
Adjacent versus remote attack in the inner-sphere ET reactions of $[Co(NH_3)_5SCN]^{2+}$ or $[Co(NH_3)_5NCS]^{2+}$ with $[Cr(H_2O)_6]^{2+}$.

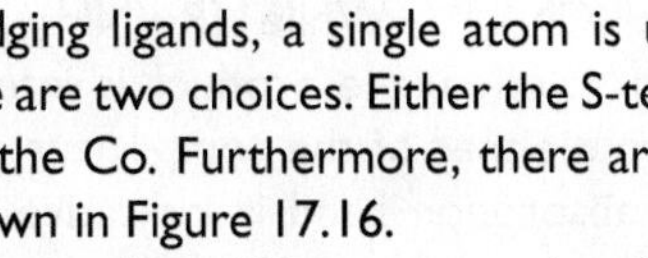

FIGURE 17.17
Example of a bridging ligand that can also be reduced to a radical.

important in the kinetics of inner-sphere ET. The rate constants for ET between $[Co(NH_3)_5L]^{3+}$ and $[Cr(H_2O)_5]^{2+}$ for a variety of different bridging ligands are listed in Table 17.15.

For most of the bridging ligands, a single atom is used as the ligand bridge. However, for SCN^-, there are two choices. Either the S-terminus or the N-terminus of the ligand can bind to the Co. Furthermore, there are two modes of attack by the bridging ligand, as shown in Figure 17.16.

In certain types of inner-sphere ET reactions, the electron may actually spend some time on the bridging ligand. For example, the bridging ligand shown in Figure 17.17 is itself reducible to form a radical intermediate. The detection of radicals can be a challenge to the experimentalist. So-called radical traps exist, but the kinetics of trapping must be faster than the kinetics of interest in order to prove the presence of the radical intermediate. If a radical mechanism is occurring, the ET occurs from the reductant to the ligand and it should not be affected much by the nature of the oxidant. On the other hand, if the ligand is only serving as

TABLE 17.16 The ratio of the rate constants for the reaction of $[Cr(H_2O)_6]^{2+}$ with either $[Co(NH_3)_5L]^{3+}$ or $[Cr(NH_3)_5L]^{3+}$ as a function of the bridging ligand.

L	k_{Co}/k_{Cr}
NCS^-	10^5
F^-	10^7
OH^-	10^9
N, O, NH_2	10
O, –O, COOH	0.4
O, –O, HOOC	50

a "wire" to connect the two metal complexes, then the nature of the oxidant is very important, as it will dictate the overall thermodynamic driving force for the reaction.

Table 17.16 lists the ratio of the rate constants for the reaction of $[Cr(H_2O)_6]^{2+}$ with either $[Co(NH_3)_5L]^{3+}$ or $[Cr(NH_3)_5L]^{3+}$. When this ratio is very large, the mechanism proceeds without a radical intermediate and the ligand bridge simply serves as a wire connecting the reductant to the oxidant. This is certainly the case for L = NCS^-, F^-, and OH^-. However, if the ratio is close to unity, as it is for the remaining ligands in Table 17.16, a radical mechanism is likely, where the electron spends some of its time on the ligand before transferring to the opposite transition metal complex.

Sometimes the successor complex of an inner-sphere ET reaction is also the final product. Consider, for example, the reaction of $[Ru(NH_3)_5H_2O]^{2+}$ with $[Fe(CN)_6]^{3-}$. The E° of the $[Ru(NH_3)_6]^{3+/2+}$ couple is −0.14 V versus SCE, while that for $[Fe(CN)_6]^{3-/4-}$ is 0.19 V versus SCE. Thus, Fe(III) is thermodynamically capable of oxidizing Ru(II). An inner-sphere reaction mechanism occurs with substitution for the aqua ligand, leading to the following ET product: $[(H_3N)_5Ru^{III}\text{-}\mu\text{-}NC\text{-}Fe^{II}(CN)_5]^-$. While $[Ru(NH_3)_6]^{3+}$ is pale yellow and $[Fe(CN)_6]^{4-}$ is colorless in solution, the product is intensely green, with a λ_{max} at 975 nm that is absent from either of the component coordination compounds. The presence of this new absorption band is indicative of a mixed-valence (MV) coordination compound.

MV compounds contain an electron donor (usually a metal in a low oxidation state) and an electron acceptor (typically a metal in a high oxidation state) within the same molecule. The electronic coupling between the donor and acceptor metals leads to new physical and spectroscopic properties for the MV compound, which are not present in either component alone. Most notable among these new properties is the presence of an intervalent transition (IT) in the UV/VIS or near-IR spectrum of the molecule. The term *IT* is usually reserved for those species where the two metals

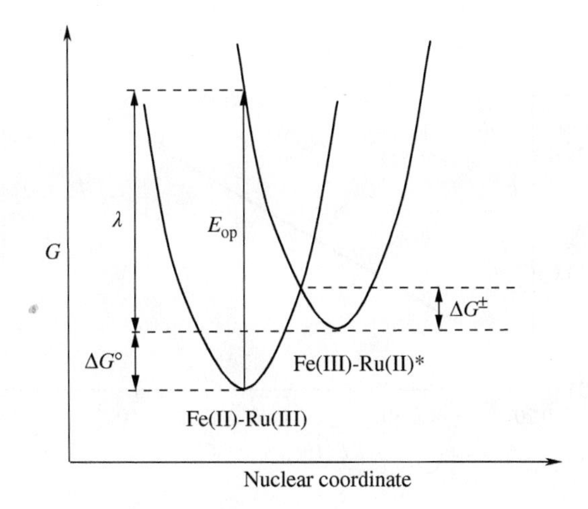

FIGURE 17.18
Free energy diagram for the MV compound $[(H_3N)_5Ru^{III}\text{-}\mu(NC)\text{-}Fe^{II}(CN)_5]^-$.

are in identical coordination environments and differ only in their formal oxidation numbers. For two different metals, the term *metal-to-metal charge transfer* (MM'CT) band is preferred.

While MV compounds can also be tight-ion pairs, such as $[Ru(NH_3)_6]^{3+}$, $[Fe(CN)_6{}^{4-}]$, most of them contain two (or more) transition metal complexes linked together by a bridging ligand(s). One of the defining characteristics of MV compounds is that the energy of the MM'CT band can be predicted using Marcus–Hush theory from the GS free energy difference (which is based on their redox potentials) and reorganization energy (λ) of the individual components. Conversely, the activation barrier for ET can be calculated from $\Delta E°$ and E_{op} (the energy of the MM'CT band). This is best illustrated by an example, with the help of the Marcus–Hush parameters illustrated in Figure 17.18.

The presence of an MM'CT band provides the molecule with an alternative way to effect the reverse ET. When light having a wavelength that falls inside the range of the MM'CT band impinges upon the MV compound, a photon can be used to photochemically transfer an electron from the reduced to the oxidized metal center (in this case, from Fe(II) → Ru(III) to generate the redox isomer in a vibrationally excited state. It should be self-evident from the definitions in Figure 17.18 that the optical energy (E_{op}) is equal to the sum of $\Delta G°$ and λ, as shown in Equation (17.53). Furthermore, because of the parabolic nature of the potential energy surfaces in the harmonic oscillator approximation, the value of $\Delta G^{\pm}$ can be calculated using the Hush relation in Equation (17.54). Thus, MV compounds provide a convenient way of relating GS thermodynamic properties to the rates of inner-sphere ET reactions, as k_{et} is related to $\Delta G^{\pm}$ through the Eyring equation.

$$E_{op} = \lambda + \Delta G° = \lambda - nF\Delta E° \quad (17.53)$$

$$\Delta G^{\pm} = \frac{E_{op}^2}{4\lambda} \quad (17.54)$$

In all of these examples, the products of the inner-sphere ET reaction had the same coordination number as the reactants. However, this is not always the case. In 1990, Zhou and Bocarsly reported the inner-sphere ET synthesis of a trinuclear MV compound, according to Equation (17.55). In this case, two ferricyanide ions are used to oxidize a square planar Pt(II) compound to form an octahedral Pt(IV), where

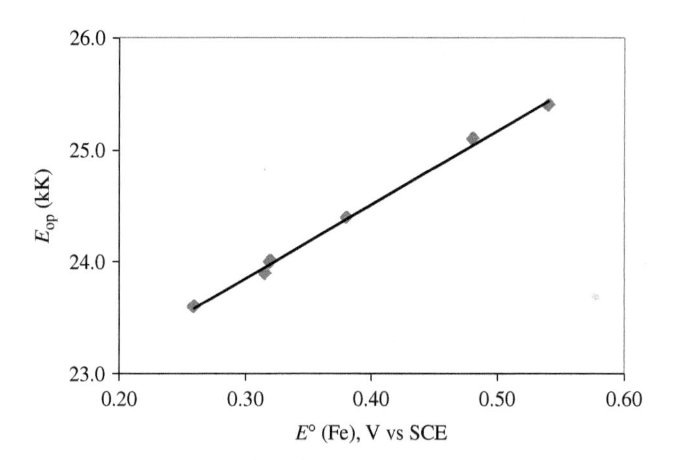

FIGURE 17.19
Linear relationship between E_{op} (MM'CT) and $E°$ (Fe) in a series of MV compounds having the formula: $[L(NC)_4Fe^{II}\text{-}\mu(CN)\text{-}Pt^{IV}(NH_3)_4\text{-}\mu(NC)\text{-}Fe^{II}(CN)_4L]^{4-}$.

the bridging cyanide ligands are retained in the MV product.

$$2[Fe^{III}(CN)_6]^{3-} + [Pt^{II}(NH_3)_4]^{2+} \rightarrow [(NC)_5Fe^{II} - \mu - (CN) - Pt^{IV}(NH_3)_4 - \mu - (NC) - Fe^{II}(CN)_5]^{4-} \qquad (17.55)$$

The MV compound has a deep red color with an MM'CT band at 424 nm. The energy of the IT band can be "tuned" in a predictable manner using Equation (17.53) by substituting $[Fe(CN)_5L]^{2-}$ for $[Fe(CN)_6]^{3-}$ in Equation (17.55). The energies of the MM'CT bands for the series of compounds having L = CN^-, 4,4′-bpy, pyrazine, 4-cyanopyridine, 2-pyridinecarboxylic acid, 2-fluoropyridine, and *N*-methylpyrazinium are shown in Figure 17.19. As expected, there is a linear correlation between E_{op} for the MM'CT band and $E°$ for the $[Fe(CN)_5L]^{2-}$ component.

Example 17-5. Given that λ_{max} for $[(H_3N)_5Ru^{III}\text{-}\mu\text{-}NC\text{-}Fe^{II}(CN)_5]^-$ is 975 nm (10, 300 cm^{-1}) and that $E° = 0.19$ V versus SCE for $[Fe(CN)_6]^{3-/4-}$ and $E° = -0.14$ V versus SCE for $[Ru(NH_3)_6]^{3+/2+}$, calculate $\Delta G^{\pm}$ for the MV compound.

Solution. Conversion of the wavelength of the IT band first into wavenumbers and then into energy units yields $E_{op} = 123$ kJ/mol. From $\Delta G° = -nF\Delta E°$, $\Delta G° = -1(96{,}485\ C/mol)(0.33\ V) = 32$ kJ/mol. By difference (from Eq. 17.53), the reorganization energy is $\lambda = 91$ kJ/mol. Using Equation (17.54) to calculate the free energy of activation, $\Delta G^{\pm}$ = 42 kJ/mol. This very large activation barrier effectively prevents reverse ET in the MV compound from reforming the original reactants by thermal ET.

The Hush relations given by Equations (17.53) and (17.54) strictly hold true only in those cases where the oxidation states in the MV compound are primarily localized on the two different metal centers. However, depending on the nature of the bridging ligand and the π-backbonding abilities of the metals, the electron might be significantly delocalized onto the ligand in the electronic ground state. In this event, the degree of electronic coupling between the two metal centers will be significant and it will affect the calculated free energy of activation. The magnitude of the electronic coupling is defined by H_{ab}, the matrix element $\langle \Phi_D | H | \Phi_A \rangle$, where H is the Hamiltonian derived from first-order perturbation theory and Φ_D and Φ_A are

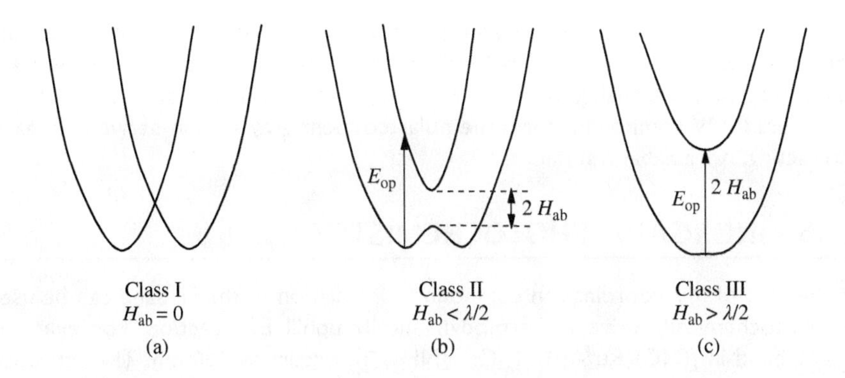

FIGURE 17.20
Robin and Day classification of the degree of electronic coupling (H_{ab}) in MV compounds. (a) Class I compounds are completely localized, (b) Class II compounds are weakly coupled, and (c) Class III compounds are delocalized.

the diabatic electronic wavefunctions for the ground and excited states, respectively. Whenever $H_{ab} \neq 0$, the ground and excited state potential energy surfaces in the MV compound are split by $2H_{ab}$, as shown in Figure 17.20. This electronic interaction manifests itself in the oscillator strength of the IT absorption band, as well as in the electrochemical and IR spectroscopic properties of the MV compound.

The Robin and Day classification of MV compounds is shown in Figure 17.20. Class I compounds have $H_{ab} = 0$. Because there is no electronic coupling at all between the metal centers, an IT band will not be observed. Class I compounds form the theoretical limit from which the Hush relations in Equations (17.53) and (17.54) were derived. Class II compounds have moderate values of H_{ab}, where $H_{ab} \leq \lambda/2$. The Hush relations can be used for Class II compounds with only the slightest of modifications. The magnitude of H_{ab} can be calculated from Equation (17.56), where ε_{max} is the molar absorptivity of the IT band at its maximum wavenumber (ν_{max}, cm^{-1}), $\Delta\nu_{1/2}$ is the full-width at half-maximum of the IR band (in cm^{-1}), and r_{ab} is the distance that the electron must transfer between the donor and acceptor metals, measured in Å.

$$H_{ab} = \frac{0.0206}{r_{ab}}(\varepsilon_{max}\nu_{max}\Delta\nu_{1/2})^{1/2} \tag{17.56}$$

For $[(H_3N)_5Ru^{III}\text{-}\mu(NC)\text{-}Fe^{II}(CN)_5]^-$, $\varepsilon_{max} = 3000\ M^{-1}\ cm^{-1}$, $\Delta\nu_{1/2} = 4900\ cm^{-1}$, and r_{ab} (estimated from crystallographic data) is 5.1 Å. Using Equation (17.56), the calculated value of $H_{ab} = 1600\ cm^{-1}$. This value is considerably smaller than the reorganization energy ($\lambda = 7600\ cm^{-1}$). Generally speaking, as long as $H_{ab} < \lambda/2$, the molecule can be classified as a Robin and Day Class II MV compound and the Hush equations will provide a reasonable estimate of the ET parameters.

Class III MV compounds are completely delocalized so that the oxidation states of the two metals are averaged. An arbitrary cutoff for Class III behavior is when $H_{ab} > \lambda/2$. One of the earliest known examples of an MV compound to exhibit Class III behavior is the Creutz–Taube ion, which is shown in Figure 17.21. The short nature of the ligand bridge and the even energy match between the d_π(Ru) and $\pi^*(p_z)$ MOs allow the transferring electron to be delocalized onto the bridging ligand, so that the oxidation states in the Creutz–Taube ion are best represented as

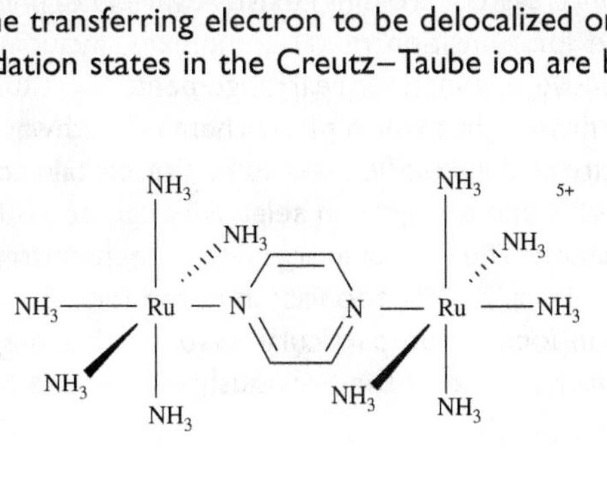

FIGURE 17.21
Structure of the Creutz–Taube ion.

+2.5 and +2.5 instead of as +2 and +3. In a sense, Class I MV compounds are the ET equivalent to ionic bonding (noninteracting, diabatic surfaces), while Class III MV compounds are analogous to covalent bonding (interacting, adiabatic surfaces) and Class II MV compounds form the polar covalent gray area in between (weakly interacting, nonadiabatic surfaces).

17.5 INORGANIC PHOTOCHEMISTRY

In the case of MV coordination compounds, irradiation of the IT band can be used to photochemically drive a thermodynamically uphill ET reaction. For example, the IT band in $[(NC)_5Ru^{II}\text{-}\mu(CN)\text{-}Co^{III}(NH_3)_5]^-$ occurs at 366 nm. The activation barrier for thermal ET in the MV compound can be calculated from spectroscopic and electrochemical data using Equations (17.53) and (17.54) and is quite large in this case (~100 kJ/mol). While the electron is thermally trapped, it can be optically transferred by irradiation of the Ru(II) → Co(III) IT band to generate a $[(NC)_5Ru^{III}\text{-}\mu(CN)\text{-}Co^{II}(NH_3)_5]^{-*}$ excited state. Because Co(II) is labile, the redox isomer in the ES rapidly decomposes into the final products: $[Ru^{III}(CN)_6]^{3-} + Co^{2+} + 5\ NH_3$ with a quantum yield of 0.46. The *quantum yield* is a measure of the efficiency of the photochemical process; a value of unity would indicate that 100% of the absorbed photons lead to the observed product. This example highlights one of the main advantages of using photochemistry: the absorption of light can be used to drive chemical reactions that are thermally inaccessible. Because photochemical excitation can be used to transfer an electron between an electron donor–acceptor pair, the electronic excited states of coordination compounds are often better oxidants and better reductants than the ground state. Thus, the ES has a wider range of chemical reactivity than the GS.

The first law of photochemistry was developed by Grotthüs and Draper and states that "only light absorbed by the reacting system can be effective in producing a chemical change." The second law of photochemistry, which was formulated by Einstein and Stark, is even more specific, stating that "each photon of light absorbed activates one molecule." The absorption of light by molecules occurs by an electronic transition between states derived from the one-electron MOs of the compound between the different term symbols describing the multielectronic wavefunctions of the molecule. However, the excitation often generates an ES electronic state that is vibrationally excited as well. This collection of Franck–Condon excited states can thermally equilibrate to a Boltzmann distribution population known as a *thexi* state, which is short for "thermally equilibrated excited state." Thexi states have an average molar enthalpy, entropy, free energy, and (in some cases) redox potential that are different from those of the GS. They also have a definite and unique structure, chemical reactivity, absorption spectrum, and IR spectrum; but they are very short-lived. The thexi state will eventually return back down to the GS unless quenching or photochemical pathways that have rate constants competitive with those for radiative decay (fluorescence or phosphorescence) or nonradiative decay (internal and external conversion and intersystem crossing) exist. A variety of photochemical pathways have been observed for transition metal complexes, including dissociation, ligand isomerization, reductive elimination, rearrangement, substitution, energy transfer, ET, and exciplex formation; but which photochemical pathway (if any) will be active depends on the nature of the specific thexi state. For certain compounds, more than one pathway is possible and a degree of selectivity can be achieved by changing the wavelength of irradiation. The field of inorganic photochemistry has burgeoned over the past several decades and is far too vast in scope for a comprehensive review in this text. Instead, the focus is on particular systems that might serve a pedagogical purpose to reinforce some of the previously introduced principles of inorganic chemistry.

17.5.1 Photochemistry of Chromium(III) Ammine Compounds

One of the more classical examples of inorganic photochemistry includes the photoaquation reactions of Cr(III) ammines. The generic reaction is shown in Equation (17.57), but a variety of N-coordinating ligands can be substituted for NH_3 in this expression and sometimes another ligand other than NH_3 is dissociated.

$$[Cr^{III}(NH_3)_5X]^{2+} + h\nu \rightarrow [Cr^{III}(NH_3)_4(H_2O)X]^{2+} + NH_3 \quad (17.57)$$

Some selected Cr(III) ammines, their photochemical products, and the quantum yields for product formation are listed in Table 17.17.

Some general observations about the data in Table 17.17 follow:

1. Although the Cr(III) ground state is substitutionally inert (d^3), the quantum yields for the photochemical processes are rather large.
2. The photochemical product differs from the thermal product in most cases and occurs $10^{14}-10^{17}$ times faster! For example, the thermal product for aquation of $[Cr(NH_3)_5Cl]^{2+}$ is $[Cr(NH_3)_5(H_2O)]^{3+}$, while the photochemical product is *cis*-$[Cr(NH_3)_4(H_2O)Cl]^{2+}$.
3. In general (although not always), the axis in the molecule with the weakest average LF strength is the one from which the ligand will be labilized. Further, if this axis contains more than one type of ligand, the one with the strongest LF strength is the one which is lost. This observation is known as *Adamson's rule.* The molecules shown in Figure 17.22 illustrate some examples of Adamson's rule.
4. The stereochemistry of the products usually follows the *edge displacement rule*, which is shown in Figure 17.23. The nucleophile enters the coordination

TABLE 17.17 Photoaquation reactions involving Cr(III) ammine compounds and their quantum yields (ϕ).

Reactant	Product	ϕ
$[Cr(NH_3)_5Cl]^{2+}$	*cis*-$[Cr(NH_3)_4(H_2O)Cl]^{2+}$	0.36
$[Cr(NH_3)_5Br]^{2+}$	*cis*- $[Cr(NH_3)_4(H_2O)Br]^{2+}$	0.35
$[Cr(NH_3)_5NCS]^{2+}$	*cis*- $[Cr(NH_3)_4(H_2O)NCS]^{2+}$	0.48
$[Cr(NH_3)_5CN]^{2+}$	*cis*- $[Cr(NH_3)_4(H_2O)CN]^{2+}$	0.33
trans-$[Cr(en)_2(NH_3)Cl]^{2+}$	*cis*-$[Cr(en)_2(H_2O)Cl]^{2+}$	0.34
trans-$[Cr(NH_3)_4Cl_2]^{+}$	*cis*-$[Cr(NH_3)_4(H_2O)Cl]^{2+}$	0.44
trans-$[Cr(NH_3)_4(H_2O^*)Cl]^{2+}$	*cis*-$[Cr(NH_3)_4(H_2O)Cl]^{2+}$	0.40
trans-$[Cr(NH_3)_4(NCS)Cl]^{+}$	*cis*-$[Cr(NH_3)_4(H_2O)Cl]^{2+}$ or *cis*-$[Cr(NH_3)_4(NCS)(H_2O)]^{2+}$	0.27 0.14
cis-$[Cr(NH_3)_4Cl_2]^{+}$	*mer*-$[Cr(NH_3)_3(H_2O)Cl_2]^{+}$	0.32
cis-$[Cr(en)_2Cl_2]^{+}$	en loss	0.13
trans-$[Cr(en)_2Cl_2]^{+}$	*cis*-$[Cr(en)_2(H_2O)Cl]^{2+}$	0.32
trans-$[Cr(en)_2Br_2]^{+}$	*cis*-$[Cr(en)_2(H_2O)Br]^{2+}$	0.16
trans-$[Cr(en)_2FCl]^{+}$	*trans*-$[Cr(en)_2(H_2O)F]^{2+}$	0.31

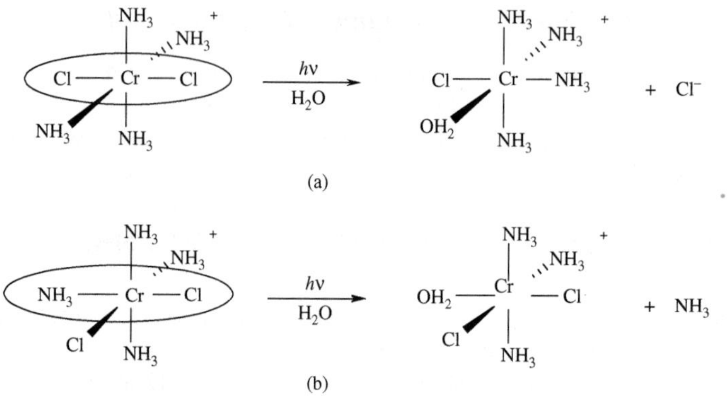

FIGURE 17.22
(a, b) Examples of Adamson's rule. The axis with the weakest overall LF strength is circled.

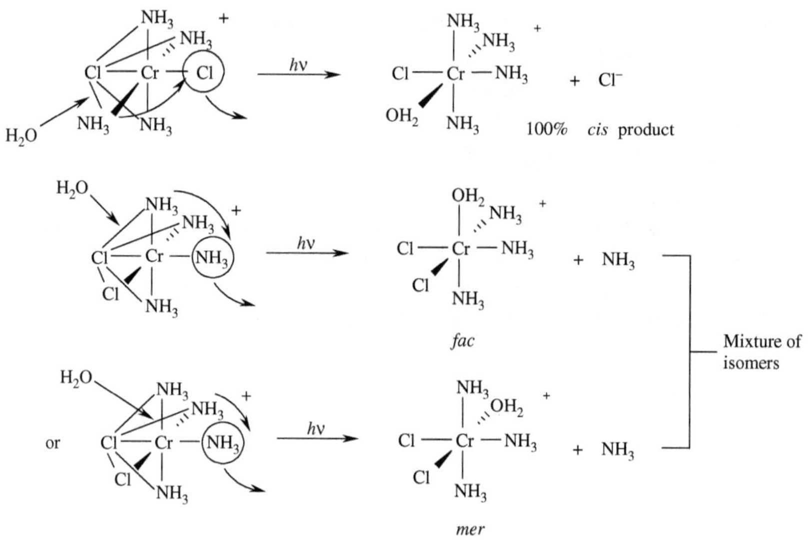

FIGURE 17.23
Examples of the edge displacement rule.

sphere of the octahedral complex along any of the four edges that are *trans* to the leaving group. The ligand in the *cis* position moves over to replace the leaving group. Some examples of the edge displacement rule are shown in Figure 17.23.

The observations listed raise more questions than they answer. First of all, which excited state(s) lead(s) to the observed photochemistry? Secondly, why do Adamson's rules seem to work? And lastly, what causes the edge displacement rule to predict the correct stereochemistry? The Tanabe–Sugano diagram for the d^3 chromium(III) ion is shown in Figure 17.24. The GS term symbol is $^4A_{2g}$. The lowest energy spin-allowed electronic transitions are $^4T_{2g} \leftarrow {}^4A_{2g}$ and $^4T_{1g} \leftarrow {}^4A_{2g}$. The lowest energy spin-forbidden electronic transitions are $^2T_{1g}$, $^2E_g \leftarrow {}^4A_{2g}$. Table 17.18 shows the quantum yield for the photochemical process as a function of the electronic transition. The data seem to indicate that internal conversion or intersystem crossing rapidly lead to population of the lowest energy quartet and doublet excited states, which are in thermal equilibrium with one another. For comparison, $[Cr(CN)_6]^{3-}$ phosphoresces from the doublet excited state and the molecule also undergoes a photoaquation reaction. The photochemical process occurs with identical efficiency even when the phosphorescence is quenched by O_2. These results imply that the photochemistry (at least in the case of $[Cr(CN)_6]^{3-}$) occurs from the ^4T_g ES.

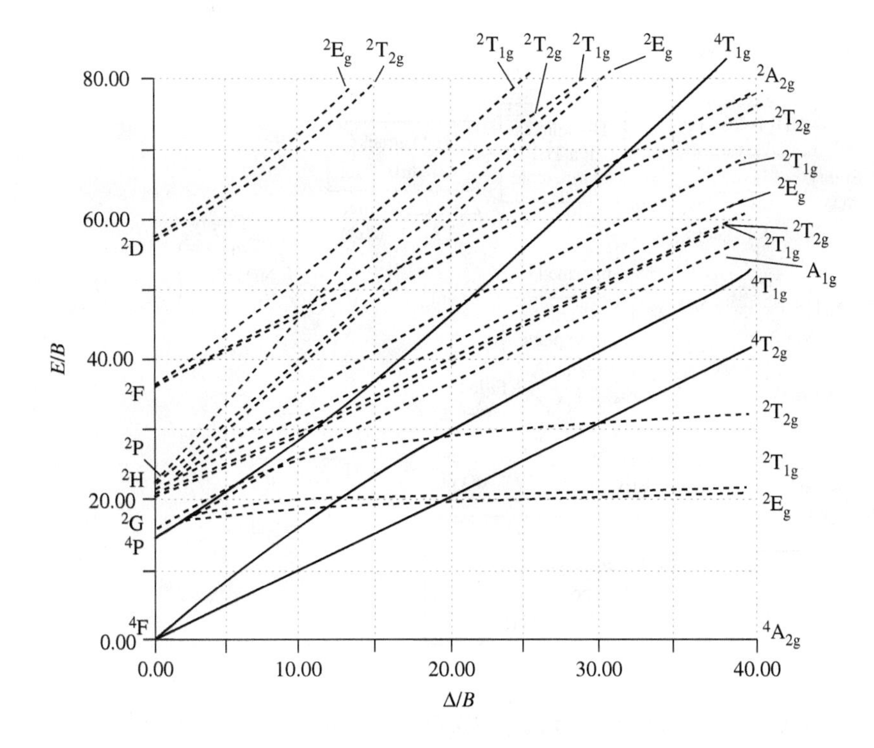

FIGURE 17.24 Tanabe–Sugano diagram for d^3 (O_h), where $B = 933\ cm^{-1}$ and $C/B = 4.5$. [Reproduced from Figgis, B. N.; Hitchman, M. A. *Ligand Field Theory and Its Applications*, Wiley-VCH: New York, 2000. This material is reproduced with permission of John Wiley & Sons, Inc.]

TABLE 17.18 The quantum yield of product formation for selected Cr(III) ammines as a function of the electronic transition excited photochemically.

Reactant	ϕ ($^4T_{1g}$ ES)	ϕ ($^4T_{2g}$ ES)	ϕ ($^2T_{2g}/^2E_g$ ES)
$[Cr(NH_3)_6]^{3+}$	0.47	0.47	0.47
cis-$[Cr(en)_2(NH_3)_2]^{3+}$	0.44	0.47	
trans-$[Cr(en)_2(NH_3)_2]^{3+}$	0.46	0.47	
$[Cr(en)_3]^{3+}$	0.37	0.37	0.40

Assuming that the photochemistry occurs from the quartet excited state, a semiempirical LF model can be used to estimate the ES bond strengths of the different ligands based on their σ- and π-donor strengths. The ligand with the weakest ES bond strength is the one that is labilized. Typically, this is the same ligand as the one predicted by Adamson's rule. The stereochemistry of the reaction can be rationalized using the one-electron MO treatment and reaction coordinate diagram for photoaquation of $[Cr(NH_3)_5Cl]^{2+}$ shown in Figure 17.25. In the octahedral geometry, the d^3 metal ion has a $t_{2g}^3e_g^0$ electron configuration. The edge displacement rule shows that all of the action occurs in a single plane of the molecule. Labeling the pertinent plane as the *xy*-plane, photochemical excitation of the molecule occurs by transferring an electron from the d_{xy} orbital in the t_{2g} level to the $d_{x^2-y^2}$ orbital in the e_g* level. Population of the antibonding $d_{x^2-y^2}$ orbital will weaken the bonding in the *xy*-plane. Because NH_3 is a stronger sigma donor than Cl^-, it will have the shorter bond length and will be the ligand that is most affected when the $d_{x^2-y^2}$ orbital is populated in the ES. Thus, it is the NH_3 ligand that is dissociated. The resulting square pyramidal TS relaxes to a trigonal bipyramidal intermediate, which

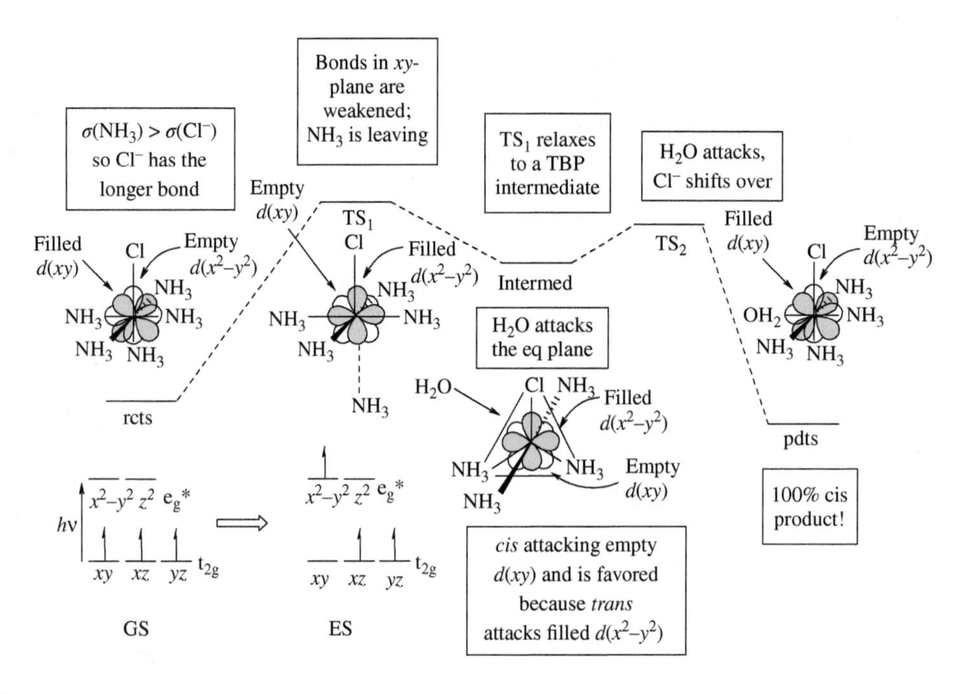

FIGURE 17.25
Proposed reaction coordinate diagram for the photoaquation of $[Cr(NH_3)_5Cl]^{2+}$ explaining the observed stereochemistry using the edge displacement rule.

is favored on the basis of sterics. In accordance with Bent's rule, the more electronegative Cl^- ligand will occupy one of the equatorial sites. In the second step, the nucleophile (water in this case) approaches the molecule along one of the edges in the equatorial plane, where there is more room for it. When the water molecule enters the plane, it will be *cis* to the Cl^- if it attacks along either of the two edges in the trigonal plane that contain the Cl^-, or it will be *trans* to the Cl^- if it attacks the opposite edge. Water is a good Lewis base, so it will head straight for an empty MO on the metal. Because the opposite edge contains the filled, antibonding $d_{x^2-y^2}$ orbital while the two adjacent edges contain the empty, lower-lying d_{xy} orbital, adjacent attack is strongly preferred on an electronic basis, leading exclusively to the *cis* product.

Example 17-6. The photoaquation of *cis*-$[Cr(NH_3)_4Cl_2]^+$ leads exclusively to the *mer*-$[Cr(NH_3)_3(H_2O)Cl]^+$ product. Using the edge displacement rule, construct a reaction coordinate diagram similar to the one in Figure 17.25 to explain the observed stereoselectivity.

Solution. See the given diagram. Assuming that the weaker Cl^- ligands occupy the *xy*-plane, photochemical excitation takes an electron from the d_{xy} orbital and places it in an antibonding $d_{x^2-y^2}$ orbital. This weakens the bonding of all the ligands in the *xy*-plane; but because the Cr–N bond length is shorter, the NH_3 ligands are especially affected. The first transition state therefore involves the loss of an ammine ligand from the plane. The square pyramidal TS relaxes to the more stable (based on sterics) trigonal bipyramidal intermediate, where the more electronegative Cl^- ligands lie in the equatorial plane. The H_2O nucleophile enters along one of the edges in the equatorial plane. Attack in the position *trans* to the equatorial ammine would be antibonding, so that *cis* attack is favored, leading to the *mer* geometric isomer. This intermediate is still in an electronic ES because the weaker Cl^- ligands should go with the antibonding orbital instead of the NH_3. Thus, the first intermediate can also undergo an electronic transition to form a second trigonal bipyramidal intermediate. In the second intermediate, *cis* attack would be antibonding, so that *trans* attack is favored, leading once again to the *mer* product.

17.5.2 Light-Induced Excited State Spin Trapping in Iron(II) Compounds

You might recall from our discussion of CFT that the metal–ligand bond length affects the magnitude of the CF splitting parameter, Δ_o. In point of fact, Δ_o varies as $1/r^6$, where r represents the M–L bond length. For the same compound, *r* can vary as a function of temperature because there are less energetic M–L bond stretches present at extremely low temperatures. If the coordination compound in question is poised near the interface between an HS and an LS complex, a crossover in the spin state can occur because of the changing magnitude of Δ_o with temperature. This is exactly what occurs when the temperature of crystalline $[Fe(ptz)_6](BF_4)_2$ is lowered from 293 K to 128 K (ptz = 1-propyltetrazole). The Fe–N bond length changes from 2.18 Å at 293 K to 1.98 Å at 128 K, thereby increasing the magnitude of Δ_o at the lower temperature. At 293 K, the complex is HS d^6 and appears white, with a single weak electronic transition at 850 nm (11.8 kK), as shown in Figure 17.26(a). If the temperature is lowered to 128 K, the compound becomes an intense red-purple color and it undergoes spin crossover to a LS d^6 configuration. As a result, its electronic spectrum now has four peaks: two very weak transitions at 980 nm (10.3 kK) and 670 nm (14.3 kK), and two stronger transitions at 549 nm (18.2 kK) and 379 nm (26.4 kK), as shown in Figure 17.26(b).

The Tanabe–Sugano diagram for a d^6 ion is shown in Figure 17.27. For the HS complex at 293 K, the ground-state term symbol is $^5T_{2g}$ and there is only one spin-allowed transition (to the ^5E_g ES). The GS term symbol for LS Fe(II) is $^1A_{1g}$ and several spin-allowed transitions to low-lying excited states are possible. The absorption bands for $[Fe(ptz)_6](BF_4)_2$ at 293 K and at 20 K are listed in Table 17.19 along with their assigned electronic transitions.

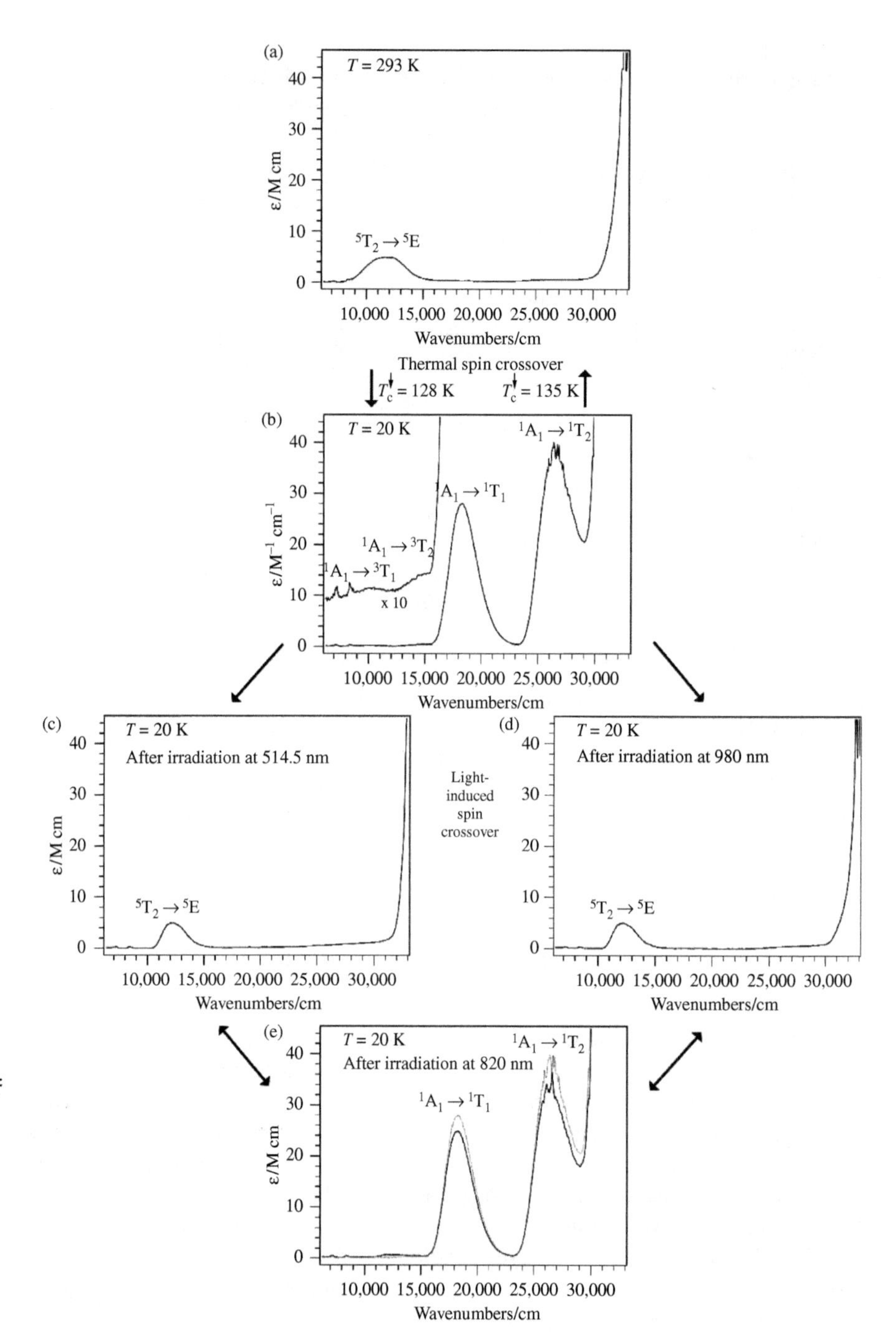

FIGURE 17.26
Electronic absorption spectra of $[Fe(ptz)_6](BF_4)_2$ (a) at 293 K, (b) after cooling to 20 K, (c) after irradiation at 514.5 nm at 20 K, (d) after irradiation of (c) at 820.0 nm at 20 K, and (e) after irradiation of (b) or (d) at 980.0 nm at 20 K. [Courtesy of Andreas Hauser.]

When the LS Fe(II) complex is irradiated at 20 K into its $^1T_{1g} \leftarrow {}^1A_{1g}$ transition using a LASER line at 514.5 nm, the molecule undergoes a photoinduced spin flip and reverts to the original HS Fe(II) compound, as shown in Figure 17.26(c). When the LS Fe(II) complex is irradiated at 20 K using 980.0 nm light, it is also converted back to the original HS compound, as shown in Figure 17.26(e). The resulting compound is metastable and can exist virtually forever when kept at 20 K. The process itself is known as the *LIESST effect*, which is short for "light-induced excited state spin trapping." Although the effect was first discovered using this specific compound, it has since been observed in a variety of spin crossover coordination complexes and is therefore a general phenomenon. Irradiation of the $^5E_g \leftarrow {}^5T_{2g}$ transition in the

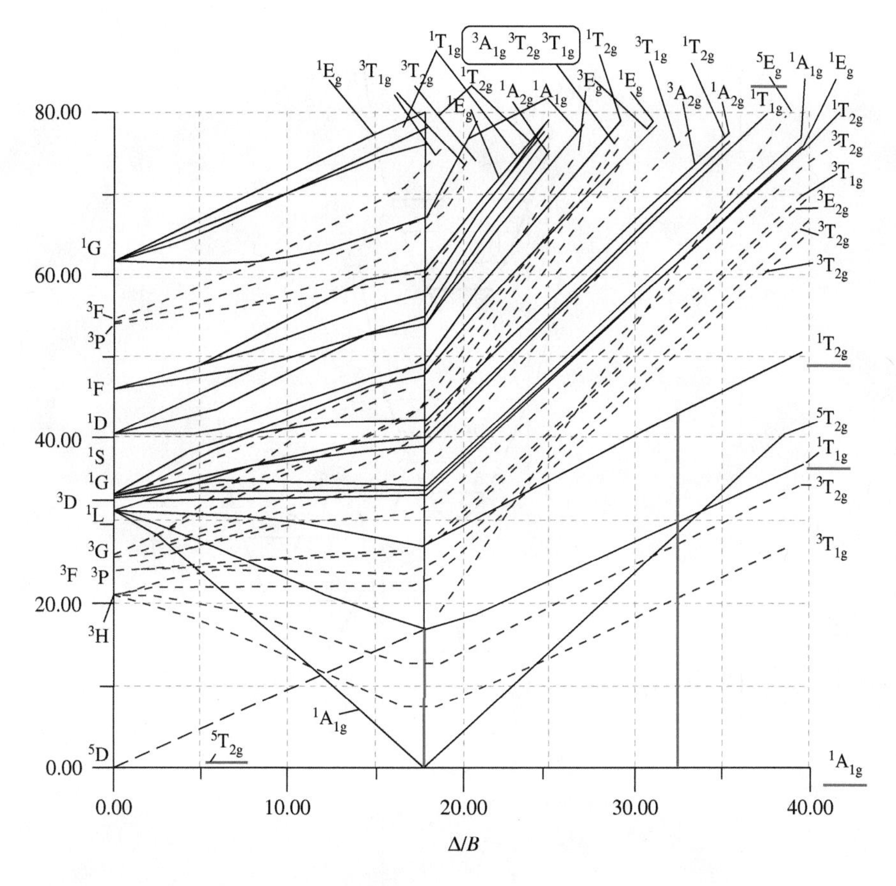

FIGURE 17.27
Tanabe–Sugano diagram for d^6 (O_h), where the diagram has been annotated to indicate the observed electronic transitions for both the HS and LS forms of $[Fe(ptz)_6](BF_4)_2$, which occur at 292 and 20 K, respectively. [Adapted from Figgis, B. N.; Hitchman, M. A. *Ligand Field Theory and Its Applications*, Wiley-VCH: New York, 2000. This material is reproduced with permission of John Wiley & Sons, Inc.]

TABLE 17.19 Absorption bands for crystalline $[Fe(ptz)_6](BF_4)_2$ and their assigned electronic transitions at two different temperatures.

293 K	HS d^6	20 K	LS d^6
850 nm (11.8 kK)	$^5E_g \leftarrow {}^5T_{2g}$	980 nm (10.3 kK)	$^3T_{1g} \leftarrow {}^1A_{1g}$
		670 nm (14.3 kK)	$^3T_{2g} \leftarrow {}^1A_{1g}$
		549 nm (18.2 kK)	$^1T_{1g} \leftarrow {}^1A_{1g}$
		379 nm (26.4 kK)	$^1T_{2g} \leftarrow {}^1A_{1g}$

HS Fe(II) compound in either (c) or (e) of Figure 17.26 with 820.0 nm light leads to a partial reverse LIESST effect, where the LS Fe(II) compound is photochemically regenerated at 20 K, as shown in Figure 17.26(d).

The fact that irradiation of both the $^3T_{1g} \leftarrow {}^1A_{1g}$ and $^1T_{1g} \leftarrow {}^1A_{1g}$ bands for the LS Fe(II) compound at 20 K lead to the observed LIESST effect implies that the lower energy $^3T_{1g}$ is an intermediate in the photochemical process. Excitation of the higher lying, spin-allowed $^1T_{1g} \leftarrow {}^1A_{1g}$ transition in the LS form of the complex is followed by intersystem crossing and internal conversion to the active $^3T_{1g}$ excited state, as shown in Figure 17.28. Likewise, irradiation of the spin-allowed $^5E_g \leftarrow {}^5T_{2g}$ in the HS complex at 20 K is followed by intersystem crossing to

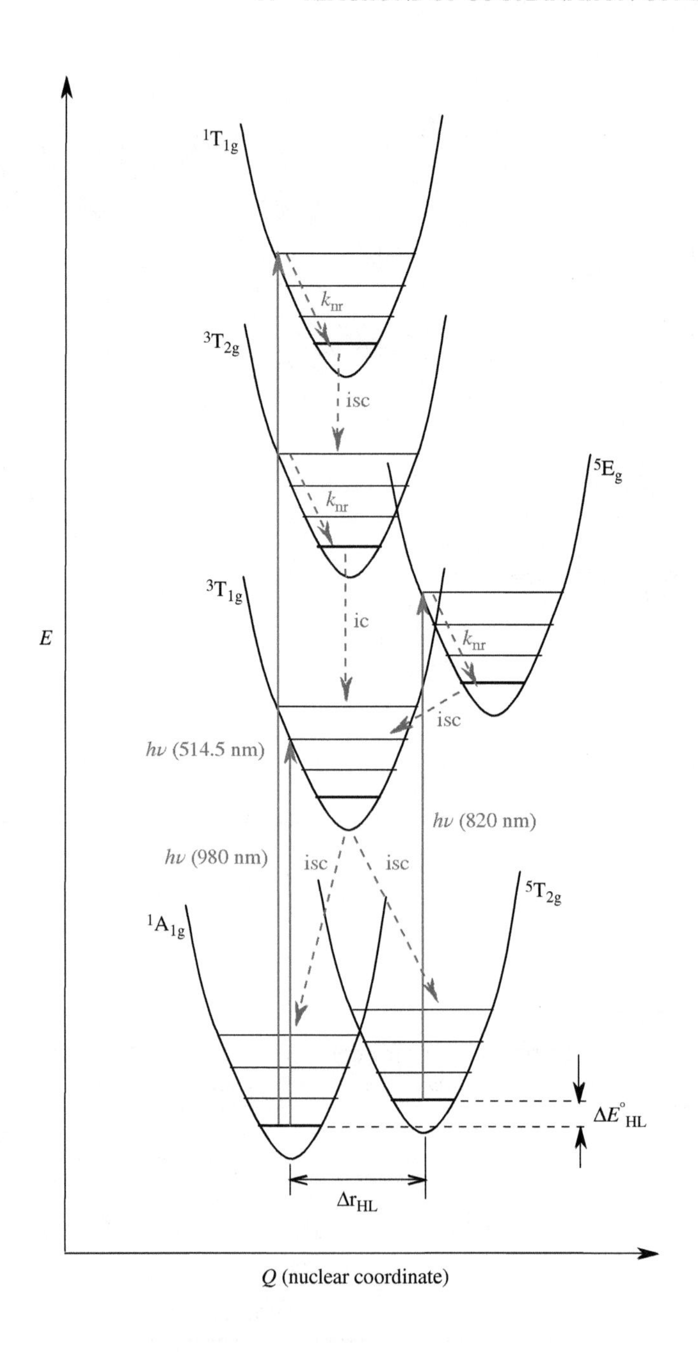

FIGURE 17.28
Jablonski diagram illustrating the role of the $^3T_{1g}$ in the LIESST and reverse-LIESST effect of spin crossover Fe(II) coordination compounds.

the same $^3T_{1g}$ excited state. This triplet state is fairly long-lived because relaxation to the $^5T_{2g}$ HS ground state is a spin-forbidden process and relaxation to the $^1A_{1g}$ LS ground state is likewise spin-forbidden. Thus, once the $^3T_{2g}$ ES is populated, it acts as a fulcrum for the photo-induced ET and allows the electron to be transferred in one direction or the other based on the wavelength of irradiation. Furthermore, the fact that the higher energy HS Fe(II) complex is metastable following LIESST implies that the thermal activation barrier for spin crossover from the $^5T_{2g}$ to the $^1A_{1g}$ is rather sizeable. The combination of these two physical properties is what makes the molecule work as a sort of optical switch. Furthermore, because the HS state is strongly paramagnetic and the LS Fe(II) complex has all of its electrons paired, the magnetic properties of the molecule can also be tuned photochemically!

Example 17-7. Using the energies of the electronic transitions for the HS and LS forms of $[Fe(ptz)_6](BF_4)_2$ listed in Table 17.19, calculate the values of Δ_o at 293 K and at 20 K.

Solution. The ratio of the energies of the $^1T_{2g} \leftarrow {}^1A_{1g}$ and $^1T_{1g} \leftarrow {}^1A_{1g}$ electronic transitions is 26.4 kK/18.2 kK = 1.45. Sliding our ruler across the horizontal axis of the Tanabe–Sugano diagram in Figure 17.27, we find that the point at which the ratio of these two transitions is 1.45 occurs at a value of $Dq/B' = 3.25$, as indicated by the vertical line on the LS side of the diagram. The corresponding values of E/B' at this point on the Tanabe–Sugano diagram are 43.0 and 30.0 kK, corresponding to a ratio of 1.5. Given that $E_2 = 26.4$ kK and (E_2/B') occurs at 43.0 kK, the calculated value of B' is 0.610 kK or 610 cm^{-1}. A quick check of this value using $E_1 = 18.2$ kK and $E_1/B' = 30.0$ kK confirms this result. Using this value of B' and the fact that $Dq/B' = 3.25$, the calculated value for the LS $Dq = 1.98$ kK, so that $\Delta_o = 10\ Dq = 19.8$ kK at 20 K, in close agreement with the value of 20.2 kK reported by Hauser. On the HS side, there is a single electronic transition ($^5E_g \leftarrow {}^5T_{2g}$) occurring at 11.8 kK. Given that the CF splitting between the T_{2g} and E_g terms in an octahedron is *defined* as Δ_o, this value will be 11.8 kK for the HS complex at 293 K.

Example 17-8. Given that the value of Δ_o goes as $1/r^6$ and that the measured Fe–N bond lengths at 293 and 20 K are 2.18 and 1.98 Å, respectively, show that the calculated ratio of $\Delta_o(LS)/\Delta_o(HS)$ from Example 17-7 agrees with the ratio predicted on the basis of the difference in metal–ligand bond lengths.

Solution. The ratio $(2.18)^6/(1.98)^6 = 1.78$ is roughly equal to the ratio $(19.8)/(11.8) = 1.68$.

17.5.3 MLCT Photochemistry in Pentaammineruthenium(II) Compounds

For the series of pentaammineruthenium(II) compounds having the general formula $[Ru(NH_3)_5\ L]^{2+}$, where L is a substituted pyridine ligand, two types of electronic transitions are observed in the UV/VIS spectra. The weaker LF transitions around 400 nm are largely obscured by the more intense Ru(II) → L MLCT absorptions, which range from 400 to 550 nm depending on the nature of the pyridine substituents. Because all of the compounds contain five Ru–NH_3 bonds and one Ru–py bond, the LF strengths of the different species should be relatively unaffected by the nature of the substituent group. On the other hand, the energy of the MLCT band is strongly dependent on the Hammett sigma parameter of the substituted pyridine ligand. The standard reduction potential of the ligand will shift more positive as electron-withdrawing groups are added to the pyridine, while the redox potential of the metal will only shift slightly. Therefore, the MLCT band will move to lower energies (longer wavelengths) for the more electron-withdrawing substituents, as shown in Figure 17.29.

The pentaammineruthenium(II) complexes undergo photosubstitution of the pyridine ligand with the quantum yields indicated in Table 17.20 when their MLCT bands are irradiated at pH = 3 and an ionic strength of 0.2 M.

For the first five compounds in the table, the quantum yields are moderately large, but they then drop off by 2–3 orders of magnitude as more electron-withdrawing groups are added to the pyridine and the MLCT band is shifted to lower and lower energies. One proposed mechanism for the observed

TABLE 17.20 Wavelengths and quantum yields for the photosubstitution of pyridine in $[Ru(NH_3)_5L]^{2+}$ as a function of the pyridine substituents.

L	λ_{max}^{MLCT}, nm	$\phi_L \times 10^3$
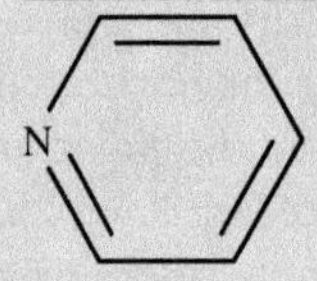 N	405	45
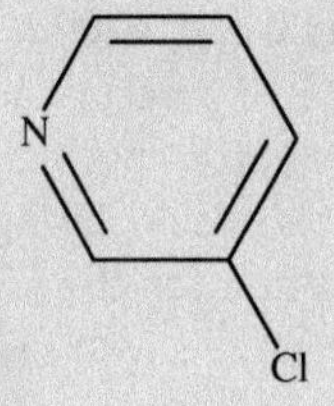 N, Cl	426	48
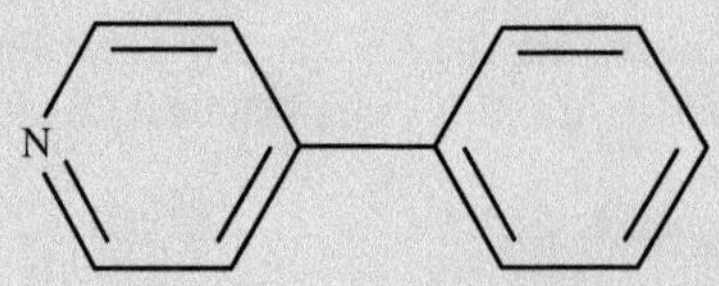 N	446	33
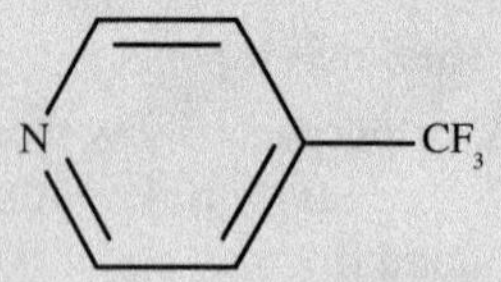 N, CF_3	454	22
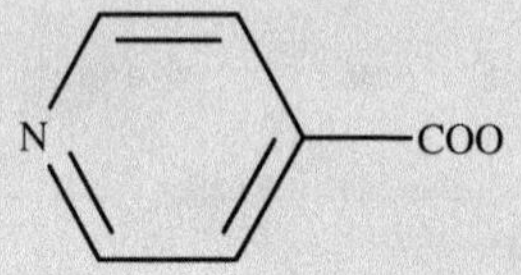 N, COO	457	26
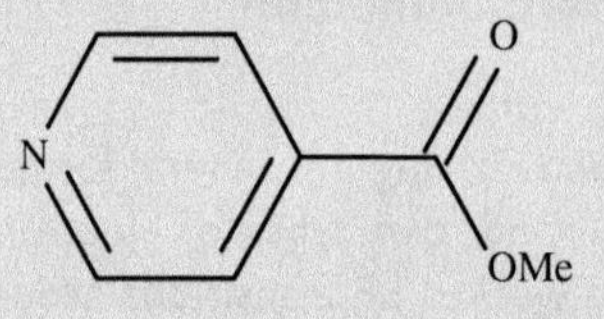 N, O, OMe	497	0.26
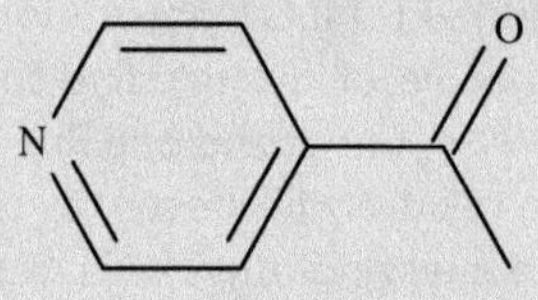 N, O	523	0.25
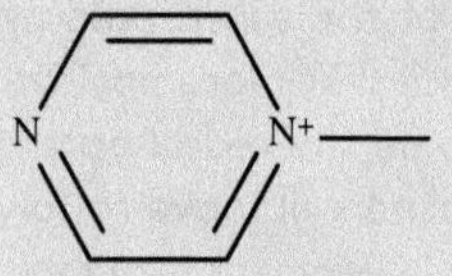 N, N^+	540	0.04
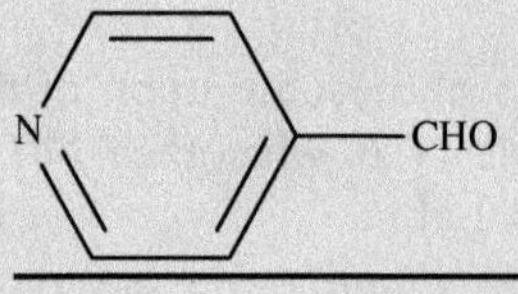 N, CHO	545	0.05

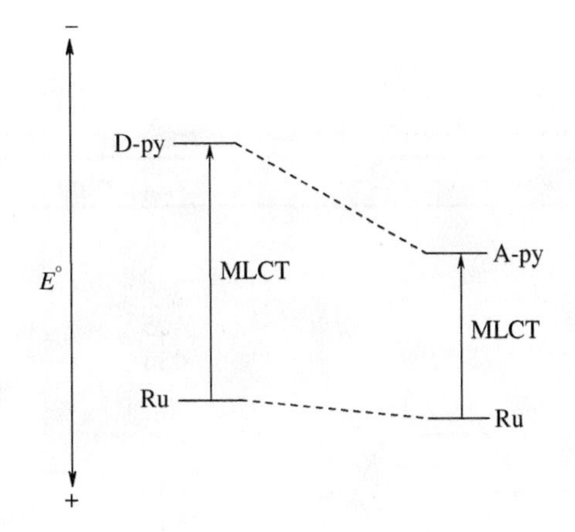

FIGURE 17.29
The MLCT band in $[Ru(NH_3)_5L]^{2+}$ shifts to lower energies as more electron-withdrawing substituents are added to the pyridine ligand because the redox potential of L shifts more positive.

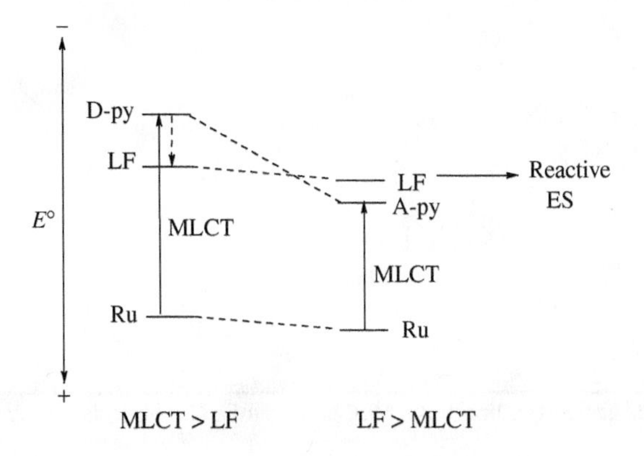

FIGURE 17.30
Proposed explanation for the dependence of the quantum yield of photosubstitution in $[Ru(NH_3)_5L]^{2+}$ on the nature of the substituents on L and the wavelength of irradiation.

photochemistry is that the photochemistry occurs from the LF, not the MLCT, excited state. Excitation of the MLCT band is followed by rapid intersystem crossing and internal conversion to populate the $^3T_{1g}$ LF excited state, which then undergoes the photosubstitution, as shown in Figure 17.30. When the more electron-withdrawing substituents are added to the pyridine ligand, the energy of the MLCT excited state falls below that of the LF excited state and the quantum yield of the reaction drops precipitously. For those compounds where the energies of the two excited states are roughly the same, a simple change of the solvent to favor a lower LF excited state over the MLCT excited state increases the quantum yield, in support of the proposed mechanism. Furthermore, the wavelength-dependence studies shown in Table 17.21 indicate that when the MLCT > LF excited state, the quantum yield of photoaquation increases at longer wavelengths as the energy of the light approaches that of the LF excited state; but when the LF > MLCT excited state, the quantum yield decreases at longer wavelengths where the LF excited state is less populated.

17.5.4 Photochemistry and Photophysics of Ruthenium(II) Polypyridyl Compounds

Literally thousands of papers have been written describing the photochemical and photophysical properties of $[Ru(bpy)_3]^{2+}$ over the past several decades, making it one of the most widely studied compounds in the history of inorganic chemistry. The compound exhibits a rich electronic spectrum, with several high-energy $\pi \rightarrow \pi^*$ ligand-centered transitions in the near-UV, weak LF absorptions around 400 nm, and a series of intense MLCT bands around 450 nm. Excitation of the MLCT bands leads

TABLE 17.21 Wavelength dependence of the quantum yield of photosubstitution in $[Ru(NH_3)_5\,L]^{2+}$ as a function of the electron-withdrawing abilities of L.

L	λ_{irrad}, nm	$\phi_L \times 10^3$
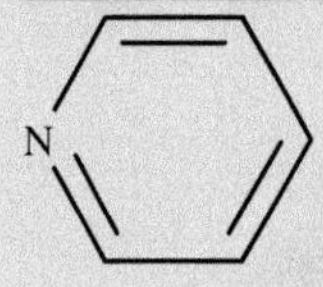	366	43
	405	45
	436	51
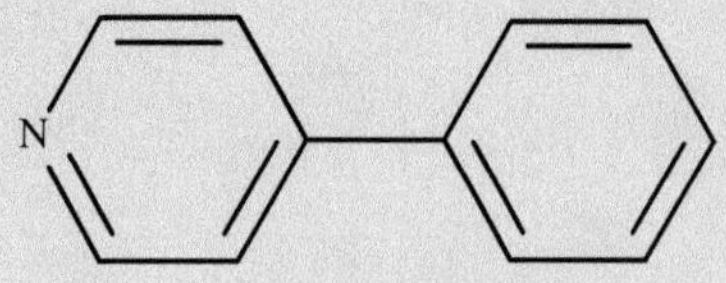	405	31
	449	33
	500	37
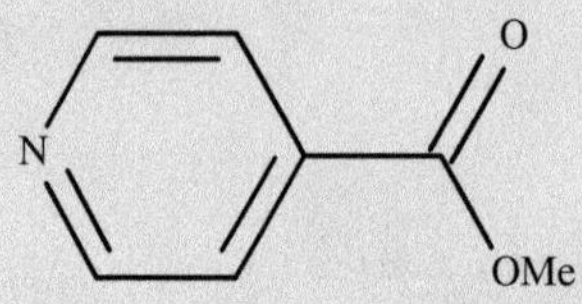	449	1.3
	479	0.72
	500	0.27
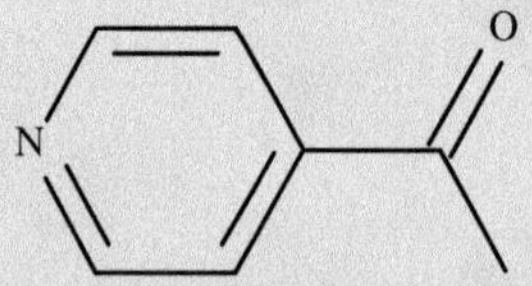	405	4.5
	449	1.4
	520	0.25

Those ligands above the line have MLCT > LF, while the ligands below the line have LF > MLCT.

to a fairly strong emission at 627 nm with a room-temperature excited-state lifetime of 620 ns in aqueous solution. The luminescent MLCT excited state is fairly stable with respect to photosubstitution, and its relatively long ES lifetime allows it to undergo bimolecular reactions with other molecules in solution.

What makes this compound so interesting is the fact that its MLCT ES is both a better oxidant and a better reductant than the GS, which imparts the ES with a wider range of chemical reactivity than the GS. Following excitation of the Ru(II) → bpy MLCT band, the electron is formally transferred from a t_{2g} orbital on the metal to a π^* MO on the bpy ligand. Although there has been some debate over the years as to whether the MLCT excited state has the electron localized on a single bipyridine or delocalized over all three ligands, it now appears that the ES electron is localized to form a $[Ru(III)\text{-}bpy^{\cdot -}]^*$ ES. The simplified MO diagram shown in Figure 17.31 provides a good overview of the relative energies of the different energy levels in $[Ru(bpy)_3]^{2+}$. The MLCT ES is a better oxidant than the GS because the complex can donate an electron more easily from the $bpy^{\cdot -}$ radical anion than it could from the filled t_{2g} level in the Ru(II) GS. The ES is also a better reductant than the GS because the vacancy in the t_{2g} level of the $[Ru(III)\text{-}bpy^{\cdot -}]^*$ ES can accept an electron more readily than the empty e_g^* in the Ru(II) GS.

The reduction potentials of the excited state can be calculated from the GS reduction potentials and knowledge of the difference in energy (E_{0-0}) between the GS and ES, according to Equations (17.58) and (17.59), as shown schematically in Figure 17.32. E_{0-0} is typically calculated from the wavelength where the MLCT

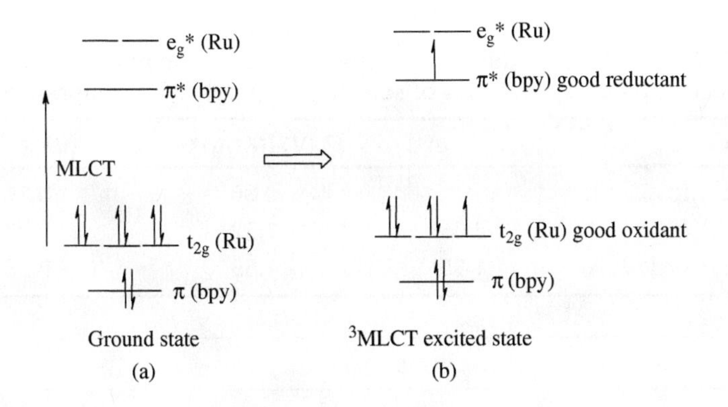

FIGURE 17.31 Simplified MO diagram for $[Ru(bpy)_3]^{2+}$ in its GS (a) and $[Ru(III)\text{-}bpy^{\cdot-}]^*$ ES (b).

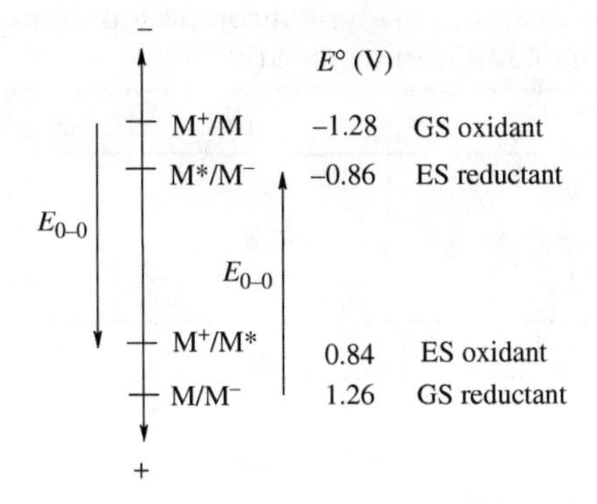

FIGURE 17.32 Schematic diagram indicating the relationships between the GS and ES potentials for $[Ru(bpy)_3]^{2+}$, where $E_{0-0} = 2.12$ eV.

absorption and emission bands overlap.

$$E°(M^+/M*) = E°(M^+/M) - E_{0-0} \tag{17.58}$$

$$E°(M*/M^-) = E°(M/M^-) + E_{0-0} \tag{17.59}$$

The relative energies of the different excited states in $[Ru(bpy)_3]^{2+}$ and related compounds can be tuned by introducing substituents to the bpy ring or by changing the nature of the chelating ligand. Because only one or all three of the ligands can be modified, literally hundreds of different combinations are possible. Consider the three examples listed in Table 17.22. The electron-donating ability of the ligand increases down the series. The stronger the σ-donating substituent on the 1,10-phenanthroline ring (phen), the easier it will be to oxidize the MLCT ES and the harder it will be to reduce the ES.

By replacing one of the chelates with two monodentate ligands, the energy of E_{0-0} can be tuned, as shown in Table 17.23. Electron donors raise the energy of the

TABLE 17.22 Effects of the ligand substituents on the photophysical and electrochemical properties of selected $[Ru(phen)_3]^{2+}$ compounds.

Compound	$E_{0-0,}$ eV	$E°$ (V) (M^+/M^*)	$E°$ (V) (M^*/M^-)
$[Ru(5\text{-}Cl\text{-}phen)_3]^{2+}$	2.13	−1.01	0.76
$[Ru(phen)_3]^{2+}$	2.13	−1.11	0.55
$[Ru(dimethyl\text{-}phen)_3]^{2+}$	2.08	−1.25	0.43

TABLE 17.23 Effects of the nonchelating ligand on the photophysical and electrochemical properties of selected $[Ru(bpy)_2X_2]^{2+}$ compounds.

Compound	E_{0-0}, eV	$E°$ (V) (M^+/M^*)	$E°$ (V) (M^*/M^-)
$[Ru(bpy)_2(py)_2]^{2+}$	>2	−0.86	0.76
$[Ru(bpy)_2(CN)_2]$	2.05	−1.32	0.37
$[Ru(bpy)_2Cl_2]$	1.65	−1.59	−0.26

TABLE 17.24 Effects of mixed chelate ligands on the photophysical properties of selected Ru(II) polypyridyl compounds, where dcb = 2,2′-bipyridine-4,4′-dicarboxylic acid.

Compound	λ_{max}, nm	λ_{em}, nm	τ, ns
$[Ru(dcb)_3]^{2+}$	464	635	800
$[Ru(dcb)_2(bpy)]^{2+}$	462	651	560
$[Ru(dcb)_2(dimethyl\text{-}bpy)]^{2+}$	474	655	485

t_{2g} MOs, lowering the energy of the MLCT band. The redox potentials of both the GS and ES will also vary with the ligand.

When using mixed chelate ligands on the Ru complexes, the lowest-energy MLCT (from which the emission occurs) is the chelate which is most easily reduced. This makes sense from the simplified MO diagram in Figure 17.32, as the more easily reduced ligand will have its π^* MO at lower energy. Several examples are shown in Table 17.24.

Notice that the lifetime (τ) of the emissive ES decreases as the energy of emission decreases. In fact, this is a general phenomenon known as the *energy gap law*. The *energy gap law* is completely analogous to the behavior observed for the Marcus inverted region of ET theory. As the energy difference between the GS and MLCT ES increases, the rate constant for nonradiative decay (k_{nr}) back to the GS decreases because of the decreased rate of intramolecular ET in the complex. When the ES is potentially emissive, a decrease in k_{nr} implies less competition for deactivation of the MLCT ES by a nonradiative pathway and therefore a longer lived emissive ES. The effect is particularly pronounced in Os(II) polypyridyl compounds, where the larger LF strength of Os increases the energy of the LF transitions so that they do not interfere with the photochemistry of the lower lying MLCT states. The emission energies, excited state lifetimes, and nonradiative decay rate constants for a series of Os(II) polypyridyl compounds are listed in Table 17.25. The definitions of the abbreviations for the different ligands can be found in the references. A plot of $\ln(k_{nr})$ versus E_{em} (kK) is linear, as shown in Figure 17.33.

Because the ES lifetimes of Ru(II) polypyridyl compounds are long-lived enough to engage in bimolecular collisions with other compounds in the solvent and considering their redox properties, it is possible to quench the emissive MLCT ES by either oxidative quenching or reductive quenching. These are shown in Equations (17.60) and (17.61), respectively.

$$[Ru(bpy)_3]^{2+*} + ArNO_2 \rightarrow [Ru(bpy)_3]^{3+} + ArNO_2^- \qquad (17.60)$$

$$[Ru(bpy)_3]^{2+*} + R_3N \rightarrow [Ru(bpy)_3]^{+} + R_3N^{+} \qquad (17.61)$$

TABLE 17.25 The emission energies, excited state lifetimes, and nonradiative decay rate constants for a series of Os(II) polypyridyl compounds.

Compound	E_{em}, kK	τ, ns	$k_{nr} \times 10^{-6}$, s^{-1}
$[Os(bpy)(dppy)_2]^{2+}$	18.62	1684	0.37
$[Os(bpy)_2(DMSO)_2]^{2+}$	17.39	1500	0.47
$[Os(bpy)_2(CNCH_2Ph)_2]^{2+}$	17.24	1228	0.61
$[Os(bpy)_2(dppy)]^{2+}$	15.87	500	1.8
$[Os(bpy)_2(dppb)]^{2+}$	15.72	344	2.8
$[Os(bpy)_2(PPh_2Me)_2]^{2+}$	15.11	261	3.7
$[Os(bpy)_2(CH_3CN)_2]^{2+}$	14.12	157	6.3
$[Os(bpy)_2(pyz)_2]^{2+}$	13.93	93	10.8
$[Os(bpy)_3]^{2+}$	13.46	60	16.6
$[Os(bpy)_2(py)_2]^{2+}$	12.99	41	24.4

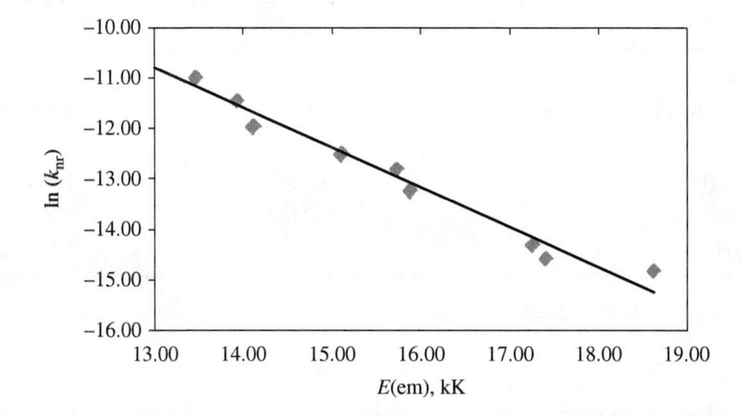

FIGURE 17.33
Linear relationship between $\ln(k_{nr})$ and E_{em} (kK) for the Os(II) polypyridyl compounds in Table 17.25.

Table 17.26 lists some of the quenching rate constants (k_q) for oxidative quenching of $[Ru(bpy)_3]^{2+*}$ with a series of $ArNO_2$ quenchers. The $E°(M^+/M^*)$ for $[Ru(bpy)_3]^{2+}$ is −0.86 V. As one proceeds down the column, the thermodynamic driving force for oxidative quenching decreases. Because the reactions all fall in the normal region for ET, as $\Delta G°$ decreases, the quenching rate constant will also decrease.

Table 17.27 lists some of the quenching rate constants (k_q) for reductive quenching of $[Ru(bpy)_3]^{2+*}$ with a series of R_3N quenchers. The $E°(M^*/M^-)$ for $[Ru(bpy)_3]^{2+}$ is +0.84 V. As one proceeds down the column, the thermodynamic driving force for reductive quenching therefore decreases. Because these reactions also lie in the normal region for ET, as $\Delta G°$ decreases, the quenching rate constant will also decrease.

The ability of the MLCT ES in Ru(II) polypyridyl compounds to undergo facile excited state ET reactions, combined with the stability of the complexes and the tunability of their ES properties, makes this class of compounds attractive candidates for a number of practical applications. Ruthenium(II) polypyridyl compounds have already found applications in the field of photovoltaics as sensitizers for wide-bandgap semiconductors (such as TiO_2) and as photocatalysts for the photochemical splitting of water. Their photophysical and photoredox properties

TABLE 17.26 Reduction potentials and quenching rate constants (k_q) for oxidative quenching of $[Ru(bpy)_3]^{2+*}$ with a series of $ArNO_2$ quenchers.

Quencher	$E_{1/2}$, V	k_q, M^{-1} s^{-1}
p-$NO_2C_6H_4NO$	−0.52	9.2×10^9
p-$NO_2C_6H_4NO_2$	−0.69	8.6×10^9
o-$NO_2C_6H_4NO_2$	−0.81	3.1×10^9
p-$NO_2C_6H_4CHO$	−0.86	2.0×10^9
m-$NO_2C_6H_4NO_2$	−0.90	1.6×10^9
p-$NO_2C_6H_4CO_2CH_3$	−0.95	6.6×10^8
4,4′-$NO_2C_6H_4C_6H_4NO_2$	−1.00	1.2×10^8

TABLE 17.27 Reduction potentials and quenching rate constants (k_q) for reductive quenching of $[Ru(bpy)_3]^{2+*}$ with a series of R_3N quenchers.

Quencher	$E_{1/2}$, V	k_q, M^{-1} s^{-1}
p-$Me_2NC_6H_4NMe_2$	0.12	1.2×10^{10}
p-$Me_2NC_6H_4C_6H_4NMe_2$	0.43	4.3×10^9
p-$Me_2NC_6H_4Me$	0.71	1.5×10^9
$Et_2NC_6H_5$	0.76	1.5×10^8
$Me_2NC_6H_5$	0.81	7.2×10^7
$N(C_6H_5)_3$	1.06	9.5×10^5

have also been exploited in molecular sensors, luminescent immunoassays, and even to probe the rate of protein-to-protein ET using a "flash-quench" technique.

EXERCISES

17.1. In the octahedral substitution reaction $[Co(NH_3)_5X]^{2+} + H_2O$, which ligand is expected to yield the larger rate constant for substitution, X = F^- or I^-? Explain.

17.2. For the octahedral substitution reaction $[M(H_2O)_6]^{2+} + H_2O^*$, which metal is expected to yield the larger rate constant for substitution, M = Mn or Ni? Explain.

17.3. The volumes of activation for the base hydrolysis of some Co(III) ammine complexes are listed in the given table. Rationalize the trends in $\Delta V^{\pm}$ for the series based on the charge on the starting complex and on the leaving group.

Reactant	$\Delta V^{\pm}$, cm^3/mol
$[Co(NH_3)_5dmso]^{3+}$	42.0
$[Co(NH_3)_5Cl]^{2+}$	33.0
$[Co(NH_3)_5(SO_4)]^+$	22.7

17.4. Predict the products of each of the following reactions (show their three-dimensional geometries) and briefly explain how each mechanism occurs:

a. $[Pt(PR_3)_4]^{2+} + 2\ Cl^- \rightarrow$

b. $[Rh(PPh_3)_3Cl] + H_2\ (g) \rightarrow$

c. *cis*-$[Cr(NH_3)_4I_2]^{2+} + Cl^- \rightarrow$

d. $[Co(NH_3)_5SCN]^{2+} + [Cr(H_2O)_6]^{2+} \rightarrow$

17.5. Explain the trend in the following rate constants for the square planar substitution reaction of *cis*-$[PtClL(PEt_3)_2]$ with H_2O: L = pyridine ($k = 8.0 \times 10^{-2}\ s^{-1}$), L = 2-methylpyridine ($k = 2.0 \times 10^{-4}\ s^{-1}$), and L = 2,6-dimethylpyridine ($k = 1.0 \times 10^{-6}\ s^{-1}$). Do these data support an associative or dissociative mechanism?

17.6. Predict the stereochemistry of the products for each of the following reactions:

a. $[Pt(NH_3)_4]^{2+} + 2\ PR_3 \rightarrow$

b. *cis*-$[Pt(NH_3)_2(py)_2]^{2+} + 2\ Br^- \rightarrow$

c. $[Pt(NH_3)Br_3]^- + NH_3 \rightarrow$

17.7. Starting with $[PtCl_4]^{2-}$, devise an experimental procedure to make each of the following isomers:

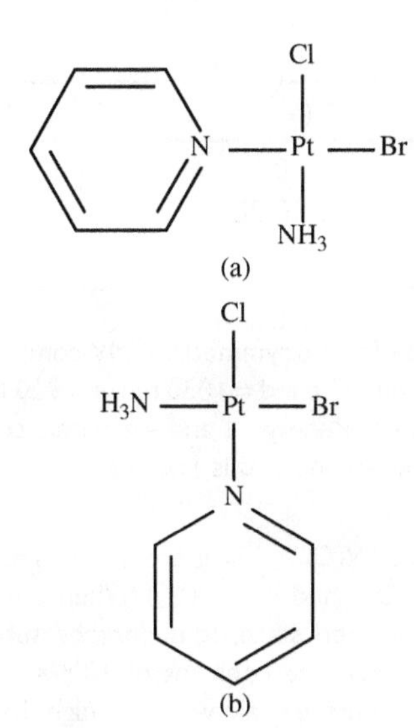

17.8. During the course of the ET reaction of $[Co(NCS)(NH_3)_5]^{2+}$ with $[Fe(H_2O)_6]^{2+}$, a small amount of $[Fe(SCN)(H_2O)_5]^{2+}$ was observed. Provide a feasible mechanism for this reaction.

17.9. Using data in Table 17.13, calculate the predicted cross exchange rate constant for the ET reaction of $[Fe(CN)_6]^{4-}$ with MnO_4^- if $\Delta E° = 0.20$ V and $f = 0.66$.

17.10. The complex $[Rh(NH_3)_5Cl]^{2+}$ undergoes photoaquation from its lowest energy LF excited state to form two products: $[Rh(NH_3)_5H_2O]^{3+}$ and *trans*-$[Rh(NH_3)4(H_2O)Cl]^{2+}$. Which photochemical product is the one predicted by Adamson's strong field ligand/weak field axis rule? Does it follow the edge displacement rule? Explain your answer.

17.11. Consider the luminescent complex $Pt_2(\mu\text{-}P_2O_5H_2)_4^{4-}$ and commonly referred to as *platinum pop*. Platinum pop absorbs light at 367 nm and emits an intense green light at 514 nm. It can also be quenched by a series of aromatic amine compounds, whose standard reduction potentials, along with their quenching rate constants at 25 °C, are listed here. Assuming that these quenching rate constants lie in the "normal" region, is this oxidative or reductive electron transfer? Explain your answer.

Quencher	$E°$, V	k_q, s^{-1}
N, *N*, *N*′, *N*′-tetramethyl-1,4-benzenediamine	0.11	1.2×10^{10}
N, *N*, *N*′, *N*′-tetramethyl-(1,1′-biphenyl)-4,4′-diamine	0.36	3.0×10^{9}
N, *N*-trimethylbenzeneamine	0.71	3.9×10^{7}
N, *N*-diphenylbenzeneamine	0.92	1.5×10^{6}

17.12. If E_{0-0} for platinum pop occurs at 475 nm and its excited state reduction potential is −1.0 V, estimate the ground-state potential for the following reaction:

$$Pt_2(\mu\text{-}P_2O_5H_2)_4^{\ 4-} \rightarrow Pt_2(\mu\text{-}P_2O_5H_2)_4^{\ 3-} + e^-$$

17.13. Is the pK_a of $[Ru(NH_3)_5pz]^{2+}$ (pz = pyrazine) larger or smaller in the excited state than in the ground state? Explain.

17.14. Consider the trinuclear MV compound: $[(NC)_5M^{II}\text{-}\mu\text{-}(CN)\text{-}Pt^{IV}(NH_3)_4\text{-}\mu\text{-}(NC)\text{-}M^{II}(CN)_5]^{4-}$, where M is a Group 8 B metal ion. Given the data in the table and assuming that Hush theory applies to this Robin and Day Class II series of compounds, determine $E°$ for the $[Os(CN)_6]^{3-/4-}$ couple.

Metal (M)	$E°$ (V vs SCE)	E_{op} (IT band, nm)
Fe	0.19	424
Ru	0.70	354
Os	*x*	380

17.15. Given the following data for the symmetrical MV compound $[(NH_3)_5Ru^{II}\text{-}\mu(4,4'\text{-bpy})\text{-}Ru^{III}(NH_3)_5]^{5+}$: $r = 113$ pm, IT band = 1030 nm, $\varepsilon = 920\ M^{-1}\ cm^{-1}$, $\Delta\nu_{1/2} = 5200\ cm^{-1}$, calculate the reorganization energy λ and electronic coupling matrix element H_{ab} in units of cm^{-1}. Then determine if this is a Robin and Day Class II or Class III MV compound.

17.16. Tungsten hexacarbonyl, $W(CO)_6$, is inert in the ground state with $k = 10^{-6}\ s^{-1}$ for substitution of a CO ligand with PPh_3. When the compound is irradiated into its lowest energy LF excited state, it undergoes substitution with PPh_3 to yield $W(CO)_5PPh_3 + CO$ with a rate constant of $10^{10}\ s^{-1}$ (or 16 orders of magnitude faster!). Using LFT arguments, explain why this might be the case. What is the lowest energy LF transition (use term symbols)?

17.17. The photochemistry of $W(CO)_5L$ compounds is quite interesting. These molecules have C_{4v} symmetry. Sketch the CF splitting diagram for the *d*-orbitals in $W(CO)_5py$ (py = pyridine) and label each with their group theoretical symbols. Fill in the correct number of electrons. According to the one-electron MO diagram, there should be two LF transitions between the highest filled *d*-orbital and each of the unfilled *d*-orbitals. Irradiation of the lower energy LF transition leads to photodissociation of py to form $W(CO)_5 + py$, while irradiation of the higher energy LF transition leads to photodissociation of CO to form $W(CO)_4py + CO$. Explain this observation based on the nature of the *d*-orbital which is populated in each case following absorption of a photon.

BIBLIOGRAPHY

1. Atkins P, Overton T, Rourke J, Weller M, Armstrong F, Hagerman M. *Shriver& Atkins' Inorganic Chemistry*. 5th ed. New York: W. H. Freeman and Company; 2010.
2. Atwood JD. *Inorganic and Organometallic Reaction Mechanisms*. 2nd ed. New York: Wiley-VCH, Inc.; 1997.

3. Creutz C. *Mixed-Valence Complexes of $^{d}5$-$^{d}6$ Metal Centers*. Prog. Inorg. Chem. 1983;30:1–74.
4. Douglas B, McDaniel D, Alexander J. *Concepts and Models of Inorganic Chemistry*. 3rd ed. New York: John Wiley & Sons, Inc.; 1994.
5. Housecroft CE, Sharpe AG. *Inorganic Chemistry*. 3rd ed. Essex, England: Pearson Education Limited; 2008.
6. Huheey JE, Keiter EA, Keiter RL. *Inorganic Chemistry: Principles of Structure and Reactivity*. 4th ed. New York: Harper Collins College Publishers; 1993.
7. Jordan RB. *Reaction Mechanisms of Inorganic and Organometallic Systems*. 2nd ed. Oxford University Press, Inc.: New York; 1998.
8. Laidler KJ. *Theories of Chemcial Reaction Rates*. New York: McGraw-Hill, Inc.; 1969.
9. Lawrence GA. *Introduction to Coordination Chemistry*. Chichester, West Sussex, England: John Wiley & Sons, Ltd; 2010.
10. Lever ABP. *Inorganic Electronic Spectroscopy*. Amsterdam, The Netherlands: Elsevier; 1985.
11. Meyer TJ, Taube H. Electron Transfer Reactions. In: Wilkinson G, Gillard RD, McCleverty JA, editors. *Comprehensive Coordination Chemistry*. Oxford, England: Pergamon Press; 1987.
12. Miessler GL, Tarr DA. *Inorganic Chemistry*. 4th ed. Upper Saddle River, NJ: Pearson Education Inc.; 2011.
13. Rodgers GE. *Descriptive Inorganic, Coordination, and Solid-State Chemistry*. 3rd ed. Belmont, CA: Brooks/Cole Cengage Learning; 2012.
14. Roundhill DM. *Photochemistry and Photophysics of Metal Complexes*. New York: Plenum Press; 1994.
15. Sykora J, Sima J. Photochemistry of Coordination Compounds. Coord. Chem. Rev. 1990; 107:1–212
16. Wrighton MS. *Inorganic and Organometallic Photochemistry*. Washington, DC: American Chemical Society; 1978.

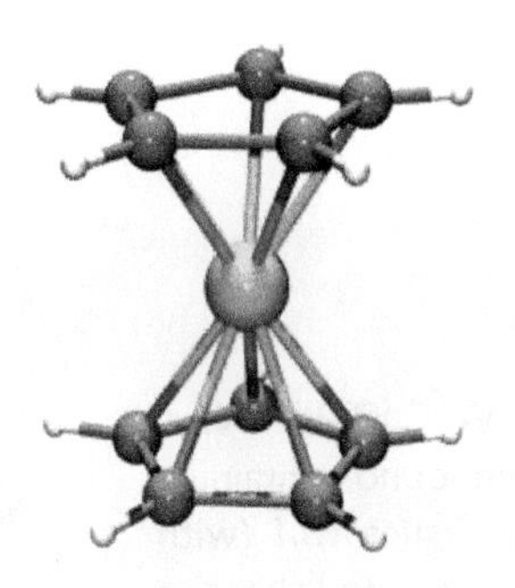

Structure and Bonding in Organometallic Compounds 18

> I can only pray that every one of you will be suffused with that most satisfying of rosy glows—the inner one of tranquility of spirit and of intellect—and that it will last you throughout your lifetime.
>
> —*Geoffrey Wilkinson*

18.1 INTRODUCTION TO ORGANOMETALLIC CHEMISTRY

This chapter's opening quote reads: "I can only pray that every one of you will be suffused with that most satisfying of rosy glows—the inner one of tranquility of spirit and of intellect—and that it will last you throughout your lifetime." Such were the words of gratitude spoken by Geoffrey Wilkinson upon the acceptance of his 1973 Nobel Prize, which he shared with Ernst Otto Fischer for their pioneering work on organometallic "sandwich compounds." The most famous of the sandwich compounds is ferrocene, whose chemical structure is also shown on the chapter's opening cover page. Ferrocene was accidentally discovered by Kealy and Pauson in 1951 when they reacted $FeCl_3$ with the Grignard reagent *cyclo*-$(C_5H_5)MgBr$ (itself an organometallic compound). The crystal structure of the product revealed (quite surprisingly) that the resulting orange compound consisted of an Fe(II) ion sandwiched in between two planar cyclopentadienyl ions. For each of the cyclopentadienyl rings, all five of the C atoms are equidistant from the metal ion, so that the two ring systems are parallel to each other just like the slices of bread in a sandwich, with the metal stuck in between them.

Although they were not the first to discover the compound, Fischer and Wilkinson were among the first to describe the nature of this rather unusual type of metal–carbon bonding. For the next several decades, just as Werner before them, their pioneering research led to the birth of an entire new field of chemistry: that of modern organometallic chemistry. Upon the acceptance of his share of the 1973 Nobel Prize, Fischer's words were no less poignant than those of Wilkinson: "May we as chemists make harmonious music in the future for the joy and benefit of mankind … and not for its detriment."

In its simplest definition, *organometallic chemistry* is defined as the chemistry of compounds that contain at least one metal–carbon bond (other than cyanide). The

Ferrocene. [The chapter opener figure was kindly provided by D. Schaarschmidt, Technische Universitat, Chemnitz.]

Principles of Inorganic Chemistry, First Edition. Brian W. Pfennig.

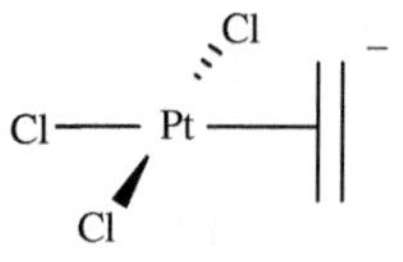

FIGURE 18.1
The structure of Zeise's salt, the original organometallic compound.

first truly organometallic compound was synthesized in 1827 by W. C. Zeise. Now known as *Zeise's salt*, this bright yellow complex anion is a Pt(II) compound containing ethylene as a ligand. The structure of the compound is shown in Figure 18.1 (with its water of crystallization not shown). Although Zeise correctly predicted that the molecule contained ethylene, this was not proven by spectroscopy until 1868, and the crystal structure of the compound was not determined until 1975. The two C atoms in the ethylene ligand lie equidistant from the metal such that the C=C double bond is perpendicular to the plane of the rest of the molecule.

The first organometallic compound containing the carbonyl ligand, $[PtCl_2(CO)_2]$ was discovered by P. Schützenberger in 1868. The first binary metal carbonyl, $Ni(CO)_4$, was discovered by L. Mond in 1890. In the late 1890s, P. Barbier and V. Grignard reported the first alkyl magnesium complexes, RMgX, which are now known as *Grignard reagents* and used extensively in organic synthesis. However, the field of modern organometallic chemistry was not officially recognized until after Kealy and Pauson's discovery of ferrocene in 1951. Fischer and Wilkinson were among the early pioneers in this blossoming new arena at the intersection of inorganic and organic chemistry. Fischer's group can also take credit for the discovery of the first metal–carbon double bond, or carbene compound, $[(CO)_5W{=}C(OCH_3)Ph]$, in 1964 and the first metal–carbon triple bond, or carbyne, $[(CO)_4XCr{\equiv}CPh]$, in 1973. Aside from his work on elucidating the structure of ferrocene, Wilkinson is perhaps best known for the hydrogenation catalyst which bears his name, Wilkinson's catalyst, $[RhCl(PPh_3)_3]$. Over the past several decades, the field of organometallic chemistry has seen enormous growth, with over six chemists receiving the Nobel Prize in chemistry for their work in organometallic chemistry since the year 2000. Organometallic compounds are primarily used as catalysts, from the early transition metal Ziegler-Natta polymerization catalysts which made the plastics industry possible to their more modern applications in enantioselective catalysis.

18.2 ELECTRON COUNTING AND THE 18-ELECTRON RULE

Because the transition metals can use their valence *d*-orbitals in their bonding, many (but not all) organometallic compounds follow the effective atomic number (EAN) rule of Sidgwick, otherwise known as the *18-electron rule*. Just as some main group compounds violate the "octet rule" (for example, BCl_3 is electron-deficient while SF_6 is hypervalent), organometallic compounds can also have less than or greater than 18 electrons in the valence shells of their central atoms. In much the same way that molecules containing main group atoms are stable when they have a filled octet, if an organometallic compound follows the 18-electron rule, it too is invariably stable. In our electron book-keeping scheme, we follow the "donor pair" formalism, which treats all of the electrons in the metal–ligand bonds as "belonging" to the ligand. The charges and numbers of electrons donated from typical organometallic ligands are listed in Table 18.1.

The symbol η is known as the *hapticity*. The *hapticity* of a ligand is defined as the number of C atoms in the ligand that are directly bonded to the metal. For the cyclopentadienyl ligand, there are three possibilities: η^1, η^3, and η^5. For η^5-Cp^-, all five of the C atoms are equidistant from the metal center, as is the case in the

TABLE 18.1 Electron counting formalism for common ligands and types of bonding in organometallic chemistry.

Ligand	Charge	e^- Count
H, CH_3, R	−1	2
X, SCN, NCS, CN, OH, N_3, etc.	−1	2
PR_3, AsR_3, SR_2, H_2O, NH_3, etc.	0	2
CO, CS	0	2
N≡O	+1	2
N=O	−1	2
Alkenes	0	2
	0	4
	0	6
	−1	6 (η^5)
	−1	4 (η^3)
H	−1	2 (η^1)
M≡N	−3	6
M–M (metal–metal bond)	0	1 ea
M=CR_2 (carbene)	0	2
M≡CR (carbyne)	+1	2

ferrocene molecule. All six π-electrons are therefore used in the bonding. For η^1, only the terminal C is bound to the metal, and this arrangement is identical to that of a bound alkyl group. Lastly, η^3 compounds have three of the C atoms equidistant from the metal and donate four electrons from two double bonds.

By convention, we will treat any valence electrons that the metal contributes to the total electron count as "*d*-electrons." We are somewhat justified in making this assumption on the basis of MOT, where the ligand electrons fill the lower lying bonding MOs and the valence electrons on the metal occupy MOs that have predominantly *d*-character. For example, the one-electron MO diagram of $Ni(CO)_4$ is shown in Figure 18.2. Each CO ligand contributes two valence electrons and the neutral metal contributes ten electrons for a total of 18 electrons, in agreement with the 18-electron rule. Although the valence electron configuration of Ni as a free atom is $[Ar]4s^23d^8$, all 10 of the valence electrons contributed from Ni occupy nonbonding MOs in the $Ni(CO)_4$ compound that are largely derived from the *d*-atomic orbitals. Thus, we will say that the Ni^0 atom in $Ni(CO)_4$ is "d^{10}."

Of course, this arbitrary assignment of the valence electrons to the *d*-orbitals on the metal or to the ligands is simply an electron counting formalism, because there is a considerable amount of metal–ligand mixing in all of the MOs in organometallic compounds. The convention is not intended to predict the true oxidation state of the metal or to describe the nature of the bonding (ionic or covalent). Its only purpose is to serve as a book-keeping procedure to account for all the valence electrons.

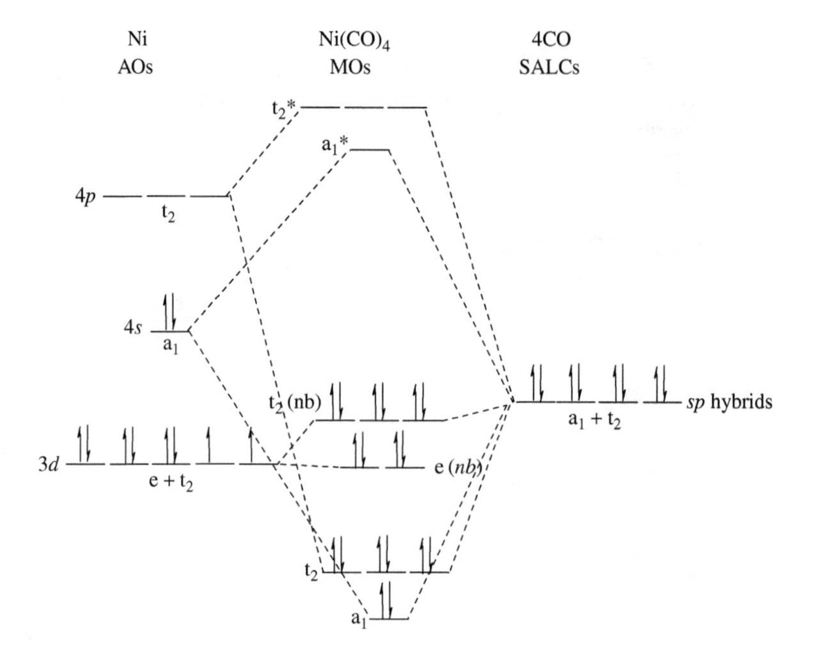

FIGURE 18.2
Simplified one-electron MO diagram for $Ni(CO)_4$.

Using the donor pair convention for the ferrocene molecule, each η^5-Cp^- ligand donates six electrons and has a charge of −1. Thus, the iron center is formally Fe(II), or d^6. One of the main reasons that this molecule is stable is because it follows the 18-electron rule. Zeise's salt, on the other hand, has 2 electrons from each Cl^- ligand, 2 electrons from the coordinated alkene, and 8 electrons from the Pt(II) center, giving it a total of only 16 valence electrons.

Organometallics that obey the 18-electron rule are said to be *saturated*, while those that have less than 18 valence electrons are *unsaturated.* The tendency of a metal to be saturated or unsaturated depends on its position in the periodic table and the types of ligands that are attached to it. In general, metals from Group 3 or 4 of the periodic table are usually unsaturated in their compounds because of the low electronegativity of the metals. These compounds are analogous to the electron-deficient elements B and Al from main group chemistry. Because they are not very electrophilic, they tend not to attract enough ligands to become saturated.

Organometallic compounds containing metals from Groups 9–12 also tend to be unsaturated, but for a very different reason. These metals are more electrophilic than those in Groups 3 or 4, but their radii are smaller so that they can only accommodate a limited number of ligands around the smaller metallic core. This is especially true for Group 9–12 metals that have large or sterically hindered ligands.

Those metals in the middle of the periodic table (Groups 5–8) are almost always saturated in their compounds because they have the perfect balance of moderate electrophilicity and size. Table 18.2 summarizes these generalities. The two examples we have already seen are Zeise's salt (which contains Pt and is unsaturated) and ferrocene (which contains Fe and is saturated). Naturally, there are exceptions to this rule of thumb. For example, many LS d^8 compounds are unsaturated because of their electronic preference for a square planar molecular geometry.

Example 18-1. For each of the following organometallic molecules, determine the formal oxidation state of the metal, count the total number of valence electrons that the compound has, and justify why any unsaturated compounds have less than 18 valence electrons.

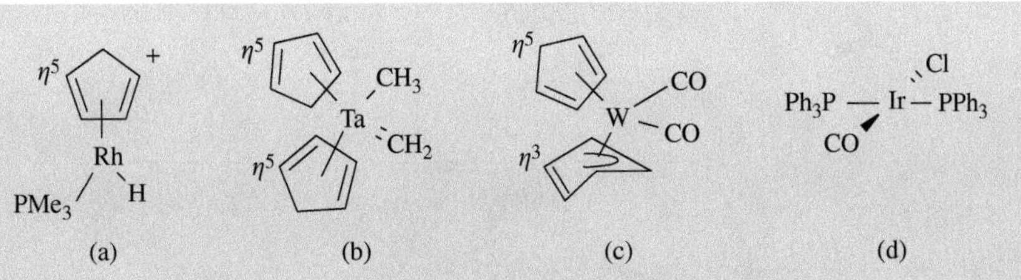

Solution. (a) Rh(III) is formally d^6. The metal therefore contributes 6 e^-, the η^5-Cp^- ligand contributes 6 e^-, H^- contributes 2 e^-, and PMe_3 contributes 2 e^-, for a total of 16 e^-. The compound is unsaturated, as predicted for Rh because of its smaller size. (b) Ta(III) is d^2. The metal contributes 2 e^-, the two η^5-Cp^- ligands contribute 12 e^- between them, the alkyl group contributes 2 e^-, and the carbene contributes 2 e^-, for a total of 18 e^-. The compound is saturated, as predicted for Ta. (c) W(II) is d^4. The metal contributes 4 e^-, the η^5-Cp^- ligand contributes 6 e^-, the η^3-Cp^- ligand contributes 4 e^-, and each CO contributes 2 e^-, for a total of 18 e^-. The compound is saturated, as predicted for W and a metal carbonyl. (d) Ir(I) is d^8 LS. The compound is square planar. The metal contributes 8 e^-, each PPh_3 ligand contributes 2 e^-, Co contributes 2 e^-, and Cl^- contributes 2 e^-, for a total of 16 e^-. The compound is unsaturated as predicted for Ir and for a LS d^8 compound.

18.3 CARBONYL LIGANDS

Metal carbonyl compounds almost always obey the 18-electron rule, regardless of the metal's identity. The main reason for this is because the CO ligand is both a good σ-donor and a good π-acceptor. The lone pair on the C-terminus of a carbonyl can donate a pair of electrons to the metal to form a coordinate covalent bond. At the same time, the π^* MOs on CO can accept electron density back from the metal as a result of π-backbonding. In an octahedral compound, such as $Cr(CO)_6$, whose MO diagram is shown in Figure 18.3, the better the σ-donating abilities of the ligand, the higher the e_g^* MOs lie, while stronger π-backbonding will lower the energy of the bonding t_{2g} orbitals.

The magnitude of Δ_o is quite large for the strong-field CO ligand. The 6 CO ligands in $Cr(CO)_6$ contribute 12 e^- to the metal-ligand bonds. The 6 electrons from the Cr^0 then occupy the π-bonding t_{2g} MOs, rounding out the 18-electron rule. Addition of an extra electron to form a 19-electron compound is highly unfavorable because the electron would have to occupy a high-lying e_g^* MO, which would necessarily weaken the M–CO bonding. Likewise, loss of an electron to form a 17-electron

TABLE 18.2 General tendencies for an organometallic compound to be saturated or unsaturated.

Sc	Ti	V	Cr	Mn	Fe	Co	Ni	Cu	Zn
Y	Zr	Nb	Mo	Tc	Ru	Rh	Pd	Ag	Cd
La	Hf	Ta	W	Re	Os	Ir	Pt	Au	Hg
Unsaturated (not χ)		Usually saturated (perfect balance of χ and r)				Unsaturated (too small r)			

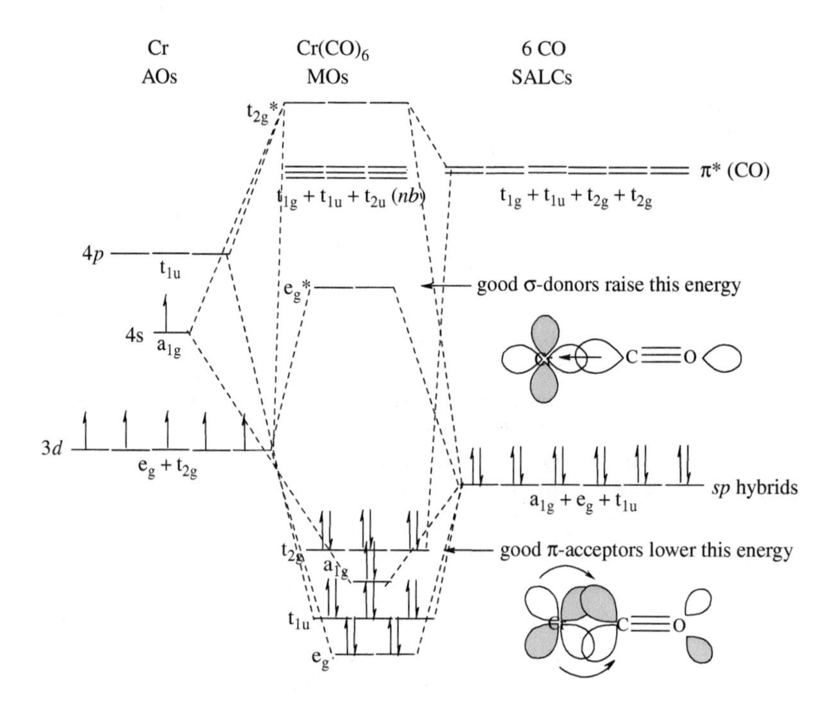

FIGURE 18.3
One-electron MO diagram for $Cr(CO)_6$.

molecule would remove an electron from a bonding t_{2g} MO and also weaken the bonding. Thus, 18 electrons is the optimum amount for forming the most stable metal carbonyl.

The presence of strong π-backbonding is of paramount importance here because CO is not a very good Lewis base. π-backbonding requires partially filled *d*-orbitals and an electropositive metal. As a result, most metal carbonyls contain metals in very low oxidation states (and are often zero-valent). As shown by the IR data in Table 18.3, the lower the valence of the metal, the more electropositive the metal will be and the more electron density it will have to engage in π-backbonding with the CO ligands. The stronger the M–CO π-backbonding, the lower the energy of ν(CO) will be because π-backbonding populates the π^* MO of CO. Notice that all of the compounds have ν(CO) lower than that for the unbound CO ligand. Likewise, the C≡O bond length in free CO (112.8 pm) lengthens upon binding to the metals (typical metal carbonyls have a C≡O bond length of ~115 pm).

The 18-electron rule is obeyed so frequently with metal carbonyls that it can often be used to predict the products of the reaction of a metal with CO (g). For example, when Ni^0 (which is d^{10}) reacts with gaseous carbon monoxide, the organometallic product must contain four CO ligands in order to satisfy the

TABLE 18.3 Infrared data for the ν(CO) stretch in a series of octahedral metal carbonyls.

Compound	ν(CO), cm^{-1}
Free CO	2143
$[Ti(CO)_6]^{2-}$	1750
$[V(CO)_6]^-$	1860
$[Cr(CO)_6]$	2000
$[Mn(CO)_6]^+$	2090

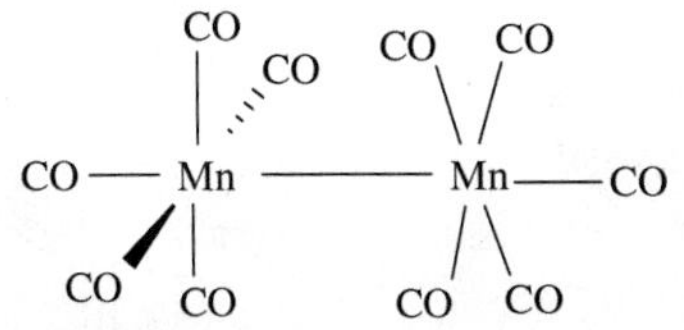

FIGURE 18.4
Structure of $Mn_2(CO)_{10}$.

18-electron rule. Indeed, the product of this reaction is the tetrahedral $Ni(CO)_4$, whose MO diagram was shown in Figure 18.2. Likewise, when Fe^0 (which is d^8) reacts with CO (g), the expected product is $Fe(CO)_5$, which has a trigonal bipyramidal molecular geometry. Because Mn^0 (d^7) has an odd number of valence electrons, it reacts with CO to form the binary compound $Mn_2(CO)_{10}$, whose structure is shown in Figure 18.4. The CO ligands on each $Mn(CO)_5$ subunit are staggered with respect to the other half of the molecule. Each Mn has 7 e^- from the metal, 1 e^- from the Mn–Mn bond, and 10 e^- from the five CO ligands, for a total of 18 valence electrons, in accordance with the 18-electron rule.

Example 18-2. Determine the symmetries of the ν(CO) stretching vibrations in $Mn_2(CO)_{10}$. Which of these vibrational modes will be IR-active and which will be Raman-active?

Solution. Using vectors to represent the CO bonds in $Mn_2(CO)_{10}$, which has the D_{4d} point group, the reducible representation for the ν(CO) stretches is given here. It reduces to: $2a_1 + 2b_2 + e_1 + e_2 + e_3$. The $2b_2$ and e_1 modes are IR-active and the $2a_1$, e_2, and e_3 modes are Raman-active. The IR-active modes occur at 2046 (b_2), 1984 (b_2), and 2015 (e_1) cm^{-1}. The Raman modes are observed at 2116 (a_1), 1997 (a_1), 2024 (e_2), and 1981 (e_3) cm^{-1}.

D_{4h}	E	$2S_8$	$2C_4$	$2S_8^3$	C_2	$4C_2'$	$4\sigma_d$	IRR's:
Γ_{CO}	10	0	2	0	2	0	4	$2a_1 + 2b_2 + e_1 + e_2 + e_3$

The carbonyl ligand can also act as a bridge between two metals, as is the case for $Co_2(CO)_8$ in the solid state, which has the molecular structure shown in Figure 18.5.

When the CO ligand acts as a symmetrical bridge, as it does in Figure 18.5, the σ_{sp} MO on the ligand can donate its pair of electrons to the *d*-orbitals on two metals simultaneously, as shown in Figure 18.6(a). Thus, in the formalism for counting electrons, only one electron per bridging ligand can "belong" to each metal. For example in $Co_2(CO)_8$ each Co atom has 9 e^- from its valence shell, 1 e^- from the Co–Co bond, 2 e^- from each of the three terminal carbonyls, and 1 e^- from each bridging CO, for a total of 18 valence electrons. The vibrational frequencies of bridging carbonyls are considerably lower in energy than those for terminal COs. For $Co_2(CO)_8$, the terminal CO stretches occur at 2075, 2064, 2047, 2035, and 2028 cm^{-1}, while

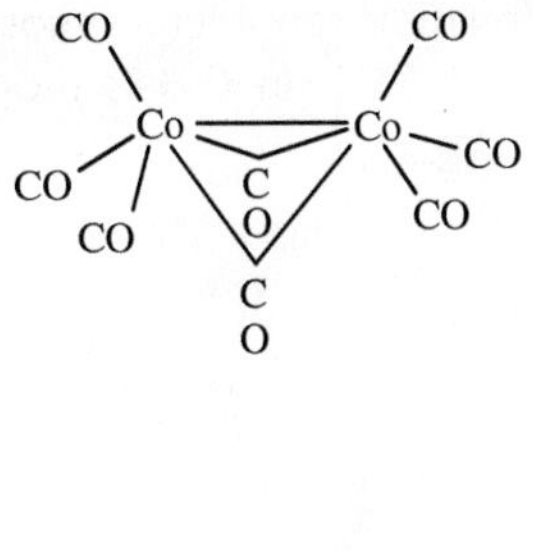

FIGURE 18.5
Structure of $Co_2(CO)_8$ in the solid state.

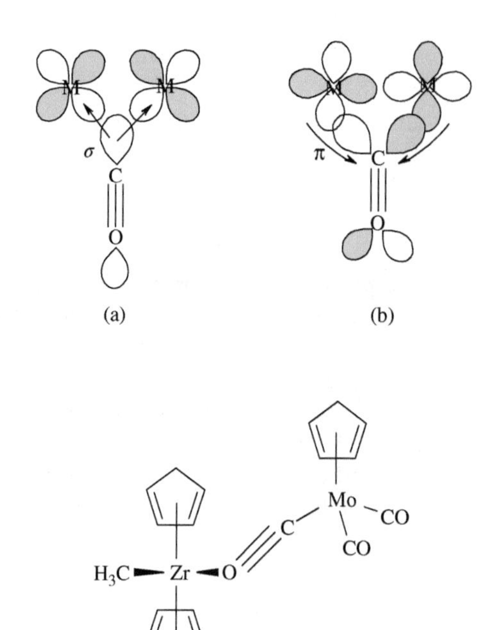

FIGURE 18.6
(a) Sigma donation and (b) π-backbonding in symmetrical bridging metal carbonyls.

FIGURE 18.7
A binary organometallic compound containing an ambidentate bridging carbonyl ligand.

the two bridging CO stretches are observed at 1867 and 1859 cm^{-1}. The reason for this bathochromic shift can be explained by the fact that the π^* MO on the carbonyl ligand is populated by *pi*-backbonding from *both* of the metal centers, as shown in Figure 18.6(b). If a carbonyl ligand were to bridge three metals, the ν(CO) stretching frequency would shift to even lower wavenumbers.

When the bridging CO ligand is asymmetrical with respect to the two metals, the term *semibridging* is employed. Occasionally, the carbonyl ligand can bridge as an ambidentate ligand (an *isocarbonyl*) between the two metals, as shown in the example in Figure 18.7. However, the bridging ligand in this case is never linear as it is for the metallocyanides. Instead, the M–C–O bond angle is bent, which significantly weakens the CO bonding and lowers the vibrational frequency of the ν(CO) stretching mode.

Example 18-3. Use the MO diagrams for CO and CN^- to show why the cyanide ligand will often form ambidentate M–C≡N–M′ bridges, and why the carbonyl ligand bridging in this manner is rare.

Solution. The MO diagrams for CO and CN^- are shown here.

The main difference between the two is the greater electronegativity difference between C and O than between C and N. This causes the σ_b HOMO in CO to be so low in energy that it has a significant percentage of C 2*s*, C 2*p*, and O 2*p* character, while the same MO in CN^- is mostly just C 2*p* and N 2*p*. Thus, the σ_b HOMO in CO has a greater percentage of C character than in CN^-, which makes it a better σ-donor on the C end than on the O end. In CN^-, σ-donation can occur almost equally through the C or N terminus. Similarly, the π^* LUMO in CO is more polarized toward the C end than it is in CN^-. In CO, π-backbonding can only occur through the C, but in CN^- it can occur through both C and N. Thus, CN^- can bond strongly with the metals through either end of the ligand, while CO can only bind strongly through the C terminus.

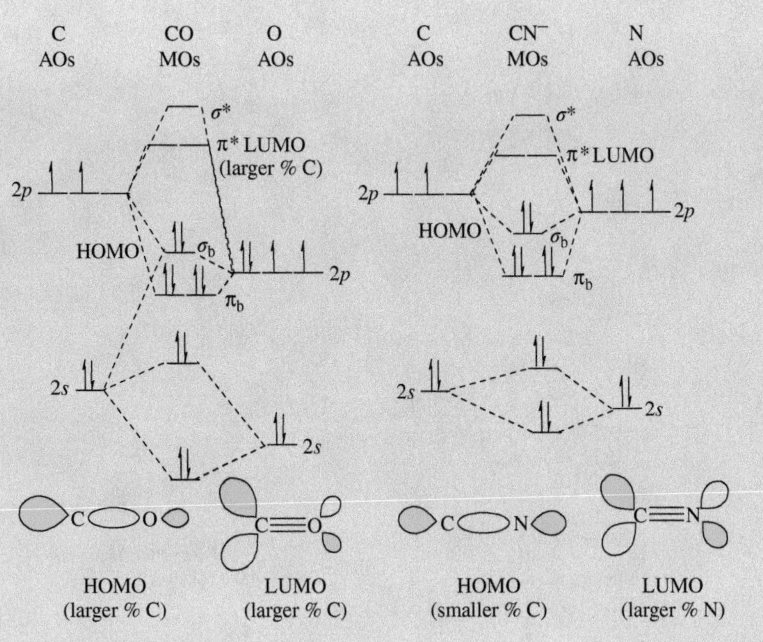

Metal carbonyls can be synthesized in the following ways:

1. Direct reaction of the zero-valent metal with CO (g) at elevated temperature:

$$\text{Fe} + 5\ \text{CO} \rightarrow \text{Fe(CO)}_5 \qquad (18.1)$$

2. Reduction of a metal precursor in the presence of CO (g):

$$\text{W(CH}_3)_6 + 9\ \text{CO} \rightarrow \text{W(CO)}_6 + 3\ \text{CH}_3\text{COCH}_3 \qquad (18.2)$$

3. Ligand substitution with CO (g):

$$[\text{Ru(NH}_3)_5\text{H}_2\text{O}]^{2+} + \text{CO} \rightarrow [\text{Ru(NH}_3)_5\text{CO}]^{2+} + \text{H}_2\text{O} \qquad (18.3)$$

4. Deinsertion:

$$[\text{RhCl(PPh}_3)_3] + \text{RCHO} \rightarrow\rightarrow\rightarrow [\text{RhCl(CO)(PPh}_3)_2 + \text{RH} + \text{PPh}_3 \quad (18.4)$$

18.4 NITROSYL LIGANDS

Similar to carbonyl, the nitrosyl ligand is a good σ-donor and a good π-acceptor. It can also serve as either a terminal ligand or a bridging ligand, as shown by the examples illustrated in Figure 18.8.

Two modes of bonding are possible. When the ligand has a +1 charge, it contains a N≡O triple bond and is *sp*-hybridized. Therefore, it tends to form a linear (165–180°) M–N≡O bond angle in its compounds, as shown in Figure 18.9(a).

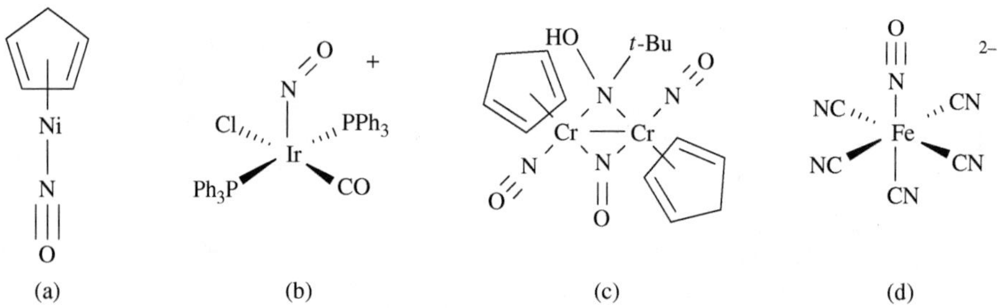

FIGURE 18.8
(a–d) Examples of organometallics containing nitrosyl ligands.

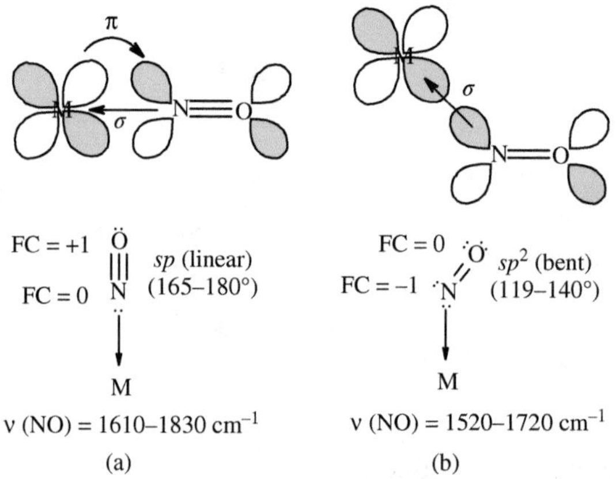

FIGURE 18.9
A comparison of the two modes of bonding for nitrosyl ligands: (a) NO^+ is *sp*-hybridized and linear, while (b) NO^- is sp^2-hybridized and bent.

Because of the triple bond, the ν(NO) stretching frequencies occur at high wavenumbers, approximately in the range of 1610–1830 cm^{-1}. On the other hand, NO^- has an N=O double bond and sp^2-hybridization, so it will bind to the metal in a bent configuration (119–140°), as shown in Figure 18.9(b). In this bonding mode, the ν(NO) stretching frequencies are redshifted to the 1520–1720 cm^{-1} range. As was the case for bridging carbonyls, the ν(NO) stretching frequencies are lower in energy than their terminal counterparts. For example, the terminal ν(NO) stretching frequencies occur at 1683 and 1625 cm^{-1} in Figure 18.8(c), while the bridging ν(NO) stretch occurs at 1499 cm^{-1}.

When CO and NO are both present as ligands, as is the case for the Ir complex in Figure 18.8(b), they prefer to be *cis* with respect to each other so as to minimize the competition for electron density from the same *d*-orbital in their π-backbonding. The compound in Figure 18.8(d) has been used as a vasodilator to lower blood pressure. The NO ligand is easily released *in vivo*, where it activates the enzyme guanylate cyclase. This enzyme is responsible for the catalytic conversion of guanosine triphosphate (GTP) into 3′,5′-cyclic guanosine monophosphate (cGMP), which causes smooth muscle relaxation from the interior walls of the arteries. The ED drug Viagra functions by blocking the enzyme PDE-5, which is responsible for the reuptake of cGMP.

Example 18-4. For each of the organometallic species in Figure 18.8, determine the oxidation number of the metal and the total number of valence electrons that the molecule has.

Solution. (a) Because the NO is linear, it has a +1 charge. The cyclopentadienyl ligand has a −1 charge. Thus, the oxidation state of Ni is formally zero. Ni^0 contributes 10 e^-, η^5-Cp^- contributes 6 e^-, and NO^+ adds another 2 e^- for a total of 18 electrons. (b) Because the NO is bent, it has a −1 charge. The chloro ligand is also −1, and the complex is +1 overall, making the oxidation number Ir(III). Ir(III) contributes 6 e^-, while each PPh_3 contributes 2 e^-, Cl^- adds another 2 e^-, and CO and NO^- both contribute 2 e^- each, for a total of 16 e^-. (c) The bridging NO ligand is −1, the N(OH)(tBu) ligand is also −1, the terminal NO ligands are +1 each, and each cyclopentadienyl ligand is −1, making the oxidation number Cr(I). Cr(I) contributes 5 e^-, the Cr–Cr bond contributes 1 e^- to each metal, the η_5-Cp^- contributes 6 e^-, the bridging ligands each contribute 1 e^- to each metal, and the terminal NO^+ ligands add another 2 e^-, making the total valence electron count 16. (d) The linear NO ligand is +1 and each cyanide ligand is −1. The complex has a 2- charge overall, making the oxidation number Fe(II). The metal contributes 6 e^-, each CN^- adds 2 more electrons, and NO^+ contributes 2 e^-, for a total of 18 valence electrons.

Example 18-5. The $[RuCl(NO)_2(PPh_3)_2]^+$ complex exhibits two ν(NO) stretches in its vibrational spectrum: one at 1845 cm^{-1} and the other at 1687 cm^{-1}. Sketch a reasonable geometry for the compound.

Solution. On the basis of the energies of the ν(NO) stretches, one of the nitrosyl ligands is NO^+ (1845 cm^{-1}) and the other is NO^- (1687 cm^{-1}). Thus, the former ligand is linear and the latter is bent. In order to minimize competition for electron density from the same *d*-orbital in their π-backbonding, the two nitrosyl ligands will prefer to occupy a *cis* geometry. The actual structure of $[RuCl(NO)_2(PPh_3)_2]^+$ is shown here (ignoring the phenyl groups on the phosphine ligands).

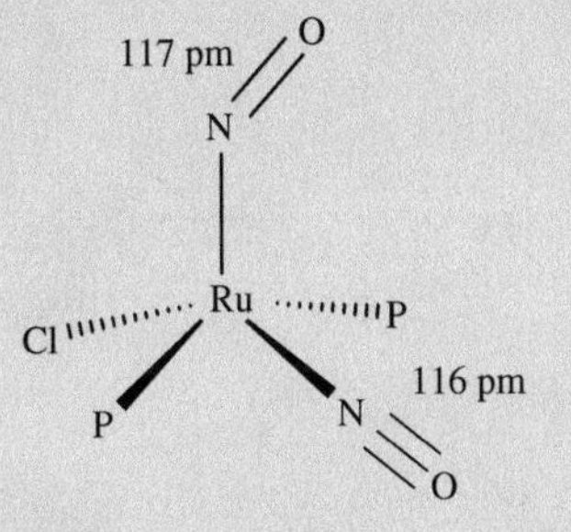

117 pm
O
N
Cl
Ru
P
P
116 pm
N
O

Nitrosyl compounds can be formed in the following ways:

1. Direct substitution of NO for the CO ligand:

$$Fe(CO)_5 + 2\,NO \rightarrow Fe(CO)_2(NO)_2 + 3\,CO \qquad (18.5)$$

2. Substitution of CO with NOCl, NO^+, or NO_2^-:

$$[Mn(CO)_5]^- + NOCl \rightarrow [Mn(CO)_4NO] + CO + Cl^- \qquad (18.6)$$

$$[Ir(PPh_3)_2(CO)Cl] + NO^+ \rightarrow [Ir(PPh_3)_2(CO)(NO)Cl]^+ \quad (18.7)$$

$$[Co(CO)_4]^- + NO_2^- + 2\,CO_2 + H_2O \rightarrow [Co(CO)_3NO] + 2\,HCO_3^- \quad (18.8)$$

18.5 HYDRIDE AND DIHYDROGEN LIGANDS

When molecular hydrogen reacts with a transition metal, it can bind to the metal in one of two ways: (i) as a hydride ligand, H^-, as shown in Figure 18.10(a); or (ii) the molecule itself can act as a η^2-ligand, as shown in Figure 18.10(b).

Although the ligand in Figure 18.10(a) is formally called a *hydride ligand* in the donor pair convention, in reality the nature of the ligand can vary from acidic to basic, depending on the electrophilicity of the metal to which it is attached. Hydrides can also act as bridging ligands, as shown in Figure 18.11. As with isocarbonyls, the M—H—M′ bond angle is never linear. In the first example shown, the M—H sigma bond itself coordinates to the second metal. In the latter example, the bonding is a three-centered, two-electron bond analogous to the bridging B–H–B bond in diborane encountered earlier in this text. Hydride ligands are easily identifiable in the ^{1}H-NMR of organometallics because any H attached to a metal is strongly shielded, with chemical shifts typically ranging from −2 to −12 ppm versus TMS. In the IR, ν(MH) stretches occur at 1600–2250 cm^{-1} (depending on the metal) for terminal hydrides and 800–1600 cm^{-1} for bridging hydrides. Hydride complexes can be made in the following ways:

1. Reduction with $LiAlH_4$ or $NaBH_4$:

$$[RuCl_2(PPh_3)_3] + \text{excess } NaBH_4 + PPh_3 \rightarrow [Ru(H)_2(PPh_3)_4] \quad (18.9)$$

2. Elimination reactions:

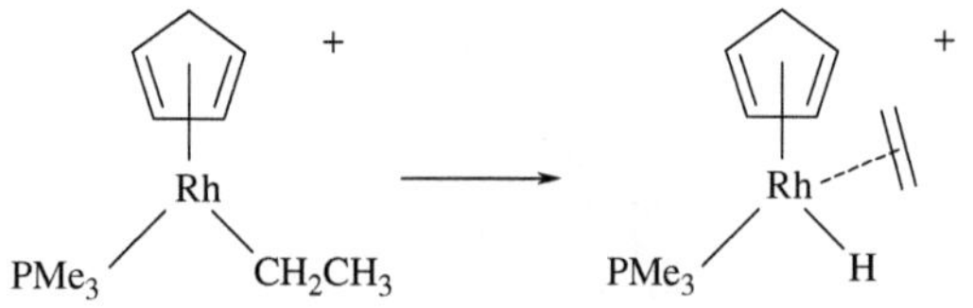

$$[(\eta^5\text{-}C_5H_5)Rh(PMe_3)(CH_2CH_3)]^+ \longrightarrow [(\eta^5\text{-}C_5H_5)Rh(PMe_3)(H)(\eta^2\text{-}C_2H_4)]^+ \quad (18.10)$$

$$[Co_2(CO)_6(\mu\text{-}CO)_2] + H\text{—}H \longrightarrow 2\,[HCo(CO)_4] \quad \text{(a)}$$

$$[Ir(CO)Cl(PPh_3)_2] + H\text{—}H \longrightarrow [Ir(\eta^2\text{-}H_2)Cl(CO)(PPh_3)_2] \quad \text{(b)}$$

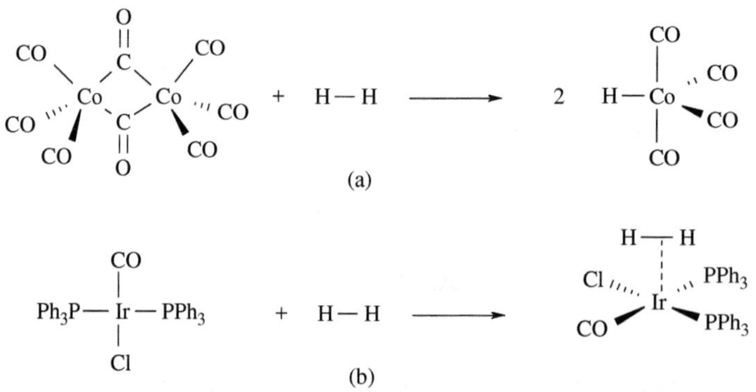

FIGURE 18.10
Reaction of molecular hydrogen with an organometallic precursor can lead to metal-hydrogen bonding either through: (a) a hydride ligand, or (b) a dihydrogen ligand.

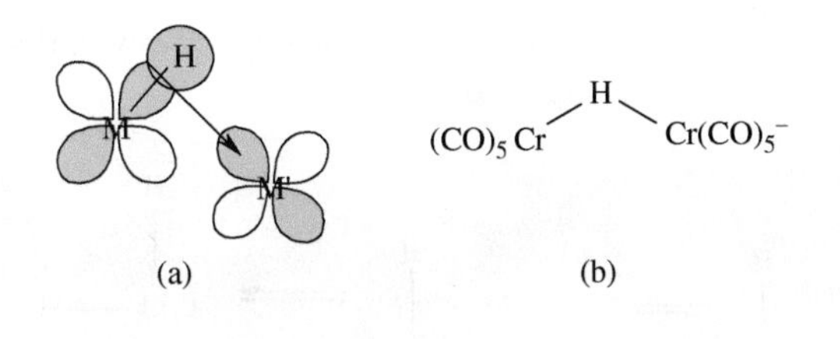

FIGURE 18.11
(a, b) Examples of bridging hydride ligands.

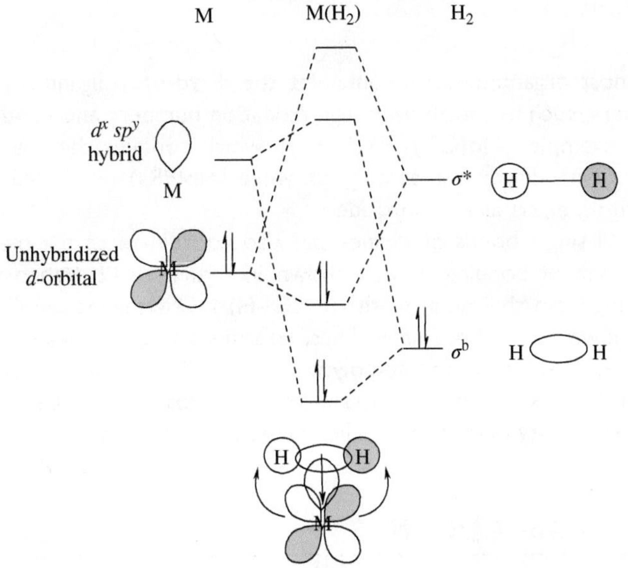

M M(H_2) H_2

$d^x sp^y$ hybrid M

Unhybridized d-orbital

σ^* H—H

σ^b H—H

FIGURE 18.12
Simplified one-electron MO diagram showing the bonding modes in a metal-dihydrogen bond.

3. Deinsertion reaction:

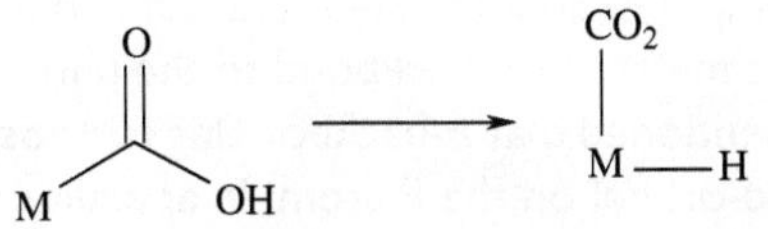

$$\mathrm{M{-}C(=O)OH} \longrightarrow \mathrm{M{-}H}\ (\mathrm{CO_2}) \qquad (18.11)$$

4. Direct reaction with H_2 (g), including oxidative addition:

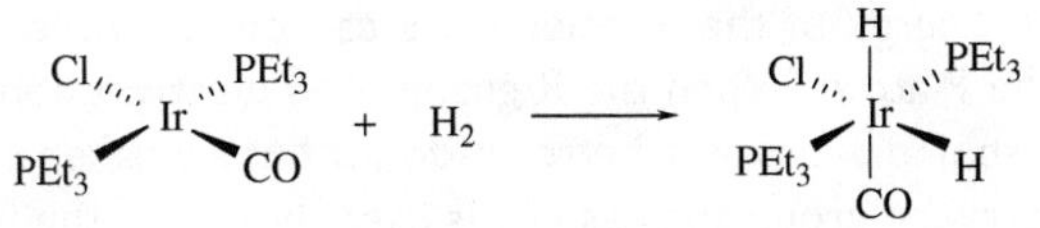

$$[\mathrm{IrCl(CO)(PEt_3)_2}] + \mathrm{H_2} \longrightarrow [\mathrm{IrCl(H)_2(CO)(PEt_3)_2}] \qquad (18.12)$$

The first definitive organometallic compound to contain dihydrogen as a ligand was not reported until 1984 when Greg Kubas published the neutron diffraction structure of [$W(CO)_3(PR_3)_2(H_2)$]. Since that time, a significant number of compounds containing coordinated H_2 have been discovered. Dihydrogen coordinates to metals using both σ-donation and π-backbonding, as shown in Figure 18.12.

The σ_b MO of H_2 donates a pair of electrons to the metal to form a three-centered, two-electron bond, while a filled d-orbital on the metal π-backbonds into the empty σ^* MO on the dihydrogen ligand. As a result of the π-backbonding, the H–H bond is weakened. Typical H–H bond lengths for coordinated H_2 are 82–90 pm, compared with a bond length of only 74.1 pm in uncoordinated H_2. If the π-backbonding is too strong, however, the H–H bond can break to form the dihydride compound:

$$L_nM + \underset{H}{\overset{H}{|}} \rightleftharpoons L_nM\cdots\underset{H}{\overset{H}{|}} \rightleftharpoons L_nM\overset{H}{\overset{|}{\ }}—H \tag{18.13}$$

Thus, most organometallics containing the dihydrogen ligand are fairly poor π-backbonders, such as metals with high oxidation numbers and good π-acceptor ligands. For example, $[Mo(CO)_3(PR_3)_2(H_2)]$, which contains the π-accepting CO ligands exists with dihydrogen as a ligand, while $[Mo(PR_3)_5(H)_2]$, which has more donating ligands, exists as the dihydride.

The C–H single bonds of alkanes can also coordinate to the metal using an analogous mode of bonding to that shown in Figure 18.12. When this occurs, π-backbonding from the metal to the σ^* (C–H) MO weakens the C–H bond in a process known as *C–H activation*. Thus, coordination of an alkane to the metal "activates" the C–H bond for homolytic cleavage. The resulting three-centered, two-electron bond is known as an *agostic complex*. Agostic intermediates have been postulated in a variety of organometallic reaction mechanisms.

18.6 PHOSPHINE LIGANDS

Phosphines are among the most prevalent ligands in organometallic chemistry. Similar to CO, PR_3 ligands and their related congeners ($P(OR)_3$, AsR_3, SbR_3, and SR_2) are good σ-donors and good π-acceptors. The sp^3-hybridized orbital on the phosphine coordinates to an empty orbital on the metal via σ-donation. At the same time, a filled *d*-orbital on the metal can π-backbond to the phosphine. In a previous chapter, we rather casually mentioned that π-backbonding to phosphine ligands occurred into an empty, low-lying *d*-orbital on the P atom. In actuality, this type of π-backbonding occurs to an MO that consists of a mixture of a low-lying *d*-orbital and a σ^* MO (which predominates) on the phosphine, as shown in Figure 18.13. Thus, for a series of isoelectronic metal phosphines, the P–R bond length increases as the charge on the metal becomes more negative (making the metal a better π-backbonder).

Phosphines are useful metals for probing organometallic reactivity for several reasons. First, the energy of the frontier MOs can be controlled by changing the substituents on the P atom. When the R group is an electron donor, as is the case for PMe_3, the phosphine becomes a better σ-donor but a weaker π-acceptor. When an electron-withdrawing group such as PF_3 is used, however, the ligand is a weaker σ-donor but a better π-acceptor. Thus, the electronic properties of organometallic compounds can be tuned by a suitable choice of the phosphine substituent groups.

The second important feature of phosphines is the tunability of their steric bulk. The "size" of a phosphine ligand is measured by its *cone angle*, which was defined by Tolman as the apex angle that originates at the metal, assuming a M–P bond length of 228 pm and including the outermost edge of the cone that encloses the van der Waals radii of all the substituent groups, as shown in Figure 18.14. The cone angles for selected phosphine ligands are listed in Table 18.4. Organometallic compounds containing the more bulky phosphine ligands seldom have room to accommodate enough ligands to satisfy the 18-electron rule. Furthermore, the steric bulk of ligands can influence the types of organometallic reactions that they undergo. Bulkier ligands, for example, can favor dissociative reaction mechanisms.

TABLE 18.4 Tolman cone angles (θ) for selected phosphine ligands.

Ligand	θ	Ligand	θ
PH_3	87°	$P(CH_3)Ph_2$	136°
PF_3	104°	$P(CF_3)_3$	137°
$P(OCH_3)_3$	107°	$P(O\text{-}o\text{-}C_6H_4CH_3)_3$	141°
$P(OCH_2CH_3)_3$	109°	PPh_3	145°
PMe_3	118°	$P(cyclo\text{-}C_6H_{11})_3$	170°
PCl_3	124°	$P(t\text{-}Bu)_3$	182°
$P(CH_3)_2Ph$	127°	$P(C_6F_5)_3$	184°
PBr_3	131°	$P(o\text{-}C_6H_4CH_3)_3$	194°
PEt_3	132°	$P(mesityl)_3$	212°

18.7 ETHYLENE AND RELATED LIGANDS

When an alkene acts as a ligand, as ethylene does in Zeise's salt, it binds to the metal in a sideways manner, where the C–C π_b MO donates electron density to an empty orbital on the metal, as shown in Figure 18.15, and the π^* MO accepts electron density back from a filled *d*-orbital on the metal. As a result, the C=C bond is weakened by both the π-donation and the π-backbonding interaction. The C=C bond length in free ethylene is 133.7 pm; in Zeise's salt it increases to 137.5 pm and it is as large as 143 pm in $[Pt(PPh_3)_2(C_2H_4)]$, where the metal is electron-rich as a result of the low oxidation number and donor ligands.

A simplified MO diagram for π-ethylene organometallics is shown in Figure 18.16. The extent of π-backbonding depends not only on the metal and its other ligands but also on the substituents on the alkene. The more electronegative the substituents, the lower the energy of the π^* MO will be and the greater the π-backbonding interaction. Thus, for example, π-backbonding is stronger to the electron-deficient C_2F_4 ligand than it is to C_2H_4. As a result, the Rh–C bond length

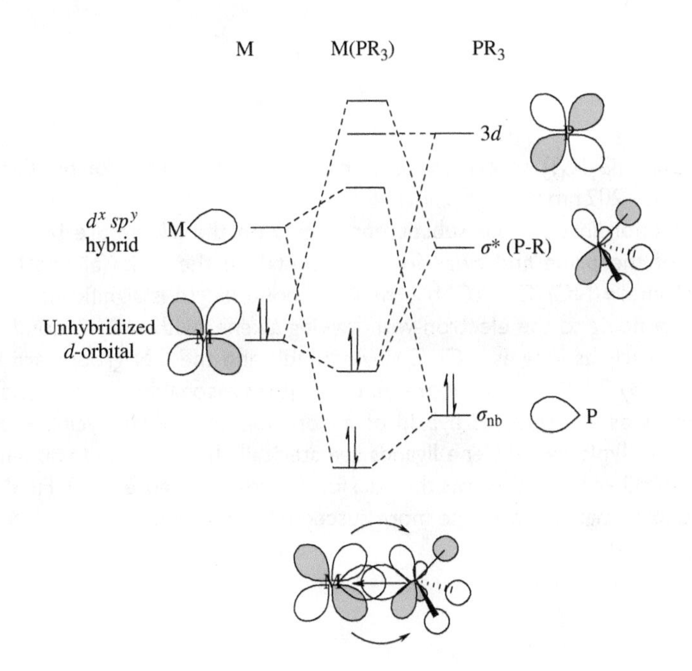

FIGURE 18.13 Metal–ligand bonding in organometallic compounds containing a phosphine ligand.

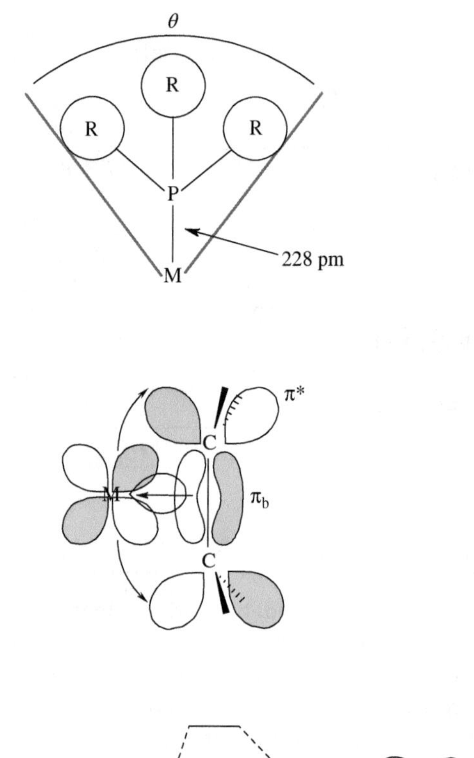

FIGURE 18.14
Definition of Tolman's cone angle for phosphine ligands.

FIGURE 18.15
Bonding modes in a π-ethylene organometallic.

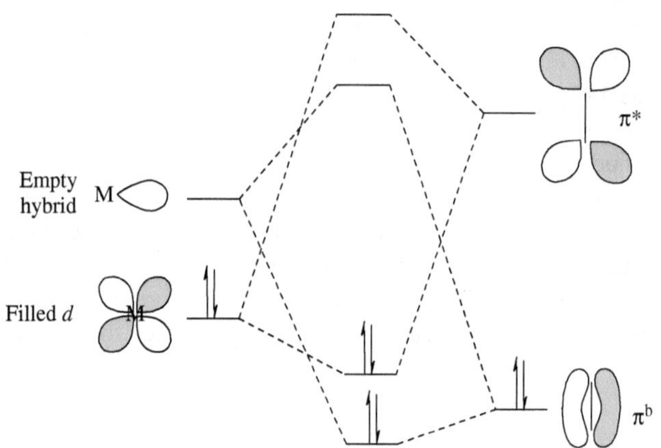

FIGURE 18.16
Simplified one-electron MO diagram for a metal–olefin bond.

in $[RhL(C_2H_4)(C_2X_4)]$, where L = acac or Cp^-, is 217–219 pm for the C_2H_4 ligand and only 201–202 pm for C_2F_4.

Upon coordination, the substituent ligands on the alkene are bent considerably out of the plane and away from the metal. In the $[NiL_2(alkene)]$ complex, where (alkene)=$(NC)_2C{=}C(CN)_2$, the C–C bond length is significantly weakened by π-backbonding to the electron-withdrawing alkene (with a C–C bond length of 147.6 pm, nearly as long as a C–C single bond!) and the CN groups are bent out of the plane by 38.4°. In such extreme cases, it is reasonable that the coordinated alkene exists as a resonance hybrid of π-donation and metallacycle, as shown in Figure 18.17. Typically, ethylene ligands are sterically hindered and will bind out of the plane of other ligands, as was the case for Zeise's salt (Figure 18.1). Furthermore, π-backbonding makes the alkene more susceptible to nucleophilic attack than when it is uncoordinated.

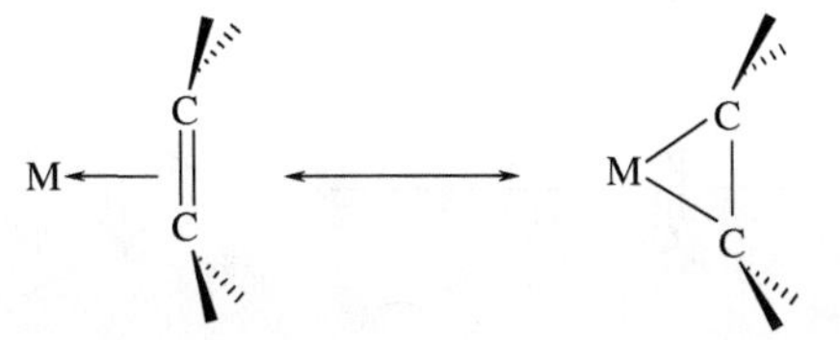

FIGURE 18.17
Canonical forms for metal–olefin bonds where the extent of π-backbonding is significant enough that the metallacycle shown at right contributes to the resonance hybrid.

Example 18-6. Use MOT to predict whether the bound ethylene ligand in $[Fe(CO)_5(C_2H_4)]$ will lie completely in the equatorial plane, as shown at left, or parallel to the two axial positions, as shown at right. Define the vertical axis of the molecule as the z-axis, so that the equatorial plane is the *xy*-plane, and let the *x*-axis bisect the C=C bond in the coordinated ethylene.

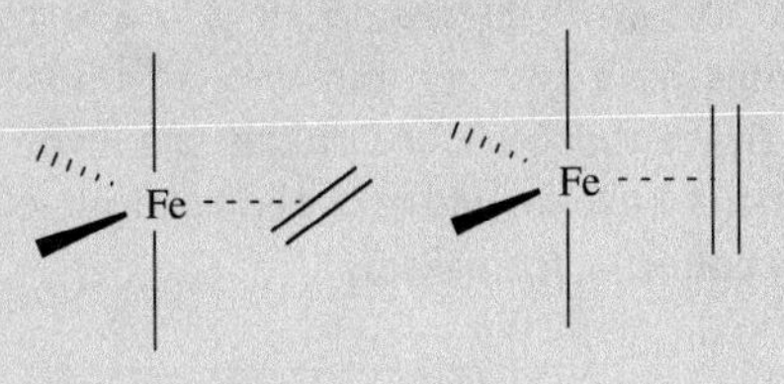

Solution. The LF diagram for a trigonal bipyramid where the CO and ethylene ligands have approximately the same degree of σ-donation is shown (center). When the π^* MO of the ethylene interacts with the *d*-orbital having appropriate symmetry for overlap, the energy of that MO will be stabilized by the π-backbonding interaction. When the ethylene ligand lies in the plane, the d_{xy} MO will have the appropriate symmetry for overlap, as shown at left. When the ethylene ligand is parallel to the principal axis of rotation, the d_{xz} MO will have the best overlap, as shown at right. Because the d_{xy} orbital is closer in energy to the π^* MO than is the d_{xz} orbital, the former will be stabilized by a greater extent than the latter. Thus, the more stable isomer is the one where the ethylene ligand lies completely in the equatorial plane (the one shown at left).

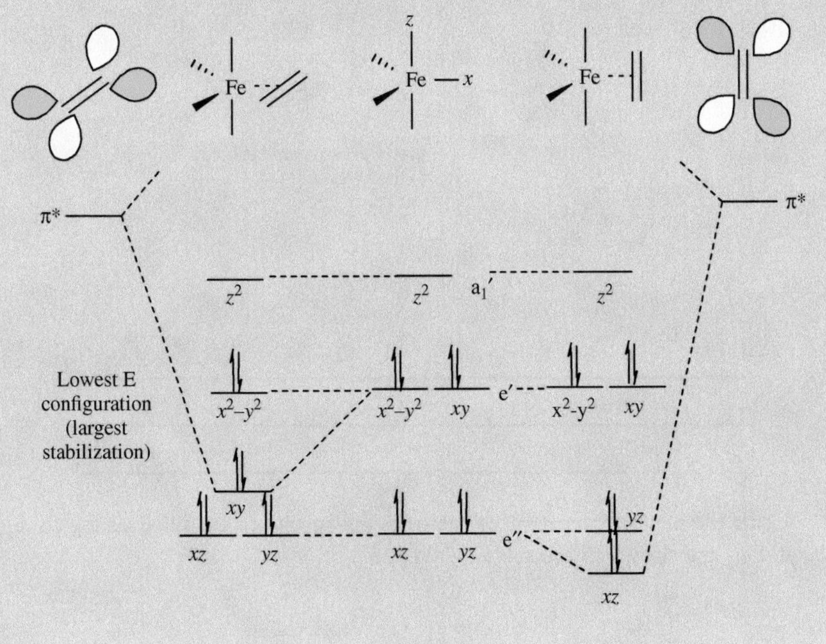

Lowest E configuration (largest stabilization)

FIGURE 18.18
Canonical forms for metal–alkyne bonds where the metallacyclopropene shown at right is the largest contributor to the resonance hybrid because of the large extent of π-backbonding in these complexes.

Metal-olefin compounds can be made by ligand substitution with the alkene, reduction of a high-valent metal in the presence of the alkene, or by beta-hydride elimination of a metal-alkyl. Organometallic compounds containing metal-alkyne bonds are also known, such as $[Pt(PPh_3)_2(C_2H_2)]$. Alkynes are better π-acceptors than their alkene counterparts. As such, the degree of π-backbonding in a metal-alkyne is stronger than that for a metal-alkene. Thus, the metallocyclopropene canonical form shown in Figure 18.18 is the largest contributor to the resonance hybrid. Alkynes have two π_b and two π^* MOs, and extended Hückel calculations seem to indicate that all four orbitals are involved in the metal-alkyne bonding. Thus, the alkyne is best viewed as a neutral, four-electron donor ligand in the donor pair electron counting formalism.

Example 18-7. The allyl ligand, $CH_2{=}CH{-}CH_2^-$, can act as a η^3-ligand using its delocalized π-MOs in its bonding. Sketch the one-electron MO diagram for π-MOs in the allyl ligand and then determine which molecular orbitals have suitable symmetry for overlap with the metal orbitals. How many electrons does the η^3-allyl ligand contribute in the donor pair electron counting formalism?

Solution. The allyl ligand belongs to the C_{2v} point group. Taking the z-axis as the principal axis and assuming that all three C atoms lie in the xz-plane, the symmetries of the π-MOs can be determined using the vectors shown here as a basis set. The reducible representation in the table reduces to $a_2 + 2b_2$.

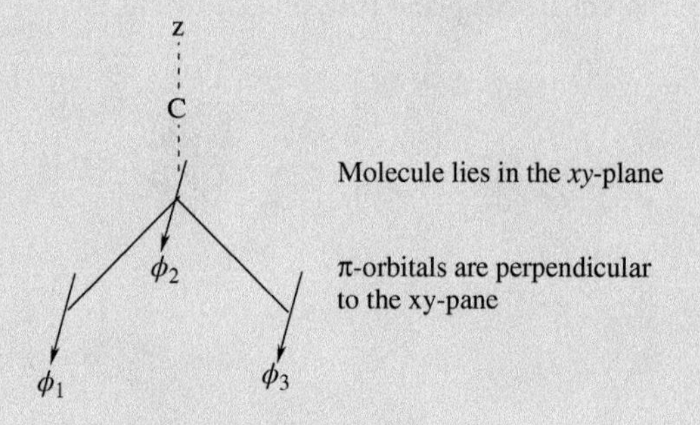

C_{2v}	E	C_2	$\sigma(xz)$	$\sigma(xy)$	IRR's:
Γ_{pi}	3	−1	−3	1	$a_2 + 2b_2$

Using the projection operator method, the results of each IRR acting on ϕ_1, ϕ_2, and ϕ_3 are shown in the table below.

C_{2v}	E	C_2	$\sigma(xz)$	$\sigma(xy)$	
ϕ_1	ϕ_1	$-\phi_3$	$-\phi_1$	ϕ_3	
ϕ_2	ϕ_2	$-\phi_2$	$-\phi_2$	ϕ_2	
ϕ_3	ϕ_3	$-\phi_1$	$-\phi_3$	ϕ_1	
P_{a2} $(\phi_1 \sim \phi_3)$	ϕ_1	$-\phi_3$	ϕ_1	$-\phi_3$	$N(\phi_1 - \phi_3)$
P_{a2} (ϕ_2)	ϕ_2	$-\phi_2$	ϕ_2	$-\phi_2$	0
P_{b2} $(\phi_1 \sim \phi_3)$	ϕ_1	ϕ_3	ϕ_1	ϕ_3	$N(\phi_1 + \phi_3)$
P_{b2} (ϕ_2)	ϕ_2	ϕ_2	ϕ_2	ϕ_2	$N(\phi_2)$

Thus, the wave function for the a_2 MO is $(1/2)^{1/2}(\phi_1 - \phi_3)$. For the two b_2 MOs, we need to take the positive and negative linear combinations P_{b2} $(\phi_1 \sim \phi_3) \pm P_{b2}$ (ϕ_2), which yield (after normalization): $(1/3)^{1/2}(\phi_1 + \phi_2 + \phi_3)$ and $(1/3)^{1/2}(\phi_1 - \phi_2 + \phi_3)$. The shapes of these MOs and the corresponding orbitals on the metal with which they can overlap are shown in the figure.

There are two filled molecular orbitals on the allyl ligand that act as good σ-donors. Thus, the allyl ligand is a four-electron donor in the donor pair electron counting formalism.

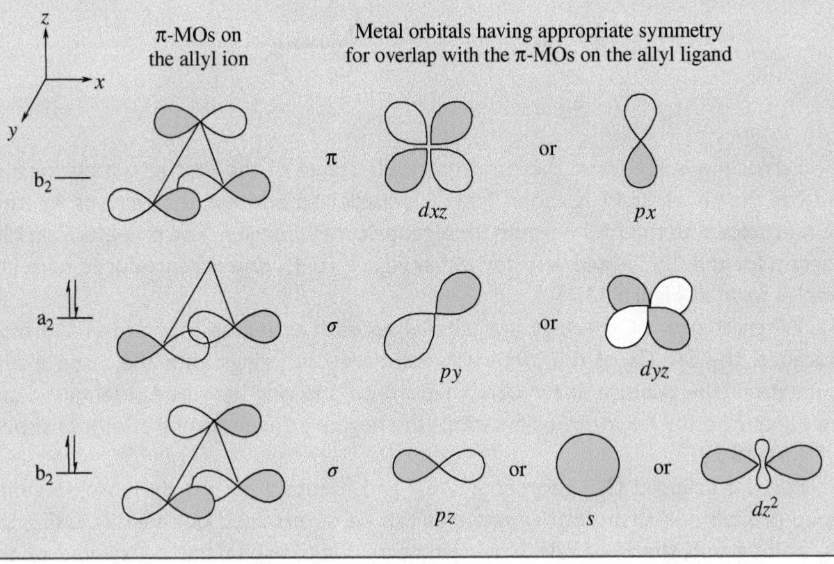

18.8 CYCLOPENTADIENE AND RELATED LIGANDS

The cyclopentadienyl ligand (C_5H_5-) is given the abbreviation Cp^-. The ligand is aromatic and contains six π-electrons. It can bind to the metal in organometallic compounds in three different ways (η^1, η^3, or η^5), as shown in Table 18.1, counting as a two, four, or six-electron donor. The discovery of ferrocene in 1951 led to the birth of the contemporary field of organometallic chemistry. As shown in the chapter heading, ferrocene has the formula $[Cp_2Fe]$ and consists of an Fe(II) ion that is "sandwiched" between two planar Cp^- ligands. As is the case in ferrocene, the normal bonding mode for the Cp^- ligand is η^5. In general, bis(cyclopentadienyl) compounds are referred to as *metallocenes*. Some properties of the first-row metallocenes are listed in Table 18.5.

TABLE 18.5 Properties of the first-row metallocenes.

	$[Cp_2V]$	$[Cp_2Cr]$	$[Cp_2Mn]$	$[Cp_2Fe]$	$[Cp_2Co]$	$[Cp_2Ni]$
Color	Purple	Red	Amber	Orange	Purple	Green
d^n count	15	16	17	18	19	20
ΔH°_{dis} (kJ/mol)				1470	1400	1320
M–C (pm)	228	217	238	206	212	220

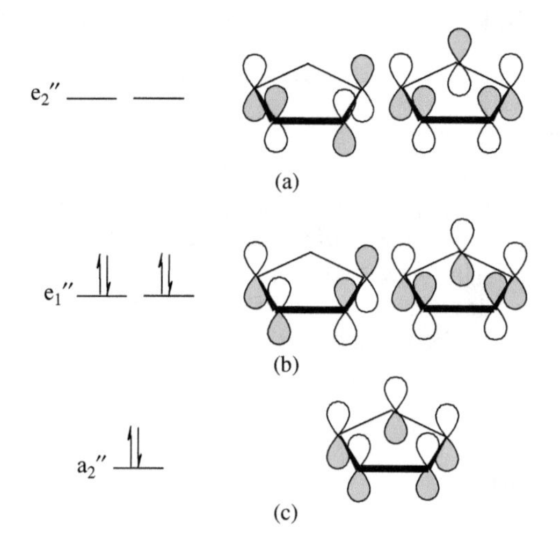

FIGURE 18.19
(a–c) Molecular orbitals and MO diagram for the Cp^- ligand.

Ferrocene is the most thermodynamically stable of the first-row metallocenes and has the shortest M–C bond length. In order to address the reasons for this, we must examine the MO diagram for a generic metallocene. The one-electron MO diagram for the Cp^- ligand was derived in Figure 10.45, and is reproduced here in a simpler form as Figure 18.19.

When the two Cp^- rings are parallel to each other, as they are in the metallocenes, the SALCs of the two rings will have both rings with the same π-MO orientation (the positive linear combination) or one ring with one orientation and the other with the opposite orientation (the negative linear combination), as shown in Figure 18.20.

For the eclipsed D_{5h} geometry, the ′ and ″ subscripts on the corresponding group orbitals and their relative energies can be determined quite easily. Using the D_{5h} point group, the s-orbital has a_1' symmetry, the p_z orbital has a_2'' symmetry, the p_x and p_y orbitals have e_1' symmetry, and the d-orbitals have the following symmetries: $d_{z^2} = a_1'$; $d_{x^2-y^2}$, $d_{xy} = e_2'$; and d_{xz}, $d_{yz} = e_1''$ Combining these atomic orbitals with the two Cp^- group orbitals, the one-electron MO diagram for the eclipsed metallocenes is shown in Figure 18.21.

Any valence electrons contributed by the metal will begin to fill in the orbitals that are inside the boxed-in region of the MO diagram. The Fe(II) ion in ferrocene contributes six valence electrons, which occupy the e_2' and a_1' bonding MOs, strengthening the M–C bonding. Those metallocenes having fewer than six valence electrons on the metal will have lower bond orders because they will have less electrons in their bonding MOs, while any metals having more than six valence electrons will begin to populate antibonding MOs. Thus, the d^6 ferrocene is the most stable of the first-row metallocenes. Odd electron compounds such as $[Cp_2Co]$ are easily oxidized or reduced. The different colors of the metallocenes are due to LF transitions between the one-electron MOs in the boxed-in region

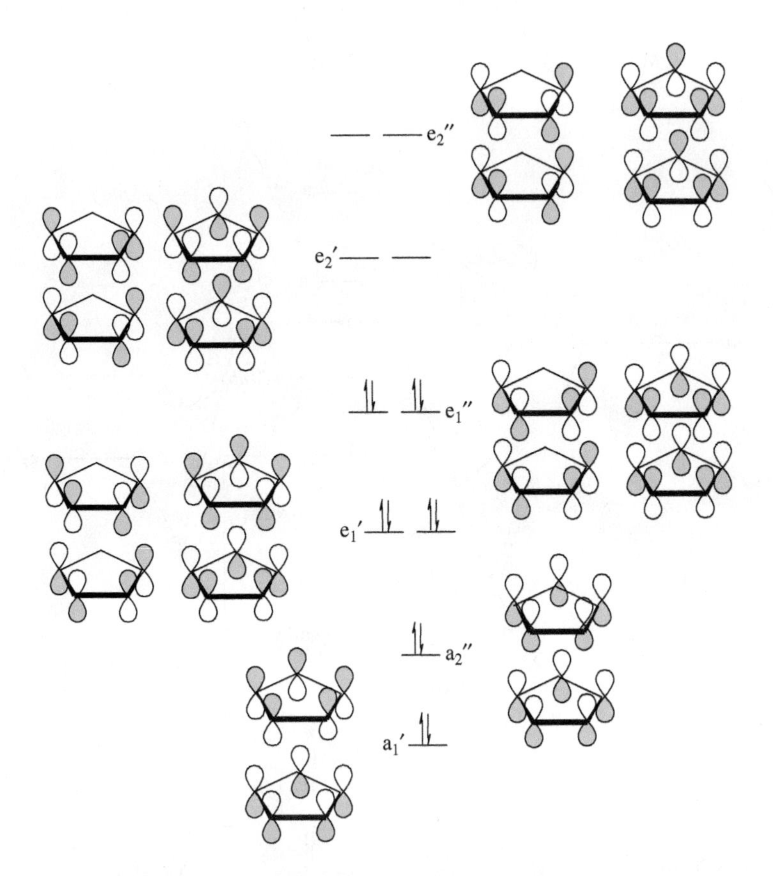

FIGURE 18.20
One-electron MO diagram for the Cp_2 group orbitals in eclipsed metallocenes.

of the MO diagram. Their energies will depend on the number of metal electrons and the magnitude of Δ_o. The transition in ferrocene, which makes the molecule orange, occurs from the filled a_1' to the empty e_1'' MO. When ferrocene is oxidized by one electron to make the Fe(III) ferrocenium ion, which is d^5, the compound changes color to an intense blue and the lowest energy transition occurs between the filled e_2' and half-filled a_a' MOs. The energy difference between these two MOs is smaller than the gap between the a_1' and e_1'' MOs in ferrocene. Thus, ferrocenium absorbs low-energy (red) light and appears blue. Despite its stability, ferrocene is still very reactive. Ferrocene can undergo electrophilic ring acylation (similar to the Friedel-Crafts acylation of benzene), sulfonation, and reaction with butyllithium. Because of π-backbonding to the Cp^- rings, the electrophilic acylation of ferrocene occurs several million times faster than it does for benzene.

The Cp^- rings in ferrocene exhibit a single resonance in the ^1H NMR, indicating that the barrier to ring rotation is small (estimated as 1–2 kcal/mol). At the two extremes are the eclipsed (D_{5h}) and staggered (D_{5d}) conformations. In dilute solution, the eclipsed conformation seems to be the more stable of the two (possibly due to van der Waals forces between the two parallel rings themselves), although stronger intermolecular forces in certain condensed phases occasionally stabilize the staggered conformation over the eclipsed form. In addition to the sandwich metallocenes, other metallocenes are bent and contain two-electron donor ligands, such as Cp_2TiCl_2. The larger size of Ti than Fe and its greater electrophilicity allow it to accept two Cl^- ligands as part of its coordination sphere.

Substituted cyclopentadienyl ligands can be used to tune the electronic properties of organometallic compounds, as well as the steric bulk of the ligand. For instance, the 1,2,3,4,5-pentamethylcyclopentadienyl ion (Cp*) is used to stabilize the Sc(III) ion in the electron-deficient Cp^*_2ScX compound. Molecular orbital calculations using the ZINDO program have shown that there is significant mixing of the

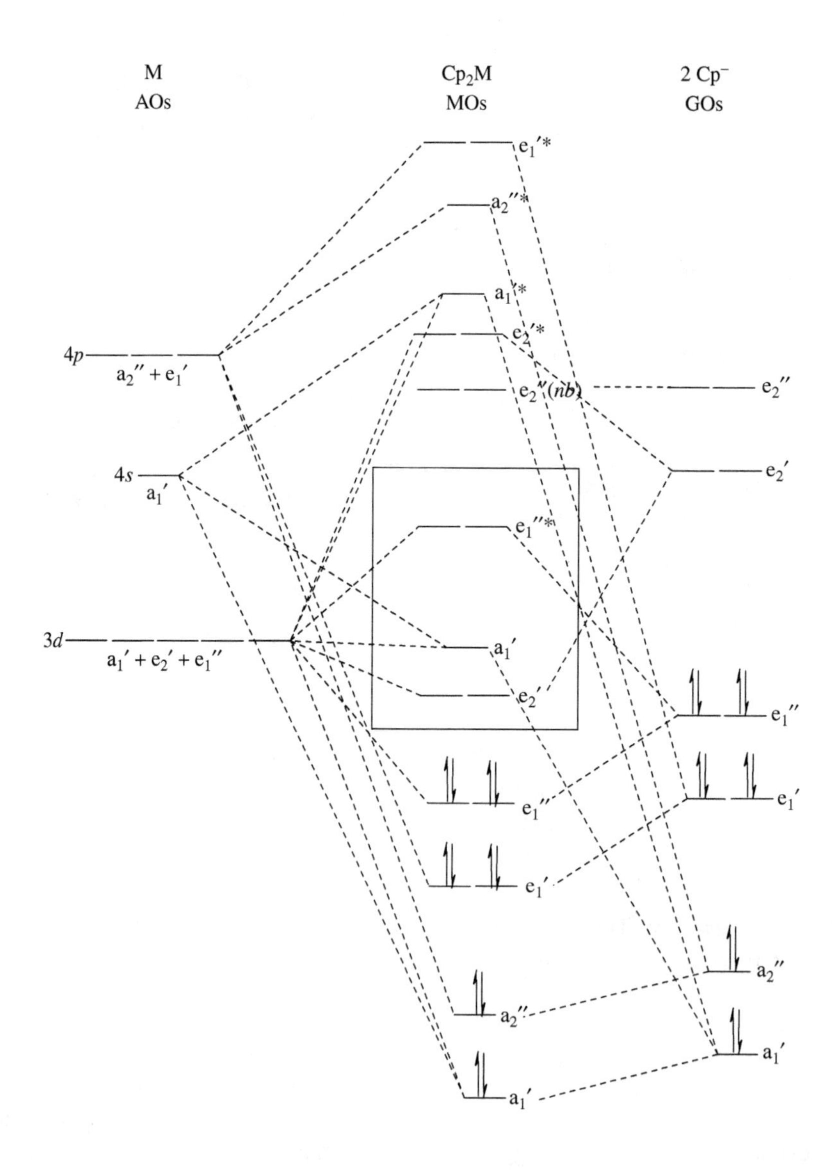

FIGURE 18.21
One-electron MO diagram for eclipsed metallocenes.

Cp* and X molecular orbitals in this molecule when X contains a lone pair of electrons. Irradiation of Cp^*_2ScX shows a room-temperature LMCT emission in fluid solution which tracks with the donor abilities of X when $X = Cl^-$, I^-, or NHPh; however, no emission is observed when $X = CH_3$. Some interesting examples of organometallics containing Cp^- ligands having different hapticities or ligands with similar pi systems are shown in Figure 18.22. Often, a Cp^- ligand will change its hapticity in the reaction mechanism in order to provide more room for coordination at a saturated metal.

18.9 CARBENES, CARBYNES, AND CARBIDOS

Carbenes (or alkylidenes in the older vernacular) are organometallic compounds that contain one or more M=C double bonds. The first carbene was discovered by Ernst Fischer in 1964. The carbene ligand can be thought of as a two-electron σ-donor and π-acceptor, as shown in Figure 18.23.

An sp^2-hybrid orbital on the C atom donates a pair of electrons to an empty orbital on the metal, while the unhybridized *p*-orbital on the C accepts electron density in a π-backbonding arrangement from a filled *d*-orbital on the metal. *Fischer-type*

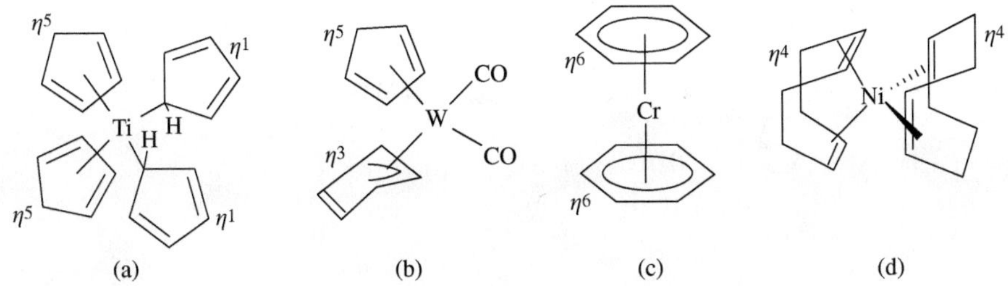

FIGURE 18.22
(a–d) Structures of some interesting organometallic molecules containing π-ligands.

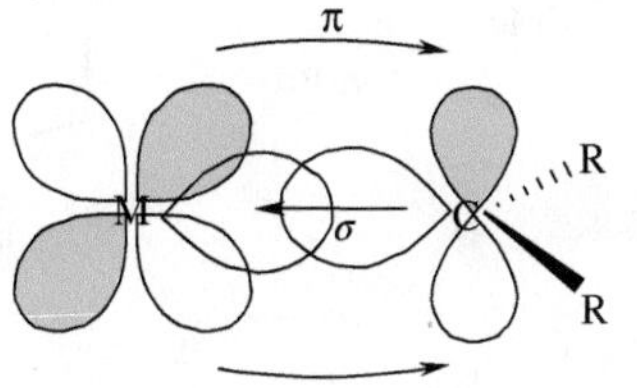

FIGURE 18.23
Sigma and pi bonding modes in a carbene.

carbenes typically contain an electronegative atom directly attached to the C atom in order to facilitate π-backbonding to the ligand, as shown in Figure 18.24(a). The electronegative O atom can stabilize the carbene using resonance with a C=O canonical structure. Most Fischer-type carbenes contain a metal in a low oxidation state and π-acceptor ligands, making them electrophilic. *Schrock-type* carbenes, on the other hand, consist of metals having high oxidation numbers, such as the example shown in Figure 18.24(b). This polarizes the M–C bond so that the C atom has a partial negative charge, making them nucleophilic. Schrock won the 2005 Nobel Prize in chemistry for his work with metal–carbon multiple bonds.

Although the C atom in carbene compounds is sp^2-hybridized, the M=C–R bond angle is typically between 160° and 170°. The α-H on the C atom can donate electron density to an empty orbital on the metal, as shown in Figure 18.25, in an agostic-type interaction that weakens the C–H bond and causes the ν(CH) stretching frequency to shift to ~2600 cm^{-1}.

Fischer was also the first to synthesize a carbyne (alkylidyne), an organometallic compound having a M≡C triple bond, as shown in Figure 18.26. A filled *sp*-hybridized orbital on the C atom donates electron density to an empty orbital on the metal. π-backbonding from filled orbitals on the metal can occur into two perpendicular sets of π^* MOs. Thus, the carbyne acts as a σ-donor and two π-acceptors. Because it has two π-acceptors and only one σ- donor, the ligand is formally

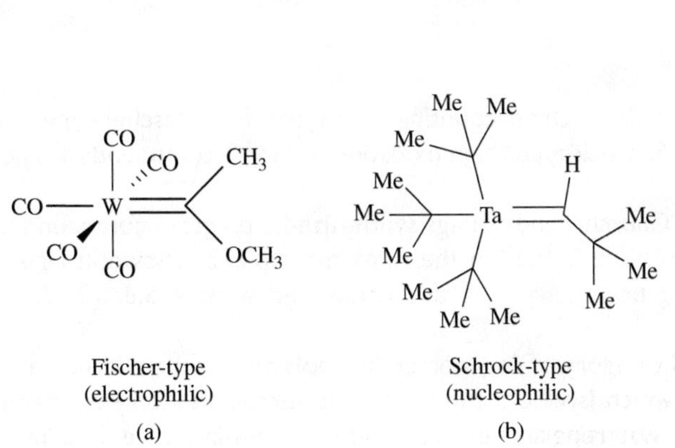

Fischer-type (electrophilic)
(a)

Schrock-type (nucleophilic)
(b)

FIGURE 18.24
(a) Fischer-type and (b) Schrock-type carbene compounds.

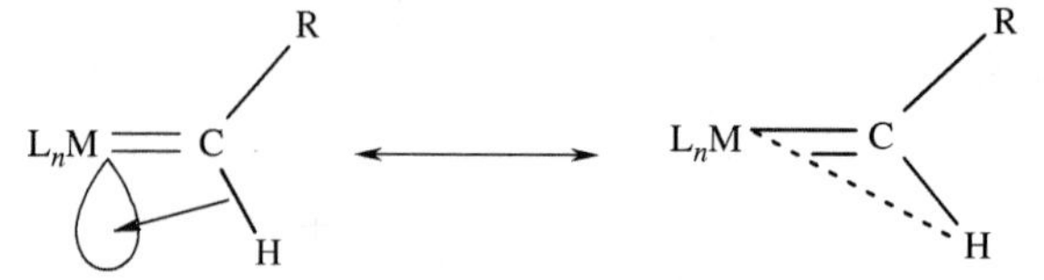

FIGURE 18.25
Agostic interaction in metal carbenes.

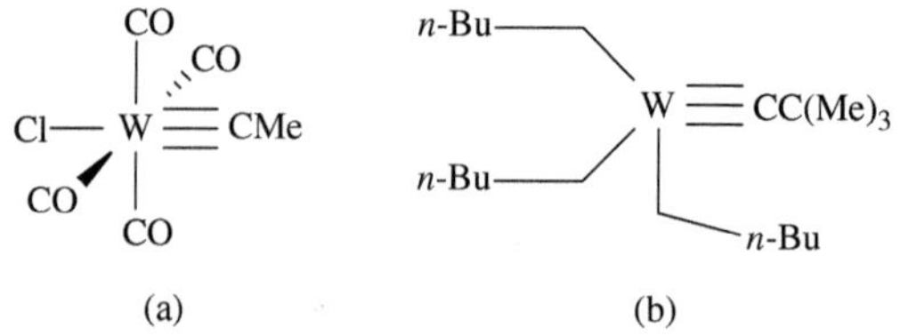

FIGURE 18.26
(a) Fischer-type and (b) Schrock-type carbynes.

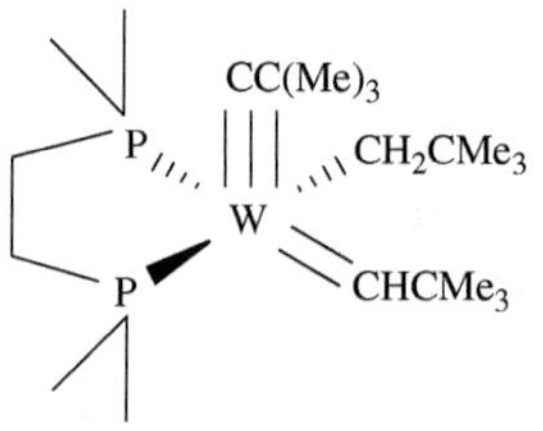

FIGURE 18.27
Example of an organometallic compound having a W–C single, double, and triple bond in the same molecule.

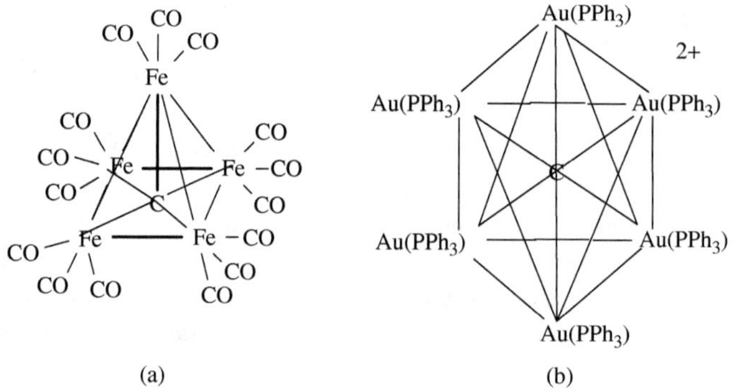

FIGURE 18.28
(a, b) Two examples of carbido compounds.

+1 in donor pair electron counting formalism. Both Fischer-type (low oxidation number) and Schrock-type (high oxidation number) compounds are now known to exist.

In 1979, Churchill and Youngs synthesized a tungsten compound having W–C, W=C, and W≡C bonds all in the same molecule, as shown in Figure 18.27! The corresponding bond lengths in this compound were 225.8, 194.2, and 178.5 pm, respectively.

The final category of organometallic molecules to be addressed is the carbido compounds, which is where the C atom is surrounded only by metal atoms. The first example was reported in 1962 and is shown in Figure 18.28(a). Never mind what you might have learned in organic chemistry about there being no such thing as a pentavalent carbon. This organometallic species contains five bonds to the same central C atom in a square pyramidal geometry. The C atom in Figure 18.28(b) is octahedral.

EXERCISES

18.1. For each of the following organometallics that obey the 18-electron rule, determine the identity of the first-row metal M (unless otherwise specified):

a. $HM(CO)_5$
b. $M(CO)_3PPh_3^-$
c. η^4-$C_8H_8M(CO)_3$
d. $[\eta^5$-$CpM(CO)_3]_2$
e. η^5-$CpM(CO)_3$ [third-row M]

18.2. For each of the following organometallic compounds, determine the formal oxidation state of the metal, the d^n count, the total number of valence electrons in the compound, and whether the metal is saturated or unsaturated:

a. η^5-CpNiNO (NO linear)
b. $W(CH_3)_6$
c. η^5-Cp_2Co^+
d. $Ir(CO)Cl(PR_3)_2$
e. $Ru(CO)(CS)(PPh_3)_2Br$

18.3. Sketch the one-electron MO diagram for the tetrahedral $Ni(CO)_4$ organometallic compound. Label each MO with its proper group theoretical symbol and fill in the correct number of valence electrons. Also, determine the number of ν(CO) stretches in the IR spectrum.

18.4. Sketch the one-electron MO diagram for the trigonal bipyramidal $Fe(CO)_5$ organometallic compound. Label each MO with its proper group theoretical symbol and fill in the correct number of valence electrons. Also, determine the number of ν(CO) stretches in the IR spectrum.

18.5. Use MOT to predict which of the following octahedral metal carbonyls will have the most positive $E°$ and explain your answer: $Cr(CO)_6$, $Mo(CO)_6$, and $W(CO)_6$.

18.6. Rationalize the following trend in the wavenumber of ν(CO):

$M(CO)_6$	ν(CO), cm^{-1}
$W(CO)_6$	1977
$Re(CO)_6^+$	2085
$Os(CO)_6^{2+}$	2190
$Ir(CO)_6^{3+}$	2254

18.7. Explain whether the following carbonyl-bridged organometallic compound obeys the 18-electron rule or not:

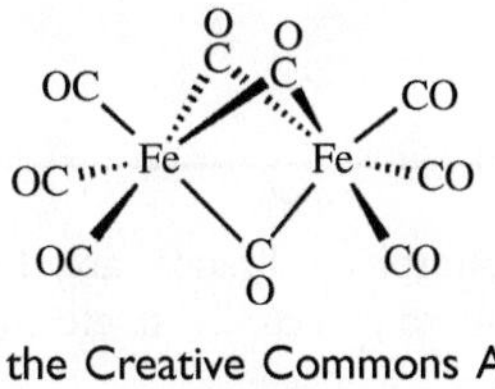

[Attributed to Smokefoot under the Creative Commons Attribution 3.0 Unported license, reproduced from http://en.wikipedia.org/wiki/Diiron_nonacarbonyl (accessed March 10, 2014).]

18.8. When $[\eta^5$-$CpMo(CO)_3]_2$ is heated, it loses two CO ligands to become $[\eta^5$-$CpMo(CO)_2]_2$ and the energies of the ν(CO) stretches change from 1960 and 1915 cm^{-1} to 1189 and 1859 cm^{-1}. The Mo–Mo bond length also shortens by 80 pm. Suggest plausible chemical structures for each of the organometallic compounds.

18.9. Which is expected to have the highest energy ν(CO), $Ni(CO)_3(PH_3)$ or $Ni(CO)_3(PPh_3)$? Explain your answer.

18.10. Rationalize why the V–C bond lengths in $V(CO)_6$ are 199.3 pm along the z-axis and 200.5 pm in the perpendicular (*xy*) plane.

18.11. Rationalize the bond angles and bond lengths of the two nitrosyl ligands in the $[Ru(PPh_3)_2(NO)_2Cl]^+$ compound shown here:

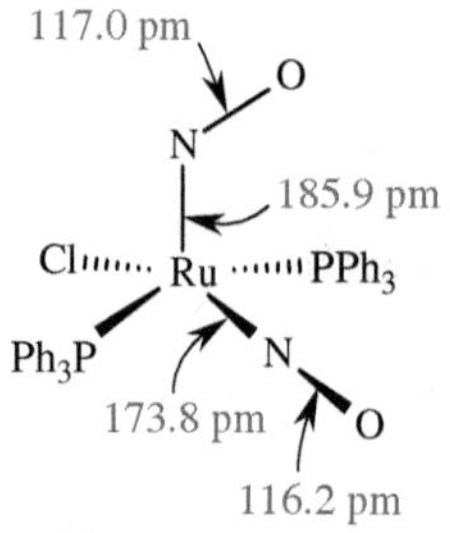

117.0 pm
185.9 pm
Cl Ru PPh_3
Ph_3P N O
173.8 pm
116.2 pm

18.12. When the compound $[Co(LL)_2NO]^{2+}$ reacts with SCN^- to make $[Co(LL)_2(NCS)NO]^+$, the NO changes from a linear to a bent geometry. Explain this finding on the basis of the 18-electron rule.

18.13. The compound $Mo(CO)_3(NCC_2H_5)_3$ exhibits two ν(CO) stretches in its IR spectrum. Is the geometry of this compound facial or meridional? Explain your answer using group theoretical arguments.

18.14. The complex cation $[Co(CO)_3(PPh_3)_2]^+$ has only a single ν(CO) stretching frequency in the IR. Suggest a plausible structure for this compound.

18.15. Coordination makes the dihydrogen ligand a much better acid than it is when it is unbound. In fact, sometimes the H_2 ligand can even be deprotonated with the base. For example, $[\eta^5\text{-CpRe(NO)(CO)(H}_2)]^+$ is a strong acid having a $pK_a = -2.5$ and can release H^+ to leave behind the hydride ligand attached to the metal. Rationalize why the dihydrogen ligand in this particular compound is so acidic.

18.16. Explain why $[Mo(CO)_3(PR_3)_2(H_2)]$ contains dihydrogen as a ligand, while $[Mo(PR_3)_5(H)_2]$ exists as the dihydride.

18.17. For the series of carbonyl compounds having the following general formula: *fac*-$[Mo(CO)_3(PX_3)_3]$, the following IR data were obtained:

Compound	ν(CO), cm^{-1}
A	2074, 2026
B	2041, 1989
C	1937, 1841

If the three different phosphine ligands are PPh_3, PF_3, and PCl_3, determine which compounds belong to which IR stretching frequencies and explain your reasoning.

18.18. The structures of two phosphine ligands are shown here. When R = methyl, the ligand reacts with $PdCl_2$ to make L_2PdCl_2; but when R = isopropyl, the ligand reacts to form the bridged species $LClPd(\mu Cl)_2PdClL$. Rationalize this result.

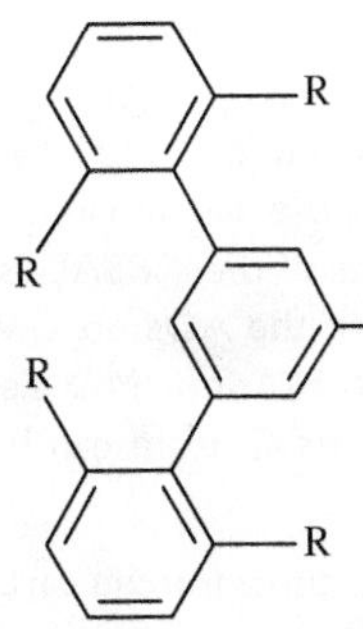

18.19. Explain why the C=C bond length increases from 133.9 pm in free ethylene to 144.5 pm in $(\eta^5\text{-Cp})Rh(\eta^2\text{-}C_2H_4)(PMe_3)$.

18.20. Predict whether the ν(C=C) stretching vibration in $Fe(CO)_4(\eta^2\text{-}C_2H_4)$ will be higher or lower in energy than the corresponding stretching frequency for free C_2H_4, which occurs at 1623 cm^{-1}. What would happen to the energy of ν(C=C) if the ligand is changed from C_2H_4 to $(CN)_2C{=}C(CN)_2$? Explain your answer.

18.21. Free N_2 has a Raman-active band at 2331 cm^{-1}. Upon coordination to a metal, would you expect this band to shift to higher or lower frequencies? Explain your answer and briefly explain what MOs are involved in the bonding of $\eta^1\text{-}N_2$ to the metal.

18.22. There are two types of C=C double bonds in the fullerene ligand, C_{60}: those that lie at the junction of two six-membered rings and those that lie at the junction of a five-membered ring with a six-membered ring. Taking into consideration the different curvature of the fullerene at these two different ring junctions, predict at which type of junction the compound $[(PPh_3)_2Pt(\eta^2\text{-}C_{60})]$ will bond. Explain your answer.

18.23. Predict the first-row metal that forms the compound $(\eta^6\text{-}C_6H_6)M(CO)_3$ and explain your answer.

18.24. Which of the following metallocenes will be easiest to oxidize: $(\eta^5\text{-Cp})_2Fe$ or $(\eta^5\text{-Cp})_2Co$? Explain your answer.

18.25. Explain why the Mn–C bond length in Table 18.5 is the longest one in the series. Hint: think about the electronic structure of HS Mn^{2+}.

18.26. When $[(\eta^5\text{-Cp})_2Co]^+$ reacts with H^-, it makes $[(\eta^5\text{-Cp})(\eta^4\text{-Cp})\ Co]$. What is the charge on the η^4-Cp ligand? Explain why the latter compound prefers η^4-hapticity instead of the more common η^3-hapticity.

18.27. Classify each of the following as Fischer-type or Schrock-type carbenes:

a. $(CO)_5Mo{=}C(OH)H$

b. $(\eta^5\text{-Cp})Cl_2Nb{=}CH_2$

c. $(\eta^5\text{-Cp})_2(CH_3)Ta{=}CH(CH_3)$

d. $(CO)_5W{=}C(OCH_3)Ph$

18.28. Match each of the items in Column B with one of the types of carbenes in Column A:

Column A	Column B
Fischer-type	Electropositive metals
Schrock-type	Electronegative metals
	Good σ-donor ligands
	Good π-acceptor ligands
	Acts as a nucleophile
	Acts as an electrophile

18.29. Determine the electron count in $Br(CO)_4Cr(\equiv CPh)$ and explain how you arrived at your answer.

18.30. The Au cluster pictured in Figure 18.28(b), which has six $Au(PPh_3)$ groups in an octahedral arrangement around a single C atom, can be modeled by the complex ion CH_6^{2+}. Construct the MO diagram for CH_6^{2+} by determining the symmetries of the C 2s and 2p AOs on one side of the diagram in the octahedral molecular point group. Then determine the symmetries of the six SALCs for the H atoms at the corners of the octahedron. Finally, combine the AOs on C with the SALCs on the 6H to form the MOs for CH_6^{2+} in the center of your MO diagram. Fill in the correct number of electrons and rationalize how the C atom can be hexavalent in this organometallic compound.

18.31. Consider the two possible structures for the carbene compound shown here.

H Cl H Pt H Cl H (a)

H Cl H Pt H Cl H (b)

y z x

For each structure:

a. Determine the molecular point group.

b. Determine the hybridization on the Pt(II) ion. Specifically, which AOs are used in hybridization?

c. Determine the CF splitting pattern of the d-orbitals based on their interactions with the ligands.

d. Determine the symmetries and shapes of the SALCs for the four σ-bonds.

e. Determine the symmetries and shapes of the SALCs for the two π-bonds.

f. Sketch the one-electron MO diagram and fill in the correct number of electrons.

g. Then use the two MO diagrams to determine which isomer will be the more stable one.

BIBLIOGRAPHY

1. Atkins P, Overton T, Rourke J, Weller M, Armstrong F, Hagerman M. *Shriver& Atkins' Inorganic Chemistry*. 5th ed. New York: W. H. Freeman and Company; 2010.
2. Crabtree RH. *The Organometallic Chemistry of the Transition Metals*. 3rd ed. John Wiley & Sons, Inc.: New York; 2001.
3. Douglas B, McDaniel D, Alexander J. *Concepts and Models of Inorganic Chemistry*. 3rd ed. New York: John Wiley & Sons, Inc.; 1994.
4. Housecroft CE, Sharpe AG. *Inorganic Chemistry*. 3rd ed. Essex, England: Pearson Education Limited; 2008.
5. Huheey JE, Keiter EA, Keiter RL. *Inorganic Chemistry: Principles of Structure and Reactivity*. 4th ed. New York: Harper Collins College Publishers; 1993.
6. Miessler GL, Tarr DA. *Inorganic Chemistry*. 4th ed. Upper Saddle River, NJ: Pearson Education Inc.; 2011.
7. Spessard GO, Miessler GL. *Organometallic Chemistry*. 2nd ed. New York: Oxford University Press; 2010.

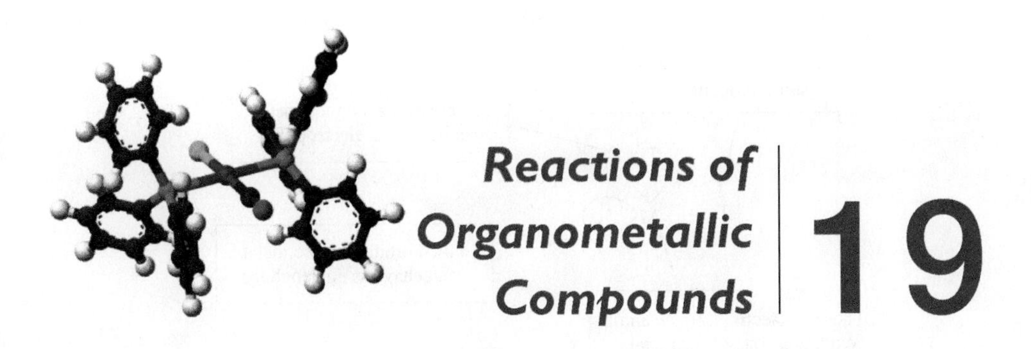

Reactions of Organometallic Compounds 19

"We never envisioned that this reaction would one day achieve the import that it has today. Our exploration has been a fascinating journey, and it is one that is ongoing."

—*Robert Grubbs*

19.1 SOME GENERAL PRINCIPLES

Organometallic compounds undergo a wide variety of chemical reactions analogous to those of organic and inorganic compounds. For example, organometallics can undergo electrophilic addition, nucleophilic substitution, and elimination reactions—just to name a few. At the same time, organometallic molecules can participate in the same types of reaction mechanisms that coordination compounds do—ligand substitution, isomerization, and electron transfer reactions. What makes organometallic chemistry fundamentally unique is the tunability of the metal's electronic properties by the appropriate choice of ligand. As shown in Figure 19.1, organometallic compounds can behave as either an electrophile (or Lewis acid) using an empty orbital on the metal or a nucleophile (or Lewis base) using a filled orbital on the metal. Electrophilic metals are characterized by electron-deficient metals, generally those from the first-row and right-hand side of the periodic table and having a high oxidation number with acceptor ligands. Nucleophilic metals, on the other hand, are electron-rich and generally consist of third-row transition metals from the left-hand side of the periodic table in low oxidation states and with donor ligands. Furthermore, the size of a metal, its electrophilicity, its d^n electron count, and the steric bulk of the ligands control whether the compound will be associative (adding a ligand) or dissociative (losing a ligand) as part of its reaction mechanism.

Vaska's complex. [Attributed to Benjah-bmm27. Reproduced from http://en.wikipedia.org/wiki/Vaska's_complex (accessed March 11, 2014).]

Principles of Inorganic Chemistry, First Edition. Brian W. Pfennig.

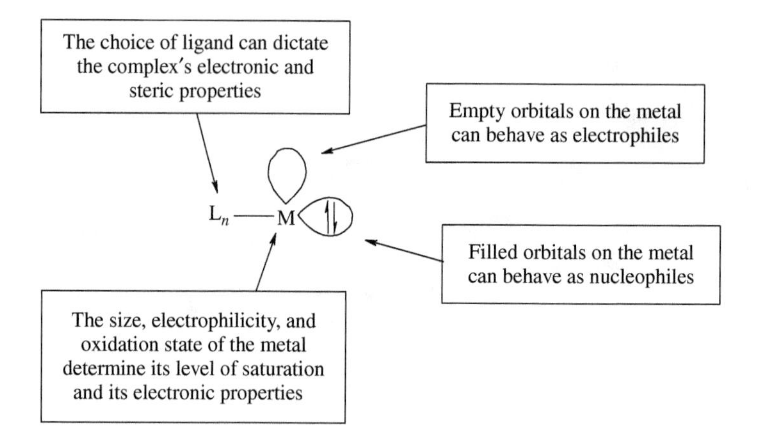

FIGURE 19.1
The ability to tune the steric and electronic properties of organometallic compounds gives them unique reaction chemistry.

19.2 ORGANOMETALLIC REACTIONS INVOLVING CHANGES AT THE METAL

19.2.1 Ligand Substitution Reactions

The generic chemical equation for a ligand exchange reaction is given by Equation (19.1), where L represents a spectator ligand, M is the metal, X is the leaving group, and Y is the entering ligand:

$$L_nM\text{-}X + Y \rightarrow L_nM\text{-}Y + X \quad (19.1)$$

As was the case for transition metal complexes, ligand substitution reactions of organometallic compounds can occur through associative (A), dissociative (D) or interchange (I) pathway. The precise type of mechanism will depend on the size and electrophilicity of the metal, its degree of saturation, the M–X bond strength, the basicity of Y, and the size and nature of the other ligands bound to the metal. As a general rule, metals from the second transition series are more reactive than those from either the first or third transition series. One underlying reason for the enhanced reactivity of the second row metals is that they strike the perfect balance between the larger steric size and effective nuclear charge (which favor third-row metals) and the diminished M–X bond strength (which are weaker for first-row metals).

Associative (A)

In an associative mechanism, the metal expands its coordination number in the rate-determining first step of the mechanism, as shown in Equation (19.2) and then rapidly dissociates the leaving group in the second step of the mechanism, as shown in Equation (19.3). Because the metal has to increase its coordination number, associative-type substitution generally occurs in complexes that are coordinatively unsaturated and are favored by large metals with smaller-sized ligands. The most common candidates for associative ligand substitution are the square planar 16-electron compounds containing Ni(II), Pd(II), Ir(I), and Au(III). The kinetics of associative ligand substitution reactions is second order, as shown in Equation (19.4):

$$L_nM\text{-}X + Y \rightarrow L_nMXY \quad k_1(\text{RDS}) \quad (19.2)$$

$$L_nMXY \rightarrow L_nM\text{-}Y + X \quad k_2(\text{fast}) \quad (19.3)$$

$$\text{Rate} = k_1[L_nM\text{-}X][Y] \quad \text{2nd order} \quad (19.4)$$

As is the case for S_N2 reactions in organic chemistry, the rate of substitution depends on the nucleophilicity of the incoming ligand because the nucleophile directly participates in the rate-determining step of an associative mechanism. The affinity of a metal for a particular nucleophile will depend primarily on the hard–soft acid–base properties of the pair. Soft metals will prefer soft ligands, whereas hard metals will favor hard ligands. Some examples of associative substitution are given in Equations (19.5)–(19.8):

$$trans\text{-}[(C_6H_5CH_3)(PEt_3)_2PtCl] + py \rightarrow trans\text{-}[(C_6H_5CH_3)(PEt_3)_2Ptpy] + Cl^- \quad (19.5)$$

$$[V(CO)_6] + PPh_3 \rightarrow [V(CO)_5PPh_3] + CO \quad (19.6)$$

$$[Co(CO)_3NO] + PR_3 \rightarrow [Co(CO)_2(NO)(PR_3)] + CO \quad (19.7)$$

$$[(\eta_6\text{-}C_6H_6)Mo(CO)_3] + 3\,PR_3 \rightarrow [(CO)_3Mo(PR_3)_3] + C_6H_6 \quad (19.8)$$

Equation (19.5) is a classic square planar substitution reaction occurring at an unsaturated (16 e^-) Pt(II) center, where the soft metal prefers the softer ligand py over Cl^-. For square planar substitution reactions, the nature of the ligand *trans* to the leaving group will affect the rate of substitution as we learned in Chapter 17. Ligands that are good σ-donors or strong π-acceptors will be strongly *trans*-directing. Square planar substitution reactions are generally believed to occur through a square pyramidal intermediate, which rearranges to trigonal bipyramidal geometry before dissociation of the leaving group and (unlike S_N2 reactions in organic chemistry) generally occurs with retention of configuration.

Equation (19.6) involves substitution of the 17-electron $[V(CO)_6]$ radical with triphenylphosphine. The −28 J/K·mol activation entropy for this reaction is clearly consistent with an associative-type mechanism. Furthermore, the rate of substitution occurs 10^{10} faster than does the same reaction with the corresponding 18-electron compound $[Cr(CO)_6]$.

As a general rule, 17-electron compounds will undergo ligand substitution reactions faster than their 18-electron counterparts. For example, oxidation of the relatively inert 18-electron $(MeCp)Mn(CO)_3$ compound by one electron to form the 17-electron complex cation significantly increases the rate of substitution. Equation (19.7) begins with the nitrosyl ligand in a linear position, giving the Co a formal oxidation state of −1 (d^{10}) and making the molecule an 18-electron species. During the course of the reaction, the NO shifts to a bent position, changing the formal oxidation number of the Co to +1 (d^8) and the total electron count to $16e^-$ in order to add the phosphine group in an associative-type substitution reaction.

In Equation (19.8), a similar mechanism occurs for the 18-electron $[(\eta_6\text{-}C_6H_6)Mo(CO)_3]$ molecule, with the coordinated benzene ring slipping to an η_4-linkage to allow room for the initial PR_3, where the intermediate $[(\eta_4\text{-}C_6H_6)Mo(CO)_3(PR_3)]$ has been isolated and characterized in the solid state. Sometimes an associative mechanism will also involve a second term in its kinetics rate law involving a competition between addition of Y and addition of solvent, as was observed in Chapter 17 for square planar substitution reactions. In this case, the solvent term will follow pseudo-first-order kinetics.

Dissociative (D)

In the dissociative mechanism, the slow step is dissociation of X, as shown in Equation (19.9) to form an intermediate with a lower coordination number. The entering ligand then adds to the metal in a rapid second step, as shown in

Equation (19.10). The kinetics of D-type mechanisms are first order as shown in Equation (19.11), as long as $[Y] \gg [X]$ or whenever k_{-1} is small.

$$L_nM{-}X \Longleftrightarrow L_nM + X \qquad K = k_1/k_{-1}(\text{RDS}) \tag{19.9}$$

$$L_nM + Y \rightarrow L_nM{-}Y \tag{19.10}$$

$$\text{rate} = \frac{kk_2[L_nM{-}X][Y]}{k_{-1}[X] + k_2[Y]} = k_{obs}[L_nM{-}X] \tag{19.11}$$

Dissociative-type mechanisms are far more common, especially with coordinatively saturated, 18-electron complexes having sterically demanding ligands. One classic example occurs for the tetrahedral, d^{10} $[Ni(CO)_4]$ molecule, which shows a first-order rate law that is independent of the incoming ligand and a positive entropy of activation. The compound is so reactive that it can replace all four CO ligands in a stepwise reaction mechanism, as shown in Equation (19.12):

$$[Ni(CO)_4] + L \rightarrow [Ni(CO)_3L] + CO$$

$$[Ni(CO)_3L] + L \rightarrow [Ni(CO)_2L_2] + CO$$

$$[Ni(CO)_2L_2] + L \rightarrow [Ni(CO)L_3] + CO$$

$$[Ni(CO)L_3] + L \rightarrow NiL_4 + CO \tag{19.12}$$

As would be expected for a dissociative-type mechanism, the rates of substitution are strongly dependent on the steric bulk of the leaving group. Larger leaving groups favor a dissociative mechanism and enhance the rate of substitution. Most ligand substitution reactions involving 19-electron compounds also proceed by a dissociative mechanism. When interchange reactions occur in organometallic chemistry, they are generally of an I_d nature.

19.2.2 Oxidative Addition and Reductive Elimination

Oxidative addition (OA) reactions involve the simultaneous oxidation of the metal by two electrons and an increase in the coordination number by the addition of a ligand or ligands, as shown in Figure 19.2. Reductive elimination (RE) is the exact opposite process.

OA requires a vacant coordination site on the metal, *d*-electrons that can be lost ($d^n \geq 2$), and suitable orbitals for bond formation (an unsaturated metal). It is favored by electropositive metals from the lower, left-hand side of the periodic table in low oxidation states and possessing electron-donating ligands. One of the first known examples of OA was the reaction of Vaska's complex (which is shown on the chapter's opening page), *trans*-$[Ir(PPh_3)_2(CO)Cl]$ with H_2, as shown in Figure 19.3. The formal oxidation state of iridium in the reactant is Ir(I), which has a d^8 electron configuration, making this a 16-electron compound. Following reaction with H_2 (which is normally considered as a reducing agent!), the formal oxidation state of iridium is Ir(III). In this prototypical OA mechanism, the molecule has been oxidized by two electrons and has added two ligands to its coordination sphere.

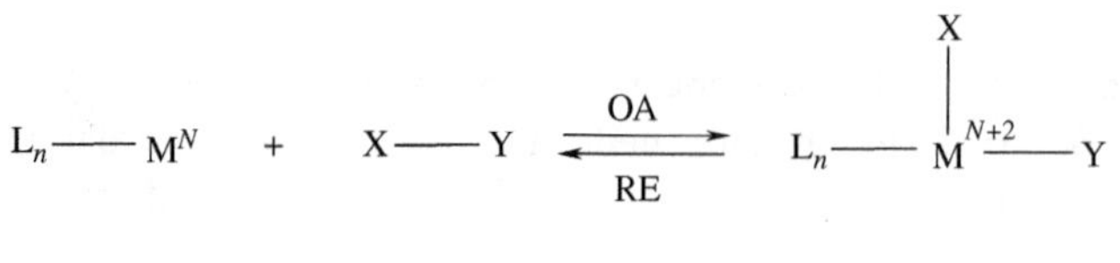

$$L_n{-}M^N + X{-}Y \underset{RE}{\overset{OA}{\rightleftharpoons}} L_n{-}M^{N+2}(X){-}Y$$

FIGURE 19.2
The processes of oxidative addition (OA) and reductive elimination (RE).

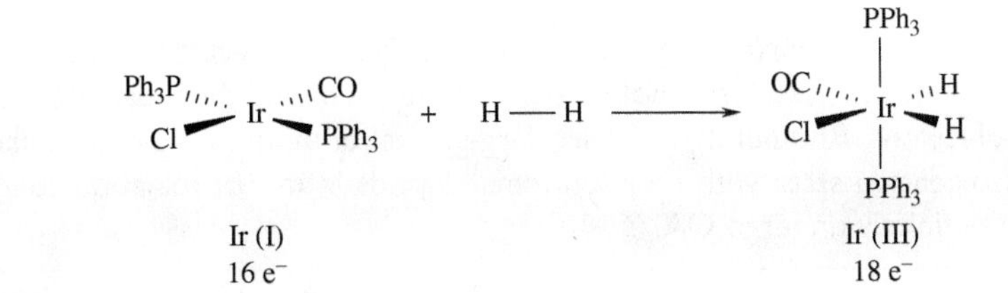

FIGURE 19.3
Oxidative addition of hydrogen to Vaska's complex.

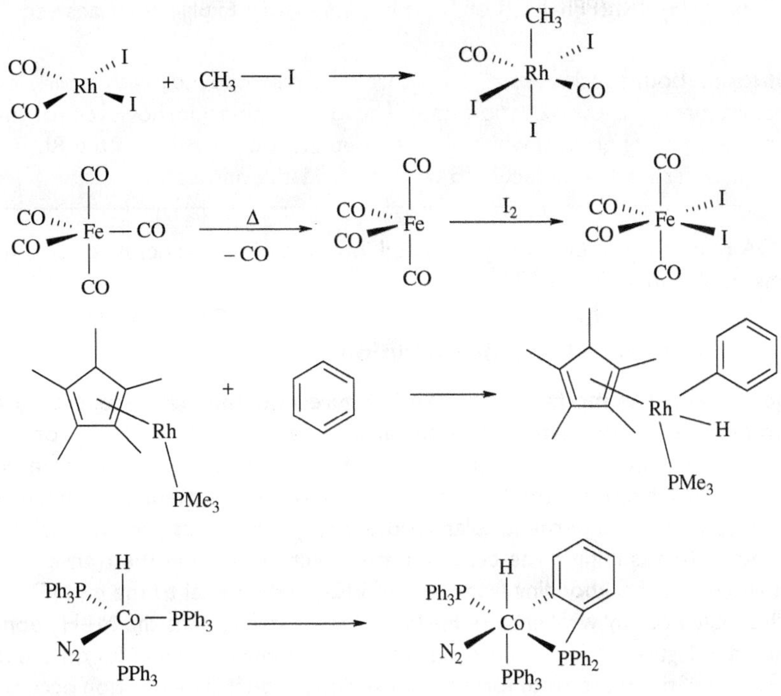

FIGURE 19.4
Examples of oxidative addition reactions.

Other examples of OA reactions are shown in Figure 19.4. The first reaction listed is an important intermediate in the Monsanto acetic acid process. In the second example, the 18-electron $[Fe(CO)_5]$ compound must first dissociate a ligand before it can undergo OA. The third reaction represents an example of C–H bond activation and the fourth reaction involves an intramolecular cyclometalation of the PPh_3 ligand.

Example 19-1. Which of the following organometallic species are incapable of undergoing an oxidative addition reaction and why not: (a) Cp_2TiCl_2, (b) $[Pt(PPh_3)_4]^{2+}$, (c) $[Ni(CO)_4]$, (d) $[Cp_2{}^*Th(tBu)_2]$, (e) $[Cp^*Ir(PMe_3)_2(H_2)]$, and (f) $[Cp^*Rh(PMe_3)_3]$.

Solution. (a) Ti(IV) is d^0 and the compound has 16 valence electrons. It cannot undergo OA because there are no oxidizable electrons on the d^0 metal. (b) Pt(II) is d^8 and the complex has 16 valence electrons and only 4 ligands. This species can undergo OA. (c) Ni(0) is d^{10}, so it has oxidizable electrons and plenty of space, but it is already an 18-electron compound, so it will be unable to participate in an OA reaction. (d) Th(IV) is d^0 and this is a 16-electron compound. It cannot undergo OA because there are no oxidizable d-electrons and the molecule is sterically crowded. (e) Ir(III) is d^6, but this is already an 18-electron compound in a high oxidation state, so OA is not feasible. (f) Rh(I) is d^8 and the molecule has 16 valence electrons. Rhodium(I) is a fairly large metal of moderate electrophilicity in a low oxidation state with only four small ligands in its coordination sphere, so it can definitely undergo OA.

Example 19-2. Which of the following will be more reactive toward oxidative addition of H_2: $[Rh(PPh_3)_3Cl]$ or $[Rh(PPh_3)_2(CO)Cl]$? Explain your answer.

Solution. Both species are d^8 LS square planar compounds with a fairly electrophilic metal in a low oxidation state. The former molecule, however, has more electron-donating groups, which will build up electron density on the Rh(I) and make it easier for the molecule to undergo oxidative addition.

OA reactions can be loosely classified into one of four general reaction mechanisms, as shown in Figure 19.5.

Nonpolar (Agnostic) Oxidative Addition

In the nonpolar OA mechanism, shown in Figure 19.6, the metal acts initially as an electrophile as it coordinates the nonpolar substrate (e.g. H_2, R–H, Y_2, or Ar–H) using a three-centered, two-electron bonded transition state known as an *agostic* complex. The dihydrogen (or Y_2) molecule approaches the metal in such a manner that its σ_b HOMO is perpendicular to the empty orbital (usually the LUMO) on the metal. The resulting three-centered, two-electron bond in the transition state is stabilized by π-backbonding from the HOMO on the metal to the σ^* MO on H_2, which simultaneously weakens the H–H bond as it strengthens the M–H_2 bonding, as shown in Figure 19.7. The reaction is completed when a filled MO on the metal attacks one of the H atoms to form the dihydrido product. The reaction occurs with retention of configuration at the α-C atom if a chiral alkane acts as the substrate and the rate law obeys complex second-order kinetics. Because of the concerted nature of the mechanism, the geometry of the product is always such that the two additional ligands are *cis* to one another in the product. As a result, this mechanism is also known as *cis addition*. The OA of H_2 to Vaska's complex, shown in Figure 19.3, is the classic example of this type of mechanism. Notice that the two hydride ligands are *cis* in the resulting octahedral compound.

When the substrate is an arene, the reaction proceeds via a Whalen intermediate, as shown in Figure 19.8. The positive charge on the benzene ring in the Whalen intermediate is stabilized by electron donor atoms in the *ortho* and *para* positions.

Polar S_N2 Oxidative Addition

The polar S_N2 OA reaction mechanism occurs with second-order reaction kinetics and large, negative entropies of activation (−40 to −50 J/K·mol), suggestive of an

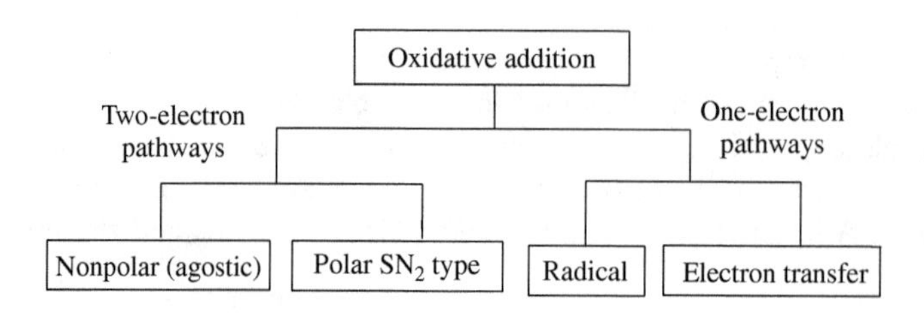

FIGURE 19.5
General types of oxidative addition reaction mechanisms.

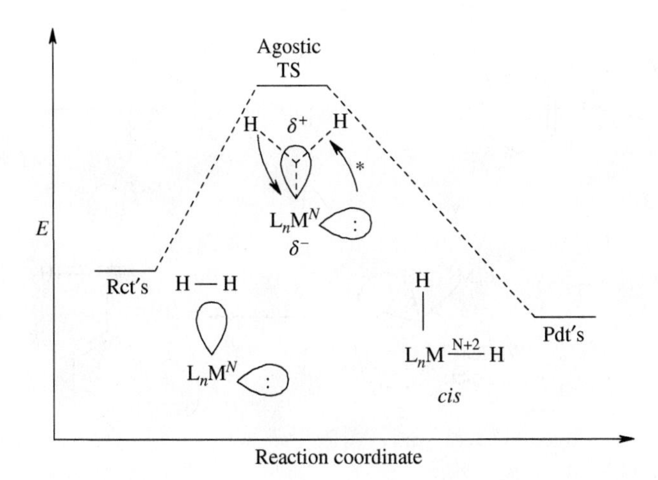

FIGURE 19.6
Agostic oxidative addition mechanism for nonpolar substrates.

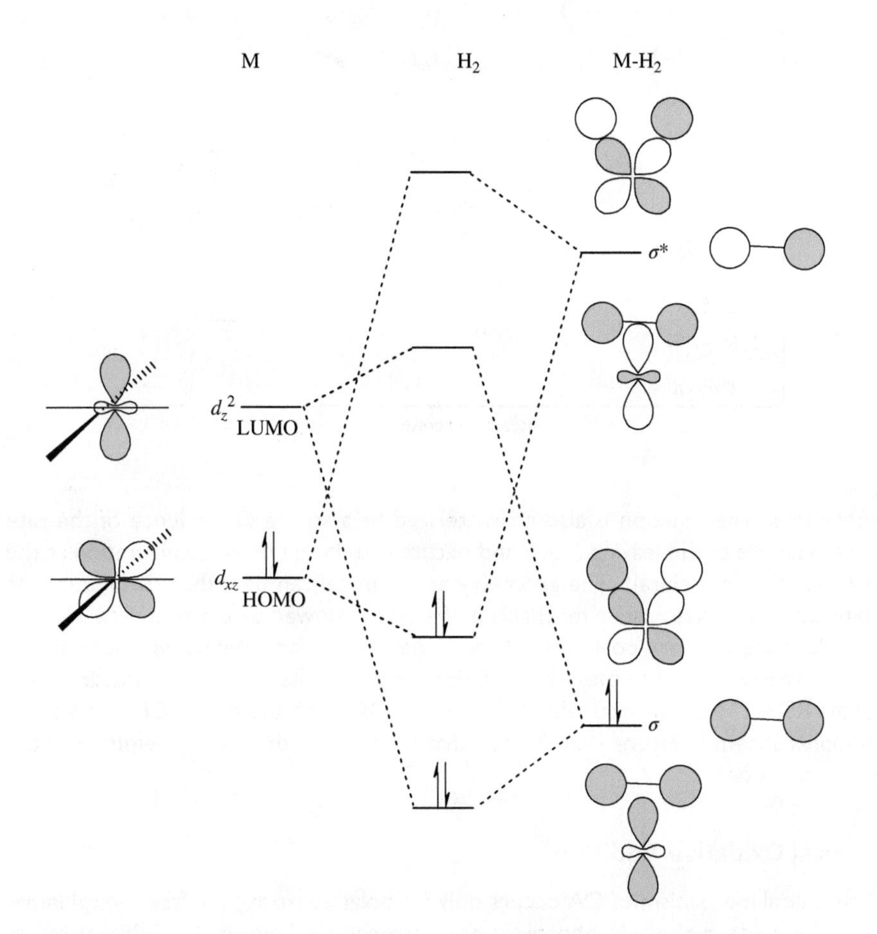

FIGURE 19.7
Frontier MO diagram for the interaction of a square planar d^8 metal compound with H_2 in a nonpolar oxidative addition reaction.

associative-type mechanism analogous to S_N2 reactions in organic chemistry. In this reaction, the metal acts as a nucleophile, attacking the R–X group in a rear-face position to form a trigonal bipyramidal transition state where the metal has a partial positive charge and the leaving group has a partial negative charge, as shown in Figure 19.9.

When the leaving group departs, its negative charge is attracted to the positive metal center, and it rapidly adds to the metal in the second step of the mechanism. The rate is enhanced by small, highly polarizable nucleophiles and soft, unsaturated

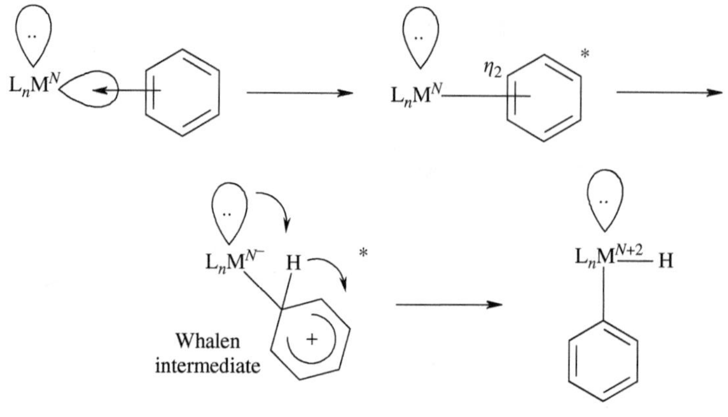

FIGURE 19.8
The Whalen intermediate in the nonpolar oxidative addition mechanism for arene substrates.

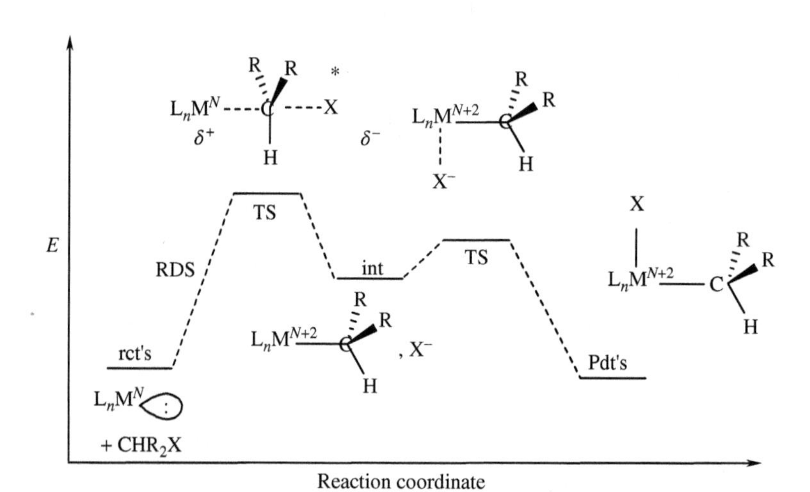

FIGURE 19.9
S_N2-type oxidative addition mechanism for polar RX compounds.

metal ions. The reaction is also characterized by a strong dependence of the rate on the nature of the leaving group and occurs with inversion of configuration at the α-C atom (if it is chiral). The geometry at the metal can be either *cis* or *trans*. As expected for an associative mechanism, the rate is slowed by the presence of bulky ligands in the auxiliary positions. At the same time, polar solvents will increase the rate of reaction because they help stabilize the partial charges in the transition state of the RDS. The classic example of a polar S_N2 OA is the addition of CH_3I to Vaska's complex shown in Figure 19.4. Notice that in that case the stereochemistry of the product is *trans*.

Radical Oxidative Addition

The radical mechanism of OA occurs only for polar substrates. A free radical initiator (I·) is made, typically by photolysis or electrochemical means. The initiator reacts with the metal complex to oxidize it by one electron, as shown in Figure 19.10. The M^{N+1} species can then react with RX to generate R·. The R· radical undergoes a chain reaction with a second metal complex to make R–M^{N+2}–X and another R· radical. This continues until chain termination by two R· radicals coupling together or by radical trapping. The propagation step in the mechanism competes with isomerization or racemization of R·, so that the product is almost always a racemic mixture of optical isomers when a chiral C atom is used. Unlike the S_N2 mechanism, the rate of the reaction is independent of steric bulk on the transition metal. Furthermore, the reaction sequence with respect to $3° > 2° > 1° > CH_3$ (which maps with the

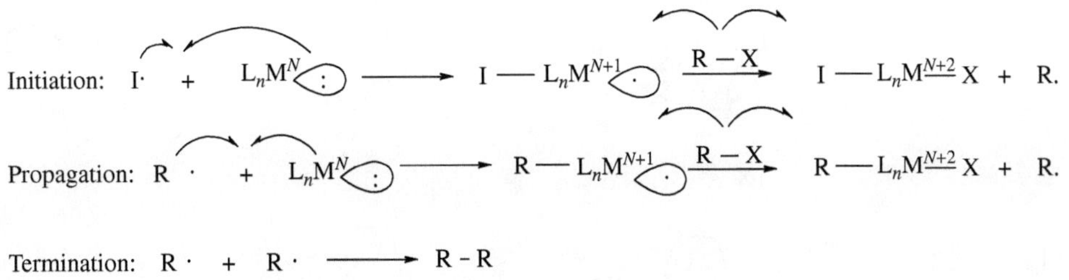

Initiation: $I\cdot + L_nM^N \longrightarrow I-L_nM^{N+1} \xrightarrow{R-X} I-L_nM^{N+2}X + R\cdot$

Propagation: $R\cdot + L_nM^N \longrightarrow R-L_nM^{N+1} \xrightarrow{R-X} R-L_nM^{N+2}X + R\cdot$

Termination: $R\cdot + R\cdot \longrightarrow R-R$

FIGURE 19.10
Radical mechanism for oxidative addition of polar substrates.

$PtL_3 \xrightarrow{\text{Fast}} PtL_2 + L$

$PtL_2 + RX \xrightarrow{\text{RDS}} [\cdot PtL_2X + R\cdot]_{\text{Cage}}$

$[\cdot PtL_2X + R\cdot]_{\text{Cage}} \xrightarrow{\text{Fast}} RPtL_2X$

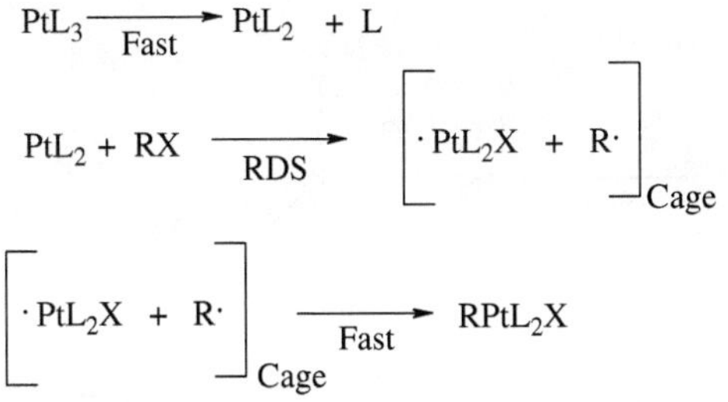

FIGURE 19.11
Electron transfer mechanism for oxidative addition of polar substrates.

stability of the alkyl radicals) is exactly opposite to that expected for an S_N2 mechanism. Metals having an odd number of valence electrons, such as Co(II), Rh(II), and Mn(0), are particularly amenable to the radical mechanism. Further evidence in support of a radical mechanism can occur if the radical can rearrange. For example, the hexenyl radical is commonly known to undergo a rapid rearrangement to the cyclopropylmethyl radical. Evidence of the latter compound in the final organometallic compound would be a strong indication of a radical pathway.

Electron Transfer Oxidative Addition

The fourth OA mechanism involves an innersphere electron transfer reaction between the organometallic compound and R–X, leading to the one-electron oxidized complex: L_nM^{N+1}–X, R· radical cage. The radical cage then undergoes collapse to form the OA product or escape to continue the radical chain mechanism shown earlier. An example of the ET/radical cage mechanism is shown in Figure 19.11.

Reductive Elimination

RE occurs when the metal needs to shed some of its ligands to assume a lower oxidation number. RE is therefore favored by (i) a metal in a high oxidation state, (ii) a saturated (18 e^-) organometallic species, (iii) steric bulk on the metal, and (iv) metals that prefer to be unsaturated (small or not very electrophilic metals). In the case of RE, the metal acts as an electrophile. It steals the electrons in an M–L bond for itself and becomes reduced in the process. The ligand winds up as part of a three-centered, two-electron agostic complex in the intermediate before dissociating into the products. The general mechanism is shown in Figure 19.12 alongside several typical examples.

Most reactions occur through the agostic intermediate, so the leaving groups have to be *cis* to one another in the complex. Sometimes the reduced metal reacts

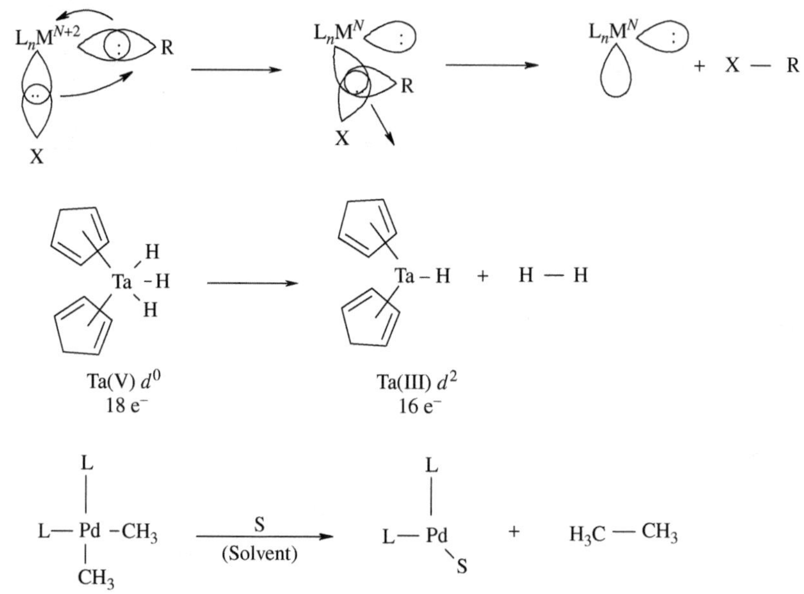

L_nM^{N+2} R X → L_nM^N R X → L_nM^N + X — R

Cp_2TaH_3 → Cp_2Ta-H + H — H

Ta(V) d^0 18 e^- Ta(III) d^2 16 e^-

$L_2Pd(CH_3)_2$ —S (Solvent)→ L_2Pd-S + H_3C-CH_3

FIGURE 19.12
Examples of reductive elimination reactions.

TABLE 19.1 Rate constants of reductive elimination for the reaction of $L_2Pd(CH_3)_2$ with solvent, as shown in Figure 19.12.

L	T, °C	k, s^{-1}
PPh_3	60	1.04×10^{-3}
PPh_2Me	60	9.63×10^{-5}
Ph_2P–CH_2CH_2–PPh_2	80	4.78×10^{-7}

with solvent in order to fill one of its vacant coordination sites following RE. The rate of reaction increases with the steric bulk of the ligand, as shown in Table 19.1. This is to be expected for any dissociation reaction. Notice how the rate constant for RE when L = PPh_3 (which is the bulkiest ligand in Table 19.1) indicates that this will be the fastest of the three reactions. Substitution of just one phenyl ring with the smaller methyl group in L = PPh_2Me decreases the rate of RE by a factor of ~10-fold. The slowest rates occur for the chelate L = dppe because the bite angle of the ligand is not conducive to the decrease in the coordination number of the metal.

19.3 ORGANOMETALLIC REACTIONS INVOLVING CHANGES AT THE LIGAND

19.3.1 Insertion and Elimination Reactions

An insertion reaction, as the name implies, involves the insertion of a ligand into another metal–ligand bond. Insertion reactions are generally classified based on the position of the inserting ligand that binds to the metal center, as shown for a generic organometallic compound in Figure 19.13(a). The inserting ligand is generally an unsaturated one, such as carbonyls, alkenes, nitrosyls, carbenes, isonitriles, or molecular oxygen. Insertion reactions can be classified by the numerical position on the ligand that attaches itself to the metal during the insertion reaction.

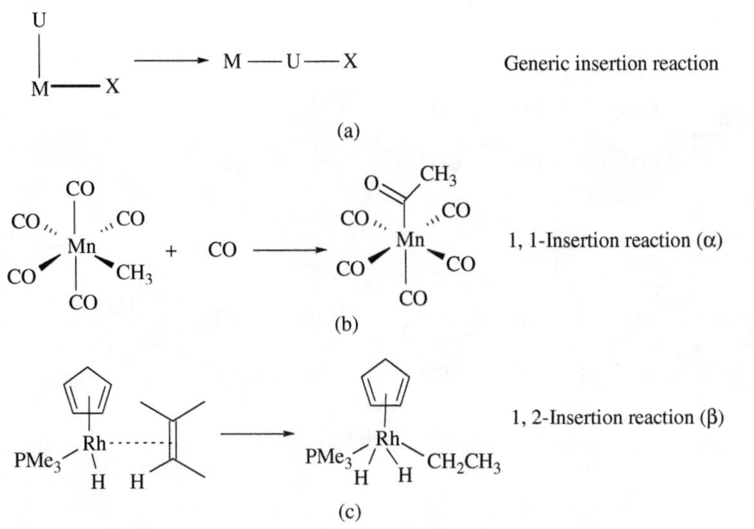

FIGURE 19.13
(a) A generic insertion, (b) 1,1-insertion, and (c) 1,2-insertion reaction, where U = an unsaturated ligand.

Alpha, or 1,1-insertion, reactions, for example, occur when the α-C atom inserts itself between the metal and another ligand, as shown in Figure 19.13(b). Beta, or 1,2-insertion reactions, occur when the β-atom inserts between the metal and an existing ligand. When the β-atom is hydrogen, the reaction is also called a *β-hydride addition reaction*, as shown in Figure 19.13(c). Gamma, delta, and even epsilon insertion reactions are also possible.

1,1-Insertion and Deinsertion

The generic chemical equation for a 1,1-insertion reaction is shown in Equation (19.13). The numbering system refers to the ligand. In the example provided in Figure 19.13(b), we will assign the C atom on the CO ligand as position 1. Following the insert, both the metal and the original alkyl group are attached to C_1. Hence, the reaction is called a *1,1-insertion*.

$$L_n\text{—}M(\text{—}Y)\text{—}X{=}Z \longrightarrow L_n\text{—}M({=}Z)\text{—}X\text{—}Y \qquad (19.13)$$

Although it appears as though the CO ligand is inserting between the metal and the alkyl group in Figure 19.13(b), the reaction actually occurs by an *alkyl migration*. In order to better elucidate the mechanism of this so-called 1,1-insertion reaction, the reverse reaction was examined: 1,1-deinsertion using radiolabeled ^{13}CO, which is readily detectable in the IR spectrum. When $[Mn(CO)_5(CH_3)]$ reacts with ^{13}CO, the radioactive isotope only occurs in the CO ligands of the compound, never in the acyl group. Furthermore, when the reaction was run with the radiolabel on one of the CO ligands: $[Mn(CO)_4(^{13}CO)(CH_3)] + CO$, the methyl group always wound up in a *cis* position with respect to the radiolabeled carbonyl. If the CO in the acyl group had moved instead of the alkyl group, then one would expect 25% of the product molecules to have no radioactive label and the other 75% to have the methyl group *cis* to the labeled carbonyl, as shown in Figure 19.14. On the other hand, if the alkyl group migrates, then 25% of the time the products will lack a radiolabel, 50% of the time the methyl group will be *cis* to the label, and 25% of the time the methyl group will be *trans* to the ^{13}CO ligand. The observed experimental ratio is 2 : 1 *cis* : *trans*, which is consistent with methyl group migration.

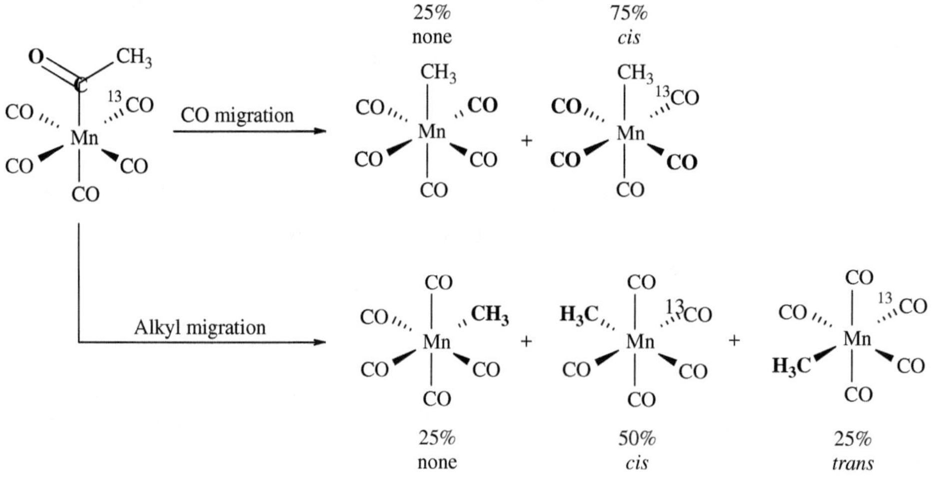

FIGURE 19.14
Deinsertion from $[Mn(CO)_4(^{13}CO)(CH_3)]$ and the predicted product distributions for CO migration versus CH_3 migration. The ligands in boldface represent the possible positions for where the migrating group might wind up.

The relative rates of alkyl migration decrease in the order: methyl > phenyl > benzyl. The above analysis assumes that the deinsertion and 1,1-insertion reactions exhibit microscopic reversibility (both the forward and the reverse reactions proceed by the same lowest-energy pathway). A chiral iron organometallic compound has also been used to address the question of whether CO insertion reactions occur by movement of the CO ligand or the alkyl ligand. This result also indicates that it is the alkyl group that actually does the migration. Of course, these results do not imply that all 1,1-insertion reactions occur by alkyl migration. In fact, there are other molecules in the literature where both mechanisms appear to be occurring simultaneously.

1,2-Insertion and Deinsertion

The generic chemical equation for a 1,1-insertion reaction is shown in Equation (19.14).

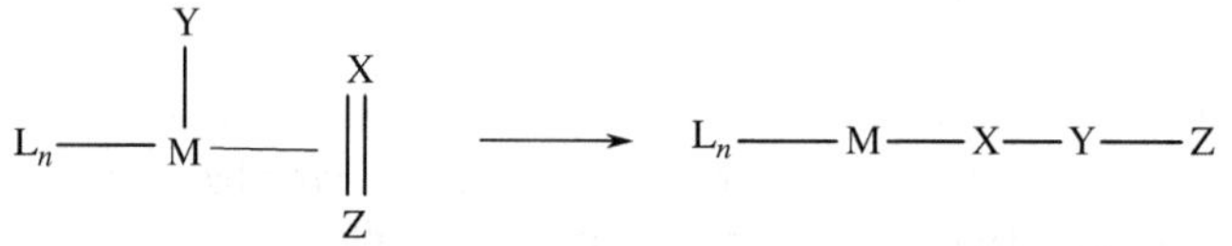

$$L_n\text{—}M(\text{—}Y)\text{—}(X{=}Z) \longrightarrow L_n\text{—}M\text{—}X\text{—}Y\text{—}Z \qquad (19.14)$$

Beta or 1,2-insertion reactions occur most frequently with alkenes or alkynes as the inserting ligand. Mechanistic studies of 1,2-insertion reactions indicate that the alkene and the M–H bond into which the alkene inserts itself must lie in the same plane so that they can form a four-centered agostic intermediate or transition state, as shown in Figure 19.15. Several four-centered agostic intermediates have been isolated and characterized by X-ray or neutron diffraction. The η_2-coordinated alkene slips into an η_1-configuration in the agostic transition state and a hydrogen atom is abstracted from the β-C atom to form the metal-alkyl product. The exact opposite reaction is called *β-elimination* or *1,2-deinsertion.* In reality, an equilibrium between the two extremes occurs. In order to shift the reaction in the forward direction (1,2-insertion), an electrophilic metal with bulky or good acceptor ligands should be used. Electropositive metals having strong donor ligands will drive the equilibrium

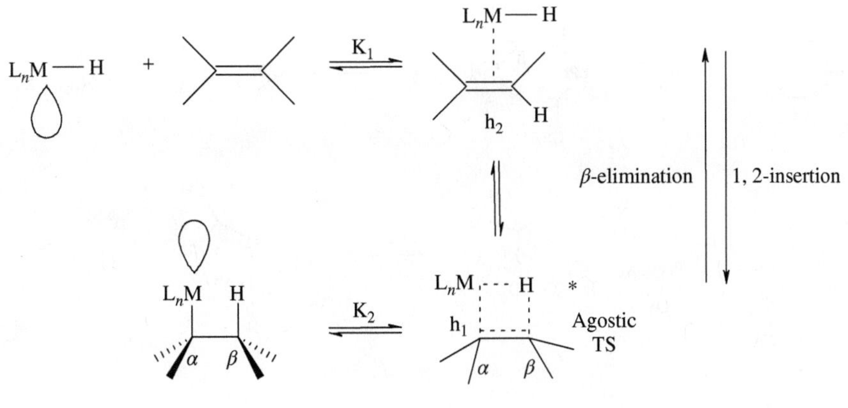

FIGURE 19.15
Proposed reaction mechanism for 1,2-insertion reactions (forward direction) and β-elimination reactions (reverse direction) through a four-centered agostic-type transition state.

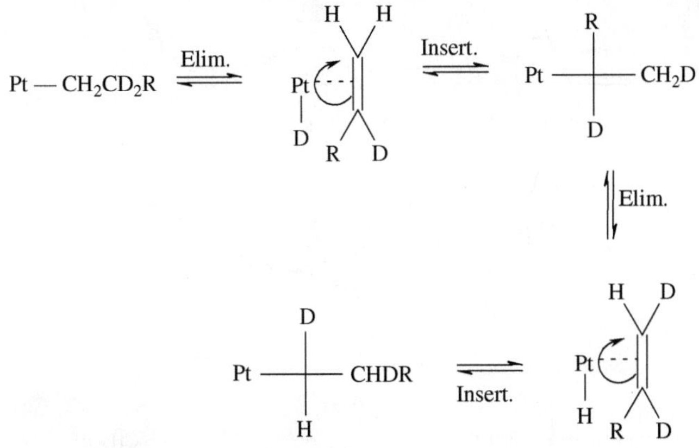

FIGURE 19.16
Process by which the H and D atoms in a metal–alkyl can scramble their positions through a series of sequential β-elimination and 1,2-insertion reactions.

in the reverse direction (β-elimination) because the increased π-backbonding to the metal–alkene intermediate weakens the C=C double bond. Because of the requirement that the four-centered agostic intermediate lie in the same plane, β-elimination reactions almost always occur with *syn* stereochemistry.

When the metal is bound to the η_2-alkene, there is free rotation around the metal–alkene bond as long as the substituent groups on the alkene are not too bulky. Thus, the dual processes of 1,2-insertion and β-elimination can be used to scramble the positions of the H atoms on metal alkyl groups, as shown in Figure 19.16.

Although much less common than α- or β-elimination reactions, elimination reactions involving the gamma, delta, or epsilon positions are also possible. Typically, these types of elimination reactions occur for the later transition metals that are coordinatively unsaturated. Two such examples are shown in Figure 19.17. As is evident from the two reactions in the figure, these types of more remote elimination reactions can be used in the synthesis of metallacycles and they are sometimes referred to as *cyclometalation reactions.*

19.3.2 Nucleophilic Attack on the Ligands

The coordination of a ligand to a metal can sometimes make the ligand itself more susceptible to nucleophilic attack. This is especially true when the metal has a high oxidation number and/or electron-withdrawing ligands. Some examples of nucleophilic attack at the ligand are shown in Figure 19.18.

When the ligand is attached to a metal that is electron withdrawing, ligands that are typically electron-rich (such as CO, alkenes, and arenes) now

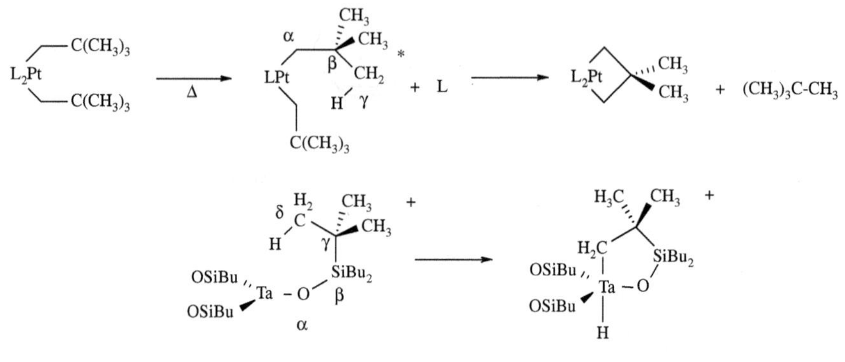

FIGURE 19.17
Cyclometalation reactions resulting from γ-elimination and δ-elimination reactions, respectively.

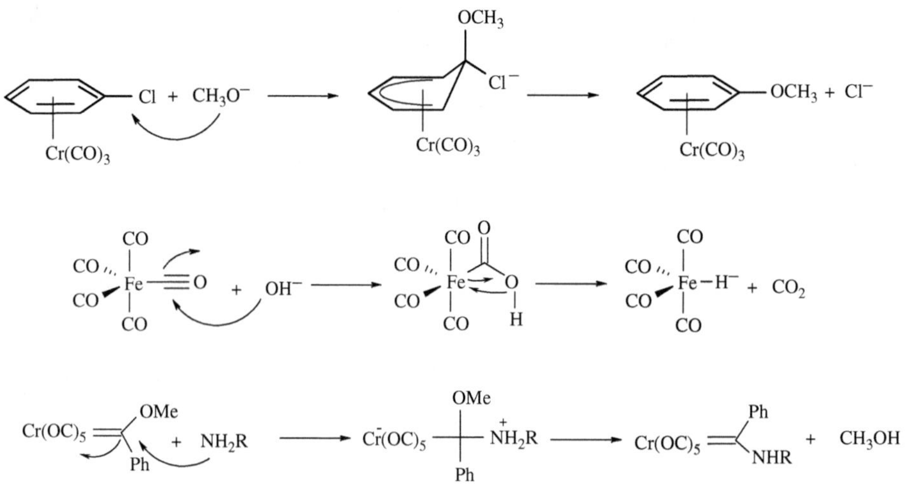

FIGURE 19.18
Examples of nucleophilic attack on the ligands.

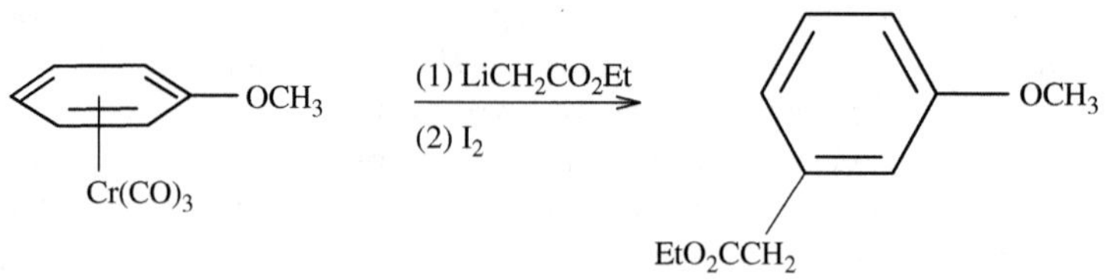

FIGURE 19.19
Attachment of the arene ring to the electron-withdrawing metal preferentially directs the nucleophilic attack on the ring to the *meta* position.

become electron-deficient. The German word for this is *umpolung*, which means "reverse polarity." Thus, for example, the methoxy group OCH_3 is usually an *ortho-/para*-directing group on arenes. However, the product distribution on the metalated arene shown in Figure 19.19 is only 4% *ortho* and 96% *meta*!

Davies, Green, and Mingos have devised a set of rules, known as the *DGM rules*, for coordinated π-systems, such as conjugated alkenes. In these π-ligands, nucleophilic attack occurs preferentially at ligands that have even-numbered hapticities and open π-systems, as shown by the sequence of ligands in Figure 19.20. Thus, for example, nucleophilic attack proceeds more rapidly at η_4-1,3-butadiene than at η_3-allyl or cyclobutadienyl ligands. Some examples are shown in Figure 19.21.

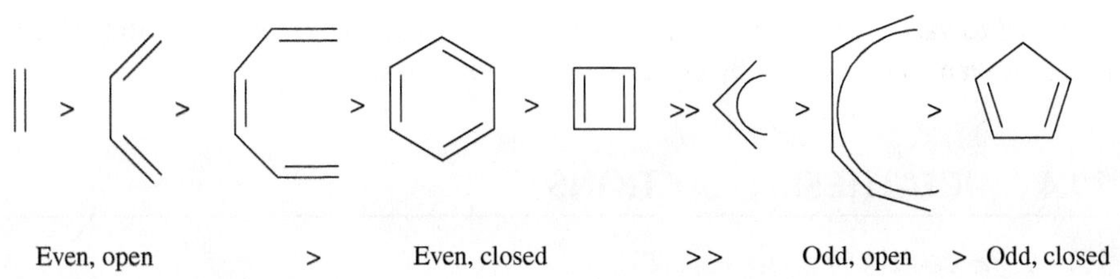

Even, open > Even, closed >> Odd, open > Odd, closed

FIGURE 19.20 DGM rules for the relative reactivity of π-ligands to nucleophilic attack.

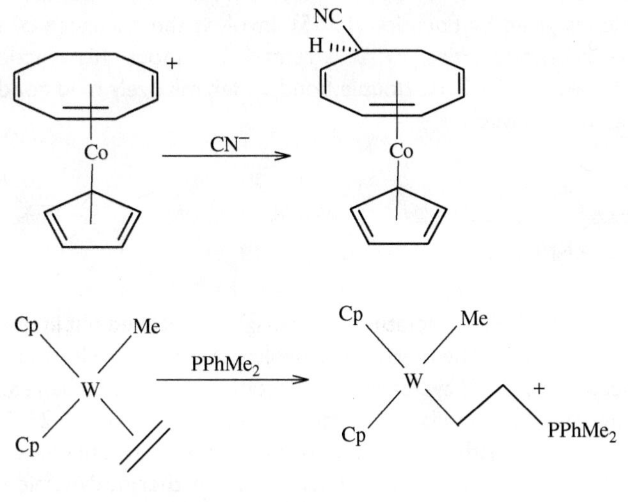

FIGURE 19.21 Examples of the DGM rules at work—nucleophilic attack occurs preferentially at even hapticity and open ligands.

19.3.3 Electrophilic Attack on the Ligands

At other times, the coordination of a ligand to a metal can sometimes make the ligand more susceptible to electrophilic attack. This is particularly true for metals having low oxidation numbers and high d^n counts. Some examples are illustrated in Figure 19.22.

The Friedel–Crafts acylation of the cyclopentadienyl rings on ferrocene occurs about a million times faster than the corresponding acylation of benzene. The $AlCl_3$ catalyst reacts with acetyl chloride to make $AlCl_4^-$ and the strong electrophile $CH_3C^+{=}O$, which then attacks the ring system. The π-MOs on the Cp^- ligand are

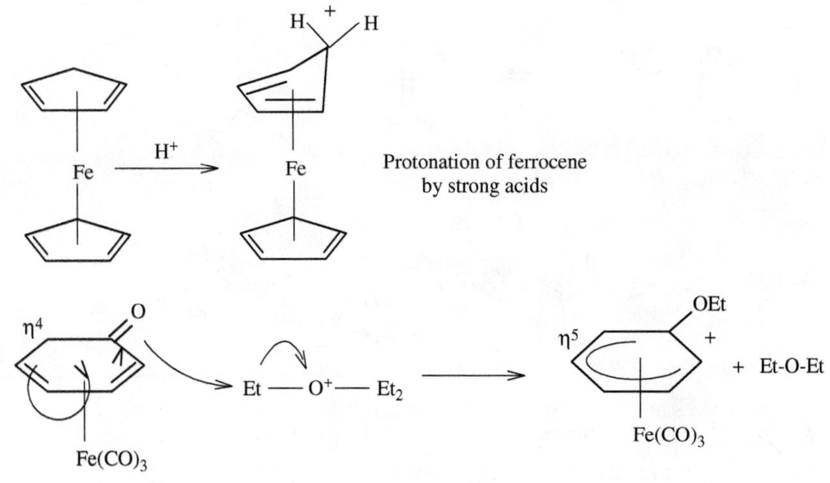

Protonation of ferrocene by strong acids

FIGURE 19.22 Examples of electrophilic attack on the ligand.

activated toward electrophilic attack by strong π-backbonding from the metal, which helps weaken the π-system by populating a π^* MO on the ring.

19.4 METATHESIS REACTIONS

19.4.1 π-Bond Metathesis

Olefin metathesis (or π-metathesis) reactions are one of the most unusual transformations in chemistry and also one of the most commercially important. The general reaction, which is given by Equation (19.15), involves the exchange of substituents between two different alkenes. What is remarkable about this reaction is that it involves the breakage of a C=C double bond under relatively mild conditions using an organometallic catalyst:

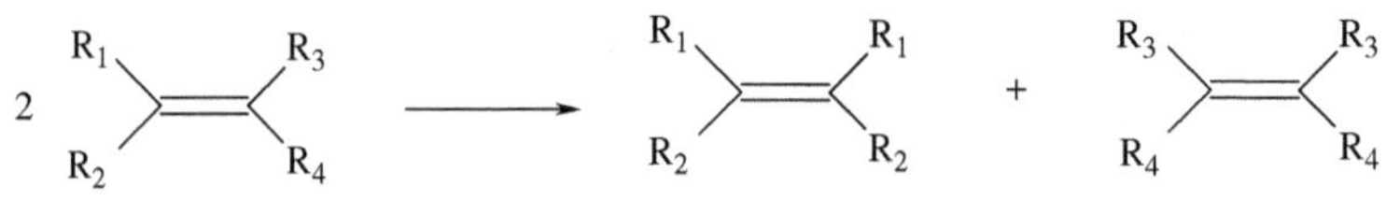

$$2\ R_1R_2C{=}CR_3R_4 \longrightarrow R_1R_2C{=}CR_1R_2 + R_3R_4C{=}CR_3R_4 \qquad (19.15)$$

The mechanism of olefin metathesis was originally worked out in the early 1970s by Hérisson and Chauvin. The mechanism involves the nonpairwise cleavage of C=C bonds that occurs in a [2+2] cycloaddition reaction between a carbene and an alkene to form an intermediate metallacyclobutane, as shown in Figure 19.23. The metallocyclobutane can open in either direction, such that an equilibrium mixture of alkenes results with the product distribution dictated by the thermodynamic stabilities of the different alkenes. Two of the more important organometallic catalysts for olefin metathesis are shown in Figure 19.24.

A variation of olefin metathesis known as *ring-opening metathesis polymerization* (ROMP) was developed by the Grubbs group in the 1990s. The reaction, which is shown in Figure 19.25, is used to open ring systems containing alkenes using an organometallic catalyst. The main difference between this mechanism and olefin metathesis is that the product remains attached to the catalyst as part of a growing

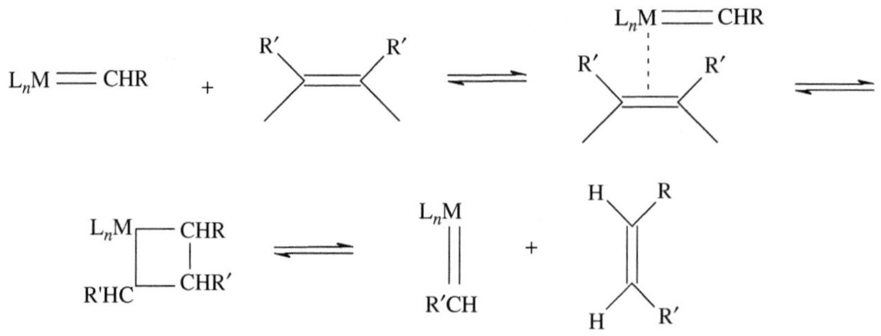

FIGURE 19.23
The Hérisson–Chauvin mechanism of olefin metathesis.

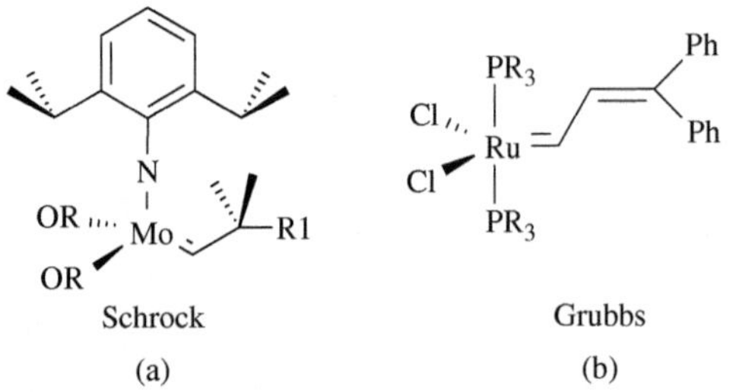

FIGURE 19.24
(a, b) Organometallic catalysts for olefin metathesis.

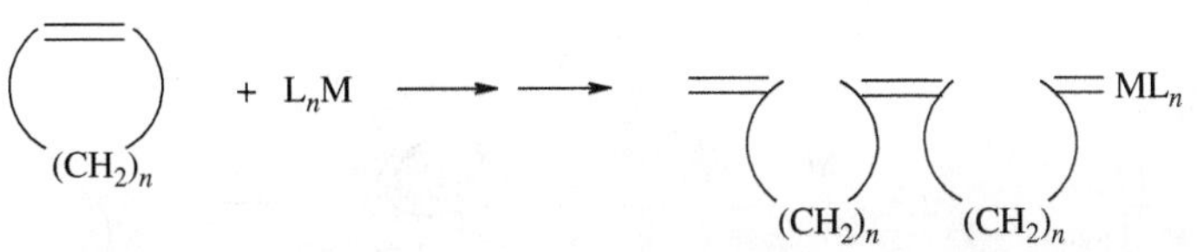

FIGURE 19.25
Ring-opening metathesis polymerization.

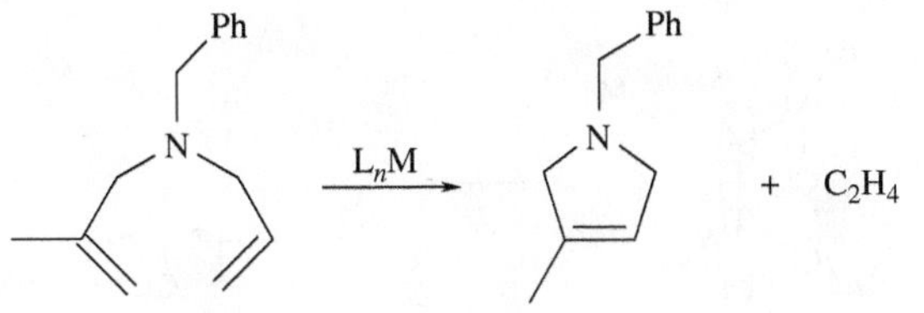

FIGURE 19.26
An example of a ring closing metathesis reaction.

polymeric chain. The reaction is largely driven by the relief of ring strain. Therefore, unstrained alkenes, such as benzene of cyclohexene, will not undergo polymerization.

Ring-closing metathesis reactions are also known, where essentially the reverse process of ROMP is used to close ring systems. However, the resulting rings cannot be appreciably strained in this case. The other product of ring-closing metathesis reactions is typically ethylene, whose very volatility helps drive the equilibrium to the right. One example of a ring-closing metathesis reaction is shown in Figure 19.26.

19.4.2 Ziegler–Natta Polymerization of Alkenes

In the 1967 cult film *The Graduate*, the following conversation ensues between the recent high school graduate Benjamin Braddock and Mr. McGuire who is about to offer him some career advice:

Mr. McGuire: "I want to say one word to you. Just one word."
Benjamin: "Yes, sir."
Mr. McGuire: "Are you listening?"
Benjamin: "Yes, I am."
Mr. McGuire: "Plastics."

Enormous quantities of the plastics polyethylene and polypropylene are manufactured by the polymerization of alkenes under ambient conditions of temperature and pressure using Ziegler–Natta catalysis. Karl Ziegler and Giulio Natta won the 1963 Nobel Prize in chemistry for their work. The catalyst is a mixture of $TiCl_3$ with $Al(C_2H_5)_2Cl$ or $TiCl_4$ with $Al(C_2H_5)_3$. According to the Cossee–Arlman mechanism, which is shown in Figure 19.27, an alkene binds at the defects (chloride vacancies) in the resulting crystalline structure, where it is converted into an alkyl group. The metal is left with a vacant coordination site, where it can bind another alkene to continue the polymerization. The polymerization essentially proceeds by a series of 1,2-insertion reactions.

19.4.3 σ-Bond Metathesis

The hallmark of any chemical transformation is the bond-breaking and bond-making that occur in reaction coordinate diagrams. One of the most important types of organometallic reactions involves C–H or C–C bond activation. We have already shown how C–H bonds can be activated through a combination of OA through a nonpolar agnostic intermediate, followed by RE. This process is reviewed in

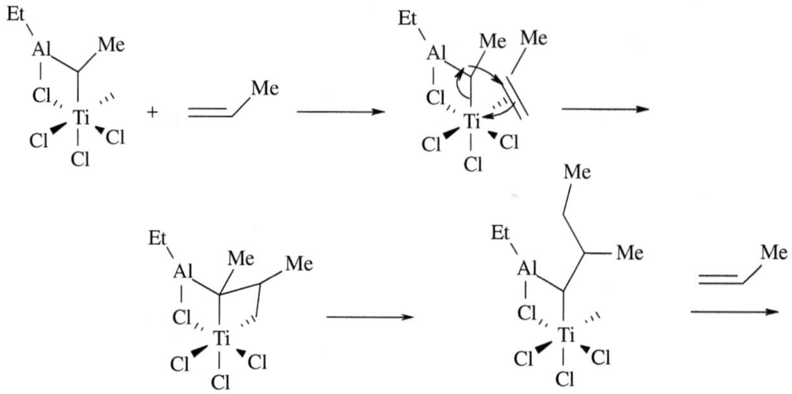

FIGURE 19.27
The Cossee–Arlman mechanism proposed for the Ziegler–Natta polymerization of alkenes.

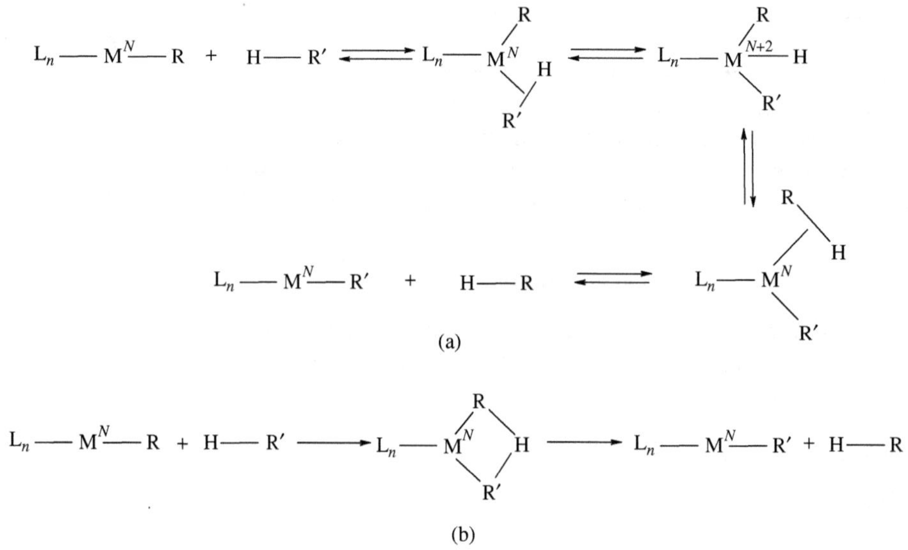

FIGURE 19.28
Two different methods of C–H bond activation: (a) oxidative addition followed by reductive elimination and (b) σ-bond metathesis.

Figure 19.28(a). An alternative mechanism of C–H bond activation occurs in σ-bond metathesis, which is shown in Figure 19.28(b)

Notice that unlike the first pathway, σ-bond metathesis does not involve a change in the oxidation number of the metal. In fact, this mechanism is actually favored by metals from the left-hand side of the periodic table having high oxidation numbers (typically Sc^{3+} or Ta^{5+}). While [2+2] cycloadditions, as the one shown in Figure 19.30(b), are disallowed in organic chemistry, they are possible in organometallic systems because of the involvement of the metal's *d*-orbitals in stabilizing the transition state. Most of the organometallics that undergo σ-bond metathesis reactions involve d^0 metals, whose empty *d*-orbitals are used to coordinate the alkanes in the cyclometalated transition state. Some representative σ-bond metathesis reactions are shown in Equations (19.16) and (19.17):

$$Cp*_2Sc\text{-}CH_3 + Ph\text{-}H \rightarrow Cp*_2Sc\text{-}Ph + CH_4 \quad (19.16)$$

$$Cp^*_2Sc\text{-}CH_3 + H_3C\text{-}C_6H_2(CH_3)_2\text{-}SiH_3 \longrightarrow H_3C\text{-}C_6H_2(CH_3)_2(Cp^*_2Sc\text{-}SiH_2) + CH_4 \quad (19.17)$$

A summary of the most common types of organometallic reactions and the conditions that favor each mechanism is listed in Table 19.2.

TABLE 19.2 Summary of the different types of organometallic reactions covered in this textbook.

Generic Reaction	Comments	Favored by
Ligand substitution		
Associative $L_nMX + Y \rightarrow [L_nMXY]^* \rightarrow L_nMY + X$	S_N2-like 2nd order follow HSAB	Good Nu^- small, unsat'd M *trans* effect
Dissociative $L_nMX \rightarrow [L_nM]^* + X(+Y) \rightarrow L_nMY$	S_N1-like 1st order if $[Y] \gg [X]$	Sat'd M ($18e^-$, $19e^-$) bulky ligands
Oxidative addition		
$L_nM^N + X\text{-}Y \Longleftrightarrow L_nM^{N+2}(X)(Y)$	Increase in M's OS and CN by a factor of 2	Electropositive M unsat'd M with low OS donor ligands vacant site
Reductive elimination		
$L_nM^{N+2}(X)(Y) \Longleftrightarrow L_nM^N + X\text{-}Y$	Decrease in M's OS and CN by a factor of 2	Electropositive M sat'd M with high OS small M bulky ligands
1,1-Insertion (alkyl migration) $L_nM(Y)(\eta^1\ X{=}Z) \Longleftrightarrow L_nM(X{=}Z)\text{-}Y$	Reverse: deinsertion (elimination) Reduction in M's CN by 1, net $2e^-$ transfer	Sat'd M with large CN unsat'd X=Z Me > Ph > Bz
1,2-Insertion $L_nM(Y)(\eta^2\ X{=}Z) \Longleftrightarrow L_nM\text{-}X\text{-}Y\text{-}Z$	Reverse: β-elimination Reduction in M's CN by 1, net $2e^-$ transfer	Sat'd, electrophilic M acceptor ligands bulky ligands
Nucleophilic addition		
$L_nM(Y)\ (\eta^2\ X{=}Z) + Nu^- \rightarrow L_nM\text{-}X\text{-}Z\text{-}Nu^-$	DGM rules (even, open)	M in high OS acceptor ligands
Electrophilic addition		
$L_nM\text{-}X{=}Z + E^+ \rightarrow L_nM{=}X\text{-}Z\text{-}E^+$		Strong π_{bb} M electropositive M low OS donor ligands
π-Bond metathesis		
$2\,R\text{-}C{=}C\text{-}R' \rightarrow R\text{-}C{=}C\text{-}R + R'\text{-}C{=}C\text{-}R'$	Chauvin mechanism (carbene [2 + 2] cycloaddition)	
σ-Bond metathesis		
$L_nMR + H\text{-}R' \rightarrow L_nMR' + H\text{-}R$	Occurs by [2 + 2] cycloaddition	Electropositive M high OS, low d^n

19.5 COMMERCIAL CATALYTIC PROCESSES

One of the main reasons why organometallic chemistry has become a major subdiscipline of inorganic chemistry is the fact that a very large number of commercially important catalytic processes involve organometallic reagents as the catalyst. A *catalyst* is a substance that is present in equal amounts before and after the reaction, but which speeds up the rate of a chemical reaction by lowering the overall activation barrier. It does so by providing an alternative, lower-energy pathway for the reaction to occur. We have already shown in the previous discussion how organometallic catalysts can "activate" molecules that otherwise might be unreactive by weakening their chemical bonding upon coordination to the metal. The C–H bond activation using OA or σ-bond metathesis reactions are but two examples. At the same time, an organometallic catalyst can sometimes be used to guide the stereochemistry of the products or to provide the correct orientation for the desired reaction to occur. The vast majority of commercially important chemical syntheses today require the use of a catalyst. Catalysts can be either *homogeneous* if they are in the same phase as the reactants (homogeneous catalysts are usually in solution) or *heterogeneous* if they are in a different phase from the reactants (heterogeneous catalysts are usually solids).

19.5.1 Catalytic Hydrogenation

One of the more important catalytic processes involves the addition of H_2 across a double bond, such as the reduction of C=C or C=O containing species. These types of reactions can occur with either a monohydride or a dihydride intermediate. We have already mentioned Wilkinson's catalyst before in the context of OA reactions. In 1965, Geoffrey Wilkinson discovered that $RhCl(PPh_3)_3$ could catalyze the hydrogenation of alkenes under ambient conditions of atmospheric pressure and 298 K. The catalyst can be used with a variety of different alkenes, although the rate constant for hydrogenation drops off with steric congestion around the C=C double bond. The proposed mechanism (determined by Halpern and Tolman) is shown in Figure 19.29. The first step involves ligand substitution of PPh_3 with a solvent ligand. In the second step, OA of H_2 occurs to form the octahedral *cis*-dihydride intermediate. Coordination of the η^2-alkene occurs with displacement of the solvent. The rate-determining step involves a combination of a 1,2-insertion reaction and *trans* → *cis* isomerization. Finally, the product is released by RE and the solvated form of the catalyst is free to begin another cycle. In the years since Wilkinson's initial discovery, other hydrogenation catalysts have been developed, including Schrock's $[Rh(dppe)S_2]^+$ (S = solvent) and Crabtree's $[Ir(COD)(py)(PCy_3)]^+$ catalysts.

A ruthenium catalyst has also been used in hydrogenation reactions that involve only a monohydride intermediate. The 16-electron catalyst $[RuCl_2(PPh_3)_3]$ adds dihydrogen as a ligand in its vacant coordination site. The presence of triethylamine in solution then abstracts HCl to form the active monohydride $[HRuCl(PPh_3)_3]$. The monohydride coordinates the alkene, which then undergoes 1,2-insertion to form the alkylruthenium compound having a vacant coordination site. Addition of dihydrogen is followed by dissociation of the alkane to regenerate the monohydride.

19.5.2 Hydroformylation

The cobalt-catalyzed hydroformylation reaction was originally discovered in 1938 by Otto Roelen and was used by the Germans in WWII to convert alkenes, H_2, and CO into aldehydes, which could then be used to make alcohols. The process is still used today, most commonly to convert propylene into butyraldehyde. Using $HCo(CO)_4$ as the catalyst, the mechanism of hydroformylation, which is shown in Figure 19.30,

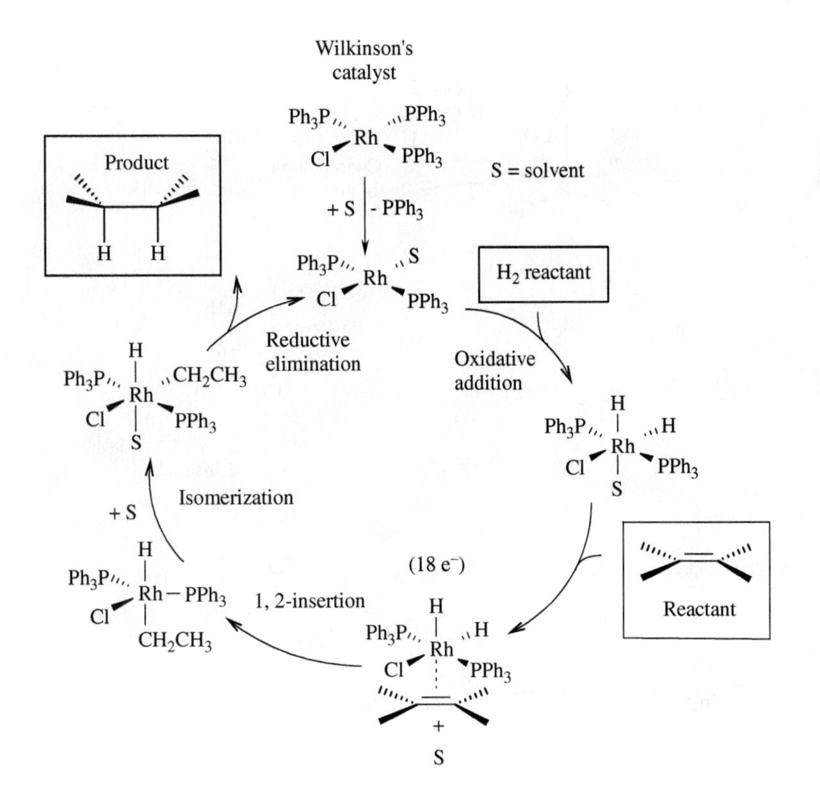

FIGURE 19.29
Proposed mechanism for the catalytic hydrogenation of alkenes using Wilkinson's catalyst.

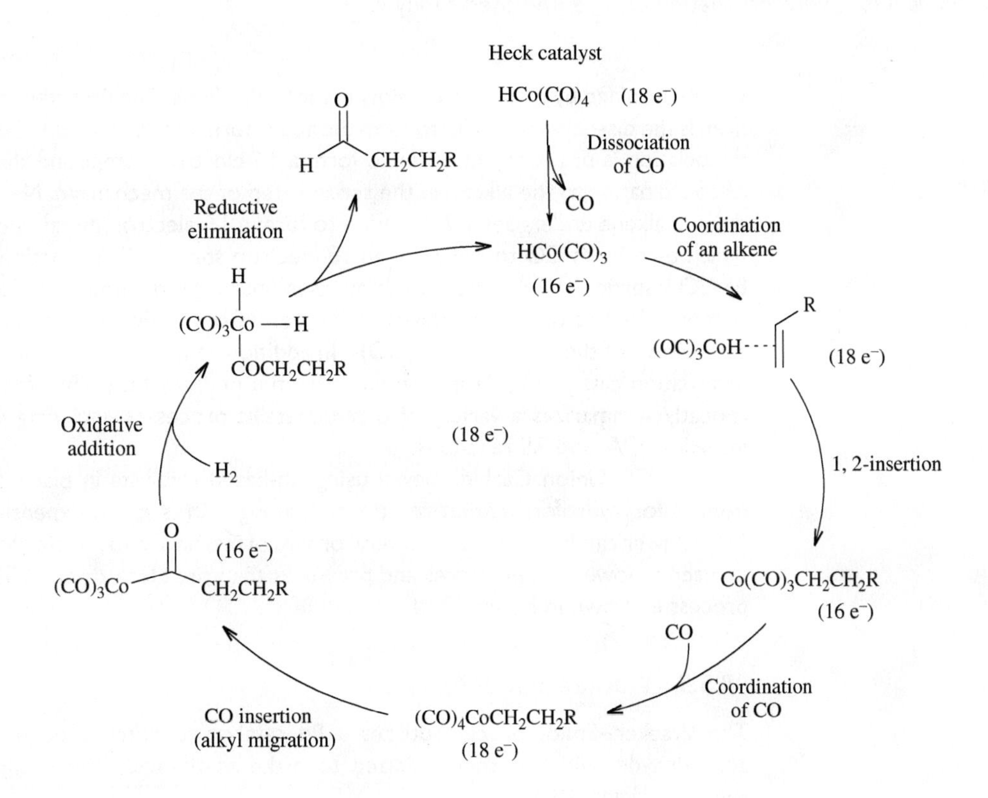

FIGURE 19.30
The Heck–Breslow catalytic cycle for hydroformylation.

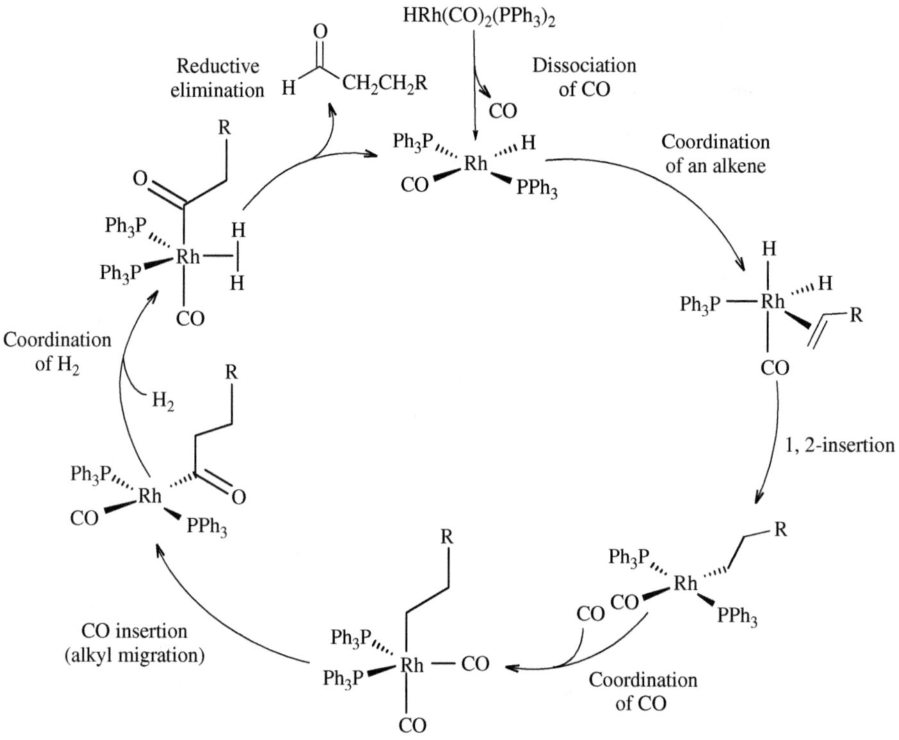

FIGURE 19.31
Proposed mechanism for hydroformylation using a Rh-based catalyst.

was determined by Heck and Breslow in the early 1960s. The first step in the mechanism is the dissociation of CO to form the active form of the catalyst $HCo(CO)_3$. The dissociation is necessary in order to form a 16-electron compound that is capable of coordinating to the alkene in the second step of the mechanism. Next, the coordinated alkene undergoes 1,2-insertion to form a 16-electron metal–alkyl complex, which then binds CO to revert to an 18-electron species. This reaction is followed by CO insertion (alkyl migration) to make an unsaturated compound. Following OA of molecular hydrogen, the catalyst undergoes RE of the aldehyde to regenerate the active form of the catalyst, $HCo(CO)_3$. In addition to its synthetic utility, the hydroformylation catalytic cycle is interesting from a pedagogical point of view because it neatly summarizes a variety of organometallic processes, including dissociation, insertion, OA, and RE reactions.

In 1976, Union Carbide began using Rh-based catalysts in place of the Heck catalyst for hydroformylation reactions. Although Rh is more expensive than Co, these newer catalysts were also several orders of magnitude more efficient and could be used at lower temperatures and pressures than the Heck catalyst. This catalytic process is shown in Figure 19.31.

19.5.3 Wacker–Smidt Process

The Wacker–Smidt process utilizes a Pd catalyst to convert ethylene gas into acetaldehyde, which is then oxidized to make acetic acid. The catalytic cycle is shown in Figure 19.32.

In the first step of the mechanism, the tetrachloropalladate ion dissociates chloride and coordinates to the alkene. In the second step, water replaces chloride. In the third step, water acts as a nucleophile and attacks the coordinated alkene (recall

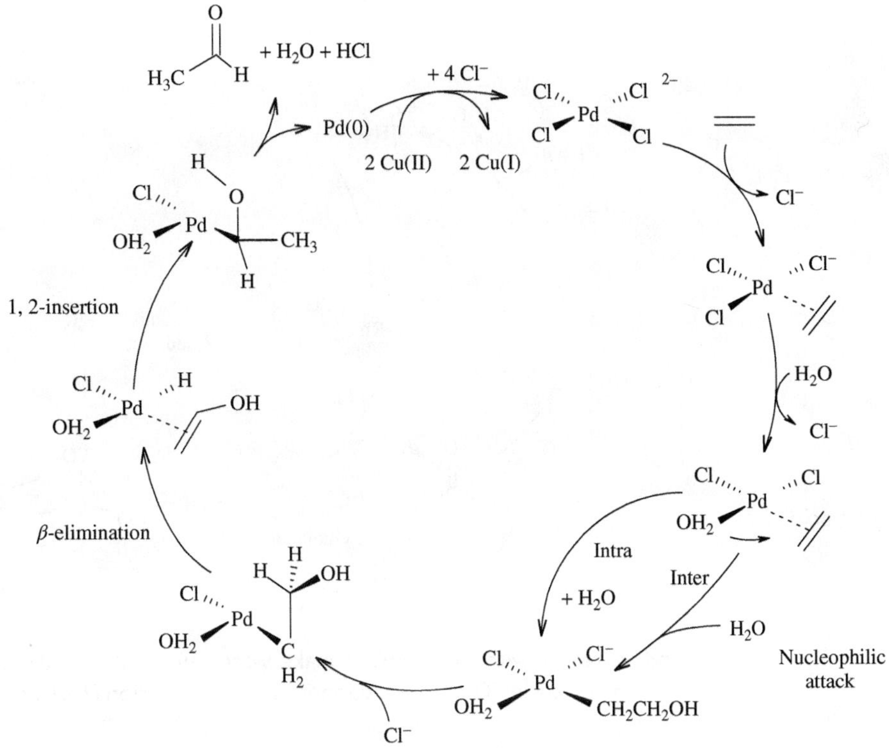

FIGURE 19.32
The Wacker–Smidt process.

that coordinated alkenes are more susceptible to nucleophilic attack than are the free alkenes). This step can occur by an intramolecular or intermolecular mechanism, as shown in the figure. The dissociation of a third equivalent of Cl^- is followed by a β-hydride elimination reaction. Following 1,2-insertion, the Pd dissociates acetaldehyde and undergoes RE of HCl to make Pd(0). The original catalyst is regenerated by oxidation with Cu(II) in the presence of chloride ion. In reality, $CuCl_2$ acts as a reprocessing catalyst. The Cu(I) by-product is reoxidized in aqueous solution by molecular oxygen. The acetaldehyde product can easily be oxidized to acetic acid. Wacker–Smidt process was used to make acetic acid for many years until it was replaced by the Monsanto acetic acid process, which uses the direct insertion of CO gas into methanol generated from synthesis gas. Synthesis gas is a mixture of CO and H_2 that can be produced from coal.

19.5.4 Monsanto Acetic Acid Process

The Monsanto acetic acid process produces acetic acid from methanol and CO gas under fairly mild conditions (180 °C, 30–40 atm). The process utilizes a square planar Rh(I) catalyst. As shown in Figure 19.33, the first step in the catalytic mechanism is the OA of methyl iodide to form an 18-electron compound. In the second step, CO insertion (alkyl migration) occurs, resulting in a 16-electron species. Carbon monoxide adds to the vacant coordination site to regenerate a saturated compound, which then undergoes RE of CH_3COI to regenerate the catalyst. The CH_3COI product is further processed by reaction with water to make acetic acid and HI. The latter

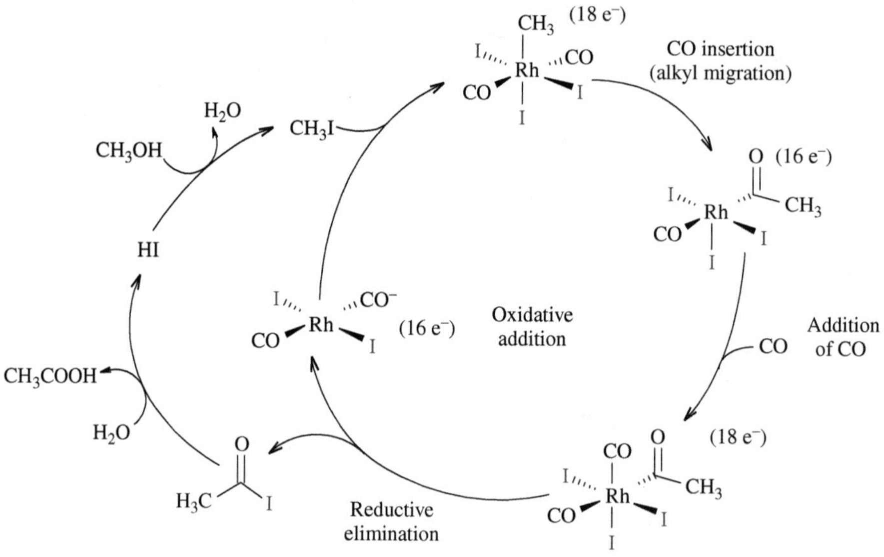

FIGURE 19.33
The Monsanto acetic acid process.

product reacts with methanol to make water and methyl iodide, which continues the catalytic cycle. The net reaction is given by Equation (19.18):

$$CH_3OH + CO \rightarrow CH_3COOH \quad (19.18)$$

19.6 ORGANOMETALLIC PHOTOCHEMISTRY

As was also the case for coordination compounds, the excited states of organometallic molecules have distinct physical and chemical properties that differ from those of the ground state. The photochemistry of organometallic compounds can often open reaction mechanisms that are thermally inaccessible.

19.6.1 Photosubstitution of CO

Because of π-backbonding, metal carbonyls are usually quite stable in their ground states. Irradiation of metal carbonyls, on the other hand, often leads to photosubstitution of CO. Several examples are shown in Figure 19.34. This process can be quite useful from a synthetic point of view when replacement of a single CO ligand is desired.

Let us consider $[W(CO)_6]$ as the prototypical example. The rate constant for CO dissociation in this compound is $10^{-6}\ s^{-1}$ at STP. Tungsten(0) is LS d^6 and has the electron configuration $t_{2g}{}^6e_g{}^0$, which makes it thermally inert. Irradiation of the spin-allowed LF transition ${}^1T_{1g} \leftarrow {}^1A_{1g}$ is followed by intersystem crossing to the lowest-energy ${}^3T_{1g}$ excited state. As shown in Figure 19.35, irradiation of the LF absorption band corresponds with the transition of an electron from a bonding t_{2g} MO to an antibonding e_g* MO, which weakens the W−C bond and labilizes CO. The excited state lifetime of the ${}^3T_{1g}$ excited state is $<10^{-10}\ s^{-1}$ and it undergoes photosubstitution with a quantum efficiency approaching unity. Thus, photosubstitution in this molecule is approximately 10^{16} times faster than the corresponding thermal reaction!

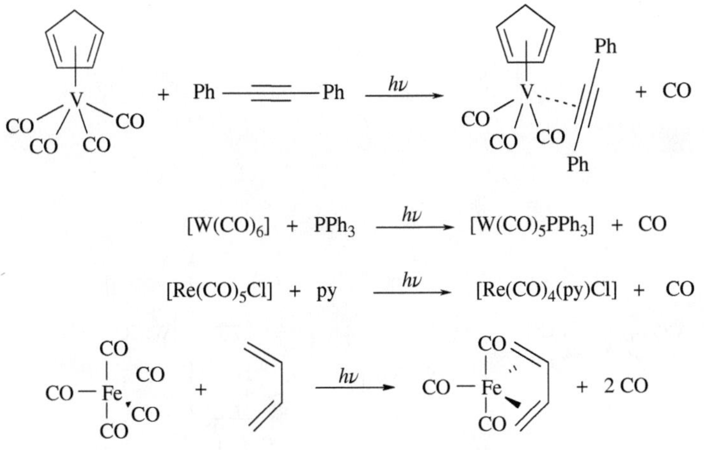

$$[W(CO)_6] + PPh_3 \xrightarrow{h\nu} [W(CO)_5PPh_3] + CO$$

$$[Re(CO)_5Cl] + py \xrightarrow{h\nu} [Re(CO)_4(py)Cl] + CO$$

FIGURE 19.34
Some examples of photosubstitution reactions of metal carbonyls.

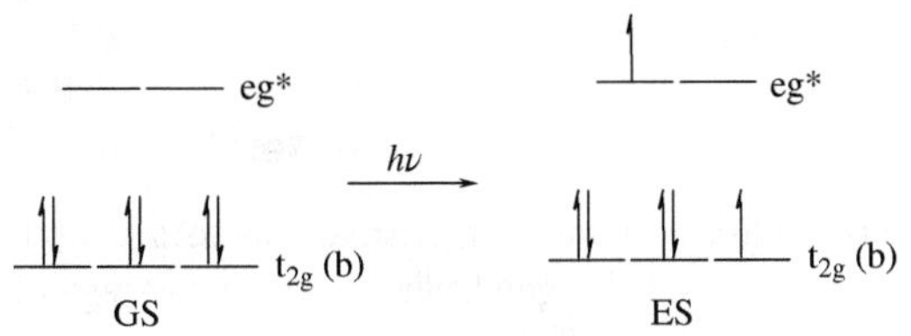

FIGURE 19.35
Photoexcitation of the ${}^1T_{1g} \leftarrow {}^1A_{1g}$ transition in $[W(CO)_6]$ is followed by intersystem crossing to the reactive ${}^3T_{1g}$ ES, which dissociates a CO ligand. The W–C bond is weakened in two ways: by removal of an electron from the π_b t_{2g} level and by population of the antibonding e_g* MO.

In order to determine whether the weakening of the M–C π-backbonding or the population of the σ-antibonding e_g* MO had the greater effect, the photochemistry of $[V(CO)_6]$ was examined. This species has a $t_{2g}{}^5e_g{}^0$ ground state, which is analogous to the removal of an electron from the π-bonding t_{2g} MO of $[W(CO)_6]$. $[V(CO)_6]$ is relatively photochemically inert. $[V(CO)_6]^-$, on the other hand, is photochemically labile. These results seem to imply that it is the population of the e_g* MO that causes the photochemical labilization of a CO ligand. The substitution of CO ligands on metal carbonyls decreases the quantum yield of CO loss. Thus, for instance, photosubstitution occurs much less efficiently in $[Fe(CO)_3(PPh_3)_2]$ than it does in $[Fe(CO)_5]$. The obvious reason for the decrease in quantum efficiency is that the PPh_3 ligand is a weaker π-backbonder than is CO. Thus, all of the Fe–C bonds are stronger in $[Fe(CO)_3(PPh_3)_2]$ than they are in $[Fe(CO)_5]$.

The photochemistry of the mono-substituted $[M(CO)_5L]$ class of organometallics is particularly interesting from a synthetic point of view. These compounds can dissociate either CO or L depending on the wavelength of irradiation. Higher energy light favors loss of CO, whereas longer wavelengths of irradiation favor the loss of L. A reasonable explanation for this behavior can be rationalized on the basis of the one-electron MO diagram for the *d*-orbitals in C_{4v}, which is shown in Figure 19.36. The lowest-energy LF transition occurs by transferring an electron from the b_2 (d_{xy}) MO to the a_1 (d_{z^2}) MO. This weakens the bonding along the z-axis of the molecule, so that either CO or L is labilized. Irradiation at higher energies transfers an electron from the b_2 (d_{xy}) MO to the b_1 ($d_{x^2-y^2}$) MO, which weakens the M–C bonds in the *xy*-plane and preferentially labilizes only the CO ligands.

If L is a strong electron acceptor, the quantum efficiency for CO loss decreases dramatically, as shown by the data in Table 19.3. The reason for this decrease in the quantum yield is explained in Figure 19.37. The acceptor ligand lowers the energy of the MLCT excited state to the point that is it now lower than the photo-active LF excited state. Thus, the quantum efficiency of CO loss decreases because there is less population of the LF excited state.

FIGURE 19.36
The wavelength dependence of photosubstitution for $[W(CO)_5L]$ is a result of the population of two different MOs. The lowest-energy electronic transition populates the a_1 MO, which preferentially labilizes the ligands along the *z*-axis, resulting primarily in dissociation of L. On the other hand, irradiation at higher energies selectively populates the b_1 Mo, which labilizes the ligands in the *xy*-plane, leading to the labilization of a CO ligand.

19.6.2 Photoinduced Cleavage of Metal–Metal Bonds

When metal carbonyls exist as dimers or in small clusters containing metal–metal bonding, irradiation of the molecules generally results in cleavage of the metal–metal bond and the formation of radicals. If the photolysis reaction is run in a halogenated solvent, the radicals can extract halogen from the solvent to form a metal–halide bond. Several examples of photoinduced cleavage are shown in Figure 19.38.

Metal–metal exchange reactions can also occur photochemically, as shown in Equation (19.19):

$$[Re_2(CO)_{10}] + [Mn_2(CO)_{10}] + h\nu \rightarrow 2\,[MnRe(CO)_5] \quad (19.19)$$

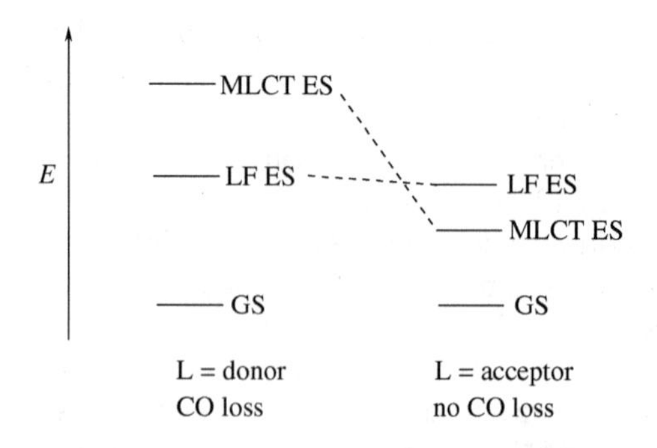

FIGURE 19.37
The quantum efficiency of photosubstitution of CO in $[M(CO)_5L]$ compounds decreases when L is a good acceptor ligand because the MLCT excited state is now longer in energy than the LF excited state.

TABLE 19.3 Quantum efficiencies for the loss of CO in a series of $[W(CO)_5L]$ compounds as a function of the electron-accepting properties of the ligand L.

L	$\phi_{CO\ loss}$
Pyridine	0.62
4-Benzoylpyridine	0.12
4-Formylpyridine	0.05

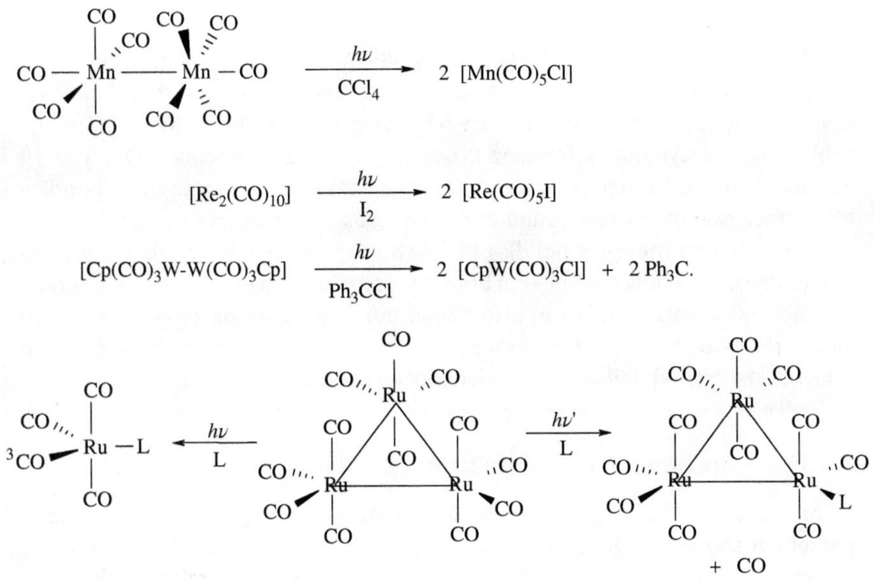

$$[(CO)_5Mn{-}Mn(CO)_5] \xrightarrow[CCl_4]{h\nu} 2\ [Mn(CO)_5Cl]$$

$$[Re_2(CO)_{10}] \xrightarrow[I_2]{h\nu} 2\ [Re(CO)_5I]$$

$$[Cp(CO)_3W{-}W(CO)_3Cp] \xrightarrow[Ph_3CCl]{h\nu} 2\ [CpW(CO)_3Cl] + 2\ Ph_3C.$$

$$[Ru(CO)_4({}^3CO)L] \xleftarrow[L]{h\nu} [Ru_3(CO)_{12}] \xrightarrow[L]{h\nu'} [Ru_3(CO)_{11}L] + CO$$

FIGURE 19.38 Photoinduced cleavage of metal–metal bonds.

Example 19-3. Sketch the one-electron MO diagram for the staggered form of $Mn_2(CO)_{10}$ pictured in Figure 19.40 using only the *d*-orbitals in the bonding. What is the lowest-energy symmetry-allowed electronic transition? Explain why irradiation at this wavelength (336 nm) leads to cleavage of the Mn–Mn bond instead of dissociation of a CO ligand.

Solution. The one-electron MO diagram of staggered $[Mn_2(CO)_{10}]$ is shown below by construction from two C_{4v} $[Mn(CO)_5]$ fragments.

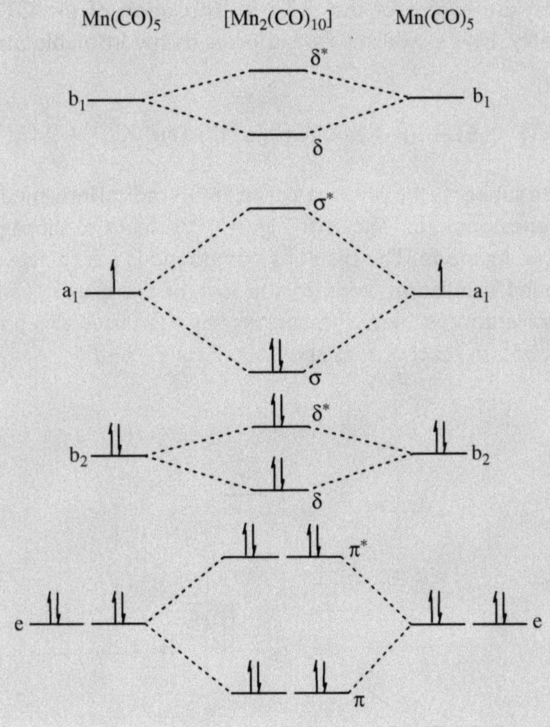

$Mn(CO)_5$ | $[Mn_2(CO)_{10}]$ | $Mn(CO)_5$

b_1 — δ^*, δ — b_1

a_1 — σ^*, σ — a_1

b_2 — δ^*, δ — b_2

e — π^*, π — e

Each Mn is d^7. The overlap of the d_{xz} and d_{yz} orbitals has π-symmetry and forms two π-bonding and two π-antibonding MOs. The d_{xy} and $d_{x^2-y^2}$ orbitals have δ-symmetry and will each form a δ-bonding and δ-antibonding MO. The d_{z^2} orbitals have σ-symmetry, forming σ-bonding and σ-antibonding MO's. The 14 valence electrons from the two Mn atoms fill the MO diagram up to the σ-bonding MO. Irradiation of the compound at 336 nm causes an electron to transfer from the σ-bonding to the σ-antibonding MO, reducing the Mn–Mn bond order from one to zero. Thus, it is the Mn–Mn bond that breaks instead of an Mn–CO bond. Photochemical loss of CO can also occur, but it does so via a radical reaction where the two $[Mn(CO)_5]$ radicals react with two equivalents of PPh_3 to make $[Mn_2(CO)_8(PPh_3)_2]$ and two equivalents of free CO.

19.6.3 Photochemistry of Metallocenes

The one-electron MO diagram for the *d*-orbitals in the staggered conformation of ferrocene is shown in Figure 19.39. The $d_{x^2-y^2}$ and d_{xy} orbitals are lowest in energy because they lie in the *xy*-plane where there are no direct interactions with the two Cp^- ligands. The d_{xz} and d_{yz} orbitals have the most direct interactions with the two ring systems and lie highest in energy. The d_{z^2} orbital falls somewhere in between these two degenerate sets of orbitals because the lobes on the z-axis point in the center of the Cp^- ring where there is not a lot of electron density and the donut part of the orbital lies in the *xy*-plane. The one-electron MO diagram predicts two LF bands in the UV/VIS: the $e_{1g} \leftarrow a_{1g}$ and $e_{1g} \leftarrow e_{2g}$ transitions, which occur at 22.7 and 30.8 kK, respectively.

Although the e_{1g} MO is antibonding, the compound is photostable in nonhalogenated solvents, such as ethyl alcohol. However, if CCl_4 is added to the solvent, a new band appears in the UV/VIS spectrum around 22.2 kK. This band has been assigned as a charge transfer to solvent (CCTS) band. The energy of the CCTS band depends on the nature of the solvent. Irradiation of the CTTS band in the presence of an alkyl halide oxidizes Fe and leads to the insoluble product shown in Equation (19.20):

$$2\,[Cp_2Fe^{II}] + h\nu + (R\text{-}X) \rightarrow [Cp_2Fe^{III}]^+,\ [Fe^{III}X_4]^- + 2\,Cp^- \quad (19.20)$$

ESR evidence supports the presence of an Fe(III) radical intermediate. Irradiation of the iron(II) arene compound shown in Figure 19.40 leads to slippage of the η_6 linkage to the η_4 bonding mode. The resulting compound is coordinatively unsaturated (16 e^-). Continued irradiation leads to the loss of the arene ligand and replacement by three solvent molecules. This species has been used as a photoinitiator for cationic polymerization reactions, as shown in Figure 19.40.

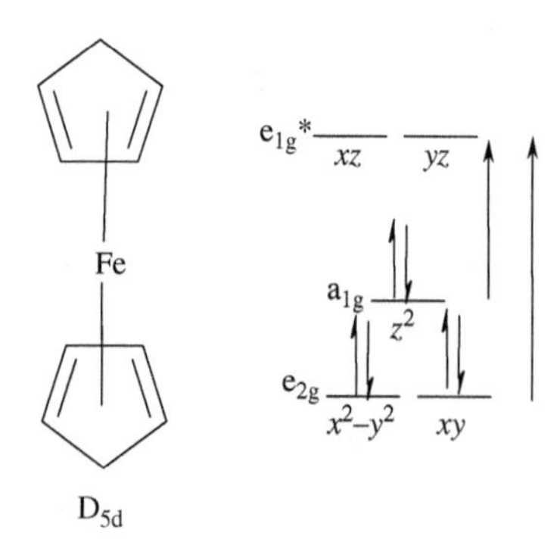

FIGURE 19.39
One-electron MO diagram for the *d*-orbitals in staggered ferrocene, showing the two possible LF electronic transitions.

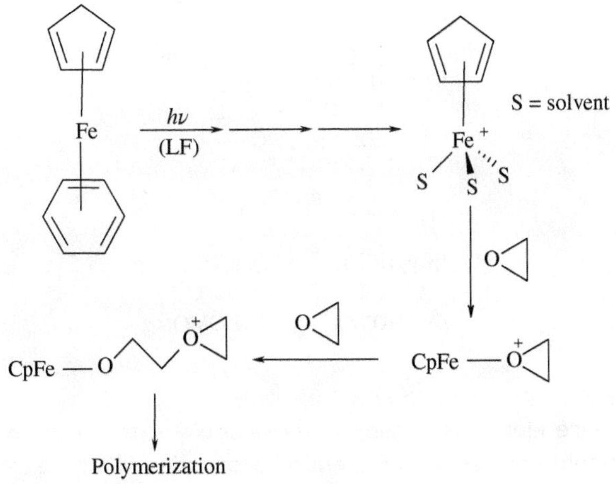

S = solvent

hv (LF)

CpFe — O

CpFe — O

Polymerization

FIGURE 19.40 Photoinduced cationic polymerization using $[CpFe(C_6H_6)]$ as the photoinitiator.

$$[Cp_2TiCl_2] + [Cp'_2TiCl_2] \xrightarrow{hv} 2\ [CpCp'TiCl_2], \text{ where } Cp' = C_5D_5^-$$

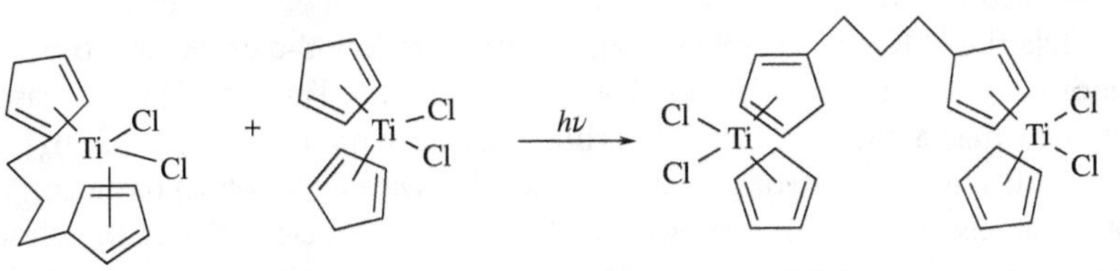

FIGURE 19.41 Photoinduced exchange reactions in bent titanocenes.

The photochemistry of the bent titanocenes $[Cp_2TiX_2]$ involves irradiation of the LMCT band instead of an LF absorption. There has been considerable debate over the years as to the nature of the HOMO in these compounds. Green classified the bent metallocenes into three categories: Class A compounds have a predominantly Cp^- HOMO, Class B have a mixture of Cp^- and X^-, and Class C compounds have primarily X^- as the HOMO. Irradiation of the LMCT band in $[Cp_2TiCl_2]$ leads to homolytic cleavage into $[CpTiCl_2]$ + Cp radical, which has been observed by radical trapping experiments. This result implies Class A behavior, as a X^- HOMO would be expected to generate the X· radical instead. Furthermore, the reactions shown in Figure 19.41 also suggest that $[Cp_2TiCl_2]$ belongs to Green's Class A categorization.

For $[Cp_2TiI_2]$, irradiation of the LMCT band leads to Ti–I bond cleavage, as shown in Equation (19.21), and is suggestive of a Class C compound. Because I^- is less electronegative than Cl^-, the lowest energy LMCT changes from Cp → Ti in $[Cp_2TiCl_2]$ to X → Ti in $[Cp_2TiI_2]$, as shown in Figure 19.42. ZINDO calculations on the corresponding Sc compounds $[Cp^*_2ScX]$ indicate that the HOMO has a mixture of Cp* and X^- character, belonging to Green's Class B category. For $[Cp^*_2ScCl]$, the HOMO is 65% Cp* and 15% Cl^-, whereas the HOMO in $[Cp^*_2ScI]$ is 15% Cp* and 78% I^-:

$$[Cp_2TiI_2] + hv \rightarrow \text{Ti-I cleavage} \qquad (19.21)$$

19.7 THE ISOLOBAL ANALOGY AND METAL–METAL BONDING IN ORGANOMETALLIC CLUSTERS

One of the underlying themes of this textbook has been the way that the electronic and molecular structure of a compound influences its chemical reactivity. Using a concept known as the *isolobal analogy*, Nobel laureate Raold Hoffmann has argued that if the symmetry, shape, and relative energy of the frontier molecular orbitals (FMOs) for two different chemical species are the same and they have the same

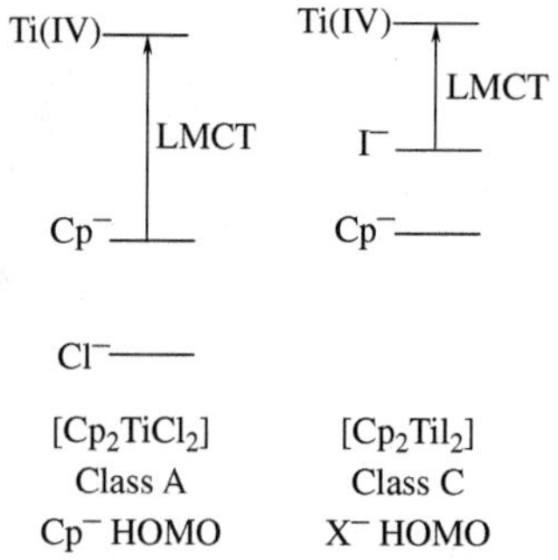

Ti(IV) — LMCT — Cp^-; Cl^- | Ti(IV) — LMCT — I^-; Cp^-

$[Cp_2TiCl_2]$ Class A Cp^- HOMO | $[Cp_2TiI_2]$ Class C X^- HOMO

FIGURE 19.42
Abbreviated MO diagram for $[Cp_2TiCl_2]$ and $[Cp_2TiI_2]$, showing how the relative energies of the Cp^- and X^- HOMOs in the two compounds lead to different photochemical reaction mechanisms.

number of valence electrons occupying those orbitals, they can be considered as isolobal and should possess similar chemical properties. We have already shown in this chapter that there are many similarities between organometallic chemistry and organic chemistry. The purpose of this section is to demonstrate how the isolobal concept can be used to draw parallels between organometallic and main group reaction chemistry as well.

While the FMOs of the main group elements follow the octet rule, those of the organometallics obey the effective atomic number rule. Thus, it might be reasonable to assume that a saturated metal carbonyl compound such as $[Cr(CO)_6]$, which contains 18 valence electrons, should be isolobal with CH_4, which has a completely filled octet. Extending this analogy, the 17-electron species $[Mn(CO)_5]$ should be isolobal with the 7-electron CH_3 radical. Indeed, the chemical reactivities of the two species are similar, as shown in Figure 19.43. Hoffmann uses a double-headed arrow

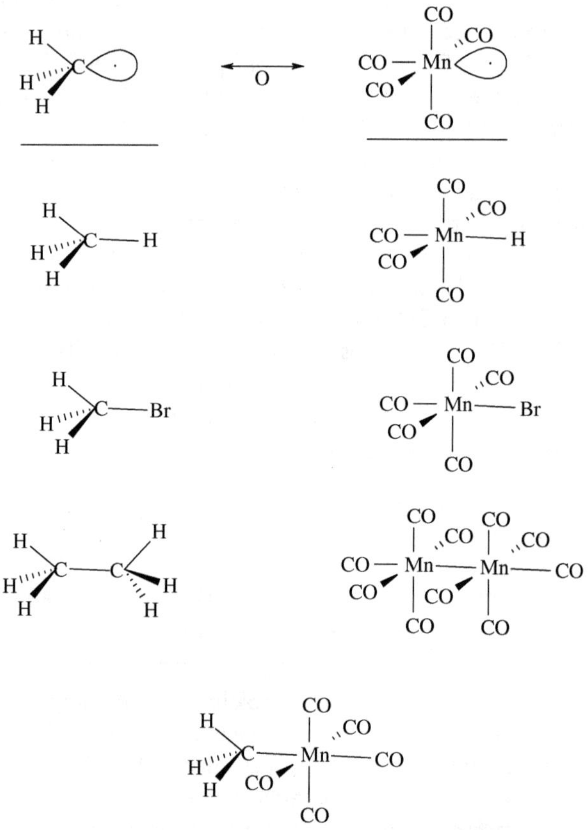

FIGURE 19.43
The isolobal fragments $CH_3\cdot$ and $[Mn(CO)_5]$ have similar reactivity.

TABLE 19.4 Examples of molecules which are isolobal.

Number of Missing Electrons	0	1	2	3
Neutral species	CH_4	CH_3	CH_2	CH
	$[Cr(CO)_6]$	$[Mn(CO)_5]$	$[Fe(CO)_4]$	$[Co(CO)_3]$
	$[Mn(CO)_6]^+$	$[Fe(CO)_5]^+$	$[Co(CO)_4]^+$	$[Ni(CO)_3]^+$
	$[Re(CO)_6]^+$	$[Cr(CO)_5]^-$	$[Mn(CO)_4]^-$	$[Fe(CO)_3]^{3-}$
		$[Os(CO)_5]^+$	$[Ir(CO)_4]^+$	$[Pt(CO)_3]^+$
	$[CpMn(CO)_3]$	$[CpFe(CO)_2]$	$[CpCo(CO)]$	$[CpNi]$
Anions	CH_3^-	CH_2^-	CH^-	
	$[Fe(CO)_5]$	$[Co(CO)_4]$	$[Ni(CO)_3]$	
Cations		CH_4^+	CH_3^+	CH_2^+
		$[V(CO)_6]$	$[Cr(CO)_5]$	$[Mn(CO)_4]$

with a loop as the symbol representing that two molecular species are isolobal, as shown in the figure.

Any organometallic compounds that are isoelectronic (and has similar FMO shapes and energies) with $[Mn(CO)_5]$ should also be isolobal; for instance, $[Fe(CO)_5]^+$. Similarly, the carbonyl ligands in $[Mn(CO)_5]$ can be replaced with any two-electron donor ligands, so that $[MnH_5]^-$, $[Mn(PR_3)_5]$, $[MnCl_5]^{5-}$, and $[Mn(NCR)_5]$ are all isolobal with $[Mn(CO)_5]$. Furthermore, as η_5-Cp^- ligands are 6-electron donors, they can replace three carbonyls, so that $[(\eta_5\text{-}Cp)Mn(CO)_2]$ and $[Mn(CO)_5]$ are also isolobal. A variety of metal carbonyl compounds and their main group analogs are listed in Table 19.4.

The isolobal analogy can apply to any molecular fragment where the FMOs have the same size, shape, energy, and occupancy. The species do not necessarily have to have the same number of electrons missing from their valence shells, as do all of the examples in Table 19.4. Thus, for example, $[Au(PPh_3)]$, which has only 13 valence electrons, and the 17-electron species $[Mn(CO)_5]$ are isolobal, as shown in Figure 19.44, because their FMOs have similar shapes and energies.

Example 19-4. For each of the organometallic compounds listed below, provide a main group fragment that is isolobal with the species: (a) $[Fe(CO)_2(PPh_3)]^-$, (b) $[Re(CO)_5]$, (c) $[(\eta_6\text{-}C_6H_6)Cr(CO)_2]$, and (d) $[(\eta_5\text{-}Cp)Fe(CO)]$.

Solution. (a) This species has 15 valence electrons, three short of the 18-electron rule, which makes it isolobal with the CH fragment. (b) This 17-electron fragment is missing a single electron and is therefore isolobal with CH_3. (c) There are 16 valence electrons in this organometallic compound, making it two short of a closed shell. The main group analog would be the CH_2 fragment. (d) This species has 16 valence electrons, making it isolobal with CH_2.

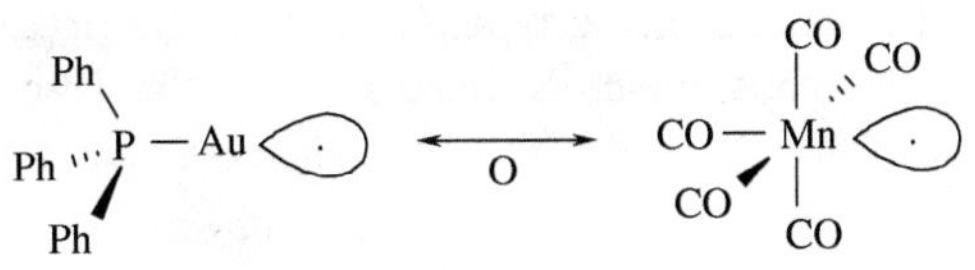

FIGURE 19.44
$[Au(PPh_3)]$ and $[Mn(CO)_5]$ are isolobal fragments.

Example 19-5. What neutral, homoleptic first-row transition metal carbonyl is isolobal with the Lewis acid BH_3?

Solution. Boron trihydride has six valence electrons, two short of a filled octet. Thus, we are looking for a 16-electron metal carbonyl. The only possibility is $[Cr(CO)_5]$.

Example 19-6. Show that the 14-electron, square planar complex $[PtCl_3]^-$ is isolobal with the 16-electron species $[Cr(CO)_5]$.

Solution. The shapes of the FMOs for both compounds are shown below. Both species have a two-electron vacancy in one of their hybrid orbitals. Because most square planar compounds are 16-electron species and most octahedral compounds obey the 18-electron rule, a square planar compound will be isolobal with an octahedral molecule that has two greater valence electrons.

Some interesting examples of compounds composed of isolobal fragments are shown in Figure 19.45.

The isolobal analogy can also be used to draw comparisons between the cluster chemistry of organometallic and the main group compounds. As we saw in Chapter 10, both boranes and carboranes form cages that are based on closed deltahedra having *n* vertices and all triangular faces. The boranes can be classified into different types, depending on the number of corners of the deltahedra that are occupied: *closo*, *nido*, *arachno*, and *hypho*. Using the isolobal analogy, it should be plausible for any organometallic fragment that is isolobal with BH can also substitute for BH to form a metallaborane cluster. The BH fragment, which has four valence electrons, is isolobal with the 14-electron species $[Fe(CO)_3]$ and $[(\eta_5\text{-Cp})Co]$. Thus, for example, $[Fe(CO)_3]$ can replace a BH group in the *nido* cluster B_5H_9 to make $[Fe(CO)_3B_4H_8]$, as shown in Figure 19.45. Similarly, anionic boranes and carboranes can also act as ligands toward transition metals. For instance, the carborane analog of ferrocene is shown in Figure 19.46.

Not surprisingly, transition metal carbonyl clusters, such as $[Ir_4(CO)_{12}]$, whose structure is shown in Figure 19.47, can also be categorized using a modification of Wade's rules, known as the *polyhedral skeletal electron pair approach*, as shown in Table 19.5. Because a transition metal atom has five additional orbitals (the *d*-orbitals) available for bonding than does a main group atom, a transition metal carbonyl cluster will contain 10 more electrons than the corresponding borane cluster. In this example, each Ir atom contributes nine valence electrons, whereas each CO group contributes two electrons each to the skeletal framework, for a total of 60 valence electrons. Using the formulas provided in Table 19.5, the $[Ir_4(CO)_{12}]$ cluster is expected to have the *nido* structure type. Given that the $[Ir(CO)_3]$ fragment has a total of 15 valence electrons, which is three short of the 18-electron rule, it will

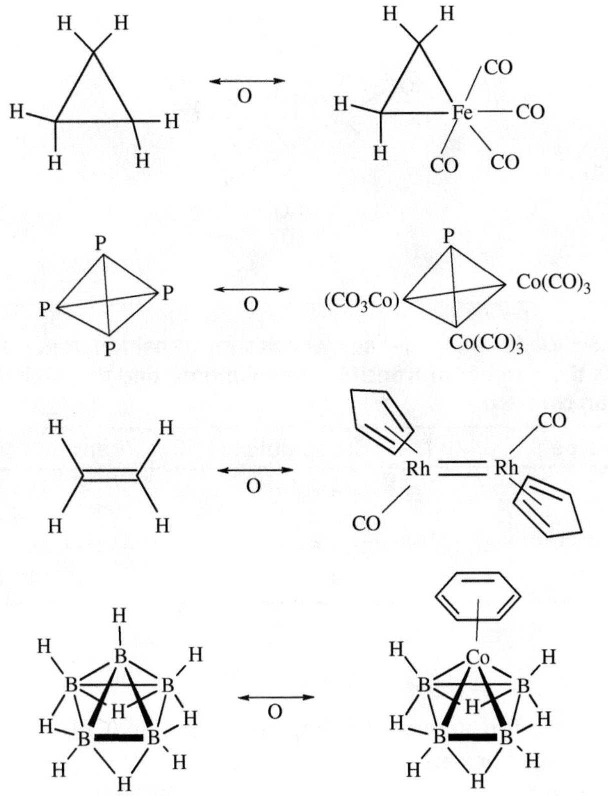

nido-B_5H_9

FIGURE 19.45
Representative main group compounds and their organometallic isolobal analogies.

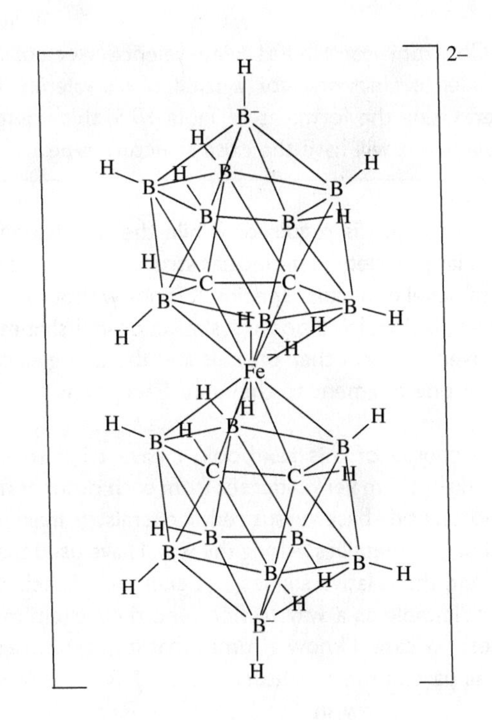

FIGURE 19.46
A carborane analog of ferrocene.

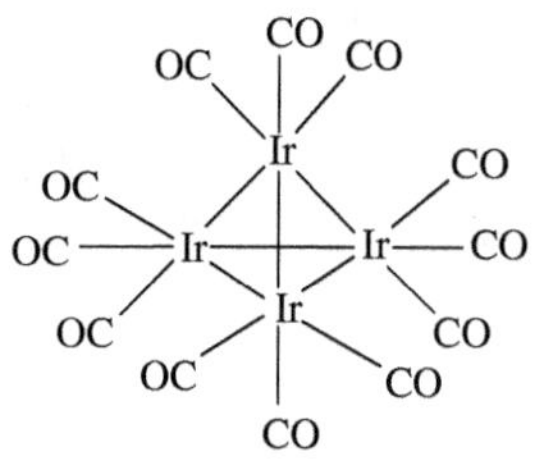

FIGURE 19.47
The structure of $[Ir_4(CO)_{12}]$.

TABLE 19.5 Wade–Mingos–Lauhrer rules for transition metal clusters, where *n* is the number of transition metal atoms, and its parallel with main group chemistry.

Structure Type	Main Group Rule	Transition Metal Rule
Closo	$4n+2$	$14n+2$
Nido	$4n+4$	$14n+4$
Arachno	$4n+6$	$14n+6$
Hypho	$4n+8$	$14n+8$

be isolobal with the five-electron BH_2 fragment. Thus, $[Ir_4(CO)_{12}]$ is isolobal with B_4H_8, which is a *nido* borane.

Example 19-7. Determine the structure type for the organometallic cluster compound $Co_6(CO)_{16}$.

Solution. Each Co atom contributes nine valence electrons and each CO contributes two valence electrons, for a total of 86 valence electrons in the $Co_6(CO)_{16}$ cluster. Using the formulas in Table 19.5, this compound will obey the $14n+2$ rule, so that it will have the *closo* structure type.

One final note of caution is required. While the isolobal analogy has proven useful for explaining a large variety of molecular structures and has even been used as a tool in the design of novel molecular species, it is not without its limitations. While two fragments that are isolobal will possess similar orbital shapes and symmetries, the extent of their overlap with other orbitals and the energies of their FMOs will differ somewhat from one fragment to the next. Exceptions to Wade's rules do in fact exist.

Throughout the course of this textbook, I have tried to draw connections between things that might seem very different from each other at first glance—ionic versus covalent bonding, acid–base versus redox chemistry, main group compounds and transition metal organometallics. Along the way, I have used the symmetry properties of molecules and the relative shapes and energies of their MOs as a unifying theme and the periodic table as a way to fine tune their chemical properties. And now at long last, I rest my case. I know at times that it might have seemed a difficult journey, but that is all part of the adventure. As the J. R .R. Tolkien character Bilbo Baggins once said to his nephew in *The Lord of the Rings*: "It is a dangerous business going out your door. You step onto the road, and if you don't keep your feet, there's no knowing where you might get swept off to." Who knows where the future of inorganic chemistry will lie? No doubt some of the principles introduced in this text may be obsolete before the book even hits the presses. Each new solution only raises a different question. It is now left for you students—the future generation of chemists, to decide where the road will lead. I wish you an enjoyable journey.

EXERCISES

19.1. Provide a rationale for how the 18-electron compound [CpRe(CO)(NO)Me] can undergo the following associative ligand substitution reaction:

$$[CpRe(CO)(NO)Me] + 2\,PMe_3 \rightarrow [CpRe(CO)(NO)(PMe_3)_2CH_3]$$

19.2. For the series of ligand substitution reactions given by the chemical equation below, values for the rate constant are given in the table. Does this reaction occur by an associative or a dissociative pathway? Explain your answer.

$$cis\text{-}[Mo(CO)_4(PR_3)_2] + CO \rightarrow cis\text{-}[Mo(CO)_5(PR_3)] + PR_3$$

Table showing the rates of ligand substitution at 70 °C in the reaction above as a function of PR_3.

PR_3	Cone Angle	k, s^{-1}
PMe_2Ph	122°	$<1.0 \times 10^{-6}$
$PMePh_2$	136°	1.33×10^{-5}
PPh_3	145°	3.16×10^{-3}
$PPh(cyclohexyl)_2$	162°	6.40×10^{-2}

19.3. Predict the product and the stereochemistry of the following reaction, which uses Wilkinson's catalyst:

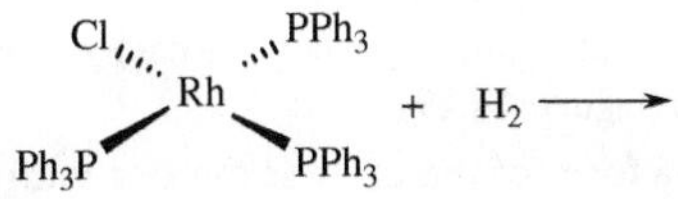

$$[RhCl(PPh_3)_3] + H_2 \longrightarrow$$

19.4. Do you think the rate of reaction in Exercise 19.3 would increase or decrease if the PPh_3 ligands on Wilkinson's catalyst were replaced by PF_3? Explain your answer.

19.5. Suggest a possible mechanism for the following reaction:

$$[(\eta^5\text{-}C_5Me_5)Ir(CO)_2] + C(CH_3)_4 \longrightarrow [(\eta^5\text{-}C_5Me_5)Ir(CO)(H)(CH_2C(CH_3)_3)]$$

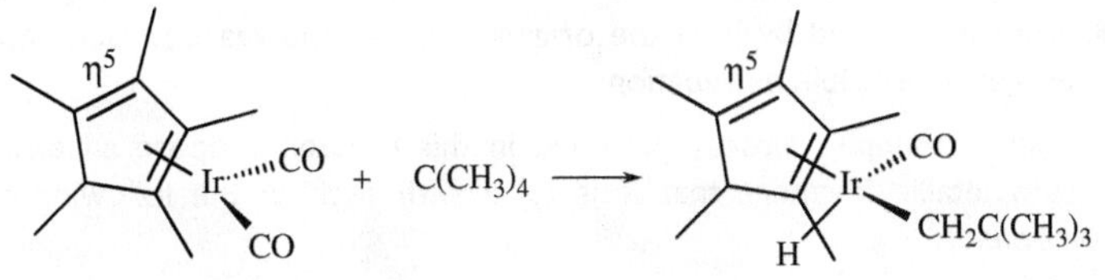

19.6. Predict the products of each of the following reductive elimination reactions:

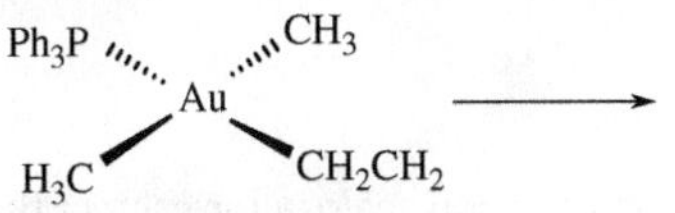

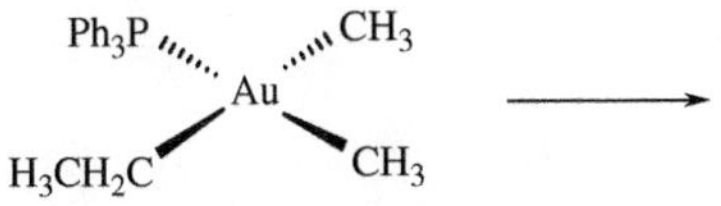

$$[Au(PPh_3)(CH_3)_2(CH_2CH_2)] \longrightarrow$$

$$[Au(PPh_3)(CH_3)_2(CH_2CH_3)] \longrightarrow$$

19.7. Predict the products of the following insertion reactions:

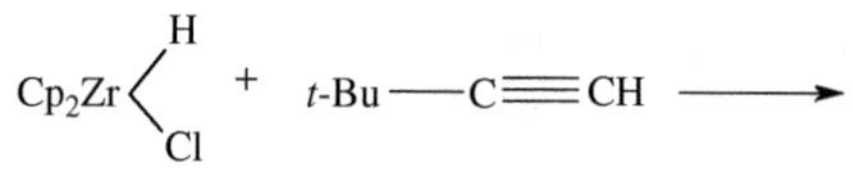

$Cp_2Zr(H)(Cl)$ + (hexene) ⟶

$Cp_2Zr(H)(Cl)$ + t-Bu—C≡CH ⟶

19.8. Would you expect the rate of elimination in the following reaction to increase or decrease if the H atoms on the α and β carbons are replaced with F atoms? Explain your answer.

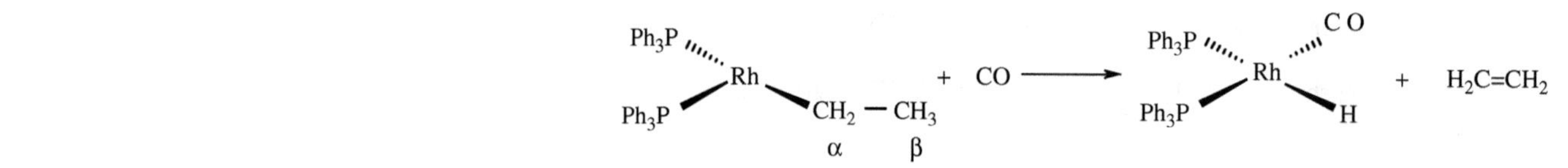

$(Ph_3P)_2Rh{-}CH_2(\alpha){-}CH_3(\beta)$ + CO ⟶ $(Ph_3P)_2Rh(CO)(H)$ + $H_2C{=}CH_2$

19.9. Predict the product of the following reaction:

(benzene)Mo(cycloheptatriene) complex (+) $\xrightarrow{MeS^-}$

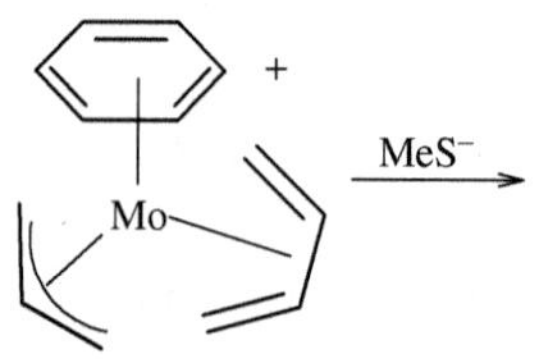

19.10. Answer the following questions regarding the catalytic hydrogenation of alkenes using Wilkinson's catalyst (Figure 19.29):

a. Why is the active form of the catalyst the one with a solvent molecule replacing a PPh_3 ligand?

b. Do you think the ligand substitution reaction occurs by an associative or dissociative mechanism? Explain.

c. Why must isomerization occur following the 1,2-insertion step before reductive elimination can occur?

d. If Rh is replaced by Ir in the original catalyst, the reaction does not proceed. Suggest a plausible explanation.

19.11. Excluding examples already provided in this chapter, propose an example of an organometallic fragment that is isolobal with each of the following main group compounds:

a. CH_3^-

b. CH_2^+

c. CH

d. CH_3^{2+}

e. CH_2^-

19.12. Provide an isolobal main group analogy for each of the following:

a. $Co(CO)_4$

b. $Ni(CO)_4$

c. $[Co(CO)_3Br]^-$

d. $Fe(CO)_3$

e. CpNi

19.13. Use the Wade–Mingos–Lauhrer rules to determine the probable structure type of each of the following organometallic cluster compounds:

a. $Co_4(CO)_{12}$
b. $Os_5(CO)_{16}$
c. $[Ni_5(CO)_{12}]^{2-}$
d. $[Rh_7(CO)_{16}]^{3-}$
e. $Rh_6(CO)_{16}$

BIBLIOGRAPHY

1. Atkins P, Overton T, Rourke J, Weller M, Armstrong F, Hagerman M. *Shriver& Atkins' Inorganic Chemistry*. 5th ed. New York: W. H. Freeman and Company; 2010.
2. Atwood JD. *Inorganic and Organometallic Reaction Mechanisms*. 2nd ed. New York: Wiley-VCH, Inc.; 1997.
3. Collman JP, Hegedus LS, Norton JR, Finke RG. *Principles and Applications of Organotransition Metal Chemistry*. Mill Valley, CA: University Science Books; 1987.
4. Crabtree RH. *The Organometallic Chemistry of the Transition Metals*. 3rd ed. John Wiley & Sons, Inc.: New York; 2001.
5. Douglas B, McDaniel D, Alexander J. *Concepts and Models of Inorganic Chemistry*. 3rd ed. New York: John Wiley & Sons, Inc.; 1994.
6. Housecroft CE, Sharpe AG. *Inorganic Chemistry*. 3rd ed. Essex, England: Pearson Education Limited; 2008.
7. Huheey JE, Keiter EA, Keiter RL. *Inorganic Chemistry: Principles of Structure and Reactivity*. 4th ed. New York: Harper Collins College Publishers; 1993.
8. Jordan RB. *Reaction Mechanisms of Inorganic and Organometallic Systems*. 2nd ed. Oxford University Press, Inc.: New York; 1998.
9. Miessler GL, Tarr DA. *Inorganic Chemistry*. 4th ed. Upper Saddle River, NJ: Pearson Education Inc.; 2011.
10. Roundhill DM. *Photochemistry and Photophysics of Metal Complexes*. New York: Plenum Press; 1994.
11. Spessard GO, Miessler GL. *Organometallic Chemistry*. 2nd ed. New York: Oxford University Press; 2010.
12. Wrighton MS. *Inorganic and Organometallic Photochemistry*. Washington, DC: American Chemical Society; 1978.

Derivation of the Classical Wave Equation | A

Consider a string of length L that is fixed at both ends, as shown here.

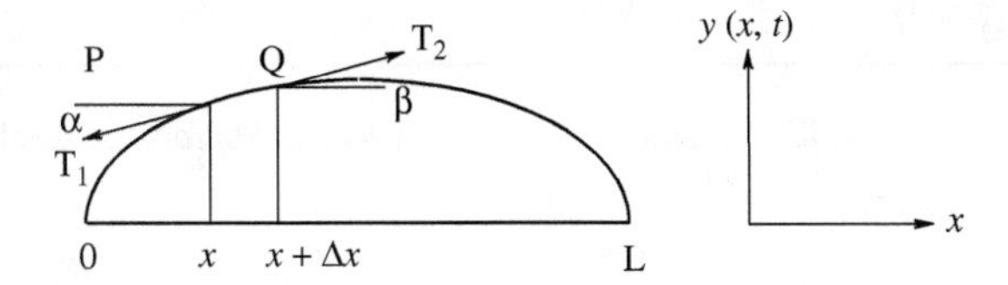

When the string is plucked, there is no net horizontal motion, so:

$$T_1 \cos\alpha = T_2 \cos\beta = T \qquad \text{(constant tension)}$$

There is a net force in the vertical direction that causes the motion of the string. This force can be calculated using Newton's second law:

$$T_2 \sin\beta - T_1 \sin\alpha = F_y = ma = \frac{\partial^2 y}{\partial t^2} m$$

Defining the mass per unit length as $\rho = m/\Delta x$ and dividing both sides by T yields the following:

$$\frac{T_2 \sin\beta}{T_2 \cos\beta} - \frac{T_1 \sin\alpha}{T_1 \cos\alpha} = \frac{\rho \Delta x}{T} \frac{\partial^2 y}{\partial t^2} = \tan\beta - \tan\alpha$$

By definition, $\tan\alpha = \partial y/\partial x$ at point x and $\tan\beta = \partial y/\partial x$ at point $x + \Delta x$. Substitution yields the following expression:

$$\frac{1}{\Delta x}\left[\frac{\partial y}{\partial x}(x + \Delta x) - \frac{\partial y}{\partial x}(x)\right] = \frac{\rho}{T}\frac{\partial^2 y}{\partial t^2}$$

Because the definition of the derivative dy/dx is the limit of $[f(x + \Delta x) - f(x)]/\Delta x$ as $\Delta x \rightarrow 0$, the equation reduces to

$$\frac{\partial^2 y}{\partial x^2} = \frac{\rho}{T}\frac{\partial^2 y}{\partial t^2}$$

Principles of Inorganic Chemistry, First Edition. Brian W. Pfennig.

The horizontal tension in the string, which is a constant, is given by Newton's second law as

$$T = F_x = ma = \frac{mv}{t} = \frac{mx}{t^2}$$

Substitution of $\rho = m/x$ and the above equation for T yields the following:

$$\frac{\rho}{T} = \frac{m/x}{mx/t^2} = \frac{t^2}{x^2} = \frac{1}{v^2}$$

Hence, the classical wave equation in one-dimension becomes:

$$\frac{\partial^2 y}{\partial x^2} = \frac{1}{v^2}\frac{\partial^2 y}{\partial t^2}$$

BIBLIOGRAPHY

1. McQuarrie DA, Simon JD. *Physical Chemistry: A Molecular Approach*. Sausalito, CA: University Science Books; 1997.

Character Tables B

Non-rotational groups

C_1	E
A	1

C_s	E	σ_h			
A′	1	1	x, y, R_z	x^2, y^2, z^2, xy	$xz^2, yz^2, x(x^2-3y^2), y(3x^2-y^2)$
A″	1	−1	z, R_z, R_y	yz, xz	$z^3, xyz, z(x^2-y^2)$

C_i	E	i			
A_g	1	1	R_x, R_y, R_z	$x^2, y^2, z^2, xy, xz, yz$	
A_u	1	−1	x, y, z		$z^3, xyz, z(x^2-y^2), xz^2, yz^2, x(x^2-3y^2), y(3x^2-y^2)$

C_n groups

C_2	E	C_2			
A	1	1	z, R_z	x^2, y^2, z^2, xy	$z^3, xyz, z(x^2-y^2)$
B	1	−1	x, y, R_x, R_y	xz, yz	$xz^2, yz^2, x(x^2-3y^2), y(3x^2-y^2)$

C_3	E	C_3	${C_3}^2$		$\varepsilon = e^{2\pi i/3}$	
A	1	1	1	z, R_z	x^2+y^2, z^2	$z^3, x(x^2-3y^2), y(3x^2-y^2)$
E	$\left\{\begin{matrix}1\\1\end{matrix}\right.$	$\begin{matrix}\varepsilon\\\varepsilon^*\end{matrix}$	$\left.\begin{matrix}\varepsilon^*\\\varepsilon\end{matrix}\right\}$	$(x, y), (R_x, R_y)$	$(x^2-y^2, xy), (xz, yz)$	$(xz^2, yz^2), [xyz, z(x^2-y^2)]$

Principles of Inorganic Chemistry, First Edition. Brian W. Pfennig.

C_4	E	C_4	C_2	C_4^3		$i = \sqrt{-1}$	
A	1	1	1	1	z, R_z	$x^2 + y^2, z^2$	z^3
B	1	−1	1	−1		$x^2 - y^2, xy$	$xyz, z(x^2 - y^2)$
E	1	i	−1	$-i$	$(x, y), (R_x, R_y)$	$(x^2 - y^2, xy), (xz, yz)$	$(xz^2, yz^2), [x(x^2 - 3y^2), y(3x^2 - y^2)]$
	1	$-i$	−1	i			

C_5	E	C_5	C_5^2	C_5^3	C_5^4		$\varepsilon = e^{2\pi i/5}$	
A	1	1	1	1	1	z, R_z	$x^2 + y^2, z^2$	z^3
E_1	1	ε	ε^2	ε^{2*}	ε^*	$(x, y), (R_x, R_y)$	(xz, yz)	(xz^2, yz^2)
	1	ε^*	ε^{2*}	ε^2	ε			
E_2	1	ε^2	ε^*	ε	ε^{2*}		$(x^2 - y^2, xy)$	$[xyz, z(x^2 - y^2)], [x(x^2 - 3y^2), y(3x^2 - y^2)]$
	1	ε^{2*}	ε	ε^*	ε^2			

C_6	E	C_6	C_3	C_2	C_3^2	C_6^5		$\varepsilon = e^{2\pi i/6}$	
A	1	1	1	1	1	1	z, R_z	$x^2 + y^2, z^2$	z^3
B	1	−1	1	−1	1	−1			$x(x^2 - 3y^2), y(3x^2 - y^2)$
E	1	ε	$-\varepsilon^*$	−1	$-\varepsilon$	ε^*	$(x, y), (R_x, R_y)$	(xz, yz)	(xz^2, yz^2)
	1	ε^*	$-\varepsilon$	−1	$-\varepsilon^*$	ε			
E	1	$-\varepsilon^*$	$-\varepsilon$	1	$-\varepsilon^*$	$-\varepsilon$		$(x^2 - y^2, xy)$	$[xyz, z(x^2 - y^2)]$
	1	$-\varepsilon$	$-\varepsilon^*$	1	$-\varepsilon$	$-\varepsilon^*$			

C_7	E	C_7	C_7^2	C_7^3	C_7^4	C_7^5	C_7^6		$\varepsilon = e^{2\pi i/7}$	
A	1	1	1	1	1	1	1	z, R_z	$x^2 + y^2, z^2$	z^3
E_1	1	ε	ε^2	ε^3	ε^{3*}	ε^{2*}	ε^*	$(x, y), (R_x, R_y)$	(xz, yz)	(xz^2, yz^2)
	1	ε^*	ε^{2*}	ε^{3*}	ε^3	ε^2	ε			
E_2	1	ε^2	ε^3	ε^*	ε	ε^3	ε^{2*}		$(x^2 - y^2, xy)$	$[xyz, z(x^2 - y^2)]$
	1	ε^{2*}	ε^{3*}	ε	ε^*	ε^{3*}	ε^2			
E_3	1	ε^3	ε^*	ε^2	ε^{2*}	ε	ε^{3*}			$[x(x^2 - 3y^2), y(3x^2 - y^2)]$
	1	ε^{3*}	ε	ε^{2*}	ε^2	ε^*	ε^3			

C_8	E	C_8	C_4	C_2	C_4^3	C_8^3	C_8^5		$\varepsilon = e^{2\pi i/8}$	
A	1	1	1	1	1	1	1	z, R_z	$x^2 + y^2, z^2$	z^3
A	1	−1	1	1	1	−1	−1			
E_1	1	ε	i	−1	$-i$	$-\varepsilon^*$	$-\varepsilon$	$(x, y), (R_x, R_y)$	(xz, yz)	(xz^2, yz^2)
	1	ε^*	$-i$	−1	i	$-\varepsilon$	$-\varepsilon^*$			
E_2	1	i	−1	1	−1	$-i$	i		$(x^2 - y^2, xy)$	$[xyz, z(x^2 - y^2)]$
	1	$-i$	−1	1	−1	i	$-i$			
E_3	1	$-\varepsilon$	i	−1	$-i$	ε^*	ε			$[x(x^2 - 3y^2), y(3x^2 - y^2)]$
	1	$-\varepsilon^*$	$-i$	−1	i	ε	ε^*			

C_{nv} groups

C_{2v}	E	C_2	σ_v (xz)	σ_v'(yz)			
A_1	1	1	1	1	z	x^2, y^2, z^2	$z^3, z(x^2-y^2)$
A_2	1	1	−1	−1	R_z	xy	xyz
B_1	1	−1	1	−1	x, R_y	xz	$xz^2, x(x^2-3y^2)$
B_2	1	−1	−1	1	y, R_x	yz	$yz^2, y(3x^2-y^2)$

C_{3v}	E	$2C_3$	$3\sigma_v$			
A_1	1	1	1	z	x^2+y^2, z^2	$z^3, x(x^2-3y^2)$
A_2	1	1	−1	R_z		$y(3x^2-y^2)$
E	2	−1	0	$(x, y), (R_x, R_y)$	$(x^2-y^2, xy), (xz, yz)$	$(xz^2, yz^2), [xyz, z(x^2-y^2)]$

C_{4v}	E	$2C_4$	C_2	$2\sigma_v$	$2\sigma_d$			
A_1	1	1	1	1	1	z	x^2+y^2, z^2	z^3
A_2	1	1	1	−1	−1	R_z		
B_1	1	−1	1	1	−1		x^2-y^2	$z(x^2-y^2)$
B_2	1	−1	1	−1	1		xy	xyz
E	2	0	−2	0	0	$(x, y), (R_x, R_y)$	(xz, yz)	$(xz^2, yz^2), [x(x^2-3y^2), y(3x^2-y^2)]$

C_{5v}	E	$2C_5$	$2C_5^2$	$5\sigma_v$			
A_1	1	1	1	1	z	x^2+y^2, z^2	z^3
A_2	1	1	1	−1	R_z		
E_1	2	$2\cos(2\pi/5)$	$2\cos(4\pi/5)$	0	$(x, y), (R_x, R_y)$	(xz, yz)	(xz^2, yz^2)
E_2	2	$2\cos(4\pi/5)$	$2\cos(2\pi/5)$	0		(x^2-y^2, xy)	$[xyz, z(x^2-y^2)], [x(x^2-3y^2), y(3x^2-y^2)]$

C_{6v}	E	$2C_6$	$2C_3$	C_2	$3\sigma_v$	$3\sigma_d$			
A_1	1	1	1	1	1	1	z	x^2+y^2, z^2	z^3
A_2	1	1	1	1	−1	−1	R_z		
B_1	1	−1	1	−1	1	−1			$x(x^2-3y^2)$
B_2	1	−1	1	−1	−1	1			$y(3x^2-y^2)$
E_1	2	1	−1	−2	0	0	$(x, y), (R_x, R_y)$	(xz, yz)	(xz^2, yz^2)

C_{7v}	E	$2C_7$	$2C_7^2$	$2C_7^3$	$7\sigma_v$			
A_1	1	1	1	1	1	z	x^2+y^2, z^2	z^3
A_2	1	1	1	1	1	R_z		
E_1	2	$2\cos(2\pi/7)$	$2\cos(4\pi/7)$	$2\cos(6\pi/7)$	0	$(x, y), (R_x, R_y)$	(xz, yz)	(xz^2, yz^2)
E_2	2	$2\cos(4\pi/7)$	$2\cos(6\pi/7)$	$2\cos(2\pi/7)$	0		(x^2-y^2, xy)	$[xyz, z(x^2-y^2)]$
E_3	2	$2\cos(6\pi/7)$	$2\cos(2\pi/7)$	$2\cos(4\pi/7)$	0			$[x(x^2-3y^2), y(3x^2-y^2)]$

C_{8v}	E	$2C_8$	$2C_4$	$2C_8^3$	C_2	$4\sigma_v$	$4\sigma_d$			
A_1	1	1	1	1	1	1	1	z	x^2+y^2, z^2	z^3
A_2	1	1	1	1	1	−1	−1	R_z		
B_1	1	−1	1	−1	1	1	−1			
B_2	1	−1	1	−1	1	−1	1			
E_1	2	$\sqrt{2}$	0	$\sqrt{2}$	−2	0	0	$(x, y), (R_x, R_y)$	(xz, yz)	(xz^2, yz^2)
E_2	2	0	−2	0	2	0	0		(x^2-y^2, xy)	$[xyz, z(x^2-y^2)]$
E_3	2	$\sqrt{2}$	0	$\sqrt{2}$	−2	0	0			$[x(x^2-3y^2), y(3x^2-y^2)]$

C_{nh} groups

C_{2h}	E	C_2	i	σ_h			
A_g	1	1	1	1	R_z	x^2, y^2, z^2, xy	
B_g	1	−1	1	−1	R_x, R_y	xz, yz	
A_u	1	1	−1	−1	z		$z^3, xyz, z(x^2-y^2)$
B_u	1	−1	−1	1	x, y		$xz^2, yz^2, x(x^2-3y^2), y(3x^2-y^2)$

C_{3h}	E	C_3	C_3^2	σ_h	S_3	S_3^5		$\varepsilon = e^{2\pi i/3}$	
A'	1	1	1	1	1	1	R_z	x^2+y^2, z^2	$[x(x^2-3y^2), y(3x^2-y^2)]$
E'	$\left\{\begin{matrix}1\\1\end{matrix}\right.$	$\begin{matrix}\varepsilon\\\varepsilon^*\end{matrix}$	$\begin{matrix}\varepsilon^*\\\varepsilon\end{matrix}$	$\begin{matrix}1\\1\end{matrix}$	$\begin{matrix}\varepsilon\\\varepsilon^*\end{matrix}$	$\left.\begin{matrix}\varepsilon^*\\\varepsilon\end{matrix}\right\}$	(x, y)	(x^2-y^2, xy)	(xz^2, yz^2)
A''	1	1	1	−1	−1	−1	z		z^3
E''	$\left\{\begin{matrix}1\\1\end{matrix}\right.$	$\begin{matrix}\varepsilon\\\varepsilon^*\end{matrix}$	$\begin{matrix}\varepsilon^*\\\varepsilon\end{matrix}$	$\begin{matrix}-1\\-1\end{matrix}$	$\begin{matrix}-\varepsilon\\-\varepsilon^*\end{matrix}$	$\left.\begin{matrix}-\varepsilon^*\\-\varepsilon\end{matrix}\right\}$	(R_x, R_y)	(xz, yz)	$[xyz, z(x^2-y^2)]$

C_{4h}	E	C_4	C_2	C_4^3	i	S_4^3	σ_h	S_4	$i = \sqrt{-1}$		
A_g	1	1	1	1	1	1	1	1	R_z	x^2+y^2, z^2	
B_g	1	−1	1	−1	1	−1	1	−1		x^2-y^2, xy	
E_g	1	i	−1	$-i$	1	i	−1	$-i$	(R_x, R_y)	(xz, yz)	
	1	$-i$	−1	i	1	$-i$	−1	i			
A_u	1	1	1	1	−1	−1	−1	−1	z		z^3
B_u	1	−1	1	−1	−1	1	−1	1			$xyz, z(x^2-y^2)$
E_u	1	i	−1	$-i$	−1	$-i$	1	i	(x, y)		$(xz^2, yz^2), [x(x^2-3y^2), y(3x^2-y^2)]$
	1	$-i$	−1	i	−1	i	1	$-i$			

C_{5h}	E	C_5	C_5^2	C_5^3	C_5^4	σ_h	S_5	S_5^7	S_5^3	S_5^9	$\varepsilon = e^{2\pi i/5}$		
A'	1	1	1	1	1	1	1	1	1	1	R_z	x^2+y^2, z^2	
E_1'	1	ε	ε^2	ε^{2*}	ε^*	1	ε	ε^2	ε^{2*}	ε^*	(x, y)		(xz^2, yz^2)
	1	ε^*	ε^{2*}	ε^2	ε	1	ε^*	ε^{2*}	ε^2	ε			
E_2'	1	ε^2	ε^*	ε	ε^{2*}	1	ε^2	ε^*	ε	ε^{2*}		(x^2-y^2, xy)	$([x(x^2-3y^2), y(3x^2-y^2)])$
	1	ε^{2*}	ε	ε^*	ε^2	1	ε^{2*}	ε	ε^*	ε^2			
A''	1	1	1	1	1	−1	−1	−1	−1	−1	z		z^3
E_1''	1	ε	ε^2	ε^{2*}	ε^*	−1	$-\varepsilon$	$-\varepsilon^2$	$-\varepsilon^{2*}$	$-\varepsilon^*$	(R_x, R_y)	(xz, yz)	
	1	ε^*	ε^{2*}	ε^2	ε	−1	$-\varepsilon^*$	$-\varepsilon^{2*}$	$-\varepsilon^2$	$-\varepsilon$			
E_2''	1	ε^2	ε^*	ε	ε^{2*}	−1	$-\varepsilon^2$	$-\varepsilon^*$	$-\varepsilon$	$-\varepsilon^{2*}$			$xyz, z(x^2-y^2)$
	1	ε^{2*}	ε	ε^*	ε^2	−1	$-\varepsilon^{2*}$	$-\varepsilon$	$-\varepsilon^*$	$-\varepsilon^2$			

C_{6h}	E	C_6	C_3	C_2	C_3^2	C_6^5	i	S_3^5	S_6^5	σ_h	S_6	S_3	$\varepsilon = e^{2\pi i/6}$		
A_g	1	1	1	1	1	1	1	1	1	1	1	1	R_z	x^2+y^2, z^2	
B_g	1	−1	1	−1	1	−1	1	−1	1	−1	1	−1			
E_{1g}	1	ε	$-\varepsilon^*$	−1	$-\varepsilon$	ε^*	1	ε	$-\varepsilon^*$	−1	$-\varepsilon$	ε^*	(R_x, R_y)	(xz, yz)	
	1	ε^*	$-\varepsilon$	−1	$-\varepsilon^*$	ε	1	ε^*	$-\varepsilon$	−1	$-\varepsilon^*$	ε			
E_{2g}	1	$-\varepsilon^*$	$-\varepsilon$	1	$-\varepsilon^*$	$-\varepsilon$	1	$-\varepsilon^*$	$-\varepsilon$	1	$-\varepsilon^*$	$-\varepsilon$		(x^2-y^2, xy)	
	1	$-\varepsilon$	$-\varepsilon^*$	1	$-\varepsilon$	$-\varepsilon^*$	1	$-\varepsilon$	$-\varepsilon^*$	1	$-\varepsilon$	$-\varepsilon^*$			
A_u	1	1	1	1	1	1	−1	−1	−1	−1	−1	−1	z		z^3
B_u	1	−1	1	−1	1	−1	−1	1	−1	1	−1	1			$([x(x^2-3y^2), y(3x^2-y^2)])$
E_{1u}	1	ε	$-\varepsilon^*$	−1	$-\varepsilon$	ε^*	−1	$-\varepsilon$	ε^*	1	ε	$-\varepsilon^*$	(x, y)		(xz^2, yz^2)
	1	ε^*	$-\varepsilon$	−1	$-\varepsilon^*$	ε	−1	$-\varepsilon^*$	ε	1	ε^*	$-\varepsilon$			
E_{2u}	1	$-\varepsilon^*$	$-\varepsilon$	1	$-\varepsilon^*$	$-\varepsilon$	−1	ε^*	ε	−1	ε^*	ε			$xyz, z(x^2-y^2)$
	1	$-\varepsilon$	$-\varepsilon^*$	1	$-\varepsilon$	$-\varepsilon^*$	−1	ε	ε^*	−1	ε	ε^*			

D_n groups

D_2	E	C_2 (z)	C_2 (y)	C_2 (x)			
A	1	1	1	1		x^2, y^2, z^2	xyz
B_1	1	1	−1	−1	z, R_z	xy	$z^3, z(x^2-y^2)$
B_2	1	−1	1	−1	y, R_y	xz	$yz^2, y(3x^2-y^2)$
B_3	1	−1	−1	1	x, R_x	yz	$xz^2, x(x^2-3y^2)$

D_3	E	$2C_3$	$3C_2$			
A_1	1	1	1		x^2+y^2, z^2	$x(x^2-3y^2)$
A_2	1	1	−1	z, R_z		$z^3, y(3x^2-y^2)$
E	2	−1	0	$(x, y), (R_x, R_y)$	$(x^2-y^2, xy), (xz, yz)$	$(xz^2, yz^2), [xyz, z(x^2-y^2)]$

D_4	E	$2C_4$	C_2	$2C_2'$	$2C_2''$			
A_1	1	1	1	1	1		x^2+y^2, z^2	
A_2	1	1	1	−1	−1	z, R_z		z^3
B_1	1	−1	1	1	−1		x^2-y^2	xyz
B_2	1	−1	1	−1	1		xy	$z(x^2-y^2)$
E	2	0	−2	0	0	$(x, y), (R_x, R_y)$	(xz, yz)	$(xz^2, yz^2), [x(x^2-3y^2), y(3x^2-y^2)]$

D_5	E	$2C_5$	$2C_5^2$	$5C_2$			
A_1	1	1	1	1		x^2, y^2, z^2	
A_2	1	1	1	−1	z, R_z		z^3
E_1	2	$2\cos(2\pi/5)$	$2\cos(4\pi/5)$	0	$(x, y), (R_x, R_y)$	(xz, yz)	(xz^2, yz^2)
E_2	2	$2\cos(4\pi/5)$	$2\cos(2\pi/5)$	0		(x^2-y^2, xy)	$[xyz, z(x^2-y^2)], [x(x^2-3y^2), y(3x^2-y^2)]$

D_6	E	$2C_6$	$2C_3$	C_2	$3C_2'$	$3C_2''$			
A_1	1	1	1	1	1	1		x^2+y^2, z^2	
A_2	1	1	1	1	−1	−1	z, R_z		z^3
B_1	1	−1	1	−1	1	−1			$x(x^2-3y^2)$
B_2	1	−1	1	−1	−1	1			$y(3x^2-y^2)$
E_1	2	1	−1	−2	0	0	$(x, y), (R_x, R_y)$	(xz, yz)	(xz^2, yz^2)
E_2	2	−1	−1	2	0	0		(x^2-y^2, xy)	$[xyz, z(x^2-y^2)]$

D_7	E	$2C_7$	$2C_7^2$	$2C_7^3$	$7C_2$			
A_1	1	1	1	1	1		x^2+y^2, z^2	
A_2	1	1	1	1	−1	z, R_z		z^3
E_1	2	$2\cos(2\pi/7)$	$2\cos(4\pi/7)$	$2\cos(6\pi/7)$	0	$(x, y), (R_x, R_y)$	(xz, yz)	(xz^2, yz^2)
E_2	2	$2\cos(4\pi/7)$	$2\cos(6\pi/7)$	$2\cos(2\pi/7)$	0		(x^2-y^2, xy)	$[xyz, z(x^2-y^2)]$
E_3	2	$2\cos(6\pi/7)$	$2\cos(2\pi/7)$	$2\cos(4\pi/7)$	0			$[x(x^2-3y^2), y(3x^2-y^2)]$

D_8	E	$2C_8$	$2C_4$	$2C_8^3$	C_2	$4C_2'$	$4C_2''$			
A_1	1	1	1	1	1	1	1		x^2+y^2, z^2	
A_2	1	1	1	1	1	−1	−1	z, R_z		z^3
B_1	1	−1	1	−1	1	1	−1			
B_2	1	−1	1	−1	1	−1	1			
E_1	2	$\sqrt{2}$	0	$\sqrt{2}$	−2	0	0	$(x, y), (R_x, R_y)$	(xz, yz)	(xz^2, yz^2)
E_2	2	0	−2	0	2	0	0		(x^2-y^2, xy)	$[xyz, z(x^2-y^2)]$
E_3	2	$-\sqrt{2}$	0	$\sqrt{2}$	−2	0	0			$[x(x^2-3y^2), y(3x^2-y^2)]$

D_{nd} groups

D_{2d}	E	$2S_4$	C_2	$2C_2'$	$2\sigma_d$			
A_1	1	1	1	1	1		x^2+y^2, z^2	xyz
A_2	1	1	1	−1	−1	R_z		$z(x^2-y^2)$
B_1	1	−1	1	1	−1		x^2-y^2	
B_2	1	−1	1	−1	1	z	xy	z^3
E	2	0	−2	0	0	$(x, y), (R_x, R_y)$	(xz, yz)	$(xz^2, yz^2), [x(x^2-3y^2), y(3x^2-y^2)]$

D_{3d}	E	$2C_3$	$3C_2$	i	$2S_6$	$3\sigma_d$			
A_{1g}	1	1	1	1	1	1		x^2+y^2, z^2	
A_{2g}	1	1	−1	1	1	−1	R_z		
E_g	2	−1	0	2	−1	0	(R_x, R_y)	$(x^2-y^2, xy), (xz, yz)$	
A_{1u}	1	1	1	−1	−1	−1			$x(x^2-3y^2)$
A_{2u}	1	1	−1	−1	−1	1	z		$z^3, y(3x^2-y^2)$
E_u	2	−1	0	−2	1	0	(x, y)		$(xz^2, yz^2), [xyz, z(x^2-y^2)]$

D_{4d}	E	$2S_8$	$2C_4$	$2S_8^3$	C_2	$4C_2'$	$4\sigma_d$			
A_1	1	1	1	1	1	1	1		x^2+y^2, z^2	
A_2	1	1	1	1	1	−1	−1	R_z		
B_1	1	−1	1	−1	1	1	−1			
B_2	1	−1	1	−1	1	−1	1	z		z^3
E_1	2	$\sqrt{2}$	0	$-\sqrt{2}$	−2	0	0	(x, y)		(xz^2, yz^2)
E_2	2	0	−2	0	2	0	0		(x^2-y^2, xy),	$[xyz, z(x^2-y^2)]$
E_3	2	$-\sqrt{2}$	0	$\sqrt{2}$	−2	0	0	(R_x, R_y)	(xz, yz)	$[x(x^2-3y^2), y(3x^2-y^2)]$

D_{5d}	E	$2C_5$	$2C_5^2$	$5C_2$	i	$2S_{10}^3$	$2S_{10}$	$5\sigma_d$	$\alpha = 2\cos(2\pi/5)$; $\beta = 2\cos(4\pi/5)$		
A_{1g}	1	1	1	1	1	1	1	1		x^2+y^2, z^2	
A_{2g}	1	1	1	−1	1	1	1	−1	R_z		
E_{1g}	2	α	β	0	2	α	β	0	(R_x, R_y)	(xz, yz)	
E_{2g}	2	β	α	0	2	β	α	0		(x^2-y^2, xy)	
A_{1u}	1	1	1	1	−1	−1	−1	−1			
A_{2u}	1	1	1	−1	−1	−1	−1	1	z		z^3
E_{1u}	2	α	β	0	−2	$-\alpha$	$-\beta$	0	(x, y)		(xz^2, yz^2)
E_{2u}	2	β	α	0	−2	$-\beta$	$-\alpha$	0			$[xyz, z(x^2-y^2)]$, $[x(x^2 3y^2), y(3x^2 y^2)]$

D_{6d}	E	$2S_{12}$	$2C_6$	$2S_4$	$2C_3$	$2S_{12}^5$	C_2	$6C_2'$	$6\sigma_d$			
A_1	1	1	1	1	1	1	1	1	1		x^2+y^2, z^2	
A_2	1	1	1	1	1	1	1	−1	−1	R_z		
B_1	1	−1	1	−1	1	−1	1	1	−1			
B_2	1	−1	1	−1	1	−1	1	−1	1	z		z^3
E_1	2	$\sqrt{3}$	1	0	−1	$-\sqrt{3}$	−2	0	0	(x, y)		(xz^2, yz^2)
E_2	2	1	−1	−2	−1	1	2	0	0		(x^2-y^2, xy)	
E_3	2	0	−2	0	2	0	−2	0	0			$[x(x^2-3y^2), y(3x^2-y^2)]$
E_4	2	−1	−1	2	−1	−1	2	0	0			$[xyz, z(x^2-y^2)]$
E_5	2	$-\sqrt{3}$	1	0	−1	$\sqrt{3}$	−2	0	0	(R_x, R_y)	(xz, yz)	

D_{nh} groups

D_{2h}	E	$C_2(z)$	$C_2(y)$	$C_2(x)$	i	$\sigma(xy)$	$\sigma(xz)$	$\sigma(yz)$			
A_g	1	1	1	1	1	1	1	1		x^2, y^2, z^2	
B_{1g}	1	1	−1	−1	1	1	−1	−1	R_z	xy	
B_{2g}	1	−1	1	−1	1	−1	1	−1	R_y	xz	
B_{3g}	1	−1	−1	1	1	−1	−1	1	R_x	yz	
A_u	1	1	1	1	−1	−1	−1	−1			xyz
B_{1u}	1	1	−1	−1	−1	−1	1	1	z		$z^3, z(x^2-y^2)$
B_{2u}	1	−1	1	−1	−1	1	−1	1	y		$yz^2, y(3x^2-y^2)$
B_{3u}	1	−1	−1	1	−1	1	1	−1	x		$xz^2, x(x^2-3y^2)$

D_{3h}	E	$2C_3$	$3C_2$	σ_h	$2S_3$	$3\sigma_v$			
A_1'	1	1	1	1	1	1		x^2+y^2, z^2	$x(x^2-3y^2)$
A_2'	1	1	−1	1	1	−1	R_z		$y(3x^2-y^2)$
E'	2	−1	0	2	−1	0	(x, y)	(x^2-y^2, xy)	(xz^2, yz^2)
A_1''	1	1	1	−1	−1	−1			
A_2''	1	1	−1	−1	−1	1	z		z^3
E''	2	−1	0	−2	1	0	(R_x, R_y)	(xz, yz)	$[xyz, z(x^2-y^2)]$

D_{4h}	E	$2C_4$	C_2	$2C_2'$	$2C_2''$	i	$2S_4$	σ_h	$2\sigma_v$	$2\sigma_d$			
A_{1g}	1	1	1	1	1	1	1	1	1	1		x^2+y^2, z^2	
A_{2g}	1	1	1	1	−1	1	1	1	−1	1	R_z		
B_{1g}	1	1	1	1	−1	1	−1	1	1	−1		x^2-y^2	
B_{2g}	1	1	1	1	1	1	−1	1	−1	1		xy	
E_g	2	0	−2	0	0	2	0	−2	0	0	(R_x, R_y)	(xz, yz)	
A_{1u}	1	1	1	1	1	−1	−1	−1	−1	−1			z^3
A_{2u}	1	1	1	−1	−1	−1	−1	−1	1	1	z		xyz
B_{1u}	1	−1	1	1	−1	−1	1	−1	−1	1			z^3
B_{2u}	1	−1	1	−1	1	−1	1	−1	1	−1			$z(x^2-y^2)$
E_u	2	0	−2	0	0	−2	0	2	0	0	(x, y)		(xz^2, yz^2), $[x(x^2-3y^2)$, $y(3x^2-y^2)]$

D_{5h}	E	$2C_5$	$2C_5^2$	$5C_2$	σ_h	$2S_5$	$2S_5^3$	$5\sigma_v$	$\alpha = 2\cos(2\pi/5)$; $\beta = 2\cos(4\pi/5)$		
A_1'	1	1	1	1	1	1	1	1		x^2+y^2, z^2	
A_2'	1	1	1	−1	1	1	1	−1	R_z		
E_1'	2	α	β	0	2	α	β	0	(x, y)		(xz^2, yz^2)
E_2'	2	β	α	0	2	β	α	0		(x^2-y^2, xy)	$[x(x^2-3y^2), y(3x^2-y^2)]$
A_1''	1	1	1	1	−1	−1	−1	−1			
A_2''	1	1	1	−1	−1	−1	−1	1	z		z^3
E_1''	2	α	β	0	−2	$-\alpha$	$-\beta$	0	(R_x, R_y)	(xz, yz)	
E_2''	2	β	α	0	−2	$-\beta$	$-\alpha$	0			$[xyz, z(x^2-y^2)]$

D_{6h}	E	$2C_6$	$2C_3$	C_2	$3C_2'$	$3C_2''$	i	$2S_3$	$2S_6$	σ_h	$3\sigma_d$	$3\sigma_v$			
A_{1g}	1	1	1	1	1	1	1	1	1	1	1	1		x^2+y^2, z^2	
A_{2g}	1	1	1	1	−1	−1	1	1	1	1	−1	−1	R_z		
B_{1g}	1	−1	1	−1	1	−1	1	−1	1	−1	1	−1			
B_{2g}	1	−1	1	−1	−1	1	1	−1	1	−1	−1	1			
E_{1g}	2	1	−1	−2	0	0	2	1	−1	−2	0	0	(R_x, R_y)	(xz, yz)	
E_{2g}	2	−1	−1	2	0	0	2	−1	−1	2	0	0		(x^2-y^2, xy)	
A_{1u}	1	1	1	1	1	1	−1	−1	−1	−1	−1	−1			
A_{2u}	1	1	1	1	−1	−1	−1	−1	−1	−1	1	1	z		z^3
B_{1u}	1	−1	1	−1	1	−1	−1	1	−1	1	−1	1			$[x(x^2-3y^2)$
B_{2u}	1	−1	1	−1	−1	1	−1	1	−1	1	1	−1			$y(3x^2-y^2)]$
E_{1u}	2	1	−1	−2	0	0	−2	−1	1	2	0	0	(x, y)		(xz^2, yz^2)
E_{2u}	2	−1	−1	2	0	0	−2	1	1	−2	0	0			$[xyz, z(x^2-y^2)]$

S_{2n} groups

S_4	E	S_4	C_2	S_4^3		$i = \sqrt{-1}$	
A	1	1	1	1	R_z	x^2, y^2, z^2	$xyz, z(x^2-y^2)$
B	1	1	1	−1	x	x^2-y^2, xy	z^3
E	$\begin{Bmatrix} 1 \\ 1 \end{Bmatrix}$	i $-i$	−1 −1	$-i$ i	$(x, y), (R_x, R_y)$	(xz, yz)	$(xz^2, yz^2), [x(x^2-3y^2), y(3x^2-y^2)]$

S_6	E	C_3	C_3^2	i	S_6^5	S_6	$\varepsilon = e^{2\pi i/3}$		
A_g	1	1	1	1	1	1	R_z	x^2+y^2, z^2	
E_g	1 1	ε ε^*	ε^* ε	1 1	ε ε^*	ε^* ε	(R_x, R_y)	$(x^2-y^2, xy), (xz, yz)$	
A_u	1	1	1	−1	−1	−1	z		$z^3, [x(x^2-3y^2), y(3x^2-y^2)]$
E_u	1 1	ε ε^*	ε^* ε	−1 −1	$-\varepsilon$ $-\varepsilon^*$	$-\varepsilon^*$ $-\varepsilon$	(x, y)		$(xz^2, yz^2), [xyz, z(x^2-y^2)]$

S_8	E	S_8	C_4	$S_8{}^3$	C_2	$S_8{}^5$	$C_4{}^3$	$S_8{}^7$	$\varepsilon = e^{2\pi i/8}$		
A	1	1	1	1	1	1	1	1	R_z	x^2+y^2, z^2	
B	1	−1	1	−1	1	−1	1	−1	z		z^3
E_1	1	ε	i	$-\varepsilon^*$	−1	$-\varepsilon$	$-i$	ε^*	(x, y)		(xz^2, yz^2)
	1	ε^*	$-i$	$-\varepsilon$	−1	$-\varepsilon^*$	i	ε			
E_2	1	i	−1	$-i$	1	i	−1	$-i$		(x^2-y^2, xy)	$xyz, z(x^2-y^2)$
	1	$-i$	−1	i	1	$-i$	−1	i			
E_3	1	$-\varepsilon^*$	$-i$	ε	−1	ε^*	i	$-\varepsilon$	(R_x, R_y)	(xz, yz)	$[x(x^2-3y^2),$
	1	$-\varepsilon$	i	ε^*	−1	ε	$-i$	$-\varepsilon^*$			$y(3x^2-y^2)]$

S_{10}	E	C_5	$C_5{}^2$	$C_5{}^3$	$C_5{}^4$	i	$S_{10}{}^7$	$S_{10}{}^9$	S_{10}	$S_{10}{}^3$	$\varepsilon = e^{2\pi i/5}$		
A_g	1	1	1	1	1	1	1	1	1	1	R_z	x^2+y^2, z^2	
E_{1g}	1	ε	ε^2	ε^{2*}	ε^*	1	ε	ε^2	ε^{2*}	ε^*	(R_x, R_y)	(xz, yz)	
	1	ε^*	ε^{2*}	ε^2	ε	1	ε^*	ε^{2*}	ε^2	ε			
E_{2g}	1	ε^2	ε^*	ε	ε^{2*}	1	ε^2	ε^*	ε	ε^{2*}		(x^2-y^2, xy)	
	1	ε^{2*}	ε	ε^*	ε^2	1	ε^{2*}	ε	ε^*	ε^2			
A_u	1	1	1	1	1	−1	−1	−1	−1	−1	z		z^3
E_{1u}	1	ε	ε^2	ε^{2*}	ε^*	−1	$-\varepsilon$	$-\varepsilon^2$	$-\varepsilon^{2*}$	$-\varepsilon^*$	(x, y)		(xz^2, yz^2)
	1	ε^*	ε^{2*}	ε^2	ε	−1	$-\varepsilon^*$	$-\varepsilon^{2*}$	$-\varepsilon^2$	$-\varepsilon$			
E_{2u}	1	ε^2	ε^*	ε	ε^{2*}	−1	$-\varepsilon^2$	$-\varepsilon^*$	$-\varepsilon$	$-\varepsilon^{2*}$			$[xyz, z(x^2-y^2)],$
	1	ε^{2*}	ε	ε^*	ε^2	−1	$-\varepsilon^{2*}$	$-\varepsilon$	$-\varepsilon^*$	$-\varepsilon^2$			$[x(x^2 3y^2),$
													$y(3x^2-y^2)]$

Cubic groups

T	E	$4C_3$	$4C_3{}^2$	$3C_2$		$\varepsilon = e^{2\pi i/3}$	
A	1	1	1	1		$x^2+y^2+z^2$	xyz
E	1	ε	ε^*	1		$(2z^2-x^2-y^2, x^2-y^2)$	
	1	ε^*	ε	1			
T	3	0	0	−1	$(x, y, z), (R_x, R_y, R_z)$	(xy, xz, yz)	$(x^3, y^3, z^3), [x(z^2-y^2),$
							$y(z^2-x^2),$
							$z(x^2-y^2)]$

T_h	E	$4C_3$	$4C_3^2$	$3C_2$	i	$4S_6$	$4S_6^5$	$3\sigma_h$		$\varepsilon = e^{2\pi i/3}$	
A_g	1	1	1	1	1	1	1	1		$x^2+y^2+z^2$	
A_u	1	1	1	1	−1	−1	−1	−1			xyz
E_g	1	ε	ε^*	1	1	ε	ε^*	1		$(2z^2-x^2-y^2, x^2-y^2)$	
	1	ε^*	ε	1	1	ε^*	ε	1			
E_u	1	ε	ε^*	1	−1	$-\varepsilon$	$-\varepsilon^*$	−1			
	1	ε^*	ε	1	−1	$-\varepsilon^*$	$-\varepsilon$	−1			
T_g	3	0	0	−1	3	0	0	−1	(R_x, R_y, R_z)	(xy, xz, yz)	
T_u	3	0	0	−1	−3	0	0	1	(x, y, z)		(x^3, y^3, z^3), $[x(z^2-y^2), y(z^2-x^2), z(x^2-y^2)]$

T_d	E	$8C_3$	$3C_2$	$6S_4$	$6\sigma_d$			
A_1	1	1	1	1	1		$x^2+y^2+z^2$	xyz
A_2	1	1	1	−1	−1			
E	2	−1	2	0	0		$(2z^2-x^2-y^2, x^2-y^2)$	
T_1	3	0	−1	1	−1	(R_x, R_y, R_z)		$[x(z^2-y^2), y(z^2-x^2), z(x^2-y^2)]$
T_2	3	0	−1	−1	1	(x, y, z)	(xy, xz, yz)	(x^3, y^3, z^3)

O	E	$6C_4$	$3C_2$	$8C_3$	$6C_2$			
A_1	1	1	1	1	1		$x^2+y^2+z^2$	
A_2	1	−1	1	1	−1			xyz
E	2	0	2	−1	0		$(2z^2-x^2-y^2, x^2-y^2)$	
T_1	3	1	−1	0	−1	(x, y, z), (R_x, R_y, R_z)		(x^3, y^3, z^3)
T_2	3	−1	−1	0	1		(xy, xz, yz)	$[x(z^2-y^2), y(z^2-x^2), z(x^2-y^2)]$

O_h	E	$8C_3$	$6C_2$	$6C_4$	$3C_2$	i	$6S_4$	$8S_6$	$3\sigma_h$	$6\sigma_d$			
A_{1g}	1	1	1	1	1	1	1	1	1	1		$x^2+y^2+z^2$	
A_{2g}	1	1	−1	−1	1	1	−1	1	1	−1			
E_g	2	−1	0	0	2	2	0	−1	2	0		$(2z^2-x^2-y^2, x^2-y^2)$	
T_{1g}	3	0	−1	1	−1	3	1	0	−1	−1	(R_x, R_y, R_z)		
T_{2g}	3	0	1	−1	−1	3	−1	0	−1	1		(xy, xz, yz)	
A_{1u}	1	1	1	1	1	−1	−1	−1	−1	−1			
A_{2u}	1	1	1	1	1	1	1	1	−1	1			xyz
E_u	2	−1	0	0	2	−2	0	1	−2	0			
T_{1u}	3	0	−1	1	−1	−3	−1	0	1	1	(x, y, z)		(x^3, y^3, z^3)
T_{2u}	3	0	1	−1	−1	−3	1	0	1	−1			$[x(z^2-y^2), y(z^2-x^2), z(x^2-y^2)]$

I	E	$12C_5$	$12C_5^2$	$20C_3$	$15C_2$	$\alpha=2\cos(2\pi/5)$; $\beta=2\cos(4\pi/5)$		
A	1	1	1	1	1		$x^2+y^2+z^2$	
T_1	3	$-\beta$	$-\alpha$	0	-1	(x, y, z), (R_x, R_y, R_z)		
T_2	3	$-\alpha$	$-\beta$	0	-1			(x^3, y^3, z^3)
G	4	-1	-1	1	0			$[x(z^2-y^2), y(z^2-x^2), z(x^2-y^2), xyz]$
H	5	0	0	-1	1		$(2z^2-x^2-y^2,$ $x^2-y^2,$ $xy, xz, yz)$	

I_h	E	$12C_5$	$12C_5^2$	$20C_3$	$15C_2$	i	$12S_{10}$	$12S_{10}^3$	$20S_6$	15σ	$\alpha=2\cos(2\pi/5)$; $\beta=2\cos(4\pi/5)$	
A_g	1	1	1	1	1	1	1	1	1	1		$x^2+y^2+z^2$
T_{1g}	3	α	β	0	-1	3	β	α	0	-1	(R_x, R_y, R_z)	
T_{2g}	3	β	α	0	-1	3	α	β	0	-1		
G_g	4	-1	-1	1	0	4	-1	-1	1	0		
H_g	5	0	0	-1	1	5	0	0	-1	1		$(2z^2-x^2-y^2,$ $x^2-y^2,$ $xy, xz, yz)$
A_u	1	1	1	1	1	-1	-1	-1	-1	-1		
T_{1u}	3	α	β	0	-1	-3	$-\beta$	$-\alpha$	0	1		(x, y, z)
T_{2u}	3	β	α	0	-1	-3	$-\alpha$	$-\beta$	0	1		(x^3, y^3, z^3)
G_u	4	-1	-1	1	0	-4	1	1	-1	0		$[x(z^2-y^2),$ $y(z^2-x^2),$ $z(x^2-y^2), xyz]$
H_u	5	0	0	-1	1	-5	0	0	1	-1		

Infinite groups

$C_{\infty v}$	E	$2C_\infty^{\phi}$	$2C_\infty^{2\phi}$	$2C_\infty^{3\phi}$	...	$15C_2$			
$A_1=\Sigma^+$	1	1	1	1	...	1	z	$x^2+y^2+z^2$	z^3
$A_2=\Sigma^-$	1	1	1	1	...	-1	R_z		
$E_1=\Pi$	2	$2\cos(\phi)$	$2\cos(2\phi)$	$2\cos(3\phi)$	...	0	(x, y), (R_x, R_y)	(xz, yz)	(xz^2, yz^2)
$E_2=\Delta$	2	$2\cos(2\phi)$	$2\cos(4\phi)$	$2\cos(6\phi)$	...	0		(x^2-y^2, xy)	$[xyz, z(x^2-y^2)]$
$E_3=\Phi$	2	$2\cos(3\phi)$	$2\cos(6\phi)$	$2\cos(2\phi)$	...	0			$[x(x^2-3y^2),$ $y(3x^2-y^2),$
:	:	:	:	:	:	:			

$D_{\infty h}$	E	$2C_{\infty}^{\phi}$	$2C_{\infty}^{2\phi}$	$2C_{\infty}^{3\phi}$	...	$\infty\sigma_v$	i	$2S_{\infty}^{\phi}$	...	∞C_2		
Σ_g^+	1	1	1	1	...	1	1	1	...	1		x^2+y^2, z^2
Σ_g^-	1	1	1	1	...	−1	1	1	...	−1	R_z	
Π_g^+	2	$2\cos(\phi)$	$2\cos(2\phi)$	$2\cos(3\phi)$	...	0	2	$-2\cos(\phi)$	...	0	(R_x, R_y)	(xz, yz)
Δ_g^+	2	$2\cos(2\phi)$	$2\cos(4\phi)$	$2\cos(6\phi)$	...	0	2	$2\cos(2\phi)$	...	0		(x^2-y^2, xy)
Φ_g^+	2	$2\cos(3\phi)$	$2\cos(6\phi)$	$2\cos(9\phi)$	...	0	2	$-2\cos(3\phi)$	...	0		
:	:	:	:	:	:	:	:	:	:	:		
Σ_u^+	1	1	1	1	...	1	−1	−1	...	−1	z	z^3
Σ_u^-	1	1	1	1	...	−1	−1	−1	...	1		
Π_u^+	2	$2\cos(\phi)$	$2\cos(2\phi)$	$2\cos(3\phi)$	...	0	−2	$2\cos(\phi)$	...	0	(x, y)	(xz^2, yz^2)
Δ_u^+	2	$2\cos(2\phi)$	$2\cos(4\phi)$	$2\cos(6\phi)$	...	0	−2	$-2\cos(2\phi)$	...	0		$[xyz, z(x^2-y^2)]$
Φ_u^+	2	$2\cos(3\phi)$	$2\cos(6\phi)$	$2\cos(9\phi)$	...	0	−2	$2\cos(3\phi)$	...	0		$[x(x^2-3y^2), y(3x^2-y^2)]$
:	:	:	:	:	:	:	:	:	:	:		

BIBLIOGRAPHY

1. Carter RL. *Molecular Symmetry and Group Theory*. New York: John Wiley & Sons, Inc.; 1998.
2. Cotton FA. *Chemical Applications of Group Theory*. 2nd ed. New York: John Wiley & Sons, Inc.; 1971.
3. Harris DC, Bertolucci MD. *Symmetry and Spectroscopy*. New York: Dover Publications, Inc.; 1978.

Direct Product Tables | C

Notes:

1. For point groups having the inversion symmetry element, add the g and u subscripts using the multiplication rules: $g \times g = g, g \times u = u \times g = u$, and $u \times u = g$.
2. The subscripts 1 and 2 are omitted if the point group has an asterisk.
3. Add the prime and double prime superscripts, when necessary, using the multiplication rules: $' \times ' = '$, $' \times '' = '' \times ' = ''$, and $'' \times '' = '$.
4. The product listed in square brackets is the antisymmetric product.
5. All the direct products commute.
6. The direct product of any IRR with the totally symmetric IRR yields back the original IRR.
7. The square of any IRR always contains the totally symmetric IRR.

C_s	A'	A''
A'	A'	A''
A''		A'

C_i	A_g	A_u
A_g	A_g	A_u
A_u		A_g

C_2, C_{2h}	A	B
A	A	B
B		A

C_{2v}	A_1	A_2	B_1	B_2
A_1	A_1	A_2	B_1	B_2
A_2		A_1	B_2	B_1
B_1			A_1	A_2
B_2				A_1

D_2, D_{2h}	A	B_1	B_2	B_3
A	A	B_1	B_2	B_3
B_1		A	B_3	B_2
B_2			A	B_1
B_3				A

C_3, C_{3h}, S_6	A	E
A	A	E
E		$[A] + A + E$

C_{3v}, D_3, D_{3d}, D_{3h}	A_1	A_2	E
A_1	A_1	A_2	E
A_2		A_1	E
E			$A_1 + [A_2] + E$

C_4, C_{4h}, S_4	A	B	E
A	A	B	E
B		A	E
E			$[A] + A + 2B$

C_{4v}, D_4, D_{2d}, D_{4h}	A_1	A_2	B_1	B_2	E
A_1	A_1	A_2	B_1	B_2	E
A_2		A_1	B_2	B_1	E
B_1			A_1	A_2	E
B_2				A_1	E
E					$A_1 + [A_2] + B_1 + B_2$

C_5, C_{5h}, S_{10}	A	E_1	E_2
A	A	E_1	E_2
E_1		$[A] + A + E_2$	$E_1 + E_2$
E_2			$[A] + A + E_1$

C_{5v}, D_5, D_{5d}, D_{5h}	A_1	A_2	E_1	E_2
A_1	A_1	A_2	E_1	E_2
A_2		A_1	E_1	E_2
E_1			$A_1 + [A_2] + E_2$	$E_1 + E_2$
E_2				$A_1 + [A_2] + E_1$

C_6, C_{6h}	A	B	E_1	E_2
A	A	B	E_1	E_2
B		A	E_2	E_1
E_1			$[A] + A + E_2$	$2B + E_1$
E_2				$[A] + A + E_2$

C_{6v}, D_6, D_{6h}	A_1	A_2	B_1	B_2	E_1	E_2
A_1	A_1	A_2	B_1	B_2	E_1	E_2
A_2		A_1	B_2	B_1	E_1	E_2
B_1			A_1	A_2	E_2	E_1
B_2				A_1	E_2	E_1
E_1					$A_1 + [A_2] + E_2$	$B_1 + B_2 + E_1$
E_2						$A_1 + [A_2] + E_2$

C_{6v}, D_6, D_{6h}	A_1	A_2	B_1	B_2	E_1	E_2	E_3	E_4	E_5
A_1	A_1	A_2	B_1	B_2	E_1	E_2	E_3	E_4	E_5
A_2		A_1	B_2	B_1	E_1	E_2	E_3	E_4	E_5
B_1			A_1	A_2	E_5	E_4	E_3	E_2	E_1
B_2				A_1	E_5	E_4	E_3	E_2	E_1
E_1					$A_1+[A_2]+E_2$	$E_1 + E_3$	$E_2 + E_4$	$E_3 + E_5$	$B_1+B_2+E_4$
E_2						$A_1+[A_2]+E_4$	$E_1 + E_5$	$B_1+B_2+E_2$	$E_3 + E_5$
E_3							$A_1+[A_2]+B_1+B_2$	$E_1 + E_5$	$E_2 + E_4$
E_4								$A_1+[A_2]+E_4$	$E_1 + E_3$
E_5									$A_1+[A_2]+E_2$

C_7, C_{7h}	A	E_1	E_2	E_3
A	A	E_1	E_2	E_3
E_1		$[A] + A + E_2$	$E_1 + E_3$	$E_2 + E_3$
E_2			$[A] + A + E_3$	$E_1 + E_2$
E_3				$[A] + A + E_1$

C_{7v}, D_7, D_{7d}, D_{7h}	A_1	A_2	E_1	E_2	E_3
A_1	A_1	A_2	E_1	E_2	E_3
A_2		A_1	E_1	E_2	E_3
E_1			$A_1 + [A_2] + E_2$	$E_1 + E_3$	$E_2 + E_3$
E_2				$A_1 + [A_2] + E_3$	$E_1 + E_2$
E_3					$A_1 + [A_2] + E_1$

C_8, C_{8h}, S_8	A	B	E_1	E_2	E_3
A	A	B	E_1	E_2	E_3
B		A	E_3	E_2	E_1
E_1			$A + [A] + E_2$	$E_1 + E_3$	$2B + E_2$
E_2				$A + [A] + 2B$	$E_1 + E_3$
E_3					$A + [A] + E_2$

C_{6v}, D_6, D_{6h}	A_1	A_2	B_1	B_2	E_1	E_2	E_3
A_1	A_1	A_2	B_1	B_2	E_1	E_2	E_3
A_2		A_1	B_2	B_1	E_1	E_2	E_3
B_1			A_1	B_2	E_3	E_2	E_1
B_2				A_1	E_3	E_2	E_1
E_1					$A_1+[A_2]+E_2$	$E_1 + E_3$	$B_1 + B_2 + E_2$
E_2						$A_1 + [A_2] + B_1 + B_2$	$E_1 + E_3$
E_3							$A_1 + [A_2] + E_2$

D_{8d}	A_1	A_2	B_1	B_2	E_1	E_2	E_3	E_4	E_5	E_6	E_7
A_1	A_1	A_2	B_1	B_2	E_1	E_2	E_3	E_4	E_5	E_6	E_7
A_2		A_1	B_2	B_1	E_1	E_2	E_3	E_4	E_5	E_6	E_7
B_1			A_1	A_2	E_7	E_6	E_5	E_4	E_3	E_2	E_1
B_2				A_1	E_7	E_6	E_5	E_4	E_3	E_2	E_1
E_1					$A_1 + [A_2] + E_2$	$E_1 + E_3$	$E_2 + E_4$	$E_3 + E_5$	$E_4 + E_6$	$E_5 + E_7$	$B_1 + B_2 + E_6$
E_2						$A_1 + [A_2] + E_4$	$E_1 + E_5$	$E_2 + E_6$	$E_3 + E_7$	$B_1 + B_2 + E_4$	$E_5 + E_7$
E_3							$A_1 + [A_2] + E_6$	$E_1 + E_7$	$B_1 + B_2 + E_2$	$E_3 + E_7$	$E_4 + E_6$
E_4								$A_1 + [A_2] + B_1 + B_2$	$E_1 + E_7$	$E_2 + E_6$	$E_3 + E_5$
E_5									$A_1 + [A_2] + E_6$	$E_1 + E_5$	$E_2 + E_4$
E_6										$A_1 + [A_2] + E_4$	$E_1 + E_3$
E_7											$A_1 + [A_2] + E_2$

O, O_h, T^*, T_d, T_h^*	A_1	A_2	E	T_1	T_2
A_1	A_1	A_2	E	T_1	T_2
A_2		A_1	E	T_2	T_1
E			$A_1 + [A_2] + E$	$T_1 + T_2$	$T_1 + T_2$
T_1				$A_1 + E + [T_1] + T_2$	$A_2 + E + T_1 + T_2$
T_2					$A_1 + E + [T_1] + T_2$

I, I_h	A	T_1	T_2	G	H
A	A	T_1	T_2	G	H
T_1		$A + [T_1] + H$	G + H	$T_2 + G + H$	$T_1 + T_2 + G + H$
T_2			$A + [T_2] + H$	$T_1 + G + H$	$T_1 + T_2 + G + H$
G				$A + [T_1] + [T_2] + G + H$	$T_1 + T_2 + G + 2H$
H					$A + [T_1] + [T_2] + [G] + G + 2H$

D_∞, $C_{\infty v}$, $D_{\infty h}$	Σ^+	Σ^-	Π	Δ	Φ	Γ
Σ^+	Σ^+	Σ^-	Π	Δ	Φ	Γ
Σ^-		Σ^+	Π	Δ	Φ	Γ
Π			$\Sigma^+ + [\Sigma^-] + \Delta$	$\Pi + \Phi$	$\Delta + \Gamma$	Φ + H
Δ				$\Sigma^+ + [\Sigma^-] + \Gamma$	Π + H	Δ + I
Φ					$\Sigma^+ + [\Sigma^-]$ + I	$\Pi + \Theta$
Γ						$\Sigma^+ + [\Sigma^-]$ + K

BIBLIOGRAPHY

1. Harris DC, Bertolucci MD. *Symmetry and Spectroscopy*. New York: Dover Publications, Inc.; 1978.

Correlation Tables D

Notes:

1. These tables list the correlations between IRRs in the parent group (boldface) and some of its subgroups.
2. The correlation tables for high-symmetry groups do not include all of the subgroups. It may be necessary to use more than one correlation table in order to obtain the symmetry connections. For instance, use $O_h \rightarrow T_d \rightarrow C_{2v}$ to connect the O_h point group with its C_{2v} subgroup.
3. When identification of the correlations between IRRs in different point groups might be ambiguous, the labels are linked by an arrow. For instance, σ_v (C_{6v}) $\rightarrow \sigma_{xz}$ (C_{2v}).
4. Paired complex conjugate IRRs are indicated in braces; for example, {E} in T_h. If the smaller subgroup does not allow degeneracy, the complex conjugate pair becomes two real-number, nondegenerate IRRs in the subgroup.

C_{2v}	C_2	$C_s\sigma(xz)$	$C_s\sigma(yz)$
A_1	A	A'	A'
A_2	A	A''	A''
B_1	B	A'	A''
B_2	B	A''	A'

C_{3v}	C_3	C_s
A_1	A	A'
A_2	A	A''
E	$\{E\}$	$A' + A''$

C_{4v}	C_4	C_{2v}, σ_v	C_{2v}, σ_d	C_2	C_s, σ_v	C_s, σ_d
A_1	A	A_1	A_1	A	A'	A'
A_2	A	A_2	A_2	A	A''	A''
B_1	B	A_1	A_2	A	A'	A''
B_2	B	A_2	A_1	A	A''	A'
E	E	$B_1 + B_2$	$B_1 + B_2$	$2B$	$A' + A''$	$A' + A''$

Principles of Inorganic Chemistry, First Edition. Brian W. Pfennig.

C_{5v}	C_5	C_s
A_1	A	A'
A_2	A	A''
E_1	$\{E_1\}$	$A' + A''$
E_2	$\{E_2\}$	$A' + A''$

C_{6v}	C_6	$C_{3v}\ \sigma_v$	$C_{3v}\ \sigma_d$	$C_{2v}\ \sigma_v \rightarrow \sigma_{xz}$	C_3	C_2	$C_s\ \sigma_v$	$C_s\ \sigma_d$
A_1	A	A_1	A_1	A_1	A	A	A'	A'
A_2	A	A_2	A_2	A_2	A	A	A''	A''
B_1	B	A_1	A_2	B_1	A	B	A'	A''
B_2	B	A_2	A_1	B_2	A	B	A''	A'
E_1	$\{E_1\}$	E	E	$B_1 + B_2$	$\{E\}$	$2B$	$A' + A''$	$A' + A''$
E_2	$\{E_2\}$	E	E	$A_1 + A_2$	$\{E\}$	$2A$	$A' + A''$	$A' + A''$

C_{2h}	C_2	Cs	C_i
A_g	A	A'	A_g
B_g	B	A''	A_g
A_u	A	A''	A_u
B_u	B	A'	A_u

C_{3h}	C_3	C_s
A'	A	A'
$\{E'\}$	$\{E\}$	$2A'$
A''	A	A''
$\{E''\}$	$\{E\}$	$2A''$

C_{4h}	C_4	S_4	C_{2h}	C_2	C_s	C_i
A_g	A	A	A_g	A	A'	A_g
B_g	B	B	A_g	A	A'	A_g
$\{E_g\}$	$\{E\}$	$\{E\}$	$2B_g$	$2B$	$2A''$	$2A_g$
A_u	A	B	A_u	A	A''	A_u
B_u	B	A	A_u	A	A''	A_u
$\{E_u\}$	$\{E\}$	$\{E\}$	$2B_u$	$2B$	$2A'$	$2A_u$

C_{5h}	C_5	C_s
A'	A	A'
$\{E_1'\}$	$\{E_1\}$	$2A'$
$\{E_2'\}$	$\{E_2\}$	$2A'$
A''	A	A''
$\{E_1''\}$	$\{E_1\}$	$2A''$
$\{E_2''\}$	$\{E_2\}$	$2A''$

C_{6h}	C_6	C_{3h}	S_6	C_{2h}	C_3	C_2	C_s	C_i
A_g	A	A'	A_g	A_g	A	A	A'	A_g
B_g	B	A''	A_g	B_g	A	B	A''	A_g
$\{E_{1g}\}$	$\{E_1\}$	$\{E''\}$	$\{E_g\}$	$2B_g$	$\{E\}$	$2B$	$2A''$	$2A_g$
$\{E_{2g}\}$	$\{E_2\}$	$\{E'\}$	$\{E_g\}$	$2A_g$	$\{E\}$	$2A$	$2A'$	$2A_g$
A_{2u}	A	A''	A_u	A_u	A	A	A''	A_u
B_u	B	A'	A_u	B_u	A	B	A'	A_u
$\{E_{1u}\}$	$\{E_1\}$	$\{E'\}$	$\{E_u\}$	$2B_u$	$\{E\}$	$2B$	$2A'$	$2A_u$
$\{E_{2u}\}$	$\{E_2\}$	$\{E''\}$	$\{E_u\}$	$2A_u$	$\{E\}$	$2A$	$2A''$	$2A_u$

		C_{2v}	C_{2v}	C_{2v}	C_{2h}	C_{2h}	C_{2h}	C_2	C_2	C_2	C_2	C_2	C_2
D_{2h}	D_2	$C_2{-}z$	$C_2{-}y$	$C_2{-}x$	$C_2{-}z$	$C_2{-}y$	$C_2{-}x$	$C_2{-}z$	$C_2{-}y$	$C_2{-}x$	σ_{xy}	σ_{xz}	σ_{yz}
Ag	A	A_1	A_1	A_1	A_g	A_g	A_g	A	A	A	A'	A'	A'
B_{1g}	B_1	A_2	B2	B_1	A_g	B_g	B_g	A	B	B	A'	A''	A''
B_{2g}	B_2	B_1	A_2	B_2	B_g	A_g	B_g	B	A	B	A''	A'	A''
B_{3g}	B_3	B_2	B_1	A_2	B_g	B_g	A_g	B	B	A	A''	A''	A'
A_u	A	A_2	A_2	A_2	A_u	A_u	A_u	A	A	A	A''	A''	A''
B_{1u}	B_1	A_1	B_1	B_2	A_u	B_u	B_u	A	B	B	A''	A'	A'
B_{2u}	B_2	B_2	A_1	B_1	B_u	A_u	B_u	B	A	B	A'	A''	A'
B_{3u}	B_3	B_1	B_2	A_1	B_u	B_u	A_u	B	B	A	A'	A'	A''

D_{3h}	D_3	C_{3v}	C_{3h}	C_{2v} $\sigma_h \rightarrow \sigma_{yz}$	C_s σ_h	C_s σ_v
A_1'	A_1	A_1	A'	A_1	A'	A'
A_2'	A_2	A_2	A'	B_2	A'	A''
E'	E	E	$\{E'\}$	$A_1 + B_2$	$2A'$	$A' + A''$
A_1''	A_1	A_2	A''	A_2	A''	A''
A_2''	A_2	A_1	A''	B_1	A''	A'
E''	E	E	$\{E''\}$	$A_2 + B_1$	$2A''$	$A' + A''$

D_{4h}	D_4	C_{4v}	C_{4h}	C_4	D_{2h} C_2'	D_{2h} C_2''	D_{2d} $C_2' \rightarrow C_2'$	D_{2d} $C_2'' \rightarrow C_2'$	S_4	D_2 C_2'
A_{1g}	A_1	A_1	A_g	A	A_g	A_g	A_1	A_1	A	A
A_{2g}	A_2	A_2	A_g	A	B_{1g}	B_{1g}	A_2	A_2	A	B_1
B_{1g}	B_1	B_1	B_g	B	A_g	B_{1g}	B_1	B_2	B	A
B_{2g}	B_2	B_2	B_g	B	B_{1g}	A_g	B_2	B_1	B	B_1
E_g	E	E	$\{E_g\}$	$\{E\}$	$B_{2g} + B_{3g}$	$B_{2g} + B_{3g}$	E	E	$\{E\}$	$B_2 + B_3$
A_{1u}	A_1	A_2	A_u	A	A_u	A_u	B_1	B_1	B	A
A_{2u}	A_2	A_1	A_u	A	B_{1u}	B_{1u}	B_2	B_2	B	B_1
B_{1u}	B_1	B_2	B_u	B	A_u	B_{1u}	A_1	A_2	A	A
B_{2u}	B_2	B_1	B_u	B	B_{1u}	A_u	A_2	A_1	A	B_1
E_u	E	E	$\{E_u\}$	$\{E\}$	$B_{2u} + B_{3u}$	$B_{2u} + B_{3u}$	E	E	$\{E\}$	$B_2 + B_3$

D_{4h} cont.	D_2 C_2''	C_{2v} C_2, σ_v	C_{2v} C_2, σ_d	C_{2v} C_2'	C_{2v} C_2''	C_{2h} C_2	C_{2h} C_2'	C_{2h} C_2''	C_s σ_h	C_s σ_v	C_s σ_d
A_{1g}	A	A_1	A_1	A_1	A_1	A_g	A_g	A_g	A'	A'	A'
A_{2g}	B_1	A_2	A_2	B_1	B_1	A_g	B_g	B_g	A'	A''	A''
B_{1g}	B_1	A_1	A_2	A_1	B_1	A_g	A_g	B_g	A'	A'	A''
B_{2g}	A	A_2	A_1	B_1	A_1	A_g	B_g	A_g	A'	A''	A'
E_g	$B_2 + B_3$	$B_1 + B_2$	$B_1 + B_2$	$A_2 + B_2$	$A_2 + B_2$	$2B_g$	$A_g + B_g$	$A_g + B_g$	$2A''$	$A' + A''$	$A' + A''$
A_{1u}	A	A_2	A_2	A_2	A_2	A_u	A_u	A_u	A''	A''	A''
A_{2u}	B_1	A_1	A_1	B_2	B_2	A_u	B_u	B_u	A''	A'	A'
B_{1u}	B_1	A_2	A_1	A_2	B_2	A_u	A_u	B_u	A''	A''	A'
B_{2u}	A	A_1	A_2	B_2	A_2	A_u	B_u	A_u	A''	A'	A''
E_u	$B_2 + B_3$	$B_1 + B_2$	$B_1 + B_2$	$A_1 + B_1$	$A_1 + B_1$	$2B_u$	$A_u + B_u$	$A_u + B_u$	$2A'$	$A' + A''$	$A' + A''$

D_{5h}	D_5	C_{5v}	C_{5h}	C_5	C_{2v} $\sigma_h \rightarrow \sigma_{xz}$
A_1'	A_1	A_1	A'	A	A_1
A_2'	A_2	A_2	A'	A	B_1
E_1'	E_1	E_1	$\{E_1'\}$	$\{E_1\}$	$A_1 + B_1$
E_2'	E_2	E_2	$\{E_2'\}$	$\{E_2\}$	$A_1 + B_1$
A_1''	A_1	A_2	A''	A	A_2
A_2''	A_2	A_1	A''	A	B_2
E_1''	E_1	E_1	$\{E_1''\}$	$\{E_1\}$	$A_2 + B_2$
E_2''	E_2	E_2	$\{E_2''\}$	$\{E_2\}$	$A_2 + B_2$

D_{6h}	D_6	C_{6v}	C_{6h}	C_6	D_{3h} C_2'	D_{3h} C_2''	D_{3d} C_2'	D_{3d} C_2''	D_{2h} $\sigma_h \rightarrow \sigma_{xy}, \sigma_v \rightarrow \sigma_{yz}$
A_{1g}	A_1	A_1	A_g	A	A_1'	A_1'	A_{1g}	A_{1g}	A_g
A_{2g}	A_2	A_2	A_g	A	A_2'	A_2'	A_{2g}	A_{2g}	B_{1g}
B_{1g}	B_1	B_2	B_g	B	A_1''	A_2''	A_{1g}	A_{2g}	B_{2g}
B_{2g}	B_2	B_1	B_g	B	A_2''	A_1''	A_{2g}	A_{1g}	B_{3g}
E_{1g}	E_1	$E1$	$\{E_{1g}\}$	$\{E\}$	E''	E''	E_g	E_g	$B_{2g} + B_{3g}$
E_{2g}	E_2	E_2	$\{E_{2g}\}$	$\{E_2\}$	E'	E'	E_g	E_g	$A_g + B_{1g}$
A_{1u}	A_1	A_2	A_u	A	A_1''	A_1''	A_{1u}	A_{1u}	A_u
A_{2u}	A_2	A_1	A_u	A	A_2''	A_2''	A_{2u}	A_{2u}	B_{1u}
B_{1u}	B_1	B_1	B_u	B	A_1'	A_2'	A_{1u}	A_{2u}	B_{2u}
B_{2u}	B_2	B_2	B_u	B	A_2'	A_1'	A_{2u}	A_{1u}	B_{3u}
E_{1u}	E_1	E_1	$\{E_{1u}\}$	$\{E_1\}$	E'	E'	E_u	E_u	$B_{2u} + B_{3u}$
E_{2u}	E_2	E_2	$\{E_{2u}\}$	$\{E_2\}$	E''	E''	E_u	E_u	$A_u + B_{1u}$

D_{2d}	S_4	D_2 $C_2 \rightarrow C_2(z)$	C_{2v}
A_1	A	A	A_1
A_2	A	B_1	A_2
B_1	B	A	A_2
B_2	B	B_1	A_1
E	$\{E\}$	$B_2 + B_3$	$B_1 + B_2$

D_{3d}	D_3	S_6	C_{3v}	C_3	C_{2h}
A_{1g}	A1	A_g	A_1	A	A_g
A_{2g}	A_2	A_g	A_2	A	B_g
E_g	E	$\{E_g\}$	E	$\{E\}$	$A_g + B_g$
A_{1u}	A_1	A_u	A_2	A	A_u
A_{2u}	A_2	A_u	A_1	A	B_u
E_u	E	$\{E_u\}$	E	$\{E\}$	$A_u + B_u$

D_{4d}	D_4	S_8	C_{4v}	C_4	C_{2v}	C_2 $C_2(z)$	C_2 C_2'	C_s
A_1	A_1	A	A_1	A	A_1	A	A	A'
A_2	A_2	A	A_2	A	A_2	A	B	A''
B_1	A_1	B	A_2	A	A_2	A	A	A''
B_2	A_2	B	A_1	A	A_1	A	B	A'
E_1	E	$\{E_1\}$	E	$\{E\}$	$B_1 + B_2$	$2B$	$A + B$	$A' + A''$
E_2	$B_1 + B_2$	$\{E_2\}$	$B_1 + B_2$	$2B$	$A_1 + A_2$	$2A$	$A + B$	$A' + A''$
E_3	E	$\{E_3\}$	E	$\{E\}$	$B_1 + B_2$	$2B$	$A + B$	$A' + A''$

D_{5d}	D_5	C_{5v}	C_5	C_2
A_{1g}	A_1	A_1	A	A
A_{2g}	A_2	A_2	A	B
E_{1g}	E_1	E_1	$\{E_1\}$	$A + B$
E_{2g}	E_2	E_2	$\{E_2\}$	$A + B$
A_{1u}	A_1	A_2	A	A
A_{2u}	A_2	A_1	A	B
E_{1u}	E_1	E_1	$\{E_1\}$	$A + B$
E_{2u}	E_2	E_2	$\{E_2\}$	$A + B$

D_{6d}	D_6	C_{6v}	D_3	D_{2d}	S_4	C_2 $C_2(z)$	C_2 C_2'
A_1	A_1	A_1	A_1	A_1	A	A	A
A_2	A_2	A_2	A_2	A_2	A	A	B
B_1	A_1	A_2	A_1	B_1	B	A	A
B_2	A_2	A_1	A_2	B_2	B	A	B
E_1	E_1	E_1	E	E	$\{E\}$	$2B$	$A + B$
E_2	E_2	E_2	E	$B_1 + B_2$	$2B$	$2A$	$A + B$
E_3	$B_1 + B_2$	$B_1 + B_2$	$A_1 + A_2$	E	$\{E\}$	$2B$	$A + B$
E_4	E_2	E_2	E	$A_1 + A_2$	$2A$	$2A$	$A + B$
E_5	E_1	E_1	E	E	$\{E\}$	$2B$	$A + B$

T	D_3	D_2	C_2
A	A	A	A
$\{E\}$	$\{E\}$	$2A$	$2A$
T	$A + \{E\}$	$B_1 + B_2 + B_3$	$A + 2B$

T_h	T	S_6	D_{2h}	D_2
A_g	A	A_g	A_g	A
$\{E_g\}$	$\{E\}$	$\{E_g\}$	$2A_g$	$2A$
T_g	T	$A_g + \{E_g\}$	$B_{1g} + B_{2g} + B_{3g}$	$B_1 + B_2 + B_3$
A_u	A	A_u	A_u	A
$\{E_u\}$	$\{E\}$	$\{E_u\}$	$2A_u$	$2A$
T_u	T	$A_u + \{E_u\}$	$B_{1u} + B_{2u} + B_{3u}$	$B_1 + B_2 + B_3$

T_d	T	C_{3v}	C_{2v}	D_{2d}
A_1	A	A_1	A_1	A_1
A_2	A	A_2	A_2	B_1
E	$\{E\}$	E	$A_1 + A_2$	$A_1 + B_1$
T_1	T	$A_2 + E$	$A_2 + B_1 + B_2$	$A_2 + E$
T_2	T	$A_1 + E$	$A_1 + B_1 + B_2$	$B_2 + E$

O	T	D_4	C_4	D_3	D_2 $3C_2$	D_2 $C_2, 2C_2'$	C_3	C_2 C_2	C_2 C_2'
A_1	A	A_1	A	A_1	A	A	A	A	A
A_2	A	B_1	B	A_2	A	B_1	A	A	B
E	$\{E\}$	$A_1 + B_1$	$A + B$	E	$2A$	$A + B$	$\{E\}$	$2A$	$A + B$
T_1	T	$A_2 + E$	$A + \{E\}$	$A_2 + E$	$B_1 + B_2 + B_3$	$B_1 + B_2 + B_3$	$A + \{E\}$	$A + 2B$	$A + 2B$
T_2	T	$B_2 + E$	$B + \{E\}$	$A_1 + E$	$B_1 + B_2 + B_3$	$A + B_2 + B_3$	$A + \{E\}$	$A + 2B$	$2A + B$

O_h	O	T_d	T_h	D_{4h}	D_{3d}
A_{1g}	A_1	A_1	A_g	A_{1g}	A_{1g}
A_{2g}	A_2	A_2	A_g	B_{1g}	A_{2g}
Eg	E	E	$\{E_g\}$	$A_{1g} + B_{1g}$	E_g
T_{1g}	T_1	T_1	T_g	$A_{2g} + E_g$	$A_{2g} + E_g$
T_{2g}	T_2	T_2	T_g	$B_{2g} + E_g$	$A_{1g} + E_g$
A_{1u}	A_1	A_2	A_u	A_{1u}	A_{1u}
A_{2u}	A_2	A_1	A_u	B_{1u}	A_{2u}
E_u	E	E	$\{E_u\}$	$A_{1u} + B_{1u}$	E_u
T_{1u}	T_1	T_2	T_u	$A_{2u} + E_u$	$A_{2u} + E_u$
T_{2u}	T_2	T_1	T_u	$B_{2u} + E_u$	$A_{1u} + E_u$

I	T	D_5	C_5	D_3	C_3	D_2	C_2
A	A	A_1	A	A_1	A	A	A
T_1	T	$A_2 + E_1$	$A + \{E_1\}$	$A_2 + E$	$A + \{E\}$	$B_1 + B_2 + B_3$	$A + 2B$
T_2	T	$A_2 + E_2$	$A + \{E_2\}$	$A_2 + E$	$A + \{E\}$	$B_1 + B_2 + B_3$	$A + 2B$
G	$A + T$	$E_1 + E_2$	$\{E_1\} + \{E_2\}$	$A_1 + A_2 + E$	$2A + \{E\}$	$A + B_1 + B_2 + B_3$	$2A + 2B$
H	$\{E\} + T$	$A_1 + E_1 + E_2$	$A + \{E_1\} + \{E_2\}$	$A_1 + 2E$	$A + 2\{E\}$	$2A + B_1 + B_2 + B_3$	$3A + 2B$

R_3	O	D_4	D_3
S	A_1	A_1	A_1
P	T_1	$A_2 + E$	$A_2 + E$
D	$E + T_2$	$A_1 + B_1 + B_2 + E$	$A_1 + 2E$
F	$A_2 + T_1 + T_2$	$A_2 + B_1 + B_2 + 2E$	$A_1 + 2A_2 + 2E$
G	$A_1 + E + T_1 + T_2$	$2A_1 + A_2 + B_1 + B_2 + 2E$	$2A_1 + A_2 + 3E$
H	$E + 2T_1 + T_2$	$A_1 + 2A_2 + B_1 + B_2 + 3E$	$A_1 + 2A_2 + 4E$

Partial correlation tables for infinite point groups:

$C_{\infty v}$	C_{2v}
$A_1 = \Sigma^+$	A_1
$A_2 = \Sigma^-$	A_2
$E_1 = \Pi$	$B_1 + B_2$
$E_2 = \Delta$	$A_1 + A_2$

$D_{\infty h}$	D_{2h}
Σ_g^+	A_g
Σ_g^-	B_{1g}
Π_g	$B_{2g} + B_{3g}$
Δ_g	$A_g + B_{1g}$
Σ_u^+	B_{1u}
Σ_u^-	A_u
Π_u	$B_{2u} + B_{3u}$
Δ_u	$A_u + B_{1u}$

BIBLIOGRAPHY

1. Carter RL. *Molecular Symmetry and Group Theory*. New York: John Wiley & Sons, Inc.; 1998.

The 230 Space Groups | E

Space Group Number	Point Group (Schoenflies)	Point Group (International)	Space Group Symbol
Triclinic			
1	$C_1{}^1$	1	*P*1
2	$C_i{}^1$	$\bar{1}$	*P*$\bar{1}$
Monoclinic			
3	$C_2{}^1$	2	P_2
4	$C_2{}^2$	2	$P2_1$
5	$C_2{}^3$	2	*C*2
6	$C_s{}^1$	*m*	*Pm*
7	$C_s{}^2$	*m*	*Pc*
8	$C_s{}^3$	*m*	*Cm*
9	$C_s{}^4$	*m*	*Cc*
10	$C_{2h}{}^1$	2/*m*	*P*2/*m*
11	$C_{2h}{}^2$	2/*m*	$P2_1/m$
12	$C_{2h}{}^3$	2/*m*	*C*2/*m*
13	$C_{2h}{}^4$	2/*m*	*P*2/*c*
14	$C_{2h}{}^5$	2/*m*	$P2_1/c$
15	$C_{2h}{}^6$	2/*m*	*C*2/*c*
Orthorhombic			
16	$D_2{}^1$	222	*P*222
17	$D_2{}^2$	222	$P222_1$
18	$D_2{}^3$	222	$P2_12_12$
19	$D_2{}^4$	222	$P2_12_12_1$
20	$D_2{}^5$	222	$C222_1$
21	$D_2{}^6$	222	*C*222
22	$D_2{}^7$	222	*F*222
23	$D_2{}^8$	222	*I*222
24	$D_2{}^9$	222	$I2_12_12_1$
25	$C_{2v}{}^1$	*mm*2	*Pmm*2

(*continued*)

Space Group Number	Point Group (Schoenflies)	Point Group (International)	Space Group Symbol
26	C_{2v}^{2}	*mm*2	$Pmc2_1$
27	C_{2v}^{3}	*mm*2	*Pcc*2
28	C_{2v}^{4}	*mm*2	*Pma*2
29	C_{2v}^{5}	*mm*2	$Pca2_1$
30	C_{2v}^{6}	*mm*2	*Pnc*2
31	C_{2v}^{7}	*mm*2	$Pmn2_1$
32	C_{2v}^{8}	*mm*2	*Pba*2
33	C_{2v}^{9}	*mm*2	$Pna2_1$
34	C_{2v}^{10}	*mm*2	*Pnn*2
35	C_{2v}^{11}	*mm*2	*Cmm*2
36	C_{2v}^{12}	*mm*2	$Cmc2_1$
37	C_{2v}^{13}	*mm*2	*Ccc*2
38	C_{2v}^{14}	*mm*2	*Amm*2
39	C_{2v}^{15}	*mm*2	*Abm*2
40	C_{2v}^{16}	*mm*2	*Ama*2
41	C_{2v}^{17}	*mm*2	*Aba*2
42	C_{2v}^{18}	*mm*2	*Fmm*2
43	C_{2v}^{19}	*mm*2	*Fdd*2
44	C_{2v}^{20}	*mm*2	*Imm*2
45	C_{2v}^{22}	*mm*2	*Iba*2
46	C_{2v}^{22}	*mm*2	*Ima*2
47	D_{2h}^{1}	*mmm*	*Pmmm*
48	D_{2h}^{2}	*mmm*	*Pnnn*
49	D_{2h}^{3}	*mmm*	*Pccm*
50	D_{2h}^{4}	*mmm*	*Pban*
51	D_{2h}^{5}	*mmm*	*Pmma*
52	D_{2h}^{6}	*mmm*	*Pnna*
53	D_{2h}^{7}	*mmm*	*Pmna*
54	D_{2h}^{8}	*mmm*	*Pcca*
55	D_{2h}^{9}	*mmm*	*Pbam*
56	D_{2h}^{10}	*mmm*	*Pccn*
57	D_{2h}^{11}	*mmm*	*Pbcm*
58	D_{2h}^{12}	*mmm*	*Pnnm*
59	D_{2h}^{13}	*mmm*	*Pmmn*
60	D_{2h}^{14}	*mmm*	*Pbcn*
61	D_{2h}^{15}	*mmm*	*Pbca*
62	D_{2h}^{16}	*mmm*	*Pnma*
63	D_{2h}^{17}	*mmm*	*Pmcm*
64	D_{2h}^{18}	*mmm*	*Cmca*
65	D_{2h}^{19}	*mmm*	*Cmmm*
66	D_{2h}^{20}	*mmm*	*Cccm*
67	D_{2h}^{21}	*mmm*	*Cmma*
68	D_{2h}^{22}	*mmm*	*Ccca*
69	D_{2h}^{23}	*mmm*	*Fmmm*
70	D_{2h}^{24}	*mmm*	*Fddd*
71	D_{2h}^{25}	*mmm*	*Immm*
72	D_{2h}^{26}	*mmm*	*Ibam*
73	D_{2h}^{27}	*mmm*	*Ibca*
74	D_{2h}^{28}	*mmm*	*Imma*

Space Group Number	Point Group (Schoenflies)	Point Group (International)	Space Group Symbol
Tetragonal			
75	C_4^1	4	*P*4
76	C_4^2	4	$P4_1$
77	C_4^3	4	$P4_2$
78	C_4^4	4	$P4_3$
79	C_4^5	4	*I*4
80	C_4^6	4	$I4_1$
81	S_4^1	$\bar{4}$	$P\bar{4}$
82	S_4^2	$\bar{4}$	$I\bar{4}$
83	C_{4h}^1	4/*m*	*P*4/*m*
84	C_{4h}^2	4/*m*	$P4_2/m$
85	C_{4h}^3	4/*m*	*P*4/*n*
86	C_{4h}^4	4/*m*	$P4_2/n$
87	C_{4h}^5	4/*m*	*I*4/*m*
88	C_{4h}^6	4/*m*	$I4_1/a$
89	D_4^1	422	*P*422
90	D_4^2	422	$P42_12$
91	D_4^3	422	$P4_122$
92	D_4^4	422	$P4_12_12$
93	D_4^5	422	$P4_222$
94	D_4^6	422	$P4_22_12$
95	D_4^7	422	$P4_322$
96	D_4^8	422	$P4_32_12$
97	D_4^9	422	*I*422
98	D_4^{10}	422	$I4_122$
99	C_{4v}^1	4*mm*	*P*4*mm*
100	C_{4v}^2	4*mm*	*P*4*bm*
101	C_{4v}^3	4*mm*	$P4_2cm$
102	C_{4v}^4	4*mm*	$P4_2nm$
103	C_{4v}^5	4*mm*	*P*4*cc*
104	C_{4v}^6	4*mm*	*P*4*nc*
105	C_{4v}^7	4*mm*	$P4_2mc$
106	C_{4v}^8	4*mm*	$P4_2bc$
107	C_{4v}^9	4*mm*	*I*4*mm*
108	C_{4v}^{10}	4*mm*	*I*4*cm*
109	C_{4v}^{11}	4*mm*	$I4_1md$
110	C_{4v}^{12}	4*mm*	$I4_1cd$
111	D_{2d}^1	$\bar{4}2m$	$P\bar{4}2m$
112	D_{2d}^2	$\bar{4}2m$	$P\bar{4}2c$
113	D_{2d}^3	$\bar{4}2m$	$P\bar{4}2_1m$
114	D_{2d}^4	$\bar{4}2m$	$P\bar{4}2_1c$
115	D_{2d}^5	$\bar{4}m2$	$P\bar{4}m2$
116	D_{2d}^6	$\bar{4}m2$	$P\bar{4}c2$
117	D_{2d}^7	$\bar{4}m2$	$P\bar{4}b2$
118	D_{2d}^8	$\bar{4}m2$	$P\bar{4}n2$
119	D_{2d}^9	$\bar{4}m2$	$I\bar{4}m2$
120	D_{2d}^{10}	$\bar{4}m2$	$I\bar{4}c2$

(*continued*)

Space Group Number	Point Group (Schoenflies)	Point Group (International)	Space Group Symbol
121	D_{2d}^{11}	$\bar{4}m2$	$I\bar{4}2m$
122	D_{2d}^{12}	$\bar{4}2m$	$I\bar{4}2d$
123	D_{4h}^{1}	$4/mmm$	$P4/mmm$
124	D_{4h}^{2}	$4/mmm$	$P4/mcc$
125	D_{4h}^{3}	$4/mmm$	$P4/nbm$
126	D_{4h}^{4}	$4/mmm$	$P4/nnc$
127	D_{4h}^{5}	$4/mmm$	$P4/mbm$
128	D_{4h}^{6}	$4/mmm$	$P4/mnc$
129	D_{4h}^{7}	$4/mmm$	$P4/nmm$
130	D_{4h}^{8}	$4/mmm$	$P4/ncc$
131	D_{4h}^{9}	$4/mmm$	$P4/mmc$
132	D_{4h}^{10}	$4/mmm$	$P4_2/mcm$
133	D_{4h}^{11}	$4/mmm$	$P4_2/nbc$
134	D_{4h}^{12}	$4/mmm$	$P4_2/nnm$
135	D_{4h}^{13}	$4/mmm$	$P4_2/mbc$
136	D_{4h}^{14}	$4/mmm$	$P4_2/mnm$
137	D_{4h}^{15}	$4/mmm$	$P4_2/nmc$
138	D_{4h}^{16}	$4/mmm$	$P4_2/ncm$
139	D_{4h}^{17}	$4/mmm$	$I4/mmm$
140	D_{4h}^{18}	$4/mmm$	$I4/mcm$
141	D_{4h}^{19}	$4/mmm$	$I4_1/amd$
142	D_{4h}^{20}	$4/mmm$	$I4_1/acd$
Trigonal			
143	C_3^{1}	3	$P3$
144	C_3^{2}	3	$P3_1$
145	C_3^{3}	3	$P3_2$
146	C_3^{4}	3	$R3$
147	C_{3i}^{1}	$\bar{3}$	$P\bar{3}$
148	C_{3i}^{2}	$\bar{3}$	$R\bar{3}$
149	D_3^{1}	312	$P312$
150	D_3^{2}	321	$P321$
151	D_3^{3}	312	$P3_112$
152	D_3^{4}	321	$P3_121$
153	D_3^{5}	312	$P3_212$
154	D_3^{6}	321	$P3_221$
155	D_3^{7}	32	$R32$
156	C_{3v}^{1}	$3m1$	$P3m1$
157	C_{3v}^{2}	$31m$	$P31m$
158	C_{3v}^{3}	$3m1$	$P3c1$
159	C_{3v}^{4}	$31m$	$P31c$
160	C_{3v}^{5}	$3m$	$R3m$
161	C_{3v}^{6}	$3m$	$R3c$
162	D_{3d}^{1}	$\bar{3}1m$	$P\bar{3}1m$
163	D_{3d}^{2}	$\bar{3}1m$	$P\bar{3}1c$
164	D_{3d}^{3}	$\bar{3}m1$	$P\bar{3}m1$
165	D_{3d}^{4}	$\bar{3}m1$	$P\bar{3}c1$
166	D_{3d}^{5}	$\bar{3}m$	$R\bar{3}m$
167	D_{3d}^{6}	$\bar{3}m$	$R\bar{3}c$

Space Group Number	Point Group (Schoenflies)	Point Group (International)	Space Group Symbol
Hexagonal			
168	C_6^1	6	$P6$
169	C_6^2	6	$P6_1$
170	C_6^3	6	$P6_5$
171	C_6^4	6	$P6_2$
172	C_6^5	6	$P6_4$
173	C_6^6	6	$P6_3$
174	C_{3h}^1	$\bar{6}$	$P\bar{6}$
175	C_{6h}^1	$6/m$	$P6/m$
176	C_{6h}^2	$6/m$	$P6_3/m$
177	D_6^1	622	$P622$
178	D_6^2	622	$P6_122$
179	D_6^3	622	$P6_522$
180	D_6^4	622	$P6_222$
181	D_6^5	622	$P6_422$
182	D_6^6	622	$P6_322$
183	C_{6v}^1	$6mm$	$P6mm$
184	C_{6v}^2	$6mm$	$P6cc$
185	C_{6v}^3	$6mm$	$P6_3cm$
186	C_{6v}^4	$6mm$	$P6_3mc$
187	D_{3h}^1	$\bar{6}m2$	$P\bar{6}m2$
188	D_{3h}^2	$\bar{6}m2$	$P\bar{6}c2$
189	D_{3h}^3	$\bar{6}2m$	$P\bar{6}2m$
190	D_{3h}^4	$\bar{6}2m$	$P\bar{6}2c$
191	D_{6h}^1	$6/mmm$	$P6/mmm$
192	D_{6h}^2	$6/mmm$	$P6/mcc$
193	D_{6h}^3	$6/mmm$	$P6_3/mcm$
194	D_{6h}^4	$6/mmm$	$P6_3/mmc$
Cubic			
195	T^1	23	$P23$
196	T^2	23	$F23$
197	T^3	23	$I23$
198	T^4	23	$P2_13$
199	T^5	23	$I2_13$
200	T_h^1	$m\bar{3}$	$Pm\bar{3}$
201	T_h^2	$m\bar{3}$	$Pn\bar{3}$
202	T_h^3	$m\bar{3}$	$Fm\bar{3}$
203	T_h^4	$m\bar{3}$	$Fd\bar{3}$
204	T_{h5}	$m\bar{3}$	$Im\bar{3}$
205	T_h^6	$m\bar{3}$	$Pa\bar{3}$
206	T_h^7	$m\bar{3}$	$Ia\bar{3}$
207	O^1	432	$P432$
208	O^2	432	$P4_232$
209	O^3	432	$F432$
210	O^4	432	$F4_132$
211	O^5	432	$I432$
212	O^6	432	$P4_332$

(continued)

Space Group Number	Point Group (Schoenflies)	Point Group (International)	Space Group Symbol
213	O^7	432	$P4_132$
214	O^8	432	$I4_132$
215	T_d^1	$\bar{4}3m$	$P\bar{4}3m$
216	T_d^2	$\bar{4}3m$	$F\bar{4}3m$
217	T_d^3	$\bar{4}3m$	$I\bar{4}3m$
218	T_d^4	$\bar{4}3m$	$P\bar{4}3n$
219	T_d^5	$\bar{4}3m$	$F\bar{4}3c$
220	T_d^6	$\bar{4}3m$	$I\bar{4}3d$
221	O_h^1	$m\bar{3}m$	$Pm\bar{3}m$
222	O_h^2	$m\bar{3}m$	$Pn\bar{3}n$
223	O_h^3	$m\bar{3}m$	$Pm\bar{3}n$
224	O_h^4	$m\bar{3}m$	$Pn\bar{3}m$
225	O_h^5	$m\bar{3}m$	$Fm\bar{3}m$
226	O_h^6	$m\bar{3}m$	$Fm\bar{3}c$
227	O_h^7	$m\bar{3}m$	$Fd\bar{3}m$
228	O_h^8	$m\bar{3}m$	$Fd\bar{3}c$
229	O_h^9	$m\bar{3}m$	$Im\bar{3}m$
230	O_h^{10}	$m\bar{3}m$	$Ia\bar{3}d$

BIBLIOGRAPHY

1. Hahn T, editor. *International Tables for Crystallography, Volume A: Space-Group Symmetry*. 4th ed. Boston: Kluwer Academic Publishers; 1995.

Index

Principles of Inorganic Chemistry, First Edition. Brian W. Pfennig.

Commonly Used Character Tables

C_{2v}	E	C_2	$\sigma_v(xz)$	$\sigma_v'(yz)$			
A_1	1	1	1	1	z	x^2, y^2, z^2	$z^3, z(x^2-y^2)$
A_2	1	1	−1	−1	R_z	xy	xyz
B_1	1	−1	1	−1	x, R_y	xz	$xz^2, x(x^2-3y^2)$
B_2	1	−1	−1	1	y, R_x	yz	$yz^2, y(3x^2-y^2)$

C_{3v}	E	$2C_3$	$3\sigma_v$			
A_1	1	1	1	z	x^2+y^2, z^2	$z^3, x(x^2-3y^2)$
A_2	1	1	−1	R_z		$y(3x^2-y^2)$
E	2	−1	0	(x, y), (R_x, R_y)	$(x^2-y^2, xy), (xz, yz)$	$(xz^2, yz^2), [xyz, z(x^2-y^2)]$

C_{4v}	E	$2C_4$	C_2	$2\sigma_v$	$2\sigma_d$			
A_1	1	1	1	1	1	z	x^2+y^2, z^2	z^3
A_2	1	1	1	−1	−1	R_z		
B_1	1	−1	1	1	−1		x^2-y^2	$z(x^2-y^2)$
B_2	1	−1	1	−1	1		xy	xyz
E	2	0	−2	0	0	$(x, y), (R_x, R_y)$	(xz, yz)	$(xz^2, yz^2), [x(x^2-3y^2), y(3x^2-y^2)]$

D_{2h}	E	$C_2(z)$	$C_2(y)$	$C_2(x)$	i	$\sigma(xy)$	$\sigma(xz)$	$\sigma(yz)$			
A_g	1	1	1	1	1	1	1	1		x^2, y^2, z^2	
B_{1g}	1	1	−1	−1	1	1	−1	−1	R_z	xy	
B_{2g}	1	−1	1	−1	1	−1	1	−1	R_y	xz	
B_{3g}	1	−1	−1	1	1	−1	−1	1	R_x	yz	
A_u	1	1	1	1	−1	−1	−1	−1			xyz
B_{1u}	1	1	−1	−1	−1	−1	1	1	z		$z^3, z(x^2-y^2)$
B_{2u}	1	−1	1	−1	−1	1	−1	1	y		$yz^2, y(3x^2-y^2)$
B_{3u}	1	−1	−1	1	−1	1	1	−1	x		$xz^2, x(x^2-3y^2)$

D_{4h}	E	$2C_4$	C_2	$2C_2'$	$2C_2''$	i	$2S_4$	σ_h	$2\sigma_v$	$2\sigma_d$			
A_{1g}	1	1	1	1	1	1	1	1	1	1		x^2+y^2, z^2	
A_{2g}	1	1	1	1	−1	1	1	1	−1	1	R_z		
B_{1g}	1	1	1	1	−1	1	−1	1	1	−1		x^2-y^2	
B_{2g}	1	1	1	1	1	1	−1	1	−1	1		xy	
E_g	2	0	−2	0	0	2	0	−2	0	0	(R_x, R_y)	(xz, yz)	
A_{1u}	1	1	1	1	1	−1	−1	−1	−1	−1			z^3
A_{2u}	1	1	1	−1	−1	−1	−1	−1	1	1	z		xyz
B_{1u}	1	−1	1	1	−1	−1	1	−1	−1	1			z^3
B_{2u}	1	−1	1	−1	1	−1	1	−1	1	−1			$z(x^2-y^2)$
E_u	2	0	−2	0	0	−2	0	2	0	0	(x, y)		$(xz^2, yz^2), [x(x^2-3y^2), y(3x^2-y^2)]$

T_d	E	$8C_3$	$3C_2$	$6S_4$	$6\sigma_d$			
A_1	1	1	1	1	1		$x^2+y^2+z^2$	xyz
A_2	1	1	1	−1	−1			
E	2	−1	2	0	0		$(2z^2-x^2-y^2, x^2-y^2)$	
T_1	3	0	−1	1	−1	(R_x, R_y, R_z)		$[x(z^2-y^2), y(z^2-x^2), z(x^2-y^2)]$
T_2	3	0	−1	−1	1	(x, y, z)	(xy, xz, yz)	(x^3, y^3, z^3)

O_h	E	$8C_3$	$6C_2$	$6C_4$	$3C_2$	i	$6S_4$	$8S_6$	$3\sigma_h$	$6\sigma_d$			
A_{1g}	1	1	1	1	1	1	1	1	1	1		$x^2+y^2+z^2$	
A_{2g}	1	1	−1	−1	1	1	−1	1	1	−1			
E_g	2	−1	0	0	2	2	0	−1	2	0		$(2z^2-x^2-y^2, x^2-y^2)$	
T_{1g}	3	0	−1	1	−1	3	1	0	−1	−1	(R_x, R_y, R_z)		
T_{2g}	3	0	1	−1	−1	3	−1	0	−1	1		(xy, xz, yz)	
A_{1u}	1	1	1	1	1	−1	−1	−1	−1	−1			
A_{2u}	1	1	1	1	1	1	1	1	−1	1			xyz
E_u	2	−1	0	0	2	−2	0	1	−2	0			
T_{1u}	3	0	−1	1	−1	−3	−1	0	1	1	(x, y, z)		(x^3, y^3, z^3)
T_{2u}	3	0	1	−1	−1	−3	1	0	1	−1			$[x(z^2-y^2), y(z^2-x^2), z(x^2-y^2)]$

Commonly Used Correlation Tables

C_{2v}	C_2	C_s, σ (xz)	C_s, σ (yz)
A_1	A	A'	A'
A_2	A	A''	A''
B_1	B	A'	A''
B_2	B	A''	A'

C_{3v}	C_3	C_s
A_1	A	A'
A_2	A	A''
E	$\{E\}$	$A' + A''$

C_{4v}	C_4	C_{2v}, σ_v	C_{2v}, σ_d	C_2	C_s, σ_v	C_s, σ_d
A_1	A	A_1	A_1	A	A'	A'
A_2	A	A_2	A_2	A	A''	A''
B_1	B	A_1	A_2	A	A'	A''
B_2	B	A_2	A_1	A	A''	A'
E	E	$B_1 + B_2$	$B_1 + B_2$	$2B$	$A' + A''$	$A' + A''$

D_{4h}	D_4	C_{4v}	C_{4h}	C_4	D_{2h} C_2'	D_{2h} C_2''	D_{2d} $C_2' \rightarrow C_2'$	D_{2d} $C_2'' \rightarrow C_2'$	S_4	D_2C_2'
A_{1g}	A_1	A_1	A_g	A	A_g	A_g	A_1	A_1	A	A
A_{2g}	A_2	A_2	A_g	A	B_{1g}	B_{1g}	A_2	A_2	A	B_1
B_{1g}	B_1	B_1	B_g	B	A_g	B_{1g}	B_1	B_2	B	A
B_{2g}	B_2	B_2	B_g	B	B_{1g}	A_g	B_2	B_1	B	B_1
E_g	E	E	$\{E_g\}$	$\{E\}$	$B_{2g} + B_{3g}$	$B_{2g} + B_{3g}$	E	E	$\{E\}$	$B_2 + B_3$
A_{1u}	A_1	A_2	A_u	A	A_u	A_u	B_1	B_1	B	A
A_{2u}	A_2	A_1	A_u	A	B_{1u}	B_{1u}	B_2	B_2	B	B_1
B_{1u}	B_1	B_2	B_u	B	A_u	B_{1u}	A_1	A_2	A	A
B_{2u}	B_2	B_1	B_u	B	B_{1u}	A_u	A_2	A_1	A	B_1
E_u	E	E	$\{E_u\}$	$\{E\}$	$B_{2u} + B_{3u}$	$B_{2u} + B_{3u}$	E	E	$\{E\}$	$B_2 + B_3$

T_d	T	C_{3v}	C_{2v}	D_{2d}
A_1	A	A_1	A_1	A_1
A_2	A	A_2	A_2	B_1
E	$\{E\}$	E	$A_1 + A_2$	$A_1 + B_1$
T_1	T	$A_2 + E$	$A_2 + B_1 + B_2$	$A_2 + E$
T_2	T	$A_1 + E$	$A_1 + B_1 + B_2$	$B_2 + E$

O_h	O	T_d	T_h	D_{4h}	D_{3d}
A_{1g}	A_1	A_1	A_g	A_{1g}	A_{1g}
A_{2g}	A_2	A_2	A_g	B_{1g}	A_{2g}
Eg	E	E	$\{E_g\}$	$A_{1g} + B_{1g}$	E_g
T_{1g}	T_1	T_1	T_g	$A_{2g} + E_g$	$A_{2g} + E_g$
T_{2g}	T_2	T_2	T_g	$B_{2g} + E_g$	$A_{1g} + E_g$
A_{1u}	A_1	A_2	A_u	A_{1u}	A_{1u}
A_{2u}	A_2	A_1	A_u	B_{1u}	A_{2u}
E_u	E	E	$\{E_u\}$	$A_{1u} + B_{1u}$	E_u
T_{1u}	T_1	T_2	T_u	$A_{2u} + E_u$	$A_{2u} + E_u$
T_{2u}	T_2	T_1	T_u	$B_{2u} + E_u$	$A_{1u} + E_u$

$C_{\infty v}$	C_{2v}
$A_1 = \Sigma^+$	A_1
$A_2 = \Sigma^-$	A_2
$E_1 = \Pi$	$B_1 + B_2$
$E_2 = \Delta$	$A_1 + A_2$

$D_{\infty h}$	D_{2h}
Σ_g^+	A_g
Σ_g^-	B_{1g}
Π_g	$B_{2g} + B_{3g}$
Δ_g	$A_g + B_{1g}$
Σ_u^+	B_{1u}
Σ_u^-	A_u
Π_u	$B_{2u} + B_{3u}$
Δ_u	$A_u + B_{1u}$